Birkhäuser

Applied and Numerical Harmonic Analysis

Gerlind Plonka • Daniel Potts • Gabriele Steidl •
Manfred Tasche

Numerical Fourier Analysis

Second Edition

Birkhäuser

Gerlind Plonka
University of Göttingen
Göttingen, Germany

Daniel Potts
Angewandte Funktionalanalysis
TU Chemnitz
Chemnitz, Sachsen, Germany

Gabriele Steidl
Technical University of Berlin
Berlin, Germany

Manfred Tasche
University of Rostock
Rostock, Germany

ISSN 2296-5009 ISSN 2296-5017 (electronic)
Applied and Numerical Harmonic Analysis
ISBN 978-3-031-35004-7 ISBN 978-3-031-35005-4 (eBook)
https://doi.org/10.1007/978-3-031-35005-4

Mathematics Subject Classification: 65-02, 65Txx, 42-02, 42Axx, 42Bxx

This book is published under the imprint Birkhäuser, www.birkhauser-science.com by the registered company Springer Nature Switzerland AG
The registered company address is: Gewerbestrasse 11, 6330 Cham, Switzerland

Paper in this product is recyclable.

ANHA Series Preface

The Applied and Numerical Harmonic Analysis (ANHA) book series aims to provide the engineering, mathematical, and scientific communities with significant developments in harmonic analysis, ranging from abstract harmonic analysis to basic applications. The title of the series reflects the importance of applications and numerical implementation, but richness and relevance of applications and implementation depend fundamentally on the structure and depth of theoretical underpinnings. Thus, from our point of view, the interleaving of theory and applications and their creative symbiotic evolution is axiomatic.

Harmonic analysis is a wellspring of ideas and applicability that has flourished, developed, and deepened over time within many disciplines and by means of creative cross-fertilization with diverse areas. The intricate and fundamental relationship between harmonic analysis and fields such as signal processing, partial differential equations (PDEs), and image processing is reflected in our state-of-the-art ANHA series.

Our vision of modern harmonic analysis includes a broad array of mathematical areas, e.g., wavelet theory, Banach algebras, classical Fourier analysis, time-frequency analysis, deep learning, and fractal geometry, as well as the diverse topics that impinge on them.

For example, wavelet theory can be considered an appropriate tool to deal with some basic problems in digital signal processing, speech and image processing, geophysics, pattern recognition, biomedical engineering, and turbulence. These areas implement the latest technology from sampling methods on surfaces to fast algorithms and computer vision methods. The underlying mathematics of wavelet theory depends not only on classical Fourier analysis, but also on ideas from abstract harmonic analysis, including von Neumann algebras and the affine group. This leads to a study of the Heisenberg group and its relationship to Gabor systems, and of the metaplectic group for a meaningful interaction of signal decomposition methods.

The unifying influence of wavelet theory in the aforementioned topics illustrates the justification for providing a means for centralizing and disseminating information from the broader, but still focused, area of harmonic analysis. This will be a key role of ANHA. We intend to publish with the scope and interaction that such a host of issues demands.

Along with our commitment to publish mathematically significant works at the frontiers of harmonic analysis, we have a comparably strong commitment to publish major advances in the following applicable topics in which harmonic analysis plays a substantial role:

*Analytic Number theory * Antenna Theory * Artificial Intelligence * Biomedical Signal Processing * Classical Fourier Analysis * Coding Theory * Communications Theory * Compressed Sensing * Crystallography and Quasi-Crystals * Data Mining * Data Science * Deep Learning * Digital Signal Processing * Dimension Reduction and Classification * Fast Algorithms * Frame Theory and Applications * Gabor Theory and Applications * Geophysics * Image Processing * Machine Learning * Manifold Learning * Numerical Partial Differential Equations * Neural Networks * Phaseless Reconstruction * Prediction Theory * Quantum Information Theory * Radar Applications * Sampling Theory (Uniform and Non-uniform) and Applications * Spectral Estimation * Speech Processing * Statistical Signal Processing * Super-resolution * Time Series * Time-Frequency and Time-Scale Analysis * Tomography * Turbulence * Uncertainty Principles *Waveform design * Wavelet Theory and Applications

The above point of view for the ANHA book series is inspired by the history of Fourier analysis itself, whose tentacles reach into so many fields.

In the last two centuries, Fourier analysis has had a major impact on the development of mathematics, on the understanding of many engineering and scientific phenomena, and on the solution of some of the most important problems in mathematics and the sciences. Historically, Fourier series were developed in the analysis of some of the classical PDEs of mathematical physics; these series were used to solve such equations. In order to understand Fourier series and the kinds of solutions they could represent, some of the most basic notions of analysis were defined, e.g., the concept of "function." Since the coefficients of Fourier series are integrals, it is no surprise that Riemann integrals were conceived to deal with uniqueness properties of trigonometric series. Cantor's set theory was also developed because of such uniqueness questions.

A basic problem in Fourier analysis is to show how complicated phenomena, such as sound waves, can be described in terms of elementary harmonics. There are two aspects of this problem: first, to find, or even define properly, the harmonics or spectrum of a given phenomenon, e.g., the spectroscopy problem in optics; second, to determine which phenomena can be constructed from given classes of harmonics, as done, for example, by the mechanical synthesizers in tidal analysis.

Fourier analysis is also the natural setting for many other problems in engineering, mathematics, and the sciences. For example, Wiener's Tauberian theorem in Fourier analysis not only characterizes the behavior of the prime numbers but is a fundamental tool for analyzing the ideal structures of Banach algebras. It also provides the proper notion of spectrum for phenomena such as white light. This latter process leads to the Fourier analysis associated with correlation functions in filtering and prediction problems. These problems, in turn, deal naturally with Hardy spaces in complex analysis, as well as inspiring Wiener to consider communications engineering in terms of feedback and stability, his cybernetics. This latter theory develops concepts to understand complex systems such as learning and cognition and neural networks, and it is arguably a precursor of deep learning and its spectacular interactions with data science and AI.

Nowadays, some of the theory of PDEs has given way to the study of Fourier integral operators. Problems in antenna theory are studied in terms of unimodular trigonometric polynomials. Applications of Fourier analysis abound in signal processing, whether with the fast Fourier transform (FFT), or filter design, or the adaptive modeling inherent in time-frequency-scale methods such as wavelet theory.

The coherent states of mathematical physics are translated and modulated Fourier transforms, and these are used, in conjunction with the uncertainty principle, for dealing with signal reconstruction in communications theory. We are back to the raison d'etre of the ANHA series!

College Park, MD, USA John Benedetto
Wojciech Czaja
Boston, MA, USA Kasso Okoudjou

Preface to the Second Edition

In the second edition of this book, we have taken the opportunity to update the monograph, thereby reflecting some recent developments in the field of Numerical Fourier Analysis. Beside minor corrections, we have included several additional comments and new references, as for example the remark on a new Fourier approach to the ANOVA decomposition of high-dimensional trigonometric polynomials in Sect. 8.5.

We have extended the first edition by adding some topics which have been missing before. A new Sect. 4.4 on the Fourier transform of measures has been incorporated, and we would like to thank Robert Beinert, who co-authored this section. We also added a short Sect. 6.3.3 on the fast two-dimensional discrete cosine transform.

Moreover, we have included recent research results on the approximation errors of the nonequispaced fast Fourier transform based on special window functions in Sect. 7.2, and extended Sect. 2.3 by introducing a new regularized Shannon sampling formula which leads to error estimates with exponential decay for bandlimited functions. Further, we added Sect. 10.2.3 to present the recently developed ESPIRA algorithm for the recovery of exponential sums.

We would like to thank all colleagues and students, who helped us to recognize typos that appeared in the first edition of the book. These have been corrected in the current edition.

Finally, we thank Springer/Birkhäuser for the excellent cooperation and for publishing this second edition of this monograph.

Göttingen, Germany	Gerlind Plonka
Chemnitz, Germany	Daniel Potts
Berlin, Germany	Gabriele Steidl
Rostock, Germany	Manfred Tasche
December 2022	

Preface to the First Edition

Fourier analysis has been grown to an essential mathematical tool with a tremendous amount of different applications in applied mathematics, engineering, physics, and other sciences. Many recent technological innovations from spectroscopy and computer tomography to speech and music signal processing are based on Fourier analysis. Fast Fourier algorithms are the heart of data processing methods, and their societal impact can hardly be overestimated.

The field of Fourier analysis is continuously developing towards the needs in applications and many topics are part of ongoing intensive research. Due to the importance of Fourier techniques there are several books on the market focusing on different aspects of Fourier theory, as e.g. [41, 75, 93, 143, 149, 156, 178, 247, 263, 265, 309, 318, 359, 399, 449, 453], or on corresponding algorithms of the discrete Fourier transform, see e.g. [52, 63, 64, 82, 197, 305, 362, 421], not counting further monographs on special applications and generalizations as wavelets [88, 101, 280].

So, why do we write a further book? Reading textbooks in Fourier analysis it appears as a shortcoming that the focus is either set only on the mathematical theory or vice versa only on the corresponding discrete Fourier and convolution algorithms, while the reader needs to consult additional references on the numerical techniques in the one case or on the analytical background in the other.

The urgent need for a unified presentation of Fourier theory and corresponding algorithms particularly emerges from new developments in function approximation using Fourier methods. It is important to understand how well a continuous signal can be approximated by employing the discrete Fourier transform to sampled spectral data. A deep understanding of function approximation by Fourier representations is even more crucial for deriving more advanced transforms as the nonequispaced fast Fourier transform, which is an approximative algorithm by nature, or sparse fast Fourier transforms on special lattices in higher dimensions.

This book encompasses the required classical Fourier theory in the first part in order to give deep insight into the construction and analysis of corresponding fast Fourier algorithms in the second part, including recent developments on nonequispaced and sparse fast Fourier transforms in higher dimensions. In the third part of the book, we present a selection of mathematical applications including recent research results on nonlinear function approximation by exponential sums.

Our book starts with two chapters on classical Fourier analysis and Chap. 3 on the discrete Fourier transform in one dimension, followed by Chap. 4 on the multivariate case. This theoretical part provides the background for all further chapters and makes the book self-contained.

Chapters 5–8 are concerned with the construction and analysis of corresponding fast algorithms in the one- and multidimensional case. While Chap. 5 covers the well-known fast Fourier transforms, Chaps. 7 and 8 are concerned with the construction of the nonequispaced fast Fourier transforms and the high-dimensional fast Fourier transforms on special lattices. Chapter 6 is devoted to discrete trigonometric transforms and Chebyshev expansions which are closely related to Fourier series.

The last part of the book contains two chapters on applications of numerical Fourier methods for improved function approximation.

Starting with Sects. 5.4 and 5.5, the book covers many recent well-recognized developments in numerical Fourier analysis which cannot be found in other books in this form, inclusive research results of the authors obtained within the last 20 years.

This includes topics as

- the analysis of the numerical stability of the radix-2 FFT in Sect. 5.5,
- fast trigonometric transforms based on orthogonal matrix factorizations and fast discrete polynomial transforms in Chap. 6,
- fast Fourier transforms and fast trigonometric transforms for nonequispaced data in space and/or frequency in Sects. 7.1–7.4, and
- fast summation at nonequispaced knots in Sect. 7.5.

More recent research results can be found on

- sparse FFT for vectors with presumed sparsity in Sect. 5.4,
- high-dimensional sparse fast FFT on rank-1 lattices in Chap. 8, and
- applications of multi-exponential analysis and Prony method for recovery of structured functions in Chap. 10.

An introductory course on Fourier analysis at the advanced undergraduate level can for example be built using Sects. 1.2–1.4, 2.1–2.2, 3.2–3.3, 4.1–4.3, and 5.1–5.2. We assume that the reader is familiar with basic knowledge on calculus of univariate and multivariate functions (including basic facts on Lebesgue integration and functional analysis) and on numerical linear algebra. Focusing a lecture on discrete fast algorithms and applications, one may consult Chaps. 3, 5, 6, and 9. Chapters 7, 8, and 10 are on an advanced level and require pre-knowledge from the Chaps. 1, 2, and 4.

Parts of the book are based a series of lectures and seminars given by the authors to students of mathematics, physics, computer science and electrical engineering. Chapters 1, 2, 3, 5, and 9 are partially based on teaching material written by G. Steidl and M. Tasche that was published in 1996 by the University of Hagen under the title "Fast Fourier Transforms—Theory and Applications" (in German). The authors wish to express their gratitude to the University of Hagen for the friendly permission to use this material for the present book.

Last but not least, the authors would like to thank Springer/Birkhäuser for publishing this book.

Göttingen, Germany — Gerlind Plonka
Chemnitz, Germany — Daniel Potts
Kaiserslautern, Germany — Gabriele Steidl
Rostock, Germany — Manfred Tasche
October 2018

Contents

1 Fourier Series 1
1.1 Fourier's Solution of Laplace Equation 1
1.2 Fourier Coefficients and Fourier Series 6
1.3 Convolution of Periodic Functions 16
1.4 Pointwise and Uniform Convergence of Fourier Series 27
1.4.1 Pointwise Convergence 30
1.4.2 Uniform Convergence 40
1.4.3 Gibbs Phenomenon 45
1.5 Discrete Signals and Linear Filters 51

2 Fourier Transform 59
2.1 Fourier Transform on $L_1(\mathbb{R})$ 59
2.2 Fourier Transform on $L_2(\mathbb{R})$ 77
2.3 Poisson Summation Formula and Shannon's Sampling Theorem . 83
2.4 Heisenberg's Uncertainty Principle 103
2.5 Fourier-Related Transforms in Time–Frequency Analysis 109
2.5.1 Windowed Fourier Transform 110
2.5.2 Fractional Fourier Transforms 116

3 Discrete Fourier Transforms 123
3.1 Motivations for Discrete Fourier Transforms 123
3.1.1 Approximation of Fourier Coefficients and Aliasing Formula 124
3.1.2 Computation of Fourier Series and Fourier Transforms .. 129
3.1.3 Trigonometric Polynomial Interpolation 130
3.2 Fourier Matrices and Discrete Fourier Transforms 135
3.2.1 Fourier Matrices 135
3.2.2 Properties of Fourier Matrices 140
3.2.3 DFT and Cyclic Convolutions 147
3.3 Circulant Matrices 154
3.4 Kronecker Products and Stride Permutations 159
3.5 Discrete Trigonometric Transforms 168

4 Multidimensional Fourier Methods ... 175
Robert Beinert, Gerlind Plonka, Daniel Potts, Gabriele Steidl, and Manfred Tasche
4.1 Multidimensional Fourier Series ... 176
4.2 Multidimensional Fourier Transform ... 183
4.2.1 Fourier Transform on $\mathscr{S}(\mathbb{R}^d)$ ... 183
4.2.2 Fourier Transforms on $L_1(\mathbb{R}^d)$ and $L_2(\mathbb{R}^d)$ ... 192
4.2.3 Poisson Summation Formula ... 194
4.2.4 Fourier Transforms of Radial Functions ... 197
4.3 Fourier Transform of Tempered Distributions ... 199
4.3.1 Tempered Distributions ... 199
4.3.2 Fourier Transforms on $\mathscr{S}'(\mathbb{R}^d)$ ... 210
4.3.3 Periodic Tempered Distributions ... 215
4.3.4 Hilbert Transform and Riesz Transform ... 222
4.4 Fourier Transform of Measures ... 229
4.4.1 Measure Spaces ... 230
4.4.2 Fourier Transform of Measures on $\mathbb{T}^d$ ... 235
4.4.3 Fourier Transform of Measures on $\mathbb{R}^d$ ... 240
4.5 Multidimensional Discrete Fourier Transforms ... 247
4.5.1 Computation of Multivariate Fourier Coefficients ... 247
4.5.2 Two-Dimensional Discrete Fourier Transforms ... 251
4.5.3 Higher-Dimensional Discrete Fourier Transforms ... 260

5 Fast Fourier Transforms ... 265
5.1 Construction Principles of Fast Algorithms ... 265
5.2 Radix–2 FFTs ... 269
5.2.1 Sande–Tukey FFT in Summation Form ... 270
5.2.2 Cooley–Tukey FFT in Polynomial Form ... 273
5.2.3 Radix–2 FFT in Matrix Form ... 276
5.2.4 Radix–2 FFT for Parallel Programming ... 281
5.2.5 Computational Cost of Radix–2 FFT ... 284
5.3 Other Fast Fourier Transforms ... 288
5.3.1 Chinese Remainder Theorem ... 288
5.3.2 Fast Algorithms for DFT of Composite Length ... 290
5.3.3 Radix–4 FFT and Split–Radix FFT ... 297
5.3.4 Rader FFT and Bluestein FFT ... 302
5.3.5 Multidimensional FFT ... 310
5.4 Sparse FFT ... 315
5.4.1 Single-Frequency Recovery ... 316
5.4.2 Recovery of Vectors with One Frequency Band ... 318
5.4.3 Recovery of Sparse Fourier Vectors ... 321
5.5 Numerical Stability of FFT ... 328

6 Chebyshev Methods and Fast DCT Algorithms 339
6.1 Chebyshev Polynomials and Chebyshev Series 339
6.1.1 Chebyshev Polynomials 340
6.1.2 Chebyshev Series 346
6.2 Fast Evaluation of Polynomials 354
6.2.1 Horner Scheme and Clenshaw Algorithm 354
6.2.2 Polynomial Evaluation and Interpolation at Chebyshev Points 357
6.2.3 Fast Evaluation of Polynomial Products 363
6.3 Fast DCT Algorithms 367
6.3.1 Fast DCT Algorithms via FFT 368
6.3.2 Fast DCT Algorithms via Orthogonal Matrix Factorizations 372
6.3.3 Fast Two-Dimensional DCT Algorithms 382
6.4 Interpolation and Quadrature Using Chebyshev Expansions 384
6.4.1 Interpolation at Chebyshev Extreme Points 384
6.4.2 Clenshaw–Curtis Quadrature 393
6.5 Discrete Polynomial Transforms 400
6.5.1 Orthogonal Polynomials 400
6.5.2 Fast Evaluation of Orthogonal Expansions 403

7 Fast Fourier Transforms for Nonequispaced Data 413
7.1 Nonequispaced Data Either in Space or Frequency Domain 413
7.2 Approximation Errors for Special Window Functions 421
7.3 Nonequispaced Data in Space and Frequency Domain 439
7.4 Nonequispaced Fast Trigonometric Transforms 442
7.5 Fast Summation at Nonequispaced Knots 448
7.6 Inverse Nonequispaced Discrete Transforms 455
7.6.1 Direct Methods for Inverse NDCT and Inverse NDFT ... 455
7.6.2 Iterative Methods for Inverse NDFT 461

8 High-Dimensional FFT 465
8.1 Fourier Partial Sums of Smooth Multivariate Functions 466
8.2 Fast Evaluation of Multivariate Trigonometric Polynomials 471
8.2.1 Rank-1 Lattices 472
8.2.2 Evaluation of Trigonometric Polynomials on Rank-1 Lattice 474
8.2.3 Evaluation of the Fourier Coefficients 475
8.3 Efficient Function Approximation on Rank-1 Lattices 478
8.4 Reconstructing Rank-1 Lattices 481
8.5 Multiple Rank-1 Lattices 486

9 Numerical Applications of DFT 493
9.1 Cardinal Interpolation by Translates 493
9.1.1 Cardinal Lagrange Function 497
9.1.2 Computation of Fourier Transforms 507

9.2 Periodic Interpolation by Translates 512
9.2.1 Periodic Lagrange Function 513
9.2.2 Computation of Fourier Coefficients 519
9.3 Quadrature of Periodic Functions 523
9.4 Accelerating Convergence of Fourier Series 529
9.4.1 Krylov–Lanczos Method 530
9.4.2 Fourier Extension 534
9.5 Fast Poisson Solvers 539
9.6 Spherical Fourier Transforms 551
9.6.1 Discrete Spherical Fourier Transforms 555
9.6.2 Fast Spherical Fourier Transforms 555
9.6.3 Fast Spherical Fourier Transforms for Nonequispaced Data 558
9.6.4 Fast Quadrature and Approximation on $\mathbb{S}^2$ 562

10 Prony Method for Reconstruction of Structured Functions 567
10.1 Prony Method 567
10.2 Recovery of Exponential Sums 574
10.2.1 MUSIC and Approximate Prony Method 576
10.2.2 ESPRIT 580
10.2.3 ESPIRA 586
10.3 Stability of Exponentials 588
10.4 Recovery of Structured Functions 602
10.4.1 Recovery from Fourier Data 603
10.4.2 Recovery from Function Samples 608
10.5 Phase Reconstruction 614

A List of Symbols and Abbreviations 621
A.1 Table of Some Fourier Series 630
A.2 Table of Some Chebyshev Series 631
A.3 Table of Some Fourier Transforms 632
A.4 Table of Some Discrete Fourier Transforms 633
A.5 Table of Some Fourier Transforms of Tempered Distributions 634

References 635

Index 653

Applied and Numerical Harmonic Analysis (106 Volumes) 661

Chapter 1
Fourier Series

Chapter 1 covers the classical theory of Fourier series of 2π-periodic functions. In the introductory Sect. 1.1, we sketch Fourier's theory on heat propagation. Section 1.2 introduces some basic notions such as Fourier coefficients and Fourier series of a 2π-periodic function. The convolution of 2π-periodic functions is handled in Sect. 1.3. Section 1.4 presents the main results on the pointwise and uniform convergence of Fourier series. For a 2π-periodic, piecewise continuously differentiable function f, a complete proof of the important convergence theorem of Dirichlet–Jordan is given. Further we describe the Gibbs phenomenon for partial sums of the Fourier series of f near a jump discontinuity. Finally, in Sect. 1.5, we apply Fourier series in digital signal processing and describe the linear filtering of discrete signals.

1.1 Fourier's Solution of Laplace Equation

In 1804, the French mathematician and Egyptologist Jean Baptiste Joseph Fourier (1768–1830) began his studies on the heat propagation in solid bodies. In 1807, he finished a first paper about heat propagation. He discovered the fundamental partial differential equation of heat propagation and developed a new method to solve this equation. The mathematical core of Fourier's idea was that each periodic function can be well approximated by a linear combination of sine and cosine terms. This theory contradicted the previous views on functions and was met with resistance by some members of the French Academy of Sciences, so that a publication was initially prevented. Later, Fourier presented these results in the famous book *The Analytical Theory of Heat* published firstly 1822 in French, cf. [149]. For an image of Fourier, see Fig. 1.1 (Image source: https://commons.wikimedia.org/wiki/File:Joseph_Fourier.jpg).

G. Plonka et al., *Numerical Fourier Analysis*, Applied and Numerical Harmonic Analysis, https://doi.org/10.1007/978-3-031-35005-4_1

Fig. 1.1 The mathematician and Egyptologist Jean Baptiste Joseph Fourier (1768–1830)

In the following, we describe Fourier's idea by a simple example. We consider the open unit disk $\Omega = \{(x, y) \in \mathbb{R}^2 : x^2 + y^2 < 1\}$ with the boundary $\Gamma = \{(x, y) \in \mathbb{R}^2 : x^2 + y^2 = 1\}$. Let $v(x, y, t)$ denote the temperature at the point $(x, y) \in \Omega$ and the time $t \geq 0$. For physical reasons, the temperature fulfills the *heat equation*:

$$\frac{\partial^2 v}{\partial x^2} + \frac{\partial^2 v}{\partial y^2} = c\,\frac{\partial v}{\partial t}, \quad (x, y) \in \Omega,\ t > 0$$

with some constant $c > 0$. At steady state, the temperature is independent of the time such that $v(x, y, t) = v(x, y)$ satisfies the *Laplace equation*:

$$\frac{\partial^2 v}{\partial x^2} + \frac{\partial^2 v}{\partial y^2} = 0, \quad (x, y) \in \Omega.$$

What is the temperature $v(x, y)$ at any point $(x, y) \in \Omega$, if the temperature at each point of the boundary Γ is known?

Using polar coordinates

$$x = r\,\cos\varphi\,,\ y = r\,\sin\varphi, \quad 0 < r < 1,\ 0 \leq \varphi < 2\pi,$$

we obtain for the temperature $u(r, \varphi) := v(r\,\cos\varphi, r\,\sin\varphi)$ by chain rule

$$\frac{\partial^2 v}{\partial x^2} + \frac{\partial^2 v}{\partial y^2} = \frac{\partial^2 u}{\partial r^2} + \frac{1}{r}\,\frac{\partial u}{\partial r} + \frac{1}{r^2}\,\frac{\partial^2 u}{\partial \varphi^2} = 0\,.$$

If we extend the variable φ periodically to the real line $\mathbb{R}$, then $u(r, \varphi)$ is 2π-periodic with respect to φ and fulfills

$$r^2\,\frac{\partial^2 u}{\partial r^2} + r\,\frac{\partial u}{\partial r} = -\frac{\partial^2 u}{\partial \varphi^2}, \quad 0 < r < 1,\ \varphi \in \mathbb{R}. \tag{1.1}$$

Since the temperature at the boundary Γ is given, we know the boundary condition:

$$u(1, \varphi) = f(\varphi), \quad \varphi \in \mathbb{R}, \tag{1.2}$$

where f is a given continuously differentiable, 2π-periodic function. Applying *separation of variables*, we seek nontrivial solutions of (1.1) of the form $u(r, \varphi) = p(r)\, q(\varphi)$, where p is bounded on $(0, 1)$ and q is 2π-periodic. From (1.1) it follows

$$\big(r^2\, p''(r) + r\, p'(r)\big)\, q(\varphi) = -p(r)\, q''(\varphi)$$

and hence

$$\frac{r^2\, p''(r) + r\, p'(r)}{p(r)} = -\frac{q''(\varphi)}{q(\varphi)} \,. \tag{1.3}$$

The variables r and φ can be independently chosen. If φ is fixed and r varies, then the left-hand side of (1.3) is a constant. Analogously, if r is fixed and φ varies, then the right-hand side of (1.3) is a constant. Let λ be the common value of both sides. Then, we obtain two linear differential equations:

$$r^2\, p''(r) + r\, p'(r) - \lambda\, p(r) = 0 \,, \tag{1.4}$$

$$q''(\varphi) + \lambda\, q(\varphi) = 0 \,. \tag{1.5}$$

Since nontrivial solutions of (1.5) must have the period 2π, we obtain the solutions $\frac{a_0}{2}$ for $\lambda = 0$ and $a_n \cos(n\varphi) + b_n \sin(n\varphi)$ for $\lambda = n^2$, $n \in \mathbb{N}$, where a_0, a_n, and b_n with $n \in \mathbb{N}$ are real constants. For $\lambda = 0$, the linear differential equation (1.4) has the linearly independent solutions 1 and $\ln r$, where only 1 is bounded on $(0, 1)$. For $\lambda = n^2$, Eq. (1.4) has the linearly independent solutions r^n and r^{-n}, where only r^n is bounded on $(0, 1)$. Thus, we see that $\frac{a_0}{2}$ and $r^n\, (a_n \cos(n\varphi) + b_n \sin(n\varphi))$, $n \in \mathbb{N}$, are the special solutions of the Laplace equation (1.1). If u_1 and u_2 are solutions of the linear equation (1.1), then $u_1 + u_2$ is a solution of (1.1) too. Using the *superposition principle*, we obtain a formal solution of (1.1) of the form

$$u(r, \varphi) = \frac{a_0}{2} + \sum_{n=1}^{\infty} r^n \big(a_n \cos(n\varphi) + b_n \sin(n\varphi)\big) \,. \tag{1.6}$$

By the boundary condition (1.2), the coefficients a_0, a_n, and b_n with $n \in \mathbb{N}$ must be chosen so that

$$u(1, \varphi) = \frac{a_0}{2} + \sum_{n=1}^{\infty} \big(a_n \cos(n\varphi) + b_n \sin(n\varphi)\big) = f(\varphi), \quad \varphi \in \mathbb{R}. \tag{1.7}$$

Fourier conjectured that this could be done for an arbitrary 2π-periodic function f. We will see that this is only the case, if f fulfills some additional conditions. As shown in the next section, from (1.7), it follows that

$$a_n = \frac{1}{\pi} \int_0^{2\pi} f(\psi)\, \cos(n\psi)\, \mathrm{d}\psi, \qquad n \in \mathbb{N}_0, \tag{1.8}$$

$$b_n = \frac{1}{\pi} \int_0^{2\pi} f(\psi)\, \sin(n\psi)\, \mathrm{d}\psi, \qquad n \in \mathbb{N}. \tag{1.9}$$

By assumption, f is bounded on $\mathbb{R}$, i.e., $|f(\psi)| \le M$. Thus, we obtain that

$$|a_n| \le \frac{1}{\pi} \int_0^{2\pi} |f(\psi)|\, \mathrm{d}\psi \le 2M, \qquad n \in \mathbb{N}_0.$$

Analogously, it holds $|b_n| \le 2M$ for all $n \in \mathbb{N}$.

Now we have to show that the constructed function (1.6) with the coefficients (1.8) and (1.9) is really a solution of (1.1) which fulfills the boundary condition (1.2). Since the 2π-periodic function f is continuously differentiable, we will see by Theorem 1.37 that

$$\sum_{n=1}^{\infty} \big(|a_n| + |b_n|\big) < \infty\,.$$

Introducing $u_n(r, \varphi) := r^n \big(a_n\, \cos(n\varphi) + b_n\, \sin(n\varphi)\big)$, we can estimate

$$|u_n(r, \varphi)| \le |a_n| + |b_n|, \qquad (r, \varphi) \in [0,\ 1] \times \mathbb{R}.$$

From Weierstrass criterion for uniform convergence, it follows that the series $\frac{a_0}{2} + \sum_{n=1}^{\infty} u_n$ converges uniformly on $[0,\ 1] \times \mathbb{R}$. Since each term u_n is continuous on $[0,\ 1] \times \mathbb{R}$, the sum u of this uniformly convergent series is continuous on $[0,\ 1] \times \mathbb{R}$, too. Note that the temperature in the origin of the closed unit disk is equal to the mean value $\frac{a_0}{2} = \frac{1}{2\pi} \int_0^{2\pi} f(\psi)\, \mathrm{d}\psi$ of the temperature f at the boundary.

Now we show that u fulfills the Laplace equation in $[0,\ 1) \times \mathbb{R}$. Let $0 < r_0 < 1$ be arbitrarily fixed. By

$$\frac{\partial^k}{\partial \varphi^k}\, u_n(r, \varphi) = r^n\, n^k \left(a_n\, \cos\left(n\varphi + \frac{k\pi}{2}\right) + b_n\, \sin(n\varphi + \frac{k\pi}{2})\right)$$

for arbitrary $k \in \mathbb{N}$, we obtain

$$\Big|\frac{\partial^k}{\partial \varphi^k}\, u_n(r, \varphi)\Big| \le 4\, r^n\, n^k\, M \le 4\, r_0^n\, n^k\, M$$

for $0 \le r \le r_0$. The series $4\, M \sum_{n=1}^{\infty} r_0^n\, n^k$ is convergent. By the Weierstrass criterion, $\sum_{n=1}^{\infty} \frac{\partial^k}{\partial \varphi^k}\, u_n$ is uniformly convergent on $[0,\ r_0] \times \mathbb{R}$. Consequently, $\frac{\partial^k}{\partial \varphi^k}\, u$ exists and

$$\frac{\partial^k}{\partial \varphi^k} u = \sum_{n=1}^{\infty} \frac{\partial^k}{\partial \varphi^k} u_n .$$

Analogously, one can show that $\frac{\partial^k}{\partial r^k} u$ exists and

$$\frac{\partial^k}{\partial r^k} u = \sum_{n=1}^{\infty} \frac{\partial^k}{\partial r^k} u_n .$$

Since all u_n are solutions of the Laplace equation (1.1) in $[0,\ 1) \times \mathbb{R}$, it follows by term by term differentiation that u is also a solution of (1.1) in $[0,\ 1) \times \mathbb{R}$.

Finally, we simplify the representation of the solution (1.6) with the coefficients (1.8) and (1.9). Since the series in (1.6) converges uniformly on $[0,\ 1] \times \mathbb{R}$, we can change the order of summation and integration such that

$$u(r, \varphi) = \frac{1}{\pi} \int_0^{2\pi} f(\psi) \left(\frac{1}{2} + \sum_{n=1}^{\infty} r^n \cos\left(n(\varphi - \psi)\right) \right) \mathrm{d}\psi .$$

Taking the real part of the geometric series

$$1 + \sum_{n=1}^{\infty} r^n \, \mathrm{e}^{\mathrm{i}n\theta} = \frac{1}{1 - r\mathrm{e}^{\mathrm{i}\theta}} .$$

it follows

$$1 + \sum_{n=1}^{\infty} r^n \cos(n\theta) = \frac{1 - r \cos\theta}{1 + r^2 - 2r \cos\theta}$$

and hence

$$\frac{1}{2} + \sum_{n=1}^{\infty} r^n \cos(n\theta) = \frac{1}{2} \, \frac{1 - r^2}{1 + r^2 - 2r \cos\theta} .$$

Thus, for $0 \le r < 1$ and $\varphi \in \mathbb{R}$, the solution of (1.6) can be represented as *Poisson integral*:

$$u(r, \varphi) = \frac{1}{2\pi} \int_0^{2\pi} f(\psi) \, \frac{1 - r^2}{1 + r^2 - 2r \cos(\varphi - \psi)} \, \mathrm{d}\psi .$$

1.2 Fourier Coefficients and Fourier Series

A complex-valued function $f : \mathbb{R} \to \mathbb{C}$ is 2π*-periodic* or *periodic with period* 2π, if $f(x + 2\pi) = f(x)$ for all $x \in \mathbb{R}$. In the following, we identify any 2π-periodic function $f : \mathbb{R} \to \mathbb{C}$ with the corresponding function $f : \mathbb{T} \to \mathbb{C}$ defined on the *torus* $\mathbb{T}$ of length 2π. The torus $\mathbb{T}$ can be considered as quotient space $\mathbb{R}/(2\pi\mathbb{Z})$ or its representatives, e.g., the interval $[0, 2\pi]$ with identified endpoints 0 and 2π. For short, one can also geometrically think of the unit circle with circumference 2π. Typical examples of 2π-periodic functions are 1, $\cos(n\cdot)$, $\sin(n\cdot)$ for each angular frequency $n \in \mathbb{N}$ and the *complex exponentials* $\mathrm{e}^{\mathrm{i}k\cdot}$ for each $k \in \mathbb{Z}$.

By $C(\mathbb{T})$ we denote the Banach space of all continuous functions $f : \mathbb{T} \to \mathbb{C}$ with the norm

$$\|f\|_{C(\mathbb{T})} := \max_{x\in\mathbb{T}} |f(x)|$$

and by $C^r(\mathbb{T})$, $r \in \mathbb{N}$ the Banach space of r-times continuously differentiable functions $f : \mathbb{T} \to \mathbb{C}$ with the norm

$$\|f\|_{C^r(\mathbb{T})} := \|f\|_{C(\mathbb{T})} + \|f^{(r)}\|_{C(\mathbb{T})} .$$

Clearly, we have $C^r(\mathbb{T}) \subset C^s(\mathbb{T})$ for $r > s$.

Let $L_p(\mathbb{T})$, $1 \le p \le \infty$ be the Banach space of measurable functions $f : \mathbb{T} \to \mathbb{C}$ with finite norm:

$$\|f\|_{L_p(\mathbb{T})} := \Big(\frac{1}{2\pi}\int_{-\pi}^{\pi} |f(x)|^p \,\mathrm{d}x\Big)^{1/p}, \quad 1 \le p < \infty ,$$

$$\|f\|_{L_\infty(\mathbb{T})} := \mathrm{ess\ sup}\,\{|f(x)| : x \in \mathbb{T}\} ,$$

where we identify almost equal functions. If a 2π-periodic function f is integrable on $[-\pi, \pi]$, then we have

$$\int_{-\pi}^{\pi} f(x)\,\mathrm{d}x = \int_{-\pi+a}^{\pi+a} f(x)\,\mathrm{d}x$$

for all $a \in \mathbb{R}$ so that we can integrate over any interval of length 2π.

Using Hölder's inequality it can be shown that the spaces $L_p(\mathbb{T})$ for $1 \le p \le \infty$ are continuously embedded as

$$L_1(\mathbb{T}) \supset L_2(\mathbb{T}) \supset \ldots \supset L_\infty(\mathbb{T}).$$

In the following we are mainly interested in the Hilbert space $L_2(\mathbb{T})$ consisting of all absolutely square-integrable functions $f : \mathbb{T} \to \mathbb{C}$ with inner product and norm:

$$\langle f, g\rangle_{L_2(\mathbb{T})} := \frac{1}{2\pi}\int_{-\pi}^{\pi} f(x)\,\overline{g(x)}\,\mathrm{d}x\,, \quad \|f\|_{L_2(\mathbb{T})} := \Big(\frac{1}{2\pi}\int_{-\pi}^{\pi} |f(x)|^2\,\mathrm{d}x\Big)^{1/2}.$$

If it is clear from the context which inner product or norm is addressed, we abbreviate $\langle f, g\rangle := \langle f, g\rangle_{L_2(\mathbb{T})}$ and $\|f\| := \|f\|_{L_2(\mathbb{T})}$. For all $f,\, g \in L_2(\mathbb{T})$, it holds the Cauchy–Schwarz inequality:

$$|\langle f, g\rangle_{L_2(\mathbb{T})}| \le \|f\|_{L_2(\mathbb{T})}\,\|g\|_{L_2(\mathbb{T})}\,.$$

Theorem 1.1 *The set of complex exponentials*

$$\big\{\mathrm{e}^{\mathrm{i}k\cdot} = \cos(k\cdot) + \mathrm{i}\,\sin(k\cdot) : k \in \mathbb{Z}\big\} \tag{1.10}$$

forms an orthonormal basis of $L_2(\mathbb{T})$.

Proof

1. By definition, an orthonormal basis is a complete orthonormal system. First we show the orthonormality of the complex exponentials in (1.10). We have:

$$\langle \mathrm{e}^{\mathrm{i}k\cdot}, \mathrm{e}^{\mathrm{i}j\cdot}\rangle = \frac{1}{2\pi}\int_{-\pi}^{\pi} \mathrm{e}^{\mathrm{i}(k-j)x}\,\mathrm{d}x\,,$$

which implies for integers $k = j$

$$\langle \mathrm{e}^{\mathrm{i}k\cdot},\, \mathrm{e}^{\mathrm{i}k\cdot}\rangle = \frac{1}{2\pi}\int_{-\pi}^{\pi} 1\,\mathrm{d}x = 1.$$

On the other hand, we obtain for distinct integers j, k:

$$\begin{aligned}\langle \mathrm{e}^{\mathrm{i}k\cdot},\, \mathrm{e}^{\mathrm{i}j\cdot}\rangle &= \frac{1}{2\pi\mathrm{i}(k-j)}\big(\mathrm{e}^{\pi\mathrm{i}(k-j)} - \mathrm{e}^{-\pi\mathrm{i}(k-j)}\big)\\ &= \frac{2\mathrm{i}\,\sin\pi(k-j)}{2\pi\mathrm{i}(k-j)} = 0\,.\end{aligned}$$

2. Now we prove the completeness of the set (1.10). We have to show that $\big\langle f, \mathrm{e}^{\mathrm{i}k\cdot}\big\rangle = 0$ for all $k \in \mathbb{Z}$ implies $f = 0$.
First we consider a continuous function $f \in C(\mathbb{T})$ having $\langle f, \mathrm{e}^{\mathrm{i}k\cdot}\rangle = 0$ for all $k \in \mathbb{Z}$. Let us denote by

$$\mathscr{T}_n := \Big\{\sum_{k=-n}^{n} c_k \mathrm{e}^{\mathrm{i}k\cdot} : c_k \in \mathbb{C}\Big\} \tag{1.11}$$

the *space of all trigonometric polynomials up to degree* n. By the approximation theorem of Weierstrass (see Theorem 1.21,) there exists for any function $f \in C(\mathbb{T})$ a sequence $(p_n)_{n\in\mathbb{N}_0}$ of trigonometric polynomials $p_n \in \mathscr{T}_n$, which converges uniformly to f, i.e.,

$$\|f - p_n\|_{C(\mathbb{T})} = \max_{x\in\mathbb{T}} \big|f(x) - p_n(x)\big| \to 0 \quad \text{for } n \to \infty\,.$$

By assumption we have:

$$\langle f, p_n\rangle = \langle f, \sum_{k=-n}^{n} c_k\, \mathrm{e}^{\mathrm{i}k\cdot}\rangle = \sum_{k=-n}^{n} \bar{c}_k\, \langle f, \mathrm{e}^{\mathrm{i}k\cdot}\rangle = 0\,.$$

Hence we conclude:

$$\|f\|^2 = \langle f,\, f\rangle - \langle f, p_n\rangle = \langle f, f - p_n\rangle \to 0 \tag{1.12}$$

as $n \to \infty$, so that $f = 0$.

3. Now let $f \in L_2(\mathbb{T})$ with $\langle f, \mathrm{e}^{\mathrm{i}k\cdot}\rangle = 0$ for all $k \in \mathbb{Z}$ be given. Then

$$h(x) := \int_0^x f(t)\,\mathrm{d}t, \quad x \in [0, 2\pi),$$

is an absolutely continuous function satisfying $h'(x) = f(x)$ almost everywhere. We have further $h(0) = h(2\pi) = 0$. For $k \in \mathbb{Z}\setminus\{0\}$, we obtain:

$$\begin{aligned}\langle h, \mathrm{e}^{\mathrm{i}k\cdot}\rangle &= \frac{1}{2\pi}\int_0^{2\pi} h(x)\,\mathrm{e}^{-\mathrm{i}kx}\,\mathrm{d}x\\ &= -\frac{1}{2\pi\mathrm{i}k}\,h(x)\,\mathrm{e}^{-\mathrm{i}kx}\Big|_0^{2\pi} + \frac{1}{2\pi\mathrm{i}k}\int_0^{2\pi} \underbrace{h'(x)}_{=f(x)}\,\mathrm{e}^{-\mathrm{i}kx}\,\mathrm{d}x = \frac{1}{\mathrm{i}k}\,\langle f, \mathrm{e}^{\mathrm{i}k\cdot}\rangle = 0\,.\end{aligned}$$

Hence, the 2π-periodically continued continuous function $h - \langle h, 1\rangle$ fulfills $\big\langle h - \langle h, 1\rangle, \mathrm{e}^{\mathrm{i}k\cdot}\big\rangle = 0$ for all $k \in \mathbb{Z}$. Using the first part of this proof, we obtain $h = \langle h, 1\rangle = \text{const}$. Since $f(x) = h'(x) = 0$ almost everywhere, this yields the assertion. ■

Once we have an orthonormal basis of a Hilbert space, we can represent its elements with respect to this basis. Let us consider the finite sum:

$$S_n f := \sum_{k=-n}^{n} c_k(f)\,\mathrm{e}^{\mathrm{i}k\cdot} \in \mathcal{T}_n\,, \quad c_k(f) := \big\langle f, \mathrm{e}^{\mathrm{i}k\cdot}\big\rangle = \frac{1}{2\pi}\int_{-\pi}^{\pi} f(x)\,\mathrm{e}^{-\mathrm{i}kx}\,\mathrm{d}x\,,$$

called *nth Fourier partial sum* of f with the *Fourier coefficients* $c_k(f)$. By definition $S_n : L_2(\mathbb{T}) \to L_2(\mathbb{T})$ is a linear operator which possesses the following important approximation property.

Lemma 1.2 *The Fourier partial sum operator $S_n : L_2(\mathbb{T}) \to L_2(\mathbb{T})$ is an orthogonal projector onto $\mathcal{T}_n$, i.e.,*

$$\|f - S_n f\| = \min\{\|f - p\| : p \in \mathcal{T}_n\}$$

for arbitrary $f \in L_2(\mathbb{T})$. In particular, it holds:

$$\|f - S_n f\|^2 = \|f\|^2 - \sum_{k=-n}^{n} |c_k(f)|^2. \tag{1.13}$$

Proof

1. For each trigonometric polynomial

$$p = \sum_{k=-n}^{n} c_k \, \mathrm{e}^{\mathrm{i}k\cdot} \tag{1.14}$$

with arbitrary $c_k \in \mathbb{C}$ and all $f \in L_2(\mathbb{T})$, we have:

$$\begin{aligned} \|f - p\|^2 &= \|f\|^2 - \langle f, p\rangle - \langle p, f\rangle + \|p\|^2 \\ &= \|f\|^2 + \sum_{k=-n}^{n} \big(-\overline{c_k}\, c_k(f) - c_k\, \overline{c_k(f)} + |c_k|^2\big) \\ &= \|f\|^2 - \sum_{k=-n}^{n} |c_k(f)|^2 + \sum_{k=-n}^{n} |c_k - c_k(f)|^2. \end{aligned}$$

Thus,

$$\|f - p\|^2 \ge \|f\|^2 - \sum_{k=-n}^{n} |c_k(f)|^2,$$

where equality holds only in the case $c_k = c_k(f)$, $k = -n, \ldots, n$, i.e., if and only if $p = S_n f$.

2. For $p \in \mathcal{T}_n$ of the form (1.14), the corresponding Fourier coefficients are $c_k(p) = c_k$ for $k = -n, \ldots, n$ and $c_k(p) = 0$ for all $|k| > n$. Thus, we have $S_n p = p$ and $S_n(S_n f) = S_n f$ for arbitrary $f \in L_2(\mathbb{T})$. Hence, $S_n : L_2(\mathbb{T}) \to L_2(\mathbb{T})$ is a projection onto $\mathcal{T}_n$. By

$$\langle S_n f, g\rangle = \sum_{k=-n}^{n} c_k(f)\, \overline{c_k(g)} = \langle f, S_n g\rangle$$

for all $f, g \in L_2(\mathbb{T})$, the Fourier partial sum operator S_n is self-adjoint, i.e., S_n is an orthogonal projection. Moreover, S_n has the operator norm $\|S_n\|_{L_2(\mathbb{T})\to L_2(\mathbb{T})} = 1$. ■

As an immediate consequence of Lemma 1.2, we obtain the following

Theorem 1.3 *Every function $f \in L_2(\mathbb{T})$ has a unique representation of the form*

$$f = \sum_{k\in\mathbb{Z}} c_k(f)\, \mathrm{e}^{\mathrm{i}k\cdot}, \quad c_k(f) := \langle f, \mathrm{e}^{\mathrm{i}k\cdot}\rangle = \frac{1}{2\pi}\int_{-\pi}^{\pi} f(x)\, \mathrm{e}^{-\mathrm{i}kx}\, \mathrm{d}x\,, \tag{1.15}$$

where the series $(S_n f)_{n=0}^{\infty}$ converges in $L_2(\mathbb{T})$ to f, i.e.,

$$\lim_{n\to\infty} \|S_n f - f\| = 0\,.$$

Further the Parseval equality *is fulfilled:*

$$\|f\|^2 = \sum_{k\in\mathbb{Z}} |\langle f, \mathrm{e}^{\mathrm{i}k\cdot}\rangle|^2 = \sum_{k\in\mathbb{Z}} |c_k(f)|^2 < \infty\,. \tag{1.16}$$

Proof By Lemma 1.2, we know that for each $n \in \mathbb{N}_0$

$$\|S_n f\|^2 = \sum_{k=-n}^{n} |c_k(f)|^2 \le \|f\|^2 < \infty\,.$$

For $n \to \infty$, we obtain *Bessel's inequality*

$$\sum_{k=-\infty}^{\infty} |c_k(f)|^2 \le \|f\|^2\,.$$

Consequently, for arbitrary $\varepsilon > 0$, there exists an index $N(\varepsilon) \in \mathbb{N}$ such that

$$\sum_{|k|>N(\varepsilon)} |c_k(f)|^2 < \varepsilon\,.$$

For $m > n \ge N(\varepsilon)$ we obtain:

$$\|S_m f - S_n f\|^2 = \left(\sum_{k=-m}^{-n-1} + \sum_{k=n+1}^{m}\right) |c_k(f)|^2 \le \sum_{|k|>N(\varepsilon)} |c_k(f)|^2 < \varepsilon\,.$$

Hence $(S_n f)_{n=0}^{\infty}$ is a Cauchy sequence. In the Hilbert space $L_2(\mathbb{T})$, each Cauchy sequence is convergent. Assume that $\lim_{n\to\infty} S_n f = g$ with $g \in L_2(\mathbb{T})$. Since

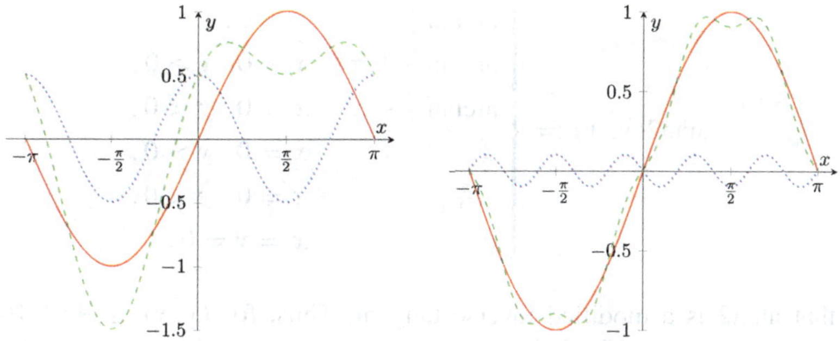

Fig. 1.2 Two 2π-periodic functions $\sin x + \frac{1}{2}\cos(2x)$ (left) and $\sin x - \frac{1}{10}\sin(4x)$ as superpositions of sine and cosine functions

$$\langle g,\, \mathrm{e}^{\mathrm{i}k\cdot}\rangle = \lim_{n\to\infty} \langle S_n f,\, \mathrm{e}^{\mathrm{i}k\cdot}\rangle = \lim_{n\to\infty} \langle f,\, S_n \mathrm{e}^{\mathrm{i}k\cdot}\rangle = \langle f,\, \mathrm{e}^{\mathrm{i}k\cdot}\rangle$$

for all $k \in \mathbb{Z}$, we conclude by Theorem 1.1 that $f = g$. Letting $n \to \infty$ in (1.13), we obtain the Parseval equality (1.16). ■

The representation (1.15) is the so-called Fourier series of f. Figure 1.2 shows 2π-periodic functions as superposition of two 2π-periodic functions.

Clearly, the partial sums of the Fourier series are the Fourier partial sums. The constant term $c_0(f) = \frac{1}{2\pi}\int_{-\pi}^{\pi} f(x)\,\mathrm{d}x$ in the Fourier series of f is the *mean value* of f.

Remark 1.4 For fixed $L > 0$, a function $f : \mathbb{R} \to \mathbb{C}$ is called *L-periodic*, if $f(x + L) = f(x)$ for all $x \in \mathbb{R}$. By substitution we see that the Fourier series of an L-periodic function f reads as follows:

$$f = \sum_{k\in\mathbb{Z}} c_k^{(L)}(f)\,\mathrm{e}^{2\pi\mathrm{i}k\cdot/L}\,, \quad c_k^{(L)}(f) := \frac{1}{L}\int_{-L/2}^{L/2} f(x)\,\mathrm{e}^{-2\pi\mathrm{i}kx/L}\,\mathrm{d}x\,. \tag{1.17}$$

□

In polar coordinates we can represent the Fourier coefficients in the form

$$c_k(f) = |c_k(f)|\,\mathrm{e}^{\mathrm{i}\,\varphi_k}, \quad \varphi_k := \mathrm{atan2}\left(\mathrm{Im}\; c_k(f),\, \mathrm{Re}\; c_k(f)\right), \tag{1.18}$$

where

$$\operatorname{atan2}(y,\,x) := \begin{cases} \arctan\frac{y}{x} & x>0\,,\\ \arctan\frac{y}{x}+\pi & x<0,\ y\ge 0\,,\\ \arctan\frac{y}{x}-\pi & x<0,\ y<0\,,\\ \frac{\pi}{2} & x=0,\ y>0\,,\\ -\frac{\pi}{2} & x=0,\ y<0\,,\\ 0 & x=y=0\,. \end{cases}$$

Note that atan2 is a modified inverse tangent. Thus, for $(x, y) \in \mathbb{R}^2 \setminus \{(0, 0)\}$, $\operatorname{atan2}(y, x) \in (-\pi, \pi]$ is defined as the angle between the vectors $(1, 0)^\top$ and $(x, y)^\top$. The sequence $\big(|c_k(f)|\big)_{k\in\mathbb{Z}}$ is called the *spectrum* or *modulus* of f and $(\varphi_k)_{k\in\mathbb{Z}}$ the *phase* of f.

For fixed $a \in \mathbb{R}$, the 2π-*periodic extension* of a function $f : [-\pi + a, \pi + a) \to \mathbb{C}$ to the whole line $\mathbb{R}$ is given by $f(x + 2\pi n) := f(x)$ for all $x \in [-\pi + a, \pi + a)$ and all $n \in \mathbb{Z}$. Often we have $a = 0$ or $a = \pi$.

Example 1.5 Consider the 2π-periodic extension of the real-valued function $f(x) = \mathrm{e}^{-x}$, $x \in (-\pi, \pi)$ with $f(\pm\pi) = \cosh\pi = \frac{1}{2}(\mathrm{e}^{-\pi} + \mathrm{e}^{\pi})$. Then, the Fourier coefficients $c_k(f)$ are given by

$$\begin{aligned} c_k(f) &= \frac{1}{2\pi}\int_{-\pi}^{\pi} \mathrm{e}^{-(1+\mathrm{i}k)x}\,\mathrm{d}x \\ &= -\frac{1}{2\pi\,(1+\mathrm{i}k)}\Big(\mathrm{e}^{-(1+\mathrm{i}k)\pi} - \mathrm{e}^{(1+\mathrm{i}k)\pi}\Big) = \frac{(-1)^k\,\sinh\pi}{(1+\mathrm{i}\,k)\,\pi}\,. \end{aligned}$$

Figure 1.3 shows both the 8th and 16th Fourier partial sums $S_8 f$ and $S_{16} f$. □

For $f \in L_2(\mathbb{T})$, it holds the Parseval equality (1.16). Thus, the Fourier coefficients $c_k(f)$ converge to zero as $|k| \to \infty$. Since

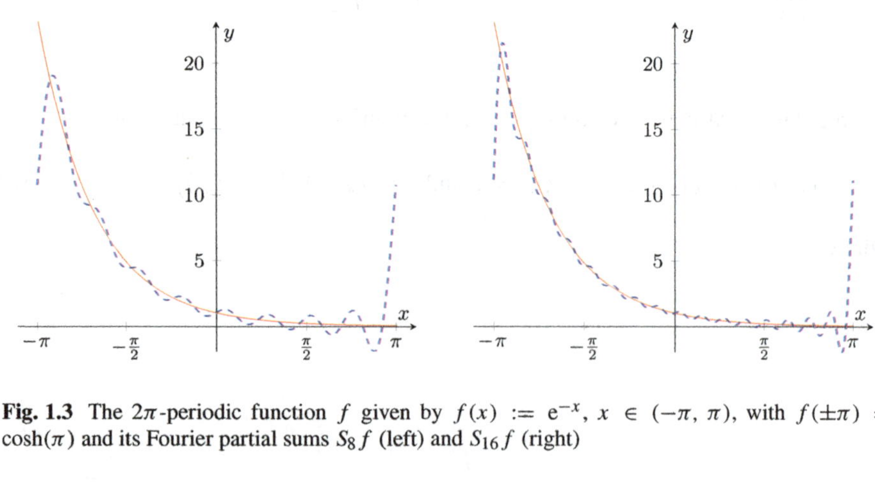

Fig. 1.3 The 2π-periodic function f given by $f(x) := \mathrm{e}^{-x}$, $x \in (-\pi, \pi)$, with $f(\pm\pi) = \cosh(\pi)$ and its Fourier partial sums $S_8 f$ (left) and $S_{16} f$ (right)

$$|c_k(f)| \le \frac{1}{2\pi} \int_{-\pi}^{\pi} |f(x)| \, \mathrm{d}x = \|f\|_{L_1(\mathbb{T})} ,$$

the integrals

$$c_k(f) = \frac{1}{2\pi} \int_{-\pi}^{\pi} f(x) \, \mathrm{e}^{-\mathrm{i}kx} \, \mathrm{d}x , \quad k \in \mathbb{Z}$$

also exist for all functions $f \in L_1(\mathbb{T})$, i.e., the Fourier coefficients are well-defined for any function of $L_1(\mathbb{T})$. The next lemma contains simple properties of Fourier coefficients.

Lemma 1.6 *The Fourier coefficients of $f, g \in L_1(\mathbb{T})$ have the following properties for all $k \in \mathbb{Z}$:*

1. Linearity*: For all $\alpha, \beta \in \mathbb{C}$,*

$$c_k(\alpha f + \beta g) = \alpha \, c_k(f) + \beta \, c_k(g) .$$

2. Translation–Modulation*: For all $x_0 \in [0, 2\pi)$ and $k_0 \in \mathbb{Z}$*

$$c_k\big(f(\cdot - x_0)\big) = \mathrm{e}^{-\mathrm{i}kx_0} \, c_k(f) ,$$

$$c_k(\mathrm{e}^{-\mathrm{i}k_0 \cdot} f) = c_{k+k_0}(f) .$$

In particular $|c_k(f(\cdot - x_0))| = |c_k(f)|$, i.e., translation does not change the spectrum of f.

3. Differentiation–Multiplication*: For absolute continuous functions $f \in L_1(\mathbb{T})$, i.e., $f' \in L_1(\mathbb{T})$, we have:*

$$c_k(f') = \mathrm{i} \, k \, c_k(f) .$$

Proof The first property follows directly from the linearity of the integral. The translation–modulation property can be seen as

$$\begin{aligned} c_k\big(f(\cdot - x_0)\big) &= \frac{1}{2\pi} \int_{-\pi}^{\pi} f(x - x_0) \, \mathrm{e}^{-\mathrm{i}kx} \, \mathrm{d}x \\ &= \frac{1}{2\pi} \int_{-\pi}^{\pi} f(y) \, \mathrm{e}^{-\mathrm{i}k(y+x_0)} \, \mathrm{d}y = \mathrm{e}^{-\mathrm{i}kx_0} \, c_k(f), \end{aligned}$$

and similarly for the modulation–translation property.

For the differentiation property, recall that an absolutely continuous function has a derivative almost everywhere. Then, we obtain by integration by parts:

$$\frac{1}{2\pi}\int_{-\pi}^{\pi} \mathrm{i}k\, f(x)\,\mathrm{e}^{-\mathrm{i}kx}\,\mathrm{d}x = \frac{1}{2\pi}\int_{-\pi}^{\pi} f'(x)\,\mathrm{e}^{-\mathrm{i}kx}\,\mathrm{d}x = c_k(f').$$

■

The complex Fourier series

$$f = \sum_{k\in\mathbb{Z}} c_k(f)\,\mathrm{e}^{\mathrm{i}k\cdot}$$

can be rewritten using Euler's formula $\mathrm{e}^{\mathrm{i}k\cdot} = \cos(k\cdot) + \mathrm{i}\,\sin(k\cdot)$ as

$$f = \frac{1}{2}\,a_0(f) + \sum_{k=1}^{\infty} \big(a_k(f)\,\cos(k\cdot) + b_k(f)\,\sin(k\cdot)\big)\,, \tag{1.19}$$

where

$$a_k(f) = c_k(f) + c_{-k}(f) = 2\,\langle f, \cos(k\cdot)\rangle\,, \quad k \in \mathbb{N}_0\,,$$
$$b_k(f) = \mathrm{i}\,\big(c_k(f) - c_{-k}(f)\big) = 2\,\langle f, \sin(k\cdot)\rangle\,, \quad k \in \mathbb{N}\,.$$

Consequently $\Big\{1, \sqrt{2}\,\cos(k\cdot)\colon k \in \mathbb{N}\Big\} \cup \Big\{\sqrt{2}\,\sin(k\cdot)\colon k \in \mathbb{N}\Big\}$ form also an orthonormal basis of $L_2(\mathbb{T})$. If $f : \mathbb{T} \to \mathbb{R}$ is a real-valued function, then $c_k(f) = \overline{c_{-k}(f)}$ and (1.19) is the *real Fourier series* of f. Using polar coordinates (1.18), the Fourier series of a real-valued function $f \in L_2(\mathbb{T})$ can be written in the form

$$f = \frac{1}{2}\,a_0(f) + \sum_{k=1}^{\infty} r_k\,\sin\Big(k\cdot + \frac{\pi}{2} + \varphi_k\Big).$$

with sine oscillations of amplitudes $r_k = 2\,|c_k(f)|$, angular frequencies k, and phase shifts $\frac{\pi}{2} + \varphi_k$. For even/odd functions, the Fourier series simplify to pure cosine/sine series.

Lemma 1.7 *If $f \in L_2(\mathbb{T})$ is* even, *i.e., $f(x) = f(-x)$ for all $x \in \mathbb{T}$, then $c_k(f) = c_{-k}(f)$ for all $k \in \mathbb{Z}$ and f can be represented as a* Fourier cosine series*:*

$$f = c_0(f) + 2\sum_{k=1}^{\infty} c_k(f)\,\cos(k\cdot) = \frac{1}{2}\,a_0(f) + \sum_{k=1}^{\infty} a_k(f)\,\cos(k\cdot)\,.$$

If $f \in L_2(\mathbb{T})$ is odd, *i.e., $f(x) = -f(-x)$ for all $x \in \mathbb{T}$, then $c_k(f) = -c_{-k}(f)$ for all $k \in \mathbb{Z}$ and f can be represented as a* Fourier sine series

$$f = 2\mathrm{i} \sum_{k=1}^{\infty} c_k(f) \sin(k\cdot) = \sum_{k=1}^{\infty} b_k(f) \sin(k\cdot).$$

The simple proof of Lemma 1.7 is left as an exercise.

Example 1.8 The 2π-periodic extension of the function $f(x) = x^2$, $x \in [-\pi, \pi)$ is even and has the Fourier cosine series:

$$\frac{\pi^2}{3} + 4 \sum_{k=1}^{\infty} \frac{(-1)^k}{k^2} \cos(k\cdot)\,.$$

□

Example 1.9 The 2π-periodic extension of the function $s(x) = \frac{\pi - x}{2\pi}$, $x \in (0, 2\pi)$, with $s(0) = 0$ is odd and has jump discontinuities at $2\pi k$, $k \in \mathbb{Z}$, of unit height. This so-called sawtooth function has the Fourier sine series:

$$\sum_{k=1}^{\infty} \frac{1}{\pi\, k} \sin(k\cdot)\,.$$

Figure 1.4 illustrates the corresponding Fourier partial sum $S_8 f$. Applying the Parseval equality (1.16), we obtain:

$$\sum_{k=1}^{\infty} \frac{1}{2\pi^2 k^2} = \|s\|^2 = \frac{1}{12}.$$

This implies

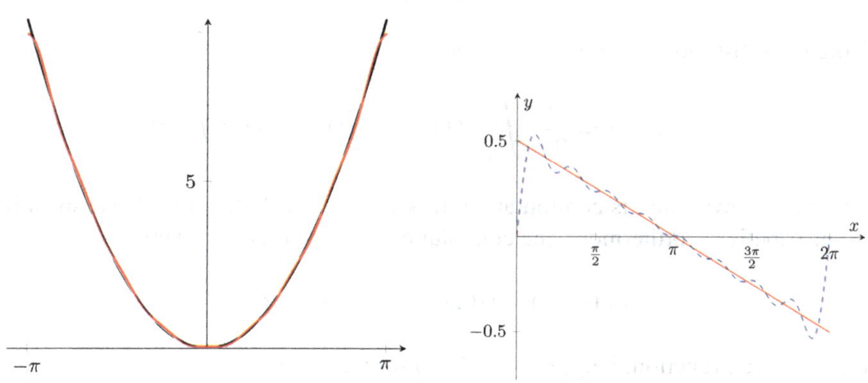

Fig. 1.4 The Fourier partial sums $S_8 f$ of the even 2π-periodic function f given by $f(x) := x^2$, $x \in [-\pi, \pi)$ (left) and of the odd 2π-periodic function f given by $f(x) = \frac{1}{2} - \frac{x}{2\pi}$, $x \in (0, 2\pi)$, with $f(0) = f(2\pi) = 0$ (right)

$$\sum_{k=1}^{\infty} \frac{1}{k^2} = \frac{\pi^2}{6}.$$

The last equation can be also obtained from the Fourier series in Example 1.8 by setting $x := \pi$ and assuming that the series converges in this point. □

Example 1.10 We consider the 2π-periodic extension of the *rectangular pulse function* $f : [-\pi, \pi) \to \mathbb{R}$ given by

$$f(x) = \begin{cases} 0 & x \in (-\pi, 0), \\ 1 & x \in (0, \pi) \end{cases}$$

and $f(-\pi) = f(0) = \frac{1}{2}$. The function $f - \frac{1}{2}$ is odd and the Fourier series of f reads

$$\frac{1}{2} + \sum_{n=1}^{\infty} \frac{2}{(2n-1)\pi} \sin\big((2n-1)\,\cdot\big).$$

□

1.3 Convolution of Periodic Functions

The *convolution* of two 2π-periodic functions $f, g \in L_1(\mathbb{T})$ is the function $h = f * g$ given by

$$h(x) := (f * g)(x) = \frac{1}{2\pi} \int_{-\pi}^{\pi} f(y)\, g(x-y)\, \mathrm{d}y\,.$$

Using the substitution $y = x - t$, we see

$$(f * g)(x) = \frac{1}{2\pi} \int_{-\pi}^{\pi} f(x-t)\, g(t)\, \mathrm{d}t = (g * f)(x)$$

so that the convolution is commutative. It is easy to check that it is also associative and distributive. Furthermore, the convolution is translation invariant:

$$(f(\cdot - t) * g)(x) = (f * g)(x - t)\,.$$

If g is an even function, i.e., $g(x) = g(-x)$ for all $x \in \mathbb{R}$, then

$$(f * g)(x) = \frac{1}{2\pi} \int_{-\pi}^{\pi} f(y)\, g(y - x)\, \mathrm{d}y.$$

Figure 1.5 shows the convolution of two special 2π-periodic functions.

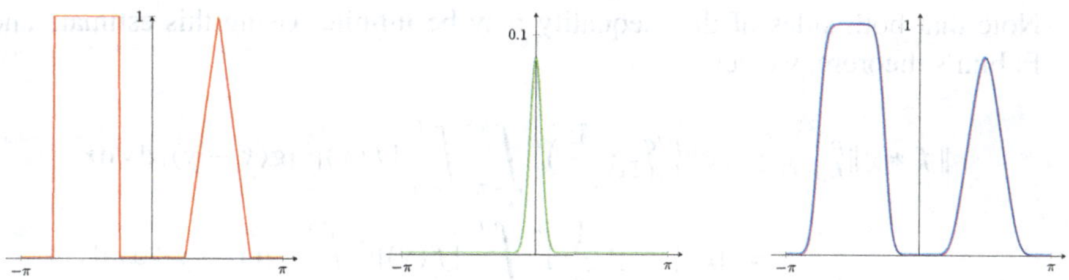

Fig. 1.5 Two 2π-periodic functions f (red) and g (green). right: The corresponding convolution $f * g$ (blue)

The following theorem shows that the convolution is well-defined for certain functions.

Theorem 1.11

1. *Let $f \in L_p(\mathbb{T})$, $1 \le p \le \infty$ and $g \in L_1(\mathbb{T})$ be given. Then, $f * g$ exists almost everywhere and $f * g \in L_p(\mathbb{T})$. Further we have the* Young inequality*:*

$$\|f * g\|_{L_p(\mathbb{T})} \le \|f\|_{L_p(\mathbb{T})}\|g\|_{L_1(\mathbb{T})}.$$

2. *Let $f \in L_p(\mathbb{T})$ and $g \in L_q(\mathbb{T})$, where $1 \le p, q \le \infty$ and $\frac{1}{p} + \frac{1}{q} = 1$. Then $(f * g)(x)$ exists for every $x \in \mathbb{T}$ and $f * g \in C(\mathbb{T})$. It holds:*

$$\|f * g\|_{C(\mathbb{T})} \le \|f\|_{L_p(\mathbb{T})}\|g\|_{L_q(\mathbb{T})}.$$

3. *Let $f \in L_p(\mathbb{T})$ and $g \in L_q(\mathbb{T})$, where $\frac{1}{p} + \frac{1}{q} = \frac{1}{r} + 1$, $1 \le p, q, r \le \infty$. Then $f * g$ exists almost everywhere and $f * g \in L_r(\mathbb{T})$. Further we have the* generalized Young inequality*:*

$$\|f * g\|_{L_r(\mathbb{T})} \le \|f\|_{L_p(\mathbb{T})}\|g\|_{L_q(\mathbb{T})}.$$

Proof

1. Let $p \in (1, \infty)$ and $\frac{1}{p} + \frac{1}{q} = 1$. Then, we obtain by Hölder's inequality:

$$\begin{aligned}
|(f * g)(x)| &\le \frac{1}{2\pi}\int_{-\pi}^{\pi} |f(y)|\, \underbrace{|g(x-y)|}_{=|g|^{1/p}\,|g|^{1/q}}\, \mathrm{d}y \\
&\le \Big(\frac{1}{2\pi}\int_{-\pi}^{\pi} |f(y)|^p |g(x-y)| \mathrm{d}y\Big)^{1/p} \Big(\frac{1}{2\pi}\int_{-\pi}^{\pi} |g(x-y)|\, \mathrm{d}x\Big)^{1/q} \\
&= \|g\|_{L_1(\mathbb{T})}^{1/q} \Big(\frac{1}{2\pi}\int_{-\pi}^{\pi} |f(y)|^p |g(x-y)| \mathrm{d}y\Big)^{1/p}.
\end{aligned}$$

Note that both sides of the inequality may be infinite. Using this estimate and Fubini's theorem, we get:

$$\begin{aligned}\|f * g\|^p_{L_p(\mathbb{T})} &\le \|g\|^{p/q}_{L_1(\mathbb{T})}\Big(\frac{1}{2\pi}\Big)^2 \int_{-\pi}^{\pi}\int_{-\pi}^{\pi} |f(y)|^p |g(x-y)|\,\mathrm{d}y\,\mathrm{d}x \\ &= \|g\|^{p/q}_{L_1(\mathbb{T})}\Big(\frac{1}{2\pi}\Big)^2 \int_{-\pi}^{\pi} |f(y)|^p \int_{-\pi}^{\pi} |g(x-y)|\,\mathrm{d}x\,\mathrm{d}y \\ &= \|g\|^{1+p/q}_{L_1(\mathbb{T})}\|f\|^p_{L_p(\mathbb{T})} = \|g\|^p_{L_1(\mathbb{T})}\|f\|^p_{L_p(\mathbb{T})}.\end{aligned}$$

The cases $p = 1$ and $p = \infty$ are straightforward and left as an exercise.

2. Let $f \in L_p(\mathbb{T})$ and $g \in L_q(\mathbb{T})$ with $\frac{1}{p} + \frac{1}{q} = 1$ and $p > 1$ be given. By Hölder's inequality it follows

$$\begin{aligned}|(f * g)(x)| &\le \Big(\frac{1}{2\pi}\int_{-\pi}^{\pi} |f(x-y)|^p\,\mathrm{d}y\Big)^{1/p}\Big(\frac{1}{2\pi}\int_{-\pi}^{\pi} |g(y)|^q\,\mathrm{d}y\Big)^{1/q} \\ &\le \|f\|_{L_p(\mathbb{T})}\,\|g\|_{L_q(\mathbb{T})}\end{aligned}$$

and consequently

$$|(f * g)(x+t) - (f * g)(x)| \le \|f(\cdot + t) - f\|_{L_p(\mathbb{T})}\|g\|_{L_q(\mathbb{T})}.$$

Now the second assertion follows, since the translation is continuous in the $L_p(\mathbb{T})$ norm (see [144, Proposition 8.5]), i.e., $\|f(\cdot+t) - f\|_{L_p(\mathbb{T})} \to 0$ as $t \to 0$. The case $p = 1$ is straightforward.

3. Finally, let $f \in L_p(\mathbb{T})$ and $g \in L_q(\mathbb{T})$ with $\frac{1}{p} + \frac{1}{q} = \frac{1}{r} + 1$ for $1 \le p, q, r \le \infty$ be given. The case $r = \infty$ is described in Part 2 so that it remains to consider $1 \le r < \infty$. Then $p \le r$ and $q \le r$, since otherwise we would get the contradiction $q < 1$ or $p < 1$. Set $s := p\,(1 - \frac{1}{q}) = 1 - \frac{p}{r} \in [0, 1)$ and $t := \frac{r}{q} \in [1, \infty)$. Define q' by $\frac{1}{q} + \frac{1}{q'} = 1$. Then, we obtain by Hölder's inequality:

$$\begin{aligned}h(x) &:= \frac{1}{2\pi}\int_{-\pi}^{\pi} |f(x-y)g(y)|\,\mathrm{d}y \\ &= \frac{1}{2\pi}\int_{-\pi}^{\pi} |f(x-y)|^{1-s}\,|g(y)|\,|f(x-y)|^s\,\mathrm{d}y \\ &\le \Big(\frac{1}{2\pi}\int_{-\pi}^{\pi} |f(x-y)|^{(1-s)q}\,|g(y)|^q\,\mathrm{d}y\Big)^{1/q} \\ &\quad\times \Big(\frac{1}{2\pi}\int_{-\pi}^{\pi} |f(x-y)|^{sq'}\,\mathrm{d}y\Big)^{1/q'}.\end{aligned}$$

Using that by definition $sq' = p$ and $q/q' = (sq)/p$, this implies

$$\begin{aligned} h^q(x) &\le \frac{1}{2\pi}\int_{-\pi}^{\pi} |f(x-y)|^{(1-s)q}\,|g(y)|^q\,\mathrm{d}y\Big(\frac{1}{2\pi}\int_{-\pi}^{\pi}|f(x-y)|^p\,\mathrm{d}y\Big)^{q/q'} \\ &= \frac{1}{2\pi}\int_{-\pi}^{\pi} |f(x-y)|^{(1-s)q}\,|g(y)|^q\,\mathrm{d}y\Big(\frac{1}{2\pi}\int_{-\pi}^{\pi}|f(x-y)|^p\,\mathrm{d}y\Big)^{(sq)/p} \\ &= \frac{1}{2\pi}\int_{-\pi}^{\pi} |f(x-y)|^{(1-s)q}\,|g(y)|^q\,\mathrm{d}y\,\|f\|_{L_p(\mathbb{T})}^{sq} \end{aligned}$$

such that

$$\begin{aligned} \|h\|_{L_r(\mathbb{T})}^q &= \Big(\frac{1}{2\pi}\int_{-\pi}^{\pi}|h(x)|^{qt}\,\mathrm{d}x\Big)^{q/(qt)} = \Big(\frac{1}{2\pi}\int_{-\pi}^{\pi}|h^q(x)|^{t}\,\mathrm{d}x\Big)^{1/t} \\ &= \|h^q\|_{L_t(\mathbb{T})} \\ &\le \|f\|_{L_p(\mathbb{T})}^{sq}\Big(\frac{1}{2\pi}\int_{-\pi}^{\pi}\Big(\frac{1}{2\pi}\int_{-\pi}^{\pi}|f(x-y)|^{(1-s)q}\,|g(y)|^q\,\mathrm{d}y\Big)^t\,\mathrm{d}x\Big)^{1/t} \end{aligned}$$

and further by $(1-s)qt = p$ and generalized Minkowski's inequality:

$$\begin{aligned} \|h\|_{L_r(\mathbb{T})}^q &\le \|f\|_{L_p(\mathbb{T})}^{sq}\,\frac{1}{2\pi}\int_{-\pi}^{\pi}\Big(\frac{1}{2\pi}\int_{-\pi}^{\pi}|f(x-y)|^{(1-s)qt}\,|g(y)|^{qt}\,\mathrm{d}x\Big)^{1/t}\,\mathrm{d}y \\ &= \|f\|_{L_p(\mathbb{T})}^{sq}\,\frac{1}{2\pi}\int_{-\pi}^{\pi}|g(y)|^q\Big(\frac{1}{2\pi}\int_{-\pi}^{\pi}|f(x-y)|^{(1-s)qt}\,\mathrm{d}x\Big)^{1/t}\,\mathrm{d}y \\ &= \|f\|_{L_p(\mathbb{T})}^{sq}\,\|f\|_{L_{(1-s)qt}(\mathbb{T})}^{(1-s)q}\,\frac{1}{2\pi}\int_{-\pi}^{\pi}|g(y)|^q\,\mathrm{d}y = \|f\|_{L_p(\mathbb{T})}^q\,\|g\|_{L_q(\mathbb{T})}^q\,. \end{aligned}$$

Taking the qth root finishes the proof. Alternatively, the third step can be proved using the Riesz–Thorin theorem. ■

The convolution of an $L_1(\mathbb{T})$ function and an $L_p(\mathbb{T})$ function with $1 \le p < \infty$ is in general not defined pointwise as the following example shows.

Example 1.12 We consider the 2π-periodic extension of $f : [-\pi, \pi) \to \mathbb{R}$ given by

$$f(y) := \begin{cases} |y|^{-3/4} & y \in [-\pi, \pi) \setminus \{0\}\,, \\ 0 & y = 0\,. \end{cases} \tag{1.20}$$

The extension denoted by f is even and belongs to $L_1(\mathbb{T})$. The convolution $(f * f)(x)$ is finite for all $x \in [-\pi, \pi) \setminus \{0\}$. However, for $x = 0$, this does not hold true, since

$$\int_{-\pi}^{\pi} f(y)\, f(-y)\, \mathrm{d}y = \int_{-\pi}^{\pi} |y|^{-3/2}\, \mathrm{d}y = \infty\,.$$

□

The following lemma describes the *convolution property of Fourier series*.

Lemma 1.13 *For $f,\ g \in L_1(\mathbb{T})$, it holds:*

$$c_k(f * g) = c_k(f)\, c_k(g)\,, \quad k \in \mathbb{Z}.$$

Proof Using Fubini's theorem, we obtain by the 2π-periodicity of $g\, \mathrm{e}^{-\mathrm{i}k\,\cdot}$ that

$$\begin{aligned} c_k(f * g) &= \frac{1}{(2\pi)^2} \int_{-\pi}^{\pi} \Big(\int_{-\pi}^{\pi} f(y)\, g(x-y)\, \mathrm{d}y \Big) \mathrm{e}^{-\mathrm{i}kx}\, \mathrm{d}x \\ &= \frac{1}{(2\pi)^2} \int_{-\pi}^{\pi} f(y)\, \mathrm{e}^{-\mathrm{i}ky} \Big(\int_{-\pi}^{\pi} g(x-y)\, \mathrm{e}^{-\mathrm{i}k(x-y)}\, \mathrm{d}x \Big)\, \mathrm{d}y \\ &= \frac{1}{(2\pi)^2} \int_{-\pi}^{\pi} f(y)\, \mathrm{e}^{-\mathrm{i}ky} \Big(\int_{-\pi}^{\pi} g(t)\, \mathrm{e}^{-\mathrm{i}kt}\, \mathrm{d}t \Big)\, \mathrm{d}y = c_k(f)\, c_k(g)\,. \end{aligned}$$

■

The convolution of functions with certain functions, so-called kernels, is of particular interest.

Example 1.14 The nth *Dirichlet kernel* for $n \in \mathbb{N}_0$ is defined by

$$D_n(x) := \sum_{k=-n}^{n} \mathrm{e}^{\mathrm{i}kx}\,, \quad x \in \mathbb{R}\,. \tag{1.21}$$

By Euler's formula it follows:

$$D_n(x) = 1 + 2 \sum_{k=1}^{n} \cos(kx)\,.$$

Obviously, $D_n \in \mathscr{T}_n$ is real-valued and even. For $x \in (0, \pi]$ and $n \in \mathbb{N}$, we can express $\big(\sin \frac{x}{2}\big)\, D_n(x)$ as telescope sum

$$\begin{aligned} \Big(\sin \frac{x}{2}\Big)\, D_n(x) &= \sin \frac{x}{2} + \sum_{k=1}^{n} 2\, \cos(kx)\, \sin \frac{x}{2} \\ &= \sin \frac{x}{2} + \sum_{k=1}^{n} \Big(\sin \frac{(2k+1)x}{2} - \sin \frac{(2k-1)x}{2} \Big) = \sin \frac{(2n+1)x}{2}\,. \end{aligned}$$

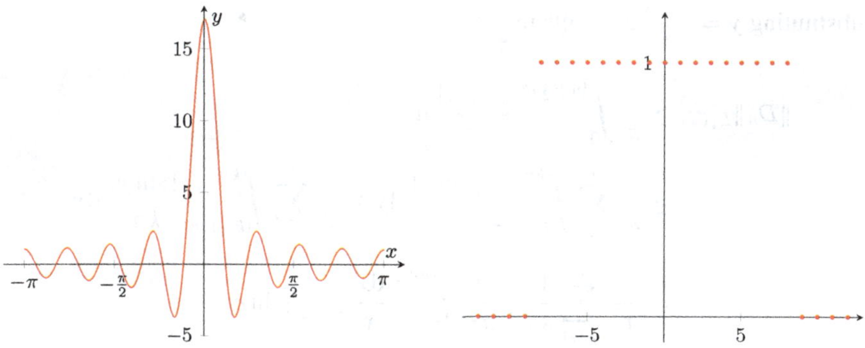

Fig. 1.6 The Dirichlet kernel D_8 (left) and its Fourier coefficients $c_k(D_8)$ (right)

Thus, the nth Dirichlet kernel can be represented as a fraction:

$$D_n(x) = \frac{\sin\frac{(2n+1)x}{2}}{\sin\frac{x}{2}}, \quad x \in [-\pi, \pi)\backslash\{0\}, \tag{1.22}$$

with $D_n(0) = 2n + 1$. Figure 1.6 depicts the Dirichlet kernel D_8. The Fourier coefficients of D_n are

$$c_k(D_n) = \begin{cases} 1 & k = -n, \ldots, n, \\ 0 & |k| > n \,. \end{cases}$$

For $f \in L_1(\mathbb{T})$ with Fourier coefficients $c_k(f)$, $k \in \mathbb{Z}$, we obtain by Lemma 1.13 that

$$f * D_n = \sum_{k=-n}^{n} c_k(f)\, \mathrm{e}^{\mathrm{i}k\cdot} = S_n f\,, \tag{1.23}$$

which is just the nth Fourier partial sum of f. By the following calculations, the Dirichlet kernel fulfills

$$\|D_n\|_{L_1(\mathbb{T})} = \frac{1}{2\pi}\int_{-\pi}^{\pi} |D_n(x)|\, \mathrm{d}x \geq \frac{4}{\pi^2}\ln n\,. \tag{1.24}$$

Note that $\|D_n\|_{L_1(\mathbb{T})}$ are called *Lebesgue constants*. Since $\sin x \leq x$ for $x \in [0, \frac{\pi}{2})$, we get by (1.22) that

$$\|D_n\|_{L_1(\mathbb{T})} = \frac{1}{\pi}\int_0^{\pi} \frac{|\sin((2n+1)x/2)|}{\sin(x/2)}\, \mathrm{d}x \geq \frac{2}{\pi}\int_0^{\pi} \frac{|\sin((2n+1)x/2)|}{x}\, \mathrm{d}x.$$

Substituting $y = \frac{2n+1}{2}\,x$ results in

$$\begin{aligned}\|D_n\|_{L_1(\mathbb{T})} &\ge \frac{2}{\pi}\int_0^{(n+\frac{1}{2})\pi} \frac{|\sin y|}{y}\,\mathrm{d}y \\ &\ge \frac{2}{\pi}\sum_{k=1}^{n}\int_{(k-1)\pi}^{k\pi} \frac{|\sin y|}{y}\,\mathrm{d}y \ge \frac{2}{\pi}\sum_{k=1}^{n}\int_{(k-1)\pi}^{k\pi} \frac{|\sin y|}{k\pi}\,\mathrm{d}y \\ &= \frac{4}{\pi^2}\sum_{k=1}^{n}\frac{1}{k} \ge \frac{4}{\pi^2}\int_1^{n+1}\frac{\mathrm{d}x}{x} \ge \frac{4}{\pi^2}\,\ln n\,.\end{aligned}$$

The Lebesgue constants fulfill

$$\|D_n\|_{L_1(\mathbb{T})} = \frac{4}{\pi^2}\,\ln n + \mathcal{O}(1)\,, \qquad n \to \infty\,.$$

□

Example 1.15 The nth *Fejér kernel* for $n \in \mathbb{N}_0$ is defined by

$$F_n := \frac{1}{n+1}\sum_{j=0}^{n} D_j \in \mathcal{T}_n\,. \tag{1.25}$$

By (1.22) and (1.25), we obtain $F_n(0) = n+1$ and for $x \in [-\pi,\,\pi)\setminus\{0\}$

$$F_n(x) = \frac{1}{n+1}\sum_{j=0}^{n}\frac{\sin\big((j+\frac{1}{2})x\big)}{\sin\frac{x}{2}}\,.$$

Multiplying the numerator and denominator of each right-hand fraction by $2\,\sin\frac{x}{2}$ and replacing the product of sines in the numerator by the differences $\cos(jx) - \cos\big((j+1)x\big)$, we find by cascade summation that F_n can be represented in the form

$$F_n(x) = \frac{1}{2(n+1)}\,\frac{1-\cos\big((n+1)x\big)}{\big(\sin\frac{x}{2}\big)^2} = \frac{1}{n+1}\Big(\frac{\sin\frac{(n+1)x}{2}}{\sin\frac{x}{2}}\Big)^2\,. \tag{1.26}$$

In contrast to the Dirichlet kernel, the Fejér kernel is nonnegative. Figure 1.7 shows the Fejér kernel F_8. The Fourier coefficients of F_n are

$$c_k(F_n) = \begin{cases} 1-\frac{|k|}{n+1} & k = -n,\ldots,n\,,\\ 0 & |k| > n\,.\end{cases}$$

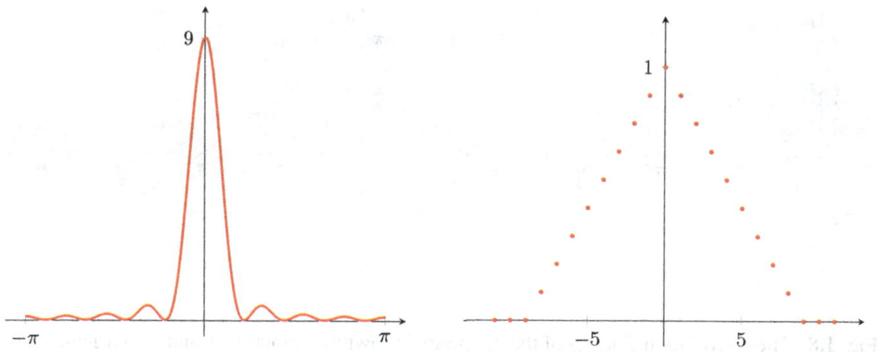

Fig. 1.7 The Fejér kernel F_8 (left) and its Fourier coefficients $c_k(F_8)$ (right)

Using the convolution property, the convolution $f * F_n$ for arbitrary $f \in L_1(\mathbb{T})$ is given by

$$\sigma_n f := f * F_n = \sum_{k=-n}^{n} \Big(1 - \frac{|k|}{n+1}\Big)\, c_k(f)\, \mathrm{e}^{\mathrm{i}k\cdot}\,. \tag{1.27}$$

Then, $\sigma_n f$ is called the *nth Fejér sum* or *nth Cesàro sum* of f. Further, we have:

$$\|F_n\|_{L_1(\mathbb{T})} = \frac{1}{2\pi} \int_{-\pi}^{\pi} F_n(x)\, \mathrm{d}x = 1.$$

□

Example 1.16 The nth *de la Vallée Poussin kernel* V_{2n} for $n \in \mathbb{N}$ is defined by

$$V_{2n} = \frac{1}{n} \sum_{j=n}^{2n-1} D_j = 2\, F_{2n-1} - F_{n-1} = \sum_{k=-2n}^{2n} c_k(V_{2n})\, \mathrm{e}^{\mathrm{i}k\cdot}$$

with the Fourier coefficients

$$c_k(V_{2n}) = \begin{cases} 2 - \frac{|k|}{n} & k = -2n, \ldots, -(n+1), n+1, \ldots, 2n\,, \\ 1 & k = -n, \ldots, n\,, \\ 0 & |k| > 2n\,. \end{cases}$$

□

By Theorem 1.11 the convolution of two $L_1(\mathbb{T})$ functions is again a function in $L_1(\mathbb{T})$. The space $L_1(\mathbb{T})$ forms together with the addition and the convolution a so-called Banach algebra. Unfortunately, there does not exist an identity element

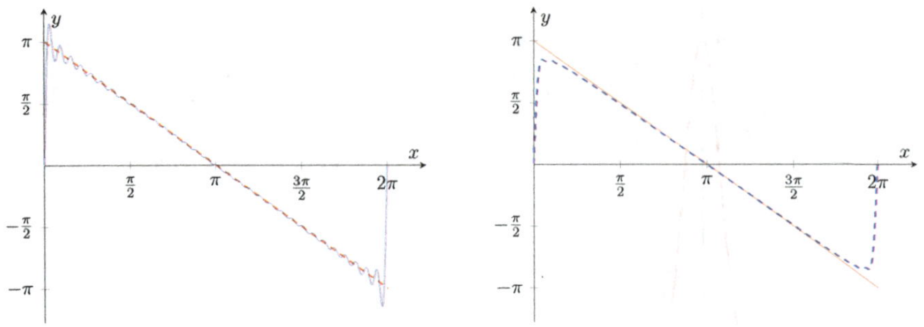

Fig. 1.8 The convolution $f * D_{32}$ of the 2π-periodic sawtooth function f and the Dirichlet kernel D_{32} approximates f quite good except at the jump discontinuities (left). The convolution $f * F_{32}$ of f and the Fejér kernel F_{32} approximates f not as good as $f * D_{32}$, but it does not oscillate near the jump discontinuities (right)

with respect to $*$, i.e., there is no function $g \in L_1(\mathbb{T})$ such that $f * g = f$ for all $f \in L_1(\mathbb{T})$. As a remedy we can define approximate identities. Figure 1.8 shows the different approximations $f * D_{32}$ and $f * F_{32}$ of the 2π-periodic sawtooth function f.

A sequence $(K_n)_{n\in\mathbb{N}}$ of functions $K_n \in L_1(\mathbb{T})$ is called an *approximate identity* or a *summation kernel* , if it satisfies the following properties:

1. $\frac{1}{2\pi}\int_{-\pi}^{\pi} K_n(x)\,\mathrm{d}x = 1$ for all $n \in \mathbb{N}$,
2. $\|K_n\|_{L_1(\mathbb{T})} = \frac{1}{2\pi}\int_{-\pi}^{\pi} |K_n(x)|\,\mathrm{d}x \le C < \infty$ for all $n \in \mathbb{N}$,
3. $\lim_{n\to\infty}\big(\int_{-\pi}^{-\delta} + \int_{\delta}^{\pi}\big)|K_n(x)|\,\mathrm{d}x = 0$ for each $0 < \delta < \pi$.

Theorem 1.17 *For an approximate identity* $(K_n)_{n\in\mathbb{N}}$*, it holds:*

$$\lim_{n\to\infty} \|K_n * f - f\|_{C(\mathbb{T})} = 0$$

for all $f \in C(\mathbb{T})$.

Proof Since a continuous function is uniformly continuous on a compact interval, for all $\varepsilon > 0$ there exists a number $\delta > 0$ so that for all $|u| < \delta$

$$\|f(\cdot - u) - f\|_{C(\mathbb{T})} < \varepsilon\,. \tag{1.28}$$

Using the first property of an approximate identity, we obtain:

$$\begin{aligned}
\|K_n * f - f\|_{C(\mathbb{T})} &= \sup_{x\in\mathbb{T}} \Big|\frac{1}{2\pi}\int_{-\pi}^{\pi} f(x-u)\,K_n(u)\,\mathrm{d}u - f(x)\Big| \\
&= \sup_{x\in\mathbb{T}} \Big|\frac{1}{2\pi}\int_{-\pi}^{\pi} \big(f(x-u) - f(x)\big)\,K_n(u)\,\mathrm{d}u\Big| \\
&\le \frac{1}{2\pi}\sup_{x\in\mathbb{T}} \int_{-\pi}^{\pi} |f(x-u) - f(x)|\,|K_n(u)|\,\mathrm{d}u
\end{aligned}$$

$$= \frac{1}{2\pi} \sup_{x\in\mathbb{T}} \Big(\int_{-\pi}^{-\delta} + \int_{-\delta}^{\delta} + \int_{\delta}^{\pi} \Big) |f(x-u) - f(x)| \, |K_n(u)| \, du \, .$$

By (1.28) the right-hand side can be estimated as

$$\frac{\varepsilon}{2\pi} \int_{-\delta}^{\delta} |K_n(u)| \, du + \frac{1}{2\pi} \sup_{x\in\mathbb{T}} \Big(\int_{-\pi}^{-\delta} + \int_{\delta}^{\pi} \Big) |f(x-u) - f(x)| \, |K_n(u)| \, du \, .$$

By the properties 2 and 3 of the reproducing kernel K_n, we obtain for sufficiently large $n \in \mathbb{N}$ that

$$\|K_n * f - f\|_{C(\mathbb{T})} \le \varepsilon C + \frac{1}{\pi} \, \|f\|_{C(\mathbb{T})} \, \varepsilon.$$

Since $\varepsilon > 0$ can be chosen arbitrarily small, this yields the assertion. ■

Example 1.18 The sequence $(D_n)_{n\in\mathbb{N}}$ of Dirichlet kernels defined in Example 1.14 is not an approximate identity, since $\|D_n\|_{L_1(\mathbb{T})}$ is not uniformly bounded for all $n \in \mathbb{N}$ by (1.24). Indeed we will see in the next section that $S_n f = D_n * f$ does in general not converge uniformly to $f \in C(\mathbb{T})$ for $n \to \infty$. A general remedy in such cases consists in considering the Cesàro mean as shown in the next example. □

Example 1.19 The sequence $(F_n)_{n\in\mathbb{N}}$ of Fejér kernels defined in Example 1.15 possesses by definition the first two properties of an approximate identity and also fulfills the third one by (1.26) and

$$\begin{aligned}
\Big(\int_{-\pi}^{-\delta} + \int_{\delta}^{\pi} \Big) F_n(x) \, \mathrm{d}x &= 2 \int_{\delta}^{\pi} F_n(x) \, \mathrm{d}x \\
&= \frac{2}{n+1} \int_{\delta}^{\pi} \Big(\frac{\sin((n+1)x/2)}{\sin(x/2)} \Big)^2 \mathrm{d}x \\
&\le \frac{2}{n+1} \int_{\delta}^{\pi} \frac{\pi^2}{x^2} \, \mathrm{d}x = \frac{2\pi}{n+1} \Big(\frac{\pi}{\delta} - 1 \Big) .
\end{aligned}$$

The right-hand side tends to zero as $n \to \infty$ so that $(F_n)_{n\in\mathbb{N}}$ is an approximate identity.

It is not hard to verify that the sequence $(V_{2n})_{n\in\mathbb{N}}$ of de la Vallée Poussin kernels defined in Example 1.16 is also an approximate identity. □

From Theorem 1.17 and Example 1.19, it follows immediately

Theorem 1.20 (Approximation Theorem of Fejér) *If $f \in C(\mathbb{T})$, then the Fejér sums $\sigma_n f$ converge uniformly to f as $n \to \infty$. If $m \le f(x) \le M$ for all $x \in \mathbb{T}$ with m, $M \in \mathbb{R}$, then $m \le (\sigma_n f)(x) \le M$ for all $n \in \mathbb{N}$.*

Proof Since $(F_n)_{n\in\mathbb{N}}$ is an approximate identity, the Fejér sums $\sigma_n f$ converge uniformly to f as $n \to \infty$. If a real-valued function $f : \mathbb{T} \to \mathbb{R}$ fulfills the estimate $m \le f(x) \le M$ for all $x \in \mathbb{T}$ with certain constants $m,\ M \in \mathbb{R}$, then

$$(\sigma_n f)(x) = \frac{1}{2\pi} \int_{-\pi}^{\pi} F_n(y)\, f(x-y)\, \mathrm{d}y$$

fulfills also $m \le (\sigma_n f)(x) \le M$ for all $x \in \mathbb{T}$, since $F_n(y) \ge 0$ and $\frac{1}{2\pi}\int_{-\pi}^{\pi} F_n(y)\,\mathrm{d}y = c_0(F_n) = 1$. ∎

The Theorem 1.20 of Fejér has many important consequences such as follows.

Theorem 1.21 (Approximation Theorem of Weierstrass) *If $f \in C(\mathbb{T})$, then for each $\varepsilon > 0$ there exists a trigonometric polynomial $p = \sigma_n f \in \mathcal{T}_n$ of sufficiently large degree n such that $\|f - p\|_{C(\mathbb{T})} < \varepsilon$. Further this trigonometric polynomial p is a weighted Fourier partial sum given by* (1.27).

Finally we present two important inequalities for any trigonometric polynomial $p \in \mathcal{T}_n$ with fixed $n \in \mathbb{N}$. The inequality of S. M. Nikolsky compares different norms of any trigonometric polynomial $p \in \mathcal{T}_n$. The inequality of S. N. Bernstein estimates the norm of the derivative p' by the norm of a trigonometric polynomial $p \in \mathcal{T}_n$.

Theorem 1.22 *Assume that $1 \le q \le r \le \infty$, where q is finite and $s := \lceil q/2 \rceil$. Then for all $p \in \mathcal{T}_n$, it holds the* Nikolsky inequality *:*

$$\|p\|_{L_r(\mathbb{T})} \le (2n\,s + 1)^{1/q - 1/r}\, \|p\|_{L_q(\mathbb{T})} \tag{1.29}$$

and the Bernstein inequality

$$\|p'\|_{L_r(\mathbb{T})} \le n\, \|p\|_{L_r(\mathbb{T})}\,. \tag{1.30}$$

Proof

1. Setting $m := n\,s$, we have $p^s \in \mathcal{T}_m$ and hence $p^s * D_m = p^s$ by (1.23). Using the Cauchy–Schwarz inequality, we can estimate

$$\begin{aligned}
|p(x)^s| &\le \frac{1}{2\pi} \int_{-\pi}^{\pi} |p(t)|^s\, |D_m(x-t)|\, \mathrm{d}t \\
&\le \|p\|_{C(\mathbb{T})}^{s-q/2}\, \frac{1}{2\pi} \int_{-\pi}^{\pi} |p(t)|^{q/2}\, |D_m(x-t)|\, \mathrm{d}t \\
&\le \|p\|_{C(\mathbb{T})}^{s-q/2}\, \|\,|p|^{q/2}\,\|_{L_2(\mathbb{T})}\, \|D_m\|_{L_2(\mathbb{T})}\,.
\end{aligned}$$

Since

$$\|\,|p|^{q/2}\,\|_{L_2(\mathbb{T})} = \|p\|_{L_q(\mathbb{T})}^{q/2}\,, \qquad \|D_m\|_{L_2(\mathbb{T})} = (2m+1)^{1/2}\,,$$

we obtain:

$$\|p\|_{C(\mathbb{T})}^{s} \leq (2m+1)^{1/2}\, \|p\|_{C(\mathbb{T})}^{s-q/2}\, \|p\|_{L_q(\mathbb{T})}^{q/2}$$

and hence the Nikolsky inequality (1.29) for $r = \infty$, i.e.,

$$\|p\|_{L_\infty(\mathbb{T})} = \|p\|_{C(\mathbb{T})} \leq (2m+1)^{1/q}\, \|p\|_{L_q(\mathbb{T})}\,. \tag{1.31}$$

For finite $r > q$, we use the inequality:

$$\begin{aligned} \|p\|_{L_r(\mathbb{T})} &= \Big(\frac{1}{2\pi}\int_{-\pi}^{\pi} |p(t)|^{r-q}\, |p(t)|^{q}\, \mathrm{d}t\Big)^{1/r} \\ &\leq \|p\|_{C(\mathbb{T})}^{1-q/r} \Big(\frac{1}{2\pi}\int_{-\pi}^{\pi} |p(t)|^{q}\, \mathrm{d}t\Big)^{1/r} = \|p\|_{C(\mathbb{T})}^{1-q/r}\, \|p\|_{L_q(\mathbb{T})}^{q/r} \end{aligned}$$

Then from (1.31) it follows the Nikolsky inequality (1.29).

2. For simplicity, we show the Bernstein inequality (1.30) only for $r = 2$. An arbitrary trigonometric polynomial $p \in \mathscr{T}_n$ has the form:

$$p(x) = \sum_{k=-n}^{n} c_k\, \mathrm{e}^{\mathrm{i}\,k\,x}$$

with certain coefficients $c_k = c_k(p) \in \mathbb{C}$ such that

$$p'(x) = \sum_{k=-n}^{n} \mathrm{i}\,k\, c_k\, \mathrm{e}^{\mathrm{i}\,k\,x}\,.$$

Thus, by the Parseval equality (1.16), we obtain:

$$\|p\|_{L_2(\mathbb{T})}^2 = \sum_{k=-n}^{n} |c_k|^2\,, \quad \|p'\|_{L_2(\mathbb{T})}^2 = \sum_{k=-n}^{n} k^2\, |c_k|^2 \leq n^2\, \|p\|_{L_2(\mathbb{T})}^2\,.$$

For a proof in the general case $1 \leq r \leq \infty$, we refer to [112, pp. 97–102]. The Bernstein inequality is best possible, since we have equality in (1.30) for $p(x) = \mathrm{e}^{\mathrm{i}\,n\,x}$. ■

1.4 Pointwise and Uniform Convergence of Fourier Series

In Sect. 1.3, it was shown that a Fourier series of an arbitrary function $f \in L_2(\mathbb{T})$ converges in the norm of $L_2(\mathbb{T})$, i.e.,

$$\lim_{n\to\infty} \|S_n f - f\|_{L_2(\mathbb{T})} = \lim_{n\to\infty} \|f * D_n - f\|_{L_2(\mathbb{T})} = 0\,.$$

In general, the pointwise or almost everywhere convergence of a sequence $(f_n)_{n\in\mathbb{N}}$ of functions $f_n \in L_2(\mathbb{T})$ does not result the convergence in $L_2(\mathbb{T})$.

Example 1.23 Let $f_n : \mathbb{T} \to \mathbb{R}$ be the 2π-extension of

$$f_n(x) := \begin{cases} n & x \in (0,\, 1/n)\,, \\ 0 & x \in \{0\} \cup [1/n,\, 2\pi)\,. \end{cases}$$

Obviously, we have $\lim_{n\to\infty} f_n(x) = 0$ for all $x \in [0,\, 2\pi]$. But it holds for $n \to \infty$,

$$\|f_n\|^2_{L_2(\mathbb{T})} = \frac{1}{2\pi} \int_0^{1/n} n^2\, \mathrm{d}x = \frac{n}{2\pi} \to \infty\,.$$

□

As known (see, e.g., [274, pp. 52–53]), if a sequence $(f_n)_{n\in\mathbb{N}}$, where $f_n \in L_p(\mathbb{T})$ with $1 \le p \le \infty$, converges to $f \in L_p(\mathbb{T})$ in the norm of $L_p(\mathbb{T})$, then there exists a subsequence $\big(f_{n_k}\big)_{k\in\mathbb{N}}$ such that for almost all $x \in [0,\, 2\pi]$,

$$\lim_{k\to\infty} f_{n_k}(x) = f(x)\,.$$

In 1966, L. Carleson proved the fundamental result that the Fourier series of an arbitrary function $f \in L_p(\mathbb{T})$, $1 < p < \infty$, converges almost everywhere. For a proof see, e.g., [178, pp. 232–233]. Kolmogoroff [245] showed that an analog result for $f \in L_1(\mathbb{T})$ is false.

A natural question is whether the Fourier series of every function $f \in C(\mathbb{T})$ converges *uniformly* or at least *pointwise* to f. From Carleson's result it follows that the Fourier series of $f \in C(\mathbb{T})$ converges almost everywhere, i.e., in all points of $[0,\, 2\pi]$ except for a set of Lebesgue measure zero. In fact, many mathematicians like Riemann, Weierstrass, and Dedekind conjectured over long time that the Fourier series of a function $f \in C(\mathbb{T})$ converges pointwise to f. But one has neither pointwise nor uniform convergence of the Fourier series of a function $f \in C(\mathbb{T})$ in general. A concrete counterexample was constructed by Du Bois–Reymond in 1876 and was a quite remarkable surprise. It was shown that there exists a real-valued function $f \in C(\mathbb{T})$ such that

$$\lim_{n\to\infty} \sup |S_n f(0)| = \infty.$$

To see that pointwise convergence fails in general, we need the following principle of uniform boundedness of sequences of linear bounded operators; see, e.g., [434, Korollar 2.4].

Theorem 1.24 (Theorem of Banach–Steinhaus) *Let X and Y be Banach spaces. Then a sequence $(T_n)_{n\in\mathbb{N}}$ of bounded linear operators from X to Y converges pointwise to a bounded linear operator $T : X \to Y$, i.e., for all $f \in X$, we have:*

$$Tf = \lim_{n\to\infty} T_n f, \tag{1.32}$$

if and only if it fulfills the following properties:

1. $\sup_{n\in\mathbb{N}} \|T_n\|_{X\to Y} \le \text{const} < \infty$, *and*
2. *there exists a dense subset* $D \subseteq X$ *such that* $\lim_{n\to\infty} T_n p = Tp$ *for all* $p \in D$.

Theorem 1.25 *There exists a function* $f \in C(\mathbb{T})$ *whose Fourier series does not converge pointwise.*

Proof Applying Theorem 1.24 of Banach–Steinhaus, we choose $X = C(\mathbb{T})$, $Y = \mathbb{C}$ and $D = \bigcup_{n=0}^{\infty} \mathscr{T}_n$. By the approximation Theorem 1.21 of Weierstrass, the set D of all trigonometric polynomials is dense in $C(\mathbb{T})$. Then, we consider the linear bounded functionals $T_n f := (S_n f)(0)$ for $n \in \mathbb{N}$ and $Tf := f(0)$ for $f \in C(\mathbb{T})$. Note that instead of 0 we can choose any fixed $x_0 \in \mathbb{T}$.

We want to show that the norms $\|T_n\|_{C(\mathbb{T})\to\mathbb{C}}$ are not uniformly bounded with respect to n. More precisely, we will deduce $\|T_n\|_{C(\mathbb{T})\to\mathbb{C}} = \|D_n\|_{L_1(\mathbb{T})}$ which are not uniformly bounded by (1.24). Then by the Banach–Steinhaus Theorem 1.24, there exists a function $f \in C(\mathbb{T})$ whose Fourier series does not converge in the point 0.

Let us determine the norm $\|T_n\|_{C(\mathbb{T})\to\mathbb{C}}$. From

$$\begin{aligned} |T_n f| = |(S_n f)(0)| &= |(D_n * f)(0)| \\ &= \Big|\frac{1}{2\pi}\int_{-\pi}^{\pi} D_n(x)\, f(x)\, \mathrm{d}x\Big| \le \|f\|_{C(\mathbb{T})}\|D_n\|_{L_1(\mathbb{T})} \end{aligned}$$

for arbitrary $f \in C(\mathbb{T})$, it follows $\|T_n\|_{C(\mathbb{T})\to\mathbb{C}} \le \|D_n\|_{L_1(\mathbb{T})}$. To verify the opposite direction, consider for an arbitrary $\varepsilon > 0$ the function

$$f_\varepsilon := \frac{D_n}{|D_n| + \varepsilon} \in C(\mathbb{T}),$$

which has $C(\mathbb{T})$ norm smaller than 1. Then,

$$\begin{aligned} |T_n f_\varepsilon| = (D_n * f_\varepsilon)(0) &= \frac{1}{2\pi}\int_{-\pi}^{\pi} \frac{|D_n(x)|^2}{|D_n(x)| + \varepsilon}\, \mathrm{d}x \\ &\ge \frac{1}{2\pi}\int_{-\pi}^{\pi} \frac{|D_n(x)|^2 - \varepsilon^2}{|D_n(x)| + \varepsilon}\, \mathrm{d}x \\ &\ge \Big(\frac{1}{2\pi}\int_{-\pi}^{\pi} |D_n(x)|\, \mathrm{d}x - \varepsilon\Big)\, \|f_\varepsilon\|_{C(\mathbb{T})} \end{aligned}$$

implies $\|T_n\|_{C(\mathbb{T})\to\mathbb{C}} \ge \|D_n\|_{L_1(\mathbb{T})} - \varepsilon$. For $\varepsilon \to 0$, we obtain the assertion. ∎

Remark 1.26 Theorem 1.25 indicates that there exists a function $f \in C(\mathbb{T})$ such that $\big(S_n f\big)_{n\in\mathbb{N}_0}$ is not convergent in $C(\mathbb{T})$. Analogously, one can show by Theorem 1.24 of Banach–Steinhaus that there exists a function $f \in L_1(\mathbb{T})$ such that $\big(S_n f\big)_{n\in\mathbb{N}_0}$ is not convergent in $L_1(\mathbb{T})$ (cf. [265, p. 52]). Later we will see that the Fourier series of any $f \in L_1(\mathbb{T})$ converges to f in the weak sense of distribution theory (see Lemma 4.57 or [156, pp. 336–337]).

1.4.1 Pointwise Convergence

In the following we will see that for frequently appearing classes of functions, stronger convergence results can be proved. A function $f\colon \mathbb{T} \to \mathbb{C}$ is called *piecewise continuously differentiable*, if there exist finitely many points $0 \le x_0 < x_1 < \ldots < x_{n-1} < 2\pi$ such that f is continuously differentiable on each subinterval (x_j, x_{j+1}), $j = 0, \ldots, n-1$ with $x_n = x_0 + 2\pi$, and the left- and right-hand limits $f(x_j \pm 0)$, $f'(x_j \pm 0)$ for $j = 0, \ldots, n$ exist and are finite. In the case $f(x_j - 0) \ne f(x_j + 0)$, the piecewise continuously differentiable function $f : \mathbb{T} \to \mathbb{C}$ has a *jump discontinuity* at x_j with jump height $|f(x_j+0) - f(x_j-0)|$. Simple examples of piecewise continuously differentiable functions $f : \mathbb{T} \to \mathbb{C}$ are the sawtooth function and the rectangular pulse function (see Examples 1.9 and 1.10). This definition is illustrated in Fig. 1.9.

The next convergence statements will use the following result of Riemann–Lebesgue.

Lemma 1.27 (Lemma of Riemann–Lebesgue) *Let $f \in L_1\big((a,b)\big)$ with $-\infty \le a < b \le \infty$ be given. Then, the following relations hold:*

$$\lim_{|v|\to\infty} \int_a^b f(x)\, \mathrm{e}^{-\mathrm{i}xv}\, \mathrm{d}x = 0\,,$$

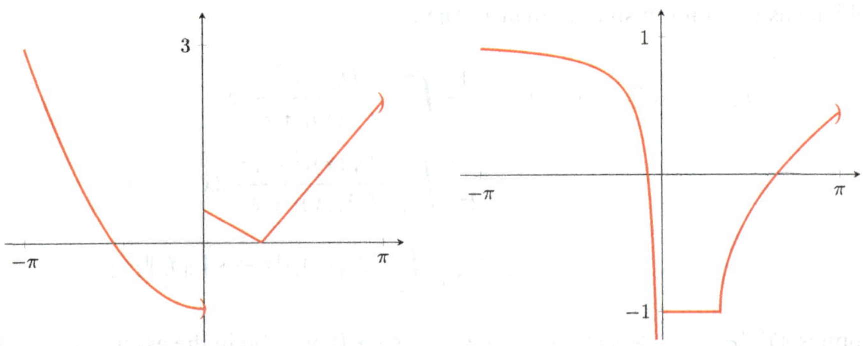

Fig. 1.9 A piecewise continuously differentiable function (left) and a function that is not piecewise continuously differentiable (right)

$$\lim_{|v|\to\infty} \int_a^b f(x)\, \sin(xv)\, \mathrm{d}x = 0\,, \qquad \lim_{|v|\to\infty} \int_a^b f(x)\, \cos(xv)\, \mathrm{d}x = 0\,.$$

Especially, for $f \in L_1(\mathbb{T})$, we have:

$$\lim_{|k|\to\infty} c_k(f) = \frac{1}{2\pi} \lim_{|k|\to\infty} \int_{-\pi}^{\pi} f(x)\, \mathrm{e}^{-\mathrm{i}xk}\, \mathrm{d}x = 0\,.$$

Proof We prove only

$$\lim_{|v|\to\infty} \int_a^b f(x)\, p(vx)\, \mathrm{d}x = 0 \tag{1.33}$$

for $p(t) = \mathrm{e}^{-\mathrm{i}t}$. The other cases $p(t) = \sin t$ and $p(t) = \cos t$ can be shown analogously.

For the characteristic function $\chi_{[\alpha,\beta]}$ of a finite interval $[\alpha, \beta] \subseteq \overline{(a, b)}$, it follows for $v \neq 0$ that

$$\Big| \int_a^b \chi_{[\alpha,\beta]}(x)\, \mathrm{e}^{-\mathrm{i}xv}\, \mathrm{d}x \Big| = \Big| -\frac{1}{\mathrm{i}v}\, (\mathrm{e}^{-\mathrm{i}v\beta} - \mathrm{e}^{-\mathrm{i}v\alpha}) \Big| \leq \frac{2}{|v|}\,.$$

This becomes arbitrarily small as $|v| \to \infty$ so that characteristic functions and also all linear combinations of characteristic functions (i.e., step functions) fulfill the assertion.

The set of all step functions is dense in $L_1\big(\overline{(a, b)}\big)$, i.e., for any $\varepsilon > 0$ and $f \in L_1\big(\overline{(a, b)}\big)$, there exists a step function φ such that

$$\|f - \varphi\|_{L_1(\overline{(a,b)})} = \int_a^b |f(x) - \varphi(x)|\, \mathrm{d}x < \varepsilon\,.$$

By

$$\begin{aligned} \Big| \int_a^b f(x)\, \mathrm{e}^{-\mathrm{i}xv}\, \mathrm{d}x \Big| &\leq \Big| \int_a^b (f(x) - \varphi(x))\, \mathrm{e}^{-\mathrm{i}xv}\, \mathrm{d}x \Big| + \Big| \int_a^b \varphi(x)\, \mathrm{e}^{-\mathrm{i}xv}\, \mathrm{d}x \Big| \\ &\leq \varepsilon + \Big| \int_a^b \varphi(x)\, \mathrm{e}^{-\mathrm{i}xv}\, \mathrm{d}x \Big| \end{aligned}$$

we obtain the assertion. ■

Next we formulate a localization principle, which states that the convergence behavior of a Fourier series of a function $f \in L_1(\mathbb{T})$ at a point x_0 depends merely on the values of f in some arbitrarily small neighborhood—despite the fact that the Fourier coefficients are determined by all function values on $\mathbb{T}$.

Theorem 1.28 (Riemann's Localization Principle) *Let $f \in L_1(\mathbb{T})$ and $x_0 \in \mathbb{R}$ be given. Then we have:*

$$\lim_{n\to\infty} (S_n f)(x_0) = c$$

for some $c \in \mathbb{R}$ if and only if for some $\delta \in (0, \pi)$

$$\lim_{n\to\infty} \int_0^{\delta} \big(f(x_0 - t) + f(x_0 + t) - 2\,c\big)\, D_n(t)\,\mathrm{d}t = 0\,.$$

Proof Since $D_n \in C(\mathbb{T})$ is even, we get:

$$\begin{aligned}(S_n f)(x_0) &= \frac{1}{2\pi}\Big(\int_{-\pi}^{0} + \int_0^{\pi}\Big) f(x_0 - t)\, D_n(t)\,\mathrm{d}t \\ &= \frac{1}{2\pi}\int_0^{\pi} \big(f(x_0 - t) + f(x_0 + t)\big)\, D_n(t)\,\mathrm{d}t\,.\end{aligned}$$

Using $\pi = \int_0^{\pi} D_n(t)\,\mathrm{d}t$, we conclude further:

$$(S_n f)(x_0) - c = \frac{1}{2\pi}\int_0^{\pi} \big(f(x_0 - t) + f(x_0 + t) - 2\,c\big)\, D_n(t)\,\mathrm{d}t.$$

By Example 1.14, we have $D_n(t) = \sin\big((n + \frac{1}{2})t\big)/\sin\frac{t}{2}$ for $t \in (0, \pi]$. By the Lemma 1.27 of Riemann–Lebesgue, we obtain:

$$\lim_{n\to\infty} \int_{\delta}^{\pi} \frac{f(x_0 - t) + f(x_0 + t) - 2\,c}{\sin\frac{t}{2}}\, \sin\big((n + \frac{1}{2})\,t\big)\,\mathrm{d}t = 0$$

and hence

$$\lim_{n\to\infty} (S_n f)(x_0) - c = \lim_{n\to\infty} \frac{1}{2\pi}\int_0^{\delta} \big(f(x_0 - t) + f(x_0 + t) - 2\,c\big)\, D_n(t)\,\mathrm{d}t\,,$$

if one of the limits exists. ■

For a complete proof of the main result on the convergence of Fourier series, we need some additional preliminaries. Here we follow mainly the ideas of [202, p. 137 and pp. 144–148].

Let a compact interval $[a, b] \subset \mathbb{R}$ with $-\infty < a < b < \infty$ be given. Then a function $\varphi : [a, b] \to \mathbb{C}$ is called a *function of bounded variation*, if

$$V_a^b(\varphi) := \sup \sum_{j=1}^{n} |\varphi(x_j) - \varphi(x_{j-1})| < \infty\,, \tag{1.34}$$

where the supremum is taken over all partitions $a = x_0 < x_1 < \ldots < x_n = b$ of $[a, b]$. The nonnegative number $V_a^b(\varphi)$ is the *total variation* of φ on $[a, b]$. We set $V_a^a(\varphi) := 0$. For instance, each monotone function $\varphi : [a, b] \to \mathbb{R}$ is a function of bounded variation with $V_a^b(\varphi) = |\varphi(b) - \varphi(a)|$. Because

$$|\varphi(x)| \le |\varphi(a)| + |\varphi(x) - \varphi(a)| \le |\varphi(a)| + V_a^x(\varphi) \le |\varphi(a)| + V_a^b(\varphi) < \infty$$

for all $x \in [a, b]$, each function of bounded variation is bounded on $[a, b]$.

Lemma 1.29 *Let $\varphi : [a, b] \to \mathbb{C}$ and $\psi : [a, b] \to \mathbb{C}$ be functions of bounded variation. Then for arbitrary $\alpha \in \mathbb{C}$ and $c \in [a, b]$, it holds:*

$$V_a^b(\alpha\,\varphi) = |\alpha|\, V_a^b(\varphi)\,,$$

$$V_a^b(\varphi + \psi) \le V_a^b(\varphi) + V_a^b(\psi)\,,$$

$$V_a^b(\varphi) = V_a^c(\varphi) + V_c^b(\varphi)\,, \tag{1.35}$$

$$\max\{V_a^b(\mathrm{Re}\,\varphi),\ V_a^b(\mathrm{Im}\,\varphi)\} \le V_a^b(\varphi) \le V_a^b(\mathrm{Re}\,\varphi) + V_a^b(\mathrm{Im}\,\varphi)\,. \tag{1.36}$$

The simple proof is omitted here. For details see, e.g., [402, pp. 159–162].

Theorem 1.30 (Jordan's Decomposition Theorem) *Let $\varphi : [a, b] \to \mathbb{C}$ be a given function of bounded variation. Then, there exist four nondecreasing functions $\varphi_j : [a, b] \to \mathbb{R}$, $j = 1, \ldots, 4$, such that φ possesses the* Jordan decomposition

$$\varphi = (\varphi_1 - \varphi_2) + \mathrm{i}\,(\varphi_3 - \varphi_4)\,,$$

where $\mathrm{Re}\,\varphi = \varphi_1 - \varphi_2$ and $\mathrm{Im}\,\varphi = \varphi_3 - \varphi_4$ are functions of bounded variation. If φ is continuous, then φ_j, $j = 1, \ldots, 4$, are continuous too.

Proof From (1.36) it follows that $\mathrm{Re}\,\varphi$ and $\mathrm{Im}\,\varphi$ are functions of bounded variation. We decompose $\mathrm{Re}\,\varphi$. Obviously,

$$\varphi_1(x) := V_a^x(\mathrm{Re}\,\varphi)\,, \quad x \in [a, b]\,,$$

is nondecreasing by (1.35). Then,

$$\varphi_2(x) := \varphi_1(x) - \mathrm{Re}\,\varphi(x)\,, \quad x \in [a, b]\,,$$

is nondecreasing too, since for $a \le x < y \le b$ it holds

$$\big|\mathrm{Re}\,\varphi(y) - \mathrm{Re}\,\varphi(x)\big| \le V_x^y(\mathrm{Re}\,\varphi) = \varphi_1(y) - \varphi_1(x)$$

and hence

$$\varphi_2(y) - \varphi_2(x) = \big(\varphi_1(y) - \varphi_1(x)\big) - \big(\mathrm{Re}\,\varphi(y) - \mathrm{Re}\,\varphi(x)\big) \ge 0\,.$$

Thus, we obtain $\operatorname{Re}\varphi = \varphi_1 - \varphi_2$. Analogously, we can decompose $\operatorname{Im}\varphi = \varphi_3 - \varphi_4$. Using $\varphi = \operatorname{Re}\varphi + \mathrm{i}\, \operatorname{Im}\varphi$, we receive the above Jordan decomposition of φ. If φ is continuous at $x \in [a,\, b]$, then, by definition, each φ_j is continuous at x. ■

A 2π-periodic function $f\,:\,\mathbb{T} \to \mathbb{C}$ with $V_0^{2\pi}(f) < \infty$ is called a *2π-periodic function of bounded variation*. By (1.35) a 2π-periodic function of bounded variation has the property $V_a^b(f) < \infty$ for each compact interval $[a,\, b] \subset \mathbb{R}$.

Example 1.31 Let $f : \mathbb{T} \to \mathbb{C}$ be a piecewise continuously differentiable function with jump discontinuities at distinct points $x_j \in [0,\, 2\pi)$, $j = 1, \ldots, n$. Assume that it holds $f(x) = \frac{1}{2}\big(f(x+0) + f(x-0)\big)$ for all $x \in [0,\, 2\pi)$. Then f is a 2π-periodic function of bounded variation, since

$$V_0^{2\pi}(f) = \sum_{j=1}^{n} |f(x_j+0) - f(x_j-0)| + \int_0^{2\pi} |f'(t)|\, \mathrm{d}t < \infty\,.$$

The functions given in Examples 1.5, 1.8, 1.9, and 1.10 are 2π-periodic functions of bounded variation. □

Lemma 1.32 *There exists a constant $c_0 > 0$ such that for all $\alpha,\, \beta \in [0,\, \pi]$ and all $n \in \mathbb{N}$ it holds*

$$\Big|\int_\alpha^\beta D_n(t)\, \mathrm{d}t\Big| \le c_0\,. \tag{1.37}$$

Proof We introduce the function $h \in C[0,\, \pi]$ by

$$h(t) := \frac{1}{\sin\frac{t}{2}} - \frac{2}{t}\,, \qquad t \in (0,\, \pi]\,,$$

and $h(0) := 0$. This continuous function h is increasing, and we have $0 \le h(t) \le h(\pi) < \frac{1}{2}$ for all $t \in [0,\, \pi]$. Using (1.22), for arbitrary $\alpha,\, \beta \in [0,\, \pi]$, we estimate

$$\begin{aligned}\Big|\int_\alpha^\beta D_n(t)\, \mathrm{d}t\Big| &\le \Big|\int_\alpha^\beta h(t)\, \sin\big(n + \frac{1}{2}\big)t\, \mathrm{d}t\Big| + 2\Big|\int_\alpha^\beta \frac{\sin\big(n+\frac{1}{2}\big)t}{t}\, \mathrm{d}t\Big| \\ &\le \frac{\pi}{2} + 2\Big|\int_\alpha^\beta \frac{\sin\big(n+\frac{1}{2}\big)t}{t}\, \mathrm{d}t\Big|\,.\end{aligned}$$

By the sine integral

$$\mathrm{Si}(x) := \int_0^x \frac{\sin t}{t}\, \mathrm{d}t\,, \qquad x \in \mathbb{R}\,,$$

it holds for all $\gamma \ge 0$ (see Lemma 1.41)

$$\Big|\int_0^{\gamma} \frac{\sin x}{x}\,\mathrm{d}x\Big| \le \mathrm{Si}(\pi) < 2\,.$$

From

$$\int_{\alpha}^{\beta} \frac{\sin\big(n+\frac{1}{2}\big)t}{t}\,\mathrm{d}t = \int_0^{(n+\frac{1}{2})\beta} \frac{\sin x}{x}\,\mathrm{d}x - \int_0^{(n+\frac{1}{2})\alpha} \frac{\sin x}{x}\,\mathrm{d}x$$

it follows that

$$\Big|\int_{\alpha}^{\beta} \frac{\sin\big(n+\frac{1}{2}\big)t}{t}\,\mathrm{d}t\Big| \le 4\,,$$

i.e., (1.37) is fulfilled for the constant $c_0 = \frac{\pi}{2} + 8$. ■

Lemma 1.33 *Assume that* $0 < a < b < 2\pi$, $\delta > 0$, *and* $b - a + 2\delta < 2\pi$ *be given. Let* $\varphi : [a - \delta - \pi,\, b + \delta + \pi] \to \mathbb{R}$ *be nondecreasing, piecewise continuous function which is continuous on* $[a - \delta,\, b + \delta]$.

Then, for each $\varepsilon > 0$, *there exists an index* $n_0(\varepsilon)$ *such that for all* $n > n_0(\varepsilon)$ *and all* $x \in [a,\, b]$

$$\Big|\int_0^{\pi} \big(\varphi(x+t) + \varphi(x-t) - 2\,\varphi(x)\big)\, D_n(t)\,\mathrm{d}t\Big| < \varepsilon\,.$$

Proof

1. For $(x, t) \in [a - \delta,\, b + \delta] \times [0,\, \pi]$, we introduce the functions:

$$\begin{aligned} g(x,t) &:= \varphi(x+t) + \varphi(x-t) - 2\,\varphi(x)\,,\\ h_1(x,t) &:= \varphi(x+t) - \varphi(x) \ge 0\,,\\ h_2(x,t) &:= \varphi(x) - \varphi(x-t) \ge 0 \end{aligned}$$

such that $g = h_1 - h_2$. For fixed $x \in [a,\, b]$, both functions $h_j(x,\cdot)$, $j = 1, 2$, are nondecreasing on $[0,\, \pi]$. Since $h_j(\cdot,\pi)$, $j = 1,\, 2$, are piecewise continuous on $[a,\, b]$, there exists a constant $c_1 > 0$ such that for all $(x, t) \in [a,\, b] \times [0,\, \pi]$

$$\big|h_j(x,t)\big| \le c_1\,. \tag{1.38}$$

Since φ is continuous on the compact interval $[a - \delta,\, b + \delta]$, the function φ is uniformly continuous on $[a - \delta,\, b + \delta]$, i.e., for each $\varepsilon > 0$ there exists $\beta \in (0,\, \delta)$ such that for all $y, z \in [a - \delta,\, b + \delta]$ with $|y - z| \le \beta$ we have

$$\big|\varphi(y) - \varphi(z)\big| < \frac{\varepsilon}{4\,c_0}\,.$$

By the proof of Lemma 1.32, we can choose $c_0 = \frac{\pi}{2} + 8$. Hence, we obtain for all $(x, t) \in [a, b] \times [0, \beta]$ and $j = 1, 2$:

$$0 \le h_j(x, t) < \frac{\varepsilon}{4\, c_0}\,. \tag{1.39}$$

2. Now we split the integral:

$$\int_0^\pi g(x, t)\, D_n(t)\, \mathrm{d}t = \int_0^\beta g(x, t)\, D_n(t)\, \mathrm{d}t + \int_\beta^\pi g(x, t)\, D_n(t)\, \mathrm{d}t \tag{1.40}$$

into a sum of two integrals, where the first integral can be written in the form

$$\int_0^\beta g(x, t)\, D_n(t)\, \mathrm{d}t = \int_0^\beta h_1(x, t)\, D_n(t)\, \mathrm{d}t - \int_0^\beta h_2(x, t)\, D_n(t)\, \mathrm{d}t\,. \tag{1.41}$$

Observing that $h_j(x, \cdot)$, $j = 1, 2$, are nondecreasing for fixed $x \in [a, b]$, we obtain by the second mean value theorem for integrals; see, e.g., [402, pp. 328–329], that for certain $\alpha_j(x) \in [0, \beta]$

$$\begin{aligned}\int_0^\beta h_j(x, t)\, D_n(t)\, \mathrm{d}t &= h_j(x, 0) \int_0^{\alpha_j(x)} D_n(t)\, \mathrm{d}t + h_j(x, \beta) \int_{\alpha_j(x)}^\beta D_n(t)\, \mathrm{d}t \\ &= 0 + h_j(x, \beta) \int_{\alpha_j(x)}^\beta D_n(t)\, \mathrm{d}t\,, \quad j = 1, 2\,.\end{aligned}$$

By (1.37) and (1.39), this integral can be estimated for all $x \in [a, b]$ by

$$\Big|\int_0^\beta h_j(x, t)\, D_n(t)\, \mathrm{d}t\Big| \le \frac{\varepsilon}{4\, c_0}\, c_0 = \frac{\varepsilon}{4}$$

such that by (1.41) for all $x \in [a, b]$

$$\Big|\int_0^\beta g(x, t)\, D_n(t)\, \mathrm{d}t\Big| \le \frac{\varepsilon}{4} + \frac{\varepsilon}{4} = \frac{\varepsilon}{2}\,. \tag{1.42}$$

3. Next we consider the second integral in (1.40) which can be written as

$$\int_\beta^\pi g(x, t)\, D_n(t)\, \mathrm{d}t = \int_\beta^\pi h_1(x, t)\, D_n(t)\, \mathrm{d}t - \int_\beta^\pi h_2(x, t)\, D_n(t)\, \mathrm{d}t\,. \tag{1.43}$$

Since $h_j(x, \cdot)$, $j = 1, 2$, are nondecreasing for fixed $x \in [a, b]$, the second mean value theorem for integrals provides the existence of certain $\gamma_j(x) \in [\beta, \pi]$ such that

$$\int_\beta^\pi h_j(x, t)\, D_n(t)\, \mathrm{d}t = h_j(x, \beta) \int_\beta^{\gamma_j(x)} D_n(t)\, \mathrm{d}t + h_j(x, \pi) \int_{\gamma_j(x)}^\pi D_n(t)\, \mathrm{d}t\,. \tag{1.44}$$

From (1.22) it follows:

$$\int_\beta^{\gamma_j(x)} D_n(t)\,\mathrm{d}t = \int_\beta^{\gamma_j(x)} \frac{1}{\sin\frac{t}{2}}\,\sin\big(n+\frac{1}{2}\big)t\,\mathrm{d}t\,.$$

Since $\big(\sin\frac{t}{2}\big)^{-1}$ is monotone on $[\beta,\,\gamma_j(x)]$, again by the second mean value theorem for integrals, there exist $\eta_j(x) \in [\beta,\,\gamma_j(x)]$ with

$$\int_\beta^{\gamma_j(x)} D_n(t)\,\mathrm{d}t = \frac{1}{\sin\frac{\beta}{2}}\int_\beta^{\eta_j(x)} \sin\big(n+\frac{1}{2}\big)t\,\mathrm{d}t + \frac{1}{\sin\frac{\gamma_j(x)}{2}}\int_{\eta_j(x)}^{\gamma_j(x)} \sin\big(n+\frac{1}{2}\big)t\,\mathrm{d}t\,. \tag{1.45}$$

Now we estimate both integrals in (1.45) such that

$$\Big|\int_\beta^{\eta_j(x)} \sin\big(n+\frac{1}{2}\big)t\,\mathrm{d}t\Big| \le \frac{4}{2n+1}\,,$$

$$\Big|\int_{\eta_j(x)}^{\gamma_j(x)} \sin\big(n+\frac{1}{2}\big)t\,\mathrm{d}t\Big| \le \frac{4}{2n+1}\,.$$

Applying the above inequalities, we see by (1.45) for all $x \in [a,\,b]$ and $j = 1,\,2$ that

$$\Big|\int_\beta^{\gamma_j(x)} D_n(t)\,\mathrm{d}t\Big| \le \frac{8}{(2n+1)\,\sin\frac{\beta}{2}}\,. \tag{1.46}$$

Analogously, one can show for all $x \in [a,\,b]$ and $j = 1,\,2$ that

$$\Big|\int_{\gamma_j(x)}^{\pi} D_n(t)\,\mathrm{d}t\Big| \le \frac{8}{(2n+1)\,\sin\frac{\beta}{2}}\,. \tag{1.47}$$

Using (1.38) and (1.44), the inequalities (1.46) and (1.47) yield for all $x \in [a,\,b]$ and $j = 1,\,2$,

$$\Big|\int_\beta^{\pi} h_j(x,t)\,D_n(t)\,\mathrm{d}t\Big| \le \frac{16\,c_1}{(2n+1)\,\sin\frac{\beta}{2}}$$

and hence by (1.43)

$$\Big|\int_\beta^{\pi} g(x,t)\,D_n(t)\,\mathrm{d}t\Big| \le \frac{32\,c_1}{(2n+1)\,\sin\frac{\beta}{2}}\,.$$

Therefore, for the chosen $\varepsilon > 0$, there exists an index $n_0(\varepsilon) \in \mathbb{N}$ such that for all $n > n_0(\varepsilon)$ and all $x \in [a,\, b]$,

$$\Big| \int_\beta^\pi g(x,t)\, D_n(t)\, \mathrm{d}t \Big| < \frac{\varepsilon}{2}\,. \tag{1.48}$$

Together with (1.40), (1.42), and (1.48), it follows for all $n > n_0(\varepsilon)$ and all $x \in [a,\, b]$,

$$\Big| \int_0^\pi g(x,t)\, D_n(t)\, \mathrm{d}t \Big| < \varepsilon\,.$$

This completes the proof. ■

Based on Riemann's localization principle and these preliminaries, we can prove the following important theorem concerning pointwise convergence of the Fourier series of a piecewise continuously differentiable function $f : \mathbb{T} \to \mathbb{C}$.

Theorem 1.34 (Convergence Theorem of Dirichlet–Jordan) *Let $f : \mathbb{T} \to \mathbb{C}$ be a piecewise continuously differentiable function. Then at every point $x_0 \in \mathbb{R}$, the Fourier series of f converges as*

$$\lim_{n\to\infty} (S_n f)(x_0) = \frac{1}{2}\big(f(x_0+0) + f(x_0-0)\big)\,.$$

In particular, if f is continuous at x_0, then

$$\lim_{n\to\infty} (S_n f)(x_0) = f(x_0).$$

Further the Fourier series of f converges uniformly on any closed interval $[a,\, b] \subset (0,\, 2\pi)$, if f is continuous on $[a - \delta,\, b + \delta]$ with certain $\delta > 0$. Especially, if $f \in C(\mathbb{T})$ is piecewise continuously differentiable, then the Fourier series of f converges uniformly to f on $\mathbb{R}$.

Proof

1. By assumption there exists $\delta \in (0, \pi)$, such that f is continuously differentiable in $[x_0 - \delta, x_0 + \delta]\backslash\{x_0\}$. Let

$$M := \max_{t\in[-\pi,\,\pi]} \{|f'(t+0)|, |f'(t-0)|\}\,.$$

By the mean value theorem, we conclude:

$$|f(x_0+t) - f(x_0+0)| \le t\, M\,, \quad |f(x_0-t) - f(x_0-0)| \le t\, M$$

for all $t \in (0, \delta]$. This implies

$$\int_0^\delta \frac{|f(x_0 - t) + f(x_0 + t) - f(x_0 + 0) - f(x_0 - 0)|}{t}\,\mathrm{d}t \le 2\,M\delta < \infty\,.$$

By $\frac{t}{\pi} \le \sin\frac{t}{2}$ for $t \in [0,\,\pi]$, the function

$$h(t) := \frac{f(x_0 - t) + f(x_0 + t) - f(x_0 + 0) - f(x_0 - 0)}{t}\,\frac{t}{\sin\frac{t}{2}}\,, \qquad t \in (0,\,\delta]\,,$$

is absolutely integrable on $[0,\,\delta]$. By Lemma 1.27 of Riemann–Lebesgue, we get:

$$\lim_{n\to\infty} \int_0^\delta h(t)\,\sin\Big(\Big(n + \frac{1}{2}\Big)\,t\Big)\,\mathrm{d}t = 0\,.$$

Using Riemann's localization principle, cf. Theorem 1.28, we obtain the assertion with $2\,c = f(x_0 + 0) + f(x_0 - 0)$.

2. By assumption and Example 1.31, the given function f is a 2π-periodic function of bounded variation. Then, it follows that $V_{a-\delta-\pi}^{b+\delta+\pi}(f) < \infty$. By the Jordan decomposition Theorem 1.30, the function f restricted on $[a-\delta-\pi,\,b+\delta+\pi]$ can be represented in the form

$$f = (\varphi_1 - \varphi_2) + \mathrm{i}\,(\varphi_3 - \varphi_4)\,,$$

where $\varphi_j : [a - \delta - \pi,\, b + \delta + \pi] \to \mathbb{R}$, $j = 1, \ldots, 4$, are nondecreasing and piecewise continuous. Since f is continuous on $[a,\,b]$, each φ_j, $j = 1, \ldots, 4$, is continuous on $[a,\,b]$ too. Applying Lemma 1.33, we obtain that for each $\varepsilon > 0$ there exists an index $N(\varepsilon) \in \mathbb{N}$ such that for $n > N(\varepsilon)$ and all $x \in [a,\,b]$,

$$|(S_n f)(x) - f(x)| = \frac{1}{2\pi}\,\Big|\int_0^\pi \big(f(x+t) + f(x-t) - 2\,f(x)\big)\,D_n(t)\,\mathrm{d}t\Big| < \varepsilon\,.$$

This completes the proof. ■

Example 1.35 The functions $f : \mathbb{T} \to \mathbb{C}$ given in Examples 1.5, 1.8, 1.9, and 1.10 are piecewise continuously differentiable. If $x_0 \in \mathbb{R}$ is a jump discontinuity of f, then the value $f(x_0)$ is equal to the mean $\frac{1}{2}\,\big(f(x_0+0)+f(x_0-0)\big)$ of right and left limits. By the convergence Theorem 1.34 of Dirichlet–Jordan, the Fourier series of f converges to f in each point of $\mathbb{R}$. On each closed interval, which does not contain any discontinuity of f, the Fourier series converges uniformly. Since the piecewise continuously differentiable function of Example 1.8 is contained in $C(\mathbb{T})$, its Fourier series converges uniformly on $\mathbb{R}$. □

Remark 1.36 The convergence Theorem 1.34 of Dirichlet–Jordan is also valid for each 2π-periodic function $f : \mathbb{T} \to \mathbb{C}$ of bounded variation (see, e.g., [402, pp. 546–547]). □

1.4.2 Uniform Convergence

A useful criterion for uniform convergence of the Fourier series of a function $f \in C(\mathbb{T})$ is the following.

Theorem 1.37 *If $f \in C(\mathbb{T})$ fulfills the condition*

$$\sum_{k\in\mathbb{Z}} |c_k(f)| < \infty\,, \tag{1.49}$$

then the Fourier series of f converges uniformly to f. Each function $f \in C^1(\mathbb{T})$ has the property (1.49).

Proof By the assumption (1.49) and

$$|c_k(f)\,\mathrm{e}^{\mathrm{i}\,k\cdot}| = |c_k(f)|\,,$$

the uniform convergence of the Fourier series follows from the Weierstrass criterion of uniform convergence. If $g \in C(\mathbb{T})$ is the sum of the Fourier series of f, then we obtain for all $k \in \mathbb{Z}$

$$c_k(g) = \langle g,\, \mathrm{e}^{\mathrm{i}\,k\cdot}\rangle = \sum_{n\in\mathbb{Z}} c_n(f)\,\langle \mathrm{e}^{\mathrm{i}\,n\cdot},\, \mathrm{e}^{\mathrm{i}\,k\cdot}\rangle = c_k(f)$$

such that $g = f$ by Theorem 1.1.

Assume that $f \in C^1(\mathbb{T})$. By the convergence Theorem 1.34 of Dirichlet–Jordan, we know already that the Fourier series of f converges uniformly to f. This could be also seen as follows: by the differentiation property of the Fourier coefficients in Lemma 1.6, we have $c_k(f) = (\mathrm{i}\,k)^{-1}\,c_k(f')$ for all $k \neq 0$ and $c_0(f') = 0$. By Parseval equality of $f' \in L_2(\mathbb{T})$, it follows:

$$\|f'\|^2 = \sum_{k\in\mathbb{Z}} |c_k(f')|^2 < \infty\,.$$

Using Cauchy–Schwarz inequality, we get finally

$$\begin{aligned}\sum_{k\in\mathbb{Z}} |c_k(f)| &= |c_0(f)| + \sum_{k\neq 0} \frac{1}{|k|}\,|c_k(f')| \\ &\le |c_0(f)| + \Big(\sum_{k\neq 0}\frac{1}{k^2}\Big)^{1/2}\Big(\sum_{k\neq 0}|c_k(f')|^2\Big)^{1/2} < \infty.\end{aligned}$$

This completes the proof. ∎

Remark 1.38 If $f \in C^1(\mathbb{T})$, then by the mean value theorem it follows that

$$|f(x+h) - f(x)| \le |h| \max_{t \in \mathbb{T}} |f'(t)|$$

for all $x, x+h \in \mathbb{T}$, which means f is *Lipschitz continuous* on $\mathbb{T}$. More generally, a function $f \in C(\mathbb{T})$ is called *Hölder continuous of order* $\alpha \in (0,\,1]$ on $\mathbb{T}$, if

$$|f(x+h) - f(x)| \le c\,|h|^{\alpha}$$

for all $x, x+h \in \mathbb{T}$ with certain constant $c \ge 0$ which depends on f. One can show that the Fourier series of a function $f \in C(\mathbb{T})$ which is Hölder continuous of order $\alpha \in (0,\,1]$ converges uniformly to f and it holds:

$$\|S_n f - f\|_{C(\mathbb{T})} = \mathcal{O}(n^{-\alpha}\,\ln n)\,, \quad n \to \infty$$

(see [453, Vol. I, p. 64]). □

In practice, the following convergence result of Fourier series for a sufficiently smooth, 2π–periodic function is very useful.

Theorem 1.39 (Bernstein) *Let $f \in C^r(\mathbb{T})$ with fixed $r \in \mathbb{N}$ be given. Then, the Fourier coefficients $c_k(f)$ have the form:*

$$c_k(f) = \frac{1}{(\mathrm{i}\,k)^r}\, c_k(f^{(r)})\,, \quad k \in \mathbb{Z} \setminus \{0\}\,. \tag{1.50}$$

Further the approximation error $f - S_n f$ can be estimated for all $n \in \mathbb{N} \setminus \{1\}$ by

$$\|f - S_n f\|_{C(\mathbb{T})} \le c\,\|f^{(r)}\|_{C(\mathbb{T})}\,\frac{\ln n}{n^r}\,, \tag{1.51}$$

where the constant $c > 0$ is independent of f and n.

Proof

1. Repeated integration by parts provides (1.50). By Lemma 1.27 of Riemann–Lebesgue, we know

$$\lim_{|k| \to \infty} c_k(f^{(r)}) = 0$$

such that

$$\lim_{|k| \to \infty} k^r\, c_k(f) = 0\,.$$

2. The nth partial sum of the Fourier series of $f^{(r)} \in C(\mathbb{T})$ can be written in the form

$$\big(S_n f^{(r)}\big)(x) = \frac{1}{\pi} \int_0^{\pi} \big(f^{(r)}(x+y) + f^{(r)}(x-y)\big)\, \frac{\sin(n+\frac{1}{2})y}{2\,\sin\frac{y}{2}}\, \mathrm{d}y\,.$$

Then, we estimate:

$$
\begin{aligned}
\big|(S_n f^{(r)})(x)\big| &\le \frac{2}{\pi}\,\|f^{(r)}\|_{C(\mathbb{T})} \int_0^{\pi} \frac{|\sin(n+\frac{1}{2})y|}{2\,\sin\frac{y}{2}}\,\mathrm{d}y \\
&< \|f^{(r)}\|_{C(\mathbb{T})} \int_0^{\pi} \frac{|\sin(n+\frac{1}{2})y|}{y}\,\mathrm{d}y \\
&= \|f^{(r)}\|_{C(\mathbb{T})} \int_0^{(n+\frac{1}{2})\pi} \frac{|\sin u|}{u}\,\mathrm{d}u \\
&< \|f^{(r)}\|_{C(\mathbb{T})} \Big(1 + \int_1^{(n+\frac{1}{2})\pi} \frac{1}{u}\,\mathrm{d}u\Big) \\
&= \|f^{(r)}\|_{C(\mathbb{T})} \Big(1 + \ln\big(n+\frac{1}{2}\big)\pi\Big)\,.
\end{aligned}
$$

For a convenient constant $c > 0$, we obtain for all $n \in \mathbb{N} \setminus \{1\}$ that

$$
\|S_n f^{(r)}\|_{C(\mathbb{T})} \le c\,\|f^{(r)}\|_{C(\mathbb{T})}\,\ln n\,. \tag{1.52}
$$

By Theorem 1.37 the Fourier series of f converges uniformly to f such that by (1.50)

$$
\begin{aligned}
f - S_n f &= \sum_{k=n+1}^{\infty} \big(c_k(f)\,\mathrm{e}^{\mathrm{i}k\cdot} + c_{-k}(f)\,\mathrm{e}^{-\mathrm{i}k\cdot}\big) \\
&= \sum_{k=n+1}^{\infty} \frac{1}{(\mathrm{i}k)^r}\big(c_k(f^{(r)})\,\mathrm{e}^{\mathrm{i}k\cdot} + (-1)^r\,c_{-k}(f^{(r)})\,\mathrm{e}^{-\mathrm{i}k\cdot}\big)\,.
\end{aligned} \tag{1.53}
$$

3. For even smoothness $r = 2s$, $s \in \mathbb{N}$, we obtain by (1.53) that

$$
\begin{aligned}
f - S_n f &= (-1)^s \sum_{k=n+1}^{\infty} \frac{1}{k^r}\big(c_k(f^{(r)})\,\mathrm{e}^{\mathrm{i}k\cdot} + c_{-k}(f^{(r)})\,\mathrm{e}^{-\mathrm{i}k\cdot}\big) \\
&= (-1)^s \sum_{k=n+1}^{\infty} \frac{1}{k^r}\big(S_k f^{(r)} - S_{k-1} f^{(r)}\big)\,.
\end{aligned}
$$

Obviously, for $N > n$, it holds the identity:

$$
\sum_{k=n+1}^{N} a_k\,(b_k - b_{k-1}) = a_N\,b_N - a_{n+1}\,b_n + \sum_{k=n+1}^{N-1} (a_k - a_{k+1})\,b_k \tag{1.54}
$$

for arbitrary complex numbers a_k and b_k. We apply (1.54) to $a_k = k^{-r}$ and $b_k = S_k f^{(r)}$. Then for $N \to \infty$ we receive

$$f - S_n f = (-1)^{s+1} \frac{1}{(n+1)^r} S_n f^{(r)} + (-1)^s \sum_{k=n+1}^{\infty} \Big(\frac{1}{k^r} - \frac{1}{(k+1)^r}\Big) S_k f^{(r)}, \tag{1.55}$$

since by (1.52)

$$\frac{1}{N^r} \|S_N f^{(r)}\|_{C(\mathbb{T})} \le c \, \|f^{(r)}\|_{C(\mathbb{T})} \frac{\ln N}{N^r} \to 0 \quad \text{as } N \to \infty .$$

Thus, we can estimate the approximation error (1.55) by

$$\|f - S_n f\|_{C(\mathbb{T})} \le c \, \|f^{(r)}\|_{C(\mathbb{T})} \Big(\frac{\ln n}{(n+1)^r} + \sum_{k=n+1}^{\infty} \Big(\frac{1}{k^r} - \frac{1}{(k+1)^r}\Big) \ln k\Big).$$

Using the identity (1.54) for $a_k = \ln k$ and $b_k = -(k+1)^{-r}$, we see that

$$\sum_{k=n+1}^{\infty} \Big(\frac{1}{k^r} - \frac{1}{(k+1)^r}\Big) \ln k = \frac{\ln(n+1)}{(n+1)^r} + \sum_{k=n+1}^{\infty} \frac{1}{(k+1)^r} \ln\Big(1 + \frac{1}{k}\Big),$$

since $(N+1)^{-k} \ln N \to 0$ as $N \to \infty$. From $\ln(1 + \frac{1}{k}) < \frac{1}{k}$, it follows that

$$\begin{aligned}\sum_{k=n+1}^{\infty} \frac{1}{(k+1)^r} \ln\Big(1 + \frac{1}{k}\Big) &< \sum_{k=n+1}^{\infty} \frac{1}{k\,(k+1)^r} < \sum_{k=n+1}^{\infty} \frac{1}{k^{r+1}} \\ &< \int_n^{\infty} \frac{1}{x^{r+1}} \, \mathrm{d}x = \frac{1}{r\,n^r}.\end{aligned}$$

Hence, for convenient constant $c_1 > 0$, we have:

$$\|f - S_n f\|_{C(\mathbb{T})} \le c_1 \, \|f^{(r)}\|_{C(\mathbb{T})} \frac{1}{n^r} (1 + \ln n).$$

This inequality implies (1.51) for even r.

4. The case of odd smoothness $r = 2s+1$, $s \in \mathbb{N}_0$, can be handled similarly as the case of even r. By (1.53) we obtain

$$\begin{aligned} f - S_n f &= (-1)^s \, \mathrm{i} \sum_{k=n+1}^{\infty} \frac{1}{k^r} \big(c_{-k}(f^{(r)})\, \mathrm{e}^{-\mathrm{i}k\cdot} - c_k(f^{(r)})\, \mathrm{e}^{\mathrm{i}k\cdot}\big) \\ &= (-1)^s \sum_{k=n+1}^{\infty} \frac{1}{k^r} \big(\tilde{S}_k f^{(r)} - \tilde{S}_{k-1} f^{(r)}\big) \end{aligned} \tag{1.56}$$

with the *nth partial sum of the conjugate Fourier series* of $f^{(r)}$

$$\tilde{S}_n f^{(r)} := \mathrm{i} \sum_{j=1}^{n} \big(c_{-j}(f^{(r)})\, \mathrm{e}^{-\mathrm{i}j\cdot} - c_j(f^{(r)})\, \mathrm{e}^{\mathrm{i}j\cdot}\big)\,.$$

From

$$\begin{aligned} \mathrm{i}\,\big(c_{-j}(f^{(r)})\, \mathrm{e}^{-\mathrm{i}jx} - c_j(f^{(r)})\, \mathrm{e}^{\mathrm{i}jx}\big) &= -\frac{1}{\pi} \int_{-\pi}^{\pi} f^{(r)}(y)\, \sin j(y-x)\, \mathrm{d}y \\ &= -\frac{1}{\pi} \int_{-\pi}^{\pi} f^{(r)}(x+y)\, \sin(jy)\, \mathrm{d}y \\ &= -\frac{1}{\pi} \int_{0}^{\pi} \big(f^{(r)}(x+y) - f^{(r)}(x-y)\big) \\ &\quad \times \sin(jy)\, \mathrm{d}y \end{aligned}$$

and

$$\sum_{j=1}^{n} \sin(jy) = \frac{\cos\frac{y}{2} - \cos(n+\frac{1}{2})y}{2\,\sin\frac{y}{2}}\,, \qquad y \in \mathbb{R} \setminus 2\pi\,\mathbb{Z}\,,$$

it follows that

$$(\tilde{S}_n f^{(r)})(x) = -\frac{1}{\pi} \int_{0}^{\pi} \big(f^{(r)}(x+y) - f^{(r)}(x-y)\big)\, \frac{\cos\frac{y}{2} - \cos(n+\frac{1}{2})y}{2\,\sin\frac{y}{2}}\, \mathrm{d}y$$

and hence

$$\begin{aligned} |(\tilde{S}_n f^{(r)})(x)| &\le \frac{2}{\pi}\, \|f^{(r)}\|_{C(\mathbb{T})} \int_0^{\pi} \frac{|\cos\frac{y}{2} - \cos(n+\frac{1}{2})y|}{2\,\sin\frac{y}{2}}\, \mathrm{d}y \\ &= \frac{4}{\pi}\, \|f^{(r)}\|_{C(\mathbb{T})} \int_0^{\pi} \frac{|\sin\frac{ny}{2}\, \sin\frac{(n+1)y}{2}|}{2\,\sin\frac{y}{2}}\, \mathrm{d}y \\ &< \frac{4}{\pi}\, \|f^{(r)}\|_{C(\mathbb{T})} \int_0^{\pi} \frac{|\sin\frac{(n+1)y}{2}|}{2\,\sin\frac{y}{2}}\, \mathrm{d}y\,. \end{aligned}$$

Similarly as in step 2, we obtain for any $n \in \mathbb{N} \setminus \{1\}$

$$\|\tilde{S}_n f^{(r)}\|_{C(\mathbb{T})} \le c\, \|f^{(r)}\|_{C(\mathbb{T})}\, \ln n$$

with some constant $c > 0$.

Now we apply the identity (1.54) to $a_k = k^{-r}$ and $b_k = \tilde{S}_k f^{(r)}$. For $N \to \infty$ it follows from (1.56) that

$$f - S_n f = (-1)^{s+1} \frac{1}{(n+1)^r} \tilde{S}_n f^{(r)} + (-1)^s \sum_{k=n+1}^{\infty} \Big(\frac{1}{k^r} - \frac{1}{(k+1)^r}\Big) \tilde{S}_k f^{(r)} .$$

Thus, we obtain the estimate:

$$\|f - S_n f\|_{C(\mathbb{T})} \le c\, \|f^{(r)}\|_{C(\mathbb{T})} \Big(\frac{\ln n}{(n+1)^r} + \sum_{k=n+1}^{\infty} \Big(\frac{1}{k^r} - \frac{1}{(k+1)^r}\Big) \ln k\Big) .$$

We proceed as in step 3 and show the estimate (1.51) for odd r. ■

Roughly speaking we can say by Bernstein's theorem 1.39: The smoother a function $f : \mathbb{T} \to \mathbb{C}$ is, the faster its Fourier coefficients $c_k(f)$ tend to zero as $|k| \to \infty$ and the faster its Fourier series converges uniformly to f.

Remark 1.40 Let $f \in C^{r-1}(\mathbb{T})$ with fixed $r \in \mathbb{N}$ be given. Assume that $f^{(r)}$ exists in $[0, 2\pi)$ without finitely many points $x_j \in [0, 2\pi)$. Suppose that both one-sided derivatives $f^{(r)}(x_j \pm 0)$ exist and are finite for each x_j and that $f^{(r)}(x_j) := \frac{1}{2}\big(f^{(r)}(x_j+0)+f^{(r)}(x_j-0)\big)$. If $f^{(r)}$ is of bounded variation $V_0^{2\pi}\big(f^{(r)}\big)$, c.f. (1.34), then for all $k \in \mathbb{Z} \setminus \{0\}$ we have:

$$|c_k(f)| \le \frac{V_0^{2\pi}\big(f^{(r)}\big)}{2\pi\, |k|^{r+1}} .$$

This upper bound can be derived by integrating $c_k(f)$ by parts r-times, followed by partial integration of a Stieltjes integral:

$$\int_0^{2\pi} f^{(r)}(x)\, \mathrm{e}^{-\mathrm{i}kx}\, \mathrm{d}x = \int_0^{2\pi} f^{(r)}(x)\, \mathrm{d}g(x) = f^{(r)}(x)\, g(x)\big|_0^{2\pi} - \int_0^{2\pi} g(x)\, \mathrm{d}f^{(r)}(x)$$

with $g(x) = \frac{1}{-\mathrm{i}k}\, \mathrm{e}^{-\mathrm{i}kx}$; see, e.g., [63, pp. 186–188], [441, Theorem 4.3]. □

1.4.3 Gibbs Phenomenon

Let $f : \mathbb{T} \to \mathbb{C}$ be a piecewise continuously differentiable function with a jump discontinuity at $x_0 \in \mathbb{R}$. Then Theorem 1.34 of Dirichlet–Jordan implies

$$\lim_{n\to\infty} (S_n f)(x_0) = \frac{f(x_0 - 0) + f(x_0 + 0)}{2} .$$

Clearly, the Fourier series of f cannot converge uniformly in any small neighborhood of x_0, because the uniform limit of the continuous functions $S_n f$ would be continuous. The *Gibbs phenomenon* describes the bad convergence behavior of the Fourier partial sums $S_n f$ in a small neighborhood of x_0. If $n \to \infty$, then $S_n f$ overshoots and undershoots f near the jump discontinuity at x_0; see the right Fig. 1.4.

First we analyze the convergence of the Fourier partial sums $S_n s$ of the sawtooth function s from Example 1.9 which is piecewise linear with $s(0) = 0$ and therefore piecewise continuously differentiable. The nth Fourier partial sum $S_n s$ reads as

$$(S_n s)(x) = \sum_{k=1}^{n} \frac{1}{\pi k} \sin(kx).$$

By the Theorem 1.34 of Dirichlet–Jordan, $(S_n s)(x)$ converges to $s(x)$ as $n \to \infty$ at each point $x \in \mathbb{R} \setminus 2\pi\,\mathbb{Z}$ such that

$$s(x) = \sum_{k=1}^{\infty} \frac{1}{\pi k} \sin(kx)\,.$$

Now we compute $S_n s$ in the neighborhood $\big(-\frac{\pi}{2}, \frac{\pi}{2}\big)$ of the jump discontinuity at $x_0 = 0$. By Example 1.14 we have:

$$\frac{1}{2} + \sum_{k=1}^{n} \cos(kt) = \frac{1}{2}\, D_n(t)\,, \quad t \in \mathbb{R}\,,$$

and hence by integration

$$\begin{aligned} \frac{x}{2\pi} + (S_n s)(x) &= \frac{1}{2\pi} \int_0^x D_n(t)\,\mathrm{d}t = \frac{1}{\pi} \int_0^{x/2} \frac{\sin\big((2n+1)t\big)}{t}\,\mathrm{d}t \\ &\quad + \frac{1}{\pi} \int_0^{x/2} h(t)\, \sin\big((2n+1)t\big)\,\mathrm{d}t\,, \end{aligned} \tag{1.57}$$

where the function

$$h(t) := \begin{cases} (\sin t)^{-1} - t^{-1} & t \in \big(-\frac{\pi}{2}, \frac{\pi}{2}\big) \setminus \{0\}, \\ 0 & t = 0 \end{cases}$$

is continuously differentiable in $\big(-\frac{\pi}{2}, \frac{\pi}{2}\big)$. Integration by parts yields:

$$\frac{1}{\pi} \int_0^{x/2} h(t)\, \sin\big((2n+1)\,t\big)\,\mathrm{d}t = \mathcal{O}(n^{-1})\,, \quad n \to \infty\,.$$

Using the *sine integral*

$$\mathrm{Si}(y) := \int_0^y \frac{\sin t}{t}\,\mathrm{d}t\,, \quad y \in \mathbb{R}\,,$$

we obtain:

$$(S_n s)(x) = \frac{1}{\pi}\,\mathrm{Si}\Big(\Big(n + \frac{1}{2}\Big)\,x\Big) - \frac{x}{2\pi} + \mathcal{O}(n^{-1})\,, \quad n \to \infty\,. \tag{1.58}$$

Lemma 1.41 *The sine integral has the property:*

$$\lim_{y\to\infty} \mathrm{Si}(y) = \int_0^\infty \frac{\sin t}{t}\,\mathrm{d}t = \frac{\pi}{2}\,.$$

Further $\mathrm{Si}(\pi)$ *is the maximum value of the sine integral.*

Proof Introducing

$$a_k := \int_{k\pi}^{(k+1)\pi} \frac{\sin t}{t}\,\mathrm{d}t\,, \quad k \in \mathbb{N}_0\,,$$

we see that $\operatorname{sgn} a_k = (-1)^k$, $|a_k| > |a_{k+1}|$ and $\lim_{k\to\infty}|a_k| = 0$. By the Leibniz criterion for alternating series, we obtain that

$$\int_0^\infty \frac{\sin t}{t}\,\mathrm{d}t = \sum_{k=0}^\infty a_k < \infty\,,$$

i.e., $\lim_{y\to\infty} \mathrm{Si}(y)$ exists. From (1.57) with $x = \pi$, it follows that

$$\frac{\pi}{2} = \int_0^{\pi/2} \frac{\sin\big((2k+1)\,t\big)}{t}\,\mathrm{d}t + \int_0^{\pi/2} h(t)\,\sin\big((2k+1)\,t\big)\,\mathrm{d}t\,.$$

By the Lemma 1.27 of Riemann–Lebesgue, we conclude for $k \to \infty$ that

$$\frac{\pi}{2} = \lim_{k\to\infty} \int_0^{\pi/2} \frac{\sin\big((2k+1)\,t\big)}{t}\,\mathrm{d}t = \lim_{k\to\infty} \int_0^{(k+\frac{1}{2})\pi} \frac{\sin x}{x}\,\mathrm{d}x\,.$$

Consequently,

$$\sum_{k=0}^\infty a_k = \frac{\pi}{2}\,, \quad \mathrm{Si}(n\pi) = \sum_{k=0}^{n-1} a_k\,, \quad n \in \mathbb{N}\,.$$

The function Si defined on $[0,\,\infty)$ is continuous, bounded, and nonnegative. Further Si increases monotonically on $[2k\pi,\,(2k+1)\pi]$ and decreases monotonically on

$[(2k+1)\pi,\ (2k+2)\pi]$ for all $k \in \mathbb{N}_0$. Thus, we have:

$$\max\{\text{Si}(y) : y \in [0,\ \infty)\} = \text{Si}(\pi) \approx 1.8519\,. \qquad \blacksquare$$

For $x = \frac{2\pi}{2n+1}$, we obtain by (1.58) and Lemma 1.41 that

$$(S_n s)\Big(\frac{2\pi}{2n+1}\Big) = \frac{1}{\pi}\,\text{Si}(\pi) - \frac{1}{2n+1} + \mathcal{O}(n^{-1})\,, \qquad n \to \infty\,,$$

where $\frac{1}{\pi}\,\text{Si}(\pi)$ is the maximum value of $\frac{1}{\pi}\,\text{Si}\big((n+\frac{1}{2})x\big)$ for all $x > 0$.

Ignoring the term $-\frac{1}{2n+1} + \mathcal{O}(n^{-1})$ for large n, we conclude that

$$\lim_{n\to\infty}(S_n s)\Big(\frac{2\pi}{2n+1}\Big) = \frac{1}{\pi}\,\text{Si}(\pi) = s(0+0) + \Big(\frac{1}{\pi}\,\text{Si}(\pi) - \frac{1}{2}\Big)\,\big(s(0+0) - s(0-0)\big)\,,$$

where $\frac{1}{\pi}\,\text{Si}(\pi) - \frac{1}{2} \approx 0.08949$. Since the sawtooth function $s : \mathbb{T} \to \mathbb{C}$ is odd, we obtain that

$$\lim_{n\to\infty}(S_n s)\Big(-\frac{2\pi}{2n+1}\Big) = -\frac{1}{\pi}\,\text{Si}(\pi) = s(0-0) - \Big(\frac{1}{\pi}\,\text{Si}(\pi) - \frac{1}{2}\Big)\,\big(s(0+0) - s(0-0)\big)\,.$$

Thus for large n, we observe an overshooting and undershooting of $S_n s$ at both sides of the jump discontinuity of approximately 9% of the jump height. This behavior does not change with growing n und is typical for the convergence of $S_n s$ near a jump discontinuity. Figure 1.10 illustrates this behavior.

A general description of the Gibbs phenomenon is given by the following.

Theorem 1.42 (Gibbs Phenomenon) *Let $f : \mathbb{T} \to \mathbb{C}$ be a piecewise continuously differentiable function with a jump discontinuity at $x_0 \in \mathbb{R}$. Assume that $f(x_0) = \frac{1}{2}\big(f(x_0 - 0) + f(x_0 + 0)\big)$. Then, it holds:*

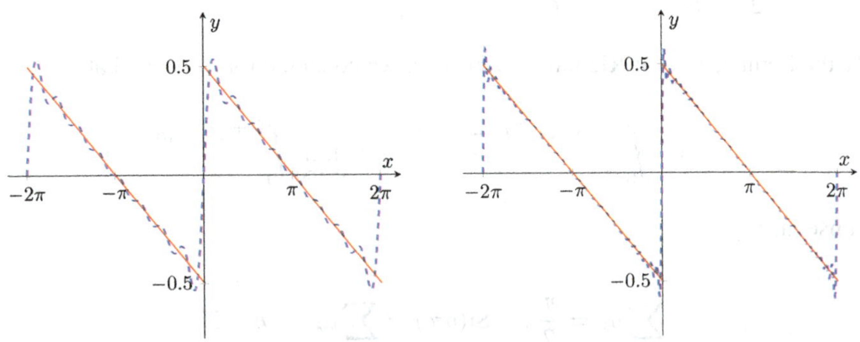

Fig. 1.10 Gibbs phenomenon for the Fourier partial sums $S_8 s$ (blue, left) and $S_{16} s$ (blue, right), where s is the 2π-periodic sawtooth function (red)

$$\lim_{n\to\infty} (S_n f)\Big(x_0 + \frac{2\pi}{2n+1}\Big) = f(x_0+0) + \Big(\frac{1}{\pi}\,\mathrm{Si}(\pi) - \frac{1}{2}\Big) \times \big(f(x_0+0) - f(x_0-0)\big)\,,$$

$$\lim_{n\to\infty} (S_n f)\Big(x_0 - \frac{2\pi}{2n+1}\Big) = f(x_0-0) - \Big(\frac{1}{\pi}\,\mathrm{Si}(\pi) - \frac{1}{2}\Big) \times \big(f(x_0+0) - f(x_0-0)\big)\,.$$

Proof Let $s : \mathbb{T} \to \mathbb{C}$ denote the sawtooth function of Example 1.9. We consider the function:

$$g := f - \big(f(x_0+0) - f(x_0-0)\big)\, s(\cdot - x_0)\,.$$

Then, $g : \mathbb{T} \to \mathbb{C}$ is also piecewise continuously differentiable and continuous in an interval $[x_0 - \delta,\, x_0 + \delta]$ with $\delta > 0$. Further we have $g(x_0) = f(x_0) = \frac{1}{2}\big(f(x_0-0) + f(x_0+0)\big)$. By the Theorem 1.34 of Dirichlet–Jordan, the Fourier series of g converges uniformly to g in $[x_0 - \delta,\, x_0 + \delta]$. By

$$(S_n f)(x) = (S_n g)(x) + \big(f(x_0+0) - f(x_0-0)\big) \sum_{k=1}^{n} \frac{1}{\pi k}\, \sin\big(k\,(x - x_0)\big)$$

it follows for $x = x_0 \pm \frac{2\pi}{2n+1}$ and $n \to \infty$ that

$$\lim_{n\to\infty} (S_n f)\Big(x_0 + \frac{2\pi}{2n+1}\Big) = g(x_0) + \frac{1}{\pi}\,\mathrm{Si}(\pi)\,\big(f(x_0+0) - f(x_0-0)\big)\,,$$

$$\lim_{n\to\infty} (S_n f)\Big(x_0 - \frac{2\pi}{2n+1}\Big) = g(x_0) - \frac{1}{\pi}\,\mathrm{Si}(\pi)\,\big(f(x_0+0) - f(x_0-0)\big)\,.$$

This completes the proof. ■

For large n, the Fourier partial sum $S_n f$ of a piecewise continuously differentiable function $f : \mathbb{T} \to \mathbb{C}$ exhibits the overshoot and undershoot at each point of jump discontinuity. If f is continuous at x_0, then $S_n f$ converges uniformly to f as $n \to \infty$ in a certain neighborhood of x_0 and the Gibbs phenomenon is absent.

Remark 1.43 Assume that $f : \mathbb{T} \to \mathbb{C}$ is a piecewise continuously differentiable function. By the Gibbs phenomenon, the truncation of Fourier series to $S_n f$ causes ripples in a neighborhood of each point of jump discontinuity. These ripples can be removed by the use of properly weighted Fourier coefficients such as by Fejér summation or Lanczos smoothing.

By the *Fejér summation* , we take the arithmetic mean $\sigma_n f$ of all Fourier partial sums $S_k f$, $k = 0, \ldots, n$, i.e.,

$$\sigma_n f = \frac{1}{n+1} \sum_{k=0}^{n} S_k f \in \mathscr{T}_n \, .$$

Then $\sigma_n f$ is the nth Fejér sum of f. With the Fejér kernel

$$F_n = \frac{1}{n+1} \sum_{k=0}^{n} D_k \in \mathscr{T}_n$$

of Example 1.15 and by $S_k f = f * D_k$, $k = 0, \ldots, n$, we obtain the representation $\sigma_n f = f * F_n$. Since

$$S_k f = \sum_{j=-k}^{k} c_j(f) \, \mathrm{e}^{\mathrm{i}j\cdot} \, ,$$

then it follows that

$$\sigma_n f = \frac{1}{n+1} \sum_{k=0}^{n} \sum_{j=-k}^{k} c_j(f) \, \mathrm{e}^{\mathrm{i}j\cdot} = \sum_{\ell=-n}^{n} \Big(1 - \frac{|\ell|}{n+1}\Big) c_\ell(f) \, \mathrm{e}^{\mathrm{i}\ell\cdot} \, .$$

Note that the positive weights

$$\omega_\ell := 1 - \frac{|\ell|}{n+1} \, , \qquad \ell = -n, \ldots, n$$

decay linearly from $\omega_0 = 1$ to $\omega_n = \omega_{-n} = (n+1)^{-1}$ as $|\ell|$ increases from 0 to n.

In contrast to the Fejér summation, the *Lanczos smoothing* uses the means of the function $S_n f$ over the intervals $[x - \frac{\pi}{n}, x - \frac{\pi}{n}]$ for each $x \in \mathbb{T}$, i.e., we form

$$(\Lambda_n f)(x) := \frac{n}{2\pi} \int_{x-\pi/n}^{x+\pi/n} (S_n f)(u) \, \mathrm{d}u \, .$$

By

$$S_n f = \sum_{k=-n}^{n} c_k(f) \, \mathrm{e}^{\mathrm{i}k\cdot} \, ,$$

we obtain the weighted Fourier partial sum:

$$(\Lambda_n f)(x) = \frac{n}{2\pi} \sum_{k=-n}^{n} c_k(f) \int_{x-\pi/n}^{x+\pi/n} \mathrm{e}^{\mathrm{i}ku} \, \mathrm{d}u$$

$$= \sum_{k=-n}^{n} \Big(\operatorname{sinc} \frac{k\pi}{n} \Big) c_k(f)\, \mathrm{e}^{\mathrm{i}kx} \,,$$

where the nonnegative weights $\omega_k := \operatorname{sinc} \frac{k\pi}{n}$, $k = -n, \ldots, n$, decay from $\omega_0 = 1$ to $\omega_n = \omega_{-n} = 0$ as $|\ell|$ increases from 0 to n. If we arrange that $\omega_k := 0$ for all $k \in \mathbb{Z}$ with $|k| > n$, then we obtain a so-called window sequence which will be considered in the next section. □

1.5 Discrete Signals and Linear Filters

In this section we apply Fourier series in the digital signal processing. The set of all bounded complex sequences $x = \big(x_k\big)_{k\in\mathbb{Z}}$ is denoted by $\ell_\infty(\mathbb{Z})$. It turns out that $\ell_\infty(\mathbb{Z})$ is a Banach space under the norm

$$\|x\|_\infty := \sup\{|x_k| : k \in \mathbb{Z}\}.$$

For $1 \le p < \infty$, we denote by $\ell_p(\mathbb{Z})$ the set of all complex sequences $x = \big(x_k\big)_{k\in\mathbb{Z}}$ such that

$$\|x\|_p := \Big(\sum_{k\in\mathbb{Z}} |x_k|^p\Big)^{1/p} < \infty \,.$$

Then $\ell_p(\mathbb{Z})$ is a Banach space. For $p = 2$, we obtain the Hilbert space $\ell_2(\mathbb{Z})$ with the inner product and the norm

$$\langle x,\, y\rangle := \sum_{k\in\mathbb{Z}} x_k\, \overline{y}_k \,, \qquad \|x\|_2 := \Big(\sum_{k\in\mathbb{Z}} |x_k|^2\Big)^{1/2}$$

for $x = \big(x_k\big)_{k\in\mathbb{Z}}$ and $y = \big(y_k\big)_{k\in\mathbb{Z}} \in \ell_2(\mathbb{Z})$. Note that

$$\|x\|_2^2 = \sum_{k\in\mathbb{Z}} |x_k|^2$$

is the so-called energy of x. The Cauchy–Schwarz inequality reads for all $x,\, y \in \ell_2(\mathbb{Z})$ as follows:

$$|\langle x,\, y\rangle| \le \|x\|_2\, \|y\|_2 \,.$$

A *discrete signal* is defined as a bounded complex sequence $x = \big(x_k\big)_{k\in\mathbb{Z}} \in \ell_\infty(\mathbb{Z})$. If $f : \mathbb{R} \to \mathbb{C}$ is a bounded function, then we obtain a discrete signal $x = \big(x_k\big)_{k\in\mathbb{Z}} \in \ell_\infty(\mathbb{Z})$ by equidistant sampling $x_k := f(k\,t_0)$ for all $k \in \mathbb{Z}$ and fixed $t_0 > 0$.

A discrete signal $x = \big(x_k\big)_{k\in\mathbb{Z}}$ is called *N-periodic* with $N \in \mathbb{N}$, if $x_k = x_{k+N}$ for all $k \in \mathbb{Z}$. Obviously, one can identify an N-periodic discrete signal $x = \big(x_k\big)_{k\in\mathbb{Z}}$ and the vector $\big(x_k\big)_{k=0}^{N-1} \in \mathbb{C}^N$. The Fourier analysis in $\mathbb{C}^N$ will be handled in Chap. 3.

Example 1.44 Special discrete signals are the *pulse sequence* $\delta := \big(\delta_k\big)_{k\in\mathbb{Z}}$ with the Kronecker symbol δ_k, the *jump sequence* $\big(u_k\big)_{k\in\mathbb{Z}}$ with $u_k := 1$ for $k \ge 0$ and $u_k := 0$ for $k < 0$, and the *exponential sequence* $\big(\mathrm{e}^{\mathrm{i}\omega_0 k}\big)_{k\in\mathbb{Z}}$ with certain $\omega_0 \in \mathbb{R}$. If $\omega_0\, N \in 2\pi\, \mathbb{Z}$ with $N \in \mathbb{N}$, the exponential sequence is N-periodic. □

A *digital filter* H is an operator from the domain $\operatorname{dom} H \subseteq \ell_\infty(\mathbb{Z})$ into $\ell_\infty(\mathbb{Z})$ that converts input signals of $\operatorname{dom} H$ into output signals in $\ell_\infty(\mathbb{Z})$ by applying a specific rule. In the following linear filters are of special interest. A *linear filter* is a linear operator $H : \operatorname{dom} H \to \ell_\infty(\mathbb{Z})$ such that for all $x,\ y \in \operatorname{dom} H$ and each $\alpha \in \mathbb{C}$

$$H(x+y) = Hx + Hy\,, \qquad H(\alpha\, x) = \alpha\, Hx\,.$$

A digital filter $H : \operatorname{dom} H \to \ell_\infty(\mathbb{Z})$ is called *shift-invariant* or *time-invariant*, if $z = Hx = \big(z_k\big)_{k\in\mathbb{Z}}$ with arbitrary input signal $x = \big(x_k\big)_{k\in\mathbb{Z}} \in \operatorname{dom} H$ implies that for each $\ell \in \mathbb{Z}$

$$\big(z_{k-\ell}\big)_{k\in\mathbb{Z}} = H\,\big(x_{k-\ell}\big)_{k\in\mathbb{Z}}\,.$$

In other words, a shift-invariant digital filter transforms each shifted input signal to a shifted output signal.

A digital filter $H : \operatorname{dom} H \to \ell_\infty(\mathbb{Z})$ is called *bounded* on $\ell_p(\mathbb{Z})$ with $1 \le p < \infty$, if $Hx \in \ell_p(\mathbb{Z})$ for any input signal $x \in \ell_p(\mathbb{Z})$.

Example 1.45 A simple digital filter is the *forward shift* $Vx := \big(x_{k-1}\big)_{k\in\mathbb{Z}}$ for any $x = \big(x_k\big)_{k\in\mathbb{Z}} \in \ell_\infty(\mathbb{Z})$. Then we obtain that $V^n x = \big(x_{k-n}\big)_{k\in\mathbb{Z}}$ for $n \in \mathbb{Z}$. In particular for $n = -1$, we have the *backward shift* $V^{-1}x = \big(x_{k+1}\big)_{k\in\mathbb{Z}}$. These filters are linear, shift-invariant, and bounded on $\ell_p(\mathbb{Z})$, $1 \le p \le \infty$.

Another digital filter is the *moving average* $Ax := \big(z_k\big)_{k\in\mathbb{Z}}$ which is defined by

$$z_k := \frac{1}{M_1 + M_2 + 1} \sum_{\ell=-M_1}^{M_2} x_{k-\ell}\,, \quad k \in \mathbb{Z}\,,$$

where M_1 and M_2 are fixed nonnegative integers. This filter is linear, shift-invariant, and bounded on $\ell_p(\mathbb{Z})$, $1 \le p \le \infty$ too.

The *modulation filter* $Mx := \big(z_k\big)_{k\in\mathbb{Z}}$ is defined by

$$z_k := \mathrm{e}^{\mathrm{i}\,(k\omega_0 + \varphi_0)}\, x_k\,, \quad k \in \mathbb{Z}\,,$$

for any $x = (x_k)_{k\in\mathbb{Z}} \in \ell_\infty(\mathbb{Z})$ and fixed $\omega_0,\ \varphi_0 \in \mathbb{R}$. Obviously, this filter is linear and bounded on $\ell_p(\mathbb{Z})$, $1 \le p \le \infty$. In the case $\omega_0 \in 2\pi\,\mathbb{Z}$, the modulation filter is shift-invariant, since then $z_k = \mathrm{e}^{\mathrm{i}\varphi_0}\, x_k$.

Finally, the *accumulator* $Sx := (z_k)_{k\in\mathbb{Z}}$ is defined on $\ell_1(\mathbb{Z})$ by

$$z_k := \sum_{\ell=-\infty}^{k} x_\ell\,, \quad k \in \mathbb{Z}\,.$$

This filter is linear and shift-invariant, but not bounded. □

A linear, shift-invariant filter or linear time-invariant filter is abbreviated as LTI *filter*. If a digital filter H has the output signal $h := H\,\delta$ of the pulse sequence $\delta = (\delta_k)_{k\in\mathbb{Z}}$ (see Example 1.44) as input signal, then h is called *impulse response*. The components h_k of the impulse response $h = (h_k)_{k\in\mathbb{Z}}$ are called *filter coefficients*.

Theorem 1.46 *Let H be an LTI filter with the impulse response $h = (h_k)_{k\in\mathbb{Z}} \in \ell_1(\mathbb{Z})$.*

Then for each input signal $x = (x_k)_{k\in\mathbb{Z}} \in \ell_p(\mathbb{Z})$ with $1 \le p < \infty$ the output signal $z = Hx = (z_k)_{k\in\mathbb{Z}}$ can be represented as discrete convolution *$z = h * x$ with*

$$z_k := \sum_{j\in\mathbb{Z}} h_j\, x_{k-j}\,.$$

Proof For the impulse response $h \in \ell_1(\mathbb{Z})$ and arbitrary input signal $x \in \ell_p(\mathbb{Z})$, the discrete convolution $h * x$ is contained in $\ell_p(\mathbb{Z})$, since by Hölder inequality we have

$$\|h * x\|_p \le \|h\|_1\, \|x\|_p\,.$$

Each input signal $x = (x_k)_{k\in\mathbb{Z}} \in \ell_p(\mathbb{Z})$ can be represented in the form

$$x = \sum_{j\in\mathbb{Z}} x_j\, V^j\delta\,,$$

because the shifted pulse sequences $V^j\delta$, $j \in \mathbb{Z}$, form a basis of $\ell_p(\mathbb{Z})$. Obviously, the shift-invariant filter H has the property $H\,V^j\,\delta = V^j\,H\,\delta = V^j\,h$ for each $j \in \mathbb{Z}$.

Since the LTI filter H is linear and shift-invariant, we obtain the following representation of the output signal:

$$\begin{aligned} z = H\,x &= H\Big(\sum_{j\in\mathbb{Z}} x_j\, V^j\delta\Big) = \sum_{j\in\mathbb{Z}} x_j\, H\, V^j\delta \\ &= \sum_{j\in\mathbb{Z}} x_j\, V^j\, H\delta = \sum_{j\in\mathbb{Z}} x_j\, V^j h \end{aligned}$$

that means

$$z_k = \sum_{j\in\mathbb{Z}} x_j\, h_{k-j} = \sum_{n\in\mathbb{Z}} h_n\, x_{k-n}\,, \quad k \in \mathbb{Z}\,.$$

In particular, the operator $H : \ell_p(\mathbb{Z}) \to \ell_p(\mathbb{Z})$ is bounded with the operator norm $\|h\|_1$. ■

Remark 1.47 In $\ell_1(\mathbb{Z})$ the discrete convolution is a commutative, associative, and distributive multiplication which has the pulse sequence δ as unit. Further we have $\|x * y\|_1 \le \|x\|_1\, \|y\|_1$ for all $x,\ y \in \ell_1(\mathbb{Z})$. □

Using Fourier series, we discuss some properties of LTI filters. Applying the exponential sequence $x = \big(\mathrm{e}^{\mathrm{i}\omega k}\big)_{k\in\mathbb{Z}}$ for $\omega \in \mathbb{R}$ as input signal of the LTI filter H with the impulse response $h \in \ell_1(\mathbb{Z})$, by Theorem 1.46 we obtain the output signal $z = \big(z_k\big)_{k\in\mathbb{Z}} = h * x$ with

$$z_k = \sum_{j\in\mathbb{Z}} h_j\, \mathrm{e}^{\mathrm{i}\omega(k-j)} = \mathrm{e}^{\mathrm{i}\omega k} \sum_{j\in\mathbb{Z}} h_j\, \mathrm{e}^{-\mathrm{i}\omega j} = \mathrm{e}^{\mathrm{i}\omega k}\, H(\omega)$$

with the so-called *transfer function* of the LTI filter H defined by

$$H(\omega) := \sum_{j\in\mathbb{Z}} h_j\, \mathrm{e}^{-\mathrm{i}\omega j}\,. \tag{1.59}$$

By Theorem 1.37 the Fourier series in (1.59) is uniformly convergent and has the variable $-\omega$ instead of ω. Thus the exponential sequence $x = \big(\mathrm{e}^{\mathrm{i}\omega k}\big)_{k\in\mathbb{Z}}$ for $\omega \in \mathbb{R}$ has the property

$$H\,x = H(\omega)\,x\,,$$

i.e., $H(\omega) \in \mathbb{C}$ is an eigenvalue of the LTI filter H with the corresponding eigensequence x.

Example 1.48 Let $a \in \mathbb{C}$ with $|a| < 1$ be given. We consider the LTI filter H with the impulse response $h = \big(h_k\big)_{k\in\mathbb{Z}}$, where $h_k = a^k$ for $k \ge 0$ and $h_k = 0$ for $k < 0$ such that $h \in \ell_1(\mathbb{Z})$. Then, the transfer function of H reads as follows:

$$H(\omega) = \sum_{k=0}^{\infty} (a\,\mathrm{e}^{-\mathrm{i}\omega})^k = \frac{1}{1 - a\,\mathrm{e}^{-\mathrm{i}\omega}} = \frac{1 - a\,\cos\omega + \mathrm{i}\,a\,\sin\omega}{1 + a^2 - 2a\,\cos\omega}\,.$$

With the corresponding *magnitude response*

$$|H(\omega)| = \big(1 + a^2 - 2a\,\cos\omega\big)^{-1/2}$$

and *phase response*

$$\arg\, H(\omega) = \arctan \frac{a\, \sin \omega}{1 - a\, \cos \omega}$$

we obtain the following representation of the transfer function:

$$H(\omega) = |H(\omega)|\, \mathrm{e}^{\mathrm{i}\, \arg\, H(\omega)}\,.$$

□

An LTI filter H with finitely many non-zero filter coefficients is called FIR *filter*, where FIR means finite impulse response. Then, the transfer function of an FIR filter H has the form

$$H(\omega) = \mathrm{e}^{\mathrm{i}\,\omega\, N_0} \sum_{k=0}^{m} h_k\, \mathrm{e}^{-\mathrm{i}\,\omega\, k}$$

with certain $m \in \mathbb{N}$ and $N_0 \in \mathbb{Z}$, where $h_0\, h_m \neq 0$. The filter coefficients h_k of an FIR filter can be chosen in a way such that the transfer function $H(\omega)$ possesses special properties with respect to the magnitude response $|H(\omega)|$ and the phase response $\arg\, H(\omega)$.

Example 1.49 We consider the so-called comb filter H with the filter coefficients $h_0 = h_m = 1$ for certain $m \in \mathbb{N}$ and $h_k = 0$ for $k \in \mathbb{Z} \setminus \{0,\, m\}$. Then the transfer function of H is given by $H(\omega) = 1 + \mathrm{e}^{-\mathrm{i}\,m\,\omega}$ so that

$$H(\omega) = \big(\mathrm{e}^{\mathrm{i}m\omega/2} + \mathrm{e}^{-\mathrm{i}m\omega/2}\big)\, \mathrm{e}^{-\mathrm{i}m\omega/2} = 2\, \cos \frac{m\omega}{2}\, \mathrm{e}^{-\mathrm{i}m\omega/2}\,.$$

For $\frac{(2k-1)\pi}{m} < \omega < \frac{(2k+1)\pi}{m}$ with $k \in \mathbb{Z}$ it holds

$$\cos \frac{m\omega}{2} = \big|\cos \frac{m\omega}{2}\big|\, (-1)^k = \big|\cos \frac{m\omega}{2}\big|\, \mathrm{e}^{\mathrm{i}\pi k}\,.$$

Hence, we obtain:

$$H(\omega) = 2\, \big|\cos \frac{m\omega}{2}\big|\, \mathrm{e}^{\mathrm{i}\,(-m\omega/2 + \pi k)}$$

so that

$$|H(\omega)| = 2\, |\cos \frac{m\omega}{2}|\,, \quad \arg\, H(\omega) = -\frac{m\omega}{2} + \pi k\,.$$

Thus the phase response $\arg\, H(\omega)$ is piecewise linearly with respect to ω. □

An FIR filter H possesses a *linear phase* , if its phase response is a linear function $\arg H(\omega) = \gamma + c\,\omega$ with parameters $\gamma \in [0,\,2\pi)$ and $c \in \mathbb{R}$.

An *ideal low-pass filter* H_{LP} with the cutoff frequency $\omega_0 \in (0,\,\pi)$ is defined by its transfer function:

$$H_{LP}(\omega) := \begin{cases} 1 & |\omega| \le \omega_0\,, \\ 0 & \omega_0 < |\omega| \le \pi\,. \end{cases}$$

The interval $(-\omega_0,\,\omega_0)$ is called *transmission band* , and the set $[-\pi,\,-\omega_0) \cup (\omega_0,\,\pi]$ is the so-called stop band of the ideal low-pass filter H_{LP}. The corresponding filter coefficients h_k of the ideal low-pass filter H_{LP} coincide with the Fourier coefficients of $H_{LP}(\omega)$ and read as follows:

$$\begin{aligned} h_k &= \frac{1}{2\pi}\int_{-\pi}^{\pi} H_{LP}(\omega)\,\mathrm{e}^{\mathrm{i}\omega k}\,\mathrm{d}\omega = \frac{1}{2\pi}\int_{-\omega_0}^{\omega_0} \mathrm{e}^{\mathrm{i}\omega k}\,\mathrm{d}\omega \\ &= \frac{\omega_0}{\pi}\,\mathrm{sinc}\,(\omega_0\,k) \end{aligned}$$

with

$$\mathrm{sinc}\;x := \begin{cases} \frac{\sin x}{x} & x \ne 0\,, \\ 1 & x = 0\,. \end{cases}$$

Thus, we obtain the Fourier expansion:

$$H_{LP}(\omega) = \sum_{k\in\mathbb{Z}} \frac{\omega_0}{\pi}\,\mathrm{sinc}\,(\omega_0\,k)\,\mathrm{e}^{-\mathrm{i}k\omega}\,, \tag{1.60}$$

i.e., the ideal low-pass filter H_{LP} is not an FIR filter and $h_k = \mathcal{O}(|k|^{-1})$ for $|k| \to \infty$.

An *ideal high-pass filter* H_{HP} with the cutoff frequency $\omega_0 \in (0,\,\pi)$ is defined by its transfer function:

$$H_{HP}(\omega) := \begin{cases} 0 & |\omega| \le \pi - \omega_0\,, \\ 1 & \pi - \omega_0 < |\omega| \le \pi\,. \end{cases}$$

We see immediately that $H_{HP}(\omega) = H_{LP}(\omega + \pi)$ and hence by (1.60)

$$H_{HP}(\omega) = \sum_{k\in\mathbb{Z}} (-1)^k\,\frac{\omega_0}{\pi}\,\mathrm{sinc}\,(\omega_0\,k)\,\mathrm{e}^{-\mathrm{i}k\omega}\,,$$

i.e., the ideal high-pass filter is not an FIR filter too.

In the following we consider the construction of low-pass FIR filters. A simple construction of a low-pass FIR filter can be realized by the nth Fourier partial sum of (1.60):

$$H_{LP,n}(\omega) := \sum_{k=-n}^{n} \frac{\omega_0}{\pi} \operatorname{sinc}(\omega_0 k) \, \mathrm{e}^{-\mathrm{i}k\omega} \tag{1.61}$$

with certain $n \in \mathbb{N}$. Then $H_{LP,n}(\omega)$ oscillates inside the transmission band or rather the stop band. Further, $H_{LP,n}(\omega)$ has the Gibbs phenomenon at $\omega = \pm\omega_0$. In order to reduce these oscillations of $H_{LP,n}(\omega)$, we apply so-called window sequences. The simplest example is the *rectangular window sequence* :

$$f_{k,n}^{R} := \begin{cases} 1 & k = -n, \ldots, n\,, \\ 0 & |k| > n \end{cases}$$

such that the related spectral function is the nth Dirichlet function

$$F_n^R(\omega) = \sum_{k=-n}^{n} 1 \cdot \mathrm{e}^{-\mathrm{i}k\omega} = D_n(\omega)$$

and hence

$$H_{LP,n}(\omega) = H_{LP}(\omega) * F_n^R(\omega) = H_{LP}(\omega) * D_n(\omega)\,.$$

The *Hann window sequence* is defined by

$$f_{k,n}^{Hn} := \begin{cases} \frac{1}{2}\left(1 + \cos\frac{\pi k}{n}\right) & k = -n, \ldots, n\,, \\ 0 & |k| > n \end{cases}$$

and has the related spectral function

$$\begin{aligned} F_n^{Hn}(\omega) &= \sum_{k=-n}^{n} \frac{1}{2}\Big(1 + \cos\frac{\pi k}{n}\Big)\, \mathrm{e}^{-\mathrm{i}k\omega} \\ &= \frac{1}{2} \sum_{k=-n}^{n} \mathrm{e}^{-\mathrm{i}k\omega} + \frac{1}{4} \sum_{k=-n}^{n} \mathrm{e}^{-\mathrm{i}k(\omega+\pi/n)} + \frac{1}{4} \sum_{k=-n}^{n} \mathrm{e}^{-\mathrm{i}k(\omega-\pi/n)} \\ &= \frac{1}{4}\Big(2\, F_n^R(\omega) + F_n^R\Big(\omega + \frac{\pi}{n}\Big) + F_n^R\Big(\omega - \frac{\pi}{n}\Big)\Big)\,. \end{aligned}$$

Thus $F_n^{Hn}(\omega)$ is the weighted mean of $F_n^R(\omega)$ and the corresponding shifts $F_n^R(\omega \pm \frac{\pi}{n})$.

The *Hamming window sequence* generalizes the Hann window sequence and is defined by

$$f_{k,n}^{Hm} := \begin{cases} 0.54 + 0.46\, \cos\frac{\pi k}{n} & k = -n, \ldots, n\,, \\ 0 & |k| > n \end{cases}$$

The filter coefficients $f_{k,n}^{Hm}$ are chosen in a way such that the first overshoot of the spectral function $F_n^R(\omega)$ is annihilated as well as possible. Let the spectral function $F_n^{Hm}(\omega)$ be of the form

$$F_n^{Hm}(\omega) = (1-\alpha)\, F_n^R(\omega) + \frac{\alpha}{2}\, F_n^R\Big(\omega + \frac{\pi}{n}\Big) + \frac{\alpha}{2}\, F_n^R\Big(\omega - \frac{\pi}{n}\Big)$$

with certain $\alpha \in (0,\ 1)$. We calculate the first side lobe of $F_n^R(\omega) = D_n(\omega)$ by considering the zeros of $D_n'(\omega)$. By considering

$$D_n'(\omega) = \Big(\frac{\sin\frac{(2n+1)\omega}{2}}{\sin\frac{\omega}{2}}\Big)' = 0\,,$$

we obtain the approximate value $\omega = \frac{5\pi}{2n+1}$ as first side lobe. From

$$\begin{aligned} F_n^{Hm}\Big(\frac{5\pi}{2n+1}\Big) &= (1-\alpha)\, F_n^R\Big(\frac{5\pi}{2n+1}\Big) + \frac{\alpha}{2}\, F_n^R\Big(\frac{5\pi}{2n+1} + \frac{\pi}{n}\Big) \\ &\quad + \frac{\alpha}{2}\, F_n^R\Big(\frac{5\pi}{2n+1} - \frac{\pi}{n}\Big) = 0 \end{aligned}$$

it follows that $\alpha = \frac{21}{46} \approx 0.46$.

A further generalization of the Hann window sequence is the *Blackman window sequence* which is defined by

$$f_{k,n}^{Bl} := \begin{cases} 0.42 + 0.5\,\cos\frac{\pi k}{n} + 0.08\,\cos\frac{2\pi k}{n} & k = -n, \ldots, n\,, \\ 0 & |k| > n\,. \end{cases}$$

The corresponding spectral function reads as follows:

$$F_n^{Bl}(\omega) = \sum_{k=-n}^{n} f_{k,n}^{Bl}\, \mathrm{e}^{-\mathrm{i}k\omega}\,.$$

Chapter 2
Fourier Transform

Fourier transforms of integrable functions defined on the whole real line $\mathbb{R}$ are studied in Chap. 2. First, in Sect. 2.1, the Fourier transform is defined on the Banach space $L_1(\mathbb{R})$. The main properties of the Fourier transform are handled, such as the Fourier inversion formula and the convolution property. Then, in Sect. 2.2, the Fourier transform is introduced as a bijective mapping of the Hilbert space $L_2(\mathbb{R})$ onto itself by the theorem of Plancherel. The Hermite functions, which form an orthogonal basis of $L_2(\mathbb{R})$, are eigenfunctions of the Fourier transform. In Sect. 2.3, we present the Poisson summation formula and Shannon's sampling theorem. Unfortunately, the practical use of Shannon's sampling theorem is limited, since it requires infinitely many samples, which is impossible in practice. To overcome this drawback, we present a regularized Shannon sampling formula, which works with localized sampling. Heisenberg's uncertainty principle is discussed in Sect. 2.4. Finally, two generalizations of the Fourier transform are sketched in Sect. 2.5, namely, the windowed Fourier transform and the fractional Fourier transform.

2.1 Fourier Transform on $L_1(\mathbb{R})$

Let $C_0(\mathbb{R})$ denote the Banach space of continuous functions $f : \mathbb{R} \to \mathbb{C}$ vanishing as $|x| \to \infty$ with norm

$$\|f\|_{C_0(\mathbb{R})} = \|f\|_\infty := \max_{x\in\mathbb{R}} |f(x)|$$

and let $C_c(\mathbb{R})$ be the subspace of continuous, compactly supported functions. By $C^r(\mathbb{R})$, $r \in \mathbb{N}$, we denote the set of r-times continuously differentiable functions on $\mathbb{R}$. Accordingly $C_0^r(\mathbb{R})$ and $C_c^r(\mathbb{R})$ are defined.

For $1 \le p \le \infty$, let $L_p(\mathbb{R})$ denote the Banach space of all measurable functions $f : \mathbb{R} \to \mathbb{C}$ with finite norm:

G. Plonka et al., *Numerical Fourier Analysis*, Applied and Numerical Harmonic Analysis, https://doi.org/10.1007/978-3-031-35005-4_2

$$\|f\|_{L_p(\mathbb{R})} := \begin{cases} \left(\int_{\mathbb{R}} |f(x)|^p \, \mathrm{d}x\right)^{1/p} & 1 \le p < \infty\,, \\ \operatorname{ess\,sup}\{|f(x)| : x \in \mathbb{R}\} & p = \infty\,, \end{cases}$$

where we identify almost equal functions. In particular, we are interested in the Hilbert space $L_2(\mathbb{R})$ with inner product and norm:

$$\langle f, g \rangle_{L_2(\mathbb{R})} := \int_{\mathbb{R}} f(x)\, \overline{g(x)}\, \mathrm{d}x\,, \quad \|f\|_{L_2(\mathbb{R})} := \left(\int_{\mathbb{R}} |f(x)|^2 \, \mathrm{d}x\right)^{1/2}.$$

If it is clear from the context which inner product or norm is addressed, we abbreviate $\langle f, g\rangle := \langle f, g \rangle_{L_2(\mathbb{R})}$ and $\|f\| := \|f\|_{L_2(\mathbb{R})}$.

Note that in contrast to the periodic setting, there is no continuous embedding of the spaces $L_p(\mathbb{R})$. We have neither $L_1(\mathbb{R}) \subset L_2(\mathbb{R})$ nor $L_1(\mathbb{R}) \supset L_2(\mathbb{R})$. For example, $f(x) := \frac{1}{x}\, \chi_{[1,\infty)}(x)$, where $\chi_{[1,\infty)}$ denotes the *characteristic function* of the interval $[1, \infty)$, is in $L_2(\mathbb{R})$, but not in $L_1(\mathbb{R})$. On the other hand, $f(x) := \frac{1}{\sqrt{x}}\, \chi_{(0,1]}(x)$ is in $L_1(\mathbb{R})$, but not in $L_2(\mathbb{R})$.

Remark 2.1 Note that each function $f \in C_0(\mathbb{R})$ is uniformly continuous on $\mathbb{R}$ by the following reason: For arbitrary $\varepsilon > 0$, there exists $L = L(\varepsilon) > 0$ such that $|f(x)| \le \varepsilon/3$ for all $|x| \ge L$. If $x, y \in [-L, L]$, then there exists $\delta > 0$ such that $|f(x) - f(y)| \le \varepsilon/3$ whenever $|x - y| \le \delta$. If $x, y \in \mathbb{R}\backslash[-L, L]$, then $|f(x) - f(y)| \le |f(x)| + |f(y)| \le 2\varepsilon/3$. If $x \in [-L, L]$ and $y \in \mathbb{R}\backslash[-L, L]$, say $y > L$ with $|x - y| \le \delta$, then $|f(x) - f(y)| \le |f(x) - f(L)| + |f(L) - f(y)| \le \varepsilon$. In summary we have then $|f(x) - f(y)| \le \varepsilon$ whenever $|x - y| \le \delta$. □

The (*continuous*) *Fourier transform* $\hat{f} = \mathcal{F} f$ of a function $f \in L_1(\mathbb{R})$ is defined by

$$\hat{f}(\omega) = (\mathcal{F} f)(\omega) := \int_{\mathbb{R}} f(x)\, \mathrm{e}^{-\mathrm{i}x\omega}\, \mathrm{d}x\,, \quad \omega \in \mathbb{R}\,. \tag{2.1}$$

Since $|f(x)\, \mathrm{e}^{-\mathrm{i}x\omega}| = |f(x)|$ and $f \in L_1(\mathbb{R})$, the integral (2.1) is well-defined. In practice, the variable x denotes mostly the time or the space and the variable ω is the frequency. Therefore, the domain of the Fourier transform is called *time domain* or *space domain*. The range of the Fourier transform is called *frequency domain*. Roughly spoken, the Fourier transform (2.1) measures how much oscillations around the frequency ω are contained in $f \in L_1(\mathbb{R})$. The function $\hat{f} = |f|\, \mathrm{e}^{\mathrm{i} \arg f}$ is also called *spectrum* of f with *modulus* $|\hat{f}|$ and *phase* $\arg f$.

Remark 2.2 In the literature, the Fourier transform is not consistently defined. For instance, other frequently applied definitions are

$$\frac{1}{\sqrt{2\pi}} \int_{\mathbb{R}} f(x)\, \mathrm{e}^{-\mathrm{i}\omega x}\, \mathrm{d}x\,, \quad \int_{\mathbb{R}} f(x)\, \mathrm{e}^{-2\pi \mathrm{i}\omega x}\, \mathrm{d}x\,.$$

□

Example 2.3 Let $L > 0$. The *rectangle function*

$$f(x) := \begin{cases} 1 & x \in (-L, L), \\ \frac{1}{2} & x \in \{-L, L\}, \\ 0 & \text{otherwise}, \end{cases}$$

has the Fourier transform

$$\hat{f}(\omega) = \int_{-L}^{L} \mathrm{e}^{-\mathrm{i}\omega x}\, \mathrm{d}x = \frac{-\mathrm{e}^{-\mathrm{i}\omega L} + \mathrm{e}^{\mathrm{i}L\omega}}{\mathrm{i}\omega} = \frac{2\mathrm{i}\,L\,\sin(\omega L)}{\mathrm{i}\,L\omega}$$
$$= \frac{2L\,\sin(L\omega)}{L\omega} = 2L\,\mathrm{sinc}(L\omega)$$

with the *cardinal sine function* or sinc *function*

$$\mathrm{sinc}(x) = \begin{cases} \frac{\sin x}{x} & x \in \mathbb{R} \setminus \{0\}, \\ 1 & x = 0. \end{cases}$$

Figure 2.1 shows the cardinal sine function. While $\mathrm{supp}\, f = [-L, L]$ is bounded, this is not the case for $\hat{f}$. Even worse, $\hat{f} \notin L_1(\mathbb{R})$, since for $n \in \mathbb{N} \setminus \{1\}$

$$\int_0^{n\pi} |\mathrm{sinc}(x)|\, \mathrm{d}x = \sum_{k=1}^{n} \int_{(k-1)\pi}^{k\pi} |\mathrm{sinc}(x)|\mathrm{d}x \geq \sum_{k=1}^{n} \frac{1}{k\pi} \int_{(k-1)\pi}^{k\pi} |\sin x|\mathrm{d}x = \frac{2}{\pi} \sum_{k=1}^{n} \frac{1}{k}$$

and the last sum becomes infinitely large as $n \to \infty$. Thus, the Fourier transform does not map $L_1(\mathbb{R})$ into itself. □

Example 2.4 For given $L > 0$, the *hat function*

Fig. 2.1 The sinc function on $[-20, 20]$

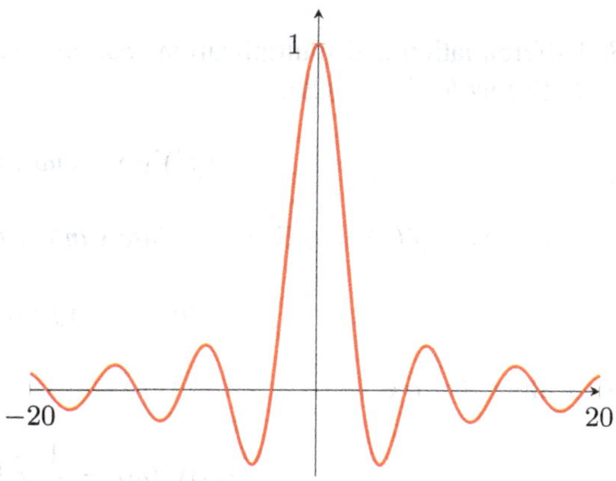

$$f(x) := \begin{cases} 1 - \frac{|x|}{L} & x \in [-L,\, L]\,, \\ 0 & \text{otherwise}\,, \end{cases}$$

has the Fourier transform

$$\begin{aligned} \hat{f}(\omega) &= 2 \int_0^L \big(1 - \frac{x}{L}\big) \cos(\omega x)\, \mathrm{d}x = \frac{2}{L\omega} \int_0^L \sin(\omega x)\, \mathrm{d}x \\ &= \frac{2}{L\omega^2} \big(1 - \cos(L\omega)\big) = L\, \big(\mathrm{sinc} \frac{L\omega}{2}\big)^2 \end{aligned}$$

for $\omega \in \mathbb{R} \setminus \{0\}$. In the case $\omega = 0$, we obtain:

$$\hat{f}(0) = 2 \int_0^L \big(1 - \frac{x}{L}\big)\, \mathrm{d}x = L\,.$$

□

The following theorem collects basic properties of the Fourier transform which can easily be proved.

Theorem 2.5 (Properties of the Fourier Transform) *Let $f,\, g \in L_1(\mathbb{R})$. Then the Fourier transform possesses the following properties:*

1. Linearity*: For all $\alpha,\ \beta \in \mathbb{C}$,*

$$(\alpha f + \beta g)\hat{} = \alpha\, \hat{f} + \beta\, \hat{g}\,.$$

2. Translation and modulation*: For each $x_0,\, \omega_0 \in \mathbb{R}$,*

$$\begin{aligned} \big(f(\cdot - x_0)\big)\hat{}(\omega) &= \mathrm{e}^{-\mathrm{i}\, x_0 \omega}\, \hat{f}(\omega), \\ \big(\mathrm{e}^{-\mathrm{i}\, \omega_0 \cdot}\, f\big)\hat{}(\omega) &= \hat{f}(\omega_0 + \omega)\,. \end{aligned}$$

3. Differentiation and multiplication*: For an absolutely continuous function $f \in L_1(\mathbb{R})$ with $f' \in L_1(\mathbb{R})$,*

$$(f')\hat{}(\omega) = \mathrm{i}\omega\, \hat{f}(\omega)\,.$$

If $h(x) := x\, f(x)$, $x \in \mathbb{R}$, is absolutely integrable, then

$$\hat{h}(\omega) = \mathrm{i}\, (\hat{f})'(\omega)\,.$$

4. Scaling*: For $c \in \mathbb{R} \setminus \{0\}$,*

$$\big(f(c\cdot)\big)\hat{}(\omega) = \frac{1}{|c|} \hat{f}\left(c^{-1}\omega\right).$$

Applying these properties we can calculate the Fourier transforms of some special functions.

Example 2.6 We consider the *normalized Gaussian function* :

$$f(x) := \frac{1}{\sqrt{2\pi\sigma^2}}\, \mathrm{e}^{-x^2/(2\sigma^2)}\,, \quad x \in \mathbb{R}\,, \tag{2.2}$$

with standard deviation $\sigma > 0$. Note that $\int_{\mathbb{R}} f(x)\,\mathrm{d}x = 1$, since for $a > 0$ we obtain using polar coordinates r and φ that

$$\Big(\int_{\mathbb{R}} \mathrm{e}^{-ax^2}\,\mathrm{d}x\Big)^2 = \Big(\int_{\mathbb{R}} \mathrm{e}^{-ax^2}\,\mathrm{d}x\Big)\Big(\int_{\mathbb{R}} \mathrm{e}^{-ay^2}\,\mathrm{d}y\Big) = \int_{\mathbb{R}}\int_{\mathbb{R}} \mathrm{e}^{-a(x^2+y^2)}\,\mathrm{d}x\,\mathrm{d}y$$
$$= \int_0^{2\pi}\Big(\int_0^{\infty} r\,\mathrm{e}^{-ar^2}\,\mathrm{d}r\Big)\,\mathrm{d}\varphi = \frac{\pi}{a}\,.$$

Now we compute the Fourier transform:

$$\hat{f}(\omega) = \frac{1}{\sqrt{2\pi\sigma^2}} \int_{\mathbb{R}} \mathrm{e}^{-x^2/(2\sigma^2)}\,\mathrm{e}^{-\mathrm{i}\omega x}\,\mathrm{d}x\,. \tag{2.3}$$

This integral can be calculated by Cauchy's integral theorem of complex function theory. Here we use another technique. Obviously, the Gaussian function (2.2) satisfies the differential equation:

$$f'(x) + \frac{x}{\sigma^2}\, f(x) = 0\,.$$

Applying Fourier transform to this differential equation, we obtain by the differentiation–multiplication property of Theorem 2.5:

$$\mathrm{i}\omega\,\hat{f}(\omega) + \frac{\mathrm{i}}{\sigma^2}\,(\hat{f})'(\omega) = 0\,.$$

This linear differential equation has the general solution:

$$\hat{f}(\omega) = C\,\mathrm{e}^{-\sigma^2\omega^2/2}\,,$$

with an arbitrary constant C. From (2.3) it follows that

$$\hat{f}(0) = C = \int_{\mathbb{R}} f(x)\,\mathrm{d}x = 1$$

and hence

$$\hat{f}(\omega) = \mathrm{e}^{-\sigma^2\omega^2/2} \tag{2.4}$$

is a non-normalized Gaussian function with standard deviation $1/\sigma$. The smaller the standard deviation is in the space domain, the larger it is in the frequency domain. In particular for $\sigma = 1$, the Gaussian function (2.2) coincides with its Fourier transform $\hat{f}$ up to the factor $1/\sqrt{2\pi}$. Note that the Gaussian function is the only function with this behavior. □

Example 2.7 Let $a > 0$ and $b \in \mathbb{R} \setminus \{0\}$ be given. We consider the *Gaussian chirp* :

$$f(x) := \mathrm{e}^{-(a-\mathrm{i}\,b)\,x^2}\,. \tag{2.5}$$

The Fourier transform of (2.5) reads as follows:

$$\hat{f}(\omega) = C\,\exp\,\frac{-(a+\mathrm{i}\,b)\omega^2}{4\,(a^2+b^2)}\,,$$

which can be calculated by a similar differential equation as in Example 2.6. Using Cauchy's integral theorem, the constant C reads as follows:

$$C = \hat{f}(0) = \int_{\mathbb{R}} \mathrm{e}^{-(a-\mathrm{i}b)x^2}\,\mathrm{d}x = (a-\mathrm{i}b)^{-1/2}\,\sqrt{\pi}$$

such that

$$(a-\mathrm{i}b)^{-1/2} = \begin{cases} \frac{1}{\sqrt{2}}\left(\sqrt{\frac{\sqrt{a^2+b^2}+a}{a^2+b^2}} + \mathrm{i}\sqrt{\frac{\sqrt{a^2+b^2}-a}{a^2+b^2}}\right) & b > 0\,, \\ \frac{1}{\sqrt{2}}\left(\sqrt{\frac{\sqrt{a^2+b^2}+a}{a^2+b^2}} - \mathrm{i}\sqrt{\frac{\sqrt{a^2+b^2}-a}{a^2+b^2}}\right) & b < 0\,. \end{cases}$$

□

Note that $(a-ib)^{-1/2}$ is the square root of $(a-ib)^{-1} = \frac{a+ib}{a^2+b^2}$ with positive real part. In Example 2.3 we have seen that the Fourier transform of a function $f \in L_1(\mathbb{R})$ is not necessarily in $L_1(\mathbb{R})$. By the following theorem, the Fourier transform of $f \in L_1(\mathbb{R})$ is a continuous function, which vanishes at $\pm\infty$.

Theorem 2.8 *The Fourier transform $\mathcal{F}$ defined by* (2.1) *is a linear, continuous operator from $L_1(\mathbb{R})$ into $C_0(\mathbb{R})$ with operator norm $\|\mathcal{F}\|_{L_1(\mathbb{R})\to C_0(\mathbb{R})} = 1$.*

More precisely $\mathcal{F}$ maps $L_1(\mathbb{R})$ onto a dense subspace of $C_0(\mathbb{R})$, such that $\mathcal{F}$: $L_1(\mathbb{R}) \to C_0(\mathbb{R})$ is not surjective (see [208, pp. 172–173]).

Proof The linearity of $\mathcal{F}$ follows from those of the integral operator. Let $f \in L_1(\mathbb{R})$. For any $\omega, h \in \mathbb{R}$, we can estimate

$$|\hat{f}(\omega+h) - \hat{f}(\omega)| = \Big|\int_{\mathbb{R}} f(x)\,\mathrm{e}^{-\mathrm{i}\omega x}\,(\mathrm{e}^{-\mathrm{i}hx} - 1)\,\mathrm{d}x\Big| \le \int_{\mathbb{R}} |f(x)|\,|\mathrm{e}^{-\mathrm{i}hx} - 1|\,\mathrm{d}x\,.$$

Since $|f(x)|\,|\mathrm{e}^{-ihx}-1| \le 2\,|f(x)| \in L_1(\mathbb{R})$ and

$$|\mathrm{e}^{-ihx}-1| = \big((\cos(hx)-1)^2+(\sin(hx))^2\big)^{1/2} = \big(2-2\cos(hx)\big)^{1/2} \to 0$$

as $h \to 0$, we obtain by the dominated convergence theorem of Lebesgue:

$$\begin{aligned}\lim_{h\to 0} |\hat f(\omega+h)-\hat f(\omega)| &\le \lim_{h\to 0}\int_{\mathbb{R}} |f(x)|\,|\mathrm{e}^{-ihx}-1|\,\mathrm{d}x\\ &= \int_{\mathbb{R}} |f(x)|\,\big(\lim_{h\to 0}|\mathrm{e}^{-ihx}-1|\big)\,\mathrm{d}x = 0\,.\end{aligned}$$

Hence, $\hat f$ is continuous. Further, we know by Lemma 1.27 of Riemann–Lebesgue that $\lim_{|\omega|\to\infty} \hat f(\omega) = 0$. Thus $\hat f = \mathscr{F} f \in C_0(\mathbb{R})$.

For $f \in L_1(\mathbb{R})$ we have:

$$|\hat f(\omega)| \le \int_{\mathbb{R}} |f(x)|\,\mathrm{d}x = \|f\|_{L_1(\mathbb{R})}\,,$$

so that

$$\|\mathscr{F} f\|_{C_0(\mathbb{R})} = \|\hat f\|_{C_0(\mathbb{R})} \le \|f\|_{L_1(\mathbb{R})}$$

and consequently $\|\mathscr{F}\|_{L_1(\mathbb{R})\to C_0(\mathbb{R})} \le 1$. In particular we obtain for $g(x) := \frac{1}{\sqrt{2\pi}}\,\mathrm{e}^{-x^2/2}$ that $\|g\|_{L_1(\mathbb{R})} = 1$ and $\hat g(\omega) = \mathrm{e}^{-\omega^2/2}$; see Example 2.6. Hence, we have $\|\hat g\|_{C_0(\mathbb{R})} = 1$ and $\|\mathscr{F}\|_{L_1(\mathbb{R})\to C_0(\mathbb{R})} = 1$. ■

Using the Theorem 2.8, we obtain following result.

Lemma 2.9 *Let $f,\, g \in L_1(\mathbb{R})$. Then, we have $\hat f\, g,\, \hat g\, f \in L_1(\mathbb{R})$ and*

$$\int_{\mathbb{R}} \hat f(x)\, g(x)\,\mathrm{d}x = \int_{\mathbb{R}} f(\omega)\,\hat g(\omega)\,\mathrm{d}\omega\,. \tag{2.6}$$

Proof By Theorem 2.8 we know that $\hat f,\, \hat g \in C_0(\mathbb{R})$ are bounded so that $\hat f g,\, f\,\hat g \in L_1(\mathbb{R})$. Taking into account that $f(\omega)\,g(x)\,\mathrm{e}^{-\mathrm{i}x\omega} \in L_1(\mathbb{R}^2)$, equality (2.6) follows by Fubini's theorem:

$$\begin{aligned}\int_{\mathbb{R}} \hat f(x)\, g(x)\,\mathrm{d}x &= \int_{\mathbb{R}}\int_{\mathbb{R}} f(\omega)\, g(x)\,\mathrm{e}^{-\mathrm{i}x\omega}\,\mathrm{d}\omega\,\mathrm{d}x\\ &= \int_{\mathbb{R}} f(\omega)\int_{\mathbb{R}} g(x)\,\mathrm{e}^{-\mathrm{i}x\omega}\,\mathrm{d}x\,\mathrm{d}\omega = \int_{\mathbb{R}} f(\omega)\,\hat g(\omega)\,\mathrm{d}\omega\,.\end{aligned}$$

■

Next we examine under which assumptions on $f \in L_1(\mathbb{R})$ the *Fourier inversion formula*:

$$f(x) = (\hat{f})^{\check{}}(x) := \frac{1}{2\pi} \int_{\mathbb{R}} \hat{f}(\omega)\, \mathrm{e}^{\mathrm{i}\omega x}\, \mathrm{d}\omega \tag{2.7}$$

holds true. Note that (2.7) is almost the same formula as (2.1), except of the plus sign in the exponential and the factor $\frac{1}{2\pi}$.

Theorem 2.10 (Fourier Inversion Formula for $L_1(\mathbb{R})$ Functions) *Let $f \in L_1(\mathbb{R})$ with $\hat{f} \in L_1(\mathbb{R})$ be given.*

Then the Fourier inversion formula (2.7) *holds true for almost every $x \in \mathbb{R}$. For $f \in L_1(\mathbb{R}) \cap C_0(\mathbb{R})$ with $\hat{f} \in L_1(\mathbb{R})$, the Fourier inversion formula holds for all $x \in \mathbb{R}$.*

In the following we give a proof for a function $f \in L_1(\mathbb{R}) \cap C_0(\mathbb{R})$ with $\hat{f} \in L_1(\mathbb{R})$. For the general setting, we refer, e.g., to [82, pp. 38–44].

Proof

1. For any $n \in \mathbb{N}$, we use the function $g_n(x) := \frac{1}{2\pi}\, \mathrm{e}^{-|x|/n}$ which has by straightforward computation the Fourier transform:

$$\hat{g}_n(\omega) = \frac{n}{\pi(1 + n^2\omega^2)}\,.$$

Both functions g_n and $\hat{g}_n$ are in $L_1(\mathbb{R})$. By (2.6) and Theorem 2.5, we deduce the relation for the functions $f(x)$ and $g_n(x)\, \mathrm{e}^{\mathrm{i}xy}$:

$$\int_{\mathbb{R}} \hat{f}(x)\, g_n(x)\, \mathrm{e}^{\mathrm{i}xy}\, \mathrm{d}x = \int_{\mathbb{R}} f(\omega)\, \hat{g}_n(\omega - y)\, \mathrm{d}\omega\,,$$

where $y \in \mathbb{R}$ is arbitrary fixed. We examine this equation as $n \to \infty$. We have $\lim_{n\to\infty} g_n(x) = \frac{1}{2\pi}$. For the left-hand side, since $|\hat{f}(x)\, g_n(x)\, \mathrm{e}^{\mathrm{i}xy}| \le \frac{1}{2\pi}\, |\hat{f}(x)|$ and $\hat{f} \in L_1(\mathbb{R})$, we can pass to the limit under the integral by the dominated convergence theorem of Lebesgue:

$$\lim_{n\to\infty} \int_{\mathbb{R}} \hat{f}(x)\, g_n(x)\, \mathrm{e}^{\mathrm{i}xy}\, \mathrm{d}x = \frac{1}{2\pi} \int_{\mathbb{R}} \hat{f}(x)\, \mathrm{e}^{\mathrm{i}xy}\, \mathrm{d}x = (\hat{f})^{\check{}}(y)\,.$$

2. It remains to show that the limit on the right-hand side is equal to

$$\lim_{n\to\infty} \int_{\mathbb{R}} f(\omega)\, \hat{g}_n(\omega - y)\, \mathrm{d}\omega = f(y)\,.$$

By assumption, $f \in L_1(\mathbb{R}) \cap C_0(\mathbb{R})$. Then $f \in C_0(\mathbb{R})$ and hence f is uniformly continuous on $\mathbb{R}$ by Remark 2.1, i.e., for every $\varepsilon > 0$, there exists $\delta = \delta(\varepsilon) > 0$ such that $|f(x) - f(y)| < \varepsilon$ if $|x - y| \le \delta$.

Note that $\hat{g}_n \in L_1(\mathbb{R})$ fulfills the relation:

$$\int_{\mathbb{R}} \hat{g}_n(\omega)\,\mathrm{d}\omega = \lim_{L\to\infty} \int_{-L}^{L} \hat{g}_n(\omega)\,\mathrm{d}\omega = \frac{2}{\pi} \lim_{L\to\infty} \arctan(nL) = 1$$

Then, we get:

$$\int_{\mathbb{R}} f(\omega)\,\hat{g}_n(\omega - y)\,\mathrm{d}\omega - f(y) = \int_{\mathbb{R}} \big(f(\omega + y) - f(y)\big)\,\hat{g}_n(\omega)\,\mathrm{d}\omega$$

$$= \int_{-\delta}^{\delta} \big(f(\omega + y) - f(y)\big)\,\hat{g}_n(\omega)\,\mathrm{d}\omega + \int_{|\omega|\geq\delta} \big(f(\omega + y) - f(y)\big)\,\hat{g}_n(\omega)\,\mathrm{d}\omega\,.$$

3. For all $n \in \mathbb{N}$, we obtain:

$$\Big|\int_{-\delta}^{\delta} \big(f(\omega + y) - f(y)\big)\,\hat{g}_n(\omega)\,\mathrm{d}\omega\Big| \leq \int_{-\delta}^{\delta} \big|f(\omega + y) - f(y)\big|\,\hat{g}_n(\omega)\,\mathrm{d}\omega$$

$$\leq \varepsilon \int_{-\delta}^{\delta} \hat{g}_n(\omega)\,\mathrm{d}\omega \leq \varepsilon\,.$$

Next we see

$$\Big|\int_{|\omega|\geq\delta} f(y)\,\hat{g}_n(\omega)\,\mathrm{d}\omega\Big| \leq |f(y)|\,\big(1 - \frac{2}{\pi}\arctan(n\delta)\big)\,. \tag{2.8}$$

Since the even function $\hat{g}_n$ is decreasing on $[0, \infty)$, we receive

$$\Big|\int_{|\omega|\geq\delta} f(\omega + y)\,\hat{g}_n(\omega)\,\mathrm{d}\omega\Big| \leq \hat{g}_n(\delta)\,\|f\|_{L_1(\mathbb{R})}\,. \tag{2.9}$$

As $n \to \infty$ the right-hand sides in (2.8) and (2.9) go to zero. This completes the proof. ■

As a corollary we obtain that the Fourier transform is one-to-one.

Corollary 2.11 *If $f \in L_1(\mathbb{R})$ fulfills $\hat{f} = 0$, then $f = 0$ almost everywhere on $\mathbb{R}$.*

We have seen that a 2π-periodic function can be reconstructed from its Fourier coefficients by the Fourier series in the $L_2(\mathbb{T})$ sense and that pointwise/uniform convergence of the Fourier series requires additional assumptions on the function.

Now we consider a corresponding problem in $L_1(\mathbb{R})$ and ask for the convergence of *Cauchy principal value* (of an improper integral) :

$$\lim_{L\to\infty} \frac{1}{2\pi} \int_{-L}^{L} \hat{f}(\omega)\,\mathrm{e}^{\mathrm{i}\,\omega x}\,\mathrm{d}\omega\,.$$

Note that for a Lebesgue integrable function on $\mathbb{R}$, the Cauchy principal value coincides with the integral over $\mathbb{R}$.

Similar to Riemann's localization principle in Theorem 1.28 in the 2π-periodic setting, we have the following result.

Theorem 2.12 (Riemann's Localization Principle) *Let $f \in L_1(\mathbb{R})$ and $x_0 \in \mathbb{R}$. Further let $\varphi(t) := f(x_0+t) + f(x_0-t) - 2f(x_0)$, $t \in \mathbb{R}$. Assume that for some $\delta > 0$*

$$\int_0^\delta \frac{|\varphi(t)|}{t}\,\mathrm{d}t < \infty\,.$$

Then, it holds:

$$f(x_0) = \lim_{L\to\infty} \frac{1}{2\pi} \int_{-L}^{L} \hat{f}(\omega)\,\mathrm{e}^{\mathrm{i}\,\omega x_0}\,\mathrm{d}\omega\,.$$

Proof It follows

$$\begin{aligned} I_L(x_0) &:= \frac{1}{2\pi}\int_{-L}^{L} \hat{f}(\omega)\,\mathrm{e}^{\mathrm{i}\,\omega x_0}\,\mathrm{d}\omega = \frac{1}{2\pi}\int_{-L}^{L}\int_{\mathbb{R}} f(u)\,\mathrm{e}^{-\mathrm{i}\,\omega u}\,\mathrm{d}u\,\mathrm{e}^{\mathrm{i}\,\omega x_0}\,\mathrm{d}\omega \\ &= \frac{1}{2\pi}\int_{-L}^{L}\int_{\mathbb{R}} f(u)\,\mathrm{e}^{\mathrm{i}\,\omega(x_0-u)}\,\mathrm{d}u\,\mathrm{d}\omega\,. \end{aligned}$$

Since $|f(u)\,\mathrm{e}^{\mathrm{i}\,\omega\,(x_0-u)}| = |f(u)|$ and $f \in L_1(\mathbb{R})$, we can change the order of integration in $I_L(x_0)$ by Fubini's theorem which results in

$$\begin{aligned} I_L(x_0) &= \frac{1}{2\pi}\int_{\mathbb{R}} f(u)\int_{-L}^{L} \mathrm{e}^{\mathrm{i}\,\omega(x_0-u)}\,\mathrm{d}\omega\,\mathrm{d}u = \frac{1}{\pi}\int_{\mathbb{R}} f(u)\,\frac{\sin\big(L\,(x_0-u)\big)}{x_0-u}\,\mathrm{d}u \\ &= \frac{1}{\pi}\int_0^\infty \big(f(x_0+t)+f(x_0-t)\big)\,\frac{\sin(L\,t)}{t}\,\mathrm{d}t\,. \end{aligned}$$

Since we have by Lemma 1.41 that

$$\int_0^\infty \frac{\sin(L\,t)}{t}\,\mathrm{d}t = \int_0^\infty \frac{\sin t}{t}\,\mathrm{d}t = \frac{\pi}{2}\,, \tag{2.10}$$

we conclude

$$\begin{aligned} I_L(x_0) - f(x_0) &= \frac{1}{\pi}\int_0^\infty \frac{\varphi(t)}{t}\,\sin(L\,t)\,\mathrm{d}t \\ &= \frac{1}{\pi}\int_0^\delta \frac{\varphi(t)}{t}\,\sin(L\,t)\mathrm{d}t + \frac{1}{\pi}\int_\delta^\infty \frac{f(x_0+t)+f(x_0-t)}{t}\,\sin(Lt)\mathrm{d}t \end{aligned}$$

$$- \frac{2}{\pi} f(x_0) \int_\delta^\infty \frac{\sin(Lt)}{t} \, dt .$$

Since $\varphi(t)/t \in L_1([0,\delta])$ by assumption, the first integral converges to zero as $L \to \infty$ by Lemma 1.27 of Riemann–Lebesgue. The same holds true for the second integral. Concerning the third integral, we use

$$\begin{aligned} \frac{\pi}{2} = \int_0^\infty \frac{\sin(L\,t)}{t} \, dt &= \int_0^\delta \frac{\sin(L\,t)}{t} \, dt + \int_\delta^\infty \frac{\sin(L\,t)}{t} \, dt \\ &= \int_0^{L\delta} \frac{\sin t}{t} \, dt + \int_\delta^\infty \frac{\sin(L\,t)}{t} \, dt . \end{aligned}$$

Since the first summand converges to $\frac{\pi}{2}$ as $L \to \infty$, the integral $\int_\delta^\infty \frac{\sin(L\,t)}{t} \, dt$ converges to zero as $L \to \infty$. This finishes the proof. ■

A function $f : \mathbb{R} \to \mathbb{C}$ is called *piecewise continuously differentiable* on $\mathbb{R}$, if there exists a finite partition of $\mathbb{R}$ determined by $-\infty < x_0 < x_1 < \ldots < x_n < \infty$ of $\mathbb{R}$ such that f is continuously differentiable on each interval $(-\infty, x_0)$, (x_0, x_1), ..., (x_{n-1}, x_n), (x_n, ∞) and the one-sided limits $\lim_{x \to x_j \pm 0} f(x)$ and $\lim_{x \to x_j \pm 0} f'(x)$, $j = 0, \ldots, n$ exist. Similarly as in the proof of Theorem 1.34 of Dirichlet–Jordan, the previous theorem can be used to prove that for a piecewise continuously differentiable function $f \in L_1(\mathbb{R})$, it holds

$$\frac{1}{2}\big(f(x+0) + f(x-0)\big) = \lim_{L \to \infty} \frac{1}{2\pi} \int_{-L}^{L} \hat{f}(\omega)\, e^{i\omega x} \, d\omega$$

for all $x \in \mathbb{R}$.

The Fourier transform is again closely related to the convolution of functions. If $f : \mathbb{R} \to \mathbb{C}$ and $g : \mathbb{R} \to \mathbb{C}$ are given functions, then their *convolution* $f * g$ is defined by

$$(f * g)(x) := \int_{\mathbb{R}} f(y)\, g(x - y)\, dy \,, \quad x \in \mathbb{R} \,, \tag{2.11}$$

provided that this integral (2.11) exists. Note that the convolution is a commutative, associative, and distributive operation. Various conditions can be imposed on f and g to ensure that (2.11) exists. For instance, if f and g are both in $L_1(\mathbb{R})$, then $(f * g)(x)$ exists for almost every $x \in \mathbb{R}$ and further $f * g \in L_1(\mathbb{R})$. In the same way as for 2π-periodic functions, we can prove the following result.

Theorem 2.13

1. *Let $f \in L_p(\mathbb{R})$ with $1 \le p \le \infty$ and $g \in L_1(\mathbb{R})$ be given. Then $f * g$ exists almost everywhere and $f * g \in L_p(\mathbb{R})$. Further we have the* Young inequality*:*

$$\|f * g\|_{L_p(\mathbb{R})} \le \|f\|_{L_p(\mathbb{R})} \, \|g\|_{L_1(\mathbb{R})} \, .$$

2. *Let $f \in L_p(\mathbb{R})$ and $g \in L_q(\mathbb{R})$, where $1 < p < \infty$ and $\frac{1}{p} + \frac{1}{q} = 1$. Then $f * g \in C_0(\mathbb{R})$ fulfills*

$$\|f * g\|_{C_0(\mathbb{R})} \le \|f\|_{L_p(\mathbb{R})} \, \|g\|_{L_q(\mathbb{R})} \, .$$

3. *Let $f \in L_p(\mathbb{R})$ and $g \in L_q(\mathbb{R})$, where $1 \le p,\, q,\, r \le \infty$ and $\frac{1}{p} + \frac{1}{q} = \frac{1}{r} + 1$. Then $f * g \in L_r(\mathbb{R})$ and we have the* generalized Young inequality*:*

$$\|f * g\|_{L_r(\mathbb{R})} \le \|f\|_{L_p(\mathbb{R})} \, \|g\|_{L_q(\mathbb{R})} \, .$$

Differentiation and convolution are related by the following.

Lemma 2.14 *Let $f \in L_1(\mathbb{R})$ and $g \in C^r(\mathbb{R})$, where $g^{(k)}$ for $k = 0, \ldots, r$ are bounded on $\mathbb{R}$. Then $f * g \in C^r(\mathbb{R})$ and*

$$(f * g)^{(k)} = f * g^{(k)} \, , \quad k = 1, \ldots, r \, .$$

Proof Since $g^{(k)} \in L_\infty(\mathbb{R})$, the first assertion follows by the first part of Theorem 2.13. The function $x \mapsto f(y)g(x-y)$ is r-times differentiable, and for $k = 0, \ldots, r$ we have:

$$\big|f(y)g^{(k)}(x-y)\big| \le |f(y)| \sup_{t \in \mathbb{R}} |g^{(k)}(t)|.$$

Since $f \in L_1(\mathbb{R})$, we can differentiate under the integral sign; see [156, Proposition 14.2.2], which results in

$$(f * g)^{(k)}(x) = \int_{\mathbb{R}} f(y)\, g^{(k)}(x-y)\, \mathrm{d}y = \big(f * g^{(k)}\big)(x)\, .$$

■

The following theorem presents the most important property of the Fourier transform.

Theorem 2.15 (Convolution Property of Fourier Transform) *Let $f, g \in L_1(\mathbb{R})$. Then, we have:*

$$(f * g)\hat{} = \hat{f}\, \hat{g}\, .$$

Proof For $f, g \in L_1(\mathbb{R})$ we have $f * g \in L_1(\mathbb{R})$ by Theorem 2.13. Using Fubini's theorem, we obtain for all $\omega \in \mathbb{R}$

$$\begin{aligned}(f*g)\hat{}(\omega) &= \int_{\mathbb{R}} (f*g)(x)\,\mathrm{e}^{-\mathrm{i}\omega x}\,\mathrm{d}x = \int_{\mathbb{R}} \Big(\int_{\mathbb{R}} f(y)\,g(x-y)\,\mathrm{d}y\Big)\,\mathrm{e}^{-\mathrm{i}\omega x}\,\mathrm{d}x \\ &= \int_{\mathbb{R}} f(y)\,\Big(\int_{\mathbb{R}} g(x-y)\,\mathrm{e}^{-\mathrm{i}\omega(x-y)}\,\mathrm{d}x\Big)\,\mathrm{e}^{-\mathrm{i}\omega y}\,\mathrm{d}y \\ &= \int_{\mathbb{R}} f(y)\,\Big(\int_{\mathbb{R}} g(t)\,\mathrm{e}^{-\mathrm{i}\omega t}\,\mathrm{d}t\Big)\,\mathrm{e}^{-\mathrm{i}\omega y}\,\mathrm{d}y = \hat{f}(\omega)\,\hat{g}(\omega)\,.\end{aligned}$$

■

Applying these properties of the Fourier transform, we can calculate the Fourier transforms of some special functions.

Example 2.16 Let $N_1 : \mathbb{R} \to \mathbb{R}$ denote the *cardinal* B-*spline of order one* defined by

$$N_1(x) := \begin{cases} 1 & x \in (0,\,1)\,,\\ 1/2 & x \in \{0,\,1\}\,,\\ 0 & \text{otherwise}\,.\end{cases}$$

For $m \in \mathbb{N}$, the convolution

$$N_{m+1}(x) := (N_m * N_1)(x) = \int_0^1 N_m(x-t)\,\mathrm{d}t\,,$$

is the *cardinal* B-*spline of order* $m+1$. Especially, for $m = 1$, we obtain the linear cardinal B-spline:

$$N_2(x) := \begin{cases} x & x \in [0,\,1)\,,\\ 2-x & x \in [1,\,2)\,,\\ 0 & \text{otherwise}\,.\end{cases}$$

Note that the support of N_m is the interval $[0,\,m]$. By

$$\hat{N}_1(\omega) = \int_0^1 \mathrm{e}^{-\mathrm{i}x\omega}\,\mathrm{d}x = \frac{1-\mathrm{e}^{-\mathrm{i}\omega}}{\mathrm{i}\,\omega}\,, \qquad \omega \in \mathbb{R}\setminus\{0\}\,,$$

and $\hat{N}_1(0) = 1$, we obtain:

$$\hat{N}_1(\omega) = \mathrm{e}^{-\mathrm{i}\omega/2}\,\operatorname{sinc}\frac{\omega}{2}\,, \qquad \omega \in \mathbb{R}\,.$$

By the convolution property of Theorem 2.15, we obtain

$$\hat{N}_{m+1}(\omega) = \hat{N}_m(\omega)\,\hat{N}_1(\omega) = \big(\hat{N}_1(\omega)\big)^{m+1}\,.$$

Hence, the Fourier transform of the cardinal B-spline N_m reads as follows:

$$\hat{N}_m(\omega) = \mathrm{e}^{-\mathrm{i}\, m\, \omega/2} \, \big(\operatorname{sinc} \frac{\omega}{2}\big)^m \,.$$

For the *centered cardinal* B-*spline of order* $m \in \mathbb{N}$ defined by

$$M_m(x) := N_m(x + \frac{m}{2})\,,$$

we obtain by the translation property of Theorem 2.5 that

$$\hat{M}_m(\omega) = \big(\operatorname{sinc} \frac{\omega}{2}\big)^m \,.$$

□

The Banach space $L_1(\mathbb{R})$ with the addition and convolution of functions is a *Banach algebra.* As for periodic functions, there is no identity element with respect to the convolution. A remedy is again to work with an approximate identity. We start with following observation.

Lemma 2.17

1. If $f \in L_p(\mathbb{R})$ with $1 \le p \le \infty$, then

$$\lim_{y \to 0} \|f(\cdot - y) - f\|_{L_p(\mathbb{R})} = 0\,.$$

2. If $f \in C_0(\mathbb{R})$, then

$$\lim_{y \to 0} \|f(\cdot - y) - f\|_{C_0(\mathbb{R})} = 0\,.$$

Proof

1. Let $f \in L_p(\mathbb{R})$ be given. Then for arbitrary $\varepsilon > 0$, there exists a step function:

$$s(x) := \sum_{j=1}^{n} a_j\, \chi_{I_j}(x)$$

with constants $a_j \in \mathbb{C}$ and pairwise disjoint intervals $I_j \subset \mathbb{R}$ such that $\|f - s\|_{L_p(\mathbb{R})} < \varepsilon/3$. Corresponding to ε, we choose $\delta > 0$ so that $\|s(\cdot - y) - s\|_{L_p(\mathbb{R})} < \varepsilon/3$ for all $|y| < \delta$. Thus, we obtain:

$$\begin{aligned}\|f(\cdot - y) - f\|_{L_p(\mathbb{R})} &\le \|f(\cdot - y) - s(\cdot - y)\|_{L_p(\mathbb{R})} \\ &\quad + \|s(\cdot - y) - s\|_{L_p(\mathbb{R})} + \|s - f\|_{L_p(\mathbb{R})}\end{aligned}$$

$$\le \frac{\varepsilon}{3} + \frac{\varepsilon}{3} + \frac{\varepsilon}{3} = \varepsilon\,.$$

2. Now let $f \in C_0(\mathbb{R})$ be given. Then by Remark 2.1, f is uniformly continuous on $\mathbb{R}$. Thus for each $\varepsilon > 0$, there exists $\delta > 0$ such that $|f(x-y) - f(x)| < \varepsilon$ for all $|y| < \delta$ and all $x \in \mathbb{R}$, i.e., $\|f(\cdot - y) - f\|_{C_0(\mathbb{R})} < \varepsilon$. ■

Theorem 2.18 *Let $\varphi \in L_1(\mathbb{R})$ with $\int_{\mathbb{R}} \varphi(x)\,\mathrm{d}x = 1$ be given and let*

$$\varphi_\sigma(x) := \frac{1}{\sigma}\varphi\Big(\frac{x}{\sigma}\Big)\,, \quad \sigma > 0\,,$$

be a so-called approximate identity. *Then the following relations hold true:*

1. *For $f \in L_p(\mathbb{R})$ with $1 \le p < \infty$, we have:*

$$\lim_{\sigma \to 0} \|f * \varphi_\sigma - f\|_{L_p(\mathbb{R})} = 0\,.$$

2. *For $f \in C_0(\mathbb{R})$, we have:*

$$\lim_{\sigma \to 0} \|f * \varphi_\sigma - f\|_{C_0(\mathbb{R})} = 0\,.$$

Proof

1. Let $f \in L_p(\mathbb{R})$, $1 \le p < \infty$, be given. Since $\int_{\mathbb{R}} \varphi_\sigma(y)\,\mathrm{d}y = 1$ for all $\sigma > 0$, we obtain:

$$(f * \varphi_\sigma)(x) - f(x) = \int_{\mathbb{R}} \big(f(x-y) - f(x)\big)\,\varphi_\sigma(y)\,\mathrm{d}y$$

and hence

$$\begin{aligned} \big|(f * \varphi_\sigma)(x) - f(x)\big| &\le \int_{\mathbb{R}} |f(x-y) - f(x)|\,|\varphi_\sigma(y)|\,\mathrm{d}y \\ &= \int_{\mathbb{R}} \big(|f(x-y) - f(x)|\,|\varphi_\sigma(y)|^{1/p}\big)\,|\varphi_\sigma(y)|^{1/q}\,\mathrm{d}y\,, \end{aligned}$$

where $\frac{1}{p} + \frac{1}{q} = 1$. Using Hölder's inequality, the above integral can be estimated by

$$\Big(\int_{\mathbb{R}} |f(x-y) - f(x)|^p\,|\varphi_\sigma(y)|\,\mathrm{d}y\Big)^{1/p}\,\Big(\int_{\mathbb{R}} |\varphi_\sigma(y)|\,\mathrm{d}y\Big)^{1/q}\,.$$

Thus, we obtain:

$$\int_{\mathbb{R}} \big|(f * \varphi_\sigma)(x) - f(x)\big|^p \,\mathrm{d}x$$
$$\leq \Big(\int_{\mathbb{R}} \int_{\mathbb{R}} |f(x-y) - f(x)|^p \,|\varphi_\sigma(y)|\,\mathrm{d}y\,\mathrm{d}x\Big) \Big(\int_{\mathbb{R}} |\varphi_\sigma(y)|\,\mathrm{d}y\Big)^{p/q}.$$

For arbitrary $\varepsilon > 0$, we choose $\delta > 0$ by Lemma 2.17 so that $\|f(\cdot - y) - f\|_{L_p(\mathbb{R})} < \varepsilon$ for $|y| < \delta$. Changing the order of integration, we see that the last integral term is bounded by

$$\|\varphi_\sigma\|_{L_1(\mathbb{R})}^{p/q} \Big(\int_{\mathbb{R}} |\varphi_\sigma(y)| \int_{\mathbb{R}} |f(x-y) - f(x)|^p \,\mathrm{d}x\,\mathrm{d}y\Big)$$
$$\leq \|\varphi_\sigma\|_{L_1(\mathbb{R})}^{p/q} \Big(\int_{-\delta}^{\delta} + \int_{|y|>\delta}\Big) |\varphi_\sigma(y)|\, \|f(\cdot - y) - f\|_{L_p(\mathbb{R})}^p \,\mathrm{d}y\,.$$

Then we receive

$$\int_{-\delta}^{\delta} |\varphi_\sigma(y)|\, \|f(\cdot - y) - f\|_{L_p(\mathbb{R})}^p \,\mathrm{d}y \leq \varepsilon \int_{-\delta}^{\delta} |\varphi_\sigma(y)|\,\mathrm{d}y \leq \varepsilon\, \|\varphi\|_{L_1(\mathbb{R})}\,.$$

Since $\|f(\cdot - y) - f\|_{L_p(\mathbb{R})} \leq \|f(\cdot - y)\|_{L_p(\mathbb{R})} + \|f\|_{L_p(\mathbb{R})} = 2\,\|f\|_{L_p(\mathbb{R})}$, we conclude that

$$\int_{|y|>\delta} |\varphi_\sigma(y)|\, \|f(\cdot - y) - f\|_{L_p(\mathbb{R})}^p \,\mathrm{d}y \leq 2^p\, \|f\|_{L_p(\mathbb{R})}^p \int_{|y|>\delta} |\varphi_\sigma(y)|\,\mathrm{d}y$$
$$= 2^p\, \|f\|_{L_p(\mathbb{R})}^p \int_{|x|>\delta/\sigma} |\varphi(x)|\,\mathrm{d}x\,.$$

Observing

$$\lim_{\sigma \to 0} \int_{|x|>\delta/\sigma} |\varphi(x)|\,\mathrm{d}x = 0$$

by $\varphi \in L_1(\mathbb{R})$, we obtain:

$$\limsup_{\sigma \to 0} \|f * \varphi_\sigma - f\|_{L_p(\mathbb{R})}^p \leq \varepsilon\, \|\varphi\|_{L_1(\mathbb{R})}^p\,.$$

2. Now we consider $f \in C_0(\mathbb{R})$. For arbitrary $\varepsilon > 0$ we choose $\delta > 0$ by Lemma 2.17 so that $\|f(\cdot - y) - f\|_{C_0(\mathbb{R})} < \varepsilon$ for $|y| < \delta$. As in the first step, we preserve

$$\big|(f * \varphi_\sigma)(x) - f(x)\big| \leq \int_{\mathbb{R}} \big|f(x-y) - f(x)\big|\, \big|\varphi_\sigma(y)\big|\,\mathrm{d}y$$

$$\le \Big(\int_{-\delta}^{\delta} + \int_{|y|>\delta}\Big) |\varphi_\sigma(y)|\, |f(x-y) - f(x)|\, \mathrm{d}y\,.$$

Now we estimate both integrals:

$$\int_{-\delta}^{\delta} |\varphi_\sigma(y)|\, |f(x-y) - f(x)|\, \mathrm{d}y \le \varepsilon \int_{-\delta}^{\delta} |\varphi_\sigma(y)|\, \mathrm{d}y \le \varepsilon\, \|\varphi\|_{L_1(\mathbb{R})}\,,$$

$$\begin{aligned}\int_{|y|>\delta} |\varphi_\sigma(y)|\, |f(x-y) - f(x)|\, \mathrm{d}y &\le 2\, \|f\|_{C_0(\mathbb{R})} \int_{|y|>\delta} |\varphi_\sigma(y)|\, \mathrm{d}y \\ &= 2\, \|f\|_{C_0(\mathbb{R})} \int_{|y|>\delta/\sigma} |\varphi(x)|\, \mathrm{d}x\end{aligned}$$

and obtain

$$\limsup_{\sigma\to 0} \|f * \varphi_\sigma - f\|_{C_0(\mathbb{R})} \le \varepsilon\, \|\varphi\|_{L_1(\mathbb{R})}\,.$$

■

Example 2.19 Let $f \in L_1(\mathbb{R})$ be given. We choose

$$\varphi(x) := \frac{1}{\sqrt{2\pi}}\, \mathrm{e}^{-x^2/2}\,, \quad x \in \mathbb{R}\,.$$

Then by Example 2.6, the approximate identity φ_σ coincides with the normalized Gaussian function (2.2) with standard deviation $\sigma > 0$. Then for each continuity point $x_0 \in \mathbb{R}$ of f, it holds:

$$\lim_{\sigma\to 0} (f * \varphi_\sigma)(x_0) = f(x_0)\,.$$

This can be seen as follows: For any $\varepsilon > 0$, there exists $\delta > 0$ such that $|f(x_0 - y) - f(x_0)| < \varepsilon$ for all $|y| \le \delta$. Since $\int_{\mathbb{R}} \varphi_\sigma(y)\, \mathrm{d}y = 1$ by Example 2.6, we get:

$$(f * \varphi_\sigma)(x_0) - f(x_0) = \int_{\mathbb{R}} \big(f(x_0 - y) - f(x_0)\big)\, \varphi_\sigma(y)\, \mathrm{d}y$$

and consequently

$$\big|(f * g_\sigma)(x_0) - f(x_0)\big| \le \Big(\int_{-\delta}^{\delta} + \int_{|y|>\delta}\Big) \varphi_\sigma(y)\, |f(x_0 - y) - f(x_0)|\, \mathrm{d}y$$

$$\le \varepsilon \int_{-\delta}^{\delta} \varphi_\sigma(y)\, \mathrm{d}y + \int_{|y|>\delta} |f(x_0 - y)|\, \varphi_\sigma(y)\, \mathrm{d}y + |f(x_0)| \int_{|y|>\delta} \varphi_\sigma(y)\, \mathrm{d}y$$

$$\leq \varepsilon + \|f\|_{L_1(\mathbb{R})}\, \varphi_\sigma(\delta) + \frac{1}{\sqrt{2\pi}}\, |f(x_0)| \int_{|x|>\delta/\sigma} \mathrm{e}^{-x^2/2}\, \mathrm{d}x\,.$$

Thus, we obtain:

$$\limsup_{\sigma\to 0} |(f * \varphi_\sigma)(x_0) - f(x_0)| \leq \varepsilon\,.$$

□

An important consequence of Theorem 2.18 is the following result.

Theorem 2.20 *The set $C_c^\infty(\mathbb{R})$ of all compactly supported, infinitely differentiable functions is dense in $L_p(\mathbb{R})$ for $1 \leq p < \infty$ and in $C_0(\mathbb{R})$.*

Proof Let

$$\varphi(x) := \begin{cases} c\, \exp\big(-\frac{1}{1-x^2}\big) & x \in (-1,\, 1)\,, \\ 0 & x \in \mathbb{R} \setminus (-1,\, 1)\,, \end{cases}$$

where the constant $c > 0$ is determined by the condition $\int_{\mathbb{R}} \varphi(x)\, \mathrm{d}x = 1$. Then φ is infinitely differentiable and has compact support $[-1,\, 1]$, i.e., $\varphi \in C_c^\infty(\mathbb{R})$. We choose φ_σ with $\sigma > 0$ as approximate identity.

Let $f \in L_p(\mathbb{R})$. For arbitrary $\varepsilon > 0$, there exists $N \in \mathbb{N}$ such that

$$\|f - f_N\|_{L_p(\mathbb{R})} < \frac{\varepsilon}{2}\,,$$

where f_N is the restricted function

$$f_N(x) := \begin{cases} f(x) & x \in [-N,\, N]\,, \\ 0 & x \in \mathbb{R} \setminus [-N,\, N]\,. \end{cases}$$

By the first part of Theorem 2.18, we know that

$$\|f_N * \varphi_\sigma - f_N\|_{L_p(\mathbb{R})} < \frac{\varepsilon}{2}$$

for sufficiently small $\sigma > 0$. Thus, $f_N * \varphi_\sigma$ is a good approximation of f, because

$$\|f_N * \varphi_\sigma - f\|_{L_p(\mathbb{R})} \leq \|f_N * \varphi_\sigma - f_N\|_{L_p(\mathbb{R})} + \|f_N - f\|_{L_p(\mathbb{R})} < \varepsilon\,.$$

By Lemma 2.14, the function $f_N * \varphi_\sigma$ is infinitely differentiable. Further this convolution product has compact support $\operatorname{supp} f_N * \varphi_\sigma \subseteq [-N - \sigma,\, N + \sigma]$. Consequently, $C_c^\infty(\mathbb{R})$ is a dense subset of $L_p(\mathbb{R})$.

Using the second part of Theorem 2.18, one can analogously prove the assertion for $f \in C_0(\mathbb{R})$. ■

2.2 Fourier Transform on $L_2(\mathbb{R})$

Up to now we have considered the Fourier transforms of $L_1(\mathbb{R})$ functions. Next we want to establish a Fourier transform on the Hilbert space $L_2(\mathbb{R})$, where the Fourier integral

$$\int_{\mathbb{R}} f(x)\,\mathrm{e}^{-\mathrm{i}x\omega}\,\mathrm{d}x$$

may not exist, i.e., it does not take a finite value for some $\omega \in \mathbb{R}$. Therefore, we define the Fourier transform of an $L_2(\mathbb{R})$ function in a different way based on the following result.

Lemma 2.21 *Let $f,\, g \in L_1(\mathbb{R})$, such that $\hat{f},\, \hat{g} \in L_1(\mathbb{R})$. Then, the following* Parseval equality *is valid:*

$$2\pi\,\langle f,\, g\rangle_{L_2(\mathbb{R})} = \langle \hat{f},\, \hat{g}\rangle_{L_2(\mathbb{R})}\,. \tag{2.12}$$

Note that $f,\, \hat{f} \in L_1(\mathbb{R})$ implies that $(\hat{f})\check{} = f$ almost everywhere by Theorem 2.10 and $(\hat{f})\check{} \in C_0(\mathbb{R})$ by Theorem 2.8. Thus, we have $f \in L_2(\mathbb{R})$, since

$$\int_{\mathbb{R}} |f(x)|^2\,\mathrm{d}x = \int_{\mathbb{R}} |(\hat{f})\check{}(x)|\,|f(x)|\,\mathrm{d}x \le \|(\hat{f})\check{}\|_{C_0(\mathbb{R})}\,\|f\|_{L_1(\mathbb{R})} < \infty\,.$$

Proof Using Fubini's theorem and Fourier inversion formula (2.7), we obtain:

$$\begin{aligned}\int_{\mathbb{R}} \hat{f}(\omega)\,\overline{\hat{g}(\omega)}\,\mathrm{d}\omega &= \int_{\mathbb{R}} \hat{f}(\omega)\,\overline{\int_{\mathbb{R}} g(x)\,\mathrm{e}^{-\mathrm{i}x\omega}\,\mathrm{d}x}\,\mathrm{d}\omega \\ &= \int_{\mathbb{R}} \overline{g(x)} \int_{\mathbb{R}} \hat{f}(\omega)\,\mathrm{e}^{\mathrm{i}x\omega}\,\mathrm{d}\omega\,\mathrm{d}x = 2\pi \int_{\mathbb{R}} \overline{g(x)}\,f(x)\,\mathrm{d}x\,.\end{aligned}$$

■

Applying Theorem 2.20, for any function $f \in L_2(\mathbb{R})$, there exists a sequence $(f_j)_{j\in\mathbb{N}}$ of functions $f_j \in C_c^{\infty}(\mathbb{R})$ such that

$$\lim_{j\to\infty} \|f - f_j\|_{L_2(\mathbb{R})} = 0\,.$$

Thus, $(f_j)_{j\in\mathbb{N}}$ is a Cauchy sequence in $L_2(\mathbb{R})$, i.e., for every $\varepsilon > 0$, there exists an index $N(\varepsilon) \in \mathbb{N}$ so that for all $j, k \ge N(\varepsilon)$

$$\|f_k - f_j\|_{L_2(\mathbb{R})} < \varepsilon\,.$$

Clearly, $f_j,\, \hat{f}_j \in L_1(\mathbb{R})$. By Parseval equality (2.12), we obtain for all $j, k \ge N(\varepsilon)$

$$\|f_k - f_j\|_{L_2(\mathbb{R})} = \frac{1}{\sqrt{2\pi}} \|\hat{f}_k - \hat{f}_j\|_{L_2(\mathbb{R})} \le \varepsilon ,$$

so that $(\hat{f}_j)_{j\in\mathbb{N}}$ is also a Cauchy sequence in $L_2(\mathbb{R})$. Since $L_2(\mathbb{R})$ is complete, this Cauchy sequence converges to some function in $L_2(\mathbb{R})$. We define the *Fourier transform* $\hat{f} = \mathscr{F} f \in L_2(\mathbb{R})$ of $f \in L_2(\mathbb{R})$ as

$$\hat{f} = \mathscr{F} f := \lim_{j\to\infty} \hat{f}_j .$$

In this way the domain of the Fourier transform is extended to $L_2(\mathbb{R})$. Note that the set $L_1(\mathbb{R}) \cap L_2(\mathbb{R})$ is dense in $L_2(\mathbb{R})$, since $C_c^\infty(\mathbb{R})$ is contained in $L_1(\mathbb{R}) \cap L_2(\mathbb{R})$ and $C_c^\infty(\mathbb{R})$ is dense in $L_2(\mathbb{R})$ by Theorem 2.20.

By the continuity of the inner product, we obtain also the Parseval equality in $L_2(\mathbb{R})$. We summarize.

Theorem 2.22 (Plancherel) *The Fourier transform truncated on $L_1(\mathbb{R}) \cap L_2(\mathbb{R})$ can be uniquely extended to a bounded linear operator of $L_2(\mathbb{R})$ onto itself which satisfies the* Parseval equalities

$$2\pi \, \langle f,\, g\rangle_{L_2(\mathbb{R})} = \langle \hat{f},\, \hat{g}\rangle_{L_2(\mathbb{R})} , \quad \sqrt{2\pi}\, \|f\|_{L_2(\mathbb{R})} = \|\hat{f}\|_{L_2(\mathbb{R})} \tag{2.13}$$

for all $f,\, g \in L_2(\mathbb{R})$.

Note that Theorem 2.5 is also true for $L_2(\mathbb{R})$ functions. Moreover, we have the following Fourier inversion formula.

Theorem 2.23 (Fourier Inversion Formula for $L_2(\mathbb{R})$ Functions)
Let $f \in L_2(\mathbb{R})$ with $\hat{f} \in L_1(\mathbb{R})$ be given. Then the Fourier inversion formula

$$f(x) = \frac{1}{2\pi} \int_{\mathbb{R}} \hat{f}(\omega)\, \mathrm{e}^{\mathrm{i}\,\omega x}\, \mathrm{d}\omega \tag{2.14}$$

holds true for almost every $x \in \mathbb{R}$. If f is in addition continuous, then the Fourier inversion formula (2.14) *holds pointwise for all $x \in \mathbb{R}$.*

Remark 2.24 Often the integral notation

$$\hat{f}(\omega) = \int_{\mathbb{R}} f(x)\, \mathrm{e}^{-\mathrm{i}x\omega}\, \mathrm{d}x$$

is also used for the Fourier transform of $L_2(\mathbb{R})$ functions, although the integral may not converge pointwise. But it may be interpreted by a limiting process. For $\varepsilon > 0$ and $f \in L_2(\mathbb{R})$, the function $g_\varepsilon : \mathbb{R} \to \mathbb{C}$ is defined by

$$g_\varepsilon(\omega) := \int_{\mathbb{R}} \mathrm{e}^{-\varepsilon^2 x^2}\, f(x)\, \mathrm{e}^{-\mathrm{i}x\omega}\, \mathrm{d}x .$$

Then, g_ε converges in the $L_2(\mathbb{R})$ norm and pointwise almost everywhere to $\hat{f}$ for $\varepsilon \to 0$. □

Finally we introduce an orthogonal basis of $L_2(\mathbb{R})$, whose elements are eigenfunctions of the Fourier transform. For $n \in \mathbb{N}_0$, the *nth Hermite polynomial* H_n is defined by

$$H_n(x) := (-1)^n \, \mathrm{e}^{x^2} \frac{\mathrm{d}^n}{\mathrm{d}x^n} \mathrm{e}^{-x^2} , \quad x \in \mathbb{R} .$$

In particular we have

$$H_0(x) = 1 , \quad H_1(x) = 2x , \quad H_2(x) = 4x^2 - 2 , \quad H_3(x) = 8x^3 - 12x .$$

The Hermite polynomials fulfill the three-term relation:

$$H_{n+1}(x) = 2x \, H_n(x) - 2n \, H_{n-1}(x) , \quad n \in \mathbb{N} , \tag{2.15}$$

and the recursion

$$H_n'(x) = 2n \, H_{n-1}(x) , \quad n \in \mathbb{N} . \tag{2.16}$$

For $n \in \mathbb{N}_0$, the *nth Hermite function* h_n is given by

$$h_n(x) := H_n(x) \, \mathrm{e}^{-x^2/2} = (-1)^n \, \mathrm{e}^{x^2/2} \frac{\mathrm{d}^n}{\mathrm{d}x^n} \mathrm{e}^{-x^2} , \quad x \in \mathbb{R} .$$

In particular, we have $h_0(x) = \mathrm{e}^{-x^2/2}$ which has the Fourier transform $\hat{h}_0(\omega) = \frac{1}{\sqrt{2\pi}} \mathrm{e}^{-\omega^2/2}$ by Example 2.6. The Hermite functions fulfill the differential equation:

$$h_n''(x) - (x^2 - 2n - 1) \, h_n(x) = 0 \tag{2.17}$$

and can be computed recursively by

$$h_{n+1}(x) = x \, h_n(x) - h_n'(x) , \quad n \in \mathbb{N}_0 .$$

Theorem 2.25 *The Hermite functions h_n, $n \in \mathbb{N}_0$, with*

$$\langle h_n, h_n \rangle_{L_2(\mathbb{R})} = \sqrt{\pi} \, 2^n n!$$

form a complete orthogonal system in $L_2(\mathbb{R})$. The Fourier transforms of the Hermite functions are given by

$$\hat{h}_n(\omega) = \sqrt{2\pi} \, (-\mathrm{i})^n \, h_n(\omega) . \tag{2.18}$$

In other words, the functions h_n are the eigenfunctions of the Fourier transform $\mathscr{F} : L_2(\mathbb{R}) \to L_2(\mathbb{R})$ with eigenvalues $\sqrt{2\pi}\,(-\mathrm{i})^n$ for all $n \in \mathbb{N}_0$.

By Theorem 2.25 we see that the Hermite polynomials are orthogonal polynomials in the weighted Hilbert space $L_{2,w}(\mathbb{R})$ with $w(x) := \mathrm{e}^{-x^2}$, $x \in \mathbb{R}$, i.e., they are orthogonal with respect to the weighted Lebesgue measure $\mathrm{e}^{-x^2}\,\mathrm{d}x$.

Proof

1. We show that $\langle h_m, h_n\rangle_{L_2(\mathbb{R})} = 0$ for $m \neq n$. By the differential equation (2.17), we obtain:

$$\begin{aligned}(h_m'' - x^2 h_m)\,h_n &= -(2m+1)\,h_m\,h_n\,,\\ (h_n'' - x^2 h_n)\,h_m &= -(2n+1)\,h_m\,h_n\,.\end{aligned}$$

Substraction yields

$$h_m''\,h_n - h_n''\,h_m = (h_m' h_n - h_n' h_m)' = 2(n-m)\,h_m\,h_n\,,$$

which results after integration in

$$\begin{aligned}2(n-m)\,\langle h_m, h_n\rangle_{L_2(\mathbb{R})} &= 2(m-n)\int_{\mathbb{R}} h_m(x)\,h_n(x)\,\mathrm{d}x\\ &= \big(h_m'(x)\,h_n(x) - h_n'(x)\,h_m(x)\big)\big|_{-\infty}^{\infty} = 0\,.\end{aligned}$$

2. Next we prove for $n \in \mathbb{N}_0$ that

$$\langle h_n, h_n\rangle_{L_2(\mathbb{R})} = \sqrt{\pi}\,2^n n!\,. \tag{2.19}$$

For $n = 0$ the relation holds true by Example 2.6. We show the recursion:

$$\langle h_{n+1}, h_{n+1}\rangle_{L_2(\mathbb{R})} = 2(n+1)\,\langle h_n, h_n\rangle_{L_2(\mathbb{R})} \tag{2.20}$$

which implies (2.19). Using (2.16), integration by parts, and step 1 of this proof, we obtain:

$$\begin{aligned}\langle h_{n+1}, h_{n+1}\rangle_{L_2(\mathbb{R})} &= \int_{\mathbb{R}} \mathrm{e}^{-x^2}\,(H_{n+1}(x))^2\,\mathrm{d}x = \int_{\mathbb{R}} \big(2x\mathrm{e}^{-x^2}\big)\big(H_n(x)H_{n+1}(x)\big)\mathrm{d}x\\ &= \int_{\mathbb{R}} \mathrm{e}^{-x^2}\,\big(H_n'(x)\,H_{n+1}(x) + H_n(x)\,H_{n+1}'(x)\big)\,\mathrm{d}x\\ &= 2\,(n+1)\int_{\mathbb{R}} \mathrm{e}^{-x^2}\,\big(H_n(x)\big)^2\,\mathrm{d}x = 2\,(n+1)\,\langle h_n, h_n\rangle_{L_2(\mathbb{R})}\,.\end{aligned}$$

3. To verify the completeness of the orthogonal system $\{h_n : n \in \mathbb{N}_0\}$, we prove that $f \in L_2(\mathbb{R})$ with $\langle f, h_n\rangle_{L_2(\mathbb{R})} = 0$ for all $n \in \mathbb{N}_0$ implies $f = 0$ almost everywhere. To this end, we consider the complex function $g\colon \mathbb{C} \to \mathbb{C}$ defined by

$$g(z) := \int_{\mathbb{R}} h_0(x)\, f(x)\, \mathrm{e}^{-\mathrm{i}xz}\, \mathrm{d}x\,, \quad z \in \mathbb{C}\,.$$

This is the holomorphic continuation of the Fourier transform of $h_0\, f$ onto whole $\mathbb{C}$. For every $m \in \mathbb{N}_0$, it holds:

$$g^{(m)}(z) = (-\mathrm{i})^m \int_{\mathbb{R}} x^m\, h_0(x)\, f(x)\, \mathrm{e}^{-\mathrm{i}xz}\, \mathrm{d}x\,, \quad z \in \mathbb{C}\,.$$

Since $g^{(m)}(0)$ is a certain linear combination of $\langle f, h_n\rangle_{L_2(\mathbb{R})}$, $n = 0, \ldots, m$, we conclude that $g^{(m)}(0) = 0$ for all $m \in \mathbb{N}_0$. Thus, we get $g = 0$ and $(h_0 f)^\vee = 0$. By Corollary 2.11 we have $h_0 f = 0$ almost everywhere and consequently $f = 0$ almost everywhere.

4. By Example 2.6 we know that

$$\hat{h}_0(\omega) = \int_{\mathbb{R}} \mathrm{e}^{-\mathrm{i}x\omega - x^2/2}\, \mathrm{d}x = \sqrt{2\pi}\, \mathrm{e}^{-\omega^2/2}\,, \quad \omega \in \mathbb{R}\,.$$

We compute the Fourier transform of h_n and obtain after n times integration by parts

$$\begin{aligned}\hat{h}_n(\omega) &= \int_{\mathbb{R}} h_n(x)\, \mathrm{e}^{-\mathrm{i}\omega x}\, \mathrm{d}x = (-1)^n \int_{\mathbb{R}} \mathrm{e}^{-\mathrm{i}\omega x + x^2/2} \Big(\frac{\mathrm{d}^n}{\mathrm{d}x^n}\, \mathrm{e}^{-x^2}\Big)\, \mathrm{d}x \\ &= \int_{\mathbb{R}} \mathrm{e}^{-x^2} \Big(\frac{\mathrm{d}^n}{\mathrm{d}x^n}\, \mathrm{e}^{-\mathrm{i}\omega x + x^2/2}\Big)\, \mathrm{d}x = \mathrm{e}^{\omega^2/2} \int_{\mathbb{R}} \mathrm{e}^{-x^2} \Big(\frac{\mathrm{d}^n}{\mathrm{d}x^n}\, \mathrm{e}^{(x-\mathrm{i}\omega)^2/2}\Big)\, \mathrm{d}x\,.\end{aligned}$$

By symmetry reasons we have

$$\frac{\mathrm{d}^n}{\mathrm{d}x^n}\, \mathrm{e}^{(x-\mathrm{i}\omega)^2/2} = \mathrm{i}^n\, \frac{\mathrm{d}^n}{\mathrm{d}\omega^n}\, \mathrm{e}^{(x-\mathrm{i}\omega)^2/2}\,,$$

so that

$$\begin{aligned}\hat{h}_n(\omega) &= \mathrm{i}^n\, \mathrm{e}^{\omega^2/2} \int_{\mathbb{R}} \mathrm{e}^{-x^2} \Big(\frac{\mathrm{d}^n}{\mathrm{d}\omega^n}\, \mathrm{e}^{(x-\mathrm{i}\omega)^2/2}\Big)\, \mathrm{d}x \\ &= \mathrm{i}^n\, \mathrm{e}^{\omega^2/2}\, \frac{\mathrm{d}^n}{\mathrm{d}\omega^n} \Big(\mathrm{e}^{-\omega^2/2} \int_{\mathbb{R}} \mathrm{e}^{-\mathrm{i}x\omega - x^2/2}\, \mathrm{d}x\Big) \\ &= \sqrt{2\pi}\, \mathrm{i}^n\, \mathrm{e}^{\omega^2/2}\, \frac{\mathrm{d}^n}{\mathrm{d}\omega^n}\, \mathrm{e}^{-\omega^2} = \sqrt{2\pi}\, (-\mathrm{i})^n\, h_n(\omega)\,.\end{aligned}$$

■

In addition to Theorem 2.15, the following relation between the convolution of functions and the Fourier transform in $L_2(\mathbb{R})$ is often useful.

Theorem 2.26 (Convolution Property of Fourier Transform) *For $f, g \in L_2(\mathbb{R})$ the following relations hold true:*

$$\frac{1}{2\pi}\,(\hat{f} * \hat{g}) = \mathcal{F}(f\,g)\,, \quad f * g = \mathcal{F}^{-1}(\hat{f}\,\hat{g})\,, \tag{2.21}$$

where $\mathcal{F} : L_1(\mathbb{R}) \to C_0(\mathbb{R})$ is given by (2.1) and $\mathcal{F}^{-1} : L_1(\mathbb{R}) \to C_0(\mathbb{R})$ is defined for $h \in L_1(\mathbb{R})$ by

$$(\mathcal{F}^{-1}h)(x) := \frac{1}{2\pi} \int_{\mathbb{R}} h(\omega)\, \mathrm{e}^{\mathrm{i}\omega x}\, \mathrm{d}\omega\,, \quad x \in \mathbb{R}\,.$$

Note that $f, g \in L_2(\mathbb{R})$ implies $\hat{f}, \hat{g} \in L_2(\mathbb{R})$ and by Theorem 2.13 then $f * g$, $\hat{f} * \hat{g} \in C_0(\mathbb{R})$. Further, by Hölder's inequality, we obtain $f\,g$, $\hat{f}\,\hat{g} \in L_1(\mathbb{R})$.

Proof By the modulation and scaling properties of the Fourier transform on $L_2(\mathbb{R})$, it holds for $f, g \in L_2(\mathbb{R})$ and $\omega \in \mathbb{R}$ that

$$\hat{f}(\omega - \cdot) = \big(f(-\cdot)\,\mathrm{e}^{\mathrm{i}\omega\,\cdot}\big)^{\wedge}, \quad \bar{\hat{g}} = \big(\bar{g}(-\cdot)\big)^{\wedge}.$$

Then, we obtain by Plancherel's theorem 2.22 for any $\omega \in \mathbb{R}$ that

$$\begin{aligned}\frac{1}{2\pi}\,(\hat{f} * \hat{g})(\omega) &= \frac{1}{2\pi}\,\langle \hat{f}(\omega - \cdot),\, \bar{\hat{g}} \rangle_{L_2(\mathbb{R})} \\ &= \frac{1}{2\pi}\,\langle \big(f(-\cdot)\mathrm{e}^{\mathrm{i}\omega\,\cdot}\big)^{\wedge},\, \big(\bar{g}(-\cdot)\big)^{\wedge} \rangle_{L_2(\mathbb{R})} \\ &= \langle f(-\cdot)\mathrm{e}^{\mathrm{i}\omega\,\cdot},\, \bar{g}(-\cdot) \rangle_{L_2(\mathbb{R})} = \mathcal{F}(f\,g)(\omega)\,.\end{aligned}$$

The second formula in (2.21) follows immediately from the first one, if we replace f by $\hat{f}$ and g by $\hat{g}$ and use the relations:

$$\hat{\hat{f}} = 2\pi\, f(-\cdot)\,, \quad \hat{\hat{g}} = 2\pi\, g(-\cdot)\,.$$

This completes the proof. ∎

Example 2.27 We consider the function:

$$f(x) := \frac{1}{2}\, \chi_{(-1,1)}(x)\,, \quad x \in \mathbb{R}\,.$$

By Example 2.3 it holds:

$$\hat{f}(\omega) = \operatorname{sinc} \omega \in L_2(\mathbb{R}) \setminus L_1(\mathbb{R})\,.$$

Using (2.21), we determine the convolution $\hat{f} * \hat{f}$ that means

$$\frac{1}{2\pi}(\hat{f} * \hat{f})(\omega) = \frac{1}{4}\big(\mathscr{F}\chi_{(-1,1)}\big)(\omega) = \frac{1}{2}\,\mathrm{sinc}\,\omega\,.$$

Thus, we obtain:

$$\frac{1}{\pi}\int_{\mathbb{R}} \mathrm{sinc}(\omega - \tau)\,\mathrm{sinc}\,\tau\,\mathrm{d}\tau = \mathrm{sinc}\,\omega\,, \qquad \omega \in \mathbb{R}\,.$$

□

2.3 Poisson Summation Formula and Shannon's Sampling Theorem

Poisson summation formula establishes an interesting relation between Fourier series and Fourier transforms. For $n \in \mathbb{N}$ and $f \in L_1(\mathbb{R})$, we consider the functions:

$$\varphi_n(x) := \sum_{k=-n}^{n} |f(x + 2k\pi)|\,,$$

which fulfill

$$\int_{-\pi}^{\pi} \varphi_n(x)\,\mathrm{d}x = \int_{-\pi}^{\pi} \sum_{k=-n}^{n} |f(x + 2k\pi)|\,\mathrm{d}x = \sum_{k=-n}^{n} \int_{-\pi}^{\pi} |f(x + 2k\pi)|\,\mathrm{d}x$$

$$= \sum_{k=-n}^{n} \int_{2k\pi-\pi}^{2k\pi+\pi} |f(x)|\,\mathrm{d}x = \int_{-2n\pi-\pi}^{2n\pi+\pi} |f(x)|\,\mathrm{d}x \le \|f\|_{L_1(\mathbb{R})} < \infty\,.$$

Since $(\varphi_n)_{n\in\mathbb{N}}$ is a monotone increasing sequence of nonnegative functions, we obtain by the monotone convergence theorem of B. Levi that the function $\varphi(x) := \lim_{n\to\infty} \varphi_n(x)$ for almost all $x \in \mathbb{R}$ is measurable and fulfills

$$\int_{-\pi}^{\pi} \varphi(x)\,\mathrm{d}x = \lim_{n\to\infty} \int_{-\pi}^{\pi} \varphi_n(x)\,\mathrm{d}x = \|f\|_{L_1(\mathbb{R})} < \infty\,.$$

We introduce the 2π-periodic function:

$$\tilde{f}(x) := \sum_{k\in\mathbb{Z}} f(x + 2k\pi)\,. \tag{2.22}$$

The 2π–periodic function $\tilde{f} \in L_1(\mathbb{T})$ is called 2π*–periodization* of $f \in L_1(\mathbb{R})$. Since

$$|\tilde{f}(x)| = \Big|\sum_{k\in\mathbb{Z}} f(x+2k\pi)\Big| \le \sum_{k\in\mathbb{Z}} |f(x+2k\pi)| = \varphi(x)\,,$$

we obtain

$$\int_{-\pi}^{\pi} |\tilde{f}(x)|\,\mathrm{d}x \le \int_{-\pi}^{\pi} |\varphi(x)|\,\mathrm{d}x = \|f\|_{L_1(\mathbb{R})} < \infty$$

so that $\tilde{f} \in L_1(\mathbb{T})$. After these preparations we can formulate the Poisson summation formula.

Theorem 2.28 (Poisson Summation Formula) *Assume that* $f \in L_1(\mathbb{R}) \cap C_0(\mathbb{R})$ *fulfills the conditions*

1. $\sum_{k\in\mathbb{Z}} \max\limits_{x\in[-\pi,\,\pi]} |f(x+2k\pi)| < \infty\,,$
2. $\sum_{k\in\mathbb{Z}} |\hat{f}(k)| < \infty\,.$

Then for all $x \in \mathbb{R}$*, the following relation is fulfilled:*

$$2\pi\,\tilde{f}(x) = 2\pi \sum_{k\in\mathbb{Z}} f(x+2k\pi) = \sum_{k\in\mathbb{Z}} \hat{f}(k)\,\mathrm{e}^{\mathrm{i}kx}\,. \tag{2.23}$$

Both series in (2.23) *converge absolutely and uniformly on* $\mathbb{R}$.

For $x = 0$ *this implies the* Poisson summation formula

$$2\pi \sum_{k\in\mathbb{Z}} f(2k\pi) = \sum_{k\in\mathbb{Z}} \hat{f}(k)\,. \tag{2.24}$$

Proof By the first assumption, we have absolute and uniform convergence of the series (2.22) by the known Weierstrass criterion of uniform convergence. Since $f \in C_0(\mathbb{R})$, we see that $\tilde{f} \in C(\mathbb{T})$. Because the uniformly convergent series (2.22) can be integrated term by term, we obtain for the Fourier coefficients of $\tilde{f}$

$$\begin{aligned} 2\pi\,c_k(\tilde{f}) &= \int_{-\pi}^{\pi} \sum_{\ell\in\mathbb{Z}} f(x+2\ell\pi)\,\mathrm{e}^{-\mathrm{i}kx}\,\mathrm{d}x = \sum_{\ell\in\mathbb{Z}} \int_{-\pi}^{\pi} f(x+2\ell\pi)\,\mathrm{e}^{-\mathrm{i}kx}\,\mathrm{d}x \\ &= \int_{\mathbb{R}} f(x)\,\mathrm{e}^{-\mathrm{i}kx}\,\mathrm{d}x = \hat{f}(k)\,. \end{aligned}$$

Thus,

$$\tilde{f}(x) = \sum_{k\in\mathbb{Z}} c_k(\tilde{f})\,\mathrm{e}^{\mathrm{i}kx} = \frac{1}{2\pi} \sum_{k\in\mathbb{Z}} \hat{f}(k)\,\mathrm{e}^{\mathrm{i}kx}$$

where by the second assumption and Theorem 1.37 the Fourier series of $\tilde{f}$ converges uniformly to $\tilde{f}$ on $\mathbb{R}$. By the second assumption, the Fourier series of $\tilde{f}$ is absolutely convergent. ■

Remark 2.29 The general Poisson summation formula (2.23) is fulfilled, if $f \in L_1(\mathbb{R}) \cap C_0(\mathbb{R})$ fulfills the conditions

$$|f(x)| \le C\,(1+|x|)^{-1-\varepsilon}\,, \quad |\hat{f}(\omega)| \le C\,(1+|\omega|)^{-1-\varepsilon}$$

for some $C > 0$ and $\varepsilon > 0$. For further details see [399, pp. 250–253] or [178, pp. 171–173]. The Poisson summation formula was generalized for slowly growing functions in [301]. □

We illustrate the performance of Poisson summation formula (2.24) by an example.

Example 2.30 For fixed $\alpha > 0$, we consider the function $f(x) := \mathrm{e}^{-\alpha|x|}$, $x \in \mathbb{R}$. Simple calculation shows that its Fourier transform reads

$$\hat{f}(\omega) = \int_0^\infty \big(\mathrm{e}^{(\alpha-\mathrm{i}\,\omega)x} + \mathrm{e}^{(\alpha+\mathrm{i}\,\omega)x}\big)\,\mathrm{d}x = \frac{2\alpha}{\alpha^2+\omega^2}\,.$$

Note that by Fourier inversion formula (2.14), the function $g(x) := (x^2+\alpha^2)^{-1}$ has the Fourier transform $\hat{g}(\omega) = \frac{\pi}{\alpha}\,\mathrm{e}^{-\alpha|\omega|}$.

The function f is contained in $L_1(\mathbb{R}) \cap C_0(\mathbb{R})$ and fulfills both conditions of Theorem 2.28. Since

$$\sum_{k\in\mathbb{Z}} f(2\pi k) = 1 + 2\sum_{k=1}^{\infty}\big(\mathrm{e}^{-2\pi\alpha}\big)^k = \frac{1+\mathrm{e}^{-2\pi\alpha}}{1-\mathrm{e}^{-2\pi\alpha}}\,,$$

we obtain by the Poisson summation formula (2.24) that

$$\sum_{k\in\mathbb{Z}} \frac{1}{\alpha^2+k^2} = \frac{\pi}{\alpha}\,\frac{1+\mathrm{e}^{-2\pi\alpha}}{1-\mathrm{e}^{-2\pi\alpha}}\,.$$

□

The following sampling theorem was discovered independently by the mathematician E.T. Whittaker [435] as well as the electrical engineers V.A. Kotelnikov [250] and C.E. Shannon [385]; see also [155, 419]. Shannon first recognized the significance of the sampling theorem in digital signal processing. The sampling theorem answers the question how to sample a function f by its values $f(nT)$, $n \in \mathbb{Z}$, for an appropriate $T > 0$ while keeping the whole information contained in f. The distance T between two successive sample points is called *sampling period.* In other words, we want to find a convenient sampling period T such that f can be recovered from its samples $f(nT)$, $n \in \mathbb{Z}$. The *sampling rate* is defined as the reciprocal value $\frac{1}{T}$ of the sampling period T. Indeed this question can be only answered for a certain class of functions.

A function $f \in L_2(\mathbb{R})$ is called *bandlimited* on $[-L,\, L]$ with some $L > 0$, if $\operatorname{supp} \hat{f} \subseteq [-L,\, L]$, i.e., if $\hat{f}(\omega) = 0$ for all $|\omega| > L$. The positive number L is the *bandwidth* of f. A typical bandlimited function on $[-L,\, L]$ is

$$h(x) = \frac{L}{\pi}\, \operatorname{sinc}(Lx)\,, \quad x \in \mathbb{R}\,.$$

Note that $h \in L_2(\mathbb{R}) \setminus L_1(\mathbb{R})$. By Example 2.3 and Theorem 2.22 of Plancherel, the Fourier transform $\hat{h}$ is equal to

$$\hat{h}(\omega) = \begin{cases} 1 & \omega \in (-L,\, L)\,, \\ \frac{1}{2} & \omega \in \{-L,\, L\}\,, \\ 0 & \omega \in \mathbb{R} \setminus [-L,\, L]\,. \end{cases} \tag{2.25}$$

Theorem 2.31 (Sampling Theorem of Shannon–Whittaker–Kotelnikov) *Let $f \in L_1(\mathbb{R}) \cap C_0(\mathbb{R})$ be bandlimited on $[-L, L]$. Let $M \geq L > 0$.*

Then f is completely determined by its values $f\big(\frac{k\pi}{M}\big)$, $k \in \mathbb{Z}$, and further f can be represented in the form:

$$f(x) = \sum_{k \in \mathbb{Z}} f\big(\frac{k\pi}{M}\big)\, \operatorname{sinc}(Mx - k\pi)\,, \tag{2.26}$$

where the series (2.26) *converges absolutely and uniformly on* $\mathbb{R}$.

Proof

1. From $f \in L_1(\mathbb{R}) \cap C_0(\mathbb{R})$, it follows that $f \in L_2(\mathbb{R})$, since

$$\|f\|^2_{L_2(\mathbb{R})} = \int_{\mathbb{R}} |f(x)|\, |f(x)|\, \mathrm{d}x \leq \|f\|_{C_0(\mathbb{R})}\, \|f\|_{L_1(\mathbb{R})} < \infty\,.$$

Let $x \in \mathbb{R}$ be an arbitrary point. Since $f \in L_2(\mathbb{R})$ is bandlimited on $[-L,\, L]$ and $M \geq L$, by Theorem 2.10, we obtain:

$$f(x) = \frac{1}{2\pi} \int_{-M}^{M} \hat{f}(\omega)\, \mathrm{e}^{\mathrm{i}\,\omega x}\, \mathrm{d}\omega\,. \tag{2.27}$$

Let $\varphi,\, \psi \in L_2(\mathbb{T})$ be the 2π-periodic extensions of

$$\varphi(\omega) := \hat{f}\big(\frac{M\omega}{\pi}\big)\,, \quad \psi(\omega) := \mathrm{e}^{-\mathrm{i}\,xM\omega/\pi}\,, \quad \omega \in [-\pi,\, \pi)\,.$$

By (2.27) these functions possess the Fourier coefficients:

$$c_k(\varphi) = \langle \varphi,\, \mathrm{e}^{\mathrm{i}k\cdot} \rangle_{L_2(\mathbb{T})} = \frac{1}{2M} \int_{-M}^{M} \hat{f}(\omega)\, \mathrm{e}^{-\mathrm{i}k\pi\omega/M}\, \mathrm{d}\omega = \frac{\pi}{M}\, f\big(-\frac{k\pi}{M}\big),$$

$$c_k(\psi) = \langle \psi,\, \mathrm{e}^{\mathrm{i}k\cdot} \rangle_{L_2(\mathbb{T})} = \frac{1}{2M} \int_{-M}^{M} \mathrm{e}^{-\mathrm{i}\,(x+\frac{k\pi}{M})\,\omega}\, \mathrm{d}\omega = \mathrm{sinc}\,(Mx + k\pi)\,.$$

From (2.27) it follows that

$$f(x) = \frac{M}{2\pi^2} \int_{-\pi}^{\pi} \hat{f}\big(\frac{M\omega}{\pi}\big)\, \mathrm{e}^{\mathrm{i}xM\omega/\pi}\, \mathrm{d}\omega = \frac{M}{\pi}\, \langle \varphi,\, \psi \rangle_{L_2(\mathbb{T})}$$

and hence by the Parseval equality (1.16)

$$\begin{aligned} f(x) &= \frac{M}{\pi} \sum_{k\in\mathbb{Z}} c_k(\varphi)\, \overline{c_k(\psi)} = \sum_{k\in\mathbb{Z}} f\big(-\frac{k\pi}{M}\big)\, \mathrm{sinc}\,(Mx + k\pi) \\ &= \sum_{k\in\mathbb{Z}} f\big(\frac{k\pi}{M}\big)\, \mathrm{sinc}\,(Mx - k\pi)\,. \end{aligned}$$

2. As shown in the first step, it holds for arbitrary $n \in \mathbb{N}$:

$$\begin{aligned} f(x) - \sum_{k=-n}^{n} f\big(\frac{k\pi}{M}\big)\, \mathrm{sinc}\,(Mx - k\pi) &= \frac{1}{2\pi} \int_{-M}^{M} \big(\hat{f}(\omega) \\ &\quad - \sum_{k=-n}^{n} c_k(\varphi)\, \mathrm{e}^{\mathrm{i}k\pi\omega/M}\big)\, \mathrm{e}^{\mathrm{i}x\omega}\, \mathrm{d}\omega \\ &= \frac{M}{2\pi^2} \int_{-\pi}^{\pi} \big(\varphi(\omega) - (S_n\varphi)(\omega)\big)\, \mathrm{e}^{\mathrm{i}xM\omega/\pi}\, \mathrm{d}\omega\,. \end{aligned}$$

Using the Cauchy–Schwarz inequality in $L_2(\mathbb{T})$, we obtain for all $x \in \mathbb{R}$

$$\begin{aligned} \big|f(x) - \sum_{k=-n}^{n} f\big(\frac{k\pi}{M}\big)\mathrm{sinc}(Mx - k\pi)\big| &= \frac{M}{2\pi^2}\big| \int_{-\pi}^{\pi} \big(\varphi(\omega) - (S_n\varphi)(\omega)\big)\mathrm{e}^{\mathrm{i}xM\omega/\pi}\,\mathrm{d}\omega\big| \\ &\le \frac{M}{\pi}\, \|\varphi - S_n\varphi\|_{L_2(\mathbb{T})} \end{aligned}$$

and thus by Theorem 1.3

$$\|f - \sum_{k=-n}^{n} f\big(\frac{k\pi}{M}\big)\, \mathrm{sinc}\,(M\cdot - k\pi)\|_{C_0(\mathbb{R})} \le \frac{M}{\pi}\, \|\varphi - S_n\varphi\|_{L_2(\mathbb{T})} \to 0 \quad \text{as } n \to \infty\,.$$

Consequently, the series (2.26) converges uniformly on $\mathbb{R}$. Note that each summand of the series (2.26) has the following interpolation property:

$$f\big(\frac{k\pi}{M}\big)\,\mathrm{sinc}\,(Mx - k\pi) = \begin{cases} f\big(\frac{k\pi}{M}\big) & x = \frac{k\pi}{M}\,, \\ 0 & x \in \frac{\pi}{M}\,(\mathbb{Z}\setminus\{k\})\,. \end{cases}$$

3. The absolute convergence of the series (2.26) is an immediate consequence of the Cauchy–Schwarz inequality in $\ell_2(\mathbb{Z})$ as well as the Parseval equalities of φ and ψ:

$$\sum_{k=-\infty}^{\infty} \big|f\big(\frac{k\pi}{M}\big)\big|\,\big|\mathrm{sinc}\,(M\,\cdot - k\pi)\big| = \frac{M}{\pi}\sum_{k=-\infty}^{\infty} |c_k(\varphi)|\,|c_k(\psi)|$$

$$\le \frac{M}{\pi}\,\Big(\sum_{k=-\infty}^{\infty} |c_k(\varphi)|^2\Big)^{1/2}\,\Big(\sum_{k=-\infty}^{\infty} |c_k(\psi)|^2\Big)^{1/2} = \frac{M}{\pi}\,\|\varphi\|_{L_2(\mathbb{T})} < \infty\,.$$

■

By the sampling Theorem 2.31, a bandlimited function f with $\mathrm{supp}\,\hat{f} \subseteq [-L,\,L]$ can be reconstructed from its equispaced samples $f\big(\frac{k\pi}{M}\big)$, $k \in \mathbb{Z}$, with $M \ge L > 0$. Hence, the sampling period $T = \frac{\pi}{L}$ is the largest, and the sampling rate $\frac{L}{\pi}$ is the smallest possible one. Then, $\frac{\pi}{L}$ is called *Nyquist rate*; see [306].

Remark 2.32 The sinc function decreases to zero only slightly as $|x| \to \infty$ so that we have to incorporate many summands in a truncation of the series (2.26) to get a good approximation of f. One can obtain a better approximation of f by so-called oversampling, i.e., by the choice of a higher sampling rate $\frac{L(1+\lambda)}{\pi}$ with some $\lambda > 0$ and corresponding sample values $f\big(\frac{k\pi}{L(1+\lambda)}\big)$, $k \in \mathbb{Z}$.

The choice of a lower sampling rate $\frac{L(1-\lambda)}{\pi}$ with some $\lambda \in (0,1)$ is called *undersampling*, which results in a reconstruction of a function f° where higher-frequency parts of f appear in lower-frequency parts of f°. This effect is called *aliasing* in signal processing or *Moiré effect* in imaging. □

Unfortunately, the practical use of the sampling Theorem 2.31 is limited, since it requires the knowledge of infinitely many samples which is impossible in practice. Further the sinc function decays very slowly such that the Shannon sampling series (2.26) has rather poor convergence. Moreover, in the presence of noise in the samples $f\big(\frac{k\pi}{M}\big)$, $k \in \mathbb{Z}$, the convergence of Shannon sampling series may even break down completely.

To overcome these drawbacks, one can use the following regularization technique (see [242, 275, 356]). Assume that $f \in L_1(\mathbb{R}) \cap C_0(\mathbb{R})$ is bandlimited on $[-L,\,L]$ with $L > 0$. Note that it holds $f \in L_2(\mathbb{R})$, since

$$\|f\|_{L_2(\mathbb{R})}^2 \le \|f\|_{C_0(\mathbb{R})}\,\|f\|_{L_1(\mathbb{R})} < \infty\,. \tag{2.28}$$

Let $M = (1+\lambda)\,L$ with $\lambda > 0$. Further let $\varphi : \mathbb{R} \to [0,\,1]$ be a given window function such as the Gaussian window function:

$$\varphi(x) := \mathrm{e}^{-x^2/(2\sigma^2)}\,, \quad x \in \mathbb{R}\,, \tag{2.29}$$

with $\sigma = \frac{1}{L}\sqrt{\frac{\pi m}{(1+\lambda)\lambda}}$ or the sinh-type window function

$$\varphi(x) := \frac{1}{\sinh \beta}\,\sinh\big(\beta\sqrt{1-\frac{M^2x^2}{m^2\pi^2}}\big)\,\chi_{[-m\pi/M,\,m\pi/M]}(x)\,, \quad x \in \mathbb{R}\,, \tag{2.30}$$

with $\beta = \frac{m\pi\lambda}{1+\lambda}$ and $m \in \mathbb{N}\setminus\{1\}$. Then, the function $\mathrm{sinc}(M\,\cdot)$ is regularized by the truncated window function:

$$\varphi_m(x) := \varphi(x)\,\chi_{[-m\pi/M,\,m\pi/M]}(x)\,, \quad x \in \mathbb{R}\,,$$

where $\chi_{[-m\pi/M,\,m\pi/M]}$ denotes the characteristic function of the interval $\big[-\frac{m\pi}{M},\,\frac{m\pi}{M}\big]$ and where $\varphi : \mathbb{R} \to [0,\,1]$ is an even window function of $L_1(\mathbb{R}) \cap L_2(\mathbb{R})$ which is decreasing on $[0,\,\infty)$ with $\varphi(0) = 1$.

Now we recover the given function $f \in L_1(\mathbb{R}) \cap C_0(\mathbb{R})$ which is bandlimited on $[-L,\,L]$ by the *regularized Shannon sampling formula* :

$$(R_{\varphi,m}f)(x) := \sum_{k\in\mathbb{Z}} f\big(\frac{k\pi}{M}\big)\,\mathrm{sinc}\big(M\,(x-\frac{k\pi}{M})\big)\,\varphi_m\big(x-\frac{k\pi}{M}\big)\,, \quad x \in \mathbb{R}\,. \tag{2.31}$$

Obviously, (2.31) is an interpolating approximation of f, since it holds

$$(R_{\varphi,m}f)\big(\frac{k\pi}{M}\big) = f\big(\frac{k\pi}{M}\big)\,, \quad k \in \mathbb{Z}\,.$$

The use of the truncated window function φ_m with the compact support $\big[-\frac{m\pi}{M},\,\frac{m\pi}{M}\big]$ leads to *localized sampling* of f, i.e., the computation of $(R_{\varphi,m}f)(x)$ for $x \in \mathbb{R}\setminus\frac{\pi}{M}\,\mathbb{Z}$ requires only $2m+1$ samples $f\big(\frac{k\pi}{M}\big)$, where $k \in \mathbb{Z}$ fulfills the condition $|\pi k - Mx| \le m\pi$. Thus, we can easily and rapidly reconstruct f by (2.31) on each open interval $\big(\frac{\ell\pi}{M},\,\frac{(\ell+1)\pi}{M}\big)$, $\ell \in \mathbb{Z}$. Especially for $x \in \big(0,\,\frac{\pi}{M}\big)$, we obtain by (2.31) that

$$(R_{\varphi,m}f)(x) = \sum_{k=-m+1}^{m} f\big(\frac{k\pi}{M}\big)\,\psi\big(x-\frac{k\pi}{M}\big)\,,$$

where $\psi(x) := \mathrm{sinc}(Mx)\,\varphi(x)$ denotes the regularized sinc function. Note that $\psi \in L_1(\mathbb{R}) \cap L_2(\mathbb{R})$ is continuous on $\mathbb{R}$. For the recovery of f on any interval $\big(\frac{\ell\pi}{M},\,\frac{(\ell+1)\pi}{M}\big)$ with $\ell \in \mathbb{Z}$, we use the simplified formula:

$$(R_{\varphi,m}f)\big(x+\frac{\ell\pi}{M}\big) = \sum_{k=-m+1}^{m} f\big(\frac{(k+\ell)\,\pi}{M}\big)\,\psi\big(x-\frac{k\pi}{M}\big)\,, \quad x \in \big(0,\,\frac{\pi}{M}\big)\,. \tag{2.32}$$

Hence, the reconstruction of f on the interval $[-\pi, \pi]$ requires only $2m + 2M$ samples $f\big(\frac{n\pi}{M}\big)$ with $n = -M - m + 1, \ldots, m + M$. Thus, the regularized Shannon sampling (2.31) is mainly based on oversampling and localized sampling.

Now we estimate the uniform approximation error:

$$\|f - R_{\varphi,m}f\|_{C_0(\mathbb{R})} := \max_{x\in\mathbb{R}} \big|f(x) - (R_{\varphi,m}f)(x)\big|\,.$$

Theorem 2.33 *Let $f \in L_1(\mathbb{R}) \cap C_0(\mathbb{R})$ be bandlimited on $[-L, L]$ with $L > 0$. Assume that $M = (1+\lambda)\,L$, $\lambda > 0$, and $m \in \mathbb{N} \setminus \{1\}$ be given. Further let $\varphi : \mathbb{R} \to [0,\,1]$ be an even window function of $L_2(\mathbb{R}) \cap C_0(\mathbb{R})$ which is decreasing on $[0,\,\infty)$ with $\varphi(0) = 1$.*

Then, the regularized Shannon sampling formula (2.31) satisfies the error estimate:

$$\|f - R_{\varphi,m}f\|_{C_0(\mathbb{R})} \le \big(E_1(m, M, L) + E_2(m, M, L)\big)\, \|f\|_{L^2(\mathbb{R})}$$

with the error constants

$$E_1(m, M, L) := \sqrt{\frac{L}{\pi}}\, \max_{\omega\in[-L,\,L]} \Big|1 - \frac{1}{2\pi}\int_{\omega-M}^{\omega+M} \hat{\varphi}(\tau)\,\mathrm{d}\tau\Big|\,, \tag{2.33}$$

$$E_2(m, M, L) := \frac{1}{\pi m}\sqrt{\frac{2M}{\pi}\Big(\varphi^2\big(\frac{\pi m}{M}\big) + \frac{M}{\pi}\int_{\pi m/M}^{\infty} \varphi^2(t)\,\mathrm{d}t\Big)}\,. \tag{2.34}$$

Proof By (2.28) it holds $f \in L_2(\mathbb{R})$. First we consider the error on the interval $\big[0,\,\frac{\pi}{M}\big]$. Here we split the approximation error:

$$f(t) - (R_{\varphi,m}f)(t) = e_1(t) + e_{2,0}(t)\,, \quad t \in \big[0,\,\frac{\pi}{M}\big]\,,$$

into the regularization error

$$e_1(t) := f(t) - \sum_{k\in\mathbb{Z}} f\big(\frac{k\pi}{M}\big)\,\psi\big(t - \frac{k\pi}{M}\big)\,, \quad t \in \mathbb{R}\,, \tag{2.35}$$

and the truncation error

$$e_{2,0}(t) := \sum_{k\in\mathbb{Z}} f\big(\frac{k\pi}{M}\big)\,\psi\big(t - \frac{k\pi}{M}\big) - (R_{\varphi,m}f)(t)\,, \quad t \in \big[0,\,\frac{\pi}{M}\big]\,. \tag{2.36}$$

By Theorem 2.26, the Fourier transform of $\psi \in L_1(\mathbb{R}) \cap L_2(\mathbb{R})$ reads as follows:

$$\hat{\psi}(\omega) = \frac{1}{2\pi} \int_{\mathbb{R}} \frac{\pi}{M} \chi_{[-M,\,M]}(\omega - \tau)\, \hat{\varphi}(\tau)\, \mathrm{d}\tau = \frac{1}{2M} \int_{\omega-M}^{\omega+M} \hat{\varphi}(\tau)\, \mathrm{d}\tau\,.$$

Using the shifting property of the Fourier transform (see Theorem 2.5), the Fourier transform of $\psi\big(\cdot - \frac{\pi k}{M}\big)$ has the form

$$\frac{1}{2M}\, \mathrm{e}^{-\pi k \mathrm{i}\omega/M} \int_{\omega-M}^{\omega+M} \hat{\varphi}(\tau)\, \mathrm{d}\tau\,.$$

Therefore, the Fourier transform of the regularization error (2.35) can be represented as

$$\hat{e}_1(\omega) = \hat{f}(\omega) - \bigg(\sum_{k\in\mathbb{Z}} f\Big(\frac{k\pi}{M}\Big) \frac{1}{2M}\, \mathrm{e}^{-\pi \mathrm{i}k\omega/M}\bigg) \int_{\omega-M}^{\omega+M} \hat{\varphi}(\tau)\, \mathrm{d}\tau\,. \tag{2.37}$$

By assumption it holds $\operatorname{supp} \hat{f} \subseteq [-L,\, L] \subset [-M,\, M]$ such that the restricted function $\hat{f}|_{[-M,\,M]}$ belongs to $L_2([-M,\, M])$. Thus, this function possesses the $(2M)$-periodic Fourier expansion:

$$\hat{f}(\omega) = \sum_{k\in\mathbb{Z}} c_k(\hat{f})\, \mathrm{e}^{-\pi \mathrm{i}k\omega/M}\,, \qquad \omega \in [-M,\, M]\,,$$

with the Fourier coefficients

$$c_k(\hat{f}) = \frac{1}{2M} \int_{-M}^{M} \hat{f}(\tau)\, \mathrm{e}^{\pi \mathrm{i}k\tau/M}\, \mathrm{d}\tau = \frac{\pi}{M}\, f\Big(\frac{\pi k}{M}\Big), \quad k \in \mathbb{Z}\,,$$

where we have used the Fourier inversion formula (2.7). Then, $\hat{f}$ can be written as

$$\hat{f}(\omega) = \hat{f}(\omega)\, \chi_{[-L,\,L]}(\omega) = \Big(\frac{\pi}{M} \sum_{k\in\mathbb{Z}} f\Big(\frac{\pi k}{M}\Big)\, \mathrm{e}^{-\pi \mathrm{i}k\omega/M}\Big)\, \chi_{[-L,\,L]}(\omega)\,. \tag{2.38}$$

Introducing the function

$$\eta(\omega) := \chi_{[-L,\,L]}(\omega) - \frac{1}{2\pi} \int_{\omega-M}^{\omega+M} \hat{\varphi}(\tau)\, \mathrm{d}\tau\,, \qquad \omega \in \mathbb{R}\,, \tag{2.39}$$

we see that $\hat{e}_1(\omega) = \hat{f}(\omega)\, \eta(\omega)$ and thereby $|\hat{e}_1(\omega)| = |\hat{f}(\omega)|\, |\eta(\omega)|$. Thus, by the Fourier inversion formula (2.7), we get:

$$|e_1(t)| = \frac{1}{2\pi}\,\Big|\int_{\mathbb{R}} \hat{e}_1(\omega)\,\mathrm{e}^{\mathrm{i}\omega t}\,\mathrm{d}\omega\Big| \le \frac{1}{2\pi}\int_{\mathbb{R}} |\hat{e}_1(\omega)|\,\mathrm{d}\omega$$
$$\le \frac{1}{2\pi}\int_{-L}^{L} |\hat{f}(\omega)|\,|\eta(\omega)|\,\mathrm{d}\omega \le \frac{1}{2\pi}\max_{\omega\in[-L,\,L]} |\eta(\omega)| \int_{-L}^{L} |\hat{f}(\omega)|\,\mathrm{d}\omega\,.$$

Applying the Cauchy–Schwarz inequality and the Parseval equality (2.13), we see that

$$\int_{-L}^{L} |\hat{f}(\omega)|\,\mathrm{d}\omega \le \Big(\int_{-L}^{L} 1^2\,\mathrm{d}\omega\Big)^{1/2}\Big(\int_{-L}^{L} |\hat{f}(\omega)|^2\,\mathrm{d}\omega\Big)^{1/2}$$
$$= \sqrt{2L}\,\|\hat{f}\|_{L_2(\mathbb{R})} = 2\sqrt{L\pi}\,\|f\|_{L_2(\mathbb{R})}\,.$$

Using the error constant (2.33), this yields:

$$\|e_1\|_{C_0(\mathbb{R})} \le E_1(m, M, L)\,\|f\|_{L_2(\mathbb{R})}\,.$$

Now we estimate the truncation error (2.36). Then, for $t \in \big(0, \frac{\pi}{M}\big)$, it holds:

$$e_{2,0}(t) = \sum_{k\in\mathbb{Z}} f\Big(\frac{k\pi}{M}\Big)\,\psi\Big(t - \frac{k\pi}{M}\Big)\Big[1 - \chi_{[-m\pi/M,m\pi/M]}\Big(t - \frac{k\pi}{M}\Big)\Big]$$
$$= \sum_{k\in\mathbb{Z}\setminus\{1-m,\ldots,m\}} f\Big(\frac{k\pi}{M}\Big)\,\psi\Big(t - \frac{k\pi}{M}\Big)\,.$$

By the non-negativity of φ, we receive

$$|e_{2,0}(t)| \le \sum_{k\in\mathbb{Z}\setminus\{1-m,\ldots,m\}} \Big|f\Big(\frac{k\pi}{M}\Big)\Big|\,\big|\mathrm{sinc}\big(Mt - k\pi\big)\big|\,\varphi\Big(t - \frac{k\pi}{M}\Big)\,.$$

For $t \in \big(0, \frac{\pi}{M}\big)$ and $k \in \mathbb{Z}\setminus\{1-m,\ldots,m\}$, it holds:

$$\big|\mathrm{sinc}\big(Mt - k\pi\big)\big| \le \frac{1}{|Mt - k\pi|} \le \frac{1}{\pi m}$$

and hence

$$|e_{2,0}(t)| \le \frac{1}{\pi m}\sum_{k\in\mathbb{Z}\setminus\{1-m,\ldots,m\}} \Big|f\Big(\frac{k\pi}{M}\Big)\Big|\,\varphi\Big(t - \frac{k\pi}{M}\Big)\,.$$

Then, the Cauchy–Schwarz inequality implies

$$|e_{2,0}(t)| \le \frac{1}{\pi m} \Big(\sum_{k\in\mathbb{Z}\setminus\{1-m,\ldots,m\}} |f(\frac{k\pi}{M})|^2\Big)^{1/2} \Big(\sum_{k\in\mathbb{Z}\setminus\{1-m,\ldots,m\}} \varphi^2(t-\frac{k\pi}{M})\Big)^{1/2}.$$

For the bandlimited function f, it holds:

$$\frac{\pi}{M}\sum_{k\in\mathbb{Z}} |f(\frac{k\pi}{M})|^2 = \|f\|^2_{L_2(\mathbb{R})} \tag{2.40}$$

and hence

$$\Big(\sum_{k\in\mathbb{Z}\setminus\{1-m,\ldots,m\}} |f(\frac{k\pi}{M})|^2\Big)^{1/2} \le \sqrt{\frac{M}{\pi}}\, \|f\|_{L_2(\mathbb{R})}\,.$$

The equality (2.40) can be shown as follows. The Fourier transform of $\operatorname{sinc}(Mt)$ is equal to $\frac{\pi}{M}\,\chi_{[-M,M]}(\omega)$ such that the Fourier transform of the shifted function $\operatorname{sinc}(Mt - k\pi)$ is equal to $\frac{\pi}{M}\,\exp(-ik\pi\omega/M)\,\chi_{[-M,M]}(\omega)$ for $k \in \mathbb{Z}$. By the Parseval equality (2.13), it follows that for all $k,\,\ell \in \mathbb{Z}$ it holds

$$2\pi\,\langle \operatorname{sinc}(Mt-k\pi),\, \operatorname{sinc}(Mt-\ell\pi)\rangle_{L_2(\mathbb{R})} = \frac{\pi^2}{M^2}\int_{-M}^{M} \mathrm{e}^{-\mathrm{i}\,(k-\ell)\pi\omega/M}\,\mathrm{d}\omega = \frac{2\pi^2}{M}\,\delta_{k-\ell}$$

and hence

$$\langle \operatorname{sinc}(Mt-k\pi),\, \operatorname{sinc}(Mt-\ell\pi)\rangle_{L_2(\mathbb{R})} = \frac{\pi}{M}\,\delta_{k-\ell}\,.$$

By the sampling Theorem 2.31, it holds (2.26) and hence (2.40).

Since φ is even and $\varphi|_{[0,\,\infty)}$ decreases monotonically by assumption, for $t \in (0,\, \frac{\pi}{M})$, we can estimate the series:

$$\begin{aligned}\sum_{k\in\mathbb{Z}\setminus\{1-m,\ldots,m\}} \varphi^2(t-\frac{\pi k}{M}) &= \Big(\sum_{k=-\infty}^{-m} + \sum_{k=m+1}^{\infty}\Big)\varphi^2(t-\frac{\pi k}{M})\\ &= \sum_{k=m}^{\infty}\varphi^2(t+\frac{\pi k}{M}) + \sum_{k=m+1}^{\infty}\varphi^2(t-\frac{\pi k}{M})\\ &\le \sum_{k=m}^{\infty}\varphi^2(\frac{\pi k}{M}) + \sum_{k=m+1}^{\infty}\varphi^2(\frac{\pi}{M}-\frac{\pi k}{M})\\ &= 2\sum_{k=m}^{\infty}\varphi^2(\frac{\pi k}{M})\,.\end{aligned}$$

Using the integral test for convergence of series, we obtain that

$$\sum_{k=m}^{\infty} \varphi^2\big(\frac{\pi k}{M}\big) = \varphi^2\big(\frac{\pi m}{M}\big) + \sum_{k=m+1}^{\infty} \varphi^2\big(\frac{\pi k}{M}\big) < \varphi^2\big(\frac{\pi m}{M}\big) + \int_m^{\infty} \varphi^2\big(\frac{\pi t}{M}\big)\,\mathrm{d}t$$
$$= \varphi^2\big(\frac{\pi m}{M}\big) + \frac{M}{\pi}\int_{\pi m/M}^{\infty} \varphi^2(t)\,\mathrm{d}t\,.$$

By the interpolation property of $R_{\varphi,m} f$, it holds $e_{2,0}(0) = e_{2,0}\big(\frac{\pi}{M}\big) = 0$. Thus, we obtain by (2.34) that

$$\max_{t\in[0,\pi/M]} |e_{2,0}(t)| \le \frac{1}{\pi m}\sqrt{\frac{2M}{\pi}}\,\|f\|_{L^2(\mathbb{R})}\Big(\varphi^2\big(\frac{\pi m}{M}\big) + \frac{M}{\pi}\int_{\pi m/M}^{\infty} \varphi^2(t)\,\mathrm{d}t\Big)^{1/2}$$
$$= E_2(m, M, L)\,\|f\|_{L^2(\mathbb{R})}\,.$$

By the same technique, this error estimate can be shown for each interval $\big[\frac{k\pi}{M}, \frac{(k+1)\pi}{M}\big]$ with arbitrary $k \in \mathbb{Z}$. On the interval $\big[\frac{k\pi}{M}, \frac{(k+1)\pi}{M}\big]$, we decompose the approximation error in the form

$$f\big(t + \frac{k\pi}{M}\big) - (R_{\varphi,m} f)\big(t + \frac{k\pi}{M}\big) = e_1\big(t + \frac{k\pi}{M}\big) + e_{2,k}(t)\,, \quad t \in \big[0, \frac{\pi}{M}\big]\,,$$

with

$$e_1\big(t + \frac{k\pi}{M}\big) = f\big(t + \frac{k\pi}{M}\big) - \sum_{\ell\in\mathbb{Z}} f\big(\frac{\ell\pi}{M}\big)\,\psi\big(t - \frac{(\ell - k)\pi}{M}\big)$$
$$= f\big(t + \frac{k\pi}{M}\big) - \sum_{\ell\in\mathbb{Z}} f\big(\frac{(\ell + k)\pi}{M}\big)\,\psi\big(t - \frac{\ell\pi}{M}\big)\,,$$
$$e_{2,k}(t) := \sum_{\ell\in\mathbb{Z}\setminus\{1-m,\ldots,m\}} f\big(\frac{(\ell + k)\pi}{M}\big)\,\psi\big(t - \frac{\ell\pi}{M}\big)\,.$$

As shown above, it holds:

$$\big\|e_1\big(\cdot + \frac{k\pi}{M}\big)\big\|_{C_0(\mathbb{R})} = \|e_1\|_{C_0(\mathbb{R})}\,,$$
$$|e_{2,k}(t)| \le E_2(m, M, L)\,\|f\|_{L_2(\mathbb{R})}\,, \quad t \in \big(0, \frac{\pi}{M}\big)\,.$$

By the interpolation property of $R_{\varphi,m} f$, we have $e_{2,k}(0) = e_{2,k}\big(\frac{\pi}{M}\big) = 0$ for each $k \in \mathbb{Z}$. Hence, it follows that

$$\max_{t\in[k\pi/M,\,(k+1)\pi/M]} |f(t) - (R_{\varphi,m} f)(t)| \le \|e_1\|_{C_0(\mathbb{R})} + \max_{t\in[0,\,\pi/M]} |e_{2,k}(t)|$$
$$\le \big(E_1(m, M, L) + E_2(m, M, L)\big)\|f\|_{L^2(\mathbb{R})}\,.$$

This completes the proof. ■

Using Theorem 2.33, we show that for the regularized Shannon sampling formula (2.31) with the Gaussian window function (2.29) the approximation error decays exponentially with respect to m.

Corollary 2.34 *Let $f \in L_1(\mathbb{R}) \cap C_0(\mathbb{R})$ be bandlimited on $[-L, L]$ with $L > 0$. Assume that $M = (1+\lambda)\,L$, $\lambda > 0$, and $m \in \mathbb{N} \setminus \{1\}$ be given. Further let φ be the Gaussian window function* (2.29) *with $\sigma = \frac{1}{L}\sqrt{\frac{\pi m}{(1+\lambda)\,\lambda}}$.*

Then, the regularized Shannon sampling formula (2.31) *satisfies the error estimate:*

$$\|f - R_{\varphi,m} f\|_{C_0(\mathbb{R})} \le C(m, M, L)\, \mathrm{e}^{\pi m\lambda/(2+2\lambda)}\, \|f\|_{L_2(\mathbb{R})}$$

with the constant

$$C(m, M, L) := \frac{\sqrt{L}}{\pi}\sqrt{\frac{2+2\lambda}{\lambda \pi m}} + \frac{1}{\pi m}\sqrt{\frac{L}{\pi}\Big(2 + 2\lambda + \frac{(1+\lambda)^2}{\pi\lambda}\Big)}\,.$$

Proof The Fourier transform of (2.29) reads as follows:

$$\hat{\varphi}(\omega) = \sqrt{2\pi}\,\sigma\, \mathrm{e}^{-\omega^2\sigma^2/2}\,, \quad \omega \in \mathbb{R}\,.$$

First we estimate the error constant:

$$E_1(m, M, L) = \sqrt{\frac{L}{\pi}} \max_{\omega \in [-L,\, L]} |\eta(\omega)|$$

for the Gaussian window function (2.29), where

$$\begin{aligned}
\eta(\omega) &= 1 - \frac{1}{2\pi}\int_{\omega-M}^{\omega+M} \hat{\varphi}(\tau)\,\mathrm{d}\tau = 1 - \frac{1}{\sqrt{\pi}}\int_{(\omega-M)\sigma/\sqrt{2}}^{(\omega+M)\sigma/\sqrt{2}} \mathrm{e}^{-t^2}\,\mathrm{d}t \\
&= \frac{1}{\sqrt{\pi}}\Big[\int_{\mathbb{R}} \mathrm{e}^{-t^2}\,\mathrm{d}t - \int_{(\omega-M)\sigma/\sqrt{2}}^{(\omega+M)\sigma/\sqrt{2}} \mathrm{e}^{-t^2}\,\mathrm{d}t\Big] \\
&= \frac{1}{\sqrt{\pi}}\Big[\int_{-\infty}^{(\omega-M)\sigma/\sqrt{2}} \mathrm{e}^{-t^2}\,\mathrm{d}t + \int_{(\omega+M)\sigma/\sqrt{2}}^{\infty} \mathrm{e}^{-t^2}\,\mathrm{d}t\Big] \\
&= \frac{1}{\sqrt{\pi}}\Big[\int_{(M-\omega)\sigma/\sqrt{2}}^{\infty} \mathrm{e}^{-t^2}\,\mathrm{d}t + \int_{(\omega+M)\sigma/\sqrt{2}}^{\infty} \mathrm{e}^{-t^2}\,\mathrm{d}t\Big]
\end{aligned}$$

for $\omega \in [-L, L]$. Since η is even, we consider only $\omega \in [0, L]$. Applying the inequality

$$\int_a^\infty e^{-t^2}\,dt = \int_0^\infty e^{-(t+a)^2}\,dt \le e^{-a^2}\int_0^\infty e^{-2at}\,dt = \frac{1}{2a}\,e^{-a^2} \tag{2.41}$$

with $a > 0$, we obtain for $\omega \in [0,\ L]$ that

$$\begin{aligned} 0 \le \eta(\omega) &\le \frac{1}{\sqrt{2\pi}}\Big(\frac{e^{-(M-\omega)^2\sigma^2/2}}{(M-\omega)\sigma} + \frac{e^{-(M+\omega)^2\sigma^2/2}}{(M+\omega)\sigma}\Big) \\ &\le \sqrt{\frac{2}{\pi}}\,\frac{e^{-(M-\omega)^2\sigma^2/2}}{(M-\omega)\sigma}\,. \end{aligned}$$

Thus, for all $\omega \in [-L,\ L]$ it holds:

$$0 \le \eta(\omega) \le \sqrt{\frac{2}{\pi}}\,\frac{e^{-(M-|\omega|)^2\sigma^2/2}}{(M-|\omega|)\sigma}$$

and hence

$$E_1(m, M, L) \le \frac{\sqrt{2L}}{(M-L)\pi\sigma}\,e^{-(M-L)^2\sigma^2/2}\,. \tag{2.42}$$

Now we examine the error constant:

$$E_2(m, M, L) = \frac{1}{\pi m}\sqrt{\frac{2M}{\pi}\Big(\varphi^2(\frac{m\pi}{M}) + \frac{M}{\pi}\int_{m\pi/M}^\infty \varphi^2(t)\,dt\Big)}$$

for the Gaussian window function (2.29). By (2.41) it follows that

$$\begin{aligned} \varphi^2(\frac{m\pi}{M}) &= e^{-\pi^2m^2/(M^2\sigma^2)}\,, \\ \frac{M}{\pi}\int_{m\pi/M}^\infty e^{-t^2/\sigma^2}\,dt &= \frac{M\sigma}{\pi}\int_{m\pi/(\sigma M)}^\infty e^{-s^2}\,ds \\ &\le \frac{\sigma^2M^2}{2\pi^2m}\,e^{-\pi^2m^2/(\sigma^2M^2)} \end{aligned}$$

and hence

$$E_2(m, M, L) \le \frac{1}{\pi m}\sqrt{\frac{2M}{\pi}\Big(1 + \frac{\sigma^2M^2}{2\pi^2m}\Big)}\,e^{-\pi^2m^2/(2\,M^2\sigma^2)}\,. \tag{2.43}$$

In the case

$$\sigma = \sqrt{\frac{\pi m}{M\,(M-L)}} = \frac{1}{L}\sqrt{\frac{\pi m}{(1+\lambda)\,\lambda}}\,,$$

both error terms (2.42) and (2.43) possess the same exponential decay by

$$\frac{\pi^2 m^2}{2\sigma^2 M^2} = \frac{(M-L)^2\sigma^2}{2}\,.$$

Then, it holds:

$$E_1(m, M, L) \le \frac{\sqrt{L}}{\pi}\sqrt{\frac{2+2\lambda}{\lambda\pi m}}\,\mathrm{e}^{-\pi m\lambda/(2+2\lambda)}\,,$$

$$E_2(m, M, L) \le \frac{1}{\pi m}\sqrt{\frac{L}{\pi}\left(2+2\lambda+\frac{(1+\lambda)^2}{\pi\lambda}\right)}\,\mathrm{e}^{-\pi m\lambda/(2+2\lambda)}\,.$$

This completes the proof. ■

Now we demonstrate that for the regularized Shannon sampling formula (2.31) with the sinh-type window function (2.30), the approximation error has a much better exponential decay.

Corollary 2.35 *Let $f \in L_1(\mathbb{R}) \cap C_0(\mathbb{R})$ be bandlimited on $[-L, L]$ with $L > 0$. Assume that $M = (1+\lambda)\,L$, $\lambda > 0$, and $m \in \mathbb{N}\setminus\{1\}$ be given. Further let φ be the* sinh-*type window function* (2.30) *with $\beta = \frac{m\pi\lambda}{1+\lambda}$.*

Then, the regularized Shannon sampling formula (2.31) *satisfies the error estimate:*

$$\|f - R_{\varphi,m} f\|_{C_0(\mathbb{R})} \le \sqrt{\frac{L}{\pi}}\,\mathrm{e}^{-\pi m\lambda/(1+\lambda)}\,\|f\|_{L_2(\mathbb{R})}$$

Proof By Theorem 2.33, we have to estimate only the error constant $E_1(m, M, L)$, since $E_2(m, M, L) = 0$. Using the scaled frequency $v = \frac{m\pi}{\beta M}\,\omega$, the Fourier transform of (2.30) reads by [307, 7.58] as follows:

$$\hat{\varphi}(\omega) = \frac{m\pi^2}{M\,\sinh\beta}\cdot\begin{cases}(1-v^2)^{-1/2}\,I_1\big(\beta\sqrt{1-v^2}\big) & |v| < 1\,,\\ (v^2-1)^{-1/2}\,J_1\big(\beta\sqrt{v^2-1}\big) & |v| > 1\,.\end{cases}$$

Then, it holds:

$$E_1(m, M, L) = \sqrt{\frac{L}{\pi}}\,\max_{\tau\in[-L,\,L]}|\eta(\tau)|$$

with the function

$$\eta(\tau) = 1 - \frac{1}{2\pi}\int_{\tau-M}^{\tau+M}\hat{\varphi}(\omega)\,\mathrm{d}\omega\,,\quad \tau\in[-L,\,L]\,.$$

Substituting $v = \frac{m\pi}{\beta M}\,\omega$, it follows that

$$\eta(\tau) = 1 - \frac{\beta M}{2\pi^2 m} \int_{-v_1(-\tau)}^{v_1(\tau)} \hat{\varphi}\big(\frac{\beta M}{m\pi}\,v\big)\,\mathrm{d}v$$

with the linear increasing function $v_1(\tau) := \frac{m\pi}{\beta M}\,(\tau + M)$. By the choice of the parameter $\beta = \frac{m\pi\lambda}{1+\lambda}$ with $\lambda > 0$, it holds $v_1(-L) = 1$ and $v_1(\tau) \ge 1$ for all $\tau \in [-L,\, L]$. We decompose $\eta(\tau)$ in the form

$$\eta(\tau) = \eta_1(\tau) - \eta_2(\tau)$$

with

$$\eta_1(\tau) := 1 - \frac{\beta}{2\,\sinh\beta} \int_{-1}^{1} \frac{I_1\big(\beta\sqrt{1-v^2}\big)}{\sqrt{1-v^2}}\,\mathrm{d}v\,,$$

$$\eta_2(\tau) := \frac{\beta}{2\,\sinh\beta}\Big(\int_{-v_1(-\tau)}^{-1} + \int_1^{v_1(\tau)}\Big)\frac{J_1\big(\beta\sqrt{v^2-1}\big)}{\sqrt{v^2-1}}\,\mathrm{d}v$$

$$= \frac{\beta}{2\,\sinh\beta}\Big(\int_1^{v_1(-\tau)} + \int_1^{v_1(\tau)}\Big)\frac{J_1\big(\beta\sqrt{v^2-1}\big)}{\sqrt{v^2-1}}\,\mathrm{d}v\,.$$

By [170, 6.681–3] and [2, 10.2.13], it holds:

$$\int_{-1}^{1} \frac{I_1\big(\beta\sqrt{1-v^2}\big)}{\sqrt{1-v^2}}\,\mathrm{d}v = \int_{-\pi/2}^{\pi/2} I_1(\beta\,\cos s)\,\mathrm{d}s$$

$$= \pi\,\Big(I_{1/2}\big(\frac{\beta}{2}\big)\Big)^2 = \frac{4}{\beta}\,\big(\sinh\frac{\beta}{2}\big)^2 \qquad (2.44)$$

and hence

$$\eta_1(\tau) = 1 - \frac{2\,\big(\sinh\frac{\beta}{2}\big)^2}{\sinh\beta} = \frac{2\,\mathrm{e}^{-\beta}}{1+\mathrm{e}^{-\beta}}\,.$$

By [170, 6.645–1] we have:

$$\int_1^{\infty} \frac{J_1\big(\beta\sqrt{v^2-1}\big)}{\sqrt{v^2-1}}\mathrm{d}v = I_{1/2}\big(\frac{\beta}{2}\big)K_{1/2}\big(\frac{\beta}{2}\big) = \frac{2}{\sqrt{\pi\beta}}\,\sinh\frac{\beta}{2}\cdot\sqrt{\frac{\pi}{\beta}}\,\mathrm{e}^{-\beta/2} = \frac{1-\mathrm{e}^{-\beta}}{\beta}\,,$$

where $I_{1/2}$ and $K_{1/2}$ are modified Bessel functions of half order (see [2, 10.2.13, 10.2.14, and 10.2.17]). Numerical experiments have shown that for all $T > 1$ it holds

$$0 < \int_1^T \frac{J_1\big(\beta\sqrt{v^2-1}\big)}{\sqrt{v^2-1}}\,\mathrm{d}v \le \frac{3\,(1-\mathrm{e}^{-\beta})}{2\,\beta}\,.$$

Thus, we obtain:

$$\begin{aligned} 0 \le \eta_2(\tau) &= \frac{\beta}{2\,\sinh\beta}\Big(\int_1^{v_1(-\tau)} + \int_1^{v_1(\tau)}\Big)\frac{J_1\big(\beta\,\sqrt{v^2-1}\big)}{\sqrt{v^2-1}}\,\mathrm{d}v \\ &\le \frac{\beta}{2\,\sinh\beta}\,\frac{3\,\big(1-\mathrm{e}^{-\beta}\big)}{\beta} = \frac{3\,\mathrm{e}^{-\beta}}{1+\mathrm{e}^{-\beta}}\,. \end{aligned}$$

Hence for all $\tau \in [-L,\, L]$, it holds the estimate:

$$|\eta(\tau)| = |\eta_1(\tau) - \eta_2(\tau)| \le \frac{\mathrm{e}^{-\beta}}{1+\mathrm{e}^{-\beta}} < \mathrm{e}^{-\beta}\,.$$

This implies that

$$E_1(m, M, L) \le \sqrt{\frac{L}{\pi}}\,\mathrm{e}^{-\beta} = \sqrt{\frac{L}{\pi}}\,\mathrm{e}^{-m\pi\lambda/(1+\lambda)}\,.$$

This completes the proof. ∎

Example 2.36 We illustrate the excellent error behavior of the regularized Shannon sampling formula (2.31). The function

$$g(t) = \frac{L^2}{2\pi}\,\big(\mathrm{sinc}\,\frac{L\,t}{2}\big)^2\,, \quad t \in \mathbb{R}\,,$$

has the Fourier transform

$$\hat{g}(\omega) = (L - |\omega|)\,\chi_{[-L,\,L]}(\omega)\,, \quad \omega \in \mathbb{R}\,,$$

because by the Fourier inversion formula (2.7) we have

$$g(t) = \frac{1}{2\pi}\int_{\mathbb{R}} \hat{g}(\omega)\,\mathrm{e}^{\mathrm{i}\,\omega t}\,\mathrm{d}\omega = \frac{1}{\pi}\int_0^L (L-\omega)\,\cos(\omega t)\,\mathrm{d}\omega\,.$$

Since by [170, 3.827–7] it holds

$$\|g\|_{L_2(\mathbb{R})} = \frac{L^2}{2\pi}\Big(\int_{\mathbb{R}}\big(\mathrm{sinc}\,\frac{L\,t}{2}\big)^4\,\mathrm{d}t\Big)^{1/2} = \sqrt{\frac{L^3}{3\pi}}\,,$$

the normed function

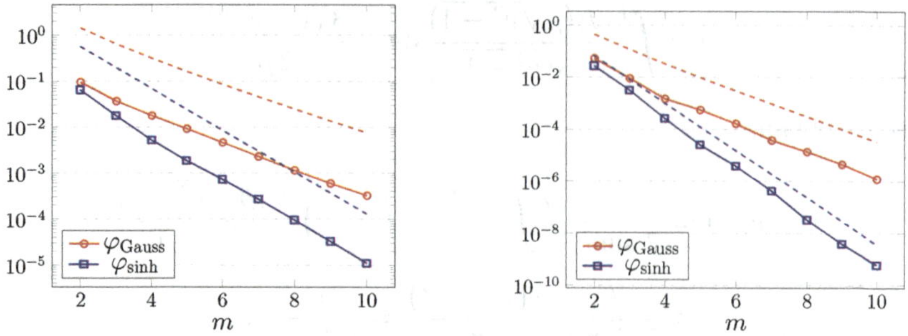

Fig. 2.2 Approximation of the test function (2.45) with $L = 64$ by the regularized Shannon sampling formula (2.31) with the Gaussian window function (red) and the sinh-type window function (blue) for various $m \in \{2, 3, \ldots, 10\}$. The actual approximation error (2.46) (solid line) is compared with the theoretical upper bound $E_1(m, M, L) + E_2(m, M, L)$ (dashed line) for $\lambda = 0.5$ (left) and $\lambda = 2$ (right)

$$f(t) = \sqrt{\frac{3L}{4\pi}}\,\Big(\mathrm{sinc}\,\frac{L\,t}{2}\Big)^2\,, \quad t \in \mathbb{R}\,, \tag{2.45}$$

is bandlimited on $[-L,\ L]$. For $L = 64$, we consider the regularized Shannon sampling formula $R_{\varphi,m} f$ with the Gaussian window function (2.29) and the sinh-type window function (2.30). In Fig. 2.2 we compare the upper bound $E_1(m, M, L) + E_2(m, M, L)$ with the approximation error:

$$\max_{t\in[-\pi,\,\pi]} \big|f(t) - (R_{\varphi,m} f)(t)\big| \tag{2.46}$$

evaluated at 10^4 equidistant points in $[-\pi,\ \pi]$. Here we choose $m \in \{2,\ 3,\ \ldots,\ 10\}$ and $\lambda \in \{0.5,\ 2\}$. □

If only erroneous samples $\tilde{f}_k := f\big(\frac{k\pi}{M}\big) + \varepsilon_k$ with bounded error terms $|\varepsilon_k| \le \varepsilon$, $k \in \mathbb{Z}$, are given, then the corresponding Shannon sampling series may differ appreciably from f. In contrast to the Shannon sampling series, the regularized Shannon sampling formula (2.31) is numerically robust. Here we denote the regularized Shannon sampling formula with erroneous samples $\tilde{f}_k$ by

$$(R_{\varphi,m} \tilde{f})(t) := \sum_{k\in\mathbb{Z}} \tilde{f}_k\, \mathrm{sinc}\big(Mt - k\pi\big)\, \varphi_m\big(t - \frac{k\pi}{M}\big)\,, \quad t \in \mathbb{R}\,. \tag{2.47}$$

The *perturbation error of the regularized Shannon sampling formula* (2.31) is measured by

$$\|R_{\varphi,m} f - R_{\varphi,m} \tilde{f}\,\|_{C_0(\mathbb{R})}\,. \tag{2.48}$$

Theorem 2.37 *Let $f \in L_1(\mathbb{R}) \cap C_0(\mathbb{R})$ be bandlimited on $[-L, L]$ with $L > 0$. Assume that $M = (1+\lambda)\,L$, $\lambda > 0$, and $m \in \mathbb{N}\setminus\{1\}$ be given. Let $\tilde{f}_k := f\big(\frac{k\pi}{M}\big)+\varepsilon_k$, $k \in \mathbb{Z}$, be noisy samples, where it holds $|\varepsilon_k| \le \varepsilon$ for all $k \in \mathbb{Z}$. Further let $\varphi : \mathbb{R} \to [0, 1]$ be an even window function of $L_2(\mathbb{R}) \cap C_0(\mathbb{R})$ which decreases on $[0, \infty)$ with $\varphi(0) = 1$.*

Then, the regularized Shannon sampling formula (2.31) *is numerically robust, and it holds:*

$$\|R_{\varphi,m} f - R_{\varphi,m} \tilde{f}\|_{C_0(\mathbb{R})} \le \varepsilon\,\Big(2 + \frac{M}{2\pi}\,\hat{\varphi}(0)\Big)\,.$$

Proof First we consider the perturbation error on the interval $\big[0, \frac{\pi}{M}\big]$. By (2.31) and (2.47), it holds:

$$\begin{aligned}\tilde{e}_0(t) &:= (R_{\varphi,m}\tilde{f})(t) - (R_{\varphi,m} f)(t)\\ &= \sum_{k=-m+1}^{m} \Big(\tilde{f}_k - f\big(\frac{k\pi}{M}\big)\Big)\,\psi\big(t - \frac{k\pi}{M}\big)\\ &= \sum_{k=-m+1}^{m} \varepsilon_k\,\psi\big(t - \frac{k\pi}{M}\big)\,, \quad t \in \big[0, \frac{\pi}{M}\big]\,.\end{aligned}$$

By the non-negativity of φ and $|\varepsilon_k| \le \varepsilon$, we receive

$$|\tilde{e}_0(t)| \le \sum_{k=-m+1}^{m} |\varepsilon_k|\,\big|\mathrm{sinc}(Mt - k\pi)\big|\,\varphi\big(t - \frac{k\pi}{M}\big) \le \varepsilon \sum_{k=-m+1}^{m} \varphi\big(t - \frac{k\pi}{M}\big)\,.$$

Since φ is even and decreases on $[0, \infty)$ by assumption, we estimate this sum for $t \in \big(0, \frac{\pi}{M}\big)$ as follows:

$$\begin{aligned}\sum_{k=-m+1}^{m} \varphi\big(t - \frac{k\pi}{M}\big) &= \Big(\sum_{k=-m+1}^{0} + \sum_{k=1}^{m}\Big)\,\varphi\big(t - \frac{k\pi}{M}\big)\\ &= \sum_{k=0}^{m-1} \varphi\big(\frac{k\pi}{M} + t\big) + \sum_{k=1}^{m} \varphi\big(\frac{k\pi}{M} - t\big)\\ &\le \sum_{k=0}^{m-1} \varphi\big(\frac{k\pi}{M}\big) + \sum_{k=1}^{m} \varphi\big(\frac{(k-1)\pi}{M}\big) = 2\sum_{k=0}^{m-1} \varphi\big(\frac{k\pi}{M}\big)\,.\end{aligned}$$

Using the integral test for convergence of series, we obtain:

$$\sum_{k=0}^{m-1} \varphi\big(\frac{k\pi}{M}\big) \le \varphi(0) + \int_0^{m-1} \varphi\big(\frac{\pi t}{M}\big)\,\mathrm{d}t = 1 + \frac{M}{\pi} \int_0^{(m-1)\pi/M} \varphi(t)\,\mathrm{d}t\,.$$

By the definition of the Fourier transform, it holds:

$$\hat{\varphi}(0) = \int_{\mathbb{R}} \varphi(t)\,\mathrm{d}t = 2 \int_0^{\infty} \varphi(t)\,\mathrm{d}t \ge 2 \int_0^{(m-1)\pi/M} \varphi(t)\,\mathrm{d}t$$

and therefore

$$|\tilde{e}_0(t)| \le 2\,\varepsilon \sum_{k=0}^{m-1} \varphi\big(\frac{k\pi}{M}\big) \le \varepsilon\,\big(2 + \frac{M}{2\pi}\,\hat{\varphi}(0)\big)\,, \quad t \in \big(0,\, \frac{\pi}{M}\big)\,.$$

By the interpolation property of (2.31), it holds $|\tilde{e}_0(0)| = |\varepsilon_0| \le \varepsilon$ as well as $\big|\tilde{e}_0\big(\frac{\pi}{M}\big)\big| = |\varepsilon_1| \le \varepsilon$. Hence, we obtain:

$$\max_{t\in[0,\pi/M]} |\tilde{e}_0(t)| \le \varepsilon\,\big(2 + \frac{M}{2\pi}\,\hat{\varphi}(0)\big)\,.$$

Analogously, this error estimate can be shown for each interval $\big[\frac{k\pi}{M},\, \frac{(k+1)\pi}{M}\big]$ with arbitrary $k \in \mathbb{Z}$. Using (2.32), we introduce the error function:

$$\begin{aligned}\tilde{e}_k(t) &:= (R_{\varphi,m}\tilde{f})\big(t + \frac{k\pi}{M}\big) - (R_{\varphi,m}f)\big(t + \frac{k\pi}{M}\big)\\ &= \sum_{\ell=-m+1}^{m} \varepsilon_{\ell+k}\,\psi\big(t - \frac{\ell\pi}{M}\big)\,, \quad t \in \big[0,\, \frac{\pi}{M}\big]\,.\end{aligned}$$

As above shown, it holds

$$|\tilde{e}_k(t)| \le \varepsilon\big(2 + \frac{M}{2\pi}\,\hat{\varphi}(0)\big)\,, \quad t \in \big(0,\, \frac{\pi}{M}\big)\,.$$

By the interpolation property, it holds

$$\big|\tilde{e}_k(0)\big| = |\varepsilon_k| \le \varepsilon\,, \quad \big|\tilde{e}_k\big(\frac{\pi}{M}\big)\big| = |\varepsilon_{k+1}| \le \varepsilon\,.$$

Thus, we obtain that

$$\max_{t\in[k\pi/M,\,(k+1)\pi/M]} \big|(R_{\varphi,m}\tilde{f})(t) - (R_{\varphi,m}f)(t)\big| \le \varepsilon\big(2 + \frac{M}{2\pi}\,\hat{\varphi}(0)\big)$$

for each $k \in \mathbb{Z}$. ■

Then, one can show that the perturbation error (2.48) for the regularized Shannon sampling formula (2.31) with the Gaussian window function (2.29) or the sinh-type window function (2.30) grows only as $\mathcal{O}(\sqrt{m})$.

Corollary 2.38 *Let $f \in L_1(\mathbb{R}) \cap C_0(\mathbb{R})$ be bandlimited on $[-L, L]$ with $L > 0$. Assume that $M = (1+\lambda)\, L$, $\lambda > 0$, and $m \in \mathbb{N}\setminus\{1\}$ be given. Let $\tilde{f}_k := f\big(\frac{k\pi}{M}\big)+\varepsilon_k$, $k \in \mathbb{Z}$, be noisy samples, where it holds $|\varepsilon_k| \le \varepsilon$ for all $k \in \mathbb{Z}$.*

Then, the regularized Shannon sampling formula (2.31) *with the Gaussian window function* (2.29) *with $\sigma = \frac{1}{L}\sqrt{\frac{\pi m}{(1+\lambda)\,\lambda}}$ is numerically robust, and it holds:*

$$\|R_{\varphi,m} f - R_{\varphi,m} \tilde{f}\|_{C_0(\mathbb{R})} \le \varepsilon \left(2 + \frac{1}{2}\sqrt{2 + \frac{2}{\lambda}}\,\sqrt{m}\right).$$

Further the regularized Shannon sampling formula (2.31) *with the* sinh-*type window function* (2.30) *with $\beta = \frac{m\pi\lambda}{1+\lambda}$ is numerically robust and it holds*

$$\|R_{\varphi,m} f - R_{\varphi,m} \tilde{f}\|_{C_0(\mathbb{R})} \le \varepsilon \left(2 + \frac{1}{2}\sqrt{2 + \frac{2}{\lambda}}\,\big(1 - \mathrm{e}^{-2m\pi\lambda/(1+\lambda)}\big)^{-1}\,\sqrt{m}\right).$$

For further details see [242].

2.4 Heisenberg's Uncertainty Principle

In this section, we consider a nonzero function $f \in L_2(\mathbb{R})$ with squared $L_2(\mathbb{R})$ norm:

$$\|f\|^2 := \|f\|^2_{L_2(\mathbb{R})} = \int_{\mathbb{R}} |f(x)|^2\, \mathrm{d}x > 0\,,$$

which is called the *energy* of f in some applications. A signal f is often measured in time. We keep the spatial variable x instead of t also when speaking about a time-dependent signal. In the following, we investigate the time–frequency locality of f and $\hat{f}$.

It is impossible to construct a nonzero compactly supported function $f \in L_2(\mathbb{R})$ whose Fourier transform $\hat{f}$ has a compact support too. More generally, we show the following result.

Lemma 2.39 *If the Fourier transform $\hat{f}$ of a nonzero function $f \in L_2(\mathbb{R})$ has compact support, then f cannot be zero on a whole interval. If a nonzero function $f \in L_2(\mathbb{R})$ has compact support, then $\hat{f}$ cannot be zero on a whole interval.*

Proof We consider $f \in L_2(\mathbb{R})$ with $\operatorname{supp} \hat{f} \subseteq [-L, L]$ with some $L > 0$. By the Fourier inversion formula (2.14), we have almost everywhere

$$f(x) = \frac{1}{2\pi} \int_{-L}^{L} \hat{f}(\omega)\, \mathrm{e}^{\mathrm{i}\,\omega x}\, \mathrm{d}\omega\,,$$

where the function on the right-hand side is infinitely differentiable. Since we identify almost everywhere equal functions in $L_2(\mathbb{R})$, we can assume that $f \in C^\infty(\mathbb{R})$.

Assume that $f(x) = 0$ for all $x \in [a, b]$ with $a < b$. For $x_0 = \frac{a+b}{2}$ we obtain by repeated differentiation with respect to x that

$$f^{(n)}(x_0) = \frac{1}{2\pi} \int_{-L}^{L} \hat{f}(\omega)\, (\mathrm{i}\,\omega)^n\, \mathrm{e}^{\mathrm{i}\,\omega x_0}\, \mathrm{d}\omega = 0\,, \quad n \in \mathbb{N}_0\,.$$

Expressing the exponential $\mathrm{e}^{\mathrm{i}\,\omega(x-x_0)}$ as power series, we see that for all $x \in \mathbb{R}$,

$$\begin{aligned} f(x) &= \frac{1}{2\pi} \int_{-L}^{L} \hat{f}(\omega)\, \mathrm{e}^{\mathrm{i}\,\omega(x-x_0)}\, \mathrm{e}^{\mathrm{i}\,\omega x_0}\, \mathrm{d}\omega \\ &= \frac{1}{2\pi} \sum_{n=0}^{\infty} \frac{(x-x_0)^n}{n!} \int_{-L}^{L} \hat{f}(\omega)\, (\mathrm{i}\,\omega)^n\, \mathrm{e}^{\mathrm{i}\,\omega x_0}\, \mathrm{d}\omega = 0\,. \end{aligned}$$

This contradicts the assumption that $f \neq 0$. Analogously, we can show the second assertion. ∎

Lemma 2.39 describes a special aspect of a general principle that says that both f and $\hat{f}$ cannot be highly localized, i.e., if $|f|^2$ vanishes or is very small outside some small interval, then $|\hat{f}|^2$ spreads out, and conversely. We measure the *dispersion of f about the time $x_0 \in \mathbb{R}$* by

$$\Delta_{x_0} f := \frac{1}{\|f\|^2} \int_{\mathbb{R}} (x - x_0)^2\, |f(x)|^2 \mathrm{d}x > 0\,.$$

Note that if $x\, f(x)$, $x \in \mathbb{R}$, is not in $L_2(\mathbb{R})$, then $\Delta_{x_0} f = \infty$ for any $x_0 \in \mathbb{R}$. The dispersion $\Delta_{x_0} f$ measures how much $|f(x)|^2$ spreads out in a neighborhood of x_0. If $|f(x)|^2$ is very small outside a small neighborhood of x_0, then the factor $(x - x_0)^2$ makes the numerator of $\Delta_{x_0} f$ small in comparison to the denominator $\|f\|^2$. Otherwise, if $|f(x)|^2$ is large far away from x_0, then the factor $(x - x_0)^2$ makes the numerator of $\Delta_{x_0} f$ large in comparison to the denominator $\|f\|^2$.

Analogously, we measure the *dispersion of $\hat{f}$ about the frequency $\omega_0 \in \mathbb{R}$* by

$$\Delta_{\omega_0} \hat{f} := \frac{1}{\|\hat{f}\|^2} \int_{\mathbb{R}} (\omega - \omega_0)^2\, |\hat{f}(\omega)|^2\, \mathrm{d}\omega > 0\,.$$

By the Parseval equality $\|\hat{f}\|^2 = 2\pi\, \|f\|^2 > 0$ we obtain:

$$\Delta_{\omega_0} \hat{f} = \frac{1}{2\pi \, \|f\|^2} \int_{\mathbb{R}} (\omega - \omega_0)^2 \, |\hat{f}(\omega)|^2 \, \mathrm{d}\omega \, .$$

If $\omega \, f(\omega)$, $\omega \in \mathbb{R}$, is not in $L_2(\mathbb{R})$, then $\Delta_{\omega_0} f = \infty$ for any $\omega_0 \in \mathbb{R}$.

Example 2.40 As in Example 2.6 we consider the normalized Gaussian function

$$f(x) := \frac{1}{\sqrt{2\pi\sigma^2}} \, \mathrm{e}^{-x^2/(2\sigma^2)}$$

with standard deviation $\sigma > 0$. Then f has $L_1(\mathbb{R})$ norm one, but the energy

$$\|f\|^2 = \frac{1}{2\pi\sigma^2} \int_{\mathbb{R}} \mathrm{e}^{-x^2/\sigma^2} \, \mathrm{d}x = \frac{1}{2\sigma\sqrt{\pi}} \, .$$

Further f has the Fourier transform

$$\hat{f}(\omega) = \mathrm{e}^{-\sigma^2\omega^2/2}$$

with the energy

$$\|\hat{f}\|^2 = \int_{\mathbb{R}} \mathrm{e}^{-\sigma^2\omega^2} \, \mathrm{d}\omega = \frac{\sqrt{\pi}}{\sigma} .$$

For small deviation σ we observe that f is highly localized near zero, but its Fourier transform $\hat{f}$ has the large deviation $\frac{1}{\sigma}$ and is not concentrated near zero. Now we measure the dispersion of f around the time $x_0 \in \mathbb{R}$ by

$$\begin{aligned} \Delta_{x_0} f &= \frac{1}{2\pi\sigma^2 \, \|f\|^2} \int_{\mathbb{R}} (x - x_0)^2 \, \mathrm{e}^{-x^2/\sigma^2} \, \mathrm{d}x \\ &= \frac{1}{2\pi\sigma^2 \, \|f\|^2} \int_{\mathbb{R}} x^2 \, \mathrm{e}^{-x^2/\sigma^2} \, \mathrm{d}x + x_0^2 = \frac{\sigma^2}{2} + x_0^2 \, . \end{aligned}$$

For the dispersion of $\hat{f}$ about the frequency $\omega_0 \in \mathbb{R}$, we obtain

$$\begin{aligned} \Delta_{\omega_0} \hat{f} &= \frac{1}{\|\hat{f}\|^2} \int_{\mathbb{R}} (\omega - \omega_0)^2 \, \mathrm{e}^{-\sigma^2\omega^2} \, \mathrm{d}\omega \\ &= \frac{1}{\|\hat{f}\|^2} \int_{\mathbb{R}} \omega^2 \, \mathrm{e}^{-\sigma^2\omega^2} \, \mathrm{d}\omega + \omega_0^2 = \frac{1}{2\sigma^2} + \omega_0^2 \, . \end{aligned}$$

Thus, for each $\sigma > 0$, we get the inequality:

$$(\Delta_{x_0} f) \, (\Delta_{\omega_0} \hat{f}) = \Big(\frac{\sigma^2}{2} + x_0^2\Big) \Big(\frac{1}{2\sigma^2} + \omega_0^2\Big) \ge \frac{1}{4}$$

with equality for $x_0 = \omega_0 = 0$. □

Heisenberg's uncertainty principle says that for any $x_0, \omega_0 \in \mathbb{R}$, both functions f and $\hat{f}$ cannot be simultaneously localized around time $x_0 \in \mathbb{R}$ and frequency $\omega_0 \in \mathbb{R}$.

Theorem 2.41 (Heisenberg's Uncertainty Principle) *For any nonzero function $f \in L_2(\mathbb{R})$, the inequality*

$$(\Delta_{x_0} f)\,(\Delta_{\omega_0} \hat{f}) \ge \frac{1}{4} \tag{2.49}$$

is fulfilled for each $x_0, \omega_0 \in \mathbb{R}$. The equality in (2.49) *holds if and only if*

$$f(x) = C\,\mathrm{e}^{\mathrm{i}\,\omega_0 x}\,\mathrm{e}^{-a\,(x-x_0)^2/2}\,, \quad x \in \mathbb{R}\,, \tag{2.50}$$

with some $a > 0$ and complex constant $C \neq 0$.

Proof

1. Without lost of generality, we can assume that both functions $x\,f(x)$, $x \in \mathbb{R}$, and $\omega\,\hat{f}(\omega)$, $\omega \in \mathbb{R}$ are contained in $L_2(\mathbb{R})$ too, since otherwise we have $(\Delta_{x_0} f)\,(\Delta_{\omega_0} \hat{f}) = \infty$ and the inequality (2.49) is true.
2. In the special case $x_0 = \omega_0 = 0$, we obtain by the definitions that

$$(\Delta_0 f)\,(\Delta_0 \hat{f}) = \frac{1}{2\pi\,\|f\|^4}\Big(\int_{\mathbb{R}} |x\,f(x)|^2\,\mathrm{d}x\Big)\Big(\int_{\mathbb{R}} |\omega\,\hat{f}(\omega)|^2\,\mathrm{d}\omega\Big)\,.$$

For simplicity we additionally assume the differentiability of f. From $\omega \hat{f}(\omega) \in L_2(\mathbb{R})$, it follows by Theorems 2.5 and 2.22 that $f' \in L_2(\mathbb{R})$. Thus, we get by $(\mathscr{F} f')(\omega) = \mathrm{i}\,\omega \hat{f}(\omega)$ and the Parseval equality (2.13) that

$$\begin{aligned}(\Delta_0 f)\,(\Delta_0 \hat{f}) &= \frac{1}{2\pi\,\|f\|^4}\Big(\int_{\mathbb{R}} |x\,f(x)|^2\,\mathrm{d}x\Big)\Big(\int_{\mathbb{R}} |(\mathscr{F} f')(\omega)|^2\,\mathrm{d}\omega\Big)\\ &= \frac{1}{\|f\|^4}\Big(\int_{\mathbb{R}} |x\,f(x)|^2\,\mathrm{d}x\Big)\Big(\int_{\mathbb{R}} |f'(x)|^2\,\mathrm{d}x\Big)\,.\end{aligned} \tag{2.51}$$

By integration by parts, we obtain:

$$\int_{\mathbb{R}} \big(x\,\overline{f(x)}\big)\,f'(x)\,\mathrm{d}x = \underbrace{x\,|f(x)|^2\Big|_{-\infty}^{\infty}}_{=0} - \int_{\mathbb{R}} \big(|f(x)|^2 + x\,f(x)\,\overline{f'(x)}\big)\,\mathrm{d}x$$

and hence

$$\|f\|^2 = -2\,\mathrm{Re}\int_{\mathbb{R}} \overline{x\,f(x)}\,f'(x)\,\mathrm{d}x\,.$$

By the Cauchy–Schwarz inequality in $L_2(\mathbb{R})$, it follows that

$$\begin{aligned}\|f\|^4 &= 4\,\big(\mathrm{Re}\int_{\mathbb{R}} \overline{x\,f(x)}\,f'(x)\,\mathrm{d}x\big)^2\\ &\le 4\,\big|\int_{\mathbb{R}} \overline{x\,f(x)}\,f'(x)\,\mathrm{d}x\big|^2\\ &\le 4\,\big(\int_{\mathbb{R}} x^2\,|f(x)|^2\,\mathrm{d}x\big)\,\big(\int_{\mathbb{R}} |f'(x)|^2\,\mathrm{d}x\big)\,. \end{aligned} \tag{2.52}$$

Then, by (2.51) and (2.52), we obtain the inequality (2.49) for $x_0 = \omega_0 = 0$.

3. Going through the previous step of the proof, we see that we have equality in (2.49) if and only if

$$\int_{\mathbb{R}} \overline{x\,f(x)}\,f'(x)\,\mathrm{d}x \in \mathbb{R} \tag{2.53}$$

and equality holds true in the Cauchy–Schwarz estimate

$$\big|\int_{\mathbb{R}} \overline{x\,f(x)}\,f'(x)\,\mathrm{d}x\big|^2 = \big(\int_{\mathbb{R}} x^2\,|f(x)|^2\,\mathrm{d}x\big)\,\big(\int_{\mathbb{R}} |f'(x)|^2\,\mathrm{d}x\big).$$

The latter is the case if and only if $x\,f(x)$ and $f'(x)$ are linearly dependent, i.e.,

$$f'(x) + a\,x f(x) = 0\,, \quad a \in \mathbb{C}\,.$$

Plugging this into (2.53), we see that the integral can become only real if $a \in \mathbb{R}$. The above ordinary differential equation has the solution $f(x) = C\,\mathrm{e}^{-a\,x^2/2}$ which belongs to $L_2(\mathbb{R})$ only for $a > 0$.

4. In the general case with any $x_0, \omega_0 \in \mathbb{R}$, we introduce the function:

$$g(x) := \mathrm{e}^{-\mathrm{i}\omega_0 x}\,f(x + x_0)\,, \quad x \in \mathbb{R}\,. \tag{2.54}$$

Obviously, $g \in L_2(\mathbb{R})$ is nonzero. By Theorem 2.5, this function g has the Fourier transform:

$$\hat{g}(\omega) = \mathrm{e}^{\mathrm{i}(\omega+\omega_0)x_0}\,\hat{f}(\omega + \omega_0)\,, \quad \omega \in \mathbb{R}\,,$$

such that

$$\begin{aligned}\Delta_0 g &= \frac{1}{\|f\|^2}\int_{\mathbb{R}} x^2\,|f(x + x_0)|^2\,\mathrm{d}x = \Delta_{x_0} f\,,\\ \Delta_0 \hat{g} &= \frac{1}{\|\hat{f}\|^2}\int_{\mathbb{R}} \omega^2\,|\hat{f}(\omega + \omega_0)|^2\,\mathrm{d}\omega = \Delta_{\omega_0} \hat{f}\,.\end{aligned}$$

Thus, we obtain by step 2 that

$$(\Delta_{x_0} f)\,(\Delta_{\omega_0} \hat{f}) = (\Delta_0 g)\,(\Delta_0 \hat{g}) \geq \frac{1}{4}\,.$$

5. From the equality $(\Delta_0 g)\,(\Delta_0 \hat{g}) = \frac{1}{4}$, it follows by step 3 that $g(x) = C\,\mathrm{e}^{-ax^2/2}$ with $C \in \mathbb{C}$ and $a > 0$. By the substitution (2.54), we see that the equality in (2.49) means that f has the form (2.50). ■

Remark 2.42 In above proof, the additional assumption that f is differentiable is motivated by the following example. The hat function $f(x) = \max_{x\in\mathbb{R}}\{1 - |x|,\, 0\}$ possesses the Fourier transform $\hat{f}(\omega) = \big(\operatorname{sinc}\frac{\omega}{2}\big)^2$ (cf. Example 2.4). Hence $x\,f(x)$ and $\omega\,\hat{f}(\omega)$ are in $L_2(\mathbb{R})$, but f is not differentiable. In Sect. 4.3.1 we will see that we have to deal indeed with functions which are differentiable in the distributional sense. The distributional derivative of the hat function f is equal to $\chi_{[-1,\,0]} - \chi_{[0,\,1]}$ (cf. Remark 4.44). □

The *average time* of a nonzero function $f \in L_2(\mathbb{R})$ is defined by

$$x^* := \frac{1}{\|f\|^2} \int_{\mathbb{R}} x\,|f(x)|^2\,\mathrm{d}x\,.$$

This value exists and is a real number, if $\int_{\mathbb{R}} |x|\,|f(x)|^2\,\mathrm{d}x < \infty$. For a nonzero function $f \in L_2(\mathbb{R})$ with $x^* \in \mathbb{R}$, the quantity $\Delta_{x^*} f$ is the so-called temporal variance of f. Analogously, the *average frequency* of the Fourier transform $\hat{f} \in L_2(\mathbb{R})$ is defined by

$$\omega^* := \frac{1}{\|\hat{f}\|^2} \int_{\mathbb{R}} \omega\,|\hat{f}(\omega)|^2\,\mathrm{d}\omega\,.$$

For a Fourier transform $\hat{f}$ with $\omega^* \in \mathbb{R}$, the quantity $\Delta_{\omega^*} \hat{f}$ is the so-called frequency variance of $\hat{f}$.

Example 2.43 The normalized Gaussian function in Example 2.40 has the average time zero and the temporal variance $\Delta_0 f = \frac{\sigma^2}{2}$, where $\sigma > 0$ denotes the standard deviation of the normalized Gaussian function (2.2). Its Fourier transform has the average frequency zero and the frequency variance $\Delta_0 \hat{f} = \frac{1}{2\sigma^2}$. □

Lemma 2.44 *For each nonzero function $f \in L_2(\mathbb{R})$ with finite average time x^*, it holds the estimate:*

$$\Delta_{x_0} f = \Delta_{x^*} f + (x^* - x_0)^2 \geq \Delta_{x^*} f$$

for any $x_0 \in \mathbb{R}$.

Similarly, for each nonzero function $f \in L_2(\mathbb{R})$ with finite average frequency ω^* of $\hat{f}$, it holds the estimate:

$$\Delta_{\omega_0} \hat{f} = \Delta_{\omega^*} \hat{f} + (\omega^* - \omega_0)^2 \geq \Delta_{\omega^*} \hat{f}$$

for any $\omega_0 \in \mathbb{R}$.

Proof From

$$(x - x_0)^2 = (x - x^*)^2 + 2\,(x - x^*)(x^* - x_0) + (x^* - x_0)^2$$

it follows immediately that

$$\int_{\mathbb{R}} (x - x_0)^2\, |f(x)|^2\, \mathrm{d}x = \int_{\mathbb{R}} (x - x^*)^2\, |f(x)|^2\, \mathrm{d}x + 0 + (x^* - x_0)^2\, \|f\|^2$$

and hence

$$\Delta_{x_0} f = \Delta_{x^*} f + (x^* - x_0)^2 \geq \Delta_{x^*} f\,.$$

Analogously, one can show the second inequality. ■

Applying Theorem 2.41 in the special case $x_0 = x^*$ and $\omega_0 = \omega^*$, we obtain the following result.

Corollary 2.45 *For any nonzero function $f \in L_2(\mathbb{R})$ with finite average time x^* and finite average frequency ω^*, the inequality*

$$(\Delta_{x^*} f)\,(\Delta_{\omega^*} \hat{f}) \geq \frac{1}{4}$$

is fulfilled. The equality in the above inequality holds if and only if

$$f(x) = C\, \mathrm{e}^{\mathrm{i}\,\omega^* x}\, \mathrm{e}^{-a\,(x-x^*)^2/2}\,, \quad x \in \mathbb{R}\,,$$

with some $a > 0$ and complex constant $C \neq 0$.

2.5 Fourier-Related Transforms in Time–Frequency Analysis

In time–frequency analysis, time-dependent functions with changing frequency characteristics appearing, e.g., in music, speech, or radar signal, are studied in time and frequency simultaneously. By the uncertainty principle, the Fourier transform does not yield a good description of the local spatial behavior of the frequency content. A standard tool in time–frequency analysis is the windowed Fourier transform which is discussed in the first subsection. In order to obtain information about local properties of $f \in L_2(\mathbb{R})$, we restrict f to small intervals and examine the resulting Fourier transforms. Another popular tool in time–frequency analysis is the fractional Fourier transform which is handled in the second subsection. Note that wavelet theory also belongs to time–frequency analysis, but is out of the scope of this book.

2.5.1 Windowed Fourier Transform

The Fourier transform $\hat{f}$ contains frequency information of the whole function $f \in L_2(\mathbb{R})$. Now we are interested in simultaneous information about time and frequency of a given function $f \in L_2(\mathbb{R})$. In time–frequency analysis, we ask for frequency information of f near certain time. Analogously, we are interested in the time information of the Fourier transform $\hat{f}$ near certain frequency. Therefore, we localize the function f and its Fourier transform $\hat{f}$ by using windows.

A real, even nonzero function $\psi \in L_2(\mathbb{R})$, where ψ and $\hat{\psi}$ are localized near zero, is called a *window function* or simply *window*. Thus, $\hat{\psi}$ is a window too.

Example 2.46 Let $L > 0$ be fixed. Frequently applied window functions are the *rectangular window* :

$$\psi(x) = \chi_{[-L,\, L]}(x)\,,$$

the *triangular window*

$$\psi(x) = \Big(1 - \frac{|x|}{L}\Big)\, \chi_{[-L,\, L]}(x)\,,$$

the *Gaussian window* with deviation $\sigma > 0$

$$\psi(x) = \frac{1}{\sqrt{2\pi\sigma^2}}\, \mathrm{e}^{-x^2/(2\sigma^2)}\,,$$

the *Hanning window*

$$\psi(x) = \frac{1}{2}\Big(1 + \cos\frac{\pi x}{L}\Big)\, \chi_{[-L,\, L]}(x)\,,$$

and the *Hamming window*

$$\psi(x) = \Big(0.54 + 0.46\,\cos\frac{\pi x}{L}\Big)\, \chi_{[-L,\, L]}(x)\,,$$

where $\chi_{[-L,\, L]}$ denotes the characteristic function of the interval $[-L,\, L]$ (Fig. 2.3). □

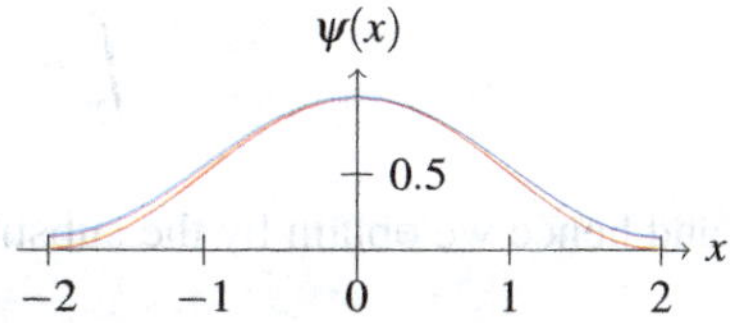

Fig. 2.3 Hanning window (red) and Hamming window (blue) for $L = 2$

Using the shifted window $\psi(\cdot - b)$, we consider the product $f\,\psi(\cdot - b)$ which is localized in some neighborhood of $b \in \mathbb{R}$. Then, we form the Fourier transform of the localized function $f\,\psi(\cdot - b)$. The mapping $\mathcal{F}_\psi \colon L_2(\mathbb{R}) \to L_2(\mathbb{R}^2)$ defined by

$$(\mathcal{F}_\psi f)(b, \omega) := \int_{\mathbb{R}} f(x)\,\psi(x-b)\,\mathrm{e}^{-\mathrm{i}\,\omega x}\,\mathrm{d}x = \langle f,\, \Psi_{b,\omega}\rangle_{L_2(\mathbb{R})} \tag{2.55}$$

with the *time–frequency atom*

$$\Psi_{b,\omega}(x) := \psi(x-b)\,\mathrm{e}^{\mathrm{i}\,\omega x}\,, \quad x \in \mathbb{R}\,,$$

is called *windowed Fourier transform* or *short-time Fourier transform* (STFT); see [185, pp. 37–58]. Note that the time–frequency atom $\Psi_{b,\omega}$ is concentrated in time b and in frequency ω. A special case of the windowed Fourier transform is the *Gabor transform* [155] which uses a Gaussian window. The squared magnitude $|(\mathcal{F}_\psi f)(b,\omega)|^2$ of the windowed Fourier transform is called *spectrogram* of f with respect to ψ.

The windowed Fourier transform $\mathcal{F}_\psi f$ can be interpreted as a joint time–frequency information of f. Thus, $(\mathcal{F}_\psi f)(b,\omega)$ can be considered as a measure for the amplitude of a frequency band near ω at time b.

Example 2.47 We choose the Gaussian window ψ with deviation $\sigma = 1$, i.e.,

$$\psi(x) := \frac{1}{\sqrt{2\pi}}\,\mathrm{e}^{-x^2/2}\,, \quad x \in \mathbb{R}\,,$$

and consider the $L_2(\mathbb{R})$ function $f(x) := \psi(x)\,\mathrm{e}^{\mathrm{i}\,\omega_0 x}$ with fixed frequency $\omega_0 \in \mathbb{R}$. We show that the frequency ω_0 can be detected by windowed Fourier transform $\mathcal{F}_\psi f$ which reads as follows:

$$\begin{aligned}(\mathcal{F}_\psi f)(b,\omega) &= \frac{1}{2\pi}\,\mathrm{e}^{-b^2/2}\int_{\mathbb{R}} \mathrm{e}^{-x^2}\,\mathrm{e}^{b\,x + \mathrm{i}\,(\omega_0-\omega)\,x}\,\mathrm{d}x \\ &= \frac{1}{2\pi}\,\mathrm{e}^{-b^2/4}\int_{\mathbb{R}} \mathrm{e}^{-(x-b/2)^2}\,\mathrm{e}^{\mathrm{i}\,(\omega_0-\omega)\,x}\,\mathrm{d}x\,.\end{aligned}$$

From Example 2.6 we know that

$$\int_{\mathbb{R}} \mathrm{e}^{-t^2} \mathrm{e}^{\mathrm{i}\omega t} \, \mathrm{d}t = \sqrt{\pi} \, \mathrm{e}^{-\omega^2/4}$$

and hence we obtain by the substitution $t = x - \frac{b}{2}$ that

$$(\mathscr{F}_\psi f)(b, \omega) = \frac{1}{2\sqrt{\pi}} \, \mathrm{e}^{-b^2/4} \, \mathrm{e}^{-(\omega_0-\omega)^2/4} \, \mathrm{e}^{\mathrm{i}\, b\, (\omega_0-\omega)/2} .$$

Thus, the spectrogram is given by

$$\big|(\mathscr{F}_\psi f)(b, \omega)\big|^2 = \frac{1}{4\pi} \, \mathrm{e}^{-b^2/2} \, \mathrm{e}^{-(\omega_0-\omega)^2/2} .$$

For each time $b \in \mathbb{R}$, the spectrogram has its maximum at the frequency $\omega = \omega_0$. In practice, one can detect ω_0 only, if $|b|$ is not too large. □

The following identity combines f and $\hat{f}$ in a joint time–frequency representation.

Lemma 2.48 *Let ψ be a window. Then, for all time–frequency locations $(b, \omega) \in \mathbb{R}^2$, we have:*

$$2\pi \, (\mathscr{F}_\psi f)(b, \omega) = \mathrm{e}^{-\mathrm{i}\, b\omega} \, (\mathscr{F}_{\hat{\psi}} \hat{f})(\omega, -b) .$$

Proof Since ψ is real and even by definition, its Fourier transform $\hat{\psi}$ is real and even too. Thus, $\hat{\psi}$ is a window too. By Theorem 2.5 and Parseval equality (2.13), we obtain:

$$2\pi \, \langle f, \, \psi(\cdot - b) \, \mathrm{e}^{\mathrm{i}\omega \cdot} \rangle_{L_2(\mathbb{R})} = \langle \hat{f}, \, \hat{\psi}(\cdot - \omega) \, \mathrm{e}^{-\mathrm{i}b(\cdot - \omega)} \rangle_{L_2(\mathbb{R})}$$

and hence

$$\begin{aligned} 2\pi \int_{\mathbb{R}} f(x) \, \psi(x - b) \, \mathrm{e}^{-\mathrm{i}\omega x} \, \mathrm{d}x &= \int_{\mathbb{R}} \hat{f}(u) \, \hat{\psi}(u - \omega) \, \mathrm{e}^{\mathrm{i}\, b\, (u-\omega)} \, \mathrm{d}u \\ &= \mathrm{e}^{-\mathrm{i}\, b\omega} \int_{\mathbb{R}} \hat{f}(u) \, \hat{\psi}(u - \omega) \, \mathrm{e}^{\mathrm{i}\, bu} \, \mathrm{d}u . \end{aligned}$$

■

Remark 2.49 Let ψ be a window function, where the functions $x\, \psi(x)$ and $\omega\, \hat{\psi}(\omega)$ are in $L_2(\mathbb{R})$ too. For all time–frequency locations $(b, \omega) \in \mathbb{R}^2$, the time–frequency atoms $\Psi_{b,\omega} = \psi(\cdot - b) \, \mathrm{e}^{\mathrm{i}\, \omega \cdot}$ and their Fourier transforms $\hat{\Psi}_{b,\omega} = \hat{\psi}(\cdot - \omega) \, \mathrm{e}^{-\mathrm{i}\, b\, (\cdot - \omega)}$ have constant energies $\|\Psi_{b,\omega}\|^2 = \|\psi\|^2$ and $\|\hat{\Psi}_{b,\omega}\|^2 = \|\hat{\psi}\|^2 = 2\pi \, \|\psi\|^2$. Then, the atom $\Psi_{b,\omega}$ has the average time $x^* = b$ and $\hat{\Psi}_{b,\omega}$ has the average frequency $\omega^* = \omega$, since

$$x^* = \frac{1}{\|\psi\|^2} \int_{\mathbb{R}} x \, |\Psi_{b,\omega}(x)|^2 \, \mathrm{d}x = \frac{1}{\|\psi\|^2} \int_{\mathbb{R}} (x+b) \, |\psi(x)|^2 \, \mathrm{d}x = b \,,$$

$$\omega^* = \frac{1}{\|\hat{\psi}\|^2} \int_{\mathbb{R}} u \, |\hat{\Psi}_{b,\omega}(u)|^2 \, \mathrm{d}u = \frac{1}{\|\hat{\psi}\|^2} \int_{\mathbb{R}} (u+\omega) \, |\hat{\psi}(u)|^2 \, \mathrm{d}u = \omega \,.$$

Further, the temporal variance of the time–frequency atom $\Psi_{b,\omega}$ is invariant for all time–frequency locations $(b, \omega) \in \mathbb{R}^2$, because

$$\Delta_b \Psi_{b,\omega} = \frac{1}{\|\psi\|^2} \int_{\mathbb{R}} (x-b)^2 \, |\Psi_{b,\omega}(x)|^2 \, \mathrm{d}x = \frac{1}{\|\psi\|^2} \int_{\mathbb{R}} x^2 \, |\psi(x)|^2 \, \mathrm{d}x = \Delta_0 \psi \,.$$

Analogously, the frequency variance of $\hat{\Psi}_{b,\omega}$ is constant for all time–frequency locations $(b, \omega) \in \mathbb{R}^2$, because

$$\Delta_\omega \hat{\Psi}_{b,\omega} = \frac{1}{\|\hat{\psi}\|^2} \int_{\mathbb{R}} (u-\omega)^2 \, |\hat{\Psi}_{b,\omega}(u)|^2 \, \mathrm{d}u = \frac{1}{\|\hat{\psi}\|^2} \int_{\mathbb{R}} u^2 \, |\hat{\psi}(u)|^2 \, \mathrm{d}u = \Delta_0 \hat{\psi} \,.$$

For arbitrary $f \in L_2(\mathbb{R})$, we obtain by Parseval equality (2.13)

$$2\pi \, (\mathscr{F}_\psi)(b, \omega) = 2\pi \, \langle f, \, \Psi_{b,\omega} \rangle_{L_2(\mathbb{R})} = \langle \hat{f}, \, \hat{\Psi}_{b,\omega} \rangle_{L_2(\mathbb{R})} \,.$$

Hence, the value $(\mathscr{F}_\psi)(b, \omega)$ contains information on f in the *time–frequency window* or *Heisenberg box* :

$$\big[b - \sqrt{\Delta_0 \psi}, \, b + \sqrt{\Delta_0 \psi}\big] \times \big[\omega - \sqrt{\Delta_0 \hat{\psi}}, \, \omega + \sqrt{\Delta_0 \hat{\psi}}\big] ,$$

since the deviation is the square root of the variance. Note that the area of the Heisenberg box cannot become arbitrary small, i.e., it holds by Heisenberg's uncertainty principle (see Corollary 2.45) that

$$\big(2 \sqrt{\Delta_0 \psi}\big) \big(2 \sqrt{\Delta_0 \hat{\psi}}\big) \ge 2.$$

The size of the Heisenberg box is independent of the time–frequency location $(b, \omega) \in \mathbb{R}^2$. This means that a windowed Fourier transform has the same resolution across the whole time–frequency plane $\mathbb{R}^2$. □

Theorem 2.50 *Let ψ be a window function. Then for $f, g \in L_2(\mathbb{R})$, the following relation holds true:*

$$\langle \mathscr{F}_\psi f, \, \mathscr{F}_\psi g \rangle_{L_2(\mathbb{R}^2)} = 2\pi \, \|\psi\|^2_{L_2(\mathbb{R})} \, \langle f, g \rangle_{L_2(\mathbb{R})} \,.$$

In particular, for $\|\psi\|_{L_2(\mathbb{R})} = 1$ the energies of $\mathscr{F}_\psi f$ and f are equal up to the factor 2π,

$$\|\mathscr{F}_\psi f\|^2_{L_2(\mathbb{R}^2)} = 2\pi \, \|f\|^2_{L_2(\mathbb{R})} \,.$$

Proof

1. First, let $\psi \in L_1(\mathbb{R}) \cap L_\infty(\mathbb{R})$. Then, we have:

$$\langle \mathscr{F}_\psi f, \mathscr{F}_\psi g\rangle_{L_2(\mathbb{R}^2)} = \int_{\mathbb{R}} \int_{\mathbb{R}} (\mathscr{F}_\psi f)(b,\omega)\, \overline{(\mathscr{F}_\psi g)(b,\omega)}\, \mathrm{d}\omega\, \mathrm{d}b \,.$$

We consider the inner integral:

$$\int_{\mathbb{R}} (\mathscr{F}_\psi f)(b,\omega)\, \overline{(\mathscr{F}_\psi g)(b,\omega)}\, \mathrm{d}\omega = \int_{\mathbb{R}} (f\, \psi(\cdot - b))\hat{\ }(\omega)\, \overline{(g\, \psi(\cdot - b))\hat{\ }(\omega)}\, \mathrm{d}\omega \,.$$

By

$$\int_{\mathbb{R}} |f(x)\, \psi(x-b)|^2\, \mathrm{d}x \le \|\psi\|^2_{L_\infty(\mathbb{R})}\, \|f\|^2_{L_2(\mathbb{R})} < \infty$$

we see that $f\, \psi(\cdot - b) \in L_2(\mathbb{R})$ such that we can apply the Parseval equality (2.13)

$$\int_{\mathbb{R}} (\mathscr{F}_\psi f)(b,\omega)\, \overline{(\mathscr{F}_\psi g)(b,\omega)}\, \mathrm{d}\omega = 2\pi \int_{\mathbb{R}} f(x)\, \overline{g(x)}\, |\psi(x-b)|^2\, \mathrm{d}x \,.$$

Using this in the above inner product results in

$$\langle \mathscr{F}_\psi f, \mathscr{F}_\psi g\rangle_{L_2(\mathbb{R}^2)} = 2\pi \int_{\mathbb{R}} \int_{\mathbb{R}} f(x)\, \overline{g(x)}\, |\psi(x-b)|^2\, \mathrm{d}x\, \mathrm{d}b \,.$$

Since $f, g \in L_2(\mathbb{R})$, we see as in the above argumentation that the absolute integral exists. Hence we can change the order of integration by Fubini's theorem which results in

$$\begin{aligned}\langle \mathscr{F}_\psi f, \mathscr{F}_\psi g\rangle_{L_2(\mathbb{R}^2)} &= 2\pi \int_{\mathbb{R}} f(x)\, \overline{g(x)} \int_{\mathbb{R}} |\psi(x-b)|^2\, \mathrm{d}b\, \mathrm{d}x \\ &= 2\pi\, \|\psi\|^2_{L_2(\mathbb{R})}\, \langle f, g\rangle_{L_2(\mathbb{R})} \,.\end{aligned}$$

2. Let $f, g \in L_2(\mathbb{R})$ be fixed. By $\psi \mapsto \langle \mathscr{F}_\psi f, \mathscr{F}_\psi g\rangle_{L_2(\mathbb{R}^2)}$ a continuous functional is defined on $L_1(\mathbb{R}) \cap L_\infty(\mathbb{R})$. Now $L_1(\mathbb{R}) \cap L_\infty(\mathbb{R})$ is a dense subspace of $L_2(\mathbb{R})$. Then this functional can be uniquely extended on $L_2(\mathbb{R})$, where $\langle f, g\rangle_{L_2(\mathbb{R})}$ is kept. ■

Remark 2.51 By Theorem 2.50 under the condition $\|\psi\|_{L_2(\mathbb{R})} = 1$, we know that

$$\int_{\mathbb{R}} |f(x)|^2 \, \mathrm{d}x = \frac{1}{2\pi} \int_{\mathbb{R}} \int_{\mathbb{R}} |(\mathscr{F}_\psi f)(b, \omega)|^2 \, \mathrm{d}b \, \mathrm{d}\omega \, .$$

Hence, the spectrogram $|(\mathscr{F}_\psi f)(b, \omega)|^2$ can be interpreted as an energy density, i.e., the time–frequency rectangle $[b,\, b + \Delta b] \times [\omega,\, \omega + \Delta\omega]$ corresponds to the energy:

$$\frac{1}{2\pi} \, |(\mathscr{F}_\psi f)(b, \omega)|^2 \, \Delta b \; \Delta\omega \, .$$

□

By Theorem 2.50 the windowed Fourier transform represents a univariate signal $f \in L_2(\mathbb{R})$ by a bivariate function $\mathscr{F}_\psi f \in L_2(\mathbb{R}^2)$. Conversely, from given windowed Fourier transform $\mathscr{F}_\psi f$, one can recover the function f:

Corollary 2.52 *Let ψ be a window function with $\|\psi\|_{L_2(\mathbb{R})} = 1$. Then for all $f \in L_2(\mathbb{R})$, it holds the representation formula:*

$$f(x) = \frac{1}{2\pi} \int_{\mathbb{R}} \int_{\mathbb{R}} (\mathscr{F}_\psi f)(b, \omega) \; \psi(x - b) \, \mathrm{e}^{\mathrm{i}\omega x} \, \mathrm{d}b \, \mathrm{d}\omega \, ,$$

where the integral is meant in the weak sense.

Proof Let

$$\tilde{f}(x) := \int_{\mathbb{R}} \int_{\mathbb{R}} (\mathscr{F}_\psi f)(b, \omega) \; \psi(x - b) \, \mathrm{e}^{\mathrm{i}\omega x} \, \mathrm{d}b \, \mathrm{d}\omega \, , \quad x \in \mathbb{R} \, .$$

By Theorem 2.50 we obtain:

$$\begin{aligned} \langle \tilde{f},\, h \rangle_{L_2(\mathbb{R})} &= \int_{\mathbb{R}} \int_{\mathbb{R}} (\mathscr{F}_\psi f)(b, \omega) \; \langle \psi(\cdot - b) \, \mathrm{e}^{\mathrm{i}\omega \cdot},\, h \rangle_{L_2(\mathbb{R})} \, \mathrm{d}b \, \mathrm{d}\omega \\ &= \langle \mathscr{F}_\psi f,\, \mathscr{F}_\psi h \rangle_{L_2(\mathbb{R}^2)} = 2\pi \, \langle f,\, h \rangle_{L_2(\mathbb{R})} \end{aligned}$$

for all $h \in L_2(\mathbb{R})$ so that $\tilde{f} = 2\pi \, f$ in $L_2(\mathbb{R})$. ■

A typical application of this time–frequency analysis consists in the following three steps:

1. For a given (noisy) signal $f \in L_2(\mathbb{R})$, compute the windowed Fourier transform $\mathscr{F}_\psi f$ with respect to a suitable window ψ.
2. Then, $(\mathscr{F}_\psi f)(b, \omega)$ is transformed into a new function $g(b, \omega)$ by so-called signal compression. Usually, $(\mathscr{F}_\psi f)(b, \omega)$ is truncated to a region of interest, where $|(\mathscr{F}_\psi f)(b, \omega)|$ is larger than a given threshold.
3. By the compressed function g, compute an approximate signal $\tilde{f}$ (of the given signal f) by a modified reconstruction formula of Corollary 2.52:

$$\tilde{f}(x) = \frac{1}{2\pi} \int_{\mathbb{R}} \int_{\mathbb{R}} g(b, \omega)\, \varphi(x - b)\, \mathrm{e}^{\mathrm{i}\omega x}\, \mathrm{d}b\, \mathrm{d}\omega\,,$$

where φ is a convenient window. Note that distinct windows ψ and φ may be used in steps 1 and 3.

For an application of the windowed Fourier transform in music analysis, we refer to [140].

2.5.2 *Fractional Fourier Transforms*

The fractional Fourier transform (FRFT) is another Fourier-related transform in time–frequency analysis. Some of its roots can be found in quantum mechanics and in optics, where the FRFT can be physically realized. For more details see [69, 71, 309] and in particular for numerical algorithms to compute the FRFT [70]. The definition of FRFT is based on the spectral decomposition of the Fourier transform on $L_2(\mathbb{R})$. To this end, we consider the normalized Fourier transform:

$$\frac{1}{\sqrt{2\pi}}\, (\mathcal{F} f)(u) = \frac{1}{\sqrt{2\pi}} \int_{\mathbb{R}} f(x)\, \mathrm{e}^{-\mathrm{i}xu}\, \mathrm{d}x$$

which is a unitary operator on $L_2(\mathbb{R})$ by Plancherel's theorem 2.22. By Theorem 2.25, the normalized Hermite functions

$$\varphi_n(x) := (2^n\, n!)^{-1/2}\, \pi^{-1/4}\, H_n(x)\, \mathrm{e}^{-x^2/2}\,, \quad n \in \mathbb{N}_0\,,$$

are eigenfunctions of $\frac{1}{\sqrt{2\pi}}\, \mathcal{F}$ related to the eigenvalues $(-\mathrm{i})^n = \mathrm{e}^{-\mathrm{i}n\pi/2}$, i.e.,

$$\frac{1}{\sqrt{2\pi}}\, \mathcal{F} \varphi_n = \mathrm{e}^{-\mathrm{i}n\pi/2}\, \varphi_n\,, \quad n \in \mathbb{N}_0\,. \tag{2.56}$$

Since $\{\varphi_n : n \in \mathbb{N}_0\}$ is an orthonormal basis of $L_2(\mathbb{R})$, every function $f \in L_2(\mathbb{R})$ can be represented in the form

$$f = \sum_{n=0}^{\infty} \langle f, \varphi_n \rangle_{L_2(\mathbb{R})}\, \varphi_n\,.$$

Then, by the Theorems 2.22 and 2.56, it follows the spectral decomposition:

$$\frac{1}{\sqrt{2\pi}}\, (\mathcal{F} f)(u) = \sum_{n=0}^{\infty} \mathrm{e}^{-\mathrm{i}n\pi/2}\, \langle f, \varphi_n \rangle_{L_2(\mathbb{R})}\, \varphi_n(u) = \int_{\mathbb{R}} K_{\pi/2}(x, u)\, f(x)\, \mathrm{d}x$$

with the kernel of the normalized Fourier transform

$$K_{\pi/2}(x, u) := \sum_{n=0}^{\infty} \mathrm{e}^{-\mathrm{i}n\pi/2}\, \varphi_n(x)\, \varphi_n(u) = \frac{1}{\sqrt{2\pi}}\, \mathrm{e}^{-\mathrm{i}xu}\,.$$

We use the spectral decomposition of $\frac{1}{\sqrt{2\pi}}\, \mathcal{F}$ to define the *fractional Fourier transform* $\mathcal{F}_\alpha$ *of order* $\alpha \in \mathbb{R}$ as the series

$$(\mathcal{F}_\alpha f)(u) := \sum_{n=0}^{\infty} \mathrm{e}^{-\mathrm{i}n\alpha}\, \langle f,\, \varphi_n \rangle_{L_2(\mathbb{R})}\, \varphi_n(u) \tag{2.57}$$

for arbitrary $f \in L_2(\mathbb{R})$. Obviously, $\mathcal{F}_\alpha$ is a continuous linear operator of $L_2(\mathbb{R})$ into itself with the property:

$$\mathcal{F}_\alpha \varphi_n = \mathrm{e}^{-\mathrm{i}n\alpha}\, \varphi_n\,, \quad n \in \mathbb{N}_0\,. \tag{2.58}$$

Since the operator $\mathcal{F}_\alpha$ is 2π-periodic with respect to α, i.e., $\mathcal{F}_{\alpha+2\pi} f = \mathcal{F}_\alpha f$ for all $f \in L_2(\mathbb{R})$, we can restrict ourselves to the case $\alpha \in [-\pi,\, \pi)$. Using (2.57) we see that $\mathcal{F}_0 f = f$, $\mathcal{F}_{\pi/2} f = \frac{1}{\sqrt{2\pi}}\, \mathcal{F} f$, and $\mathcal{F}_{-\pi/2} f = \frac{1}{\sqrt{2\pi}}\, \mathcal{F}^{-1} f$ for all $f \in L_2(\mathbb{R})$. Applying $H_n(-x) = (-1)^n\, H_n(x)$, we obtain:

$$(\mathcal{F}_{-\pi} f)(u) = \sum_{n=0}^{\infty} (-1)^n\, \langle f,\, \varphi_n \rangle_{L_2(\mathbb{R})}\, \varphi_n = \sum_{n=0}^{\infty} \langle f(-\cdot),\, \varphi_n \rangle_{L_2(\mathbb{R})}\, \varphi_n(u) = f(-u)\,.$$

Roughly speaking, the fractional Fourier transform can be interpreted by a rotation through an angle $\alpha \in [-\pi,\, \pi)$ in the time–frequency plane. Let u and v be the new rectangular coordinates in the time–frequency plane, i.e.,

$$\begin{pmatrix} u \\ v \end{pmatrix} = \begin{pmatrix} \cos\alpha & \sin\alpha \\ -\sin\alpha & \cos\alpha \end{pmatrix} \begin{pmatrix} x \\ \omega \end{pmatrix}.$$

Using (2.57) and setting

$$u = x\, \cos\alpha + \omega\, \sin\alpha\,, \tag{2.59}$$

we obtain the following connections.

α	$u = x \cos\alpha + \omega \sin\alpha$	$(\mathscr{F}_\alpha f)(u)$
$-\pi$	$u = -x$	$(\mathscr{F}_{-\pi} f)(-x) = f(x)$
$-\frac{\pi}{2}$	$u = -\omega$	$(\mathscr{F}_{-\pi/2} f)(-\omega) = \frac{1}{\sqrt{2\pi}}\,(\mathscr{F}^{-1} f)(-\omega)$
0	$u = x$	$(\mathscr{F}_0 f)(x) = f(x)$
$\frac{\pi}{2}$	$u = \omega$	$(\mathscr{F}_{\pi/2} f)(\omega) = \frac{1}{\sqrt{2\pi}}\,(\mathscr{F} f)(\omega)$
π	$u = -x$	$(\mathscr{F}_\pi f)(-x) = (\mathscr{F}_{-\pi} f)(-x) = f(x)$

As seen in the table, the FRFT is essentially a rotation in the time–frequency plane. By (2.59) the u-axis rotates around the origin such that, e.g., for increasing $\alpha \in [0, \frac{\pi}{2}]$, the FRFT $(\mathscr{F}_\alpha f)(u)$ describes the change of $f(x)$ toward the normalized Fourier transform $(2\pi)^{-1/2}\,\hat{f}(\omega)$.

For $0 < |\alpha| < \pi$, Mehler's formula [267, p. 61] implies for $x, u \in \mathbb{R}$ that

$$\begin{aligned} K_\alpha(x, u) &:= \sum_{n=0}^{\infty} \mathrm{e}^{-\mathrm{i}n\alpha}\, \varphi_n(x)\, \varphi_n(u) \\ &= \sqrt{\frac{1 - \mathrm{i}\cot\alpha}{2\pi}}\, \exp\Big(\frac{\mathrm{i}}{2}\,(x^2 + u^2)\cot\alpha - \frac{\mathrm{i}\,xu}{\sin\alpha}\Big)\,, \end{aligned}$$

where the square root $\sqrt{1 - \mathrm{i}\cot\alpha}$ has positive real part. For computational purposes, it is more practical to use the *integral representation* of the FRFT:

$$\begin{aligned} (\mathscr{F}_\alpha f)(u) &= \int_{\mathbb{R}} f(x)\, K_\alpha(x, u)\, \mathrm{d}x \\ &= \sqrt{\frac{1 - \mathrm{i}\cot\alpha}{2\pi}} \int_{\mathbb{R}} f(x)\, \exp\Big(\frac{\mathrm{i}}{2}\,(x^2 + u^2)\cot\alpha - \frac{\mathrm{i}\,xu}{\sin\alpha}\Big)\, \mathrm{d}x\,. \end{aligned}$$

Remark 2.53 The FRFT $\mathscr{F}_\alpha f$ exists if the Fourier transform $\mathscr{F} f$ exists, in particular for $f \in L_2(\mathbb{R})$ or $f \in L_1(\mathbb{R})$. Similarly to the Fourier transform of tempered distributions introduced in Sect. 4.3, the FRFT can be extended to tempered distributions. □

In the most interesting case $\alpha \in (-\pi, \pi) \setminus \{-\frac{\pi}{2}, 0, \frac{\pi}{2}\}$, the FRFT $\mathscr{F}_\alpha f$ can be formed in three steps:

1. multiplication of the given function $f(x)$ with the linear chirp $\exp(\frac{\mathrm{i}}{2}\, x^2 \cot\alpha)$,
2. Fourier transform of the product for the scaled argument $\frac{u}{\sin\alpha}$, and
3. multiplication of the intermediate result with the linear chirp

$$\sqrt{\frac{1 - \mathrm{i}\cot\alpha}{2\pi}}\, \exp\Big(\frac{\mathrm{i}}{2}\, u^2 \cot\alpha\Big)\,.$$

Similarly as the complex exponentials are the basic functions for the Fourier transform, linear chirps are basic functions for the FRFT $\mathscr{F}_\alpha f$ for $\alpha \in (-\pi, \pi) \setminus \{-\frac{\pi}{2}, 0, \frac{\pi}{2}\}$.

From the definition of FRFT, we obtain the following properties of the FRFT.

Lemma 2.54 *For all $\alpha,\ \beta \in \mathbb{R}$, the* FRFT *has the following properties:*

1. *$\mathcal{F}_0$ is the identity operator and $\mathcal{F}_{\pi/2}$ coincides with the normalized Fourier transform $\frac{1}{\sqrt{2\pi}}\,\mathcal{F}$.*
2. *$\mathcal{F}_\alpha\,\mathcal{F}_\beta = \mathcal{F}_{\alpha+\beta}$.*
3. *$\mathcal{F}_{-\alpha}$ is the inverse of $\mathcal{F}_\alpha$.*
4. *For all $f,\, g \in L_2(\mathbb{R})$, it holds the Parseval equality:*

$$\langle f,\, g\rangle_{L_2(\mathbb{R})} = \int_{\mathbb{R}} f(x)\,\overline{g(x)}\,\mathrm{d}x = \langle \mathcal{F}_\alpha f,\, \mathcal{F}_\alpha g\rangle_{L_2(\mathbb{R})}\,.$$

Proof The first property follows immediately from the definition of the FRFT $\mathcal{F}_\alpha$ for $\alpha = 0$ and $\alpha = \frac{\pi}{2}$. The second property can be seen as follows. For arbitrary $f \in L_2(\mathbb{R})$, we have:

$$\mathcal{F}_\beta f = \sum_{n=0}^{\infty} \mathrm{e}^{-\mathrm{i}n\beta}\,\langle f, \varphi_n\rangle_{L_2(\mathbb{R})}\,\varphi_n\,.$$

Since $\mathcal{F}_\alpha$ is a continuous linear operator with the property (2.58), we conclude

$$\mathcal{F}_\alpha(\mathcal{F}_\beta f) = \sum_{n=0}^{\infty} \mathrm{e}^{-\mathrm{i}n(\alpha+\beta)}\,\langle f, \varphi_n\rangle_{L_2(\mathbb{R})}\,\varphi_n = \mathcal{F}_{\alpha+\beta} f\,.$$

The third property is a simple consequence of the first and second properties.

Finally, the Parseval equality holds for all $f,\, g \in L_2(\mathbb{R})$, since

$$\begin{aligned}\langle \mathcal{F}_\alpha f,\, \mathcal{F}_\alpha g\rangle_{L_2(\mathbb{R})} &= \sum_{n=0}^{\infty}\sum_{m=0}^{\infty} \mathrm{e}^{-\mathrm{i}(n-m)\alpha}\,\langle f,\, \varphi_n\rangle_{L_2(\mathbb{R})}\,\overline{\langle g,\, \varphi_m\rangle_{L_2(\mathbb{R})}}\,\delta_{n-m}\\ &= \sum_{n=0}^{\infty} \langle f,\, \varphi_n\rangle_{L_2(\mathbb{R})}\,\overline{\langle g,\, \varphi_n\rangle_{L_2(\mathbb{R})}} = \langle f,\, g\rangle_{L_2(\mathbb{R})}\,.\end{aligned}$$

This completes the proof. ■

The following theorem collects further basic properties of the FRFT which can easily be proved.

Theorem 2.55 *Let $\alpha \in (-\pi,\, \pi) \setminus \{-\frac{\pi}{2},\, 0,\, \frac{\pi}{2}\}$ be given. Then, the* FRFT *of order α has the following properties:*

1. Linearity*: For all $c,\, d \in \mathbb{C}$ and $f,\, g \in L_2(\mathbb{R})$,*

$$\mathcal{F}_\alpha(c\,f + d\,g) = c\,\mathcal{F}_\alpha f + d\,\mathcal{F}_\alpha\,.$$

2. Translation and modulation*: For all $b \in \mathbb{R}$ and $f \in L_2(\mathbb{R})$,*

$$\big(\mathscr{F}_\alpha f(\cdot - b)\big)(u) = \exp\big(\frac{\mathrm{i}}{2}\, b^2 \sin\alpha\, \cos\alpha - \mathrm{i}\, ub\, \sin\alpha\big)\, (\mathscr{F}_\alpha f)(u - b\, \cos\alpha)\,,$$

$$\big(\mathscr{F}_\alpha \mathrm{e}^{-\mathrm{i}b\,\cdot} f\big)(u) = \exp\big(-\frac{\mathrm{i}}{2}\, b^2 \sin\alpha\, \cos\alpha - \mathrm{i}\, ub\, \cos\alpha\big)\, (\mathscr{F}_\alpha f)(u + b\, \sin\alpha)\,.$$

3. Differentiation and multiplication*: For $f \in L_2(\mathbb{R})$ with $f' \in L_2(\mathbb{R})$,*

$$(\mathscr{F}_\alpha f')(u) = \cos\alpha\, \frac{\mathrm{d}}{\mathrm{d}u}\, (\mathscr{F}_\alpha f)(u) + \mathrm{i}\, u\, \sin\alpha\, (\mathscr{F}_\alpha f)(u)\,.$$

If f and $g(x) := x\, f(x)$, $x \in \mathbb{R}$, are contained in $L_2(\mathbb{R})$, then

$$(\mathscr{F}_\alpha g)(u) = u\, \cos\alpha\, (\mathscr{F}_\alpha f)(u) + \mathrm{i}\, \sin\alpha\, \frac{\mathrm{d}}{\mathrm{d}u}\, (\mathscr{F}_\alpha f)(u)\,.$$

4. Scaling*: For $b \in \mathbb{R} \setminus \{0\}$, $\big(\mathscr{F}_\alpha f(b\,\cdot)\big)(u)$ reads as follows:*

$$\frac{1}{|b|}\, \frac{\sqrt{1 - \mathrm{i}\, \cot\alpha}}{\sqrt{1 - \mathrm{i}\, \cot\beta}}\, \exp\Big(\frac{\mathrm{i}}{2} u^2 \cot\alpha\, \big(1 - \frac{(\cos\beta)^2}{(\cos\alpha)^2}\big)\Big)\, (\mathscr{F}_\beta f)\big(\frac{u\, \sin\beta}{b\, \sin\alpha}\big)$$

with $\beta := \arctan(b^2\, \tan\alpha)$.

Example 2.56 The function $f(x) = (4x^2 - 10)\, \mathrm{e}^{-x^2/2} = \big(H_2(x) - 8\big)\, \mathrm{e}^{-x^2/2}$ is contained in $L_2(\mathbb{R})$, where $H_2(x) = 4x^2 - 2$ is the second Hermite polynomial. Since the FRFT $\mathscr{F}_\alpha$ is a linear operator, we obtain the FRFT:

$$(\mathscr{F}_\alpha f)(u) = (\mathrm{e}^{-2\mathrm{i}\alpha}\, H_2(u) - 8)\, \mathrm{e}^{-u^2/2} = (4\, \mathrm{e}^{-2\mathrm{i}\alpha}\, u^2 - 2\, \mathrm{e}^{-2\mathrm{i}\alpha} - 8)\, \mathrm{e}^{-u^2/2}\,.$$

Using the scaling property of the FRFT in Theorem 2.55, the function $g(x) = \mathrm{e}^{-b^2 x^2/2}$, $x \in \mathbb{R}$, with $b \in \mathbb{R} \setminus \{0\}$ possesses the FRFT

$$\frac{1}{|b|}\, \frac{\sqrt{1 - \mathrm{i}\, \cot\alpha}}{\sqrt{1 - \mathrm{i}\, \cot\beta}}\, \exp\Big(\frac{\mathrm{i}}{2} u^2\, \cot\alpha\, \big(1 - \frac{(\cos\beta)^2}{(\cos\alpha)^2}\big)\Big)\, \exp\big(-\frac{u^2}{2}\, \frac{(\sin\beta)^2}{b^2\, (\sin\alpha)^2}\big)$$

with $\beta := \arctan(b^2\, \tan\alpha)$. □

The linear canonical transform (see [197]) is a generalization of the FRFT. As shown, the FRFT of order α is related to the rotation matrix:

$$\mathbf{R}_\alpha := \begin{pmatrix} \cos\alpha & \sin\alpha \\ -\sin\alpha & \cos\alpha \end{pmatrix}.$$

If

$$\mathbf{A} := \begin{pmatrix} a & b \\ c & d \end{pmatrix}$$

is a real matrix with determinant 1 and $b > 0$, then the *linear canonical transform* $\mathscr{L}_{\mathbf{A}} : L_2(\mathbb{R}) \to L_2(\mathbb{R})$ is defined by

$$(\mathscr{L}_{\mathbf{A}} f)(u) := \int_{\mathbb{R}} k_{\mathbf{A}}(u, x)\, f(x)\, \mathrm{d}x\,, \quad f \in L_2(\mathbb{R})\,,$$

with the kernel

$$k_{\mathbf{A}}(u, x) := \frac{1}{\sqrt{2\pi b}} \exp\Big(\mathrm{i}\,\big(\frac{a\,x^2}{2b} + \frac{d\,u^2}{2b} - \frac{u\,x}{b} - \frac{\pi}{4}\big)\Big)\,.$$

For $\mathbf{A} = \mathbf{R}_\alpha$ with $\sin\alpha > 0$, i.e., $\alpha \in (0, \pi)$, the linear canonical transform $\mathscr{L}_{\mathbf{A}}$ coincides with a scaled FRFT $\mathscr{F}_\alpha$, since for arbitrary $f \in L_2(\mathbb{R})$ it holds

$$\begin{aligned} \mathscr{L}_{\mathbf{A}} f)(u) &= \frac{\mathrm{e}^{-\mathrm{i}\pi/4}}{\sqrt{2\pi\, \sin\alpha}} \int_{\mathbb{R}} f(x)\, \exp\Big(\frac{\mathrm{i}}{2}\,(x^2 + u^2)\,\cot\alpha - \frac{\mathrm{i}xu}{\sin\alpha}\Big)\,\mathrm{d}x \\ &= \frac{\mathrm{e}^{-\mathrm{i}\pi/4}}{\sqrt{\sin\alpha - \mathrm{i}\,\cos\alpha}}\,(\mathscr{F}_\alpha f)(u)\,, \end{aligned}$$

where $\sqrt{\sin\alpha - \mathrm{i}\,\cos\alpha}$ means the square root with positive real part.

The *special affine Fourier transform* (SAFT) was introduced in [1] for the study of certain operations on optical wave functions. The SAFT is formally defined as

$$(\mathscr{F}_A f)(u) := \int_{\mathbb{R}} k_A(u, x)\, f(x)\, \mathrm{d}x\,, \quad f \in L_2(\mathbb{R})\,,$$

where $A = (a, b, c, d, p, q)$ is a real vector with $ad - bc = 1$ and $b \neq 0$. The kernel k_A has the form

$$k_A(u, x) := \frac{1}{\sqrt{2\pi\,|b|}} \exp\Big(\mathrm{i}\,\big(\frac{a\,x^2}{2b} + \frac{d\,u^2}{2b} - \frac{u\,x}{b} + \frac{p\,x}{b} + \frac{(bq - dp)\,u}{b}\big)\Big)\,.$$

For example, $A = (0, 1, -1, 0, 0, 0)$ gives the normed Fourier transform and $A = (0, -1, 1, 0, 0, 0)$ its inverse. Note that the vector $A = (1, \lambda, 0, 1, 0, 0)$ with $\lambda \neq 0$ produces the *Fresnel transform*. For a comprehensive study of the properties of SAFT (such as inverse of $\mathscr{F}_A$, translation for $\mathscr{F}_A$, shift property of $\mathscr{F}_A$, related convolution, Poisson-type summation formula, and Shannon-type sampling theorem), we refer to [47].

Chapter 3
Discrete Fourier Transforms

This chapter deals with the discrete Fourier transform (DFT). In Sect. 3.1, we show that numerical realizations of Fourier methods, such as the computation of Fourier coefficients, Fourier transforms, or trigonometric interpolation, lead to the DFT. We also present barycentric formulas for interpolating trigonometric polynomials. In Sect. 3.2, we study the basic properties of the Fourier matrix and of the DFT. In particular, we consider the eigenvalues of the Fourier matrix with their multiplicities. Further, we present the intimate relations between cyclic convolutions and the DFT. In Sect. 3.3, we show that cyclic convolutions and circulant matrices are closely related and that circulant matrices can be diagonalized by the Fourier matrix. Section 3.4 presents the properties of Kronecker products and stride permutations, which we will need later in Chap. 5 for the factorization of a Fourier matrix. We show that block circulant matrices can be diagonalized by Kronecker products of Fourier matrices. Finally, Sect. 3.5 addresses real versions of the DFT, such as the discrete cosine transform (DCT) and the discrete sine transform (DST). These linear transforms are generated by orthogonal matrices.

3.1 Motivations for Discrete Fourier Transforms

Discrete Fourier methods can be traced back to the eighteenth and nineteenth centuries, where they have been used already for determining the orbits of celestial bodies. The corresponding data contain periodic patterns that can be well interpolated by trigonometric polynomials. In order to calculate the coefficients of trigonometric polynomials, we need to employ the so-called discrete Fourier transform (DFT). Clairaut, Lagrange, and later Gauss already considered the DFT to solve the problem of fitting astronomical data. In 1754, Clairaut published a first formula for a discrete Fourier transform. For historical remarks see [63, pp. 2–6].

G. Plonka et al., *Numerical Fourier Analysis*, Applied and Numerical Harmonic Analysis, https://doi.org/10.1007/978-3-031-35005-4_3

We start with introducing the discrete Fourier transform. For a given vector $\mathbf{a} = (a_j)_{j=0}^{N-1} \in \mathbb{C}^N$, we call the vector $\hat{\mathbf{a}} = (\hat{a}_k)_{k=0}^{N-1} \in \mathbb{C}^N$ the *discrete Fourier transform of* **a** *with length* N, if

$$\hat{a}_k = \sum_{j=0}^{N-1} a_j \, \mathrm{e}^{-2\pi \mathrm{i}\, jk/N} = \sum_{j=0}^{N-1} a_j \, w_N^{jk} \,, \qquad k = 0, \ldots, N-1 \,, \tag{3.1}$$

where

$$w_N := \mathrm{e}^{-2\pi\, \mathrm{i}/N} = \cos \frac{2\pi}{N} - \mathrm{i} \, \sin \frac{2\pi}{N} \,. \tag{3.2}$$

Obviously, $w_N \in \mathbb{C}$ is a *primitive Nth root of unity*, because $w_N^N = 1$ and $w_N^k \neq 1$ for $k = 1, \ldots, N-1$. Since

$$(w_N^k)^N = \left(\mathrm{e}^{-2\pi \mathrm{i}\, k/N}\right)^N = \mathrm{e}^{-2\pi \mathrm{i} k} = 1 \,,$$

all numbers w_N^k, $k = 0, \ldots, N-1$ are Nth roots of unity and form the vertices of a regular N-gon inscribed in the complex unit circle.

In this section we will show that the discrete Fourier transform naturally comes into play for the numerical solution of following fundamental problems:

- computation of Fourier coefficients of a function $f \in C(\mathbb{T})$,
- computation of the values of a trigonometric polynomial on a uniform grid of the interval $[0, 2\pi)$,
- calculation of the Fourier transform of a function $f \in L_1(\mathbb{R}) \cap C(\mathbb{R})$ on a uniform grid of an interval $[-n\pi, n\pi)$ with certain $n \in \mathbb{N}$,
- interpolation by trigonometric polynomials on a uniform grid of the interval $[0, 2\pi)$.

3.1.1 Approximation of Fourier Coefficients and Aliasing Formula

First we describe a numerical approach to compute the Fourier coefficients $c_k(f)$, $k \in \mathbb{Z}$, of a given function $f \in C(\mathbb{T})$, where f is given by its values sampled on the uniform grid $\{\frac{2\pi j}{N} : j = 0, \ldots, N-1\}$. Assume that $N \in \mathbb{N}$ is even. Using the trapezoidal rule for numerical integration, we can compute $c_k(f)$ for each $k \in \mathbb{Z}$ approximately. By $f(0) = f(2\pi)$, we find that

$$
\begin{aligned}
c_k(f) &= \frac{1}{2\pi}\int_0^{2\pi} f(t)\,\mathrm{e}^{-\mathrm{i}kt}\,\mathrm{d}t \\
&\approx \frac{1}{2N}\sum_{j=0}^{N-1}\Big[f\big(\frac{2\pi j}{N}\big)\,\mathrm{e}^{-2\pi\mathrm{i}\,jk/N} + f\big(\frac{2\pi (j+1)}{N}\big)\,\mathrm{e}^{-2\pi\mathrm{i}\,(j+1)k/N}\Big] \\
&= \frac{1}{2N}\sum_{j=0}^{N-1} f\big(\frac{2\pi j}{N}\big)\,\mathrm{e}^{-2\pi\mathrm{i}\,jk/N} + \frac{1}{2N}\sum_{j=1}^{N} f\big(\frac{2\pi j}{N}\big)\,\mathrm{e}^{-2\pi\mathrm{i}\,jk/N} \\
&= \frac{1}{N}\sum_{j=0}^{N-1} f\big(\frac{2\pi j}{N}\big)\,\mathrm{e}^{-2\pi\mathrm{i}\,jk/N}\,, \quad k\in\mathbb{Z}\,.
\end{aligned}
$$

Thus, we obtain:

$$
\hat{f}_k := \frac{1}{N}\sum_{j=0}^{N-1} f\big(\frac{2\pi j}{N}\big)\,w_N^{jk} \tag{3.3}
$$

as approximate values of $c_k(f)$. If f is real-valued, then we observe the symmetry relation:

$$
\hat{f}_k = \overline{\hat{f}_{-k}}\,, \quad k\in\mathbb{Z}\,.
$$

Obviously, the values $\hat{f}_k$ are N-periodic, i.e., $\hat{f}_{k+N} = \hat{f}_k$ for all $k\in\mathbb{Z}$, since $w_N^N = 1$. However, by the Lemma 1.27 of Riemann–Lebesgue, we know that $c_k(f)\to 0$ as $|k|\to\infty$. Therefore, $\hat{f}_k$ is only an acceptable approximation of $c_k(f)$ for small $|k|$, i.e.,

$$
\hat{f}_k \approx c_k(f)\,, \quad k = -\frac{N}{2},\ldots,\frac{N}{2}-1\,.
$$

Example 3.1 Let f be the 2π-periodic extension of the pulse function:

$$
f(x) := \begin{cases} 1 & x\in(-\frac{\pi}{2},\,\frac{\pi}{2})\,,\\ \frac{1}{2} & x\in\{-\frac{\pi}{2},\,\frac{\pi}{2}\}\,,\\ 0 & x\in[-\pi,\,-\frac{\pi}{2})\cup(\frac{\pi}{2},\,\pi)\,. \end{cases}
$$

Note that f is even. Then, its Fourier coefficients read for $k\in\mathbb{Z}\setminus\{0\}$ as follows:

$$
c_k(f) = \frac{1}{2\pi}\int_{-\pi}^{\pi} f(x)\,\mathrm{e}^{-\mathrm{i}kx}\,\mathrm{d}x = \frac{1}{\pi}\int_0^{\pi/2}\cos(kx)\,\mathrm{d}x = \frac{1}{\pi k}\,\sin\frac{\pi k}{2}
$$

and $c_0(f) = \frac{1}{2}$. For fixed $N\in 4\,\mathbb{N}$, we obtain the related approximate values:

$$\hat{f}_k = \frac{1}{N} \sum_{j=-N/2}^{N/2-1} f\big(\frac{2\pi j}{N}\big)\, w_N^{jk}$$
$$= \frac{1}{N} \Big(\cos\frac{\pi k}{2} + 1 + 2 \sum_{j=1}^{N/4-1} \cos\frac{2\pi jk}{N}\Big) \quad k \in \mathbb{Z}\,.$$

Hence, we have $\hat{f}_k = \frac{1}{2}$ for $k \in N\,\mathbb{Z}$. Using the Dirichlet kernel $D_{N/4-1}$ with (1.22), it follows that for $k \in \mathbb{Z} \setminus (N\,\mathbb{Z})$

$$\hat{f}_k = \frac{1}{N} \Big(\cos\frac{\pi k}{2} + D_{N/4-1}\big(\frac{2\pi k}{N}\big)\Big) = \frac{1}{N} \sin\frac{\pi k}{2} \cot\frac{\pi k}{N}\,.$$

This example illustrates the different asymptotic behavior of the Fourier coefficients $c_k(f)$ and its approximate values $\hat{f}_k$ for $|k| \to \infty$. □

To see this effect more clearly, we will derive a so-called aliasing formula for Fourier coefficients. To this end we use the following notations. As usual, δ_j, $j \in \mathbb{Z}$, denotes the *Kronecker symbol* with

$$\delta_j := \begin{cases} 1 & j = 0\,, \\ 0 & j \neq 0\,. \end{cases}$$

For $j \in \mathbb{Z}$, we denote the *nonnegative residue modulo* $N \in \mathbb{N}$ by $j \bmod N$, where $j \bmod N \in \{0, \ldots, N-1\}$ and N is a divisor of $j - (j \bmod N)$. Note that we have for all $j,\, k \in \mathbb{Z}$

$$(j\,k) \bmod N = \big((j \bmod N)\, k\big) \bmod N\,. \tag{3.4}$$

Lemma 3.2 *Let $N \in \mathbb{N}$ be given. For each $j \in \mathbb{Z}$, the primitive Nth root of unity w_N has the property:*

$$\sum_{k=0}^{N-1} w_N^{jk} = N\, \delta_{j \bmod N}\,, \tag{3.5}$$

where

$$\delta_{j \bmod N} := \begin{cases} 1 & j \bmod N = 0\,, \\ 0 & j \bmod N \neq 0 \end{cases}$$

denotes the N-periodic Kronecker symbol.

Proof In the case $j \bmod N = 0$, we have $j = \ell N$ with certain $\ell \in \mathbb{Z}$ and hence $w_N^j = (w_N^N)^\ell = 1$. This yields (3.5) for $j \bmod N = 0$.

In the case $j \bmod N \neq 0$ we have $j = \ell N + m$ with certain $\ell \in \mathbb{Z}$ and $m \in \{1, \ldots, N-1\}$ such that $w_N^j = (w_N^N)^\ell \, w_N^m = w_N^m \neq 1$. For arbitrary $x \neq 1$, it holds:

$$\sum_{k=0}^{N-1} x^k = \frac{x^N - 1}{x - 1} .$$

For $x = w_N^j$, we obtain (3.5) for $j \bmod N \neq 0$. ■

Lemma 3.2 can be used to prove the following aliasing formula, which describes the relation between the Fourier coefficients $c_k(f)$ and their approximate values $\hat{f}_k$.

Theorem 3.3 (Aliasing Formula for Fourier Coefficients) *Let $f \in C(\mathbb{T})$ be given. Assume that the Fourier coefficients of f satisfy the condition $\sum_{k\in\mathbb{Z}}|c_k(f)| < \infty$. Then the* aliasing formula

$$\hat{f}_k = \sum_{\ell\in\mathbb{Z}} c_{k+\ell N}(f) , \quad k \in \mathbb{Z} , \tag{3.6}$$

holds.

Proof Using Theorem 1.37, the Fourier series of f converges uniformly to f. Hence for each $x \in \mathbb{T}$,

$$f(x) = \sum_{\ell\in\mathbb{Z}} c_\ell(f)\, \mathrm{e}^{\mathrm{i}\,\ell x} .$$

For $x = \frac{2\pi j}{N}$, $j = 0, \ldots, N-1$, we obtain that

$$f\big(\frac{2\pi j}{N}\big) = \sum_{\ell\in\mathbb{Z}} c_\ell(f)\, \mathrm{e}^{2\pi\mathrm{i}\, j\ell/N} = \sum_{\ell\in\mathbb{Z}} c_\ell(f)\, w_N^{-\ell j} .$$

Hence due to (3.3) and the convergence of the Fourier series

$$\hat{f}_k = \frac{1}{N} \sum_{j=0}^{N-1} \Big(\sum_{\ell\in\mathbb{Z}} c_\ell(f)\, w_N^{-j\ell} \Big)\, w_N^{jk} = \frac{1}{N} \sum_{\ell\in\mathbb{Z}} c_\ell(f) \sum_{j=0}^{N-1} w_N^{j(k-\ell)} ,$$

which yields by (3.5) the aliasing formula (3.6). ■

By Theorem 3.3 we have no aliasing effect, if f is a trigonometric polynomial of degree $< \frac{N}{2}$, i.e., for

$$f = \sum_{k=-N/2+1}^{N/2-1} c_k(f)\, \mathrm{e}^{2\pi\mathrm{i}\, k \cdot}$$

we have $\hat{f}_k = c_k(f)$, $k = -N/2+1, \ldots, N/2-1$.

Corollary 3.4 *Under the assumptions of Theorem 3.3, the error estimate*

$$|\hat{f}_k - c_k(f)| \le \sum_{\ell \in \mathbb{Z}\setminus\{0\}} |c_{k+\ell N}(f)| \tag{3.7}$$

holds for $k = -\frac{N}{2}, \ldots, \frac{N}{2} - 1$. *Especially for* $f \in C^r(\mathbb{T})$, $r \in \mathbb{N}$, *with the property*

$$|c_k(f)| \le \frac{c}{|k|^{r+1}}, \quad k \in \mathbb{Z} \setminus \{0\}, \tag{3.8}$$

where $c > 0$ *is a constant, we have the error estimate:*

$$|\hat{f}_k - c_k(f)| \le \frac{c}{r\, N^{r+1}} \Big(\big(\frac{1}{2} + \frac{k}{N}\big)^{-r} + \big(\frac{1}{2} - \frac{k}{N}\big)^{-r} \Big) \tag{3.9}$$

for $|k| < \frac{N}{2}$.

Proof The estimate (3.7) immediately follows from the aliasing formula (3.6) by triangle inequality. With the assumption (3.8), formula (3.7) implies that

$$\begin{aligned} |\hat{f}_k - c_k(f)| &\le \sum_{\ell=1}^{\infty} \Big(|c_{k+\ell N}(f)| + |c_{k-\ell N}(f)| \Big) \\ &\le \frac{c}{N^{r+1}} \sum_{\ell=1}^{\infty} \Big(\big|\ell + \frac{k}{N}\big|^{-r-1} + \big|\ell - \frac{k}{N}\big|^{-r-1} \Big). \end{aligned}$$

For $|s| < \frac{1}{2}$ and $\ell \in \mathbb{N}$, it holds by the Hermite–Hadamard inequality that

$$\big(\ell + s\big)^{-r-1} \le \int_{\ell-1/2}^{\ell+1/2} (x+s)^{-r-1}\, \mathrm{d}x,$$

since the function $g(x) = (x+s)^{-r-1}$, $x \in \big[\frac{1}{2}, \infty\big)$, is convex. Hence,

$$\sum_{\ell=1}^{\infty} \big(\ell + s\big)^{-r-1} \le \int_{1/2}^{\infty} (x+s)^{-r-1}\, \mathrm{d}x = \frac{1}{r} \big(\frac{1}{2} + s\big)^{-r},$$

since for $s = \pm \frac{k}{N}$ with $|k| < \frac{N}{2}$ we have $|s| < \frac{1}{2}$. This completes the proof of (3.9). ■

3.1.2 Computation of Fourier Series and Fourier Transforms

First we study the computation of a trigonometric polynomial $p \in \mathscr{T}_n$, $n \in \mathbb{N}$, on a uniform grid of $[0, 2\pi)$. Choosing $N \in \mathbb{N}$ with $N \geq 2n + 1$, we want to calculate the value of $p = \sum_{j=-n}^{n} c_j \, \mathrm{e}^{\mathrm{i}\, j \cdot}$ at all grid points $\frac{2\pi k}{N}$ for $k = 0, \ldots, N-1$, where the coefficients $c_j \in \mathbb{C}$ are given. Using (3.2) we have:

$$p\big(\frac{2\pi k}{N}\big) = \sum_{j=-n}^{n} c_j \, \mathrm{e}^{2\pi \mathrm{i}\, jk/N} = \sum_{j=-n}^{n} c_j \, w_N^{-jk} = \sum_{j=0}^{n} c_{-j} \, w_N^{jk} + \sum_{j=1}^{n} c_j \, w_N^{(N-j)k}$$
$$= \sum_{j=0}^{n} c_{-j} \, w_N^{jk} + \sum_{j=N-n}^{N-1} c_{N-j} \, w_N^{jk} \,. \tag{3.10}$$

Introducing the entries

$$d_j := \begin{cases} c_{-j} & j = 0, \ldots, n\,, \\ 0 & j = n+1, \ldots, N-n-1\,, \\ c_{N-j} & j = N-n, \ldots, N-1\,, \end{cases}$$

we obtain:

$$p\big(\frac{2\pi k}{N}\big) = \sum_{j=0}^{N-1} d_j \, w_N^{jk}\,, \quad k = 0, \ldots, N-1\,, \tag{3.11}$$

which can be interpreted as a discrete Fourier transform of length N.

Now, in order to evaluate a Fourier series on a uniform grid of an interval of length 2π, we use their partial sum $p = S_n f$ as an approximation. For smooth functions, the Fourier series converges rapidly; see Theorem 1.39, such that we can approximate the Fourier series very accurately by proper choosing the polynomial degree n.

Next we sketch the computation of the Fourier transform $\hat{f}$ of a given function $f \in L_1(\mathbb{R}) \cap C(\mathbb{R})$. Since $f(x) \to 0$ for $|x| \to \infty$, we obtain for sufficiently large $n \in \mathbb{N}$ that

$$\hat{f}(\omega) = \int_{\mathbb{R}} f(x) \, \mathrm{e}^{-\mathrm{i} x\omega} \, \mathrm{d}x \approx \int_{-n\pi}^{n\pi} f(x) \, \mathrm{e}^{-\mathrm{i} x\omega} \, \mathrm{d}x\,, \quad \omega \in \mathbb{R}\,.$$

Using the uniform grid $\{\frac{2\pi j}{N} : j = -\frac{nN}{2}, \ldots, \frac{nN}{2} - 1\}$ of the interval $[-n\pi, n\pi)$ for even $n \in \mathbb{N}$, we approximate the integral by the rectangle rule:

$$\int_{-n\pi}^{n\pi} f(x) \, \mathrm{e}^{-\mathrm{i} x\omega} \, \mathrm{d}x \approx \frac{2\pi}{N} \sum_{j=-nN/2}^{nN/2-1} f\big(\frac{2\pi j}{N}\big) \, \mathrm{e}^{-2\pi \mathrm{i}\, j\omega/N}\,.$$

For $\omega = \frac{k}{n}$ with $k = -\frac{nN}{2}, \ldots, \frac{nN}{2} - 1$, we find the following approximate value of $\hat{f}\big(\frac{k}{n}\big)$:

$$\hat{f}\big(\frac{k}{n}\big) \approx \frac{2\pi}{N} \sum_{j=-nN/2}^{nN/2-1} f\big(\frac{2\pi j}{N}\big)\, w_{nN}^{jk} . \tag{3.12}$$

This is indeed a discrete Fourier transform of length nN, when we shift the summation index similarly as in (3.10). Here, as before when evaluating the Fourier coefficients, the approximation is only acceptable for the $|k| \le \frac{nN}{2}$, since the approximate values of $\hat{f}\big(\frac{k}{n}\big)$ are nN-periodic, while the Fourier transform decays with $\lim_{|\omega|\to\infty} |\hat{f}(\omega)| = 0$.

Remark 3.5 In Sects. 9.1 and 9.3, we will present more accurate methods for the computation of Fourier transforms and Fourier coefficients. The sampling of trigonometric polynomials on a nonuniform grid will be considered in Chap. 7. □

3.1.3 Trigonometric Polynomial Interpolation

Finally we consider the *interpolation by a trigonometric polynomial* on a uniform grid of $[0,\, 2\pi)$. First we discuss the trigonometric interpolation with an *odd* number of equidistant nodes $x_k := \frac{2\pi k}{2n+1} \in [0,\, 2\pi)$, $k = 0, \ldots, 2n$.

Lemma 3.6 *Let $n \in \mathbb{N}$ be given and $N = 2n + 1$. For arbitrary $p_k \in \mathbb{C}$, $k = 0, \ldots, N-1$, there exists a unique trigonometric polynomial of degree n:*

$$p = \sum_{\ell=-n}^{n} c_\ell\, \mathrm{e}^{\mathrm{i}\ell\,\cdot} \in \mathscr{T}_n \tag{3.13}$$

satisfying the interpolation conditions

$$p(x_k) = p\big(\frac{2\pi k}{2n+1}\big) = p_k\,, \quad k = 0, \ldots, 2n\,. \tag{3.14}$$

The coefficients $c_\ell \in \mathbb{C}$ of (3.13) *are given by*

$$c_\ell = \frac{1}{2n+1} \sum_{k=0}^{2n} p_k\, w_N^{\ell k}\,, \quad \ell = -n, \ldots, n\,. \tag{3.15}$$

Using the Dirichlet kernel D_n, the interpolating trigonometric polynomial (3.13) *can be written in the form*

$$p = \frac{1}{2n+1} \sum_{k=0}^{2n} p_k \, D_n(\cdot - x_k)\,. \tag{3.16}$$

Proof

1. From the interpolation conditions (3.14), it follows by (3.2) that solving the trigonometric interpolation problem is equivalent to solving the system of linear equations:

$$p(x_k) = \sum_{\ell=-n}^{n} c_\ell \, w_N^{-\ell k} = p_k\,, \quad k = 0, \ldots, 2n\,. \tag{3.17}$$

Assume that $c_\ell \in \mathbb{C}$ solve (3.17). Then, by Lemma 3.2 we obtain:

$$\sum_{k=0}^{2n} p_k \, w_N^{jk} = \sum_{k=0}^{2n} \Big(\sum_{\ell=-n}^{n} c_\ell \, w_N^{-k\ell} \Big) w_N^{jk}$$
$$= \sum_{\ell=-n}^{n} c_\ell \Big(\sum_{k=0}^{2n} w_N^{(j-\ell)k} \Big) = (2n+1)\, c_j\,.$$

Hence, any solution of (3.17) has to be of the form (3.15).

On the other hand, for c_ℓ given by (3.15), we find by Lemma 3.2 that for $k = 0, \ldots, 2n$

$$p\big(\frac{2\pi k}{2n+1}\big) = p(x_k) = \sum_{\ell=-n}^{n} c_\ell \, w_N^{-\ell k} = \frac{1}{2n+1} \sum_{\ell=-n}^{n} \Big(\sum_{j=0}^{2n} p_j \, w_N^{j\ell} \Big) w_N^{-\ell k}$$
$$= \frac{1}{2n+1} \sum_{j=0}^{2n} p_j \Big(\sum_{\ell=-n}^{n} w_N^{(j-k)\ell} \Big) = p_k\,.$$

Thus, the linear system (3.17) is uniquely solvable.

2. From (3.13) and (3.15), it follows by $c_{-\ell} = c_{N-\ell}$, $\ell = 1, \ldots, n$ that

$$p(x) = c_0 + \sum_{\ell=1}^{n} \big(c_\ell \, \mathrm{e}^{\mathrm{i}\,\ell x} + c_{N-\ell} \, \mathrm{e}^{-\mathrm{i}\,\ell x} \big)$$
$$= \frac{1}{2n+1} \sum_{k=0}^{2n} p_k \Big(1 + \sum_{\ell=1}^{n} \big(\mathrm{e}^{\mathrm{i}\,\ell(x-x_k)} + \mathrm{e}^{-\mathrm{i}\,\ell(x-x_k)} \big) \Big)$$

and we conclude (3.16) by the definition (1.21) of the Dirichlet kernel D_n. ∎

Formula (3.16) particularly implies that the *trigonometric Lagrange polynomials* $\ell_k \in \mathscr{T}_n$ with respect to the uniform grid $\{x_k = \frac{2\pi k}{2n+1} : k = 0, \ldots, 2n\}$ are given by

$$\ell_k := \frac{1}{2n+1}\, D_n(\cdot - x_k)\,, \quad k = 0, \ldots, 2n\,.$$

By Lemma 3.6 the trigonometric Lagrange polynomials ℓ_k, $k = 0, \ldots, N-1$, form a basis of $\mathscr{T}_n$ and satisfy the interpolation conditions:

$$\ell_k(x_j) = \delta_{j-k}\,, \quad j,\, k = 0, \ldots, 2n\,.$$

Further, the trigonometric Lagrange polynomials generate a *partition of unity*, since (3.16) yields for $p = 1$ that

$$1 = \frac{1}{2n+1} \sum_{k=0}^{2n} p_k\, D_n(\cdot - x_k) = \sum_{k=0}^{2n} \ell_k\,. \tag{3.18}$$

Now we consider the trigonometric interpolation for an *even* number of equidistant nodes $x_k^* := \frac{\pi k}{n} \in [0,\, 2\pi)$, $k = 0, \ldots, 2n-1$.

Lemma 3.7 *Let $n \in \mathbb{N}$ be given and $N = 2n$. For arbitrary $p_k^* \in \mathbb{C}$, $k = 0, \ldots, 2n-1$, there exists a unique trigonometric polynomial of the special form:*

$$p^* = \sum_{\ell=1-n}^{n-1} c_\ell^*\, \mathrm{e}^{\mathrm{i}\ell\,\cdot} + \frac{1}{2}\, c_n^* \left(\mathrm{e}^{\mathrm{i}n\,\cdot} + \mathrm{e}^{-\mathrm{i}n\,\cdot}\right) \in \mathscr{T}_n \tag{3.19}$$

satisfying the interpolation conditions

$$p^*\Big(\frac{2\pi k}{2n}\Big) = p_k^*\,, \quad k = 0, \ldots, 2n-1\,. \tag{3.20}$$

The coefficients $c_\ell^* \in \mathbb{C}$ of (3.19) *are given by*

$$c_\ell^* = \frac{1}{2n} \sum_{k=0}^{2n-1} p_k^*\, w_N^{\ell k}\,, \quad \ell = 1-n, \ldots, n\,. \tag{3.21}$$

The interpolating trigonometric polynomial (3.19) *can be written in the form*

$$p^* = \frac{1}{2n} \sum_{k=0}^{2n-1} p_k^*\, D_n^*(\cdot - x_k^*)\,, \tag{3.22}$$

where $D_n^* := D_n - \cos(n\,\cdot)$ denotes the modified nth Dirichlet kernel.

A proof of Lemma 3.7 is omitted here, since this result can be similarly shown as Lemma 3.6.

Remark 3.8 By $\sin(nx_k^*) = \sin(\pi k) = 0$ for $k = 0, \ldots, 2n-1$, each trigonometric polynomial $p^* + c\,\sin(n\,\cdot)$ with arbitrary $c \in \mathbb{C}$ is a solution of the trigonometric interpolation problem (3.20). Therefore, the restriction to trigonometric polynomials of the special form (3.19) is essential for the unique solvability of the trigonometric interpolation problem (3.20). □

Formula (3.22) implies that the *trigonometric Lagrange polynomials* $\ell_k^* \in \mathcal{T}_n$ with respect to the uniform grid $\{x_k^* = \frac{\pi k}{n} : k = 0, \ldots, 2n-1\}$ are given by

$$\ell_k^* := \frac{1}{2n}\, D_n^*(\cdot - x_k^*)\,, \quad k = 0, \ldots, 2n-1\,.$$

By Lemma 3.7 the $2n$ trigonometric Lagrange polynomials ℓ_k^* are linearly independent, but they do not form a basis of $\mathcal{T}_n$, since $\dim \mathcal{T}_n = 2n+1$.

Finally, we study efficient and numerically stable representations of the interpolating trigonometric polynomials (3.16) and (3.22). For that purpose we employ the *barycentric formulas for interpolating trigonometric polynomials* introduced by P. Henrici [201]. For a survey on barycentric interpolation formulas, we refer to [44] and [415, pp. 33–41].

Theorem 3.9 (Barycentric Formulas for Trigonometric Interpolation) *Let $n \in \mathbb{N}$ be given. For odd integer $N = 2n+1$ and $x_k = \frac{2\pi k}{2n+1}$, $k = 0, \ldots, 2n$, the interpolating trigonometric polynomial in* (3.16) *satisfies the* barycentric formula*:*

$$p(x) = \begin{cases} \dfrac{\displaystyle\sum_{k=0}^{2n} (-1)^k\, p_k \,\operatorname{cosec} \frac{x - x_k}{2}}{\displaystyle\sum_{k=0}^{2n} (-1)^k \,\operatorname{cosec} \frac{x - x_k}{2}} & x \in \mathbb{R} \setminus \bigcup_{k=0}^{2n} \left(\{x_k\} + 2\pi\mathbb{Z}\right), \\ p_j & x \in \{x_j\} + 2\pi\mathbb{Z}\,,\ j = 0, \ldots, 2n\,. \end{cases}$$

For even integer $N = 2n$ and $x_k^* = \frac{\pi k}{n}$, $k = 0, \ldots, 2n-1$, the interpolating trigonometric polynomial (3.22) *satisfies the* barycentric formula*:*

$$p^*(x) = \begin{cases} \dfrac{\displaystyle\sum_{k=0}^{2n-1} (-1)^k\, p_k^* \,\cot \frac{x - x_k^*}{2}}{\displaystyle\sum_{k=0}^{2n-1} (-1)^k \,\cot \frac{x - x_k^*}{2}} & x \in \mathbb{R} \setminus \bigcup_{k=0}^{2n-1} \left(\{x_k^*\} + 2\pi\mathbb{Z}\right), \\ p_j^* & x \in \{x_j^*\} + 2\pi\mathbb{Z}\,,\ j = 0, \ldots, 2n-1\,. \end{cases}$$

Proof

1. Let $N = 2n+1$ be odd. We consider $x \in \mathbb{R} \setminus \bigcup_{k=0}^{2n} \big(\{x_k\} + 2\pi\mathbb{Z}\big)$. From (3.16) and (1.22) it follows for all x_k, $k = 0, \ldots, 2n$, that

$$p(x) = \frac{1}{2n+1} \sum_{k=0}^{2n} p_k \frac{\sin \frac{(2n+1)(x-x_k)}{2}}{\sin \frac{x-x_k}{2}} = \frac{\sin\big(n+\frac{1}{2}\big)x}{2n+1} \sum_{k=0}^{2n} (-1)^k\, p_k \,\frac{1}{\sin \frac{x-x_k}{2}}$$

$$= \frac{\sin\big(n+\frac{1}{2}\big)x}{2n+1} \sum_{k=0}^{2n} (-1)^k\, p_k \,\operatorname{cosec} \frac{x-x_k}{2}\,. \tag{3.23}$$

Especially for $p = 1$, we obtain:

$$1 = \frac{\sin\big(n+\frac{1}{2}\big)x}{2n+1} \sum_{k=0}^{2n} (-1)^k \operatorname{cosec} \frac{x-x_k}{2}\,. \tag{3.24}$$

Dividing (3.23) by (3.24) and canceling the common factor, we find the first barycentric formula.

2. For even $N = 2n$, we consider $x \in \mathbb{R} \setminus \bigcup_{k=0}^{2n-1} \big(\{x_k^*\} + 2\pi\mathbb{Z}\big)$. By (1.22) the modified nth Dirichlet kernel can be written in the form

$$D_n^*(x) = D_n(x) - \cos(nx) = \frac{\sin\big(n+\frac{1}{2}\big)x}{\sin\frac{x}{2}} - \cos(nx) = \sin(nx)\,\cot\frac{x}{2}\,.$$

Then from (3.22) it follows that

$$p^*(x) = \frac{\sin(nx)}{2n} \sum_{k=0}^{2n-1} (-1)^k\, p_k^* \,\cot \frac{x-x_k^*}{2}\,. \tag{3.25}$$

Especially for $p^* = 1$ we receive

$$1 = \frac{\sin(nx)}{2n} \sum_{k=0}^{2n-1} (-1)^k \cot \frac{x-x_k^*}{2}\,. \tag{3.26}$$

Dividing (3.25) by (3.26) and canceling the common factor, we get the second barycentric formula. ∎

For an efficient numerical realization of these barycentric formulas, one can apply the fast summation technique presented in Sect. 7.5.

The results of the four problems presented in (3.3), (3.11), (3.12), and (3.15) have almost the same structure and motivate the detailed study of the DFT in the next section. For fast algorithms for the DFT, we refer to Chap. 5.

3.2 Fourier Matrices and Discrete Fourier Transforms

In this section we present the main properties of Fourier matrices and discrete Fourier transforms.

3.2.1 Fourier Matrices

For fixed $N \in \mathbb{N}$, we consider the vectors $\mathbf{a} = (a_j)_{j=0}^{N-1}$ and $\mathbf{b} = (b_j)_{j=0}^{N-1}$ with components $a_j,\, b_j \in \mathbb{C}$. As usual, the inner product and the Euclidean norm in the vector space $\mathbb{C}^N$ are defined by

$$\langle \mathbf{a}, \mathbf{b} \rangle := \mathbf{a}^\top \overline{\mathbf{b}} = \sum_{j=0}^{N-1} a_j\, \overline{b}_j\,, \quad \|\mathbf{a}\|_2 := \sqrt{\sum_{j=0}^{N-1} |a_j|^2}\,.$$

Lemma 3.10 *Let $N \in \mathbb{N}$ be given and $w_N := \mathrm{e}^{-2\pi\,\mathrm{i}/N}$. Then, the set of the* exponential vectors $\mathbf{e}_k := \big(w_N^{jk}\big)_{j=0}^{N-1}$, $k = 0, \ldots, N-1$, *forms an orthogonal basis of $\mathbb{C}^N$, where $\|\mathbf{e}_k\|_2 = \sqrt{N}$ for each $k = 0, \ldots, N-1$. Any $\mathbf{a} \in \mathbb{C}^N$ can be represented in the form*

$$\mathbf{a} = \frac{1}{N} \sum_{k=0}^{N-1} \langle \mathbf{a},\, \mathbf{e}_k \rangle\, \mathbf{e}_k\,. \tag{3.27}$$

The set of complex conjugate exponential vectors $\overline{\mathbf{e}}_k = \big(w_N^{-jk}\big)_{j=0}^{N-1}$, $k = 0, \ldots, N-1$, forms also an orthogonal basis of $\mathbb{C}^N$.

Proof For $k, \ell \in \{0, \ldots, N-1\}$, the inner product $\langle \mathbf{e}_k,\, \mathbf{e}_\ell \rangle$ can be calculated by Lemma 3.2 such that

$$\langle \mathbf{e}_k,\, \mathbf{e}_\ell \rangle = \sum_{j=0}^{N-1} w_N^{(k-\ell)j} = N\, \delta_{(k-\ell) \bmod N}\,.$$

Thus $\{\mathbf{e}_k : k = 0, \ldots, N-1\}$ is an orthogonal basis of $\mathbb{C}^N$, because the N exponential vectors $\mathbf{e}_k$ are linearly independent and $\dim \mathbb{C}^N = N$. Consequently, each vector $\mathbf{a} \in \mathbb{C}^N$ can be expressed in the form (3.27). Analogously, the vectors $\overline{\mathbf{e}}_k$, $k = 0, \ldots, N-1$, form an orthogonal basis of $\mathbb{C}^N$. ■

The N-by-N *Fourier matrix* is defined by

$$\mathbf{F}_N := \big(w_N^{jk}\big)_{j,k=0}^{N-1} = \begin{pmatrix} 1 & 1 & \ldots & 1 \\ 1 & w_N & \ldots & w_N^{N-1} \\ \vdots & \vdots & & \vdots \\ 1 & w_N^{N-1} & \ldots & w_N \end{pmatrix} .$$

Due to the properties of the primitive Nth root of unity w_N, the Fourier matrix $\mathbf{F}_N$ consists of only N distinct entries. Obviously, $\mathbf{F}_N$ is symmetric, $\mathbf{F}_N = \mathbf{F}_N^\top$, but $\mathbf{F}_N$ is not Hermitian for $N > 2$. The columns of $\mathbf{F}_N$ are the vectors $\mathbf{e}_k$ of the orthogonal basis of $\mathbb{C}^N$ such that by Lemma 3.10

$$\mathbf{F}_N^\top \overline{\mathbf{F}}_N = N\,\mathbf{I}_N\,, \tag{3.28}$$

where $\mathbf{I}_N$ denotes the N-by-N identity matrix. Hence, the scaled Fourier matrix $\frac{1}{\sqrt{N}}\,\mathbf{F}_N$ is unitary.

The linear map from $\mathbb{C}^N$ to $\mathbb{C}^N$, which is represented as the matrix vector product

$$\hat{\mathbf{a}} = \mathbf{F}_N\,\mathbf{a} = \big(\langle \mathbf{a},\, \overline{\mathbf{e}}_k\rangle\big)_{k=0}^{N-1}\,, \quad \mathbf{a} \in \mathbb{C}^N\,,$$

is called *discrete Fourier transform of length N* and abbreviated by DFT(N). The transformed vector $\hat{\mathbf{a}} = (\hat{a}_k)_{k=0}^{N-1}$ is called the discrete Fourier transform (DFT) of $\mathbf{a} = (a_j)_{j=0}^{N-1}$, and we have:

$$\hat{a}_k = \langle \mathbf{a},\, \overline{\mathbf{e}}_k\rangle = \sum_{j=0}^{N-1} a_j\, w_N^{jk}\,, \quad k = 0, \ldots, N-1\,. \tag{3.29}$$

In practice, one says that the DFT(N) maps from *time domain* $\mathbb{C}^N$ to *frequency domain* $\mathbb{C}^N$.

The main importance of the DFT arises from the fact that there exist fast and numerically stable algorithms for its computation; see Chap. 5.

Example 3.11 For $N \in \{2,\, 3,\, 4\}$, we obtain the Fourier matrices:

$$\mathbf{F}_2 = \begin{pmatrix} 1 & 1 \\ 1 & -1 \end{pmatrix}, \quad \mathbf{F}_3 = \begin{pmatrix} 1 & 1 & 1 \\ 1 & w_3 & \overline{w}_3 \\ 1 & \overline{w}_3 & w_3 \end{pmatrix}, \quad \mathbf{F}_4 = \begin{pmatrix} 1 & 1 & 1 & 1 \\ 1 & -\mathrm{i} & -1 & \mathrm{i} \\ 1 & -1 & 1 & -1 \\ 1 & \mathrm{i} & -1 & -\mathrm{i} \end{pmatrix}$$

with $w_3 = -\frac{1}{2} - \frac{\sqrt{3}}{2}\,\mathrm{i}$. Figure 3.1 displays both real and imaginary parts of the Fourier matrix $\mathbf{F}_{16}$ and a plot of the second row of both below. In the grayscale images, white corresponds to the value 1 and black corresponds to -1. □

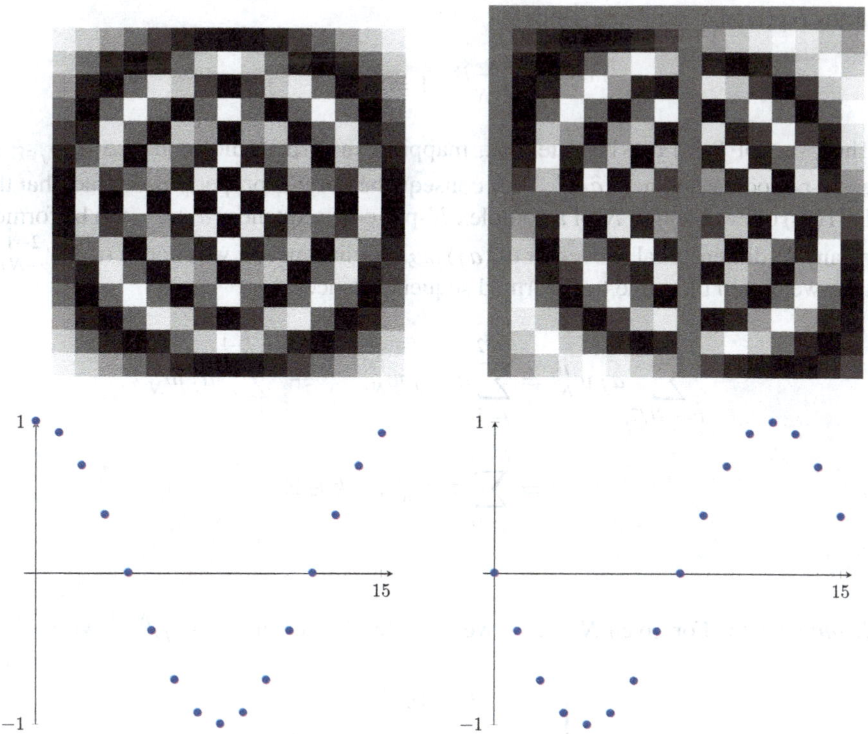

Fig. 3.1 Grayscale images of real and imaginary part of the Fourier matrix $\mathbf{F}_{16}$ (top left and right) and the values of the corresponding second rows (bottom)

Remark 3.12 Let $N \in \mathbb{N}$ with $N > 1$ be given. Obviously we can compute the values

$$\hat{a}_k = \sum_{j=0}^{N-1} a_j\, w_N^{jk} \tag{3.30}$$

for all $k \in \mathbb{Z}$. From

$$w_N^{j(k+N)} = w_N^{jk} \cdot 1 = w_N^{jk}\,, \quad k \in \mathbb{Z}\,,$$

we observe that the resulting sequence $(\hat{a}_k)_{k\in\mathbb{Z}}$ is N-periodic. The same is true for the inverse DFT(N). For a given vector $(\hat{a}_k)_{k=0}^{N-1}$ the sequence $(a_j)_{j\in\mathbb{Z}}$ with

$$a_j = \frac{1}{N} \sum_{k=0}^{N-1} \hat{a}_k\, w_N^{-jk}\,, \quad j \in \mathbb{Z}\,,$$

is an N-periodic sequence, since

$$w_N^{-(j+N)k} = w_N^{-jk} \cdot 1 = w_N^{-jk}, \quad j \in \mathbb{Z}.$$

Thus, the DFT(N) can be extended, mapping an N-periodic sequence $(a_j)_{j\in\mathbb{Z}}$ to an N-periodic sequence $(\hat{a}_k)_{k=0}^{N-1}$. A consequence of this property is the fact that the DFT(N) of even length N of a complex N-periodic sequence $(a_j)_{j\in\mathbb{Z}}$ can be formed by any N-dimensional subvector of $(a_j)_{j\in\mathbb{Z}}$. For instance, if we choose $(a_j)_{j=-N/2}^{N/2-1}$, then we obtain the same transformed sequence, since

$$\begin{aligned}\sum_{j=-N/2}^{N/2-1} a_j\, w_N^{jk} &= \sum_{j=1}^{N/2} a_{N-j}\, w_N^{(N-j)k} + \sum_{j=0}^{N/2-1} a_j\, w_N^{jk} \\ &= \sum_{j=0}^{N-1} a_j\, w_N^{jk}, \quad k \in \mathbb{Z}.\end{aligned}$$

□

Example 3.13 For given $N \in 2\,\mathbb{N}$, we consider the vector $\mathbf{a} = (a_j)_{j=0}^{N-1}$ with

$$a_j = \begin{cases} 0 & j \in \{0, \frac{N}{2}\}, \\ 1 & j = 1, \ldots, \frac{N}{2} - 1, \\ -1 & j = \frac{N}{2} + 1, \ldots, N-1. \end{cases}$$

We determine the DFT(N) of $\mathbf{a}$, i.e., $\hat{\mathbf{a}} = (\hat{a}_k)_{k=0}^{N-1}$. Obviously, we have $\hat{a}_0 = 0$. For $k \in \{1, \ldots, N-1\}$, we obtain:

$$\hat{a}_k = \sum_{j=1}^{N/2-1} w_N^{jk} - \sum_{j=N/2+1}^{N-1} w_N^{jk} = \big(1 - (-1)^k\big) \sum_{j=1}^{N/2-1} w_N^{jk}$$

and hence $\hat{a}_k = 0$ for even k. Using

$$\sum_{j=1}^{N/2-1} x^j = \frac{x - x^{N/2}}{1-x}, \quad x \neq 1,$$

it follows for $x = w_N^k$ with odd k that

$$\hat{a}_k = 2\,\frac{w_N^k - w_N^{kN/2}}{1 - w_N^k} = 2\,\frac{w_N^k + 1}{1 - w_N^k} = 2\,\frac{w_{2N}^k + w_{2N}^{-k}}{w_{2N}^{-k} - w_{2N}^k} = -2\mathrm{i}\,\cot\frac{\pi k}{N}.$$

Thus, we receive

$$\hat{a}_k = \begin{cases} 0 & k = 0, 2, \ldots, N-2, \\ -2\mathrm{i} \cot \frac{\pi k}{N} & k = 1, 3, \ldots, N-1. \end{cases}$$

□

Example 3.14 For given $N \in \mathbb{N} \setminus \{1\}$, we consider the vector $\mathbf{a} = (a_j)_{j=0}^{N-1}$ with

$$a_j = \begin{cases} \frac{1}{2} & j = 0, \\ \frac{j}{N} & j = 1, \ldots, N-1. \end{cases}$$

Note that the related N-periodic sequence $(a_j)_{j \in \mathbb{Z}}$ with $a_j = a_{j \bmod N}$, $j \in \mathbb{Z}$, is a sawtooth sequence. Now we calculate the DFT(N) of $\mathbf{a}$, i.e., $\hat{\mathbf{a}} = (\hat{a}_k)_{k=0}^{N-1}$. Obviously, we have:

$$\hat{a}_0 = \frac{1}{2} + \frac{1}{N} \sum_{j=1}^{N-1} j = \frac{1}{2} + \frac{N\,(N-1)}{2\,N} = \frac{N}{2}\,.$$

Using the sum formula

$$\sum_{j=1}^{N-1} j\,x^j = -\frac{(N-1)\,x^N}{1-x} + \frac{x - x^N}{(1-x)^2}\,, \qquad x \neq 1\,,$$

we obtain for $x = w_N^k$ with $k \in \{1, \ldots, N-1\}$ that

$$\sum_{j=1}^{N-1} j\,w_N^{jk} = \frac{-(N-1)}{1-w_N^k} + \frac{w_N^k - 1}{(1-w_N^k)^2} = -\frac{N}{1-w_N^k}$$

and hence

$$\hat{a}_k = \frac{1}{2} + \frac{1}{N} \sum_{j=1}^{N-1} j\,w_N^{jk} = \frac{1}{2} - \frac{1}{1-w_N^k} = -\frac{1+w_N^k}{2\,(1-w_N^k)} = \frac{\mathrm{i}}{2} \cot \frac{\pi k}{N}\,.$$

□

Remark 3.15 In the literature, the Fourier matrix is not consistently defined. In particular, the normalization constants differ, and one finds, for example, $\big(w_N^{-jk}\big)_{j,k=0}^{N-1}$, $\frac{1}{\sqrt{N}} \big(w_N^{jk}\big)_{j,k=0}^{N-1}$, $\frac{1}{N} \big(w_N^{jk}\big)_{j,k=0}^{N-1}$, and $\big(w_N^{jk}\big)_{j,k=1}^{N}$. Consequently, there exist different forms of the DFT(N). For the sake of clarity, we emphasize that the DFT(N)

is differently defined in the respective package documentations. For instance, *Mathematica* uses the DFT(N) of the form

$$\hat{a}_k = \frac{1}{\sqrt{N}} \sum_{j=1}^{N} a_j\, w_N^{-(j-1)(k-1)}\,, \quad k = 1, \ldots, N\,.$$

In *Matlab*, the DFT(N) is defined by

$$\hat{a}_{k+1} = \sum_{j=0}^{N-1} a_{j+1}\, w_N^{jk}\,, \quad k = 0, \ldots, N-1\,.$$

In *Maple*, the definition of DFT(N) reads as follows:

$$\hat{a}_k = \frac{1}{\sqrt{N}} \sum_{j=1}^{N} a_j\, w_N^{(j-1)(k-1)}\,, \quad k = 1, \ldots, N\,.$$

□

3.2.2 Properties of Fourier Matrices

Now we describe the main properties of Fourier matrices.

Theorem 3.16 *The Fourier matrix* $\mathbf{F}_N$ *is invertible and its inverse reads as follows:*

$$\mathbf{F}_N^{-1} = \frac{1}{N}\, \overline{\mathbf{F}}_N = \frac{1}{N}\, \big(w_N^{-jk}\big)_{j,k=0}^{N-1}\,. \tag{3.31}$$

The corresponding DFT *is a bijective map on* $\mathbb{C}^N$. *The* inverse DFT of length N *is given by the matrix-vector product:*

$$\mathbf{a} = \mathbf{F}_N^{-1}\, \hat{\mathbf{a}} = \frac{1}{N}\, \big(\langle \hat{\mathbf{a}},\, \mathbf{e}_k\rangle\big)_{k=0}^{N-1}\,, \quad \hat{\mathbf{a}} \in \mathbb{C}^N$$

such that

$$a_j = \frac{1}{N}\, \langle \hat{\mathbf{a}},\, \mathbf{e}_j\rangle = \frac{1}{N} \sum_{k=0}^{N-1} \hat{a}_k\, w_N^{-jk}\,, \quad j = 0, \ldots, N-1\,. \tag{3.32}$$

Proof Relation (3.31) follows immediately from (3.28). Consequently, the DFT(N) is bijective on $\mathbb{C}^N$. ■

Lemma 3.17 *The Fourier matrix* $\mathbf{F}_N$ *satisfies*

$$\mathbf{F}_N^2 = N\,\mathbf{J}'_N\,, \quad \mathbf{F}_N^4 = N^2\,\mathbf{I}_N\,, \tag{3.33}$$

with the flip matrix

$$\mathbf{J}'_N := \big(\delta_{(j+k)\bmod N}\big)_{j,k=0}^{N-1} = \begin{pmatrix} 1 & & & \\ & & & 1 \\ & & \iddots & \\ & 1 & & \end{pmatrix}.$$

Further we have:

$$\mathbf{F}_N^{-1} = \frac{1}{N}\,\mathbf{J}'_N\,\mathbf{F}_N = \frac{1}{N}\,\mathbf{F}_N\,\mathbf{J}'_N\,. \tag{3.34}$$

Proof Let $\mathbf{F}_N^2 = \big(c_{j,\ell}\big)_{j,\ell=0}^{N-1}$. Using Lemma 3.2, we find

$$c_{j,\ell} = \sum_{k=0}^{N-1} w_N^{jk}\,w_N^{k\ell} = \sum_{k=0}^{N-1} w_N^{(j+\ell)k} = N\,\delta_{(j+\ell)\bmod N}\,.$$

and hence $\mathbf{F}_N^2 = N\,\mathbf{J}'_N$. From $(\mathbf{J}'_N)^2 = \mathbf{I}_N$, it follows that

$$\mathbf{F}_N^4 = \mathbf{F}_N^2\,\mathbf{F}_N^2 = (N\,\mathbf{J}'_N)\,(N\,\mathbf{J}'_N) = N^2\,(\mathbf{J}'_N)^2 = N^2\,\mathbf{I}_N\,.$$

By $N\,\mathbf{F}_N\,\mathbf{J}'_N = N\,\mathbf{J}'_N\,\mathbf{F}_N = \mathbf{F}_N^3$ and $\mathbf{F}_N^4 = N^2\,\mathbf{I}_N$, we finally obtain:

$$\mathbf{F}_N^{-1} = \frac{1}{N^2}\,\mathbf{F}_N^3 = \frac{1}{N}\,\mathbf{F}_N\,\mathbf{J}'_N = \frac{1}{N}\,\mathbf{J}'_N\,\mathbf{F}_N\,.$$

This completes the proof. ■

Using (3.34), the inverse DFT(N) can be computed by the *same* algorithm as the DFT(N) employing a reordering and a scaling.

Remark 3.18 The application of the flip matrix $\mathbf{J}'_N$ to a vector $\mathbf{a} = (a_k)_{k=0}^{N-1}$ provides the vector:

$$\mathbf{J}'_N\,\mathbf{a} = \big(a_{(N-j)\bmod N}\big)_{j=0}^{N-1} = (a_0,\,a_{N-1},\,\ldots,\,a_1)^\top\,,$$

i.e., the components of $\mathbf{a}$ are "flipped." From Lemma 3.17 it follows immediately that

$$\mathbf{F}_N\hat{\mathbf{a}} = \mathbf{F}_N^2\mathbf{a} = N\,\mathbf{J}'_N\,\mathbf{a} = N\,\big(a_{(N-j)\bmod N}\big)_{j=0}^{N-1}\,.$$

This equality can be used as a simple test for correctness of a DFT algorithm. □

Now we want to study the spectral properties of the Fourier matrix in a more detailed manner. For that purpose, let the *counter-identity matrix* $\mathbf{J}_N$ be defined by

$$\mathbf{J}_N := \big(\delta_{(j+k+1) \bmod N}\big)_{j,k=0}^{N-1} = \begin{pmatrix} & & 1 \\ & \cdot^{\cdot^{\cdot}} & \\ 1 & & \end{pmatrix}$$

having nonzero entries 1 only on the main counter-diagonal. Then, $\mathbf{J}_N\mathbf{a}$ provides the reversed vector:

$$\mathbf{J}_N\mathbf{a} = \big(a_{(-j-1) \bmod N}\big)_{j=0}^{N-1} = (a_{N-1},\, a_{N-2},\, \ldots,\, a_1,\, a_0)^\top .$$

First we obtain the following result about the eigenvalues of $\mathbf{F}_N$.

Lemma 3.19 *For $N \in \mathbb{N} \setminus \{1\}$, the Fourier matrix $\mathbf{F}_N$ possesses at most the four distinct eigenvalues $\sqrt{N}$, $-\sqrt{N}$, $-\mathrm{i}\sqrt{N}$, and $\mathrm{i}\sqrt{N}$.*

Proof Let $\lambda \in \mathbb{C}$ be an eigenvalue of $\mathbf{F}_N$ with the corresponding eigenvector $\mathbf{a} \in \mathbb{C}^N$, i.e., $\mathbf{F}_N\mathbf{a} = \lambda\,\mathbf{a}$, $\mathbf{a} \neq \mathbf{0}$. Hence by (3.33) we obtain $N^2\,\mathbf{a} = \mathbf{F}_N^4\mathbf{a} = \lambda^4\,\mathbf{a}$ such that $\lambda^4 - N^2 = 0$. Hence, possible eigenvalues of $\mathbf{F}_N$ are $\sqrt{N}$, $-\sqrt{N}$, $-\mathrm{i}\sqrt{N}$, and $\mathrm{i}\sqrt{N}$. ■

Now, we want to determine the exact multiplicities of the distinct eigenvalues of the Fourier matrix $\mathbf{F}_N$. We start by considering the characteristic polynomial of the matrix $\mathbf{F}_N^2$.

Lemma 3.20 *For $N \in \mathbb{N}$ with $N \geq 4$, we have:*

$$\det(\lambda\mathbf{I}_N - \mathbf{F}_N^2) = \begin{cases} (\lambda - N)^{(N+2)/2}\,(\lambda + N)^{(N-2)/2} & N \text{ even}, \\ (\lambda - N)^{(N+1)/2}\,(\lambda + N)^{(N-1)/2} & N \text{ odd}. \end{cases}$$

Proof

1. For $n \in \mathbb{N}$ we consider the matrix $\mathbf{T}_n(\lambda) := \lambda\,\mathbf{I}_n - N\,\mathbf{J}_n$. For even n, the matrix is of the form

$$\mathbf{T}_n(\lambda) = \begin{pmatrix} \lambda & & & & & -N \\ & \ddots & & & \cdot^{\cdot^{\cdot}} & \\ & & \lambda & -N & & \\ & & -N & \lambda & & \\ & \cdot^{\cdot^{\cdot}} & & & \ddots & \\ -N & & & & & \lambda \end{pmatrix}.$$

We show for even n by induction that

$$\det \mathbf{T}_n(\lambda) = (\lambda - N)^{n/2}\,(\lambda + N)^{n/2}\,. \tag{3.35}$$

Indeed, for $n = 2$ we have:

$$\det \mathbf{T}_2(\lambda) = \det \begin{pmatrix} \lambda & -N \\ -N & \lambda \end{pmatrix} = (\lambda - N)\,(\lambda + N)\,.$$

Assume now that (3.35) is true for an even $n \in \mathbb{N}$. Expanding $\det \mathbf{T}_{n+2}(\lambda)$ with respect to the 0-th column, we obtain:

$$\begin{aligned} \det \mathbf{T}_{n+2}(\lambda) &= \lambda \,\det \begin{pmatrix} \mathbf{T}_n(\lambda) & \\ & \lambda \end{pmatrix} + N \,\det \begin{pmatrix} & -N \\ \mathbf{T}_n(\lambda) & \end{pmatrix} \\ &= (\lambda^2 - N^2)\,\det \mathbf{T}_n(\lambda) \\ &= (\lambda - N)^{(n+2)/2}\,(\lambda + N)^{(n+2)/2}\,. \end{aligned}$$

2. By (3.33) we obtain:

$$\det(\lambda \mathbf{I}_N - \mathbf{F}_N^2) = \det(\lambda \mathbf{I}_N - N\,\mathbf{J}_N') = \det \begin{pmatrix} \lambda - N & \\ & T_{N-1}(\lambda) \end{pmatrix}.$$

For odd N, we find

$$\det(\lambda \mathbf{I}_N - N\,\mathbf{J}_N') = (\lambda - N)\det T_{N-1}(\lambda) = (\lambda - N)^{(N+1)/2}(\lambda + N)^{(N-1)/2}.$$

For even N we expand $T_{N-1}(\lambda)$ with respect to the $\frac{(N-1)}{2}$-th column that contains only one nonzero value $\lambda - N$ in the center. We obtain:

$$\det(\lambda \mathbf{I}_N - N\,\mathbf{J}_N') = (\lambda - N)^2 \det T_{N-2}(\lambda) = (\lambda - N)^{(N+2)/2}(\lambda + N)^{(N-2)/2}.$$

This completes the proof. ∎

Since $\det(\lambda \mathbf{I}_N - \mathbf{F}_N^2)$ is the characteristic polynomial of $\mathbf{F}_N^2$, we can conclude already the multiplicities of the eigenvalues of $\mathbf{F}_N^2$. We denote the multiplicities of the eigenvalues $\sqrt{N}$, $-\sqrt{N}$, $-\mathrm{i}\sqrt{N}$, and $\mathrm{i}\sqrt{N}$ of $\mathbf{F}_N$ by m_1, m_2, m_3, and m_4. Thus, the eigenvalue N of $\mathbf{F}_N^2$ possesses the multiplicity $m_1 + m_2$, and the eigenvalue $-N$ has the multiplicity $m_3 + m_4$. Lemma 3.20 implies

$$m_1 + m_2 = \begin{cases} (N+2)/2 & N \text{ even}\,, \\ (N+1)/2 & N \text{ odd}\,, \end{cases} \tag{3.36}$$

$$m_3 + m_4 = \begin{cases} (N-2)/2 & N \text{ even}\,, \\ (N-1)/2 & N \text{ odd}\,. \end{cases} \tag{3.37}$$

In order to deduce m_1, m_2, m_3, and m_4, we also consider the trace and the determinant of $\mathbf{F}_N$. We recall that the *trace* of a square matrix $\mathbf{A}_N = \big(a_{j,k}\big)_{j,k=0}^{N-1} \in \mathbb{C}^{N\times N}$ is equal to the sum of its eigenvalues and that the determinant $\det \mathbf{A}_N$ is the product of its eigenvalues, i.e.,

$$\operatorname{tr} \mathbf{A}_N = \sum_{j=0}^{N-1} a_{j,j} = \sum_{j=0}^{N-1} \lambda_j , \quad \det \mathbf{A}_N = \prod_{j=0}^{N-1} \lambda_j . \tag{3.38}$$

For the Fourier matrix $\mathbf{F}_N$, we obtain:

$$\operatorname{tr} \mathbf{F}_N = \sqrt{N}\,(m_1 - m_2) + \mathrm{i}\,\sqrt{N}\,(m_4 - m_3) . \tag{3.39}$$

Now we calculate the trace of $\mathbf{F}_N$:

$$\operatorname{tr} \mathbf{F}_N = \sum_{j=0}^{N-1} \mathrm{e}^{-2\pi \mathrm{i} j^2/N} .$$

The above sum is called *quadratic Gauss sum*. The following computation of the quadratic Gauss sum is based on ideas of Dirichlet and is a nice application of 1-periodic Fourier series.

Lemma 3.21 *For $N \in \mathbb{N} \setminus \{1\}$, we have:*

$$\operatorname{tr} \mathbf{F}_N = \sqrt{N}\,(1 + \mathrm{i}^N)(1 - \mathrm{i}) . \tag{3.40}$$

Proof

1. We consider the 1-periodic function h, which is given on $[0,\ 1)$ by

$$h(x) := \sum_{j=0}^{N-1} \mathrm{e}^{-2\pi \mathrm{i}\,(x+j)^2/N} , \quad x \in [0,\ 1) .$$

Then, we obtain:

$$\begin{aligned} \frac{1}{2}\big(h(0+0) + h(0-0)\big) &= \frac{1}{2}\big(h(0+0) + h(1-0)\big) \\ &= \frac{1}{2} \sum_{j=0}^{N-1} \big(\mathrm{e}^{-2\pi \mathrm{i}\, j^2/N} + \mathrm{e}^{-2\pi \mathrm{i}\,(j+1)^2/N}\big) \\ &= \frac{1}{2} + \sum_{j=1}^{N-1} \mathrm{e}^{-2\pi \mathrm{i}\, j^2/N} + \frac{1}{2} = \operatorname{tr} \mathbf{F}_N . \end{aligned}$$

The function h is piecewise continuously differentiable and can be represented by its 1-periodic Fourier series:

$$h(x) = \sum_{k \in \mathbb{Z}} c_k^{(1)}(h)\, \mathrm{e}^{2\pi \mathrm{i} k x} .$$

By Theorem 1.34 of Dirichlet–Jordan, this Fourier series converges at the point $x = 0$ to

$$\sum_{k\in\mathbb{Z}} c_k^{(1)}(h) = \frac{1}{2}\big(h(0+0) + h(0-0)\big) = \operatorname{tr} \mathbf{F}_N\,.$$

2. Now we calculate the Fourier coefficients:

$$\begin{aligned} c_k^{(1)}(h) &= \sum_{j=0}^{N-1} \int_0^1 \mathrm{e}^{-2\pi\mathrm{i}\,(u+j)^2/N}\,\mathrm{e}^{-2\pi\mathrm{i}ku}\,\mathrm{d}u = \int_0^N \mathrm{e}^{-2\pi\mathrm{i}\,y^2/N}\,\mathrm{e}^{-2\pi\mathrm{i}ky}\,\mathrm{d}y \\ &= \mathrm{e}^{\pi\mathrm{i}Nk^2/2} \int_0^N \mathrm{e}^{-2\pi\mathrm{i}\,(y+kN/2)^2/N}\,\mathrm{d}y\,. \end{aligned}$$

Thus, we obtain for even $k = 2r$, $r \in \mathbb{Z}$,

$$c_{2r}^{(1)}(h) = \int_0^N \mathrm{e}^{-2\pi\mathrm{i}\,(y+rN)^2/N}\,\mathrm{d}y\,,$$

and for odd $k = 2r+1$, $r \in \mathbb{Z}$,

$$c_{2r+1}^{(1)}(h) = \mathrm{e}^{\pi\mathrm{i}\,N/2} \int_0^N \mathrm{e}^{-2\pi\mathrm{i}\,(y+rN+N/2)^2/N}\,\mathrm{d}y = \mathrm{i}^N \int_{N/2}^{3N/2} \mathrm{e}^{-2\pi\mathrm{i}\,(y+rN)^2/N}\,\mathrm{d}y\,.$$

Consequently, it holds:

$$\begin{aligned} \operatorname{tr} \mathbf{F}_N &= \sum_{r\in\mathbb{Z}} c_{2r}^{(1)}(h) + \sum_{r\in\mathbb{Z}} c_{2r+1}^{(1)}(h) \\ &= (1+\mathrm{i}^N) \int_{\mathbb{R}} \mathrm{e}^{-2\pi\mathrm{i}\,y^2/N}\,\mathrm{d}y = 2\,(1+\mathrm{i}^N) \int_0^\infty \mathrm{e}^{-2\pi\mathrm{i}\,y^2/N}\,\mathrm{d}y \\ &= (1+\mathrm{i}^N)\sqrt{\frac{2N}{\pi}} \int_0^\infty \mathrm{e}^{-\mathrm{i}\,v^2}\,\mathrm{d}v\,. \end{aligned}$$

3. The integral

$$\int_0^\infty \mathrm{e}^{-\mathrm{i}\,v^2}\,\mathrm{d}v = \frac{1}{2}\sqrt{\frac{\pi}{2}}\,(1-\mathrm{i})$$

can be computed by Cauchy's integral theorem. For this we consider the sector with arbitrary radius R and angle $-\frac{\pi}{4}$. Then for the holomorphic function $\mathrm{e}^{-\mathrm{i}z^2}$, $z \in \mathbb{C}$, it holds:

$$\int_0^R \mathrm{e}^{-\mathrm{i}v^2}\,\mathrm{d}v = I_1(R) + I_2(R)\,,$$

where

$$I_1(R) := \int_{\Gamma_1} \mathrm{e}^{-\mathrm{i}\,z^2}\,\mathrm{d}z\,, \quad \Gamma_1 := \{z = \frac{1}{\sqrt{2}}\,(1-\mathrm{i})\,t \,:\, 0 \le t \le R\}\,,$$

$$I_2(R) := \int_{\Gamma_2} \mathrm{e}^{-\mathrm{i}\,z^2}\,\mathrm{d}z\,, \quad \Gamma_2 := \{z = R\,\mathrm{e}^{\mathrm{i}t} \,:\, -\frac{\pi}{4} \le t \le 0\}\,.$$

The integral $I_2(R)$ tends to zero as $R \to \infty$, since

$$|I_2(R)| \le \int_0^{\pi/4} R\,\mathrm{e}^{-R^2 \sin t}\,\mathrm{d}t \le R \int_0^{\pi/4} \mathrm{e}^{-2R^2 t/\pi}\,\mathrm{d}t = -\frac{\pi}{R}\,\mathrm{e}^{-R^2/2} + \frac{\pi}{R}\,.$$

The integral

$$I_1(R) = \frac{1}{\sqrt{2}}\,(1-\mathrm{i}) \int_0^R \mathrm{e}^{-t^2}\,\mathrm{d}t$$

tends to $\frac{1}{\sqrt{2}}\,(1-\mathrm{i})\,\frac{\sqrt{\pi}}{2}$ as $R \to \infty$ by Example 2.6. Note that $\frac{1}{\sqrt{2}}\,(1-\mathrm{i})$ is the square root of $-\mathrm{i}$ with positive real part.

Hence, it follows that

$$\operatorname{tr}\mathbf{F}_N = (1+\mathrm{i}^N)\,\sqrt{\frac{2N}{\pi}}\,\frac{1}{\sqrt{2}}\,(1-\mathrm{i})\,\frac{\sqrt{\pi}}{2} = \sqrt{N}\,(1+\mathrm{i}^N)(1-\mathrm{i})\,.$$

This completes the proof. ∎

Theorem 3.22 *For $N \in \mathbb{N}$ with $N > 4$, the Fourier matrix $\mathbf{F}_N$ has four distinct eigenvalues $\sqrt{N}$, $-\sqrt{N}$, $-\mathrm{i}\,\sqrt{N}$, and $\mathrm{i}\,\sqrt{N}$ with corresponding multiplicities m_1, m_2, m_3, and m_4 given in the table.*

N	m_1	m_2	m_3	m_4	$\det \mathbf{F}_N$
$4n$	$n+1$	n	n	$n-1$	$\mathrm{i}\,(-1)^{n+1}\,N^{N/2}$
$4n+1$	$n+1$	n	n	n	$(-1)^n\,N^{N/2}$
$4n+2$	$n+1$	$n+1$	n	n	$(-1)^{n+1}\,N^{N/2}$
$4n+3$	$n+1$	$n+1$	$n+1$	n	$\mathrm{i}\,(-1)^n\,N^{N/2}$

Proof Each integer $N > 4$ can be represented in the form $N = 4n + k$ with $n \in \mathbb{N}$ and $k \in \{0,\ 1,\ 2,\ 3\}$. From (3.39) and (3.40),

$$\begin{aligned}\operatorname{tr} \mathbf{F}_N &= \sqrt{N}\,(1+\mathrm{i}^N)(1-\mathrm{i}) = \frac{\sqrt{N}}{2}\,\big(1-\mathrm{i}+\mathrm{i}^N-\mathrm{i}^{N+1}\big)\\ &= \sqrt{N}\,(m_1-m_2)+\mathrm{i}\sqrt{N}\,(m_4-m_3)\,,\end{aligned}$$

it follows that

$$m_1 - m_2 = \begin{cases} 1 & N=4n\,,\\ 1 & N=4n+1\,,\\ 0 & N=4n+2\,,\\ 0 & N=4n+3\,,\end{cases} \tag{3.41}$$

$$m_4 - m_3 = \begin{cases} -1 & N=4n\,,\\ 0 & N=4n+1\,,\\ 0 & N=4n+2\,,\\ -1 & N=4n+3\,.\end{cases} \tag{3.42}$$

Using the linear equations (3.36) and (3.41), we compute m_1 and m_2. Analogously we determine m_3 and m_4 by solving the linear equations (3.37) and (3.42). Finally, we conclude

$$\det \mathbf{F}_N = (-1)^{m_2+m_3}\,\mathrm{i}^{m_3+m_4}\,N^{N/2}$$

as the product of all N eigenvalues of $\mathbf{F}_N$. ■

Remark 3.23 For the computation of the eigenvectors of the Fourier matrix $\mathbf{F}_N$, we refer to [286, 294]. □

3.2.3 DFT and Cyclic Convolutions

The *cyclic convolution* of the vectors $\mathbf{a} = (a_k)_{k=0}^{N-1}$, $\mathbf{b} = (b_k)_{k=0}^{N-1} \in \mathbb{C}^N$ is defined as the vector $\mathbf{c} = (c_n)_{n=0}^{N-1} := \mathbf{a} * \mathbf{b} \in \mathbb{C}^N$ with the components:

$$c_n = \sum_{k=0}^{N-1} a_k\, b_{(n-k) \bmod N} = \sum_{k=0}^{n} a_k\, b_{n-k} + \sum_{k=n+1}^{N-1} a_k\, b_{N+n-k}\,, \quad n=0,\ldots,N-1\,.$$

The cyclic convolution in $\mathbb{C}^N$ is a commutative, associative, and distributive operation with the unity $\mathbf{b}_0 = \big(\delta_{j \bmod N}\big)_{j=0}^{N-1} = (1,0,\ldots,0)^\top$ which is the so-called pulse vector.

The *forward-shift matrix* $\mathbf{V}_N$ is defined by

$$\mathbf{V}_N := \big(\delta_{(j-k-1) \bmod N}\big)_{j,k=0}^{N-1} = \begin{pmatrix} & & & 1 \\ 1 & & & \\ & \ddots & & \\ & & 1 & \end{pmatrix}.$$

The application of $\mathbf{V}_N$ to a vector $\mathbf{a} = (a_k)_{k=0}^{N-1}$ provides the forward-shifted vector:

$$\mathbf{V}_N \mathbf{a} = \big(a_{(j-1) \bmod N}\big)_{j=0}^{N-1} = (a_{N-1},\, a_0,\, a_1,\, \ldots,\, a_{N-2})^\top .$$

Hence, we obtain:

$$\mathbf{V}_N^2 := \big(\delta_{(j-k-2) \bmod N}\big)_{j,k=0}^{N-1} = \begin{pmatrix} & & & 1 & \\ & & & & 1 \\ 1 & & & & \\ & \ddots & & & \\ & & 1 & & \end{pmatrix}$$

and

$$\mathbf{V}_N^2 \mathbf{a} = \big(a_{(j-2) \bmod N}\big)_{j=0}^{N-1} = (a_{N-2},\, a_{N-1},\, a_0,\, \ldots,\, a_{N-3})^\top .$$

Further we have $\mathbf{V}_N^N = \mathbf{I}_N$ and

$$\mathbf{V}_N^\top = \mathbf{V}_N^{-1} = \mathbf{V}_N^{N-1} = \begin{pmatrix} & 1 & & \\ & & \ddots & \\ & & & 1 \\ 1 & & & \end{pmatrix},$$

which is called *backward-shift matrix*, since

$$\mathbf{V}_N^{-1} \mathbf{a} = \big(a_{(j+1) \bmod N}\big)_{j=0}^{N-1} = (a_1,\, a_2,\, \ldots,\, a_{N-1},\, a_0)^\top .$$

is the backward-shifted vector of $\mathbf{a}$.

The matrix $\mathbf{I}_N - \mathbf{V}_N$ is the *cyclic difference matrix*, since

$$(\mathbf{I}_N - \mathbf{V}_N)\, \mathbf{a} = \big(a_j - a_{(j-1) \bmod N}\big)_{j=0}^{N-1} = (a_0 - a_{N-1},\, a_1 - a_0,\, \ldots,\, a_{N-1} - a_{N-2})^\top .$$

We observe that

$$\mathbf{I}_N + \mathbf{V}_N + \mathbf{V}_N^2 + \ldots + \mathbf{V}_N^{N-1} = \big(1\big)_{j,k=0}^{N-1}.$$

We want to characterize all linear maps $\mathbf{H}_N$ from $\mathbb{C}^N$ to $\mathbb{C}^N$ which are *shift-invariant*, i.e., satisfy

$$\mathbf{H}_N\,(\mathbf{V}_N\,\mathbf{a}) = \mathbf{V}_N\,(\mathbf{H}_N\,\mathbf{a})$$

for all $\mathbf{a} \in \mathbb{C}^N$. Thus, we have $\mathbf{H}_N\,\mathbf{V}_N^k = \mathbf{V}_N^k\,\mathbf{H}_N$, $k = 0, \ldots, N-1$. Shift-invariant maps play an important role for signal filtering. We show that any shift-invariant, linear map $\mathbf{H}_N$ can be represented by a cyclic convolution.

Lemma 3.24 *Each shift-invariant, linear map $\mathbf{H}_N$ from $\mathbb{C}^N$ to $\mathbb{C}^N$ can be represented in the form*

$$\mathbf{H}_N\,\mathbf{a} = \mathbf{a} * \mathbf{h}\,, \quad \mathbf{a} \in \mathbb{C}^N\,,$$

where $\mathbf{h} := \mathbf{H}_N\,\mathbf{b}_0$ is the impulse response vector *of the pulse vector $\mathbf{b}_0$.*

Proof Let $\mathbf{b}_k := \big(\delta_{(j-k) \bmod N}\big)_{j=0}^{N-1}$, $k = 0, \ldots, N-1$, be the standard basis vectors of $\mathbb{C}^N$. Then $\mathbf{b}_k = \mathbf{V}_N^k\,\mathbf{b}_0$, $k = 0, \ldots, N-1$. An arbitrary vector $\mathbf{a} = (a_k)_{k=0}^{N-1} \in \mathbb{C}^N$ can be represented in the standard basis as

$$\mathbf{a} = \sum_{k=0}^{N-1} a_k\,\mathbf{b}_k = \sum_{k=0}^{N-1} a_k\,\mathbf{V}_N^k\,\mathbf{b}_0\,.$$

Applying the linear, shift-invariant map $\mathbf{H}_N$ to this vector $\mathbf{a}$, we get:

$$\begin{aligned}\mathbf{H}_N\,\mathbf{a} &= \sum_{k=0}^{N-1} a_k\,\mathbf{H}_N(\mathbf{V}_N^k\,\mathbf{b}_0) = \sum_{k=0}^{N-1} a_k\,\mathbf{V}_N^k\,(\mathbf{H}_N\,\mathbf{b}_0) = \sum_{k=0}^{N-1} a_k\,\mathbf{V}_N^k\,\mathbf{h} \\ &= \big(\mathbf{h} \mid \mathbf{V}_N\,\mathbf{h} \mid \ldots \mid \mathbf{V}_N^{N-1}\,\mathbf{h}\big)\,\mathbf{a}\end{aligned}$$

that means

$$\begin{aligned}\mathbf{H}_N\,\mathbf{a} &= \begin{pmatrix} h_0 & h_{N-1} & \ldots & h_2 & h_1 \\ h_1 & h_0 & \ldots & h_3 & h_2 \\ \vdots & \vdots & & \vdots & \vdots \\ h_{N-1} & h_{N-2} & \ldots & h_1 & h_0 \end{pmatrix} \begin{pmatrix} a_0 \\ a_1 \\ \vdots \\ a_{N-1} \end{pmatrix} \\ &= \Big(\sum_{k=0}^{N-1} a_k\,h_{(n-k) \bmod N}\Big)_{n=0}^{N-1} = \mathbf{a} * \mathbf{h}\,.\end{aligned}$$

This completes the proof. ■

Now we present the basic properties of DFT(N) and start with an example.

Example 3.25 Let $\mathbf{b}_k = \big(\delta_{(j-k) \bmod N}\big)_{j=0}^{N-1}$, $k = 0, \ldots, N-1$, be the standard basis vectors of $\mathbb{C}^N$ and let $\mathbf{e}_k = \big(w_N^{jk}\big)_{j=0}^{N-1}$, $k = 0, \ldots, N-1$, be the exponential vectors in Lemma 3.10 that form the columns of $\mathbf{F}_N$. Then, we obtain for $k = 0, \ldots, N-1$ that

$$\mathbf{F}_N\, \mathbf{b}_k = \mathbf{e}_k\,, \quad \mathbf{F}_N\, \mathbf{e}_k = \mathbf{F}_N^2\, \mathbf{b}_k = N\, \mathbf{J}_N'\, \mathbf{b}_k = N\, \mathbf{b}_{(-k) \bmod N}\,.$$

In particular, we observe that the sparse vectors $\mathbf{b}_k$ are transformed into non-sparse vectors $\mathbf{e}_k$, since all components of $\mathbf{e}_k$ are nonzero. Further we obtain that for all $k = 0, \ldots, N-1$

$$\mathbf{F}_N\, \mathbf{V}_N\, \mathbf{b}_k = \mathbf{F}_N\, \mathbf{b}_{(k+1) \bmod N} = \mathbf{e}_{(k+1) \bmod N} = \mathbf{M}_N\, \mathbf{F}_N\, \mathbf{b}_k\,,$$

where $\mathbf{M}_N := \operatorname{diag} \mathbf{e}_1$ is the so-called *modulation matrix* which generates a modulation or frequency shift by the property $\mathbf{M}_N\, \mathbf{e}_k = \mathbf{e}_{(k+1) \bmod N}$. Consequently, we have

$$\mathbf{F}_N\, \mathbf{V}_N = \mathbf{M}_N\, \mathbf{F}_N \tag{3.43}$$

and more generally $\mathbf{F}_N\, \mathbf{V}_N^k = \mathbf{M}_N^k\, \mathbf{F}_N$, $k = 1, \ldots, N-1$. Transposing the last equation for $k = N-1$, we obtain:

$$\mathbf{V}_N^\top\, \mathbf{F}_N = \mathbf{V}_N^{-1}\, \mathbf{F}_N = \mathbf{F}_N\, \mathbf{M}_N, \quad \mathbf{V}_N\, \mathbf{F}_N = \mathbf{F}_N\, \mathbf{M}_N^{-1}\,. \tag{3.44}$$

□

Theorem 3.26 (Properties of DFT(N)) *The* DFT(N) *possesses the following properties:*

1. Linearity: *For all* $\mathbf{a}, \mathbf{b} \in \mathbb{C}^N$ *and* $\alpha \in \mathbb{C}$, *we have:*

$$(\mathbf{a} + \mathbf{b})\hat{} = \hat{\mathbf{a}} + \hat{\mathbf{b}}\,, \quad (\alpha\, \mathbf{a})\hat{} = \alpha\, \hat{\mathbf{a}}\,.$$

2. Inversion: *For all* $\mathbf{a} \in \mathbb{C}^N$, *we have:*

$$\mathbf{a} = \mathbf{F}_N^{-1}\, \hat{\mathbf{a}} = \frac{1}{N}\, \overline{\mathbf{F}}_N\, \hat{\mathbf{a}} = \frac{1}{N}\, \mathbf{J}_N'\, \mathbf{F}_N\, \hat{\mathbf{a}}\,.$$

3. Flipping property: *For all* $\mathbf{a} \in \mathbb{C}^N$, *we have:*

$$(\mathbf{J}_N'\, \mathbf{a})\hat{} = \mathbf{J}_N'\, \hat{\mathbf{a}}\,, \quad (\overline{\mathbf{a}})\hat{} = \mathbf{J}_N'\, \overline{\hat{\mathbf{a}}}\,.$$

4. Shifting in time and frequency domain: *For all* $\mathbf{a} \in \mathbb{C}^N$, *we have:*

$$(\mathbf{V}_N\, \mathbf{a})\hat{} = \mathbf{M}_N\, \hat{\mathbf{a}}\,, \quad (\mathbf{M}_N^{-1}\, \mathbf{a})\hat{} = \mathbf{V}_N\, \hat{\mathbf{a}}\,.$$

5. Cyclic convolution in time and frequency domain: *For all* $\mathbf{a}$, $\mathbf{b} \in \mathbb{C}^N$, *we have:*

$$(\mathbf{a} * \mathbf{b})\hat{} = \hat{\mathbf{a}} \circ \hat{\mathbf{b}}\,, \quad N\,(\mathbf{a} \circ \mathbf{b})\hat{} = \hat{\mathbf{a}} * \hat{\mathbf{b}}\,,$$

where $\mathbf{a} \circ \mathbf{b} := (a_k\, b_k)_{k=0}^{N-1}$ *denotes the* componentwise product *of the vectors* $\mathbf{a} = (a_k)_{k=0}^{N-1}$ *and* $\mathbf{b} = (b_k)_{k=0}^{N-1}$.

6. Parseval equality: *For all* $\mathbf{a}$, $\mathbf{b} \in \mathbb{C}^N$ *we have*

$$\frac{1}{N}\,\langle \hat{\mathbf{a}},\, \hat{\mathbf{b}}\rangle = \langle \mathbf{a},\, \mathbf{b}\rangle\,, \quad \frac{1}{N}\,\|\hat{\mathbf{a}}\|_2^2 = \|\mathbf{a}\|_2^2\,.$$

7. Difference property in time and frequency domain: *For all* $\mathbf{a} \in \mathbb{C}^N$ *we have*

$$\big((\mathbf{I}_N - \mathbf{V}_N)\,\mathbf{a}\big)\hat{} = (\mathbf{I}_N - \mathbf{M}_N)\,\hat{\mathbf{a}}\,, \quad \big((\mathbf{I}_N - \mathbf{M}_N^{-1})\,\mathbf{a}\big)\hat{} = (\mathbf{I}_N - \mathbf{V}_N)\,\hat{\mathbf{a}}\,.$$

8. Permutation property: *Let* $p \in \mathbb{Z}$ *and* N *be relatively prime. Assume that* $q \in \mathbb{Z}$ *satisfies the condition* $(p\,q)$ mod $N = 1$ *and that the* DFT(N) *of* $(a_j)_{j=0}^{N-1} \in \mathbb{C}^N$ *is equal to* $(\hat{a}_k)_{k=0}^{N-1}$. *Then the* DFT(N) *of the permuted vector* $\big(a_{(pj) \bmod N}\big)_{j=0}^{N-1}$ *is equal to the permuted vector* $\big(\hat{a}_{(qk) \bmod N}\big)_{k=0}^{N-1}$.

Proof

1. The linearity follows from the definition of the DFT(N).
2. The second property is obtained from (3.31) and (3.34).
3. By (3.31) and (3.34), we have $\mathbf{F}_N\,\mathbf{J}'_N = \mathbf{J}'_N\,\mathbf{F}_N = \overline{\mathbf{F}}_N$ and hence

$$\begin{aligned}(\mathbf{J}'_N\,\mathbf{a})\hat{} &= \mathbf{F}_N\,\mathbf{J}'_N\,\mathbf{a} = \mathbf{J}'_N\,\mathbf{F}_N\,\mathbf{a} = \mathbf{J}'_N\,\hat{\mathbf{a}}\,,\\ (\overline{\mathbf{a}})\hat{} &= \mathbf{F}_N\,\overline{\mathbf{a}} = \overline{\overline{\mathbf{F}}_N\,\mathbf{a}} = \overline{\mathbf{J}'_N\,\mathbf{F}_N\,\mathbf{a}} = \mathbf{J}'_N\,\overline{\hat{\mathbf{a}}}\,.\end{aligned}$$

4. From (3.43) and (3.44), it follows that

$$\begin{aligned}(\mathbf{V}_N\,\mathbf{a})\hat{} &= \mathbf{F}_N\,\mathbf{V}_N\,\mathbf{a} = \mathbf{M}_N\,\mathbf{F}_N\,\mathbf{a} = \mathbf{M}_N\,\hat{\mathbf{a}}\,,\\ (\mathbf{M}_N^{-1}\,\mathbf{a})\hat{} &= \mathbf{F}_N\,\mathbf{M}_N^{-1}\,\mathbf{a} = \mathbf{V}_N\,\mathbf{F}_N\,\mathbf{a} = \mathbf{V}_N\,\hat{\mathbf{a}}\,.\end{aligned}$$

5. Let $\mathbf{c} = \mathbf{a} * \mathbf{b}$ be the cyclic convolution of $\mathbf{a}$ and $\mathbf{b}$ with the components

$$c_j = \sum_{n=0}^{N-1} a_n\, b_{(j-n) \bmod N}\,, \quad j = 0, \ldots, N-1\,.$$

We calculate the components of $\hat{\mathbf{c}} = \big(\hat{c}_k\big)_{k=0}^{N-1}$ and obtain for $k = 0, \ldots, N-1$:

$$\begin{aligned}\hat{c}_k &= \sum_{j=0}^{N-1}\Big(\sum_{n=0}^{N-1} a_n\, b_{(j-n) \bmod N}\Big)\, w_N^{jk} \\ &= \sum_{n=0}^{N-1} a_n\, w_N^{nk}\Big(\sum_{j=0}^{N-1} b_{(j-n) \bmod N}\, w_N^{((j-n) \bmod N)\,k}\Big) \\ &= \Big(\sum_{n=0}^{N-1} a_n\, w_N^{nk}\Big)\, \hat{b}_k = \hat{a}_k\, \hat{b}_k\,.\end{aligned}$$

Now let $\mathbf{c} = \mathbf{a} \circ \mathbf{b} = (a_j\, b_j)_{j=0}^{N-1}$. Using the second property, we get:

$$a_j = \frac{1}{N}\sum_{k=0}^{N-1} \hat{a}_k\, w_N^{-jk}\,, \quad b_j = \frac{1}{N}\sum_{\ell=0}^{N-1} \hat{b}_\ell\, w_N^{-j\ell}\,, \quad j = 0, \ldots, N-1\,.$$

Thus, we obtain that for $j = 0, \ldots, N-1$

$$\begin{aligned}c_j = a_j\, b_j &= \frac{1}{N^2}\Big(\sum_{k=0}^{N-1} \hat{a}_k\, w_N^{-jk}\Big)\Big(\sum_{\ell=0}^{N-1} \hat{b}_\ell\, w_N^{-j\ell}\Big) \\ &= \frac{1}{N^2}\sum_{k=0}^{N-1}\sum_{\ell=0}^{N-1} \hat{a}_k\, \hat{b}_\ell\, w_N^{-j(k+\ell)} \\ &= \frac{1}{N^2}\sum_{n=0}^{N-1}\Big(\sum_{k=0}^{N-1} \hat{a}_k\, \hat{b}_{(n-k) \bmod N}\Big)\, w_N^{-jn}\,,\end{aligned}$$

i.e., $\mathbf{c} = \frac{1}{N}\,\mathbf{F}_N^{-1}(\hat{\mathbf{a}} * \hat{\mathbf{b}})$ and hence $N\,\hat{\mathbf{c}} = \hat{\mathbf{a}} * \hat{\mathbf{b}}$.

6. For arbitrary $\mathbf{a}, \mathbf{b} \in \mathbb{C}^N$, we conclude

$$\langle \hat{\mathbf{a}},\, \hat{\mathbf{b}}\rangle = \mathbf{a}^\top\, \mathbf{F}_N\, \overline{\mathbf{F}}_N\, \overline{\mathbf{b}} = N\, \mathbf{a}^\top\, \overline{\mathbf{b}} = N\, \langle \mathbf{a},\, \mathbf{b}\rangle\,.$$

7. The difference properties follow directly from the shift properties.
8. Since $p \in \mathbb{Z}$ and N are relatively prime, the greatest common divisor of p and N is one. Then, there exist $q,\ M \in \mathbb{Z}$ with $p\,q + M\,N = 1$ (see [8, p. 21]). By the Euler–Fermat theorem (see [8, p. 114]), the (unique modulo N) solution of the linear congruence $p\,q \equiv 1 \pmod N$ is given by $q \equiv p^{\varphi(N)-1} \pmod N$, where $\varphi(N)$ denotes the Euler totient function.

Now we compute the DFT(N) of the permuted vector $\big(a_{(pj) \bmod N}\big)_{j=0}^{N-1}$. Then, the kth component of the transformed vector reads:

$$\sum_{j=0}^{N-1} a_{(pj) \bmod N}\, w_N^{jk}\,. \tag{3.45}$$

The value (3.45) does not change, if the sum is reordered and the summation index $j = 0, \ldots, N-1$ is replaced by $(q\,\ell) \bmod N$ with $\ell = 0, \ldots, N-1$. Indeed, by $p\,q \equiv 1 \pmod N$ and (3.4), we have:

$$\ell = (p\,q\,\ell) \bmod N = \big[\big((q\,\ell) \bmod N\big)\,p\big] \bmod N$$

and furthermore

$$w_N^{[(q\,\ell) \bmod N]k} = w_N^{q\,\ell\,k} = w_N^{\ell\,[(q\,k) \bmod N]}\,.$$

Thus, we obtain

$$\begin{aligned}\sum_{j=0}^{N-1} a_{(pj) \bmod N}\, w_N^{jk} &= \sum_{\ell=0}^{N-1} a_{(p\,q\,\ell) \bmod N}\, w_N^{q\,\ell\,k} \\ &= \sum_{j=0}^{N-1} a_\ell\, w_N^{\ell\,[(q\,k) \bmod N]} = \hat{a}_{(qk) \bmod N}\,.\end{aligned}$$

For example, in the special case $p = q = -1$, the flipped vector $\big(a_{(-j) \bmod N}\big)_{j=0}^{N-1}$ is transformed to the flipped vector $\big(\hat{a}_{(-k) \bmod N}\big)_{k=0}^{N-1}$. ■

Now we analyze the symmetry properties of DFT(N). A vector $\mathbf{a} = (a_j)_{j=0}^{N-1} \in \mathbb{C}^N$ is called *even*, if $\mathbf{a} = \mathbf{J}'_N\,\mathbf{a}$, i.e., $a_j = a_{(N-j) \bmod N}$ for all $j = 0, \ldots, N-1$, and it is called *odd*, if $\mathbf{a} = -\mathbf{J}'_N\,\mathbf{a}$, i.e., $a_j = -a_{(N-j) \bmod N}$ for all $j = 0, \ldots, N-1$. For $N = 6$ the vector $(a_0,\, a_1,\, a_2,\, a_3,\, a_2,\, a_1)^\top$ is even and $(0,\, a_1,\, a_2,\, 0,\, -a_2,\, -a_1)^\top$ is odd.

Corollary 3.27 *For* $\mathbf{a} \in \mathbb{R}^N$ *and* $\hat{\mathbf{a}} = \mathbf{F}_N\mathbf{a} = \big(\hat{a}_j\big)_{j=0}^{N-1}$, *we have:*

$$\overline{\hat{\mathbf{a}}} = \mathbf{J}'_N\,\hat{\mathbf{a}}\,,$$

i.e., $\overline{\hat{a}}_j = \hat{a}_{(N-j) \bmod N}$, $j = 0, \ldots, N-1$. *In other words,* Re $\hat{\mathbf{a}}$ *is even and* Im $\hat{\mathbf{a}}$ *is odd.*

Proof By $\mathbf{a} = \overline{\mathbf{a}} \in \mathbb{R}^N$ and $\overline{\mathbf{F}}_N = \mathbf{J}'_N\,\mathbf{F}_N$, it follows that

$$\mathbf{J}'_N\,\hat{\mathbf{a}} = \mathbf{J}'_N\,\mathbf{F}_N\,\mathbf{a} = \overline{\mathbf{F}}_N\,\mathbf{a} = \overline{\mathbf{F}}_N\,\overline{\mathbf{a}} = \overline{\hat{\mathbf{a}}}\,.$$

For $\hat{\mathbf{a}} = \text{Re}\ \hat{\mathbf{a}} + \text{i}\ \text{Im}\ \hat{\mathbf{a}}$, we obtain:

$$\overline{\hat{\mathbf{a}}} = \text{Re}\,\hat{\mathbf{a}} - \text{i}\,\text{Im}\,\hat{\mathbf{a}} = \mathbf{J}'_N\,\hat{\mathbf{a}} = \mathbf{J}'_N\,(\text{Re}\,\hat{\mathbf{a}}) + \text{i}\,\mathbf{J}'_N\,(\text{Im}\,\hat{\mathbf{a}})$$

and hence $\text{Re}\,\hat{\mathbf{a}} = \mathbf{J}'_N\,(\text{Re}\,\hat{\mathbf{a}})$ and $\text{Im}\,\hat{\mathbf{a}} = -\mathbf{J}'_N\,(\text{Im}\,\hat{\mathbf{a}})$. ■

Corollary 3.28 *If $\mathbf{a} \in \mathbb{C}^N$ is even/odd, then $\hat{\mathbf{a}} = \mathbf{F}_N \mathbf{a}$ is even/odd.*
If $\mathbf{a} \in \mathbb{R}^N$ is even, then $\hat{\mathbf{a}} = \text{Re}\,\hat{\mathbf{a}} \in \mathbb{R}^N$ is even.
If $\mathbf{a} \in \mathbb{R}^N$ is odd, then $\hat{\mathbf{a}} = \text{i}\,\text{Im}\,\hat{\mathbf{a}} \in \text{i}\,\mathbb{R}^N$ is odd.

Proof From $\mathbf{a} = \pm\mathbf{J}'_N\,\mathbf{a}$, it follows that

$$\hat{\mathbf{a}} = \mathbf{F}_N\,\mathbf{a} = \pm\mathbf{F}_N\,\mathbf{J}'_N\,\mathbf{a} = \pm\mathbf{J}'_N\,\mathbf{F}_N\,\mathbf{a} = \pm\mathbf{J}'_N\,\hat{\mathbf{a}}\,.$$

For even $\mathbf{a} \in \mathbb{R}^N$, we obtain by Corollary 3.27 that $\overline{\hat{\mathbf{a}}} = \mathbf{J}'_N\,\hat{\mathbf{a}} = \hat{\mathbf{a}}$, i.e., $\hat{\mathbf{a}} \in \mathbb{R}^N$ is even. Analogously we can show the assertion for odd $\mathbf{a} \in \mathbb{R}^N$. ■

3.3 Circulant Matrices

An N-by-N matrix

$$\text{circ}\,\mathbf{a} := \big(a_{(j-k)\bmod N}\big)_{j,k=0}^{N-1} = \begin{pmatrix} a_0 & a_{N-1} & \ldots & a_2 & a_1 \\ a_1 & a_0 & \ldots & a_3 & a_2 \\ \vdots & \vdots & & \vdots & \vdots \\ a_{N-1} & a_{N-2} & \ldots & a_1 & a_0 \end{pmatrix} \tag{3.46}$$

is called *circulant matrix* generated by $\mathbf{a} = (a_k)_{k=0}^{N-1} \in \mathbb{C}^N$. The first column of $\text{circ}\,\mathbf{a}$ is equal to $\mathbf{a}$. A circulant matrix is a special Toeplitz matrix in which the diagonals wrap around. Remember that a *Toeplitz matrix* is a structured matrix $\big(a_{j-k}\big)_{j,k=0}^{N-1}$ for given $\big(a_k\big)_{k=1-N}^{N-1} \in \mathbb{C}^{2N-1}$ such that the entries along each diagonal are constant.

Example 3.29 If $\mathbf{b}_k = (\delta_{j-k})_{j=0}^{N-1}$, $k = 0, \ldots, N-1$, denote the standard basis vectors of $\mathbb{C}^N$, then the forward-shifted matrix $\mathbf{V}_N$ is a circulant matrix, since $\mathbf{V}_N = \text{circ}\,\mathbf{b}_1$. More generally, we obtain that

$$\mathbf{V}_N^k = \text{circ}\,\mathbf{b}_k\,, \quad k = 0, \ldots, N-1\,.$$

with $\mathbf{V}_N^0 = \text{circ}\,\mathbf{b}_0 = \mathbf{I}_N$ and $\mathbf{V}_N^{N-1} = \mathbf{V}_N^{-1} = \text{circ}\,\mathbf{b}_{N-1}$. The cyclic difference matrix is also a circulant matrix, since $\mathbf{I}_N - \mathbf{V}_N = \text{circ}\,(\mathbf{b}_0 - \mathbf{b}_1)$. □

Remark 3.30 In the literature, a circulant matrix is not consistently defined. For instance in [102, p. 66] and [205, p. 33], a circulant matrix of $\mathbf{a} \in \mathbb{C}^N$ is defined by $\big(a_{(k-j)\bmod N}\big)_{j,k=0}^{N-1} = (\text{circ}\,\mathbf{a})^\top$ such that the first row is equal to $\mathbf{a}^\top$. □

Circulant matrices and cyclic convolutions of vectors in $\mathbb{C}^N$ are closely related. From Lemma 3.24 it follows that for arbitrary vectors $\mathbf{a}, \mathbf{b} \in \mathbb{C}^N$

$$(\text{circ}\,\mathbf{a})\,\mathbf{b} = \mathbf{a} * \mathbf{b}\,.$$

Using the cyclic convolution property of DFT(N) (see property 5 of Theorem 3.26), we obtain that a circulant matrix can be diagonalized by Fourier matrices.

Theorem 3.31 *For each* $\mathbf{a} \in \mathbb{C}^N$, *the circulant matrix* circ $\mathbf{a}$ *can be diagonalized by the Fourier matrix* $\mathbf{F}_N$. *We have:*

$$\text{circ}\,\mathbf{a} = \mathbf{F}_N^{-1}\,\big(\text{diag}\,(\mathbf{F}_N\mathbf{a})\big)\,\mathbf{F}_N\,. \tag{3.47}$$

Proof For any $\mathbf{b} \in \mathbb{C}^N$, we form the cyclic convolution of $\mathbf{a}$ and $\mathbf{b}$. Then by the cyclic convolution property of Theorem 3.26, we obtain that

$$\mathbf{F}_N\,\mathbf{c} = (\mathbf{F}_N\,\mathbf{a}) \circ (\mathbf{F}_N\,\mathbf{b}) = \big(\text{diag}\,(\mathbf{F}_N\,\mathbf{a})\big)\,\mathbf{F}_N\,\mathbf{b}\,.$$

and hence

$$\mathbf{c} = \mathbf{F}_N^{-1}\,\big(\text{diag}\,(\mathbf{F}_N\,\mathbf{a})\big)\,\mathbf{F}_N\,\mathbf{b}\,.$$

On the other hand, we have $\mathbf{c} = (\text{circ}\,\mathbf{a})\,\mathbf{b}$ such that for all $\mathbf{b} \in \mathbb{C}^N$

$$(\text{circ}\,\mathbf{a})\,\mathbf{b} = \mathbf{F}_N^{-1}\,\big(\text{diag}\,(\mathbf{F}_N\,\mathbf{a})\big)\,\mathbf{F}_N\,\mathbf{b}\,.$$

This completes the proof of (3.47). ■

Remark 3.32 Using the decomposition (3.47), the matrix-vector product (circ $\mathbf{a}$) $\mathbf{b}$ can be realized by employing three DFT(N) and one componentwise vector multiplication. We compute

$$(\text{circ}\,\mathbf{a})\,\mathbf{b} = \mathbf{F}_N^{-1}\,\big(\text{diag}\,(\mathbf{F}_N\mathbf{a})\big)\,\mathbf{F}_N\,\mathbf{b} = \mathbf{F}_N^{-1}\,(\text{diag}\,\hat{\mathbf{a}})\,\hat{\mathbf{b}} = \mathbf{F}_N^{-1}\,(\hat{\mathbf{a}} \circ \hat{\mathbf{b}})\,.$$

As we will see in Chap. 5, one DFT(N) of radix-2 length can be realized by $\mathcal{O}(N \log N)$ arithmetical operations such that $(\text{circ}\,\mathbf{a})\,\mathbf{b} = \mathbf{a} * \mathbf{b}$ can be computed by $\mathcal{O}(N \log N)$ arithmetical operations too. □

Corollary 3.33 *For arbitrary* $\mathbf{a} \in \mathbb{C}^N$, *the eigenvalues of* circ $\mathbf{a}$ *coincide with the components* $\hat{a}_j$, $j = 0, \ldots, N-1$, *of* $(\hat{a}_j)_{j=0}^{N-1} = \mathbf{F}_N\,\mathbf{a}$. *A right eigenvector related to the eigenvalue* $\hat{a}_j$, $j = 0, \ldots, N-1$, *is the complex conjugate exponential vector* $\bar{\mathbf{e}}_j = (w_N^{-jk})_{k=0}^{N-1}$ *and a left eigenvector of* $\hat{a}_j$ *is* $\mathbf{e}_j^\top$, *i.e.*,

$$(\text{circ}\,\mathbf{a})\,\bar{\mathbf{e}}_j = \hat{a}_j\,\bar{\mathbf{e}}_j\,, \quad \mathbf{e}_j^\top\,(\text{circ}\,\mathbf{a}) = \hat{a}_j\,\mathbf{e}_j^\top\,. \tag{3.48}$$

Proof Using (3.47), we obtain that

$$\big(\operatorname{circ}\mathbf{a}\big)\,\mathbf{F}_N^{-1} = \mathbf{F}_N^{-1}\,\operatorname{diag}\,(\hat{a}_j)_{j=0}^{N-1}\,, \quad \mathbf{F}_N\,\operatorname{circ}\mathbf{a} = \big(\operatorname{diag}\,(\hat{a}_j)_{j=0}^{N-1}\big)\,\mathbf{F}_N$$

with

$$\mathbf{F}_N = \begin{pmatrix} \mathbf{e}_0^\top \\ \mathbf{e}_1^\top \\ \vdots \\ \mathbf{e}_{N-1}^\top \end{pmatrix}, \quad \mathbf{F}_N^{-1} = \frac{1}{N}\,\big(\bar{\mathbf{e}}_0\,|\,\bar{\mathbf{e}}_1\,|\,\ldots\,|\,\bar{\mathbf{e}}_{N-1}\big)\,. \tag{3.49}$$

Hence, we conclude that (3.48) holds. Note that the eigenvalues $\hat{a}_j$ of circ **a** need not be distinct. ■

By the definition of the forward-shifted matrix $\mathbf{V}_N$, each circulant matrix (3.46) can be written in the form

$$\operatorname{circ}\mathbf{a} = \sum_{k=0}^{N-1} a_k\,\mathbf{V}_N^k\,, \tag{3.50}$$

where $\mathbf{V}_N^0 = \mathbf{V}_N^N = \mathbf{I}_N$. Therefore, $\mathbf{V}_N$ is called *basic circulant matrix.*

The representation (3.50) reveals that N-by-N circulant matrices form a commutative algebra. Linear combinations and products of circulant matrices are also circulant matrices, and products of any two circulant matrices commute. The inverse of a nonsingular circulant matrix is again a circulant matrix. The following result is very useful for the computation with circulant matrices.

Theorem 3.34 (Properties of Circulant Matrices) *For arbitrary* **a**, **b** $\in \mathbb{C}^N$ *and* $\alpha \in \mathbb{C}$, *we have:*

1. $(\operatorname{circ}\mathbf{a})^\top = \operatorname{circ}(\mathbf{J}_N'\,\mathbf{a})\,,$
2. $(\operatorname{circ}\mathbf{a}) + (\operatorname{circ}\mathbf{b}) = \operatorname{circ}(\mathbf{a}+\mathbf{b})\,, \quad \alpha\,(\operatorname{circ}\mathbf{a}) = \operatorname{circ}(\alpha\,\mathbf{a})\,,$
3. $(\operatorname{circ}\mathbf{a})\,(\operatorname{circ}\mathbf{b}) = (\operatorname{circ}\mathbf{b})\,(\operatorname{circ}\mathbf{a}) = \operatorname{circ}(\mathbf{a} * \mathbf{b})\,,$
4. circ **a** *is a normal matrix with the* spectral decomposition (3.47),
5. $\det(\operatorname{circ}\mathbf{a}) = \prod_{j=0}^{N-1}\hat{a}_j$ *with* $\big(\hat{a}_j\big)_{j=0}^{N-1} = \mathbf{F}_N\,\mathbf{a}\,.$
6. *The Moore–Penrose pseudo-inverse of* circ **a** *has the form*

$$(\operatorname{circ}\mathbf{a})^{+} = \mathbf{F}_N^{-1}\,\big(\operatorname{diag}\,(\hat{a}_j^{+})_{j=0}^{N-1}\big)\,\mathbf{F}_N\,,$$

   where $\hat{a}_j^{+} := \hat{a}_j^{-1}$ *if* $\hat{a}_j \neq 0$ *and* $\hat{a}_j^{+} := 0$ *if* $\hat{a}_j = 0$.
7. circ **a** *is invertible if and only if* $\hat{a}_j \neq 0$ *for all* $j = 0, \ldots, N-1$. *Under this condition,* $(\operatorname{circ}\mathbf{a})^{-1}$ *is the circulant matrix:*

$$(\operatorname{circ}\mathbf{a})^{-1} = \mathbf{F}_N^{-1}\,\big(\operatorname{diag}\,(\hat{a}_j^{-1})_{j=0}^{N-1}\big)\,\mathbf{F}_N\,.$$

Proof

1. Using $\mathbf{V}_N^\top = \mathbf{V}_N^{-1}$ and $\mathbf{V}_N^N = \mathbf{I}_N$, we obtain for $\mathbf{a} = (a_k)_{k=0}^{N-1} \in \mathbb{C}^N$ by (3.50) that

$$\begin{aligned}(\text{circ }\mathbf{a})^\top &= \sum_{k=0}^{N-1} a_k \left(\mathbf{V}_N^k\right)^\top = \sum_{k=0}^{N-1} a_k \left(\mathbf{V}_N^\top\right)^k = \sum_{k=0}^{N-1} a_k \mathbf{V}_N^{-k} = \sum_{k=0}^{N-1} a_k \mathbf{V}_N^{N-k} \\ &= a_0 \mathbf{I}_N + a_{N-1} \mathbf{V}_N + \ldots + a_1 \mathbf{V}_N^{N-1} = \text{circ}\,(\mathbf{J}_N' \mathbf{a})\,.\end{aligned}$$

2. The two relations follow from the definition (3.46).
3. Let $\mathbf{a} = (a_k)_{k=0}^{N-1}$, $\mathbf{b} = (b_\ell)_{\ell=0}^{N-1} \in \mathbb{C}^N$ be given. Using $\mathbf{V}_N^N = \mathbf{I}_N$, we conclude that by (3.50)

$$(\text{circ }\mathbf{a})\,(\text{circ }\mathbf{b}) = \Big(\sum_{k=0}^{N-1} a_k \mathbf{V}_N^k\Big)\Big(\sum_{\ell=0}^{N-1} b_\ell \mathbf{V}_N^\ell\Big) = \sum_{n=0}^{N-1} c_n \mathbf{V}_N^n$$

with the entries

$$c_n = \sum_{j=0}^{N-1} a_j\, b_{(n-j) \bmod N}\,, \quad n = 0, \ldots, N-1\,.$$

By $(c_n)_{n=0}^{N-1} = \mathbf{a} * \mathbf{b}$, we obtain $(\text{circ }\mathbf{a})\,(\text{circ }\mathbf{b}) = \text{circ}\,(\mathbf{a} * \mathbf{b})$. Since the cyclic convolution is commutative, the product of circulant matrices is also commutative.

4. By property 1, the conjugate transpose of circ $\mathbf{a}$ is again a circulant matrix. Since circulant matrices commute by property 3, circ $\mathbf{a}$ is a normal matrix. By (3.47) we obtain the spectral decomposition of the normal matrix:

$$\text{circ }\mathbf{a} = \frac{1}{\sqrt{N}}\,\overline{\mathbf{F}}_N \left(\text{diag}\,(\mathbf{F}_N \mathbf{a})\right) \frac{1}{\sqrt{N}}\,\mathbf{F}_N\,, \tag{3.51}$$

because $\frac{1}{\sqrt{N}}\,\mathbf{F}_N$ is unitary.

5. The determinant det (circ $\mathbf{a}$) of the matrix product (3.50) can be computed by

$$\det(\text{circ }\mathbf{a}) = \left(\det \mathbf{F}_N\right)^{-1} \Big(\prod_{j=0}^{N-1} \hat{a}_j\Big) \det \mathbf{F}_N = \prod_{j=0}^{N-1} \hat{a}_j\,.$$

6. The *Moore–Penrose pseudo-inverse* $\mathbf{A}_N^+$ of an N-by-N matrix $\mathbf{A}_N$ is uniquely determined by the properties:

$$\mathbf{A}_N\,\mathbf{A}_N^+\,\mathbf{A}_N = \mathbf{A}_N\,, \quad \mathbf{A}_N^+\,\mathbf{A}_N\,\mathbf{A}_N^+ = \mathbf{A}_N^+\,,$$

where $\mathbf{A}_N \mathbf{A}_N^+$ and $\mathbf{A}_N^+ \mathbf{A}_N$ are Hermitian. From the spectral decomposition (3.51) of circ $\mathbf{a}$, it follows that

$$(\mathrm{circ}\ \mathbf{a})^+ = \frac{1}{\sqrt{N}} \overline{\mathbf{F}}_N \left(\mathrm{diag}\,(\hat{a}_j)_{j=0}^{N-1}\right)^+ \frac{1}{\sqrt{N}} \mathbf{F}_N = \mathbf{F}_N^{-1} \left(\mathrm{diag}\,(\hat{a}_j^+)_{j=0}^{N-1}\right) \mathbf{F}_N \,.$$

The matrix circ $\mathbf{a}$ is invertible if and only if $\det(\mathrm{circ}\ \mathbf{a}) \neq 0$, i.e., if $\hat{a}_j \neq 0$ for all $j = 0, \ldots, N-1$. In this case,

$$\mathbf{F}_N^{-1} \left(\mathrm{diag}\,(\hat{a}_j^{-1})_{j=0}^{N-1}\right) \mathbf{F}_N$$

is the inverse of circ $\mathbf{a}$. ■

Circulant matrices can be characterized by the following property.

Lemma 3.35 *An N-by-N matrix $\mathbf{A}_N$ is a circulant matrix if and only if $\mathbf{A}_N$ and the basic circulant matrix $\mathbf{V}_N$ commute, i.e.,*

$$\mathbf{V}_N \mathbf{A}_N = \mathbf{A}_N \mathbf{V}_N \,. \tag{3.52}$$

Proof Each circulant matrix circ $\mathbf{a}$ with $\mathbf{a} \in \mathbb{C}^N$ can be represented in the form (3.50). Hence circ $\mathbf{a}$ commutes with $\mathbf{V}_N$.

Let $\mathbf{A}_N = \left(a_{j,k}\right)_{j,k=0}^{N-1}$ be an arbitrary N-by-N matrix with the property (3.52) such that $\mathbf{V}_N \mathbf{A}_N \mathbf{V}_N^{-1} = \mathbf{A}_N$. From

$$\mathbf{V}_N \mathbf{A}_N \mathbf{V}_N^{-1} = \left(a_{(j-1)\ \mathrm{mod}\ N, (k-1)\ \mathrm{mod}\ N}\right)_{j,k=0}^{N-1}$$

it follows for all $j,\ k = 0, \ldots, N-1$

$$a_{(j-1)\ \mathrm{mod}\ N,\ (k-1)\ \mathrm{mod}\ N} = a_{j,k} \,.$$

Setting $a_j := a_{j,0}$ for $j = 0, \ldots, N-1$, we conclude that $a_{j,k} = a_{(j-k)\ \mathrm{mod}\ N}$ for $j,\ k = 0, \ldots, N-1$, i.e., $\mathbf{A}_N = \mathrm{circ}\,(a_j)_{j=0}^{N-1}$. ■

Remark 3.36 For arbitrarily given $t_k \in \mathbb{C}$, $k = 1-N, \ldots, N-1$, we consider the N-by-N *Toeplitz matrix* :

$$\mathbf{T}_N := \left(t_{j-k}\right)_{j,k=0}^{N-1} = \begin{pmatrix} t_0 & t_{-1} & \ldots & t_{2-N} & t_{1-N} \\ t_1 & t_0 & \ldots & t_{3-N} & t_{2-N} \\ \vdots & \vdots & & \vdots & \vdots \\ t_{N-1} & t_{N-2} & \ldots & t_1 & t_0 \end{pmatrix}.$$

In general, $\mathbf{T}_N$ is not a circulant matrix. But $\mathbf{T}_N$ can be extended to a $2N$-by-$2N$ circulant matrix $\mathbf{C}_{2N}$. We define

$$\mathbf{C}_{2N} := \begin{pmatrix} \mathbf{T}_N & \mathbf{E}_N \\ \mathbf{E}_N & \mathbf{T}_N \end{pmatrix}$$

with

$$\mathbf{E}_N := \begin{pmatrix} 0 & t_{N-1} & \ldots & t_2 & t_1 \\ t_{1-N} & 0 & \ldots & t_3 & t_2 \\ \vdots & \vdots & & \vdots & \vdots \\ t_{-1} & t_{-2} & \ldots & t_{1-N} & 0 \end{pmatrix}.$$

Then, $\mathbf{C}_{2N} = \operatorname{circ} \mathbf{c}$ with the vector

$$\mathbf{c} := (t_0,\, t_1,\, \ldots,\, t_{N-1},\, 0,\, t_{1-N}, \ldots,\, t_{-1})^\top \in \mathbb{C}^{2N}.$$

Thus, for an arbitrary vector $\mathbf{a} \in \mathbb{C}^N$, the matrix-vector product $\mathbf{T}_N\,\mathbf{a}$ can be computed using the circulant matrix vector product:

$$\mathbf{C}_{2N} \begin{pmatrix} \mathbf{a} \\ \mathbf{0} \end{pmatrix} = \begin{pmatrix} \mathbf{T}_N\,\mathbf{a} \\ \mathbf{E}_N\,\mathbf{a} \end{pmatrix},$$

where $\mathbf{0} \in \mathbb{C}^N$ denotes the zero vector. Applying a fast Fourier transform of Chap. 5, this matrix-vector product can therefore be realized with only $\mathcal{O}(N \log N)$ arithmetical operations. □

3.4 Kronecker Products and Stride Permutations

In this section we consider special block matrices that are obtained by employing the Kronecker product of two matrices. These special matrices often occur by reshaping linear equations with matrix-valued unknowns to matrix-vector representations. We will also show that block circulant matrices can again be simply diagonalized using Kronecker products of Fourier matrices.

For arbitrary matrices $\mathbf{A}_{M,N} = \big(a_{j,k}\big)_{j,k=0}^{M-1,N-1} \in \mathbb{C}^{M\times N}$ and $\mathbf{B}_{P,Q} \in \mathbb{C}^{P\times Q}$, the *Kronecker product* of $\mathbf{A}_{M,N}$ and $\mathbf{B}_{P,Q}$ is defined as the block matrix:

$$\begin{aligned} \mathbf{A}_{M,N} \otimes \mathbf{B}_{P,Q} &:= \big(a_{j,k}\,\mathbf{B}_{P,Q}\big)_{j,k=0}^{M-1,N-1} \\ &= \begin{pmatrix} a_{0,0}\,\mathbf{B}_{P,Q} & \ldots & a_{0,N-1}\,\mathbf{B}_{P,Q} \\ \vdots & & \vdots \\ a_{M-1,0}\,\mathbf{B}_{P,Q} & \ldots & a_{M-1,N-1}\,\mathbf{B}_{P,Q} \end{pmatrix} \in \mathbb{C}^{MP\times NQ}. \end{aligned}$$

In particular, for $\mathbf{a} = (a_j)_{j=0}^{M-1} \in \mathbb{C}^M$ and $\mathbf{b} \in \mathbb{C}^P$, we obtain the Kronecker product:

$$\mathbf{a} \otimes \mathbf{b} = \big(a_j\,\mathbf{b}\big)_{j=0}^{M-1} = \begin{pmatrix} a_0\,\mathbf{b} \\ \vdots \\ a_{M-1}\,\mathbf{b} \end{pmatrix} \in \mathbb{C}^{MP} .$$

The Kronecker product of the identity matrix $\mathbf{I}_M$ and the square matrix $\mathbf{B}_P$ is equal to the *block diagonal matrix* :

$$\mathbf{I}_M \otimes \mathbf{B}_P = \big(\delta_{j-k}\,\mathbf{B}_P\big)_{j,k=0}^{M-1} = \begin{pmatrix} \mathbf{B}_P & & \\ & \ddots & \\ & & \mathbf{B}_P \end{pmatrix} \in \mathbb{C}^{MP \times MP} .$$

Example 3.37 For the Fourier matrix $\mathbf{F}_2$, we obtain that

$$\mathbf{I}_2 \otimes \mathbf{F}_2 = \begin{pmatrix} 1 & 1 & 0 & 0 \\ 1 & -1 & 0 & 0 \\ 0 & 0 & 1 & 1 \\ 0 & 0 & 1 & -1 \end{pmatrix}, \quad \mathbf{F}_2 \otimes \mathbf{I}_2 = \begin{pmatrix} 1 & 0 & 1 & 0 \\ 0 & 1 & 0 & 1 \\ 1 & 0 & -1 & 0 \\ 0 & 1 & 0 & -1 \end{pmatrix} .$$

□

By definition, the Kronecker product $\mathbf{A}_{M,N} \otimes \mathbf{B}_{P,Q}$ has the entry $a_{j,k}\, b_{m,n}$ in the $(j\,P+m)$th row and $(k\,Q+n)$th column for $j = 0, \ldots, M-1$, $k = 0, \ldots, N-1$, $m = 0, \ldots, P-1$, and $n = 0, \ldots, Q-1$. Further it follows from the definition that the Kronecker product is associative, i.e., for arbitrary matrices $\mathbf{A}_{M,N} \in \mathbb{C}^{M\times N}$, $\mathbf{B}_{P,Q} \in \mathbb{C}^{P\times Q}$, and $\mathbf{C}_{R,S} \in \mathbb{C}^{R\times S}$, we have:

$$\big(\mathbf{A}_{M,N} \otimes \mathbf{B}_{P,Q}\big) \otimes \mathbf{C}_{R,S} = \mathbf{A}_{M,N} \otimes \big(\mathbf{B}_{P,Q} \otimes \mathbf{C}_{R,S}\big) . \tag{3.53}$$

In many applications, we especially consider Kronecker products of square matrices and of vectors. For square matrices $\mathbf{A}_M$ and $\mathbf{B}_P$, we simply observe by blockwise multiplication that

$$\mathbf{A}_M \otimes \mathbf{B}_P = \big(\mathbf{A}_M \otimes \mathbf{I}_P\big)\big(\mathbf{I}_M \otimes \mathbf{B}_P\big) = \big(\mathbf{I}_M \otimes \mathbf{B}_P\big)\big(\mathbf{A}_M \otimes \mathbf{I}_P\big) . \tag{3.54}$$

As we can see in Example 3.37, the Kronecker product is not commutative. In order to understand the relation between the Kronecker products $\mathbf{A}_M \otimes \mathbf{B}_P$ and $\mathbf{B}_P \otimes \mathbf{A}_M$, we introduce special permutation matrices. An N-by-N *permutation matrix* is obtained by permuting the columns of the N-by-N identity matrix $\mathbf{I}_N$. For instance, the counter-identity matrix $\mathbf{J}_N$, the flip matrix $\mathbf{J}'_N$, and the forward-shifted matrix $\mathbf{V}_N$ are permutation matrices. For $N \in \mathbb{N}$ with $N = L\,M$, where $L,\, M \geq 2$ are integers, the *L-stride permutation matrix* $\mathbf{P}_N(L) \in \mathbb{C}^{N\times N}$ is defined by the property

$$\begin{aligned}\mathbf{P}_N(L)\,\mathbf{a} &= \big((a_{Lk+\ell})_{k=0}^{M-1}\big)_{\ell=0}^{L-1}\\ &= \big(a_0, \ldots, a_{L(M-1)}\,|\,a_1, \ldots, a_{L(M-1)+1}\,|\ldots|a_{L-1}, \ldots, a_{L(M-1)+L-1}\big)^\top\end{aligned}$$

for arbitrary $\mathbf{a} = \big(a_j\big)_{j=0}^{N-1} \in \mathbb{C}^N$. Note that

$$\mathbf{P}_N(L) = \big(\delta_{(j\,L-k) \bmod (N-1)}\big)_{j,k=0}^{N-1}\,.$$

For even $N \in \mathbb{N}$, the 2-stride permutation matrix or *even-odd permutation matrix* $\mathbf{P}_N(2)$ is of special interest in the fast computation of DFT(N) (see Chap. 5). Then, we have:

$$\mathbf{P}_N(2)\,\mathbf{a} = \big((a_{2k+\ell})_{k=0}^{N/2-1}\big)_{\ell=0}^{1} = \big(a_0, a_2, \ldots, a_{N-2}\,|\,a_1, a_3, \ldots, a_{N-1}\big)^\top\,.$$

Example 3.38 For $N = 6$, $L = 2$, and $M = 3$, we get that

$$\mathbf{P}_6(2) = \begin{pmatrix} 1\,0\,0\,0\,0\,0\\ 0\,0\,1\,0\,0\,0\\ 0\,0\,0\,0\,1\,0\\ 0\,1\,0\,0\,0\,0\\ 0\,0\,0\,1\,0\,0\\ 0\,0\,0\,0\,0\,1 \end{pmatrix}, \quad \mathbf{P}_6(2)^\top = \mathbf{P}_6(3) = \begin{pmatrix} 1\,0\,0\,0\,0\,0\\ 0\,0\,0\,1\,0\,0\\ 0\,1\,0\,0\,0\,0\\ 0\,0\,0\,0\,1\,0\\ 0\,0\,1\,0\,0\,0\\ 0\,0\,0\,0\,0\,1 \end{pmatrix}.$$

Then, we have $\mathbf{P}_6(2)\,\mathbf{P}_6(3) = \mathbf{I}_6$ and

$$\mathbf{P}_6(2)\,\mathbf{a} = (a_0,\, a_2,\, a_4,\, a_1,\, a_3,\, a_5)^\top\,, \quad \mathbf{P}_6(3)\,\mathbf{a} = (a_0,\, a_3,\, a_1,\, a_4,\, a_2,\, a_5)^\top$$

for $\mathbf{a} = \big(a_j\big)_{j=0}^{5}$. □

The following property of stride-permutation matrices can be simply shown using Kronecker products.

Lemma 3.39 *If $N \in \mathbb{N}\setminus\{1\}$ can be factorized in the form $N = K\,L\,M$ with $K \in \mathbb{N}$ and $L,\, M \in \mathbb{N}\setminus\{1\}$, then*

$$\mathbf{P}_N(L\,M) = \mathbf{P}_N(L)\,\mathbf{P}_N(M) = \mathbf{P}_N(M)\,\mathbf{P}_N(L)\,. \tag{3.55}$$

In particular, for $N = L\,M$, we have:

$$\mathbf{P}_N(L)^{-1} = \mathbf{P}_N(L)^\top = \mathbf{P}_N(M)\,. \tag{3.56}$$

Proof For arbitrary vectors $\mathbf{a} \in \mathbb{C}^K$, $\mathbf{b} \in \mathbb{C}^L$, and $\mathbf{c} \in \mathbb{C}^M$, we obtain from the definitions on the one hand that

$$\mathbf{P}_N(L\,M)\,(\mathbf{a} \otimes \mathbf{b} \otimes \mathbf{c}) = \mathbf{P}_N(LM)\,\big(\mathbf{a} \otimes (\mathbf{b} \otimes \mathbf{c})\big) = \mathbf{b} \otimes \mathbf{c} \otimes \mathbf{a}\,.$$

On the other hand,

$$\begin{aligned}\mathbf{P}_N(L)\,\mathbf{P}_N(M)\,(\mathbf{a} \otimes \mathbf{b} \otimes \mathbf{c}) &= \mathbf{P}_N(L)\,\big[\mathbf{P}_N(M)\,\big((\mathbf{a} \otimes \mathbf{b}) \otimes \mathbf{c}\big)\big] \\ &= \mathbf{P}_N(L)\,\big(\mathbf{c} \otimes (\mathbf{a} \otimes \mathbf{b})\big) = \mathbf{P}_N(L)\,\big((\mathbf{c} \otimes \mathbf{a}) \otimes \mathbf{b}\big) \\ &= \mathbf{b} \otimes \mathbf{c} \otimes \mathbf{a}\,.\end{aligned}$$

The two equations are true for all vectors $\mathbf{a} \otimes \mathbf{b} \otimes \mathbf{c} \in \mathbb{C}^N$, which span $\mathbb{C}^N$. Consequently, $\mathbf{P}_N(L\,M) = \mathbf{P}_N(L)\,\mathbf{P}_N(M)$. Analogously, one can show that $\mathbf{P}_N(L\,M) = \mathbf{P}_N(M)\,\mathbf{P}_N(L)$.

If $N = L\,M$, then $\mathbf{P}_N(L\,M) = \mathbf{P}_N(N) = \mathbf{I}_N$ and hence $\mathbf{P}_N(M)\,\mathbf{P}_N(L) = \mathbf{I}_N$. Since $\mathbf{P}_N(M)$ and $\mathbf{P}_N(L)$ are orthogonal matrices, we conclude (3.56). ∎

Corollary 3.40 *Let $p \in \mathbb{N}$ be a prime. For $N = p^n$ with $n \in \mathbb{N}$, the set $\{\mathbf{P}_N(p^k) : k = 0, \ldots, n-1\}$ is a cyclic group generated by the p-stride permutation matrix $\mathbf{P}_N(p)$ with $\mathbf{P}_N(p)^k = \mathbf{P}_N(p^k)$ for $k = 0, \ldots, n-1$.*

This corollary follows immediately from Lemma 3.39.

We are interested in the properties of Kronecker products. For simplicity, we restrict ourselves to square matrices.

Lemma 3.41 *For arbitrary matrices $\mathbf{A}_L \in \mathbb{C}^{L\times L}$ and $\mathbf{B}_M \in \mathbb{C}^{M\times M}$ and for all vectors $\mathbf{a} \in \mathbb{C}^L$ and $\mathbf{b} \in \mathbb{C}^M$, we have:*

$$(\mathbf{A}_L \otimes \mathbf{B}_M)\,(\mathbf{a} \otimes \mathbf{b}) = (\mathbf{A}_L\,\mathbf{a}) \otimes (\mathbf{B}_M\,\mathbf{b})\,.$$

Proof Assume that $\mathbf{A}_L = \big(a_{j,k}\big)_{j,k=0}^{L-1}$ and $\mathbf{a} = \big(a_k\big)_{k=0}^{L-1}$. From

$$\mathbf{A}_L \otimes \mathbf{B}_M = \big(a_{j,k}\,\mathbf{B}_M\big)_{j,k=0}^{L-1}\,, \quad \mathbf{a} \otimes \mathbf{b} = \big(a_k\,\mathbf{b}\big)_{k=0}^{L-1}$$

it follows by blockwise computation that

$$(\mathbf{A}_L \otimes \mathbf{B}_M)\,(\mathbf{a} \otimes \mathbf{b}) = \Big(\big(\sum_{k=0}^{L-1} a_{j,k}\,a_k\big)\,\mathbf{B}_M\,\mathbf{b}\Big) = (\mathbf{A}_L\,\mathbf{a}) \otimes (\mathbf{B}_M\,\mathbf{b})\,,$$

since $\sum_{k=0}^{L-1} a_{j,k}\,a_k$ is the jth component of $\mathbf{A}_L\,\mathbf{a}$. ∎

We denote with $\mathrm{vec}\,\mathbf{X}$ the *vectorization of a matrix* $\mathbf{X} = (\mathbf{x}_0|\ldots|\mathbf{x}_{M-1}) \in \mathbb{C}^{L\times M}$ into the vector $(\mathbf{x}_0^\top, \ldots, \mathbf{x}_{M-1}^\top)^\top = (\mathbf{x}_k)_{k=0}^{M-1} \in \mathbb{C}^{LM}$.

Theorem 3.42 *Let $\mathbf{A}_L$, $\mathbf{C}_L \in \mathbb{C}^{L\times L}$, and $\mathbf{B}_M$, $\mathbf{D}_M \in \mathbb{C}^{M\times M}$ be arbitrary square matrices. Then the Kronecker product possesses the following properties:*

1. $(\mathbf{A}_L \otimes \mathbf{B}_M)\,(\mathbf{C}_L \otimes \mathbf{D}_M) = (\mathbf{A}_L\,\mathbf{C}_L) \otimes (\mathbf{B}_M\,\mathbf{D}_M)$,
2. $(\mathbf{A}_L \otimes \mathbf{B}_M)^\top = \mathbf{A}_L^\top \otimes \mathbf{B}_M^\top$,
3. $\mathbf{P}_N(M)\,(\mathbf{A}_L \otimes \mathbf{B}_M)\,\mathbf{P}_N(L) = \mathbf{B}_M \otimes \mathbf{A}_L$ *with* $N = L\,M$,
4. $\det(\mathbf{A}_L \otimes \mathbf{B}_M) = (\det \mathbf{A}_L)^M\,(\det \mathbf{B}_M)^L$,
5. $(\mathbf{A}_L \otimes \mathbf{B}_M)^{-1} = \mathbf{A}_L^{-1} \otimes \mathbf{B}_M^{-1}$, *if* $\mathbf{A}_L$ *and* $\mathbf{B}_M$ *are invertible.*
6. *Let* $\mathbf{X} \in \mathbb{C}^{L\times M}$, *then*

$$\mathrm{vec}(\mathbf{A}_L\,\mathbf{X}\,\mathbf{B}_M) = (\mathbf{B}_M^\top \otimes \mathbf{A}_L)\ \mathrm{vec}\,\mathbf{X}.$$

Proof

1. For arbitrary matrices $\mathbf{A}_L = \big(a_{j,k}\big)_{j,k=0}^{L-1}$ and $\mathbf{C}_L = \big(c_{k,\ell}\big)_{k,\ell=0}^{L-1} \in \mathbb{C}^{L\times L}$, we obtain that

$$\mathbf{A}_L \otimes \mathbf{B}_M := \big(a_{j,k}\,\mathbf{B}_M\big)_{j,k=0}^{L-1}, \quad \mathbf{C}_L \otimes \mathbf{D}_M := \big(c_{k,\ell}\,\mathbf{D}_M\big)_{k,\ell=0}^{L-1}.$$

By blockwise multiplication of the two matrices, we conclude that

$$(\mathbf{A}_L \otimes \mathbf{B}_M)\,(\mathbf{C}_L \otimes \mathbf{D}_M) = \Big(\sum_{k=0}^{L-1} a_{j,k}\,c_{k,\ell}\,\mathbf{B}_M\,\mathbf{D}_M\Big)_{j,\ell=0}^{L-1} = (\mathbf{A}_L\,\mathbf{C}_L) \otimes (\mathbf{B}_M\,\mathbf{D}_M),$$

since $\sum_{k=0}^{L-1} a_{j,k}\,c_{k,\ell}$ is the (j,ℓ)th entry of the matrix product $\mathbf{A}_L\,\mathbf{C}_L$. The first property of the Kronecker product is of high importance for efficient multiplication of block matrices, since the multiplication of two matrices of large size $L\,M$-by-$L\,M$ can be transferred to the multiplication of matrices of lower sizes L-by-L and M-by-M and the realization of one Kronecker product, being computed by elementwise multiplication.
2. The second property follows immediately from the definition of the Kronecker product.
3. For arbitrary vectors $\mathbf{a} \in \mathbb{C}^L$ and $\mathbf{b} \in \mathbb{C}^M$, we find

$$\begin{aligned}\mathbf{P}_N(M)\,(\mathbf{A}_L \otimes \mathbf{B}_M)\,\big(\mathbf{P}_N(L)\,(\mathbf{b} \otimes \mathbf{a})\big) &= \mathbf{P}_N(M)\,(\mathbf{A}_L \otimes \mathbf{B}_M)\,(\mathbf{a} \otimes \mathbf{b})\\ &= \mathbf{P}_N(M)\,\big((\mathbf{A}_L\,\mathbf{a}) \otimes (\mathbf{B}_M\,\mathbf{b})\big)\\ &= (\mathbf{B}_M\,\mathbf{b}) \otimes (\mathbf{A}_L\,\mathbf{a}) = (\mathbf{B}_M \otimes (\mathbf{A}_L)\,(\mathbf{b} \otimes \mathbf{a}).\end{aligned}$$

Since arbitrary vectors of the form $\mathbf{b} \otimes \mathbf{a}$ span the vector space $\mathbb{C}^{L\,M}$, the so-called commutation property is shown.
4. The Kronecker product $\mathbf{A}_L \otimes \mathbf{B}_M$ can be factorized in the form

$$\mathbf{A}_L \otimes \mathbf{B}_M = (\mathbf{A}_L \otimes \mathbf{I}_M)\,(\mathbf{I}_L \otimes \mathbf{B}_M).$$

For the block diagonal matrix $\mathbf{I}_L \otimes \mathbf{B}_M$, we obtain that

$$\det(\mathbf{I}_L \otimes \mathbf{B}_M) = (\det \mathbf{B}_M)^L\,.$$

By the commutation property, it follows that with $N = L\,M$

$$\mathbf{P}_N(M)\,(\mathbf{A}_L \otimes \mathbf{I}_M)\,\mathbf{P}_N(L) = \mathbf{P}_N(M)\,(\mathbf{A}_L \otimes \mathbf{I}_M)\,\mathbf{P}_N(M)^{-1} = \mathbf{I}_M \otimes \mathbf{A}_L$$

and hence

$$\det(\mathbf{A}_L \otimes \mathbf{I}_M) = \det(\mathbf{I}_M \otimes \mathbf{A}_L) = (\det \mathbf{A}_L)^M\,.$$

Thus, we have:

$$\det(\mathbf{A}_L \otimes \mathbf{B}_M) = \big(\det(\mathbf{A}_L \otimes \mathbf{I}_M)\big)\,\big(\det(\mathbf{I}_L \otimes \mathbf{B}_M)\big) = (\det \mathbf{A}_L)^M\,(\det \mathbf{B}_M)^L\,.$$

5. By property 4, the Kronecker product $\mathbf{A}_L \otimes \mathbf{B}_M$ is invertible if and only if $\mathbf{A}_L$ and $\mathbf{B}_M$ are invertible. By property 1, the inverse of $\mathbf{A}_L \otimes \mathbf{B}_M$ reads $\mathbf{A}_L^{-1} \otimes \mathbf{B}_M^{-1}$, if $\mathbf{A}_L$ and $\mathbf{B}_M$ are invertible.
6. Let the columns of $\mathbf{X}$ be given by $\mathbf{X} = (\mathbf{x}_0\,|\,\ldots\,|\,\mathbf{x}_{M-1})$, and let $\mathbf{B}_M = (b_{j,k})_{j,k=0}^{M-1} = (\mathbf{b}_0\,|\,\ldots\,|\,\mathbf{b}_{M-1})$. On the one hand, we find

$$\operatorname{vec}(\mathbf{A}_L\,\mathbf{X}\,\mathbf{B}_M) = \operatorname{vec}(\mathbf{A}_L\,\mathbf{X}\,\mathbf{b}_0\,|\,\ldots\,|\,\mathbf{A}_L\,\mathbf{X}\,\mathbf{b}_{M-1})\,.$$

On the other hand,

$$(\mathbf{B}_M^\top \otimes \mathbf{A}_L)\,\operatorname{vec}\mathbf{X} = \begin{pmatrix} b_{0,0}\,\mathbf{A}_L & \ldots & b_{M-1,0}\,\mathbf{A}_L \\ b_{0,1}\,\mathbf{A}_L & \ldots & b_{M-1,1}\,\mathbf{A}_L \\ \vdots & & \vdots \\ b_{0,M-1}\,\mathbf{A}_L & \ldots & b_{M-1,M-1}\,\mathbf{A}_L \end{pmatrix} \begin{pmatrix} \mathbf{x}_0 \\ \vdots \\ \mathbf{x}_{M-1} \end{pmatrix}$$

$$= \begin{pmatrix} b_{0,0}\,\mathbf{A}_L\,\mathbf{x}_0 + \ldots + b_{M-1,0}\,\mathbf{A}_L\,\mathbf{x}_{M-1} \\ b_{0,1}\,\mathbf{A}_L\,\mathbf{x}_0 + \ldots + b_{M-1,1}\,\mathbf{A}_L\,\mathbf{x}_{M-1} \\ \vdots \\ b_{0,M-1}\,\mathbf{A}_L\,\mathbf{x}_0 + \ldots + b_{M-1,M-1}\,\mathbf{A}_L\,\mathbf{x}_{M-1} \end{pmatrix}$$

$$= \begin{pmatrix} \mathbf{A}_L\,\mathbf{X}\,\mathbf{b}_0 \\ \mathbf{A}_L\,\mathbf{X}\,\mathbf{b}_1 \\ \vdots \\ \mathbf{A}_L\,\mathbf{X}\,\mathbf{b}_{M-1} \end{pmatrix}.$$

Thus, the assertion follows. ∎

Now we use Kronecker products of Fourier matrices in order to diagonalize generalized circulant matrices. Assume that vectors $\mathbf{c}_k \in \mathbb{C}^M$, $k = 0, \ldots, L-1$, are given and let $\mathbf{A}(k) := \text{circ}\ \mathbf{c}_k$ be the corresponding M-by-M circulant matrices. Then the L-by-L block matrix

$$\mathbf{A}_{L\,M} := \big(\mathbf{A}((j-k) \bmod L)\big)_{j,k=0}^{L-1} = \begin{pmatrix} \mathbf{A}(0) & \mathbf{A}(L-1) \dots & \mathbf{A}(1) \\ \mathbf{A}(1) & \mathbf{A}(0) \quad \dots & \mathbf{A}(2) \\ \vdots & \vdots & \vdots \\ \mathbf{A}(L-1) & \mathbf{A}(L-2) \dots & \mathbf{A}(0) \end{pmatrix} \in \mathbb{C}^{LM \times LM}$$

is called L-by-L *block circulant matrix with* M-by-M *circulant blocks*. We observe that L-by-L block circulant matrices with M-by-M circulant blocks commute, since circulant matrices commute by Theorem 3.34. Note that $\mathbf{A}_{LM}$ is already determined by the vectors $\mathbf{c}_k$, $k = 0, \dots, L-1$, or equivalently by

$$\mathbf{C}_{M,L} := \big(\mathbf{c}_0 \,|\, \dots \,|\, \mathbf{c}_{L-1}\big) \in \mathbb{C}^{M \times L}.$$

Example 3.43 Let $\mathbf{A}_L$ be an L-by-L circulant matrix and let $\mathbf{B}_M$ be an M-by-M circulant matrix. Obviously, the Kronecker product $\mathbf{A}_L \otimes \mathbf{B}_M$ is an L-by-L block circulant matrix with M-by-M circulant blocks $\mathbf{B}_M$. In particular, $\mathbf{I}_L \otimes \mathbf{B}_M$ is a block diagonal matrix with circulant blocks. The so-called Kronecker sum of $\mathbf{A}_L$ and $\mathbf{B}_M$ defined by

$$(\mathbf{A}_L \otimes \mathbf{I}_M) + (\mathbf{I}_L \otimes \mathbf{B}_M) \in \mathbb{C}^{LM \times LM} \tag{3.57}$$

is also a block matrix with circulant blocks. □

Lemma 3.44 *Each L-by-L block circulant matrix $\mathbf{A}_{L\,M} \in \mathbb{C}^{LM \times LM}$ with M-by-M circulant blocks $\mathbf{A}(k) = \text{circ}\ \mathbf{c}_k$ with $\mathbf{c}_k \in \mathbb{C}^M$, $k = 0, \dots, L-1$ can be diagonalized by the Kronecker product $\mathbf{F}_L \otimes \mathbf{F}_M$ of the Fourier matrices $\mathbf{F}_L$ and $\mathbf{F}_M$ in the following form:*

$$\mathbf{A}_{L\,M} = \big(\mathbf{F}_L \otimes \mathbf{F}_M\big)^{-1} \big(\text{diag}\,(\text{vec}\ \hat{\mathbf{C}}_{M,L})\big) \big(\mathbf{F}_L \otimes \mathbf{F}_M\big),$$

where $\hat{\mathbf{C}}_{M,L} := \mathbf{F}_M\, \mathbf{C}_{M,L}\, \mathbf{F}_L$. Moreover we have:

$$\big(\mathbf{F}_L \otimes \mathbf{F}_M\big)\ \text{vec}\ \mathbf{C}_{M,L} = \text{vec}\ \hat{\mathbf{C}}_{M,L}\,.$$

Proof First we compute the product $\big(\mathbf{F}_L \otimes \mathbf{I}_M\big)\, \mathbf{A}_{L\,M}\, \big(\mathbf{F}_L^{-1} \otimes \mathbf{I}_M\big)$ by blockwise multiplication:

$$\frac{1}{L}\, \big(w_L^{jk}\, \mathbf{I}_M\big)_{j,k=0}^{L-1} \big(\mathbf{A}((k-\ell) \bmod L)\big)_{k,\ell=0}^{L-1} \big(w_L^{-\ell n}\, \mathbf{I}_M\big)_{\ell,n=0}^{L-1}.$$

The result is a block matrix $\big(\mathbf{B}(j,n)\big)_{j,n=0}^{L-1}$ with the blocks

$$\mathbf{B}(j,n) = \frac{1}{L}\sum_{k=0}^{L-1}\sum_{\ell=0}^{L-1} w_L^{jk-\ell n}\,\mathbf{A}((k-\ell) \bmod L) \in \mathbb{C}^{M\times M}\,.$$

In the case $j = n$, we obtain after substitution $m = (k-\ell) \bmod L$ that

$$\mathbf{B}(j,j) = \sum_{m=0}^{L-1} w_L^{jm}\,\mathbf{A}(m)\,. \tag{3.58}$$

In the case $j \neq n$, we see by Lemma 3.2 that $\mathbf{B}(j,n)$ is equal to the M-by-M zero matrix. Hence $\big(\mathbf{F}_L \otimes \mathbf{I}_M\big)\,\mathbf{A}_{L\,M}\,\big(\mathbf{F}_L^{-1} \otimes \mathbf{I}_M\big)$ is a block diagonal matrix with the diagonal blocks in (3.58). By Theorem 3.42, the block circulant matrix $\mathbf{A}_{L\,M}$ with circulant blocks can be represented in the form

$$\mathbf{A}_{L\,M} = \big(\mathbf{F}_L^{-1} \otimes \mathbf{I}_M\big)\Big[\operatorname{diag}\big(\mathbf{B}(j,j)\big)_{j=0}^{L-1}\Big]\big(\mathbf{F}_L \otimes \mathbf{I}_M\big)\,.$$

Since $\mathbf{A}(m) = \operatorname{circ}\mathbf{c}_m$ are circulant matrices by assumption, we obtain by Theorem 3.31 that

$$\mathbf{A}(m) = \mathbf{F}_M^{-1}\,\big(\operatorname{diag}(\mathbf{F}_M\,\mathbf{c}_m)\big)\,\mathbf{F}_M$$

and hence by (3.58)

$$\mathbf{B}(j,j) = \mathbf{F}_M^{-1}\,\Big(\sum_{m=0}^{L-1} w_L^{jm}\,\operatorname{diag}(\mathbf{F}_M\,\mathbf{c}_m)\Big)\,\mathbf{F}_M\,.$$

Thus by Theorem 3.42 we conclude

$$\begin{aligned}
\mathbf{A}_{L\,M} &= \big(\mathbf{F}_L^{-1} \otimes \mathbf{I}_M\big)\big(\mathbf{I}_L \otimes \mathbf{F}_M^{-1}\big)\Big[\operatorname{diag}\Big(\sum_{m=0}^{L-1} w_L^{jm}\,\mathbf{F}_M\,\mathbf{c}_m\Big)_{j=0}^{L-1}\Big] \\
&\quad\times\big(\mathbf{I}_L \otimes \mathbf{F}_M\big)\big(\mathbf{F}_L \otimes \mathbf{I}_M\big) \\
&= \big(\mathbf{F}_L^{-1} \otimes \mathbf{F}_M^{-1}\big)\Big[\operatorname{diag}\Big(\sum_{m=0}^{L-1} w_L^{jm}\,\mathbf{F}_M\,\mathbf{c}_m\Big)_{j=0}^{L-1}\Big]\big(\mathbf{F}_L \otimes \mathbf{F}_M\big) \\
&= \big(\mathbf{F}_L \otimes \mathbf{F}_M\big)^{-1}\Big[\operatorname{diag}\Big(\sum_{m=0}^{L-1} w_L^{jm}\,\mathbf{F}_M\,\mathbf{c}_m\Big)_{j=0}^{L-1}\Big]\big(\mathbf{F}_L \otimes \mathbf{F}_M\big)\,.
\end{aligned}$$

Finally, we show that

$$\Big(\sum_{m=0}^{L-1} w_L^{jm}\,\mathbf{F}_M\,\mathbf{c}_m\Big)_{j=0}^{L-1} = \text{vec}\,\hat{\mathbf{C}}_{M,L}\,.$$

We recall that $\mathbf{C}_{M,L} = (c_{j,k})_{j,k=0}^{M-1,L-1}$ has the columns $\mathbf{c}_m = (c_{j,m})_{j=0}^{M-1}$, $m = 0,\ldots,L-1$, and $\hat{\mathbf{C}}_{M,L} := \mathbf{F}_M\,\mathbf{C}_{M,L}\,\mathbf{F}_L = (\hat{c}_{\ell,k})_{\ell,k=0}^{M-1,L-1}$ has the entries

$$\hat{c}_{\ell,k} = \sum_{j=0}^{M-1}\sum_{m=0}^{L-1} c_{j,m}\,w_M^{j\ell}\,w_L^{km}\,.$$

From $\mathbf{F}_L \otimes \mathbf{F}_M = (w_L^{km}\,\mathbf{F}_M)_{k,m=0}^{L-1}$ and $\text{vec}\,\mathbf{C}_{M,L} = (\mathbf{c}_m)_{m=0}^{L-1}$, it follows by blockwise multiplication:

$$\begin{aligned}(\mathbf{F}_L \otimes \mathbf{F}_M)\,\text{vec}\,\mathbf{C}_{M,L} &= \Big(\sum_{m=0}^{L-1} w_L^{km}\,\mathbf{F}_M\,\mathbf{c}_m\Big)_{k=0}^{L-1}\\ &= \Big(\Big(\sum_{m=0}^{L-1}\sum_{j=0}^{M-1} c_{j,m}\,w_M^{j\ell}\,w_L^{km}\Big)_{\ell=0}^{M-1}\Big)_{k=0}^{L-1}\\ &= \big((\hat{c}_{\ell,k})_{\ell=0}^{M-1}\big)_{k=0}^{L-1} = \text{vec}\,\hat{\mathbf{C}}_{M,L}\,.\end{aligned}$$

This completes the proof. ■

Corollary 3.45 *Let* $\mathbf{a} \in \mathbb{C}^L$ *and* $\mathbf{b} \in \mathbb{C}^M$ *be arbitrary given vectors and let* $\mathbf{A}_L = \text{circ}\,\mathbf{a}$ *and* $\mathbf{B}_M = \text{circ}\,\mathbf{b}$ *be the corresponding circulant matrices. Then, both the Kronecker product* $\mathbf{A}_L \otimes \mathbf{B}_M$ *and the Kronecker sum* (3.57) *of* $\mathbf{A}_L$ *and* $\mathbf{B}_M$ *can be diagonalized by the Kronecker product* $\mathbf{F}_L \otimes \mathbf{F}_M$.

If $\lambda \in \mathbb{C}$ *be an eigenvalue of* $\mathbf{A}_L$ *with corresponding eigenvector* $\mathbf{x} \in \mathbb{C}^L$ *and if* $\mu \in \mathbb{C}$ *be an eigenvalue of* $\mathbf{B}_M$ *with corresponding eigenvector* $\mathbf{y} \in \mathbb{C}^M$, *then* $\lambda + \mu$ *is an eigenvalue of the Kronecker sum* (3.57) *of* $\mathbf{A}_L$ *and* $\mathbf{B}_M$ *with an eigenvector* $\mathbf{x} \otimes \mathbf{y} \in \mathbb{C}^{LM}$.

Proof From Lemma 3.44 and Example 3.43, it follows immediately that the Kronecker product $\mathbf{A}_L \otimes \mathbf{B}_M$ and the Kronecker sum (3.57) can be diagonalized by $\mathbf{F}_L \otimes \mathbf{F}_M$.

If λ be an eigenvalue of $\mathbf{A}_L$ and if μ be an eigenvalue of $\mathbf{B}_M$, then we see by Lemma 3.41 that

$$\begin{aligned}\big[(\mathbf{A}_L \otimes \mathbf{I}_M) + (\mathbf{I}_L \otimes \mathbf{B}_M)\big]\,(\mathbf{x} \otimes \mathbf{y}) &= \big((\mathbf{A}_L\,\mathbf{x}) \otimes \mathbf{y}\big) + \big((\mathbf{x} \otimes (\mathbf{B}_M\,\mathbf{y})\big)\\ &= (\lambda + \mu)\,(\mathbf{x} \otimes \mathbf{y})\,.\end{aligned}$$

■

3.5 Discrete Trigonometric Transforms

In this section we consider some real versions of the DFT. Discrete trigonometric transforms are widely used in applied mathematics, digital signal processing, and image compression. Examples of such real transforms are a discrete cosine transform (DCT), discrete sine transform (DST), and discrete Hartley transform (DHT). We will introduce these transforms and show that they are generated by orthogonal matrices.

Lemma 3.46 *Let $N \ge 2$ be a given integer. Then the set of* cosine vectors of type I

$$\mathbf{c}_k^{\mathrm{I}} := \sqrt{\frac{2}{N}}\, \varepsilon_N(k)\, \Big(\varepsilon_N(j)\, \cos\frac{jk\pi}{N}\Big)_{j=0}^{N}, \quad k = 0, \ldots, N,$$

forms an orthonormal basis of $\mathbb{R}^{N+1}$, where $\varepsilon_N(0) = \varepsilon_N(N) := \frac{\sqrt{2}}{2}$ and $\varepsilon_N(j) := 1$ for $j = 1, \ldots, N-1$. The $(N+1)$-by-$(N+1)$ cosine matrix of type I *is defined by*

$$\mathbf{C}_{N+1}^{\mathrm{I}} := \sqrt{\frac{2}{N}}\, \Big(\varepsilon_N(j)\, \varepsilon_N(k)\, \cos\frac{jk\pi}{N}\Big)_{j,k=0}^{N}, \tag{3.59}$$

i.e., it has the cosine vectors of type I *as columns. Then $\mathbf{C}_{N+1}^{\mathrm{I}}$ is symmetric and orthogonal, i.e., $\big(\mathbf{C}_{N+1}^{\mathrm{I}}\big)^{-1} = \mathbf{C}_{N+1}^{\mathrm{I}}$.*

Proof By Example 1.14 we know that for $x \in \mathbb{R} \setminus 2\pi\,\mathbb{Z}$

$$\sum_{j=1}^{N-1} \cos(jx) = \frac{\sin\frac{(2N-1)x}{2}}{2\,\sin\frac{x}{2}} - \frac{1}{2}.$$

In particular, it follows for $x = \frac{2\pi k}{N}$,

$$\sum_{j=1}^{N-1} \cos\frac{2kj\pi}{N} = -1, \quad k \in \mathbb{Z} \setminus N\,\mathbb{Z}, \tag{3.60}$$

and for $x = \frac{(2k+1)\pi}{N}$,

$$\sum_{j=1}^{N-1} \cos\frac{(2k+1)\,j\pi}{N} = 0, \quad k \in \mathbb{Z}. \tag{3.61}$$

For $k,\, \ell \in \{0, \ldots, N\}$, the inner product $\langle \mathbf{c}_k^{\mathrm{I}},\, \mathbf{c}_\ell^{\mathrm{I}}\rangle$ can be calculated as follows:

$$\begin{aligned}\langle \mathbf{c}_k^{\mathrm{I}}, \mathbf{c}_\ell^{\mathrm{I}}\rangle &= \frac{2}{N}\,\varepsilon_N(k)\,\varepsilon_N(\ell)\sum_{j=0}^{N}\varepsilon_N(j)^2\cos\frac{jk\pi}{N}\,\cos\frac{j\ell\pi}{N}\\ &= \frac{1}{N}\,\varepsilon_N(k)\,\varepsilon_N(\ell)\Big(1+(-1)^{k+\ell}+\sum_{j=1}^{N-1}\cos\frac{(k-\ell)j\pi}{N}\\ &\quad + \sum_{j=1}^{N-1}\cos\frac{(k+\ell)j\pi}{N}\Big)\,. \end{aligned} \tag{3.62}$$

In the case $k \neq \ell$ with even $k+\ell$, the integer $k-\ell$ is also even. Hence by (3.60) and (3.62), we obtain $\langle \mathbf{c}_k^{\mathrm{I}}, \mathbf{c}_\ell^{\mathrm{I}}\rangle = 0$. In the case $k \neq \ell$ with odd $k+\ell$, the integer $k-\ell$ is odd such that $\langle \mathbf{c}_k^{\mathrm{I}}, \mathbf{c}_\ell^{\mathrm{I}}\rangle = 0$ by (3.61) and (3.62). For $k = \ell \in \{1, \ldots, N-1\}$, it follows from (3.60) and (3.62) that

$$\langle \mathbf{c}_k^{\mathrm{I}}, \mathbf{c}_\ell^{\mathrm{I}}\rangle = \frac{1}{N}\Big(2+(N-1)+\sum_{j=1}^{N-1}\cos\frac{2kj\pi}{N}\Big) = 1\,.$$

For $k = \ell \in \{0, N\}$, we get:

$$\langle \mathbf{c}_k^{\mathrm{I}}, \mathbf{c}_\ell^{\mathrm{I}}\rangle = \frac{1}{2N}\big(2+(N-1)+(N-1)\big) = 1\,.$$

Thus, the set $\{\mathbf{c}_k^{\mathrm{I}} : k = 0, \ldots, N\}$ is an orthonormal basis of $\mathbb{R}^{N+1}$, because the $N+1$ cosine vectors of type I are linearly independent and $\dim \mathbb{R}^{N+1} = N+1$. ∎

The linear map from $\mathbb{R}^{N+1}$ to $\mathbb{R}^{N+1}$, which is represented by the matrix-vector product $\mathbf{C}_{N+1}^{\mathrm{I}}\mathbf{a} = \big(\langle \mathbf{a}, \mathbf{c}_k^{\mathrm{I}}\rangle\big)_{k=0}^{N}$ for arbitrary $\mathbf{a} \in \mathbb{R}^{N+1}$, is called *discrete cosine transform of type* I *and length* $N+1$ and will be abbreviated by DCT-I $(N+1)$. As we will show in Sect. 6.2 (see, e.g., Algorithm 6.22), this transform plays an important role for evaluating expansions of Chebyshev polynomials.

Lemma 3.47 *Let $N \geq 2$ be a given integer. Then the set of* cosine vectors of type II

$$\mathbf{c}_k^{\mathrm{II}} := \sqrt{\frac{2}{N}}\,\Big(\varepsilon_N(j)\,\cos\frac{(2k+1)j\pi}{2N}\Big)_{j=0}^{N-1}\,,\quad k = 0, \ldots, N-1\,,$$

forms an orthonormal basis of $\mathbb{R}^N$. The N-by-N cosine matrix of type II *is defined by the column vectors $\mathbf{c}_k^{\mathrm{II}}$:*

$$\mathbf{C}_N^{\mathrm{II}} := \sqrt{\frac{2}{N}}\,\Big(\varepsilon_N(j)\,\cos\frac{(2k+1)j\pi}{2N}\Big)_{j,k=0}^{N-1}\,. \tag{3.63}$$

The matrix $\mathbf{C}_N^{\mathrm{II}}$ is orthogonal, but it is not symmetric. We have $\big(\mathbf{C}_N^{\mathrm{II}}\big)^{-1} = \big(\mathbf{C}_N^{\mathrm{II}}\big)^{\top}$.

Proof For $k,\ \ell \in \{0, \ldots, N-1\}$, we calculate the inner product:

$$\langle \mathbf{c}_k^{\mathrm{II}}, \mathbf{c}_\ell^{\mathrm{II}} \rangle = \frac{2}{N} \sum_{j=0}^{N-1} \varepsilon_N(j)^2 \cos\frac{(2k+1)j\pi}{2N} \cos\frac{(2\ell+1)j\pi}{2N}$$

$$= \frac{1}{N}\Big(1 + \sum_{j=1}^{N-1} \cos\frac{(k-\ell)j\pi}{N} + \sum_{j=1}^{N-1} \cos\frac{(k+\ell+1)j\pi}{N}\Big). \tag{3.64}$$

In the case $k \neq \ell$, this inner product vanishes by (3.60), (3.61), and (3.64). For $k = \ell$ we obtain by (3.61) and (3.64) that

$$\langle \mathbf{c}_k^{\mathrm{II}}, \mathbf{c}_k^{\mathrm{II}} \rangle = \frac{1}{N}\Big(1 + (N-1) + \sum_{j=1}^{N-1} \cos\frac{(2k+1)j\pi}{N}\Big) = 1\,.$$

■

The linear map from $\mathbb{R}^N$ to $\mathbb{R}^N$, represented by $\mathbf{C}_N^{\mathrm{II}}\mathbf{a} = \big(\langle \mathbf{a}, \mathbf{c}_k^{\mathrm{II}} \rangle\big)_{k=0}^{N-1}$ for arbitrary $\mathbf{a} \in \mathbb{R}^N$, is called *discrete cosine transform of type* II *and length* N and will be abbreviated by DCT-II (N). The DCT-II plays a special role for decorrelation of digital signals and images.

The N-by-N *cosine matrix of type* III is defined by

$$\mathbf{C}_N^{\mathrm{III}} := \sqrt{\frac{2}{N}}\,\big(\varepsilon_N(k) \cos\frac{(2j+1)k\pi}{2N}\big)_{j,k=0}^{N-1}. \tag{3.65}$$

Obviously, $\mathbf{C}_N^{\mathrm{III}} = \big(\mathbf{C}_N^{\mathrm{II}}\big)^\top = \big(\mathbf{C}_N^{\mathrm{II}}\big)^{-1}$. The columns

$$\mathbf{c}_k^{\mathrm{III}} := \sqrt{\frac{2}{N}}\,\varepsilon_N(k)\,\big(\cos\frac{(2j+1)k\pi}{2N}\big)_{j=0}^{N-1}, \quad k = 0, \ldots, N-1,$$

of $\mathbf{C}_N^{\mathrm{III}}$ form an orthonormal basis of $\mathbb{R}^N$ and are called *cosine vectors of type* III. The matrix $\mathbf{C}_N^{\mathrm{III}}$ also generates a linear orthogonal map from $\mathbb{R}^N$ to $\mathbb{R}^N$, which is called *discrete cosine transform of type* III *and length* N, abbreviated by DCT-III (N). In particular, the DCT-III (N) is the inverse DCT-II (N).

Lemma 3.48 *Let $N \geq 2$ be a given integer. Then the set of* cosine vectors of type IV

$$\mathbf{c}_k^{\mathrm{IV}} := \sqrt{\frac{2}{N}}\,\big(\cos\frac{(2j+1)(2k+1)\pi}{4N}\big)_{j=0}^{N-1}, \quad k = 0, \ldots, N-1,$$

forms an orthonormal basis of $\mathbb{R}^N$. *The* N*-by-*N cosine matrix of type IV, *defined by the columns* $\mathbf{c}_k^{\mathrm{IV}}$,

$$\mathbf{C}_N^{\mathrm{IV}} := \sqrt{\frac{2}{N}} \Big(\cos \frac{(2j+1)(2k+1)\pi}{4N}\Big)_{j,k=0}^{N-1}, \tag{3.66}$$

is symmetric and orthogonal, i.e., $\big(\mathbf{C}_N^{\mathrm{IV}}\big)^{-1} = \mathbf{C}_N^{\mathrm{IV}}$.

Proof From

$$2 \sin x \sum_{j=0}^{N-1} \cos(2j+1)x = \sum_{j=0}^{N-1} \big(\sin(2j+2)x - \sin(2jx)\big) = \sin(2Nx)$$

it follows for $x \in \mathbb{R} \setminus \pi\mathbb{Z}$

$$\sum_{j=0}^{N-1} \cos(2j+1)x = \frac{\sin(2Nx)}{2 \sin x}$$

and hence

$$\sum_{j=0}^{N-1} \cos \frac{(2j+1)k\pi}{N} = 0\,, \quad k \in \mathbb{Z} \setminus N\mathbb{Z}\,, \tag{3.67}$$

$$\sum_{j=0}^{N-1} \cos \frac{(2j+1)(2k+1)\pi}{2N} = 0\,, \quad k \in \mathbb{Z}\,. \tag{3.68}$$

For $k,\, \ell \in \{0, \ldots, N-1\}$, we calculate the inner product:

$$\langle \mathbf{c}_k^{\mathrm{IV}}, \mathbf{c}_\ell^{\mathrm{IV}} \rangle = \frac{2}{N} \sum_{j=0}^{N-1} \cos \frac{(2j+1)(2k+1)\pi}{4N} \cos \frac{(2j+1)(2\ell+1)\pi}{4N}$$

$$= \frac{1}{N} \Big(\sum_{j=0}^{N-1} \cos \frac{(2j+1)(k-\ell)\pi}{2N} + \sum_{j=0}^{N-1} \cos \frac{(2j+1)(k+\ell+1)\pi}{2N} \Big)\,. \tag{3.69}$$

In the case $k \neq \ell$, this inner product vanishes by (3.67), (3.68), and (3.69). For $k = \ell$ we obtain by (3.69) and (3.68) that

$$\langle \mathbf{c}_k^{\mathrm{IV}}, \mathbf{c}_\ell^{\mathrm{IV}} \rangle = \frac{1}{N} \Big(N + \sum_{j=0}^{N-1} \cos \frac{(2j+1)(2k+1)\pi}{2N} \Big) = 1\,.$$

■

The linear map from $\mathbb{R}^N$ to $\mathbb{R}^N$, given by $\mathbf{C}_N^{\mathrm{IV}}\,\mathbf{a} = \big(\langle \mathbf{a},\, \mathbf{c}_k^{\mathrm{IV}}\rangle\big)_{k=0}^{N-1}$ for arbitrary $\mathbf{a} \in \mathbb{R}^N$, is called *discrete cosine transform of type* IV *and length N* and will be abbreviated by DCT-IV (N). We will study the interplay between DCT-II (N) and DCT-IV (N) in Sect. 6.3.

Analogously to DCT's of types I–IV, one can introduce discrete sine transforms.

Lemma 3.49 *Let $N \ge 2$ be a given integer. Then the set of* sine vectors of type I

$$\mathbf{s}_k^{\mathrm{I}} := \sqrt{\frac{2}{N}}\,\Big(\sin\frac{(j+1)(k+1)\pi}{N}\Big)_{j=0}^{N-2}\,, \qquad k = 0, \ldots, N-2\,,$$

forms an orthonormal basis of $\mathbb{R}^{N-1}$. The $(N-1)$-by-$(N-1)$ sine matrix of type I, *defined by*

$$\mathbf{S}_{N-1}^{\mathrm{I}} := \sqrt{\frac{2}{N}}\,\Big(\sin\frac{(j+1)(k+1)\pi}{N}\Big)_{j,k=0}^{N-2}\,, \tag{3.70}$$

is symmetric and orthogonal, i.e., $\big(\mathbf{S}_{N-1}^{\mathrm{I}}\big)^{-1} = \mathbf{S}_{N-1}^{\mathrm{I}}$.

Proof For $k,\, \ell \in \{0, \ldots, N-1\}$, we calculate the inner product:

$$\begin{aligned}\langle \mathbf{s}_k^{\mathrm{I}},\, \mathbf{s}_\ell^{\mathrm{I}}\rangle &= \frac{2}{N}\sum_{j=0}^{N-2} \sin\frac{(j+1)(k+1)\pi}{N}\,\sin\frac{(j+1)(\ell+1)\pi}{N} \\ &= \frac{1}{N}\Big(\sum_{j=1}^{N-1}\cos\frac{(k-\ell)j\pi}{N} - \sum_{j=1}^{N-1}\cos\frac{(k+\ell+2)j\pi}{N}\Big)\,. \end{aligned} \tag{3.71}$$

In the case $k \ne \ell$, we observe that $k-\ell$ and $k+\ell+2$ are either both even or both odd. Hence, $\langle \mathbf{s}_k^{\mathrm{I}},\, \mathbf{s}_\ell^{\mathrm{I}}\rangle$ vanishes by (3.60), (3.61), and (3.71). For $k = \ell$ we obtain by (3.71) and (3.61) that

$$\langle \mathbf{s}_k^{\mathrm{I}},\, \mathbf{s}_k^{\mathrm{I}}\rangle = \frac{1}{N}\Big((N-1) - \sum_{j=1}^{N-1}\cos\frac{(2k+2)j\pi}{N}\Big) = 1\,.$$

■

The linear map from $\mathbb{R}^{N-1}$ to $\mathbb{R}^{N-1}$ generated by $\mathbf{S}_{N-1}^{\mathrm{I}}\mathbf{a} = \big(\langle \mathbf{a},\, \mathbf{s}_k^{\mathrm{I}}\rangle\big)_{k=0}^{N-2}$ for arbitrary $\mathbf{a} \in \mathbb{R}^{N-1}$ is called *discrete sine transform of type* I *and length* $N-1$ and will be abbreviated by DST-I $(N-1)$.

Let $N \ge 2$ be a given integer. The N-by-N *sine matrices of type* II – IV are defined by

$$\mathbf{S}_N^{\mathrm{II}} := \sqrt{\frac{2}{N}} \left(\varepsilon_N(j+1) \sin \frac{(j+1)(2k+1)\pi}{2N}\right)_{j,k=0}^{N-1}, \tag{3.72}$$

$$\mathbf{S}_N^{\mathrm{III}} := \left(\mathbf{S}_N^{\mathrm{II}}\right)^{\top}, \tag{3.73}$$

$$\mathbf{S}_N^{\mathrm{IV}} := \sqrt{\frac{2}{N}} \left(\sin \frac{(2j+1)(2k+1)\pi}{4N}\right)_{j,k=0}^{N-1}. \tag{3.74}$$

The *discrete sine transform of types* II–IV *and length* N is the linear mapping from $\mathbb{R}^N$ to $\mathbb{R}^N$, which is generated by the matrix-vector product with the N-by-N sine matrix of types II–IV. For these discrete sine transforms, we use the abbreviations DST-II (N), DST-III (N), and DST-IV (N).

In the following lemma, we recall the intertwining relations of above cosine and sine matrices.

Lemma 3.50 *For each integer $N \geq 2$, the cosine and sine matrices satisfy the following intertwining relations:*

$$\begin{aligned}
\mathbf{C}_{N+1}^{\mathrm{I}} \mathbf{J}_{N+1} &= \mathbf{D}_{N+1} \mathbf{C}_{N+1}^{\mathrm{I}}, \quad \mathbf{S}_{N-1}^{\mathrm{I}} \mathbf{J}_{N-1} = \mathbf{D}_{N-1} \mathbf{S}_{N-1}^{\mathrm{I}}, \\
\mathbf{C}_N^{\mathrm{II}} \mathbf{J}_N &= \mathbf{D}_N \mathbf{C}_N^{\mathrm{II}}, \quad \mathbf{S}_N^{\mathrm{II}} \mathbf{J}_N = \mathbf{D}_N \mathbf{S}_N^{\mathrm{II}}, \\
\mathbf{J}_N \mathbf{C}_N^{\mathrm{III}} &= \mathbf{C}_N^{\mathrm{III}} \mathbf{D}_N, \quad \mathbf{J}_N \mathbf{S}_N^{\mathrm{III}} = \mathbf{S}_N^{\mathrm{III}} \mathbf{D}_N, \\
(-1)^{N-1} \mathbf{C}_N^{\mathrm{IV}} \mathbf{J}_N \mathbf{D}_N &= \mathbf{J}_N \mathbf{D}_N \mathbf{C}_N^{\mathrm{IV}}, \quad (-1)^{N-1} \mathbf{S}_N^{\mathrm{IV}} \mathbf{J}_N \mathbf{D}_N = \mathbf{J}_N \mathbf{D}_N \mathbf{S}_N^{\mathrm{IV}}
\end{aligned}$$

and further

$$\mathbf{J}_N \mathbf{C}_N^{\mathrm{II}} = \mathbf{S}_N^{\mathrm{II}} \mathbf{D}_N, \quad \mathbf{C}_N^{\mathrm{III}} \mathbf{J}_N = \mathbf{D}_N \mathbf{S}_N^{\mathrm{III}}, \quad \mathbf{C}_N^{\mathrm{IV}} \mathbf{J}_N = \mathbf{D}_N \mathbf{S}_N^{\mathrm{IV}}, \tag{3.75}$$

where $\mathbf{J}_N$ denotes the counter-identity matrix and where $\mathbf{D}_N := \operatorname{diag}\left((-1)^k\right)_{k=0}^{N-1}$ is the diagonal sign matrix.

The proof is straightforward and is omitted here.

Lemma 3.51 *For each integer $N \geq 2$, the N-by-N sine matrices of type* II – IV *are orthogonal. The columns of $\mathbf{S}_N^{\mathrm{II}}$, $\mathbf{S}_N^{\mathrm{III}}$, or $\mathbf{S}_N^{\mathrm{IV}}$ form an orthonormal basis of $\mathbb{R}^N$.*

Proof As shown by Lemmas 3.47 and 3.48, the cosine matrices of types II, III, and IV are orthogonal. Obviously, the matrices $\mathbf{J}_N$ and $\mathbf{D}_N$ are orthogonal. By (3.75), the sine matrices of types II–IV can be represented as products of orthogonal matrices:

$$\mathbf{S}_N^{\mathrm{II}} = \mathbf{J}_N \mathbf{C}_N^{\mathrm{II}} \mathbf{D}_N, \quad \mathbf{S}_N^{\mathrm{III}} = \left(\mathbf{S}_N^{\mathrm{II}}\right)^{\top} = \mathbf{D}_N \mathbf{C}_N^{\mathrm{III}} \mathbf{J}_N, \quad \mathbf{S}_N^{\mathrm{IV}} = \mathbf{D}_N \mathbf{C}_N^{\mathrm{IV}} \mathbf{J}_N.$$

Hence, the sine matrices $\mathbf{S}_N^{\mathrm{II}}$, $\mathbf{S}_N^{\mathrm{III}}$, and $\mathbf{S}_N^{\mathrm{IV}}$ are orthogonal, i.e., the corresponding columns of these sine matrices form orthonormal bases of $\mathbb{R}^N$. ■

Remark 3.52 First cosine and sine matrices appeared in connection with trigonometric approximation (see [198] and [371]). In signal processing, cosine matrices of types II and III were introduced in [4]. The above classification of cosine and sine matrices was given in [426] (cf. [363, pp. 12–21]).

Other proofs for the orthogonality of the cosine matrices $\mathbf{C}_{N+1}^{\mathrm{I}}$, $\mathbf{C}_N^{\mathrm{II}}$, $\mathbf{C}_N^{\mathrm{III}}$, and $\mathbf{C}_N^{\mathrm{IV}}$ can be found in [363, pp. 12–16] and [436, pp. 85–90]. G. Strang [401] pointed out that the cosine vectors of each type are eigenvectors of a symmetric second difference matrix and therefore orthogonal.

We will study discrete trigonometric transforms and their applications in more detail in Chap. 6. In Sect. 6.3, we will derive fast algorithms for these discrete trigonometric transforms with computational costs $\mathcal{O}(N \log N)$, if the transform length N is a power of two. □

Remark 3.53 The N-by-N *Hartley matrix*

$$\mathbf{H}_N := \frac{1}{N}\big(\operatorname{cas}\frac{jk\pi}{N}\big)_{j,k=0}^{N-1}$$

with $\operatorname{cas} x := \cos x + \sin x$ for $x \in \mathbb{R}$, where "cas" is an abbreviation of the expression "cosine and sine." The historical roots of this matrix go back to the introduction of continuous Hartley transform by R. Hartley in 1942. The need to sample signals and approximate the continuous Hartley transform by a matrix-vector product led to the Hartley matrix introduced by R.N. Bracewell [61]. The Hartley matrix is symmetric and orthogonal, since the *Hartley vectors*

$$\mathbf{h}_k := \frac{1}{N}\big(\operatorname{cas}\frac{jk\pi}{N}\big)_{j=0}^{N-1}, \quad k = 0, \ldots, N-1$$

form an orthonormal basis of $\mathbb{R}^N$. The linear map from $\mathbb{R}^N$ to $\mathbb{R}^N$ generated by the matrix vector product $\mathbf{H}_N\,\mathbf{a} = \big(\langle \mathbf{a}, \mathbf{h}_k\rangle\big)_{k=0}^{N-1}$ for arbitrary $\mathbf{a} \in \mathbb{R}^N$ is called *discrete Hartley transform of length N* and will be abbreviated by DHT(N). Note that the basic properties of DHT are discussed in [9] that also presents fast and numerically stable algorithms for DHT of radix-2 length N. □

Chapter 4
Multidimensional Fourier Methods

Robert Beinert, Gerlind Plonka, Daniel Potts, Gabriele Steidl, and Manfred Tasche

In this chapter, we consider d-dimensional Fourier methods for fixed $d \in \mathbb{N}$. We start with Fourier series of d-variate, 2π-periodic functions $f : \mathbb{T}^d \to \mathbb{C}$ in Sect. 4.1, where we follow the lines of Chap. 1. In particular, we present basic properties of the Fourier coefficients and learn about their decay for smooth functions.

Then, in Sect. 4.2, we deal with Fourier transforms of functions defined on $\mathbb{R}^d$. Here we follow another path than in the case $d = 1$ considered in Chap. 2. We show that the Fourier transform is a linear, bijective operator on the Schwartz space $\mathscr{S}(\mathbb{R}^d)$ of rapidly decaying functions. Using the density of $\mathscr{S}(\mathbb{R}^d)$ in $L_1(\mathbb{R}^d)$ and $L_2(\mathbb{R}^d)$, the Fourier transform on these spaces is discussed. The Poisson summation formula and the Fourier transforms of radial functions are also addressed.

In Sect. 4.3, we introduce tempered distributions as linear, continuous functionals on the Schwartz space $\mathscr{S}(\mathbb{R}^d)$. We consider Fourier transforms of tempered distributions and Fourier series of periodic tempered distributions. Further, we

---

Robert Beinert is the co-author for this chapter.

R. Beinert (✉)
Technische Universität Berlin, Institut für Mathematik, Berlin, Germany
e-mail: robert.beinert@tu-berlin.de

G. Plonka
University of Göttingen, Göttingen, Germany

D. Potts
Angewandte Funktionalanalysis, TU Chemnitz, Chemnitz, Sachsen, Germany

G. Steidl
Technical University of Berlin, Berlin, Germany

M. Tasche
University of Rostock, Rostock, Germany

G. Plonka et al., *Numerical Fourier Analysis*, Applied and Numerical Harmonic Analysis, https://doi.org/10.1007/978-3-031-35005-4_4

introduce the Hilbert transform and its multidimensional generalization, the Riesz transform.

In Sect. 4.4, we deal with Fourier transforms of measures. We show that there is a one-to-one relation between positive definite functions and Fourier transforms of positive measures. As in the case $d = 1$, any numerical application of d-dimensional Fourier series or Fourier transforms leads to d-dimensional discrete Fourier transforms handled in Sect. 4.5. We present the basic properties of the two-dimensional and higher dimensional DFT, including the convolution property and the aliasing formula.

4.1 Multidimensional Fourier Series

We consider *d-variate, 2π-periodic functions* $f : \mathbb{R}^d \to \mathbb{C}$, i.e., functions fulfilling $f(\mathbf{x}) = f(\mathbf{x} + 2\pi\,\mathbf{k})$ for all $\mathbf{x} = (x_j)_{j=1}^d \in \mathbb{R}^d$ and all $\mathbf{k} = (k_j)_{j=1}^d \in \mathbb{Z}^d$. Note that the function f is 2π-periodic in each variable x_j, $j = 1, \ldots, d$, and that f is uniquely determined by its restriction to the hypercube $[0, 2\pi)^d$. Hence f can be considered as a function defined on the d-dimensional torus $\mathbb{T}^d = \mathbb{R}^d/(2\pi\,\mathbb{Z}^d)$. For fixed $\mathbf{n} = (n_j)_{j=1}^d \in \mathbb{Z}^d$, the *d-variate complex exponential*

$$\mathrm{e}^{\mathrm{i}\,\mathbf{n}\cdot\mathbf{x}} = \prod_{j=1}^{d} \mathrm{e}^{\mathrm{i}\,n_j x_j}\,, \quad \mathbf{x} \in \mathbb{R}^d\,,$$

is 2π-periodic, where $\mathbf{n} \cdot \mathbf{x} := n_1 x_1 + \ldots + n_d x_d$ is the inner product of $\mathbf{n} \in \mathbb{Z}^d$ and $\mathbf{x} \in \mathbb{R}^d$. Further, we use the Euclidean norm $\|\mathbf{x}\|_2 := (\mathbf{x} \cdot \mathbf{x})^{1/2}$ of $\mathbf{x} \in \mathbb{R}^d$. For a multi-index $\boldsymbol{\alpha} = (\alpha_k)_{k=1}^d \in \mathbb{N}_0^d$ with $|\boldsymbol{\alpha}| = \alpha_1 + \ldots + \alpha_d$, we use the notation

$$\mathbf{x}^{\boldsymbol{\alpha}} := \prod_{k=1}^{d} x_k^{\alpha_k}\,.$$

Let $C(\mathbb{T}^d)$ be the Banach space of continuous functions $f : \mathbb{T}^d \to \mathbb{C}$ equipped with the norm

$$\|f\|_{C(\mathbb{T}^d)} := \max_{\mathbf{x}\in\mathbb{T}^d} |f(\mathbf{x})|\,.$$

By $C^r(\mathbb{T}^d)$, $r \in \mathbb{N}$, we denote the Banach space of r-times continuously differentiable functions with the norm

$$\|f\|_{C^r(\mathbb{T}^d)} := \sum_{|\boldsymbol{\alpha}|\le r} \max_{\mathbf{x}\in\mathbb{T}^d} |D^{\boldsymbol{\alpha}} f(\mathbf{x})|\,,$$

where

$$D^{\boldsymbol{\alpha}} f(\mathbf{x}) := \frac{\partial^{\alpha_1}}{\partial x_1^{\alpha_1}} \cdots \frac{\partial^{\alpha_d}}{\partial x_d^{\alpha_d}} f(\mathbf{x})$$

denotes the partial derivative with the multi-index $\boldsymbol{\alpha} = (\alpha_j)_{j=1}^d \in \mathbb{N}_0^d$ and $|\boldsymbol{\alpha}| \le r$. For $1 \le p \le \infty$, let $L_p(\mathbb{T}^d)$ denote the Banach space of all measurable functions $f : \mathbb{T}^d \to \mathbb{C}$ with finite norm

$$\|f\|_{L_p(\mathbb{T}^d)} := \begin{cases} \left(\frac{1}{(2\pi)^d} \int_{[0,\,2\pi]^d} |f(\mathbf{x})|^p \, \mathrm{d}\mathbf{x} \right)^{1/p} & 1 \le p < \infty\,, \\ \operatorname{ess\,sup} \{|f(\mathbf{x})| : \mathbf{x} \in [0,\, 2\pi]^d\} & p = \infty\,, \end{cases}$$

where almost everywhere equal functions are identified. The spaces $L_p(\mathbb{T}^d)$ with $1 < p < \infty$ are continuously embedded as

$$L_1(\mathbb{T}^d) \supset L_p(\mathbb{T}^d) \supset L_\infty(\mathbb{T}^d)\,.$$

By the periodicity of $f \in L_1(\mathbb{T}^d)$, we have

$$\int_{[0,\,2\pi]^d} f(\mathbf{x}) \, \mathrm{d}\mathbf{x} = \int_{[-\pi,\,\pi]^d} f(\mathbf{x}) \, \mathrm{d}\mathbf{x}\,.$$

For $p = 2$, we obtain the Hilbert space $L_2(\mathbb{T}^d)$ with the inner product and norm

$$\langle f,\, g\rangle_{L_2(\mathbb{T}^d)} := \frac{1}{(2\pi)^d} \int_{[0,\,2\pi]^d} f(\mathbf{x})\, \overline{g(\mathbf{x})} \, \mathrm{d}\mathbf{x}\,, \quad \|f\|_{L_2(\mathbb{T}^d)} := \sqrt{\langle f,\, f\rangle_{L_2(\mathbb{T}^d)}}$$

for arbitrary $f,\, g \in L_2(\mathbb{T}^d)$. For all $f,\, g \in L_2(\mathbb{T}^d)$, it holds the Cauchy–Schwarz inequality

$$|\langle f,\, g\rangle_{L_2(\mathbb{T}^d)}| \le \|f\|_{L_2(\mathbb{T}^d)} \, \|g\|_{L_2(\mathbb{T}^d)}\,.$$

The set of all complex exponentials $\{\mathrm{e}^{\mathrm{i}\,\mathbf{k}\cdot\mathbf{x}} : \mathbf{k} \in \mathbb{Z}^d\}$ forms an orthonormal basis of $L_2(\mathbb{T}^d)$. A linear combination of complex exponentials

$$p(\mathbf{x}) = \sum_{\mathbf{k}\in\mathbb{Z}^d} a_{\mathbf{k}}\, \mathrm{e}^{\mathrm{i}\,\mathbf{k}\cdot\mathbf{x}}$$

with only finitely many coefficients $a_{\mathbf{k}} \in \mathbb{C} \setminus \{0\}$ is called *d-variate, 2π-periodic trigonometric polynomial*. The *degree* of p is the largest number $\|\mathbf{k}\|_1 = |k_1| + \ldots + |k_d|$ such that $a_{\mathbf{k}} \ne 0$ with $\mathbf{k} = (k_j)_{j=1}^d \in \mathbb{Z}^d$. The set of all trigonometric polynomials is dense in $L_p(\mathbb{T}^d)$ for $1 \le p < \infty$ (see [178, p. 168]).

For $f \in L_1(\mathbb{T}^d)$ and arbitrary $\mathbf{k} \in \mathbb{Z}^d$, the $\mathbf{k}$th *Fourier coefficient* of f is defined as

$$c_{\mathbf{k}}(f) := \langle f(\mathbf{x}), \mathrm{e}^{\mathrm{i}\,\mathbf{k}\cdot\mathbf{x}}\rangle_{L_2(\mathbb{T}^d)} = \frac{1}{(2\pi)^d} \int_{[0,\,2\pi]^d} f(\mathbf{x})\,\mathrm{e}^{-\mathrm{i}\mathbf{k}\cdot\mathbf{x}}\,\mathrm{d}\mathbf{x}\,.$$

As in the univariate case, the $\mathbf{k}$th *modulus* and *phase* of f are defined by $|c_{\mathbf{k}}(f)|$ and $\arg c_{\mathbf{k}}(f)$, respectively. Obviously, we have

$$|c_{\mathbf{k}}(f)| \le \frac{1}{(2\pi)^d} \int_{[0,\,2\pi]^d} |f(\mathbf{x})|\,\mathrm{d}\mathbf{x} = \|f\|_{L_1(\mathbb{T}^d)}\,.$$

The Fourier coefficients possess similar properties as in the univariate setting (cf. Lemma 1.6 and Lemma 1.13).

Lemma 4.1 *The Fourier coefficients of any functions $f,\, g \in L_1(\mathbb{T}^d)$ have the following properties for all $\mathbf{k} = (k_j)_{j=1}^d \in \mathbb{Z}^d$:*

1. Uniqueness: *If $c_{\mathbf{k}}(f) = c_{\mathbf{k}}(g)$ for all $\mathbf{k} \in \mathbb{Z}^d$, then $f = g$ almost everywhere.*
2. Linearity: *For all $\alpha,\, \beta \in \mathbb{C}$,*

$$c_{\mathbf{k}}(\alpha\, f + \beta\, g) = \alpha\, c_{\mathbf{k}}(f) + \beta\, c_{\mathbf{k}}(g)\,.$$

3. Translation and modulation: *For all $\mathbf{x}_0 \in [0,\, 2\pi)^d$ and $\mathbf{k}_0 \in \mathbb{Z}^d$,*

$$c_{\mathbf{k}}\big(f(\mathbf{x} - \mathbf{x}_0)\big) = \mathrm{e}^{-\mathrm{i}\,\mathbf{k}\cdot\mathbf{x}_0}\, c_{\mathbf{k}}(f)\,,$$
$$c_{\mathbf{k}}\big(\mathrm{e}^{-\mathrm{i}\,\mathbf{k}_0\cdot\mathbf{x}}\, f(\mathbf{x})\big) = c_{\mathbf{k}+\mathbf{k}_0}(f)\,.$$

4. Differentiation: *For $f \in L_1(\mathbb{T}^d)$ with partial derivative $\frac{\partial f}{\partial x_j} \in L_1(\mathbb{T}^d)$*

$$c_{\mathbf{k}}\Big(\frac{\partial f}{\partial x_j}\Big) = \mathrm{i}\, k_j\, c_{\mathbf{k}}(f)\,.$$

5. Convolution: *For $f, g \in L_1(\mathbb{T}^d)$, the d*-variate convolution

$$(f * g)(\mathbf{x}) := \frac{1}{(2\pi)^d} \int_{[0,\,2\pi]^d} f(\mathbf{y})\, g(\mathbf{x} - \mathbf{y})\,\mathrm{d}\mathbf{y}\,, \quad \mathbf{x} \in \mathbb{R}^d\,,$$

is contained in $L_1(\mathbb{T}^d)$ and we have

$$c_{\mathbf{k}}(f * g) = c_{\mathbf{k}}(f)\, c_{\mathbf{k}}(g)\,.$$

The proof of Lemma 4.1 can be given similarly as in the univariate case and is left to the reader.

Remark 4.2 The differentiation property 4 of Lemma 4.1 can be generalized. Assume that $f \in L_1(\mathbb{R}^d)$ possesses partial derivatives $D^{\boldsymbol{\alpha}} f \in L_1(\mathbb{T}^d)$ for all multi-indices $\boldsymbol{\alpha} \in \mathbb{N}_0^d$ with $|\boldsymbol{\alpha}| \le r$, where $r \in \mathbb{N}$ is fixed. Repeated application of the differentiation property 4 of Lemma 4.1 provides

$$c_{\mathbf{k}}(D^{\boldsymbol{\alpha}} f) = (\mathrm{i}\,\mathbf{k})^{\boldsymbol{\alpha}}\, c_{\mathbf{k}}(f) \tag{4.1}$$

for all $\mathbf{k} \in \mathbb{Z}^d$, where $(\mathrm{i}\,\mathbf{k})^{\boldsymbol{\alpha}}$ denotes the product $(\mathrm{i}\,k_1)^{\alpha_1} \ldots (\mathrm{i}\,k_d)^{\alpha_d}$ with the convention $0^0 = 1$. □

Remark 4.3 If the 2π-periodic function

$$f(\mathbf{x}) = \prod_{j=1}^{d} f_j(x_j)$$

is the product of univariate functions $f_j \in L_1(\mathbb{T})$, $j = 1, \ldots, d$, then we have for all $\mathbf{k} = (k_j)_{j=1}^d \in \mathbb{Z}^d$

$$c_{\mathbf{k}}(f) = \prod_{j=1}^{d} c_{k_j}(f_j)\,.$$

□

Example 4.4 Let $n \in \mathbb{N}_0$ be given. The nth *Dirichlet kernel* $D_n : \mathbb{T}^d \to \mathbb{C}$

$$D_n(\mathbf{x}) := \sum_{k_1=-n}^{n} \ldots \sum_{k_d=-n}^{n} \mathrm{e}^{\mathrm{i}\,\mathbf{k}\cdot\mathbf{x}}$$

is a trigonometric polynomial of degree $d\,n$. It is the product of univariate nth Dirichlet kernels

$$D_n(\mathbf{x}) = \prod_{j=1}^{d} D_n(x_j)\,.$$

□

For arbitrary $n \in \mathbb{N}_0$, the nth *Fourier partial sum* of $f \in L_1(\mathbb{T}^d)$ is defined by

$$(S_n f)(\mathbf{x}) := \sum_{k_1=-n}^{n} \ldots \sum_{k_d=-n}^{n} c_{\mathbf{k}}(f)\, \mathrm{e}^{\mathrm{i}\,\mathbf{k}\cdot\mathbf{x}}\,. \tag{4.2}$$

Using the nth Dirichlet kernel D_n, the nth Fourier partial sum $S_n f$ can be represented as convolution $S_n f = f * D_n$.

For $f \in L_1(\mathbb{T}^d)$, the d-dimensional *Fourier series*

$$\sum_{\mathbf{k} \in \mathbb{Z}^d} c_{\mathbf{k}}(f)\, \mathrm{e}^{\mathrm{i}\,\mathbf{k}\cdot\mathbf{x}} \tag{4.3}$$

is called *convergent* to f in $L_2(\mathbb{T}^d)$, if the sequence of Fourier partial sums (4.2) converges to f, i.e.,

$$\lim_{n\to\infty} \|f - S_n f\|_{L_2(\mathbb{T}^d)} = 0\,.$$

Then it holds the following result on convergence in $L_2(\mathbb{T}^d)$ (cf. Lemma 1.3 for $d = 1$):

Theorem 4.5 *Every function $f \in L_2(\mathbb{T}^d)$ can be expanded into the Fourier series* (4.3) *which converges to f in $L_2(\mathbb{T}^d)$. Further the* Parseval equality

$$\|f\|^2_{L_2(\mathbb{T}^d)} = \frac{1}{(2\pi)^d} \int_{[0,\,2\pi]^d} |f(\mathbf{x})|^2\, \mathrm{d}\mathbf{x} = \sum_{\mathbf{k}\in\mathbb{Z}^d} |c_{\mathbf{k}}(f)|^2 \tag{4.4}$$

is fulfilled.

Now we investigate the relation between the smoothness of the function $f : \mathbb{T}^d \to \mathbb{C}$ and the decay of its Fourier coefficients $c_{\mathbf{k}}(f)$ as $\|\mathbf{k}\|_2 \to \infty$. We show that the smoother a function $f : \mathbb{T}^d \to \mathbb{C}$ is, the faster its Fourier coefficients $c_{\mathbf{k}}(f)$ tend to zero as $\|\mathbf{k}\|_2 \to \infty$ (cf. Lemma 1.27 and Theorem 1.39 for $d = 1$).

Lemma 4.6 *1. For $f \in L_1(\mathbb{T}^d)$ we have*

$$\lim_{\|\mathbf{k}\|_2\to\infty} c_{\mathbf{k}}(f) = 0\,. \tag{4.5}$$

2. *Let $r \in \mathbb{N}$ be given. If f and its partial derivatives $D^{\boldsymbol{\alpha}} f$ are contained in $L_1(\mathbb{T}^d)$ for all multi-indices $\boldsymbol{\alpha} \in \mathbb{N}_0^d$ with $|\boldsymbol{\alpha}| \le r$, then*

$$\lim_{\|\mathbf{k}\|_2\to\infty} (1 + \|\mathbf{k}\|_2^r)\, c_{\mathbf{k}}(f) = 0\,. \tag{4.6}$$

Proof

1. If $f \in L_2(\mathbb{T}^d)$, then (4.5) is a consequence of the Parseval equality (4.4). For all $\varepsilon > 0$, any function $f \in L_1(\mathbb{T}^d)$ can be approximated by a trigonometric polynomial p of degree n such that $\|f - p\|_{L_1(\mathbb{T}^d)} < \varepsilon$. Then the Fourier coefficients of $r := f - p \in L_1(\mathbb{T}^d)$ fulfill $|c_{\mathbf{k}}(r)| \le \|r\|_{L_1(\mathbb{T}^d)} < \varepsilon$ for all $\mathbf{k} \in \mathbb{Z}^d$. Further we have $c_{\mathbf{k}}(p) = 0$ for all $\mathbf{k} \in \mathbb{Z}^d$ with $\|\mathbf{k}\|_1 > n$, since the trigonometric polynomial p has the degree n. By the linearity of the Fourier coefficients and by $\|\mathbf{k}\|_1 \ge \|\mathbf{k}\|_2$, we obtain for all $\mathbf{k} \in \mathbb{Z}^d$ with $\|\mathbf{k}\|_2 > n$ that

$$|c_{\mathbf{k}}(f)| = |c_{\mathbf{k}}(p) + c_{\mathbf{k}}(r)| = |c_{\mathbf{k}}(r)| < \varepsilon\,.$$

2. We consider a fixed multi-index $\mathbf{k} \in \mathbb{Z}^d \setminus \{\mathbf{0}\}$ with $|k_\ell| = \max_{j=1,\ldots,d} |k_j| > 0$. From (4.1) it follows that

$$(\mathrm{i}\, k_\ell)^r\, c_{\mathbf{k}}(f) = c_{\mathbf{k}}\Big(\frac{\partial^r f}{\partial x_\ell^r}\Big)\,.$$

Using $\|\mathbf{k}\|_2 \le \sqrt{d}\, |k_\ell|$, we obtain the estimate

$$\|\mathbf{k}\|_2^r\, |c_{\mathbf{k}}(f)| \le d^{r/2}\, \Big|c_{\mathbf{k}}\Big(\frac{\partial^r f}{\partial x_\ell^r}\Big)\Big| \le d^{r/2} \max_{|\alpha|=r} |c_{\mathbf{k}}(D^\alpha f)|\,.$$

Then from (4.5) it follows the assertion (4.6). ∎

Now we consider the uniform convergence of d-dimensional Fourier series.

Theorem 4.7 *If $f \in C(\mathbb{T}^d)$ has the property*

$$\sum_{\mathbf{k}\in\mathbb{Z}^d} |c_{\mathbf{k}}(f)| < \infty\,, \tag{4.7}$$

then the d-dimensional Fourier series (4.3) *converges uniformly to f on $\mathbb{T}^d$, i.e.,*

$$\lim_{n\to\infty} \|f - S_n f\|_{C(\mathbb{T}^d)} = 0\,.$$

Proof By (4.7), the Weierstrass criterion of uniform convergence ensures that the Fourier series (4.3) converges uniformly to a continuous function

$$g(\mathbf{x}) := \sum_{\mathbf{k}\in\mathbb{Z}^d} c_{\mathbf{k}}(f)\, \mathrm{e}^{\mathrm{i}\,\mathbf{k}\cdot\mathbf{x}}.$$

Since f and g have the same Fourier coefficients, the uniqueness property in Lemma 4.1 gives $f = g$ on $\mathbb{T}^d$. ∎

Now we want to show that a sufficiently smooth function $f : \mathbb{T}^d \to \mathbb{C}$ fulfills condition (4.7). We need the following result.

Lemma 4.8 *If $2\,r > d$, then it holds*

$$\sum_{\mathbf{k}\in\mathbb{Z}^d\setminus\{\mathbf{0}\}} \|\mathbf{k}\|_2^{-2r} < \infty\,. \tag{4.8}$$

Proof For all $\mathbf{k} = (k_j)_{j=1}^d \in \mathbb{Z}^d \setminus \{\mathbf{0}\}$, we have $\|\mathbf{k}\|_2 \ge 1$. Using the inequality of arithmetic and geometric means, it follows

$$(d+1)\, \|\mathbf{k}\|_2^2 \ge d + \|\mathbf{k}\|_2^2 = \sum_{j=1}^d (1 + k_j^2) \ge d\, \Big(\prod_{j=1}^d (1+k_j^2)\Big)^{1/d}$$

and hence

$$\|\mathbf{k}\|_2^{-2r} \le \Big(\frac{d+1}{d}\Big)^r \prod_{j=1}^{d} (1+k_j^2)^{-r/d}\,.$$

Consequently, we obtain

$$\sum_{\mathbf{k}\in\mathbb{Z}^d\setminus\{\mathbf{0}\}} \|\mathbf{k}\|_2^{-2r} \le \Big(\frac{d+1}{d}\Big)^r \sum_{k_1\in\mathbb{Z}} (1+k_1^2)^{-r/d} \ldots \sum_{k_d\in\mathbb{Z}} (1+k_d^2)^{-r/d}$$

$$= \Big(\frac{d+1}{d}\Big)^r \Big(\sum_{k\in\mathbb{Z}} (1+k^2)^{-r/d}\Big)^d < \Big(\frac{d+1}{d}\Big)^r \Big(1+2\sum_{k=1}^{\infty} k^{-2r/d}\Big)^d < \infty.$$

■

Theorem 4.9 *If $f \in C^r(\mathbb{T}^d)$ with $2r > d$, then the condition* (4.7) *is fulfilled and the d-dimensional Fourier series* (4.3) *converges uniformly to f on $\mathbb{T}^d$.*

Proof By assumption, each partial derivative $D^{\boldsymbol{\alpha}} f$ with $|\boldsymbol{\alpha}| \le r$ is continuous on $\mathbb{T}^d$. Hence we have $D^{\boldsymbol{\alpha}} f \in L_2(\mathbb{T}^d)$ such that by (4.1) and the Parseval equality (4.4),

$$\sum_{|\boldsymbol{\alpha}|=r} \sum_{\mathbf{k}\in\mathbb{Z}^d} |c_{\mathbf{k}}(f)|^2\, \mathbf{k}^{2\boldsymbol{\alpha}} < \infty\,,$$

where $\mathbf{k}^{2\boldsymbol{\alpha}}$ denotes the product $k_1^{2\alpha_1} \ldots k_d^{2\alpha_d}$ with $0^0 := 1$. Then there exists a positive constant c, depending only on the dimension d and on r, such that

$$\sum_{|\boldsymbol{\alpha}|=r} \mathbf{k}^{2\boldsymbol{\alpha}} \ge c\, \|\mathbf{k}\|_2^{2r}\,.$$

By the Cauchy–Schwarz inequality in $\ell_2(\mathbb{Z}^d)$ and by Lemma 4.8, we obtain

$$\sum_{\mathbf{k}\in\mathbb{Z}^d\setminus\{\mathbf{0}\}} |c_{\mathbf{k}}(f)| \le \sum_{\mathbf{k}\in\mathbb{Z}^d\setminus\{\mathbf{0}\}} |c_{\mathbf{k}}(f)|\, \Big(\sum_{|\boldsymbol{\alpha}|=r} \mathbf{k}^{2\boldsymbol{\alpha}}\Big)^{1/2} c^{-1/2}\, \|\mathbf{k}\|_2^{-r}$$

$$\le \Big(\sum_{|\boldsymbol{\alpha}|=r} \sum_{\mathbf{k}\in\mathbb{Z}^d} |c_{\mathbf{k}}(f)|^2\, \mathbf{k}^{2\boldsymbol{\alpha}}\Big)^{1/2} \Big(\sum_{\mathbf{k}\in\mathbb{Z}^d\setminus\{\mathbf{0}\}} \|\mathbf{k}\|_2^{-2r}\Big)^{1/2} c^{-1/2} < \infty\,.$$

■

For further discussion on the theory of multidimensional Fourier series, see [399, pp. 245–250], [178, pp. 161–248], and [386, pp. 1–137].

4.2 Multidimensional Fourier Transform

Let $C_b(\mathbb{R}^d)$ denote the Banach space of all bounded, continuous functions $f : \mathbb{R}^d \to \mathbb{C}$ and $C_0(\mathbb{R}^d)$ the Banach space of all continuous functions vanishing as $\|\mathbf{x}\|_2 \to \infty$, with norm

$$\|f\|_\infty := \sup_{\mathbf{x}\in\mathbb{R}^d} |f(\mathbf{x})| .$$

Let $C_c(\mathbb{R}^d)$ be the subspace of all continuous functions with compact supports. By $C^r(\mathbb{R}^d)$, $r \in \mathbb{N} \cup \{\infty\}$, we denote the set of r-times continuously differentiable functions and by $C_c^r(\mathbb{R}^d)$ the set of r-times continuously differentiable functions with compact supports.

For $1 \le p \le \infty$, let $L_p(\mathbb{R}^d)$ be the Banach space of all measurable functions $f : \mathbb{R}^d \to \mathbb{C}$ with finite norm

$$\|f\|_{L_p(\mathbb{R}^d)} := \begin{cases} \big(\int_{\mathbb{R}^d} |f(\mathbf{x})|^p \, \mathrm{d}\mathbf{x}\big)^{1/p} & 1 \le p < \infty , \\ \operatorname{ess\,sup} \{|f(\mathbf{x})| : \mathbf{x} \in \mathbb{R}^d\} & p = \infty , \end{cases}$$

where almost everywhere equal functions are identified. In particular, we are interested in the Hilbert space $L_2(\mathbb{R}^d)$ with inner product and norm

$$\langle f, g\rangle_{L_2(\mathbb{R}^d)} := \int_{\mathbb{R}^d} f(\mathbf{x}) \, \overline{g(\mathbf{x})} \, \mathrm{d}\mathbf{x} , \quad \|f\|_{L_2(\mathbb{R}^d)} := \Big(\int_{\mathbb{R}^d} |f(\mathbf{x})|^2 \, \mathrm{d}\mathbf{x}\Big)^{1/2}.$$

4.2.1 Fourier Transform on $\mathcal{S}(\mathbb{R}^d)$

By $\mathcal{S}(\mathbb{R}^d)$, we denote the set of all functions $\varphi \in C^\infty(\mathbb{R}^d)$ with the property $\mathbf{x}^{\boldsymbol{\alpha}} D^{\boldsymbol{\beta}} \varphi(\mathbf{x}) \in C_0(\mathbb{R}^d)$ for all multi-indices $\boldsymbol{\alpha}, \boldsymbol{\beta} \in \mathbb{N}_0^d$. We define the *convergence in* $\mathcal{S}(\mathbb{R}^d)$ as follows: A sequence $(\varphi_k)_{k\in\mathbb{N}}$ of functions $\varphi_k \in \mathcal{S}(\mathbb{R}^d)$ converges to $\varphi \in \mathcal{S}(\mathbb{R}^d)$, if for all multi-indices $\boldsymbol{\alpha}, \boldsymbol{\beta} \in \mathbb{N}_0^d$, the sequences $\big(\mathbf{x}^{\boldsymbol{\alpha}} D^{\boldsymbol{\beta}} \varphi_k\big)_{k\in\mathbb{N}}$ converge uniformly to $\mathbf{x}^{\boldsymbol{\alpha}} D^{\boldsymbol{\beta}} \varphi$ on $\mathbb{R}^d$. We will write $\varphi_k \underset{\mathcal{S}}{\longrightarrow} \varphi$ as $k \to \infty$. Then the linear space $\mathcal{S}(\mathbb{R}^d)$ with this convergence is called *Schwartz space* or *space of rapidly decreasing functions*. The name is in honor of the French mathematician L. Schwartz (1915–2002).

Any function $\varphi \in \mathcal{S}(\mathbb{R}^d)$ is *rapidly decreasing* in the sense that for all multi-indices $\boldsymbol{\alpha}, \boldsymbol{\beta} \in \mathbb{N}_0^d$

$$\lim_{\|\mathbf{x}\|_2\to\infty} \mathbf{x}^{\boldsymbol{\alpha}} D^{\boldsymbol{\beta}} \varphi(\mathbf{x}) = 0 .$$

Introducing

$$\|\varphi\|_m := \max_{|\boldsymbol{\beta}|\le m} \|(1 + \|\mathbf{x}\|_2)^m \, D^{\boldsymbol{\beta}} \varphi(\mathbf{x})\|_{C_0(\mathbb{R}^d)} , \quad m \in \mathbb{N}_0 , \tag{4.9}$$

we see that $\|\varphi\|_0 \le \|\varphi\|_1 \le \|\varphi\|_2 \le \ldots$ are finite for all $\varphi \in \mathscr{S}(\mathbb{R}^d)$. We can describe the convergence in the Schwartz space by the help of (4.9).

Lemma 4.10 *For $\varphi_k, \varphi \in \mathscr{S}(\mathbb{R}^d)$, we have $\varphi_k \underset{\mathscr{S}}{\longrightarrow} \varphi$ as $k \to \infty$ if and only if for all $m \in \mathbb{N}_0$*

$$\lim_{k\to\infty} \|\varphi_k - \varphi\|_m = 0\,. \tag{4.10}$$

Proof

1. Let (4.10) be fulfilled for all $m \in \mathbb{N}_0$. Then for all $\boldsymbol{\alpha} = (\alpha_j)_{j=1}^d \in \mathbb{N}_0^d \setminus \{\mathbf{0}\}$ with $|\boldsymbol{\alpha}| \le m$, we get by the relation between geometric and quadratic means that

$$|\mathbf{x}^{\boldsymbol{\alpha}}| \le \left(\frac{\alpha_1 x_1^2 + \ldots + \alpha_d x_d^2}{|\boldsymbol{\alpha}|}\right)^{|\boldsymbol{\alpha}|/2} \le \left(x_1^2 + \ldots + x_d^2\right)^{|\boldsymbol{\alpha}|/2} \le (1 + \|\mathbf{x}\|_2)^m$$

so that

$$|\mathbf{x}^{\boldsymbol{\alpha}} D^{\boldsymbol{\beta}}(\varphi_k - \varphi)(\mathbf{x})| \le (1 + \|\mathbf{x}\|_2)^m\, |D^{\boldsymbol{\beta}}(\varphi_k - \varphi)(\mathbf{x})|.$$

Hence, for all $\boldsymbol{\beta} \in \mathbb{N}_0^d$ with $|\boldsymbol{\beta}| \le m$, it holds

$$\|\mathbf{x}^{\boldsymbol{\alpha}} D^{\boldsymbol{\beta}}(\varphi_k - \varphi)(\mathbf{x})\|_{C_0(\mathbb{R}^d)} \le \sup_{\mathbf{x}\in\mathbb{R}^d} (1 + \|\mathbf{x}\|_2)^m\, |D^{\boldsymbol{\beta}}(\varphi_k - \varphi)(\mathbf{x})| \le \|\varphi_k - \varphi\|_m\,.$$

2. Assume that $\varphi_k \underset{\mathscr{S}}{\longrightarrow} \varphi$ as $k \to \infty$, i.e., for all $\boldsymbol{\alpha}, \boldsymbol{\beta} \in \mathbb{N}_0^d$, we have

$$\lim_{k\to\infty} \|\mathbf{x}^{\boldsymbol{\alpha}} D^{\boldsymbol{\beta}}(\varphi_k - \varphi)(\mathbf{x})\|_{C_0(\mathbb{R}^d)} = 0\,.$$

We consider multi-indices $\boldsymbol{\alpha}, \boldsymbol{\beta} \in \mathbb{N}_0^d$ with $|\boldsymbol{\alpha}| \le m$ and $|\boldsymbol{\beta}| \le m$ for $m \in \mathbb{N}$. Since norms in $\mathbb{R}^n$ are equivalent, we obtain

$$\begin{aligned}(1 + \|\mathbf{x}\|_2)^m &\le C_1(1 + \|\mathbf{x}\|_2^m) \le C_1\Big(1 + C_2 \sum_{j=1}^d |x_j|^m\Big)\\ &\le C \sum_{|\boldsymbol{\alpha}|\le m} |\mathbf{x}^{\boldsymbol{\alpha}}|\,.\end{aligned}$$

This implies

$$\|(1 + \|\mathbf{x}\|_2)^m\, D^{\boldsymbol{\beta}}(\varphi_k - \varphi)(\mathbf{x})\|_{C_0(\mathbb{R}^d)} \le C \sum_{|\boldsymbol{\alpha}|\le m} \|\mathbf{x}^{\boldsymbol{\alpha}}\, D^{\boldsymbol{\beta}}(\varphi_k - \varphi)(\mathbf{x})\|_{C_0(\mathbb{R}^d)}$$

and hence

$$\|\varphi_k - \varphi\|_m \leq C \max_{|\boldsymbol{\beta}|\leq m} \sum_{|\boldsymbol{\alpha}|\leq m} \|\mathbf{x}^{\boldsymbol{\alpha}} D^{\boldsymbol{\beta}}(\varphi_k - \varphi)(\mathbf{x})\|_{C_0(\mathbb{R}^d)}$$

such that $\lim_{k\to\infty} \|\varphi_k - \varphi\|_m = 0$. ■

Remark 4.11 Using Lemma 4.10 it is not hard to check that the convergence in the Schwartz space $\mathscr{S}(\mathbb{R}^d)$ is induced by the *metric*

$$\rho(\varphi, \psi) := \sum_{m=0}^{\infty} \frac{1}{2^m} \frac{\|\varphi - \psi\|_m}{1 + \|\varphi - \psi\|_m}, \quad \varphi, \psi \in \mathscr{S}(\mathbb{R}^d),$$

i.e., the convergence $\varphi_k \underset{\mathscr{S}}{\longrightarrow} \varphi$ as $k \to \infty$ is equivalent to

$$\lim_{k\to\infty} \rho(\varphi_k, \varphi) = 0.$$

Moreover, the metric space is complete by the following reason: Let $(\varphi_k)_{k\in\mathbb{N}}$ be a Cauchy sequence with respect to ρ. Then, for every $\boldsymbol{\alpha}, \boldsymbol{\beta} \in \mathbb{N}_0^d$, $\left(x^{\boldsymbol{\alpha}} D^{\boldsymbol{\beta}} \varphi_k\right)_{k\in\mathbb{N}}$ is a Cauchy sequence in Banach space $C_0(\mathbb{R}^d)$ and converges uniformly to a function $\psi_{\boldsymbol{\alpha},\boldsymbol{\beta}}$. Then, by definition of $\mathscr{S}(\mathbb{R}^d)$, it follows $\psi_{\boldsymbol{\alpha},\boldsymbol{\beta}}(\mathbf{x}) = \mathbf{x}^{\boldsymbol{\alpha}} D^{\boldsymbol{\beta}} \psi_{\mathbf{0},\mathbf{0}}(\mathbf{x})$ with $\psi_{\mathbf{0},\mathbf{0}} \in \mathscr{S}(\mathbb{R}^d)$ and hence $\varphi_k \underset{\mathscr{S}}{\longrightarrow} \psi_{\mathbf{0},\mathbf{0}}$ as $k \to \infty$.

Note that the metric ρ is not generated by a norm, since $\rho(c\,\varphi, 0) \neq |c|\,\rho(\varphi, 0)$ for all $c \in \mathbb{C} \setminus \{0\}$ with $|c| \neq 1$ and non-vanishing $\varphi \in \mathscr{S}(\mathbb{R}^d)$. The Schwartz space is a so-called locally convex space. □

Clearly, it holds $\mathscr{S}(\mathbb{R}^d) \subset C_0(\mathbb{R}^d) \subset L_\infty(\mathbb{R}^d)$ and $\mathscr{S}(\mathbb{R}^d) \subset L_p(\mathbb{R}^d)$, $1 \leq p < \infty$, by the following argument: For each $\varphi \in \mathscr{S}(\mathbb{R}^d)$, we have by (4.9)

$$|\varphi(\mathbf{x})| \leq \|\varphi\|_{d+1}\,(1 + \|\mathbf{x}\|_2)^{-d-1}$$

for all $\mathbf{x} \in \mathbb{R}^d$. Then, using polar coordinates with $r = \|\mathbf{x}\|_2$, we obtain

$$\begin{aligned}\int_{\mathbb{R}^d} |\varphi(\mathbf{x})|^p\,\mathrm{d}\mathbf{x} &\leq \|\varphi\|_{d+1}^p \int_{\mathbb{R}^d} (1 + \|\mathbf{x}\|_2)^{-p(d+1)}\,\mathrm{d}\mathbf{x} \\ &\leq C \int_0^\infty \frac{r^{d-1}}{(1+r)^{p(d+1)}}\,\mathrm{d}r \leq C \int_0^\infty \frac{1}{(1+r)^2}\,\mathrm{d}r < \infty\end{aligned}$$

with some constant $C > 0$. Hence the Schwartz space $\mathscr{S}(\mathbb{R}^d)$ is contained in $L_1(\mathbb{R}^d) \cap L_2(\mathbb{R}^d)$.

Obviously, $C_c^\infty(\mathbb{R}^d) \subset \mathscr{S}(\mathbb{R}^d)$. Since $C_c^\infty(\mathbb{R}^d)$ is dense in $L_p(\mathbb{R}^d)$, $p \in [1, \infty)$ (see, e.g., [416, Satz 3.6]), we also have that $\mathscr{S}(\mathbb{R}^d)$ is dense in $L_p(\mathbb{R}^d)$, $p \in [1, \infty)$. Summarizing, it holds

$$C_c^\infty(\mathbb{R}^d) \subset \mathscr{S}(\mathbb{R}^d) \subset C_0^\infty(\mathbb{R}^d) \subset C^\infty(\mathbb{R}^d). \tag{4.11}$$

Example 4.12 A typical function in $C_c^\infty(\mathbb{R}^d) \subset \mathscr{S}(\mathbb{R}^d)$ is the test function

$$\varphi(\mathbf{x}) := \begin{cases} \exp\big(-\frac{1}{1-\|\mathbf{x}\|_2^2}\big) & \|\mathbf{x}\|_2 < 1\,, \\ 0 & \|\mathbf{x}\|_2 \geq 1\,. \end{cases} \tag{4.12}$$

The compact support of φ is the unit ball $\{\mathbf{x} \in \mathbb{R}^d : \|\mathbf{x}\|_2 \leq 1\}$.

Any Gaussian function $\mathrm{e}^{-a\,\|\mathbf{x}\|_2^2}$ with $a > 0$ is contained in $\mathscr{S}(\mathbb{R}^d)$, but it is not in $C_c^\infty(\mathbb{R}^d)$.

For any $n \in \mathbb{N}$, the function

$$f(\mathbf{x}) := (1 + \|\mathbf{x}\|_2^2)^{-n} \in C_0^\infty(\mathbb{R}^d)$$

does not belong to $\mathscr{S}(\mathbb{R}^d)$, since $\|\mathbf{x}\|_2^{2n}\, f(\mathbf{x})$ does not tend to zero as $\|\mathbf{x}\|_2 \to \infty$. □

Example 4.13 In the univariate case, each product of a polynomial and the Gaussian function $\mathrm{e}^{-x^2/2}$ is a rapidly decreasing function. By Theorem 2.25, the Hermite functions $h_n(x) = H_n(x)\,\mathrm{e}^{-x^2/2}$, $n \in \mathbb{N}_0$, are contained in $\mathscr{S}(\mathbb{R})$ and form an orthogonal basis of $L_2(\mathbb{R})$. Here H_n denotes the nth Hermite polynomial. Thus $\mathscr{S}(\mathbb{R})$ is dense in $L_2(\mathbb{R})$. For each multi-index $\mathbf{n} = (n_j)_{j=1}^d \in \mathbb{N}_0^d$, the function $\mathbf{x^n}\,\mathrm{e}^{-\|\mathbf{x}\|_2^2/2}$, $\mathbf{x} = (x_j)_{j=1}^d \in \mathbb{R}^d$, is a rapidly decreasing function. The set of all functions

$$h_{\mathbf{n}}(\mathbf{x}) := \mathrm{e}^{-\|\mathbf{x}\|_2^2/2} \prod_{j=1}^{d} H_{n_j}(x_j) \in \mathscr{S}(\mathbb{R}^d)\,, \quad \mathbf{n} \in \mathbb{N}_0^d\,,$$

is an orthogonal basis of $L_2(\mathbb{R}^d)$. Further $\mathscr{S}(\mathbb{R}^d)$ is dense in $L_2(\mathbb{R}^d)$. □

For $f \in L_1(\mathbb{R}^d)$ we define its *Fourier transform* at $\boldsymbol{\omega} \in \mathbb{R}^d$ by

$$\mathscr{F} f(\boldsymbol{\omega}) = \hat{f}(\boldsymbol{\omega}) := \int_{\mathbb{R}^d} f(\mathbf{x})\,\mathrm{e}^{-\mathrm{i}\,\mathbf{x}\cdot\boldsymbol{\omega}}\,\mathrm{d}\mathbf{x}\,, \tag{4.13}$$

where $\mathbf{x}\cdot\boldsymbol{\omega} := x_1\omega_1 + \ldots + x_d\omega_d$. Since

$$|\hat{f}(\boldsymbol{\omega})| \leq \int_{\mathbb{R}^d} |f(\mathbf{x})|\,\mathrm{d}\mathbf{x} = \|f\|_{L_1(\mathbb{R}^d)}\,,$$

the Fourier transform (4.13) exists for all $\boldsymbol{\omega} \in \mathbb{R}^d$ and is bounded on $\mathbb{R}^d$.

Example 4.14 Let $L > 0$ be given. The characteristic function $f(\mathbf{x})$ of the hypercube $[-L,\,L]^d \subset \mathbb{R}^d$ is the product $\prod_{j=1}^d \chi_{[-L,\,L]}(x_j)$ of univariate characteristic functions. By Example 2.3, the related Fourier transform reads as follows:

$$\hat{f}(\boldsymbol{\omega}) = (2L)^d \prod_{j=1}^{d} \mathrm{sinc}(L\omega_j)\,.$$

□

Example 4.15 The Gaussian function $f(\mathbf{x}) := \mathrm{e}^{-\|\sigma \mathbf{x}\|_2^2/2}$ with fixed $\sigma > 0$ is the product of the univariate functions $f(x_j) = \mathrm{e}^{-\sigma^2 x_j^2/2}$ such that by Example 2.6,

$$\hat{f}(\boldsymbol{\omega}) = \left(\frac{2\pi}{\sigma^2}\right)^{d/2} \mathrm{e}^{-\|\boldsymbol{\omega}\|_2^2/(2\sigma^2)} .$$

□

By the following theorem, the Fourier transform maps the Schwartz space $\mathscr{S}(\mathbb{R}^d)$ into itself.

Theorem 4.16 *For every $\varphi \in \mathscr{S}(\mathbb{R}^d)$, it holds $\mathscr{F}\varphi \in \mathscr{S}(\mathbb{R}^d)$, i.e., $\mathscr{F} : \mathscr{S}(\mathbb{R}^d) \to \mathscr{S}(\mathbb{R}^d)$. Furthermore, we have $D^{\boldsymbol{\alpha}}(\mathscr{F}\varphi) \in \mathscr{S}(\mathbb{R}^d)$ and $\mathscr{F}(D^{\boldsymbol{\alpha}}\varphi) \in \mathscr{S}(\mathbb{R}^d)$ with*

$$D^{\boldsymbol{\alpha}}(\mathscr{F}\varphi) = (-\mathrm{i})^{|\boldsymbol{\alpha}|}\, \mathscr{F}(\mathbf{x}^{\boldsymbol{\alpha}}\,\varphi)\,, \tag{4.14}$$

$$\boldsymbol{\omega}^{\boldsymbol{\alpha}}\,(\mathscr{F}\varphi) = (-\mathrm{i})^{|\boldsymbol{\alpha}|}\, \mathscr{F}(D^{\boldsymbol{\alpha}}\,\varphi)\,, \tag{4.15}$$

for all $\boldsymbol{\alpha} \in \mathbb{N}_0^d$, where the partial derivative $D^{\boldsymbol{\alpha}}$ in (4.14) acts on $\boldsymbol{\omega}$ and in (4.15) on $\mathbf{x}$.

Proof

1. We consider $\boldsymbol{\alpha} = \mathbf{e}_1$. Using Fubini's theorem, we obtain

$$\begin{aligned}\frac{\partial}{\partial\omega_1}(\mathscr{F}\varphi)(\boldsymbol{\omega}) &= \lim_{h\to 0}\frac{1}{h}\big((\mathscr{F}\varphi)(\boldsymbol{\omega}+h\mathbf{e}_1) - (\mathscr{F}\varphi)(\boldsymbol{\omega})\big)\\ &= \lim_{h\to 0}\int_{\mathbb{R}^d}\varphi(\mathbf{x})\frac{1}{h}\left(\mathrm{e}^{-\mathrm{i}\mathbf{x}\cdot(\boldsymbol{\omega}+h\mathbf{e}_1)} - \mathrm{e}^{-\mathrm{i}\mathbf{x}\cdot\boldsymbol{\omega}}\right)\mathrm{d}\mathbf{x} \qquad (4.16)\\ &= \lim_{h\to 0}\int_{\mathbb{R}^{d-1}}\mathrm{e}^{-\mathrm{i}\tilde{\mathbf{x}}\cdot\tilde{\boldsymbol{\omega}}}\int_{\mathbb{R}}\varphi(\mathbf{x})\frac{1}{h}\left(\mathrm{e}^{-2x_1(\omega_1+h)} - \mathrm{e}^{-\mathrm{i}x_1\omega_1}\right)\mathrm{d}x_1\,\mathrm{d}\tilde{\mathbf{x}}\end{aligned}$$

where $\tilde{\mathbf{x}} := (x_2, \ldots, x_d)$. Now $g(\omega) := \mathrm{e}^{-\mathrm{i}x\omega}$ is Lipschitz continuous with Lipschitz constant $\sup_\omega |g'(\omega)| = \sup_\omega |-\mathrm{i}x\,\mathrm{e}^{-\mathrm{i}x\omega}| = |x|$ so that

$$\frac{1}{h}\left|\mathrm{e}^{-\mathrm{i}x_1(\omega_1+h)} - \mathrm{e}^{-\mathrm{i}x_1\omega_1}\right| \le |x_1|.$$

Since $\varphi \in \mathscr{S}(\mathbb{R}^d)$, we conclude that $|\varphi(\mathbf{x})\,x_1|$ is an integrable upper bound of the sequence in the integrand of (4.16). Lebesque's dominated convergence theorem can change the order of differentiation and integration which results in

$$\begin{aligned}\frac{\partial}{\partial\omega_1}(\mathscr{F}\varphi)(\boldsymbol{\omega}) &= \int_{\mathbb{R}^d}\varphi(\mathbf{x})\frac{\partial}{\partial\omega_1}\mathrm{e}^{-\mathrm{i}\mathbf{x}\cdot\boldsymbol{\omega}}\,\mathrm{d}\mathbf{x}\\ &= -\mathrm{i}\int_{\mathbb{R}^d}\varphi(\mathbf{x})\,x_1\,\mathrm{e}^{-\mathrm{i}\,\mathbf{x}\cdot\boldsymbol{\omega}}\,\mathrm{d}\mathbf{x} = -\mathrm{i}\,(\mathscr{F}(x_1\varphi)(\boldsymbol{\omega})).\end{aligned}$$

For arbitrary $\boldsymbol{\alpha} \in \mathbb{N}_0^d$ the assertion follows by induction.

2. We start by considering $\boldsymbol{\alpha} = \mathbf{e}_1$. From the theorem of Fubini, it follows

$$\omega_1\,(\mathcal{F}\varphi)(\boldsymbol{\omega}) = \int_{\mathbb{R}^d} \omega_1\,\mathrm{e}^{-\mathrm{i}\,\mathbf{x}\cdot\boldsymbol{\omega}}\,\varphi(\mathbf{x})\,\mathrm{d}\mathbf{x}$$
$$= \int_{\mathbb{R}^{d-1}} \mathrm{e}^{-\mathrm{i}\,\tilde{\mathbf{x}}\cdot\tilde{\boldsymbol{\omega}}} \int_{\mathbb{R}} \mathrm{i}\omega_1\,\mathrm{e}^{-\mathrm{i}\,x_1\omega_1}\,\varphi(\mathbf{x})\,\mathrm{d}x_1\,\mathrm{d}x_2\ldots\mathrm{d}x_d\,.$$

For the inner integral, integration by parts yields

$$\int_{\mathbb{R}} \mathrm{i}\omega_1\,\mathrm{e}^{-\mathrm{i}\omega_1 x_1}\,\varphi(\mathbf{x})\,\mathrm{d}x_1 = \int_{\mathbb{R}} \mathrm{e}^{-\mathrm{i}\,x_1\omega_1}\,\frac{\partial}{\partial x_1}\varphi(\mathbf{x})\,\mathrm{d}x_1\,.$$

Thus we obtain

$$\omega_1\,(\mathcal{F}\varphi)(\boldsymbol{\omega}) = -\mathrm{i}\,\mathcal{F}\big(\frac{\partial}{\partial x_1}\varphi\big)(\boldsymbol{\omega})\,.$$

For an arbitrary multi-index $\boldsymbol{\alpha} \in \mathbb{N}_0^d$, formula (4.15) follows by induction.

3. From (4.14) and (4.15), it follows for all multi-indices $\boldsymbol{\alpha},\ \boldsymbol{\beta} \in \mathbb{N}_0^d$ and each $\varphi \in \mathcal{S}(\mathbb{R}^d)$

$$\boldsymbol{\omega}^{\boldsymbol{\alpha}}\,[D^{\boldsymbol{\beta}}(\mathcal{F}\varphi)] = (-\mathrm{i})^{|\boldsymbol{\beta}|}\,\boldsymbol{\omega}^{\boldsymbol{\alpha}}\,\mathcal{F}(\mathbf{x}^{\boldsymbol{\beta}}\varphi) = (-\mathrm{i})^{|\boldsymbol{\alpha}|+|\boldsymbol{\beta}|}\,\mathcal{F}[D^{\boldsymbol{\alpha}}(\mathbf{x}^{\boldsymbol{\beta}}\varphi)]\,. \tag{4.17}$$

Since

$$|\boldsymbol{\omega}^{\boldsymbol{\alpha}}\,[D^{\boldsymbol{\beta}}(\mathcal{F}\varphi)](\boldsymbol{\omega})| = |\mathcal{F}[D^{\boldsymbol{\alpha}}(\mathbf{x}^{\boldsymbol{\beta}}\varphi)](\boldsymbol{\omega})| \le \int_{\mathbb{R}^d} |D^{\boldsymbol{\alpha}}(\mathbf{x}^{\boldsymbol{\beta}}\varphi)|\,\mathrm{d}\mathbf{x} < \infty\,,$$

we conclude that $\boldsymbol{\omega}^{\boldsymbol{\alpha}}\,[D^{\boldsymbol{\beta}}(\mathcal{F}\varphi)](\boldsymbol{\omega})$ is uniformly bounded on $\mathbb{R}^d$, so that $\mathcal{F}\varphi \in \mathcal{S}(\mathbb{R}^d)$. ■

Remark 4.17 The *Leibniz product rule* for the partial differentiation of the product of two functions reads as follows:

$$D^{\boldsymbol{\alpha}}(\varphi\,\psi) = \sum_{\boldsymbol{\beta}\le\boldsymbol{\alpha}} \binom{\boldsymbol{\alpha}}{\boldsymbol{\beta}}\,(D^{\boldsymbol{\beta}}\varphi)\,(D^{\boldsymbol{\alpha}-\boldsymbol{\beta}}\psi) \tag{4.18}$$

with $\boldsymbol{\alpha} = (\alpha_j)_{j=1}^d \in \mathbb{N}_0^d$, where the sum runs over all $\boldsymbol{\beta} = (\beta_j)_{j=1}^d \in \mathbb{N}_0^d$ with $\beta_j \le \alpha_j$ for $j = 1, \ldots, d$, and where

$$\binom{\boldsymbol{\alpha}}{\boldsymbol{\beta}} := \frac{\alpha_1!\,\ldots\,\alpha_d!}{\beta_1!\,\ldots\,\beta_d!\,(\alpha_1-\beta_1)!\,\ldots\,(\alpha_d-\beta_d)!}\,.$$

□

Based on Theorem 4.16, we can show that the Fourier transform is indeed a bijection on $\mathscr{S}(\mathbb{R}^d)$.

Theorem 4.18 *The Fourier transform* $\mathscr{F} : \mathscr{S}(\mathbb{R}^d) \to \mathscr{S}(\mathbb{R}^d)$ *is a linear, bijective mapping. Further, the Fourier transform is continuous with respect to the convergence in* $\mathscr{S}(\mathbb{R}^d)$*, i.e., for* $\varphi_k, \varphi \in \mathscr{S}(\mathbb{R}^d)$*,* $\varphi_k \underset{\mathscr{S}}{\longrightarrow} \varphi$ *as* $k \to \infty$ *implies* $\mathscr{F}\varphi_k \underset{\mathscr{S}}{\longrightarrow} \mathscr{F}\varphi$ *as* $k \to \infty$*.*

For all $\varphi \in \mathscr{S}(\mathbb{R}^d)$*, the* inverse Fourier transform $\mathscr{F}^{-1} : \mathscr{S}(\mathbb{R}^d) \to \mathscr{S}(\mathbb{R}^d)$ *is given by*

$$(\mathscr{F}^{-1}\varphi)(\mathbf{x}) := \frac{1}{(2\pi)^d} \int_{\mathbb{R}^d} \varphi(\boldsymbol{\omega})\, \mathrm{e}^{\mathrm{i}\,\mathbf{x}\cdot\boldsymbol{\omega}}\, \mathrm{d}\boldsymbol{\omega}\,. \tag{4.19}$$

The inverse Fourier transform is also a linear, bijective mapping on $\mathscr{S}(\mathbb{R}^d)$ *which is continuous with respect to the convergence in* $\mathscr{S}(\mathbb{R}^d)$*. Further, for all* $\varphi \in \mathscr{S}(\mathbb{R}^d)$ *and all* $\mathbf{x} \in \mathbb{R}^d$*, it holds the* Fourier inversion formula

$$\varphi(\mathbf{x}) = \frac{1}{(2\pi)^d} \int_{\mathbb{R}^d} (\mathscr{F}\varphi)(\boldsymbol{\omega})\, \mathrm{e}^{\mathrm{i}\,\mathbf{x}\cdot\boldsymbol{\omega}}\, \mathrm{d}\boldsymbol{\omega}\,.$$

Proof

1. By Theorem 4.16 the Fourier transform $\mathscr{F}$ maps the Schwartz space $\mathscr{S}(\mathbb{R}^d)$ into itself. The linearity of the Fourier transform $\mathscr{F}$ follows from those of the integral operator (4.13). For arbitrary $\varphi \in \mathscr{S}(\mathbb{R}^d)$, for all $\boldsymbol{\alpha}, \boldsymbol{\beta} \in \mathbb{N}_0^d$ with $|\boldsymbol{\alpha}| \le m$ and $|\boldsymbol{\beta}| \le m$, and for all $\boldsymbol{\omega} \in \mathbb{R}^d$, we obtain by (4.17) and Leibniz product rule

$$\begin{aligned} |\boldsymbol{\omega}^{\boldsymbol{\beta}}\, D^{\boldsymbol{\alpha}}(\mathscr{F}\varphi)(\boldsymbol{\omega})| &= |\mathscr{F}\big(D^{\boldsymbol{\beta}}(\mathbf{x}^{\boldsymbol{\alpha}}\, \varphi(\mathbf{x}))\big)(\boldsymbol{\omega})| \le \int_{\mathbb{R}^d} |D^{\boldsymbol{\beta}}(\mathbf{x}^{\boldsymbol{\alpha}}\, \varphi(\mathbf{x}))|\, \mathrm{d}\mathbf{x} \\ &\le C \int_{\mathbb{R}^d} (1 + \|\mathbf{x}\|_2)^m \sum_{|\boldsymbol{\gamma}| \le m} |D^{\boldsymbol{\gamma}}\varphi(\mathbf{x})|\, \mathrm{d}\mathbf{x} \\ &\le C \int_{\mathbb{R}^d} \frac{(1 + \|\mathbf{x}\|_2)^{m+d+1}}{(1 + \|\mathbf{x}\|_2)^{d+1}} \sum_{|\boldsymbol{\gamma}| \le m} |D^{\boldsymbol{\gamma}}\varphi(\mathbf{x})|\, \mathrm{d}\mathbf{x} \\ &\le C \int_{\mathbb{R}^d} \frac{\mathrm{d}\mathbf{x}}{(1 + \|\mathbf{x}\|_2)^{d+1}}\, \|\varphi\|_{m+d+1}\,. \end{aligned}$$

By

$$\|\mathscr{F}\varphi\|_m = \max_{|\boldsymbol{\gamma}| \le m} \|(1 + \|\boldsymbol{\omega}\|_2)^m\, D^{\boldsymbol{\gamma}}\, \mathscr{F}\varphi(\boldsymbol{\omega})\|_{C_0(\mathbb{R}^d)}$$

we see that

$$\|\mathscr{F}\varphi\|_m \le C' \, \|\varphi\|_{m+d+1} \tag{4.20}$$

for all $\varphi \in \mathscr{S}(\mathbb{R}^d)$ and each $m \in \mathbb{N}_0$.

Assume that $\varphi_k \underset{\mathscr{S}}{\longrightarrow} \varphi$ as $k \to \infty$ for $\varphi_k,\, \varphi \in \mathscr{S}(\mathbb{R}^d)$. Applying the inequality (4.20) to $\varphi_k - \varphi$, we obtain for all $m \in \mathbb{N}_0$

$$\|\mathscr{F}\varphi_k - \mathscr{F}\varphi\|_m \le C' \, \|\varphi_k - \varphi\|_{m+d+1} \,.$$

From Lemma 4.10, it follows that $\mathscr{F}\varphi_k \underset{\mathscr{S}}{\longrightarrow} \mathscr{F}\varphi$ as $k \to \infty$.

2. The mapping

$$(\tilde{\mathscr{F}}\varphi)(\mathbf{x}) := \frac{1}{(2\pi)^d} \int_{\mathbb{R}^d} \varphi(\boldsymbol{\omega})\, \mathrm{e}^{\mathrm{i}\,\mathbf{x}\cdot\boldsymbol{\omega}}\, \mathrm{d}\boldsymbol{\omega}\,, \quad \varphi \in \mathscr{S}(\mathbb{R}^d)\,,$$

is linear and continuous from $\mathscr{S}(\mathbb{R}^d)$ into itself by the first step of this proof, since $(\tilde{\mathscr{F}}\varphi)(\mathbf{x}) = \frac{1}{(2\pi)^d}\, (\mathscr{F}\varphi)(-\mathbf{x})$.

Now we show that $\tilde{\mathscr{F}}$ is the inverse mapping of $\mathscr{F}$. For arbitrary $\varphi,\, \psi \in \mathscr{S}(\mathbb{R}^d)$, it holds by Fubini's theorem

$$\begin{aligned}
\int_{\mathbb{R}^d} (\mathscr{F}\varphi)(\boldsymbol{\omega})\, \psi(\boldsymbol{\omega})\, \mathrm{e}^{\mathrm{i}\,\boldsymbol{\omega}\cdot\mathbf{x}}\, \mathrm{d}\boldsymbol{\omega} &= \int_{\mathbb{R}^d} \Big(\int_{\mathbb{R}^d} \varphi(\mathbf{y})\, \mathrm{e}^{-\mathrm{i}\,\boldsymbol{\omega}\cdot\mathbf{y}}\, \mathrm{d}\mathbf{y}\Big)\, \psi(\boldsymbol{\omega})\, \mathrm{e}^{\mathrm{i}\,\boldsymbol{\omega}\cdot\mathbf{x}}\, \mathrm{d}\boldsymbol{\omega} \\
&= \int_{\mathbb{R}^d} \varphi(\mathbf{y}) \Big(\int_{\mathbb{R}^d} \psi(\boldsymbol{\omega})\, \mathrm{e}^{\mathrm{i}\,(\mathbf{x}-\mathbf{y})\cdot\boldsymbol{\omega}}\, \mathrm{d}\boldsymbol{\omega}\Big) \mathrm{d}\mathbf{y} \\
&= \int_{\mathbb{R}^d} \varphi(\mathbf{y})\, (\mathscr{F}\psi)(\mathbf{y}-\mathbf{x})\, \mathrm{d}\mathbf{y} \\
&= \int_{\mathbb{R}^d} \varphi(\mathbf{z}+\mathbf{x})\, (\mathscr{F}\psi)(\mathbf{z})\, \mathrm{d}\mathbf{z}\,.
\end{aligned}$$

For the Gaussian function $\psi(\mathbf{x}) := \mathrm{e}^{-\|\varepsilon\mathbf{x}\|_2^2/2}$ with $\varepsilon > 0$, we have by Example 4.15 that $(\mathscr{F}\psi)(\boldsymbol{\omega}) = \big(\frac{2\pi}{\varepsilon^2}\big)^{d/2}\, \mathrm{e}^{-\|\boldsymbol{\omega}\|_2^2/(2\varepsilon^2)}$ and consequently

$$\begin{aligned}
\int_{\mathbb{R}^d} (\mathscr{F}\varphi)(\boldsymbol{\omega})\, \mathrm{e}^{-\|\varepsilon\boldsymbol{\omega}\|_2^2/2} \mathrm{e}^{\mathrm{i}\,\boldsymbol{\omega}\cdot\mathbf{x}}\, \mathrm{d}\boldsymbol{\omega} &= \Big(\frac{2\pi}{\varepsilon^2}\Big)^{d/2} \int_{\mathbb{R}^d} \varphi(\mathbf{z}+\mathbf{x})\, \mathrm{e}^{-\|\mathbf{z}\|_2^2/(2\varepsilon^2)}\, \mathrm{d}\mathbf{z} \\
&= (2\pi)^{d/2} \int_{\mathbb{R}^d} \varphi(\varepsilon\,\mathbf{y}+\mathbf{x})\, \mathrm{e}^{-\|\mathbf{y}\|_2^2/2}\, \mathrm{d}\mathbf{y}\,.
\end{aligned}$$

Since $|(\mathscr{F}\varphi)(\boldsymbol{\omega})\, \mathrm{e}^{-\|\varepsilon\boldsymbol{\omega}\|_2^2/2}| \le |\mathscr{F}\varphi(\boldsymbol{\omega})|$ for all $\boldsymbol{\omega} \in \mathbb{R}^d$ and $\mathscr{F}\varphi \in \mathscr{S}(\mathbb{R}^d) \subset L_1(\mathbb{R}^d)$, we obtain by Lebesgue's dominated convergence theorem

$$\big(\tilde{\mathscr{F}}(\mathscr{F}\varphi)\big)(\mathbf{x}) = \frac{1}{(2\pi)^d} \lim_{\varepsilon\to 0} \int_{\mathbb{R}^d} (\mathscr{F}\varphi)(\boldsymbol{\omega})\, \mathrm{e}^{-\|\varepsilon\boldsymbol{\omega}\|_2^2/2}\, \mathrm{e}^{\mathrm{i}\,\boldsymbol{\omega}\cdot\mathbf{x}}\, \mathrm{d}\boldsymbol{\omega}$$

$$= (2\pi)^{-d/2} \lim_{\varepsilon \to 0} \int_{\mathbb{R}^d} \varphi(\mathbf{x} + \varepsilon \mathbf{y})\, \mathrm{e}^{-\|\mathbf{y}\|_2^2/2}\, \mathrm{d}\mathbf{y}$$

$$= (2\pi)^{-d/2} \varphi(\mathbf{x}) \int_{\mathbb{R}^d} \mathrm{e}^{-\|\mathbf{y}\|_2^2/2}\, \mathrm{d}\mathbf{y} = \varphi(\mathbf{x})\,,$$

since by Example 2.6

$$\int_{\mathbb{R}^d} \mathrm{e}^{-\|\mathbf{y}\|_2^2/2}\, \mathrm{d}\mathbf{y} = \Big(\int_{\mathbb{R}} \mathrm{e}^{-y^2/2}\, \mathrm{d}y\Big)^d = (2\pi)^{d/2}\,.$$

From $\tilde{\mathscr{F}}(\mathscr{F}\varphi) = \varphi$, it follows immediately that $\mathscr{F}(\tilde{\mathscr{F}}\varphi) = \varphi$ for all $\varphi \in \mathscr{S}(\mathbb{R}^d)$. Hence, $\tilde{\mathscr{F}} = \mathscr{F}^{-1}$ and $\mathscr{F}$ are bijective. ■

The *convolution* $f * g$ of two d-variate functions $f, g \in L_1(\mathbb{R}^d)$ is defined by

$$(f * g)(\mathbf{x}) := \int_{\mathbb{R}^d} f(\mathbf{y})\, g(\mathbf{x} - \mathbf{y})\, \mathrm{d}\mathbf{y}\,.$$

Theorem 2.13 carries over to our multivariate setting. Moreover, by the following lemma, the product and the convolution of two rapidly decreasing functions are again rapidly decreasing.

Lemma 4.19 *For arbitrary $\varphi, \psi \in \mathscr{S}(\mathbb{R}^d)$, the product $\varphi\, \psi$ and the convolution $\varphi * \psi$ are in $\mathscr{S}(\mathbb{R}^d)$, and it holds $\mathscr{F}(\varphi * \psi) = \hat{\varphi}\, \hat{\psi}$.*

Proof By the Leibniz product rule (4.18), we obtain that $\mathbf{x}^{\boldsymbol{\gamma}}\, D^{\boldsymbol{\alpha}}\big(\varphi(\mathbf{x})\, \psi(\mathbf{x})\big) \in C_0(\mathbb{R}^d)$ for all $\boldsymbol{\alpha}, \boldsymbol{\gamma} \in \mathbb{N}_0^d$, i.e., $\varphi\, \psi \in \mathscr{S}(\mathbb{R}^d)$.

By Theorem 4.18, we know that $\hat{\varphi}, \hat{\psi} \in \mathscr{S}(\mathbb{R}^d)$ and hence $\hat{\varphi}\, \hat{\psi} \in \mathscr{S}(\mathbb{R}^d)$ by the first step. Using Theorem 4.18, we obtain that $\mathscr{F}(\hat{\varphi}\, \hat{\psi}) \in \mathscr{S}(\mathbb{R}^d)$. On the other hand, we conclude by Fubini's theorem

$$\mathscr{F}(\varphi * \psi)(\boldsymbol{\omega}) = \int_{\mathbb{R}^d} \Big(\int_{\mathbb{R}^d} \varphi(\mathbf{y})\, \psi(\mathbf{x} - \mathbf{y})\, \mathrm{d}\mathbf{y}\Big)\, \mathrm{e}^{-\mathrm{i}\,\mathbf{x}\cdot\boldsymbol{\omega}}\, \mathrm{d}\mathbf{x}$$

$$= \int_{\mathbb{R}^d} \varphi(\mathbf{y})\, \mathrm{e}^{-\mathrm{i}\,\mathbf{y}\cdot\boldsymbol{\omega}} \Big(\int_{\mathbb{R}^d} \psi(\mathbf{x} - \mathbf{y})\, \mathrm{e}^{-\mathrm{i}\,(\mathbf{x}-\mathbf{y})\cdot\boldsymbol{\omega}}\, \mathrm{d}\mathbf{x}\Big)\, \mathrm{d}\mathbf{y}$$

$$= \Big(\int_{\mathbb{R}^d} \varphi(\mathbf{y})\, \mathrm{e}^{-\mathrm{i}\,\mathbf{y}\cdot\boldsymbol{\omega}}\, \mathrm{d}\mathbf{y}\Big)\, \hat{\psi}(\boldsymbol{\omega}) = \hat{\varphi}(\boldsymbol{\omega})\, \hat{\psi}(\boldsymbol{\omega})\,.$$

Therefore $\varphi * \psi = \mathscr{F}^{-1}(\hat{\varphi}\, \hat{\psi}) \in \mathscr{S}(\mathbb{R}^d)$. ■

The basic properties of the d-variate Fourier transform on $\mathscr{S}(\mathbb{R}^d)$ can be proved similarly as in Theorems 2.5 and 2.15. The following properties 1, 3, and 4 hold also true for functions in $L_1(\mathbb{R}^d)$, whereas property 2 holds only under additional smoothness assumptions.

Theorem 4.20 (Properties of the Fourier Transform on $\mathscr{S}(\mathbb{R}^d)$) *The Fourier transform of a function $\varphi \in \mathscr{S}(\mathbb{R}^d)$ has the following properties:*

1. Translation and modulation: *For fixed* $\mathbf{x}_0, \boldsymbol{\omega}_0 \in \mathbb{R}^d$,

$$\big(\varphi(\mathbf{x}-\mathbf{x}_0)\big)\hat{}(\boldsymbol{\omega}) = \mathrm{e}^{-\mathrm{i}\,\mathbf{x}_0\cdot\boldsymbol{\omega}}\,\hat{\varphi}(\boldsymbol{\omega})\,,$$

$$\big(\mathrm{e}^{-\mathrm{i}\,\boldsymbol{\omega}_0\cdot\mathbf{x}}\,\varphi(\mathbf{x})\big)\hat{}(\boldsymbol{\omega}) = \hat{\varphi}(\boldsymbol{\omega}+\boldsymbol{\omega}_0)\,.$$

2. Differentiation and multiplication: *For* $\boldsymbol{\alpha} \in \mathbb{N}_0^d$,

$$\big(D^{\boldsymbol{\alpha}}\varphi(\mathbf{x})\big)\hat{}(\boldsymbol{\omega}) = \mathrm{i}^{|\boldsymbol{\alpha}|}\,\boldsymbol{\omega}^{\boldsymbol{\alpha}}\,\hat{\varphi}(\boldsymbol{\omega})\,,$$

$$\big(\mathbf{x}^{\boldsymbol{\alpha}}\varphi(\mathbf{x})\big)\hat{}(\boldsymbol{\omega}) = \mathrm{i}^{|\boldsymbol{\alpha}|}\,(D^{\boldsymbol{\alpha}}\hat{\varphi})(\boldsymbol{\omega})\,.$$

3. Scaling: *For* $c \in \mathbb{R}\setminus\{0\}$,

$$\big(\varphi(c\,\mathbf{x})\big)\hat{}(\boldsymbol{\omega}) = \frac{1}{|c|^d}\,\hat{\varphi}(c^{-1}\,\boldsymbol{\omega})\,.$$

4. Convolution: *For* $\varphi,\,\psi \in \mathscr{S}(\mathbb{R}^d)$,

$$(\varphi * \psi)\hat{}(\boldsymbol{\omega}) = \hat{\varphi}(\boldsymbol{\omega})\,\hat{\psi}(\boldsymbol{\omega})\,.$$

4.2.2 Fourier Transforms on $L_1(\mathbb{R}^d)$ and $L_2(\mathbb{R}^d)$

Similar to the univariate case (see Theorem 2.8), we obtain the following theorem for the Fourier transform on $L_1(\mathbb{R}^d)$.

Theorem 4.21 *The Fourier transform $\mathscr{F}$ defined by* (4.13) *is a linear continuous operator from $L_1(\mathbb{R}^d)$ into $C_0(\mathbb{R}^d)$ with the operator norm* $\|\mathscr{F}\|_{L_1(\mathbb{R}^d)\to C_0(\mathbb{R}^d)} = 1$.

Proof By (4.11) there exists for any $f \in L_1(\mathbb{R}^d)$ a sequence $(\varphi_k)_{k\in\mathbb{N}}$ with $\varphi_k \in \mathscr{S}(\mathbb{R}^d)$ such that $\lim_{k\to\infty}\|f-\varphi_k\|_{L_1(\mathbb{R}^d)} = 0$. Then the $C_0(\mathbb{R}^d)$ norm of $\mathscr{F}f - \mathscr{F}\varphi_k$ can be estimated by

$$\|\mathscr{F}f - \mathscr{F}\varphi_k\|_{C_0(\mathbb{R}^d)} = \sup_{\boldsymbol{\omega}\in\mathbb{R}^d}|\mathscr{F}(f-\varphi_k)(\boldsymbol{\omega})| \le \|f-\varphi_k\|_{L_1(\mathbb{R}^d)}\,,$$

i.e., $\lim_{k\to\infty}\mathscr{F}\varphi_k = \mathscr{F}f$ in the norm of $C_0(\mathbb{R}^d)$. By $\mathscr{S}(\mathbb{R}^d) \subset C_0(\mathbb{R}^d)$ and the completeness of $C_0(\mathbb{R}^d)$, we conclude that $\mathscr{F}f \in C_0(\mathbb{R}^d)$. The operator norm of $\mathscr{F} : L_1(\mathbb{R}^d) \to C_0(\mathbb{R}^d)$ can be deduced as in the univariate case, where we have just to use the d-variate Gaussian function. ■

Theorem 4.22 (Fourier Inversion Formula for $L_1(\mathbb{R}^d)$ Functions) *Let $f \in L_1(\mathbb{R}^d)$ and $\hat{f} \in L_1(\mathbb{R}^d)$. Then the* Fourier inversion formula

$$f(\mathbf{x}) = \frac{1}{(2\pi)^d} \int_{\mathbb{R}^d} \hat{f}(\boldsymbol{\omega})\, \mathrm{e}^{\mathrm{i}\boldsymbol{\omega}\cdot\mathbf{x}}\, \mathrm{d}\boldsymbol{\omega} \tag{4.21}$$

holds true for almost all $\mathbf{x} \in \mathbb{R}^d$.

The proof follows similar lines as those of Theorem 2.10 in the univariate case. Another proof of Theorem 4.22 is sketched in Remark 4.49.

The following lemma is related to the more general Lemma 2.21 proved in the univariate case.

Lemma 4.23 *For arbitrary $\varphi,\, \psi \in \mathscr{S}(\mathbb{R}^d)$, the following* Parseval equality *is valid:*

$$(2\pi)^d\, \langle \varphi,\, \psi \rangle_{L_2(\mathbb{R}^d)} = \langle \mathscr{F}\varphi,\, \mathscr{F}\psi \rangle_{L_2(\mathbb{R}^d)}\,.$$

In particular, we have $(2\pi)^{d/2}\, \|\varphi\|_{L_2(\mathbb{R}^d)} = \|\mathscr{F}\varphi\|_{L_2(\mathbb{R}^d)}$.

Proof By Theorem 4.18 we have $\varphi = \mathscr{F}^{-1}(\mathscr{F}\varphi)$ for $\varphi \in \mathscr{S}(\mathbb{R}^d)$. Then Fubini's theorem yields

$$\begin{aligned}
(2\pi)^d\, \langle \varphi,\, \psi \rangle_{L_2(\mathbb{R}^d)} &= (2\pi)^d \int_{\mathbb{R}^d} \varphi(\mathbf{x})\, \overline{\psi(\mathbf{x})}\, \mathrm{d}\mathbf{x} \\
&= \int_{\mathbb{R}^d} \overline{\psi(\mathbf{x})} \Big(\int_{\mathbb{R}^d} (\mathscr{F}\varphi)(\boldsymbol{\omega})\, \mathrm{e}^{\mathrm{i}\,\mathbf{x}\cdot\boldsymbol{\omega}}\, \mathrm{d}\boldsymbol{\omega} \Big)\, \mathrm{d}\mathbf{x} \\
&= \int_{\mathbb{R}^d} (\mathscr{F}\varphi)(\boldsymbol{\omega})\, \overline{\int_{\mathbb{R}^d} \psi(\mathbf{x})\, \mathrm{e}^{-\mathrm{i}\,\mathbf{x}\cdot\boldsymbol{\omega}}\, \mathrm{d}\mathbf{x}}\, \mathrm{d}\boldsymbol{\omega} \\
&= \int_{\mathbb{R}^d} \mathscr{F}\varphi(\boldsymbol{\omega})\, \overline{\mathscr{F}\psi(\boldsymbol{\omega})}\, \mathrm{d}\boldsymbol{\omega} = \langle \mathscr{F}\varphi,\, \mathscr{F}\psi \rangle_{L_2(\mathbb{R}^d)}\,.
\end{aligned}$$

■

We will use the following extension theorem of bounded linear operator (see, e.g., [12, Theorem 2.4.1]) to extend the Fourier transform from $\mathscr{S}(\mathbb{R}^d)$ to $L_2(\mathbb{R}^d)$.

Theorem 4.24 (Extension of a Bounded Linear Operator) *Let H be a Hilbert space and let $D \subset H$ be a linear subset which is dense in H. Further let $F : D \to H$ be a linear bounded operator. Then F admits a unique* extension *to a bounded linear operator $\tilde{F} : H \to H$, and it holds*

$$\|F\|_{D\to H} = \|\tilde{F}\|_{H\to H}\,.$$

For each $f \in H$ with $f = \lim_{k\to\infty} f_k$, where $f_k \in D$, it holds $\tilde{F} f = \lim_{k\to\infty} F f_k$.

Theorem 4.25 (Plancherel) *The Fourier transform* $\mathscr{F} : \mathscr{S}(\mathbb{R}^d) \to \mathscr{S}(\mathbb{R}^d)$ *can be uniquely extended to a linear continuous bijective transform* $\mathscr{F} : L_2(\mathbb{R}^d) \to L_2(\mathbb{R}^d)$, *which fulfills the* Parseval equality

$$(2\pi)^d \, \langle f,\, g\rangle_{L_2(\mathbb{R}^d)} = \langle \mathscr{F} f,\, \mathscr{F} g\rangle_{L_2(\mathbb{R}^d)} \tag{4.22}$$

for all $f,\, g \in L_2(\mathbb{R}^d)$. *In particular, it holds* $(2\pi)^{d/2}\, \|f\|_{L_2(\mathbb{R}^d)} = \|\mathscr{F} f\|_{L_2(\mathbb{R}^d)}$.

The above extension is also called *Fourier transform* on $L_2(\mathbb{R}^d)$ or sometimes *Fourier–Plancherel transform.*

Proof Applying Theorem 4.24, we consider $D = \mathscr{S}(\mathbb{R}^d)$ as linear, dense subspace of the Hilbert space $H = L_2(\mathbb{R}^d)$. By Lemma 4.23 we know that $\mathscr{F}$ and $\mathscr{F}^{-1}$ are bounded linear operators from D to H with the operator norms $(2\pi)^{d/2}$ and $(2\pi)^{-d/2}$. Therefore both operators admit unique extensions $\mathscr{F} : L_2(\mathbb{R}^d) \to L_2(\mathbb{R}^d)$ and $\mathscr{F}^{-1} : L_2(\mathbb{R}^d) \to L_2(\mathbb{R}^d)$ and (4.22) is fulfilled. ■

Theorem 4.26 (Convolution Property of Fourier Transform) *For the Fourier transform* $\mathscr{F} : L_1(\mathbb{R}^d) \to C_0(\mathbb{R}^d)$, *it holds*

$$\mathscr{F}(f * g) = (\mathscr{F} f)(\mathscr{F} g)$$

for all $f,\, g \in L_1(\mathbb{R}^d)$.

For the Fourier transform $\mathscr{F} : L_2(\mathbb{R}^d) \to L_2(\mathbb{R}^d)$, *it holds*

$$f * g = \mathscr{F}^{-1}\big[(\mathscr{F} f)(\mathscr{F} g)\big]$$

for all $f,\, g \in L_2(\mathbb{R}^d)$, *where* $\mathscr{F}^{-1} : L_1(\mathbb{R}^d) \to C_0(\mathbb{R}^d)$ *is defined by*

$$(\mathscr{F}^{-1} h)(\mathbf{x}) := \frac{1}{(2\pi)^d} \int_{\mathbb{R}^d} h(\boldsymbol{\omega})\, \mathrm{e}^{\mathrm{i}\boldsymbol{\omega}\cdot\mathbf{x}}\, \mathrm{d}\boldsymbol{\omega}\,, \quad \mathbf{x} \in \mathbb{R}^d\,,$$

for $h \in L_1(\mathbb{R}^d)$.

The proof of Theorem 4.26 follows similar lines as the proofs of Theorems 2.15 and 2.26.

4.2.3 Poisson Summation Formula

Now we generalize the one-dimensional Poisson summation formula (see Theorem 2.28). For $f \in L_1(\mathbb{R}^d)$ we introduce its 2π*–periodization* by

$$\tilde{f}(\mathbf{x}) := \sum_{\mathbf{k}\in\mathbb{Z}^d} f(\mathbf{x} + 2\pi\mathbf{k})\,, \quad \mathbf{x} \in \mathbb{R}^d\,. \tag{4.23}$$

First we prove the existence of the 2π–periodization $\tilde{f} \in L_1(\mathbb{T}^d)$ of $f \in L_1(\mathbb{R}^d)$.

Lemma 4.27 *For given $f \in L_1(\mathbb{R}^d)$, the series in (4.23) converges absolutely for almost all $\mathbf{x} \in \mathbb{R}^d$, and $\tilde{f}$ is contained in $L_1(\mathbb{T}^d)$.*

Proof At first we show that the 2π-periodization φ of $|f|$ belongs to $L_1(\mathbb{T}^d)$, i.e.,

$$\varphi(\mathbf{x}) := \sum_{\mathbf{k}\in\mathbb{Z}^d} |f(\mathbf{x} + 2\pi\,\mathbf{k})|\,.$$

For each $n \in \mathbb{N}$, we form the nonnegative function

$$\varphi_n(\mathbf{x}) := \sum_{k_1=-n}^{n-1} \ldots \sum_{k_d=-n}^{n-1} |f(\mathbf{x} + 2\pi\,\mathbf{k})|\,.$$

Then we obtain

$$\begin{aligned}\int_{[0,\,2\pi]^d} \varphi_n(\mathbf{x})\,\mathrm{d}\mathbf{x} &= \sum_{k_1=-n}^{n-1} \ldots \sum_{k_d=-n}^{n-1} \int_{[0,\,2\pi]^d} |f(\mathbf{x} + 2\pi\,\mathbf{k})|\,\mathrm{d}\mathbf{x} \\ &= \sum_{k_1=-n}^{n-1} \ldots \sum_{k_d=-n}^{n-1} \int_{2\pi\mathbf{k}+[0,\,2\pi]^d} |f(\mathbf{x})|\,\mathrm{d}\mathbf{x} \\ &= \int_{[-2\pi n,\,2\pi n]^d} |f(\mathbf{x})|\,\mathrm{d}\mathbf{x}\end{aligned}$$

and hence

$$\lim_{n\to\infty} \int_{[0,\,2\pi]^d} \varphi_n(\mathbf{x})\,\mathrm{d}\mathbf{x} = \int_{\mathbb{R}^d} |f(\mathbf{x})|\,\mathrm{d}\mathbf{x} = \|f\|_{L_1(\mathbb{R}^d)} < \infty\,. \tag{4.24}$$

Since $(\varphi_n)_{n\in\mathbb{N}}$ is a monotone increasing sequence of nonnegative integrable functions with the property (4.24), we receive by the monotone convergence theorem of B. Levi that $\lim_{n\to\infty} \varphi_n(\mathbf{x}) = \varphi(\mathbf{x})$ for almost all $\mathbf{x} \in \mathbb{R}^d$ and $\varphi \in L_1(\mathbb{T}^d)$, where it holds

$$\int_{[0,\,2\pi]^d} \varphi(\mathbf{x})\,\mathrm{d}\mathbf{x} = \lim_{n\to\infty} \int_{[0,\,2\pi]^d} \varphi_n(\mathbf{x})\,\mathrm{d}\mathbf{x} = \|f\|_{L_1(\mathbb{R}^d)}\,.$$

In other words, the series in (4.23) converges absolutely for almost all $\mathbf{x} \in \mathbb{R}^d$. From

$$|\tilde{f}(\mathbf{x})| = \Big|\sum_{\mathbf{k}\in\mathbb{Z}^d} f(\mathbf{x} + 2\pi\mathbf{k})\Big| \le \sum_{\mathbf{k}\in\mathbb{Z}^d} |f(\mathbf{x} + 2\pi\mathbf{k})| = \varphi(\mathbf{x})\,,$$

it follows that $\tilde{f} \in L_1(\mathbb{T}^d)$ with

$$\|\tilde{f}\|_{L_1(\mathbb{T}^d)} = \int_{[0,\,2\pi]^d} |\tilde{f}(\mathbf{x})|\,\mathrm{d}\mathbf{x} \le \int_{[0,\,2\pi]^d} \varphi(\mathbf{x})\,\mathrm{d}\mathbf{x} = \|f\|_{L_1(\mathbb{R})^d}\,.$$

■

The d-dimensional Poisson summation formula describes an interesting connection between the values $\hat{f}(\mathbf{n})$, $\mathbf{n} \in \mathbb{Z}^d$, of the Fourier transform $\hat{f}$ of a given function $f \in L_1(\mathbb{R}^d) \cap C_0(\mathbb{R}^d)$ and the Fourier series of the 2π-periodization $\tilde{f}$.

Theorem 4.28 *Let* $f \in C_0(\mathbb{R}^d)$ *be a given function which fulfills the decay conditions*

$$|f(\mathbf{x})| \le \frac{c}{1 + \|\mathbf{x}\|_2^{d+\varepsilon}}\,, \qquad |\hat{f}(\boldsymbol{\omega})| \le \frac{c}{1 + \|\boldsymbol{\omega}\|_2^{d+\varepsilon}} \tag{4.25}$$

for all $\mathbf{x}$, $\boldsymbol{\omega} \in \mathbb{R}^d$ *with some constants* $\varepsilon > 0$ *and* $c > 0$.

Then for all $\mathbf{x} \in \mathbb{R}^d$, *it holds the* Poisson summation formula

$$(2\pi)^d\, \tilde{f}(\mathbf{x}) = (2\pi)^d \sum_{\mathbf{k}\in\mathbb{Z}^d} f(\mathbf{x} + 2\pi\,\mathbf{k}) = \sum_{\mathbf{n}\in\mathbb{Z}^d} \hat{f}(\mathbf{n})\,\mathrm{e}^{\mathrm{i}\,\mathbf{n}\cdot\mathbf{x}}\,, \tag{4.26}$$

where both series in (4.26) *converge absolutely and uniformly on* $\mathbb{R}^d$. *In particular, for* $\mathbf{x} = \mathbf{0}$ *it holds*

$$(2\pi)^d \sum_{\mathbf{k}\in\mathbb{Z}^d} f(2\pi\,\mathbf{k}) = \sum_{\mathbf{n}\in\mathbb{Z}^d} \hat{f}(\mathbf{n})\,.$$

Proof From the decay conditions (4.25), it follows that $f, \hat{f} \in L_1(\mathbb{R}^d)$ such that $\tilde{f} \in L_1(\mathbb{T}^d)$ by Lemma 4.27. Then we obtain

$$\begin{aligned} c_{\mathbf{n}}(\tilde{f}) &= \frac{1}{(2\pi)^d} \int_{[0,\,2\pi]^d} \tilde{f}(\mathbf{x})\,\mathrm{e}^{-\mathrm{i}\,\mathbf{n}\cdot\mathbf{x}}\,\mathrm{d}\mathbf{x} \\ &= \frac{1}{(2\pi)^d} \int_{[0,\,2\pi]^d} \Big(\sum_{\mathbf{k}\in\mathbb{Z}^d} f(\mathbf{x} + 2\pi\,\mathbf{k})\,\mathrm{e}^{-\mathrm{i}\,\mathbf{n}\cdot(\mathbf{x}+2\pi\mathbf{k})}\Big)\,\mathrm{d}\mathbf{x} \\ &= \frac{1}{(2\pi)^d} \int_{\mathbb{R}^d} f(\mathbf{x})\,\mathrm{e}^{-\mathrm{i}\,\mathbf{n}\cdot\mathbf{x}}\,\mathrm{d}\mathbf{x} = \frac{1}{(2\pi)^d}\,\hat{f}(\mathbf{n})\,. \end{aligned}$$

From the second decay condition and Lemma 4.8, it follows that $\sum_{\mathbf{n}\in\mathbb{Z}^d} |\hat{f}(\mathbf{n})| < \infty$. Thus, by Theorem 4.7, the 2π-periodization $\tilde{f} \in C(\mathbb{T}^d)$ possesses the uniformly convergent Fourier series

$$\tilde{f}(\mathbf{x}) = \frac{1}{(2\pi)^d} \sum_{\mathbf{n}\in\mathbb{Z}^d} \hat{f}(\mathbf{n})\, \mathrm{e}^{\mathrm{i}\,\mathbf{n}\cdot\mathbf{x}}\,.$$

Further we have $\tilde{f} \in C(\mathbb{T}^d)$ such that (4.26) is valid for all $\mathbf{x} \in \mathbb{R}^d$. ■

Remark 4.29 The decay conditions (4.25) on f and $\hat{f}$ are needed only for the absolute and uniform convergence of both series and the pointwise validity of (4.26). Obviously, any $f \in \mathscr{S}(\mathbb{R}^d)$ fulfills the decay conditions (4.25). Note that the Poisson summation formula (4.26) holds pointwise or almost everywhere under much weaker conditions on f and $\hat{f}$; see [184]. □

4.2.4 Fourier Transforms of Radial Functions

A function $f : \mathbb{R}^d \to \mathbb{C}$ is called a *radial function*, if $f(\mathbf{x}) = f(\mathbf{y})$ for all $\mathbf{x}, \mathbf{y} \in \mathbb{R}^d$ with $\|\mathbf{x}\|_2 = \|\mathbf{y}\|_2$. Thus a radial function f can be written in the form $f(\mathbf{x}) = F(\|\mathbf{x}\|_2)$ with certain univariate function $F : [0, \infty) \to \mathbb{C}$. A radial function f is characterized by the property $f(\mathbf{A}\,\mathbf{x}) = f(\mathbf{x})$ for all orthogonal matrices $\mathbf{A} \in \mathbb{R}^{d\times d}$. The Gaussian function in Example 4.15 is a typical example of a radial function. Extended material on radial functions can be found in [433].

Lemma 4.30 *Let* $\mathbf{A} \in \mathbb{R}^{d\times d}$ *be invertible and let* $f \in L_1(\mathbb{R}^d)$. *Then we have*

$$\big(f(\mathbf{A}\,\mathbf{x})\big)^{\wedge}(\boldsymbol{\omega}) = \frac{1}{|\det \mathbf{A}|}\, \hat{f}(\mathbf{A}^{-\top}\boldsymbol{\omega})\,.$$

In particular, for an orthogonal matrix $\mathbf{A} \in \mathbb{R}^{d\times d}$, *we have the relation*

$$\big(f(\mathbf{A}\,\mathbf{x})\big)^{\wedge}(\boldsymbol{\omega}) = \hat{f}(\mathbf{A}\boldsymbol{\omega})\,.$$

Proof Substituting $\mathbf{y} := \mathbf{A}\,\mathbf{x}$, it follows

$$\begin{aligned}\big(f(\mathbf{A}\,\mathbf{x})\big)^{\wedge}(\boldsymbol{\omega}) &= \int_{\mathbb{R}^d} f(\mathbf{A}\mathbf{x})\, \mathrm{e}^{-\mathrm{i}\,\boldsymbol{\omega}\cdot\mathbf{x}}\, \mathrm{d}\mathbf{x} \\ &= \frac{1}{|\det \mathbf{A}|} \int_{\mathbb{R}^d} f(\mathbf{y})\, \mathrm{e}^{-\mathrm{i}\,(\mathbf{A}^{-\top}\boldsymbol{\omega})\cdot\mathbf{y}}\, \mathrm{d}\mathbf{y} = \frac{1}{|\det \mathbf{A}|}\, \hat{f}(\mathbf{A}^{-\top}\boldsymbol{\omega})\,.\end{aligned}$$

If $\mathbf{A}$ is orthogonal, then $\mathbf{A}^{-\top} = \mathbf{A}$ and $|\det \mathbf{A}| = 1$. ■

Corollary 4.31 *Let* $f \in L_1(\mathbb{R}^d)$ *be a radial function of the form* $f(\mathbf{x}) = F(r)$ *with* $r := \|\mathbf{x}\|_2$. *Then its Fourier transform* $\hat{f}$ *is also a radial function. In the case* $d = 2$, *we have*

$$\hat{f}(\boldsymbol{\omega}) = 2\pi \int_0^\infty F(r)\, J_0(r\,\|\boldsymbol{\omega}\|_2)\, r\, \mathrm{d}r\,, \tag{4.27}$$

where J_0 *denotes the* Bessel function of order zero

$$J_0(x) := \sum_{k=0}^{\infty} \frac{(-1)^k}{(k!)^2} \left(\frac{x}{2}\right)^{2k}.$$

Proof The first assertion is an immediate consequence of Lemma 4.30. Let $d = 2$. Using polar coordinates $(r,\, \varphi)$ and $(\rho,\, \psi)$ with $r = \|\mathbf{x}\|_2$, $\rho = \|\boldsymbol{\omega}\|_2$, and $\varphi,\, \psi \in [0,\, 2\pi)$ such that

$$\mathbf{x} = (r\, \cos\varphi,\, r\, \sin\varphi)^\top\,, \quad \boldsymbol{\omega} = (\rho\, \cos\psi,\, \rho\, \sin\psi)^\top\,,$$

we obtain

$$\begin{aligned}\hat{f}(\boldsymbol{\omega}) &= \int_{\mathbb{R}^2} f(\mathbf{x})\, \mathrm{e}^{-\mathrm{i}\,\mathbf{x}\cdot\boldsymbol{\omega}}\, \mathrm{d}\mathbf{x} \\ &= \int_0^\infty \int_0^{2\pi} F(r)\, \mathrm{e}^{-\mathrm{i}\,r\rho\, \cos(\varphi-\psi)}\, r\, \mathrm{d}\varphi\, \mathrm{d}r\,.\end{aligned}$$

The inner integral with respect to φ is independent of ψ, since the integrand is 2π-periodic. For $\psi = -\frac{\pi}{2}$ we conclude by Bessel's integral formula

$$\int_0^{2\pi} \mathrm{e}^{-\mathrm{i}\,r\rho\, \cos(\varphi+\pi/2)}\, \mathrm{d}\varphi = \int_0^{2\pi} \mathrm{e}^{\mathrm{i}\,r\rho\, \sin\varphi}\, \mathrm{d}\varphi = 2\pi\, J_0(r\rho)\,.$$

This yields the integral representation (4.27), which is called *Hankel transform of order zero* of F. ■

Remark 4.32 The *Hankel transform of order zero* $\mathscr{H} : L_2\big((0,\, \infty)\big) \to L_2\big((0,\, \infty)\big)$ is defined by

$$(\mathscr{H} F)(\rho) := \int_0^\infty F(r)\, J_0(r\,\rho)\, r\, \mathrm{d}r\,.$$

□

Remark 4.33 In the case $d = 3$, we can use spherical coordinates for the computation of the Fourier transform of a radial function $f \in L_1(\mathbb{R}^3)$, where $f(\mathbf{x}) = F(\|\mathbf{x}\|_2)$. This results in

$$\hat{f}(\boldsymbol{\omega}) = \frac{4\pi}{\|\boldsymbol{\omega}\|_2} \int_0^\infty F(r)\, r\, \sin(r\,\|\boldsymbol{\omega}\|_2)\, \mathrm{d}r\,, \quad \boldsymbol{\omega} \in \mathbb{R}^3 \setminus \{\mathbf{0}\}\,. \tag{4.28}$$

For an arbitrary dimension $d \in \mathbb{N} \setminus \{1\}$, we obtain

$$\hat{f}(\boldsymbol{\omega}) = (2\pi)^{d/2} \, \|\boldsymbol{\omega}\|_2{}^{1-d/2} \int_0^\infty F(r)\, r^{d/2}\, J_{d/2-1}(r\, \|\boldsymbol{\omega}\|_2)\, \mathrm{d}r\,, \quad \boldsymbol{\omega} \in \mathbb{R}^d \setminus \{\mathbf{0}\}\,,$$

where

$$J_\nu(x) := \sum_{k=0}^{\infty} \frac{(-1)^k}{k!\,\Gamma(k+\nu+1)} \left(\frac{x}{2}\right)^{2k+\nu}$$

denotes the *Bessel function of order* $\nu \geq 0$; see [399, p. 155]. □

Example 4.34 Let $f : \mathbb{R}^2 \to \mathbb{R}$ be the characteristic function of the unit disk, i.e., $f(\mathbf{x}) := 1$ for $\|\mathbf{x}\|_2 \leq 1$ and $f(\mathbf{x}) := 0$ for $\|\mathbf{x}\|_2 > 1$. By (4.27) it follows for $\boldsymbol{\omega} \in \mathbb{R}^2 \setminus \{\mathbf{0}\}$ that

$$\hat{f}(\boldsymbol{\omega}) = 2\pi \int_0^1 J_0(r\, \|\boldsymbol{\omega}\|_2)\, r\, \mathrm{d}r = \frac{2\pi}{\|\boldsymbol{\omega}\|_2}\, J_1(\|\boldsymbol{\omega}\|_2)$$

and $\hat{f}(\mathbf{0}) = \pi$.

Let $f : \mathbb{R}^3 \to \mathbb{R}$ be the characteristic function of the unit ball. Then from (4.28) it follows for $\boldsymbol{\omega} \in \mathbb{R}^3 \setminus \{\mathbf{0}\}$ that

$$\hat{f}(\boldsymbol{\omega}) = \frac{4\pi}{\|\boldsymbol{\omega}\|_2^3} \left(\sin \|\boldsymbol{\omega}\|_2 - \|\boldsymbol{\omega}\|_2 \cos \|\boldsymbol{\omega}\|_2\right),$$

and in particular $\hat{f}(\mathbf{0}) = \frac{4\pi}{3}$. □

4.3 Fourier Transform of Tempered Distributions

Now we show that the Fourier transform can be generalized to so-called tempered distributions which are linear continuous functionals on the Schwartz space $\mathcal{S}(\mathbb{R}^d)$; see [382]. The simplest tempered distribution, which cannot be described just by integrating the product of some function with functions from $\mathcal{S}(\mathbb{R}^d)$, is the Dirac distribution δ defined by $\langle \delta, \varphi \rangle := \varphi(\mathbf{0})$ for all $\varphi \in \mathcal{S}(\mathbb{R}^d)$.

4.3.1 Tempered Distributions

A *tempered distribution* T is a continuous linear functional on $\mathcal{S}(\mathbb{R}^d)$. In other words, a tempered distribution $T : \mathcal{S}(\mathbb{R}^d) \to \mathbb{C}$ fulfills the following conditions:

(i) Linearity: For all $\alpha_1, \alpha_2 \in \mathbb{C}$ and all $\varphi_1, \varphi_2 \in \mathscr{S}(\mathbb{R}^d)$,

$$\langle T, \alpha_1 \varphi_1 + \alpha_2 \varphi_2 \rangle = \alpha_1 \langle T, \varphi_1 \rangle + \alpha_2 \langle T, \varphi_2 \rangle .$$

(ii) Continuity: If $\varphi_j \underset{\mathscr{S}}{\longrightarrow} \varphi$ as $j \to \infty$ with $\varphi_j, \varphi \in \mathscr{S}(\mathbb{R}^d)$, then

$$\lim_{j\to\infty} \langle T,\, \varphi_j \rangle = \langle T,\, \varphi \rangle .$$

The set of tempered distributions is denoted by $\mathscr{S}'(\mathbb{R}^d)$. Defining for $T_1,\, T_2 \in \mathscr{S}'(\mathbb{R}^d)$ and all $\varphi \in \mathscr{S}(\mathbb{R}^d)$ the operation

$$\langle \alpha_1 T_1 + \alpha_2 T_2, \varphi \rangle := \alpha_1 \langle T_1, \varphi \rangle + \alpha_2 \langle T_2,\, \varphi \rangle,$$

the set $\mathscr{S}'(\mathbb{R}^d)$ becomes a linear space. We say that a sequence $(T_k)_{k\in\mathbb{N}}$ of tempered distributions $T_k \in \mathscr{S}'(\mathbb{R}^d)$ *converges in* $\mathscr{S}'(\mathbb{R}^d)$ to $T \in \mathscr{S}'(\mathbb{R}^d)$, if for all $\varphi \in \mathscr{S}(\mathbb{R}^d)$,

$$\lim_{k\to\infty} \langle T_k,\, \varphi \rangle = \langle T,\, \varphi \rangle .$$

We will use the notation $T_k \underset{\mathscr{S}'}{\longrightarrow} T$ as $k \to \infty$.

Lemma 4.35 (Schwartz) *A linear functional $T : \mathscr{S}(\mathbb{R}^d) \to \mathbb{C}$ is a tempered distribution if and only if there exist constants $m \in \mathbb{N}_0$ and $C \ge 0$ such that for all $\varphi \in \mathscr{S}(\mathbb{R}^d)$*

$$|\langle T, \varphi \rangle| \le C\, \|\varphi\|_m . \tag{4.29}$$

Proof

1. Assume that (4.29) holds true. Let $\varphi_j \underset{\mathscr{S}}{\longrightarrow} \varphi$ as $j \to \infty$, i.e., by Lemma 4.10, $\lim_{j\to\infty} \|\varphi_j - \varphi\|_m = 0$ for all $m \in \mathbb{N}_0$. From (4.29) it follows
$$|\langle T, \varphi_j - \varphi \rangle| \le C\, \|\varphi_j - \varphi\|_m$$
for some $m \in \mathbb{N}_0$ and $C \ge 0$. Thus $\lim_{j\to\infty} \langle T, \varphi_j - \varphi \rangle = 0$ and hence $\lim_{j\to\infty} \langle T,\, \varphi_j \rangle = \langle T,\, \varphi \rangle$.
2. Conversely, let $T \in \mathscr{S}'(\mathbb{R}^d)$. Then $\varphi_j \underset{\mathscr{S}}{\longrightarrow} \varphi$ as $j \to \infty$ implies $\lim_{j\to\infty} \langle T, \varphi_j \rangle = \langle T, \varphi \rangle$.
Assume that for all $m \in \mathbb{N}$ and $C > 0$, there exists $\varphi_{m,C} \in \mathscr{S}(\mathbb{R}^d)$ such that
$$|\langle T,\, \varphi_{m,C} \rangle| > C\, \|\varphi_{m,C}\|_m .$$

Choose $C = m$ and set $\varphi_m := \varphi_{m,m}$. Then it follows $|\langle T, \varphi_m\rangle| > m\|\varphi_m\|_m$ and hence

$$1 = |\langle T, \frac{\varphi_m}{\langle T, \varphi_m\rangle}\rangle| > m\,\|\frac{\varphi_m}{\langle T, \varphi_m\rangle}\|_m\,.$$

We introduce the function

$$\psi_m := \frac{\varphi_m}{\langle T, \varphi_m\rangle} \in \mathscr{S}(\mathbb{R}^d)$$

which has the properties $\langle T, \psi_m\rangle = 1$ and $\|\psi_m\|_m < \frac{1}{m}$. Thus, $\psi_m \underset{\mathscr{S}}{\longrightarrow} 0$ as $m \to \infty$. On the other hand, we have by assumption $T \in \mathscr{S}'(\mathbb{R}^d)$ that $\lim_{m\to\infty}\langle T, \psi_m\rangle = 0$. This contradicts $\langle T, \psi_m\rangle = 1$. ■

A measurable function $f : \mathbb{R}^d \to \mathbb{C}$ is called *slowly increasing*, if there exist $C > 0$ and $N \in \mathbb{N}_0$ such that it holds almost everywhere

$$|f(\mathbf{x})| \le C\,(1 + \|\mathbf{x}\|_2)^N\,. \tag{4.30}$$

These functions grow at most polynomial as $\|\mathbf{x}\|_2 \to \infty$. In particular, polynomials and complex exponential functions $\mathrm{e}^{\mathrm{i}\,\boldsymbol{\omega}\cdot\mathbf{x}}$ are slowly increasing functions. But the reciprocal Gaussian function $f(\mathbf{x}) := \mathrm{e}^{\|\mathbf{x}\|_2^2}$ is not a slowly increasing function.

For each slowly increasing function f, we can form the linear functional $T_f : \mathscr{S}(\mathbb{R}^d) \to \mathbb{C}$,

$$\langle T_f, \varphi\rangle := \int_{\mathbb{R}^d} f(\mathbf{x})\,\varphi(\mathbf{x})\,\mathrm{d}\mathbf{x}\,, \quad \varphi \in \mathscr{S}(\mathbb{R}^d)\,. \tag{4.31}$$

By Lemma 4.35 we obtain $T_f \in \mathscr{S}'(\mathbb{R}^d)$, because for every $\varphi \in \mathscr{S}(\mathbb{R}^d)$,

$$\begin{aligned}
|\langle T_f, \varphi\rangle| &\le \int_{\mathbb{R}^d} \frac{|f(\mathbf{x})|}{(1 + \|\mathbf{x}\|_2)^{N+d+1}}\,(1 + \|\mathbf{x}\|_2)^{N+d+1}\,|\varphi(\mathbf{x})|\,\mathrm{d}\mathbf{x}\\
&\le C\int_{\mathbb{R}^d} \frac{\mathrm{d}\mathbf{x}}{(1 + \|\mathbf{x}\|_2)^{d+1}}\,\sup_{\mathbf{x}\in\mathbb{R}^d}\big((1 + \|\mathbf{x}\|_2)^{N+d+1}\,|\varphi(\mathbf{x})|\big)\\
&\le C\int_{\mathbb{R}^d} \frac{\mathrm{d}\mathbf{x}}{(1 + \|\mathbf{x}\|_2)^{d+1}}\,\|\varphi\|_{N+d+1}\,.
\end{aligned}$$

A function in $L_p(\mathbb{R}^d)$ is not slowly increasing; however these functions give also rise to tempered distributions as the following example shows.

Example 4.36 Every function $f \in L_p(\mathbb{R}^d)$, $1 \le p \le \infty$, is in $\mathscr{S}'(\mathbb{R}^d)$ by Lemma 4.35. For $p = 1$ we have

$$|\langle T_f, \varphi\rangle| \le \int_{\mathbb{R}^d} |f(\mathbf{x})||\varphi(\mathbf{x})|\,\mathrm{d}\mathbf{x} \le \|f\|_{L_1(\mathbb{R}^d)}\|\varphi\|_0 < \infty\,.$$

For $1 < p \le \infty$, let q be given by $\frac{1}{p} + \frac{1}{q} = 1$, where $q = 1$ if $p = \infty$. Then we obtain for $m \in \mathbb{N}_0$ with $m\,q \ge d+1$ by Hölder's inequality

$$\begin{aligned}|\langle T_f, \varphi\rangle| &\le \int_{\mathbb{R}^d} |f(\mathbf{x})|(1+\|\mathbf{x}\|_2)^{-m}(1+\|\mathbf{x}\|_2)^m\,|\varphi(\mathbf{x})|\,\mathrm{d}\mathbf{x}\\ &\le \|\varphi\|_m \int_{\mathbb{R}^d} |f(\mathbf{x})|(1+\|\mathbf{x}\|_2)^{-m}\,\mathrm{d}\mathbf{x}\\ &\le \|\varphi\|_m\,\|f\|_{L_p(\mathbb{R}^d)}\Big(\int_{\mathbb{R}^d}(1+\|\mathbf{x}\|_2)^{-qm}\,\mathrm{d}\mathbf{x}\Big)^{1/q}\,.\end{aligned}$$

□

If a distribution $T \in \mathscr{S}'(\mathbb{R}^d)$ arises from a function in the sense that $\langle T, \varphi\rangle = \int_{\mathbb{R}^d} f(\mathbf{x})\varphi(\mathbf{x})\,\mathrm{d}\mathbf{x}$ is well-defined for all $\varphi \in \mathscr{S}(\mathbb{R}^d)$, then we speak about a *regular tempered distribution*. The following example describes a distribution which is not regular.

Example 4.37 The *Dirac distribution* δ is defined by

$$\langle \delta, \varphi\rangle := \varphi(\mathbf{0})$$

for all $\varphi \in \mathscr{S}(\mathbb{R}^d)$. Clearly, the Dirac distribution δ is a continuous linear functional with $|\langle\delta, \varphi\rangle| \le \|\varphi\|_0$ for all $\varphi \in \mathscr{S}(\mathbb{R}^d)$ so that $\delta \in \mathscr{S}'(\mathbb{R}^d)$. By the following argument, the Dirac distribution is not regular: Assume in contrary that there exists a function f such that

$$\varphi(\mathbf{0}) = \int_{\mathbb{R}^d} f(\mathbf{x})\,\varphi(\mathbf{x})\,\mathrm{d}\mathbf{x}$$

for all $\varphi \in \mathscr{S}(\mathbb{R}^d)$. By (4.30) this function f is integrable over the unit ball. Let φ be the compactly supported test function (4.12) and $\varphi_n(\mathbf{x}) := \varphi(n\,\mathbf{x})$ for $n \in \mathbb{N}$. Then we obtain the contradiction

$$\begin{aligned}\mathrm{e}^{-1} = |\varphi_n(\mathbf{0})| = \Big|\int_{\mathbb{R}^d} f(\mathbf{x})\,\varphi_n(\mathbf{x})\,\mathrm{d}\mathbf{x}\Big| &\le \int_{B_{1/n}(\mathbf{0})} |f(\mathbf{x})|\,|\varphi(n\mathbf{x})|\,\mathrm{d}\mathbf{x}\\ &\le \mathrm{e}^{-1}\int_{B_{1/n}(\mathbf{0})} |f(\mathbf{x})|\,\mathrm{d}\mathbf{x} \to 0 \quad \text{as} \quad n \to \infty\,,\end{aligned}$$

where $B_{1/n}(\mathbf{0}) = \{\mathbf{x} \in \mathbb{R}^d : \|\mathbf{x}\|_2 \le 1/n\}$. □

Remark 4.38 In quantum mechanics, the distribution δ was introduced by the physicist Paul Dirac. It is used to model the density of an idealized point mass as a "generalized function" which is equal to zero everywhere except for zero and whose integral over $\mathbb{R}$ is equal to one. Since there does not exist a function with these properties, the Dirac distribution was defined by L. Schwartz [382, p. 19] as a continuous linear functional that maps every test function $\varphi \in \mathscr{S}(\mathbb{R})$ to its value $\varphi(0)$. In signal processing, the Dirac distribution is also known as the *unit impulse signal*. The Kronecker symbol which is usually defined on $\mathbb{Z}$ is a discrete analogon of the Dirac distribution. □

Important operations on tempered distributions are translations, dilations, and multiplications with smooth, sufficiently fast-decaying functions and derivations. In the following, we consider these operations.

The *translation* by $\mathbf{x}_0 \in \mathbb{R}^d$ of a tempered distribution $T \in \mathscr{S}'(\mathbb{R}^d)$ is the tempered distribution $T(\cdot - \mathbf{x}_0)$ defined for all $\varphi \in \mathscr{S}(\mathbb{R}^d)$ by

$$\langle T(\cdot - \mathbf{x}_0), \varphi \rangle := \langle T, \varphi(\cdot + \mathbf{x}_0) \rangle .$$

The *scaling* with $c \in \mathbb{R} \setminus \{0\}$ of $T \in \mathscr{S}'(\mathbb{R}^d)$ is the tempered distribution $T(c\,\cdot)$ given for all $\varphi \in \mathscr{S}(\mathbb{R}^d)$ by

$$\langle T(c\,\cdot), \varphi \rangle := \frac{1}{|c|^d}\, \langle T, \varphi(c^{-1}\,\cdot) \rangle .$$

In particular for $c = -1$, we obtain the *reflection* of $T \in \mathscr{S}'(\mathbb{R}^d)$, namely,

$$\langle T(-\,\cdot), \varphi \rangle := \langle T, \tilde{\varphi} \rangle$$

for all $\varphi \in \mathscr{S}(\mathbb{R}^d)$, where $\tilde{\varphi}(\mathbf{x}) := \varphi(-\mathbf{x})$ denotes the reflection of $\varphi \in \mathscr{S}(\mathbb{R}^d)$.

Assume that $\psi \in C^\infty(\mathbb{R}^d)$ fulfills

$$|D^{\boldsymbol{\alpha}} \psi(\mathbf{x})| \le C_{\boldsymbol{\alpha}}\, (1 + \|\mathbf{x}\|_2)^{N_{\boldsymbol{\alpha}}} \tag{4.32}$$

for all $\boldsymbol{\alpha} \in \mathbb{N}_0^d$ and positive constants $C_{\boldsymbol{\alpha}}$ and $N_{\boldsymbol{\alpha}}$, i.e., $D^{\boldsymbol{\alpha}}\psi$ has at most polynomial growth at infinity for all $\boldsymbol{\alpha} \in \mathbb{N}_0^d$. Then the *product* of ψ with a tempered distribution $T \in \mathscr{S}'(\mathbb{R}^d)$ is the tempered distribution $\psi\, T$ defined as

$$\langle \psi\, T, \varphi \rangle := \langle T, \psi\, \varphi \rangle , \quad \varphi \in \mathscr{S}(\mathbb{R}^d) .$$

Note that the product of an arbitrary $C^\infty(\mathbb{R}^d)$ function with a tempered distribution is not defined.

Example 4.39 For a regular tempered distribution $T_f \in \mathscr{S}'(\mathbb{R}^d)$ and $c \neq 0$, we obtain

$$T_f(\cdot - \mathbf{x}_0) = T_{f(\cdot - \mathbf{x}_0)} , \quad T_f(c\cdot) = T_{f(c\cdot)} , \quad \psi T_f = T_{\psi f} .$$

For the Dirac distribution δ, we have

$$\langle \delta(\cdot - \mathbf{x}_0), \varphi \rangle = \langle \delta, \varphi(\cdot + \mathbf{x}_0) \rangle = \varphi(\mathbf{x}_0),$$
$$\langle \delta(c\cdot), \varphi \rangle = \frac{1}{|c|^d} \langle \delta, \varphi\big(\frac{\cdot}{c}\big) \rangle = \frac{1}{|c|^d} \varphi(\mathbf{0}),$$
$$\langle \psi\, \delta, \varphi \rangle = \langle \delta, \psi\, \varphi \rangle = \psi(\mathbf{0})\, \varphi(\mathbf{0})$$

for all $\varphi \in \mathscr{S}(\mathbb{R}^d)$, where $\psi \in C^\infty(\mathbb{R}^d)$ fulfills (4.32) for all $\boldsymbol{\alpha} \in \mathbb{N}_0^d$. □

Example 4.40 The distribution T_f arising from the function $f(x) := \ln(|x|)$ for $x \neq 0$ and $f(x) = 0$ for $x = 0$ is in $\mathscr{S}'(\mathbb{R})$ by the following reason: For all $\varphi \in \mathscr{S}(\mathbb{R})$ we have

$$\begin{aligned}\langle \ln(|x|), \varphi(x) \rangle &= \int_{-\infty}^{0} \ln(-x)\, \varphi(x)\, \mathrm{d}x + \int_{0}^{\infty} \ln(x)\, \varphi(x)\, \mathrm{d}x \\ &= \int_{0}^{\infty} \ln(x)\, \varphi(-x)\, \mathrm{d}x + \int_{0}^{\infty} \ln(x)\, \varphi(x)\, \mathrm{d}x .\end{aligned}$$

Since $\ln(x) \leq x$ for $x \geq 1$, we obtain

$$\begin{aligned}\int_{0}^{\infty} \ln(x)\, \varphi(x)\, \mathrm{d}x &= \int_{0}^{1} \ln(x)\, \varphi(x)\, \mathrm{d}x + \int_{1}^{\infty} \ln(x)\, \varphi(x)\, \mathrm{d}x \\ &\leq \|\varphi\|_{C_0(\mathbb{R})} \int_{0}^{1} \ln(x)\, \mathrm{d}x + \int_{1}^{\infty} x \varphi(x)\, \mathrm{d}x \\ &= \|\varphi\|_{C_0(\mathbb{R})} \lim_{\varepsilon \to 0} \int_{\varepsilon}^{1} \ln(x)\, \mathrm{d}x + \int_{1}^{\infty} x \varphi(x)\, \mathrm{d}x\end{aligned}$$

and similarly for $\varphi(-x)$. Since $\varphi \in \mathscr{S}(\mathbb{R})$, the second integral exists. For the first integral, we get by integration by parts

$$\int_{\varepsilon}^{1} \ln(x)\, \mathrm{d}x = x \ln(x)|_{\varepsilon}^{1} - \int_{\varepsilon}^{1} \frac{1}{x}\, x\, \mathrm{d}x = -\varepsilon \ln(\varepsilon) - (1 - \varepsilon)$$

and by l'Hospital's rule

$$\lim_{\varepsilon \to 0} \int_{\varepsilon}^{1} \ln(x)\, \mathrm{d}x = \lim_{\varepsilon \to 0} (-\varepsilon \ln(\varepsilon) - (1 - \varepsilon)) = -1 .$$

Therefore $\langle \ln(|x|), \varphi(x) \rangle$ is well-defined for all $\varphi \in \mathscr{S}(\mathbb{R})$ and a tempered distribution of function type. Similarly as above we can conclude that $\ln(|x|)$ is absolutely integrable on any compact set. □

Another important operation on tempered distributions is the differentiation. For $\boldsymbol{\alpha} \in \mathbb{N}_0^d$, the *derivative* $D^{\boldsymbol{\alpha}} T$ of a distribution $T \in \mathscr{S}'(\mathbb{R}^d)$ is defined for all $\varphi \in \mathscr{S}(\mathbb{R}^d)$ by

$$\langle D^{\boldsymbol{\alpha}} T, \varphi\rangle := (-1)^{|\boldsymbol{\alpha}|} \langle T, D^{\boldsymbol{\alpha}} \varphi\rangle . \tag{4.33}$$

Assume that $f \in C^r(\mathbb{R}^d)$ with $r \in \mathbb{N}$ possesses slowly increasing partial derivatives $D^{\boldsymbol{\alpha}} f$ for all $|\boldsymbol{\alpha}| \le r$. Thus $T_{D^{\boldsymbol{\alpha}} f} \in \mathscr{S}'(\mathbb{R}^d)$. Then we see by integration by parts that $T_{D^{\boldsymbol{\alpha}} f} = D^{\boldsymbol{\alpha}} T_f$ for all $\boldsymbol{\alpha} \in \mathbb{N}_0^d$ with $|\boldsymbol{\alpha}| \le r$, i.e., the distributional derivatives and the classical derivatives coincide.

Lemma 4.41 *Let $T, T_k \in \mathscr{S}'(\mathbb{R}^d)$ with $k \in \mathbb{N}$ be given. For $\lambda_1, \lambda_2 \in \mathbb{R}$, and $\boldsymbol{\alpha}, \boldsymbol{\beta} \in \mathbb{N}_0^d$, the following relations hold true:*

1. $D^{\boldsymbol{\alpha}} T \in \mathscr{S}'(\mathbb{R}^d)$,
2. $D^{\boldsymbol{\alpha}} (\lambda_1 T_1 + \lambda_2 T_2) = \lambda_1 D^{\boldsymbol{\alpha}} T_1 + \lambda_2 D^{\boldsymbol{\alpha}} T_2$,
3. $D^{\boldsymbol{\alpha}} (D^{\boldsymbol{\beta}} T) = D^{\boldsymbol{\beta}} (D^{\boldsymbol{\alpha}} T) = D^{\boldsymbol{\alpha}+\boldsymbol{\beta}} T$.
4. $T_k \underset{\mathscr{S}'}{\longrightarrow} T$ *as* $k \to \infty$ *implies* $D^{\boldsymbol{\alpha}} T_k \underset{\mathscr{S}'}{\longrightarrow} D^{\boldsymbol{\alpha}} T$ *as* $k \to \infty$.

Proof The properties 1–3 follow directly from the definition of the derivative of tempered distributions. Property 4 can be derived by

$$\lim_{k\to\infty} \langle D^{\boldsymbol{\alpha}} T_k, \varphi\rangle = \lim_{k\to\infty} (-1)^{|\boldsymbol{\alpha}|} \langle T_k, D^{\boldsymbol{\alpha}} \varphi\rangle = (-1)^{|\boldsymbol{\alpha}|} \langle T, D^{\boldsymbol{\alpha}} \varphi\rangle = \langle D^{\boldsymbol{\alpha}} T, \varphi\rangle$$

for all $\varphi \in \mathscr{S}(\mathbb{R}^d)$. ■

Example 4.42 For the slowly increasing univariate function

$$f(x) := \begin{cases} 0 & x \le 0 , \\ x & x > 0 , \end{cases}$$

we obtain

$$\begin{aligned} \langle D\, T_f, \varphi\rangle &= -\langle f, \varphi'\rangle = -\int_{\mathbb{R}} f(x)\, \varphi'(x)\, \mathrm{d}x \\ &= -\int_0^\infty x\, \varphi'(x)\, \mathrm{d}x = -x\, \varphi(x)\big|_0^\infty + \int_0^\infty \varphi(x)\, \mathrm{d}x = \int_0^\infty \varphi(x)\, \mathrm{d}x \end{aligned}$$

so that

$$D\, T_f(x) = H(x) := \begin{cases} 0 & x \le 0 , \\ 1 & x > 0 . \end{cases}$$

The function H is called *Heaviside function*. Further we get

$$\langle D^2\, T_f,\, \varphi\rangle = -\langle D\, T_f,\, \varphi'\rangle = -\int_0^\infty \varphi'(x)\, \mathrm{d}x = -\varphi(x)\big|_0^\infty = \varphi(0) = \langle \delta,\, \varphi\rangle$$

so that $D^2\, T_f = D\, T_H = \delta$. Thus the distributional derivative of the Heaviside function is equal to the Dirac distribution. □

Example 4.43 We are interested in the distributional derivative of regular tempered distribution T_f of Example 4.40. For all $\varphi \in \mathscr{S}(\mathbb{R})$, we get by integration by parts

$$\begin{aligned}\langle D\ln(|x|),\, \varphi(x)\rangle &= -\langle \ln(|x|),\, \varphi'(x)\rangle \\ &= -\int_0^\infty \ln(x)\big(\varphi'(x) + \varphi'(-x)\big)\, \mathrm{d}x \\ &= -\ln(x)\big(\varphi(x) - \varphi(-x)\big)\big|_0^\infty + \int_0^\infty \frac{1}{x}\big(\varphi(x) - \varphi(-x)\big)\, \mathrm{d}x \\ &= \lim_{\varepsilon\to 0} \ln(\varepsilon)\big(\varphi(\varepsilon) - \varphi(-\varepsilon)\big) + \lim_{\varepsilon\to 0}\int_\varepsilon^\infty \frac{1}{x}\big(\varphi(x) - \varphi(-x)\big)\, \mathrm{d}x\end{aligned}$$

Taylor expansion yields

$$\varphi(\varepsilon) = \varphi(0) + \varepsilon\, \varphi'(\xi_\varepsilon)\,, \quad \xi_\varepsilon \in (0,\, \varepsilon)$$

so that by the mean value theorem

$$\varphi(\varepsilon) - \varphi(-\varepsilon) = \varepsilon\big(\varphi'(\xi_\varepsilon) + \varphi'(\xi_{-\varepsilon})\big) = 2\varepsilon\, \varphi'(\xi)\,, \quad |\xi| < \varepsilon\,.$$

Thus,

$$\begin{aligned}\langle D\ln(|x|), \varphi(x)\rangle &= \lim_{\varepsilon\to 0}\int_\varepsilon^\infty \frac{1}{x}\big(\varphi(x) - \varphi(-x)\big)\, \mathrm{d}x \\ &= \lim_{\varepsilon\to 0}\Big(\int_\varepsilon^\infty + \int_{-\infty}^{-\varepsilon}\Big)\frac{1}{x}\varphi(x)\, \mathrm{d}x \\ &= \mathrm{pv}\int_{\mathbb{R}} \frac{1}{x}\, \varphi(x)\, \mathrm{d}x,\end{aligned}$$

where pv denotes the Cauchy principle value integral. We see that the tempered distribution $\mathrm{pv}\big(\frac{1}{x}\big)$ defined for all $\varphi \in \mathscr{S}(\mathbb{R})$ by

$$\langle \mathrm{pv}\big(\frac{1}{x}\big), \varphi(x)\rangle := \mathrm{pv}\int_{\mathbb{R}} \frac{1}{x}\, \varphi(x)\, \mathrm{d}x \tag{4.34}$$

fulfills $D\ln(|x|) = \mathrm{pv}\big(\frac{1}{x}\big)$. Note that the integral $\int_{\mathbb{R}} \frac{1}{x}\, \varphi(x)\, \mathrm{d}x$ does not exist for all $\varphi \in \mathscr{S}(\mathbb{R})$. □

Remark 4.44 Let $f \in C^1(\mathbb{R}\setminus\{x_1, \ldots, x_n\})$ be given, where $x_k \in \mathbb{R}$, $k = 1, \ldots, n$, are distinct jump discontinuities of f. Then the distributional derivative of T_f reads as follows:

$$D\,T_f = f' + \sum_{k=1}^{n} \big(f(x_k + 0) - f(x_k - 0)\big)\,\delta(\cdot\, - x_k)\,.$$

For example, the distributional derivative of the characteristic function $f = \chi_{[a,\,b]}$, where $[a,\, b] \subset \mathbb{R}$ is a compact interval, is equal to

$$D\,T_f = \delta(\cdot\, - a) - \delta(\cdot\, - b)\,.$$

If $f = N_2$ is the cardinal B-spline of order 2 (cf. Example 2.16), then the first and second distributional derivatives of T_f are

$$D\,T_f = \chi_{[0,\,1]} - \chi_{[1,\,2]}\,, \quad D^2\,T_f = \delta - 2\,\delta(\cdot\, - 1) + \delta(\cdot\, - 2)\,.$$

□

For arbitrary $\psi \in \mathscr{S}(\mathbb{R}^d)$ and $T \in \mathscr{S}'(\mathbb{R}^d)$, the *convolution* $\psi * T$ is defined as

$$\langle \psi * T, \varphi\rangle := \langle T,\, \tilde{\psi} * \varphi\rangle\,, \quad \varphi \in \mathscr{S}(\mathbb{R}^d)\,, \tag{4.35}$$

where $\tilde{\psi}$ denotes the reflection of ψ, i.e., $\tilde{\psi}(\mathbf{x}) := \psi(-\mathbf{x})$ for $\mathbf{x} \in \mathbb{R}^d$.

Example 4.45 Let f be a slowly increasing function. For the regular tempered distribution $T_f \in \mathscr{S}'(\mathbb{R}^d)$ and $\psi \in \mathscr{S}(\mathbb{R}^d)$, we have by Fubini's theorem for all $\varphi \in \mathscr{S}(\mathbb{R}^d)$

$$\begin{aligned}\langle \psi * T_f,\, \varphi\rangle &= \langle T_f,\, \tilde{\psi} * \varphi\rangle = \int_{\mathbb{R}^d} f(\mathbf{y})\,(\tilde{\psi} * \varphi)(\mathbf{y})\,\mathrm{d}\mathbf{y}\\ &= \int_{\mathbb{R}^d} f(\mathbf{y})\,\Big(\int_{\mathbb{R}^d} \psi(\mathbf{x} - \mathbf{y})\,\varphi(\mathbf{x})\,\mathrm{d}\mathbf{x}\Big)\,\mathrm{d}\mathbf{y} = \int_{\mathbb{R}^d} (\psi * f)(\mathbf{x})\,\varphi(\mathbf{x})\,\mathrm{d}\mathbf{x}\,,\end{aligned}$$

i.e., $\psi * T_f = T_{\psi * f}$ is a regular tempered distribution generated by the $C^\infty(\mathbb{R}^d)$ function

$$\int_{\mathbb{R}^d} \psi(\mathbf{x} - \mathbf{y})\,f(\mathbf{y})\,\mathrm{d}\mathbf{y} = \langle T_f,\, \psi(\mathbf{x} -\, \cdot)\rangle\,.$$

For the Dirac distribution δ and $\psi \in \mathscr{S}(\mathbb{R}^d)$, we get for all $\varphi \in \mathscr{S}(\mathbb{R}^d)$

$$\langle \psi * \delta,\, \varphi\rangle = \langle \delta,\, \tilde{\psi} * \varphi\rangle = (\tilde{\psi} * \varphi)(\mathbf{0}) = \int_{\mathbb{R}^d} \psi(\mathbf{x})\,\varphi(\mathbf{x})\,\mathrm{d}\mathbf{x}$$

i.e., $\psi * \delta = \psi$. □

The convolution $\psi * T$ of $\psi \in \mathcal{S}(\mathbb{R}^d)$ and $T \in \mathcal{S}'(\mathbb{R}^d)$ possesses the following properties.

Theorem 4.46 *For all $\psi \in \mathcal{S}(\mathbb{R}^d)$ and $T \in \mathcal{S}'(\mathbb{R}^d)$, the convolution $\psi * T$ is a regular tempered distribution generated by the slowly increasing $C^\infty(\mathbb{R}^d)$ function $\langle T, \psi(\mathbf{x} - \cdot)\rangle$, $\mathbf{x} \in \mathbb{R}^d$. For all $\boldsymbol{\alpha} \in \mathbb{N}_0^d$, it holds*

$$D^{\boldsymbol{\alpha}}(\psi * T) = (D^{\boldsymbol{\alpha}}\psi) * T = \psi * (D^{\boldsymbol{\alpha}}T). \tag{4.36}$$

Proof

1. For arbitrary $\varphi \in \mathcal{S}(\mathbb{R}^d)$, $T \in \mathcal{S}'(\mathbb{R}^d)$, and $\boldsymbol{\alpha} \in \mathbb{N}_0^d$, we obtain by (4.33) and (4.35)

$$\langle D^{\boldsymbol{\alpha}}(\psi * T), \varphi\rangle = (-1)^{|\boldsymbol{\alpha}|}\, \langle \psi * T,\, D^{\boldsymbol{\alpha}}\varphi\rangle = (-1)^{|\boldsymbol{\alpha}|}\, \langle T,\, \tilde{\psi} * D^{\boldsymbol{\alpha}}\varphi\rangle\,,$$

where $\tilde{\psi}(\mathbf{x}) = \psi(-\mathbf{x})$ and

$$(\tilde{\psi} * D^{\boldsymbol{\alpha}}\varphi)(\mathbf{x}) = \int_{\mathbb{R}^d} \tilde{\psi}(\mathbf{y})\, D^{\boldsymbol{\alpha}}\varphi(\mathbf{x} - \mathbf{y})\, \mathrm{d}\mathbf{y}\,.$$

Now we have

$$\begin{aligned}(\tilde{\psi} * D^{\boldsymbol{\alpha}}\varphi)(\mathbf{x}) &= \int_{\mathbb{R}^d} \tilde{\psi}(\mathbf{y})\, D^{\boldsymbol{\alpha}}\varphi(\mathbf{x} - \mathbf{y})\, \mathrm{d}\mathbf{y} = D^{\boldsymbol{\alpha}}\,(\tilde{\psi} * \varphi)(\mathbf{x}) \\ &= D^{\boldsymbol{\alpha}} \int_{\mathbb{R}^d} \tilde{\psi}(\mathbf{x} - \mathbf{y})\, \varphi(\mathbf{y})\, \mathrm{d}\mathbf{y} \\ &= \int_{\mathbb{R}^d} D^{\boldsymbol{\alpha}}\, \tilde{\psi}(\mathbf{x} - \mathbf{y})\, \varphi(\mathbf{y})\, \mathrm{d}\mathbf{y} = (D^{\boldsymbol{\alpha}}\, \tilde{\psi} * \varphi)(\mathbf{x})\,,\end{aligned}$$

since the interchange of differentiation and integration in above integrals is justified, because $\tilde{\psi}$ and φ belong to $\mathcal{S}(\mathbb{R}^d)$. From

$$D^{\boldsymbol{\alpha}}\, \tilde{\psi} = (-1)^{|\boldsymbol{\alpha}|}\, \widetilde{D^{\boldsymbol{\alpha}}\, \psi}$$

it follows that

$$\begin{aligned}\langle D^{\boldsymbol{\alpha}}(\psi * T), \varphi\rangle &= (-1)^{|\boldsymbol{\alpha}|}\, \langle \psi * T,\, D^{\boldsymbol{\alpha}}\varphi\rangle = \langle D^{\boldsymbol{\alpha}}\, T,\, \tilde{\psi} * \varphi\rangle = \langle \psi * D^{\boldsymbol{\alpha}}\, T,\, \varphi\rangle \\ &= (-1)^{|\boldsymbol{\alpha}|}\, \langle T,\, D^{\boldsymbol{\alpha}}\tilde{\psi} * \varphi\rangle = \langle T,\, \widetilde{D^{\boldsymbol{\alpha}}\psi} * \varphi\rangle = \langle (D^{\boldsymbol{\alpha}}\psi) * T, \varphi\rangle.\end{aligned}$$

Thus we have shown (4.36).

2. Now we prove that the convolution $\psi * T$ is a regular tempered distribution generated by the complex-valued function $\langle T,\, \psi(\mathbf{x}-\cdot)\rangle$ for $\mathbf{x} \in \mathbb{R}^d$. In Example 4.45, we have seen that this is true for each regular tempered distribution.
Let $\psi,\, \varphi \in \mathscr{S}(\mathbb{R}^d)$, and $T \in \mathscr{S}'(\mathbb{R}^d)$ be given. By Lemma 4.19 we know that $\tilde{\psi} * \varphi \in \mathscr{S}(\mathbb{R}^d)$. We represent $(\tilde{\psi} * \varphi)(\mathbf{y})$ for arbitrary $\mathbf{y} \in \mathbb{R}^d$ as a limit of Riemann sums

$$(\tilde{\psi} * \varphi)(\mathbf{y}) = \int_{\mathbb{R}^d} \psi(\mathbf{x}-\mathbf{y})\, \varphi(\mathbf{x})\, \mathrm{d}\mathbf{x} = \lim_{j \to \infty} \sum_{\mathbf{k} \in \mathbb{Z}^d} \psi(\mathbf{x_k}-\mathbf{y})\psi(\mathbf{x_k})\, \frac{1}{j^d}\,,$$

where $\mathbf{x_k} := \frac{\mathbf{k}}{j}$, $\mathbf{k} \in \mathbb{Z}^d$, is the midpoint of a hypercube with side length $\frac{1}{j}$. Indeed, since $\tilde{\psi} * \varphi \in \mathscr{S}(\mathbb{R}^d)$, it is not hard to check that the above Riemann sums converge in $\mathscr{S}(\mathbb{R}^d)$. Since T is a continuous linear functional, we get

$$\begin{aligned}\langle T,\, \tilde{\psi} * \varphi\rangle &= \lim_{j \to \infty} \langle T,\, \sum_{\mathbf{k} \in \mathbb{Z}^d} \varphi(\mathbf{x_k} - \cdot)\, \psi(\mathbf{x_k})\, \frac{1}{j^d}\rangle \\ &= \lim_{j \to \infty} \sum_{\mathbf{k} \in \mathbb{Z}^d} \varphi(\mathbf{x_k})\, \frac{1}{j^d}\, \langle T,\, \psi(\mathbf{x_k} - \cdot)\rangle = \int_{\mathbb{R}^d} \langle T,\, \psi(\mathbf{x} - \cdot)\rangle\, \varphi(\mathbf{x})\, \mathrm{d}\mathbf{x}\,,\end{aligned}$$

i.e., the convolution $\psi * T$ is a regular tempered distribution generated by the function $\langle T,\, \psi(\mathbf{x} - \cdot)\rangle$ which belongs to $C^\infty(\mathbb{R}^d)$ by (4.36).
3. Finally, we show that the $C^\infty(\mathbb{R}^d)$ function $\langle T,\, \psi(\mathbf{x} - \cdot)\rangle$ is slowly increasing. Here we use the simple estimate

$$1 + \|\mathbf{x} - \mathbf{y}\|_2 \le 1 + \|\mathbf{x}\|_2 + \|\mathbf{y}\|_2 \le (1 + \|\mathbf{x}\|_2)\, (1 + \|\mathbf{y}\|_2)$$

for all $\mathbf{x},\, \mathbf{y} \in \mathbb{R}^d$.
For arbitrary fixed $\mathbf{x}_0 \in \mathbb{R}^d$ and every $m \in \mathbb{N}_0$, we obtain for $\psi \in \mathscr{S}(\mathbb{R}^d)$,

$$\begin{aligned}\|\psi(\mathbf{x}_0 - \cdot)\|_m &= \max_{|\boldsymbol{\beta}| \le m} \|(1 + \|\mathbf{x}\|_2)^m\, D^{\boldsymbol{\beta}} \psi(\mathbf{x}_0 - \mathbf{x})\|_{C_0(\mathbb{R}^d)} \\ &= \max_{|\boldsymbol{\beta}| \le m}\, \max_{\mathbf{x} \in \mathbb{R}^d} (1 + \|\mathbf{x}\|_2)^m\, |D^{\boldsymbol{\beta}}\, \psi(\mathbf{x}_0 - \mathbf{x})| \\ &= \max_{|\boldsymbol{\beta}| \le m}\, \max_{\mathbf{y} \in \mathbb{R}^d} (1 + \|\mathbf{x}_0 - \mathbf{y}\|_2)^m\, |D^{\boldsymbol{\beta}} \psi(\mathbf{y})| \\ &\le (1 + \|\mathbf{x}_0\|_2)^m \sup_{|\boldsymbol{\beta}| \le m}\, \sup_{\mathbf{y} \in \mathbb{R}^d} (1 + \|\mathbf{y}\|_2)^m\, |D^{\boldsymbol{\beta}}\, \psi(\mathbf{y})| \\ &= (1 + \|\mathbf{x}_0\|_2)^m\, \|\psi\|_m\,.\end{aligned}$$

Since $T \in \mathscr{S}'(\mathbb{R}^d)$, by Lemma 4.35 of Schwartz, there exist constants $m \in \mathbb{N}_0$ and $C > 0$, so that $|\langle T,\, \varphi\rangle| \le C\, \|\varphi\|_m$ for all $\varphi \in \mathscr{S}(\mathbb{R}^d)$. Then we conclude

$$|\langle T,\ \psi(\mathbf{x}-\cdot)\rangle| \le C\,\|\psi(\mathbf{x}-\cdot)\|_m \le C\,(1+\|\mathbf{x}\|_2)^m\,\|\psi\|_m\,.$$

Hence $\langle T,\ \psi(\mathbf{x}-\cdot)\rangle$ is a slowly increasing function. ■

4.3.2 Fourier Transforms on $\mathscr{S}'(\mathbb{R}^d)$

The *Fourier transform* $\mathscr{F}T = \hat{T}$ of a tempered distribution $T \in \mathscr{S}'(\mathbb{R}^d)$ is defined by

$$\langle \mathscr{F}T, \varphi\rangle = \langle \hat{T}, \varphi\rangle := \langle T, \mathscr{F}\varphi\rangle = \langle T, \hat{\varphi}\rangle \tag{4.37}$$

for all $\varphi \in \mathscr{S}(\mathbb{R}^d)$. Indeed $\hat{T}$ is again a continuous linear functional on $\mathscr{S}(\mathbb{R}^d)$, since by Theorem 4.18, the expression $\langle T, \mathscr{F}\varphi\rangle$ defines a linear functional on $\mathscr{S}(\mathbb{R}^d)$. Further, $\varphi_k \underset{\mathscr{S}}{\longrightarrow} \varphi$ as $k \to \infty$, implies $\mathscr{F}\varphi_k \underset{\mathscr{S}}{\longrightarrow} \mathscr{F}\varphi$ as $k \to \infty$ so that for $T \in \mathscr{S}'(\mathbb{R}^d)$, it follows

$$\lim_{k\to\infty} \langle \hat{T}, \varphi_k\rangle = \lim_{k\to\infty} \langle T,\ \mathscr{F}\varphi_k\rangle = \langle T,\ \mathscr{F}\varphi\rangle = \langle \hat{T}, \varphi\rangle\,.$$

Example 4.47 Let $f \in L_1(\mathbb{R}^d)$. Then we obtain for an arbitrary $\varphi \in \mathscr{S}(\mathbb{R}^d)$ by Fubini's theorem

$$\begin{aligned}\langle \mathscr{F}T_f,\ \varphi\rangle = \langle T_f,\ \hat{\varphi}\rangle &= \int_{\mathbb{R}^d}\Big(\int_{\mathbb{R}^d}\varphi(\mathbf{x})\,\mathrm{e}^{-\mathrm{i}\,\mathbf{x}\cdot\boldsymbol{\omega}}\,\mathrm{d}\mathbf{x}\Big)\,f(\boldsymbol{\omega})\,\mathrm{d}\boldsymbol{\omega}\\ &= \int_{\mathbb{R}^d}\hat{f}(\mathbf{x})\,\varphi(\mathbf{x})\,\mathrm{d}\mathbf{x} = \langle T_{\hat{f}},\ \varphi\rangle\,,\end{aligned}$$

i.e., $\mathscr{F}T_f = T_{\mathscr{F}f}$.

Let $\mathbf{x}_0 \in \mathbb{R}^d$ be fixed. For the shifted Dirac distribution $\delta(\cdot - \mathbf{x}_0)$, we have

$$\begin{aligned}\langle \mathscr{F}\delta(\cdot-\mathbf{x}_0),\ \varphi\rangle = \langle \delta(\cdot-\mathbf{x}_0),\ \hat{\varphi}\rangle &= \langle \delta(\cdot-\mathbf{x}_0),\ \int_{\mathbb{R}^d}\varphi(\boldsymbol{\omega})\mathrm{e}^{-\mathrm{i}\,\boldsymbol{\omega}\cdot\mathbf{x}}\,\mathrm{d}\boldsymbol{\omega}\rangle\\ &= \int_{\mathbb{R}^d}\varphi(\boldsymbol{\omega})\,\mathrm{e}^{-\mathrm{i}\,\boldsymbol{\omega}\cdot\mathbf{x}_0}\,\mathrm{d}\boldsymbol{\omega} = \langle \mathrm{e}^{-\mathrm{i}\,\boldsymbol{\omega}\cdot\mathbf{x}_0},\ \varphi(\boldsymbol{\omega})\rangle\,,\end{aligned}$$

so that $\mathscr{F}\delta(\cdot - \mathbf{x}_0) = \mathrm{e}^{-\mathrm{i}\,\boldsymbol{\omega}\cdot\mathbf{x}_0}$ and in particular, for $\mathbf{x}_0 = \mathbf{0}$ we obtain $\mathscr{F}\delta = 1$. □

Theorem 4.48 *The Fourier transform on $\mathscr{S}'(\mathbb{R}^d)$ is a linear, bijective operator $\mathscr{F} : \mathscr{S}'(\mathbb{R}^d) \to \mathscr{S}'(\mathbb{R}^d)$. The Fourier transform on $\mathscr{S}'(\mathbb{R}^d)$ is continuous in the sense that for $T_k,\ T \in \mathscr{S}'(\mathbb{R}^d)$ the convergence $T_k \underset{\mathscr{S}'}{\longrightarrow} T$ as $k \to \infty$ implies $\mathscr{F}T_k \underset{\mathscr{S}'}{\longrightarrow} \mathscr{F}T$ as $k \to \infty$. The* inverse Fourier transform *is given by*

$$\langle \mathscr{F}^{-1}T,\, \varphi\rangle = \langle T,\, \mathscr{F}^{-1}\varphi\rangle \tag{4.38}$$

for all $\varphi \in \mathscr{S}(\mathbb{R}^d)$ *which means*

$$\mathscr{F}^{-1}T := \frac{1}{(2\pi)^d}\,\mathscr{F}\,T(-\,\cdot)\,.$$

For all $T \in \mathscr{S}'(\mathbb{R}^d)$, *it holds the* Fourier inversion formula

$$\mathscr{F}^{-1}(\mathscr{F}\,T) = \mathscr{F}\,(\mathscr{F}^{-1}T) = T\,.$$

Proof By definition (4.37), the Fourier transform $\mathscr{F}$ maps $\mathscr{S}'(\mathbb{R}^d)$ into itself. Obviously, $\mathscr{F}$ is a linear operator. We show that $\mathscr{F}$ is a continuous linear operator of $\mathscr{S}'(\mathbb{R}^d)$ onto $\mathscr{S}'(\mathbb{R}^d)$. Assume that $T_k \underset{\mathscr{S}'}{\longrightarrow} T$ as $k \to \infty$. Then, we get by (4.37),

$$\lim_{k\to\infty} \langle \mathscr{F}\,T_k,\, \varphi\rangle = \lim_{k\to\infty} \langle T_k,\, \mathscr{F}\,\varphi\rangle = \langle T,\, \mathscr{F}\,\varphi\rangle = \langle \mathscr{F}\,T,\, \varphi\rangle$$

for all $\varphi \in \mathscr{S}(\mathbb{R}^d)$. This means that $\mathscr{F}\,T_k \underset{\mathscr{S}'}{\longrightarrow} \mathscr{F}\,T$ as $k \to \infty$, i.e., the operator $\mathscr{F} : \mathscr{S}'(\mathbb{R}^d) \to \mathscr{S}'(\mathbb{R}^d)$ is continuous.

Next we show that (4.38) is the inverse Fourier transform, i.e.,

$$\mathscr{F}^{-1}\,(\mathscr{F}\,T) = T\,, \quad \mathscr{F}\,(\mathscr{F}^{-1}\,T) = T \tag{4.39}$$

for all $T \in \mathscr{S}'(\mathbb{R}^d)$. By Theorem 4.18, we find that for all $\varphi \in \mathscr{S}(\mathbb{R}^d)$,

$$\begin{aligned}
\langle \mathscr{F}^{-1}\,(\mathscr{F}\,T),\, \varphi\rangle &= \frac{1}{(2\pi)^d}\,\langle \mathscr{F}\,\big(\mathscr{F}\,T(-\,\cdot)\big),\, \varphi\rangle \\
&= \frac{1}{(2\pi)^d}\,\langle \mathscr{F}\,T(-\,\cdot),\, \mathscr{F}\,\varphi\rangle = \frac{1}{(2\pi)^d}\,\langle \mathscr{F}\,T,\, (\mathscr{F}\,\varphi)(-\,\cdot)\rangle \\
&= \langle \mathscr{F}\,T,\, \mathscr{F}^{-1}\,\varphi\rangle = \langle T,\, \mathscr{F}(\mathscr{F}^{-1}\,\varphi)\rangle = \langle T,\, \varphi\rangle\,.
\end{aligned}$$

By (4.39), each $T \in \mathscr{S}'(\mathbb{R}^d)$ is the Fourier transform of the tempered distribution $S = \mathscr{F}^{-1}\,T$, i.e., $T = \mathscr{F}\,S$. Thus both $\mathscr{F}$ and $\mathscr{F}^{-1}$ map $\mathscr{S}'(\mathbb{R}^d)$ one-to-one onto $\mathscr{S}'(\mathbb{R}^d)$. ■

Remark 4.49 From Theorem 4.48 it follows immediately Theorem 4.22. If $f \in L_1(\mathbb{R}^d)$ with $\hat{f} \in L_1(\mathbb{R}^d)$ is given, then T_f and $T_{\hat{f}}$ are regular tempered distributions by Example 4.36. By Theorem 4.48 and Example 4.47, we have

$$T_{\mathscr{F}^{-1}\hat{f}} = \mathscr{F}^{-1}\,T_{\hat{f}} = \mathscr{F}^{-1}(\mathscr{F}T_f) = T_f$$

so that the functions f and

$$(\mathscr{F}^{-1}\hat{f})(\mathbf{x}) = \frac{1}{(2\pi)^d}\int_{\mathbb{R}^d}\hat{f}(\boldsymbol{\omega})\,\mathrm{e}^{\mathrm{i}\,\mathbf{x}\cdot\boldsymbol{\omega}}\,\mathrm{d}\boldsymbol{\omega}$$

are equal almost everywhere. □

The following theorem summarizes properties of Fourier transform on $\mathscr{S}'(\mathbb{R}^d)$.

Theorem 4.50 (Properties of the Fourier Transform on $\mathscr{S}'(\mathbb{R}^d)$) *The Fourier transform of a tempered distribution $T \in \mathscr{S}'(\mathbb{R}^d)$ has the following properties:*

1. Translation and modulation: *For fixed* $\mathbf{x}_0,\ \boldsymbol{\omega}_0 \in \mathbb{R}^d$,

$$\mathscr{F}T(\cdot - \mathbf{x}_0) = \mathrm{e}^{-\mathrm{i}\,\boldsymbol{\omega}\cdot\mathbf{x}_0}\,\mathscr{F}T\,,$$
$$\mathscr{F}\big(\mathrm{e}^{-\mathrm{i}\,\boldsymbol{\omega}_0\cdot\mathbf{x}}\,T\big) = \mathscr{F}T(\cdot + \boldsymbol{\omega}_0)\,.$$

2. Differentiation and multiplication: *For* $\boldsymbol{\alpha} \in \mathbb{N}_0^d$,

$$\mathscr{F}(D^{\boldsymbol{\alpha}}T) = \mathrm{i}^{|\boldsymbol{\alpha}|}\,\boldsymbol{\omega}^{\boldsymbol{\alpha}}\,\mathscr{F}T\,,$$
$$\mathscr{F}(\mathbf{x}^{\boldsymbol{\alpha}}T) = \mathrm{i}^{|\boldsymbol{\alpha}|}\,D^{\boldsymbol{\alpha}}\mathscr{F}T\,.$$

3. Scaling: *For* $c \in \mathbb{R}\setminus\{0\}$,

$$\mathscr{F}T(c\,\cdot) = \frac{1}{|c|^d}\,\mathscr{F}T(c^{-1}\,\cdot)\,.$$

4. Convolution: *For* $\varphi \in \mathscr{S}(\mathbb{R}^d)$,

$$\mathscr{F}(T * \varphi) = (\mathscr{F}T)\,(\mathscr{F}\varphi)\,.$$

The proof follows in a straightforward way from the definitions of corresponding operators, in particular the Fourier transform (4.37) on $\mathscr{S}'(\mathbb{R}^d)$ and Theorem 4.20.

Finally, we present some additional examples of Fourier transforms of tempered distributions.

Example 4.51 In Example 4.47 we have seen that for fixed $\mathbf{x}_0 \in \mathbb{R}^d$

$$\mathscr{F}\delta(\cdot - \mathbf{x}_0) = \mathrm{e}^{-\mathrm{i}\,\boldsymbol{\omega}\cdot\mathbf{x}_0}\,, \quad \mathscr{F}\delta = 1\,.$$

Now we determine $\mathscr{F}^{-1}1$. By Theorem 4.48, we obtain

$$\mathscr{F}^{-1}1 = \frac{1}{(2\pi)^d}\,\mathscr{F}1(-\,\cdot) = \frac{1}{(2\pi)^d}\,\mathscr{F}1,$$

since the reflection $1(-\,\cdot)$ is equal to 1. Thus, we have $\mathscr{F}1 = (2\pi)^d\,\delta$. From Theorem 4.50, it follows for any $\boldsymbol{\alpha} \in \mathbb{N}_0^d$

$$\mathscr{F}(D^{\alpha}\,\delta) = (\mathrm{i}\,\omega)^{\alpha}\,\mathscr{F}\,\delta = (\mathrm{i}\,\omega)^{\alpha}\,1 = (\mathrm{i}\,\omega)^{\alpha}\,,$$

$$\mathscr{F}(\mathbf{x}^{\alpha}) = \mathscr{F}(\mathbf{x}^{\alpha}\,1) = \mathrm{i}^{|\alpha|}\,D^{\alpha}\,\mathscr{F}\,1 = (2\pi)^d\,\mathrm{i}^{|\alpha|}\,D^{\alpha}\,\delta\,.$$

□

Example 4.52 We are interested in the Fourier transform of the distribution $\mathrm{pv}\left(\frac{1}{x}\right)$ from Example 4.40. First it is not hard to check that for any $T \in \mathscr{S}'(\mathbb{R})$

$$xT = 1 \qquad \Longleftrightarrow \qquad T = \mathrm{pv}\left(\frac{1}{\cdot}\right) + C\delta$$

with a constant $C \in \mathbb{R}$. Similarly as in Example 4.42, the derivative of the sign function

$$\mathrm{sgn}(x) := \begin{cases} 1 & x > 0\,, \\ 0 & x = 0\,, \\ -1 & x < 0\,, \end{cases}$$

is $D\,\mathrm{sgn} = 2\delta$. Then we obtain by Theorem 4.50 for all $\varphi \in \mathscr{S}(\mathbb{R})$

$$\begin{aligned} \langle \mathscr{F}(D\,\mathrm{sgn}), \varphi \rangle &= \langle D\,\mathrm{sgn}, \hat{\varphi} \rangle = 2\hat{\varphi}(0) = \langle 2\hat{\delta}, \varphi \rangle = \langle 2, \varphi \rangle \\ &= \langle \mathrm{i}\omega\,\mathrm{sgn}\hat{\ }(\omega), \varphi(\omega) \rangle \end{aligned}$$

so that $\mathrm{i}\omega\,\mathrm{sgn}\hat{\ }(\omega) = 2$ and

$$\mathrm{sgn}\hat{\ } = \frac{2}{\mathrm{i}}\mathrm{pv}\left(\frac{1}{\cdot}\right) + C\delta = \frac{2}{\mathrm{i}}\mathrm{pv}\left(\frac{1}{\cdot}\right),$$

where the last equality, i.e., $C = 0$, can be seen using the Gaussian φ. Hence it follows

$$\mathrm{pv}\left(\frac{1}{\cdot}\right)\hat{\ } = -\mathrm{i}\pi\;\mathrm{sgn}.$$

□

Remark 4.53 Using Theorem 4.48, we can simplify the d-dimensional Poisson summation formula 4.26. For arbitrary $\varphi \in \mathscr{S}(\mathbb{R}^d)$, we introduce the 2π-*periodization operator* $P_{2\pi} : \mathscr{S}(\mathbb{R}^d) \to C^{\infty}(\mathbb{T}^d)$ by

$$P_{2\pi}\,\varphi := \sum_{\mathbf{k}\in\mathbb{Z}^d} \varphi(\cdot + 2\pi\,\mathbf{k})\,.$$

Note that this series converges absolutely and uniformly on $\mathbb{R}^d$. Then $P_{2\pi}\varphi \in C^{\infty}(\mathbb{T}^d)$ can be represented as uniformly convergent Fourier series

$$(P_{2\pi}\varphi)(\mathbf{x}) = \sum_{\mathbf{k}\in\mathbb{Z}^d} c_{\mathbf{k}}(P_{2\pi}\varphi)\, \mathrm{e}^{\mathrm{i}\,\mathbf{k}\cdot\mathbf{x}}\,,$$

where

$$\begin{aligned} c_{\mathbf{k}}(P_{2\pi}\varphi) &= \frac{1}{(2\pi)^d}\int_{[0,\,2\pi]^d} (P_{2\pi}\varphi)(\mathbf{x})\, \mathrm{e}^{-\mathrm{i}\,\mathbf{k}\cdot\mathbf{x}}\,\mathrm{d}\mathbf{x} \\ &= \frac{1}{(2\pi)^d}\int_{\mathbb{R}^d} \varphi(\mathbf{x})\, \mathrm{e}^{-\mathrm{i}\,\mathbf{k}\cdot\mathbf{x}}\,\mathrm{d}\mathbf{x} = \frac{1}{(2\pi)^d}\,\hat{\varphi}(\mathbf{k})\,. \end{aligned}$$

Hence we obtain the d-dimensional Poisson summation formula

$$(P_{2\pi}\varphi)(\mathbf{x}) = \frac{1}{(2\pi)^d}\sum_{\mathbf{k}\in\mathbb{Z}^d} \hat{\varphi}(\mathbf{k})\, \mathrm{e}^{\mathrm{i}\,\mathbf{k}\cdot\mathbf{x}}\,,$$

where the Fourier series converges absolutely and uniformly on $\mathbb{R}^d$ too. For $\hat{\varphi} \in \mathscr{S}(\mathbb{R}^d)$, we can form the *uniform sampling operator* $S_1 : \mathscr{S}(\mathbb{R}^d) \to \mathscr{S}'(\mathbb{R}^d)$ by

$$S_1\hat{\varphi} = \sum_{\mathbf{k}\in\mathbb{Z}^d} \hat{\varphi}(\mathbf{k})\,\delta(\cdot - \mathbf{k})\,.$$

Obviously, $S_1\hat{\varphi}$ is a 1-periodic tempered distribution. For the distributional inverse Fourier transform $\mathscr{F}^{-1}$, we have $\mathscr{F}^{-1}\,\mathrm{e}^{-\mathrm{i}\,\mathbf{k}\cdot\boldsymbol{\omega}} = \delta(\cdot\; - \mathbf{k})$. Thus for arbitrary $\varphi \in \mathscr{S}(\mathbb{R}^d)$, we obtain by Theorem 4.48 the equation

$$P_{2\pi}\varphi = \mathscr{F}^{-1}\, S_1\, \mathscr{F}\,\varphi\,.$$

In [301], the Poisson summation formula is generalized for regular tempered distributions generated by continuous, slowly increasing functions. □

The spaces $\mathscr{S}(\mathbb{R}^d)$, $L_2(\mathbb{R}^d)$, and $\mathscr{S}'(\mathbb{R}^d)$ are a typical example of a so-called Gelfand triple named after the mathematician I.M. Gelfand (1913–2009).To obtain a Gelfand triple $(B,\ H,\ B')$, we equip a Hilbert space H with a dense topological vector subspace B of test functions carrying a finer topology than H such that the natural (injective) inclusion $B \subset H$ is continuous. Let B' be the dual space of all linear continuous functionals on B with its (weak-$*$) topology. Then the embedding of H' in B' is injective and continuous. Applying the Riesz representation theorem, we can identify H with H' leading to the Gelfand triple

$$B \subset H \cong H' \subset B'.$$

We are interested in

$$\mathscr{S}(\mathbb{R}^d) \subset L_2(\mathbb{R}^d) \cong L_2(\mathbb{R}^d)' \subset \mathscr{S}'(\mathbb{R}^d). \tag{4.40}$$

Note that we already know that $\mathscr{S}(\mathbb{R}^d)$ is dense in $L_2(\mathbb{R}^d)$. Moreover, the natural embedding is indeed continuous, since $\varphi_k \underset{\mathscr{S}}{\longrightarrow} \varphi$ as $k \to \infty$ implies

$$\begin{aligned}\|\varphi_k - \varphi\|_{L_2(\mathbb{R}^d)}^2 &= \int_{\mathbb{R}^d} (1 + \|\mathbf{x}\|_2)^{-d-1} (1 + \|\mathbf{x}\|_2)^{d+1} |\varphi_k(\mathbf{x}) - \varphi(\mathbf{x})|^2 \,\mathrm{d}\mathbf{x} \\ &\le \sup_{\mathbf{x}\in\mathbb{R}^d} (1 + \|\mathbf{x}\|_2)^{d+1} |\varphi_k(\mathbf{x}) - \varphi(\mathbf{x})|^2 \int_{\mathbb{R}^d} \frac{\mathrm{d}\mathbf{y}}{(1 + \|\mathbf{y}\|_2)^{d+1}} \\ &\le C \sup_{\mathbf{x}\in\mathbb{R}^d} (1 + \|\mathbf{x}\|_2)^{d+1} |\varphi_k(\mathbf{x}) - \varphi(\mathbf{x})|^2 \to 0\end{aligned}$$

as $k \to \infty$.

Corollary 4.54 *If we identify $f \in L_2(\mathbb{R}^d)$ with $T_f \in \mathscr{S}'(\mathbb{R}^d)$, then the Fourier transforms on $L_2(\mathbb{R}^d)$ and $\mathscr{S}'(\mathbb{R}^d)$ coincide in the sense $\mathscr{F} T_f = T_{\mathscr{F} f}$.*

Proof For any sequence $(f_k)_{k\in\mathbb{N}}$ of functions $f_k \in \mathscr{S}(\mathbb{R}^d)$ converging to f in $L_2(\mathbb{R}^d)$, we obtain

$$\lim_{k\to\infty} \langle \mathscr{F}\varphi,\, \bar{f}_k \rangle_{L_2(\mathbb{R}^d)} = \langle \mathscr{F}\varphi,\, \bar{f} \rangle_{L_2(\mathbb{R}^d)} = \langle T_f,\, \mathscr{F}\varphi \rangle = \langle \mathscr{F} T_f,\, \varphi \rangle$$

for all $\varphi \in \mathscr{S}(\mathbb{R}^d)$. On the other hand, we conclude by definition of $\mathscr{F}$ on $L_2(\mathbb{R}^d)$ that

$$\lim_{k\to\infty} \langle \mathscr{F}\varphi,\, \bar{f}_k \rangle_{L_2(\mathbb{R}^d)} = \lim_{k\to\infty} \langle \varphi,\, \overline{\mathscr{F} f_k} \rangle_{L_2(\mathbb{R}^d)} = \langle \varphi,\, \overline{\mathscr{F} f} \rangle_{L_2(\mathbb{R}^d)} = \langle T_{\mathscr{F} f},\, \varphi \rangle$$

for all $\varphi \in \mathscr{S}(\mathbb{R}^d)$. Thus, $\mathscr{F} T_f = T_{\mathscr{F} f}$ and we are done. ■

4.3.3 Periodic Tempered Distributions

Next we describe the connection between 2π-periodic tempered distributions and Fourier series. We restrict our attention to the case $d = 1$.

A tempered distribution $T \in \mathscr{S}'(\mathbb{R})$ with the property $T = T(\cdot - 2\pi)$, i.e.,

$$\langle T, \varphi \rangle = \langle T, \varphi(\cdot + 2\pi) \rangle$$

for all $\varphi \in \mathscr{S}(\mathbb{R})$ is called *2π-periodic tempered distribution*.

Example 4.55 We consider the so-called Dirac comb

$$\Delta_{2\pi} := \sum_{k\in\mathbb{Z}} \delta(\cdot - 2\pi k) = \lim_{n\to\infty} \sum_{k=-n}^{n} \delta(\cdot - 2\pi k)\,,$$

which is meant as follows: For every $\varphi \in \mathscr{S}(\mathbb{R}^d)$ and $T_n := \sum_{k=-n}^{n} \delta(\cdot - 2\pi k) \in \mathscr{S}'(\mathbb{R})$, we have

$$|T_n(\varphi)| = \Big| \sum_{k=-n}^{n} \frac{(1+4\pi^2 k^2)\,\varphi(2\pi k)}{1+4\pi^2 k^2} \Big|$$

$$\le \sum_{k=-n}^{n} \frac{1}{1+4\pi^2 k^2} \sup_{x\in\mathbb{R}} (1+|x|)^2\, |\varphi(x)| \le C\, \|\varphi\|_m\,,$$

where $m = 2$. Hence the absolute sum is bounded for all $n \in \mathbb{N}$ and thus converges for $n \to \infty$.The limit is $\langle \Delta_{2\pi}, \varphi \rangle$. □

Lemma 4.56 (Poisson Summation Formula for Dirac Comb) *In $\mathscr{S}'(\mathbb{R})$ it holds*

$$2\pi\, \Delta_{2\pi} = \sum_{k\in\mathbb{Z}} \mathrm{e}^{-\mathrm{i}k\cdot}\,.$$

Proof Since $\mathscr{F}$ is continuous on $\mathscr{S}'(\mathbb{R})$, we have

$$\mathscr{F}\Big(\sum_{k\in\mathbb{Z}} \delta(\cdot - k)\Big) = \sum_{k\in\mathbb{Z}} \mathrm{e}^{\mathrm{i}k\cdot} \in \mathscr{S}'(\mathbb{R})\,. \tag{4.41}$$

The functions from $\mathscr{S}(\mathbb{R})$ fulfill the assumptions of Poisson summation formula (4.26). Using this formula and (4.41), we obtain for all $\varphi \in \mathscr{S}(\mathbb{R})$

$$2\pi \Big\langle \sum_{k\in\mathbb{Z}} \delta(\cdot - 2\pi k), \varphi \Big\rangle = 2\pi \sum_{k\in\mathbb{Z}} \varphi(2\pi k) = \sum_{k\in\mathbb{Z}} \hat{\varphi}(k) = \Big\langle \sum_{k\in\mathbb{Z}} \mathrm{e}^{-\mathrm{i}k\cdot}, \varphi \Big\rangle\,.$$

This yields the assertion. ■

By Theorem 1.3, we have known that every function $f \in L_2(\mathbb{T})$ possesses a convergent Fourier series in $L_2(\mathbb{R})$. On the other hand, we know from Sect. 1.4 that there exists $f \in L_1(\mathbb{T})$ such that the sequence of Fourier partial sums $(S_n f)(x)$ is not convergent in $L_1(\mathbb{T})$ as $n \to \infty$; see [265, p. 52]. But every $f \in L_1(\mathbb{T})$ generates a regular tempered distribution T_f which is 2π-periodic. Next we show that the Fourier series of any function $f \in L_1(\mathbb{T})$ converges to f in $\mathscr{S}'(\mathbb{R})$.

Lemma 4.57 *For $f \in L_1(\mathbb{T})$, the Fourier series*

$$T := \sum_{k\in\mathbb{Z}} c_k(f)\,\mathrm{e}^{\mathrm{i}k\cdot}\,, \quad c_k(f) := \frac{1}{2\pi}\int_0^{2\pi} f(t)\,\mathrm{e}^{-\mathrm{i}kt}\,\mathrm{d}t\,,$$

is a 2π-periodic tempered distribution which coincides with T_f in $\mathscr{S}'(\mathbb{R})$. The Fourier transform $\mathscr{F}\,T_f$ reads as

$$\mathscr{F} T_f = 2\pi \sum_{k\in\mathbb{Z}} c_k(f)\,\delta(\cdot - k)$$

and the distributional derivative $D\,T_f$ as

$$D\,T_f = \sum_{k\in\mathbb{Z}\setminus\{0\}} \mathrm{i}\,k\,c_k(f)\,\mathrm{e}^{\mathrm{i}\,k\,\cdot}\,.$$

Proof Consider the nth Fourier partial sum

$$S_n\,f = \sum_{k=-n}^{n} c_k(f)\,\mathrm{e}^{\mathrm{i}\,k\,\cdot}\,.$$

For arbitrary $\varphi \in \mathscr{S}(\mathbb{R})$, we obtain by Fubini's theorem

$$\begin{aligned}\langle S_n\,f, \varphi\rangle &= \int_{\mathbb{R}} (S_n\,f)(x)\,\varphi(x)\,\mathrm{d}x \\ &= \int_0^{2\pi} \Big(\int_{\mathbb{R}} \frac{1}{2\pi} \sum_{k=-n}^{n} \mathrm{e}^{\mathrm{i}\,k\,(x-t)}\,\varphi(x)\,\mathrm{d}x\Big)\,f(t)\,\mathrm{d}t\,.\end{aligned}$$

By Lemma 4.56, we know that $\frac{1}{2\pi}\sum_{k=-n}^{n} \mathrm{e}^{\mathrm{i}\,k\,(\cdot-t)}$ converges to $\Delta_{2\pi}(\cdot - t)$ in $\mathscr{S}'(\mathbb{R})$. Taking the limes for $n \to \infty$ and using Lebesgue's dominated convergence theorem, it follows

$$\begin{aligned}\langle T, \varphi\rangle &= \int_0^{2\pi} \langle \Delta_{2\pi}(\cdot - t), \varphi\rangle\,f(t)\,\mathrm{d}t = \int_0^{2\pi} \Big(\sum_{k\in\mathbb{Z}} \varphi(t + 2\pi\,k)\Big)\,f(t)\,\mathrm{d}t \\ &= \int_{\mathbb{R}} \varphi(t)\,f(t)\,\mathrm{d}t = \langle f, \varphi\rangle\end{aligned}$$

for all $\varphi \in \mathscr{S}(\mathbb{R})$. Thus, $T = T_f$. Since $\mathscr{F}$ and D are linear continuous operations in $\mathscr{S}'(\mathbb{R})$, the Fourier transform $\mathscr{F}\,T_f$ and the derivative $D\,T_f$ can be formed term-by-term by Theorem 4.48 and Lemma 4.41. ■

Example 4.58 As in Example 1.9, we consider the 2π-periodic sawtooth function f given by $f(x) = \frac{1}{2} - \frac{x}{2\pi}$, $x \in (0, 2\pi)$, and $f(0) = 0$. This function possesses in $L_2(\mathbb{R})$ the pointwise convergent Fourier expansion

$$\sum_{k\in\mathbb{Z}\setminus\{0\}} \frac{1}{2\pi\,\mathrm{i}\,k}\,\mathrm{e}^{\mathrm{i}\,k\,x}\,.$$

By Lemma 4.57 we obtain the representation

$$T_f = \sum_{k \in \mathbb{Z}\setminus\{0\}} \frac{1}{2\pi \mathrm{i}\, k}\, \mathrm{e}^{\mathrm{i}\, k\, \cdot}$$

in $\mathscr{S}'(\mathbb{R})$. Using Lemma 4.41, this Fourier series can be termwise differentiated in $\mathscr{S}'(\mathbb{R})$, so that

$$\begin{aligned} D\, T_f &= \frac{1}{2\pi} \sum_{k \in \mathbb{Z}\setminus\{0\}} \mathrm{e}^{\mathrm{i}\, k\, \cdot} \\ &= -\frac{1}{2\pi} + \sum_{k \in \mathbb{Z}} \delta(\cdot - 2\pi k) = -\frac{1}{2\pi} + \varDelta_{2\pi}\,. \end{aligned}$$

□

Finally, we want to extend Lemma 4.57 to 2π-periodic tempered distributions. Since the function $\mathrm{e}^{\mathrm{i}\, k\, \cdot}$ is not in $\mathscr{S}(\mathbb{R})$ the expression $\langle T, \mathrm{e}^{\mathrm{i}\, k\, \cdot}\rangle$ is not defined. Therefore, we choose a nonnegative compactly supported function $\theta \in C_c^\infty(\mathbb{R})$ which generates a *partition of unity*

$$\sum_{k \in \mathbb{Z}} \theta(x + 2\pi k) = 1\,, \quad x \in \mathbb{R}\,. \tag{4.42}$$

Note that the series (4.42) is *locally finite*, i.e., on every compact interval, only a finite number of terms $\theta(x + 2\pi k)$ does not vanish identically.

Example 4.59 The even function $\theta \in C_c^\infty(\mathbb{R})$ with $\theta(x) := 1$ for $0 \le x \le \frac{2\pi}{3}$, $\theta(x) := 0$ for $x \ge \frac{4\pi}{3}$ and

$$\theta(x) := \Big(1 + \frac{\mathrm{e}^{1/(4\pi/3 - x)}}{\mathrm{e}^{1/(x - 2\pi/3)}}\Big)^{-1}\,, \quad \frac{2\pi}{3} < x < \frac{4\pi}{3}$$

supported in $\big[-\frac{4\pi}{3}, \frac{4\pi}{3}\big]$ fulfills (4.42). This follows immediately from the facts that for $x \in \big(\frac{2\pi}{3}, \frac{4\pi}{3}\big)$, we have

$$\sum_{k \in \mathbb{Z}} \theta(x + 2\pi k) = \theta(x) + \theta(x - 2\pi) = 1$$

by definition of θ. For $x \in \big[0, \frac{2\pi}{3}\big]$, we have

$$\sum_{k \in \mathbb{Z}} \theta(x + 2\pi k) = \theta(x) = 1$$

and for $x \in \big[\frac{4\pi}{3}, 2\pi\big]$

$$\sum_{k\in\mathbb{Z}} \theta(x + 2\pi k) = \theta(x - 2\pi) = 1\,.$$

□

Then $\theta\,\mathrm{e}^{-\mathrm{i}k\,\cdot} \in \mathscr{S}(\mathbb{R})$, so that the *Fourier coefficients of a 2π-periodic tempered distribution $T \in \mathscr{S}'(\mathbb{R})$* given by

$$c_k(T) := \frac{1}{2\pi}\,\langle T,\, \theta\,\mathrm{e}^{-\mathrm{i}k\,\cdot}\rangle\,, \quad k \in \mathbb{Z} \tag{4.43}$$

are well-defined. By the following reason, the Fourier coefficients are independent of the chosen $\theta \in C_c^\infty(\mathbb{R})$: Let $\theta_1 \in C_c^\infty(\mathbb{R})$ be another function with property (4.42). Then, by the 2π-periodicity of T, we obtain

$$\begin{aligned}\langle T,\, \theta_1\,\mathrm{e}^{-\mathrm{i}k\,\cdot}\rangle &= \langle T,\, \sum_{\ell\in\mathbb{Z}} \theta(\cdot + 2\ell\pi)\,\theta_1\,\mathrm{e}^{-\mathrm{i}k\,\cdot}\rangle \\ &= \langle T,\, \sum_{\ell\in\mathbb{Z}} \theta\,\theta_1(\cdot - 2\ell\pi)\,\mathrm{e}^{-\mathrm{i}k\,\cdot}\rangle = \langle T,\, \theta\,\mathrm{e}^{-\mathrm{i}k\,\cdot}\rangle\,.\end{aligned}$$

Theorem 4.60 *Let $T \in \mathscr{S}'(\mathbb{R})$ be a 2π-periodic tempered distribution. Then its Fourier coefficients* (4.43) *are slowly increasing, i.e., there exist $m \in \mathbb{N}$ and $C > 0$ such that for all $k \in \mathbb{Z}$*

$$|c_k(T)| \le C\,(1 + |k|)^m \tag{4.44}$$

and T has the distributional Fourier series

$$T = \sum_{k\in\mathbb{Z}} c_k(T)\,\mathrm{e}^{\mathrm{i}k\,\cdot} \tag{4.45}$$

which converges in $\mathscr{S}'(\mathbb{R})$. Further, we have for all $\varphi \in \mathscr{S}(\mathbb{R})$

$$\langle T, \varphi\rangle = \sum_{k\in\mathbb{Z}} c_k(T)\,\hat{\varphi}(-k)\,.$$

The Fourier transform $\mathscr{F}\,T$ reads as

$$\mathscr{F}\,T = 2\pi \sum_{k\in\mathbb{Z}} c_k(T)\,\delta(\cdot - k)\,. \tag{4.46}$$

Proof By the Lemma 4.35 we know that there exists $C_1 > 0$ and $m \in \mathbb{N}_0$ such that

$$|c_k(T)| = \frac{1}{2\pi}\,|\langle T,\, \theta\,\mathrm{e}^{-\mathrm{i}k\,\cdot}\rangle| \le C_1\,\|\theta\,\mathrm{e}^{-\mathrm{i}k\,\cdot}\|_m\,.$$

Then, we get by the Leibniz product rule

$$\|\theta\, \mathrm{e}^{-\mathrm{i}k\,\cdot}\|_m = \max_{|\beta|\le m} \|(1+|x|)^m\, D^\beta\big(\theta(x)\, \mathrm{e}^{-\mathrm{i}kx}\big)\|_{C(\mathrm{supp}\,\varphi)} \le C\,(1+|k|)^m$$

which gives (4.44).

Next we show that for arbitrary $\varphi \in \mathscr{S}(\mathbb{R})$, the series

$$\sum_{k\in\mathbb{Z}} c_k(T)\, \hat{\varphi}(-k) \tag{4.47}$$

converges. By $\varphi \in \mathscr{S}(\mathbb{R})$ we have $\hat{\varphi} \in \mathscr{S}(\mathbb{R})$, so that for some $C_1 > 0$ and m from above

$$(1+|k|)^{m+2}\, |\hat{\varphi}(-k)| \le C_1\,.$$

Together with (4.44), this implies

$$|c_k(T)\, \hat{\varphi}(-k)| \le C\, C_1 (1+|k|)^{-2}$$

for all $k \in \mathbb{Z}$, so that the series (4.47) converges absolutely.

For $n \in \mathbb{N}$, we have

$$\langle \sum_{k=-n}^{n} c_k(T)\, \mathrm{e}^{\mathrm{i}k\,\cdot},\, \varphi\rangle = \sum_{k=-n}^{n} c_k(T) \int_{\mathbb{R}} \varphi(x)\, \mathrm{e}^{\mathrm{i}kx}\, \mathrm{d}x = \sum_{k=-n}^{n} c_k(T)\, \hat{\varphi}(-k)$$

and letting n go to infinity, we obtain

$$\lim_{n\to\infty} \langle \sum_{k=-n}^{n} c_k(T)\, \mathrm{e}^{\mathrm{i}k\,\cdot},\, \varphi\rangle = \sum_{k\in\mathbb{Z}} c_k(T)\, \hat{\varphi}(-k)\,.$$

Define

$$\langle \sum_{k\in\mathbb{Z}} c_k(T)\, \mathrm{e}^{\mathrm{i}k\,\cdot},\, \varphi\rangle := \sum_{k\in\mathbb{Z}} c_k(T)\, \hat{\varphi}(-k)$$

for all $\varphi \in \mathscr{S}(\mathbb{R})$. By definition of the Fourier coefficients , we see that

$$\sum_{k=-n}^{n} c_k(T)\, \hat{\varphi}(-k) = \langle T,\, \theta \int_{\mathbb{R}} \frac{1}{2\pi} \sum_{k=-n}^{n} \mathrm{e}^{\mathrm{i}k\,(x-\cdot)}\, \varphi(x)\, \mathrm{d}x\rangle$$

and for $n \to \infty$ by Poisson summation formula (2.28)

$$\sum_{k\in\mathbb{Z}} c_k(T)\,\hat{\varphi}(-k) = \langle T,\, \theta \sum_{k\in\mathbb{Z}} \varphi(\cdot - 2\pi k)\rangle\,.$$

Now the 2π-periodicity of T and (4.42) implies

$$\langle T,\, \theta \sum_{k\in\mathbb{Z}} \varphi(\cdot + 2\pi k)\rangle = \langle T,\, \varphi \sum_{k\in\mathbb{Z}} \theta(\cdot - 2\pi k)\rangle = \langle T,\, \varphi\rangle$$

for all $\varphi \in \mathscr{S}(\mathbb{R})$ and consequently

$$T = \sum_{k\in\mathbb{Z}} c_k(T)\,\mathrm{e}^{\mathrm{i}k\,\cdot}\,.$$

Since the Fourier transform is continuous on $\mathscr{S}'(\mathbb{R})$, we obtain (4.46). ■

Example 4.61 The Fourier coefficients of the 2π-periodic Dirac comb $\Delta_{2\pi} \in \mathscr{S}'(\mathbb{R})$ are given by

$$c_k(\Delta_{2\pi}) = \frac{1}{2\pi}\,\langle \Delta_{2\pi},\, \theta\,\mathrm{e}^{-\mathrm{i}k\,\cdot}\rangle = \frac{1}{2\pi}\sum_{k\in\mathbb{Z}} \theta(2\pi k) = \frac{1}{2\pi}\,.$$

Thus, by Theorem 4.60, the 2π-periodic Dirac comb $\Delta_{2\pi}$ can be represented as distributional Fourier series

$$\Delta_{2\pi} = \frac{1}{2\pi}\sum_{k\in\mathbb{Z}} \mathrm{e}^{\mathrm{i}k\,\cdot}$$

which is in agreement with Lemma 4.56. By (4.46) the distributional Fourier transform of $\Delta_{2\pi}$ is equal to the 1-periodic Dirac comb

$$\mathscr{F}\,\Delta_{2\pi} = \Delta_1 := \sum_{k\in\mathbb{Z}} \delta(\cdot - k)\,.$$

□

Remark 4.62 As known, the asymptotic behavior of the Fourier coefficients $c_k(f)$, $k \in \mathbb{Z}$, of a given 2π-periodic function f reflects the smoothness of f. By the Lemma 1.27, the Fourier coefficients $c_k(f)$ of $f \in L_1(\mathbb{T})$ tend to zero as $|k| \to \infty$. With increasing smoothness of f, the decay of $|c_k(f)|$ becomes faster; see Theorem 1.39. In contrast to that, Fourier coefficients $c_k(T)$ of a 2π-periodic tempered distribution T possess another asymptotic behavior. By (4.44), the values $|c_k(T)|$ may possibly grow infinitely, but the growth is at most polynomial. For example, the Fourier coefficients $c_k(\Delta_{2\pi}) = \frac{1}{2\pi}$ of the 2π-periodic Dirac comb $\Delta_{2\pi}$ are constant. □

4.3.4 Hilbert Transform and Riesz Transform

In this subsection, we introduce the Hilbert transform and a generalization thereof to higher dimensions, the Riesz transform. Both are closely related to so-called quadrature operators [162, 398] which will be not considered here. The transforms have many applications in signal and image processing, and we will refer to some of them in the following. Concerning further information on the Hilbert transform, the reader may consult the books [189, 237, 238].

The *Hilbert transform* $\mathscr{H} : L_2(\mathbb{R}) \to L_2(\mathbb{R})$ is defined by

$$\mathscr{H} f = \mathscr{F}^{-1}(-\mathrm{i}\,\mathrm{sgn}(\cdot)\,\hat{f}) . \tag{4.48}$$

In the Fourier domain, it reads

$$\widehat{\mathscr{H} f}(\omega) = -\mathrm{i}\,\mathrm{sgn}(\omega)\,\hat{f}(\omega) = \begin{cases} -\mathrm{i}\hat{f}(\omega) & \omega > 0\,, \\ 0 & \omega = 0\,, \\ \mathrm{i}\hat{f}(\omega) & \omega < 0\,. \end{cases}$$

Since we have by Example 4.52 that $\mathrm{pv}\left(\frac{1}{\cdot}\right)^{\wedge} = -\mathrm{i}\pi\,\mathrm{sgn}$, we expect by formally applying the convolution property of the Fourier transform that

$$\mathscr{H} f(x) = \frac{1}{\pi}\,\big(f * \mathrm{pv}\big(\frac{1}{\cdot}\big)\big)(x) = \frac{1}{\pi}\,\mathrm{pv}\int_{\mathbb{R}} \frac{f(y)}{x-y}\,\mathrm{d}y\,.$$

However, the convolution of a tempered distribution and a function in $L_2(\mathbb{R})$ is in general not defined so that we have to verify that the above integral is indeed defined almost everywhere.

Theorem 4.63 *The Hilbert transform* (4.48) *can be expressed as*

$$\mathscr{H} f(x) = \frac{1}{\pi}\,\big(f * \mathrm{pv}\big(\frac{1}{\cdot}\big)\big)(x) = \frac{1}{\pi}\,\mathrm{pv}\int_{\mathbb{R}} \frac{f(y)}{x-y}\,\mathrm{d}y = \frac{1}{\pi}\,\mathrm{pv}\int_{\mathbb{R}} \frac{f(x-y)}{y}\,\mathrm{d}y\,.$$

Proof For $\varepsilon > 0$ we define the $L_2(\mathbb{R})$ function

$$g_\varepsilon(x) := \begin{cases} \frac{1}{x} & |x| > \varepsilon\,, \\ 0 & |x| \le \varepsilon\,. \end{cases}$$

By the convolution Theorem 2.26 of the Fourier transform, we obtain for all $f \in L_2(\mathbb{R})$ that

$$\int_{|y|>\varepsilon} \frac{f(x-y)}{y}\,\mathrm{d}y = (f * g_\varepsilon)(x) = \mathscr{F}^{-1}\left(\hat{f}\,\hat{g}_\varepsilon\right)(x)\,.$$

We have

$$\hat{g}_\varepsilon(\omega) = \int_{|x|>\varepsilon} \frac{1}{x}\, \mathrm{e}^{-\mathrm{i}x\omega}\, \mathrm{d}x = \int_\varepsilon^\infty \frac{1}{x}\, (\mathrm{e}^{-\mathrm{i}x\omega} - \mathrm{e}^{\mathrm{i}x\omega})\, \mathrm{d}x$$
$$= -2\mathrm{i} \int_\varepsilon^\infty \frac{\sin(x\omega)}{x}\, \mathrm{d}x = -2\mathrm{i}\, \mathrm{sgn}(\omega) \int_{\omega\varepsilon}^\infty \frac{\sin y}{y}\, \mathrm{d}y\,.$$

Using Lemma 1.41 we conclude

$$\lim_{\varepsilon\to 0} \hat{g}_\varepsilon(\omega) = -\mathrm{i}\pi\ \mathrm{sgn}(\omega)\,.$$

For all $\omega \neq 0$ we know that $|\hat{g}_\varepsilon(\omega)| \le 2\ \sup_{0<\tau<t} \big|\int_\tau^t \frac{\sin(y)}{y}\, \mathrm{d}y\big| < \infty$ so that

$$\lim_{\varepsilon\to 0} \|\hat{f}\hat{g}_\varepsilon + \mathrm{i}\,\pi\ \mathrm{sgn}\,\hat{f}\|_{L_2(\mathbb{R})} = 0$$

and by continuity of the Fourier transform on $L_2(\mathbb{R})$

$$\frac{1}{\pi} \lim_{\varepsilon\to 0} \int_{|y|>\varepsilon} \frac{f(x-y)}{y}\, \mathrm{d}y = \frac{1}{\pi} \lim_{\varepsilon\to 0} \mathcal{F}^{-1}\Big(\hat{f}\,\hat{g}_\varepsilon\Big)(x)$$
$$= \mathcal{F}^{-1}\big(-\mathrm{i}\,\mathrm{sgn}(\omega)\, \hat{f}(\omega)\big)(x)\,.$$

■

In particular the Theorem 4.63 shows that the Hilbert transform of a real-valued function in $L_2(\mathbb{R})$, denoted by $L_2(\mathbb{R}, \mathbb{R})$, is again real-valued. The Hilbert transform has various useful properties.

Theorem 4.64 (Properties of Hilbert Transform) *The Hilbert transform* $\mathcal{H} : L_2(\mathbb{R}) \to L_2(\mathbb{R})$

1. *multiplied by* $\sqrt{2\pi}$ *is an isometry,*
2. *commutes with translations,*
3. *commutes with positive dilations,*
4. *is self-inverting, i.e.,* $(\mathrm{i}\mathcal{H})^2$ *is the identity,*
5. *is anti-selfadjoint on* $L_2(\mathbb{R}, \mathbb{R})$*, that is,* $\mathcal{H}^* = -\mathcal{H}$*, and*
6. *anti-commutes with reflections on* $L_2(\mathbb{R}, \mathbb{R})$*, that is,* $\mathcal{H}\,(f(-\,\cdot)) = -\mathcal{H} f$ *for all* $f \in L_2(\mathbb{R}, \mathbb{R})$.

Proof

1. The first property follows by the Parseval equality and since $|-\mathrm{i}\,\mathrm{sgn}(\omega)| = 1$ for $\omega \neq 0$.
2. By the property 1 of Theorem 4.50, we obtain

$$\mathcal{H}\big(f(\cdot - x_0)\big)(x) = \mathcal{F}^{-1}\big(-\mathrm{i}\,\mathrm{sgn}\,\mathcal{F}(f(\cdot - x_0))\big)(x)$$

$$= \mathcal{F}^{-1}\big(-\mathrm{i}\,\mathrm{sgn}(\omega)\,\mathrm{e}^{-\mathrm{i}x_0\omega}\,\hat{f}(\omega)\big)(x)$$
$$= \mathcal{F}^{-1}\big(-\mathrm{i}\,\mathrm{sgn}\,\hat{f}\big)(x-x_0) = \mathcal{H} f(x-x_0)\,.$$

3. For $c > 0$, it follows

$$\mathcal{H}\big(f(c\cdot)\big)(x) = \frac{1}{\pi}\,\mathrm{pv}\int_{\mathbb{R}} \frac{f(cy)}{x-y}\,\mathrm{d}y = \frac{1}{\pi}\,\mathrm{pv}\int_{\mathbb{R}} \frac{f(s)}{cx-s}\,\mathrm{d}s = \mathcal{H} f(cx)\,.$$

4. For all $f \in L_2(\mathbb{R})$, we get by (4.48)

$$\mathrm{i}\,\mathcal{H} f = \mathrm{i}\,\mathcal{F}^{-1}(-\mathrm{i}\,\mathrm{sgn}\,\hat{f}) = \mathcal{F}^{-1}(\mathrm{sgn}\,\hat{f})$$

and hence

$$(\mathrm{i}\,\mathcal{H})^2 f = \mathcal{F}^{-1}\big(\mathrm{sgn}\,\mathcal{F}\mathcal{F}^{-1}(\mathrm{sgn}\,\hat{f})\big) = \mathcal{F}^{-1}(\mathrm{sgn})^2\,\hat{f} = f\,.$$

5. By the Parseval equality, we conclude for all real-valued functions $f,\, g \in L_2(\mathbb{R}, \mathbb{R})$ that

$$\begin{aligned}\langle \mathcal{H} f,\, g\rangle_{L_2(\mathbb{R})} &= \int_{\mathbb{R}} \mathcal{F}^{-1}\left(-\mathrm{i}\,\mathrm{sgn}\,\hat{f}\right)(x)\,g(x)\,\mathrm{d}x\\ &= \frac{1}{2\pi}\int_{\mathbb{R}} \big(-\mathrm{i}\,\mathrm{sgn}(\omega)\,\hat{f}(\omega)\big)\,\overline{\hat{g}(\omega)}\,\mathrm{d}\omega\\ &= -\frac{1}{2\pi}\int_{\mathbb{R}} \hat{f}(\omega)\,\overline{(-\mathrm{i}\,\mathrm{sgn}(\omega)\,\hat{g}(\omega))}\,\mathrm{d}\omega = -\,\langle f,\, \mathcal{H} g\rangle_{L_2(\mathbb{R})}\,.\end{aligned}$$

6. For all $f \in L_2(\mathbb{R}, \mathbb{R})$, we have $\big(f(-\,\cdot)\big)\hat{} = \bar{\hat{f}}$ so that

$$\mathcal{H}\big(f(-\,\cdot)\big) = \mathcal{F}^{-1}\big(-\mathrm{i}\,\mathrm{sgn}\,\bar{\hat{f}}\big) = -\overline{\mathcal{F}^{-1}(-\mathrm{i}\,\mathrm{sgn}\,\hat{f})} = -\mathcal{H} f\,.$$

■

Note that up to a constant, the Hilbert transform is the only operator on $L_2(\mathbb{R}, \mathbb{R})$ with properties 1–3 and 6 and up to the sign, the only operator which fulfills properties 1–5; see [398].

The Hilbert transform can be used to construct functions which Fourier transform is only supported on the positive interval. For a real-valued function $f \in L_2(\mathbb{R}, \mathbb{R})$, the function

$$f_a(x) := f(x) + \mathrm{i}\,\mathcal{H} f(x)$$

with

$$\hat{f}_a(\omega) = \hat{f}(\omega) + \mathrm{i}\widehat{\mathscr{H} f}(\omega) = \begin{cases} 2\hat{f}(\omega) & \omega > 0\,, \\ 0 & \omega < 0 \end{cases}$$

is called *analytic signal* of f with *amplitude* $|f_a(x)| := \big(f(x)^2 + \mathscr{H} f(x)^2\big)^{1/2}$, *phase* $\phi(x) := \operatorname{atan2}(\mathscr{H} f(x), f(x))$, and *instantaneous phase* $\nu(x) = \phi'(x)$. Note that any complex-valued function f can be written as

$$f(x) = A(x)\,\mathrm{e}^{\mathrm{i}\phi(x)} = A(x)\,\cos\big(\phi(x)\big) + \mathrm{i}\,A(x)\,\sin\big(\phi(x)\big)$$

with a nonnegative function $A(x) = |f(x)|$ and $\phi(x) = \operatorname{atan2}\big(\operatorname{Im}\phi(x), \operatorname{Re}\phi(x)\big)$. If f is real-valued, we have only the cosine part representation. For applications of analytic signals in time-frequency analysis, see, e.g., [280, Chapter 4.4].

Example 4.65 For the function

$$f(x) := A(x)\,\cos(\omega_0\,x + \zeta_0) = \frac{1}{2}\,A(x)\,\big(\mathrm{e}^{\mathrm{i}(\omega_0\,x+\zeta_0)} + \mathrm{e}^{-\mathrm{i}(\omega_0\,x+\zeta_0)}\big)$$

with a nonnegative, continuous function $A \in L_2(\mathbb{R})$ and $\omega_0 > 0$, we are interested in finding its amplitude $A(x)$ and phase $\phi(x) = \omega_0\,x + \zeta_0$ which is an affine function. This would be easy if we could compute somehow

$$g(x) := A(x)\,\sin(\omega_0\,x + \zeta_0) = \frac{1}{2\mathrm{i}}\,A(x)\,\big(\mathrm{e}^{\mathrm{i}(\omega_0\,x+\zeta_0)} - \mathrm{e}^{-\mathrm{i}(\omega_0\,x+\zeta_0)}\big)$$

because of $f(x) + \mathrm{i}\,g(x) = A(x)\,\mathrm{e}^{\mathrm{i}\,(\omega_0\,x+\zeta_0)}$ so that

$$A(x) = |f(x) + \mathrm{i}\,g(x)|\,, \quad \phi(x) = \operatorname{atan2}\big(g(x), f(x)\big)\,.$$

Indeed g can sometimes be computed by the Hilbert transform of f: By the translation-modulation property of the Fourier transform, we obtain

$$\begin{aligned}
\hat{f}(\omega) &= \frac{1}{2}\int_{\mathbb{R}} \big(A(x)\,\mathrm{e}^{-\mathrm{i}\,(\omega-\omega_0)\,x}\,\mathrm{e}^{\mathrm{i}\,\zeta_0} + A(x)\,\mathrm{e}^{-\mathrm{i}\,(\omega+\omega_0)\,x}\,\mathrm{e}^{-\mathrm{i}\,\zeta_0}\big)\,\mathrm{d}x \\
&= \frac{1}{2}\big(\mathrm{e}^{\mathrm{i}\,\zeta_0}\,\hat{A}(\omega-\omega_0) + \mathrm{e}^{-\mathrm{i}\,\zeta_0}\,\hat{A}(\omega+\omega_0)\big)\,, \\
\hat{g}(\omega) &= \frac{1}{2\mathrm{i}}\int_{\mathbb{R}} \big(A(x)\,\mathrm{e}^{-\mathrm{i}\,(\omega-\omega_0)\,x}\,\mathrm{e}^{\mathrm{i}\,\zeta_0} - A(x)\,\mathrm{e}^{-\mathrm{i}\,(\omega+\omega_0)\,x}\,\mathrm{e}^{-\mathrm{i}\,\zeta_0}\big)\,\mathrm{d}x \\
&= -\frac{\mathrm{i}}{2}\,\big(\mathrm{e}^{\mathrm{i}\,\zeta_0}\,\hat{A}(\omega-\omega_0) - \mathrm{e}^{-\mathrm{i}\,\zeta_0}\,\hat{A}(\omega+\omega_0)\big)
\end{aligned}$$

and

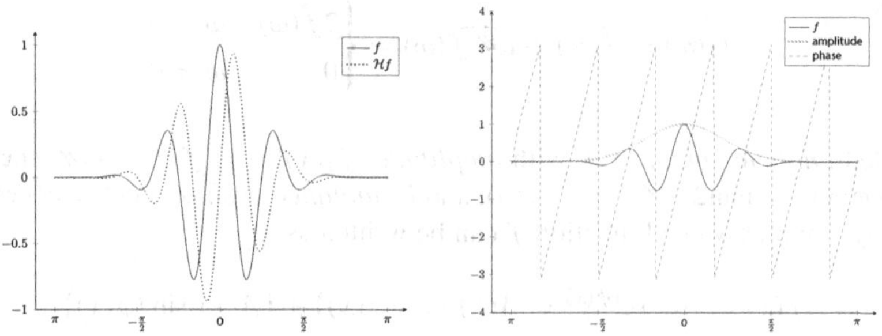

Fig. 4.1 Left: function (4.49) and its Hilbert transform. Right: amplitude $|f_a|$ and the phase ϕ of its analytical signal f_a

$$\widehat{\mathscr{H}f}(\omega) = -\frac{\mathrm{i}}{2}\cdot\begin{cases}\big(\mathrm{e}^{\mathrm{i}\zeta_0}\,\hat{A}(\omega-\omega_0)+\mathrm{e}^{-\mathrm{i}\zeta_0}\,\hat{A}(\omega+\omega_0)\big) & \omega>0\,,\\ \big(-\mathrm{e}^{\mathrm{i}\zeta_0}\,\hat{A}(\omega-\omega_0)-\mathrm{e}^{-\mathrm{i}\zeta_0}\,\hat{A}(\omega+\omega_0)\big) & \omega<0\,.\end{cases}$$

We see that $\widehat{\mathscr{H}f}(\omega)=\hat{g}(\omega)$ if and only if almost everywhere

$$\hat{A}(\omega+\omega_0)=0,\quad \omega>0 \quad\text{and}\quad \hat{A}(\omega-\omega_0)=0,\quad \omega<0$$

which is fulfilled if and only if $\hat{A}$ is supported (up to a set of measure zero) in $[-\omega_0,\,\omega_0]$. The function

$$f(x) := \begin{cases}\mathrm{e}^{-x^2}\cos(6x) & x\in[-\pi,\,\pi]\,,\\ 0 & x\in\mathbb{R}\setminus[-\pi,\,\pi]\end{cases} \tag{4.49}$$

together with its Hilbert transform as well as the amplitude and phase of its analytical signal f_a are shown in Fig. 4.1.

The example can be also seen using the so-called Bedrosian theorem which formally says that

$$\mathscr{H}(f\,g) = f\,\mathscr{H}(g),$$

if either supp $\hat{f}\subseteq(0,\,\infty)$ and supp $\hat{g}\subseteq(0,\,\infty)$ or supp $\hat{f}\subseteq[-a,a]$ and supp $\hat{g}\subseteq\mathbb{R}\backslash[-a,a]$. For the theorem of Bedrosian and corresponding extensions, see, e.g. [31, 66, 444].

There are several ways to generalize the Hilbert transform to higher dimensions. In the following, we concentrate on the most frequently used generalizations, namely, the partial Hilbert transform and the Riesz transform.

Let $\boldsymbol{\omega}_0\in\mathbb{R}^d\setminus\{\mathbf{0}\}$ be given. The *partial Hilbert transform with respect to* $\boldsymbol{\omega}_0$ is defined for $f\in L_2(\mathbb{R}^d)$ by

$$\mathscr{H}_{\boldsymbol{\omega}_0}f := \mathscr{F}^{-1}\big(-\mathrm{i}\,\mathrm{sgn}(\boldsymbol{\omega}\cdot\boldsymbol{\omega}_0)\,\hat{f}(\boldsymbol{\omega})\big)\,.$$

The partial Hilbert transform occurs in the context of functions whose Fourier transform is supported in one halfspace; see [68] or [280, Chapter 4].

Let $L_2(\mathbb{R}^d, \mathbb{R}^d)$ denote the (Bochner) space of functions $F : \mathbb{R}^d \to \mathbb{R}^d$ such that $\|F\|_2 \in L_2(\mathbb{R}^d)$. Since all norms in $\mathbb{R}^d$ are equivalent, this is the same as saying that $F_j \in L_2(\mathbb{R}^d)$, $j = 1, \ldots, d$, where $F = (F_j)_{j=1}^d$.

The *Riesz transform* $\mathscr{R} : L_2(\mathbb{R}^d, \mathbb{R}) \to L_2(\mathbb{R}^d, \mathbb{R}^d)$ is defined by

$$\mathscr{R}f = \mathscr{F}^{-1}\big(- \mathrm{i}\,\frac{\boldsymbol{\omega}}{\|\boldsymbol{\omega}\|_2}\,\hat{f}(\boldsymbol{\omega})\big)\,,$$

where the inverse Fourier transform is taken componentwise, i.e., $\mathscr{R}f = (\mathscr{R}_j f)_{j=1}^d$ with

$$\mathscr{R}_j f := \mathscr{F}^{-1}\big(- \mathrm{i}\,\frac{\omega_j}{\|\boldsymbol{\omega}\|_2}\,(\mathscr{F}f)(\boldsymbol{\omega})\big)\,, \quad \boldsymbol{\omega} = (\omega_j)_{j=1}^d \in \mathbb{R}^d\,.$$

Note that for $f \in L_2(\mathbb{R}^d)$, we have $\hat{f} \in L_2(\mathbb{R}^d)$ and $\frac{\omega_j}{\|\boldsymbol{\omega}\|_2}\,\hat{f} \in L_2(\mathbb{R}^d)$, $j = 1, \ldots, d$, where we set $\frac{\omega_j}{\|\boldsymbol{\omega}\|_2} := 0$ if $\boldsymbol{\omega} = \mathbf{0}$. Therefore the inverse Fourier transform is well-defined. For $d = 1$ the Riesz transform coincides with the Hilbert transform.

In the Fourier domain, it reads

$$\widehat{\mathscr{R}f}(\boldsymbol{\omega}) = -\mathrm{i}\,\frac{\boldsymbol{\omega}}{\|\boldsymbol{\omega}\|_2}\,\hat{f}(\boldsymbol{\omega})\,.$$

Similarly as for the Hilbert transform, it can be shown that the Riesz transform can be rewritten as

$$\begin{aligned}\mathscr{R}f(\mathbf{x}) &= C_d\,\left(f * \mathrm{pv}\left(\frac{\cdot}{\|\cdot\|_2^{d+1}}\right)\right)(\mathbf{x})\\ &= C_d\,\lim_{\varepsilon\to 0}\int_{\mathbb{R}^d\setminus B_\varepsilon(\mathbf{0})}\frac{\mathbf{y}}{\|\mathbf{y}\|_2^{d+1}}\,f(\mathbf{x}-\mathbf{y})\,\mathrm{d}\mathbf{y}\end{aligned}$$

where

$$C_d := \pi^{-(d+1)/2}\,\Gamma\big(\frac{d+1}{2}\big)$$

with the Gamma function $\Gamma(z) := \int_0^\infty t^{z-1}\mathrm{e}^{-t}\,\mathrm{d}t$. In particular we have

$$C_1 = \frac{1}{\pi}\,, \quad C_2 = \frac{1}{2\pi}\,, \quad C_3 = \frac{1}{\pi^2}\,.$$

The Riesz transform was introduced in [365] and arises in the study of differentiability properties of harmonic potentials.

Theorem 4.66 (Properties of Riesz Transform) *The Riesz transform* $\mathscr{R} : L_2(\mathbb{R}^d, \mathbb{R}) \to L_2(\mathbb{R}^d, \mathbb{R}^d)$

1. *multiplied by* $(2\pi)^{d/2}$ *is an isometry,*
2. *commutes with translations,*
3. *commutes with positive dilations,*
4. *fulfills for each orthogonal matrix* $\mathbf{U} \in \mathbb{R}^{d\times d}$ *the relation*

$$\mathscr{R}\big(f(\mathbf{U}^{-1}\cdot)\big) = \mathbf{U}\,\mathscr{R} f(\mathbf{U}^{-1}\cdot)\,,$$

5. *is anti-selfadjoint, that is,* $\mathscr{R}^* g = -\sum_{j=1}^{d} \mathscr{R}_j\, g_j$ *for all* $g = (g_j)_{j=1}^{d} \in L_2(\mathbb{R}^d, \mathbb{R}^d)$*, and*
6. *anti-commutes with reflections on* $L_2(\mathbb{R}^d, \mathbb{R})$.

Proof We only prove the fourth property since the other ones follow similarly as for the Riesz transform. By Lemma 4.30 we obtain

$$\begin{aligned}\mathscr{R}\Big(f(\mathbf{U}^{-1}\cdot)\Big)(\mathbf{x}) &= \mathscr{F}^{-1}\Big(-\mathrm{i}\,\frac{\boldsymbol{\omega}}{\|\boldsymbol{\omega}\|_2}\,\hat{f}(\mathbf{U}^{-1}\boldsymbol{\omega})\Big)(\mathbf{x})\\ &= \frac{1}{(2\pi)^d}\int_{\mathbb{R}}\Big(-\mathrm{i}\,\frac{\boldsymbol{\omega}}{\|\boldsymbol{\omega}\|_2}\,\hat{f}(\mathbf{U}^{-1}\boldsymbol{\omega})\Big)\,\mathrm{e}^{\mathrm{i}\,\boldsymbol{\omega}\cdot\mathbf{x}}\,\mathrm{d}\boldsymbol{\omega}\\ &= \frac{1}{(2\pi)^d}\int_{\mathbb{R}}\Big(-\mathrm{i}\,\frac{\mathbf{U}\mathbf{v}}{\|\mathbf{v}\|_2}\,\hat{f}(\mathbf{v})\,\mathrm{e}^{\mathrm{i}\,\mathbf{v}\cdot\mathbf{U}^{-1}\mathbf{x}}\Big)\,\mathrm{d}\mathbf{v}\\ &= \mathbf{U}\mathscr{F}^{-1}\Big(-\mathrm{i}\,\frac{\mathbf{v}}{\|\mathbf{v}\|_2}\,\hat{f}\Big)(\mathbf{U}^{-1}\mathbf{x})\,.\end{aligned}$$

■

The multidimensional counterpart of an analytic signal is the monogenic signal. Let $d = 2$ and $\mathscr{R} f = (\mathscr{R}_1 f, \mathscr{R}_2 f)^\top$. Then the *monogenic signal* of a function $f \in L_2(\mathbb{R}^2, \mathbb{R})$ is defined by

$$f_m := (f,\, \mathscr{R}_1 f,\, \mathscr{R}_2 f)^\top\,.$$

The monogenic signal was introduced in image processing by Felsberg and Sommer [138] and in the context of optics by Larkin et al. [264]. It has the *amplitude*

$$A := \sqrt{f^2 + (\mathscr{R}_1 f)^2 + (\mathscr{R}_2 f)^2}.$$

Its *local orientation* $\theta \in (-\pi, \pi]$ and *instantaneous phase* $\xi \in [0, \pi]$ are determined by

$$f = A\,\cos\xi\,,\quad \mathscr{R}_1 f = A\,\sin\xi\,\cos\theta\,,\quad \mathscr{R}_2 f = A\,\sin\xi\,\sin\theta\,.$$

The instantaneous phase can be recovered by

$$\xi = \arccos \frac{f}{A}.$$

With $r := \sqrt{(\mathscr{R}_1 f)^2 + (\mathscr{R}_2 f)^2} = A \sin \xi$, we obtain the *local orientation vector* by $(\frac{\mathscr{R}_1 f}{r}, \frac{\mathscr{R}_2 f}{r})^\top = (\cos\theta, \sin\theta)^\top$ and the *local orientation* by

$$\theta := \operatorname{atan2}(\mathscr{R}_2 f, \mathscr{R}_1 f).$$

Thus, the Riesz transform of a two-dimensional signal provides information about the amplitude, the instantaneous phase, and the local orientation (of the phase) of the signal. Therefore it contains, in contrast to, e.g., the directional Hilbert transform where the desired direction has to be addressed in advance, an "automatic" orientation component. The Riesz transform can replace the (smoothed) gradient in structure tensors as those of Förstner and Gülch [145] to make them more robust; see, e.g., [251]. It was used in the context of steerable wavelets [420], curvelets [400], and shearlets [193].

4.4 Fourier Transform of Measures

The space of tempered distributions provides a quite general setting for defining the Fourier transform. However, these distributions are defined with respect to the topology of the Schwartz space, which is not a normed space; see Remark 4.11. In this section, we return to normed spaces and deal with the Fourier transform on the Banach space of finite, complex-valued measures $\mathscr{M}(\mathbb{X})$, $\mathbb{X} \in \{\mathbb{T}^d, \mathbb{R}^d\}$. Indeed by the Riesz representation theorem, the measure space $\mathscr{M}(\mathbb{X})$ is isometrically isomorphic to the dual space of continuous complex-valued functions. We will see that every measure can be considered as a tempered distribution, but not conversely. In contrast to general tempered distribution, the convolution of measures is well-defined and related to their Fourier transforms by multiplication. Moreover, we will see that there is a one-to-one relation between positive definite functions and the Fourier transform of positive measures, which is known as Herglotz' theorem for $\mathbb{X} = \mathbb{T}^d$ and Bochner's theorem for $\mathbb{X} = \mathbb{R}^d$. Fourier-transformed probability measures are known as *characteristic functions*; see, e.g. [137]. Besides probability theory and statistics, the Fourier transform of measures has recently found applications in signal and image processing, in particular in image approximation [125, 268] and image dithering [177, 412], sketching for large-scale learning [236], and compressive learning for image restoration [387] as well as superresolution [33, 79, 81, 92] closely related to the Prony-like method considered in Chap. 10, to name only a few.

We start by recalling basic facts on measure spaces. Then we consider the spaces $\mathbb{X} = \mathbb{T}^d$ and $\mathbb{X} = \mathbb{R}^d$ separately. For further material on the topic, we refer to [203] and [284, Chapter 6].

4.4.1 Measure Spaces

Let $\mathbb{X}$ be a locally compact, separable, and complete metric space, e.g., a locally compact Polish space. In the following subsections, we will focus on $\mathbb{X} = \mathbb{T}^d$ and $\mathbb{X} = \mathbb{R}^d$. By $\mathscr{B}(\mathbb{X})$, we denote the Borel σ-algebra of $\mathbb{X}$, which is the σ-algebra generated by the open sets of $\mathbb{X}$. A finite, complex-valued Borel measure is a mapping $\mu : \mathscr{B}(\mathbb{X}) \to \mathbb{C}$ with $\mu(\emptyset) = 0$, $|\mu(B)| < \infty$ for all $B \in \mathscr{B}(\mathbb{X})$ and

$$\mu\Big(\bigcup_{k=1}^{\infty} B_k\Big) = \sum_{k=1}^{\infty} \mu(B_k)$$

for any pairwise disjoint sets $B_k \in \mathscr{B}(\mathbb{X})$. To each finite, complex-valued Borel measure μ, we can introduce the *total variation measure* $|\mu|$ by

$$|\mu|(B) := \sup\Big\{\sum_{k=1}^{\infty} |\mu(B_k)| : \bigcup_{k=1}^{\infty} B_k = B\Big\}, \qquad B \in \mathscr{B}(\mathbb{X}),$$

where the sets B_k are pairwise disjoint. In Polish spaces, finite Borel measures like the total variation measure are automatically *Radon measures*, meaning that they satisfy

$$|\mu|(B) = \inf\{\, |\mu|(U) : B \subset U,\ U \text{ open}\} \qquad \text{(outer regularity)},$$
$$|\mu|(B) = \sup\{\, |\mu(K)| : K \subset B,\ K \text{ compact}\} \qquad \text{(inner regularity)}$$

for all $B \in \mathscr{B}(\mathbb{X})$; see [144, p. 222].

Equipped with the *total variation norm*

$$\|\mu\|_{\mathrm{TV}} := |\mu|(\mathbb{X}),$$

the linear space of finite Borel measures on $\mathbb{X}$ becomes a Banach space, which is denoted by $\mathscr{M}(\mathbb{X})$. When working with measures, it is important that $\mathscr{M}(\mathbb{X})$ can be considered as dual space of continuous functions by the Riesz representation theorem. More precisely, we have the following isometric isomorphisms between measures μ and linear functionals T_μ on spaces of continuous functions:

- $C_c(\mathbb{X})' \cong \mathscr{M}(\mathbb{X})$ with $T_\mu(\varphi) = \langle \mu, \varphi \rangle := \int_{\mathbb{X}} \varphi \, d\mu$ for all $\varphi \in C_c(\mathbb{X})$.
- $C_0(\mathbb{X})' \cong \mathscr{M}(\mathbb{X})$ with $T_\mu(\varphi) = \langle \mu, \varphi \rangle := \int_{\mathbb{X}} \varphi \, d\mu$ for all $\varphi \in C_0(\mathbb{X})$.

Note that we prefer to use the same order in the dual pairing as for tempered distributions. The dual norm on $\mathcal{M}(\mathbb{R}^d)$ coincides with the total variation norm, i.e.,

$$\|\mu\|_{\mathrm{TV}} = \sup_{\substack{\varphi \in C_c(\mathbb{X}),\\ \|\varphi\|_\infty \le 1}} |\langle \mu, \varphi\rangle| = \sup_{\substack{\varphi \in C_0(\mathbb{X}),\\ \|\varphi\|_\infty \le 1}} |\langle \mu, \varphi\rangle|.$$

Here are two examples of measures that are related to Dirac distributions and regular tempered distributions by Remark 4.69.

Example 4.67 The *Dirac measure* $\delta_{\mathbf{x}}$, $\mathbf{x} \in \mathbb{X}$, is given by

$$\delta_{\mathbf{x}}(B) := \begin{cases} 1 & \text{if } \mathbf{x} \in B, \\ 0 & \text{if } \mathbf{x} \notin B \end{cases}$$

for all $B \in \mathscr{B}(\mathbb{X})$. An *atomic measure* is a linear combination $\mu = \sum_{j=1}^n \mu_j \delta_{\mathbf{x}_j}$ of Dirac measures $\delta_{\mathbf{x}_j}$ and weights $\mu_j \in \mathbb{C}$. Its total variation is given by

$$\|\mu\|_{\mathrm{TV}} = \sup_{\substack{\varphi \in C_c(\mathbb{X}),\\ \|\varphi\|_\infty \le 1}} \left| \sum_{j=1}^n \mu_j \, \varphi(\mathbf{x}_j) \right| = \sum_{j=1}^n |\mu_j|.$$

□

Example 4.68 Assume that the measure μ is absolutely continuous with respect to the Lebesgue measure λ, which means that $\lambda(B) = 0$ implies $\mu(B) = 0$ for $B \in \mathscr{B}(\mathbb{X})$. As a consequence of the Radon–Nikodym theorem, the measure μ has the form $\mu = f\lambda$, where $f \in L_1(\mathbb{X})$ is the so-called Radon–Nikodym derivative $\mathrm{d}\mu / \mathrm{d}\lambda$.Then we obtain

$$\|\mu\|_{\mathrm{TV}} = \sup_{\substack{\varphi \in C_c(\mathbb{X}),\\ \|\varphi\|_\infty \le 1}} \left| \int_{\mathbb{X}} \varphi(\mathbf{x}) \, f(\mathbf{x}) \, \mathrm{d}\mathbf{x} \right| = \|f\|_{L_1(\mathbb{X})}.$$

□

By the following remark, the measures on $\mathbb{X} = \mathbb{R}^d$ can be considered as a subspace of $\mathscr{S}'(\mathbb{R}^d)$.

Remark 4.69 Since $\mathscr{S}(\mathbb{R}^d)$ is a dense subspace of $\big(C_0(\mathbb{R}^d), \|\cdot\|_\infty\big)$, we know by the Riesz representation theorem that $\mu \in \mathcal{M}(\mathbb{R}^d)$ can be identified as above with a linear, continuous mapping $T_\mu : \mathscr{S}(\mathbb{R}^d) \to \mathbb{C}$, which acts on any $\varphi \in \mathscr{S}(\mathbb{R}^d)$ by

$$T_\mu(\varphi) = \langle T_\mu, \varphi\rangle := \langle \mu, \varphi\rangle = \int_{\mathbb{R}^d} \varphi \, \mathrm{d}\mu.$$

Moreover, the mapping T_μ is also continuous with respect to the convergence in the Schwartz space. This follows from the bound

$$|\langle T_\mu, \varphi\rangle| \le \|\mu\|_{\mathrm{TV}}\, \|\varphi\|_\infty = \|\mu\|_{\mathrm{TV}}\, \|\varphi\|_0$$

with $\|\cdot\|_0$ from (4.9). Hence Lemma 4.35 implies that T_μ is actually a tempered distribution, i.e., $T_\mu \in \mathscr{S}'(\mathbb{R}^d)$ for all $\mu \in \mathscr{M}(\mathbb{R}^d)$. In this sense, the space of measures $\mathscr{M}(\mathbb{R}^d)$ can be considered a subspace of $\mathscr{S}'(\mathbb{R}^d)$.

On the other hand, not every tempered distribution is induced by a measure. For instance, assume that the derivative of the Delta distribution $\delta' \in \mathscr{S}'(\mathbb{R}^d)$ can be written as T_μ for some $\mu \in \mathscr{M}(\mathbb{R}^d)$. This would imply

$$|\varphi'(0)| = |\langle \delta', \varphi\rangle| = |\langle T_\mu, \varphi\rangle| \le \|\mu\|_{\mathrm{TV}}\, \|\varphi\|_\infty \quad \text{for all } \varphi \in \mathscr{S}(\mathbb{R}^d).$$

Since the derivative of a test function φ with $\|\varphi\|_\infty \le 1$ can be arbitrary steep, the above inequality cannot hold in general, which contradicts the assumption. Thus, δ' cannot be represented by a measure. □

Clearly, for compact $\mathbb{X}$ like $\mathbb{X} = \mathbb{T}^d$, the spaces $C_c(\mathbb{X})$ and $C_0(\mathbb{X})$ are the same, and we will just write $C(\mathbb{X})$ in the next subsection. However, for $\mathbb{X} = \mathbb{R}^d$, we have to be careful since weak-* convergence of measures with respect to $C_c(\mathbb{X})'$ and $C_0(\mathbb{X})'$ might be different as the following example shows.

Example 4.70 Take $\mathbb{X} = \mathbb{R}$ and $\mu_n := n\delta_n$. Then, for all $\varphi \in C_c(\mathbb{R})$

$$\int_{\mathbb{R}} \varphi \,\mathrm{d}\mu_n = n \int_{\mathbb{R}} \varphi \,\mathrm{d}\delta_n \to 0 \quad \text{as } n \to \infty.$$

Hence $\mu_n \overset{*}{\rightharpoonup} 0$ in $C_c(\mathbb{R})'$. But for the $\varphi(x) := \frac{\sin(x)}{x}$ in $C_0(\mathbb{R})$, we obtain

$$\int_{\mathbb{R}} \varphi \,\mathrm{d}\mu_n = \sin(n),$$

which diverges as $n \to \infty$. Hence $(\mu_n)_{n\in\mathbb{N}}$ does not converge weak-* in $C_0(\mathbb{R})'$. □

The problem in the above example is that $(\mu_n)_{n\in\mathbb{N}}$ is unbounded. For bounded sequences, the Banach–Steinhaus theorem together with $\overline{C_c(\mathbb{X})} = C_0(\mathbb{X})$ gives immediately the following equivalence.

Lemma 4.71 *Let $(\mu_n)_{n\in\mathbb{N}} \subset \mathscr{M}(\mathbb{X})$ be bounded. Then*

$$\mu_n \overset{*}{\rightharpoonup} \mu \text{ in } C_c(\mathbb{X})' \quad \text{if and only if} \quad \mu_n \overset{*}{\rightharpoonup} \mu \text{ in } C_0(\mathbb{X})'.$$

In the next remark, we have a look at the *space of bounded, continuous functions* $C_b(\mathbb{X})$ with the supremum norm. Let rba($\mathbb{X}$) denote the space of finitely additive set functions. These measures are not σ-additive in general.

Remark 4.72 It holds

- $C_b(\mathbb{X})' \cong \text{rba}(\mathbb{X})$ with $T_\mu(\varphi) = \langle \varphi, \mu \rangle := \int_{\mathbb{X}} \varphi \, \mathrm{d}\mu$ for all $\varphi \in C_b(\mathbb{X})$.

Note that $C_0(\mathbb{X}) \subset C_b(\mathbb{X})$, but $C_b(\mathbb{X})' \cong \text{rba}(\mathbb{X}) \not\subseteq \mathscr{M}(\mathbb{X}) \cong C_0(\mathbb{X})'$. Consider $\mathbb{X} = \mathbb{R}$ and the bounded sequence of measures $(\mu_n)_{n\in\mathbb{N}} := (\delta_n)_{n\in\mathbb{N}}$. Then

$$\int_{\mathbb{X}} \varphi \, \mathrm{d}\mu_n = \varphi(n) \to 0 \quad \text{for all } \varphi \in C_0(\mathbb{R}).$$

Hence $\mu_n \xrightarrow{*} 0$ in $C_0(\mathbb{R})'$. But for $\varphi(x) := \sin(x)$ in $C_b(\mathbb{R})$, we obtain $\int_{\mathbb{X}} \varphi \, \mathrm{d}\mu_n = \sin(n)$ which diverges. Hence $(\mu_n)_{n\in\mathbb{N}}$ is not weak-* convergent in $C_b(\mathbb{R})'$. The different behavior appears because the mass of $(\mu_n)_{n\in\mathbb{N}}$ "wanders off to infinity." To solve the issue, $(\mu_n)_{n\in\mathbb{N}}$ must be *tight*, i.e., for all $\epsilon > 0$, there exists a compact set $K \subset \mathbb{X}$ such that $|\mu_n|(\mathbb{X} \setminus K) < \epsilon$ for all $n \in \mathbb{N}$. If $(\mu_n)_{n\in\mathbb{N}} \subset \mathscr{M}(\mathbb{X})$ be bounded and tight, then it also holds

$$\mu_n \xrightarrow{*} \mu \text{ in } C_c(\mathbb{X})' \quad \text{if and only if} \quad \mu_n \xrightarrow{*} \mu \text{ in } C_b(\mathbb{X})'.$$

While the implication from right to left is clear, the opposite direction can be seen as follows: Let $f \in C_b(\mathbb{X})$ and $\epsilon > 0$. By the tightness of $(\mu_n)_{n\in\mathbb{N}}$, there exists a compact set $K \subset \mathbb{X}$, such that $|\mu_n|(X \setminus K) < \epsilon$ for all $n \in \mathbb{N}$ and $|\mu|(\mathbb{X} \setminus K) < \epsilon$. Since $\mathbb{X}$ is a metric space, exploiting *Urysohn's lemma*, we can construct $\tilde{f} \in C_c(\mathbb{X})$ such that $\tilde{f}_{|K} = f_{|K}$ and $\|\tilde{f}\|_\infty \le \|f\|_\infty$. Now it holds

$$\begin{aligned}
\left| \int_{\mathbb{X}} f \, d\mu - \int_{\mathbb{X}} f \, \mathrm{d}\mu_n \right| &\le \left| \int_{\mathbb{X}} (f - \tilde{f}) \, \mathrm{d}\mu \right| + \left| \int_{\mathbb{X}} \tilde{f} \, d\mu - \int_{\mathbb{X}} \tilde{f} \, \mathrm{d}\mu_n \right| \\
&\quad + \left| \int_{\mathbb{X}} (\tilde{f} - f) \, \mathrm{d}\mu_n \right| \\
&\le 2\|f\|_\infty |\mu|(\mathbb{X} \setminus K) + \underbrace{\left| \int_{\mathbb{X}} \tilde{f} \, \mathrm{d}\mu - \int_{\mathbb{X}} \tilde{f} \, \mathrm{d}\mu_n \right|}_{\text{small by weak-* conv. in } C_c'(\mathbb{X})} \\
&\quad + 2\|f\|_\infty |\mu_n|(\mathbb{X} \setminus K) \le (4\|f\|_\infty + 1)\epsilon
\end{aligned}$$

for sufficiently large $n \in \mathbb{N}$. This shows the weak-* convergence in $C_b(\mathbb{X})'$. □

When dealing with the Herglotz theorem and the Bochner theorem, we will be concerned with positive measures. A finite, real-valued (signed) Borel measure $\mu : \mathscr{B}(\mathbb{X}) \to \mathbb{R}$ in $\mathscr{M}(\mathbb{X})$ is called *positive* if $\mu(B) \ge 0$ for all $B \in \mathscr{B}(\mathbb{X})$. We denote the subset of positive measures by $\mathscr{M}_+(\mathbb{X}) \subset \mathscr{M}(\mathbb{X})$. By the Hahn–Jordan decomposition, we have the unique decomposition $\mu = \mu^+ - \mu^-$ with positive measures μ^+, μ^- for every real-valued $\mu \in \mathscr{M}(\mathbb{X})$. The total variation measure then fulfills $|\mu| = \mu^+ + \mu^-$, where the three positive measures are again Radon measures. Analogously, there exists a unique Hahn–Jordan decomposition for every complex-valued measure $\mu \in \mathscr{M}(\mathbb{X})$, where the decomposition is separately applied to the real and imaginary part of μ. We have the following proposition.

Theorem 4.73 *A measure $\mu \in \mathscr{M}(\mathbb{X})$ is positive if and only if*

$$\langle \mu, \varphi \rangle = \int_{\mathbb{X}} \varphi \, \mathrm{d}\mu \geq 0 \tag{4.50}$$

for all $\varphi \in C_c(\mathbb{X})$ with $\varphi : \mathbb{X} \to \mathbb{R}_+$.

Proof By definition of the integral it follows immediately that (4.50) holds true for positive measures. The opposite direction can be shown by contradiction. Assume that there exists $B \in \mathscr{B}(\mathbb{X})$ such that $\mu(B) < 0$. More precisely, we may assume without loss of generality that μ has a Hahn–Jordan decomposition $\mu^-(B) > 0$ and $\mu^+(B) = 0$. By the regularity of $|\mu|$, we find a sequence of compact sets $K_n \in \mathscr{B}(\mathbb{X})$ and open sets $U_n \in \mathscr{B}(\mathbb{X})$ with $K_n \subset B \subset U_n$ such that $\lim_{n\to\infty} |\mu|(K_n) = |\mu|(B)$ and $\lim_{n\to\infty} |\mu|(U_n) = |\mu|(B)$. In particular, we have $\mu^-(K_n) = |\mu|(K_n) \to |\mu|(B) = \mu^-(B)$. As a consequence of Urysohn's lemma, there exist $f_n \in C_c(\mathbb{X})$ such that

$$\mathbf{1}_{K_n}(x) \leq f_n(x) \leq \mathbf{1}_{U_n}(x) \quad \text{for all } x \in \mathbb{X}.$$

Exploiting the upper and lower bound of f_n, we obtain

$$\begin{aligned} 0 \leq \int_{\mathbb{X}} f_n \, \mathrm{d}\mu &= \int_{\mathbb{X}} f_n \, \mathrm{d}\mu^+ - \int_{\mathbb{X}} f_n \, \mathrm{d}\mu^- \\ &\leq \mu^+(U_n \setminus B) - \mu^-(K_n) \leq |\mu|(U_n \setminus B) - \mu^-(K_n). \end{aligned}$$

Considering the right-hand side for $n \to \infty$, we realize that $0 \leq -\mu^-(B) = \mu(B) < 0$, which yields the contradiction. ■

As a consequence, the limit of positive measures is positive too.

Corollary 4.74 *Let $(\mu_n)_{n\in\mathbb{N}}$ be a sequence of positive measures, and let $\mu \in \mathscr{M}(\mathbb{X})$ be the weak-* limit $\mu_n \overset{*}{\rightharpoonup} \mu$ in $C_c(\mathbb{X})'$. Then μ is positive.*

Besides weak-* convergence known from functional analysis, there is the notation of weak and vague convergence of measures in measure theory. A sequence of positive measures $\mu_n \in \mathscr{M}_+(\mathbb{X})$ converges *vaguely* to $\mu \in \mathscr{M}(\mathbb{X})$ if

$$\int_{\mathbb{X}} \varphi \, \mathrm{d}\mu_n \to \int_{\mathbb{X}} \varphi \, \mathrm{d}\mu \quad \text{for all } \varphi \in C_0(\mathbb{X})$$

(sometimes the definition is also given with respect to $C_c(\mathbb{X})$). A sequence of positive measures $\mu_n \in \mathrm{rba}(\mathbb{X})$ converges *weakly* to $\mu \in \mathrm{rba}(\mathbb{X})$ if

$$\int_{\mathbb{X}} \varphi \, \mathrm{d}\mu_n \to \int_{\mathbb{X}} \varphi \, \mathrm{d}\mu \quad \text{for all } \varphi \in C_b(\mathbb{X}).$$

Therefore a sequence $(\mu_n)_{n\in\mathbb{N}} \subset \mathscr{M}(\mathbb{X})$ converges vaguely to $\mu \subset \mathscr{M}(\mathbb{X})$ if and only if $\mu_n \overset{*}{\rightharpoonup} \mu$ in $C_0(\mathbb{X})'$ and μ_n are positive measures.

Finally, let us consider probability measures $\mathscr{P}(\mathbb{X}) \subset \mathscr{M}_+(\mathbb{X})$ having the additional property that $\mu(\mathbb{X}) = 1$. Clearly, we have for $\mu \in \mathscr{P}(\mathbb{X})$ that $\|\mu\|_{\mathrm{TV}} = 1$. Now let $(\mu_n)_{n\in\mathbb{N}}$ be a sequence of probability measures such that $\mu_n \overset{*}{\rightharpoonup} \mu$ in $C_b(\mathbb{X})'$ with $\mu \in \mathscr{M}(\mathbb{X})$. By Corollary 4.74, we know that μ is a positive measure. It is even a probability measure, since for $\varphi \equiv 1 \in C_b(\mathbb{X})$, we have

$$1 = \mu_n(\mathbb{X}) = \int_{\mathbb{X}} 1 \, \mathrm{d}\mu_n \to \int_{\mathbb{X}} 1 \, \mathrm{d}\mu = \mu(\mathbb{X}). \tag{4.51}$$

Finally, we like to state the following important theorem on the weak convergence of sequences of probability measures.

Theorem 4.75 (Prokhorov) *Let $(\mu_n)_{n\in\mathbb{N}}$ be a tight sequence of probability measures. Then there exists a subsequence $(\mu_{n_k})_{k\in\mathbb{N}}$ which converges weakly to a probability measure μ.*

Proof Since $\|\mu_n\|_{\mathrm{TV}} = 1$ for all $n \in \mathbb{N}$, the sequence $(\mu_n)_{n\in\mathbb{N}}$ is bounded in $C_0(\mathbb{X})'$. Since $\mathbb{X}$ is separable, $C_0(\mathbb{X})$ is separable. By the sequential Banach–Alaoglu theorem, there exists a subsequence $(\mu_{n_k})_{k\in\mathbb{Z}}$ with $\mu_{n_k} \overset{*}{\rightharpoonup} \mu \in \mathscr{M}(\mathbb{X})$ in $C_0(\mathbb{X})'$. Since $(\mu_n)_{n\in\mathbb{N}}$ is tight, Remark 4.72 shows that $\mu_{n_k} \overset{*}{\rightharpoonup} \mu \in C_b(\mathbb{X})'$. By (4.51), we have that μ is also a probability measure. ■

4.4.2 Fourier Transform of Measures on $\mathbb{T}^d$

In this subsection, we deal with $\mathbb{X} = \mathbb{T}^d$. We use the complex harmonics to introduce Fourier coefficients similarly to those for periodic functions and periodic tempered distributions. The *Fourier coefficients* of a measure $\mu \in \mathscr{M}(\mathbb{T}^d)$ are defined by

$$c_{\mathbf{k}}(\mu) := \frac{1}{(2\pi)^d} \langle \mu, \mathrm{e}^{-\mathrm{i}\mathbf{k}\cdot} \rangle, \qquad \mathbf{k} \in \mathbb{Z}^d.$$

By $|\langle \mu, \mathrm{e}^{-\mathrm{i}\mathbf{k}\cdot} \rangle| \le \|\mu\|_{\mathrm{TV}}$, we conclude that $(c_{\mathbf{k}}(\mu))_{\mathbf{k}\in\mathbb{Z}^d} \in \ell^\infty(\mathbb{Z}^d)$. By the following theorem, a measure $\mu \in \mathscr{M}(\mathbb{T}^d)$ is uniquely determined by its Fourier coefficients.

Theorem 4.76 (Unique Fourier Representation of Measures) *Let $\mu \in \mathscr{M}(\mathbb{T}^d)$ be a measure with $c_{\mathbf{k}}(\mu) = 0$ for all $\mathbf{k} \in \mathbb{Z}^d$. Then μ is the zero measure.*

Proof We restrict our attention to $d = 1$. The general case follows by taking tensor products. Let $\varphi \in C(\mathbb{T})$ with Fourier series

$$\varphi = \sum_{k \in \mathbb{Z}} c_k(\varphi)\, \mathrm{e}^{\mathrm{i}k\,\cdot}.$$

Exploiting that the Fejér kernel in Example 1.15 is an approximate identity, we know that

$$F_n * \varphi = \sum_{k=-n}^{n} \frac{n+1-|k|}{n+1}\, c_k(\varphi)\, \mathrm{e}^{\mathrm{i}k\,\cdot}$$

converges uniformly to φ as $n \to \infty$. Since μ acts as continuous operator on $C(\mathbb{T})$, we obtain by assumption on μ that $\langle \mu, \varphi \rangle = 0$. Since $\varphi \in C(\mathbb{T})$ was chosen arbitrarily, this implies that μ is the zero measure. ■

Similarly to distributions, we can define the convolution of a complex-valued measure and a continuous function. For $\mu \in \mathscr{M}(\mathbb{T}^d)$ and $\varphi \in C(\mathbb{T}^d)$, the *convolution* $\mu * \varphi$ is defined by

$$(\mu * \varphi)(\mathbf{x}) := \frac{1}{(2\pi)^d}\, \langle \mu, \varphi(\mathbf{x} - \cdot) \rangle = \frac{1}{(2\pi)^d} \int_{\mathbb{T}^d} \varphi(\mathbf{x} - \mathbf{y})\, \mathrm{d}\mu(\mathbf{y}).$$

Clearly, the convolution is pointwise well-defined. Moreover, since it is the composition of the two continuous operators $\mathbf{x} \mapsto \varphi(\mathbf{x} - \cdot)$ from $\mathbb{T}^d$ to $C(\mathbb{T}^d)$ and $\langle \mu, \cdot \rangle$ from $C(\mathbb{T}^d)$ to $\mathbb{C}$, we conclude $\mu * \varphi \in C(\mathbb{T}^d)$. Further, we have the relations

$$\|\mu * \varphi\|_\infty = \frac{1}{(2\pi)^d} \max_{\mathbf{x} \in \mathbb{T}^d} |\langle \mu, \varphi(\mathbf{x} - \cdot) \rangle| \le \frac{1}{(2\pi)^d}\, \|\mu\|_{\mathrm{TV}}\, \|\varphi\|_\infty$$

and

$$c_{\mathbf{k}}(\mu * \varphi) = \frac{1}{(2\pi)^{2d}} \int_{\mathbb{T}^d} \int_{\mathbb{T}^d} \varphi(\mathbf{x} - \mathbf{y}) \mathrm{e}^{-\mathrm{i}\mathbf{k} \cdot \mathbf{x}}\, \mathrm{d}\mu(\mathbf{y})\, \mathrm{d}\mathbf{x} = c_{\mathbf{k}}(\mu)\, c_{\mathbf{k}}(\varphi).$$

In contrast to general tempered distributions, we can also define a convolution of two measures. For $\mu, \nu \in \mathscr{M}(\mathbb{T}^d)$, the *convolution* $\mu * \nu \in \mathscr{M}(\mathbb{T}^d)$ is defined by

$$\langle \mu * \nu, \varphi \rangle := \frac{1}{(2\pi)^d}\, \langle \mu, \langle \nu, \varphi(\mathbf{x} + \mathbf{y}) \rangle \rangle = \frac{1}{(2\pi)^d} \int_{\mathbb{T}^d} \int_{\mathbb{T}^d} \varphi(\mathbf{x} + \mathbf{y})\, \mathrm{d}\nu(\mathbf{y})\, \mathrm{d}\mu(\mathbf{x}),$$

where $\varphi \in C(\mathbb{T}^d)$. With an analogous argumentation as above, the inner integral $\mathbf{x} \mapsto \int_{\mathbb{T}^d} \varphi(\mathbf{x} + \mathbf{y})\, \mathrm{d}\nu(\mathbf{y})$ is again a continuous function such that the convolution is well defined. Moreover, since

$$|\langle \mu * \nu, \varphi \rangle| \le \|\mu\|_{\mathrm{TV}} \|\nu\|_{\mathrm{TV}} \|\varphi\|_\infty,$$

we obtain that $\|\mu * \nu\|_{\mathrm{TV}} \le \|\mu\|_{\mathrm{TV}} \|\nu\|_{\mathrm{TV}}$. The convolution of measures fits to the notations of the convolution of signals and functions as can be seen in the following examples.

Example 4.77 We consider the two atomic measures $\mu := \sum_{k=0}^{N-1} \mu_k \delta_{2\pi k/N}$ and $\nu := \sum_{k=0}^{N-1} \nu_k \delta_{2\pi k/N}$ with $\mu_k, \nu_k \in \mathbb{C}$ on $\mathbb{T}$. The convolution is given by

$$\begin{aligned}\langle \mu * \nu, \varphi \rangle &= \frac{1}{2\pi} \Big\langle \mu, \sum_{k=0}^{N-1} \nu_k \, \varphi(\cdot + 2\pi k/N) \Big\rangle \\ &= \frac{1}{2\pi} \sum_{k,\ell=0}^{N-1} \mu_\ell \nu_k \, \varphi(2\pi k/N + 2\pi \ell/N) \\ &= \frac{1}{2\pi} \Big\langle \sum_{k,\ell=0}^{N-1} \mu_\ell \nu_k \, \delta_{2\pi k/N + 2\pi \ell/N}, \varphi \Big\rangle\end{aligned}$$

implying that

$$\mu * \nu = \sum_{k=0}^{N-1} \eta_k \, \delta_{2\pi k/N} \qquad \text{with} \qquad \eta_k := \frac{1}{2\pi} \sum_{\ell=0}^{N-1} \mu_{(k-\ell) \bmod N} \, \nu_\ell .$$

Note that $(\eta_k)_k$ is exactly the cyclic convolution of $(\mu_k)_k$ and $(\nu_k)_k$. □

Example 4.78 If the measures μ and ν are absolutely continuous with respect to the Lebesgue measure, then the theorem of Radon–Nikodym allows us to write them in the form $\mu = f_\mu \lambda$ and $\nu = f_\nu \lambda$, where $f_\mu, f_\nu \in L_1(\mathbb{T})$. Now the convolution reads as

$$\begin{aligned}\langle \mu * \nu, \varphi \rangle &= \frac{1}{2\pi} \Big\langle \mu, \int_{\mathbb{T}} \varphi(y + \cdot) f_\nu(y) \, \mathrm{d}y \Big\rangle \\ &= \frac{1}{2\pi} \int_{\mathbb{T}} \int_{\mathbb{T}} \varphi(y + x) f_\mu(x) f_\nu(y) \, \mathrm{d}y \, \mathrm{d}x \\ &= \frac{1}{2\pi} \int_{\mathbb{T}} \varphi(z) \int_{\mathbb{T}} f_\mu(x) f_\nu(z - x) \, \mathrm{d}x \, \mathrm{d}z = \int_{\mathbb{T}} \varphi(z) (f_\mu * f_\nu)(z) \, \mathrm{d}z\end{aligned}$$

so that $\mu * \nu = (f_\mu * f_\nu)\lambda$. □

The Fourier transform of the convolution of two measures becomes a pointwise product of their Fourier coefficients.

Theorem 4.79 (Convolution Property of Fourier Transform) *For $\mu, \nu \in \mathcal{M}(\mathbb{T}^d)$, the Fourier coefficients of $\mu * \nu \in \mathcal{M}(\mathbb{T}^d)$ fulfill*

$$c_{\mathbf{k}}(\mu * \nu) = c_{\mathbf{k}}(\mu)\, c_{\mathbf{k}}(\nu), \qquad \mathbf{k} \in \mathbb{Z}^d.$$

Proof The statement follows immediately by

$$\begin{aligned}\frac{1}{(2\pi)^d} \langle \mu * \nu, \mathrm{e}^{-\mathrm{i}\mathbf{k}\cdot} \rangle &= \frac{1}{(2\pi)^{2d}} \Big\langle \mu, \int_{\mathbb{T}^d} \mathrm{e}^{-\mathrm{i}\mathbf{k}(\mathbf{y}+\cdot)}\, \mathrm{d}\nu(\mathbf{y}) \Big\rangle \\ &= \frac{1}{(2\pi)^{2d}} \Big\langle \mu, \mathrm{e}^{-\mathrm{i}\mathbf{k}\cdot} \int_{\mathbb{T}^d} \mathrm{e}^{-\mathrm{i}\mathbf{k}\cdot\mathbf{y}}\, \mathrm{d}\nu(\mathbf{y}) \Big\rangle \\ &= \frac{1}{(2\pi)^{2d}} \langle \mu, \mathrm{e}^{-\mathrm{i}\mathbf{k}\cdot} \rangle \langle \nu, \mathrm{e}^{-\mathrm{i}\mathbf{k}\cdot} \rangle = c_{\mathbf{k}}(\mu)\, c_{\mathbf{k}}(\nu).\end{aligned}$$

■

Finally, we are interested in positive measures and their relation to so-called positive definite sequences. A sequence $(a_{\mathbf{k}})_{\mathbf{k}\in\mathbb{Z}^d}$ of complex numbers is *positive definite* if

$$\sum_{\mathbf{k},\mathbf{l}\in\mathbb{Z}^d} a_{\mathbf{k}-\mathbf{l}}\, \xi_{\mathbf{k}}\, \bar{\xi}_{\mathbf{l}} \ge 0.$$

for any sequence $(\xi_{\mathbf{k}})_{\mathbf{k}\in\mathbb{Z}^d}$ of complex numbers with only finitely many nonzero components.

Remark 4.80 By the following reason, positive definite sequences are always bounded: We consider the sequence $e_{\mathbf{n}} \in \mathbb{Z}^d$ which has zero components except for the $\mathbf{n}$-th component which is 1. Choosing $\xi = e_{\mathbf{0}}$ in the definition of positive definiteness, we have $a_{\mathbf{0}} \ge 0$. Using $\xi = e_{\mathbf{0}} \pm e_{\mathbf{n}}$ and $\xi = e_{\mathbf{0}} \pm \mathrm{i} e_{\mathbf{n}}$ for an $\mathbf{n} \in \mathbb{Z}^d \setminus \{\mathbf{0}\}$, we obtain $-2a_{\mathbf{0}} \le a_{\mathbf{n}} + a_{-\mathbf{n}} \le 2a_{\mathbf{0}}$ and $-2a_{\mathbf{0}} \le \mathrm{i}(a_{\mathbf{n}} - a_{-\mathbf{n}}) \le 2a_{\mathbf{0}}$ such that the real and imaginary parts of $a_{\mathbf{n}}$ are bounded independently of $\mathbf{n}$. Thus $(a_{\mathbf{k}})_{\mathbf{k}\in\mathbb{Z}^d}$ is contained in $\ell^\infty(\mathbb{Z}^d)$. □

By the following theorem, the Fourier coefficients of a positive measure form a positive definite sequence.

Theorem 4.81 *Let $\mu \in \mathcal{M}(\mathbb{T}^d)$ be a positive measure. Then $\{c_{\mathbf{k}}(\mu)\}_{\mathbf{k}\in\mathbb{Z}^d}$ is a positive definite sequence.*

Proof Let $\{\xi_{\mathbf{k}}\}_{\mathbf{k}\in\mathbb{Z}^d}$ be any sequence of complex numbers with only finitely many nonzero components. We consider the trigonometric polynomial

$$\varphi := \sum_{\mathbf{k}\in\mathbb{Z}^d} \xi_{\mathbf{k}}\, \mathrm{e}^{\mathrm{i}\mathbf{k}\cdot}.$$

Now $|\varphi|^2$ is a continuous nonnegative function. Integrating with respect to μ, we obtain

$$0 \le \langle \mu, |\varphi|^2 \rangle = \Big\langle \mu, \sum_{\mathbf{k},\mathbf{l} \in \mathbb{Z}^d} \xi_{\mathbf{k}} \bar{\xi}_{\mathbf{l}} \mathrm{e}^{-\mathrm{i}(\mathbf{l}-\mathbf{k})\cdot} \Big\rangle = (2\pi)^d \sum_{\mathbf{k},\mathbf{l} \in \mathbb{Z}^d} \xi_{\mathbf{k}} \bar{\xi}_{\mathbf{l}} c_{\mathbf{l}-\mathbf{k}}(\mu)$$

■

Interestingly, every positive definite sequence defines also a positive measure via its Fourier coefficients such that we have a one-to-one correspondence between positive measures and positive definite sequences.

Theorem 4.82 (Herglotz) *Let $(a_{\mathbf{k}})_{\mathbf{k}\in\mathbb{Z}^d}$ be a positive definite sequence. Then there exists a unique positive measure $\mu \in \mathcal{M}(\mathbb{T}^d)$ such that $c_{\mathbf{k}}(\mu) = a_{\mathbf{k}}$ for all $\mathbf{k} \in \mathbb{Z}^d$.*

Proof For ease of notation, we restrict our attention to the case $d = 1$.

1. For the positive definite sequence $(a_k)_{k\in\mathbb{Z}}$, we define the associated trigonometric polynomials $\varphi_n(x) := \sum_{k=-n}^{n} a_k \, \mathrm{e}^{\mathrm{i}kx}$ and show that their convolution $F_{2n} * \varphi_{2n}$ with the Fejér kernel is a nonnegative function. To this end, define for any $x \in \mathbb{T}$ the sequence $(\xi_k)_{k\in\mathbb{Z}}$ by

$$\xi_k := \begin{cases} \mathrm{e}^{\mathrm{i}kx} & \text{if } |k| \le n, \\ 0 & \text{otherwise.} \end{cases}$$

Exploiting the positive definiteness, we get

$$\begin{aligned} \sum_{k,\ell\in\mathbb{Z}} a_{k-\ell}\, \xi_k \, \bar{\xi}_\ell &= \sum_{k,\ell=-n}^{n} a_{k-\ell} \, \mathrm{e}^{\mathrm{i}(k-\ell)x} = \sum_{k=-2n}^{2n} (2n+1-|k|) \, a_k \, \mathrm{e}^{\mathrm{i}kx} \\ &= (2n+1)\,(F_{2n} * \varphi_{2n})(x) \ge 0. \end{aligned}$$

In particular, the function $F_{2n} * \varphi_{2n}$ is real-valued, which is only possible if $(a_k)_{k\in\mathbb{Z}}$ is conjugate symmetric, i.e., $a_{-k} = \bar{a}_k$. Further, we notice for any $n \in \mathbb{N}$ that

$$\frac{1}{2\pi} \int_{-\pi}^{\pi} (F_{2n} * \varphi_{2n})(x) \, \mathrm{d}x = a_0. \tag{4.52}$$

2. Next we define the linear functional T_μ on the vector space $\mathcal{T}$ of trigonometric polynomials by

$$T_\mu(p) := 2\pi \sum_{k=-M}^{M} \bar{a}_k \, c_k \qquad \text{for} \qquad p = \sum_{k=-M}^{M} c_k \, \mathrm{e}^{\mathrm{i}k\cdot}.$$

By definition we have

$$T_\mu(\mathrm{e}^{-\mathrm{i}k\cdot}) = \bar{a}_{-k} = a_k,$$

so that this operator is our candidate to determine the unique measure μ. We can rewrite T_μ as

$$T_\mu(p) = 2\pi \lim_{n\to\infty} \sum_{k=-M}^{M} \frac{2n+1-|k|}{2n+1}\, \bar{a}_k\, c_k = \lim_{n\to\infty} \int_{-\pi}^{\pi} p(x)\,(F_{2n} * \varphi_{2n})(x)\,\mathrm{d}x.$$

Exploiting the nonnegativity of the convolution shown in the first part of the proof and (4.52), we obtain

$$\left|\int_{-\pi}^{\pi} p(x)\,(F_{2n} * \varphi_{2n})(x)\,\mathrm{d}x\right| \le 2\pi\,|a_0|\,\|p\|_\infty.$$

Since this bound is independent of n, the functional T_μ is bounded on $\mathscr{T}$. Since the trigonometric polynomials are dense in $C(\mathbb{T})$, the functional T_μ has a unique continuous extension to $C(\mathbb{T})$. By Riesz's representation theorem, T_μ corresponds to a complex-valued measure $\mu \in \mathscr{M}(\mathbb{T})$.

3. It remains to show that $\mu \in \mathscr{M}(\mathbb{T})$ is positive. Let $\psi \in C(\mathbb{T})$ be a nonnegative function. Since the Fejér kernel defines an approximate identity, the trigonometric polynomials $F_{2n} * \psi$ converge uniformly to ψ. Since T_μ is continuous on $C(\mathbb{T})$, we obtain

$$\langle \mu, \psi\rangle = \lim_{n\to\infty} \langle \mu, F_{2n} * \psi\rangle = \lim_{n\to\infty} \int_{-\pi}^{\pi} \psi(x)\,(F_{2n} * \varphi_{2n})(x)\,\mathrm{d}x.$$

Since each term in the last limit is nonnegative, this implies that μ is positive. ■

4.4.3 Fourier Transform of Measures on $\mathbb{R}^d$

In this subsection, we focus on $\mathbb{X} = \mathbb{R}^d$. The *Fourier transform* on $\mathscr{M}(\mathbb{R}^d)$ is defined by $\mathscr{F} : \mathscr{M}(\mathbb{R}^d) \to C_b(\mathbb{R}^d)$ with

$$\mathscr{F}\mu(\boldsymbol{\omega}) = \hat{\mu}(\boldsymbol{\omega}) := \langle \mu, \mathrm{e}^{-\mathrm{i}\boldsymbol{\omega}\cdot}\rangle, \qquad \boldsymbol{\omega} \in \mathbb{R}^d.$$

We have the following theorem.

Theorem 4.83 *The Fourier transform on $\mathscr{M}(\mathbb{R}^d)$ is a linear, bounded operator mapping into $C_b(\mathbb{R}^d)$ with norm $\|\mathscr{F}\|_{\mathscr{M}(\mathbb{R}^d)\to C_b(\mathbb{R}^d)} = 1$.*

Proof In order to show that the Fourier-transformed measure is indeed continuous, we exploit the pointwise convergence $\mathrm{e}^{-\mathrm{i}\omega_2\cdot\mathbf{x}} \to \mathrm{e}^{-\mathrm{i}\omega_1\cdot\mathbf{x}}$ for $\omega_2 \to \omega_1$ and the uniform bound $|\mathrm{e}^{-\mathrm{i}\omega_1\cdot\mathbf{x}} - \mathrm{e}^{-\mathrm{i}\omega_2\cdot\mathbf{x}}| \le 2$. Then, since the constant function is integrable with respect to any finite measure, Lebesgue's dominated convergence theorem implies

$$\begin{aligned}\lim_{\omega_2\to\omega_1} |\hat{\mu}(\omega_1) - \hat{\mu}(\omega_2)| &= \lim_{\omega_2\to\omega_1} |\langle \mu, \mathrm{e}^{-\mathrm{i}\omega_1\cdot} - \mathrm{e}^{-\mathrm{i}\omega_2\cdot}\rangle| \\ &\le \lim_{\omega_2\to\omega_1} \langle |\mu|, |\mathrm{e}^{-\mathrm{i}\omega_1\cdot} - \mathrm{e}^{-\mathrm{i}\omega_2\cdot}|\rangle \\ &= \langle |\mu|, \lim_{\omega_2\to\omega_1} |\mathrm{e}^{-\mathrm{i}\omega_1\cdot} - \mathrm{e}^{-\mathrm{i}\omega_2\cdot}|\rangle = 0.\end{aligned}$$

Thus $\hat{\mu}$ is continuous. The operator norm simply follows from the estimation $\|\hat{\mu}\|_\infty = \sup_{\omega\in\mathbb{R}^d} |\langle \mu, \mathrm{e}^{-\mathrm{i}\omega\cdot}\rangle| \le \|\mu\|_{\mathrm{TV}}$, where we have equality for the atomic measure $\delta_{\mathbf{0}}$. ■

The next remark shows the relation between the Fourier transform of measures and those of the corresponding tempered distributions.

Remark 4.84 Since every measure defines a tempered distribution T_μ in the sense of Remark 4.69, we can calculate the distributional Fourier transform

$$\begin{aligned}\langle \hat{T}_\mu, \varphi\rangle &= \langle T_\mu, \hat{\varphi}\rangle = \int_{\mathbb{R}^d} \hat{\varphi}(\boldsymbol{\omega})\,\mathrm{d}\mu(\boldsymbol{\omega}) = \int_{\mathbb{R}^d}\int_{\mathbb{R}^d} \varphi(\mathbf{x})\,\mathrm{e}^{-\mathrm{i}\mathbf{x}\cdot\boldsymbol{\omega}}\,\mathrm{d}\boldsymbol{\omega}\,\mathrm{d}\mu(\mathbf{x}) \\ &= \int_{\mathbb{R}^d} \varphi(\mathbf{x})\,\hat{\mu}(\mathbf{x})\,\mathrm{d}\mathbf{x} = \langle T_{\hat{\mu}}, \varphi\rangle\end{aligned}$$

for all $\varphi \in \mathscr{S}(\mathbb{R}^d)$. This implies $\hat{T}_\mu = T_{\hat{\mu}}$. □

Example 4.85 For $\mu := \sum_{k=1}^N \mu_k \delta_{\mathbf{x}_k}$ we have

$$\hat{\mu}(\boldsymbol{\omega}) = \sum_{k=1}^N \mu_k\,\mathrm{e}^{-\mathrm{i}\boldsymbol{\omega}\cdot\mathbf{x}_k}.$$

For an absolutely continuous measure $\mu = f\lambda$ with density $f \in L_1(\mathbf{R}^d)$, we obtain that $\hat{\mu} = \hat{f}\lambda$. □

By the following theorem, a measure $\mu \in \mathscr{M}(\mathbb{T}^d)$ is uniquely determined by its Fourier transform.

Theorem 4.86 (Unique Fourier Representation of Measures) *Let $\mu \in \mathscr{M}(\mathbb{R}^d)$ be a measure with $\hat{\mu} \equiv 0$. Then μ is the zero measure.*

Proof Assume that there exists a nonzero measure μ with $\hat{\mu} \equiv 0$. Then we can find $\varphi \in C_c(\mathbb{R}^d)$ with $|\langle \mu, \varphi\rangle| > 0$. Next, we approximate φ using the approximate identity

$$\psi_n(\mathbf{x}) := n^d \psi(n\mathbf{x}) \qquad \text{with} \qquad \psi(\mathbf{x}) := \Big(\prod_{j=1}^d \operatorname{sinc}(x_j/2)\Big)^2 .$$

Its Fourier transform $\hat{\psi}(\boldsymbol{\omega}) = \prod_{j=1}^d M_2(\omega_j)$ has compact support, cf. Example 2.16. Since $\varphi_n := \psi_n * \varphi$ converges uniformly to φ by a multidimensional version of Theorem 2.18, we conclude that for sufficiently large $n \in \mathbb{N}$, it also holds $|\langle \mu, \varphi_n\rangle| > 0$. By the Fourier convolution theorem, we obtain

$$\hat{\varphi}_n(\boldsymbol{\omega}) = \mathscr{F}(\varphi * \psi_n)(\boldsymbol{\omega}) = \hat{\varphi}(\boldsymbol{\omega})\, \hat{\psi}(\boldsymbol{\omega}/n)$$

and by the Fourier inversion in Theorem 2.10 further

$$\varphi_n(\mathbf{x}) = \frac{1}{(2\pi)^d} \int_{[-n,n]^d} \hat{\varphi}_n(\boldsymbol{\omega})\, \mathrm{e}^{\mathrm{i}\mathbf{x}\cdot\boldsymbol{\omega}}\, \mathrm{d}\boldsymbol{\omega}.$$

The integral can be uniformly approximated on every ball $B_R := \{\mathbf{x} \in \mathbb{R}^d : \|\mathbf{x}\|_2 \le R\}$ by a Riemann sum

$$\rho(\mathbf{x}) = \frac{1}{(2\pi)^d} \sum_{k=1}^K \hat{\varphi}_n(\boldsymbol{\omega}_k)\, \mathrm{e}^{\mathrm{i}\boldsymbol{\omega}_k\cdot\mathbf{x}}\, w_k, \qquad \sum_{k=1}^N w_k = (2n)^d,$$

so that $|\varphi_n(\mathbf{x}) - \rho(\mathbf{x})| \le \epsilon$ for any given $\epsilon > 0$. We can estimate

$$\|\rho\|_\infty \le \frac{(2n)^d}{(2\pi)^d}\, \|\hat{\varphi}_n\|_\infty.$$

Finally, splitting the integral over the ball B_R and its complement, we obtain

$$\begin{aligned} |\langle \mu, \varphi_n\rangle - \langle \mu, \rho\rangle| &= \Big| \int_{B_R} \varphi_n - \rho \,\mathrm{d}\mu \Big| + \Big| \int_{\mathbb{R}^d \setminus B_R} \varphi_n - \rho\, \mathrm{d}\mu \Big| \\ &\le \epsilon\, |\mu|(B_R) + \Big(\|\varphi_n\|_\infty + \frac{(2n)^d}{(2\pi)^d}\, \|\hat{\varphi}_n\|_\infty\Big)\, |\mu|(\mathbb{R}^d \setminus B_R). \end{aligned}$$

Exploiting that

$$|\mu|(B_R) \to \|\mu\|_{\mathrm{TV}} \quad \text{and} \quad |\mu|(\mathbb{R}^d \setminus B_R) \to 0 \tag{4.53}$$

as $R \to \infty$, we can make the right-hand side arbitrary small for suitable ϵ and R. Consequently, $|\langle \mu, \rho\rangle| > 0$. However, since ρ is an exponential sum, we have by assumption on $\hat{\mu}$ that $|\langle \mu, \rho\rangle| = 0$, which is a contradiction. ■

Next we consider convolutions. For $\mu \in \mathcal{M}(\mathbb{R}^d)$ and $\varphi \in C_0(\mathbb{R}^d)$, the *convolution* $\mu * \varphi$ is defined by

$$(\mu * \varphi)(\mathbf{x}) := \langle \mu, \varphi(\mathbf{x} - \cdot) \rangle = \int_{\mathbb{R}^d} \varphi(\mathbf{x} - \mathbf{y})\, \mathrm{d}\mu(\mathbf{y}).$$

Again, $\mu * \varphi$ is a continuous function. Moreover, we have

$$\Big| \int_{\mathbb{R}^d} \varphi(\mathbf{x} - \mathbf{y})\, \mathrm{d}\mu(\mathbf{y}) \Big| \le \int_{B_{\|\mathbf{x}\|_2/2}} |\varphi(\mathbf{x} - \mathbf{y})|\, \mathrm{d}|\mu|(\mathbf{y}) + \|\varphi\|_\infty |\mu|(\mathbb{R}^d \setminus B_{\|\mathbf{x}\|_2/2})$$

which approaches zero as $\|\mathbf{x}\| \to \infty$ by the decrease of φ and (4.53). Consequently, it holds $\mu * \varphi \in C_0(\mathbb{R}^d)$.

For $\mu, \nu \in \mathcal{M}(\mathbb{R}^d)$, the *convolution* $\mu * \nu \in \mathcal{M}(\mathbb{R}^d)$ is defined by

$$\langle \mu * \nu, \varphi \rangle := \langle \mu, \langle \nu, \varphi(\mathbf{x} + \mathbf{y}) \rangle \rangle = \int_{\mathbb{R}^d} \int_{\mathbb{R}^d} \varphi(\mathbf{x} + \mathbf{y})\, \mathrm{d}\nu(\mathbf{y})\, \mathrm{d}\mu(\mathbf{x}),$$

where $\varphi \in C_0(\mathbb{R}^d)$. Then we have the Fourier convolution theorem.

Theorem 4.87 (Convolution Property of Fourier Transform) *For $\mu, \nu \in \mathcal{M}(\mathbb{R}^d)$, the Fourier-transformed measures fulfill*

$$\widehat{\mu * \nu} = \hat{\mu} \cdot \hat{\nu}.$$

Proof The statement follows by

$$\langle \mu * \nu, \mathrm{e}^{-\mathrm{i}\boldsymbol{\omega}\cdot} \rangle = \Big\langle \mu, \int_{\mathbb{R}^d} \mathrm{e}^{-\mathrm{i}\boldsymbol{\omega}\cdot(\mathbf{y}+\cdot)}\, \mathrm{d}\nu(y) \Big\rangle = \Big\langle \mu, \mathrm{e}^{-\mathrm{i}\boldsymbol{\omega}\cdot} \int_{\mathbb{R}^d} \mathrm{e}^{-\mathrm{i}\boldsymbol{\omega}\cdot\mathbf{y}}\, \mathrm{d}\nu(y) \Big\rangle$$
$$= \langle \mu, \mathrm{e}^{-\mathrm{i}\boldsymbol{\omega}\cdot} \rangle \langle \nu, \mathrm{e}^{-\mathrm{i}\boldsymbol{\omega}\cdot} \rangle = \hat{\mu}(\boldsymbol{\omega})\, \hat{\nu}(\boldsymbol{\omega}).$$

■

The convolution of two absolutely continuous measures is again an absolutely continuous measure, whose density is just the convolution of the densities of the two initial measures. Here is another example with atomic measures.

Example 4.88 For $\mu := \sum_{k=1}^N \mu_k \delta_{\mathbf{x}_k}$ and $\nu := \sum_{l=1}^M \nu_l \delta_{\mathbf{y}_l}$, we obtain

$$\mu * \nu = \sum_{k=1}^{N} \sum_{l=1}^{M} \mu_k \nu_l \delta_{\mathbf{x}_k + \mathbf{y}_l}.$$

□

Finally, we deal again with positive measures. A function $\varphi \in C(\mathbb{R}^d)$ is *positive definite* if

$$\sum_{k,\ell\in\mathbb{Z}} \varphi(\mathbf{x}_k - \mathbf{x}_\ell)\, \xi_k \bar{\xi}_\ell \geq 0$$

for any sequence $(\mathbf{x}_k)_{k\in\mathbb{Z}}$ in $\mathbb{R}^d$ and any sequence $(\xi_k)_{k\in\mathbb{Z}}$ of complex numbers with only finitely many nonzero elements. Using the argumentation from Remark 4.80, we can show that every positive definite function has to be bounded. By the following theorem, the Fourier transform of a positive measure is a positive definite function.

Theorem 4.89 *Let $\mu \in \mathscr{M}(\mathbb{R}^d)$ be a positive measure. Then $\hat{\mu}$ is a positive definite function.*

Proof Let $(\boldsymbol{\omega}_k)_{k\in\mathbb{Z}}$ be a sequence in $\mathbb{R}^d$ and $(\xi_k)_{k\in\mathbb{Z}}$ a sequence of complex numbers with only finitely many nonzero elements. We define the nonnegative auxiliary function

$$\varphi(\mathbf{x}) = \Big|\sum_{k\in\mathbb{Z}} \xi_k\, \mathrm{e}^{-\mathrm{i}\boldsymbol{\omega}_k\cdot\mathbf{x}}\Big|^2 \geq 0.$$

Expanding the square, we can rewrite the auxiliary function as

$$\varphi(\mathbf{x}) = \Big(\sum_{k\in\mathbb{Z}} \xi_k\, \mathrm{e}^{-\mathrm{i}\boldsymbol{\omega}_k\cdot\mathbf{x}}\Big)\overline{\Big(\sum_{\ell\in\mathbb{Z}} \xi_\ell\, \mathrm{e}^{-\mathrm{i}\boldsymbol{\omega}_\ell\cdot\mathbf{x}}\Big)} = \sum_{k,\ell\in\mathbb{Z}} \xi_k\bar{\xi}_\ell\, \mathrm{e}^{-\mathrm{i}(\boldsymbol{\omega}_k-\boldsymbol{\omega}_\ell)\cdot\mathbf{x}}.$$

Since the measure μ is positive, the corresponding integral is also positive, which implies

$$\sum_{k,\ell\in\mathbb{Z}} \xi_k\bar{\xi}_\ell\, \langle \mu, \mathrm{e}^{-\mathrm{i}(\boldsymbol{\omega}_k-\boldsymbol{\omega}_\ell)\cdot}\rangle = \sum_{k,\ell\in\mathbb{Z}} \xi_k\bar{\xi}_\ell\, \hat{\mu}(\boldsymbol{\omega}_k - \boldsymbol{\omega}_\ell) \geq 0.$$

■

Salomon Bochner [53] has shown the converse of the above theorem, which can be seen as counterpart to Herglotz' theorem for measures on the torus.

Theorem 4.90 (Bochner) *Let $\varphi \in C_b(\mathbb{R}^d)$ be a positive definite function. Then there exists a unique positive measure $\mu \in \mathscr{M}_+(\mathbb{R}^d)$ such that $\hat{\mu} = \varphi$.*

The proof of Bochner's theorem requires the following two lemmata.

Lemma 4.91 *Let $\varphi \in C_b(\mathbb{R}^d)$ be a positive definite function. If $\psi \in L_1(\mathbb{R}^d)$ is nonnegative almost everywhere, then $\varphi\hat{\psi}$ is positive definite.*

Proof Let $(\boldsymbol{\omega}_k)_{k\in\mathbb{Z}}$ be a sequence in $\mathbb{R}^d$ and $(\xi_k)_{k\in\mathbb{Z}}$ be a sequence of complex numbers with only finitely many nonzero elements. Then we conclude

$$\begin{aligned}
&\sum_{k,\ell\in\mathbb{Z}} \varphi(\omega_k-\omega_\ell)\,\hat{\psi}(\omega_k-\omega_\ell)\,\xi_k\bar{\xi}_\ell \\
&= \sum_{k,\ell\in\mathbb{Z}} \varphi(\omega_k-\omega_\ell)\,\xi_k\bar{\xi}_\ell \int_{\mathbb{R}^d} \psi(\mathbf{x})\,\mathrm{e}^{-\mathrm{i}(\omega_k-\omega_\ell)\cdot\mathbf{x}}\,\mathrm{d}\mathbf{x} \\
&= \int_{\mathbb{R}^d} \Big(\sum_{k,\ell\in\mathbb{Z}} \varphi(\omega_k-\omega_\ell)\,\xi_k \mathrm{e}^{-\mathrm{i}\omega_k\cdot\mathbf{x}}\,\overline{\xi_\ell \mathrm{e}^{-\mathrm{i}\omega_\ell\cdot\mathbf{x}}}\Big)\,\psi(\mathbf{x})\,\mathrm{d}\mathbf{x} \ge 0.
\end{aligned}$$

■

Lemma 4.92 *Let $\varphi \in C_b(\mathbb{R}^d) \cap L_1(\mathbb{R}^d)$ be positive definite. Then the inverse Fourier transform $\check{\varphi}$ is nonnegative and absolutely integrable.*

Proof

1. If the inverse Fourier transform $\check{\varphi}$ is not nonnegative, then there exists $\psi \in C_c(\mathbb{R}^d)$ such that

$$\int_{\mathbb{R}^d} \check{\varphi}(\mathbf{x})\,|\psi(\mathbf{x})|^2\,\mathrm{d}\mathbf{x} < 0. \tag{4.54}$$

Applying the definition of the inverse Fourier transform, Fubini's theorem, and Parseval's identity, we can rewrite the integral

$$\begin{aligned}
\int_{\mathbb{R}^d} \check{\varphi}(\mathbf{x})\,|\psi(\mathbf{x})|^2\,\mathrm{d}\mathbf{x} &= \frac{1}{(2\pi)^d}\int_{\mathbb{R}^d}\int_{\mathbb{R}^d} \varphi(\omega)\,|\psi(\mathbf{x})|^2\,\mathrm{e}^{\mathrm{i}\omega\cdot\mathbf{x}}\,\mathrm{d}\omega\,\mathrm{d}\mathbf{x} \\
&= \frac{1}{(2\pi)^d}\int_{\mathbb{R}^d}\int_{\mathbb{R}^d} \varphi(\omega)\,\psi(\mathbf{x})\,\overline{\psi(\mathbf{x})\,\mathrm{e}^{-\mathrm{i}\omega\cdot\mathbf{x}}}\,\mathrm{d}\mathbf{x}\,\mathrm{d}\omega \\
&= \frac{1}{(2\pi)^d}\int_{\mathbb{R}^d} \varphi(\omega) \int_{\mathbb{R}^d} \psi(\mathbf{x})\,\overline{\psi(\mathbf{x})\,\mathrm{e}^{-\mathrm{i}\omega\cdot\mathbf{x}}}\,\mathrm{d}\mathbf{x}\,\mathrm{d}\omega \\
&= \frac{1}{(2\pi)^{2d}}\int_{\mathbb{R}^d} \varphi(\omega) \int_{\mathbb{R}^d} \hat{\psi}(\eta)\overline{\hat{\psi}(\omega+\eta)}\,\mathrm{d}\eta\,\mathrm{d}\omega \\
&= \frac{1}{(2\pi)^{2d}}\int_{\mathbb{R}^d}\int_{\mathbb{R}^d} \varphi(\omega-\eta)\,\hat{\psi}(\eta)\,\overline{\hat{\psi}(\omega)}\,\mathrm{d}\eta\,\mathrm{d}\omega.
\end{aligned}$$

Since φ and $\hat{\psi}$ are continuous, we can approximate the integral arbitrary well by a Riemann sum with respect to an equispaced grid on a sufficiently large cube C. In this manner, we get the approximation

$$\frac{w}{(2\pi)^{2d}} \sum_{k,\ell=1}^{N} \varphi(\omega_k-\omega_\ell)\,\hat{\psi}(\omega_\ell)\,\overline{\hat{\psi}(\omega_k)},$$

where w is the uniform weight and where N is the number of equispaced points $\boldsymbol{\omega}_k$ in C. Since φ is positive definite, this Riemann sum is nonnegative, which is a contradiction to the negativity in (4.54).

2. To show the integrability, we consider the function

$$\psi(\mathbf{x}) := \prod_{j=1}^{d} M_2(x_j) \qquad \text{with} \qquad \hat{\psi}(\boldsymbol{\omega}) = \Big(\prod_{j=1}^{d} \operatorname{sinc}(\omega_j/2)\Big)^2,$$

see Example 2.16. Setting $\psi_n(\mathbf{x}) := \psi(\mathbf{x}/n)$ with $\hat{\psi}_n(\boldsymbol{\omega}) = n^d\, \hat{\psi}(n\boldsymbol{\omega})$, we obtain

$$\begin{aligned}\int_{\mathbb{R}^d} \psi_n(\mathbf{x})\, \check{\varphi}(\mathbf{x})\, \mathrm{d}\mathbf{x} &= \frac{1}{(2\pi)^d} \int_{\mathbb{R}^d} \int_{\mathbb{R}^d} \psi_n(\mathbf{x})\, \varphi(\boldsymbol{\omega})\, \mathrm{e}^{\mathrm{i}\boldsymbol{\omega}\cdot\mathbf{x}}\, \mathrm{d}\boldsymbol{\omega}\, \mathrm{d}\mathbf{x} \\ &= \frac{1}{(2\pi)^d} \int_{\mathbb{R}^d} \hat{\psi}_n(-\boldsymbol{\omega})\, \varphi(\boldsymbol{\omega})\, \mathrm{d}\boldsymbol{\omega} = \frac{1}{(2\pi)^d}\, (\hat{\psi}_n * \varphi)(\mathbf{0}).\end{aligned}$$

Notice that $\hat{\psi}_n$ is an approximate identity, meaning that the last term converges to $\varphi(\mathbf{0})/(2\pi)^d$ by Theorem 2.18. Due to Levi's monotone convergence theorem, the first integral converges to $\int_{\mathbb{R}^d} \check{\varphi}(\mathbf{x})\, \mathrm{d}\mathbf{x}$ establishing the second assertion. ■

***Proof of Theorem* 4.90** Analogously to the previous proof, we employ the Fourier-transformed approximation identity

$$\psi(\mathbf{x}) := \prod_{j=1}^{d} M_2(x_j) \qquad \text{with} \qquad \check{\psi}(\boldsymbol{\omega}) = \Big(\prod_{j=1}^{d} \operatorname{sinc}(\omega_j/2)\Big)^2,$$

and define $\psi_n(\mathbf{x}) := \psi(\mathbf{x}/n)$ with $\check{\psi}_n(\boldsymbol{\omega}) = n^d\, \check{\psi}(n\boldsymbol{\omega})$. Lemma 4.91 ensures that the product $\varphi\psi_n \in L_1(\mathbb{R}^d)$ is positive definite and Lemma 4.92 that $f_n := \mathcal{F}^{-1}(\varphi\psi_n)$ is nonnegative and absolutely integrable. The Fourier transform of the absolutely continuous measure $\mu_n := f_n\lambda$ is $\varphi\psi_n$. Further we have

$$\|\mu_n\|_{\mathrm{TV}} = \int_{\mathbb{R}^d} f_n(\mathbf{x})\, \mathrm{d}\mathbf{x} = \hat{f}_n(\mathbf{0}) = \varphi(\mathbf{0}),$$

i.e., $\mu_n \in \mathcal{M}(\mathbb{R}^d)$. By the sequential Banach–Alaoglu theorem, $(\mu_n)_{n\in\mathbb{N}}$ contains a weak-* convergent subsequence $(\mu_{n_k})_{k\in\mathbb{N}}$ meaning that there exists $\mu \in \mathcal{M}(\mathbb{R}^d)$ such that

$$\langle \mu_{n_k}, \phi\rangle \to \langle \mu, \phi\rangle \qquad \text{for all} \qquad \phi \in C_0(\mathbb{R}^d). \tag{4.55}$$

The measure μ is positive since it is the weak-* limit of positive measures; see Corollary 4.74. To verify that μ has the wanted Fourier transform, we take an arbitrary $g \in L_1(\mathbb{R}^d)$ and consider the integral

$$\int_{\mathbb{R}^d} g(\boldsymbol{\omega})\, \hat{\mu}(\boldsymbol{\omega})\, \mathrm{d}\boldsymbol{\omega} = \int_{\mathbb{R}^d} \int_{\mathbb{R}^d} g(\boldsymbol{\omega})\, \mathrm{e}^{-\mathrm{i}\boldsymbol{\omega}\cdot\mathbf{x}}\, \mathrm{d}\mu(\mathbf{x})\, \mathrm{d}\boldsymbol{\omega} = \int_{\mathbb{R}^d} \hat{g}(\mathbf{x})\, \mathrm{d}\mu(\mathbf{x}),$$

where we have applied Fubini's theorem. Since $\hat{g} \in C_0(\mathbb{R}^d)$, we exploit the weak-* convergence (4.55) and revert the above rearrangements to obtain

$$\int_{\mathbb{R}^d} g(\boldsymbol{\omega})\, \hat{\mu}(\boldsymbol{\omega})\, \mathrm{d}\boldsymbol{\omega} = \lim_{k\to\infty} \int_{\mathbb{R}^d} \hat{g}(\mathbf{x})\, \mathrm{d}\mu_{n_k}(\mathbf{x}) = \lim_{k\to\infty} \int_{\mathbb{R}^d} g(\boldsymbol{\omega})\, \hat{\mu}_{n_k}(\boldsymbol{\omega})\, \mathrm{d}\boldsymbol{\omega}.$$

Finally, the Fourier inversion theorem and Lebesgue's dominated convergence theorem together with the pointwise convergence of $(\psi_{n_k})_{k\in\mathbb{Z}}$ to one yield

$$\int_{\mathbb{R}^d} g(\boldsymbol{\omega})\, \hat{\mu}(\boldsymbol{\omega})\, \mathrm{d}\boldsymbol{\omega} = \lim_{k\to\infty} \int_{\mathbb{R}^d} g(\boldsymbol{\omega})\, \varphi(\boldsymbol{\omega})\, \psi_{n_k}(\boldsymbol{\omega})\, \mathrm{d}\boldsymbol{\omega} = \int_{\mathbb{R}^d} g(\boldsymbol{\omega})\, \varphi(\boldsymbol{\omega})\, \mathrm{d}\boldsymbol{\omega}.$$

Since the equality holds for all $g \in L_1(\mathbb{R}^d)$ and $\hat{\mu}, \varphi \in C_b(\mathbb{R}^d)$, this implies $\hat{\mu} \equiv \varphi$. This finishes the proof. ■

Remark 4.93 Owing to the close relation between the Fourier transform of measure spaces and tempered distributions, the Bochner theorem immediately applies to regular tempered distributions. More precisely, if T_φ is a regular tempered distribution associated with a positive definite function $\varphi \in C_b(\mathbb{R}^d)$, then there exists a measure $\mu \in \mathscr{M}(\mathbb{R}^d)$ such that $T_\varphi = \hat{T}_\mu$. □

4.5 Multidimensional Discrete Fourier Transforms

The multidimensional DFT is necessary for the computation of Fourier coefficients of a function $f \in C(\mathbb{T}^d)$ as well as for the calculation of the Fourier transform of a function $f \in L_1(\mathbb{R}^d) \cap C(\mathbb{R}^d)$. Further the two-dimensional DFT finds numerous applications in image processing. The properties of the one-dimensional DFT (see Chap. 3) can be extended to the multidimensional DFT in a straightforward way.

4.5.1 Computation of Multivariate Fourier Coefficients

We describe the computation of Fourier coefficients $c_{\mathbf{k}}(f)$, $\mathbf{k} = (k_j)_{j=1}^d \in \mathbb{Z}^d$, of a given function $f \in C(\mathbb{T}^d)$, where f is sampled on the uniform grid $\{\frac{2\pi}{N}\,\mathbf{n} : \mathbf{n} \in I_N^d\}$, where $N \in \mathbb{N}$ is even, $I_N := \{0, \ldots, N-1\}$, and $I_N^d := \{\mathbf{n} = (n_j)_{j=1}^d : n_j \in I_N,\ j = 1, \ldots, d\}$. Using the rectangle rule of numerical integration, we can compute $c_{\mathbf{k}}(f)$ for $\mathbf{k} \in \mathbb{Z}^d$ approximately. Since $[0,\, 2\pi]^d$ is equal to the union of the N^d hypercubes $\frac{2\pi}{N}\,\mathbf{n} + [0,\, \frac{2\pi}{N}]^d$, $\mathbf{n} \in I_N^d$, we obtain

$$c_{\mathbf{k}}(f) = \frac{1}{(2\pi)^d} \int_{[0,\,2\pi]^d} f(\mathbf{x})\,\mathrm{e}^{-\mathrm{i}\,\mathbf{k}\cdot\mathbf{x}}\,\mathrm{d}\mathbf{x} \approx \frac{1}{N^d} \sum_{\mathbf{n}\in I_N^d} f\Big(\frac{2\pi}{N}\,\mathbf{n}\Big)\,\mathrm{e}^{-2\pi\mathrm{i}\,(\mathbf{k}\cdot\mathbf{n})/N}$$

$$= \frac{1}{N^d} \sum_{\mathbf{n}\in I_N^d} f\Big(\frac{2\pi}{N}\,\mathbf{n}\Big)\,w_N^{\mathbf{k}\cdot\mathbf{n}}$$

with $w_N = \mathrm{e}^{-2\pi\mathrm{i}/N}$. The expression

$$\sum_{\mathbf{n}\in I_N^d} f\Big(\frac{2\pi}{N}\,\mathbf{n}\Big)\,w_N^{\mathbf{k}\cdot\mathbf{n}}$$

is called the *d-dimensional discrete Fourier transform of size* $N_1 \times \ldots \times N_d$ of the d-dimensional array $\big(f(\frac{2\pi}{N}\,\mathbf{n})\big)_{\mathbf{n}\in I_N^d}$, where $N_1 = \ldots = N_d := N$. Thus we obtain the approximate Fourier coefficients

$$\hat{f}_{\mathbf{k}} := \frac{1}{N^d} \sum_{\mathbf{n}\in I_N^d} f\big(\frac{2\pi}{N}\,\mathbf{n}\big)\,w_N^{\mathbf{k}\cdot\mathbf{n}}\,. \tag{4.56}$$

Obviously, the values $\hat{f}_{\mathbf{k}}$ are N-periodic, i.e., for all $\mathbf{k}, \mathbf{m} \in \mathbb{Z}^d$ we have

$$\hat{f}_{\mathbf{k}+N\,\mathbf{m}} = \hat{f}_{\mathbf{k}}\,.$$

But by Lemma 4.6 we know that $\lim_{\|\mathbf{k}\|_2\to\infty} c_{\mathbf{k}}(f) = 0$. Therefore we can only expect that

$$\hat{f}_{\mathbf{k}} \approx c_{\mathbf{k}}(f)\,, \quad k_j = -\frac{N}{2}, \ldots, \frac{N}{2} - 1;\ j = 1, \ldots, d\,.$$

To see this effect more clearly, we will derive a multidimensional aliasing formula. By $\delta_{\mathbf{m}}$, $\mathbf{m} \in \mathbb{Z}^d$, we denote the *d-dimensional Kronecker symbol*

$$\delta_{\mathbf{m}} := \begin{cases} 1 & \mathbf{m} = \mathbf{0}\,, \\ 0 & \mathbf{m} \in \mathbb{Z}^d \setminus \{\mathbf{0}\}\,. \end{cases}$$

First we present a generalization of Lemma 3.2.

Lemma 4.94 *Let* $N_j \in \mathbb{N} \setminus \{1\}$, $j = 1, \ldots, d$, *be given. Then for each* $\mathbf{m} = (m_j)_{j=1}^d \in \mathbb{Z}^d$, *we have*

$$\sum_{k_1=0}^{N_1-1} \cdots \sum_{k_d=0}^{N_d-1} w_{N_1}^{m_1 k_1} \ldots w_{N_d}^{m_d k_d} = \prod_{j=1}^{d} (N_j\,\delta_{m_j \bmod N})$$

$$= \begin{cases} \prod_{j=1}^{d} N_j & \mathbf{m} \in N_1 \mathbb{Z} \times \ldots \times N_d \mathbb{Z}, \\ 0 & \mathbf{m} \in \mathbb{Z}^d \setminus (N_1 \mathbb{Z} \times \ldots \times N_d \mathbb{Z}). \end{cases}$$

If $N_1 = \ldots = N_d = N$, then for each $\mathbf{m} \in \mathbb{Z}^d$,

$$\sum_{\mathbf{k} \in I_N^d} w_N^{\mathbf{m} \cdot \mathbf{k}} = N^d \, \delta_{\mathbf{m} \bmod N} = \begin{cases} N^d & \mathbf{m} \in N \mathbb{Z}^d, \\ 0 & \mathbf{m} \in \mathbb{Z}^d \setminus (N \mathbb{Z}^d), \end{cases}$$

where the vector $\mathbf{m} \bmod N := (m_j \bmod N)_{j=1}^d$ *denotes the* nonnegative residue of $\mathbf{m} \in \mathbb{Z}^d$ modulo N, *and*

$$\delta_{\mathbf{m} \bmod N} = \prod_{j=1}^{d} \delta_{m_j \bmod N}.$$

Proof This result is an immediate consequence of Lemma 3.2, since

$$\sum_{k_1=0}^{N_1-1} \ldots \sum_{k_d=0}^{N_d-1} w_{N_1}^{m_1 k_1} \ldots w_{N_d}^{m_d k_d} = \prod_{j=1}^{d} \Big(\sum_{k_j=0}^{N_j-1} w_{N_j}^{m_j k_j} \Big) = \prod_{j=1}^{d} (N_j \, \delta_{m_j \bmod N_j}).$$

■

The following aliasing formula describes a close relation between the Fourier coefficients $c_{\mathbf{k}}(f)$ and the approximate values $\hat{f}_{\mathbf{k}}$.

Theorem 4.95 (Aliasing Formula for d-Variate Fourier Coefficients) *Let $N \in \mathbb{N}$ be even and let $f \in C(\mathbb{T}^d)$ be given. Assume that the Fourier coefficients $c_{\mathbf{k}}(f)$ satisfy the condition $\sum_{\mathbf{k} \in \mathbb{Z}^d} |c_{\mathbf{k}}(f)| < \infty$.*

Then we have the aliasing formula

$$\hat{f}_{\mathbf{k}} = \sum_{\mathbf{m} \in \mathbb{Z}^d} c_{\mathbf{k}+N\,\mathbf{m}}(f). \tag{4.57}$$

Thus for $k_j = -\frac{N}{2}, \ldots, \frac{N}{2} - 1$ and $j = 1, \ldots, d$, we have the error estimate

$$|\hat{f}_{\mathbf{k}} - c_{\mathbf{k}}(f)| \le \sum_{\mathbf{m} \in \mathbb{Z}^d \setminus \{\mathbf{0}\}} |c_{\mathbf{k}+N\,\mathbf{m}}(f)|.$$

Proof By Theorem 4.7, the d-dimensional Fourier series of f converges uniformly to f. Hence for all $\mathbf{x} \in \mathbb{T}^d$, we have

$$f(\mathbf{x}) = \sum_{\mathbf{m}\in\mathbb{Z}^d} c_{\mathbf{m}}(f)\, \mathrm{e}^{\mathrm{i}\,\mathbf{m}\cdot\mathbf{x}}\,.$$

In particular for $\mathbf{x} = \frac{2\pi}{N}\,\mathbf{n}$, $\mathbf{n} \in I_N^d$, we obtain

$$f\Big(\frac{2\pi}{N}\,\mathbf{n}\Big) = \sum_{\mathbf{m}\in\mathbb{Z}^d} c_{\mathbf{m}}(f)\, \mathrm{e}^{2\pi\,\mathrm{i}\,(\mathbf{m}\cdot\mathbf{n})/N} = \sum_{\mathbf{m}\in\mathbb{Z}^d} c_{\mathbf{m}}(f)\, w_N^{-\mathbf{m}\cdot\mathbf{n}}\,.$$

Hence due to (4.56) and the pointwise convergence of the Fourier series,

$$\begin{aligned} \hat{f}_{\mathbf{k}} &= \frac{1}{N^d} \sum_{\mathbf{n}\in\mathbf{I}_N} \Big(\sum_{\mathbf{m}\in\mathbb{Z}^d} c_{\mathbf{m}}(f)\, w_N^{-\mathbf{m}\cdot\mathbf{n}} \Big)\, w_N^{\mathbf{k}\cdot\mathbf{n}} \\ &= \frac{1}{N^d} \sum_{\mathbf{m}\in\mathbb{Z}^d} c_{\mathbf{m}}(f) \sum_{\mathbf{n}\in I_N^d} w_N^{(\mathbf{k}-\mathbf{m})\cdot\mathbf{n}}\,, \end{aligned}$$

which yields the aliasing formula (4.57) by Lemma 4.94. ■

Now we sketch the computation of the Fourier transform $\hat{f}$ of a given function $f \in L_1(\mathbb{R}^d) \cap C_0(\mathbb{R}^d)$. Since $f(\mathbf{x}) \to 0$ as $\|\mathbf{x}\|_2 \to \infty$, we obtain for sufficiently large $n \in \mathbb{N}$ that

$$\hat{f}(\boldsymbol{\omega}) = \int_{\mathbb{R}^d} f(\mathbf{x})\, \mathrm{e}^{-\mathrm{i}\,\mathbf{x}\cdot\boldsymbol{\omega}}\, \mathrm{d}\mathbf{x} \approx \int_{[-n\pi,\, n\pi]^d} f(\mathbf{x})\, \mathrm{e}^{-\mathrm{i}\,\mathbf{x}\cdot\boldsymbol{\omega}}\, \mathrm{d}\mathbf{x}\,, \qquad \boldsymbol{\omega} \in \mathbb{R}^d\,.$$

Using the uniform grid $\{\frac{2\pi}{N}\,\mathbf{k} : k_j = -\frac{nN}{2}, \ldots, \frac{nN}{2} - 1;\ j = 1, \ldots, d\}$ of the hypercube $[-n\pi,\, n\pi)^d$ for even $N \in \mathbb{N}$, we receive by the rectangle rule of numerical integration

$$\begin{aligned} &\int_{[-n\pi,\, n\pi]^d} f(\mathbf{x})\, \mathrm{e}^{-\mathrm{i}\,\mathbf{x}\cdot\boldsymbol{\omega}}\, \mathrm{d}\mathbf{x} \\ &\quad\approx \Big(\frac{2\pi}{N}\Big)^d \sum_{k_1=-nN/2}^{nN/2-1} \cdots \sum_{k_d=-nN/2}^{nN/2-1} f\Big(\frac{2\pi}{N}\,\mathbf{k}\Big)\, \mathrm{e}^{-2\pi\mathrm{i}\,(\mathbf{k}\cdot\boldsymbol{\omega})/N}\,. \end{aligned}$$

For $\boldsymbol{\omega} = \frac{1}{n}\,\mathbf{m}$ with $m_j = -\frac{nN}{2}, \ldots, \frac{nN}{2} - 1$ and $j = 1, \ldots, d$, we obtain the following values:

$$\Big(\frac{2\pi}{N}\Big)^d \sum_{k_1=-nN/2}^{nN/2-1} \cdots \sum_{k_d=-nN/2}^{nN/2-1} f\Big(\frac{2\pi}{N}\,\mathbf{k}\Big)\, w_{nN}^{\mathbf{k}\cdot\boldsymbol{\omega}} \approx \hat{f}\Big(\frac{1}{n}\,\mathbf{m}\Big)\,,$$

which can be considered as d-dimensional DFT$(N_1 \times \ldots \times N_d)$ with $N_1 = \ldots = N_d = n\,N$.

4.5.2 Two-Dimensional Discrete Fourier Transforms

Let $N_1,\, N_2 \in \mathbb{N} \setminus \{1\}$ be given, and let $I_{N_j} := \{0, \ldots, N_j - 1\}$ for $j = 1,\, 2$ be the corresponding index sets. The linear map which maps any matrix $\mathbf{A} = (a_{k_1,k_2})_{k_1,k_2=0}^{N_1-1,N_2-1} \in \mathbb{C}^{N_1 \times N_2}$ to the matrix

$$\hat{\mathbf{A}} = (\hat{a}_{n_1,n_2})_{n_1,n_2=0}^{N_1-1,N_2-1} := \mathbf{F}_{N_1}\,\mathbf{A}\,\mathbf{F}_{N_2} \in \mathbb{C}^{N_1 \times N_2}\,,$$

is called *two-dimensional discrete Fourier transform of size* $N_1 \times N_2$ and abbreviated by DFT$(N_1 \times N_2)$. The entries of the transformed matrix $\hat{\mathbf{A}}$ read as follows

$$\hat{a}_{n_1,n_2} = \sum_{k_1=0}^{N_1-1} \sum_{k_2=0}^{N_2-1} a_{k_1,k_2}\, w_{N_1}^{k_1 n_1}\, w_{N_2}^{k_2 n_2}\,, \qquad n_j \in I_{N_j};\; j = 1,\, 2\,. \tag{4.58}$$

If we form the entries (4.58) for all $n_1,\, n_2 \in \mathbb{Z}$, then we observe the *periodicity* of DFT$(N_1 \times N_2)$, i.e., for all $\ell_1,\, \ell_2 \in \mathbb{Z}$, one has

$$\hat{a}_{n_1,n_2} = \hat{a}_{n_1+\ell_1 N_1,n_2+\ell_2 N_2}\,, \qquad n_j \in I_{N_j}\,,\; j = 1,\, 2\,.$$

Remark 4.96 The two-dimensional DFT is of great importance for digital image processing. The light intensity measured by a camera is generally sampled over a rectangular array of pictures elements, so-called pixels. Thus a *digital grayscale image* is a matrix $\mathbf{A} = (a_{k_1,k_2})_{k_1,k_2=0}^{N_1-1,N_2-1}$ of $N_1\,N_2$ pixels $(k_1, k_2) \in I_{N_1} \times I_{N_2}$ and corresponding grayscale values $a_{k_1,k_2} \in \{0,\, 1,\, \ldots,\, 255\}$, where zero means black and 255 is white. Typically, $N_1,\, N_2 \in \mathbb{N}$ are relatively large, for instance, $N_1 = N_2 = 512$.

The *modulus* of the transformed matrix $\hat{\mathbf{A}}$ is given by $|\hat{\mathbf{A}}| := (|\hat{a}_{n_1,n_2}|)_{n_1,n_2=0}^{N_1-1,N_2-1}$ and its *phase* by

$$\operatorname{atan2}(\operatorname{Im}\hat{\mathbf{A}}, \operatorname{Re}\hat{\mathbf{A}}) := \big(\operatorname{atan2}(\operatorname{Im}\hat{a}_{n_1,n_2}, \operatorname{Re}\hat{a}_{n_1,n_2})\big)_{n_1,n_2=0}^{N_1-1,N_2-1}\,,$$

where atan2 is defined in Remark 1.4. In natural images the phase contains important structure information as illustrated in Fig. 4.2. For image sources, we refer to [59] and the databank of the Signal and Image Processing Institute of the University of Southern California (USA). □

For the computation of DFT$(N_1 \times N_2)$ the following simple relation to one-dimensional DFTs is very useful. If the data a_{k_1,k_2} can be factorized as

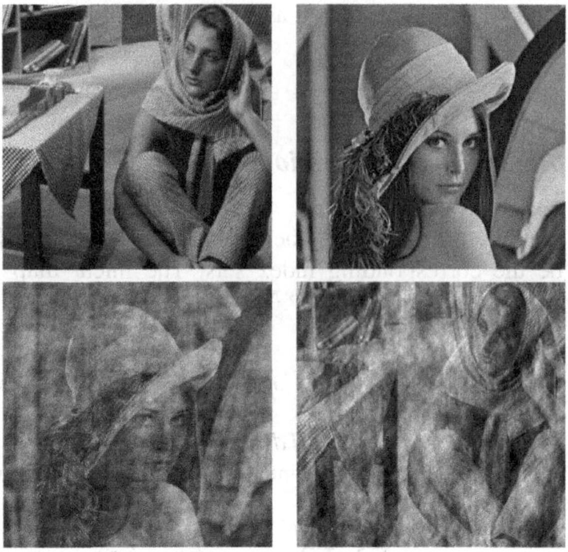

Fig. 4.2 Top: images *Barbara* (left) and *Lena* (right). Bottom: images reconstructed with modulus of *Barbara* and phase of *Lena* (left) and, conversely, with modulus of *Lena* and phase of *Barbara* (right). The phase appears to be dominant with respect to structures

$$a_{k_1,k_2} = b_{k_1}\, c_{k_2}\,, \quad k_j \in I_{N_j};\; j = 1,\, 2\,,$$

then the DFT($N_1 \times N_2$) of $\mathbf{A} = (a_{k_1,k_2})_{k_1,k_2=0}^{N_1-1,N_2-1} = \mathbf{b}\,\mathbf{c}^\top$ reads as follows:

$$\hat{\mathbf{A}} = \mathbf{F}_{N_1} \mathbf{b}\mathbf{c}^\top \mathbf{F}_{N_2}^\top = (\hat{b}_{n_1}\, \hat{c}_{n_2})_{n_1,n_2=0}^{N_1-1,N_2-1}\,, \tag{4.59}$$

where $(\hat{b}_{n_1})_{n_1=0}^{N_1-1}$ is the one-dimensional DFT(N_1) of $\mathbf{b} = (b_{k_1})_{k_1=0}^{N_1-1}$ and $(\hat{c}_{n_2})_{n_2=0}^{N_2-1}$ is the one-dimensional DFT(N_2) of $\mathbf{c} = (c_{k_2})_{k_2=0}^{N-1}$.

Example 4.97 For fixed $s_j \in I_{N_j}$, $j = 1,\, 2$, the sparse matrix

$$\mathbf{A} := \big(\delta_{(k_1-s_1) \bmod N_1}\, \delta_{(k_2-s_2) \bmod N_2}\big)_{k_1,k_2=0}^{N_1-1,N_2-1}$$

is transformed to $\hat{\mathbf{A}} = \big(w_{N_1}^{n_1 s_1}\, w_{N_2}^{n_2 s_2}\big)_{n_1,n_2=0}^{N_1-1,N_2-1}$. Thus we see that a sparse matrix (i.e., a matrix with few nonzero entries) is not transformed to a sparse matrix.

Conversely, the matrix $\mathbf{B} = \big(w_{N_1}^{-s_1 k_1}\, w_{N_2}^{-s_2 k_2}\big)_{k_1,k_2=0}^{N_1-1,N_2-1}$ is mapped to

$$\hat{\mathbf{B}} := N_1\, N_2\, \big(\delta_{(n_1-s_1) \bmod N_1}\, \delta_{(n_2-s_2) \bmod N_2}\big)_{n_1,n_2=0}^{N_1-1,N_2-1}\,.$$

□

Example 4.98 Let $N_1 = N_2 = N \in \mathbb{N} \setminus \{1\}$. We consider the matrix $\mathbf{A} = \big(a_{k_1}\, a_{k_2}\big)_{k_1,k_2=0}^{N-1}$, where a_{k_j} is defined as in Example 3.13 by

$$a_{k_j} := \begin{cases} \frac{1}{2} & k_j = 0\,, \\ \frac{k_j}{N} & k_j = 1, \ldots, N-1\,. \end{cases}$$

Thus by (4.59) and Example 3.13, we obtain the entries of the transformed matrix $\hat{\mathbf{A}}$ by

$$\hat{a}_{n_1,n_2} = \hat{a}_{n_1}\,\hat{a}_{n_2} = -\frac{1}{4}\,\cot\frac{\pi n_1}{N}\,\cot\frac{\pi n_2}{N}\,, \qquad n_j \in I_{N_j};\; j = 1, 2\,.$$

□

By Theorem 3.16, the DFT($N_1 \times N_2$) maps $\mathbb{C}^{N_1 \times N_2}$ one-to-one onto itself. The *inverse* DFT($N_1 \times N_2$) *of size* $N_1 \times N_2$ is given by

$$\mathbf{A} = \mathbf{F}_{N_1}^{-1}\,\hat{\mathbf{A}}\,\mathbf{F}_{N_2}^{-1} = \frac{1}{N_1 N_2}\,\mathbf{J}'_{N_1}\,\mathbf{F}_{N_1}\,\hat{\mathbf{A}}\,\mathbf{F}_{N_2}\,\mathbf{J}'_{N_2}$$

such that

$$a_{k_1,k_2} = \frac{1}{N_1 N_2}\sum_{n_1=0}^{N_1-1}\sum_{n_2=0}^{N_2-1} \hat{a}_{n_1,n_2}\, w_{N_1}^{-k_1 n_1}\, w_{N_2}^{-k_2 n_2}\,, \qquad k_j \in I_{N_j};\; j = 1, 2\,.$$

In practice, one says that the DFT($N_1 \times N_2$) is defined on the *time domain* or *space domain* $\mathbb{C}^{N_1 \times N_2}$. The range of the DFT($N_1 \times N_2$) is called *frequency domain* which is $\mathbb{C}^{N_1 \times N_2}$ too.

In the linear space $\mathbb{C}^{N_1 \times N_2}$, we introduce the *inner product* of two complex matrices $\mathbf{A} = \big(a_{k_1,k_2}\big)_{k_1,k_2=0}^{N_1-1,N_2-1}$ and $\mathbf{B} = \big(b_{k_1,k_2}\big)_{k_1,k_2=0}^{N_1-1,N_2-1}$ by

$$\langle \mathbf{A},\, \mathbf{B}\rangle := \sum_{k_1=0}^{N_1-1}\sum_{k_2=0}^{N_2-1} a_{k_1,k_2}\,\overline{b}_{k_1,k_2}$$

as well as the *Frobenius norm* of $\mathbf{A}$ by

$$\|\mathbf{A}\|_{\mathrm{F}} := \langle \mathbf{A},\, \mathbf{A}\rangle^{1/2} = \Big(\sum_{k_1=0}^{N_1-1}\sum_{k_2=0}^{N_2-1} |a_{k_1,k_2}|^2\Big)^{1/2}\,.$$

Lemma 4.99 *For given* $N_1,\, N_2 \in \mathbb{N} \setminus \{1\}$, *the set of* exponential matrices

$$\mathbf{E}_{m_1,m_2} := \big(w_{N_1}^{-k_1 m_1}\, w_{N_2}^{-k_2 m_2}\big)_{k_1,k_2=0}^{N_1-1,N_2-1}$$

forms an orthogonal basis of $\mathbb{C}^{N_1 \times N_2}$, *where* $\|\mathbf{E}_{m_1,m_2}\|_{\mathrm{F}} = \sqrt{N_1\,N_2}$ *for all* $m_j \in I_{N_j}$ *and* $j = 1, 2$. *Any matrix* $\mathbf{A} \in \mathbb{C}^{N_1 \times N_2}$ *can be represented in the form*

$$\mathbf{A} = \frac{1}{N_1\,N_2} \sum_{m_1=0}^{N_1-1} \sum_{m_2=0}^{N_2-1} \langle \mathbf{A},\, \mathbf{E}_{m_1,m_2}\rangle\, \mathbf{E}_{m_1,m_2}\,,$$

and we have

$$\hat{\mathbf{A}} = \big(\langle \mathbf{A},\, \mathbf{E}_{m_1,m_2}\rangle\big)_{m_1,m_2=0}^{N_1-1,N_2-1}\,.$$

Proof From Lemma 4.94 it follows that for $p_j \in I_{N_j}$, $j = 1,\, 2$,

$$\begin{aligned}
\langle \mathbf{E}_{m_1,m_2},\, \mathbf{E}_{p_1,p_2}\rangle &= \sum_{k_1=0}^{N_1-1} \sum_{k_2=0}^{N_2-1} w_{N_1}^{k_1\,(p_1-m_1)}\, w_{N_2}^{k_2\,(p_2-m_2)} \\
&= N_1 N_2\, \delta_{(m_1-p_1) \bmod N_1}\, \delta_{(m_2-p_2) \bmod N_2} \\
&= \begin{cases} N_1\,N_2 & (m_1,\, m_2) = (p_1,\, p_2)\,, \\ 0 & (m_1,\, m_2) \neq (p_1,\, p_2)\,. \end{cases}
\end{aligned}$$

Further we see that $\|\mathbf{E}_{m_1,m_2}\|_{\mathrm{F}} = \sqrt{N_1\,N_2}$. Since $\dim \mathbb{C}^{N_1\times N_2} = N_1\,N_2$, the set of the $N_1\,N_2$ exponential matrices forms an orthogonal basis of $\mathbb{C}^{N_1\times N_2}$. ■

In addition, we introduce the *cyclic convolution*

$$\mathbf{A} * \mathbf{B} := \bigg(\sum_{k_1=0}^{N_1-1} \sum_{k_2=0}^{N_2-1} a_{k_1,k_2}\, b_{(m_1-k_1) \bmod N_1,\, (m_2-k_2) \bmod N_2}\bigg)_{m_1,m_2=0}^{N_1-1,N_2-1}$$

and the *entrywise product*

$$\mathbf{A} \circ \mathbf{B} := \big(a_{k_1,k_2}\, b_{k_1,k_2}\big)_{k_1,k_2=0}^{N_1-1,N_2-1}\,.$$

In the case $N_1 = N_2 = N$, the cyclic convolution in $\mathbb{C}^{N\times N}$ is a commutative, associative, and distributive operation with the unity $(\delta_{k_1 \bmod N}\, \delta_{k_2 \bmod N})_{k_1,k_2=0}^{N-1}$.

Remark 4.100 The cyclic convolution is of great importance for digital image filtering. Assume that $\mathbf{A} = \big(a_{k_1,k_2}\big)_{k_1,k_2=0}^{N_1-1,\, N_2-1}$ is a given grayscale image. The matrix $\mathbf{G} = \big(g_{k_1,k_2}\big)_{k_1,k_2=0}^{N_1-1,N_2-1}$ with

$$g_{k_1,k_2} := \begin{cases} \frac{1}{4} & (k_1,\, k_2) = (0,\, 0)\,, \\ \frac{1}{8} & (k_1,\, k_2) \in \{((N_1 \pm 1) \bmod N_1,\, 0),\, (0,\, (N_2 \pm 1) \bmod N_2)\}\,, \\ \frac{1}{16} & (k_1,\, k_2) = ((N_1 \pm 1) \bmod N_1,\, (N_2 \pm 1) \bmod N_2)\,, \\ 0 & \text{otherwise} \end{cases}$$

is called *discrete Gaussian filter*. Then $\mathbf{A} * \mathbf{G}$ is the filtered image. Gaussian filtering is used to blur images and to remove noise or details. The matrix $\mathbf{L} = (\ell_{k_1,k_2})_{k_1,k_2=0}^{N_1-1,N_2-1}$ with

$$\ell_{k_1,k_2} := \begin{cases} 4 & (k_1,\, k_2) = (0,\, 0)\,, \\ -2 & (k_1,\, k_2) \in \{((N_1 \pm 1) \bmod N_1,\, 0),\; (0,\, (N_2 \pm 1) \bmod N_2)\}\,, \\ 1 & (k_1,\, k_2) = ((N_1 \pm 1) \bmod N_1,\, (N_2 \pm 1) \bmod N_2)\,, \\ 0 & \text{otherwise} \end{cases}$$

is called *discrete Laplacian filter*. Then $\mathbf{A}*\mathbf{L}$ is the filtered image. Laplacian filtering distinguishes that regions of $\mathbf{A}$ with rapid intensity change and can be used for edge detection. □

The properties of DFT($N_1 \times N_2$) are immediate generalizations of the properties of one-dimensional DFT; see Theorem 3.26.

Theorem 4.101 (Properties of Two-Dimensional DFT($N_1 \times N_2$)) *The* DFT($N_1 \times N_2$) *possesses the following properties:*

1. Linearity: *For all* $\mathbf{A}$, $\mathbf{B} \in \mathbb{C}^{N_1 \times N_2}$ *and* $\alpha \in \mathbb{C}$ *we have*

$$(\mathbf{A} + \mathbf{B})\hat{} = \hat{\mathbf{A}} + \hat{\mathbf{B}}\,, \quad (\alpha\, \mathbf{A})\hat{} = \alpha\, \hat{\mathbf{A}}\,.$$

2. Inversion: *For all* $\mathbf{A} \in \mathbb{C}^{N_1 \times N_2}$ *we have*

$$\mathbf{A} = \mathbf{F}_{N_1}^{-1}\, \hat{\mathbf{A}}\, \mathbf{F}_{N_2}^{-1} = \frac{1}{N_1\, N_2}\, \overline{\mathbf{F}}_{N_1}\, \hat{\mathbf{A}}\, \overline{\mathbf{F}}_{N_2} = \frac{1}{N_1\, N_2}\, \mathbf{J}'_{N_1}\, \mathbf{F}_{N_1}\, \hat{\mathbf{A}}\, \mathbf{F}_{N_2}\, \mathbf{J}'_{N_2}\,.$$

3. Flipping property: *For all* $\mathbf{A} \in \mathbb{C}^{N_1 \times N_2}$ *we have*

$$(\mathbf{J}'_{N_1}\, \mathbf{A}\, \mathbf{J}'_{N_2})\hat{} = \mathbf{J}'_{N_1}\, \hat{\mathbf{A}}\, \mathbf{J}'_{N_2}\,, \quad (\overline{\mathbf{A}})\hat{} = \mathbf{J}'_{N_1}\, \overline{\hat{\mathbf{A}}}\, \mathbf{J}'_{N_2}\,,$$

where $\mathbf{J}'_{N_j}$ *are the flip matrices. Note that*

$$\begin{aligned} \mathbf{J}'_{N_1}\, \mathbf{A}\, \mathbf{J}'_{N_2} &= \big(a_{(N_1-k_1) \bmod N_1,(N_2-k_2) \bmod N_2}\big)_{k_1,k_2=0}^{N_1-1,N_2-1}\,, \\ \mathbf{J}'_{N_1}\, \overline{\mathbf{A}}\, \mathbf{J}'_{N_2} &= \big(\bar{a}_{(N_1-n_1) \bmod N_1,(N_2-n_2) \bmod N_2}\big)_{n_1,n_2=0}^{N_1-1,N_2-1}\,, \\ (\hat{\mathbf{A}})\hat{} = \mathbf{F}_{N_1}\, \hat{\mathbf{A}}\, \mathbf{F}_{N_2} &= N_1 N_2\, \mathbf{J}'_{N_1}\, \mathbf{A}\, \mathbf{J}'_{N_2} \\ &= \big(a_{(N_1-k_1) \bmod N_1,(N_2-k_2) \bmod N_2}\big)_{k_1,k_2=0}^{N_1-1,N_2-1}\,. \end{aligned}$$

4. Shifting in time and frequency domain: *For all* $\mathbf{A} \in \mathbb{C}^{N_1 \times N_2}$ *and fixed* $s_j \in I_{N_j}$, $j = 1,\, 2$, *we have*

$$(\mathbf{V}_{N_1}^{s_1}\,\mathbf{A}\,\mathbf{V}_{N_2}^{s_2})\hat{}= \mathbf{M}_{N_1}^{s_1}\,\hat{\mathbf{A}}\,\mathbf{M}_{N_2}^{s_2}\,, \quad (\mathbf{M}_{N_1}^{-s_1}\,\mathbf{A}\,\mathbf{M}_{N_2}^{-s_2})\hat{}= \mathbf{V}_{N_1}^{s_1}\,\hat{\mathbf{A}}\,\mathbf{V}_{N_2}^{s_2}\,,$$

where $\mathbf{V}_{N_j}$ *are forward-shift matrices and* $\mathbf{M}_{N_j}$ *are modulation matrices. Note that*

$$\begin{aligned}
\mathbf{V}_{N_1}^{s_1}\,\mathbf{A}\,\mathbf{V}_{N_2}^{s_2} &= \big(a_{(k_1-s_1)\bmod N_1,(k_2-s_2)\bmod N_2}\big)_{k_1,k_2=0}^{N_1-1,N_2-1}\,,\\
\mathbf{M}_{N_1}^{s_1}\,\hat{\mathbf{A}}\,\mathbf{M}_{N_2}^{s_2} &= \big(w_{N_1}^{s_1 n_1}\,w_{N_2}^{s_2 n_2}\,\hat{a}_{n_1,n_2}\big)_{n_1,n_2=0}^{N_1-1,N_2-1}\,.
\end{aligned}$$

5. Cyclic convolution in time and frequency domain: *For all* $\mathbf{A},\,\mathbf{B}\in\mathbb{C}^{N_1\times N_2}$ *we have*

$$(\mathbf{A}*\mathbf{B})\hat{}= \hat{\mathbf{A}}\circ\hat{\mathbf{B}}\,, \quad N_1\,N_2\,(\mathbf{A}\circ\mathbf{B})\hat{}= \hat{\mathbf{A}}*\hat{\mathbf{B}}\,.$$

6. Parseval equality: *For all* $\mathbf{A},\,\mathbf{B}\in\mathbb{C}^{N_1\times N_2}$ *we have*

$$\langle\hat{\mathbf{A}},\,\hat{\mathbf{B}}\rangle = N_1\,N_2\,\langle\mathbf{A},\,\mathbf{B}\rangle$$

such that

$$\|\hat{\mathbf{A}}\|_{\mathrm{F}} = \sqrt{N_1\,N_2}\,\|\mathbf{A}\|_{\mathrm{F}}\,.$$

Proof

1. The linearity follows immediately from the definition of DFT$(N_1\times N_2)$.
2. By (3.31) and (3.34) we obtain the inversion property.
3. By (3.31) and (3.34) we have $\mathbf{F}_{N_j}\,\mathbf{J}'_{N_j} = \mathbf{J}'_{N_j}\,\mathbf{F}_{N_j} = \overline{\mathbf{F}}_{N_j}$ for $j=1,\,2$ and hence

$$\begin{aligned}
(\mathbf{J}'_{N_1}\,\mathbf{A}\,\mathbf{J}'_{N_2})\hat{} &= \mathbf{F}_{N_1}\,\mathbf{J}'_{N_1}\,\mathbf{A}\,\mathbf{J}'_{N_2}\,\mathbf{F}_{N_2}\\
&= \mathbf{J}'_{N_1}\,\mathbf{F}_{N_1}\,\mathbf{A}\,\mathbf{F}_{N_2}\,\mathbf{J}'_{N_2} = \mathbf{J}'_{N_1}\,\hat{\mathbf{A}}\,\mathbf{J}'_{N_2}\,.
\end{aligned}$$

Analogously, we receive that

$$(\overline{\mathbf{A}})\hat{}= \mathbf{F}_{N_1}\,\overline{\mathbf{A}}\,\mathbf{F}_{N_2} = \overline{\overline{\mathbf{F}}_{N_1}\,\mathbf{A}\,\overline{\mathbf{F}}_{N_2}} = \overline{\mathbf{J}'_{N_1}\,\mathbf{F}_{N_1}\,\mathbf{A}\,\mathbf{F}_{N_2}\,\mathbf{J}'_{N_2}} = \mathbf{J}'_{N_1}\,\overline{\hat{\mathbf{A}}}\,\mathbf{J}'_{N_2}\,.$$

From Lemma 3.17 it follows immediately that

$$(\hat{\mathbf{A}})\hat{}= \mathbf{F}_{N_1}\,\hat{\mathbf{A}}\,\mathbf{F}_{N_2} = \mathbf{F}_{N_1}^2\,\mathbf{A}\,\mathbf{F}_{N_2}^2 = N_1 N_2\,\mathbf{J}'_{N_1}\,\mathbf{A}\,\mathbf{J}'_{N_2}\,.$$

4. From (3.43) and (3.44) it follows that $\mathbf{F}_{N_j}\,\mathbf{V}_{N_j}^{s_j} = \mathbf{M}_{N_j}^{s_j}\,\mathbf{F}_{N_j}$ and $\mathbf{V}_{N_j}^{s_j}\,\mathbf{F}_{N_j} = \mathbf{F}_{N_j}\,\mathbf{M}_{N_j}^{-s_j}$ and hence

$$\begin{aligned}(\mathbf{V}_{N_1}^{s_1}\,\mathbf{A}\,\mathbf{V}_{N_2}^{s_2})\hat{} &= \mathbf{F}_{N_1}\,\mathbf{V}_{N_1}^{s_1}\,\mathbf{A}\,\mathbf{V}_{N_2}^{s_2}\,\mathbf{F}_{N_2}\\ &= \mathbf{M}_{N_1}^{s_1}\,\mathbf{F}_{N_1}\,\mathbf{A}\,\mathbf{F}_{N_2}\,\mathbf{M}_{N_2}^{s_2} = \mathbf{M}_{N_1}^{s_1}\,\hat{\mathbf{A}}\,\mathbf{M}_{N_2}^{s_2}\,.\end{aligned}$$

The transformed matrix of $\mathbf{M}_{N_1}^{-s_1}\,\mathbf{A}\,\mathbf{M}_{N_2}^{-s_2}$ can be similarly determined.

5. Let $\mathbf{C} = \big(c_{m_1,m_2}\big)_{m_1,m_2=0}^{N_1-1,N_2-1} = \mathbf{A} * \mathbf{B}$ be the cyclic convolution of the matrices $\mathbf{A} = \big(a_{k_1,k_2}\big)_{k_1,k_2=0}^{N_1-1,N_2-1}$ and $\mathbf{B} = \big(b_{k_1,k_2}\big)_{k_1,k_2=0}^{N_1-1,N_2-1}$ with the entries

$$c_{m_1,m_2} = \sum_{k_1=0}^{N_1-1}\sum_{k_2=0}^{N_2-1} a_{k_1,k_2}\, b_{(m_1-k_1) \bmod N_1,\,(m_2-k_2) \bmod N_2}\,.$$

Then we calculate the transformed matrix $\hat{\mathbf{C}} = (\hat{c}_{n_1,n_2})_{n_1,n_2=0}^{N_1-1,N_2-1}$ by

$$\begin{aligned}\hat{c}_{n_1,n_2} &= \sum_{m_1=0}^{N_1-1}\sum_{m_2=0}^{N_2-1}\Big(\sum_{k_1=0}^{N_1-1}\sum_{k_2=0}^{N_2-1} a_{k_1,k_2}\, b_{(m_1-k_1) \bmod N_1,\,(m_2-k_2) \bmod N_2}\Big)\, w_{N_1}^{m_1 n_1}\, w_{N_2}^{m_2 n_2}\\ &= \Big(\sum_{k_1=0}^{N_1-1}\sum_{k_2=0}^{N_2-1} a_{k_1,k_2}\, w_{N_1}^{k_1 n_1}\, w_{N_2}^{k_2 n_2}\Big)\,\hat{\mathbf{b}}_{n_1,n_2} = \hat{\mathbf{a}}_{n_1,n_2}\,\hat{\mathbf{b}}_{n_1,n_2}\,.\end{aligned}$$

The equation $N_1\,N_2\,(\mathbf{A}\circ\mathbf{B})\hat{} = \hat{\mathbf{A}} * \hat{\mathbf{B}}$ can be similarly shown.

6. The entries of the transformed matrices $\hat{\mathbf{A}}$ and $\hat{\mathbf{B}}$ read as follows:

$$\hat{a}_{n_1,n_2} = \sum_{k_1=0}^{N_1-1}\sum_{k_2=0}^{N_2-1} a_{k_1,k_2}\, w_{N_1}^{k_1 n_1}\, w_{N_2}^{k_2 n_2}\,, \qquad \hat{b}_{n_1,n_2} = \sum_{\ell_1=0}^{N_1-1}\sum_{\ell_2=0}^{N_2-1} b_{\ell_1,\ell_2}\, w_{N_1}^{\ell_1 n_1}\, w_{N_2}^{\ell_2 n_2}\,.$$

Applying Lemma 4.94, we obtain

$$\begin{aligned}\langle\hat{\mathbf{A}},\,\hat{\mathbf{B}}\rangle &= \sum_{n_1=0}^{N_1-1}\sum_{n_2=0}^{N_2-1} \hat{a}_{n_1,n_2}\,\overline{\hat{b}}_{n_1,n_2}\\ &= \sum_{k_1=0}^{N_1-1}\sum_{k_2=0}^{N_2-1}\sum_{\ell_1=0}^{N_1-1}\sum_{\ell_2=0}^{N_2-1} a_{k_1,k_2}\,\overline{b}_{\ell_1,\ell_2}\Big(\sum_{n_1=0}^{N_1-1}\sum_{n_2=0}^{N_2-1} w_{N_1}^{(k_1-\ell_1)\,n_1}\, w_{N_2}^{(k_2-\ell_2)\,n_2}\Big)\\ &= N_1 N_2 \sum_{k_1=0}^{N_1-1}\sum_{k_2=0}^{N_2-1} a_{k_1,k_2}\,\overline{b}_{k_1,k_2} = N_1 N_2\,\langle\mathbf{A},\,\mathbf{B}\rangle\,.\end{aligned}$$

■

Remark 4.102 Let $\mathbf{A} = \big(a_{k_1,k_2}\big)_{k_1,k_2=0}^{N_1-1,N_2-1} \in \mathbb{C}^{N_1 \times N_2}$ be an arbitrary matrix. Then from the property (3.33) of the Fourier matrix, it follows immediately that the DFT($N_1 \times N_2$) of the transformed matrix $\hat{\mathbf{A}} = \mathbf{F}_{N_1}\,\mathbf{A}\,\mathbf{F}_{N_2}$ is equal to the scaled flipped matrix, since it holds

$$\begin{aligned}(\hat{\mathbf{A}})\hat{} &= \mathbf{F}_{N_1}\,\hat{\mathbf{A}}\,\mathbf{F}_{N_2} = \mathbf{F}_{N_1}^2\,\mathbf{A}\,\mathbf{F}_{N_2}^2 = N_1 N_2\,\mathbf{J}'_{N_1}\,\mathbf{A}\,\mathbf{J}'_{N_2} \\ &= N_1 N_2\,\big(a_{(N_1-k_1) \bmod N_1,\,(N_2-k_2) \bmod N_2}\big)_{k_1,k_2=0}^{N_1-1,N_2-1}\,.\end{aligned}$$

This property can be used as simple test for correctness. □

Now we analyze the symmetry properties of DFT($N_1 \times N_2$). A matrix $\mathbf{A} = \big(a_{k_1,k_2}\big)_{k_1,k_2=0}^{N_1-1,N_2-1}$ is called *even*, if $\mathbf{A} = \mathbf{J}'_{N_1}\,\mathbf{A}\,\mathbf{J}'_{N_2}$, i.e., for all $k_j \in I_{N_j}$, $j = 1, 2$,

$$a_{k_1,k_2} = a_{(N_1-k_1) \bmod N_1,\,(N_2-k_2) \bmod N_2}\,.$$

A matrix $\mathbf{A}$ is called *odd*, if $\mathbf{A} = -\mathbf{J}'_{N_1}\,\mathbf{A}\,\mathbf{J}'_{N_2}$, i.e., for all $k_j \in I_{N_j}$, $j = 1, 2$,

$$a_{k_1,k_2} = -\,a_{(N_1-k_1) \bmod N_1,\,(N_2-k_2) \bmod N_2}\,.$$

Corollary 4.103 *If* $\mathbf{A} \in \mathbb{R}^{N_1 \times N_2}$ *is real, then* $\hat{\mathbf{A}}$ *has the symmetry property* $\overline{\hat{\mathbf{A}}} = \mathbf{J}'_{N_1}\,\hat{\mathbf{A}}\,\mathbf{J}'_{N_2}$*, i.e., for all* $n_j \in I_{N_j}$*,* $j = 1, 2$*,*

$$\overline{\hat{a}}_{n_1,n_2} = \hat{a}_{(N_1-n_1) \bmod N_1,\,(N_2-n_2) \bmod N_2}\,.$$

In other words, Re $\hat{\mathbf{A}}$ *is even and* Im $\hat{\mathbf{A}}$ *is odd.*

Proof Using $\overline{\mathbf{A}} = \mathbf{A}$ and the flipping property of Theorem 4.101, we obtain

$$\overline{\hat{\mathbf{A}}} = \mathbf{J}'_{N_1}\,\hat{\mathbf{A}}\,\mathbf{J}'_{N_2}\,.$$

From $\hat{\mathbf{A}} = \mathrm{Re}\,\hat{\mathbf{A}} + \mathrm{i}\,\mathrm{Im}\,\hat{\mathbf{A}}$ it follows that

$$\mathrm{Re}\,\hat{\mathbf{A}} - \mathrm{i}\,\mathrm{Im}\,\hat{\mathbf{A}} = \mathbf{J}'_{N_1}\,(\mathrm{Re}\,\hat{\mathbf{A}})\,\mathbf{J}'_{N_2} + \mathrm{i}\,\mathbf{J}'_{N_1}\,(\mathrm{Im}\,\hat{\mathbf{A}})\,\mathbf{J}'_{N_2}\,,$$

i.e., $\mathrm{Re}\,\hat{\mathbf{A}} = \mathbf{J}'_{N_1}\,(\mathrm{Re}\,\hat{\mathbf{A}})\,\mathbf{J}'_{N_2}$ and $\mathrm{Im}\,\hat{\mathbf{A}} = -\mathbf{J}'_{N_1}\,(\mathrm{Im}\,\hat{\mathbf{A}})\,\mathbf{J}'_{N_2}$. ■

Corollary 4.104 *If* $\mathbf{A} \in \mathbb{C}^{N_1 \times N_2}$ *is even/odd, then* $\hat{\mathbf{A}}$ *is even/odd.*

If $\mathbf{A} \in \mathbb{R}^{N_1 \times N_2}$ *is even, then* $\hat{\mathbf{A}} = \mathrm{Re}\,\hat{\mathbf{A}}$ *is even. If* $\mathbf{A} \in \mathbb{R}^{N_1 \times N_2}$ *is odd, then* $\hat{\mathbf{A}} = \mathrm{i}\,\mathrm{Im}\,\hat{\mathbf{A}}$ *is odd.*

Proof From $\mathbf{A} = \pm\mathbf{J}'_{N_1}\,\mathbf{A}\,\mathbf{J}'_{N_2}$ and (3.34) it follows that

$$\hat{\mathbf{A}} = \pm\mathbf{F}_{N_1}\,\mathbf{J}'_{N_1}\,\mathbf{A}\,\mathbf{J}'_{N_2}\,\mathbf{F}_{N_2} = \pm\mathbf{J}'_{N_1}\,\mathbf{F}_{N_1}\,\mathbf{A}\,\mathbf{F}_{N_2}\,\mathbf{J}'_{N_2} = \pm\mathbf{J}'_{N_1}\,\hat{\mathbf{A}}\,\mathbf{J}'_{N_2}\,.$$

For even $\mathbf{A} \in \mathbb{R}^{N_1 \times N_2}$, we obtain by Corollary 4.103 that $\hat{\mathbf{A}} = \mathbf{J}'_{N_1}\, \hat{\mathbf{A}}\, \mathbf{J}'_{N_2}$, i.e., $\hat{\mathbf{A}} \in \mathbb{R}^{N_1 \times N_2}$ is even. Analogously, we can show the assertion for odd $\mathbf{A} \in \mathbb{R}^{N_1 \times N_2}$. ■

Example 4.105 For fixed $s_j \in I_{N_j}$, $j = 1, 2$, we consider the real even matrix

$$\mathbf{A} = \left(\cos 2\pi \left(\frac{s_1 k_1}{N_1} + \frac{s_2 k_2}{N_2}\right)\right)_{k_1,k_2=0}^{N_1-1,N_2-1}.$$

Using Example 4.97 and Euler's formula $\mathrm{e}^{\mathrm{i}x} = \cos x + \mathrm{i}\, \sin x$, we obtain for $\mathbf{A} = \left(\mathrm{Re}\,(w_{N_1}^{-s_1 k_1}\, w_{N_2}^{-s_2 k_2})\right)_{k_1,k_2=0}^{N_1-1,N_2-1}$ the real even transformed matrix

$$\hat{\mathbf{A}} = \frac{N_1\, N_2}{2} \left(\delta_{(n_1-s_1) \bmod N_1}\, \delta_{(n_2-s_2) \bmod N_2} + \delta_{(n_1+s_1) \bmod N_1}\, \delta_{(n_2+s_2) \bmod N_2}\right)_{n_1,n_2=0}^{N_1-1,N_2-1}.$$

Analogously, the real odd matrix

$$\mathbf{B} = \left(\sin 2\pi \left(\frac{s_1 k_1}{N_1} + \frac{s_2 k_2}{N_2}\right)\right)_{k_1,k_2=0}^{N_1-1,N_2-1}$$

possesses the transformed matrix

$$\hat{\mathbf{B}} = \frac{N_1\, N_2}{2\mathrm{i}} \left(\delta_{(n_1-s_1) \bmod N_1}\, \delta_{(n_2-s_2) \bmod N_2} - \delta_{(n_1+s_1) \bmod N_1}\, \delta_{(n_2+s_2) \bmod N_2}\right)_{n_1,n_2=0}^{N_1-1,N_2-1}$$

which is imaginary and odd. □

Finally we describe two simple methods for the *computation of the two-dimensional* DFT *via one-dimensional transforms*. The *first method* reads as follows. If $\mathbf{A} \in \mathbb{C}^{N_1 \times N_2}$ has rank 2 and can be decomposed in the form

$$\mathbf{A} = \mathbf{b}_1\, \mathbf{c}_1^\top + \mathbf{b}_2\, \mathbf{c}_2^\top$$

with $\mathbf{b}_\ell \in \mathbb{C}^{N_1}$ and $\mathbf{c}_\ell \in \mathbb{C}^{N_2}$ for $\ell = 1, 2$, then by (4.59) and the linearity of DFT$(N_1 \times N_2)$, the transformed matrix is equal to

$$\hat{\mathbf{A}} = \hat{\mathbf{b}}_1\, \hat{\mathbf{c}}_1^\top + \hat{\mathbf{b}}_2\, \hat{\mathbf{c}}_2^\top\,,$$

where $\hat{\mathbf{b}}_\ell = \mathbf{F}_{N_1} \mathbf{b}_\ell$ and $\hat{\mathbf{c}}_\ell = \mathbf{F}_{N_2} \mathbf{c}_\ell$ for $\ell = 1, 2$.

In the *second method*, we reshape a matrix $\mathbf{A} = \left(a_{k_1,\,k_2}\right)_{k_1,\,k_2=0}^{N_1-1,N_2-1} \in \mathbb{C}^{N_1 \times N_2}$ into a vector $\mathbf{a} = \left(a_k\right)_{k=0}^{N_1 N_2-1} \in \mathbb{C}^{N_1 N_2}$ by *vectorization* vec : $\mathbb{C}^{N_1 \times N_2} \to \mathbb{C}^{N_1 N_2}$ by $a_{k_1+N_1 k_2} := a_{k_1,\,k_2}$ for $k_j \in I_{N_j}$, $j = 1, 2$. Obviously, vec : $\mathbb{C}^{N_1 \times N_2} \to \mathbb{C}^{N_1 N_2}$ is a linear transform which maps $\mathbb{C}^{N_1 \times N_2}$ one-to-one onto $\mathbb{C}^{N_1 N_2}$. Hence vec is invertible and its inverse map reads $\mathrm{vec}^{-1}\,(a_k)_{k=0}^{N_1N_2-1} = (a_{k_1,k_2})_{k_1,k_2=0}^{N_1-1,N_2-1}$ with $k_1 := k \bmod N_1$ and $k_2 := (k-k_1)/N_1$. For $N_1 = N_2 = 2$, we have

$$\mathrm{vec}\begin{pmatrix} a_{0,0} & a_{0,1} \\ a_{1,0} & a_{1,1} \end{pmatrix} = (a_{0,0}\; a_{1,0}\; a_{0,1}\; a_{1,1})^\top ,$$

$$\mathrm{vec}^{-1}\,(a_0\; a_1\; a_2\; a_3)^\top = \begin{pmatrix} a_0 & a_2 \\ a_1 & a_3 \end{pmatrix} .$$

By Lemma 3.44 we obtain

$$\hat{\mathbf{A}} = \mathrm{vec}^{-1}\,\big((\mathbf{F}_{N_2} \otimes \mathbf{F}_{N_1})\,\mathrm{vec}\,\mathbf{A}\big) .$$

Unfortunately, the one-dimensional transform with transform matrix $\mathbf{F}_{N_2} \otimes \mathbf{F}_{N_1}$ is not a one-dimensional DFT. However, applying Theorem 3.42, we can write

$$\mathbf{F}_{N_2} \otimes \mathbf{F}_{N_1} = (\mathbf{F}_{N_2} \otimes \mathbf{I}_{N_1})(\mathbf{I}_{N_2} \otimes \mathbf{F}_{N_1}).$$

This factorization leads to the *row–column method* ; see Sect. 5.3.5.

4.5.3 Higher-Dimensional Discrete Fourier Transforms

Assume that $d \in \mathbb{N} \setminus \{1, 2\}$ and that $N_j \in \mathbb{N} \setminus \{1\}$ for $j = 1, \ldots, d$ are given. For simplification, we shall use multi-index notations. We introduce the vector $\mathbf{N} := (N_j)_{j=1}^d$, the product $P := N_1 \ldots N_d$, and the d-dimensional index set $I_\mathbf{N} := I_{N_1} \times \cdots \times I_{N_d}$ with $I_{N_j} = \{0, \ldots, N_j - 1\}$. For $\mathbf{k} = (k_j)_{j=1}^d \in \mathbb{Z}^d$, the multiplication $\mathbf{k} \circ \mathbf{N} := (k_j\, N_j)_{j=1}^d$ and the division $\mathbf{k}/\mathbf{N} := (k_j/N_j)_{j=1}^d$ are performed elementwise. Further we set

$$\mathbf{k} \bmod \mathbf{N} := (k_j \bmod N_j)_{j=1}^d .$$

Let $\mathbb{C}^{N_1 \times \ldots \times N_d}$ denote the set of all d-dimensional arrays $\mathbf{A} = (a_\mathbf{k})_{\mathbf{k} \in I_\mathbf{N}}$ of size $N_1 \times \ldots \times N_d$ with $a_\mathbf{k} \in \mathbb{C}$. Note that two-dimensional arrays are matrices.

The linear map from $\mathbb{C}^{N_1 \times \ldots \times N_d}$ into itself, which transforms any d-dimensional array $\mathbf{A} = (a_\mathbf{k})_{\mathbf{k} \in I_\mathbf{N}}$ to an array $\hat{\mathbf{A}} = (\hat{a}_\mathbf{n})_{\mathbf{n} \in I_\mathbf{N}}$ with

$$\hat{a}_\mathbf{n} := \sum_{k_1=0}^{N_1-1} \ldots \sum_{k_d=0}^{N_d-1} a_\mathbf{k}\, w_{N_1}^{k_1 n_1} \ldots w_{N_d}^{k_d n_d} = \sum_{\mathbf{k} \in I_\mathbf{N}} a_\mathbf{k}\, \mathrm{e}^{-2\pi \mathrm{i}\, \mathbf{n} \cdot (\mathbf{k}/\mathbf{N})} \tag{4.60}$$

is called *d-dimensional discrete Fourier transform* of size $N_1 \times \ldots \times N_d$ and abbreviated by DFT($N_1 \times \ldots \times N_d$). If we form the entries (4.60) for all $\mathbf{n} \in \mathbb{Z}^d$, then we observe the *periodicity* of DFT($N_1 \times \ldots \times N_d$), namely, for all $\mathbf{p} \in \mathbb{Z}^d$ and $\mathbf{m} \in I_{\mathbf{N}}$,

$$\hat{a}_{\mathbf{m}} = \hat{a}_{\mathbf{m}+\mathbf{p}\circ\mathbf{N}} \,.$$

The *inverse* DFT($N_1 \times \ldots \times N_d$) maps each d-dimensional array $\mathbf{A} = (a_{\mathbf{n}})_{\mathbf{n}\in I_{\mathbf{N}}}$ to an array $\check{\mathbf{A}} = (\check{a}_{\mathbf{k}})_{\mathbf{k}\in I_{\mathbf{N}}}$ with

$$\check{a}_{\mathbf{k}} := \frac{1}{P} \sum_{n_1=0}^{N_1-1} \ldots \sum_{n_d=0}^{N_d-1} \hat{a}_{\mathbf{n}} w_{N_1}^{-n_1 k_1} \ldots w_{N_d}^{-n_d k_d} = \frac{1}{P} \sum_{\mathbf{n}\in I_{\mathbf{N}}} a_{\mathbf{n}}\, \mathrm{e}^{2\pi \mathrm{i}\, \mathbf{n}\cdot(\mathbf{k}/\mathbf{N})} \,.$$

From Lemma 4.94 it follows that for all $\mathbf{m} \in \mathbb{Z}^d$

$$\sum_{\mathbf{k}\in I_{\mathbf{N}}} \mathrm{e}^{2\pi \mathrm{i}\, \mathbf{m}\cdot(\mathbf{k}/\mathbf{N})} = \begin{cases} P & \mathbf{m}/\mathbf{N} \in \mathbb{Z}^d \,, \\ 0 & \mathbf{m}/\mathbf{N} \notin \mathbb{Z}^d \,. \end{cases}$$

Hence we obtain that for all $\mathbf{A} \in \mathbb{C}^{N_1\times\ldots\times N_d}$

$$\mathbf{A} = (\hat{\mathbf{A}})\check{}= (\check{\mathbf{A}})\hat{}\,.$$

As usually, we define entrywise the addition and the multiplication by a scalar in $\mathbb{C}^{N_1\times\ldots\times N_d}$. Further for d-dimensional arrays $\mathbf{A} = (a_{\mathbf{k}})_{\mathbf{k}\in I_{\mathbf{N}}}$ and $\mathbf{B} = (b_{\mathbf{k}})_{\mathbf{k}\in I_{\mathbf{N}}}$, we consider the *inner product*

$$\langle \mathbf{A},\, \mathbf{B}\rangle := \sum_{\mathbf{k}\in I_{\mathbf{N}}} a_{\mathbf{k}}\, \bar{b}_{\mathbf{k}}$$

and the related norm

$$\|\mathbf{A}\| := \langle \mathbf{A},\, \mathbf{A}\rangle^{1/2} = \Big(\sum_{\mathbf{k}\in I_{\mathbf{N}}} |a_{\mathbf{k}}|^2 \Big)^{1/2} .$$

Then the set of *exponential arrays*

$$\mathbf{E}_{\mathbf{m}} := \big(\mathrm{e}^{2\pi \mathrm{i}\, \mathbf{m}\cdot(\mathbf{k}/\mathbf{N})}\big)_{\mathbf{k}\in I_{\mathbf{N}}} \,, \quad \mathbf{m} \in I_{\mathbf{N}}$$

forms an orthogonal basis of $\mathbb{C}^{N_1\times\ldots\times N_d}$, where $\|\mathbf{E}_{\mathbf{m}}\| = \sqrt{P}$ for all $\mathbf{m} \in I_{\mathbf{N}}$. Any array $\mathbf{A} \in \mathbb{C}^{N_1\times\ldots\times N_d}$ can be represented in the form

$$\mathbf{A} = \frac{1}{P} \sum_{\mathbf{m}\in I_{\mathbf{N}}} \langle \mathbf{A},\, \mathbf{E}_{\mathbf{m}}\rangle\, \mathbf{E}_{\mathbf{m}} \,.$$

Hence $\hat{\mathbf{A}}$ is equal to the array

$$\big(\langle \mathbf{A},\, \mathbf{E_m}\rangle\big)_{\mathbf{m}\in I_{\mathbf{N}}}\,.$$

In addition, we introduce the *cyclic convolution*

$$\mathbf{A} * \mathbf{B} := \Big(\sum_{\mathbf{k}\in I_{\mathbf{N}}} a_{\mathbf{k}}\, b_{(\mathbf{m}-\mathbf{k})\bmod \mathbf{N}}\Big)_{\mathbf{m}\in I_{\mathbf{N}}}$$

and the *entrywise product*

$$\mathbf{A} \circ \mathbf{B} := (a_{\mathbf{k}}\, b_{\mathbf{k}})_{\mathbf{m}\in I_{\mathbf{N}}}$$

in $\mathbb{C}^{N_1\times\ldots\times N_d}$.

The properties of DFT($N_1 \times \ldots \times N_d$) are natural generalizations of Theorem 4.101.

Theorem 4.106 (Properties of d-Dimensional DFT($N_1 \times \ldots \times N_d$)) *The* DFT($N_1 \times \ldots \times N_d$) *possesses the following properties:*

1. Linearity: *For all* $\mathbf{A}, \mathbf{B} \in \mathbb{C}^{N_1\times\ldots\times N_d}$ *and* $\alpha \in \mathbb{C}$, *we have*

$$(\mathbf{A}+\mathbf{B})\hat{} = \hat{\mathbf{A}} + \hat{\mathbf{B}}\,, \quad (\alpha\, \mathbf{A})\hat{} = \alpha\, \hat{\mathbf{A}}\,.$$

2. Inversion: *For all* $\mathbf{A} \in \mathbb{C}^{N_1\times\ldots\times N_d}$, *we have*

$$\mathbf{A} = (\hat{\mathbf{A}})\check{} = (\check{\mathbf{A}})\hat{}.$$

3. Flipping property: *For all* $\mathbf{A} \in \mathbb{C}^{N_1\times\ldots\times N_d}$, *the* DFT($N_1 \times \ldots \times N_d$) *of the flipped array*

$$\big(a_{(\mathbf{N}-\mathbf{k})\bmod \mathbf{N}}\big)_{\mathbf{k}\in I_{\mathbf{N}}}$$

is equal to

$$\big(\hat{a}_{(\mathbf{N}-\mathbf{n})\bmod \mathbf{N}}\big)_{\mathbf{n}\in I_{\mathbf{N}}}\,.$$

The DFT($N_1 \times \ldots \times N_d$) *of the conjugate complex array* $\big(\bar{a}_{\mathbf{k}}\big)_{\mathbf{k}\in I_{\mathbf{N}}}$ *is equal to*

$$\big(\bar{\hat{a}}_{(\mathbf{N}-\mathbf{n})\bmod \mathbf{N}}\big)_{\mathbf{n}\in I_{\mathbf{N}}}\,.$$

4. Shifting in time and frequency domain: *For each* $\mathbf{A} \in \mathbb{C}^{N_1\times\ldots\times N_d}$ *and fixed* $\mathbf{s} \in I_{\mathbf{N}}$, *the* DFT($N_1 \times \ldots \times N_d$) *of the shifted array*

$$\big(a_{(\mathbf{k}-\mathbf{s}) \bmod \mathbf{N}}\big)_{\mathbf{k}\in I_{\mathbf{N}}}$$

is equal to the modulated array

$$\mathbf{E}_{-\mathbf{s}} \circ \hat{\mathbf{A}} = \big(\mathrm{e}^{-2\pi \mathrm{i}\, \mathbf{n}\cdot(\mathbf{s}/\mathbf{N})}\, \hat{a}_{\mathbf{n}}\big)_{\mathbf{n}\in I_{\mathbf{N}}} .$$

Further the DFT($N_1 \times \ldots \times N_d$) *of the modulated array*

$$\mathbf{E}_{\mathbf{s}} \circ \mathbf{A} = \big(\mathrm{e}^{2\pi \mathrm{i}\, \mathbf{k}\cdot(\mathbf{s}/\mathbf{N})}\, a_{\mathbf{k}}\big)_{\mathbf{k}\in I_{\mathbf{N}}}$$

is equal to the shifted array

$$\big(\hat{a}_{(\mathbf{n}-\mathbf{s}) \bmod \mathbf{N}}\big)_{\mathbf{n}\in I_{\mathbf{N}}} .$$

5. Cyclic convolution in time and frequency domain: *For all* $\mathbf{A}$, $\mathbf{B} \in \mathbb{C}^{N_1\times\ldots\times N_d}$ *we have*

$$(\mathbf{A} * \mathbf{B})\hat{} = \hat{\mathbf{A}} \circ \hat{\mathbf{B}} , \quad P\,(\mathbf{A} \circ \mathbf{B})\hat{} = \hat{\mathbf{A}} * \hat{\mathbf{B}} .$$

6. Parseval equality: *For all* $\mathbf{A}$, $\mathbf{B} \in \mathbb{C}^{N_1\times\ldots\times N_d}$ *we have*

$$\langle \hat{\mathbf{A}}, \hat{\mathbf{B}} \rangle = P\, \langle \mathbf{A}, \mathbf{B} \rangle$$

such that

$$\|\hat{\mathbf{A}}\| = \sqrt{P}\, \|\mathbf{A}\| .$$

The proof is similar to that of Theorem 4.101.

Now we describe the symmetry properties of DFT($N_1 \times \ldots \times N_d$). An array $\mathbf{A} = (a_{\mathbf{k}})_{\mathbf{k}\in I_{\mathbf{N}}} \in \mathbb{C}^{N_1\times\ldots\times N_d}$ is called *even*, if we have

$$a_{\mathbf{k}} = a_{(\mathbf{N}-\mathbf{k}) \bmod \mathbf{N}}$$

for all $\mathbf{k} \in I_{\mathbf{N}}$. Analogously, an array $\mathbf{A} = (a_{\mathbf{k}})_{\mathbf{k}\in I_{\mathbf{N}}} \in \mathbb{C}^{N_1\times\ldots\times N_d}$ is called *odd*, if for all $\mathbf{k} \in I_{\mathbf{N}}$,

$$a_{\mathbf{k}} = -a_{(\mathbf{N}-\mathbf{k}) \bmod \mathbf{N}} .$$

Corollary 4.107 *If* $\mathbf{A} \in \mathbb{R}^{N_1\times\ldots\times N_d}$ *is a real array, then the entries of* $\hat{\mathbf{A}} = (\hat{a}_{\mathbf{n}})_{\mathbf{n}\in I_{\mathbf{N}}}$ *possess the symmetry property*

$$\overline{\hat{a}_{\mathbf{n}}} = \hat{a}_{(\mathbf{N}-\mathbf{n}) \bmod \mathbf{N}} , \quad \mathbf{n} \in I_{\mathbf{N}} .$$

In other words, Re $\hat{\mathbf{A}}$ *is even and* Im $\hat{\mathbf{A}}$ *is odd. If* $\mathbf{A} \in \mathbb{C}^{N_1 \times \ldots \times N_d}$ *is even/odd, then* $\hat{\mathbf{A}}$ *is even/odd too.*

This corollary can be similarly shown as Corollary 4.103.

As for the two-dimensional DFT, we can compute the d-dimensional DFT using only one-dimensional transforms. If the entries of the array $\mathbf{A} \in \mathbb{C}^{N_1 \times \ldots \times N_d}$ can be factorized in the form

$$a_{\mathbf{k}} = b_{k_1} \ldots c_{k_d}\,, \quad \mathbf{k} = (k_j)_{j=1}^d \in I_{\mathbf{N}}\,,$$

then the entries the transformed matrix $\hat{\mathbf{A}}$ read as follows

$$\hat{a}_{\mathbf{n}} = \hat{b}_{n_1} \ldots \hat{c}_{n_d}\,, \quad \mathbf{n} = (n_j)_{j=1}^d \in I_{\mathbf{N}}\,,$$

where $(\hat{b}_n)_{n=0}^{N_1-1} = \mathbf{F}_{N_1} (b_k)_{k=0}^{N_1-1}, \ldots, (\hat{c}_n)_{n=0}^{N_d-1} = \mathbf{F}_{N_d} (c_k)_{k=0}^{N_d-1}$ are one-dimensional DFTs.

We can also reshape an array $\mathbf{A} = (a_{\mathbf{k}})_{\mathbf{k} \in I_{\mathbf{N}}} \in \mathbb{C}^{N_1 \times \ldots \times N_d}$ into a vector $\mathbf{a} = (a_k)_{k=0}^{P-1} \in \mathbb{C}^P$ by *vectorization* vec : $\mathbb{C}^{N_1 \times \ldots \times N_d} \to \mathbb{C}^P$ by

$$a_{k_1 + N_1 k_2 + N_1 N_2 k_3 + \ldots + N_1 \ldots N_{d-1} k_d} := a_{\mathbf{k}}\,, \quad \mathbf{k} = (k_j)_{j=1}^d \in I_{\mathbf{N}}\,.$$

Obviously, vec : $\mathbb{C}^{N_1 \times \ldots \times N_d} \to \mathbb{C}^P$ is a linear transform which maps $\mathbb{C}^{N_1 \times \ldots \times N_d}$ one-to-one onto $\mathbb{C}^P$. Hence vec is invertible and its inverse map reads $\text{vec}^{-1} (a_k)_{k=0}^{P-1} = (a_{\mathbf{k}})_{\mathbf{k} \in I_{\mathbf{N}}}$ with $k_1 := k \bmod N_1$ and

$$k_j = \frac{k - k_1 N_1 - \ldots - k_{j-1} N_1 \ldots N_{j-1}}{N_1 \ldots N_{j-1}} \bmod N_j\,, \quad j = 2, \ldots, d\,.$$

By extension of Lemma 3.44, we obtain

$$\hat{\mathbf{A}} = \text{vec}^{-1} \big((\mathbf{F}_{N_d} \otimes \ldots \otimes \mathbf{F}_{N_1}) \,\text{vec}\, \mathbf{A} \big)\,.$$

Thus, the d-dimensional DFT is converted into a matrix-vector product with a P-by-P matrix. This matrix is however different from the Fourier matrix $\mathbf{F}_P$. The Kronecker product of Fourier matrices can be rewritten into a product of d matrices, where each of the matrix factors corresponds to a one-dimensional DFT that has to be applied to subvectors. This approach leads to the generalized row–column method in Sect. 5.3.5.

Chapter 5
Fast Fourier Transforms

As shown in Chap. 3, any application of Fourier methods leads to the evaluation of a discrete Fourier transform of length N (DFT(N)). Thus the efficient computation of DFT(N) is very important. Therefore this chapter treats fast Fourier transforms. A *fast Fourier transform* (FFT) is an algorithm for computing the DFT(N) which needs only a relatively low number of arithmetic operations.

In Sect. 5.1, we summarize the essential construction principles for fast algorithms. Section 5.2 deals with radix–2 FFTs, where N is a power of 2. Here we show three different representations of these algorithms in order to give more insight into their structures. In Sect. 5.3, we derive some further FFTs. In particular, we consider the decomposition approach to reduce a DFT($N_1\, N_2$) to N_1 DFT(N_2) and N_2 DFT(N_1). We also study the radix–4 FFT and the split–radix FFT. For prime N or arbitrary $N \in \mathbb{N}$, the Rader FFT and the Bluestein FFT is considered, being based on the representation of the DFT using cyclic convolutions.

The FFTs considerably reduce the computational cost for computing the DFT(N) from $2\,N^2$ to $\mathcal{O}(N\, \log N)$ arithmetic operations. In Sect. 5.5, we examine the numerical stability of the derived FFT. Note there exists no linear algorithm that can realize the DFT(N) with a smaller computational cost than $\mathcal{O}(N\, \log N)$; see [293]. Faster algorithms can be only derived, if some a priori information on the resulting vector are available. We will consider such approaches in Sect. 5.4.

5.1 Construction Principles of Fast Algorithms

One of the main reasons for the great importance of Fourier methods is the existence of fast algorithms for the implementation of DFT. Nowadays, the FFT is one of the most well-known and mostly applied fast algorithms. Many applications in mathematical physics, engineering, and signal processing were just not possible without FFT.

G. Plonka et al., *Numerical Fourier Analysis*, Applied and Numerical Harmonic Analysis, https://doi.org/10.1007/978-3-031-35005-4_5

A frequently applied FFT is due to J.W. Cooley and J.W. Tukey [94]. Indeed an earlier fast algorithm by I.J. Good [169] used for statistical computations did not find further attention.

Around 1800, C.F. Gauss was very interested in astronomy. Using his least squares method, he determined the orbit of the asteroid Ceres with great accuracy. Later he fitted 12 equidistant data points for the position of the asteroid Pallas by a trigonometric polynomial. The solution of this interpolation problem leads to a DFT of length 12. In order to reduce arithmetical operations, Gauss had been the first who decomposed the DFT of length 12 into DFTs of shorter lengths 3 and 4. This splitting process is the main idea of FFT.

Being interested in trigonometric interpolation problems, C. Runge [370] developed in 1903 fast methods for discrete sine transforms of certain lengths.

But only the development of the computer technology heavily enforced the development of fast algorithms. After deriving the Cooley–Tukey FFT in 1965, many further FFTs emerged being mostly based on similar strategies. We especially mention the Sande–Tukey FFT as another radix–2 FFT, the radix–4 FFT, and the split–radix FFT. While these FFT methods are only suited for length $N = 2^t$ or even $N = 4^t$, other approaches employ cyclic convolutions and can be generalized to other lengths N. For the history of FFT, see [198] or [354, pp. 77–83].

First we want to present five aspects being important for the evaluation and comparison of fast algorithms, namely, computational cost, storage cost, numerical stability, suitability for parallel programming, and needed number of data rearrangements.

1. *Computational cost*

The *computational cost* of an algorithm is determined by the number of floating point operations (flops), i.e., of single (real/complex) additions and (real/complex) multiplications to perform the algorithm. For the considered FFT, we will separately give the number of required additions and multiplications.

Usually, one is only interested in the order of magnitude of the computational cost of an algorithm in dependence of the number of input values and uses the big O notation. For two functions $f,\, g : \mathbb{N} \to \mathbb{R}$ with $f(N) \neq 0$ for all $N \in \mathbb{N}$, we write $g(N) = \mathcal{O}(f(N))$ for $N \to \infty$, if there exists a constant $c > 0$ such that $|g(N)/f(N)| \le c$ holds for all $N \in \mathbb{N}$. By

$$\log_a N = (\log_a b)(\log_b N)\,, \qquad a,\, b > 1\,,$$

we have

$$\mathcal{O}(\log_a N) = \mathcal{O}(\log_b N)\,.$$

Therefore it is usual to write simply $\mathcal{O}(\log N)$ without fixing the base of the logarithm.

2. *Storage cost*

While memory capacities got tremendously cheaper within the last years, it is desired that algorithms require only a memory capacity being in the same order as the size of input data. Therefore we prefer so-called in-place algorithms, where the needed intermediate and final results can be stored by overwriting the input data. Clearly, these algorithms have to be carefully derived, since a later access to the input data or intermediate data is then impossible. Most algorithms that we consider in this chapter can be written as in-place algorithms.

3. *Numerical stability*

Since the evaluations are performed in floating point arithmetic, rounding errors can accumulate essentially during a computation leading to an inaccurate result. In Sect. 5.5 we will show that the FFTs accumulate smaller rounding errors than the direct computation of the DFT using a matrix-vector multiplication.

4. *Parallel programming*

In order to increase the speed of computation, it is of great interest to decompose the algorithm into independent subprocesses such that execution can be carried out simultaneously using multiprocessor systems. The results of these independent evaluations have to be combined afterward upon completion.

The FFT has been shown to be suitable for parallel computing. One approach to efficiently implement the FFT and to represent the decomposition of the FFT into subprocesses is to use signal flow graphs; see Sect. 5.2.

5. *Rearrangements of data*

The computation time of an algorithm mainly depends on the computational cost of the algorithm but also on the data structure as, e.g., the number and complexity of needed data rearrangements.

In practical applications, the simplicity of the implementation of an algorithm plays an important role. Therefore FFTs with a simple and clear data structure are preferred to FFTs with slightly smaller computational cost but requiring more complex data arrangements.

Basic principles for the construction of fast algorithms are:

- the application of recursions,
- the divide-and-conquer technique, and
- parallel programming.

All three principles are applied for the construction of FFTs.

Recursions can be used, if the computation of the final result can be decomposed into consecutive steps, where in the nth step only the intermediate results from the previous r steps are required. Optimally, we need only the information of one previous step to perform the next intermediate result such that an in-place processing is possible.

The *divide-and-conquer technique* is a suitable tool to reduce the execution time of an algorithm. The original problem is decomposed into several subproblems of smaller size but with the same structure. This decomposition is then iteratively applied to decrease the subproblems even further. Obviously, this technique is closely related to the recursion approach. In order to apply the divide-and-conquer technique to construct FFTs, a suitable indexing of the data is needed.

The FFTs can be described in different forms. We will especially consider the sum representation, the representation based on polynomials, and the matrix representation. The original derivation of the FFT by J.W. Cooley and J.W. Tukey [94] applied the sum representation of the DFT(N). For a vector $\mathbf{a} = (a_j)_{j=0}^{N-1} \in \mathbb{C}^N$, the DFT is given by $\hat{\mathbf{a}} = (\hat{a}_k)_{k=0}^{N-1} \in \mathbb{C}^N$ with the *sum representation*

$$\hat{a}_k := \sum_{j=0}^{N-1} a_j\, w_N^{jk}\,, \quad k = 0, \ldots, N-1\,, \quad w_N := \mathrm{e}^{-2\pi \mathrm{i}/N}\,. \tag{5.1}$$

Applying the divide-and-conquer technique, the FFT performs the above summation by iterative evaluation of partial sums.

Employing the *polynomial representation* of the FFT, we interpret the DFT(N) as evaluation of the polynomial

$$a(z) := a_0 + a_1 z + \ldots + a_{N-1} z^{N-1} \in \mathbb{C}[z]$$

at the N knots w_N^k, $k = 0, \ldots, N-1$, i.e.,

$$\hat{a}_k := a(w_N^k)\,, \quad k = 0, \ldots, N-1\,. \tag{5.2}$$

This approach to the DFT is connected with trigonometric interpolation. The FFT is now based on the fast polynomial evaluation by reducing it to the evaluation of polynomials of smaller degrees.

Besides the polynomial arithmetic, the *matrix representation* has been shown to be appropriate for representing fast DFT algorithms. Starting with the matrix representation of the DFT

$$\hat{\mathbf{a}} := \mathbf{F}_N\, \mathbf{a}\,, \tag{5.3}$$

the Fourier matrix $\mathbf{F}_N := (w_N^{jk})_{j,k=0}^{N-1}$ is factorized into a product of sparse matrices. Then the FFT is performed by successive matrix-vector multiplications. This method requires essentially less arithmetical operations than a direct multiplication with the full matrix $\mathbf{F}_N$. The obtained algorithm is recursive, where at the nth step, only the intermediate vector obtained in the previous step is employed.

Beside the three possibilities to describe the FFTs, one tool to show the data structures of the algorithm and to simplify the programming is the signal flow graph. The *signal flow graph* is a directed graph whose vertices represent the intermediate

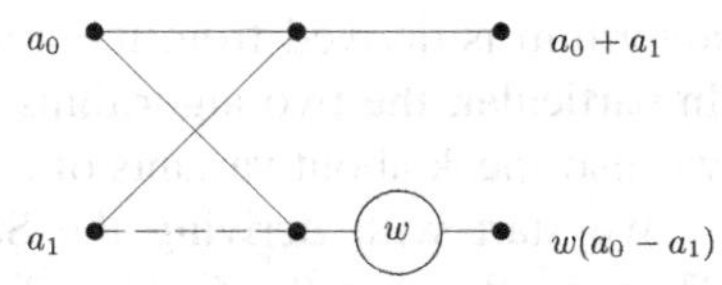

Fig. 5.1 Butterfly signal flow graph

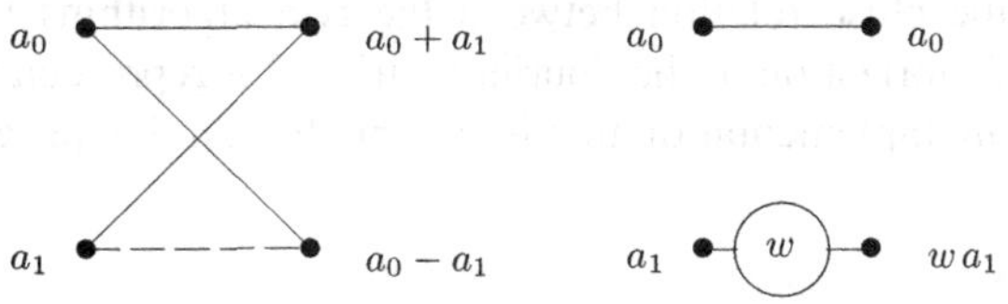

Fig. 5.2 Signal flow graphs of $\mathbf{F}_2\mathbf{a}$ and diag $(1, w)\,\mathbf{a}$

results and whose edges illustrate the arithmetical operations. In this chapter, all signal flow graphs are composed of butterfly forms as presented in Fig. 5.1.

The direction of evaluation in signal flow graphs is always from left to right. In particular, the factorization of the Fourier matrix into sparse matrices with at most two nonzero entries per row and per column can be simply transferred to a signal flow graph. For example, the matrix-vector multiplications

$$\mathbf{F}_2\mathbf{a} = \begin{pmatrix} 1 & 1 \\ 1 & -1 \end{pmatrix} \begin{pmatrix} a_0 \\ a_1 \end{pmatrix}, \quad \mathrm{diag}\,(1, w)\,\mathbf{a} = \begin{pmatrix} 1 & 0 \\ 0 & w \end{pmatrix} \begin{pmatrix} a_0 \\ a_1 \end{pmatrix}$$

with fixed $w \in \mathbb{C}$ can be transferred to the signal flow graphs in Fig. 5.2.

As seen in Chap. 3, most applications use beside the DFT also the inverse DFT such that we need also a fast algorithm for the inverse transform. However, since

$$\mathbf{F}_N^{-1} = \frac{1}{N}\,\mathbf{J}'_N\,\mathbf{F}_N$$

with the flip matrix $\mathbf{J}'_N := \big(\delta_{(j+k) \bmod N}\big)_{j,k=0}^{N-1}$ in Lemma 3.17, each fast algorithm for the DFT(N) also provides a fast algorithm for the inverse DFT(N), and we do not need to consider this case separately.

5.2 Radix–2 FFTs

Radix–2 FFTs are based on the iterative divide-and-conquer technique for computing the DFT(N), if N is a power of 2. The most well-known radix–2 FFTs are the Cooley–Tukey FFT and the Sande–Tukey FFT [94]. These algorithms can be also adapted for parallel processing. The two radix–2 FFTs only differ regarding the order of components of the input and output vector and the order of the multiplication with twiddle factors. As we will see from the corresponding factorization of the Fourier matrix into a product of sparse matrices, the one algorithm is derived from the other by using the transpose of the matrix product. In particular, the two algorithms possess the same computational costs. Therefore we also speak about variants of only one radix–2 FFT.

We start with deriving the Sande–Tukey FFT using the sum representation. Then we develop the Cooley–Tukey FFT in polynomial form. Finally we show the close relation between the two algorithms by examining the corresponding factorization of the Fourier matrix. This representation will be also applied to derive an implementation that is suitable for parallel programming.

5.2.1 *Sande–Tukey FFT in Summation Form*

Assume that $N = 2^t$, $t \in \mathbb{N} \setminus \{1\}$, is given. Then (5.1) implies

$$\begin{aligned} \hat{a}_k &= \sum_{j=0}^{N/2-1} a_j\, w_N^{jk} + \sum_{j=0}^{N/2-1} a_{N/2+j}\, w_N^{(N/2+j)k} \\ &= \sum_{j=0}^{N/2-1} \big(a_j + (-1)^k a_{N/2+j}\big)\, w_N^{jk}\,, \quad k = 0, \ldots, N-1\,. \end{aligned} \tag{5.4}$$

Considering the components of the output vector with even and odd indices, respectively, we obtain

$$\hat{a}_{2k} = \sum_{j=0}^{N/2-1} (a_j + a_{N/2+j})\, w_{N/2}^{jk}\,, \tag{5.5}$$

$$\hat{a}_{2k+1} = \sum_{j=0}^{N/2-1} (a_j - a_{N/2+j})\, w_N^{j}\, w_{N/2}^{jk}\,, \quad k = 0, \ldots, N/2-1\,. \tag{5.6}$$

Thus, using the divide-and-conquer technique, the DFT(N) is obtained by computing

- $N/2$ DFT(2) of the vectors $(a_j, a_{N/2+j})^\top$, $j = 0, \ldots, N/2-1$,
- $N/2$ multiplications with the twiddle factors w_N^j, $j = 0, \ldots, N/2-1$,
- 2 DFT($N/2$) of the vectors $(a_j + a_{N/2+j})_{j=0}^{N/2-1}$ and $\big((a_j - a_{N/2+j})\, w_N^j\big)_{j=0}^{N/2-1}$.

However, we do not evaluate the two DFT($N/2$) directly but apply the decomposition in (5.4) again to the two sums. We iteratively continue this procedure and obtain the desired output vector after t decomposition steps. At each iteration step, we require $N/2$ DFT(2) and $N/2$ multiplications with twiddle factors. As we will show in Sect. 5.2.4, this procedure reduces the computational cost to perform

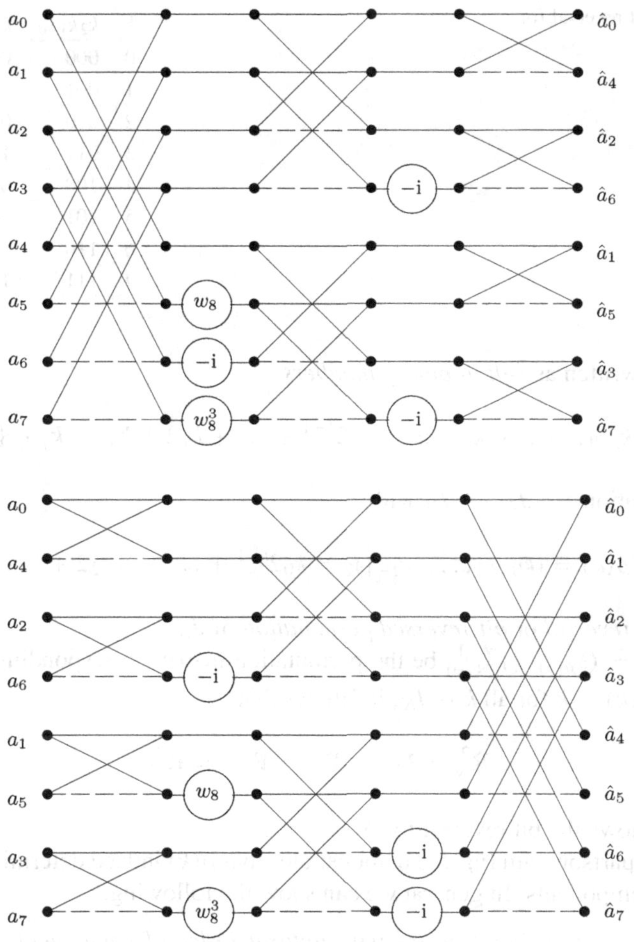

Fig. 5.3 Sande–Tukey algorithm for DFT(8) with input values in natural order (above) and in bit reversal order (below)

the DFT(N) to $\mathcal{O}(N \log N)$. This is an essential improvement! For example, for $N = 512 = 2^9$, the computation cost is reduced by more than 50 times.

The above algorithm is called *Sande–Tukey* FFT. In Fig. 5.3 we show the corresponding signal flow graph of the DFT(8).

The evaluation of $\hat{a}_0 = \sum_{j=0}^{N-1} a_j$ in the Sande–Tukey FFT is obviously executed by *cascade summation*. The signal flow graph well illustrates how to implement an in-place algorithm. Note that the output components are obtained in a different order, which can be described by a permutation of indices.

All indices are in the set

$$J_N := \{0, \ldots, N-1\} = \{0, \ldots, 2^t - 1\}$$

Table 5.1 Bit reversal for $N = 8 = 2^3$

k	$k_2k_1k_0$	$k_0k_1k_2$	$\rho(k)$
0	000	000	0
1	001	100	4
2	010	010	2
3	011	110	6
4	100	001	1
5	101	101	5
6	110	011	3
7	111	111	7

and can be written as *t-digit binary numbers*

$$k = (k_{t-1}, \ldots, k_1, k_0)_2 := k_{t-1}2^{t-1} + \ldots + k_1 2 + k_0 \,, \quad k_j \in \{0, 1\}\,.$$

The permutation $\rho : J_N \to J_N$ with

$$\rho(k) = (k_0, k_1, \ldots, k_{t-1})_2 = k_0 2^{t-1} + \ldots + k_{t-2}2 + k_{t-1}$$

is called *bit reversal* or *bit-reversed permutation* of J_N.

Let $\mathbf{R}_N := (\delta_{\rho(j)-k})_{j,k=0}^{N-1}$ be the permutation matrix corresponding to ρ. Since we have $\rho^2(k) = k$ for all $k \in J_N$, it follows that

$$\mathbf{R}_N^2 = \mathbf{I}_N \,, \quad \mathbf{R}_N = \mathbf{R}_N^{-1} = \mathbf{R}_N^{\top}\,. \tag{5.7}$$

Table 5.1 shows the bit reversal for $N = 8 = 2^3$.

The comparison with Fig. 5.3 demonstrates that $\rho(k)$ indeed determines the order of output components. In general we can show the following.

Lemma 5.1 *For an input vector with natural order of components, the Sande–Tukey* FFT *computes the output components in bit-reversed order.*

Proof We show by induction with respect to t that for $N = 2^t$ with $t \in \mathbb{N} \setminus \{1\}$, the kth value of the output vector is $\hat{a}_{\rho(k)}$.

For $t = 1$, the assertion is obviously correct. Assuming that the assertion holds for $N = 2^t$, we consider now the DFT of length $2N = 2^{t+1}$.

The first step of the algorithm decomposes the DFT($2N$) into two DFT(N), where for $k = 0, \ldots, N-1$, the values $\hat{a}_{2k}$ are computed at the kth position and $\hat{a}_{2k+1}$ at the $(N+k)$th position of the output vector. Afterward the two DFT(N) are independently computed using further decomposition steps of the Sande–Tukey FFT. By induction assumption, we thus obtain after executing the complete algorithm the values $\hat{a}_{2\rho(k)}$ at the kth position and $\hat{a}_{2\rho(k)+1}$ at the $(N+k)$th position of the output vector. The permutation $\pi : J_{2N} \to J_{2N}$ with

$$\pi(k) = 2\,\rho(k)\,, \quad \pi(k+N) = 2\,\rho(k) + 1\,, \quad k = 0, \ldots, N-1\,,$$

is by

$$\begin{aligned}\pi(k) &= \pi((0, k_{t-1}, \ldots, k_0)_2) = 2\,(0, k_0, \ldots, k_{t-2}, k_{t-1})_2 \\ &= (k_0, \ldots, k_{t-1}, 0)_2\,, \\ \pi(N+k) &= \pi((1, k_{t-1}, \ldots, k_0)_2) = 2\,(0, k_0, \ldots, k_{t-2}, k_{t-1})_2 + 1 \\ &= (k_0, \ldots, k_{t-1}, 1)_2\end{aligned}$$

indeed equivalent to the bit reversal of J_{2N}. ■

We summarize the pseudo-code for the Sande–Tukey FFT as follows.

Algorithm 5.2 (Sande–Tukey FFT)
Input: $N = 2^t$ *with* $t \in \mathbb{N} \setminus \{1\}$, $a_j \in \mathbb{C}$ *for* $j = 0, \ldots, N-1$.

```
for n := 1 to t do
  begin m := 2^(t-n+1)
   for ℓ := 0 to 2^(n-1) - 1 do
     begin
      for r := 0 to m/2 - 1 do
       begin j := r + ℓm ,
            s := a_j + a_(m/2+j) ,
            a_(m/2+j) := (a_j - a_(m/2+j)) w_m^r ,
            a_j := s
      end
     end
  end.
```

Output: $\hat{a}_k := a_{\rho(k)} \in \mathbb{C}$, $k = 0, \ldots, N-1$.

5.2.2 *Cooley–Tukey FFT in Polynomial Form*

Next, we derive the *Cooley–Tukey* FFT in polynomial form. In the presentation of the algorithm, we use multi-indices for a better illustration of the order of data. We consider the polynomial $a(z) := a_0 + a_1 z + \ldots + a_{N-1} z^{N-1}$ that has to be evaluated at the N knots $z = w_N^k$, $k = 0, \ldots, N-1$. We decompose the polynomial $a(z)$ as follows

$$a(z) = \sum_{j=0}^{N/2-1} a_j\, z^j + \sum_{j=0}^{N/2-1} a_{N/2+j}\, z^{N/2+j} = \sum_{j=0}^{N/2-1} (a_j + z^{N/2}\, a_{N/2+j})\, z^j\,.$$

By $w_N^{kN/2} = (-1)^k = (-1)^{k_0}$ for all $k \in \{0, \ldots, N-1\}$ with

$$k = (k_{t-1}, \ldots, k_0)_2\,, \quad k_j \in \{0,\ 1\}\,,$$

the term $z^{N/2} = (-1)^k$ can be only 1 or -1. Thus the evaluation of $a(z)$ at $z = w_N^k$, $k = 0, \ldots, N-1$, can be reduced to the evaluation of the two polynomials

$$a^{(i_0)}(z) := \sum_{j=0}^{N/2-1} a_j^{(i_0)} z^j , \quad i_0 = 0,\ 1 ,$$

with the coefficients

$$a_j^{(i_0)} := a_j + (-1)^{i_0}\, a_{N/2+j} , \quad j = 0, \ldots, N/2-1 ,$$

at the $N/2$ knots w_N^k with $k = (k_{t-1}, \ldots, k_1, i_0)_2$. In the first step of the algorithm, we compute the coefficients of the new polynomials $a^{(i_0)}(z)$, $i_0 = 0,\ 1$. Then we apply the method again separately to the two polynomials $a^{(i_0)}(z)$, $i_0 = 0,\ 1$. By

$$a^{(i_0)}(z) := \sum_{j=0}^{N/4-1} \big(a_j^{(i_0)} + z^{N/4} a_{N/4+j}^{(i_0)}\big)\, z^j$$

and $w_N^{kN/4} = (-1)^{k_1}\, (-\mathrm{i})^{k_0}$, this polynomial evaluation is equivalent to the evaluating the four polynomials

$$a^{(i_0,i_1)}(z) := \sum_{j=0}^{N/4-1} a_j^{(i_0,i_1)}\, z^j , \quad i_0,\ i_1 \in \{0,\ 1\} ,$$

with the coefficients

$$a_j^{(i_0,i_1)} := a_j^{(i_0)} + (-1)^{i_1}\, (-\mathrm{i})^{i_0}\, a_{N/4+j}^{(i_0)} , \quad j = 0, \ldots, N/4-1 ,$$

at the $N/4$ knots w_N^k with $k = (k_{t-1}, \ldots, k_2, i_1, i_0)_2$. Therefore, at the second step, we compute the coefficients of $a^{(i_0,i_1)}(z)$, $i_0,\ i_1 \in \{0,\ 1\}$. We iteratively continue the method and obtain after t steps N polynomials of degree 0, i.e., constants that yield the desired output values. At the $(i_0, \ldots, i_{t-1})_2$th position of the output vector, we get

$$a^{(i_0,\ldots,i_{t-1})}(z) = a_0^{(i_0,\ldots,i_{t-1})} = a(w_N^k) = \hat{a}_k , \qquad i_0, \ldots, i_{t-1} \in \{0, 1\},$$

with $k = (i_{t-1}, \ldots, i_0)_2$. Thus, the output values are again in bit-reversed order. Figure 5.4 shows the signal flow graph of the described Cooley–Tukey FFT for $N = 8$.

Remark 5.3 In the Sande–Tukey FFT, the number of output values that can be independently computed doubles at each iteration step, i.e., the sampling rate is

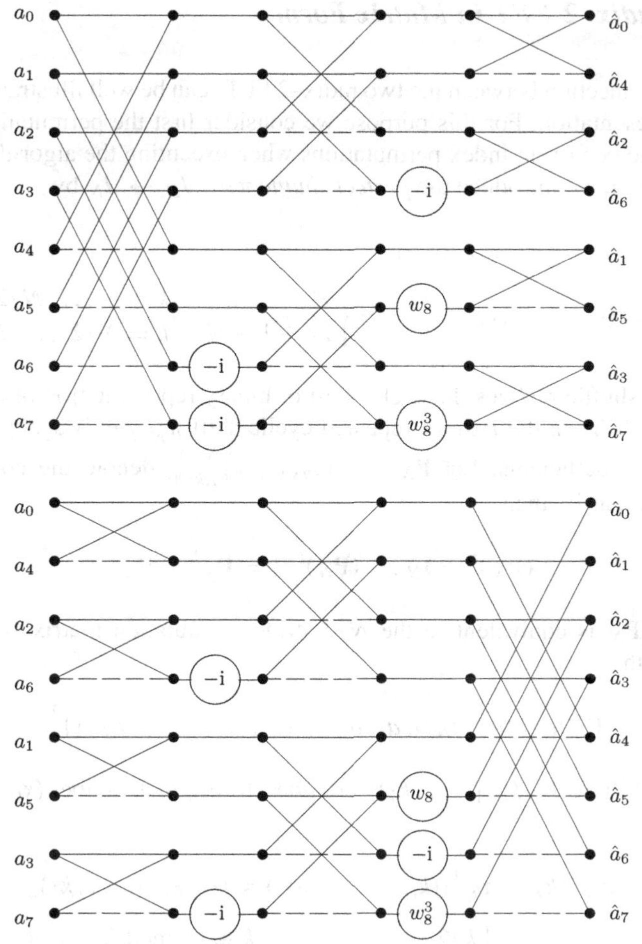

Fig. 5.4 Cooley–Tukey FFT for $N = 8$ with input values in natural order (above) and in bit reversal order (below)

iteratively reduced in frequency domain. Therefore this algorithm is also called *decimation-in-frequency* FFT; see Fig. 5.3. Analogously, the Cooley–Tukey FFT corresponds to reduction of sampling rate in time and is therefore called *decimation-in-time* FFT; see Fig. 5.4. □

5.2.3 Radix–2 FFT in Matrix Form

The close connection between the two radix–2 FFTs can be well illustrated using the matrix representation. For this purpose we consider first the permutation matrices that yield the occurring index permutations when executing the algorithms. Beside the bit reversal, we introduce the *perfect shuffle* $\pi_N : J_N \to J_N$ by

$$\begin{aligned}\pi_N(k) &= \pi_N((k_{t-1}, \ldots, k_0)_2) \\ &= (k_{t-2}, \ldots, k_0, k_{t-1})_2 = \begin{cases} 2k & k = 0, \ldots, N/2 - 1\,, \\ 2k + 1 - N & k = N/2, \ldots, N - 1\,. \end{cases}\end{aligned}$$

The perfect shuffle realizes the cyclic shift of binary representation of the numbers $0, \ldots, N-1$. Then the t-times repeated cyclic shift π_N^t yields again the original order of the coefficients. Let $\mathbf{P}_N := \big(\delta_{\pi_N(j)-k}\big)_{j,k=0}^{N-1}$ denote the corresponding permutation matrix, then

$$(\mathbf{P}_N)^t = \mathbf{I}_N\,, \quad (\mathbf{P}_N)^{t-1} = \mathbf{P}_N^{-1} = \mathbf{P}_N^\top\,. \tag{5.8}$$

Obviously, $\mathbf{P}_N$ is equivalent to the $N/2$-stride permutation matrix considered in Sect. 3.4 with

$$\mathbf{P}_N\,\mathbf{a} = \big(a_0, a_{N/2}, a_1, a_{N/2+1}, \ldots, a_{N/2-1}, a_{N-1}\big)^\top\,.$$

The cyclic shift of $(k_0, k_{t-1}, \ldots, k_1)_2$ provides the original number $(k_{t-1}, \ldots, k_0)_2$, i.e.,

$$\begin{aligned}\pi_N^{-1}(k) &= \pi_N^{-1}((k_{t-1}, \ldots, k_0)_2) = (k_0, k_{t-1}, \ldots, k_1)_2 \\ &= \begin{cases} k/2 & k \equiv 0 \mod 2\,, \\ N/2 + (k-1)/2 & k \equiv 1 \mod 2\,. \end{cases}\end{aligned}$$

Hence, at the first step of the algorithm, $\mathbf{P}_N^{-1} = \mathbf{P}_N^\top$ yields the desired rearrangement of output components $\hat{a}_k$ taking first all even and then all odd indices. Thus, $\mathbf{P}_N^{-1}$ coincides with the even–odd permutation matrix in Sect. 3.4.

Example 5.4 For $N = 8$, i.e., $t = 3$, we obtain

$$\mathbf{P}_8 = \begin{pmatrix} 1\,0\,0\,0\,0\,0\,0\,0 \\ 0\,0\,0\,0\,1\,0\,0\,0 \\ 0\,1\,0\,0\,0\,0\,0\,0 \\ 0\,0\,0\,0\,0\,1\,0\,0 \\ 0\,0\,1\,0\,0\,0\,0\,0 \\ 0\,0\,0\,0\,0\,0\,1\,0 \\ 0\,0\,0\,1\,0\,0\,0\,0 \\ 0\,0\,0\,0\,0\,0\,0\,1 \end{pmatrix}, \quad \mathbf{P}_8 \begin{pmatrix} c_0 \\ c_1 \\ c_2 \\ c_3 \\ c_4 \\ c_5 \\ c_6 \\ c_7 \end{pmatrix} = \begin{pmatrix} c_0 \\ c_4 \\ c_1 \\ c_5 \\ c_2 \\ c_6 \\ c_3 \\ c_7 \end{pmatrix}$$

and

$$\mathbf{P}_8^\top = \begin{pmatrix} 1\,0\,0\,0\,0\,0\,0\,0 \\ 0\,0\,1\,0\,0\,0\,0\,0 \\ 0\,0\,0\,0\,1\,0\,0\,0 \\ 0\,0\,0\,0\,0\,0\,1\,0 \\ 0\,1\,0\,0\,0\,0\,0\,0 \\ 0\,0\,0\,1\,0\,0\,0\,0 \\ 0\,0\,0\,0\,0\,1\,0\,0 \\ 0\,0\,0\,0\,0\,0\,0\,1 \end{pmatrix}, \quad \mathbf{P}_8^\top \begin{pmatrix} c_0 \\ c_1 \\ c_2 \\ c_3 \\ c_4 \\ c_5 \\ c_6 \\ c_7 \end{pmatrix} = \begin{pmatrix} c_0 \\ c_2 \\ c_4 \\ c_6 \\ c_1 \\ c_3 \\ c_5 \\ c_7 \end{pmatrix}.$$

□

The first step of the Sande–Tukey FFT in Algorithm 5.2 is now by (5.5) and (5.6) equivalent to the matrix factorization

$$\mathbf{F}_N = \mathbf{P}_N \, (\mathbf{I}_2 \otimes \mathbf{F}_{N/2}) \, \mathbf{D}_N \, (\mathbf{F}_2 \otimes \mathbf{I}_{N/2}) \tag{5.9}$$

with the diagonal matrix $\mathbf{W}_{N/2} := \operatorname{diag}\, (w_N^j)_{j=0}^{N/2-1}$ and the block diagonal matrix

$$\mathbf{D}_N := \operatorname{diag}\, (\mathbf{I}_{N/2}, \mathbf{W}_{N/2}) = \begin{pmatrix} \mathbf{I}_{N/2} & \\ & \mathbf{W}_{N/2} \end{pmatrix}$$

which is a diagonal matrix too. In (5.9), by $\otimes$ we denote the Kronecker product introduced in Sect. 3.4. At the second step of the decomposition, the factorization is again applied to $\mathbf{F}_{N/2}$. Thus we obtain

$$\mathbf{F}_N = \mathbf{P}_N \left(\mathbf{I}_2 \otimes \left[\mathbf{P}_{N/2} \, (\mathbf{I}_2 \otimes \mathbf{F}_{N/4}) \, \mathbf{D}_{N/2} \, (\mathbf{F}_2 \otimes \mathbf{I}_{N/4})\right]\right) \mathbf{D}_N \, (\mathbf{F}_2 \otimes \mathbf{I}_{N/2})$$

with the diagonal matrices

$$\mathbf{D}_{N/2} := \operatorname{diag}\, (\mathbf{I}_{N/4}, \mathbf{W}_{N/4})\,, \quad \mathbf{W}_{N/4} := \operatorname{diag}\, (w_{N/2}^j)_{j=0}^{N/4-1}\,.$$

Application of Theorem 3.42 yields

$$\mathbf{F}_N = \mathbf{P}_N\,(\mathbf{I}_2 \otimes \mathbf{P}_{N/2})\,(\mathbf{I}_4 \otimes \mathbf{F}_{N/4})\,(\mathbf{I}_2 \otimes \mathbf{D}_{N/2})\,(\mathbf{I}_2 \otimes \mathbf{F}_2 \otimes \mathbf{I}_{N/4})\,\mathbf{D}_N\,(\mathbf{F}_2 \otimes \mathbf{I}_{N/2})\,.$$

After t steps, we thus obtain the factorization of the Fourier matrix $\mathbf{F}_N$ into sparse matrices for the *Sande–Tukey* FFT *with natural order of input components*

$$\begin{aligned}\mathbf{F}_N &= \mathbf{R}_N\,(\mathbf{I}_{N/2} \otimes \mathbf{F}_2)\,(\mathbf{I}_{N/4} \otimes \mathbf{D}_4)\,(\mathbf{I}_{N/4} \otimes \mathbf{F}_2 \otimes \mathbf{I}_2) \\ &\quad \times (\mathbf{I}_{N/8} \otimes \mathbf{D}_8)\,\ldots\,\mathbf{D}_N\,(\mathbf{F}_2 \otimes \mathbf{I}_{N/2}) \\ &= \mathbf{R}_N \prod_{n=1}^{t} \mathbf{T}_n\,(\mathbf{I}_{N/2^n} \otimes \mathbf{F}_2 \otimes \mathbf{I}_{2^{n-1}}) \end{aligned} \tag{5.10}$$

with the permutation matrix $\mathbf{R}_N = \mathbf{P}_N\,(\mathbf{I}_2 \otimes \mathbf{P}_{N/2})\,\ldots\,(\mathbf{I}_{N/4} \otimes \mathbf{P}_4)$ and the diagonal matrices

$$\begin{aligned}\mathbf{T}_n &:= \mathbf{I}_{N/2^n} \otimes \mathbf{D}_{2^n}\,, \\ \mathbf{D}_{2^n} &:= \operatorname{diag}\,(\mathbf{I}_{2^{n-1}}, \mathbf{W}_{2^{n-1}})\,, \quad \mathbf{W}_{2^{n-1}} := \operatorname{diag}\,(w_{2^n}^j)_{j=0}^{2^{n-1}-1}\,.\end{aligned}$$

Note that $\mathbf{T}_1 = \mathbf{I}_N$.

From Lemma 5.1 and by (5.7), we know already that $\mathbf{R}_N$ in (5.10) is the permutation matrix corresponding to the bit reversal. We want to illustrate this fact taking a different view.

Remark 5.5 For distinct indices $j_1, \ldots, j_n \in J_t := \{0, \ldots, t-1\}$, let $(j_1, j_2, \ldots, j_n)$ with $1 \le n < t$ be that permutation of J_t that maps j_1 onto j_2, j_2 onto j_3, …, j_{n-1} onto j_n, and j_n onto j_1. Such a permutation is called *n-cycle*. For $N = 2^t$, the permutations of the index set J_N occurring in a radix–2 FFT can be represented by permutations of the indices in its binary presentation, i.e., $\pi : J_N \to J_N$ can be written as

$$\pi(k) = \pi((k_{t-1}, \ldots, k_0)_2) = (k_{\pi_t(k-1)}, \ldots, k_{\pi_t(0)})_2$$

with a certain permutation $\pi_t : J_t \to J_t$. The perfect shuffle $\pi_N : J_N \to J_N$ corresponds to the t-cycle

$$\pi_{N,t} := (0, \ldots, t-1)$$

and the bit reversal $\rho : J_N \to J_N$ to the permutation

$$\rho_t := \begin{cases} (0, t-1)(1, t-2)\ldots(t/2-1, t/2+1) & t \equiv 0 \mod 2\,, \\ (0, t-1)(1, t-2)\ldots((t-1)/2, (t+1)/2) & t \equiv 1 \mod 2\,. \end{cases}$$

Let $\pi_{N,n} : J_t \to J_t$ with $1 \le n \le t$ be given by the n-cycle

$$\pi_{N,n} := (0, \ldots, n-1).$$

Then we can prove by induction that

$$\rho_t = \pi_{N,t}\,\pi_{N,t-1}\,\dots\,\pi_{N,2}\,.$$

Using the matrix representation, we obtain now the desired relation

$$\mathbf{R}_N = \mathbf{P}_N\,(\mathbf{I}_2 \otimes \mathbf{P}_{N/2})\,\dots\,(\mathbf{I}_{N/4} \otimes \mathbf{P}_4)\,.$$

□

Example 5.6 The factorization of $\mathbf{F}_8$ in (5.10) has the form

$$\mathbf{F}_8 = \mathbf{R}_8\,(\mathbf{I}_4 \otimes \mathbf{F}_2)\,(\mathbf{I}_2 \otimes \mathbf{D}_4)\,(\mathbf{I}_2 \otimes \mathbf{F}_2 \otimes \mathbf{I}_2)\,\mathbf{D}_8\,(\mathbf{F}_2 \otimes \mathbf{I}_4)\,,$$

i.e.,

$$\mathbf{F}_8 = \begin{pmatrix} 1&0&0&0&0&0&0&0\\ 0&0&0&0&1&0&0&0\\ 0&0&1&0&0&0&0&0\\ 0&0&0&0&0&0&1&0\\ 0&1&0&0&0&0&0&0\\ 0&0&0&0&0&1&0&0\\ 0&0&0&1&0&0&0&0\\ 0&0&0&0&0&0&0&1 \end{pmatrix} \begin{pmatrix} 1&1&0&0&0&0&0&0\\ 1&-1&0&0&0&0&0&0\\ 0&0&1&1&0&0&0&0\\ 0&0&1&-1&0&0&0&0\\ 0&0&0&0&1&1&0&0\\ 0&0&0&0&1&-1&0&0\\ 0&0&0&0&0&0&1&1\\ 0&0&0&0&0&0&1&-1 \end{pmatrix}$$

$$\cdot \begin{pmatrix} 1&0&0&0&0&0&0&0\\ 0&1&0&0&0&0&0&0\\ 0&0&1&0&0&0&0&0\\ 0&0&0&-\mathrm{i}&0&0&0&0\\ 0&0&0&0&1&0&0&0\\ 0&0&0&0&0&1&0&0\\ 0&0&0&0&0&0&1&0\\ 0&0&0&0&0&0&0&-\mathrm{i} \end{pmatrix} \begin{pmatrix} 1&0&1&0&0&0&0&0\\ 0&1&0&1&0&0&0&0\\ 1&0&-1&0&0&0&0&0\\ 0&1&0&-1&0&0&0&0\\ 0&0&0&0&1&0&1&0\\ 0&0&0&0&0&1&0&1\\ 0&0&0&0&1&0&-1&0\\ 0&0&0&0&0&1&0&-1 \end{pmatrix}$$

$$\cdot \begin{pmatrix} 1&0&0&0&0&0&0&0\\ 0&1&0&0&0&0&0&0\\ 0&0&1&0&0&0&0&0\\ 0&0&0&1&0&0&0&0\\ 0&0&0&0&1&0&0&0\\ 0&0&0&0&0&w_8&0&0\\ 0&0&0&0&0&0&-\mathrm{i}&0\\ 0&0&0&0&0&0&0&w_8^3 \end{pmatrix} \begin{pmatrix} 1&0&0&0&1&0&0&0\\ 0&1&0&0&0&1&0&0\\ 0&0&1&0&0&0&1&0\\ 0&0&0&1&0&0&0&1\\ 1&0&0&0&-1&0&0&0\\ 0&1&0&0&0&-1&0&0\\ 0&0&1&0&0&0&-1&0\\ 0&0&0&1&0&0&0&-1 \end{pmatrix}.$$

This factorization of $\mathbf{F}_8$ yields the signal flow graph of the Sande–Tukey FFT in Fig. 5.3. □

Using (5.10), we can now derive further factorizations of the Fourier matrix $\mathbf{F}_N$ and obtain corresponding radix–2 FFTs. A new factorization is, e.g., obtained by taking the transpose of (5.10), where we use that $\mathbf{F}_N = \mathbf{F}_N^\top$. Further, we can employ the identity matrix as a new factor that is written as a product of a permutation matrix and its transpose. We finish this subsection by deriving the matrix factorizations of $\mathbf{F}_N$ for the Sande–Tukey FFT with bit-reversed order of input values and for the Cooley–Tukey FFT. In the next subsection, we will show how these slight manipulations of the found Fourier matrix factorization can be exploited for deriving a radix–2 FFT that is suitable for parallel programming.

We recall that by Theorem 3.42

$$\mathbf{P}_N\,(\mathbf{A}\otimes\mathbf{B})\,\mathbf{P}_N^\top = \mathbf{P}_N(N/2)\,(\mathbf{A}\otimes\mathbf{B})\,\mathbf{P}_N(2) = \mathbf{B}\otimes\mathbf{A}\,,$$

where $\mathbf{P}_N(2)$ denotes the even-odd permutation matrix and $\mathbf{P}_N(N/2)$ is the $N/2$-stride permutation matrix. Thus we conclude:

Corollary 5.7 *Let $N = 2^t$. Then we have*

$$\begin{aligned}
\mathbf{P}_N^n\,(\mathbf{I}_{N/2}\otimes\mathbf{F}_2)\,\mathbf{P}_N^{-n} &= \mathbf{I}_{N/2^{n+1}}\otimes\mathbf{F}_2\otimes\mathbf{I}_{2^n}\,, \quad n = 0,\ldots,t-1\,,\\
\mathbf{R}_N\,(\mathbf{I}_{N/2^n}\otimes\mathbf{F}_2\otimes\mathbf{I}_{2^{n-1}})\,\mathbf{R}_N &= \mathbf{I}_{2^{n-1}}\otimes\mathbf{F}_2\otimes\mathbf{I}_{N/2^n}\,, \quad n = 1,\ldots,t\,.
\end{aligned}$$

From (5.10) and Corollary 5.7, we conclude the factorization of $\mathbf{F}_N$ corresponding to the Sande–Tukey FFT with bit-reversed order of input values

$$\mathbf{F}_N = \Big(\prod_{n=1}^{t}\mathbf{T}_n^\rho\,(\mathbf{I}_{2^{n-1}}\otimes\mathbf{F}_2\otimes\mathbf{I}_{N/2^n})\Big)\,\mathbf{R}_N\,, \quad \mathbf{T}_n^\rho := \mathbf{R}_N\,\mathbf{T}_n\,\mathbf{R}_N\,.$$

The matrix factorization corresponding to the Cooley–Tukey FFT is obtained from (5.10) by taking the transpose. From $\mathbf{F}_N = \mathbf{F}_N^\top$ it follows that

$$\mathbf{F}_N = \Big(\prod_{n=1}^{t}(\mathbf{I}_{2^{n-1}}\otimes\mathbf{F}_2\otimes\mathbf{I}_{N/2^n})\,\mathbf{T}_{t-n+1}\Big)\,\mathbf{R}_N\,.$$

This factorization equates the Cooley–Tukey FFT with bit reversal order of input values. By Corollary 5.7 we finally observe that

$$\mathbf{F}_N = \mathbf{R}_N\,\prod_{n=1}^{t}(\mathbf{I}_{N/2^n}\otimes\mathbf{F}_2\otimes\mathbf{I}_{2^{n-1}})\,\mathbf{T}_{t-n+1}^\rho$$

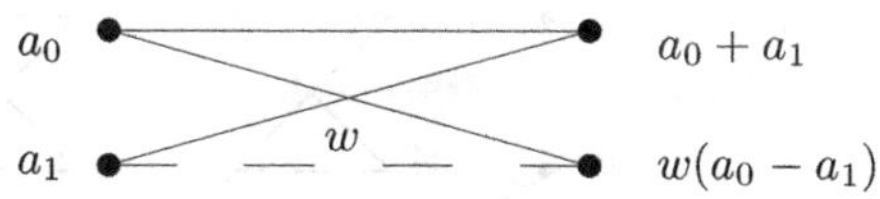

Fig. 5.5 Butterfly signal flow graph

is the matrix factorization of the Cooley–Tukey FFT with natural order of input values.

5.2.4 Radix–2 FFT *for Parallel Programming*

Now we want to consider a radix–2 FFT with respect to parallel programming. For parallel execution of the algorithm, the iteration steps should have the same structure. The common structure of the different steps of the radix–2 FFT consists in the applying $N/2$ butterfly operations that are realized with a convenient Nth root of unity w; see Fig. 5.1. We present the signal flow graph in the following simplified form in Fig. 5.5.

In the factorization of $\mathbf{F}_N$, the $N/2$ butterfly operations correspond to the product of $\tilde{\mathbf{T}}_n\,(\mathbf{I}_{2^{n-1}} \otimes \mathbf{F}_2 \otimes \mathbf{I}_{N/2^n})$ with one intermediate vector. Here, $\tilde{\mathbf{T}}_n$ denotes a suitable diagonal matrix. Since each time two components of the evaluated intermediate vector in one step depend only on two components of the previously computed vector, the $N/2$ butterfly operations can be realized independently. Therefore, these operations can be evaluated in parallel. A radix–2 FFT for parallel programming should satisfy the following requirements:

1. Uniform location of the butterfly operations at each step of the algorithm. In matrix representation, this means that the nth step corresponds to the multiplication of an intermediate vector with a matrix of the form $\tilde{\mathbf{T}}_n\,(\mathbf{I}_{N/2} \otimes \mathbf{F}_2)$. The individual steps should differ only with respect to the diagonal matrices $\tilde{\mathbf{T}}_n$.
2. Uniform data flow between the steps of the algorithm. In matrix representation, this means that the products $\tilde{\mathbf{T}}_n\,(\mathbf{I}_{N/2} \otimes \mathbf{F}_2)$ are always connected by the same permutation matrix.

Now we derive a Sande–Tukey FFT for parallel programming such that its structure satisfies the above requirements. Then the corresponding factorization of the Fourier matrix $\mathbf{F}_N$ is of the form

$$\mathbf{F}_N = \mathbf{Q} \prod_{n=1}^{t} \big(\tilde{\mathbf{T}}_n\,(\mathbf{I}_{N/2} \otimes \mathbf{F}_2)\,\mathbf{P}\big) \tag{5.11}$$

with suitable permutation matrices $\mathbf{P}$ and $\mathbf{Q}$ as well as diagonal matrices $\tilde{\mathbf{T}}_n$. We restrict ourselves to the Sande–Tukey FFT in order to illustrate the essential ideas.

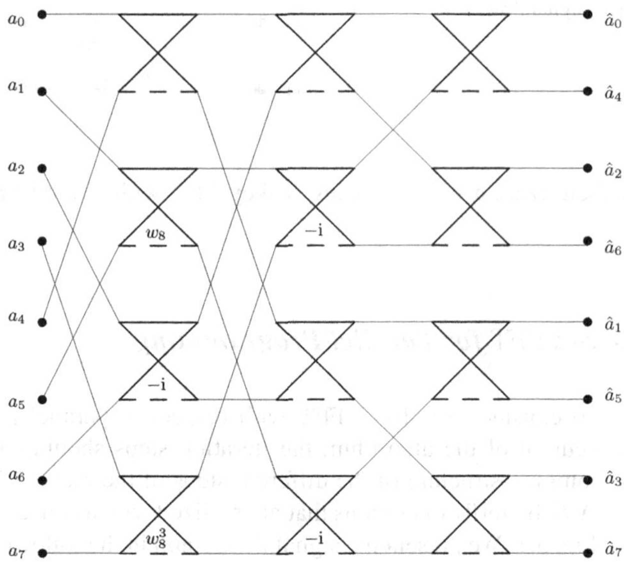

Fig. 5.6 Radix–2 FFT with uniform positions of the butterflies for DFT(8)

The Cooley–Tukey FFT and other FFTs can be treated similarly for parallelization, where we only have to take care of the appropriate diagonal matrices.

Example 5.8 We want to illustrate the approach for the Sande–Tukey FFT for the DFT(8). From the signal flow graph in Fig. 5.3 and the matrix factorization in Example 5.6, it can be seen that the algorithm does not satisfy the two requirements for parallel programming. We reorganize the wanted uniform location of the butterfly operations and obtain the algorithm as illustrated in Fig. 5.6.

The corresponding factorization of the Fourier matrix $\mathbf{F}_8$ is of the form

$$\mathbf{F}_8 = \mathbf{R}_8\,(\mathbf{I}_4 \otimes \mathbf{F}_2)\,(\mathbf{I}_2 \otimes \mathbf{P}_4)\,\mathbf{T}_2\,(\mathbf{I}_4 \otimes \mathbf{F}_2)\,\mathbf{R}_8\,\mathbf{T}_3^{\mathrm{PS}}\,(\mathbf{I}_4 \otimes \mathbf{F}_2)\,\mathbf{P}_8$$

with $\mathbf{T}_3^{\mathrm{PS}} := \mathbf{P}_8\,\mathbf{T}_3\,\mathbf{P}_8^{\top}$. This algorithm still does not satisfy the second requirement of a uniform data flow between the steps of the algorithm. A completely parallelized Sande–Tukey FFT is presented in Fig. 5.7. This algorithm corresponds to the factorization

$$\mathbf{F}_8 = \mathbf{R}_8\,(\mathbf{I}_4 \otimes \mathbf{F}_2)\,\mathbf{P}_8\,\mathbf{T}_2^{\mathrm{PS}}\,(\mathbf{I}_4 \otimes \mathbf{F}_2)\,\mathbf{P}_8\,\mathbf{T}_3^{\mathrm{PS}}\,(\mathbf{I}_4 \otimes \mathbf{F}_2)\,\mathbf{P}_8$$

with $\mathbf{T}_n^{\mathrm{PS}} := \mathbf{P}_8^{-(n-1)}\,\mathbf{T}_3\,\mathbf{P}_8^{n-1}$. The uniform data flow between the algorithm steps is realized by the perfect shuffle permutation matrix $\mathbf{P}_8$. A parallelized algorithm is well suited for hardware implementation with VLSI technology. □

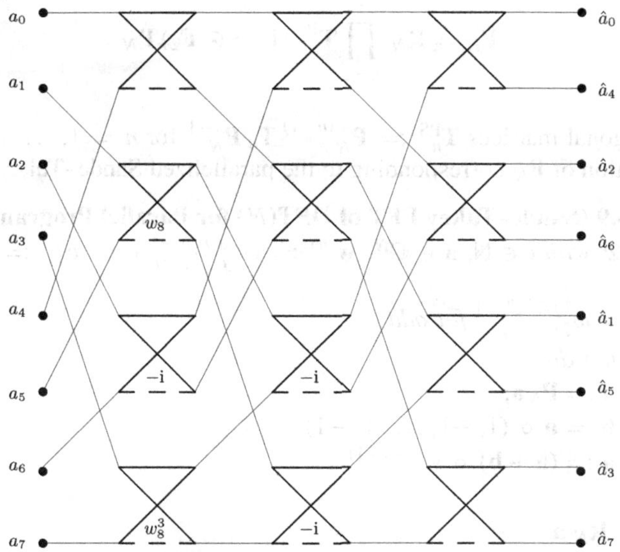

Fig. 5.7 Radix–2 FFT of DFT(8) for parallel programming

Generally, it follows from the factorization (5.10) of the Fourier matrix $\mathbf{F}_N$ and from Corollary 5.7 that

$$\begin{aligned}\mathbf{F}_N &= \mathbf{R}_N \prod_{n=1}^{t} \mathbf{T}_n\,(\mathbf{I}_{N/2^n} \otimes \mathbf{F}_2 \otimes \mathbf{I}_{2^{n-1}})\\ &= \mathbf{R}_N \prod_{n=1}^{t} \mathbf{T}_n\,\mathbf{P}_N^{n-1}\,(\mathbf{I}_{N/2} \otimes \mathbf{F}_2)\,\mathbf{P}_N^{-(n-1)}\\ &= \mathbf{R}_N\,(\mathbf{I}_{N/2} \otimes \mathbf{F}_2)\,\mathbf{T}_2\,\mathbf{P}_N\,(\mathbf{I}_{N/2} \otimes \mathbf{F}_2)\,\ldots\\ &\quad (\mathbf{I}_{N/2} \otimes \mathbf{F}_2)\,\mathbf{P}_N^{-(t-2)}\,\mathbf{T}_t\,\mathbf{P}_N^{t-1}\,(\mathbf{I}_{N/2} \otimes \mathbf{F}_2)\,\mathbf{P}_N^{-(t-1)}\end{aligned}$$

and further with (5.8)

$$\begin{aligned}\mathbf{F}_N &= \mathbf{R}_N\,(\mathbf{I}_{N/2} \otimes \mathbf{F}_2)\,\mathbf{P}_N\,\mathbf{P}_N^{-1}\,\mathbf{T}_2\,\mathbf{P}_N\,(\mathbf{I}_{N/2} \otimes \mathbf{F}_2)\,\ldots\\ &\quad (\mathbf{I}_{N/2} \otimes \mathbf{F}_2)\,\mathbf{P}_N\,\mathbf{P}_N^{-(t-1)}\,\mathbf{T}_t\,\mathbf{P}_N^{t-1}\,(\mathbf{I}_{N/2} \otimes \mathbf{F}_2)\,\mathbf{P}_N\\ &= \mathbf{R}_N \prod_{n=1}^{t} \mathbf{P}_N^{-(n-1)}\,\mathbf{T}_n\,\mathbf{P}_N^{n-1}\,(\mathbf{I}_{N/2} \otimes \mathbf{F}_2)\,\mathbf{P}_N\,,\end{aligned}$$

i.e.,

$$\mathbf{F}_N = \mathbf{R}_N \prod_{n=1}^{t} \mathbf{T}_n^{\mathrm{PS}} \, (\mathbf{I}_{N/2} \otimes \mathbf{F}_2) \, \mathbf{P}_N$$

with the diagonal matrices $\mathbf{T}_n^{\mathrm{PS}} := \mathbf{P}_N^{-(n-1)} \, \mathbf{T}_n \, \mathbf{P}_N^{n-1}$ for $n = 1, \ldots, t$. This yields the factorization of $\mathbf{F}_N$ corresponding to the parallelized Sande–Tukey FFT.

Algorithm 5.9 (Sande–Tukey FFT of DFT(N) for Parallel Programming)
Input: $N = 2^t$ *with* $t \in \mathbb{N}$, $\mathbf{a} \in \mathbb{C}^N$, $\mathbf{w}^{(n)} := (w_j^{(n)})_{j=0}^{N-1}$ *with* $w_j^{(n)} := 1$ *for even* j *and*

$w_j^{(n)} := w_{2^n}^{\lfloor j/2^{t-n+1} \rfloor}$ *for odd* j.

for $n := 1$ *to* t *do*

begin $\mathbf{a} := \mathbf{P}_N \mathbf{a}$;

$\mathbf{b} := \mathbf{a} \circ (1, -1, \ldots, 1, -1)^{\top}$;

$\mathbf{a} := (\mathbf{a} + \mathbf{b}) \circ \mathbf{w}^{(t-n+1)}$

end.

Output: $\hat{\mathbf{a}} := \mathbf{R}_N \, \mathbf{a}$.

For more information, we refer to the subroutine library FFTW; see [153, 154]. A software library for computing FFTs on massively parallel, distributed memory architectures based on the message passing interface (MPI) standard based on [321] is available; see [319].

5.2.5 Computational Cost of Radix–2 FFT

Finally, we consider the computational cost of the radix–2 FFT. We want to evaluate the number of nontrivial real multiplications and real additions.

Remark 5.10 As usually, the product of an arbitrary complex number $a = \alpha_0 + \mathrm{i}\alpha_1$ with $\alpha_0, \alpha_1 \in \mathbb{R} \setminus \{0\}$ and a known complex number $w = \omega_0 + \mathrm{i}\omega_1$ with $\omega_0, \omega_1 \in \mathbb{R} \setminus \{0\}$ requires four real multiplications and two real additions. If the values $\omega_0 \pm \omega_1$ are precomputed, then by

$$\mathrm{Re}(a\,w) = (\alpha_0 + \alpha_1)\,\omega_0 - \alpha_1\,(\omega_0 + \omega_1)\,, \quad \mathrm{Im}(a\,w) = (\alpha_0 + \alpha_1)\,\omega_0 - \alpha_0\,(\omega_0 - \omega_1)$$

we need three real multiplications and three real additions. □

We start with the following observations:

1. Multiplications with ± 1 and $\pm \mathrm{i}$ are trivial and are not taken into account.
2. The multiplication with a primitive 8th root of unity $(\pm 1 \pm \mathrm{i})/\sqrt{2}$ requires two real multiplications and two real additions.

3. The multiplication with the nth primitive root of unity for $n \in \mathbb{N} \setminus \{1, 2, 4, 8\}$ can be performed with an algorithm requiring three real multiplications and three real additions.
4. The addition of two complex numbers requires two real additions.

Let $\mu(t)$ denote the number of real multiplications and $\alpha(t)$ the number of real additions that are needed for executing the radix–2 FFT of the DFT(2^t). Then, by Fig. 5.3, we observe

$$\begin{aligned} \mu(1) &= 0\,, \quad & \alpha(1) &= 2 \cdot 2 & &= 4\,,\\ \mu(2) &= 0\,, \quad & \alpha(2) &= 2 \cdot 2 \cdot 4 & &= 16\,,\\ \mu(3) &= 4\,, \quad & \alpha(3) &= 4 + 2 \cdot 3 \cdot 8 & &= 52\,. \end{aligned}$$

Let now $N = 2^t$ with $t \in \mathbb{N} \setminus \{1\}$ be given. For evaluating the DFT($2N$) with the radix–2 FFT, we have to execute two DFT(N), $2N$ complex additions, and N complex multiplications with the twiddle factors w_{2N}^j, $j = 0, \ldots, N-1$. Among the multiplications with twiddle factors, there are two trivial for $j = 0,\ N/2$ and two multiplications with primitive 8th roots of unity for $j = N/4,\ 3N/4$. Thus for $t \ge 2$, it follows that

$$\mu(t+1) = 2\mu(t) + 3 \cdot 2^t - 8\,, \tag{5.12}$$

$$\alpha(t+1) = 2\alpha(t) + 3 \cdot 2^t - 8 + 2 \cdot 2^{t+1} = 2\alpha(t) + 7 \cdot 2^t - 8\,. \tag{5.13}$$

We want to transfer these recursions into explicit statements. For that purpose, we shortly summarize the theory of linear difference equations, since this method will give us a general way for obtaining explicit numbers for the computational cost.

In the following, we solve a *linear difference equation of order n with constant coefficients* $a_j \in \mathbb{R}$, $j = 0, \ldots, n-1$, where $a_0 \neq 0$, of the form

$$f(t+n) + a_{n-1}\, f(t+n-1) + \ldots + a_1\, f(t+1) + a_0\, f(t) = g(t)\,, \quad t \in \mathbb{N}\,. \tag{5.14}$$

Here $g : \mathbb{N} \to \mathbb{R}$ is a given sequence and $f : \mathbb{N} \to \mathbb{R}$ is the wanted sequence. If $g(t) \equiv 0$, then (5.14) is a *homogeneous* difference equation. We introduce the difference operator

$$Lf(t) := f(t+n) + a_{n-1}\, f(t+n-1) + \ldots + a_1\, f(t+1) + a_0\, f(t)\,, \quad t \in \mathbb{N}\,,$$

and the corresponding *characteristic polynomial*

$$p(\lambda) := \lambda^n + a_{n-1}\, \lambda^{n-1} + \ldots + a_1\, \lambda + a_0\,, \quad \lambda \in \mathbb{C}\,.$$

Obviously, a solution of (5.14) is uniquely determined by the initial values $f(1)$, $f(2), \ldots, f(n)$ or, more generally, by n consecutive sequence components. In the first step, we determine all solutions of the homogeneous difference equation

$Lf(t) = 0$. The ansatz $f(t) := \lambda_1^t$ with $\lambda_1 \neq 0$ provides by

$$L\lambda_1^t = p(\lambda_1)\,\lambda_1^t$$

a nontrivial solution of $Lf(t) = 0$, if and only if λ_1 is a zero of $p(\lambda)$. Let now $\lambda_j \in \mathbb{C}$, $j = 1, \ldots, s$, be distinct zeros of the characteristic polynomial with multiplicities r_j. Taking the kth derivative with regard to λ, we obtain

$$\frac{\mathrm{d}^k}{\mathrm{d}\lambda^k}\lambda^t = k!\binom{t}{k}\lambda^{t-k}\,, \quad 1 \le k \le t\,, \tag{5.15}$$

and the Leibniz product rule yields

$$L\big(\frac{\mathrm{d}^k}{\mathrm{d}\lambda^k}\lambda^t\big) = \frac{\mathrm{d}^k}{\mathrm{d}\lambda^k}(L\lambda^t) = \frac{\mathrm{d}^k}{\mathrm{d}\lambda^k}\big(p(\lambda)\,\lambda^t\big) = \sum_{\ell=0}^{k}\binom{k}{\ell}p^{(\ell)}(\lambda)\,\frac{\mathrm{d}^{k-\ell}}{\mathrm{d}\lambda^{k-\ell}}\lambda^t\,. \tag{5.16}$$

For the r_1-fold zero λ_1, we conclude

$$p(\lambda_1) = p'(\lambda_1) = \ldots = p^{(r_1-1)}(\lambda_1) = 0\,, \quad p^{(r_1)}(\lambda_1) \neq 0$$

and by (5.15) and (5.16), it follows that

$$L\binom{t}{k}\lambda_1^t = 0\,, \quad k = 0, \ldots, r_1 - 1\,.$$

Thus,

$$L(t^k\lambda_1^t) = 0\,, \quad k = 0, \ldots, r_1 - 1\,.$$

If λ_1 is a real number, then $t^k\lambda_1^t$, $k = 0, \ldots, r_1 - 1$, are r_1 real, linearly independent solutions of $Lf(t) = 0$. If $\lambda_1 = \rho_1\,\mathrm{e}^{\mathrm{i}\varphi_1}$ with $\rho_1 > 0$, $0 < \varphi_1 < 2\pi$, and $\varphi_1 \neq \pi$ is a complex r_1-fold zero of $p(\lambda)$, then $\bar{\lambda}_1$ is also an r_1–fold zero. Hence,

$$\mathrm{Re}\,(t^j\lambda_1^t) = t^j\rho_1^t\,\cos(t\varphi_1)\,, \quad \mathrm{Im}\,(t^j\lambda_1^t) = t^j\rho_1^t\,\sin(t\varphi_1)\,, \quad j = 0, \ldots, r_1 - 1,$$

are the $2r_1$ real, linearly independent solutions of $Lf(t) = 0$. In this way we obtain in any case n real, linearly independent solutions of $Lf(t) = 0$. The general solution of $Lf(t) = 0$ is an arbitrary linear combination of these n solutions; see [43].

Using superposition we find the general solution of (5.14) as the sum of the general solution of the homogeneous difference equation $Lf(t) = 0$ and one special solution of (5.14). Often, a special solution of the *inhomogeneous* difference equation $Lf(t) = g(t)$ can be found by the following method. For $g(t) = \alpha\,a^t$ with $\alpha \neq 0$ and $a \neq 1$, we choose in the case $p(a) \neq 0$ the ansatz $f(t) = c\,a^t$ and determine c. If $p(a) = p'(a) = \ldots = p^{(r-1)}(a) = 0$ and $p^{(r)}(a) \neq 0$, then the ansatz $f(t) = c\, t^r\, a^t$ leads to the desired special solution.

If $g(t)$ is a polynomial with $p(1) \neq 0$, then we choose an ansatz with a polynomial of the same degree as $g(t)$. If $p(1) = p'(1) = \ldots = p^{(r-1)}(1) = 0$, $p^{(r)}(1) \neq 0$, then this polynomial is to multiply by t^r.

Example 5.11 We consider the linear difference equation (5.12) of first order,

$$\mu(t+1) - 2\mu(t) = 3 \cdot 2^t - 8\,, \quad t \in \mathbb{N} \setminus \{1\}\,,$$

with the initial condition $\mu(2) = 0$. The corresponding characteristic polynomial $p(\lambda) := \lambda - 2$ possesses the zero $\lambda_1 = 2$, such that $\mu(t) = c\, 2^t$ with arbitrary $c \in \mathbb{R}$ is the general solution of $\mu(t+1) - 2\mu(t) = 0$. To find a special solution of the inhomogeneous difference equation, we set $\mu(t) = c_1\, t\, 2^t + c_2$ and obtain $c_1 = \frac{3}{2}$ and $c_2 = 8$. Thus, the general solution of (5.12) reads

$$\mu(t) = c\, 2^t + \frac{3}{2}\, t\, 2^t + 8\,, \quad c \in \mathbb{R}\,.$$

From the initial condition $\mu(2) = 0$, it follows that $c = -5$. □

To compute now the DFT(N) of length $N = 2^t$, $t \in \mathbb{N}$, with the Cooley–Tukey or Sande–Tukey FFT, we thus require

$$\mu(t) = \frac{3}{2}\, 2^t\, t - 5 \cdot 2^t + 8 = \frac{3}{2}\, N \log_2 N - 5\, N + 8 \tag{5.17}$$

nontrivial real multiplications. Similarly, we conclude from (5.13) the number of real additions

$$\alpha(t) = \frac{7}{2}\, 2^t\, t - 5 \cdot 2^t + 8 = \frac{7}{2}\, N \log_2 N - 5\, N + 8\,. \tag{5.18}$$

We summarize:

Theorem 5.12 *Let $N = 2^t$, $t \in \mathbb{N}$, be given. Then the computational costs of the Cooley–Tukey and Sande–Tukey* FFT *for the* DFT(N) *are equal and amount to*

$$\alpha(t) + \mu(t) = 5\, N \log_2 N - 10\, N + 16$$

real arithmetic operations.

5.3 Other Fast Fourier Transforms

In this section we want to study some further FFTs. On the one hand, we consider techniques that possess even less computational costs than described radix–2 FFTs, and on the other hand we study FFTs for DFT(N), if N is not a power of two.

5.3.1 Chinese Remainder Theorem

Efficient algorithms for computing the DFT can be deduced using the Chinese remainder theorem which was already applied in China about 1700 years ago. The first proof of this theorem was given by L. Euler in 1734. The theorem can be generalized for rings with identity element. In the following we restrict our attention to the ring of integers.

Theorem 5.13 (Chinese Remainder Theorem in $\mathbb{Z}$) *Let $N := N_1 \ldots N_d$ be the product of pairwise coprime numbers $N_j \in \mathbb{N} \setminus \{1\}$, $j = 1, \ldots, d$. Let $r_j \in \mathbb{N}_0$ with $r_j < N_j$, $j = 1, \ldots, d$ be given. Then there exists a uniquely determined integer $r \in \mathbb{N}_0$, $r < N$, with residuals*

$$r \bmod N_j = r_j, \quad j = 1, \ldots, d\,. \tag{5.19}$$

This number can be computed by one of the following methods:

1. *Lagrangian method:*

$$r = \sum_{j=1}^{d} \frac{N}{N_j} t_j r_j \bmod N_j, \quad t_j := \left(\frac{N}{N_j}\right)^{-1} \bmod N_j. \tag{5.20}$$

2. *Newton's method:*

$$r = [r_1] + [r_1 r_2]\, N_1 + \ldots + [r_1 \ldots r_d]\, N_1 \ldots N_{d-1} \tag{5.21}$$

with modular divided differences

$$\begin{aligned}
[r_j] &:= r_j \quad j = 1, \ldots, d\,,\\
\left[r_{j_1} r_{j_2}\right] &:= \frac{[r_{j_2}] - [r_{j_1}]}{N_{j_1}} \bmod N_{j_2}, \quad 1 \le j_1 < j_2 \le d\,,\\
\left[r_{j_1} \ldots r_{j_m}\right] &:= \frac{[r_{j_1} \ldots r_{j_{m-2}} r_{j_m}] - [r_{j_1} \ldots r_{j_{m-2}} r_{j_{m-1}}]}{N_{j_{m-1}}} \bmod N_{j_m},\\
&\qquad 1 \le j_1 < \ldots < j_m \le d\,.
\end{aligned}$$

Note that $(N/N_j)^{-1} \bmod N_j$ and $N_j^{-1} \bmod N_k$ for $j \neq k$ exist, since N_j and N_k, $j \neq k$, are coprime by assumption.

Proof First we show that the numbers in (5.20) and (5.21) fulfill property (5.19). Let $r \in \mathbb{N}_0$ be given by (5.20). Then we have by definition of r for any $j \in \{1, \ldots, d\}$ that

$$r \bmod N_j = \frac{N}{N_j} t_j r_j \bmod N_j = r_j \bmod N_j .$$

Next let $r \in \mathbb{N}_0$ be given (5.21). To show the assertion, we apply induction on the number d of factors of N. The case $d = 1$ is obvious.

Assume that the assertion is true, if N is the product of $d - 1$ pairwise coprime numbers. Then we have by assumption for $N := N_1 \ldots N_{d-2} N_{d-1}$ and $N := N_1 \ldots N_{d-2} N_d$ that

$$\begin{aligned} s := \ & [r_1] + [r_1 r_2]\, N_1 + \ldots + [r_1 \ldots r_{d-2}]\, N_1 \ldots N_{d-3} \\ & + [r_1 \ldots r_{d-2} r_{d-1}]\, N_1 \ldots N_{d-2} , \\ t := \ & [r_1] + [r_1 r_2]\, N_1 + \ldots + [r_1 \ldots r_{d-2}]\, N_1 \ldots N_{d-3} \\ & + [r_1 \ldots r_{d-2} r_d]\, N_1 \ldots N_{d-2} \end{aligned}$$

satisfy

$$s \bmod N_j = r_j , \quad j = 1, \ldots, d-2, d-1 , \tag{5.22}$$

$$t \bmod N_j = r_j , \quad j = 1, \ldots, d-2, d . \tag{5.23}$$

Now let $N := N_1 \ldots N_d$ and $r \in \mathbb{N}_0$ be defined by (5.21). Consider

$$\begin{aligned} \tilde{r} &:= s + (t - s)\, (N_{d-1}^{-1} \bmod N_d)\, N_{d-1} \\ &= s + ([r_1 \ldots r_{d-2} r_d] - [r_1 \ldots r_{d-2} r_{d-1}])\, (N_{d-1}^{-1} \bmod N_d)\, N_1 \ldots N_{d-1} \end{aligned}$$

By definition of the forward difference, we see that $\tilde{r} = r$. Further we obtain by (5.22) that $\tilde{r} \bmod N_j = r_j$, $j = 1, \ldots, d-1$, and by (5.23) that

$$\begin{aligned} \tilde{r} \bmod N_d = \ & s \bmod N_d \\ & + (t \bmod N_d - s \bmod N_d)\, (N_{d-1}^{-1} \bmod N_d)\, (N_{d-1} \bmod N_d) \\ = \ & t \bmod N_d = r_d . \end{aligned}$$

Hence $r \bmod N_j = \tilde{r} \bmod N_j = r_j$, $j = 1, \ldots, d$.

It remains to show that $r \in \mathbb{N}_0$ with $0 \le r < N$ is uniquely determined by its residues r_j, $j = 1, \ldots, d$. Assume that there exists another number $s \in \mathbb{N}_0$ with $0 \le s < N$ and $s \bmod N_j = r_j$ for all $j = 1, \ldots, d$. Then it holds $(r - s) \bmod N_j = 0$,

$j = 1, \ldots, d$. Since the numbers N_j, $j = 1, \ldots, d$ are pairwise coprime, this implies $N \mid r - s$ and consequently $r - s = 0$. ■

Example 5.14 We are searching for the smallest number $r \in \mathbb{N}_0$ with the property

$$r \bmod 4 = 1, \quad r \bmod 9 = 1, \quad r \bmod 25 = 4 .$$

We set $N := 4 \cdot 9 \cdot 25 = 900$. Since

$$(9 \cdot 25)^{-1} \bmod 4 = 1 , \quad (4 \cdot 25)^{-1} \bmod 9 = 1 , \quad (4 \cdot 9)^{-1} \bmod 25 = 16 ,$$

we obtain by the Lagrangian method

$$r = (9 \cdot 25 \cdot 1 \cdot 1 + 4 \cdot 25 \cdot 1 \cdot 1 + 4 \cdot 9 \cdot 16 \cdot 4) \bmod 900 = 829 .$$

Using the following scheme to compute the divided differences:

$$\begin{array}{l|llll} N_1 & [r_1] & [r_1 r_2] = \frac{r_2 - r_1}{N_1} \bmod N_2 & [r_1 r_2 r_3] = \frac{[r_1 r_3] - [r_1 r_2]}{N_2} \bmod N_3 \\ N_2 & [r_2] & [r_1 r_3] = \frac{r_3 - r_1}{N_1} \bmod N_3 & \\ N_3 & [r_3] & & \end{array}$$

that means in our case

$$\begin{array}{r|lll} 4 & 1 & 0 & 23 \\ 9 & 1 & 7 & \\ 25 & 4 & & \end{array}$$

we get by Newton's method

$$\begin{aligned} r &= [r_1] + [r_1 r_2] \, N_1 N_2 + [r_1 r_2 r_3] \, N_1 N_2 \\ &= 1 + 0 \cdot 4 + 23 \cdot 4 \cdot 9 = 829 . \end{aligned}$$

□

The Chinese remainder theorem can be generalized to polynomial rings. In this form, it can be used to design fast algorithms for DFTs; see, e.g., [438]. One can employ the Chinese remainder theorem for index permutations in higher-dimensional DFT algorithms.

5.3.2 *Fast Algorithms for DFT of Composite Length*

First we present the Coley–Tukey FFT for DFT(N), if $N = N_1 N_2$ with $N_r \in \mathbb{N} \setminus \{1\}$, $r = 1, 2$. This algorithm is also called *Gentleman–Sande* FFT; see [159]. As before, the basic idea consists in evaluating the DFT(N) by splitting it into the computation of DFTs of smaller lengths N_1 and N_2 using the divide-and-conquer technique. For a suitable indexing of the input and output components, we employ again a permutation of the index set $J_N := \{0, \ldots, N-1\}$. Let $j_1 := j \mod N_1$ denote the *nonnegative residue modulo* N_1 of $j \in J_N$ and let $j_2 := \lfloor j/N_1 \rfloor$ be the *largest integer being smaller than or equal to* j/N_1. Then we have

$$j = j_1 + j_2\, N_1 \,. \tag{5.24}$$

Analogously, let $k_1 := k \mod N_1$ and $k_2 := \lfloor k/N_1 \rfloor$ for $k \in J_N$ such that $k = k_1 + k_2\, N_1$. We introduce the permutation $\pi : J_N \to J_N$ with

$$\pi(k) = k_1\, N_2 + k_2 \tag{5.25}$$

that we will apply for the new indexing during the evaluation of the DFT(N). Let $\mathbf{P}_N$ be the permutation matrix corresponding to the permutation π of J_N, i.e.,

$$\mathbf{P}_N := \big(\delta_{\pi(j)-k}\big)_{j,k=0}^{N-1}$$

with the Kronecker symbol δ_j. Then, for $\mathbf{a} := (a_j)_{j=0}^{N-1} \in \mathbb{C}^N$ we obtain

$$\mathbf{P}_N\, \mathbf{a} = \big(a_{\pi(j)}\big)_{j=0}^{N-1} \,.$$

Example 5.15 For $N = 6$ with $N_1 = 3$ and $N_2 = 2$, the permutation π of $J_6 := \{0, \ldots, 5\}$ is given by

$$\pi(0) = 0\,,\ \pi(1) = 2\,,\ \pi(2) = 4\,,\ \pi(3) = 1\,,\ \pi(4) = 3\,,\ \pi(5) = 5$$

and corresponds to the permutation matrix

$$\mathbf{P}_6 := \begin{pmatrix} 1\,0\,0\,0\,0\,0 \\ 0\,0\,1\,0\,0\,0 \\ 0\,0\,0\,0\,1\,0 \\ 0\,1\,0\,0\,0\,0 \\ 0\,0\,0\,1\,0\,0 \\ 0\,0\,0\,0\,0\,1 \end{pmatrix} . \tag{5.26}$$

□

Index permutations play an important role in all FFTs. They are the key for applying the divide-and-conquer technique. Further, they form an essential tool for a precise presentation of the component order in input vectors, intermediate vectors, and output vectors of an FFT.

As in the previous section, we can present the FFT in a sum representation, polynomial representation, and matrix factorization. We start by considering the *sum representation* of the Cooley–Tukey FFT. From

$$\hat{a}_k := \sum_{j=0}^{N-1} a_j\, w_N^{jk}\,, \quad k = 0, \ldots, N-1\,,$$

it follows by inserting the indices as in (5.24) and (5.25) that

$$\hat{a}_{k_1 N_2 + k_2} = \sum_{j_1=0}^{N_1-1} \sum_{j_2=0}^{N_2-1} a_{j_1 + j_2 N_1}\, w_N^{(j_1 + j_2 N_1)(k_1 N_2 + k_2)}\,, \quad k_r = 0, \ldots, N_r - 1,\ r = 1,\, 2\,,$$

and by

$$w_N^{(j_1 + j_2 N_1)(k_1 N_2 + k_2)} = w_{N_1}^{j_1 k_1}\, w_N^{j_1 k_2}\, w_{N_2}^{j_2 k_2}$$

further

$$\hat{a}_{k_1 N_2 + k_2} = \sum_{j_1=0}^{N_1-1} w_{N_1}^{j_1 k_1}\, w_N^{j_1 k_2} \sum_{j_2=0}^{N_2-1} a_{j_1 + j_2 N_1}\, w_{N_2}^{j_2 k_2}\,, \quad k_r = 0, \ldots, N_r - 1,\ r = 1,\, 2\,. \tag{5.27}$$

For fixed j_1, the inner sum is equal to a DFT(N_2)

$$b_{j_1 + k_2 N_1} := \sum_{j_2=0}^{N_2-1} a_{j_1 + j_2 N_1}\, w_{N_2}^{j_2 k_2}\,, \quad k_2 = 0, \ldots, N_2 - 1\,. \tag{5.28}$$

Therefore, for each fixed $j_1 = 0, \ldots, N_1 - 1$, we first compute this DFT(N_2). It remains to evaluate

$$\hat{a}_{k_1 N_2 + k_2} = \sum_{j_1=0}^{N_1-1} b_{j_1 + k_2 N_1} w_N^{j_1 k_2} w_{N_1}^{j_1 k_1}\,, \quad k_r = 0, \ldots, N_r - 1,\ r = 1,\, 2\,. \tag{5.29}$$

Now, we first multiply the obtained intermediate values $b_{j_1 + k_2 N_1}$ with the twiddle factors $w_N^{j_1 k_2}$

$$c_{j_1 + k_2 N_1} := b_{j_1 + k_2 N_1}\, w_{N_1 N_2}^{j_1 k_2}\,, \quad j_1 = 0, \ldots N_1 - 1,\ k_2 = 0, \ldots, N_2 - 1\,,$$

and then compute for each fixed $k_2 = 0, \ldots, N_2 - 1$ the DFT(N_1) of the form

$$\hat{a}_{k_1 N_2 + k_2} = \sum_{j_1=0}^{N_1-1} c_{j_1+k_2 N_1}\, w_{N_1}^{j_1 k_1}\,, \quad k_1 = 0, \ldots, N_1 - 1\,.$$

Thus, the original problem to evaluate the DFT($N_1\, N_2$) has been decomposed into evaluating N_1 DFT(N_2) and N_2 DFT(N_1) according to the divide-and-conquer technique. We summarize the algorithm as follows.

Algorithm 5.16 (Fast Algorithm for DFT($N_1\, N_2$))
Input: $N_1, N_2 \in \mathbb{N} \setminus \{1\}$, $a_j \in \mathbb{C}$, $j = 0, \ldots, N_1 N_2 - 1$.

1. *Compute for each* $j_1 = 0, \ldots, N_1 - 1$ *the* DFT(N_2)

$$b_{j_1+k_2 N_1} := \sum_{j_2=0}^{N_2-1} a_{j_1+j_2 N_1}\, w_{N_2}^{j_2 k_2}\,, \quad k_2 = 0, \ldots, N_2 - 1\,.$$

2. *Compute the* $N_1\, N_2$ *products*

$$c_{j_1+k_2 N_1} := b_{j_1+k_2 N_1}\, w_{N_1 N_2}^{j_1 k_2}\,, \quad j_1 = 0, \ldots, N_1 - 1,\ k_2 = 0, \ldots, N_2 - 1\,.$$

3. *Compute for* $k_2 = 0, \ldots, N_2 - 1$ *the* DFT(N_1)

$$\hat{a}_{k_1 N_2 + k_2} := \sum_{j_1=0}^{N_1-1} c_{j_1+k_2 N_1}\, w_{N_1}^{j_1 k_1}\,, \quad k_1 = 0, \ldots, N_1 - 1\,.$$

Output: $\hat{a}_k \in \mathbb{C}$, $k = 0, \ldots, N_1\, N_2 - 1$.

Using the above method, we indeed save arithmetical operations. While the direct computation of the DFT($N_1\, N_2$) requires $N_1^2\, N_2^2$ complex multiplications and $N_1\, N_2\, (N_1\, N_2 - 1)$ complex additions, the application of Algorithm 5.16 needs $N_1\, N_2\, (N_1 + N_2 + 1)$ complex multiplications and $N_1\, N_2\, (N_1 + N_2 - 2)$ complex additions.

If the numbers N_1 and/or N_2 can be further factorized, then the method can be recursively applied to the DFTs of length N_1 and N_2 in Step 1 and Step 3 up to remaining prime numbers. In the special case $N_1\, N_2 = 2^t$ with $t \in \mathbb{N} \setminus \{1\}$, we can choose $N_1 = 2^{t-1}$, $N_2 = 2$. Splitting recursively the first factor again, a radix–2 FFT is obtained in the end.

Let us now derive the *polynomial representation* of (5.27)–(5.29). The computation of the DFT($N_1\, N_2$) is by (5.24) and (5.25) equivalent to evaluating the polynomial

$$a(z) = \sum_{j_1=0}^{N_1-1} \sum_{j_2=0}^{N_2-1} a_{j_1+j_2 N_1}\, z^{j_1+j_2 N_1} = \sum_{j_1=0}^{N_1-1} z^{j_1} \sum_{j_2=0}^{N_2-1} a_{j_1+j_2 N_1}\, z^{j_2 N_1}\,, \quad z \in \mathbb{C}\,,$$

of degree N_1N_2-1 at the N_1N_2 knots $w_{N_1N_2}^{k_1N_2+k_2}$ for $k_r = 0, \ldots, N_r-1, r = 1, 2$. By $w_{N_1N_2}^{(k_1N_2+k_2)j_2N_1} = w_{N_2}^{k_2j_2}$, the term $z^{j_2N_1}$ can take for all N_1N_2 knots at most N_2 different values. Therefore, evaluating $a(z)$ can be reduced to the evaluation of the N_2 polynomials of degree $N_1 - 1$

$$b^{(k_2)}(z) := \sum_{j_1=0}^{N_1-1} b_{j_1}^{(k_2)} z^{j_1}, \quad k_2 = 0, \ldots, N_2 - 1,$$

with the coefficients

$$b_{j_1}^{(k_2)} := \sum_{j_2=0}^{N_2-1} a_{j_1+j_2N_1}\, w_{N_2}^{k_2j_2}, \quad j_1 = 0, \ldots, N_1 - 1,$$

at the N_1 knots $w_{N_1N_2}^{k_1N_2+k_2}$, $k_1 = 0, \ldots, N_1 - 1$. To compute the coefficients using (5.28), i.e., $b_{j_1}^{(k_2)} = b_{j_1+k_2N_1}$, we have to evaluate the N_2 polynomials $b^{(k_2)}(z)$ at each of the N_1 knots (5.29). We summarize this procedure as follows.

Algorithm 5.17 (FFT of DFT(N_1N_2) in Polynomial Representation)
Input: $N_1, N_2 \in \mathbb{N} \setminus \{1\}$, $a_j \in \mathbb{C}$, $j = 0, \ldots, N_1N_2 - 1$.

1. *Compute for each* $j_1 = 0, \ldots, N_1 - 1$ *the* DFT(N_2)

$$b_{j_1}^{(k_2)} := \sum_{j_2=0}^{N_2-1} a_{j_1+j_2N_1}\, w_{N_2}^{j_2k_2}, \quad k_2 = 0, \ldots, N_2 - 1\,.$$

2. *Evaluate each of the* N_2 *polynomials*

$$b^{(k_2)}(z) := \sum_{j_1=0}^{N_1-1} b_{j_1}^{(k_2)}\, z^{j_1}, \quad k_2 = 0, \ldots, N_2 - 1\,,$$

at the N_1 *knots* $w_{N_1N_2}^{k_1N_2+k_2}$, $k_1 = 0, \ldots, N_1 - 1$, *by* DFT(N_1) *and set*

$$\hat{a}_{k_1N_2+k_2} := b^{(k_2)}(w_{N_1N_2}^{k_1N_2+k_2})\,.$$

Output: $\hat{a}_k \in \mathbb{C}$, $k = 0, \ldots N_1N_2 - 1$.

As before, if N_1 or N_2 can be further factorized, we can apply the method recursively and obtain a radix–2 FFT in the special case $N_1N_2 = 2^t$, $t \in \mathbb{N} \setminus \{1\}$.

Finally, we study the *matrix representation* of the Algorithm 5.16 by showing that the three steps of the algorithm correspond to a factorization of the Fourier matrix $\mathbf{F}_{N_1N_2}$ into a product of four sparse matrices. In the first step, we compute the N_1 DFT(N_2) for the N_1 partial vectors $(a_{j_1+j_2N_1})_{j_2=0}^{N_2-1}$, $j_1 = 0, \ldots, N_1 - 1$, of the input vector $\mathbf{a} = (a_j)_{j=0}^{N_1N_2-1} \in \mathbb{C}^{N_1N_2}$. This is equivalent to the matrix-vector multiplication

$$\mathbf{b} = (b_k)_{k=0}^{N_1N_2-1} := (\mathbf{F}_{N_2} \otimes \mathbf{I}_{N_1})\,\mathbf{a}\,.$$

The multiplication of the components of the intermediate vector $\mathbf{b} \in \mathbb{C}^N$ with the twiddle factors in the second step can be equivalently represented by a multiplication with a diagonal matrix $\mathbf{D}_{N_1N_2}$, i.e.,

$$\mathbf{c} = (c_k)_{k=0}^{N_1N_2-1} := \mathbf{D}_{N_1N_2}\,\mathbf{b}\,,$$

where

$$\mathbf{D}_{N_1N_2} := \mathrm{diag}\,(\mathbf{I}_{N_1}, \mathbf{W}_{N_1}, \ldots, \mathbf{W}_{N_1}^{N_2-1}) = \begin{pmatrix} \mathbf{I}_{N_1} & & & \\ & \mathbf{W}_{N_1} & & \\ & & \ddots & \\ & & & \mathbf{W}_{N_1}^{N_2-1} \end{pmatrix}$$

with $\mathbf{W}_{N_1} := \mathrm{diag}\,\big(w_{N_1N_2}^r\big)_{r=0}^{N_1-1}$.

Finally in the third step, we apply N_2 DFT(N_1) to the partial vectors $(c_{j_1+k_2N_1})_{j_1=0}^{N_1-1}$, $k_2 = 0, \ldots, N_2 - 1$, of $\mathbf{c} \in \mathbb{C}^{N_1N_2}$. This can be described by

$$\mathbf{P}_{N_1N_2}\,\hat{\mathbf{a}} = \big(\hat{a}_{\pi(\ell)}\big)_{\ell=0}^{N_1N_2-1} := (\mathbf{I}_{N_2} \otimes \mathbf{F}_{N_1})\,\mathbf{c}\,.$$

Here, π denotes the permutation in (5.25) of output indices and $\mathbf{P}_{N_1N_2}$ is the corresponding permutation matrix.

In summary, the Algorithm 5.16 corresponds to the following factorization of the Fourier matrix:

$$\mathbf{F}_{N_1N_2} = \mathbf{P}_{N_1N_2}^{\top}\,(\mathbf{I}_{N_2} \otimes \mathbf{F}_{N_1})\,\mathbf{D}_{N_1N_2}\,(\mathbf{F}_{N_2} \otimes \mathbf{I}_{N_1})\,. \tag{5.30}$$

Example 5.18 We consider the case $N = 6$ with $N_1 = 3$ and $N_2 = 2$. Then it follows from (5.30) that

$$\mathbf{F}_6 = \mathbf{P}_6^{\top}\,(\mathbf{I}_2 \otimes \mathbf{F}_3)\,\mathbf{D}_6\,(\mathbf{F}_2 \otimes \mathbf{I}_3) =$$

$$\mathbf{P}_6^\top \begin{pmatrix} 1 & 1 & 1 & 0 & 0 & 0 \\ 1 & w_3 & w_3^2 & 0 & 0 & 0 \\ 1 & w_3^2 & w_3 & 0 & 0 & 0 \\ 0 & 0 & 0 & 1 & 1 & 1 \\ 0 & 0 & 0 & 1 & w_3 & w_3^2 \\ 0 & 0 & 0 & 1 & w_3^2 & w_3 \end{pmatrix} \begin{pmatrix} 1 & 0 & 0 & 0 & 0 & 0 \\ 0 & 1 & 0 & 0 & 0 & 0 \\ 0 & 0 & 1 & 0 & 0 & 0 \\ 0 & 0 & 0 & 1 & 0 & 0 \\ 0 & 0 & 0 & 0 & w_6 & 0 \\ 0 & 0 & 0 & 0 & 0 & w_3 \end{pmatrix} \begin{pmatrix} 1 & 0 & 0 & 1 & 0 & 0 \\ 0 & 1 & 0 & 0 & 1 & 0 \\ 0 & 0 & 1 & 0 & 0 & 1 \\ 1 & 0 & 0 & -1 & 0 & 0 \\ 0 & 1 & 0 & 0 & -1 & 0 \\ 0 & 0 & 1 & 0 & 0 & -1 \end{pmatrix}$$

with the permutation matrix $\mathbf{P}_6$ in (5.26). The factorization of $\mathbf{F}_6$ yields the signal flow graph in Fig. 5.8. □

We want to illustrate the presented fast algorithm of DFT$(N_1 N_2)$ from a different point of view. For that purpose, we order the components of the input and output vectors in N_1-by-N_2 matrices $\mathbf{A} := (a_{j_1,j_2})_{j_1,j_2=0}^{N_1-1,N_2-1}$ and $\hat{\mathbf{A}} := (\hat{a}_{k_1,k_2})_{k_1,k_2=0}^{N_1-1,N_2-1}$ using the following procedure:

$$a_{j_1,j_2} := a_{j_1+j_2 N_1}\,, \quad \hat{a}_{k_1,k_2} := \hat{a}_{k_1 N_2+k_2}\,, \quad k_r,\, j_r = 0, \ldots, N_r - 1\,,\ r = 1,\, 2\,.$$

Then the first step of Algorithm 5.16 corresponds to N_1 DFT(N_2) of the row vectors of $\mathbf{A}$ and the third step to N_2 DFT(N_1) of the column vectors of the matrix $\mathbf{C} := (c_{j_1,j_2})_{j_1,j_2=0}^{N_1-1,N_2-1}$ with the intermediate values $c_{j_1,j_2} := c_{j_1+j_2N_1}$ from Step 2 as components. Figure 5.9 illustrates this representation of the DFT(6).

There exist more efficient algorithms for realizing DFT$(N_1 N_2)$, if $(N_1,\, N_2) = 1$, i.e., if N_1 and N_2 are coprime, see, e.g., [169, 305].

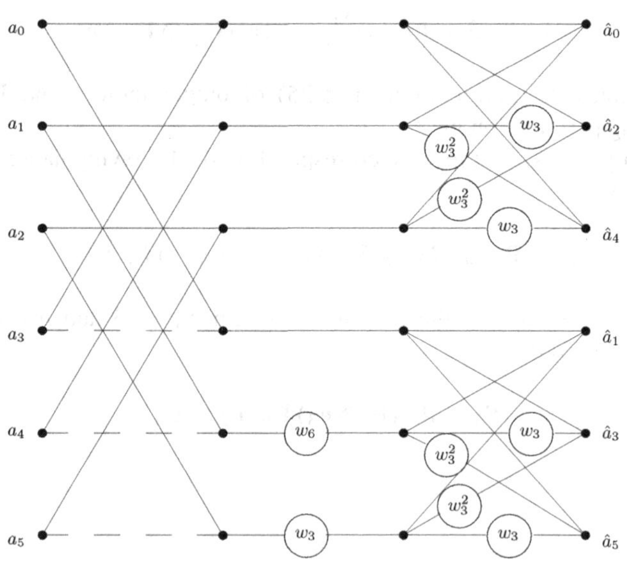

Fig. 5.8 Signal flow graph of a fast algorithm of DFT(6)

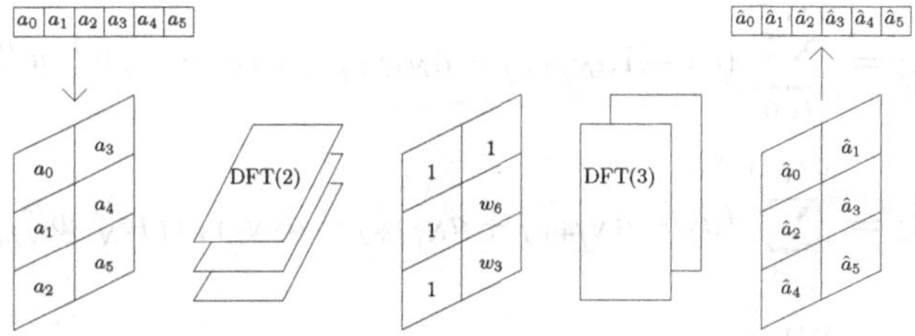

Fig. 5.9 Realization of a fast algorithm for DFT(6)

5.3.3 *Radix–4 FFT and Split–Radix FFT*

In this subsection we present a radix–4 FFT and a split–radix FFT. The advantage of these algorithms compared to radix–2 FFTs consists in lower computational costs.

The radix–4 FFT works for DFT(N) with $N = 4^t$, $t \in \mathbb{N}\backslash\{1\}$ and can be seen as a special case of the decomposition in the last subsection by taking $N_1 = 4$ and $N_2 = N/4$, where N_2 is decomposed iteratively into smaller powers of 4. The split–radix FFT uses a coupling of the radix–4 FFT and the radix–2 FFT. We restrict ourselves to only one form of the radix–2 FFT and the split–radix FFT. Similar algorithms can be derived by variations of ordering of the multiplication with twiddle factors and by changing the order of components in input and output vectors; see also Sect. 5.2. Since both algorithms are again based on butterfly operations, one can also derive a version that is suitable for parallel programming similarly as in Sect. 5.2.4.

We start with the *radix*–4 FFT. Let $N = 4^t$ with $t \in \mathbb{N} \setminus \{1\}$. We decompose the sum in (5.1) into the four partial sums

$$\begin{aligned}\hat{a}_k &= \sum_{j=0}^{N/4-1} \big(a_j\, w_N^{jk} + a_{N/4+j}\, w_N^{(N/4+j)k} + a_{N/2+j}\, w_N^{(N/2+j)k} \\ &\quad + a_{3N/4+j}\, w_N^{(3N/4+j)k}\big) \\ &= \sum_{j=0}^{N/4-1} \big(a_j + (-\mathrm{i})^k a_{N/4+j} + (-1)^k a_{N/2+j} + \mathrm{i}^k a_{3N/4+j}\big)\, w_N^{jk}\end{aligned}$$

and consider the output values with respect to the remainders of their indices modulo 4

$$\hat{a}_{4k} = \sum_{j=0}^{N/4-1} \big(a_j + a_{N/4+j} + a_{N/2+j} + a_{3N/4+j}\big)\, w_{N/4}^{jk}\,,$$

$$\hat{a}_{4k+1} = \sum_{j=0}^{N/4-1} \big(a_j - \mathrm{i}\, a_{N/4+j} - a_{N/2+j} + \mathrm{i}\, a_{3N/4+j}\big)\, w_N^j\, w_{N/4}^{jk}\,,$$

$$\hat{a}_{4k+2} = \sum_{j=0}^{N/4-1} \big(a_j - a_{N/4+j} + a_{N/2+j} - a_{3N/4+j}\big)\, w_N^{2j}\, w_{N/4}^{jk}\,,$$

$$\hat{a}_{4k+3} = \sum_{j=0}^{N/4-1} \big(a_j + \mathrm{i}\, a_{N/4+j} - a_{N/2+j} - \mathrm{i}\, a_{3N/4+j}\big)\, w_N^{3j}\, w_{N/4}^{jk}\,,$$

$$k = 0, \ldots, N/4-1\,.$$

In this way, the DFT(N) is decomposed into

- $N/4$ DFT(4) of the vectors $(a_j, a_{N/4+j}, a_{N/2+j}, a_{3N/4+j})^\top$, $j = 0, \ldots, N/4-1$,
- $3N/4$ complex multiplications with the twiddle factors w_N^{jr}, $j = 0, \ldots, N/4-1$, $r = 1, 2, 3$,
- 4 DFT($N/4$) of the vectors $\big(a_j + (-\mathrm{i})^r a_{N/4+j} + (-1)^r a_{N/2+j} + \mathrm{i}^r a_{3N/4+j}\big)_{j=0}^{N/4-1}$, $r = 0,\ 1,\ 2,\ 3$.

The $N/4$ DFT(4) as well as the multiplications with the twiddle factors are now executed in the first step of the algorithm, while the 4 DFT($N/4$) are individually decomposed using the above approach in a recursive manner. After t reduction steps, we obtain the transformed vector $\hat{\mathbf{a}}$. With this procedure, the DFT(N) is realized using only DFT(4) and multiplications with twiddle factors.

We now modify the algorithm by computing the DFT(4) with the radix–2 FFT thereby reducing the required number of additions. Then again, the algorithm only consists of butterfly operations. Figure 5.11 shows the signal flow graph of the radix–4 FFT for the DFT(16). The matrix representation of the radix–4 FFT can be obtained as follows. Let $N = 4^t = 2^{2t}$. Then the first step of the radix–4 FFT corresponds to the factorization

$$\mathbf{F}_N = \mathbf{Q}_N\, (\mathbf{I}_4 \otimes \mathbf{F}_{N/4})\, \tilde{\mathbf{D}}_N\, (\mathbf{F}_4 \otimes \mathbf{I}_{N/4})$$

with

$$\mathbf{Q}_N := \mathbf{P}_N\, (\mathbf{I}_2 \otimes \mathbf{P}_{N/2})\, (\mathbf{P}_4 \otimes \mathbf{I}_{N/4})\,,$$

$$\tilde{\mathbf{D}}_N := \mathrm{diag}\,\big(\mathbf{I}_{N/4}, \tilde{\mathbf{W}}_{N/4}, \tilde{\mathbf{W}}_{N/4}^2, \tilde{\mathbf{W}}_{N/4}^3\big)\,, \quad \tilde{\mathbf{W}}_{N/4} := \mathrm{diag}\,(w_N^j)_{j=0}^{N/4-1}\,.$$

Computing the DFT(4) with the radix–2 FFT of Algorithm 5.2 yields by (5.9) that

$$\mathbf{F}_4 = \mathbf{P}_4\, (\mathbf{I}_2 \otimes \mathbf{F}_2)\, \mathbf{D}_4\, (\mathbf{F}_2 \otimes \mathbf{I}_2)$$

and thus

$$\begin{aligned}\mathbf{F}_N &= \mathbf{Q}_N\,(\mathbf{I}_4\otimes\mathbf{F}_{N/4})\,\tilde{\mathbf{D}}_N\left[\mathbf{P}_4\,(\mathbf{I}_2\otimes\mathbf{F}_2)\,\mathbf{D}_4\,(\mathbf{F}_2\otimes\mathbf{I}_2)\right]\otimes\mathbf{I}_{N/4}\\ &= \mathbf{Q}_N\,(\mathbf{I}_4\otimes\mathbf{F}_{N/4})\,(\mathbf{P}_4\otimes\mathbf{I}_{N/4})\,\tilde{\mathbf{T}}_{2t}\,(\mathbf{I}_2\otimes\mathbf{F}_2\otimes\mathbf{I}_{N/4})\,(\mathbf{D}_4\otimes\mathbf{I}_{N/4})\,(\mathbf{F}_2\otimes\mathbf{I}_{N/2})\\ &= \mathbf{P}_N\,(\mathbf{I}_2\otimes\mathbf{P}_{N/2})\,(\mathbf{I}_4\otimes\mathbf{F}_{N/4})\,\tilde{\mathbf{T}}_{2t}\,(\mathbf{I}_2\otimes\mathbf{F}_2\otimes\mathbf{I}_{N/4})\,(\mathbf{D}_4\otimes\mathbf{I}_{N/4})\,(\mathbf{F}_2\otimes\mathbf{I}_{N/2})\end{aligned}$$

with $\mathbf{D}_4 := \operatorname{diag}(1,\,1,\,1,\,-\mathrm{i})^\top$ and

$$\tilde{\mathbf{T}}_{2t} := (\mathbf{P}_4^\top\otimes\mathbf{I}_{N/4})\,\tilde{\mathbf{D}}_N\,(\mathbf{P}_4\otimes\mathbf{I}_{N/4}) = \operatorname{diag}(\mathbf{I}_{N/4},\,\tilde{\mathbf{W}}_{N/4}^2,\,\tilde{\mathbf{W}}_{N/4},\,\tilde{\mathbf{W}}_{N/4}^3)\,.$$

The iterative application of the above factorization finally leads to the following matrix factorization that corresponds to the radix–4 FFT for the DFT(N) with $N = 4^t$

$$\mathbf{F}_N = \mathbf{R}_N\prod_{n=1}^{2t}\tilde{\mathbf{T}}_n\,(\mathbf{I}_{N/2^n}\otimes\mathbf{F}_2\otimes\mathbf{I}_{2^{n-1}}) \tag{5.31}$$

with the bit reversal matrix $\mathbf{R}_N$ and

$$\tilde{\mathbf{T}}_n := \begin{cases}\mathbf{I}_{N/2^n}\otimes\mathbf{D}_4\otimes\mathbf{I}_{2^{n-2}} & n\equiv 0 \bmod 2\,,\\ \mathbf{I}_{N/2^{n+1}}\otimes\tilde{\mathbf{D}}_{2^{n+1}} & n\equiv 1 \bmod 2\,,\end{cases}\qquad n = 1,\ldots,2t\,,$$

where

$$\tilde{\mathbf{D}}_{2^{n+1}} := \operatorname{diag}(\mathbf{I}_{2^{n-1}},\,\tilde{\mathbf{W}}_{2^{n-1}}^2,\,\tilde{\mathbf{W}}_{2^{n-1}},\,\tilde{\mathbf{W}}_{2^{n-1}}^3)\,.$$

A comparison with the factorization of $\mathbf{F}_N$ corresponding to the radix–2 FFT of Algorithm 5.2 shows that the two algorithms differ with regard to the twiddle factors. The output values are again in bit-reversed order.

We determine the numbers $\mu(t)$ and $\alpha(t)$ of nontrivial real multiplications and additions needed for executing the radix–4 FFT for the DFT(N) with $N = 4^t$, $t \in \mathbb{N}\setminus\{1\}$.

For computing the DFT($4N$) using the radix–4 algorithm, we have to execute 4 DFT(N), $8N$ complex additions, as well as $3N$ complex multiplications with twiddle factors w_{4N}^{rj}, $j = 0,\ldots,N-1$, $r = 1, 2, 3$. Among the multiplications with twiddle factors, there are *four* trivial multiplications for $(j,r) = (0,1)$, $(0,2)$, $(0,3)$, $(N/2,2)$ and 4 multiplications with 8th primitive roots of unity for $(j,r) = (N/2,1)$, $(N/2,3)$, $(N/4,2)$, $(3N/4,2)$. With the considerations in Sect. 5.2.5, we conclude

$$\begin{aligned}\mu(t+1) &= 4\,\mu(t) + 9\cdot 4^t - 16\,,\\ \alpha(t+1) &= 4\,\alpha(t) + 9\cdot 4^t - 16 + 16\cdot 4^t = 4\,\alpha(t) + 25\cdot 4^t - 16\,,\qquad t\in\mathbb{N}\,,\end{aligned}$$

with initial values $\mu(1) = 0$, $\alpha(1) = 16$. The explicit solutions of these linear difference equations are of the form

$$\begin{aligned} \mu(t) &= \tfrac{9}{4}\,t\,4^t - \tfrac{43}{12}\,4^t + \tfrac{16}{3} = \tfrac{9}{4}\,N\,\log_4 N - \tfrac{43}{12}\,N + \tfrac{16}{3}\,, \\ \alpha(t) &= \tfrac{25}{4}\,t\,4^t - \tfrac{43}{12}\,4^t + \tfrac{16}{3} = \tfrac{25}{4}\,N\,\log_4 N - \tfrac{43}{12}\,N + \tfrac{16}{3}\,, \quad t \in \mathbb{N}\,. \end{aligned} \tag{5.32}$$

A comparison of (5.32) with (5.17) and (5.18) shows that the application of the radix–4 FFT saves approximately 25% of nontrivial arithmetical operations. This saving is achieved only by a more advantageous choice of twiddle factors. Now, among the twiddle factors, there are more primitive 2^rth roots of unity with $r = 0, 1, 2, 3$.

The idea can be similarly used to construct radix–8, radix–16 FFTs, etc. Here, a transfer from a radix–2^r FFT to a radix–2^{r+1} FFT further reduces the computational cost, while at the same time this makes the algorithm more and more complex.

In the following, we derive the *split–radix* FFT. This algorithm is due to Yavne [448] and became popular under the name "split–radix FFT" in [118]. Compared to the radix–4 FFT, it reduces the computational cost further.

Let now $N = 2^t$ with $t \in \mathbb{N} \setminus \{1, 2\}$. From (5.1) we obtain

$$\hat{a}_{2k} = \sum_{j=0}^{N/2-1} (a_j + a_{N/2+j})\, w_{N/2}^{jk}\,, \quad k = 0, \ldots, N/2 - 1\,,$$

$$\hat{a}_{4k+1} = \sum_{j=0}^{N/4-1} \big((a_j - a_{N/2+j}) - \mathrm{i}\,(a_{N/4+j} - a_{3N/4+j})\big)\, w_N^j\, w_{N/4}^{jk}\,,$$
$$k = 0, \ldots, N/4 - 1\,,$$

$$\hat{a}_{4k+3} = \sum_{j=0}^{N/4-1} \big((a_j - a_{N/2+j}) + \mathrm{i}\,(a_{N/4+j} - a_{3N/4+j})\big)\, w_N^{3j}\, w_{N/4}^{jk}\,,$$
$$k = 0, \ldots, N/4 - 1\,.$$

In this way the DFT(N) is decomposed into

- $N/2$ DFT(2) of the vectors $(a_j, a_{N/2+j})^\top$, $j = 0, \ldots, N/2 - 1$,
- $N/2$ complex additions to compute the sums in the outer brackets,
- $N/2$ complex multiplications with the twiddle factors w_N^{jr}, $j = 0, \ldots, N/4 - 1$, $r = 1, 3$,
- 1 DFT($N/2$) and 2 DFT($N/4$).

This decomposition is then again applied to the DFT($N/2$) and to the two DFT($N/4$). We iteratively continue until we finally have to compute $N/2$ DFT(2) to obtain the output values which are again in bit-reversed order. Figure 5.12 shows the signal flow graph of the split–radix FFT for the DFT(16).

Let again $\mu(t)$ and $\alpha(t)$ be the numbers of needed real multiplications and additions for a transform length $N = 2^t$, $t \in \mathbb{N} \setminus \{1\}$. To evaluate the DFT(2$N$), the split–radix FFT requires one DFT(N), two DFT($N/2$), $3N$ complex additions, and N complex multiplications with the twiddle factors w_{2N}^{rj}, $j = 0, \ldots, N/2 - 1$, $r = 1, 3$. Among the multiplications with twiddle factors, there are two trivial multiplications for $(j, r) = (0, 1)$, $(0, 3)$ and two multiplications with primitive 8th roots of unity for $(j, r) = (N/4, 1)$, $(N/4, 3)$. Thus we obtain

$$\begin{aligned} \mu(t+1) &= \mu(t) + 2\,\mu(t-1) + 3 \cdot 2^t - 8\,, \\ \alpha(t+1) &= \alpha(t) + 2\,\alpha(t-1) + 3 \cdot 2^t - 8 + 6 \cdot 2^t\,. \end{aligned} \tag{5.33}$$

With the initial values

$$\begin{aligned} \mu(2) = 0\,, \quad \alpha(2) = 16\,, \\ \mu(3) = 4\,, \quad \alpha(3) = 52\,, \end{aligned}$$

we conclude that

$$\begin{aligned} \mu(t) &= t\,2^t - 3 \cdot 2^t + 4 &&= N \log_2 N - 3\,N + 4\,, \\ \alpha(t) &= 3\,t\,2^t - 3 \cdot 2^t + 4 &&= 3\,N \log_2 N - 3\,N + 4\,, \quad t \in \mathbb{N} \setminus \{1\}\,. \end{aligned} \tag{5.34}$$

We summarize:

Theorem 5.19 *For $N = 4^t$, $t \in \mathbb{N}$, the computational cost of the radix–4* FFT *for* DFT(N) *amounts*

$$\alpha(t) + \mu(t) = \frac{17}{4}\,N \log_4 N - \frac{43}{6}\,N + \frac{32}{3}$$

real arithmetical operations.

For $N = 2^t$, $t \in \mathbb{N} \setminus \{1\}$, the computational cost of the split–radix FFT *for* DFT(N) *adds up to*

$$\alpha(t) + \mu(t) = 4\,N \log_2 N - 6\,N + 8\,.$$

Note that comparing the computational cost of the radix–4 FFT with that of the radix–2 FFT or the split-radix FFT, one needs to keep in mind that $N = 4^t = 2^{2t}$, i.e., for $N = 4^t$ one needs to compare $\alpha(t)+\mu(t)$ for radix–4 FFT with $\alpha(2t)+\mu(2t)$ for radix–2 FFT and split–radix FFT.

In Tables 5.2 and 5.3, we present the number of required nontrivial real multiplications and additions for the radix–2 FFT, the radix–4 FFT, the radix–8 FFT, and the split–radix FFT. For comparison of the algorithm structures, we also present the signal flow graphs of the radix–2 FFT, the radix–4 FFT, and the split–radix FFT for the DFT(16) in Figs. 5.10, 5.11, and 5.12.

A modification of the split–radix FFT in [215] is based on the decomposition

Table 5.2 Number of real multiplications required by various FFTs for DFT(N)

N	radix–2	radix–4	radix–8	split–radix
16	24	20		20
32	88			68
64	264	208	204	196
128	712			516
256	1800	1392		1284
512	4360		13204	3076
1024	10248	7856		7172

Table 5.3 Number of real additions required by various FFTs for DFT(N)

N	radix–2	radix–4	radix–8	split–radix
16	152	148		148
32	408			388
64	1032	976	972	964
128	2504			2308
256	5896	5488		5380
512	13566		12420	12292
1024	30728	28336		27652

$$\hat{a}_k = \sum_{j=0}^{N/2-1} a_{2j}\, w_{N/2}^{jk} + w_N^k \sum_{j=0}^{N/4-1} a_{4j+1}\, w_{N/4}^{jk} + w_N^{-k} \sum_{j=0}^{N/4-1} a_{4j-1}\, w_{N/4}^{jk},$$

where $a_{-1} := a_{N-1}$. Here, we get a conjugate complex pair of twiddle factors instead of w_N^k and w_N^{3k}. The corresponding algorithm succeeds to reduce the twiddle factor load by rescaling and achieves a reduction of flops by further 5.6% compared to the usual split–radix FFT. A direct generalization of the split–radix FFT for $N = p^t$ with $t \in \mathbb{N}$ and a small prime (e.g., $p = 3$) has been considered in [422]. We remark that it is not known up to now how many flops are at least needed for an FFT of length N and whether there exist an FFT algorithm that needs even less operations than the split–radix algorithm in [215]. On the other hand, it has been shown in [293] that there exists no linear algorithm to compute the DFT(N) with less than $\mathcal{O}(N \log N)$ arithmetical operations.

5.3.4 Rader FFT and Bluestein FFT

Most previous FFTs are suitable for a special length $N = 2^t$ or even $N = 4^t$ with $t \in \mathbb{N}\backslash\{1\}$. For DFT applications with a different length, one can surely enlarge them by adding zero entries in the data vector to achieve the next radix-2 length. However, these longer data vectors have a changed structure and are not always desirable. In this subsection, we want to consider FFTs that can work with different lengths N and still achieve a computational cost of $\mathcal{O}(N \log N)$ arithmetical operations.

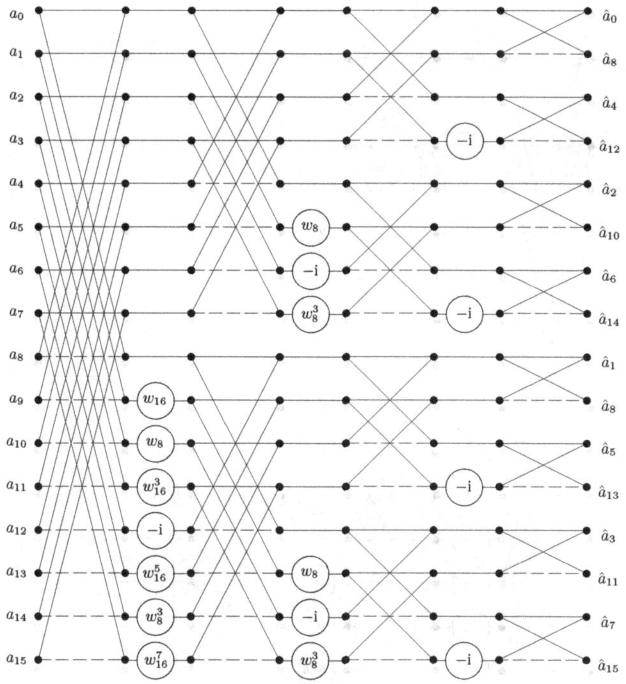

Fig. 5.10 Signal flow graph of the radix–2 FFT for DFT(16)

We start with the *Rader* FFT [358] that can be used to evaluate a DFT(p), where $p \in \mathbb{N}$ is a prime number. Again, the permutation of input and output values will play here an essential role. But the basic idea to realize the DFT is now completely different from the previously considered radix FFTs. The idea of the Rader FFT is that the DFT(p) can be rewritten using a cyclic convolution of length $p - 1$, which can then be realized efficiently by an FFT described in the previous subsections.

The Rader FFT is frequently applied to prime lengths $p \le 13$. For larger p, the Bluestein FFT is usually preferred because of its simpler structure. However, the Rader FFT is mathematically interesting, since it requires a small number of multiplications.

Let now $p \ge 3$ be a prime number. The transformed vector $\hat{\mathbf{a}} \in \mathbb{C}^p$ of $\mathbf{a} \in \mathbb{C}^p$ is given by

$$\hat{a}_k := \sum_{j=0}^{p-1} a_j\, w_p^{j\,k}\,, \quad k = 0, \ldots, p-1\,,\ w_p := \mathrm{e}^{-2\pi \mathrm{i}/p}. \tag{5.35}$$

Since p is a prime number, the index set $\{1, 2, \ldots, p-1\}$ forms the multiplicative group $(\mathbb{Z}/p\mathbb{Z})^*$ of integers modulo p. This group is cyclic of order $\varphi(p) = p - 1$, where φ denotes Euler's totient function. If g is a generating element of $(\mathbb{Z}/p\mathbb{Z})^*$,

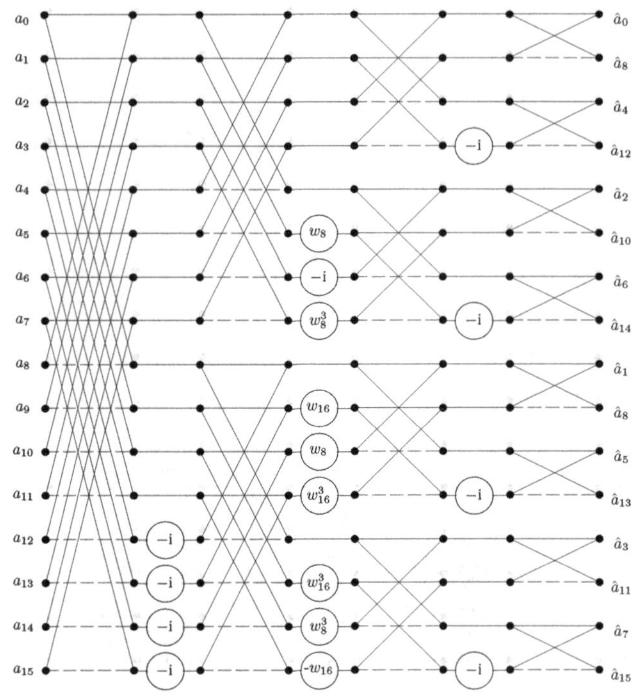

Fig. 5.11 Signal flow graph of the radix–4 FFT for DFT(16)

then each index $j \in \{1, 2, \ldots, p-1\}$ can be uniquely represented in the form

$$j = g^u \bmod p\,, \quad u \in \{0, \ldots, p-2\}\,.$$

For example, for $p = 5$ we can choose $g = 2$ as generating element of $(\mathbb{Z}/5\mathbb{Z})^*$ and find

$$1 = 2^0\,, \quad 2 = 2^1\,, \quad 4 = 2^2\,, \quad 3 = 2^3 \bmod 5\,. \tag{5.36}$$

In (5.35) we now consider the two indices $j = 0$ and $k = 0$ separately and replace $j, k \in \{1, \ldots, p-1\}$ by

$$j = g^u \bmod p\,, \quad k = g^v \bmod p, \quad u,\, v = 0, \ldots, p-2\,.$$

Then

$$\hat{a}_0 = c_0^0 + c_1^0\,, \tag{5.37}$$

$$\hat{a}_{g^v} = c_0^0 + c_v^1\,, \quad v = 0, \ldots, p-2\,, \tag{5.38}$$

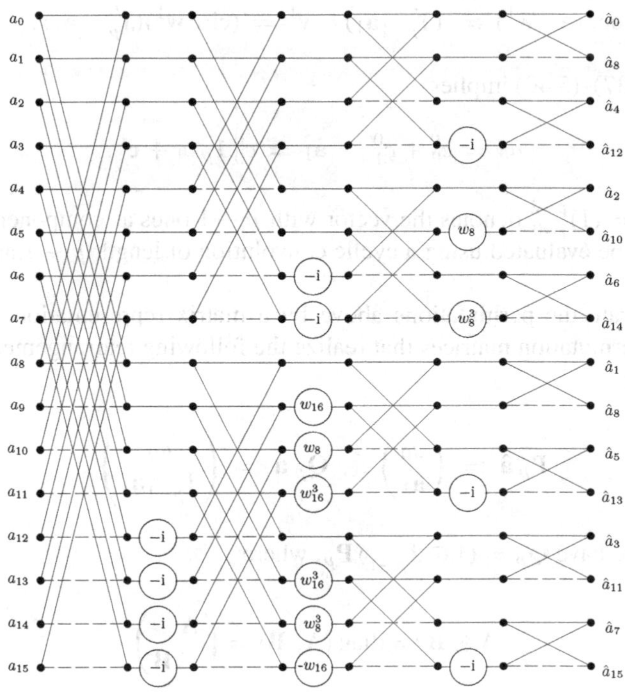

Fig. 5.12 Signal flow graph of the split–radix FFT for DFT(16)

with

$$c_0^0 := a_0, \quad c_1^0 := \sum_{u=0}^{p-2} a_{g^u},$$

$$c_v^1 := \sum_{u=0}^{p-2} a_{g^u}\, w_p^{g^{u+v}}, \quad v = 0, \ldots, p-2. \tag{5.39}$$

Obviously, (5.39) describes a *cyclic correlation* of the $(p-1)$-dimensional vectors

$$\mathbf{a}_1 := (a_{g^u})_{u=0}^{p-2}, \quad \mathbf{w}^1 := (w_p^{g^u})_{u=0}^{p-2}.$$

The cyclic correlation is closely related to the cyclic convolution considered in Sects. 3.2.3 and 3.3. Employing the flip matrix

$$\mathbf{J}'_{p-1} = \left(\delta_{(j+k)\mathrm{mod}(p-1)}\right)_{j,k=0}^{p-2} \in \mathbb{R}^{(p-1)\times(p-1)}$$

and the vector $\mathbf{c}^1 := (c_v^1)_{v=0}^{p-2}$, Eq. (5.39) can be written in the form

$$\mathbf{c}^1 = \operatorname{cor}(\mathbf{a}_1, \overline{\mathbf{w}}^1) := (\mathbf{J}'_{p-1}\mathbf{a}_1) * \mathbf{w}^1 = (\operatorname{circ}\mathbf{w}^1)(\mathbf{J}'_{p-1}\,\mathbf{a}_1)\,, \tag{5.40}$$

such that (5.37)–(5.40) implies

$$\hat{a}_0 = c_0^0 + c_1^0\,, \quad \hat{\mathbf{a}}_1 = c_0^0\,\mathbf{1}_{p-1} + \mathbf{c}^1\,.$$

Here $\mathbf{1}_{p-1} := (1)_{j=0}^{p-2}$ denotes the vector with $p-1$ ones as components. Thus the DFT(p) can be evaluated using a cyclic convolution of length $p-1$ and $2\,(p-1)$ additions.

We illustrate the permutations above by a matrix representation. Let $\mathbf{P}_p$ and $\mathbf{Q}_p$ be the permutation matrices that realize the following rearrangements of vector components:

$$\mathbf{P}_p\,\hat{\mathbf{a}} := \begin{pmatrix} \hat{a}_0 \\ \hat{\mathbf{a}}_1 \end{pmatrix}, \quad \mathbf{Q}_p\,\mathbf{a} := \begin{pmatrix} a_0 \\ \mathbf{J}'_{p-1}\mathbf{a}_1 \end{pmatrix}.$$

Obviously we have $\mathbf{Q}_p = (1 \oplus \mathbf{J}'_{p-1})\,\mathbf{P}_p$, where

$$\mathbf{A} \oplus \mathbf{B} := \operatorname{diag}(\mathbf{A},\ \mathbf{B}) = \begin{pmatrix} \mathbf{A} & \\ & \mathbf{B} \end{pmatrix}$$

denotes the block diagonal matrix of two square matrices $\mathbf{A}$ und $\mathbf{B}$. For example, for $p = 5$ we obtain with (5.36)

$$\underbrace{\begin{pmatrix} 1 & & & & \\ & 1 & & & \\ & & 1 & & \\ & & & & 1 \\ & & & 1 & \end{pmatrix}}_{\mathbf{P}_5 :=} \begin{pmatrix} \hat{a}_0 \\ \hat{a}_1 \\ \hat{a}_2 \\ \hat{a}_3 \\ \hat{a}_4 \end{pmatrix} = \begin{pmatrix} \hat{a}_0 \\ \hat{a}_1 \\ \hat{a}_2 \\ \hat{a}_4 \\ \hat{a}_3 \end{pmatrix}, \quad \underbrace{\begin{pmatrix} 1 & & & & \\ & 1 & & & \\ & & & 1 & \\ & & & & 1 \\ & & 1 & & \end{pmatrix}}_{\mathbf{Q}_5 :=} \begin{pmatrix} a_0 \\ a_1 \\ a_2 \\ a_3 \\ a_4 \end{pmatrix} = \begin{pmatrix} a_0 \\ a_1 \\ a_3 \\ a_4 \\ a_2 \end{pmatrix}.$$

From $\hat{\mathbf{a}} = \mathbf{F}_p\,\mathbf{a}$ it follows by (5.37)–(5.40) now

$$\mathbf{P}_p\,\hat{\mathbf{a}} = \mathbf{P}_p\,\mathbf{F}_p\,\mathbf{Q}_p^\top\,\mathbf{Q}_p\,\mathbf{a} = \tilde{\mathbf{F}}_p\,\mathbf{Q}_p\,\mathbf{a} \tag{5.41}$$

with the matrix $\tilde{\mathbf{F}}_p$ being composed by row and column permutations of $\mathbf{F}_p$

$$\tilde{\mathbf{F}}_p := \mathbf{P}_p\,\mathbf{F}_p\,\mathbf{Q}_p^\top = \left(\begin{array}{c|c} 1 & \mathbf{1}_{p-1}^\top \\ \hline \mathbf{1}_{p-1} & \operatorname{circ}\mathbf{w}^1 \end{array}\right). \tag{5.42}$$

A simple computation shows that $\tilde{\mathbf{F}}_p = \mathbf{A}_p\,(1 \oplus \operatorname{circ}\mathbf{w}^1)$ with

$$\mathbf{A}_p := \left(\begin{array}{c|c} 1 & -\mathbf{1}_{p-1}^\top \\ \hline \mathbf{1}_{p-1} & \mathbf{I}_{p-1} \end{array} \right). \tag{5.43}$$

For $p = 5$ we particularly obtain

$$\begin{pmatrix} \hat{a}_0 \\ \hat{a}_1 \\ \hat{a}_2 \\ \hat{a}_4 \\ \hat{a}_3 \end{pmatrix} = \underbrace{\left(\begin{array}{c|cccc} 1 & 1 & 1 & 1 & 1 \\ \hline 1 & w_5 & w_5^3 & w_5^4 & w_5^2 \\ 1 & w_5^2 & w_5 & w_5^3 & w_5^4 \\ 1 & w_5^4 & w_5^2 & w_5 & w_5^3 \\ 1 & w_5^3 & w_5^4 & w_5^2 & w_5 \end{array} \right)}_{\tilde{\mathbf{F}}_5 :=} \begin{pmatrix} a_0 \\ a_1 \\ a_3 \\ a_4 \\ a_2 \end{pmatrix}.$$

The essential part of the Rader FFT, the cyclic convolution of length $p-1$ in (5.40), can now be computed by employing a fast algorithm for cyclic convolutions based on Theorem 3.31. The multiplication with a circulant matrix can be realized with 3 DFT$(p-1)$ and $p-1$ multiplications. Indeed, (3.47) implies

$$(\text{circ}\, \mathbf{w}^1)\,(\mathbf{J}'_{p-1}\,\mathbf{a}_1) = \mathbf{F}_{p-1}^{-1}\left(\text{diag}\,(\mathbf{F}_{p-1}\mathbf{w}^1)\right)\mathbf{F}_{p-1}\,(\mathbf{J}_{p-1}\mathbf{a}_1)\,.$$

Assuming that $p-1$ can be factorized into powers of small prime factors, we may use an FFT as described in Sect. 5.3.2. For small integers $p-1$, there exist different efficient convolution algorithms; see, e.g., [52, 305], based on the Chinese remainder theorem.

Example 5.20 In the case $p = 5$, we particularly have

$$\begin{aligned} \text{circ}\, \mathbf{w}^1 &= \mathbf{F}_4^{-1}\, \text{diag}\,(\mathbf{F}_4\mathbf{w}^1)\,\mathbf{F}_4 \\ &= \frac{1}{4}\,\mathbf{P}_4 \begin{pmatrix} \mathbf{F}_2 & \\ & \mathbf{F}_2 \end{pmatrix} \bar{\mathbf{D}}_4 \begin{pmatrix} \mathbf{I}_2 & \mathbf{I}_2 \\ \mathbf{I}_2 & -\mathbf{I}_2 \end{pmatrix} (\text{diag}\,\hat{\mathbf{w}}^1)\,\mathbf{P}_4 \begin{pmatrix} \mathbf{F}_2 & \\ & \mathbf{F}_2 \end{pmatrix} \mathbf{D}_4 \begin{pmatrix} \mathbf{I}_2 & \mathbf{I}_2 \\ \mathbf{I}_2 & -\mathbf{I}_2 \end{pmatrix} \end{aligned}$$

with the even–odd permutation matrix $\mathbf{P}_4$ and $\mathbf{D}_4 = \text{diag}\,(1,\ 1,\ 1,\ -\mathrm{i})^\top$. Here the factorization (5.9) of $\mathbf{F}_4$ is used. □

A generalization of the Rader FFT is the *Winograd* FFT that can be applied for fast computation of DFT(p^t), where $p \in \mathbb{N}$ is a prime and $t \in \mathbb{N} \setminus \{1\}$. It employs the special group structure of $(\mathbb{Z}/p^t\mathbb{Z})^*$; for details, see [438].

The *Bluestein* FFT is also based on the idea to write the DFT(N) as a convolution. The obtained circulant N-by-N matrix can in turn be embedded into a circulant M-by-M matrix, where $M = 2^t$, $t \in \mathbb{N}$, satisfies $2N - 2 \le M < 4N$. To compute the obtained convolution of length M, a radix–2 FFT or split–radix FFT can be employed in order to end up with an $\mathcal{O}(N \log N)$ algorithm.

Let now $N \in \mathbb{N} \setminus \{1,\ 2\}$ be given, where N is not a power of 2. With

$$k\,j = \frac{1}{2}\left(k^2 + j^2 - (k-j)^2\right)$$

we can rewrite (5.1) as

$$\hat{a}_k = \sum_{j=0}^{N-1} a_j\, w_N^{k\,j} = w_N^{k^2/2} \sum_{j=0}^{N-1} a_j\, w_N^{j^2/2}\, w_N^{-(k-j)^2/2}\,, \quad k = 0, \ldots, N-1\,.$$

Multiplication with $w_N^{-k^2/2}$ on both sides gives

$$z_k := w_N^{-k^2/2}\hat{a}_k = \sum_{j=0}^{N-1} \left(a_j\, w_N^{j^2/2}\right) w_N^{-(k-j)^2/2} = \sum_{j=0}^{N-1} b_j\, h_{k-j}, \quad k = 0, \ldots, N-1\,, \tag{5.44}$$

with

$$\mathbf{b} := (b_j)_{j=0}^{N-1} = \left(a_j\, w_N^{j^2/2}\right)_{j=0}^{N-1}, \quad \mathbf{h} := (h_j)_{j=0}^{N-1} = \left(w_N^{-j^2/2}\right)_{j=0}^{N-1}.$$

We observe that

$$h_{k-j} = w_N^{-(k-j)^2/2} = h_{j-k}\,,$$

such that the circulant matrix circ $\mathbf{h} = (h_{(j-k) \bmod N})_{j,k=0}^{N-1}$ is symmetric. With $\mathbf{z} := (z_k)_{k=0}^{N-1}$, Eq. (5.44) can now be rewritten as

$$\mathbf{z} = (\mathrm{diag}\,\mathbf{h})\,\mathbf{F}_N\,\mathbf{a} = \mathbf{b} * \mathbf{h} = (\mathrm{circ}\,\mathbf{h})\,\mathbf{b} = (\mathrm{circ}\,\mathbf{h})\,(\mathrm{diag}\,\overline{\mathbf{h}})\,\mathbf{a}\,,$$

where $\overline{\mathbf{h}} := (\overline{h}_j)_{j=0}^{N-1}$ is the conjugate complex vector. Thus, we obtain the matrix factorization

$$\mathbf{F}_N = (\mathrm{diag}\,\mathbf{h})^{-1}\,(\mathrm{circ}\,\mathbf{h})\,(\mathrm{diag}\,\overline{\mathbf{h}})\,.$$

This representation of $\mathbf{F}_N$ is not yet efficient. But the idea is to embed circ $\mathbf{h}$ into a circulant M-by-M matrix circ $\mathbf{h}^1$ with $M = 2^t$, $t \in \mathbb{N}$, and $2N - 2 \le M < 4N$. We determine $\mathbf{h}^1 = (h_j^1)_{j=0}^{M-1} \in \mathbb{C}^M$ with

$$h_j^1 := \begin{cases} h_j & 0 \le j \le N-1\,, \\ 0 & N \le j \le M-N\,, \\ h_{M-j} & M-N+1 \le j \le M-1\,. \end{cases} \tag{5.45}$$

For example, for $N = 7$, we have to choose $M = 16 = 2^4$ such that $12 \le M < 28$, and $\mathbf{h}^1$ is of the form

$$\mathbf{h}^1 = \big(h_0, h_1, h_2, h_3, h_4, h_5, h_6, 0, 0, 0, h_6, h_5, h_4, h_3, h_2, h_1\big)^\top$$

with $h_j = w_7^{-j^2/2}$. Observe that circ $\mathbf{h}^1$ contains circ $\mathbf{h}$ as a submatrix at the left upper corner. In order to compute the convolution $\mathbf{h} * \mathbf{b} = (\text{circ}\,\mathbf{h})\,\mathbf{b}$, we therefore consider the enlarged vector $\mathbf{b}^1 = (b_j^1)_{j=0}^{M-1}$ with

$$b_j^1 := \begin{cases} b_j & 0 \le j \le N-1\,, \\ 0 & N \le j \le M-1\,, \end{cases} \tag{5.46}$$

such that the computation of $(\text{circ}\,\mathbf{h})\,\mathbf{b}$ is equivalent with evaluating the first N components of $(\text{circ}\,\mathbf{h}^1)\,\mathbf{b}^1$.

We summarize the Bluestein FFT as follows.

Algorithm 5.21 (Bluestein FFT)
Input: $N \in \mathbb{N} \setminus \{1\}$, $\mathbf{a} = (a_j)_{j=0}^{N-1} \in \mathbb{C}^N$.

1. *Determine* $M := 2^t$ *with* $t := \lfloor \log_2(4N-1) \rfloor$.
2. *Compute* $\mathbf{b} := (a_j\, w_N^{j^2/2})_{j=0}^{N-1}$ *and* $\mathbf{h} := (w_N^{-j^2/2})_{j=0}^{N-1}$.
3. *Enlarge* $\mathbf{h}$ *to* $\mathbf{h}^1$ *and* $\mathbf{b}$ *to* $\mathbf{b}^1$ *according to* (5.45) *and* (5.46).
4. *Compute* $\hat{\mathbf{h}}^1 = \mathbf{F}_M \mathbf{h}^1$ *and* $\hat{\mathbf{b}}^1 = \mathbf{F}_M \mathbf{b}^1$ *using a radix–*2 FFT *of length* M.
5. *Compute* $\hat{\mathbf{z}} := \hat{\mathbf{h}}^1 \circ \hat{\mathbf{b}}^1 = (\hat{h}_k^1\, \hat{b}_k^1)_{k=0}^{M-1}$.
6. *Compute* $\mathbf{z} = \mathbf{F}_M^{-1} \hat{\mathbf{z}}$ *using a radix–*2 FFT *of length* M.
7. *Calculate* $\hat{\mathbf{a}} := \big(w_N^{j^2/2}\, z_j\big)_{j=0}^{N-1}$.

Output: $\hat{\mathbf{a}} \in \mathbb{C}^N$.

The numerical effort for the Bluestein FFT is governed by the 3 DFT(M), but since $M < 4N$, it easily follows that the computational cost of the Bluestein FFT is still $\mathcal{O}(N \log N)$.

Remark 5.22 Let us give some further notes on FFTs. This field has been intensively studied within the last 60 years, and a lot of extensions have been suggested that we are not able to present in this chapter. Therefore we only want to give a few further ideas that can be found in the literature and refer, e.g., to [64, 119, 198, 421].

1. An early attempt to obtain a fast DFT algorithm is due to Goerzel [163], who applied a recursive scheme to the simultaneous computation of $c(x) = \sum_{k=0}^{N-1} a_k \cos kx$ and $s(x) = \sum_{k=0}^{N-1} a_k \sin kx$. The Goerzel algorithm has a higher complexity than FFTs, but it is of interest for computing only a small number of selected values of $c(x)$ and $s(x)$.
2. Bruun's FFT [67] uses z-transform filters to reduce the number of complex multiplications compared to the usual FFT.

3. Winograd developed a theory of multiplicative complexity of bilinear forms that can be exploited for fast convolution algorithms using a minimal number of multiplications; see [437, 439].
4. The FFT has also been generalized to finite fields, where the notion "cyclotomic FFT" has been coined. It is again based on a transfer to several circular convolutions; see, e.g., [442] and references therein. □

5.3.5 Multidimensional FFT

For fixed $N_1, N_2 \in \mathbb{N} \setminus \{1\}$, we consider the *two-dimensional* DFT$(N_1 \times N_2)$ of the form

$$\hat{\mathbf{A}} = \mathbf{F}_{N_1}\, \mathbf{A}\, \mathbf{F}_{N_2}\,,$$

where $\mathbf{A} = \big(a_{k_1,k_2}\big)_{k_1,k_2=0}^{N_1-1,N_2-1}$ and $\hat{\mathbf{A}} = \big(\hat{a}_{n_1,n_2}\big)_{n_1,n_2=0}^{N_1-1,N_2-1}$ are complex N_1-by-N_2 matrices as in Sect. 4.5.2. For the entries $\hat{a}_{n_1,n_2}$, we obtain from (4.58)

$$\hat{a}_{n_1,n_2} = \sum_{k_1=0}^{N_1-1} \sum_{k_2=0}^{N_2-1} a_{k_1,k_2}\, w_{N_1}^{k_1 n_1}\, w_{N_2}^{k_2 n_2}\,, \quad n_1 = 0, \ldots, N_1-1\,;\; n_2 = 0, \ldots, N_2-1\,. \tag{5.47}$$

The fast evaluation of $\hat{\mathbf{A}}$ can be performed using only one-dimensional FFTs.

First we present the *row–column method* for the two-dimensional DFT$(N_1 \times N_2)$. Let $\mathbf{A}^\top = \big(\mathbf{a}_0 \,|\, \mathbf{a}_1 \,|\, \ldots \,|\, \mathbf{a}_{N_1-1}\big)$, where $\mathbf{a}_{k_1} \in \mathbb{C}^{N_2}$, $k_1 = 0, \ldots, N_1 - 1$, denote the N_1 rows of $\mathbf{A}$. Then the product

$$\mathbf{F}_{N_2}\, \mathbf{A}^\top = \big(\mathbf{F}_{N_2}\mathbf{a}_0 \,|\, \mathbf{F}_{N_2}\mathbf{a}_1 \,|\, \ldots \,|\, \mathbf{F}_{N_2}\mathbf{a}_{N_1-1}\big)$$

can be performed by applying an FFT of length N_2 to each row of $\mathbf{A}$ separately. We obtain $\mathbf{B}^\top = (\mathbf{A}\,\mathbf{F}_{N_2})^\top = \big(\hat{\mathbf{a}}_0 \,|\, \hat{\mathbf{a}}_1 \,|\, \ldots \,|\, \hat{\mathbf{a}}_{N_1-1}\big)$ and therefore $\hat{\mathbf{A}} = \mathbf{F}_{N_1}\, \mathbf{B}$. Let now $\mathbf{B} = (\mathbf{b}_0 \,|\, \mathbf{b}_1 \,|\, \ldots \,|\, \mathbf{b}_{N_2-1})$ with columns $\mathbf{b}_{k_2} \in \mathbb{C}^{N_1}$, $k_2 = 0, \ldots, N_2 - 1$. Then

$$\hat{\mathbf{A}} = \mathbf{F}_{N_1}\, \mathbf{B} = \big(\mathbf{F}_{N_1}\mathbf{b}_0 \,|\, \mathbf{F}_{N_1}\mathbf{b}_1 \,|\, \ldots \,|\, \mathbf{F}_{N_1}\mathbf{b}_{N_2-1}\big)$$

can be performed by applying an FFT of length N_1 to each column of $\mathbf{B}$. Obviously we can also compute $\tilde{\mathbf{B}} = \mathbf{F}_{N_1}\, \mathbf{A}$ by applying a one-dimensional DFT(N_1) to each column of $\mathbf{A}$ in the first step and then compute $\tilde{\mathbf{B}}\,\mathbf{F}_{N_2}$ by applying a DFT(N_2) to each row of $\tilde{\mathbf{B}}$ in the second step.

By reshaping the matrix $\mathbf{A} = \big(a_{k_1,k_2}\big)_{k_1,k_2=0}^{N_1-1,N_2-1}$ into a vector $\mathbf{a} = \operatorname{vec} \mathbf{A} \in \mathbb{C}^{N_1 N_2}$ with $a_{k_1+N_1 k_2} = a_{k_1,k_2}$, we can transfer the two-dimensional DFT into a matrix-vector product. Applying Theorem 3.42, we obtain

$$\text{vec}\,\hat{\mathbf{A}} = (\mathbf{F}_{N_2} \otimes \mathbf{F}_{N_1})\text{vec}\,\mathbf{A} = (\mathbf{F}_{N_2} \otimes \mathbf{I}_{N_1})(\mathbf{I}_{N_2} \otimes \mathbf{F}_{N_1})\,\text{vec}\,\mathbf{A}\,.$$

The multiplication $\text{vec}\,\mathbf{B} = (\mathbf{I}_{N_2} \otimes \mathbf{F}_{N_1})\,\text{vec}\,\mathbf{A}$ is equivalent with applying the one-dimensional FFT of length N_2 to each row of $\mathbf{A}$, and the multiplication $(\mathbf{F}_{N_2} \otimes \mathbf{I}_{N_1})\,\text{vec}\,\mathbf{B}$ is equivalent with applying the one-dimensional FFT of length N_1 to each column of $\mathbf{B}$.

We also present the sum representation of the row–column method for the two-dimensional DFT$(N_1 \times N_2)$ of $\mathbf{A} = \big(a_{k_1,k_2}\big)_{k_1,k_2=0}^{N_1-1,N_2-1}$. We rewrite the double sum in (5.47)

$$\hat{a}_{n_1,n_2} = \sum_{k_1=0}^{N_1-1} w_{N_1}^{k_1 n_1} \underbrace{\Big(\sum_{k_2=0}^{N_2-1} a_{k_1,k_2}\, w_{N_2}^{k_2 n_2}\Big)}_{b_{k_1,n_2} :=}.$$

Now, for each $k_1 \in I_{N_1}$, we first compute the vectors $(b_{k_1,n_2})_{n_2=0}^{N_2-1}$ using a one-dimensional DFT(N_2) applied to the k_1th row of $\mathbf{A}$. Then we compute

$$\hat{a}_{n_1,n_2} = \sum_{k_1=0}^{N_1-1} b_{k_1,n_2}\, w_{N_1}^{k_1 n_1}\,,$$

i.e., for each $n_2 \in I_{N_2}$ we calculate the one-dimensional DFT(N_1) of the n_2th column of the intermediate array $(b_{k_1,n_2})_{k_1,n_2=0}^{N_1-1,N_2-1}$. In summary, we can compute a general DFT$(N_1 \times N_2)$ by means of N_1 DFT(N_2) and N_2 DFT(N_1).

Algorithm 5.23 (Row–Column Method for DFT$(N_1 \times N_2)$)
Input: $N_1,\, N_2 \in \mathbb{N} \setminus \{1\}$, $\mathbf{A} \in \mathbb{C}^{N_1 \times N_2}$.

1. *Compute the* DFT(N_2) *for each of the* N_1 *rows of* $\mathbf{A}$ *by one-dimensional* FFTs *to obtain*

$$\mathbf{B}^\top = (\mathbf{A}\,\mathbf{F}_{N_2})^\top = \big(\hat{\mathbf{a}}_0 \,|\, \hat{\mathbf{a}}_1 \,|\, \ldots \,|\, \hat{\mathbf{a}}_{N_1-1}\big)\,.$$

2. *Compute the* DFT(N_1) *for each of the* N_2 *columns of* $\mathbf{B} = \big(\mathbf{b}_0 \,|\, \mathbf{b}_1 \,|\, \ldots \,|\, \mathbf{b}_{N_2-1}\big)$ *by one-dimensional* FFTs *to obtain*

$$\hat{\mathbf{A}} = \mathbf{F}_{N_1}\,\mathbf{B} = \big(\hat{\mathbf{b}}_0 \,|\, \hat{\mathbf{b}}_1 \,|\, \ldots \,|\, \hat{\mathbf{b}}_{N_2-1}\big)\,.$$

Output: $\hat{\mathbf{A}} \in \mathbb{C}^{N_1 \times N_2}$.

The computational cost to apply Algorithm 5.23 is $\mathcal{O}\big(N_1\, N_2\, (\log N_1)(\log N_2)\big)$ assuming that a one-dimensional FFT of length N requires $\mathcal{O}(N\, \log N)$ floating point operations.

Now we describe the *nesting method* for the two-dimensional DFT($N_1 \times N_2$). Compared to the row–column method considered above, we can reduce the computational cost of the two-dimensional DFT using the known factorization of the Fourier matrix $\mathbf{F}_N$ that we have found to derive the one-dimensional FFTs. As shown in (5.30), for $N = M_1 M_2$, we have the matrix factorization

$$\mathbf{F}_N = \mathbf{P}_N \left(\mathbf{I}_{M_2} \otimes \mathbf{F}_{M_1}\right) \mathbf{D}_{M_1 M_2} \left(\mathbf{F}_{M_2} \otimes \mathbf{I}_{M_1}\right)$$

with the block diagonal matrix

$$\mathbf{D}_{M_1 M_2} = \operatorname{diag}\left(\mathbf{I}_{M_1}, \mathbf{W}_{M_1}, \ldots, \mathbf{W}_{M_1}^{M_2-1}\right) = \begin{pmatrix} \mathbf{I}_{M_1} & & & \\ & \mathbf{W}_{M_1} & & \\ & & \ddots & \\ & & & \mathbf{W}_{M_1}^{M_2-1} \end{pmatrix},$$

where $\mathbf{W}_{M_1} = \operatorname{diag}\,(w_N^r)_{r=0}^{M_1-1}$. Assuming now that we have the factorizations $N_1 = K_1 K_2$ and $N_2 = L_1 L_2$ with $K_1, K_2, L_1, L_2 \in \mathbb{N} \setminus \{1\}$, the two-dimensional DFT($N_1 \times N_2$) can be rewritten as

$$\hat{\mathbf{A}} = \mathbf{P}_{N_1}(\mathbf{I}_{K_2} \otimes \mathbf{F}_{K_1})\mathbf{D}_{K_1 K_2}(\mathbf{F}_{K_2} \otimes \mathbf{I}_{K_1})\mathbf{A}(\mathbf{F}_{L_2} \otimes \mathbf{I}_{L_1})\mathbf{D}_{L_1 L_2}(\mathbf{I}_{L_2} \otimes \mathbf{F}_{L_1})\mathbf{P}_{N_2}^{\top}, \tag{5.48}$$

where we have used that $\mathbf{F}_{N_2}$ and all matrix factors in the factorization up to $\mathbf{P}_{N_2}$ are symmetric. Now the computation of $\hat{\mathbf{A}}$ can be performed as follows.

Algorithm 5.24 (Nesting Method for DFT($N_1 \times N_2$))
Input: $N_1, N_2 \in \mathbb{N} \setminus \{1\}$, $\mathbf{A} \in \mathbb{C}^{N_1 \times N_2}$.

1. *Compute* $\mathbf{B} := \left(\mathbf{F}_{K_2} \otimes \mathbf{I}_{K_1}\right) \mathbf{A} \left(\mathbf{F}_{L_2} \otimes \mathbf{I}_{L_1}\right)$.
2. *Compute* $\mathbf{C} := \mathbf{D}_{K_1 K_2}\, \mathbf{B}\, \mathbf{D}_{L_1 L_2}$ *by*

$$c_{n_1,n_2} = b_{n_1,n_2}\, d_{n_1,n_2}, \quad n_1 = 0, \ldots, N_1 - 1;\; n_2 = 0, \ldots N_2 - 1,$$

   where $d_{n_1,n_2} := \mathbf{D}_{K_1 K_2}(n_1, n_1)\, \mathbf{D}_{L_1 L_2}(n_2, n_2)$ *is the product of the* n_1*th diagonal element of* $\mathbf{D}_{K_1 K_2}$ *and the* n_2*th diagonal element of* $\mathbf{D}_{L_1 L_2}$.
3. *Compute* $\hat{\mathbf{A}} := \mathbf{P}_{N_1}\left(\mathbf{I}_{K_2} \otimes \mathbf{F}_{K_1}\right) \mathbf{C} \left(\mathbf{I}_{L_2} \otimes \mathbf{F}_{L_1}\right)\mathbf{P}_{N_2}^{\top}$.

Output: $\hat{\mathbf{A}} \in \mathbb{C}^{N_1 \times N_2}$.

By reshaping the matrices to vectors using the vectorization as in Theorem 3.42, the factorization (5.48) can be rewritten as

$$\begin{aligned} \operatorname{vec} \hat{\mathbf{A}} = {} & \left(\mathbf{P}_{N_2}(\mathbf{I}_{L_2} \otimes \mathbf{F}_{L_1})\, \mathbf{D}_{L_1 L_2}(\mathbf{F}_{L_2} \otimes \mathbf{I}_{L_1})\right) \\ & \otimes \left(\mathbf{P}_{N_1}(\mathbf{I}_{K_1} \otimes \mathbf{F}_{K_1})\mathbf{D}_{K_1 K_2}(\mathbf{F}_{K_2} \otimes \mathbf{I}_{K_1})\right) \operatorname{vec} \mathbf{A} \end{aligned}$$

$$= \big(\mathbf{P}_{N_2} \otimes \mathbf{P}_{N_1}\big)\big((\mathbf{I}_{L_2} \otimes \mathbf{F}_{L_1}) \otimes (\mathbf{I}_{K_1} \otimes \mathbf{F}_{K_1})\big)\big(\mathbf{D}_{L_1 L_2} \otimes \mathbf{D}_{K_1 K_2}\big)$$
$$\cdot \big((\mathbf{F}_{L_2} \otimes \mathbf{I}_{L_1}) \otimes (\mathbf{F}_{K_2} \otimes \mathbf{I}_{K_1})\big)\, \mathrm{vec}\, \mathbf{A}\,. \tag{5.49}$$

Hence successive multiplication with these matrices corresponds to the three steps of Algorithm 5.24. Compared to the application of the row–column method connected with the considered DFT of composite length, we save multiplications assuming that the diagonal values d_{n_1,n_2} of the matrix $\mathbf{D}_{L_1L_2} \otimes \mathbf{D}_{K_1K_2}$ in Step 2 of Algorithm 5.24 are precomputed beforehand. The structure of the diagonal matrices implies that the multiplication with $\mathbf{D}_{K_1K_2}$ from the left and with $\mathbf{D}_{L_1L_2}$ from the right that needs to be performed using the row–column method requires $(N_1 - K_1)N_2 + (N_2 - L_1)N_1$ multiplications, while Step 2 of Algorithm 5.23 needs $N_1N_2 - L_1K_1$ multiplications with precomputed values $d_{n_1n_2}$. Here we have taken into consideration that $d_{n_1,n_2} = 1$ for $n_1 = 0, \ldots, K_1 - 1$ and $n_2 = 0, \ldots, L_1 - 1$. In the special case $N = N_1 = N_2$ and $L_1 = K_1$, we save $(N - K_1)^2$ multiplications. If N_1 and N_2 are powers of two, then the nesting approach can also be applied using the full factorization of the Fourier matrices $\mathbf{F}_{N_1}$ and $\mathbf{F}_{N_2}$ as given in (5.10), and we can save multiplications at each level if applying the nesting method instead of multiplying with the diagonal matrices of twiddle factors from left and right. If particularly $N_1 = N_2 = N$, then we have $\log_2 N$ levels and save $(\log_2 N)(\frac{N}{2})^2$ multiplications compared to the row–column method.

Now we consider *higher-dimensional* FFTs, i.e., $d \in \mathbb{N} \setminus \{1, 2\}$, is of moderate size. We generalize the row–column method and the nesting method to compute the d-dimensional DFT. Let $\mathbf{N} = (N_j)_{j=1}^d \in (\mathbb{N} \setminus \{1\})^d$ and the index set

$$I_{\mathbf{N}} := I_{N_1} \times I_{N_2} \ldots \times I_{N_d}$$

with $I_{N_j} := \{0, \ldots, N_j - 1\}$ be given. Recall that for a d-dimensional array $\mathbf{A} = (a_{\mathbf{k}})_{\mathbf{k} \in I_{\mathbf{N}}}$ of size $N_1 \times \ldots \times N_d$, the d-dimensional DFT $\hat{\mathbf{A}} = (\hat{a}_{\mathbf{n}})_{n \in I_{\mathbf{N}}}$ is given by (4.60), i.e.,

$$\hat{a}_{\mathbf{n}} := \sum_{k_1=0}^{N_1-1} \ldots \sum_{k_d=0}^{N_d-1} a_{\mathbf{k}}\, w_{N_1}^{k_1n_1} \ldots w_{N_d}^{k_dn_d} = \sum_{\mathbf{k} \in I_{\mathbf{N}}} a_{\mathbf{k}}\, \mathrm{e}^{-2\pi \mathrm{i}\, \mathbf{n} \cdot (\mathbf{k}/\mathbf{N})}.$$

For moderate dimension d, a generalized *row–column method* is often used for the computation of the d-dimensional DFT of $\mathbf{A} = (a_{\mathbf{k}})_{\mathbf{k} \in I_{\mathbf{N}}}$. We rewrite the multiple sum above

$$\hat{a}_{n_1,n_2,\ldots,n_d} = \sum_{k_1=0}^{N_1-1} w_{N_1}^{k_1n_1} \underbrace{\Big(\sum_{k_2=0}^{N_2-1} \ldots \sum_{k_d=0}^{N_d-1} a_{\mathbf{k}}\, w_{N_2}^{k_2n_2} \ldots w_{N_d}^{k_dn_d}\Big)}_{b_{k_1,n_2,\ldots,n_d} :=}\,.$$

Thus for a given array $(b_{k_1,n_2,\ldots,n_d})$ of size $N_1 \times \ldots \times N_d$, we compute

$$\hat{a}_{n_1,n_2,\ldots,n_d} = \sum_{k_1=0}^{N_1-1} b_{k_1,n_2,\ldots,n_d}\, w_{N_1}^{k_1 n_1}\,,$$

i.e., for each $(n_2, \ldots, n_d)^\top \in I_{N_2} \times \ldots \times I_{N_d}$, we calculate a one-dimensional DFT(N_1). The arrays $\mathbf{B}_{k_1} = (b_{k_1,n_2,\ldots,n_d})_{(n_2,\ldots,n_d)^\top \in I_{N_2} \times \ldots \times I_{N_d}}$ are obtained by a $(d-1)$-dimensional DFT($N_2 \times \ldots \times N_d$). The computational costs to compute the d-dimensional DFT are therefore

$$N_2 \ldots N_d \ \mathrm{DFT}(N_1) + N_1 \ \mathrm{DFT}(N_2 \times \ldots \times N_d)\,.$$

Recursive application of this idea with regard to each dimension thus requires for the d-dimensional DFT($N_1 \times \ldots \times N_d$)

$$N_1 \cdots N_d \Big(\frac{1}{N_1}\ \mathrm{DFT}(N_1) + \frac{1}{N_2}\ \mathrm{DFT}(N_2) + \ldots + \frac{1}{N_d}\ \mathrm{DFT}(N_d)\Big) \tag{5.50}$$

with computational cost of $\mathcal{O}\big(N_1 N_2 \ldots N_d \log_2(N_1 N_2 \ldots N_d)\big)$. If we apply the mapping $\mathrm{vec} : \mathbb{C}^{N_1 \times \ldots \times N_d} \to \mathbb{C}^P$ with $P = N_1 N_2 \cdots N_d$ introduced in Sect. 4.5.3 and reshape the array $\mathbf{A} = (a_{\mathbf{k}})_{\mathbf{k} \in I_{\mathbf{N}}} \in \mathbb{C}^{N_1 \times \ldots \times N_d}$ into a vector $\mathrm{vec}\,\mathbf{A} = \mathbf{a} = (a_k)_{k=0}^{P-1}$ by

$$a_{k_1 + N_1 k_2 + N_1 N_2 k_3 + \ldots + N_1 \ldots N_{d-1} k_d} := a_{\mathbf{k}}\,, \quad \mathbf{k} = (k_j)_{j=1}^d \in I_{\mathbf{N}}\,,$$

then we can rewrite the d-dimensional DFT as a matrix-vector product

$$\begin{aligned} \mathrm{vec}\,\hat{\mathbf{A}} &= (\mathbf{F}_{N_d} \otimes \cdots \otimes \mathbf{F}_{N_1})\,\mathrm{vec}\,\mathbf{A} \\ &= (\mathbf{F}_{N_d} \otimes \mathbf{I}_{P/N_d}) \ldots (\mathbf{I}_{P/N_1 N_2} \otimes \mathbf{F}_{N_2} \otimes \mathbf{I}_{N_1})(\mathbf{I}_{P/N_1} \otimes \mathbf{F}_{N_1})\,\mathrm{vec}\,\mathbf{A} \end{aligned}$$

Thus the row–column method can be reinterpreted as the application of the one-dimensional DFT(N_j) to subvectors of $\mathrm{vec}\,\mathbf{A}$. Similarly as for the two-dimensional FFT, we can now again employ the factorization of the Fourier matrices $\mathbf{F}_{N_j}$ and reorder the multiplications to save operations by the nesting method. Using for example a similar factorization as in (5.49), we arrive for $d = 3$ with $N_1 = K_1 K_2$, $N_2 = L_1 L_2$, and $N_3 = M_1 M_2$ that

$$\begin{aligned} \mathrm{vec}\,\hat{\mathbf{A}} = {} & \big(\mathbf{P}_{N_3} \otimes \mathbf{P}_{N_2} \otimes \mathbf{P}_{N_1}\big)\big((\mathbf{I}_{M_2} \otimes \mathbf{F}_{M_1}) \otimes (\mathbf{I}_{L_2} \otimes \mathbf{F}_{L_1}) \otimes (\mathbf{I}_{K_1} \otimes \mathbf{F}_{K_1})\big) \\ & \cdot \big(\mathbf{D}_{M_1 M_2} \otimes \mathbf{D}_{L_1 L_2} \otimes \mathbf{D}_{K_1 K_2}\big)\big((\mathbf{F}_{M_2} \otimes \mathbf{I}_{M_1}) \otimes (\mathbf{F}_{L_2} \otimes \mathbf{I}_{L_1}) \otimes (\mathbf{F}_{K_2} \otimes \mathbf{I}_{K_1})\big)\,\mathrm{vec}\,\mathbf{A}\,. \end{aligned}$$

As for the two-dimensional DFT, we can save multiplications by precomputing the diagonal matrix $\mathbf{D}_{M_1 M_2} \otimes \mathbf{D}_{L_1 L_2} \otimes \mathbf{D}_{K_1 K_2}$ and multiplying it to the vectorized array at once. For comparison, using the row–column method, we multiply with the three matrices $\mathbf{D}_{M_1 M_2} \otimes \mathbf{I}_{N_2} \otimes \mathbf{I}_{N_1}$, $\mathbf{I}_{N_3} \otimes \mathbf{D}_{L_1 L_2} \otimes \mathbf{I}_{N_1}$, and $\mathbf{I}_{N_3} \otimes \mathbf{I}_{N_2} \otimes \mathbf{D}_{K_1 K_2}$ separately.

Table 5.4 Numbers μ, α of complex multiplications and additions required by various multidimensional FFTs based on radix–2 FFT

Size	Row-column method, radix–2		Nesting method, radix–2	
	μ	α	μ	α
8×8	192	384	144	384
32×32	5120	10240	3840	10240
64×64	24576	49152	18432	49152
128×128	114688	229378	86096	229378
256×256	524288	1048576	393216	1048576
$8 \times 8 \times 8$	2304	4608	1344	4608
$64 \times 64 \times 64$	2359296	4718592	1376256	4718592

Generally, if we assume that the one-dimensional radix–2 FFT requires $\frac{N}{2}\log_2 N$ complex multiplications and $N\log_2 N$ complex additions, then the row–column method using this radix–2 FFT for the d-dimensional $\mathrm{DFT}(N_1 \times \ldots \times N_d)$ with $N_1 = \ldots = N_d = N = 2^t$ requires $(\log_2 N)^d \, \frac{N^d}{2}$ complex multiplications by (5.50). Applying the nesting method, we need only $(\log_2 N)\big(N^d - (\frac{N}{2})^d\big)$ multiplications by performing the multiplication with the diagonal matrix of precomputed twiddle factors; see Table 5.4. Here we have taken into account for both methods that the matrices of twiddle factors $\mathbf{D}_N$ possess $N/2$ ones such that the multiplication with $\mathbf{D}_N$ requires $N/2$ multiplications.

5.4 Sparse FFT

In the previous sections, we have derived fast algorithms to execute the $\mathrm{DFT}(N)$ for arbitrary input vectors $\mathbf{a} \in \mathbb{C}^N$. All fast algorithms possess the computational costs $\mathcal{O}(N \log N)$, where the split–radix FFT is one of the most efficient known FFTs up to now.

However, in recent years there has been some effort to derive so-called *sparse* FFTs with sublinear computational costs. These methods exploit a priori knowledge on the vector $\hat{\mathbf{a}}$ to be recovered. Frequently used assumptions are that $\hat{\mathbf{a}} \in \mathbb{C}^N$ is sparse or has only a small amount of significant frequencies. Often, further assumptions regard the recovery of frequencies in a certain quantized range. One can distinguish between probabilistic and deterministic algorithms on the one hand and between approximate and exact algorithms on the other hand. Further, many sparse FFTs employ special input data being different from the components of a given vector $\mathbf{a} \in \mathbb{C}^N$ to compute $\hat{\mathbf{a}} = \mathbf{F}_N\, \mathbf{a}$.

In this section, we restrict ourselves to *exact deterministic sparse* FFTs that use only the given components of $\mathbf{a} \in \mathbb{C}^N$ to evaluate $\hat{\mathbf{a}}$.

5.4.1 Single-Frequency Recovery

In the simplest case, where $\mathbf{a} \in \mathbb{C}^N$ possesses only one frequency, i.e., $\hat{\mathbf{a}} \in \mathbb{C}^N$ is 1-*sparse*. Let us assume that $\hat{\mathbf{a}} = \big(\hat{a}_k\big)_{k=0}^{N-1}$ has one nonzero component $|\hat{a}_{k_0}| \ge \theta > 0$ and $\hat{a}_k = 0$ for $k \in \{0, \ldots, N-1\} \setminus \{k_0\}$. In order to compute $\hat{\mathbf{a}}$, we only need to recover the index k_0 and the value $\hat{a}_{k_0} \in \mathbb{C}$.

Considering the inverse DFT(N), this assumption leads to

$$a_j = \frac{1}{N} \sum_{k=0}^{N-1} \hat{a}_k\, w_N^{-jk} = \frac{1}{N}\, \hat{a}_{k_0}\, w_N^{-jk_0}\,, \qquad j = 0, \ldots, N-1\,.$$

In particular,

$$a_0 = \frac{1}{N}\, \hat{a}_{k_0}\,, \quad a_1 = \frac{1}{N}\, \hat{a}_{k_0}\, w_N^{-k_0}\,.$$

Thus, only two components of $\mathbf{a}$ are sufficient to recover $\hat{\mathbf{a}}$, where

$$\hat{a}_{k_0} = N\, a_0\,, \quad w_N^{-k_0} = \frac{a_1}{a_0}\,.$$

More generally, two arbitrary components a_{j_1}, a_{j_2} of $\mathbf{a}$ with $j_1 \ne j_2$ yield

$$w_N^{-k_0(j_2-j_1)} = \frac{a_{j_2}}{a_{j_1}}\,, \quad \hat{a}_{k_0} = N\, a_{j_1}\, w_N^{k_0 j_1}\,,$$

where k_0 can be extracted from the first term. However, the above procedure is numerically instable for large N. Small perturbations in a_1 and a_0 may lead to a wrong result for k_0, since the values w_N^k lie denser on the unit circle for larger N.

We want to derive a numerically stable algorithm for the recovery of $\hat{\mathbf{a}}$. For simplicity, we assume that $N = 2^t$, $t \in \mathbb{N}\setminus\{1\}$. We introduce the *periodizations* $\hat{\mathbf{a}}^{(\ell)}$ of $\hat{\mathbf{a}}$ by

$$\hat{\mathbf{a}}^{(\ell)} := \left(\sum_{r=0}^{2^{t-\ell}-1} \hat{a}_{k+2^\ell r} \right)_{k=0}^{2^\ell-1}\,, \qquad \ell = 0, \ldots, t\,. \tag{5.51}$$

In particular, $\hat{\mathbf{a}}^{(t)} := \hat{\mathbf{a}}$, and the recursion

$$\hat{\mathbf{a}}^{(\ell)} = \big(\hat{a}_k^{(\ell+1)}\big)_{k=0}^{2^\ell-1} + \big(\hat{a}_{k+2^\ell}^{(\ell+1)}\big)_{k=0}^{2^\ell-1}\,, \qquad \ell = 0, \ldots, t-1\,, \tag{5.52}$$

is satisfied. The following lemma shows the close connection between the vector $\mathbf{a}$ being the inverse DFT of $\hat{\mathbf{a}}$ and the inverse DFTs of $\hat{\mathbf{a}}^{(\ell)}$ for $\ell = 0, \ldots, t-1$.

Lemma 5.25 *For the vectors* $\hat{\mathbf{a}}^{(\ell)} \in \mathbb{C}^{2^\ell}$, $\ell = 0, \ldots, t$, *in* (5.51), *we have*

$$\mathbf{a}^{(\ell)} := \mathbf{F}_{2^\ell}^{-1}\, \hat{\mathbf{a}}^{(\ell)} = 2^{t-\ell}\, \big(a_{2^{t-\ell}j}\big)_{j=0}^{2^\ell-1},$$

where $\mathbf{a} = (a_j)_{j=0}^{N-1} = \mathbf{F}_N^{-1}\, \hat{\mathbf{a}} \in \mathbb{C}^N$ *is the inverse* DFT *of* $\hat{\mathbf{a}} \in \mathbb{C}^N$.

Proof We obtain by (5.51) that

$$\begin{aligned} a_j^{(\ell)} &= \frac{1}{2^\ell} \sum_{k=0}^{2^\ell-1} \hat{a}_k^{(\ell)}\, w_{2^\ell}^{-jk} = \frac{1}{2^\ell} \sum_{k=0}^{2^\ell-1} \Big(\sum_{r=0}^{2^{t-\ell}-1} \hat{a}_{k+2^\ell r}\Big)\, w_{2^\ell}^{-jk} \\ &= \frac{1}{2^\ell} \sum_{k=0}^{2^t-1} \hat{a}_k\, w_{2^\ell}^{-jk} = \frac{1}{2^\ell} \sum_{k=0}^{N-1} \hat{a}_k\, w_N^{-(2^{t-\ell}j)k} = 2^{t-\ell}\, a_{2^{t-\ell}j} \end{aligned}$$

for all $j = 0, \ldots, 2^\ell - 1$. ■

We observe that all periodizations $\hat{\mathbf{a}}^{(\ell)}$ of $\hat{\mathbf{a}}^{(t)} = \hat{\mathbf{a}}$ are again 1-sparse, where the index of the nonzero frequency may change according to (5.52), while the magnitude of the nonzero frequency is always the same. For example, for $\hat{\mathbf{a}} = \hat{\mathbf{a}}^{(3)} = (0, 0, 0, 0, 0, 0, 1, 0)^\top$, we find

$$\hat{\mathbf{a}}^{(2)} = (0, 0, 1, 0)^\top, \quad \hat{\mathbf{a}}^{(1)} = (1, 0)^\top, \quad \hat{\mathbf{a}}^{(0)} = (1).$$

Denoting the index of the nonzero entry of $\hat{\mathbf{a}}^{(\ell)}$ by $k_0^{(\ell)}$, the recursion (5.52) implies that

$$k_0^{(\ell)} = \begin{cases} k_0^{(\ell+1)} & 0 \le k_0^{(\ell+1)} \le 2^\ell - 1, \\ k_0^{(\ell+1)} - 2^\ell & 2^\ell \le k_0^{(\ell+1)} \le 2^{\ell+1} - 1, \end{cases} \tag{5.53}$$

while $\hat{a}_{k_0} = \hat{a}^{(t)}_{k_0^{(t)}} = \hat{a}^{(t-1)}_{k_0^{(t-1)}} = \ldots = \hat{a}^{(0)}_{k_0^{(0)}}$. Thus, fixing $\hat{a}_{k_0} = N\, a_0$, a robust recovery of k_0 can be achieved by recursive computation of the indices $k_0^{(0)}, \ldots, k_0^{(t)}$.

Set $k_0^{(0)} := 0$, since obviously $\hat{\mathbf{a}}^{(0)} = \hat{a}_0^{(0)} = N\, a_0$. Now, we want to evaluate $k_0^{(1)}$. Using Lemma 5.25, we consider now $a_1^{(1)} = 2^{t-1}\, a_{2^{t-1}} = 2^{t-1}\, a_{N/2}$. By assumption, we find

$$a_1^{(1)} = \frac{1}{2} \sum_{k=0}^{1} \hat{a}_k^{(1)}\, w_2^{-k} = \frac{1}{2}\, \hat{a}_{k_0^{(1)}}\, (-1)^{-k_0^{(1)}} = \frac{1}{2}\, \hat{a}_{k_0}\, (-1)^{-k_0^{(1)}},$$

while $a_0^{(1)} = 2^{t-1} a_0 = \frac{1}{2} \hat{a}_{k_0}$. It follows that $k_0^{(1)} = k_0^{(0)} = 0$ if $a_0^{(1)} = a_1^{(1)}$, i.e., if $a_0 = a_{N/2}$ and $k_0^{(1)} = 1$ if $a_0 = -a_{N/2}$. Hence we set $k_0^{(1)} = k_0^{(0)}$ if $|a_0 - a_{N/2}| \le |a_0 + a_{N/2}|$ and $k_0^{(1)} = k_0^{(0)} + 1$ otherwise.

Generally, assuming that $k_0^{(\ell)}$ is known, by (5.53) we have only to decide whether $k_0^{(\ell+1)} = k_0^{(\ell)}$ or $k_0^{(\ell+1)} = k_0^{(\ell)} + 2^\ell$. We consider $a_1^{(\ell+1)} = 2^{t-(\ell+1)} a_{2^{t-(\ell+1)}}$ and $a_0^{(\ell+1)} = 2^{t-(\ell+1)} a_0$ and conclude that

$$a_1^{(\ell+1)} = \frac{1}{2^{\ell+1}} \hat{a}_{k_0}^{(\ell+1)} w_{2^{\ell+1}}^{-k_0^{(\ell+1)}} = \frac{1}{2^{\ell+1}} \hat{a}_{k_0} w_{2^{\ell+1}}^{-k_0^{(\ell+1)}}, \quad a_0^{(\ell+1)} = \frac{1}{2^{\ell+1}} \hat{a}_{k_0} .$$

Thus we choose $k_0^{(\ell+1)} = k_0^{(\ell)}$ if $|a_1^{(\ell+1)} - a_0^{(\ell+1)} w_{2^{\ell+1}}^{-k_0^{(\ell)}}| \le |a_1^{(\ell+1)} + a_0^{(\ell+1)} w_{2^{\ell+1}}^{-k_0^{(\ell)}}|$, or equivalently, if

$$|a_{2^{t-\ell-1}} - a_0 w_{2^{\ell+1}}^{-k_0^{(\ell)}}| \le |a_{2^{t-\ell-1}} + a_0 w_{2^{\ell+1}}^{-k_0^{(\ell)}}|$$

and $k_0^{(\ell+1)} = k_0^{(\ell)} + 2^\ell$ otherwise. Proceeding in this way, we compute a_{k_0} and $k_0 = k_0^{(t)}$ employing the $t+1$ values $a_0, a_{2^{t-1}}, a_{2^{t-2}}, \ldots, a_2, a_1$ with computational cost of $\mathcal{O}(\log N)$.

Algorithm 5.26 (Robust Sparse FFT for Single-Frequency Recovery)
Input: $N = 2^t$ *with* $t \in \mathbb{N} \setminus \{1\}$, *components* $a_0, a_{2^\ell} \in \mathbb{C}$, $\ell = 0, \ldots, 2^{t-1}$, *of* $\mathbf{a} = (a_j)_{j=0}^{N-1} \in \mathbb{C}^N$.

1. *Compute* $\hat{a} := N a_0$.
2. *Set* $k_0 := 0$.
3. *For* $\ell = 0, \ldots, t-1$ *if* $|a_{2^{t-\ell-1}} - a_0 w_{2^{\ell+1}}^{-k_0}| > |a_{2^{t-\ell-1}} + a_0 w_{2^{\ell+1}}^{-k_0}|$, *then* $k_0 := k_0 + 2^\ell$.

Output: *index* $k_0 \in \{0, \ldots, N-1\}$, $\hat{a}_{k_0} := \hat{a}$.
Computational cost: $\mathcal{O}(\log N)$.

5.4.2 Recovery of Vectors with One Frequency Band

The above idea can be simply transferred to the fast recovery of vectors $\hat{\mathbf{a}}$ possessing only one *frequency band of short support with given length*; see also [326]. Assume that $\hat{\mathbf{a}} = (\hat{a}_k)_{k=0}^{N-1}$ possesses a short support of given length M, i.e., we assume that there exists an index $\mu \in \{0, \ldots, N-1\}$ such that all non-vanishing components of $\hat{\mathbf{a}}$ have their indices in the support set

$$I_M := \{\mu, (\mu+1) \bmod N, \ldots, (\mu+M-1) \bmod N\},$$

while $\hat{a}_k = 0$ for $k \in \{0, \ldots, N-1\} \setminus I_M$. We assume that the frequency band is chosen of minimal size and that $|\hat{a}_\mu| \geq \theta > 0$ and $|\hat{a}_{(\mu+M-1) \bmod N}| \geq \theta > 0$ are significant frequencies; then M is called *support length* of $\hat{\mathbf{a}}$.

For example, both vectors $(0, 0, 0, 1, 2, 0, 2, 0)^\top$ and $(2, 0, 0, 0, 0, 1, 2, 0)^\top$ have a support length $M = 4$, where $\mu = 3$, $I_4 = \{3, 4, 5, 6\}$ for the first vector and $\mu = 5$ and $I_4 = \{5, 6, 7, 0\}$ for the second vector.

In order to recover $\hat{\mathbf{a}} \in \mathbb{C}^N$ with support length M, we determine $L \in \mathbb{N}$ such that $2^{L-1} < M \leq 2^L$. Then we compute in a first step the $(L+1)$-periodization $\hat{\mathbf{a}}^{(L+1)}$ of $\hat{\mathbf{a}}$ applying a DFT(2^{L+1}). More precisely, by Lemma 5.25 we have to compute

$$\hat{\mathbf{a}}^{(L+1)} = 2^{t-L-1}\, \mathbf{F}_{2^{L+1}} \big(a_{2^{t-L-1}j}\big)_{j=0}^{2^{L+1}-1}\,.$$

By (5.51) it follows that $\hat{\mathbf{a}}^{(L+1)}$ also possesses a support of length M, where the nonzero components corresponding to the support set are already the desired nonzero components that occur also in $\hat{\mathbf{a}}$. Moreover, the first support index $\mu^{(L+1)}$ of the support set of $\hat{\mathbf{a}}^{(L+1)}$ is uniquely determined. Thus, to recover $\hat{\mathbf{a}}$, we only need to compute the correct first support index $\mu = \mu^{(t)}$ in order to shift the found nonzero coefficients to their right places. This problem is now very similar to the problem to find the single support index k_0 for single-frequency recovery. Let $\mu^{(\ell)}$ denote the first support indices of $\hat{\mathbf{a}}^{(\ell)}$ for $\ell = L+1, \ldots, t$. As before, (5.52) implies that

$$\mu^{(\ell)} = \begin{cases} \mu^{(\ell+1)} & 0 \leq \mu^{(\ell+1)} \leq 2^\ell - 1\,, \\ \mu^{(\ell+1)} - 2^\ell & 2^\ell \leq \mu^{(\ell+1)} \leq 2^{\ell+1} - 1\,. \end{cases}$$

Conversely, for given $\mu^{(\ell)}$ the next index $\mu^{(\ell+1)}$ can only take the values $\mu^{(\ell)}$ or $\mu^{(\ell)} + 2^\ell$.

Example 5.27 Assume that we have a given vector $\mathbf{a} \in \mathbb{C}^{16}$ and a priori knowledge that $\mathbf{a}$ possesses a frequency band of support length $M = 3$. We want to recover $\hat{\mathbf{a}} = (\hat{a}_k)_{k=0}^{15} \in \mathbb{C}^{16}$ with $\hat{a}_{13} = \hat{a}_{14} = \hat{a}_{15} = 1$ and $\hat{a}_k = 0$ for $k = 0, \ldots, 12$.

From $M = 3$ we obtain $L = 2$ and $2^{L+1} = 8$. In a first step, we compute the periodized vector $\hat{\mathbf{a}}^{(3)} \in \mathbb{C}^8$ by applying a DFT(8) to the vector $(a_{2j})_{j=0}^7$. This gives

$$\hat{\mathbf{a}}^{(3)} = (0, 0, 0, 0, 0, 1, 1, 1)^\top\,.$$

Obviously, $\hat{\mathbf{a}}^{(3)}$ has also support length 3 and the first support index $\mu^{(3)} = 5$. In the last step, we need to recover $\hat{\mathbf{a}} = \hat{\mathbf{a}}^{(4)}$ from $\hat{\mathbf{a}}^{(3)}$. By (5.52), we only need to find out whether $\mu^{(4)} = 5$ or $\mu^{(4)} = 5 + 8 = 13$, where in the second case the already computed nonzero components of $\hat{\mathbf{a}}^{(3)}$ only need an index shift of 8. □

Generally, if $\mu^{(\ell+1)} = \mu^{(\ell)}$ is true, then each component of

$$\mathbf{F}_{2^{\ell+1}}^{-1}\, \hat{\mathbf{a}}^{(\ell+1)} = 2^{t-\ell-1} \big(a_{2^{t-\ell-1}j}\big)_{j=0}^{2^{\ell+1}-1}$$

satisfies

$$2^{t-\ell-1} a_{2^{t-\ell-1}j} = \sum_{k=\mu^{(\ell)}}^{\mu^{(\ell)}+M-1} \hat{a}^{(\ell+1)}_{k \bmod 2^{\ell+1}} \, w_{2^{\ell+1}}^{-jk} = w_{2^{\ell+1}}^{-j\mu^{(\ell)}} \sum_{k=0}^{M-1} \hat{a}^{(L+1)}_{(k+\mu^{(L+1)}) \bmod 2^{L+1}} \, w_{2^{\ell+1}}^{-jk},$$

where we have used that $\hat{\mathbf{a}}^{(L+1)}$ already contains the correct component values. Similarly, if $\mu^{(\ell+1)} = \mu^{(\ell)} + 2^\ell$, then it follows that

$$\begin{aligned} 2^{t-\ell-1} a_{2^{t-\ell-1}j} &= \sum_{k=\mu^{(\ell)}+2^\ell}^{\mu^{(\ell)}+2^\ell+M-1} \hat{a}^{(\ell+1)}_{k \bmod 2^{\ell+1}} \, w_{2^{l+1}}^{-jk} \\ &= (-1)^j \, w_{2^{\ell+1}}^{-j\mu^{(\ell)}} \sum_{k=0}^{M-1} \hat{a}^{(L+1)}_{(k+\mu^{(L+1)}) \bmod 2^{L+1}} \, w_{2^{\ell+1}}^{-jk} . \end{aligned}$$

We choose now j_ℓ as an odd integer in $\{1, 3, 5, \ldots, 2^{\ell+1}-1\}$ such that $a_{2^{t-\ell-1}j_\ell} \neq 0$. This is always possible, since if the vector $(a_{2^{t-\ell-1}(2j+1)})_{j=0}^{2^\ell-1}$ had M or more zero components, the equations above would imply that $\hat{\mathbf{a}}^{(L+1)} = \mathbf{0}$, contradicting the assumption. Now, taking $j = j_\ell$, we need to compare $a_{2^{t-\ell-1}j_\ell}$ with

$$A_\ell := 2^{\ell+1-t} \, w_{2^{\ell+1}}^{-\mu^{(\ell)} j_\ell} \sum_{k=0}^{M-1} \hat{a}^{(L+1)}_{k+\mu^{(L+1)} \bmod 2^{L+1}} \, w_{2^{\ell+1}}^{-kj_\ell} .$$

If

$$|A_\ell - a_{2^{t-\ell-1}j_\ell}| \le |A_\ell + a_{2^{t-\ell-1}j_\ell}|,$$

then we have to take $\mu^{(\ell+1)} = \mu^{(\ell)}$, and we take $\mu^{(\ell+1)} = \mu^{(\ell)} + 2^\ell$ otherwise.

Algorithm 5.28 (Sparse FFT for a Vector with Small Frequency Band)
Input: $N = 2^t$ *with* $t \in \mathbb{N} \setminus \{1\}$, $\mathbf{a} \in \mathbb{C}^N$ *vector with small frequency band, upper bound* N *of support length* M *of* $\hat{\mathbf{a}}$.

1. *Compute* $L := \lceil \log_2 M \rceil$.
2. *If* $L \ge t-1$ *compute* $\hat{\mathbf{a}} := \mathbf{F}_N \, \mathbf{a}$ *by* FFT *of length* N.
3. *If* $L < t-1$ *then*

   3.1. *Set* $\mathbf{a}^{(L+1)} := (a_{2^{t-L-1}j})_{j=0}^{2^{L+1}-1}$ *and compute* $\hat{\mathbf{a}}^{(L+1)} := \mathbf{F}_{2^{L+1}} \, \mathbf{a}^{(L+1)}$ *by* FFT *of length* 2^{L+1}.
   3.2. *Determine the first support index* μ *of* $\hat{\mathbf{a}}^{(L+1)}$ *as follows:*
   Compute

$$e_0 := \sum_{k=0}^{M-1} |\hat{a}_k^{(L+1)}|^2.$$

For $k = 1, \ldots, 2^{L+1} - 1$ *compute*

$$e_k := e_{k-1} - |\hat{a}^{(L+1)}_{(k-1) \bmod 2^{L+1}}|^2 + |\hat{a}^{(L+1)}_{(k+M-1) \bmod 2^{L+1}}|^2.$$

Compute $\mu := \arg\max \{e_k : k = 0, \ldots, 2^{L+1} - 1\}$ *and set* $\mu_0 := \mu$.

3.3. *For* $\ell = L+1, \ldots, t-1$ *choose* $j \in \{1, 3, \ldots, 2^\ell - 1\}$ *such that* $|a_{2^{t-\ell-1}j}| > \theta$. *Compute*

$$A := 2^{\ell+1-t}\, w_{2^{\ell+1}}^{-\mu j} \sum_{k=0}^{M-1} \hat{a}^{(L+1)}_{k+\mu \bmod 2^{L+1}}\, w_{2^{\ell+1}}^{-kj}\,.$$

If $|A - a_{2^{t-\ell-1}j}| > |A + a_{2^{t-\ell-1}j}|$, *then* $\mu := \mu + 2^\ell$.

3.4. *Set* $\hat{\mathbf{a}} := \mathbf{0} \in \mathbb{C}^N$.

3.5. *For* $r = 0, \ldots, M - 1$ *set*

$$\hat{a}_{(\mu+r) \bmod N} := \hat{a}^{(L+1)}_{(\mu_0+r) \bmod 2^{L+1}}\,.$$

Output: $\hat{\mathbf{a}} \in \mathbb{C}^N$, *first support index* $\mu \in \{0, \ldots, N - 1\}$.
Computational cost: $\mathscr{O}(M \log N)$.

Let us shortly study the computational cost to execute the sparse FFT in Algorithm 5.28. If $M \geq N/4$, then the usual FFT of length N should be used to recover $\hat{\mathbf{a}}$ with $\mathscr{O}(N \log N)$ flops. For $M < N/4$, Step 3.1 requires $\mathscr{O}((L + 1)2^{L+1}) = \mathscr{O}(M \log M)$ flops, since $2^{L+1} < 4M$. Step 3.2 involves the computation of energies e_k with $\mathscr{O}(2^{L+1}) = \mathscr{O}(M)$ flops. In Step 3.3 we have to perform $t-L-1$ scalar products of length M and $t-L-1$ comparisons requiring computational costs of $\mathscr{O}(M(\log N - \log M))$. Finding j requires at most $M(t-L-1)$ comparisons. The complete algorithm is governed by the DFT(2^{L+1}) and the computations in Step 3.5 with overall computational cost of $\mathscr{O}(M \log N)$.

5.4.3 Recovery of Sparse Fourier Vectors

Assume now that $\hat{\mathbf{a}} = (\hat{a}_j)_{j=0}^{N-1} \in (\mathbb{R}_+ + \mathrm{i}\,\mathbb{R}_+)^N$, i.e., $\mathrm{Re}\,\hat{a}_j \geq 0$ and $\mathrm{Im}\,\hat{a}_j \geq 0$ for all $j = 0, \ldots, N - 1$, is *M-sparse*, i.e., $\hat{\mathbf{a}}$ possesses M nonzero components, where $M \in \mathbb{N}_0$ with $M \leq N = 2^t$, $t \in \mathbb{N}$, is not a priori known. Let $\mathbf{a} = \mathbf{F}_N^{-1}\,\hat{\mathbf{a}} = (a_k)_{k=0}^{N-1}$ be the given vector of length N. We follow the ideas in [328] and want to derive a fast and numerically stable algorithm to reconstruct $\hat{\mathbf{a}}$ from adaptively chosen components of $\mathbf{a}$. For that purpose, we again use the periodized vectors $\hat{\mathbf{a}}^{(\ell)} \in (\mathbb{R}_+ + \mathrm{i}\,\mathbb{R}_+)^{2^\ell}$ as defined in (5.51).

The basic idea consists in recursive evaluation of the vectors $\hat{\mathbf{a}}^{(\ell)}$ in (5.51) for $\ell = 0, \ldots, t$, using the fact that the sparsities M_ℓ of $\hat{\mathbf{a}}^{(\ell)}$ satisfy

$$M_0 \le M_1 \le M_2 \le \ldots \le M_t = M.$$

In particular, no cancelations can occur, and the components of $\hat{\mathbf{a}}^{(\ell)}$ are contained in the first quadrant of the complex plane, i.e., $\mathrm{Re}\,\hat{a}_j^{(\ell)} \ge 0$ and $\mathrm{Im}\,\hat{a}_j^{(\ell)} \ge 0$ for $j = 0, \ldots, 2^\ell - 1$.

We start by considering $\hat{\mathbf{a}}^{(0)}$. Obviously,

$$\hat{\mathbf{a}}^{(0)} = \sum_{k=0}^{N-1} \hat{a}_k = N\, a_0.$$

Since $\hat{\mathbf{a}}$ possesses all components in the first quadrant, we can conclude that for $a_0 = 0$, the vector $\hat{\mathbf{a}}$ is the zero vector, i.e., it is 0-sparse.

Having found $\hat{\mathbf{a}}^{(0)} = N\, a_0 > 0$, we proceed and consider $\hat{\mathbf{a}}^{(1)}$. By (5.52), we find $\hat{\mathbf{a}}^{(1)} = (\hat{a}_0^{(1)}, \hat{a}_1^{(1)})^\top$, where $\hat{a}_0^{(1)} + \hat{a}_1^{(1)} = \hat{\mathbf{a}}^{(0)} = N\, a_0$ is already known. Applying Lemma 5.25, we now choose the component $\frac{N}{2} a_{N/2} = \hat{a}_0^{(1)} - \hat{a}_1^{(1)}$. Hence, with $\hat{a}_1^{(1)} = \hat{\mathbf{a}}^{(0)} - \hat{a}_0^{(1)}$, we obtain $\frac{N}{2}\, a_{N/2} = 2\,\hat{a}_0^{(1)} - \hat{\mathbf{a}}^{(0)}$, i.e.,

$$\hat{a}_0^{(1)} = \frac{1}{2}\,\big(\hat{\mathbf{a}}^{(0)} + \frac{N}{2} a_{N/2}\big)\,, \quad \hat{a}_1^{(1)} = \hat{\mathbf{a}}^{(0)} - \hat{a}_0^{(1)}\,.$$

If $\hat{a}_0^{(1)} = 0$, we can conclude that all even components of $\hat{\mathbf{a}}$ vanish, and we do not need to consider them further. If $\hat{a}_1^{(1)} = 0$, it follows analogously that all odd components of $\hat{\mathbf{a}}$ are zero.

Generally, having computed $\hat{\mathbf{a}}^{(\ell)}$ at the ℓth level of iteration, let $M_\ell \le 2^\ell$ be the obtained sparsity of $\hat{\mathbf{a}}^{(\ell)}$, and let

$$0 \le n_1^{(\ell)} < n_2^{(\ell)} < \ldots < n_{M_\ell}^{(\ell)} \le 2^\ell - 1$$

be the indices of the corresponding nonzero components of $\hat{\mathbf{a}}^{(\ell)}$. From (5.52) we can conclude that $\hat{\mathbf{a}}^{(\ell+1)}$ possesses at most $2\, M_\ell$ nonzero components, and we only need to consider $\hat{a}_{n_k}^{(\ell+1)}$ and $\hat{a}_{n_k+2^\ell}^{(\ell+1)}$ for $k = 1, \ldots, M_\ell$ as candidates for nonzero entries, while all other components of $\hat{\mathbf{a}}^{(\ell+1)}$ can be assumed to be zero. Moreover, (5.52) provides already M_ℓ conditions on these values

$$\hat{a}_{n_k}^{(\ell+1)} + \hat{a}_{n_k+2^\ell}^{(\ell+1)} = \hat{a}_{n_k}^{(\ell)}\,, \quad k = 1, \ldots, M_\ell\,.$$

Therefore, we need only M_ℓ further data to recover $\hat{\mathbf{a}}^{(\ell+1)}$. In particular, we can show the following result; see [328].

Theorem 5.29 *Let $\hat{\mathbf{a}}^{(\ell)}$, $\ell = 0, \ldots, t$, be the vectors defined in* (5.51). *Then for each $\ell = 0, \ldots, t-1$, we have:*

If $\hat{\mathbf{a}}^{(\ell)}$ is M_ℓ-sparse with support indices $0 \le n_1^{(\ell)} < n_2^{(\ell)} < \ldots < n_{M_\ell}^{(\ell)} \le 2^\ell - 1$, then the vector $\hat{\mathbf{a}}^{(\ell+1)}$ can be uniquely recovered from $\hat{\mathbf{a}}^{(\ell)}$ and M_ℓ components $a_{j_1}, \ldots, a_{j_{M_\ell}}$ of $\mathbf{a} = \mathbf{F}_N^{-1}\,\hat{\mathbf{a}}$, where the indices $j_1, \ldots, j_{M_\ell}$ are taken from the set $\{2^{t-\ell-1}(2j+1) : j = 0, \ldots, 2^\ell - 1\}$ such that the matrix

$$\big(w_N^{j_p n_r^{(\ell)}}\big)_{p,r=1}^{M_\ell} = \big(\mathrm{e}^{-2\pi\mathrm{i}\, j_p n_r^{(\ell)}/N}\big)_{p,r=1}^{M_\ell} \in \mathbb{C}^{M_\ell \times M_\ell}$$

is invertible.

Proof Using the vector notation $\hat{\mathbf{a}}_0^{(\ell+1)} := (\hat{a}_k^{(\ell+1)})_{k=0}^{2^\ell-1}$ and $\hat{\mathbf{a}}_1^{(\ell+1)} := (\hat{a}_k^{(\ell+1)})_{k=2^\ell}^{2^{\ell+1}-1}$, the recursion (5.52) yields

$$\hat{\mathbf{a}}^{(\ell)} = \hat{\mathbf{a}}_0^{(\ell+1)} + \hat{\mathbf{a}}_1^{(\ell+1)}. \tag{5.54}$$

Therefore, for given $\hat{\mathbf{a}}^{(\ell)}$, we only need to compute $\hat{\mathbf{a}}_0^{(\ell+1)}$ to recover $\hat{\mathbf{a}}^{(\ell+1)}$. By Lemma 5.25, we find

$$\begin{aligned}\big(a_{2^{t-\ell-1}j}\big)_{j=0}^{2^{\ell+1}-1} &= \mathbf{a}^{(\ell+1)} = \mathbf{F}_{2^{\ell+1}}^{-1}\begin{pmatrix}\hat{\mathbf{a}}_0^{(\ell+1)}\\ \hat{\mathbf{a}}_1^{(\ell+1)}\end{pmatrix} = \mathbf{F}_{2^{\ell+1}}^{-1}\begin{pmatrix}\hat{\mathbf{a}}_0^{(\ell+1)}\\ \hat{\mathbf{a}}^{(\ell)} - \hat{\mathbf{a}}_0^{(\ell+1)}\end{pmatrix}\\ &= 2^{-\ell-1}\big(w_{2^{\ell+1}}^{-jk}\big)_{j,k=0}^{2^{\ell+1}-1,2^\ell-1}\,\hat{\mathbf{a}}_0^{(\ell+1)} + 2^{-\ell-1}\big((-1)^j w_{2^{\ell+1}}^{-jk}\big)_{j,k=0}^{2^{\ell+1}-1,2^\ell-1}\big(\hat{\mathbf{a}}^{(\ell)} - \hat{\mathbf{a}}_0^{(\ell+1)}\big).\end{aligned} \tag{5.55}$$

Let now $0 \le n_1^{(\ell)} < n_1^{(\ell)} < \ldots < n_{M_\ell}^{(\ell)} \le 2^\ell - 1$ be the indices of the nonzero entries of $\hat{\mathbf{a}}^{(\ell)}$. Then by (5.54) also $\hat{\mathbf{a}}_0^{(\ell+1)}$ can have nonzero entries only at these components. We restrict the vectors accordingly to

$$\tilde{\hat{\mathbf{a}}}_0^{(\ell+1)} := \big(\hat{a}_{n_r^{(\ell)}}^{(\ell+1)}\big)_{r=1}^{M_\ell}, \quad \tilde{\hat{\mathbf{a}}}^{(\ell)} := \big(\hat{a}_{n_r^{(\ell)}}^{(\ell)}\big)_{r=1}^{M_\ell}.$$

Further, let $j_1, \ldots, j_{M_\ell}$ be distinct indices from $\{2^{t-\ell-1}(2r+1) : r = 0, \ldots, 2^\ell - 1\}$, i.e., we have $j_p := 2^{t-\ell-1}(2\kappa_p + 1)$ with $\kappa_p \in \{0, \ldots, 2^\ell - 1\}$ for $p = 1, \ldots, M_\ell$. We now restrict the linear system (5.55) to the M_ℓ equations corresponding to these indices $j_1, \ldots, j_{M_\ell}$ and find

$$\mathbf{b}^{(\ell+1)} := \begin{pmatrix} a_{j_1} \\ \vdots \\ a_{j_{M_\ell}} \end{pmatrix} = \begin{pmatrix} a^{(\ell+1)}_{2\kappa_1+1} \\ \vdots \\ a^{(\ell+1)}_{2\kappa_{M_\ell}+1} \end{pmatrix} = \mathbf{A}^{(\ell+1)}\, \widetilde{\hat{\mathbf{a}}}_0^{(\ell+1)} - \mathbf{A}^{(\ell+1)}\, \big(\widetilde{\hat{\mathbf{a}}}^{(\ell)} - \widetilde{\hat{\mathbf{a}}}_0^{(\ell+1)}\big)\,, \tag{5.56}$$

where

$$\begin{aligned} \mathbf{A}^{(\ell+1)} &= 2^{-\ell-1}\,\big(w_N^{-j_p n_r^{(\ell)}}\big)_{p,r=1}^{M_\ell} \\ &= 2^{-\ell-1}\,\big(w_{2^\ell}^{-\kappa_p n_r^{(\ell)}}\big)_{p,r=1}^{M_\ell}\, \mathrm{diag}\big(\omega_{2^{\ell+1}}^{-n_1^{(\ell)}}, \ldots, \omega_{2^{\ell+1}}^{-n_{M_\ell}^{(\ell)}}\big)^\top . \end{aligned} \tag{5.57}$$

If $\mathbf{A}^{(\ell+1)}$ and $\big(w_{2^\ell}^{-\kappa_p n_r^{(\ell)}}\big)_{p,r=1}^{M_\ell}$, respectively, is invertible, it follows from (5.56) that

$$\mathbf{A}^{(\ell+1)}\, \widetilde{\hat{\mathbf{a}}}_0^{(\ell+1)} = \frac{1}{2}\,\big(\mathbf{b}^{(\ell+1)} + \mathbf{A}^{(\ell+1)}\, \widetilde{\hat{\mathbf{a}}}^{(\ell)}\big)\,. \tag{5.58}$$

Thus, to recover $\widetilde{\hat{\mathbf{a}}}_0^{(\ell+1)}$ we have to solve this system of M_ℓ linear equations in M_ℓ unknowns, and the components of $\hat{\mathbf{a}}^{(\ell+1)}$ are given by

$$\hat{a}_k^{(\ell+1)} = \begin{cases} \big(\widetilde{\hat{\mathbf{a}}}_0^{(\ell+1)}\big)_r & k = n_r^{(\ell)}\,, \\ \big(\widetilde{\hat{\mathbf{a}}}^{(\ell)}\big)_r - \big(\widetilde{\hat{\mathbf{a}}}_0^{(\ell+1)}\big)_r & k = n_r^{(\ell)} + 2^\ell\,, \\ 0 & \text{otherwise.} \end{cases}$$

This completes the proof. ∎

Theorem 5.29 yields that we essentially have to solve the linear system (5.58) in order to compute $\hat{\mathbf{a}}^{(\ell+1)}$ from $\hat{\mathbf{a}}^{(\ell)}$. We summarize this approach in the following algorithm, where the conventional FFT is used at each step as long as this is more efficient than solving the linear system (5.58).

Algorithm 5.30 (Recovery of Sparse Fourier Transformed Vector $\hat{\mathbf{a}} \in (\mathbb{R}_+ + \mathrm{i}\,\mathbb{R}_+)^N$)
Input: $N = 2^t$ *with* $t \in \mathbb{N}$, $\mathbf{a} = (a_j)_{j=0}^{N-1} \in \mathbb{C}^N$, $\theta > 0$ *shrinkage constant.*

1. *Set* $M := 0$ *and* $K := \{0\}$.
2. *If* $a_0 < \theta$, *then* $\hat{\mathbf{a}} = \mathbf{0}$ *and* $I^{(t)} = \emptyset$.
3. *If* $a_0 \geq \theta$, *then*

   3.1. *Set* $M := 1$, $I^{(0)} := \{0\}$, $\hat{\mathbf{a}}^{(0)} := N\,a_0$, *and* $\widetilde{\hat{\mathbf{a}}}^{(0)} := \hat{\mathbf{a}}^{(0)}$.
   3.2. *For* $\ell = 0$ *to* $t-1$ *do*

If $M^2 \geq 2^\ell$, *then choose* $\mathbf{b}^{(\ell+1)} := \big(a_{2p+1}^{(\ell+1)}\big)_{p=0}^{2^\ell-1} = \big(a_{2^{t-\ell-1}(2p+1)}\big)_{p=0}^{2^\ell-1} \in \mathbb{C}^M$ *and solve the linear system*

$$\mathbf{F}_{2^\ell}^{-1}\left(\operatorname{diag}(w_{2^{\ell+1}}^{-k})_{k=0}^{2^\ell-1}\right)\hat{\mathbf{a}}_0^{(\ell+1)} = \frac{1}{2}\left(2\,\mathbf{b}^{(\ell+1)} + \mathbf{F}_{2^\ell}^{-1}\left(\operatorname{diag}(w_{2^{\ell+1}}^{-k})_{k=0}^{2^\ell-1}\right)\hat{\mathbf{a}}^{(\ell)}\right)$$

using an FFT *of length* 2^ℓ.
Find the index set I_ℓ *of components in* $\hat{\mathbf{a}}^{(\ell)}$ *with* $\hat{a}_k^{(\ell)} \geq \theta$ *for* $k \in I^{(\ell)}$ *and set its cardinality* $M := |I_\ell|$,
else

- *Choose* M *indices* $j_p = 2^{t-\ell-1}(2\kappa_p + 1)$ *with* $\kappa_p \in \{0, \ldots, 2^\ell - 1\}$ *for* $p = 1, \ldots, M$ *such that*

$$\mathbf{A}^{(\ell+1)} := 2^{-\ell-1}\left(w_N^{-j_p r}\right)_{p=1,\ldots,M;\, r\in I^{(\ell)}}$$

is well-conditioned and set $K := K \cup \{j_1, \ldots, j_M\}$.
- *Choose the vector* $\mathbf{b}^{(\ell+1)} := (a_{j_p})_{p=1}^{M} \in \mathbb{C}^M$ *and solve the linear system*

$$\mathbf{A}^{(\ell+1)}\widetilde{\mathbf{a}}_0^{(\ell+1)} = \frac{1}{2}\left(\mathbf{b}^{(\ell+1)} + \mathbf{A}^{(\ell+1)}\,\widetilde{\mathbf{a}}^{(\ell)}\right).$$

- *Set* $\widetilde{\mathbf{a}}_1^{(\ell+1)} := \widetilde{\mathbf{a}}^{(\ell)} - \widetilde{\mathbf{a}}_0^{(\ell+1)}$ *and* $\widetilde{\mathbf{a}}^{(\ell+1)} := \left(\left(\widetilde{\mathbf{a}}_0^{(\ell+1)}\right)^\top, \left(\widetilde{\mathbf{a}}_1^{(\ell+1)}\right)^\top\right)^\top$.
- *Determine the index set* $I^{(\ell+1)} \subset \left(I^{(\ell)} \cup (I^{(\ell)} + 2^\ell)\right)$ *such that* $\hat{a}_k^{(\ell+1)} \geq \theta$ *for* $k \in I^{(\ell+1)}$. *Set* $M := |I^{(j+1)}|$.

Output: $I^{(t)}$ *is the set of active indices in* $\hat{\mathbf{a}}$ *with* $M = |I^{(t)}|$ *and* $\widetilde{\mathbf{a}} = \widetilde{\mathbf{a}}^{(t)} = (a_k)_{k\in I^{(t)}}$.
K *is the index set of used components of* $\mathbf{a}$.

Note that the matrices $\mathbf{A}^{(\ell+1)}$ are scaled partial Fourier matrices, namely, the restrictions of the Fourier matrix $\mathbf{F}_N^{-1}$ to the columns $n_1^{(\ell)}, \ldots, n_{M_\ell}^{(\ell)}$ and the rows $k_1, \ldots, k_{M_\ell}$. For given M_ℓ column indices $n_1^{(\ell)}, \ldots, n_{M_\ell}^{(\ell)}$, we are allowed to choose the row indices in a way such that a good condition of $\mathbf{A}^{(\ell+1)}$ is ensured. Observe that we can always choose $k_p = 2^{t-\ell-1}(2\kappa_p + 1)$ with $\kappa_p = p - 1$, for $p = 1, \ldots, M_\ell$ to ensure invertibility. This means, we can just take the first M_ℓ rows of $\mathbf{F}_{2^\ell}^{-1}$ in the product representation (5.57). However, the condition of this matrix can get very large for larger ℓ.

Example 5.31 Assume that we want to recover the 4-sparse vector $\hat{\mathbf{a}} \in \mathbb{C}^{128}$ with $\hat{a}_k = 1$ for $k \in I^{(7)} := \{1, 6, 22, 59\}$. For the periodizations of $\hat{\mathbf{a}}$, we find the sparsity and the index sets

$$I^{(0)} = \{0\}\,, \qquad M_0 = 1\,,$$

$$I^{(1)} = \{0, 1\}\,, \qquad M_1 = 2\,,$$

$$I^{(2)} = \{1, 2, 3\}\,, \qquad M_2 = 3\,,$$

$$\begin{aligned} I^{(3)} &= \{1, 3, 6\}\,, & M_3 &= 3\,, \\ I^{(4)} &= \{1, 6, 11\}\,, & M_4 &= 3\,, \\ I^{(5)} &= \{1, 6, 22, 27\}\,, & M_5 &= 4\,, \\ I^{(6)} &= \{1, 6, 22, 59\}\,, & M_6 &= 4\,. \end{aligned}$$

For $\ell = 0,\ 1,\ 2,\ 3$ we have $M_\ell^2 \geq 2^\ell$ and therefore just apply the FFT of length 2^ℓ as described in the first part of the algorithm to recover

$$\hat{\mathbf{a}}^{(4)} = (0, 1, 0, 0, 0, 0, 2, 0, 0, 0, 0, 1, 0, 0, 0, 0)^\top .$$

For $\ell = 4$, we have $M_4^2 < 2^4$ and apply the restricted Fourier matrix for the recovery of $\hat{\mathbf{a}}^{(5)}$. The index set $I^{(5)}$ of nonzero components of $\hat{\mathbf{a}}^{(5)}$ is a subset of $I^{(4)} \cup (I^{(4)} + 16) = \{1, 6, 11, 17, 22, 27\}$. We simply choose $k_p^{(4)} = 2^{7-4-1}(2\kappa_p + 1)$ with $\kappa_p = p - 1$ for $p = 1, \ldots, M_4$, i.e., $(k_1^{(4)}, k_2^{(4)}, k_3^{(4)}) = 2^2(1, 3, 5) = (4, 12, 20)$. The matrix $A^{(5)}$ reads

$$\mathbf{A}^{(5)} = \frac{1}{32}\,\big(w_{16}^{-(p-1)n_r^{(4)}}\big)_{p,r=1}^{3}\,\mathrm{diag}(w_{32}^{-1},\ w_{32}^{-6},\ w_{32}^{-11})^\top$$

and possesses the condition number $\|\mathbf{A}^{(5)}\|_2 \|(\mathbf{A}^{(5)})^{-1}\|_2 = 1.1923$. Solving a system with three linear equations in three unknowns yields $\hat{\mathbf{a}}^{(5)}$ and $I^{(5)} = \{1,\ 6,\ 22,\ 27\}$. Similarly, we find at the next iteration steps 5 and 6 the 4-by-4 coefficient matrices

$$\mathbf{A}^{(6)} = \frac{1}{64}\,\big(w_{32}^{-(p-1)n_r^{(5)}}\big)_{p,r=1}^{4}\,\mathrm{diag}(w_{64}^{-1},\ w_{64}^{-6},\ w_{64}^{-22},\ w_{64}^{-27})^\top$$

with condition number 4.7150 to recover $\hat{\mathbf{a}}^{(6)}$ and

$$\mathbf{A}^{(7)} = \frac{1}{128}\,\big(w_{64}^{-(p-1)n_r^{(6)}}\big)_{p,r=1}^{4}\,\mathrm{diag}(w_{128}^{-1},\ w_{128}^{-6},\ w_{128}^{-22},\ w_{128}^{-59})^\top$$

with condition number 21.2101 to recover $\hat{\mathbf{a}}^{(7)}$.

Thus, we have employed only the components a_{8k}, $k = 0, \ldots, 15$, in the first four iteration steps (for $\ell = 0, 1, 2, 3$) to recover $\hat{\mathbf{a}}^{(4)}$, the entries $a_{4(2k+1)}$, $k = 0, 1, 2,$ at level $\ell = 4$; $\widehat{x}_{2(2k+1)}$, $k = 0, 1, 2, 3$, at level $\ell = 5$; and $\widehat{x}_{2k+1}$, $k = 0, 1, 2, 3$, at level $\ell = 6$. Summing up, we have to employ 27 of the 127 Fourier components to recover $\hat{\mathbf{a}}$, while the computational cost is governed by solving the FFT of length 16 for getting $\hat{\mathbf{a}}$ (up to level 3) the 3 linear systems with the coefficient matrices $\mathbf{A}^{(5)}$, $\mathbf{A}^{(6)}$, and $\mathbf{A}^{(7)}$, respectively. □

While the condition numbers of $\mathbf{A}^{(\ell+1)}$ in the above example are still moderate, the choice $\kappa_p = p - 1$, $p = 1, \ldots, M_\ell$, does not always lead to good condition numbers if N is large. For example, for $N = 1024$, the recovery of a 4-sparse vector with $I^{(10)} = I^{(9)} = \{1, 6, 22, 59\}$ employs a 4-by-4 matrix $\mathbf{A}^{(10)}$ at the last iteration level $\ell = 9$ possessing the condition number 22742.

In order to be able to efficiently compute the linear system (5.58), we want to preserve a Vandermonde structure for the matrix factor $\big(w_{2^\ell}^{\kappa_p n_r^{(\ell)}}\big)_{p,r=1}^{M_\ell}$ of $\mathbf{A}^{(\ell+1)}$ and at the same time ensure a good condition of this matrix. Therefore, we only consider matrices of the form $\big(w_{2^\ell}^{(p-1)\sigma^{(\ell)} n_r^{(\ell)}}\big)_{p,r=1}^{M_\ell}$, i.e., we set $\kappa_p := (p-1)\sigma^{(\ell)}$ for $p = 1, \ldots, M_\ell$. The parameter $\sigma^{(\ell)} \in \{1, 2, \ldots, 2^\ell - 1\}$ has to be chosen in a way such that $\big(w_{2^\ell}^{(p-1)\sigma^{(\ell)} n_r^{(\ell)}}\big)_{p,r=1}^{M_\ell}$ and thus $\mathbf{A}^{(\ell+1)}$ is well-conditioned.

In [328], it has been shown that the condition of $\mathbf{A}^{(\ell+1)}$ mainly depends on the distribution of the knots $w_{2^\ell}^{\sigma^{(\ell)} n_r^{(\ell)}}$ on the unit circle or equivalently on the distribution of $\sigma^{(\ell)} n_r^{(\ell)} \bmod 2^\ell$ in the interval $[0, 2^\ell - 1]$. One simple heuristic approach suggested in [328] to choose $\sigma^{(\ell)}$ is the following procedure. Note that this procedure is only needed if $M_\ell^2 < 2^\ell$.

Algorithm 5.32 (Choice of $\sigma^{(\ell)}$ to Compute $\mathbf{A}^{(\ell+1)}$)
Input: $\ell \geq 1$, $M_{\ell-1}$, M_ℓ, $I^{(\ell)} = \{n_1^{(\ell)}, \ldots, n_{M_\ell}^{(\ell)}\}$, $\sigma^{(\ell-1)}$ *if available.*
If $\ell \leq 3$ *or* $M_\ell = 1$, *choose* $\sigma^{(\ell)} := 1$
else
if $M_{\ell-1} = M_\ell$ *then*
if $\sigma^{(\ell-1)}$ *is given, then* $\sigma^{(\ell)} := 2\sigma^{(\ell-1)}$
else set $\ell := \ell - 1$ *and start the algorithm again to compute* $\sigma^{(\ell-1)}$ *first.*
else

1. *Fix* Σ *as the set of* $M^{(\ell)}$ *largest prime numbers being smaller than* $2^{\ell-1}$.
2. *For all* $\sigma \in \Sigma$
   2.1. *Compute the set* $\sigma I^{(\ell)} := \{\sigma n_1^{(\ell)} \bmod 2^\ell, \ldots, \sigma n_{M_\ell}^{(\ell)} \bmod 2^\ell\}$.
   2.2. *Order the elements of* $\sigma I^{(\ell)}$ *by size and compute the smallest distance* L_σ *between neighboring values.*
3. *Choose* $\sigma^{(\ell)} := \arg\max\{L_\sigma : \sigma \in \Sigma\}$. *If several parameters* σ *achieve the same distance* L_σ, *choose from this subset one* $\sigma^{(\ell)} = \sigma$ *that minimizes* $|\sum_{k=1}^{M_\ell} w_{2^\ell}^{\sigma n_k^{(\ell)}}|$.

Output: $\sigma^{(\ell)} \in \{1, \ldots, 2^{\ell-1}\}$.
Computational cost: at most $\mathcal{O}(M_\ell^2)$ *flops disregarding the computation of* Σ.

The set Σ can be simply precomputed and is not counted as computational cost. Step 2 of the Algorithm 5.32 requires $\mathcal{O}(M_\ell^2)$ flops and Step 3 only $\mathcal{O}(M_\ell)$ flops.

Employing Algorithm 5.32 for computing $\mathbf{A}^{(\ell+1)}$ of the form

$$\mathbf{A}^{(\ell+1)} = 2^{-\ell-1}\,\big(w_{2^\ell}^{(p-1)\sigma^{(\ell)} n_r^{(\ell)}}\big)_{p,r=1}^{M_\ell}\,\mathrm{diag}\,\big(w_{2^{\ell+1}}^{-n_1^{(\ell)}}, \ldots, w_{2^{\ell+1}}^{-n_{M_\ell}^{(\ell)}}\big)^\top$$

in Algorithm 5.30 (for $M_\ell^2 < 2^\ell$), we can solve the linear system (5.58) with $\mathcal{O}(M_\ell^2)$ flops using the Vandermonde structure, see, e.g., [108]. Altogether, since $M_\ell \le M$ for all $\ell = 0, \ldots, t$, the computational cost of Algorithm 5.30 is at most $\mathcal{O}(2^\ell \log 2^\ell) \le \mathcal{O}(M^2 \log M^2)$ to execute all levels $\ell = 0$ to $\lfloor \log_2 M^2 \rfloor$, and $\mathcal{O}((t-\ell)M^2) = \mathcal{O}((\log N - \log M^2)\, M^2)$ for the remaining steps. Thus we obtain overall computational cost of $\mathcal{O}(M^2 \log N)$, and the algorithm is more efficient than the usual FFT of length N if $M^2 < N$. Note that the sparsity M needs not to be known in advance, and the Algorithm 5.30 in fact falls back automatically to an FFT with $\mathcal{O}(N \log N)$ arithmetical operations, if $M^2 \ge N$.

Remark 5.33 The additional strong condition that $\hat{\mathbf{a}}$ satisfies $\operatorname{Re} \hat{a}_j \ge 0$ and $\operatorname{Im} \hat{a}_j \ge 0$ is only needed in order to avoid cancelations. The approach similarly applies if, e.g., the components of $\hat{\mathbf{a}}$ are all in only one of the four quadrants. Moreover, it is very unlikely that full cancelations occur in the periodized vectors $\hat{\mathbf{a}}^{(\ell)}$, such that the idea almost always works for arbitrary sparse vectors $\hat{\mathbf{a}}$.

In [325], the Algorithm 5.30 has been modified to admit rectangular Vandermonde matrices $\mathbf{A}^{(\ell+1)}$ at each iteration step in order to improve the numerical stability of the method.

In [327], the sparse FFT for vectors with small frequency band has been considered. But differently from our considerations in Sect. 5.4.2, no a priori knowledge on the size of the frequency band is needed, but a possible band size is automatically exploited during the algorithm. This improvement goes along with the drawback that the range of the Fourier components needs to be restricted similarly as in this subsection in order to avoid cancellations.

The consideration of sublinear DFT algorithms goes back to the 1990s; we refer to [161] for a review of these randomized methods that possess a constant error probability. In the last years, the research on sparse FFT has been intensified; see, e.g., [192, 312] and the survey [160]. While randomized algorithms suffer from the fact that they provide not always correct results, there exist also recent deterministic approaches that make no errors. Beside the results that have been presented in this subsection, we refer to deterministic algorithms in [5, 48, 211, 212, 266] based on arithmetical progressions and the Chinese remainder theorem. In contrast to the results given here, these algorithms need access to special signal values in an adaptive way, and these values are usually different from the components of the given input vector $\mathbf{a} \in \mathbb{C}^N$. A further class of algorithms employs the Prony method [199, 314, 350]; see also Chap. 10, where however, the emerging Hankel matrices can have very large condition numbers. For various applications of sparse FFT, we refer to [191]. □

5.5 Numerical Stability of FFT

In this section, we show that an FFT of length N that is based on a unitary factorization of the unitary Fourier matrix $\frac{1}{\sqrt{N}} \mathbf{F}_N$ is numerically stable in practice, if $N \in \mathbb{N}$ is a power of 2. We will employ the model of floating point arithmetic and study the normwise forward and backward stability of the FFT.

Assume that we work with a *binary floating point number system* $\mathbb{F} \subset \mathbb{R}$; see [204, pp. 36–40]. This is a subset of real numbers of the form

$$\widetilde{x} = \pm m \times 2^{e-t} = \pm 2^e \left(\frac{d_1}{2} + \frac{d_2}{4} + \ldots + \frac{d_t}{2^t}\right)$$

with *precision* t, *exponent range* $e_{\min} \le e \le e_{\max}$, and $d_1, \ldots, d_t \in \{0, 1\}$. The *mantissa* m satisfies $0 \le m \le 2^t - 1$, where for each $\widetilde{x} \in \mathbb{F} \setminus \{0\}$ it is assumed that $m \ge 2^{t-1}$, i.e., $d_1 = 1$. In case of *single precision* (i.e., 24 bits for the mantissa and 8 bits for the exponent), the so-called *unit roundoff* u is given by $u = 2^{-24} \approx 5.96 \times 10^{-8}$. For *double precision* (i.e., 53 bits for the mantissa and 11 bits for the exponent), we have $u = 2^{-53} \approx 1.11 \times 10^{-16}$; see [204, p. 41].

We use the *standard model of binary floating point arithmetic* in $\mathbb{R}$; see [166, pp. 60–61] or [204, p. 40], that is based on the following assumptions:

- If $a \in \mathbb{R}$ is represented by a floating point number $\mathrm{fl}(a) \in \mathbb{F}$, then

$$\mathrm{fl}(a) = a\,(1 + \epsilon)\,, \quad |\epsilon| \le u\,,$$

  where u is the unit roundoff.
- For arbitrary floating point numbers $a, b \in \mathbb{F}$ and any arithmetical operation $\circ \in \{+, -, \times, /\}$, we assume that

$$\mathrm{fl}(a \circ b) = (a \circ b)\,(1 + \epsilon)\,, \quad |\epsilon| \le u\,.$$

For complex arithmetic that is implemented by real operations, we can conclude the following estimates; see also [86].

Lemma 5.34 *For $x = a + \mathrm{i}\,b$, $y = c + \mathrm{i}\,d \in \mathbb{C}$ with $a, b, c, d \in \mathbb{F}$, we have*

$$|\mathrm{fl}(x + y) - (x + y)| \le |x + y|\,u\,,$$
$$|\mathrm{fl}(x\,y) - x\,y| \le (1 + \sqrt{2})\,|x\,y|\,u\,.$$

Proof Using the assumptions, we obtain

$$\begin{aligned}\mathrm{fl}(x + y) &= \mathrm{fl}(a + c) + \mathrm{i}\,\mathrm{fl}(b + d)\\ &= (a + c)\,(1 + \epsilon_1) + \mathrm{i}\,(b + d)\,(1 + \epsilon_2)\\ &= (x + y) + \epsilon_1(a + c) + \mathrm{i}\,\epsilon_2(b + d)\end{aligned}$$

with $|\epsilon_1|, |\epsilon_2| \le u$ and thus

$$|\mathrm{fl}(x + y) - (x + y)|^2 \le |\epsilon_1|^2\,|a + c|^2 + |\epsilon_2|^2\,|b + d|^2 \le u^2\,|x + y|^2\,.$$

The estimate for complex multiplication can be similarly shown; see [86, Lemma 2.5.3]. ■

We study now the influence of floating point arithmetic for the computation of the DFT(N), where $N \in \mathbb{N} \setminus \{1\}$. Let $\mathbf{x} \in \mathbb{C}^N$ be an arbitrary input vector and

$$\frac{1}{\sqrt{N}}\,\mathbf{F}_N = \frac{1}{\sqrt{N}}\,\big(w_N^{jk}\big)_{j,k=0}^{N-1}\,, \qquad w_N := \mathrm{e}^{-2\pi \mathrm{i}/N}\,.$$

We denote with $\mathbf{y} := \frac{1}{\sqrt{N}}\,\mathbf{F}_N\,\mathbf{x}$ the exact output vector and with $\tilde{\mathbf{y}} \in \mathbb{C}^N$ the vector that is computed by floating point arithmetic with unit roundoff u. Then, there exists a vector $\Delta\mathbf{x} \in \mathbb{C}^N$ such that $\tilde{\mathbf{y}}$ can be written as

$$\tilde{\mathbf{y}} = \frac{1}{\sqrt{N}}\,\mathbf{F}_N\,(\mathbf{x} + \Delta\mathbf{x})\,.$$

According to [204, pp. 129–130], we say that an algorithm computing the matrix-vector product $\frac{1}{\sqrt{N}}\,\mathbf{F}_N\,\mathbf{x}$ is *normwise backward stable*, if for all vectors $\mathbf{x} \in \mathbb{C}^N$ there exists a positive constant k_N such that

$$\|\Delta\mathbf{x}\|_2 \le \big(k_N\,u + \mathcal{O}(u^2)\big)\,\|\mathbf{x}\|_2 \tag{5.59}$$

holds with $k_N\,u \ll 1$. Here, $\|\mathbf{x}\|_2 := \big(\sum_{k=0}^{N-1} |x_k|^2\big)^{1/2}$ denotes the Euclidean norm of $\mathbf{x}$. Observe that the size of k_N is a measure for the numerical stability of the algorithm. Since by (3.31), the matrix $\frac{1}{\sqrt{N}}\,\mathbf{F}_N$ is unitary, we conclude that

$$\|\Delta\mathbf{x}\|_2 = \|\frac{1}{\sqrt{N}}\,\mathbf{F}_N\,\Delta\mathbf{x}\|_2 = \|\tilde{\mathbf{y}} - \mathbf{y}\|_2\,, \quad \|\mathbf{x}\|_2 = \|\frac{1}{\sqrt{N}}\,\mathbf{F}_N\,\mathbf{x}\|_2 = \|\mathbf{y}\|_2\,.$$

Thus, the inequality (5.59) implies

$$\|\tilde{\mathbf{y}} - \mathbf{y}\|_2 = \|\Delta\mathbf{x}\|_2 \le \big(k_N\,u + \mathcal{O}(u^2)\big)\,\|\mathbf{x}\|_2 = \big(k_N\,u + \mathcal{O}(u^2)\big)\,\|\mathbf{y}\|_2\,, \tag{5.60}$$

i.e., we also have the *normwise forward stability*.

In the following, we will derive worst-case estimates for the backward stability constant k_N for the Sande–Tukey FFT and compare it to the constant obtained by employing a direct computation of $\frac{1}{\sqrt{N}}\,\mathbf{F}_N\,\mathbf{x}$.

Let us assume that the Nth roots of unity w_N^k, $k = 0, \ldots, N-1$, are precomputed by a *direct call*, i.e.,

$$\tilde{w}_N^k := \mathrm{fl}\big(\cos\frac{2k\pi}{N}\big) - \mathrm{i}\,\mathrm{fl}\big(\sin\frac{2k\pi}{N}\big)$$

using quality library routines, such that

$$|\mathrm{fl}\left(\cos\frac{2k\pi}{N}\right) - \cos\frac{2k\pi}{N}| \le u\,, \quad |\mathrm{fl}\left(\sin\frac{2k\pi}{N}\right) - \sin\frac{2k\pi}{N}| \le u\,.$$

Then it follows that

$$|\widetilde{w}_N^k - w_N^k|^2 \le |\mathrm{fl}\left(\cos\frac{2k\pi}{N}\right) - \cos\frac{2k\pi}{N} - \mathrm{i}\,\mathrm{fl}\left(\sin\frac{2k\pi}{N}\right) + \mathrm{i}\,\sin\frac{2k\pi}{N}|^2 \le 2\,u^2\,,$$

i.e., $|\widetilde{w}_N^k - w_N^k| \le \sqrt{2}\,u$. The direct call is most accurate but more time-consuming than other precomputations of w_N^k using recursions; see, e.g., [86, 341].

Next, we study how floating point errors accumulate. Assume that $\widetilde{x}_1$ and $\widetilde{x}_2$ have been obtained from previous floating point computations with discrepancies $|\widetilde{x}_j - x_j| = \delta(x_j)\,u + \mathcal{O}(u^2)$ for $j = 1, 2$. Then $\widetilde{x} = \mathrm{fl}(\widetilde{x}_1 \circ \widetilde{x}_2)$ has a new discrepancy $\delta(x)$, where $x = x_1 \circ x_2$.

Lemma 5.35 *Let $x_1, x_2 \in \mathbb{C}$ with $|\widetilde{x}_j - x_j| \le \delta(x_j)\,u + \mathcal{O}(u^2)$ for $j = 1, 2$ be given. Then we have*

$$|\mathrm{fl}(\widetilde{x}_1 + \widetilde{x}_2) - (x_1 + x_2)| \le \big(|x_1 + x_2| + \delta(x_1) + \delta(x_2)\big)\,u + \mathcal{O}(u^2)\,,$$
$$|\mathrm{fl}(\widetilde{x}_1\,\widetilde{x}_2) - (x_1\,x_2)| \le \big((1+\sqrt{2})\,|x_1\,x_2| + \delta(x_1)\,|x_2| + \delta(x_2)\,|x_1|\big)\,u + \mathcal{O}(u^2)\,.$$

Proof We obtain by Lemma 5.34 with floating point numbers $\widetilde{x}_1, \widetilde{x}_2 \in \mathbb{F}$,

$$|\mathrm{fl}(\widetilde{x}_1 + \widetilde{x}_2) - (x_1 + x_2)| \le |\mathrm{fl}(\widetilde{x}_1 + \widetilde{x}_2) - (\widetilde{x}_1 + \widetilde{x}_2)| + |(\widetilde{x}_1 + \widetilde{x}_2) - (x_1 + x_2)|$$
$$\le \big(|\widetilde{x}_1 + \widetilde{x}_2| + \delta(x_1) + \delta(x_2)\big)\,u + \mathcal{O}(u^2) = \big(|x_1 + x_2| + \delta(x_1) + \delta(x_2)\big)\,u + \mathcal{O}(u^2)\,,$$

where we have used $|\widetilde{x}_1 + \widetilde{x}_2| = |x_1 + x_2| + \mathcal{O}(u)$. Similarly, for the multiplication, we obtain

$$|\mathrm{fl}(\widetilde{x}_1\,\widetilde{x}_2) - (x_1\,x_2)| \le |\mathrm{fl}(\widetilde{x}_1\,\widetilde{x}_2) - (\widetilde{x}_1\,\widetilde{x}_2)| + |(\widetilde{x}_1\,\widetilde{x}_2) - (x_1\,x_2)|\,.$$

Lemma 5.34 implies for the first term

$$|\mathrm{fl}(\widetilde{x}_1\,\widetilde{x}_2) - (\widetilde{x}_1\,\widetilde{x}_2)| \le (1+\sqrt{2})\,|\widetilde{x}_1\,\widetilde{x}_2|\,u + \mathcal{O}(u^2)$$

and for the second term

$$|(\widetilde{x}_1\,\widetilde{x}_2) - (x_1\,x_2)| \le |\widetilde{x}_2\,(\widetilde{x}_1 - x_1)| + |x_1\,(\widetilde{x}_2 - x_2)|$$
$$\le \big(|x_2| + \delta(x_2)\,u\big)\,\delta(x_1)\,u + |x_1|\,\delta(x_2)\,u + \mathcal{O}(u^2)\,.$$

In particular, $|\widetilde{x}_1\,\widetilde{x}_2| = |x_1\,x_2| + \mathcal{O}(u)$. Thus the assertion follows. ∎

In order to estimate the stability constant k_N for a DFT(N) algorithm, we first recall the roundoff error for scalar products of vectors.

Lemma 5.36 *Let $N \in \mathbb{N}$ be fixed. Let $\mathbf{x} = (x_j)_{j=0}^{N-1}$, $\mathbf{y} = (y_j)_{j=0}^{N-1} \in \mathbb{C}^N$ be arbitrary vectors and let $\tilde{\mathbf{x}} = (\tilde{x}_j)_{j=0}^{N-1}$, $\tilde{\mathbf{y}} = (\tilde{y}_j)_{j=0}^{N-1} \in (\mathbb{F} + \mathrm{i}\,\mathbb{F})^N$ be their floating point representations such that $|\tilde{x}_j - x_j| \le \sqrt{2}\,|x_j|\,u + \mathcal{O}(u^2)$ and $|\tilde{y}_j - y_j| \le |y_j|\,u + \mathcal{O}(u^2)$ for $j = 0, \ldots, N-1$.*

Then we have

$$|\mathrm{fl}(\tilde{\mathbf{x}}^\top \tilde{\mathbf{y}}) - (\mathbf{x}^\top \mathbf{y})| \le \big((N + 1 + 2\sqrt{2})\,|\mathbf{x}|^\top |\mathbf{y}|\big)\,u + \mathcal{O}(u^2)$$

for recursive summation and

$$|\mathrm{fl}(\tilde{\mathbf{x}}^\top \tilde{\mathbf{y}}) - (\mathbf{x}^\top \mathbf{y})| \le \big((t + 2 + 2\sqrt{2})|\mathbf{x}|^\top |\mathbf{y}|\big)u + \mathcal{O}(u^2)$$

for cascade summation, where $|\mathbf{x}| := (|x_j|)_{j=0}^{N-1}$ and $t := \lceil \log_2 N \rceil$ for $N \in \mathbb{N} \setminus \{1\}$.

Proof

1. We show the estimate for recursive estimation using induction with respect to N. For $N = 1$ it follows by Lemma 5.35 with $\delta(x_0) \le \sqrt{2}\,|x_0|$ and $\delta(y_0) \le |y_0|$ that

$$\begin{aligned}|\mathrm{fl}(\tilde{x}_0\, \tilde{y}_0) - x_0\, y_0| &\le \big((1 + \sqrt{2})\,|x_0|\,|y_0| + |x_0|\,|y_0| + \sqrt{2}\,|y_0|\,|x_0|\big)\,u + \mathcal{O}(u^2)\\ &= \big((2 + 2\sqrt{2})\,|x_0|\,|y_0|\big)\,u + \mathcal{O}(u^2)\,.\end{aligned}$$

Assume now that $\tilde{z}$ is the result of the computation of $\tilde{\mathbf{x}}_1^\top \tilde{\mathbf{y}}_1$ in floating point arithmetic for $\tilde{\mathbf{x}}_1, \tilde{\mathbf{y}}_1 \in \mathbb{C}^N$. Let $\tilde{\mathbf{x}} := (\tilde{\mathbf{x}}_1^\top, \tilde{x}_{N+1})^\top \in \mathbb{C}^{N+1}$ and $\tilde{\mathbf{y}} := (\tilde{\mathbf{y}}_1^\top, \tilde{y}_{N+1})^\top \in \mathbb{C}^{N+1}$. Using the intermediate result $\tilde{z}$, we find by Lemma 5.35 and the induction hypothesis with discrepancies $\delta(z) \le (N + 1 + 2\sqrt{2})\,|z|$ and $\delta(x_{N+1}\, y_{N+1}) \le (2 + 2\sqrt{2})\,|x_{N+1}\, y_{N+1}|$

$$\begin{aligned}&|\mathrm{fl}(\tilde{\mathbf{x}}^\top \tilde{\mathbf{y}}) - (\mathbf{x}^\top \mathbf{y})| = |\mathrm{fl}\big(\tilde{z} + \mathrm{fl}(\tilde{x}_{N+1}\, \tilde{y}_{N+1})\big) - (\mathbf{x}^\top \mathbf{y})|\\ &\quad\le \big(|z + x_{N+1}\, y_{N+1}| + \delta(z) + \delta(x_{N+1}\, y_{N+1})\big)\,u + \mathcal{O}(u^2)\\ &\quad\le \big(|\mathbf{x}|^\top |\mathbf{y}| + (N + 1 + 2\sqrt{2})\,|\mathbf{x}_1|^\top |\mathbf{y}_1| + (2 + 2\sqrt{2})\,|x_{N+1}\, y_{N+1}|\big)u + \mathcal{O}(u^2)\\ &\quad\le \big((N + 2 + 2\sqrt{2})\,|\mathbf{x}|^\top |\mathbf{y}|\big)\,u + \mathcal{O}(u^2)\,.\end{aligned}$$

2. For cascade summation, we also proceed by induction, this time over t, where $t = \lceil \log_2 N \rceil$ and $N \in \mathbb{N} \setminus \{1\}$. For $t = 1$, i.e., $N = 2$, it follows from the recursive summation that

$$|\mathrm{fl}(\tilde{\mathbf{x}}^\top \tilde{\mathbf{y}}) - (\mathbf{x}^\top \mathbf{y})| \le \big((3 + 2\sqrt{2})\,|\mathbf{x}|^\top |\mathbf{y}|\big)\,u + \mathcal{O}(u^2)\,.$$

Assume now that $\tilde{z}_1$ is the result of the computation of $\tilde{\mathbf{x}}_1^\top \tilde{\mathbf{y}}_1$ in floating point arithmetic for $\tilde{\mathbf{x}}_1, \tilde{\mathbf{y}}_1 \in \mathbb{C}^N$, and $\tilde{z}_2$ is the result of the computation of $\tilde{\mathbf{x}}_2^\top \tilde{\mathbf{y}}_2$ in floating point arithmetic for $\tilde{\mathbf{x}}_2, \tilde{\mathbf{y}}_2 \in \mathbb{C}^N$ using cascade summation. Let $\tilde{\mathbf{x}} := (\tilde{\mathbf{x}}_1^\top, \tilde{\mathbf{x}}_2^\top)^\top \in \mathbb{C}^{2N}$ and $\tilde{\mathbf{y}} := (\tilde{\mathbf{y}}_1^\top, \tilde{\mathbf{y}}_2^\top)^\top \in \mathbb{C}^{2N}$. Using the intermediate results $\tilde{z}_1$, $\tilde{z}_2$ for the scalar products $z_1 = \mathbf{x}_1^\top \mathbf{y}_1$ and $z_2 = \mathbf{x}_2^\top \mathbf{y}_2$, we obtain by Lemma 5.35 and the induction hypothesis

$$\begin{aligned}
&|\mathrm{fl}(\tilde{\mathbf{x}}^\top \tilde{\mathbf{y}}) - (\mathbf{x}^\top \mathbf{y})| \\
&\quad = |\mathrm{fl}(\tilde{z}_1 + \tilde{z}_2) - (z_1 + z_2)| \le \big(|z_1 + z_2| + \delta(z_1) + \delta(z_2)\big)\, u + \mathcal{O}(u^2) \\
&\quad \le \big(|\mathbf{x}|^\top |\mathbf{y}| + (t + 2 + 2\sqrt{2})\, |\mathbf{x}_1|^\top |\mathbf{y}_1| + (t + 2 + 2\sqrt{2})\, |\mathbf{x}_2|^\top |\mathbf{y}_2|\big)\, u + \mathcal{O}(u^2) \\
&\quad \le \big((t + 3 + 2\sqrt{2})\, |\mathbf{x}|^\top |\mathbf{y}|\big)\, u + \mathcal{O}(u^2)\,.
\end{aligned}$$

Thus, the assertion follows. ∎

We are now ready to estimate the backward stability constant for the direct computation of the DFT(N) of radix–2 length N.

Theorem 5.37 *Let $N = 2^t$, $t \in \mathbb{N}$, be given. Assume that the entries of the floating point representation of the Fourier matrix $\tilde{\mathbf{F}}_N$ satisfy $|\tilde{w}_N^k - w_N^k| \le \sqrt{2}\, u$ for $k = 0, \ldots, N-1$. Further let $\tilde{\mathbf{x}} = (\tilde{x}_j)_{j=0}^{N-1}$ be the vector of floating point numbers representing $\mathbf{x} = (x_j)_{j=0}^{N-1} \in \mathbb{C}^N$ with $|\tilde{x}_j - x_j| \le |x_j|\, u$.*

Then the direct computation of $\frac{1}{\sqrt{N}} \mathbf{F}_N \mathbf{x}$ is normwise backward stable, and we have

$$\|\mathrm{fl}(\frac{1}{\sqrt{N}} \tilde{\mathbf{F}}_N \tilde{\mathbf{x}}) - \frac{1}{\sqrt{N}} \mathbf{F}_N \mathbf{x}\|_2 \le \big(k_N\, u + \mathcal{O}(u^2)\big)\, \|\mathbf{x}\|_2$$

with the constant

$$k_N = \begin{cases} \sqrt{N}\,(N + 1 + 2\sqrt{2}) & \text{for recursive summation}\,, \\ \sqrt{N}\,(\log_2 N + 2 + 2\sqrt{2}) & \text{for cascade summation}\,. \end{cases}$$

Proof We apply Lemma 5.36 to each component of the matrix-vector product $\frac{1}{\sqrt{N}} \mathbf{F}_N \mathbf{x}$, not counting the factor $\frac{1}{\sqrt{N}}$ which is for even t only a shift in binary arithmetic. For the jth component, we obtain for recursive summation

$$\begin{aligned}
&|\mathrm{fl}(\frac{1}{\sqrt{N}} \tilde{\mathbf{F}}_N \tilde{\mathbf{x}})_j - (\frac{1}{\sqrt{N}} \mathbf{F}_N \mathbf{x})_j| \le \frac{1}{\sqrt{N}} \big((N + 1 + 2\sqrt{2})\, ((1)_{k=0}^{N-1})^\top |\mathbf{x}|\big)\, u + \mathcal{O}(u^2) \\
&= \big(\frac{1}{\sqrt{N}} (N + 1 + 2\sqrt{2})\, \|\mathbf{x}\|_1\big)\, u + \mathcal{O}(u^2) \le \big((N + 1 + 2\sqrt{2})\, \|\mathbf{x}\|_2\big)\, u + \mathcal{O}(u^2)\,.
\end{aligned}$$

Here we have used that $\|\mathbf{x}\|_1 \le \sqrt{N}\, \|\mathbf{x}\|_2$. Taking now the Euclidean norm, it follows that

$$\Big(\sum_{j=0}^{N-1} |\mathrm{fl}(\frac{1}{\sqrt{N}}\,\tilde{\mathbf{F}}_N\,\tilde{\mathbf{x}})_j - (\frac{1}{\sqrt{N}}\,\mathbf{F}_N\,\mathbf{x})_j|^2\Big)^{1/2} \le \big(\sqrt{N}\,(N+1+2\sqrt{2})\,\|\mathbf{x}\|_2\big)\,u + \mathcal{O}(u^2)\,.$$

The result for cascade summation follows analogously. ∎

In comparison, we estimate now the worst-case backward stability constant for a radix–2 FFT considered in Sect. 5.2. Particularly, we employ the matrix factorization (5.10) related to the Sande–Tukey FFT, i.e.,

$$\mathbf{F}_N = \mathbf{R}_N \prod_{n=1}^{t} \mathbf{T}_n\,(\mathbf{I}_{N/2^n} \otimes \mathbf{F}_2 \otimes \mathbf{I}_{2^{n-1}}) = \mathbf{R}_N\,\mathbf{M}_N^{(t)}\,\mathbf{M}_N^{(t-1)}\,\ldots\,\mathbf{M}_N^{(1)}\,,$$

where

$$\mathbf{M}_N^{(j)} := \mathbf{T}_{t-j}\,(\mathbf{I}_{2^j} \otimes \mathbf{F}_2 \otimes \mathbf{I}_{2^{t-j-1}})\,. \tag{5.61}$$

Recall that $\mathbf{R}_N$ is a permutation matrix and

$$\mathbf{T}_{t-j} := \mathbf{I}_{2^j} \otimes \mathbf{D}_{2^{t-j}}\,,$$
$$\mathbf{D}_{2^{t-j}} := \mathrm{diag}\,(\mathbf{I}_{2^{t-j-1}},\,\mathbf{W}_{2^{t-j-1}})\,, \quad \mathbf{W}_{2^{t-j-1}} := \mathrm{diag}\,(w^j_{2^{t-j}})_{j=0}^{2^{t-j-1}-1}\,.$$

In particular, $\mathbf{T}_1 = \mathbf{I}_N$. The matrices $\mathbf{I}_{2^j} \otimes \mathbf{F}_2 \otimes \mathbf{I}_{2^{t-j-1}}$ are sparse with only two nonzero entries per row, and these entries are either 1 or -1. Multiplication with these matrices just means one addition or one subtraction per component.

Theorem 5.38 *Let $N = 2^t$, $t \in \mathbb{N}$. Assume that $|\tilde{w}_N^k - w_N^k| \le \sqrt{2}\,u$ for $k = 0, \ldots, N-1$. Further let $\tilde{\mathbf{x}} = (\tilde{x}_j)_{j=0}^{N-1}$ be the vector of floating point numbers representing $\mathbf{x} = (x_j)_{j=0}^{N-1}$ with $|\tilde{x}_j - x_j| \le |x_j|\,u$.*

Then the Sande–Tukey FFT *is normwise backward stable with the constant*

$$k_N = (2 + 3\sqrt{2})\,\log_2 N + 1\,.$$

Proof

1. Let $\tilde{\mathbf{x}}^{(0)} := \tilde{\mathbf{x}}$ such that $\|\tilde{\mathbf{x}}^{(0)} - \mathbf{x}^{(0)}\|_2 \le u\,\|\mathbf{x}\|_2$. The Sande–Tukey FFT is employed by successive multiplication with the sparse matrices $\mathbf{M}_N^{(j)}$, $j = 1, \ldots, t$, and the permutation $\mathbf{R}_N$ in (5.61). We introduce the vectors

$$\tilde{\mathbf{x}}^{(j)} := \mathrm{fl}\big(\tilde{\mathbf{M}}_N^{(j)}\,\tilde{\mathbf{x}}^{(j-1)}\big)\,, \quad j = 1, \ldots, t\,,$$

while $\mathbf{x}^{(j)} := \mathbf{M}_N^{(j)} \ldots \mathbf{M}_N^{(1)}\mathbf{x}$ is the exact result after j steps. Here,

$$\tilde{\mathbf{M}}_N^{(j)} := \tilde{\mathbf{T}}_{t-j}\,(\mathbf{I}_{2^j} \otimes \mathbf{F}_2 \otimes \mathbf{I}_{2^{t-j-1}})$$

denotes the floating point representation of $\mathbf{M}_N^{(j)}$ using $\widetilde{w}_N^k$. Note that $\frac{1}{\sqrt{N}}\, \mathbf{R}_N\, \widetilde{\mathbf{x}}^{(t)}$ is the result of the algorithm in floating point arithmetic, where we do not take into account errors caused by the multiplication with $\frac{1}{\sqrt{N}}$ as before. We consider the errors e_j of the form

$$e_j := \|\widetilde{\mathbf{x}}^{(j)} - \widetilde{\mathbf{M}}_N^{(j)}\, \widetilde{\mathbf{x}}^{(j-1)}\|_2\,, \quad j = 1, \ldots, t\,.$$

Then we can estimate the floating point error as follows:

$$\begin{aligned}\|\widetilde{\mathbf{x}}^{(j)} - \mathbf{x}^{(j)}\|_2 &\le \|\widetilde{\mathbf{x}}^{(j)} - \widetilde{\mathbf{M}}_N^{(j)}\, \widetilde{\mathbf{x}}^{(j-1)}\|_2 + \|\widetilde{\mathbf{M}}_N^{(j)}\, \widetilde{\mathbf{x}}^{(j-1)} - \widetilde{\mathbf{M}}_N^{(j)}\, \mathbf{x}^{(j-1)}\|_2 \\ &\quad + \|\widetilde{\mathbf{M}}_N^{(j)}\, \mathbf{x}^{(j-1)} - \mathbf{M}_N^{(j)}\, \mathbf{x}^{(j-1)}\|_2 \\ &\le e_j + \|\widetilde{\mathbf{M}}_N^{(j)}\|_2\, \|\widetilde{\mathbf{x}}^{(j-1)} - \mathbf{x}^{(j-1)}\|_2 + \|\widetilde{\mathbf{M}}_N^{(j)} - \mathbf{M}_N^{(j)}\|_2\, \|\mathbf{x}^{(j-1)}\|_2\,.\end{aligned}$$

Observing that

$$\|\widetilde{\mathbf{M}}_N^{(j)}\|_2 = \|\widetilde{\mathbf{T}}_{t-j}\|_2\, \|\mathbf{I}_{2^j} \otimes \mathbf{F}_2 \otimes \mathbf{I}_{2^{t-j-1}}\|_2 = \sqrt{2}\, \|\widetilde{\mathbf{T}}_{t-j}\|_2 \le \sqrt{2}\, (1 + \sqrt{2}\, u)$$

and that

$$\begin{aligned}\|\widetilde{\mathbf{M}}_N^{(j)} - \mathbf{M}_N^{(j)}\|_2 &= \|(\widetilde{\mathbf{T}}_{t-j} - \mathbf{T}_{t-j})\, (\mathbf{I}_{2^j} \otimes \mathbf{F}_2 \otimes \mathbf{I}_{2^{t-j-1}})\|_2 \\ &\le \sqrt{2}\, \|\widetilde{\mathbf{T}}_{t-j} - \mathbf{T}_{t-j}\|_2 < 2\, u\,,\end{aligned}$$

we obtain

$$\|\widetilde{\mathbf{x}}^{(j)} - \mathbf{x}^{(j)}\|_2 \le e_j + \sqrt{2}\, \|\widetilde{\mathbf{x}}^{(j-1)} - \mathbf{x}^{(j-1)}\|_2 + 2\, \|\mathbf{x}^{(j-1)}\|_2\, u + \mathscr{O}(u^2)\,. \qquad (5.62)$$

2. We show now that $e_j \le 2\sqrt{2}\, (1 + \sqrt{2})\, \|\mathbf{x}^{(j-1)}\|_2\, u$ for all $j = 1, \ldots, t$. Introducing the intermediate vectors

$$\widetilde{\mathbf{y}}^{(j)} := \mathrm{fl}\big((\mathbf{I}_{2^j} \otimes \mathbf{F}_2 \otimes \mathbf{I}_{2^{t-j-1}})\, \widetilde{\mathbf{x}}^{(j-1)}\big)\,, \quad \mathbf{y}^{(j)} := (\mathbf{I}_{2^j} \otimes \mathbf{F}_2 \otimes \mathbf{I}_{2^{t-j-1}})\, \widetilde{\mathbf{x}}^{(j-1)}\,,$$

we conclude that $\widetilde{\mathbf{x}}^{(j)} = \mathrm{fl}\big(\widetilde{\mathbf{T}}_{t-j}\, \widetilde{\mathbf{y}}^{(j)}\big)$. By Lemma 5.35 with $\delta(\widetilde{x}_k^{(j)}) = 0$ for all k, we find for each component

$$|\widetilde{y}_k^{(j)} - y_k^{(j)}| \le |y_k^{(j)}|\, u + \mathscr{O}(u^2) \qquad (5.63)$$

and thus

$$\|\widetilde{\mathbf{y}}^{(j)} - \mathbf{y}^{(j)}\|_2 \le \|\mathbf{y}^{(j)}\|_2\, u = \sqrt{2}\, \|\widetilde{\mathbf{x}}^{(j-1)}\|_2\, u\,,$$

where we have used that $\|\mathbf{I}_{2^j} \otimes \mathbf{F}_2 \otimes \mathbf{I}_{2^{t-j-1}}\|_2 = \sqrt{2}$. Next, the multiplication with the diagonal matrix $\widetilde{\mathbf{T}}_{t-j}$ implies by Lemma 5.35

$$
\begin{aligned}
|\widetilde{x}_k^{(j)} - (\mathbf{M}_N^{(j)}\, \widetilde{\mathbf{x}}^{(j-1)})_k| &= |\big(\mathrm{fl}(\widetilde{\mathbf{T}}_{t-j}\, \widetilde{\mathbf{y}}^{(j)})\big)_k - \big(\mathbf{T}_{t-j}\, \mathbf{y}^{(j)}\big)_k| \\
&\le \big((1+\sqrt{2})\,|y_k^{(j)}| + \sqrt{2}\,|y_k^{(j)}| + \delta(y_k^{(j)})\big)\, u + \mathcal{O}(u^2) \\
&\le \big(2\,(1+\sqrt{2})\,|y_k^{(j)}|\big)\, u + \mathcal{O}(u^2)\,,
\end{aligned}
$$

where we have used the assumption $|\widetilde{w}_N^k - w_N^k| \le \sqrt{2}\, u$ and that (5.63) implies $\delta(y_k^{(j)}) = |y_k^{(j)}|$. Thus, we conclude

$$
e_j = \|\widetilde{\mathbf{x}}^{(j)} - \widetilde{\mathbf{M}}_N^{(j)}\, \widetilde{\mathbf{x}}^{(j-1)}\|_2 \le 2(1+\sqrt{2})\, \|\mathbf{y}^{(j)}\|_2\, u = 2\sqrt{2}\,(1+\sqrt{2})\, \|\mathbf{x}^{(j-1)}\|_2\, u\,.
$$

3. We recall that $\|\mathbf{x}^{(j)}\|_2 = 2^{j/2}\, \|\mathbf{x}\|_2$. Thus, the relation (5.62) can be written as

$$
\begin{aligned}
\|\widetilde{\mathbf{x}}^{(j)} - \mathbf{x}^{(j)}\|_2 \le{}& 2\sqrt{2}\,(1+\sqrt{2})\, \|\mathbf{x}^{(j-1)}\|_2\, u + \sqrt{2}\, \|\widetilde{\mathbf{x}}^{(j-1)} - \mathbf{x}^{(j-1)}\|_2 \\
&+ 2\, \|\mathbf{x}^{(j-1)}\|_2\, u + \mathcal{O}(u^2)\,.
\end{aligned}
$$

Starting with $\|\widetilde{\mathbf{x}}^{(0)} - \mathbf{x}^{(0)}\|_2 \le u\, \|\mathbf{x}\|_2$, we show by induction over j that

$$
\|\widetilde{\mathbf{x}}^{(j)} - \mathbf{x}^{(j)}\|_2 \le 2^{j/2}\, \big((2j+1) + 3\sqrt{2}j\big)\, \|\mathbf{x}\|_2\, u + \mathcal{O}(u^2)
$$

is true for $j = 1, \ldots, t$. For $j = 1$,

$$
\begin{aligned}
&\|\widetilde{\mathbf{x}}^{(1)} - \mathbf{x}^{(1)}\|_2 \\
&\le 2\sqrt{2}\,(1+\sqrt{2})\, \|\mathbf{x}^{(0)}\|_2\, u + \sqrt{2}\, \|\widetilde{\mathbf{x}}^{(0)} - \mathbf{x}^{(0)}\|_2 + 2\, \|\mathbf{x}^{(0)}\|_2\, u + \mathcal{O}(u^2) \\
&= \sqrt{2}\,(3 + 3\sqrt{2})\, \|\mathbf{x}\|_2\,,
\end{aligned}
$$

and the assertion is correct. Assume now that the assertion is true for some $j \in \{1, \ldots, t-1\}$. Then

$$
\begin{aligned}
&\|\widetilde{\mathbf{x}}^{(j+1)} - \mathbf{x}^{(j+1)}\|_2 \le 2\,(\sqrt{2}+3)\, \|\mathbf{x}^{(j)}\|_2\, u + \sqrt{2}\, \|\widetilde{\mathbf{x}}^{(j)} - \mathbf{x}^{(j)}\|_2 + \mathcal{O}(u^2) \\
&= 2\,(\sqrt{2}+3)\, 2^{j/2}\, \|\mathbf{x}\|_2\, u + 2^{(j+1)/2}\, \big((2j+1) + 3\sqrt{2}j\big)\, \|\mathbf{x}\|_2\, u + \mathcal{O}(u^2) \\
&= 2^{(j+1)/2}\, \big((2j+3) + 3\sqrt{2}(j+1)\big)\, \|\mathbf{x}\|_2\, u + \mathcal{O}(u^2)\,.
\end{aligned}
$$

Finally, it follows with $\widetilde{\mathbf{F}}_N = \mathbf{R}_N\, \widetilde{\mathbf{M}}_N^{(t)} \ldots \widetilde{\mathbf{M}}_N^{(1)}$ and $t = \log_2 N$ that

$$\|\text{fl}(\frac{1}{\sqrt{N}}\,\tilde{\mathbf{F}}_N\,\tilde{\mathbf{x}}) - \frac{1}{\sqrt{N}}\,\mathbf{F}_N\,\mathbf{x}\|_2 = \frac{1}{\sqrt{N}}\,\|\tilde{\mathbf{x}}^{(t)} - \mathbf{x}^{(t)}\|_2$$
$$\leq \big((2+3\sqrt{2})\,\log_2 N + 1\big)\,\|\mathbf{x}\|_2\,u + \mathcal{O}(u^2)\,.$$

■

Comparing the constants of backward stability for the usual matrix-vector multiplication and the Sande–Tukey FFT, we emphasize that *the* FFT *not only saves computational effort but also provides much more accurate results than direct computation.*

Remark 5.39 In [204, pp. 452–454], the numerical stability of the Cooley–Tukey radix–2 FFT is investigated. The obtained result $k_N = \mathcal{O}(\log_2 N)$ for various FFTs has been shown in different papers; see [10, 86, 360, 445] under the assumption that $\mathbf{x}$ is contained in $\mathbb{F}^N$ and all twiddle factors are either exactly known or precomputed by direct call. In [86, 341, 377, 409], special attention was put on the influence of the recursive precomputation of twiddle factors that can essentially deteriorate the final result.

Beside worst-case estimates, also the average case backward numerical stability of FFT has been studied; see [77, 408] with the result $k_N = \mathcal{O}(\sqrt{\log_2 N})$. □

Chapter 6
Chebyshev Methods and Fast DCT Algorithms

This chapter is concerned with Chebyshev methods and fast algorithms for the discrete cosine transform (DCT). Chebyshev methods are fundamental for the approximation and integration of real-valued functions defined on a compact interval. In Sect. 6.1, we introduce the Chebyshev polynomials of first kind and study their properties. Further, we consider the close connection between Chebyshev expansions and Fourier expansions of even 2π-periodic functions, the convergence of Chebyshev series, and the properties of Chebyshev coefficients. Section 6.2 addresses the efficient evaluation of polynomials, which are given in the orthogonal basis of Chebyshev polynomials. We present fast DCT algorithms in Sect. 6.3. These fast DCT algorithms are based either on the FFT or on the orthogonal factorization of the related cosine matrix.

In Sect. 6.4, we describe the polynomial interpolation at Chebyshev extreme points (together with a barycentric interpolation formula) and the Clenshaw–Curtis quadrature. Fast algorithms for the evaluation of polynomials at Chebyshev extreme points, for computing products of polynomials as well as for interpolation and quadrature, involve different types of the DCT. In Sect. 6.5, we consider the discrete polynomial transform which is a far-reaching generalization of the DCT.

6.1 Chebyshev Polynomials and Chebyshev Series

The basis of Chebyshev polynomials possesses a lot of favorable properties and is therefore of high interest as an alternative to the monomial basis for representing polynomials and polynomial expansions.

G. Plonka et al., *Numerical Fourier Analysis*, Applied and Numerical Harmonic Analysis, https://doi.org/10.1007/978-3-031-35005-4_6

6.1.1 Chebyshev Polynomials

We consider the interval $I := [-1, 1]$ and define the functions

$$T_k(x) := \cos(k \arccos(x)) \tag{6.1}$$

for all $k \in \mathbb{N}_0$ and all $x \in I$. Applying the substitution $x = \cos(t)$, $t \in [0, \pi]$, we observe that

$$T_k(\cos(t)) = \cos(kt), \quad k \in \mathbb{N}_0, \tag{6.2}$$

for all $t \in [0, \pi]$ and hence for all $t \in \mathbb{R}$. Formula (6.2) implies that the graph of T_5 on I is a "distorted" harmonic oscillation (see Fig. 6.1).

The trigonometric identity

$$\cos((k + 1)t) + \cos((k - 1)t) = 2 \cos(t) \cos(kt), \quad k \in \mathbb{N},$$

provides the important *recursion formula*

$$T_{k+1}(x) = 2x\, T_k(x) - T_{k-1}(x), \quad k \in \mathbb{N}, \tag{6.3}$$

with initial polynomials $T_0(x) = 1$ and $T_1(x) = x$. Thus, T_k is an algebraic polynomial of degree k with leading coefficient 2^{k-1}. Clearly, the polynomials T_k can be extended to $\mathbb{R}$ such that the recursion formula (6.3) holds for all $x \in \mathbb{R}$. The polynomial $T_k : \mathbb{R} \to \mathbb{R}$ of degree $k \in \mathbb{N}_0$ is called the *kth Chebyshev polynomial of first kind.*

Remark 6.1 Originally, these polynomials were investigated in 1854 by the Russian mathematician P. L. Chebyshev (1821–1894) (see Fig. 6.2) (image source: [405]). Note that the Russian name has several transliterations (such as Tschebyscheff, Tschebyschew, and Tschebyschow). We emphasize that the Chebyshev polynomials are of similar importance as the complex exponentials (1.10) for the approximation

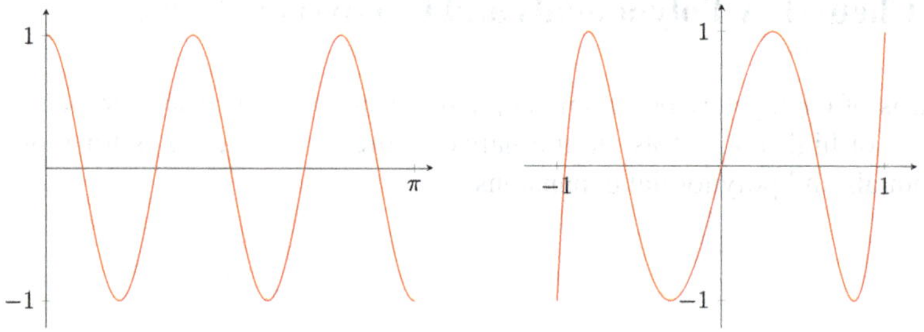

Fig. 6.1 Comparison between $\cos(5\cdot)$ restricted on $[0, \pi]$ (left) and T_5 restricted on I (right)

Fig. 6.2 The Russian mathematician Pafnuty Lvovich Chebyshev (1821–1894)

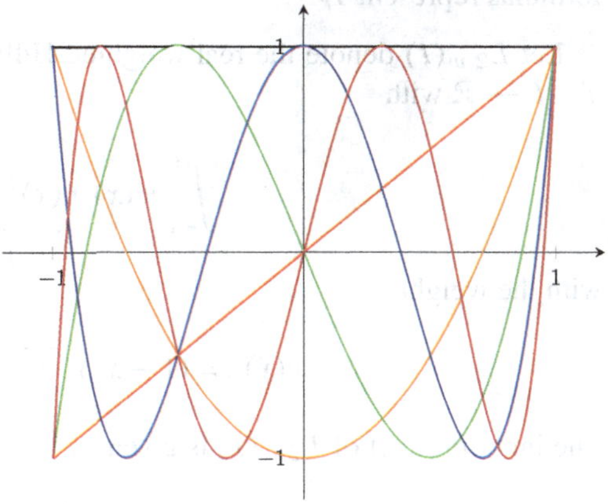

Fig. 6.3 The Chebyshev polynomials T_0 (black), T_1 (red), T_2 (orange), T_3 (green), T_4 (blue), and T_5 (violet) restricted on I

of 2π-periodic functions. There exist several excellent publications [285, 311, 366, 415] on Chebyshev polynomials. □

For $k = 2, \ldots, 5$, the recursion formula (6.3) yields

$$T_2(x) = 2x^2 - 1\,, \qquad T_3(x) = 4x^3 - 3x\,,$$
$$T_4(x) = 8x^4 - 8x^2 + 1\,,\ T_5(x) = 16x^5 - 20x^3 + 5x\,.$$

These polynomials are represented in Fig. 6.3. From

$$\arccos(-x) = \pi - \arccos(x)\,, \quad x \in I\,,$$

it follows by (6.1) that for all $k \in \mathbb{N}_0$ and all $x \in I$

$$T_k(-x) = \cos\big(k \arccos(-x)\big) = \cos(k\pi - k \arccos(x)) = (-1)^k\, T_k(x)\,. \qquad (6.4)$$

Hence, T_{2k}, $k \in \mathbb{N}_0$, is even and T_{2k+1}, $k \in \mathbb{N}_0$, is odd. Further, we have for all $k \in \mathbb{N}_0$

$$T_k(1) = 1\,, \quad T_k(-1) = (-1)^k\,, \quad T_{2k}(0) = (-1)^{k+1}\,, \quad T_{2k+1}(0) = 0\,.$$

Lemma 6.2 *For each $k \in \mathbb{N}_0$ the Chebyshev polynomial T_k possesses the explicit representation*

$$T_k(x) = \begin{cases} \frac{1}{2}\big[(x + \mathrm{i}\sqrt{1-x^2})^k + (x - \mathrm{i}\sqrt{1-x^2})^k\big] & x \in I\,, \\ \frac{1}{2}\big[(x - \sqrt{x^2-1})^k + (x + \sqrt{x^2-1})^k\big] & x \in \mathbb{R} \setminus I\,. \end{cases}$$

Proof For $k = 0$ and $k = 1$ these explicit expressions yield $T_0(x) = 1$ and $T_1(x) = x$ for all $x \in \mathbb{R}$. Simple calculation shows that for arbitrary $k \in \mathbb{N}$ the explicit expressions fulfill the recursion formula (6.3) for all $x \in \mathbb{R}$. Hence, these explicit formulas represent T_k. ∎

Let $L_{2,w}(I)$ denote the real weighted Hilbert space of all measurable functions $f : I \to \mathbb{R}$ with

$$\int_{-1}^{1} w(x)\, f(x)^2\, \mathrm{d}x < \infty$$

with the weight

$$w(x) := (1 - x^2)^{-1/2}, \quad x \in (-1,\, 1).$$

The inner product of $L_{2,w}(I)$ is given by

$$\langle f,\, g\rangle_{L_{2,w}(I)} := \frac{1}{\pi} \int_{-1}^{1} w(x)\, f(x)\, g(x)\, \mathrm{d}x$$

for all $f, g \in L_{2,w}(I)$, and the related norm of $f \in L_{2,w}(I)$ is equal to

$$\|f\|_{L_{2,w}(I)} := \langle f,\, f\rangle_{L_{2,w}(I)}^{1/2}\,.$$

As usual, almost equal functions are identified in $L_{2,w}(I)$. The following result shows that the Chebyshev polynomials satisfy similar orthogonality relations as the complex exponentials (1.10) in the weighted Hilbert space.

Theorem 6.3 *The Chebyshev polynomials T_k, $k \in \mathbb{N}_0$, form a complete orthogonal system in $L_{2,w}(I)$. For all $k,\, \ell \in \mathbb{N}_0$ we have*

$$\langle T_k,\, T_\ell\rangle_{L_{2,w}(I)} = \begin{cases} 1 & k = \ell = 0\,, \\ \frac{1}{2} & k = \ell > 0\,, \\ 0 & k \neq \ell\,. \end{cases}$$

Proof The orthogonality of the Chebyshev polynomials follows immediately from the identity

$$\langle T_k, T_\ell\rangle_{L_{2,w}(I)}=\frac{1}{\pi}\int_0^\pi \cos(kt)\,\cos(\ell t)\mathrm{d}t=\frac{1}{2\pi}\int_0^\pi \big(\cos((k-\ell)t)+\cos((k+\ell)t)\big)\mathrm{d}t.$$

The completeness of the orthogonal system $\{T_k : k \in \mathbb{N}_0\}$ is a consequence of Theorem 1.1: For $f \in L_{2,w}(I)$ with $a_k[f] = 0$ for all $k \in \mathbb{N}_0$ we can conclude that

$$0 = a_k[f] = 2\,\langle f,\, T_k\rangle_{L_{2,w}(I)} = \frac{1}{\pi}\int_{-\pi}^{\pi} \varphi(t)\,\cos(kt)\,\mathrm{d}t$$

with $\varphi = f(\cos(\cdot))$. Since φ is even, we obtain for all $k \in \mathbb{Z}$

$$\int_{-\pi}^{\pi} \varphi(t)\,\mathrm{e}^{-\mathrm{i}kt}\,\mathrm{d}t = 0\,.$$

Hence, $\varphi = 0$ almost everywhere on $\mathbb{R}$ by Theorem 1.1 and thus $f = 0$ almost everywhere on I. ■

We summarize some further useful properties of the Chebyshev polynomials on I.

Lemma 6.4 *The Chebyshev polynomials* (6.1) *possess the following properties:*

1. *For all* $k \in \mathbb{N}_0$ *we have* $|T_k(x)| \le 1$ *for* $x \in I$.
2. *The Chebyshev polynomial* T_k, $k \in \mathbb{N}$, *has exactly* $k + 1$ *extreme points*

$$x_j^{(k)} := \cos\Big(\frac{j\pi}{k}\Big) \in I\,, \quad j = 0, \ldots, k\,,$$

with $T_k(x_j^{(k)}) = (-1)^j$.

3. *The Chebyshev polynomial* T_k, $k \in \mathbb{N}$, *possesses* k *simple zeros*

$$z_j^{(k)} := \cos\Big(\frac{(2j+1)\,\pi}{2k}\Big)\,, \quad j = 0, \ldots, k-1\,.$$

Between two neighboring zeros of T_{k+1} *there is exactly one zero of* T_k.

4. *For all* k, $\ell \in \mathbb{N}_0$ *we have*

$$2\,T_k\,T_\ell = T_{k+\ell} + T_{|k-\ell|} \quad \text{and} \quad T_k(T_\ell) = T_{k\ell}\,. \tag{6.5}$$

The proof of this lemma results immediately from the representation (6.1). For fixed $N \in \mathbb{N} \setminus \{1\}$, the points $x_j^{(N)}$, $j = 0, \ldots, N$, are called *Chebyshev extreme points* or sometimes just *Chebyshev points*, and the points $z_j^{(N)}$, $j = 0, \ldots, N-1$, are called *Chebyshev zero points*.

A polynomial of the form $p(x) = p_0 + p_1 x + \ldots + p_n x^n$ with the leading coefficient $p_n = 1$ is called *monic*. For instance, $2^{-k}\,T_{k+1} \in \mathscr{P}_{k+1}$, $k \in \mathbb{N}_0$, is monic. We will show that the polynomial $2^{-k}\,T_{k+1}$ has minimal norm among all monic polynomials of $\mathscr{P}_{k+1}$ in $C(I)$.

Lemma 6.5 *Let $k \in \mathbb{N}_0$ be given. For each monic polynomial $p \in \mathscr{P}_{k+1}$, we have*

$$2^{-k} = \max_{x\in I} 2^{-k}\,|T_{k+1}(x)| \le \max_{x\in I} |p(x)| = \|p\|_{C(I)}\,.$$

Proof

1. For $x_j^{(k+1)} := \cos\big(\frac{j\pi}{k+1}\big)$, $j = 0, \ldots, k+1$, the monic polynomial $2^{-k}\,T_{k+1}$ possesses the extreme value $(-1)^j\,2^{-k}$. Hence, we see that

$$2^{-k} = \max_{x\in I} 2^{-k}\,|T_{k+1}(x)|\,.$$

2. Assume that there exists a monic polynomial $p \in \mathscr{P}_{k+1}$ with $|p(x)| < 2^{-k}$ for all $x \in I$. Then the polynomial $q := 2^{-k}\,T_{k+1} - p \in \mathscr{P}_k$ has alternating positive and negative values at the $k+2$ points $x_j^{(k+1)}$, $j = 0, \ldots, k+1$. Thus, by the intermediate value theorem, q possesses at least $k+1$ distinct zeros such that $q = 0$. Consequently, we receive $p = 2^{-k}\,T_{k+1}$ contradicting our assumption. ■

Remark 6.6 The *Chebyshev polynomials of second kind* can be defined by the recursion formula

$$U_{k+1}(x) = 2x\,U_k(x) - U_{k-1}(x)\,, \quad k \in \mathbb{N}\,,$$

starting with $U_0(x) = 1$ and $U_1(x) = 2x$ for all $x \in \mathbb{R}$. For $x \in (-1, 1)$, the Chebyshev polynomials of second kind can be represented in the form

$$U_k(x) = \frac{\sin\big((k+1)\,\arccos(x)\big)}{\sin(\arccos(x))}\,, \quad k \in \mathbb{N}_0\,. \tag{6.6}$$

Comparing this formula with (6.1), we conclude that

$$(k+1)\,U_k = T'_{k+1}\,, \quad k \in \mathbb{N}_0\,. \tag{6.7}$$

Note that for all $k \in \mathbb{N}_0$ we have

$$U_k(-1) = (-1)^k\,(k+1)\,, \quad U_k(1) = k+1\,, \quad U_{2k}(0) = (-1)^k\,, \quad U_{2k+1}(0) = 0\,. \tag{6.8}$$

Further, the $n+1$ polynomials U_k, $k = 0, \ldots, n$, form an orthonormal basis of $\mathscr{P}_n$ with respect to the inner product

$$\frac{2}{\pi}\int_{-1}^{1}\sqrt{1-x^2}\,f(x)\,g(x)\,\mathrm{d}x\,.$$

□

Remark 6.7 Chebyshev polynomials of first and second kind are special *Jacobi polynomials* , which are orthogonal polynomials related to the inner product

$$\int_{-1}^{1}(1-x)^{\alpha}\,(1+x)^{\beta}\,f(x)\,g(x)\,\mathrm{d}x$$

with certain parameters $\alpha > -1$ and $\beta > -1$. □

Finally, we consider the recursive computation of derivatives and integrals of Chebyshev polynomials.

Lemma 6.8 *The derivative of the Chebyshev polynomial T_k fulfills the recursion formula*

$$T_k' = 2k\,T_{k-1} + \frac{k}{k-2}\,T_{k-2}'\,, \quad k = 3,\, 4\,\ldots\,, \tag{6.9}$$

starting with $T_0' = 0$, $T_1' = T_0$, and $T_2' = 4\,T_1$.

The integral of the Chebyshev polynomial T_k satisfies the formula

$$\int_{-1}^{x}T_k(t)\,\mathrm{d}t = \frac{1}{2\,(k+1)}\,T_{k+1}(x) - \frac{1}{2\,(k-1)}\,T_{k-1}(x) + \frac{(-1)^{k-1}}{k^2-1}\,, \quad k = 2,\,3\,\ldots\,, \tag{6.10}$$

and

$$\int_{-1}^{x}T_0(t)\,\mathrm{d}t = T_1(x)+1\,, \quad \int_{-1}^{x}T_1(t)\,\mathrm{d}t = \frac{1}{4}\,T_2(x) - \frac{1}{4}\,.$$

Particularly, it follows for all $k \in \mathbb{N}_0$ that

$$\int_{-1}^{1}T_{2k}(t)\,\mathrm{d}t = \frac{-2}{4k^2-1}\,, \quad \int_{-1}^{1}T_{2k+1}(t)\,\mathrm{d}t = 0\,. \tag{6.11}$$

Proof

1. Let $k \in \mathbb{N}\setminus\{1\}$. Substituting $x = \cos(t)$, $t \in [0,\,\pi]$, we obtain $T_k(x) = \cos(kt)$ by (6.1). Differentiation with respect to t provides

$$\frac{1}{k}\,\frac{\mathrm{d}}{\mathrm{d}t}T_k(\cos(t)) = \frac{\sin(kt)}{\sin(t)}\,, \quad t \in (0,\,\pi)\,.$$

Thus, for $t \in (0,\,\pi)$, i.e., $x \in (-1,\,1)$, it follows the equation

$$\frac{1}{k+1}\frac{\mathrm{d}}{\mathrm{d}t}T_{k+1}(\cos(t)) - \frac{1}{k-1}\frac{\mathrm{d}}{\mathrm{d}t}T_{k-1}(\cos(t)) = \frac{\sin((k+1)t) - \sin((k-1)t)}{\sin(t)}$$
$$= 2\,\cos(kt) = 2\,T_k(\cos(t))\,.$$

We conclude that the polynomial identity

$$\frac{1}{k+1}\,T'_{k+1} - \frac{1}{k-1}\,T'_{k-1} = 2\,T_k \tag{6.12}$$

is valid on $\mathbb{R}$.

2. Integration of (6.12) yields that

$$\int_{-1}^{x} T_k(t)\,\mathrm{d}t = \frac{1}{2\,(k+1)}\,T_{k+1}(x) - \frac{1}{2\,(k-1)}\,T_{k-1}(x) + c_k$$

with some integration constant c_k. Especially for $x = -1$ we obtain by $T_k(-1) = (-1)^k$ that

$$0 = \frac{(-1)^{k+1}}{2\,(k+1)} - \frac{(-1)^{k-1}}{2\,(k-1)} + c_k$$

and hence

$$c_k = \frac{(-1)^{k-1}}{k^2-1}\,.$$

Thus, we have shown (6.10). For $x = 1$ we conclude now (6.11) from $T_k(1) = 1$. ■

6.1.2 Chebyshev Series

In this subsection we consider real-valued functions defined on the compact interval $I := [-1,\,1]$. By the substitution $x = \cos(t)$, $t \in [0,\,\pi]$, the interval $[0,\,\pi]$ can be one-to-one mapped onto I. Conversely, the inverse function $t = \arccos(x)$, $x \in I$, maps I one-to-one onto $[0,\,\pi]$ (Fig. 6.4).

Let $f : I \to \mathbb{R}$ be an arbitrary real-valued function satisfying

$$\int_{-1}^{1} \frac{1}{\sqrt{1-x^2}}\,f(x)^2\,\mathrm{d}x < \infty\,.$$

Since

$$\int_{-1}^{1} \frac{1}{\sqrt{1-x^2}}\,\mathrm{d}x = \pi\,, \tag{6.13}$$

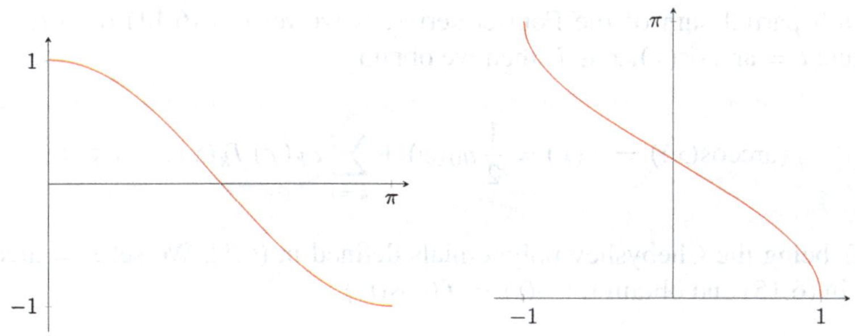

Fig. 6.4 The cosine function restricted on $[0, \pi]$ (left) and its inverse function arccos (right)

each continuous function $f : I \to \mathbb{R}$ fulfills the above condition. Now we form $f(\cos(\cdot)) : [0, \pi] \to \mathbb{R}$ and extend this function on $\mathbb{R}$ by

$$\varphi(t) := f(\cos(t)), \quad t \in \mathbb{R}.$$

Obviously, φ is a 2π-periodic, even function with

$$\int_0^{\pi} \varphi(t)^2 \,\mathrm{d}t = \int_{-1}^{1} \frac{1}{\sqrt{1-x^2}}\, f(x)^2 \,\mathrm{d}x < \infty .$$

We denote the subspace of all even, real-valued functions of $L_2(\mathbb{T})$ by $L_{2,\text{even}}(\mathbb{T})$. Recall that by Theorem 1.3 each function $\varphi \in L_{2,\text{even}}(\mathbb{T})$ can be represented as a convergent real Fourier series

$$\varphi(t) = \frac{1}{2}\, a_0(\varphi) + \sum_{k=1}^{\infty} a_k(\varphi)\, \cos(kt), \quad t \in \mathbb{R}, \tag{6.14}$$

with the Fourier coefficients

$$a_k(\varphi) := \frac{1}{\pi} \int_{-\pi}^{\pi} \varphi(t)\, \cos(kt)\,\mathrm{d}t = \frac{2}{\pi} \int_0^{\pi} \varphi(t)\, \cos(kt)\,\mathrm{d}t . \tag{6.15}$$

Here, convergence in $L_2(\mathbb{T})$ means that

$$\lim_{n\to\infty} \|\varphi - S_n\varphi\|_{L_2(\mathbb{T})} = 0, \tag{6.16}$$

where

$$S_n\varphi := \frac{1}{2}\, a_0(\varphi) + \sum_{k=1}^{n} a_k(\varphi)\, \cos(k\cdot) \tag{6.17}$$

is the nth partial sum of the Fourier series. If we restrict (6.14) onto $[0, \pi]$ and substitute $t = \arccos(x)$, $x \in I$, then we obtain

$$\varphi(\arccos(x)) = f(x) = \frac{1}{2}\, a_0(\varphi) + \sum_{k=1}^{\infty} a_k(\varphi)\, T_k(x)\,, \quad x \in I\,,$$

with T_k being the Chebyshev polynomials defined in (6.1). We set $t = \arccos(x)$, $x \in I$, in (6.15) and obtain for $\varphi(t) = f(\cos(t))$

$$a_k[f] := a_k(\varphi) = \frac{2}{\pi} \int_{-1}^{1} w(x)\, f(x)\, T_k(x)\, \mathrm{d}x\,, \quad k \in \mathbb{N}_0\,, \tag{6.18}$$

with the weight

$$w(x) := (1 - x^2)^{-1/2}\,, \quad x \in (-1,\, 1)\,.$$

The coefficient $a_k[f]$ in (6.18) is called *kth Chebyshev coefficient* of $f \in L_{2,w}(I)$.

Remark 6.9 For sufficiently large $N \in \mathbb{N}$, the numerical computation of the Chebyshev coefficients $a_k[f]$, $k = 0, \ldots, N-1$, is based on the DCT introduced in Sect. 3.5. We have

$$a_k[f] = a_k(\varphi) = \frac{2}{\pi} \int_0^{\pi} f(\cos(t))\, \cos(kt)\, \mathrm{d}t\,.$$

Analogously to the computation of the Fourier coefficients in Sect. 3.1, we split the interval $[0, \pi]$ into N subintervals of equal length and use the related midpoint rule such that

$$a_k[f] \approx \frac{2}{N} \sum_{j=0}^{N-1} f\Big(\cos\Big(\frac{(2j+1)\pi}{2N}\Big)\Big)\, \cos\Big(\frac{(2j+1)k\pi}{2N}\Big)\,, \quad k = 0, \ldots, N-1\,.$$

These sums can be calculated by the DCT-II(N) (see Sect. 6.3). □

For $f(\cos(t)) = \varphi(t)$, the Fourier series (6.14) of the transformed function $\varphi \in L_{2,\mathrm{even}}(\mathbb{T})$ transfers to the so-called Chebyshev series of $f \in L_{2,w}(I)$ which has the form

$$f = \frac{1}{2}\, a_0[f] + \sum_{k=1}^{\infty} a_k[f]\, T_k\,.$$

The *nth partial sum of the Chebyshev series* is denoted by $C_n f$.

Theorem 6.10 *Let $f \in L_{2,w}(I)$ be given. Then the sequence $(C_n f)_{n=0}^{\infty}$ of partial sums of the Chebyshev series converges to f in the norm of $L_{2,w}(I)$, i.e.,*

$$\lim_{n\to\infty} \|f - C_n f\|_{L_{2,w}(I)} = 0\,.$$

Further, for all $f,\, g \in L_{2,w}(I)$ the following Parseval equalities *are satisfied:*

$$2\,\|f\|^2_{L_{2,w}(I)} = \frac{1}{2}\,a_0[f]^2 + \sum_{k=1}^{\infty} a_k[f]^2\,,$$

$$2\,\langle f,\, g\rangle_{L_{2,w}(I)} = \frac{1}{2}\,a_0[f]\,a_0[g] + \sum_{k=1}^{\infty} a_k[f]\,a_k[g]\,. \tag{6.19}$$

Proof From $f \in L_{2,w}(I)$ it follows that $\varphi = f(\cos(\cdot)) \in L_2(\mathbb{T})$. Therefore, the Fourier partial sum $(S_n\varphi)(t)$ coincides with the partial sum $(C_n f)(x)$ of the Chebyshev series, if $x = \cos(t) \in I$ for $t \in [0,\, \pi]$. By Theorem 1.3, we know that

$$\lim_{n\to\infty} \|\varphi - S_n\varphi\|_{L_2(\mathbb{T})} = 0\,.$$

Since

$$\|f - C_n f\|_{L_{2,w}(I)} = \|\varphi - S_n\varphi\|_{L_2(\mathbb{T})}\,,$$

we obtain the convergence of the Chebyshev series of f in $L_{2,w}(I)$.

The Parseval equalities for the Chebyshev coefficients are now a consequence of the Parseval equalities for the Fourier coefficients. ■

A simple criterion for the uniform convergence of the Chebyshev series can be given as follows:

Lemma 6.11 *Let $f \in C(I)$ with*

$$\sum_{k=0}^{\infty} |a_k[f]| < \infty$$

be given. Then the Chebyshev series of f converges absolutely and uniformly on I to f.

Proof By (6.1) it holds $|T_k(x)| \le 1$ for all $x \in I$. Thus, using the Weierstrass criterion of uniform convergence, the Chebyshev series converges absolutely and uniformly on I. The limit g is continuous on I. From the completeness of the orthogonal system $\{T_k : k \in \mathbb{N}_0\}$, it follows that $f = g$ almost everywhere on I, since their Chebyshev coefficients coincide for all $k \in \mathbb{N}_0$. Observing that f, $g \in C(I)$, we conclude that the functions f and g are identical. ■

As usual, by $C(I)$ we denote the Banach space of all continuous functions $f : I \to \mathbb{R}$ with the norm

$$\|f\|_{C(I)} := \max_{x \in I} |f(x)|\,.$$

Let $C^r(I)$, $r \in \mathbb{N}$, be the set of all r-times continuously differentiable functions $f : I \to \mathbb{R}$, i.e., for each $j = 0, \ldots, r$ the derivative $f^{(j)}$ is continuous on $(-1,\,1)$ and the one-sided derivatives $f^{(j)}(-1+0)$ as well as $f^{(j)}(1-0)$ exist and fulfill the conditions

$$f^{(j)}(-1+0) = \lim_{x \to -1+0} f^{(j)}(x)\,, \quad f^{(j)}(1-0) = \lim_{x \to 1-0} f^{(j)}(x)\,.$$

Theorem 6.12 *For $f \in C^1(I)$, the corresponding Chebyshev series converges absolutely and uniformly on I to f. If $f \in C^r(I)$, $r \in \mathbb{N}$, then we have*

$$\lim_{n \to \infty} n^{r-1}\,\|f - C_n f\|_{C(I)} = 0\,.$$

Proof If $f \in C^r(I)$ with $r \in \mathbb{N}$, then the even function $\varphi = f(\cos(\cdot))$ is contained in $C^r(\mathbb{T})$. By Theorem 1.39 we have

$$\lim_{n \to \infty} n^{r-1}\,\|\varphi - S_n \varphi\|_{C(\mathbb{T})} = 0\,.$$

From $\varphi - S_n\varphi = f - C_n f$ the assertion follows. ■

Example 6.13 We consider $f(x) := |x|$, $x \in I$ (Fig. 6.5). Then $f \in C(I)$ is even. The related Chebyshev coefficients of f are for $k \in \mathbb{N}_0$ of the form

$$a_{2k}[f] = \frac{4}{\pi} \int_0^1 w(x)\, x\, T_{2k}(x)\,\mathrm{d}x\,, \quad a_{2k+1}[f] = 0\,.$$

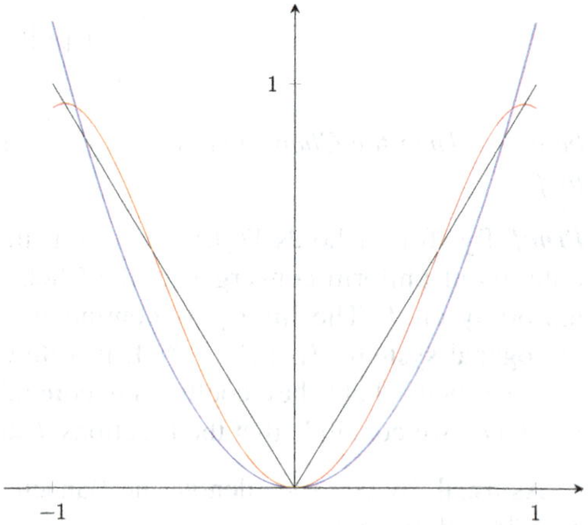

Fig. 6.5 The function $f(x) := |x|$, $x \in I$ (black) and the partial sums $C_n f$ of the Chebyshev series for $n = 2$ (blue) and $n = 4$ (red)

The substitution $x = \cos(t)$, $t \in [0, \frac{\pi}{2}]$, provides for each $k \in \mathbb{N}_0$

$$a_{2k}[f] = \frac{4}{\pi} \int_0^{\pi/2} \cos(t)\, \cos(2kt)\, \mathrm{d}t = -\frac{(-1)^k\, 4}{(4\,k^2 - 1)\pi}\,.$$

By Theorem 6.11 the Chebyshev series of f converges absolutely and uniformly on I to f, i.e.,

$$|x| = \frac{2}{\pi} - \frac{4}{\pi} \sum_{k=1}^{\infty} \frac{(-1)^k\, 4}{(4\,k^2 - 1)\pi}\, T_{2k}(x)\,, \quad x \in I\,.$$

□

Example 6.14 We consider the sign function

$$f(x) = \operatorname{sgn} x := \begin{cases} 1 & x \in (0,\, 1]\,, \\ 0 & x = 0\,, \\ -1 & x \in [-1,\, 0)\,. \end{cases}$$

Since f is odd, we find for all $k \in \mathbb{N}_0$

$$a_{2k}[f] = 0\,, \quad a_{2k+1}[f] = \frac{4}{\pi} \int_0^1 w(x)\, T_{2k+1}(x)\, \mathrm{d}x\,.$$

Substituting $x = \cos(t)$, $t \in [0, \frac{\pi}{2}]$, we obtain

$$a_{2k+1}[f] = \frac{4}{\pi} \int_0^{\pi/2} \cos((2k+1)t)\, \mathrm{d}t = \frac{(-1)^k\, 4}{(2k+1)\,\pi}\,, \quad k \in \mathbb{N}_0\,.$$

Then the Chebyshev series of f converges pointwise to f, since the even 2π-periodic function $\varphi = f(\cos(\cdot))$ is piecewise continuously differentiable and hence the Fourier series of φ converges pointwise to φ by Theorem 1.34. The jump discontinuity at $x = 0$ leads to the Gibbs phenomenon (see Sect. 1.4.3). Each partial sum $C_n f$ of the Chebyshev series oscillates with overshoot and undershoot near $x = 0$ (Fig. 6.6). □

In the following theorem, we summarize some simple properties of the Chebyshev coefficients.

Theorem 6.15 (Properties of Chebyshev Coefficients) *For all* $k \in \mathbb{N}_0$, *the Chebyshev coefficients of* $f,\, g \in L_{2,w}(I)$ *possess the following properties:*

1. Linearity: *For all* $\alpha, \beta \in \mathbb{C}$,

$$a_k[\alpha f + \beta g] = \alpha\, a_k[f] + \beta\, a_k[g]\,.$$

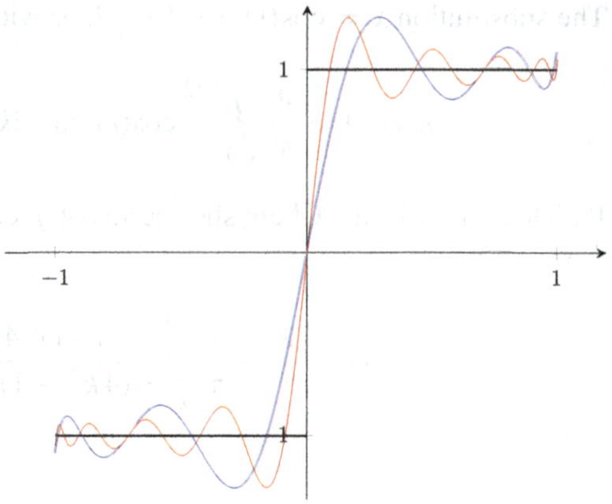

Fig. 6.6 The function $f(x) := \operatorname{sgn} x$, $x \in I$ (black) and the partial sums $C_n f$ of the Chebyshev series for $n = 8$ (blue) and $n = 16$ (red)

2. Translation: *For all* $\ell \in \mathbb{N}_0$,

$$a_k[T_\ell\, f] = \frac{1}{2}\,\big(a_{k+\ell}[f] + a_{|k-\ell|}[f]\big)\,.$$

3. Symmetry:

$$a_k[f(-\,\cdot)] = (-1)^k\, a_k[f]\,.$$

4. Differentiation: *If additionally* $f' \in L_{2,w}(I)$*, then for all* $k \in \mathbb{N}$

$$a_k[f] = \frac{1}{2k}\,\big(a_{k-1}[f'] - a_{k+1}[f']\big)\,.$$

Proof The linearity follows immediately from the definition of the Chebyshev coefficients. Using relation in (6.5), we conclude that

$$\begin{aligned} a_k[T_\ell\, f] &= \frac{2}{\pi}\int_{-1}^{1} w(x)\, f(x)\, T_\ell(x)\, T_k(x)\,\mathrm{d}x \\ &= \frac{1}{2}\,\big(a_{k+\ell}[f] + a_{|k-\ell|}[f]\big)\,. \end{aligned}$$

The symmetry is a simple consequence of (6.4). Using integration by parts, we find that

$$\begin{aligned} a_k[f] &= \frac{2}{\pi}\int_0^{\pi} f(\cos(t))\,\cos(kt)\,\mathrm{d}t \\ &= \frac{2}{k\pi}\, f(\cos(t))\,\sin(kt)\big|_0^{\pi} + \frac{2}{k\pi}\int_0^{\pi} f'(\cos(t))\,\sin(t)\,\sin(kt)\,\mathrm{d}t \end{aligned}$$

$$= \frac{1}{k\pi} \int_0^\pi f'(\cos(t)) \big(\cos((k-1)t) - \cos((k+1)t)\big)\, \mathrm{d}t$$

$$= \frac{1}{2k} \big(a_{k-1}[f'] - a_{k+1}[f']\big).$$

This completes the proof. ■

We finish this section by studying the decay properties of the Chebyshev coefficients and the Chebyshev series for expansions of smooth functions.

Theorem 6.16 *For fixed $r \in \mathbb{N}_0$, let $f \in C^{r+1}(I)$ be given. Then for all $n > r$, the Chebyshev coefficients of f satisfy the inequality*

$$|a_n[f]| \le \frac{2}{n\,(n-1)\,\ldots\,(n-r)}\, \|f^{(r+1)}\|_{C(I)}\,. \tag{6.20}$$

Further, for all $n > r$, the partial sum $C_n f$ of the Chebyshev series satisfies

$$\|f - C_n f\|_{C(I)} \le \frac{2}{r\,(n-r)^r}\, \|f^{(r+1)}\|_{C(I)}\,. \tag{6.21}$$

Proof Using $|T_n(x)| \le 1$ for all $x \in I$ and (6.13), we can estimate

$$|a_n[f^{(r+1)}]| = \frac{2}{\pi}\, \Big| \int_{-1}^1 \frac{1}{\sqrt{1-x^2}}\, f^{(r+1)}(x)\, T_n(x)\, \mathrm{d}x \Big| \le 2\, \|f^{(r+1)}\|_{C(I)}\,.$$

By the differentiation property of the Chebyshev coefficients in Theorem 6.15, we conclude that

$$|a_n[f^{(r)}]| \le \frac{1}{2n} \big(|a_{n-1}[f^{(r+1)}]| + |a_{n+1}[f^{(r+1)}]|\big) \le \frac{2}{n}\, \|f^{(r+1)}\|_{C(I)}\,.$$

Analogously, we receive

$$|a_n[f^{(r-1)}]| \le \frac{1}{2n} \big(|a_{n-1}[f^{(r)}]| + |a_{n+1}[f^{(r)}]|\big) \le \frac{2}{n\,(n-1)}\, \|f^{(r+1)}\|_{C(I)}\,.$$

If we continue in this way, we obtain (6.20) such that

$$|a_n[f]| \le \frac{2}{n(n-1)\ldots(n-r)}\, \|f^{(r+1)}\|_{C(I)} \le \frac{2}{(n-r)^{r+1}}\, \|f^{(r+1)}\|_{C(I)}\,.$$

By Theorem 6.12 the Chebyshev series of $f \in C^{r+1}(I)$ converges uniformly on I. Using $|T_k(x)| \le 1$ for all $x \in I$, the remainder

$$f - C_n f = \sum_{k=n+1}^{\infty} a_k[f]\, T_k$$

can be estimated by

$$\|f - C_n f\|_{C(I)} \le \sum_{k=n+1}^{\infty} |a_k[f]| \le 2\, \|f^{(r+1)}\|_{C(I)} \sum_{k=n+1}^{\infty} \frac{1}{(k-r)^{r+1}}$$

$$\le 2\, \|f^{(r+1)}\|_{C(I)} \int_n^{\infty} \frac{1}{(t-r)^{r+1}}\, \mathrm{d}t = 2\, \|f^{(r+1)}\|_{C(I)} \frac{1}{r\,(n-r)^r}\,.$$

■

Remark 6.17 Similar estimates of the Chebyshev coefficients $a_k[f]$ and of the remainder $f - C_n f$ are shown in [415, pp. 52–54] and [279] under the weaker assumption that $f, f', \ldots, f^{(r)}$ are absolutely continuous on I and that

$$\int_{-1}^{1} \frac{|f^{(r+1)}(x)|}{\sqrt{1-x^2}}\, \mathrm{d}x < \infty\,.$$

□

Summing up we can say by Theorems 6.12 and 6.16:
The smoother a function $f : I \to \mathbb{R}$, the faster its Chebyshev coefficients $a_k[f]$ tend to zero as $n \to \infty$ and the faster its Chebyshev series converges uniformly to f.

6.2 Fast Evaluation of Polynomials

The goal of the following considerations is the efficient evaluation of algebraic polynomials and of polynomial operations.

6.2.1 *Horner Scheme and Clenshaw Algorithm*

Let $\mathscr{P}_n$ denote the set of all real algebraic polynomials up to degree $n \in \mathbb{N}_0$

$$p(x) := p_0 + p_1\, x + \ldots + p_n\, x^n\,, \quad x \in [a,\, b]\,, \tag{6.22}$$

where $[a,\, b] \subset \mathbb{R}$ is a compact interval. We want to compute a polynomial (6.22) with real coefficients p_k, $k = 0, \ldots, n$, at one point $x_0 \in [a,\, b]$ by a low number of arithmetic operations. In order to reduce the number of needed multiplications, we write $p(x_0)$ in the form of nested multiplications

$$p(x_0) = p_0 + x_0 \Big(p_1 + x_0 \big(p_2 + x_0 (\ldots (p_{n-1} + x_0\, p_n) \ldots)\big)\Big)\,.$$

This simple idea leads to the well-known *Horner scheme*.

Algorithm 6.18 (Horner Scheme)
Input: $n \in \mathbb{N} \setminus \{1\}$, $x_0 \in [a,\, b]$, $p_k \in \mathbb{R}$ *for* $k = 0, \ldots, n$.

1. *Set* $q_{n-1} := p_n$ *and calculate recursively for* $j = 2, \ldots, n$

$$q_{n-j} := p_{n-j+1} + x_0\, q_{n-j+1}\,. \tag{6.23}$$

2. *Form* $p(x_0) := p_0 + x_0\, q_0$.

Output: $p(x_0) \in \mathbb{R}$.
Computational cost: $\mathcal{O}(n)$.

Performing n real multiplications and n real additions, we arrive at the value $p(x_0)$. But this is not the complete story of the Horner scheme. Introducing the polynomial

$$q(x) := q_0 + q_1\, x + \ldots + q_{n-1}\, x^{n-1}\,,$$

we obtain by comparing coefficient method and (6.23) that

$$p(x) = q(x)\,(x - x_0) + p(x_0)\,.$$

Hence, the Horner scheme describes also the division of the polynomial in (6.22) by the linear factor $x - x_0$. Therefore, by repeated application of the Horner scheme, we can also calculate the derivatives of the polynomial (6.22) at the point $x_0 \in [a,\, b]$. For simplicity, we only sketch the computation of $p'(x_0)$ and $p''(x_0)$. Using the Horner scheme, we divide $q(x)$ by $x - x_0$ and obtain

$$q(x) = r(x)\,(x - x_0) + q(x_0)\,.$$

Then we divide the polynomial $r(x)$ by $x - x_0$ such that

$$r(x) = s(x)\,(x - x_0) + r(x_0)\,.$$

This implies that

$$\begin{aligned} p(x) &= r(x)(x - x_0)^2 + q(x_0)(x - x_0) + p(x_0) \\ &= s(x)\,(x - x_0)^3 + r(x_0)\,(x - x_0)^2 + q(x_0)\,(x - x_0) + p(x_0) \end{aligned}$$

and hence

$$q(x_0) = p'(x_0)\,, \quad r(x_0) = \frac{1}{2}\, p''(x_0)\,.$$

As known, the monomials x^k, $k = 0, \ldots, n$, form a simple basis of $\mathcal{P}_n$. Unfortunately, the monomial basis is unfavorable from a numerical point of view. Therefore, we are interested in another basis of $\mathcal{P}_n$ which is more convenient for numerical calculations. Using the Chebyshev polynomials (6.1), such a basis of $\mathscr{P}_n$ can be formed by the polynomials

$$T_k^{[a,\,b]}(x) := T_k\Big(\frac{2x-a-b}{b-a}\Big)\,, \quad k = 0, \ldots, n\,.$$

For the interval [0, 1] we obtain the *shifted Chebyshev polynomials*

$$T_k^{[0,\,1]}(x) := T_k(2x-1)\,.$$

For the properties of shifted Chebyshev polynomials, see [311, pp. 20–21].

We restrict our considerations to polynomials on $I := [-1,\,1]$ and want to use the Chebyshev polynomials T_k, $k = 0, \ldots, n$, as orthogonal basis of $\mathscr{P}_n$. An arbitrary polynomial $p \in \mathscr{P}_n$ can be uniquely represented in the form

$$p = \frac{1}{2}\,a_0 + \sum_{k=1}^{n} a_k\,T_k \tag{6.24}$$

with some coefficients $a_k \in \mathbb{R}$, $k = 0, \ldots, n$. For an efficient computation of the polynomial value $p(x_0)$ for fixed $x_0 \in I$, we apply the *Clenshaw algorithm*. To this end, we iteratively reduce the degree of p by means of the recursion formula (6.3). Assume that $n \ge 5$ and $a_n \ne 0$. Applying (6.3) to T_n in (6.24), we obtain

$$p(x_0) = \frac{1}{2}\,a_0 + \sum_{k=1}^{n-3} a_k\,T_k(x_0) + (a_{n-2} - b_n)\,T_{n-2}(x_0) + b_{n-1}\,T_{n-1}(x_0)$$

with $b_n := a_n$ and $b_{n-1} := 2x_0\,b_n + a_{n-1}$. Next, with $b_{n-2} := 2x_0\,b_{n-1} - b_n + a_{n-2}$ it follows by (6.3) that

$$p(x_0) = \frac{1}{2}\,a_0 + \sum_{k=1}^{n-4} a_k\,T_k(x_0) + (a_{n-3} - b_{n-1})\,T_{n-3}(x_0) + b_{n-2}\,T_{n-2}(x_0)\,.$$

In this way we can continue. Thus, the Clenshaw algorithm can be considered as analogous to Algorithm 6.18 (see [90]).

Algorithm 6.19 (Clenshaw Algorithm)
Input: $n \in \mathbb{N} \setminus \{1\}$, $x_0 \in I$, $a_k \in \mathbb{R}$ *for* $k = 0, \ldots, n$.

1. *Set* $b_{n+2} = b_{n+1} := 0$ *and calculate recursively for* $j = 0, \ldots, n$

$$b_{n-j} := 2x_0\,b_{n-j+1} - b_{n-j+2} + a_{n-j}\,. \tag{6.25}$$

2. *Form* $p(x_0) := \frac{1}{2}\,(b_0 - b_2)$.

Output: $p(x_0) \in \mathbb{R}$.

The Clenshaw algorithm needs $\mathcal{O}(n)$ arithmetic operations and is convenient for the computation of few values of the polynomial (6.24). The generalization to polynomials with arbitrary three-term recurrence relation is straightforward.

6.2.2 Polynomial Evaluation and Interpolation at Chebyshev Points

Now we want to compute simultaneously all values of an arbitrary polynomial in (6.24) of high degree n on the grid of all Chebyshev zero points

$$z_k^{(N)} := \cos\Big(\frac{(2k+1)\,\pi}{2N}\Big), \quad k = 0, \ldots, N-1, \tag{6.26}$$

with an integer $N \ge n+1$. Setting $a_j := 0$, $j = n+1, \ldots, N-1$, and forming the vectors

$$\mathbf{a} := \big(\frac{\sqrt{2}}{2}\, a_0,\, a_1,\, \ldots,\, a_{N-1}\big)^\top, \quad \mathbf{p} := \big(p(z_k^{(N)})\big)_{k=0}^{N-1},$$

we obtain

$$\mathbf{p} = \sqrt{\frac{N}{2}}\, \mathbf{C}_N^{\mathrm{III}}\, \mathbf{a} \tag{6.27}$$

with the orthogonal cosine matrix of type III

$$\mathbf{C}_N^{\mathrm{III}} = \sqrt{\frac{2}{N}}\, \Big(\varepsilon_N(j)\, \cos\Big(\frac{(2k+1)j\pi}{2N}\Big)\Big)_{k,j=0}^{N-1}.$$

If N is a power of two, the vector $\mathbf{p}$ can be rapidly computed by a fast DCT-III (N) algorithm (see Sect. 6.3).

Algorithm 6.20 (Polynomial Values at Chebyshev Zero Points)
Input: $n \in \mathbb{N} \setminus \{1\}$, $N := 2^t \ge n+1$ *with* $t \in \mathbb{N} \setminus \{1\}$, $a_k \in \mathbb{R}$ *for* $k = 0, \ldots, n$.

1. *Set* $a_j := 0$, $j = n+1, \ldots, N-1$, *and* $\mathbf{a} := (\frac{\sqrt{2}}{2}\, a_0,\, a_1,\, \ldots,\, a_{N-1})^\top$.
2. *Compute* (6.27) *by Algorithm* 6.30 *or Algorithm* 6.37.

Output: $p(z_k^{(N)}) \in \mathbb{R}$, $k = 0, \ldots, N-1$.
Computational cost: $\mathcal{O}(N \log N)$.

The simultaneous computation of N values of an arbitrary polynomial of degree n with $n \le N-1$ requires only $\mathcal{O}(N \log N)$ arithmetic operations. This is an important advantage compared to the Clenshaw Algorithm 6.19, since this method would require $\mathcal{O}(n\, N)$ arithmetic operations.

From (6.27) and Lemma 3.47, it follows that

$$\mathbf{a} = \sqrt{\frac{2}{N}}\,\big(\mathbf{C}_N^{\mathrm{III}}\big)^{-1}\,\mathbf{p} = \sqrt{\frac{2}{N}}\,\mathbf{C}_N^{\mathrm{II}}\,\mathbf{p}\,. \tag{6.28}$$

In other words, the coefficients a_k, $k = 0, \ldots, N-1$, of the polynomial

$$p = \frac{1}{2}\,a_0 + \sum_{j=1}^{N-1} a_j\,T_j \tag{6.29}$$

are obtained by interpolation at Chebyshev zero points $z_k^{(N)}$, $k = 0, \ldots, N-1$ in (6.26). Thus, we obtain

Lemma 6.21 *Let $N \in \mathbb{N} \setminus \{1\}$ be given. For arbitrary $p_j \in \mathbb{R}$, $j = 0, \ldots, N-1$, there exists a unique polynomial $p \in \mathcal{P}_{N-1}$ of the form* (6.29) *which solves the interpolation problem*

$$p(z_j^{(N)}) = p_j\,, \quad j = 0, \ldots, N-1\,. \tag{6.30}$$

The coefficients of (6.29) *can be computed by* (6.28), *i.e.,*

$$a_k = \frac{2}{N}\sum_{j=0}^{N-1} p_j\,\cos\Big(\frac{(2j+1)k\pi}{2N}\Big)\,, \quad k = 0, \ldots, N-1\,.$$

The same idea of simultaneous computation of polynomial values and of polynomial interpolation can be used for the nonequispaced grid of Chebyshev extreme points $x_j^{(N)} = \cos\big(\frac{j\pi}{N}\big)$, $j = 0, \ldots, N$. In this case we represent an arbitrary polynomial $p \in \mathcal{P}_N$ in the form

$$p = \frac{1}{2}\,a_0 + \sum_{k=1}^{N-1} a_k\,T_k + \frac{1}{2}\,a_N\,T_N \tag{6.31}$$

with real coefficients a_k. For the simultaneous computation of the values $p(x_j^{(N)})$, $j = 0, \ldots, N$, we obtain that

$$\mathbf{p} = \sqrt{\frac{N}{2}}\,\mathbf{C}_{N+1}^{\mathrm{I}}\mathbf{a}\,, \tag{6.32}$$

where

$$\mathbf{p} := \big(\varepsilon_N(j)\,p(x_j^{(N)})\big)_{j=0}^{N}\,, \quad \mathbf{a} = \big(\varepsilon_N(k)\,a_k\big)_{k=0}^{N} \tag{6.33}$$

with $\varepsilon_N(0) = \varepsilon_N(N) := \frac{\sqrt{2}}{2}$ and $\varepsilon_N(j) := 1$, $j = 1, \ldots, N-1$. Here,

$$\mathbf{C}_{N+1}^{\mathrm{I}} = \sqrt{\frac{2}{N}} \left(\varepsilon_N(j)\, \varepsilon_N(k)\, \cos\left(\frac{jk\pi}{N}\right)\right)_{j,k=0}^{N}$$

denotes the orthogonal cosine matrix of type I (see Lemma 3.46). If N is a power of two, the vector $\mathbf{p}$ can be rapidly computed by a fast DCT-I $(N+1)$ algorithm (see Sect. 6.3).

Algorithm 6.22 (Polynomial Values at Chebyshev Extreme Points)
Input: $N := 2^t$ *with* $t \in \mathbb{N} \setminus \{0\}$, $a_k \in \mathbb{R}$ *for* $k = 0, \ldots, N$.

1. *Form the vector* $\mathbf{a} := (\varepsilon_N(k)\, a_k)_{k=0}^{N}$.
2. *Compute* $(p_j)_{j=0}^{N} := \sqrt{\frac{N}{2}}\, \mathbf{C}_{N+1}^{\mathrm{I}} \mathbf{a}$ *by fast* DCT-I $(N+1)$ *using Algorithm* 6.28 *or Algorithm 6.35.*
3. *Form* $p(x_j^{(N)}) := \varepsilon_N(j)^{-1}\, p_j$, $j = 0, \ldots, N$.

Output: $p(x_j^{(N)}) \in \mathbb{R}$, $j = 0, \ldots, N$.
Computational cost: $\mathcal{O}(N \log N)$.

From (6.32) and Lemma 3.46, it follows that

$$\mathbf{a} = \sqrt{\frac{2}{N}}\, \big(\mathbf{C}_{N+1}^{\mathrm{I}}\big)^{-1} \mathbf{p} = \sqrt{\frac{2}{N}}\, \mathbf{C}_{N+1}^{\mathrm{I}}\, \mathbf{p}\,. \tag{6.34}$$

In other words, the coefficients a_k, $k = 0, \ldots, N$, of the polynomial (6.31) are obtained by interpolation at Chebyshev extreme points $x_k^{(N)} = \cos\big(\frac{\pi k}{N}\big)$, $k = 0, \ldots, N$. Thus, we get the following.

Lemma 6.23 *Let* $N \in \mathbb{N} \setminus \{1\}$ *be given. For arbitrary* $p_j \in \mathbb{R}$, $j = 0, \ldots, N$, *there exists a unique polynomial* $p \in \mathcal{P}_N$ *of the form* (6.31) *which solves the interpolation problem*

$$p(x_j^{(N)}) = p_j\,, \quad j = 0, \ldots, N\,, \tag{6.35}$$

with $x_j^{(N)} = \cos\big(\frac{\pi j}{N}\big)$. *The coefficients of the polynomial in* (6.31) *can be computed by* (6.34), *i.e.,*

$$a_k = \frac{2}{N} \left(\frac{1}{2}\, p_0 + \sum_{j=1}^{N-1} p_j \cos\left(\frac{jk\pi}{N}\right) + \frac{1}{2}\, (-1)^k\, p_N \right), \quad k = 0, \ldots, N\,.$$

Now we derive an efficient and numerically stable representation of the interpolating polynomial (6.31) based on the so-called barycentric formula for interpolating polynomials introduced by H. E. Salzer [373] (see also [44] and [415, pp. 33–41]).

Theorem 6.24 (Barycentric Interpolation at Chebyshev Extreme Points) *Let $N \in \mathbb{N} \setminus \{1\}$ be given. The polynomial* (6.31) *which interpolates the real data p_j at the Chebyshev extreme points $x_j^{(N)} = \cos\big(\frac{\pi j}{N}\big)$, $j = 0, \ldots, N$, satisfies the barycentric formula*

$$p(x) = \frac{\dfrac{p_0}{2\,(x-1)} + \displaystyle\sum_{j=1}^{N-1} \frac{(-1)^j\, p_j}{x - x_j^{(N)}} + \dfrac{(-1)^N\, p_N}{2\,(x+1)}}{\dfrac{1}{2\,(x-1)} + \displaystyle\sum_{j=1}^{N-1} \frac{(-1)^j}{x - x_j^{(N)}} + \dfrac{(-1)^N}{2\,(x+1)}} \tag{6.36}$$

for all $x \in \mathbb{R} \setminus \{x_j^{(N)} : j = 0, \ldots, N\}$ and $p(x_j^{(N)}) = p_j$ for $x = x_j^{(N)}$, $j = 0, \ldots, N$.

Proof

1. Using the *node polynomial*

$$\ell(x) := \prod_{j=0}^{N} (x - x_j^{(N)})\,,$$

we form the *kth Lagrange basis polynomial*

$$\ell_k(x) := \frac{\ell(x)}{\ell'(x_k^{(N)})\,(x - x_k^{(N)})}\,, \tag{6.37}$$

where

$$\ell'(x_k^{(N)}) = \prod_{\substack{j=0\\ j\neq k}}^{N} (x_k^{(N)} - x_j^{(N)})\,.$$

The Lagrange basis polynomials possess the interpolation property

$$\ell_k(x_j^{(N)}) = \delta_{j-k}\,, \quad j, k = 0, \ldots, N\,. \tag{6.38}$$

Then the interpolation problem $p(x_j^{(N)}) = p_j$, $j = 0, \ldots, N$, has the solution

$$p(x) = \sum_{k=0}^{N} p_k\, \ell_k(x) = \ell(x) \sum_{k=0}^{N} \frac{p_k}{\ell'(x_k^{(N)})\,(x - x_k^{(N)})} \tag{6.39}$$

which is uniquely determined in $\mathscr{P}_N$. Particularly, for the constant polynomial $p \equiv 1$ we have $p_k = 1$, $k = 0, \ldots, N$, and obtain

$$1 = \sum_{k=0}^{N} \ell_k(x) = \ell(x) \sum_{k=0}^{N} \frac{1}{\ell'(x_k^{(N)})\,(x - x_k^{(N)})}\,. \tag{6.40}$$

Dividing (6.39) by (6.40), we get the barycentric formula

$$p(x) = \frac{\displaystyle\sum_{k=0}^{N} \frac{p_k}{\ell'\big(x_k^{(N)}\big)\,\big(x - x_k^{(N)}\big)}}{\displaystyle\sum_{k=0}^{N} \frac{1}{\ell'\big(x_k^{(N)}\big)\,\big(x - x_k^{(N)}\big)}}\,. \tag{6.41}$$

2. Now we calculate $\ell'(x_k^{(N)})$. Using the substitution $x = \cos(t)$, we simply observe that the monic polynomial of degree $N+1$

$$\begin{aligned} 2^{-N}\,(T_{N+1}(x) - T_{N-1}(x)) &= 2^{-N}\,(\cos((N+1)t) - \cos((N-1)t)) \\ &= -2^{1-N}\,\sin(Nt)\,\sin(t) \end{aligned}$$

possesses the $N+1$ distinct zeros $x_k^{(N)} = \cos\big(\frac{k\pi}{N}\big)$, $k = 0, \ldots, N$, such that the node polynomial reads

$$\ell(x) = 2^{-N}\,\big(T_{N+1}(x) - T_{N-1}(x)\big)\,.$$

Consequently, we obtain by (6.7) that

$$\begin{aligned} \ell'(x_k^{(N)}) &= 2^{-N}\,\big(T'_{N+1}(x_k^{(N)}) - T'_{N-1}(x_k^{(N)})\big) \\ &= 2^{-N}\,\big((N+1)\,U_N(x_k^{(N)}) - (N-1)\,U_{N-2}(x_k^{(N)})\big)\,. \end{aligned}$$

Applying (6.6) and (6.8), we find

$$(N+1)\,U_N(x_k^{(N)}) - (N-1)\,U_{N-2}(x_k^{(N)}) = \begin{cases} 4N & k = 0\,, \\ 2N\,(-1)^k & k = 1, \ldots, N-1\,, \\ 4N\,(-1)^N & k = N \end{cases}$$

and hence

$$\ell'(x_k^{(N)}) = \begin{cases} 2^{2-N}\,N & k = 0\,, \\ 2^{1-N}\,N\,(-1)^k & k = 1, \ldots, N-1\,, \\ 2^{2-N}\,N\,(-1)^N & k = N\,. \end{cases}$$

By (6.41) the above result completes the proof of (6.36). ∎

The barycentric formula (6.36) is very helpful for interpolation at Chebyshev extreme points. By (6.36) the interpolation polynomial is expressed as a weighted average of the given values p_j. This expression can be efficiently computed by the fast summation method (see Sect. 7.6).

Remark 6.25 A similar barycentric formula can be derived for the interpolation at Chebyshev zero points $z_j^{(N)} = \cos\big(\frac{(2j+1)\pi}{2N}\big)$ for $j = 0, \ldots, N-1$. Then the corresponding node polynomial has the form

$$\ell(x) = \prod_{j=0}^{N-1} (x - z_j^{(N)}) = 2^{1-N}\, T_N(x)\,.$$

By (6.7) we obtain

$$\ell'(z_j^{(N)}) = \frac{2^{1-N} N \sin\left(j + \frac{\pi}{2}\right)}{\sin\left(\frac{(2j+1)\pi}{2N}\right)} = \frac{2^{1-N}\, N\, (-1)^j}{\sin\left(\frac{(2j+1)\pi}{2N}\right)}\,, \qquad j = 0, \ldots, N-1\,.$$

Similarly to (6.41), the polynomial $p \in \mathscr{P}_{N-1}$ which interpolates the real data p_j at the Chebyshev zero points $z_j^{(N)}$, $j = 0, \ldots, N-1$, satisfies the barycentric formula

$$p(x) = \frac{\displaystyle\sum_{k=0}^{N-1} \frac{p_k}{\ell'(z_k^{(N)})\,(x - z_k^{(N)})}}{\displaystyle\sum_{k=0}^{N-1} \frac{1}{\ell'(z_k^{(N)})\,(x - z_k^{(N)})}} = \frac{\displaystyle\sum_{k=0}^{N-1} \frac{(-1)^k\, p_k \sin\left(\frac{(2k+1)\pi}{2N}\right)}{x - z_k^{(N)}}}{\displaystyle\sum_{k=0}^{N-1} \frac{(-1)^k \sin\left(\frac{(2k+1)\pi}{2N}\right)}{x - z_k^{(N)}}}$$

for all $x \in \mathbb{R} \setminus \{z_j^{(N)} : j = 0, \ldots, N-1\}$ and $p(z_j^{(N)}) = p_j$ for $x = z_j^{(N)}$, $j = 0, \ldots, N-1$. □

Next, we describe the differentiation and integration of polynomials being given in the basis of Chebyshev polynomials.

Theorem 6.26 *For fixed $n \in \mathbb{N} \setminus \{1\}$, let an arbitrary polynomial $p \in \mathscr{P}_n$ be given in the form $p = \frac{a_0}{2} + \sum_{j=1}^{n} a_j T_j$. Then the derivative p' has the form*

$$p' = \frac{1}{2}\, d_0 + \sum_{j=1}^{n-1} d_j\, T_j\,, \tag{6.42}$$

where the coefficients d_j satisfy the recursion

$$d_{n-1-j} := d_{n+1-j} + 2\,(n-j)\, a_{n-j}\,, \quad j = 0,\, 1,\, \ldots\,, n-1\,, \tag{6.43}$$

with $d_{n+1} = d_n := 0$. Further, the integral of the polynomial p can be calculated by

$$\int_{-1}^{x} p(t)\,\mathrm{d}t = \frac{1}{2}\,c_0 + \sum_{j=1}^{n+1} c_j\,T_j(x) \tag{6.44}$$

with the recursion formula

$$c_j := \frac{1}{2j}\,(a_{j-1} - a_{j+1})\,, \quad j = 1, 2, \ldots, n+1\,, \tag{6.45}$$

starting with

$$c_0 := a_0 - \frac{1}{2}\,a_1 + 2\sum_{j=2}^{n}(-1)^{j+1}\,\frac{a_j}{j^2-1}\,,$$

where we set $a_{n+1} = a_{n+2} := 0$.

Proof The integration formulas (6.44)–(6.45) are direct consequences of (6.10). For proving the differentiation formulas (6.42)–(6.43), we apply the integration formulas (6.44)–(6.45). Let $p \in \mathscr{P}_n$ be given in the form (6.24). Obviously, $p' \in \mathscr{P}_{n-1}$ can be represented in the form (6.42) with certain coefficients $d_j \in \mathbb{R}$, $j = 0, \ldots, n-1$. Then it follows that

$$\int_{-1}^{x} p'(t)\,\mathrm{d}t = p(x) - p(-1) = \big(\frac{1}{2}\,a_0 - p(-1)\big) + \sum_{j=1}^{n} a_j\,T_j(x)\,.$$

By (6.44)–(6.45) we obtain

$$a_j = \frac{1}{2j}\,(d_{j-1} - d_{j+1})\,, \quad j = 1, \ldots, n\,,$$

where we fix $d_n = d_{n+1} := 0$. Hence, the coefficients $d_{n-1}, \ldots, d_0$ can be recursively computed by (6.43). ■

6.2.3 *Fast Evaluation of Polynomial Products*

Assume that two polynomials $p, q \in \mathscr{P}_n$ are given in the monomial basis, i.e.,

$$p(x) = p_0 + p_1\,x + \ldots + p_n\,x^n\,,$$
$$q(x) = q_0 + q_1\,x + \ldots + q_n\,x^n$$

with real coefficients p_k and q_k. Then the related product $r := p\,q \in \mathscr{P}_{2n}$ possesses the form

$$r(x) = r_0 + r_1\,x + \ldots + r_{2n}\,x^{2n}$$

with real coefficients

$$r_k = \begin{cases} \sum_{j=0}^{k} p_j\,q_{k-j} & k = 0, \ldots, n\,, \\ \sum_{j=k-n}^{n} p_j\,q_{k-j} & k = n+1, \ldots, 2n\,. \end{cases}$$

This product can be efficiently calculated by cyclic convolution of the corresponding coefficient vectors, see Sect. 3.2. Let $N \geq 2n + 2$ be a fixed power of two. We introduce the corresponding coefficient vectors

$$\mathbf{p} := (p_0\,, p_1\,, \ldots\,, p_n\,, 0, \ldots\,, 0)^\top \in \mathbb{R}^N\,,$$
$$\mathbf{q} := (q_0\,, q_1\,, \ldots\,, q_n\,, 0, \ldots\,, 0)^\top \in \mathbb{R}^N\,,$$
$$\mathbf{r} := (r_0\,, r_1\,, \ldots\,, r_n\,, r_{n+1}\,, \ldots\,, r_{2n}\,, 0, \ldots\,, 0)^\top \in \mathbb{R}^N\,,$$

then it follows that $\mathbf{r} = \mathbf{p} * \mathbf{q}$. Applying the convolution property of the DFT in Theorem 3.26, we find

$$\mathbf{r} = \mathbf{F}_N^{-1}\,\big((\mathbf{F}_N\mathbf{p}) \circ (\mathbf{F}_N\mathbf{q})\big) = \frac{1}{N}\,\mathbf{J}_N'\,\mathbf{F}_N\,\big((\mathbf{F}_N\mathbf{p}) \circ (\mathbf{F}_N\mathbf{q})\big)\,,$$

with the Fourier matrix $\mathbf{F}_N$, the flip matrix $\mathbf{J}_N'$, and the component-wise product $\circ$. Using the FFT, we can thus calculate the coefficient vector $\mathbf{r}$ by $\mathscr{O}(N\,\log N)$ arithmetic operations.

Now we assume that $p, q \in \mathscr{P}_n$ are given in the basis of Chebyshev polynomials T_k, $k = 0, \ldots, n$. How can we efficiently calculate the product $p\,q \in \mathscr{P}_{2n}$ in the corresponding basis of Chebyshev polynomials?

Theorem 6.27 *For fixed $n \in \mathbb{N} \setminus \{1\}$, let $p, q \in \mathscr{P}_n$ be given polynomials of the form*

$$p = \frac{1}{2}\,a_0 + \sum_{k=1}^{n} a_k\,T_k\,, \qquad q = \frac{1}{2}\,b_0 + \sum_{\ell=1}^{n} b_\ell\,T_\ell\,,$$

where $a_k, b_\ell \in \mathbb{R}$, $k, \ell = 0, \ldots, n$.
Then the product $r := p\,q \in \mathscr{P}_{2n}$ possesses the form

$$r = \frac{1}{2}\,c_0 + \sum_{k=1}^{2n} c_k\,T_k$$

with the coefficients

$$2\,c_k := \begin{cases} a_0\,b_0 + 2\,\sum_{\ell=1}^{n} a_\ell\,b_\ell & k = 0\,,\\ \sum_{\ell=0}^{k} a_{k-\ell}\,b_\ell + \sum_{\ell=1}^{n-k}(a_\ell\,b_{\ell+k} + a_{\ell+k}\,b_\ell) & k = 1,\ldots,n-1\,,\\ \sum_{\ell=k-n}^{n} a_{k-\ell}\,b_\ell & k = n,\ldots,2n\,. \end{cases}$$

Proof

1. First, we calculate the special products $2\,p\,T_\ell$ for $\ell = 1,\ldots,n$ by means of (6.5)

$$2\,p\,T_\ell = a_0\,T_\ell + \sum_{k=1}^{n} a_k\,(2\,T_k\,T_\ell) = a_0\,T_\ell + \sum_{k=1}^{n} a_k\,T_{k+\ell} + \sum_{k=1}^{n} a_k\,T_{|k-\ell|}$$

$$= \sum_{k=\ell}^{n+\ell} a_{k-\ell}\,T_k + \sum_{k=1}^{\ell-1} a_{\ell-k}\,T_k + a_\ell + \sum_{k=1}^{n-\ell} a_{k+\ell}\,T_k\,.$$

Hence, it follows that

$$2\,p\,b_\ell\,T_\ell = \sum_{k=\ell}^{n+\ell} a_{k-\ell}\,b_\ell\,T_k + \sum_{k=1}^{\ell-1} a_{\ell-k}\,b_\ell\,T_k + a_\ell\,b_\ell + \sum_{k=1}^{n-\ell} a_{k+\ell}\,b_\ell\,T_k\,. \tag{6.46}$$

Further, we observe that

$$p\,b_0 = \frac{1}{2}\,a_0\,b_0 + \sum_{k=1}^{n} a_k\,b_0\,T_k\,. \tag{6.47}$$

2. If we sum up all Eqs. (6.46) for $\ell = 1,\ldots,n$ and Eq. (6.47), then we obtain

$$2\,p\,q = \big(\frac{1}{2}\,a_0\,b_0 + \sum_{\ell=1}^{n} a_\ell\,b_\ell\big) + \sum_{k=1}^{n} a_k\,b_0\,T_k + \sum_{\ell=1}^{n}\sum_{k=\ell}^{n+\ell} a_{k-\ell}\,b_\ell\,T_k$$

$$+ \sum_{\ell=2}^{n}\sum_{k=1}^{\ell-1} a_{\ell-k}\,b_\ell\,T_k + \sum_{\ell=1}^{n-1}\sum_{k=1}^{n-\ell} a_{k+\ell}\,b_\ell\,T_k\,.$$

We change the order of summation in the double sums

$$\sum_{\ell=1}^{n}\sum_{k=\ell}^{n+\ell} a_{k-\ell}\,b_\ell\,T_k = \Big(\sum_{k=1}^{n}\sum_{\ell=1}^{k} + \sum_{k=n+1}^{2n}\ \sum_{\ell=k-n}^{n}\Big)\,a_{k-\ell}\,b_\ell\,T_k\,,$$

$$\sum_{\ell=2}^{n}\sum_{k=1}^{\ell-1} a_{\ell-k}\, b_\ell\, T_k = \sum_{k=1}^{n-1}\sum_{\ell=k+1}^{n} a_{\ell-k}\, b_\ell\, T_k\,,$$

$$\sum_{\ell=1}^{n-1}\sum_{k=1}^{n-\ell} a_{k+\ell}\, b_\ell\, T_k = \sum_{k=1}^{n-1}\sum_{\ell=1}^{n-k} a_{k+\ell}\, b_\ell\, T_k$$

such that we receive

$$2\,p\,q = \Big(\frac{1}{2}\,a_0\,b_0 + \sum_{\ell=1}^{n} a_\ell\, b_\ell\Big) + \sum_{k=1}^{n}\Big(\sum_{\ell=0}^{k} a_{k-\ell}\, b_\ell\Big)\, T_k + \sum_{k=n+1}^{2n}\Big(\sum_{\ell=k-n}^{n} a_{k-\ell}\, b_\ell\Big)\, T_k$$

$$+ \sum_{k=1}^{n-1}\Big(\sum_{\ell=k+1}^{n} a_{\ell-k}\, b_\ell\Big)\, T_k + \sum_{k=1}^{n-1}\Big(\sum_{\ell=1}^{n-k} a_{k+\ell}\, b_\ell\Big)\, T_k\,.$$

Taking into account that

$$\sum_{\ell=k+1}^{n} a_{\ell-k}\, b_\ell = \sum_{\ell=1}^{n-k} a_\ell\, b_{\ell+k}\,,$$

we obtain the assertion. ■

The numerical computation of the coefficients c_k, $k = 0, \ldots, 2n$, of the polynomial multiplication $r = p\,q \in \mathscr{P}_{2n}$ can be efficiently realized by means of DCT. For this purpose, we choose $N \ge 2n + 2$ as a power of two, and we form the corresponding coefficient vectors

$$\mathbf{a} := \Big(\frac{\sqrt{2}}{2}\,a_0\,, a_1\,, \ldots, a_n\,, 0\,, \ldots, 0\Big)^\top \in \mathbb{R}^N\,,$$

$$\mathbf{b} := \Big(\frac{\sqrt{2}}{2}\,b_0\,, b_1\,, \ldots, b_n\,, 0\,, \ldots, 0\Big)^\top \in \mathbb{R}^N\,,$$

$$\mathbf{c} := \Big(\frac{\sqrt{2}}{2}\,c_0\,, c_1\,, \ldots, c_n\,, c_{n+1}\,, \ldots, c_{2n}\,, 0\,, \ldots, 0\Big)^\top \in \mathbb{R}^N\,.$$

From $r(z) = p(z)q(z)$, we particularly conclude for the Chebyshev zero points $z_k^{(N)} = \cos\big(\frac{(2k+1)\pi}{2N}\big)$, $k = 0, \ldots, N-1$, that

$$p(z_k^{(N)})\, q(z_k^{(N)}) = r(z_k^{(N)})\,, \quad k = 0, \ldots, N-1\,.$$

Recalling (6.27), the vectors of polynomial values and the corresponding coefficient vectors are related by

$$\mathbf{p} := \big(p(z_k^{(N)})\big)_{k=0}^{N-1} = \sqrt{\frac{N}{2}}\, \mathbf{C}_N^{\mathrm{III}} \mathbf{a}\,,$$

$$\mathbf{q} := \big(q(z_k^{(N)})\big)_{k=0}^{N-1} = \sqrt{\frac{N}{2}}\, \mathbf{C}_N^{\mathrm{III}} \mathbf{b}\,,$$

$$\mathbf{r} := \big(r(z_k^{(N)})\big)_{k=0}^{N-1} = \sqrt{\frac{N}{2}}\, \mathbf{C}_N^{\mathrm{III}} \mathbf{c}\,.$$

Since $\mathbf{r}$ is equal to the component-wise product of $\mathbf{p}$ and $\mathbf{q}$, it follows that

$$\sqrt{\frac{N}{2}}\, \mathbf{C}_N^{\mathrm{III}} \mathbf{c} = \mathbf{r} = \mathbf{p} \circ \mathbf{q} = \frac{N}{2}\, (\mathbf{C}_N^{\mathrm{III}} \mathbf{a}) \circ (\mathbf{C}_N^{\mathrm{III}} \mathbf{b})\,.$$

Hence, we obtain by Lemma 3.47 that

$$\mathbf{c} = \sqrt{\frac{N}{2}}\, \mathbf{C}_N^{\mathrm{II}} \left((\mathbf{C}_N^{\mathrm{III}} \mathbf{a}) \circ (\mathbf{C}_N^{\mathrm{III}} \mathbf{b}) \right).$$

Using fast DCT algorithms, we can thus calculate the coefficient vector $\mathbf{c}$ by $\mathscr{O}(N \log N)$ arithmetic operations (see Sect. 6.3).

6.3 Fast DCT Algorithms

In this section, we want to derive fast algorithms for the discrete cosine transform (DCT) and the discrete sine transform (DST), respectively. These discrete trigonometric transforms have been considered already in Sect. 3.5. As we have seen, for example, in Sect. 6.2, these transforms naturally occur, if we want to evaluate polynomials efficiently. Other applications relate to polynomial interpolation in Sect. 6.4 and to data decorrelation. For simplicity, we shortly recall the matrices related to DCT and DST from Sect. 3.5. Let $N \geq 2$ be a given integer. In the following, we consider *cosine* and *sine matrices of types* I – IV which are defined by

$$\begin{aligned}
\mathbf{C}_{N+1}^{\mathrm{I}} &:= \sqrt{\tfrac{2}{N}} \left(\epsilon_N(j)\, \epsilon_N(k)\, \cos\left(\tfrac{jk\pi}{N}\right) \right)_{j,k=0}^{N}, \\
\mathbf{C}_N^{\mathrm{II}} &:= \sqrt{\tfrac{2}{N}} \left(\epsilon_N(j)\, \cos\left(\tfrac{j(2k+1)\pi}{2N}\right) \right)_{j,k=0}^{N-1}, \qquad \mathbf{C}_N^{\mathrm{III}} := (\mathbf{C}_N^{\mathrm{II}})^{\top}, \\
\mathbf{C}_N^{\mathrm{IV}} &:= \sqrt{\tfrac{2}{N}} \left(\cos\left(\tfrac{(2j+1)(2k+1)\pi}{4N}\right) \right)_{j,k=0}^{N-1}, \\
\mathbf{S}_{N-1}^{\mathrm{I}} &:= \sqrt{\tfrac{2}{N}} \left(\sin\left(\tfrac{(j+1)(k+1)\pi}{N}\right) \right)_{j,k=0}^{N-2},
\end{aligned} \tag{6.48}$$

$$\mathbf{S}_N^{\mathrm{II}} := \sqrt{\tfrac{2}{N}} \left(\epsilon_N(j+1) \sin\left(\tfrac{(j+1)(2k+1)\pi}{2N}\right)\right)_{j,k=0}^{N-1}, \qquad \mathbf{S}_N^{\mathrm{III}} := (\mathbf{S}_N^{\mathrm{II}})^\top,$$

$$\mathbf{S}_N^{\mathrm{IV}} := \sqrt{\tfrac{2}{N}} \left(\sin\left(\tfrac{(2j+1)(2k+1)\pi}{4N}\right)\right)_{j,k=0}^{N-1}.$$

Here we set $\epsilon_N(0) = \epsilon_N(N) := \sqrt{2}/2$ and $\epsilon_N(j) := 1$ for $j \in \{1, \ldots, N-1\}$. In our notation, a subscript of a matrix denotes the corresponding order, while a superscript signifies the "type" of the matrix.

As shown in Sect. 3.5, the cosine and sine matrices of types I–IV are orthogonal. We say that a *discrete trigonometric transform of length M* is a linear transform that maps each vector $\mathbf{x} \in \mathbb{R}^M$ to $\mathbf{T}\mathbf{x} \in \mathbb{R}^M$, where the matrix $\mathbf{T} \in \mathbb{R}^{M\times M}$ is a cosine or sine matrix in (6.48) and $M \in \{N-1,\ N,\ N+1\}$.

There exists a large variety of fast algorithms to evaluate these matrix-vector products. We want to restrict ourselves here to two different approaches. The first method is based on the close connection between trigonometric functions and the complex exponentials by Euler's formula. Therefore, we can always employ the FFT to compute the DCT and DST. The second approach involves a direct matrix factorization of an orthogonal trigonometric matrix into a product of sparse real matrices such that the discrete trigonometric transform can be performed in real arithmetic. In particular, if this matrix factorization is additionally orthogonal, the corresponding algorithms possess excellent numerical stability (see [324]).

6.3.1 Fast DCT Algorithms via FFT

The DCT and the DST of length N (or $N+1$ and $N-1$, respectively, for DCT-I and DST-I) can always be reduced to a DFT of length $2N$ such that we can apply an FFT with computational cost of $\mathcal{O}(N \log N)$. We exemplarily show the idea for the DCTs and give the algorithms for all transforms.

Let us start with the DCT-I. In order to compute the components of $\hat{\mathbf{a}} = \mathbf{C}_{N+1}^{\mathrm{I}}\mathbf{a}$ with $\mathbf{a} = (a_k)_{k=0}^N$, we introduce the vector $\mathbf{y} \in \mathbb{R}^{2N}$ of the form

$$\mathbf{y} := (\sqrt{2}\,a_0,\ a_1, \ldots, a_{N-1},\ \sqrt{2}\,a_N,\ a_{N-1},\ a_{N-2}, \ldots, a_1)^\top.$$

Then, we obtain with $w_{2N} := \mathrm{e}^{-2\pi\mathrm{i}/(2N)} = \mathrm{e}^{-\pi\mathrm{i}/N}$ that

$$\begin{aligned}\hat{a}_j &= \sqrt{\tfrac{2}{N}}\, \epsilon_N(j) \sum_{k=0}^{N} a_k\, \epsilon_N(k) \cos\left(\tfrac{2\pi jk}{2N}\right)\\ &= \sqrt{\tfrac{2}{N}}\, \epsilon_N(j) \Big(\tfrac{\sqrt{2}}{2} a_0 + (-1)^j \tfrac{\sqrt{2}}{2} a_N + \tfrac{1}{2} \sum_{k=1}^{N-1} a_k \left(w_{2N}^{jk} + w_{2N}^{-jk}\right)\Big)\end{aligned}$$

$$= \sqrt{\tfrac{2}{N}}\, \epsilon_N(j) \Big(\tfrac{\sqrt{2}}{2} a_0 + (-1)^j \tfrac{\sqrt{2}}{2} a_N + \tfrac{1}{2} \sum_{k=1}^{N-1} a_k\, w_{2N}^{jk} + \tfrac{1}{2} \sum_{k=1}^{N-1} a_k\, w_{2N}^{j(2N-k)} \Big)$$

$$= \sqrt{\tfrac{2}{N}}\, \epsilon_N(j) \Big(\tfrac{\sqrt{2}}{2} a_0 + (-1)^j \tfrac{\sqrt{2}}{2} a_N + \tfrac{1}{2} \sum_{k=1}^{N-1} a_k\, w_{2N}^{jk} + \tfrac{1}{2} \sum_{k=N+1}^{2N-1} a_{2N-k}\, w_{2N}^{jk} \Big)$$

$$= \tfrac{1}{\sqrt{2N}}\, \epsilon_N(j) \sum_{k=0}^{2N-1} y_k\, w_{2N}^{jk}$$

for $j = 0, \ldots, N$. Thus, $\sqrt{2N}\,\big(\epsilon_N(j)^{-1}\big)_{j=0}^{N} \circ \hat{\mathbf{a}}$ is the partial vector formed by the first $N+1$ components of $\hat{\mathbf{y}} := \mathbf{F}_{2N}\,\mathbf{y}$. This observation implies the following.

Algorithm 6.28 (DCT-I ($N+1$) via DFT ($2N$))
Input: $N \in \mathbb{N} \setminus \{1\}$, $\mathbf{a} = (a_j)_{j=0}^{N} \in \mathbb{R}^{N+1}$.

1. *Determine* $\mathbf{y} \in \mathbb{R}^{2N}$ *with*

$$y_k := \begin{cases} \sqrt{2}\, a_k & k = 0,\, N\,, \\ a_k & k = 1, \ldots, N-1\,, \\ a_{2N-k} & k = N+1, \ldots, 2N-1\,. \end{cases}$$

2. *Compute* $\hat{\mathbf{y}} = \mathbf{F}_{2N}\,\mathbf{y}$ *using an* FFT *of length* $2N$.
3. *Set*

$$\hat{a}_j := \tfrac{1}{\sqrt{2N}}\, \epsilon_N(j)\, \mathrm{Re}\, \hat{y}_j\,, \quad j = 0, \ldots, N\,.$$

Output: $\hat{\mathbf{a}} = (\hat{a}_j)_{j=0}^{N} = \mathbf{C}_{N+1}^{\mathrm{I}}\, \mathbf{a} \in \mathbb{R}^{N+1}$.
Computational cost: $\mathscr{O}(N \log N)$.

For the DCT-II we proceed similarly. Let now $\hat{\mathbf{a}} := \mathbf{C}_N^{\mathrm{II}}\, \mathbf{a}$. Defining the vector

$$\mathbf{y} := (a_0,\, a_1, \ldots, a_{N-1}, a_{N-1}, \ldots, a_0)^\top \in \mathbb{R}^{2N}\,,$$

we find for $j = 0, \ldots, N-1$,

$$\hat{a}_j = \sqrt{\tfrac{2}{N}}\, \epsilon_N(j) \sum_{k=0}^{N-1} a_k \cos\big(\tfrac{2\pi j(2k+1)}{4N}\big) = \tfrac{\epsilon_N(j)}{\sqrt{2N}} \sum_{k=0}^{N-1} a_k \big(w_{4N}^{j(2k+1)} + w_{4N}^{-j(2k+2-1)}\big)$$

$$= \tfrac{\epsilon_N(j)}{\sqrt{2N}}\, w_{4N}^{j} \Big(\sum_{k=0}^{N-1} a_k\, w_{2N}^{jk} + \sum_{k=N}^{2N-1} a_{2N-k-1}\, w_{2N}^{jk} \Big) = \tfrac{\epsilon_N(j)}{\sqrt{2N}}\, w_{4N}^{j} \Big(\sum_{k=0}^{2N-1} y_k\, w_{2N}^{jk} \Big)\,,$$

implying the following.

Algorithm 6.29 (DCT-II (N) via DFT $(2N)$)
Input: $N \in \mathbb{N} \setminus \{1\}$, $\mathbf{a} = (a_j)_{j=0}^{N-1} \in \mathbb{R}^N$.

1. *Determine* $\mathbf{y} \in \mathbb{R}^{2N}$ *with*

$$y_k := \begin{cases} a_k & k = 0, \ldots, N-1\,, \\ a_{2N-k-1} & k = N, \ldots, 2N-1\,. \end{cases}$$

2. *Compute* $\hat{\mathbf{y}} = \mathbf{F}_{2N}\,\mathbf{y}$ *using an* FFT *of length* $2N$.
3. *Set*

$$\hat{a}_j := \frac{\epsilon_N(j)}{\sqrt{2N}}\,\mathrm{Re}\,(w_{4N}^j\,\hat{y}_j), \quad j = 0, \ldots, N-1\,.$$

Output: $\hat{\mathbf{a}} = (\hat{a}_j)_{j=0}^{N-1} = \mathbf{C}_N^{\mathrm{II}}\,\mathbf{a} \in \mathbb{R}^N$.
Computational cost: $\mathcal{O}(N \log N)$.

The DCT-III can be implemented using the following observation. Let now $\hat{\mathbf{a}} := \mathbf{C}_N^{\mathrm{III}}\,\mathbf{a}$. We determine

$$\mathbf{y} := (\sqrt{2}a_0,\, w_{4N}^1 a_1,\, w_{4N}^2 a_2, \ldots, w_{4N}^{N-1} a_{N-1}, 0, w_{4N}^{-N+1} a_{N-1}, \ldots, w_{4N}^{-1} a_1)^\top \in \mathbb{C}^{2N},$$

and obtain for $j = 0, \ldots, N-1$,

$$\begin{aligned} \hat{a}_j &= \sqrt{\tfrac{2}{N}} \sum_{k=0}^{N-1} \epsilon_N(k) a_k \cos\big(\tfrac{2\pi k(2j+1)}{4N}\big) \\ &= \tfrac{1}{\sqrt{2N}}\Big(\sqrt{2}a_0 + \sum_{k=1}^{N-1} a_k (w_{4N}^{(2j+1)k} + w_{4N}^{-(2j+1)k})\Big) \\ &= \tfrac{1}{\sqrt{2N}}\Big(\sqrt{2}a_0 + \sum_{k=1}^{N-1} (a_k w_{4N}^k) w_{2N}^{jk} + \sum_{k=N+1}^{2N-1} (a_{2N-k} w_{4N}^{2N+k}) w_{2N}^{jk}\Big) \\ &= \tfrac{1}{\sqrt{2N}} \sum_{k=0}^{2N-1} y_k w_{2N}^{jk}\,. \end{aligned}$$

Thus, we derive the following algorithm for the DCT-III.

Algorithm 6.30 (DCT-III (N) via DFT $(2N)$)
Input: $N \in \mathbb{N} \setminus \{1\}$, $\mathbf{a} = (a_j)_{j=0}^{N-1} \in \mathbb{R}^N$.

1. *Determine* $\mathbf{y} \in \mathbb{C}^{2N}$ *with*

$$y_k := \begin{cases} \sqrt{2}\,a_k & k = 0\,, \\ w_{4N}^k\,a_k & k = 1, \ldots, N-1\,, \\ 0 & k = N, \\ w_{4N}^{2N+k}\,a_{2N-k} & k = N+1, \ldots, 2N-1\,. \end{cases}$$

2. *Compute* $\hat{\mathbf{y}} = \mathbf{F}_{2N}\,\mathbf{y}$ *using an* FFT *of length* $2N$.
3. *Set*

$$\hat{a}_j := \frac{1}{\sqrt{2N}}\,\mathrm{Re}\,\hat{y}_j\,, \quad j = 0, \ldots, N-1\,.$$

Output: $\hat{\mathbf{a}} = (\hat{a}_j)_{j=0}^{N-1} = \mathbf{C}_N^{\mathrm{III}}\,\mathbf{a} \in \mathbb{R}^N$.
Computational cost: $\mathcal{O}(N \log N)$.

Finally, we consider the DCT-IV (N). Let $\hat{\mathbf{a}} := \mathbf{C}_N^{\mathrm{IV}}\,\mathbf{a}$. This time we employ the vector

$$\mathbf{y} := \big(w_{4N}^0\, a_0,\, w_{4N}^1 a_1, \ldots, w_{4N}^{N-1} a_{N-1}, w_{4N}^{-N} a_{N-1}, w_{4N}^{-N+1} a_{N-2}, \ldots, w_{4N}^{-1} a_0\big)^\top \in \mathbb{C}^{2N}\,.$$

Using that

$$\cos\Big(\frac{(2j+1)(2k+1)\pi}{4N}\Big) = \frac{1}{2}\Big(w_{8N}^{(2j+1)(2k+1)} + w_{8N}^{-(2j+1)(2k+1)}\Big)$$
$$= \frac{1}{2}\, w_{8N}^{(2j+1)}\Big(w_{4N}^{k}\, w_{2N}^{jk} + w_{4N}^{-(k+1)}\, w_{2N}^{-j(k+1)}\Big)\,,$$

we obtain for $j = 0, \ldots, N-1$,

$$\hat{a}_j = \sqrt{\frac{2}{N}} \sum_{k=0}^{N-1} a_k \cos\Big(\frac{2\pi(2j+1)(2k+1)}{8N}\Big)$$
$$= \frac{1}{\sqrt{2N}} \sum_{k=0}^{N-1} a_k\, w_{8N}^{(2j+1)} \big(w_{4N}^{k}\, w_{2N}^{jk} + w_{4N}^{-(k+1)}\, w_{2N}^{-j(k+1)}\big)$$
$$= \frac{1}{\sqrt{2N}}\, w_{8N}^{(2j+1)} \Big(\sum_{k=0}^{N-1} (a_k\, w_{4N}^{k})\, w_{2N}^{jk} + \sum_{k=0}^{N-1} (a_k\, w_{4N}^{-(k+1)})\, w_{2N}^{j(2N-k-1)}\Big)$$
$$= \frac{1}{\sqrt{2N}}\, w_{8N}^{(2j+1)} \Big(\sum_{k=0}^{N-1} (a_k\, w_{4N}^{k})\, w_{2N}^{jk} + \sum_{k=N}^{2N-1} (a_{2N-k-1}\, w_{4N}^{-2N+k})\, w_{2N}^{jk}\Big)$$
$$= \frac{1}{\sqrt{2N}}\, w_{8N}^{(2j+1)} \sum_{k=0}^{2N-1} y_k\, w_{2N}^{jk}\,.$$

Thus, we conclude the following algorithm for the DCT-IV.

Algorithm 6.31 (DCT-IV (N) via DFT $(2N)$)
Input: $N \in \mathbb{N} \setminus \{1\}$, $\mathbf{a} = (a_j)_{j=0}^{N-1} \in \mathbb{R}^N$.

1. *Determine* $\mathbf{y} \in \mathbb{C}^{2N}$ *with*

$$y_k := \begin{cases} w_{4N}^{k}\, a_k & k = 0, \ldots, N-1\,, \\ w_{4N}^{2N+k}\, a_{2N-k-1} & k = N, \ldots, 2N-1\,. \end{cases}$$

Table 6.1 DST algorithms of lengths $N-1$ and N, respectively, based on an FFT of length $2N$

DST	Vector $\mathbf{y}$	Vector $\hat{\mathbf{a}}$
$\hat{\mathbf{a}} = \mathbf{S}_{N-1}^{\mathrm{I}}\,\mathbf{a}$	$y_k := \begin{cases} 0 & k = 0,\, N, \\ -w_{2N}^k\, a_{k-1} & k = 1, \ldots, N-1, \\ w_{2N}^k\, a_{2N-k-1} & k = N+1, \ldots, N. \end{cases}$	$\hat{a}_j := \frac{1}{\sqrt{2N}}\,\mathrm{Re}\big((-\mathrm{i})\,\hat{y}_j\big)$, $j = 0, \ldots, N-2.$
$\hat{\mathbf{a}} = \mathbf{S}_N^{\mathrm{II}}\,\mathbf{a}$	$y_k := \begin{cases} -w_{2N}^k\, a_k & k = 0, \ldots, N-1\,, \\ w_{2N}^k\, a_{2N-k-1} & k = N, \ldots, 2N-1\,. \end{cases}$	$\hat{a}_j := \frac{1}{\sqrt{2N}}\,\epsilon_N(j+1)\,\mathrm{Re}\big((-\mathrm{i})\,w_{4N}^{j+1}\,\hat{y}_j\big)$, $j = 0, \ldots, N-1\,.$
$\hat{\mathbf{a}} = \mathbf{S}_N^{\mathrm{III}}\,\mathbf{a}$	$y_k := \begin{cases} -w_{4N}^{k+1}\, a_k & k = 0, \ldots, N-2\,, \\ \sqrt{2}\mathrm{i}\, a_{N-1} & k = N-1\,, \\ -w_{4N}^{k+1}\, a_{2N-k-2} & k = N, \ldots, 2N-2\,, \\ 0 & k = N-1\,. \end{cases}$	$\hat{a}_j := \frac{1}{\sqrt{2N}}\,\mathrm{Re}\big((-\mathrm{i})\,w_{2N}^{j}\,\hat{y}_j\big)$, $j = 0, \ldots, N-1\,.$
$\hat{\mathbf{a}} = \mathbf{S}_N^{\mathrm{IV}}\,\mathbf{a}$	$y_k := \begin{cases} -w_{4N}^k\, a_k & k = 0, \ldots, N-1\,, \\ -w_{4N}^k\, a_{2N-k-1} & k = N, \ldots, 2N-1\,. \end{cases}$	$\hat{a}_j := \frac{1}{\sqrt{2N}}\,\mathrm{Re}\big((-\mathrm{i})\,w_{8N}^{2j+1}\,\hat{y}_j\big)$, $j = 0, \ldots, N-1\,.$

2. *Compute* $\hat{\mathbf{y}} = \mathbf{F}_{2N}\,\mathbf{y}$ *using an* FFT *of length* $2N$.
3. *Set*

$$\hat{a}_j := \frac{1}{\sqrt{2N}}\,\mathrm{Re}\,(w_{8N}^{2j+1}\,\hat{y}_j)\,, \quad j = 0, \ldots, N-1\,.$$

Output: $\hat{\mathbf{a}} = (\hat{a}_j)_{j=0}^{N-1} = \mathbf{C}_N^{\mathrm{IV}}\,\mathbf{a} \in \mathbb{R}^N$.
Computational cost: $\mathcal{O}(N \log N)$.

The DST algorithms can be similarly derived from the FFT of length $2N$. We summarize them in Table 6.1, where we use the vectors $\mathbf{y} = (y_k)_{k=0}^{2N-1} \in \mathbb{C}^{2N}$ and $\hat{\mathbf{y}} = (\hat{y}_j)_{j=0}^{2N-1} = \mathbf{F}_{2N}\,\mathbf{y}$.

6.3.2 Fast DCT Algorithms via Orthogonal Matrix Factorizations

Based on the considerations in [324, 427, 428], we want to derive numerically stable fast DCT algorithms which are based on real factorizations of the corresponding cosine and sine matrices into products of sparse (almost) orthogonal matrices of simple structure. These algorithms are completely recursive and simple to implement and use only permutations, scaling with $\sqrt{2}$, butterfly operations, and plane rotations or rotation–reflections.

In order to present the sparse factorizations of the cosine and sine matrices, we first introduce a collection of special sparse matrices that we will need later. Recall that $\mathbf{I}_N$ and $\mathbf{J}_N$ denote the identity and counter-identity matrices of order N. Further, $\mathbf{P}_N = \mathbf{P}_N(2)$ denotes the 2-stride permutation matrix as in Sect. 3.4. We use the notation $\mathbf{A} \oplus \mathbf{B} = \mathrm{diag}(\mathbf{A},\, \mathbf{B})$ for block diagonal matrices, where the square matrices $\mathbf{A}$ and $\mathbf{B}$ can have different orders. Let

$$\mathbf{D}_N := \operatorname{diag}\big((-1)^k\big)_{k=0}^{N-1}$$

be the diagonal sign matrix. For even $N \ge 4$ let $N_1 := \frac{N}{2}$. We introduce the sparse orthogonal matrices

$$\mathbf{A}_N := \left(1 \oplus \frac{1}{\sqrt{2}} \begin{pmatrix} \mathbf{I}_{N_1-1} & \mathbf{I}_{N_1-1} \\ \mathbf{I}_{N_1-1} & -\mathbf{I}_{N_1-1} \end{pmatrix} \oplus (-1)\right) (\mathbf{I}_{N_1} \oplus \mathbf{D}_{N_1}\, \mathbf{J}_{N_1}),$$

and

$$\mathbf{B}_N := \frac{1}{\sqrt{2}} \begin{pmatrix} \mathbf{I}_{N_1} & \mathbf{J}_{N_1} \\ \mathbf{I}_{N_1} & -\mathbf{J}_{N_1} \end{pmatrix}, \qquad \mathbf{B}_{N+1} := \frac{1}{\sqrt{2}} \begin{pmatrix} \mathbf{I}_{N_1} & & \mathbf{J}_{N_1} \\ & \sqrt{2} & \\ \mathbf{I}_{N_1} & & -\mathbf{J}_{N_1} \end{pmatrix},$$

$$\tilde{\mathbf{B}}_N := (\mathbf{I}_{N_1} \oplus \mathbf{D}_{N_1}) \begin{pmatrix} \operatorname{diag} \mathbf{c}_{N_1} & (\operatorname{diag} \mathbf{s}_{N_1})\, \mathbf{J}_{N_1} \\ -\mathbf{J}_{N_1} \operatorname{diag} \mathbf{s}_{N_1} & \operatorname{diag}(\mathbf{J}_{N_1}\, \mathbf{c}_{N_1}) \end{pmatrix},$$

where

$$\mathbf{c}_{N_1} := \left(\cos\left(\tfrac{(2k+1)\pi}{4N}\right)\right)_{k=0}^{N_1-1}, \quad \mathbf{s}_{N_1} := \left(\sin\left(\tfrac{(2k+1)\pi}{4N}\right)\right)_{k=0}^{N_1-1}.$$

All these sparse "butterfly" matrices possess at most two nonzero components in each row and each column. The modified identity matrices are denoted by

$$\mathbf{I}'_N := \sqrt{2} \oplus \mathbf{I}_{N-1}, \quad \mathbf{I}''_N := \mathbf{I}_{N-1} \oplus \sqrt{2}.$$

Finally, let $\mathbf{V}_N$ be the forward shift matrix as in Sect. 3.2. Now, we can show the following factorizations of the cosine matrices of types I–IV.

Theorem 6.32 *Let $N \ge 4$ be an even integer and $N_1 := N/2$.*

(i) The cosine matrix $\mathbf{C}^{\mathrm{I}}_{N+1}$ can be orthogonally factorized in the form

$$\mathbf{C}^{\mathrm{I}}_{N+1} = \mathbf{P}^{\top}_{N+1}\, (\mathbf{C}^{\mathrm{I}}_{N_1+1} \oplus \mathbf{C}^{\mathrm{III}}_{N_1})\, \mathbf{B}_{N+1}\,. \tag{6.49}$$

(ii) The cosine matrix $\mathbf{C}^{\mathrm{II}}_N$ satisfies the orthogonal factorization

$$\mathbf{C}^{\mathrm{II}}_N = \mathbf{P}^{\top}_N\, (\mathbf{C}^{\mathrm{II}}_{N_1} \oplus \mathbf{C}^{\mathrm{IV}}_{N_1})\, \mathbf{B}_N\,. \tag{6.50}$$

Proof In order to show (6.50), we first permute the rows of $\mathbf{C}^{\mathrm{II}}_N$ by multiplying with $\mathbf{P}_N$ and write the result as a block matrix

$$\mathbf{P}_N \mathbf{C}_N^{\mathrm{II}} = \frac{1}{\sqrt{N_1}} \begin{pmatrix} \left(\epsilon_N(2j)\cos\left(\frac{2j(2k+1)\pi}{2N}\right)\right)_{j,k=0}^{N_1-1} & \left(\epsilon_N(2j)\cos\left(\frac{2j(N+2k+1)\pi}{2N}\right)\right)_{j,k=0}^{N_1-1} \\ \left(\cos\left(\frac{(2j+1)(2k+1)\pi}{2N}\right)\right)_{j,k=0}^{N_1-1} & \left(\cos\left(\frac{(2j+1)(N+2k+1)\pi}{2N}\right)\right)_{j,k=0}^{N_1-1} \end{pmatrix}.$$

Recalling the definition of $\mathbf{C}_N^{\mathrm{IV}}$ and using

$$\cos\left(\frac{j(N+2k+1)\pi}{N}\right) = \cos\left(\frac{j(N-2k-1)\pi}{N}\right),$$
$$\cos\left(\frac{(2j+1)(N+2k+1)\pi}{2N}\right) = -\cos\left(\frac{(2j+1)(N-2k-1)\pi}{2N}\right),$$

it follows immediately that the four blocks of $\mathbf{P}_N\,\mathbf{C}_N^{\mathrm{II}}$ can be represented by $\mathbf{C}_{N_1}^{\mathrm{II}}$ and $\mathbf{C}_{N_1}^{\mathrm{IV}}$

$$\begin{aligned}\mathbf{P}_N\,\mathbf{C}_N^{\mathrm{II}} &= \frac{1}{\sqrt{2}}\begin{pmatrix}\mathbf{C}_{N_1}^{\mathrm{II}} & \mathbf{C}_{N_1}^{\mathrm{II}}\,\mathbf{J}_{N_1}\\ \mathbf{C}_{N_1}^{\mathrm{IV}} & -\mathbf{C}_{N_1}^{\mathrm{IV}}\,\mathbf{J}_{N_1}\end{pmatrix} = (\mathbf{C}_{N_1}^{\mathrm{II}} \oplus \mathbf{C}_{N_1}^{\mathrm{IV}})\,\frac{1}{\sqrt{2}}\begin{pmatrix}\mathbf{I}_{N_1} & \mathbf{J}_{N_1}\\ \mathbf{I}_{N_1} & -\mathbf{J}_{N_1}\end{pmatrix}\\ &= (\mathbf{C}_{N_1}^{\mathrm{II}} \oplus \mathbf{C}_{N_1}^{\mathrm{IV}})\,\mathbf{B}_N.\end{aligned}$$

Since $\mathbf{P}_N^{-1} = \mathbf{P}_N^{\top}$ and $\mathbf{B}_N\,\mathbf{B}_N^{\top} = \mathbf{I}_N$, the matrices $\mathbf{P}_N$ and $\mathbf{B}_N$ are orthogonal. The proof of (6.49) follows similarly. ■

From (6.50) we also obtain a factorization of $\mathbf{C}_N^{\mathrm{III}}$

$$\mathbf{C}_N^{\mathrm{III}} = \mathbf{B}_N^{\top}\,(\mathbf{C}_{N_1}^{\mathrm{III}} \oplus \mathbf{C}_{N_1}^{\mathrm{IV}})\,\mathbf{P}_N. \tag{6.51}$$

The next theorem provides an orthogonal factorization of $\mathbf{C}_N^{\mathrm{IV}}$ for even $N \ge 4$.

Theorem 6.33 *For even $N \ge 4$, the cosine matrix $\mathbf{C}_N^{\mathrm{IV}}$ can be orthogonally factorized in the form*

$$\mathbf{C}_N^{\mathrm{IV}} = \mathbf{P}_N^{\top}\,\mathbf{A}_N\,(\mathbf{C}_{N_1}^{\mathrm{II}} \oplus \mathbf{C}_{N_1}^{\mathrm{II}})\,\tilde{\mathbf{B}}_N. \tag{6.52}$$

Proof We permute the rows of $\mathbf{C}_N^{\mathrm{IV}}$ by multiplying with $\mathbf{P}_N$ and write the result as a block matrix

$$\mathbf{P}_N\,\mathbf{C}_N^{\mathrm{IV}} = \frac{1}{\sqrt{N_1}}\begin{pmatrix}\left(\cos\left(\frac{(4j+1)(2k+1)\pi}{4N}\right)\right)_{j,k=0}^{N_1-1} & \left(\cos\left(\frac{(4j+1)(N+2k+1)\pi}{4N}\right)\right)_{j,k=0}^{N_1-1}\\ \left(\cos\left(\frac{(4j+3)(2k+1)\pi}{4N}\right)\right)_{j,k=0}^{N_1-1} & \left(\cos\left(\frac{(4j+3)(N+2k+1)\pi}{4N}\right)\right)_{j,k=0}^{N_1-1}\end{pmatrix}.$$

Now we consider the single blocks of $\mathbf{P}_N\,\mathbf{C}_N^{\mathrm{IV}}$ and represent every block by $\mathbf{C}_{N_1}^{\mathrm{II}}$ and $\mathbf{S}_{N_1}^{\mathrm{II}}$. By

$$\cos\Big(\frac{(4j+1)(2k+1)\pi}{4N}\Big)=\cos\Big(\frac{j(2k+1)\pi}{N}\Big)\cos\Big(\frac{(2k+1)\pi}{4N}\Big)-\sin\Big(\frac{j(2k+1)\pi}{N}\Big)\sin\Big(\frac{(2k+1)\pi}{4N}\Big)$$

it follows that

$$\frac{1}{\sqrt{N_1}}\Big(\cos\Big(\frac{(4j+1)(2k+1)\pi}{4N}\Big)\Big)_{j,k=0}^{N_1-1}=\frac{1}{\sqrt{2}}\Big(\mathbf{I}'_{N_1}\,\mathbf{C}^{\mathrm{II}}_{N_1}\,\mathrm{diag}\,\mathbf{c}_{N_1}-\mathbf{V}_{N_1}\,\mathbf{S}^{\mathrm{II}}_{N_1}\,\mathrm{diag}\,\mathbf{s}_{N_1}\Big)$$
$$=\frac{1}{\sqrt{2}}\big(\mathbf{I}'_{N_1}\,\mathbf{C}^{\mathrm{II}}_{N_1}\,\mathrm{diag}\,\mathbf{c}_{N_1}-\mathbf{V}_{N_1}\,\mathbf{D}_{N_1}\,\mathbf{S}^{\mathrm{II}}_{N_1}\,\mathbf{J}_{N_1}\,\mathrm{diag}\,\mathbf{s}_{N_1}\big)\,. \tag{6.53}$$

Further, with

$$\cos\Big(\frac{(4j+3)(2k+1)\pi}{4N}\Big)=\cos\Big(\frac{(j+1)(2k+1)\pi}{N}\Big)\cos\Big(\frac{(2k+1)\pi}{4N}\Big)$$
$$+\sin\Big(\frac{(j+1)(2k+1)\pi}{N}\Big)\sin\Big(\frac{(2k+1)\pi}{4N}\Big)$$

we obtain

$$\frac{1}{\sqrt{N_1}}\Big(\cos\Big(\frac{(4j+3)(2k+1)\pi}{4N}\Big)\Big)_{j,k=0}^{N_1-1}=\frac{1}{\sqrt{2}}\big(\mathbf{V}^{\top}_{N_1}\,\mathbf{C}^{\mathrm{II}}_{N_1}\,\mathrm{diag}\,\mathbf{c}_{N_1}+\mathbf{I}''_{N_1}\,\mathbf{S}^{\mathrm{II}}_{N_1}\,\mathrm{diag}\,\mathbf{s}_{N_1}\big)$$
$$=\frac{1}{\sqrt{2}}\big(\mathbf{V}^{\top}_{N_1}\,\mathbf{C}^{\mathrm{II}}_{N_1}\,\mathrm{diag}\,\mathbf{c}_{N_1}+\mathbf{I}''_{N_1}\,\mathbf{D}_{N_1}\,\mathbf{S}^{\mathrm{II}}_{N_1}\,\mathbf{J}_{N_1}\,\mathrm{diag}\,\mathbf{s}_{N_1}\big)\,. \tag{6.54}$$

From

$$\cos\Big(\frac{(4j+1)(N+2k+1)\pi}{4N}\Big)=(-1)^j\cos\Big(\frac{j(2k+1)\pi}{N}+\frac{(N+2k+1)\pi}{4N}\Big)$$
$$=(-1)^j\cos\Big(\frac{j(2k+1)\pi}{N}\Big)\sin\Big(\frac{(N-2k-1)\pi}{4N}\Big)-(-1)^j\sin\Big(\frac{j(2k+1)\pi}{N}\Big)\cos\Big(\frac{(N-2k-1)\pi}{4N}\Big)$$

we conclude

$$\frac{1}{\sqrt{N_1}}\Big(\cos\Big(\frac{(4j+1)(N+2k+1)\pi}{4N}\Big)\Big)_{j,k=0}^{N_1-1}$$
$$=\frac{1}{\sqrt{2}}\big(\mathbf{D}_{N_1}\,\mathbf{I}'_{N_1}\,\mathbf{C}^{\mathrm{II}}_{N_1}\,\mathrm{diag}\,(\mathbf{J}_{N_1}\,\mathbf{s}_{N_1})-\mathbf{D}_{N_1}\,\mathbf{V}_{N_1}\,\mathbf{S}^{\mathrm{II}}_{N_1}\,\mathrm{diag}\,(\mathbf{J}_{N_1}\,\mathbf{c}_{N_1})\big)$$
$$=\frac{1}{\sqrt{2}}\big(\mathbf{I}'_{N_1}\,\mathbf{C}^{\mathrm{II}}_{N_1}\,\mathbf{J}_{N_1}\mathrm{diag}\,(\mathbf{J}_{N_1}\,\mathbf{s}_{N_1})+\mathbf{V}_{N_1}\,\mathbf{D}_{N_1}\,\mathbf{S}^{\mathrm{II}}_{N_1}\,\mathrm{diag}\,(\mathbf{J}_{N_1}\,\mathbf{c}_{N_1})\big)\,. \tag{6.55}$$

Here we have used that $\mathbf{D}_{N_1}\,\mathbf{I}'_{N_1}=\mathbf{I}'_{N_1}\,\mathbf{D}_{N_1}$ and $-\mathbf{D}_{N_1}\mathbf{V}_{N_1}=\mathbf{V}_{N_1}\mathbf{D}_{N_1}$. Finally,

$$\cos\Big(\frac{(4j+3)(N+2k+1)\pi}{4N}\Big)=(-1)^{j+1}\cos\Big(\frac{(j+1)(2k+1)\pi}{N}-\frac{(N+2k+1)\pi}{4N}\Big)$$
$$=(-1)^{j+1}\cos\Big(\frac{(j+1)(2k+1)\pi}{N}\Big)\sin\Big(\frac{(N-2k-1)\pi}{4N}\Big)$$
$$+(-1)^{j+1}\sin\Big(\frac{(j+1)(2k+1)\pi}{N}\Big)\cos\Big(\frac{(N-2k-1)\pi}{4N}\Big)$$

implies that

$$\begin{aligned}
&\frac{1}{\sqrt{N_1}}\left(\cos\left(\frac{(4j+3)(N+2k+1)\pi}{4N}\right)\right)_{j,k=0}^{N_1-1}\\
&= -\frac{1}{\sqrt{2}}\big(\mathbf{D}_{N_1}\,\mathbf{V}_{N_1}^{\top}\,\mathbf{C}_{N_1}^{\mathrm{II}}\,\mathrm{diag}\,(\mathbf{J}_{N_1}\,\mathbf{s}_{N_1}) + \mathbf{D}_{N_1}\,\mathbf{I}''_{N_1}\,\mathbf{S}_{N_1}^{\mathrm{II}}\,\mathrm{diag}\,(\mathbf{J}_{N_1}\,\mathbf{c}_{N_1})\big)\\
&= \frac{1}{\sqrt{2}}\big(\mathbf{V}_{N_1}^{\top}\,\mathbf{D}_{N_1}\,\mathbf{C}_{N_1}^{\mathrm{II}}\,\mathrm{diag}\,(\mathbf{J}_{N_1}\,\mathbf{s}_{N_1}) - \mathbf{I}''_{N_1}\,\mathbf{D}_{N_1}\,\mathbf{S}_{N_1}^{\mathrm{II}}\,\mathrm{diag}\,(\mathbf{J}_{N_1}\,\mathbf{c}_{N_1})\big)\\
&= \frac{1}{\sqrt{2}}\big(\mathbf{V}_{N_1}^{\top}\,\mathbf{C}_{N_1}^{\mathrm{II}}\,\mathbf{J}_{N_1}\,\mathrm{diag}\,(\mathbf{J}_{N_1}\,\mathbf{s}_{N_1}) - \mathbf{I}''_{N_1}\,\mathbf{D}_{N_1}\,\mathbf{S}_{N_1}^{\mathrm{II}}\,\mathrm{diag}\,(\mathbf{J}_{N_1}\,\mathbf{c}_{N_1})\big)\,. \qquad (6.56)
\end{aligned}$$

Using the relations (6.53)–(6.56), we find the following factorization:

$$\mathbf{P}_N\,\mathbf{C}_N^{\mathrm{IV}} = \frac{1}{\sqrt{2}}\begin{pmatrix}\mathbf{I}'_{N_1} & \mathbf{V}_{N_1}\mathbf{D}_{N_1}\\ \mathbf{V}_{N_1}^{\top} & -\mathbf{I}''_{N_1}\,\mathbf{D}_{N_1}\end{pmatrix}(\mathbf{C}_{N_1}^{\mathrm{II}}\oplus\mathbf{S}_{N_1}^{\mathrm{II}})\begin{pmatrix}\mathrm{diag}\,\mathbf{c}_{N_1} & (\mathrm{diag}\,\mathbf{s}_{N_1})\,\mathbf{J}_{N_1}\\ -\mathbf{J}_{N_1}\,\mathrm{diag}\,\mathbf{s}_{N_1} & \mathrm{diag}\,(\mathbf{J}_{N_1}\mathbf{c}_{N_1})\end{pmatrix},$$

where

$$\begin{pmatrix}\mathbf{I}'_{N_1} & \mathbf{V}_{N_1}\mathbf{D}_{N_1}\\ \mathbf{V}_{N_1}^{\top} & -\mathbf{I}''_{N_1}\,\mathbf{D}_{N_1}\end{pmatrix} = \left(\sqrt{2}\oplus\begin{pmatrix}\mathbf{I}_{N_1-1} & \mathbf{I}_{N_1-1}\\ \mathbf{I}_{N_1-1} & -\mathbf{I}_{N_1-1}\end{pmatrix}\oplus(-\sqrt{2})\right)(\mathbf{I}_{N_1}\oplus\mathbf{D}_{N_1}).$$

Thus, (6.52) follows by the intertwining relation $\mathbf{S}_{N_1}^{\mathrm{II}} = \mathbf{J}_{N_1}\,\mathbf{C}_{N_1}^{\mathrm{II}}\,\mathbf{D}_{N_1}$. The orthogonality of $\mathbf{A}_N$ and $\tilde{\mathbf{B}}_N$ can be simply observed. Note that $\tilde{\mathbf{B}}_N$ consists only of N_1 plane rotations or rotation–reflections. ■

Since $\mathbf{C}_N^{\mathrm{IV}}$ is symmetric, we also obtain the factorization

$$\mathbf{C}_N^{\mathrm{IV}} = \tilde{\mathbf{B}}_N^{\top}\,(\mathbf{C}_{N_1}^{\mathrm{III}}\oplus\mathbf{C}_{N_1}^{\mathrm{III}})\,\mathbf{A}_N^{\top}\,\mathbf{P}_N. \qquad (6.57)$$

Example 6.34 For $N = 4$ we find

$$\mathbf{C}_4^{\mathrm{II}} = \mathbf{P}_4^{\top}(\mathbf{C}_2^{\mathrm{II}}\oplus\mathbf{C}_2^{\mathrm{IV}})\,\mathbf{B}_4 = \frac{1}{2}\,\mathbf{P}_4^{\top}(\sqrt{2}\,\mathbf{C}_2^{\mathrm{II}}\oplus\sqrt{2}\,\mathbf{C}_2^{\mathrm{IV}})\,\sqrt{2}\,\mathbf{B}_4$$

with

$$\mathbf{C}_2^{\mathrm{II}} = \frac{1}{\sqrt{2}}\begin{pmatrix}1 & 1\\ 1 & -1\end{pmatrix},\quad \mathbf{C}_2^{\mathrm{IV}} = \begin{pmatrix}\cos\left(\frac{\pi}{8}\right) & \sin\left(\frac{\pi}{8}\right)\\ \sin\left(\frac{\pi}{8}\right) & -\cos\left(\frac{\pi}{8}\right)\end{pmatrix},\quad \mathbf{B}_4 = \begin{pmatrix}\mathbf{I}_2 & \mathbf{J}_2\\ \mathbf{I}_2 & -\mathbf{J}_2\end{pmatrix}.$$

Thus, $\mathbf{C}_4^{\mathrm{II}}\,\mathbf{a}$ with $\mathbf{a}\in\mathbb{R}^4$ can be computed with 8 additions and 4 multiplications (not counting the scaling by $\frac{1}{2}$). Similarly,

$$\mathbf{C}_4^{\mathrm{III}} = \mathbf{B}_4^{\top}(\mathbf{C}_2^{\mathrm{III}}\oplus\mathbf{C}_2^{\mathrm{IV}})\,\mathbf{P}_4 = \tfrac{1}{2}\,(\sqrt{2}\,\mathbf{B}_4^{\top})(\sqrt{2}\,\mathbf{C}_2^{\mathrm{III}}\oplus\sqrt{2}\,\mathbf{C}_2^{\mathrm{IV}})\,\mathbf{P}_4$$

with $\mathbf{C}_2^{\mathrm{III}} = \mathbf{C}_2^{\mathrm{II}}$. Further, we find

$$\mathbf{C}_4^{\mathrm{IV}} = \mathbf{P}_4^\top \mathbf{A}_4 \, (\mathbf{C}_2^{\mathrm{II}} \oplus \mathbf{C}_2^{\mathrm{II}}) \, \tilde{\mathbf{B}}_4 = \tfrac{1}{2} \, \mathbf{P}_4^\top \sqrt{2} \, \mathbf{A}_4 \, (\sqrt{2} \, \mathbf{C}_2^{\mathrm{II}} \oplus \sqrt{2} \, \mathbf{C}_2^{\mathrm{II}}) \, \tilde{\mathbf{B}}_4$$

with

$$\mathbf{A}_4 = \frac{1}{\sqrt{2}} \begin{pmatrix} \sqrt{2} & & & \\ & 1 & 1 & \\ & 1 & -1 & \\ & & & \sqrt{2} \end{pmatrix}, \quad \tilde{\mathbf{B}}_4 = \begin{pmatrix} \cos\left(\frac{\pi}{16}\right) & & & \sin(\left(\frac{\pi}{16}\right) \\ & \cos\left(\frac{3\pi}{16}\right) & \sin\left(\frac{3\pi}{16}\right) & \\ & -\sin\left(\frac{3\pi}{16}\right) & \cos\left(\frac{3\pi}{16}\right) & \\ \sin\left(\frac{\pi}{16}\right) & & & -\cos\left(\frac{\pi}{16}\right) \end{pmatrix},$$

such that $\mathbf{C}_4^{\mathrm{IV}} \mathbf{a}$ with $\mathbf{a} \in \mathbb{R}^4$ can be computed with 10 additions and 10 multiplications. Finally,

$$\mathbf{C}_5^{\mathrm{I}} = \mathbf{P}_5^\top \, (\mathbf{C}_3^{\mathrm{I}} \oplus \mathbf{C}_2^{\mathrm{III}}) \, \mathbf{B}_5 = \frac{1}{2} \, \mathbf{P}_5^\top \, (\sqrt{2} \, \mathbf{C}_3^{\mathrm{I}} \oplus \sqrt{2} \, \mathbf{C}_2^{\mathrm{III}}) \, \sqrt{2} \, \mathbf{B}_5 \, ,$$

where particularly

$$\sqrt{2} \, \mathbf{C}_3^{\mathrm{I}} = \mathbf{P}_3^\top \, (\mathbf{C}_2^{\mathrm{II}} \oplus 1) \, \sqrt{2} \, \mathbf{B}_3$$

with

$$\mathbf{B}_5 = \frac{1}{\sqrt{2}} \begin{pmatrix} \mathbf{I}_2 & & \mathbf{J}_2 \\ & \sqrt{2} & \\ \mathbf{I}_2 & & -\mathbf{J}_2 \end{pmatrix}, \qquad \mathbf{B}_3 = \frac{1}{\sqrt{2}} \begin{pmatrix} 1 & & 1 \\ & \sqrt{2} & \\ 1 & & -1 \end{pmatrix}.$$

Thus, the computation of $\mathbf{C}_5^{\mathrm{I}} \mathbf{a}$ with $\mathbf{a} \in \mathbb{R}^5$ requires 10 additions and 4 multiplications. □

The derived factorizations of the cosine matrices of types I–IV imply the following recursive fast algorithms. We compute $\hat{\mathbf{a}} = \sqrt{N} \, \mathbf{C}_N^{\mathrm{X}} \, \mathbf{a}$ for $\mathrm{X} \in \{\mathrm{II}, \mathrm{III}, \mathrm{IV}\}$ with $\mathbf{a} \in \mathbb{R}^N$ and $\hat{\mathbf{a}} = \sqrt{N} \, \mathbf{C}_{N+1}^{\mathrm{I}} \, \mathbf{a}$ with $\mathbf{a} \in \mathbb{R}^{N+1}$. The corresponding recursive procedures are called $\mathsf{cos-I}\,(\mathbf{a}, N+1)$, $\mathsf{cos-II}\,(\mathbf{a}, N)$, $\mathsf{cos-III}\,(\mathbf{a}, N)$, and $\mathsf{cos-IV}\,(\mathbf{a}, N)$, respectively.

Algorithm 6.35 ($\mathsf{cos-I}\,(\mathbf{a}, N+1)$ **via Matrix Factorization)**
Input: $N = 2^t$, $t \in \mathbb{N}$, $N_1 = N/2$, $\mathbf{a} \in \mathbb{R}^{N+1}$.

1. *If* $N = 2$, *then*

$$\hat{\mathbf{a}} = \mathbf{P}_3^\top \, (\mathbf{C}_2^{\mathrm{II}} \oplus 1) \, \sqrt{2} \, \mathbf{B}_3 \, \mathbf{a} \, .$$

2. *If* $N \geq 4$, *then*

$$\begin{aligned}
(u_j)_{j=0}^{N} &:= \sqrt{2}\, \mathbf{B}_{N+1}\, \mathbf{a}\,, \\
\mathbf{v}' &:= \mathsf{cos-I}\big((u_j)_{j=0}^{N_1},\, N_1+1\big)\,, \\
\mathbf{v}'' &:= \mathsf{cos-III}\big((u_j)_{j=N_1+1}^{N},\, N_1\big)\,, \\
\hat{\mathbf{a}} &:= \mathbf{P}_{N+1}^{\top}\big((\mathbf{v}')^{\top},\, (\mathbf{v}'')^{\top}\big)^{\top}.
\end{aligned}$$

Output: $\hat{\mathbf{a}} = \sqrt{N}\, \mathbf{C}_{N+1}^{I}\, \mathbf{a} \in \mathbb{R}^{N+1}$.
Computational cost: $\mathcal{O}(N \log N)$.

Algorithm 6.36 ($\mathsf{cos-II}$ **(a,** N**) via Matrix Factorization)**
Input: $N = 2^t$, $t \in \mathbb{N}$, $N_1 = N/2$, $\mathbf{a} \in \mathbb{R}^N$.
1. *If* $N = 2$, *then*

$$\hat{\mathbf{a}} = \sqrt{2}\, \mathbf{C}_2^{\mathrm{II}}\, \mathbf{a}.$$

2. *If* $N \ge 4$, *then*

$$\begin{aligned}
(u_j)_{j=0}^{N-1} &:= \sqrt{2}\, \mathbf{B}_{N}\, \mathbf{a}\,, \\
\mathbf{v}' &:= \mathsf{cos-II}\big((u_j)_{j=0}^{N_1-1},\, N_1\big)\,, \\
\mathbf{v}'' &:= \mathsf{cos-IV}\big((u_j)_{j=N_1}^{N-1},\, N_1\big)\,, \\
\hat{\mathbf{a}} &:= \mathbf{P}_{N}^{\top}\big((\mathbf{v}')^{\top},\, (\mathbf{v}'')^{\top}\big)^{\top}.
\end{aligned}$$

Output: $\hat{\mathbf{a}} = \sqrt{N}\, \mathbf{C}_{N}^{\mathrm{II}}\, \mathbf{a} \in \mathbb{R}^{N}$.
Computational cost: $\mathcal{O}(N \log N)$.

Algorithm 6.37 ($\mathsf{cos-III}$ **(a,** N**) via Matrix Factorization)**
Input: $N = 2^t$, $t \in \mathbb{N}$, $N_1 = N/2$, $\mathbf{a} \in \mathbb{R}^N$.
1. *If* $N = 2$, *then*

$$\hat{\mathbf{a}} = \sqrt{2}\, \mathbf{C}_2^{\mathrm{III}}\, \mathbf{a}\,.$$

2. *If* $N \ge 4$, *then*

$$\begin{aligned}
(u_j)_{j=0}^{N-1} &:= \mathbf{P}_{N}\, \mathbf{a}\,, \\
\mathbf{v}' &:= \mathsf{cos-III}\big((u_j)_{j=0}^{N_1-1},\, N_1\big)\,, \\
\mathbf{v}'' &:= \mathsf{cos-IV}\big((u_j)_{j=N_1}^{N-1},\, N_1\big)\,, \\
\hat{\mathbf{a}} &:= \sqrt{2}\, \mathbf{B}_{N}^{\top}\big((\mathbf{v}')^{\top},\, (\mathbf{v}'')^{\top}\big)^{\top}.
\end{aligned}$$

Output: $\hat{\mathbf{a}} = \sqrt{N}\,\mathbf{C}_N^{\mathrm{III}}\,\mathbf{a} \in \mathbb{R}^N$.
Computational cost: $\mathcal{O}(N \log N)$.

Algorithm 6.38 ($\mathsf{cos-IV}\,(\mathbf{a}, N)$ **via Matrix Factorization)**
Input: $N = 2^t$, $t \in \mathbb{N}$, $N_1 = N/2$, $\mathbf{a} \in \mathbb{R}^N$.

1. *If* $N = 2$, *then*

$$\hat{\mathbf{a}} = \sqrt{2}\,\mathbf{C}_2^{\mathrm{IV}}\,\mathbf{a}\,.$$

2. *If* $N \geq 4$, *then*

$$\begin{aligned}
(u_j)_{j=0}^{N-1} &:= \sqrt{2}\,\tilde{\mathbf{B}}_N\,\mathbf{a}\,,\\
\mathbf{v}' &:= \mathsf{cos-II}\left((u_j)_{j=0}^{N_1-1},\, N_1\right),\\
\mathbf{v}'' &:= \mathsf{cos-II}\left((u_j)_{j=N_1}^{N-1},\, N_1\right),\\
\mathbf{w} &:= \mathbf{A}_N\left((\mathbf{v}')^\top, (\mathbf{v}'')^\top\right)^\top,\\
\hat{\mathbf{a}} &:= \mathbf{P}_N^\top\,\mathbf{w}\,.
\end{aligned}$$

Output: $\hat{\mathbf{a}} = \sqrt{N}\,\mathbf{C}_N^{\mathrm{IV}}\,\mathbf{a} \in \mathbb{R}^N$.
Computational cost: $\mathcal{O}(N \log N)$.

Let us consider the computational costs of these algorithms in real arithmetic. Here, we do not count permutations and multiplications with ± 1 or 2^k for $k \in \mathbb{Z}$. Let $\alpha(\mathsf{cos-II}, N)$ and $\mu(\mathsf{cos-II}, N)$ denote the number of additions and multiplications of Algorithm 6.36. For the other algorithms, we employ analogous notations. The following result is due to [324].

Theorem 6.39 *Let* $N = 2^t$, $t \in \mathbb{N} \setminus \{1\}$, *be given. Then the recursive Algorithms* 6.35–6.38 *require the following numbers of additions and multiplications*

$$\begin{aligned}
\alpha(\mathsf{cos-II}, N) = \alpha(\mathsf{cos-III}, N) &= \tfrac{4}{3}\,Nt - \tfrac{8}{9}\,N - \tfrac{1}{9}\,(-1)^t + 1\,,\\
\mu(\mathsf{cos-II}, N) = \mu(\mathsf{cos-III}, N) &= Nt - \tfrac{4}{3}\,N + \tfrac{1}{3}\,(-1)^t + 1\,,\\
\alpha(\mathsf{cos-IV}, N) &= \tfrac{4}{3}\,Nt - \tfrac{2}{9}\,N + \tfrac{2}{9}\,(-1)^t\,,\\
\mu(\mathsf{cos-IV}, N) &= Nt + \tfrac{2}{3}\,N - \tfrac{2}{3}\,(-1)^t\,,\\
\alpha(\mathsf{cos-I}, N+1) &= \tfrac{4}{3}\,Nt - \tfrac{14}{9}\,N + \tfrac{1}{2}\,t + \tfrac{7}{2} + \tfrac{1}{18}\,(-1)^t\,,\\
\mu(\mathsf{cos-I}, N+1) &= Nt - \tfrac{4}{3}\,N + \tfrac{5}{2} - \tfrac{1}{6}\,(-1)^t\,.
\end{aligned}$$

Proof
1. We compute $\alpha(\mathsf{cos-II}, N)$ and $\alpha(\mathsf{cos-IV}, N)$. From Example 6.34, it follows that

$$\alpha(\text{cos} - \text{II},\, 2) = 2\,, \qquad \alpha(\text{cos} - \text{II},\, 4) = 8\,, \tag{6.58}$$

$$\alpha(\text{cos} - \text{IV},\, 2) = 2\,, \qquad \alpha(\text{cos} - \text{II},\, 4) = 10\,. \tag{6.59}$$

Further, Algorithms 6.36 and 6.38 imply the recursions

$$\alpha(\text{cos} - \text{II},\, N) = \alpha(\sqrt{2}\,\mathbf{B}_N) + \alpha(\text{cos} - \text{II},\, N_1) + \alpha(\text{cos} - \text{IV},\, N_1)\,,$$
$$\alpha(\text{cos} - \text{IV},\, N) = \alpha(\sqrt{2}\,\tilde{\mathbf{B}}_N) + 2\alpha(\text{cos} - \text{II},\, N_1) + \alpha(\mathbf{A}_N)\,,$$

where $\alpha(\sqrt{2}\,\mathbf{B}_N)$ denotes the number of additions required for the product $\sqrt{2}\,\mathbf{B}_N\,\mathbf{a}$ for an arbitrary vector $\mathbf{a} \in \mathbb{R}^N$. Analogously, $\alpha(\sqrt{2}\,\tilde{\mathbf{B}}_N)$ and $\alpha(\mathbf{A}_N)$ are determined. From the definitions of $\mathbf{B}_N$, $\tilde{\mathbf{B}}_N$, and $\mathbf{A}_N$ it follows that

$$\alpha(\sqrt{2}\,\mathbf{B}_N) = \alpha(\sqrt{2}\,\tilde{\mathbf{B}}_N) = N\,, \quad \alpha(\sqrt{2}\,\mathbf{A}_N) = N - 2\,.$$

Thus, we obtain the linear difference equation of order 2 (with respect to $t \geq 3$),

$$\alpha(\text{cos} - \text{II},\, 2^t) = \alpha(\text{cos} - \text{II},\, 2^{t-1}) + 2\,\alpha(\text{cos} - \text{II},\, 2^{t-2}) + 2^{t+1} - 2\,.$$

With the initial conditions in (6.58), we find the unique solution

$$\alpha(\text{cos} - \text{II},\, N) = \alpha(\text{cos} - \text{III},\, N) = \tfrac{4}{3}\,Nt - \tfrac{8}{9}\,N - \tfrac{1}{9}\,(-1)^t + 1$$

which can be simply verified by induction with respect to t. Thus,

$$\alpha(\text{cos} - \text{IV},\, N) = \tfrac{4}{3}\,Nt - \tfrac{2}{9}\,N + \tfrac{2}{9}\,(-1)^t\,.$$

2. The computational cost for $\text{cos} - \text{III}(\mathbf{a},\, N)$ is the same as for $\text{cos} - \text{II}(\mathbf{a},\, N)$.
3. Finally, for $\text{cos} - \text{I}\,(\mathbf{a},\, N)$, we conclude

$$\alpha(\text{cos} - \text{I},\, N+1) = \alpha(\sqrt{2}\,\mathbf{B}_{N+1}) + \alpha(\text{cos} - \text{I},\, N_1 + 1) + \alpha(\text{cos} - \text{III},\, N_1)$$

and hence by $\alpha(\sqrt{2}\,\mathbf{B}_{N+1}) = N$ that

$$\begin{aligned}\alpha(\text{cos} - \text{I},\, 2^t + 1) &= 2^t + \alpha(\text{cos} - \text{I},\, 2^{t-1} + 1) + \tfrac{2}{3}\,2^t\,(t-1) - \tfrac{4}{9}\,2^t - \tfrac{1}{9}\,(-1)^{t-1} + 1 \\ &= \alpha(\text{cos} - \text{I},\, 2^{t-1} + 1) + \tfrac{2}{3}\,2^t t - \tfrac{1}{9}\,2^t + \tfrac{1}{9}\,(-1)^t + 1\,.\end{aligned}$$

Together with the initial condition $\alpha(\text{cos} - \text{I},\, 3) = 4$, we conclude

$$\alpha(\text{cos} - \text{I},\, 2^t) = \tfrac{4}{3}\,Nt - \tfrac{14}{9}\,N + \tfrac{1}{2}\,t + \tfrac{7}{2} + \tfrac{1}{18}\,(-1)^t\,.$$

The results for the required number of multiplications can be derived analogously. ∎

For all four types of the discrete cosine transform, the corresponding *fast inverse* DCT *algorithms* can be simply derived since we have shown the orthogonality relations

$$(\mathbf{C}_N^{\mathrm{I}})^{-1} = \mathbf{C}_N^{\mathrm{I}}, \qquad (\mathbf{C}_N^{\mathrm{II}})^{-1} = (\mathbf{C}_N^{\mathrm{II}})^{\top} = \mathbf{C}_N^{\mathrm{III}},$$
$$(\mathbf{C}_N^{\mathrm{III}})^{-1} = (\mathbf{C}_N^{\mathrm{III}})^{\top} = \mathbf{C}_N^{\mathrm{II}}, \qquad (\mathbf{C}_N^{\mathrm{IV}})^{-1} = \mathbf{C}_N^{\mathrm{IV}}$$

in Sect. 3.5. Therefore, the obtained fast DCT-I(N) algorithms, Algorithm 6.28 and 6.35, are at the same time also fast algorithms for the inverse DCT-I(N). Similarly, Algorithm 6.31 and 6.38 for the DCT-IV(N) are at the same time inverse DCT-IV(N) algorithms. The inverse DCT-II(N) coincides with the DCT-III(N); therefore, Algorithm 6.30 and 6.37 can be taken for the inverse DCT-II, and vice versa, Algorithms 6.29 and 6.36 are algorithms for the inverse DCT-III(N).

Remark 6.40

1. Comparing the computational costs of the DCT algorithms based on orthogonal factorization in real arithmetic with the FFT-based algorithms in the previous subsection, we gain a factor larger than 4. Taking, e.g., Algorithm 6.29 using the Sande–Tukey Algorithm for FFT of length $2N$ in the second step, we have by Theorem 5.12 computational costs of $10\,N\,\log_2(2N) - 20\,N + 16 = 10\,N\,\log_2 N - 10\,N + 16$ and further N multiplications to evaluate the needed vector $(w_{4N}^j\,\hat{y}_j)_{j=0}^{N-1}$. In comparison, Theorem 6.39 shows computational costs of at most $\frac{7}{3}\,N\,\log_2 N - \frac{20}{9}\,N + \frac{22}{9}$ for Algorithm 6.36.
2. A detailed analysis of the roundoff errors for the fast DCT Algorithms 6.35–6.38 shows their excellent numerical stability (see [324]).
3. Besides the proposed fast trigonometric transforms based on FFT or on orthogonal matrix factorization, there exist further fast DCT algorithms in real arithmetic via polynomial arithmetic with Chebyshev polynomials (see, e.g., [135, 136, 355, 395, 397]). These DCT algorithms generate non-orthogonal matrix factorizations of the cosine and sine matrices and therefore are inferior regarding numerical stability (see [23, 324, 381, 408]).
4. Similar orthogonal matrix factorizations and corresponding recursive algorithms can be also derived for the sine matrices (see [324]). □

Remark 6.41 Similarly, as shown in Sect. 5.4, one can derive *sparse* fast DCT algorithms with sublinear computational costs, where one exploits a priori knowledge that the vector $\hat{\mathbf{a}}$ to be recovered is sparse. These algorithms are either based on the FFT-based fast DCT (see [50]) or on the DCT algorithms via orthogonal matrix factorization (see [49]).

6.3.3 Fast Two-Dimensional DCT Algorithms

We restrict ourselves to the two-dimensional DCT-II which is mostly used in image processing. For fixed $N_1, N_2 \in \mathbb{N} \setminus \{1\}$, we consider the *two-dimensional* DCT-II$(N_1 \times N_2)$ of the form

$$\hat{\mathbf{A}} = \mathbf{C}_{N_1}^{\mathrm{II}}\, \mathbf{A}\, (\mathbf{C}_{N_2}^{\mathrm{II}})^\top , \tag{6.60}$$

where $\mathbf{A} = \big(a_{k_1,k_2}\big)_{k_1,k_2=0}^{N_1-1,N_2-1}$ and $\hat{\mathbf{A}} = \big(\hat{a}_{n_1,n_2}\big)_{n_1,n_2=0}^{N_1-1,N_2-1}$ are real N_1-by-N_2 matrices. For the entries $\hat{a}_{n_1,n_2}$, we obtain

$$\hat{a}_{n_1,n_2} = \frac{2\epsilon_{N_1}(n_1)\epsilon_{N_2}(n_2)}{\sqrt{N_1 N_2}} \sum_{k_1=0}^{N_1-1}\sum_{k_2=0}^{N_2-1} a_{k_1,k_2} \cos\Big(\frac{(2k_1+1)n_1\pi}{2N_1}\Big) \cos\Big(\frac{(2k_2+1)n_2\pi}{2N_2}\Big) \tag{6.61}$$

for $n_1 = 0, \ldots, N_1 - 1;\ n_2 = 0, \ldots, N_2 - 1$. For the fast evaluation of $\hat{\mathbf{A}}$, we can employ a fast algorithm for the one-dimensional DCT-II, similarly as it has been done for the two-dimensional FFT in Sect. 5.3.5. We employ the *row–column method* for the two-dimensional DCT-II$(N_1 \times N_2)$. Let $\mathbf{A}^\top = \big(\tilde{\mathbf{a}}_0 \,|\, \tilde{\mathbf{a}}_1 \,|\, \ldots \,|\, \tilde{\mathbf{a}}_{N_1-1}\big)$, where $\tilde{\mathbf{a}}_{k_1} \in \mathbb{R}^{N_2}$ denote the N_1 rows of $\mathbf{A}$ for $k_1 = 0, \ldots, N_1 - 1$. Then the product

$$\mathbf{B}^\top := \mathbf{C}_{N_2}^{\mathrm{II}}\, \mathbf{A}^\top = \big(\mathbf{C}_{N_2}^{\mathrm{II}} \tilde{\mathbf{a}}_0 \,|\, \mathbf{C}_{N_2}^{\mathrm{II}} \tilde{\mathbf{a}}_1 \,|\, \ldots \,|\, \mathbf{C}_{N_2}^{\mathrm{II}} \tilde{\mathbf{a}}_{N_1-1}\big)$$

can be performed by applying a fast algorithm for the DCT-II of length N_2, as, e.g., Algorithm 6.29 or 6.36, separately to each row. Then we obtain $\mathbf{B} = \mathbf{A}\, (\mathbf{C}_{N_2}^{\mathrm{II}})^\top$ and therefore $\hat{\mathbf{A}} = \mathbf{C}_{N_1}^{\mathrm{II}}\, \mathbf{B}$. Let now $\mathbf{B} = (\mathbf{b}_0 \,|\, \mathbf{b}_1 \,|\, \ldots \,|\, \mathbf{b}_{N_2-1})$ with columns $\mathbf{b}_{k_2} \in \mathbb{R}^{N_1}$, $k_2 = 0, \ldots, N_2 - 1$. Then

$$\hat{\mathbf{A}} = \mathbf{C}_{N_1}^{\mathrm{II}}\, \mathbf{B} = \big(\mathbf{C}_{N_1}^{\mathrm{II}} \mathbf{b}_0 \,|\, \mathbf{C}_{N_1}^{\mathrm{II}} \mathbf{b}_1 \,|\, \ldots \,|\, \mathbf{C}_{N_1}^{\mathrm{II}} \mathbf{b}_{N_2-1}\big)$$

can be performed by applying the DCT-II of length N_1 to each column of $\mathbf{B}$, where we again apply Algorithm 6.29 or 6.36.
Of course, we can also compute first $\tilde{\mathbf{B}} = \mathbf{C}_{N_1}^{\mathrm{II}}\, \mathbf{A}$ by applying a one-dimensional DCT-II(N_1) to each column of $\mathbf{A}$ and then compute $\tilde{\mathbf{B}}\, (\mathbf{C}_{N_2}^{\mathrm{II}})^\top$ by applying a DCT-II(N_2) to each row of $\tilde{\mathbf{B}}$ in the second step.

The matrix product (6.60) can also be rewritten as a matrix-vector product using the Kronecker product of matrices in Sect. 3.4. Denoting with $\mathrm{vec}\,\mathbf{A}$ the vectorization of $\mathbf{A} = (\mathbf{a}_0| \ldots |\mathbf{a}_{N_2-1}) \in \mathbb{R}^{N_1 \times N_2}$ into the vector $(\mathbf{a}_0^\top, \ldots, \mathbf{a}_{N_2-1}^\top)^\top = (\mathbf{a}_k)_{k=0}^{N_1} \in \mathbb{R}^{N_1 N_2}$, we obtain by Theorem 3.42 that

$$\mathrm{vec}\,\hat{\mathbf{A}} = \mathrm{vec}\big(\mathbf{C}_{N_1}^{\mathrm{II}}\, \mathbf{A}\, (\mathbf{C}_{N_2}^{\mathrm{II}})^\top\big) = (\mathbf{C}_{N_2}^{\mathrm{II}} \otimes \mathbf{C}_{N_1}^{\mathrm{II}})\mathrm{vec}\,\mathbf{A} = (\mathbf{C}_{N_2}^{\mathrm{II}} \otimes \mathbf{I}_{N_1})(\mathbf{I}_{N_2} \otimes \mathbf{C}_{N_1}^{\mathrm{II}})\,\mathrm{vec}\,\mathbf{A}\,.$$

Now, the multiplication $\operatorname{vec}\mathbf{B} = (\mathbf{I}_{N_2} \otimes \mathbf{C}^{\mathrm{II}}_{N_1}) \operatorname{vec}\mathbf{A}$ is equivalent with applying the one-dimensional DCT-II of length N_2 to each row of $\mathbf{A}$, and the multiplication $(\mathbf{C}^{\mathrm{II}}_{N_2} \otimes \mathbf{I}_{N_1}) \operatorname{vec}\mathbf{B}$ is equivalent with applying the one-dimensional DCT-II of length N_1 to each column of $\mathbf{B}$.

Using the sum representation (6.61) of the row–column method for the two-dimensional DCT-II($N_1 \times N_2$) of $\mathbf{A} = \big(a_{k_1,k_2}\big)_{k_1,k_2=0}^{N_1-1,N_2-1}$, we rewrite the row–column method as

$$\hat{a}_{n_1,n_2} = \sqrt{\tfrac{2}{N_1}}\,\epsilon_{N_1}(n_1) \sum_{k_1=0}^{N_1-1} \cos\big(\tfrac{(2k_1+1)n_1\pi}{2N_1}\big) \underbrace{\Big(\sqrt{\tfrac{2}{N_2}}\,\epsilon_{N_2}(n_2) \sum_{k_2=0}^{N_2-1} a_{k_1,k_2} \cos\big(\tfrac{(2k_2+1)n_2\pi}{2N_2}\big)\Big)}_{b_{k_1,n_2}:=} .$$

Now, for each $k_1 \in \{0, \ldots, N_1 - 1\}$ we first compute the vectors $(b_{k_1,n_2})_{n_2=0}^{N_2-1}$ using a one-dimensional DCT-II(N_2). Then we compute

$$\hat{a}_{n_1,n_2} = \sqrt{\tfrac{2}{N_1}}\,\epsilon_{N_1}(n_1) \sum_{k_1=0}^{N_1-1} b_{k_1,n_2} \cos\Big(\tfrac{(2k_1+1)n_1\pi}{2N_1}\Big) ,$$

using a DCT-II(N_1) for each $n_2 \in \{0, \ldots, N_2 - 1\}$. In summary, the row–column method requires N_1 DCT-II(N_2) and N_2 DCT-II(N_1).

Algorithm 6.42 (Row–Column Method for DCT-II($N_1 \times N_2$))
Input: $N_1, N_2 \in \mathbb{N} \setminus \{1\}$, $\mathbf{A} \in \mathbb{R}^{N_1 \times N_2}$.

1. *Apply the* DCT-II(N_2) *to each of the* N_1 *rows of* $\mathbf{A}$ *using Algorithm 6.29 or 6.36 to obtain*

$$\mathbf{B}^\top = \mathbf{C}^{\mathrm{II}}_{N_2}\, \mathbf{A}^\top .$$

2. *Apply the* DCT-II(N_1) *to each of the* N_2 *columns of* $\mathbf{B}$ *using Algorithm 6.29 or 6.36 to obtain*

$$\hat{\mathbf{A}} = \mathbf{C}^{\mathrm{II}}_{N_1}\, \mathbf{B}.$$

Output: $\hat{\mathbf{A}} \in \mathbb{R}^{N_1 \times N_2}$.

The computational cost to apply Algorithm 6.42 is $\mathcal{O}\big(N_1 N_2 (\log N_1)(\log N_2)\big)$ since the one-dimensional fast DCT-II(N) algorithms require $\mathcal{O}(N \log N)$ floating point operations.

The *inverse two-dimensional* DCT-II($N_1 \times N_2$) is given by

$$\mathbf{A} = (\mathbf{C}^{\mathrm{II}}_{N_1})^\top\, \hat{\mathbf{A}}\, \mathbf{C}^{\mathrm{II}}_{N_1} = \mathbf{C}^{\mathrm{III}}_{N_1}\, \hat{\mathbf{A}}\, (\mathbf{C}^{\mathrm{III}}_{N_1})^\top ,$$

since $(\mathbf{C}^{\mathrm{II}}_N)^{-1} = (\mathbf{C}^{\mathrm{II}}_N)^\top = \mathbf{C}^{\mathrm{III}}_N$ as shown in Lemma 3.47. We can therefore employ the row–column method also for the inverse two-dimensional DCT-II, where we replace the fast DCT-II algorithm by a fast DCT-III algorithm, e.g., Algorithm 6.30 or 6.37.

6.4 Interpolation and Quadrature Using Chebyshev Expansions

Now we show that interpolation at Chebyshev extreme points has excellent numerical properties. Further, we describe the efficient Clenshaw–Curtis quadrature which is an interpolatory quadrature rule at Chebyshev extreme points.

6.4.1 Interpolation at Chebyshev Extreme Points

Let $N \in \mathbb{N} \setminus \{1\}$ be fixed and let $I := [-1,\, 1]$. Then the nonequispaced Chebyshev extreme points $x_j^{(N)} = \cos\big(\frac{j\pi}{N}\big) \in I$, $j = 0, \ldots, N$, are denser near the endpoints ± 1 (see Fig. 6.7). We want to interpolate an arbitrary function $f \in C(I)$ at the Chebyshev extreme points $x_j^{(N)}$, $j = 0, \ldots, N$, by a polynomial $p_N \in \mathscr{P}_N$. Then the interpolation conditions

$$p_N(x_j^{(N)}) = f(x_j^{(N)})\,, \quad j = 0, \ldots, N\,, \tag{6.62}$$

have to be satisfied. Since the Chebyshev polynomials T_j, $j = 0, \ldots, N$, form a basis of $\mathscr{P}_N$, the polynomial p_N can be expressed as a Chebyshev expansion

$$p_N = \tfrac{1}{2}\, a_0^{(N)}[f] + \textstyle\sum_{k=1}^{N-1} a_k^{(N)}[f]\, T_k + \tfrac{1}{2}\, a_N^{(N)}[f]\, T_N = \sum_{k=0}^{N} \varepsilon_N(k)^2\, a_k^{(N)}[f]\, T_k \tag{6.63}$$

with certain coefficients $a_k^{(N)}[f] \in \mathbb{R}$, where $\varepsilon_N(0) = \varepsilon_N(N) := \frac{\sqrt{2}}{2}$ and $\varepsilon_N(j) := 1$, $j = 1, \ldots, N-1$. The interpolation conditions in (6.62) imply the linear system

$$f(x_j^{(N)}) = \sum_{k=0}^{N} \varepsilon_N(k)^2\, a_k^{(N)}[f]\, \cos\Big(\frac{jk\pi}{N}\Big)\,, \quad j = 0, \ldots, N\,.$$

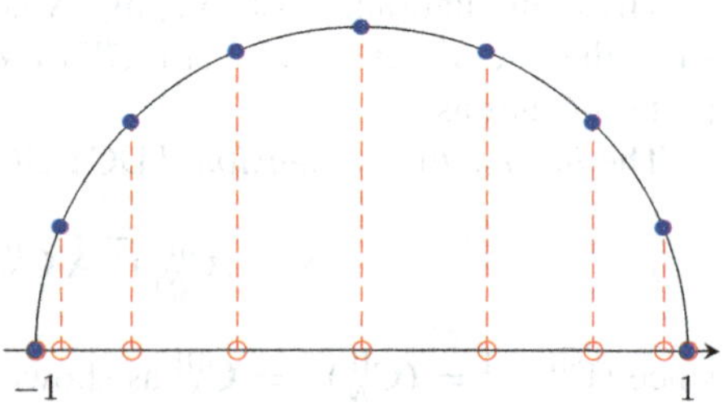

Fig. 6.7 The nonequispaced Chebyshev extreme points $x_j^{(8)} = \cos\big(\frac{j\pi}{8}\big) \in [-1,\, 1]$, $j = 0, \ldots, 8$, and the equispaced points $\mathrm{e}^{\mathrm{i}\, j\pi/8}$, $j = 0, \ldots, 8$, on the upper unit semicircle

This linear system can be written in the matrix–vector form

$$\big(\varepsilon_N(j)\, f(x_j^{(N)})\big)_{j=0}^N = \sqrt{\tfrac{N}{2}}\, \mathbf{C}_{N+1}^{\mathrm{I}}\, \big(\varepsilon_N(k)\, a_k^{(N)}[f]\big)_{k=0}^N\,, \tag{6.64}$$

where $\mathbf{C}_{N+1}^{\mathrm{I}}$ in (3.59) is the cosine matrix of type I. Recall that the symmetric cosine matrix of type I is orthogonal by Lemma 3.46, i.e., $(\mathbf{C}_{N+1}^{\mathrm{I}})^{-1} = \mathbf{C}_{N+1}^{\mathrm{I}}$. Therefore, the linear system (6.64) possesses the unique solution

$$\big(\varepsilon_N(k)\, a_k^{(N)}[f]\big)_{k=0}^N = \sqrt{\tfrac{2}{N}}\, \mathbf{C}_{N+1}^{\mathrm{I}}\, \big(\varepsilon_N(j)\, f(x_j^{(N)})\big)_{j=0}^N\,.$$

If N is a power of two, we can apply a fast algorithm for the DCT-I$(N+1)$ from Sect. 6.3 to compute the matrix–vector product above. We summarize as follows.

Lemma 6.43 *Let $N \in \mathbb{N} \setminus \{1\}$ be fixed and let $f \in C(I)$ be given. Then the interpolation problem* (6.62) *at Chebyshev extreme points $x_j^{(N)}$, $j = 0, \ldots, N$, has a unique solution of the form* (6.63) *in $\mathscr{P}_N$ with the coefficients*

$$a_k^{(N)}[f] = \tfrac{2}{N} \sum_{j=0}^{N} \varepsilon_N(j)^2\, f(x_j^{(N)})\, \cos\Big(\tfrac{jk\pi}{N}\Big)\,, \quad k = 0, \ldots, N\,. \tag{6.65}$$

If $f \in C(I)$ is even, then p_N is even and $a_{2k+1}^{(N)}[f] = 0$ for $k = 0, \ldots, \lfloor (N-1)/2 \rfloor$. If $f \in C(I)$ is odd, then p_N is odd and $a_{2k}^{(N)}[f] = 0$ for $k = 0, \ldots, \lfloor N/2 \rfloor$.

The Chebyshev coefficients $a_j[f]$, $j \in \mathbb{N}_0$, of a given function f in (6.18) and the coefficients $a_k^{(N)}[f]$, $k = 0, \ldots, N$, of the corresponding interpolation polynomial (6.65) are closely related. This connection can be described by the *aliasing formulas for Chebyshev coefficients* (see [89]).

Lemma 6.44 (Aliasing Formulas for Chebyshev Coefficients) *Let $N \in \mathbb{N}\setminus\{1\}$ be fixed. Assume that the Chebyshev coefficients of a given function $f \in C(I)$ satisfy the condition*

$$\sum_{j=0}^{\infty} |a_j[f]| < \infty\,. \tag{6.66}$$

Then the aliasing formulas

$$a_k^{(N)}[f] = a_k[f] + \sum_{\ell=1}^{\infty} \big(a_{2\ell N+k}[f] + a_{2\ell N-k}[f]\big) \tag{6.67}$$

hold for $k = 1, \ldots, N-1$, and for $k = 0$ and $k = N$ we have

$$a_0^{(N)}[f] = a_0[f] + 2 \sum_{\ell=1}^{\infty} a_{2\ell N}[f]\,, \tag{6.68}$$

$$a_N^{(N)}[f] = 2\,a_N[f] + 2\sum_{\ell=1}^{\infty} a_{(2\ell+1)\,N}[f]\,. \tag{6.69}$$

Proof By assumption (6.66) and Lemma 6.11, the Chebyshev series

$$\frac{1}{2}\,a_0[f] + \sum_{\ell=1}^{\infty} a_\ell[f]\,T_\ell$$

converges absolutely and uniformly on I to f. Thus, we obtain the function values

$$f(x_j^{(N)}) = \frac{1}{2}\,a_0[f] + \sum_{\ell=1}^{\infty} a_\ell[f]\,\cos\Big(\frac{j\ell\pi}{N}\Big)\,, \quad j = 0, \ldots, N\,,$$

at the Chebyshev extreme points $x_j^{(N)} = \cos\big(\frac{j\pi}{N}\big)$. By (6.65), the interpolation polynomial in (6.63) possesses the coefficients

$$\begin{aligned} a_k^{(N)}[f] &= \frac{2}{N}\sum_{j=0}^{N} \varepsilon_N(j)^2\, f(x_j^{(N)})\,\cos\Big(\frac{jk\pi}{N}\Big) \\ &= a_0[f]\,\frac{1}{N}\sum_{j=0}^{N} \varepsilon_N(j)^2 \cos\Big(\frac{jk\pi}{N}\Big) + \sum_{\ell=1}^{\infty} a_\ell[f]\,\frac{2}{N}\sum_{j=0}^{N} \varepsilon_N(j)^2 \cos\Big(\frac{j\ell\pi}{N}\Big)\cos\Big(\frac{jk\pi}{N}\Big)\,. \end{aligned}$$

Using (3.60) and (3.61), we see that

$$\frac{1}{N}\sum_{j=0}^{N} \varepsilon_N(j)^2 \cos\Big(\frac{jk\pi}{N}\Big) = \frac{1}{N}\Big(\frac{1}{2} + \sum_{j=1}^{N-1} \cos\Big(\frac{jk\pi}{N}\Big) + \frac{1}{2}\,(-1)^k\Big) = \begin{cases} 1 & k=0\,, \\ 0 & k=1,\ldots,N\,. \end{cases}$$

Analogously, we evaluate the sum

$$\begin{aligned} &\frac{2}{N}\sum_{j=0}^{N} \varepsilon_N(j)^2 \cos\Big(\frac{j\ell\pi}{N}\Big)\cos\Big(\frac{jk\pi}{N}\Big) \\ &= \frac{1}{N}\Big(1 + (-1)^{\ell+k} + \sum_{j=1}^{N-1} \cos\Big(\frac{j(\ell-k)\pi}{N}\Big) + \textstyle\sum_{j=1}^{N-1} \cos\Big(\frac{j(\ell+k)\pi}{N}\Big)\Big) \\ &= \begin{cases} 2 & k = 0,\ \ell = 2s\,N,\ s \in \mathbb{N}\,, \\ 1 & k = 1, \ldots, N-1\,, \ell = 2s\,N + k,\ s \in \mathbb{N}_0\,, \\ 1 & k = 1, \ldots, N-1\,, \ell = 2s\,N - k,\ s \in \mathbb{N}\,, \\ 2 & k = N,\ \ell = (2s+1)\,N,\ s \in \mathbb{N}_0\,, \\ 0 & \text{otherwise}\,. \end{cases} \end{aligned}$$

This completes the proof of the aliasing formulas for Chebyshev coefficients. ■

The aliasing formulas (6.67)–(6.69) for Chebyshev coefficients immediately provide a useful estimate of the interpolation error.

Theorem 6.45 *Let $N \in \mathbb{N} \setminus \{1\}$ be fixed. Assume that the Chebyshev coefficients of a given function $f \in C(I)$ satisfy the condition* (6.66).

Then the polynomial $p_N \in \mathcal{P}_N$ which interpolates f at the Chebyshev extreme points $x_j^{(N)}$, $j = 0, \ldots, N$, satisfies the error estimate

$$\|f - p_N\|_{C(I)} \le 2 \sum_{k=N+1}^{\infty} |a_k[f]| \,. \tag{6.70}$$

If $f \in C^{r+1}(I)$ for fixed $r \in \mathbb{N}$ and $N > r$, then

$$\|f - p_N\|_{C(I)} \le \frac{4}{r\,(N-r)^r}\, \|f^{(r+1)}\|_{C(I)} \,. \tag{6.71}$$

Proof

1. By Lemma 6.43, the interpolation polynomial p_N possesses the form (6.63) with the coefficients in (6.65). If $C_N f$ denotes the Nth partial sum of the Chebyshev series of f, then we have

$$\|f - p_N\|_{C(I)} \le \|f - C_N f\|_{C(I)} + \|C_N f - p_N\|_{C(I)} \,.$$

Obviously, we see by $|T_k(x)| \le 1$ for $x \in I$ that

$$\|f - C_N f\|_{C(I)} = \Big\| \sum_{k=N+1}^{\infty} a_k[f]\, T_k \Big\|_{C(I)} \le \sum_{k=N+1}^{\infty} |a_k[f]| \,.$$

Using the aliasing formulas (6.67)–(6.69), we obtain

$$\begin{aligned} \|C_N f - p_N\|_{C(I)} &\le \sum_{k=0}^{N-1} \varepsilon_N(k)^2\, |a_k[f] - a_k^{(N)}[f]| + |a_N[f] - \tfrac{1}{2}\, a_N^{(N)}[f]| \\ &\le \sum_{\ell=1}^{\infty} \big(|a_{2\ell N}[f]| + |a_{(2\ell+1)N}[f]|\big) + \sum_{k=1}^{N-1} \sum_{\ell=1}^{\infty} \big(|a_{2\ell N+k}[f]| + |a_{2\ell N-k}[f]|\big) \\ &= \sum_{k=N+1}^{\infty} |a_k[f]| \,. \end{aligned}$$

Thus (6.70) is shown.

2. Let $f \in C^{r+1}(I)$ for fixed $r \in \mathbb{N}$ be given. Assume that $N \in \mathbb{N}$ with $N > r$. Then for any $k > r$ the Chebyshev coefficients can be estimated by (6.20) such that

$$|a_k[f]| \le \frac{2}{(k-r)^{r+1}}\, \|f^{(r+1)}\|_{C(I)} \,.$$

Thus, from (6.70) it follows that

$$\|f - p_N\|_{C(I)} \le 4\,\|f^{(r+1)}\|_{C(I)} \sum_{k=N+1}^{\infty} \frac{1}{(k-r)^{r+1}}$$

$$\le 4\,\|f^{(r+1)}\|_{C(I)} \int_N^{\infty} \frac{1}{(x-r)^{r+1}}\,\mathrm{d}x = 4\,\|f^{(r+1)}\|_{C(I)}\,\frac{1}{r\,(N-r)^r}\,.$$

■

Example 6.46 We interpolate the function $f(x) := \mathrm{e}^x$ for $x \in I$ at Chebyshev extreme points $x_j^{(N)}$, $j = 0, \ldots, N$. Choosing $r = 10$, by (6.71) the interpolation error can be estimated for any $N > 10$ by

$$\|f - p_N\|_{C(I)} \le \frac{2\,\mathrm{e}}{5\,(N-10)^{10}}\,.$$

□

We emphasize that the polynomial interpolation at Chebyshev extreme points $x_j^{(N)} \in I$, $j = 0, \ldots, N$, has excellent properties:

The coefficients of the interpolation polynomial p_N can be rapidly computed by a fast algorithm for the DCT-I $(N+1)$.

The interpolation polynomial p_N can be stable evaluated by the barycentric formula (6.36).

The smoother the given function $f : I \to \mathbb{R}$, the faster the interpolation error $\|f - p_N\|_{C(I)}$ tends to zero for $N \to \infty$.

This situation completely changes for interpolation at equispaced points $y_j^{(N)} := -1 + \frac{2j}{N} \in I$, $j = 0, \ldots, N$. We illustrate the essential influence of the chosen interpolation points by the famous example of C. Runge [369].

Example 6.47 The *Runge phenomenon* shows that equispaced polynomial interpolation of a continuous function can be troublesome. We interpolate the rational function $f(x) := (25\,x^2 + 1)^{-1}$, $x \in I$, at the equispaced points $y_j^{(N)} := -1 + \frac{2j}{N} \in I$, $j = 0, \ldots, N$. We observe that the corresponding interpolation polynomial $q_N \in \mathcal{P}_N$ with

$$q_N(y_j^{(N)}) = f(y_j^{(N)})\,, \quad j = 0, \ldots, N\,,$$

oscillates near the endpoints ± 1 such that the interpolation error $\|f - q_N\|_{C(I)}$ increases for growing N. Thus, the interpolation polynomial q_N does not converge uniformly on I to f as $N \to \infty$. Figure 6.8 shows the graphs of f and of the related interpolation polynomials q_N for $N = 9$ and $N = 15$.

On the other hand, if we interpolate f at the nonequispaced Chebyshev extreme points $x_j^{(N)} \in I$, $j = 0, \ldots, N$, then by Theorem 6.45 the corresponding interpolation polynomials p_N converge uniformly on I to f as $N \to \infty$. Figure 6.9 illustrates the nice approximation behavior of the interpolation polynomial p_{15}. □

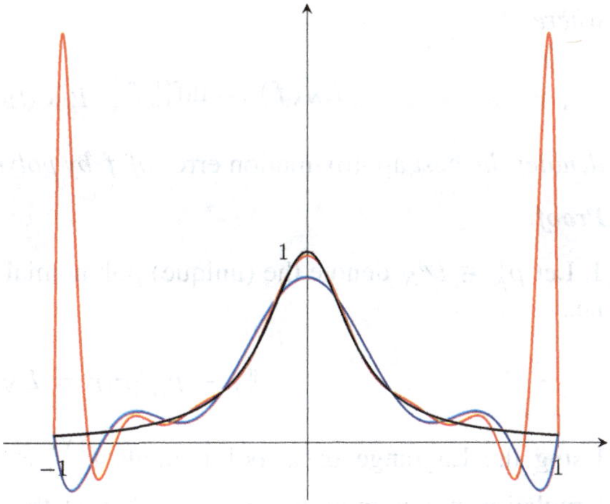

Fig. 6.8 The function $f(x) := (25\,x^2 + 1)^{-1}$ for $x \in [-1,\, 1]$ and the related interpolation polynomials q_N with equispaced nodes $y_j^{(N)}$, $j = 0, \ldots, N$, for $N = 9$ (blue) and $N = 15$ (red)

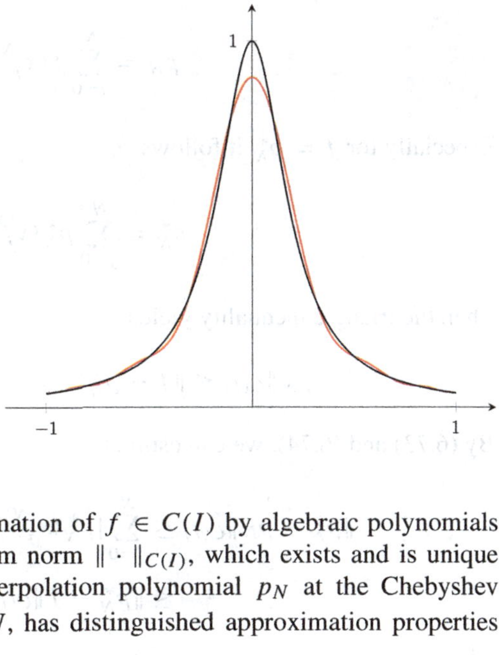

Fig. 6.9 The function $f(x) := (25\,x^2 + 1)^{-1}$ for $x \in [-1,\, 1]$ and the related interpolation polynomial p_N with nonequispaced Chebyshev extreme points $x_j^{(N)}$, $j = 0, \ldots, N$, for $N = 15$ (red)

Compared with the best approximation of $f \in C(I)$ by algebraic polynomials in $\mathscr{P}_N$ with respect to the maximum norm $\|\cdot\|_{C(I)}$, which exists and is unique on the compact interval I, the interpolation polynomial p_N at the Chebyshev extreme points $x_j^{(N)}$, $j = 0, \ldots, N$, has distinguished approximation properties for sufficiently large N.

Theorem 6.48 *Let $f \in C^r(I)$ with $r \in \mathbb{N} \setminus \{1\}$ be given. Further, let $N \in \mathbb{N}$ with $N > r$. Then the interpolation polynomial $p_N \in \mathscr{P}_N$ of f at the Chebyshev extreme points $x_j^{(N)}$, $j = 0, \ldots, N$, satisfies the inequality*

$$\|f - p_N\|_{C(I)} \le \big(5 + \tfrac{2}{\pi}\, \ln{(2N-1)}\big)\, E_N(f)\,,$$

where

$$E_N(f) := \inf\{\|f - p\|_{C(I)} : p \in \mathscr{P}_N\}$$

denotes the best approximation error *of f by polynomials in $\mathscr{P}_N$.*

Proof

1. Let $p_N^* \in \mathscr{P}_N$ denote the (unique) polynomial of best approximation of f on I, i.e.,

$$\|f - p_N^*\|_{C(I)} = E_N(f)\,. \tag{6.72}$$

Using the Lagrange basis polynomials $\ell_j^{(N)} \in \mathscr{P}_N$ defined by (6.37), the interpolation polynomial $p_N \in \mathscr{P}_N$ of f at the Chebyshev extreme points $x_j^{(N)}$, $j = 0, \ldots, N$, can be expressed as

$$p_N = \sum_{j=0}^{N} f(x_j^{(N)})\,\ell_j^{(N)}\,. \tag{6.73}$$

Especially for $f = p_N^*$ it follows

$$p_N^* = \sum_{j=0}^{N} p_N^*(x_j^{(N)})\,\ell_j^{(N)}\,. \tag{6.74}$$

Then the triangle inequality yields

$$\|f - p_N\|_{C(I)} \le \|f - p_N^*\|_{C(I)} + \|p_N^* - p_N\|_{C(I)}\,. \tag{6.75}$$

By (6.73) and (6.74), we can estimate

$$\begin{aligned}\|p_N^* - p_N\|_{C(I)} &\le \sum_{j=0}^{N} |p_N^*(x_j^{(N)}) - f(x_j^{(N)})|\,\|\ell_j^{(N)}\|_{C(I)}\\ &\le \|p_N^* - f\|_{C(I)}\,\lambda_N = E_N(f)\,\lambda_N\,,\end{aligned} \tag{6.76}$$

where

$$\lambda_N := \sum_{j=0}^{N} \|\ell_j^{(N)}\|_{C(I)} = \max_{x\in I} \sum_{j=0}^{N} |\ell_j^{(N)}(x)|$$

denotes the *Lebesgue constant for polynomial interpolation at Chebyshev extreme points*. By (6.76) the Lebesgue constant measures the distance between the interpolation polynomial p_N and the best approximation polynomial p_N^* subject to $E_N(f)$. From (6.75), (6.76), and (6.72), it follows

$$\|f - p_N\|_{C(I)} \le \|f - p_N^*\|_{C(I)} \, (1 + \lambda_N) = (1 + \lambda_N)\, E_N(f)\,.$$

2. Now we estimate the Lebesgue constant λ_N for interpolation at Chebyshev extreme points $x_j^{(N)}$, $j = 0, \ldots, N$. Using the *modified Dirichlet kernel*

$$D_N^*(t) = \tfrac{1}{2} + \sum_{j=1}^{N-1} \cos(jt) + \tfrac{1}{2}\,\cos(Nt) = \begin{cases} \tfrac{1}{2}\,\sin(Nt)\,\cot\big(\tfrac{t}{2}\big) & t \in \mathbb{R} \setminus 2\pi\,\mathbb{Z}\,, \\ N & t \in 2\pi\,\mathbb{Z}\,, \end{cases}$$

we observe that

$$D_N^*\Big(\frac{j\pi}{N}\Big) = \begin{cases} N & j \equiv 0 \bmod (2N)\,, \\ 0 & j \not\equiv 0 \bmod (2N)\,. \end{cases}$$

Thus, for $t \in [0,\, \pi]$ and $j = 0, \ldots, N$ we find

$$\ell_j^{(N)}(\cos(t)) = \tfrac{1}{N}\, D_N^*\Big(t - \tfrac{j\pi}{N}\Big) = \begin{cases} \frac{(-1)^j}{2N}\,\sin(Nt)\,\cot\big(\tfrac{t}{2} - \tfrac{j\pi}{2N}\big) & t \in [0,\,\pi] \setminus \{\tfrac{j\pi}{N}\}\,, \\ 1 & t = \tfrac{j\pi}{N}\,. \end{cases}$$

Consequently, we have to estimate the function

$$s(t) := \begin{cases} \frac{1}{2N} \sum\limits_{j=0}^{N} |\sin(Nt)\,\cot\big(\tfrac{t}{2} - \tfrac{j\pi}{2N}\big)| & t \in [0,\,\pi] \setminus \{\tfrac{j\pi}{N} : j = 0, \ldots, N\}\,, \\ 1 & t \in \{\tfrac{j\pi}{N} : j = 0, \ldots, N\}\,. \end{cases}$$

3. First, we observe that $s(t)$ is $\frac{\pi}{N}$-periodic. We consider $\sin(Nt)\,\cot\big(\frac{t}{2}\big)$ on the set $\big[\frac{-(2N+1)\pi}{2N},\, \frac{(2N+1)\pi}{2N}\big] \setminus \{0\}$. From

$$\tfrac{2}{\pi}\,|x| \le |\sin(x)| \le |x|\,, \quad x \in \big[-\tfrac{\pi}{2},\, \tfrac{\pi}{2}\big]\,,$$

and $|\cos(x)| \le 1$ we obtain the inequality

$$\Big|\sin(Nt)\,\cot\Big(\frac{t}{2}\Big)\Big| \le \frac{N\pi\,|t|}{|t|} = N\pi\,, \quad t \in \Big[-\frac{\pi}{2N},\, \frac{\pi}{2N}\Big] \setminus \{0\}\,.$$

For $t \in \big[-\frac{(2j+1)\pi}{2N},\, -\frac{(2j-1)\pi}{2N}\big] \cup \big[\frac{(2j-1)\pi}{2N},\, \frac{(2j+1)\pi}{2N}\big]$, $j = 1, \ldots, N$, we conclude

$$\Big|\sin(Nt)\,\cot\Big(\frac{t}{2}\Big)\Big| \le \frac{1}{\big|\sin\big(\frac{t}{2}\big)\big|} \le \frac{1}{\sin\Big(\frac{(2j-1)\pi}{4N}\Big)}\,,$$

since $\sin(t)$ is monotonically increasing in $[0,\, \frac{\pi}{2})$. Thus, for $t \in [0,\, \pi] \setminus \{\frac{j\pi}{N} : j = 0, \ldots, N\}$, it follows the estimate

$$s(t) \le \frac{1}{2N}\Big(N\pi + 2\sum_{j=1}^{N} \frac{1}{\sin\big(\frac{(2j-1)\pi}{4N}\big)}\Big).$$

We introduce the increasing function $h \in C^1[0, \frac{\pi}{2}]$ by

$$h(t) := \begin{cases} \frac{1}{\sin(t)} - \frac{1}{t} & t \in (0, \frac{\pi}{2}], \\ 0 & t = 0, \end{cases}$$

and obtain the following estimate for $s(t)$,

$$\begin{aligned} s(t) &\le \tfrac{\pi}{2} + \tfrac{1}{N}\sum_{j=1}^{N} \tfrac{4N}{(2j-1)\pi} + \tfrac{1}{N}\sum_{j=1}^{N} h\big(\tfrac{(2j-1)\pi}{4N}\big) \\ &= \tfrac{\pi}{2} + \tfrac{2}{\pi}\sum_{j=1}^{N} \tfrac{2}{(2j-1)} + \tfrac{2}{\pi}\,\tfrac{\pi}{2N}\sum_{j=1}^{N} h\big(\tfrac{(2j-1)\pi}{4N}\big). \end{aligned}$$

Interpreting the two sums as Riemann sums of corresponding definite integrals, this provides

$$\sum_{j=1}^{N} \tfrac{2}{(2j-1)} = 2 + \sum_{j=2}^{N} \tfrac{1}{(j-1/2)} < 2 + \int_{1/2}^{N-1/2} \tfrac{\mathrm{d}t}{t} = 2 + \ln(2N-1),$$

$$\tfrac{\pi}{2N}\sum_{j=1}^{N} h\big(\tfrac{(2j-1)\pi}{4N}\big) < \tfrac{\pi}{2}\,h(\tfrac{\pi}{2}) = \tfrac{\pi}{2} - 1.$$

Therefore, we obtain

$$s(t) \le \tfrac{\pi}{2} + \tfrac{2}{\pi}\big(2 + \ln(2N-1) + \tfrac{\pi}{2} - 1\big) < 4 + \tfrac{2}{\pi}\ln(2N-1).$$

and hence

$$\lambda_N \le 4 + \tfrac{2}{\pi}\ln(2N-1).$$

■

Remark 6.49

1. For the interpolation at the Chebyshev zero points $z_j^{(N+1)} = \cos\big(\frac{(2j+1)\pi}{2N+2}\big)$, $j = 0, \ldots, N$, one obtains similar results as for the described interpolation at Chebyshev extreme points $x_j^{(N)} = \cos\big(\frac{j\pi}{N}\big)$, $j = 0, \ldots, N$ (see [443]).
2. Let a sufficiently smooth function $f \in C^r(I)$ with fixed $r \in \mathbb{N} \setminus \{1\}$ be given. Then the *simultaneous approximation* of f and its derivatives by polynomial

interpolation can be investigated. If $p_N \in \mathcal{P}_N$ denotes the interpolation polynomial of $f \in C^r(I)$ at the Chebyshev extreme points $x_j^{(N)}$, $j = 0, \ldots, N$, then R. Haverkamp [194, 195] pointed out that

$$\|f' - p_N'\|_{C(I)} \le (2 + 2 \ln N)\, E_{N-1}(f'),$$

$$\|f'' - p_N''\|_{C(I)} \le \frac{\pi^2}{3}\, N\, E_{N-2}(f'').$$

The numerical computation of p_N' and p_N'' can be performed by Lemma 6.8.

3. Interpolation at Chebyshev nodes is also used in collocation methods for the Cauchy singular integral equation

$$a(x)\, u(x) + \frac{b(x)}{\pi} \int_{-1}^{1} \frac{u(y)}{y-x}\, \mathrm{d}y = f(x), \quad x \in (-1,\, 1),$$

where the functions $a, b, f : [-1, 1] \to \mathbb{C}$ are given and $u : (-1, 1) \to \mathbb{C}$ is the unknown solution (see, e.g., [216]). An efficient solution of the collocation equations is based on the application of fast algorithms of discrete trigonometric transforms (see [218]), and for methods based on fast summation, see [217]. □

6.4.2 Clenshaw–Curtis Quadrature

Now we will apply the interpolation polynomial (6.63) of a given function $f \in C(I)$ to numerical integration. We consider the quadrature problem, where we want to calculate an approximate value of the integral

$$I(f) := \int_{-1}^{1} f(x)\, \mathrm{d}x.$$

We obtain the famous *Clenshaw–Curtis quadrature* (see [89]), where the function f in the integral is replaced by its interpolation polynomial (6.63) at the Chebyshev extreme points $x_j^{(N)}$, $j = 0, \ldots, N$, such that

$$I(f) \approx Q_N(f) := \int_{-1}^{1} p_N(x)\, \mathrm{d}x.$$

By Lemma 6.8, the integrals of the Chebyshev polynomials possess the exact values

$$\int_{-1}^{1} T_{2j}(x)\, \mathrm{d}x = \frac{2}{1-4j^2}, \quad \int_{-1}^{1} T_{2j+1}(x)\, \mathrm{d}x = 0, \quad j \in \mathbb{N}_0.$$

Thus, we obtain the *Clenshaw–Curtis quadrature formula* from (6.63)

$$Q_N(f) = \begin{cases} a_0^{(N)}[f] + \sum\limits_{j=1}^{N/2-1} \frac{2}{1-4j^2}\, a_{2j}^{(N)}[f] + \frac{1}{1-N^2}\, a_N^{(N)}[f] & N \text{ even}, \\ a_0^{(N)}[f] + \sum\limits_{j=1}^{(N-1)/2-1} \frac{2}{1-4j^2}\, a_{2j}^{(N)}[f] & N \text{ odd} \end{cases} \tag{6.77}$$

with the corresponding *quadrature error*

$$R_N(f) := I(f) - Q_N(f)\,.$$

Note that T_{N+1} is odd for even N and hence

$$\int_{-1}^{1} T_{N+1}(x)\,\mathrm{d}x = 0\,.$$

Consequently, $R_N(p) = 0$ for all polynomials $p \in \mathscr{P}_{N+1}$, if N is even, and for all $p \in \mathscr{P}_N$, if N is odd. Therefore, all polynomials up to degree N are exactly integrated by the Clenshaw–Curtis rule.

Example 6.50 The simplest Clenshaw–Curtis quadrature formulas read as follows:

$$\begin{aligned} Q_1(f) &= f(-1) + f(1)\,, \\ Q_2(f) &= \tfrac{1}{3}\, f(-1) + \tfrac{4}{3}\, f(0) + \tfrac{1}{3}\, f(1)\,, \\ Q_3(f) &= \tfrac{1}{9}\, f(-1) + \tfrac{8}{9}\, f\big(-\tfrac{1}{2}\big) + \tfrac{8}{9}\, f\big(\tfrac{1}{2}\big) + \tfrac{1}{9}\, f(1)\,, \\ Q_4(f) &= \tfrac{1}{15}\, f(-1) + \tfrac{8}{15}\, f\big(-\tfrac{\sqrt{2}}{2}\big) + \tfrac{4}{5}\, f(0) + \tfrac{8}{15}\, f\big(\tfrac{\sqrt{2}}{2}\big) + \tfrac{1}{15}\, f(1)\,. \end{aligned}$$

Note that $Q_1(f)$ coincides with the trapezoidal rule and that $Q_2(f)$ is equal to Simpson's rule. □

Assume that $N \in \mathbb{N}$ is given. Using the explicit coefficients of p_N in (6.65) and changing the order of summations, the quadrature formula (6.77) possesses the form

$$Q_N(f) = \sum_{k=0}^{N} \varepsilon_N(k)^2\, q_k^{(N)}\, f(x_k^{(N)})$$

with the *quadrature weights*

$$q_k^{(N)} := \begin{cases} \frac{2}{N} \sum\limits_{j=0}^{N/2} \varepsilon_N(2j)^2\, \frac{2}{1-4j^2}\, \cos\Big(\frac{2jk\pi}{N}\Big) & N \text{ even}, \\ \frac{2}{N} \sum\limits_{j=0}^{(N-1)/2} \varepsilon_N(2j)^2\, \frac{2}{1-4j^2}\, \cos\Big(\frac{2jk\pi}{N}\Big) & N \text{ odd}, \end{cases} \tag{6.78}$$

for $k = 0, \ldots, N$.

Theorem 6.51 *Let $N \in \mathbb{N}$ be given. All weights $q_k^{(N)}$, $k = 0, \ldots, N$, of the Clenshaw–Curtis quadrature are positive. In particular,*

$$q_0^{(N)} = q_N^{(N)} = \begin{cases} \frac{2}{N^2-1} & N \text{ even}\,, \\ \frac{2}{N^2} & N \text{ odd}\,, \end{cases}$$

and for $k = 1, \ldots, N-1$,

$$q_k^{(N)} = q_{N-k}^{(N)} \ge \begin{cases} \frac{2}{N^2-1} & N \text{ even}\,, \\ \frac{2}{N^2} & N \text{ odd}\,. \end{cases}$$

Further,

$$\sum_{k=0}^{N} \varepsilon_N(j)^2\, q_k^{(N)} = 2\,.$$

For each $f \in C(I)$ we have

$$\lim_{N\to\infty} Q_N(f) = I(f) = \int_{-1}^{1} f(x)\, \mathrm{d}x\,. \tag{6.79}$$

If $f \in C(I)$ is odd, then $I(f) = Q_N(f) = 0$.

Proof

1. Assume that N is even. Using (6.78) we will show the inequality

$$\frac{N}{2} q_k^{(N)} = \sum_{j=0}^{N/2} \varepsilon_N(2j)^2\, \frac{2}{1-4j^2}\, \cos\Big(\frac{2jk\pi}{N}\Big) \ge \frac{N}{N^2-1}\,, \quad k = 0, \ldots, N\,.$$

Since $|\cos(x)| \le 1$ for all $x \in \mathbb{R}$, it follows by the triangle inequality that

$$\begin{aligned} \sum_{j=0}^{N/2} \varepsilon_N(2j)^2\, \frac{2}{1-4j^2}\, \cos\Big(\frac{2jk\pi}{N}\Big) &\ge 1 - \Big(\sum_{j=1}^{N/2-1} \frac{2}{4j^2-1} + \frac{1}{N^2-1}\Big) \\ &= 1 + \sum_{j=1}^{N/2-1} \Big(\frac{1}{2j+1} - \frac{1}{2j-1}\Big) - \frac{1}{N^2-1} \\ &= 1 + \Big(\frac{1}{N-1} - 1\Big) - \frac{1}{N^2-1} = \frac{N}{N^2-1}\,. \end{aligned}$$

For $k = 0$ and $k = N$ we have $\cos\big(\frac{2\pi jk}{N}\big) = 1$ for $j = 0, \ldots, \frac{N}{2}$, and therefore find the equality

$$q_0^{(N)} = q_N^{(N)} = \frac{2}{N}\Big(1 - \sum_{j=1}^{N/2-1} \frac{2}{4j^2-1} - \frac{1}{N^2-1}\Big)$$
$$= \frac{2}{N}\Big(1 + \frac{1}{N-1} - 1 - \frac{1}{N^2-1}\Big) = \frac{2}{N^2-1}\,.$$

For odd N the assertions can be shown analogously.

The Clenshaw–Curtis quadrature is exact for the constant function $f \equiv 1$, i.e.,

$$Q_N(1) = \sum_{k=0}^{N} \varepsilon_N(j)^2\, q_k^{(N)} = \int_{-1}^{1} 1\,\mathrm{d}x = 2\,.$$

2. Formula (6.79) follows from Theorem 1.24 of Banach–Steinhaus using the fact that

$$\lim_{N\to\infty} Q_N(p) = I(p) = \int_{-1}^{1} p(x)\,\mathrm{d}x$$

is satisfied for *each* polynomial p. ■

Employing Theorem 6.45, we obtain a useful estimate for the error of the Clenshaw–Curtis quadrature.

Theorem 6.52 *Let $N \in \mathbb{N}\setminus\{1\}$ and let $f \in C^{r+1}(I)$ with $r \in \mathbb{N}$ be given. Then for any $N > r+1$, the quadrature error of the Clenshaw–Curtis quadrature can be estimated by*

$$|I(f) - Q_N(f)| \le \frac{4}{r\,(N-r-1)^r}\, \|f^{(r+1)}\|_{C(I)}\,. \tag{6.80}$$

Proof

1. First, we express $f \in C^{r+1}(I)$ in the form $f = f_0 + f_1$ with

$$f_0(x) := \frac{1}{2}\big(f(x) + f(-x)\big)\,, \quad f_1(x) := \frac{1}{2}\big(f(x) - f(-x)\big)\,, \quad x \in I\,.$$

For the odd function f_1, we see that $I(f_1) = Q_N(f_1) = 0$ and hence

$$I(f) = I(f_0)\,, \quad Q_N(f) = Q_N(f_0)\,.$$

Therefore, we can replace f by its even part $f_0 \in C^{r+1}(I)$, where

$$\|f_0^{(r+1)}\|_{C(I)} \le \|f^{(r+1)}\|_{C(I)}\,.$$

2. Let $p_N \in \mathscr{P}_N$ denote the interpolation polynomial of f_0 at the Chebyshev extreme points $x_j^{(N)}$, $j = 0, \ldots, N$. Using Theorem 6.45 we estimate

$$|I(f_0) - Q_N(f_0)| = \Big| \int_{-1}^{1} \big(f_0(x) - p_N(x)\big)\,\mathrm{d}x \Big| \le 2\|f_0 - p_N\|_{C(I)}$$

$$\le 4 \sum_{\ell=\lfloor N/2\rfloor+1}^{\infty} |a_{2\ell}[f_0]|\,,$$

since $a_k[f_0] = 0$ for all odd $k \in \mathbb{N}$. By (6.20) we know for all $2\ell > r$ that

$$|a_{2\ell}[f_0]| \le \tfrac{2}{(2\ell-r)^{r+1}}\, \|f_0^{(r+1)}\|_{C(I)}\,.$$

Therefore, we obtain

$$|I(f_0) - Q_N(f_0)| \le 8\, \|f_0^{(r+1)}\|_{C(I)} \sum_{\ell=\lfloor N/2\rfloor+1}^{\infty} \tfrac{1}{(2\ell-r)^{r+1}}$$

$$\le 8\, \|f_0^{(r+1)}\|_{C(I)} \int_{\lfloor N/2\rfloor}^{\infty} \tfrac{\mathrm{d}x}{(2x-r)^{r+1}} \le \tfrac{4}{r\,(N-r-1)^r}\, \|f_0^{(r+1)}\|_{C(I)}\,.$$

■

Remark 6.53 The inequality (6.80) is not sharp. For better error estimates, we refer to [414, 443]. It is very remarkable that the Clenshaw–Curtis quadrature gives results nearly as accurate as the Gauss quadrature for most integrands (see [414] and [415, Chapter 19]). □

For the numerical realization of $Q_N[f]$, we suggest to use (6.77). By

$$a_{2k}^{(N)}[f] = \tfrac{2}{N} \sum_{j=0}^{N} \varepsilon_N(j)^2\, f(x_j^{(N)}) \cos\big(\tfrac{2jk\pi}{N}\big)$$

$$= \tfrac{2}{N} \sum_{j=0}^{N/2} \varepsilon_{N/2}(j)^2 \big(f(x_j^{(N)}) + f(x_{N-j}^{(N)})\big) \cos\big(\tfrac{2jk\pi}{N}\big)$$

we can calculate $a_{2k}^{(N)}[f]$ by means of the DCT-I $(N/2+1)$,

$$\big(\varepsilon_{N/2}(k)\, a_{2k}^{(N)}[f]\big)_{k=0}^{N/2} = \tfrac{1}{\sqrt{N}}\, \mathbf{C}_{N/2+1}^{\mathrm{I}}\, \big(\varepsilon_{N/2}(j)\, f_j\big)_{j=0}^{N/2}\,, \tag{6.81}$$

where we set $f_j := f(x_j^{(N)}) + f(x_{N-j}^{(N)})$, $j = 0, \ldots, N/2$. Thus, we obtain the following efficient algorithm for numerical integration.

Algorithm 6.54 (Clenshaw–Curtis Quadrature)
Input: $t \in \mathbb{N}\setminus\{1\}$, $N := 2^t$, $f(x_j^{(N)}) \in \mathbb{R}$, $j = 0, \ldots, N$, *for given* $f \in C(I)$, *where* $x_j^{(N)} := \cos\big(\frac{j\pi}{N}\big)$.

1. *For* $j = 0, \ldots, N/2$ *form*

$$\varepsilon_{N/2}(j)\, f_j := \varepsilon_{N/2}(j)\, \big(f(x_j^{(N)}) + f(x_{N-j}^{(N)})\big)\,.$$

2. *Compute* (6.81) *by Algorithm* 6.28 *or* 6.35.
3. *Compute*

$$Q_N(f) := \sum_{k=0}^{N/2} \varepsilon_{N/2}(k)^2\, \frac{2}{1-4\,k^2}\, a_{2k}^{(N)}[f]\,.$$

Output: $Q_N(f) \in \mathbb{R}$ *approximate value of the integral* $I(f)$.
Computational cost: $\mathscr{O}(N \log N)$.

Example 6.55 The rational function $f(x) := (x^4+x^2+\frac{9}{10})^{-1}$, $x \in I$, possesses the exact integral value $I(f) = 1.582233\ldots$. Algorithm 6.54 provides the following approximate integral values for $N = 2^t$, $t = 2, \ldots, 6$:

N	$Q_N(f)$
4	1.548821
8	1.582355
16	1.582233
32	1.582233
64	1.582233

□

Example 6.56 We consider the needle-shaped function $f(x) := (10^{-4} + x^2)^{-1}$, $x \in I$. The integral of f over I has the exact value

$$\int_{-1}^{1} \frac{\mathrm{d}x}{10^{-4}+x^2} = 200 \arctan 100 = 312.159332\ldots\,.$$

For $N = 2^t$, $t = 7, \ldots, 12$, Algorithm 6.54 provides the following results:

N	$Q_N(f)$
128	364.781238
256	315.935656
512	314.572364
1024	312.173620
2048	312.159332
4096	312.159332

The convergence of this quadrature formula can be improved by a simple trick. Since f is even, we obtain by substitution $x = \frac{t-1}{2}$

$$\int_{-1}^{1} \frac{\mathrm{d}x}{10^{-4}+x^2} = 2 \int_{-1}^{0} \frac{\mathrm{d}x}{10^{-4}+x^2} = 4 \int_{-1}^{1} \frac{\mathrm{d}t}{4 \cdot 10^{-4}+(t-1)^2} \,.$$

such that the function $g(t) = 4\,(4 \cdot 10^{-4} + (t-1)^2)^{-1}$, $t \in I$, possesses a needle at the endpoint $t = 1$. Since the Chebyshev extreme points are clustered near the endpoints of I, we obtain much faster convergence of the quadrature rule.

N	$Q_N(g)$
8	217.014988
16	312.154705
32	312.084832
64	312.159554
128	312.159332
256	312.159332

□

Summarizing we can say:
The Clenshaw–Curtis quadrature formula $Q_N(f)$ for $f \in C(I)$ is an interpolatory quadrature rule with explicitly given nodes $x_j^{(N)} = \cos\big(\frac{j\pi}{N}\big)$, $j = 0, \ldots, N$, and positive weights. For even N, the terms $a_{2j}^{(N)}[f]$, $j = 0, \ldots, \frac{N}{2}$, in $Q_N(f)$ can be efficiently and stably computed by a fast algorithm of DCT-I $(\frac{N}{2}+1)$. *Each polynomial $p \in \mathscr{P}_N$ is exactly integrated over I. For sufficiently smooth functions, the Clenshaw–Curtis quadrature gives similarly accurate results as the Gauss quadrature.*

Remark 6.57 The popular Clenshaw–Curtis quadrature for nonoscillatory integrals can be generalized to highly oscillatory integrals, i.e., integrals of highly oscillating integrands, which occur in fluid dynamics, acoustic, and electromagnetic scattering. The excellent book [107] presents efficient algorithms for computing *highly oscillatory integrals* such as

$$I_\omega[f] := \int_{-1}^{1} f(x)\, \mathrm{e}^{\mathrm{i}\omega x}\, \mathrm{d}x \,,$$

where $f \in C^s(I)$ is sufficiently smooth and $\omega \gg 1$. Efficient quadrature methods for highly oscillatory integrals use the asymptotic behavior of $I_\omega[f]$ for large ω. In the *Filon–Clenshaw–Curtis quadrature*, one interpolates f by a polynomial

$$p(x) := \sum_{j=0}^{2s+N} p_j\, T_j(x)$$

such that

$$p^{(\ell)}(-1) = f^{(\ell)}(-1)\,, \quad p^{(\ell)}(1) = f^{(\ell)}(1)\,, \quad \ell = 0, \ldots, s\,,$$
$$p\big(x_k^{(N)}\big) = f\big(x_k^{(N)}\big)\,, \quad k = 1, \ldots, N-1\,,$$

where the coefficients p_j can be calculated by DCT-I (for details see [107, pp. 62–66]). Then we determine

$$I_\omega[p] = \sum_{j=0}^{2s+N} p_j\, b_j(\omega)$$

with

$$b_j(\omega) := \int_{-1}^{1} T_j(x)\, \mathrm{e}^{\mathrm{i}\omega x}\, \mathrm{d}x\,, \quad j = 0, \ldots, 2s+N\,,$$

which can be explicitly computed,

$$b_0(\omega) = \frac{2\,\sin(\omega)}{\omega}\,, \quad b_1(\omega) = -\frac{2\mathrm{i}\,\cos(\omega)}{\omega} + \frac{2\mathrm{i}\,\sin(\omega)}{\omega^2}\,,$$
$$b_2(\omega) = \frac{8\,\cos(\omega)}{\omega^2} + \Big(\frac{2}{\omega} - \frac{8}{\omega^3}\Big)\,\sin(\omega)\,, \;\ldots\,.$$

□

6.5 Discrete Polynomial Transforms

We show that orthogonal polynomials satisfy a three-term recursion formula. Furthermore, we derive a fast algorithm to evaluate an arbitrary linear combination of orthogonal polynomials on a nonuniform grid of Chebyshev extreme points.

6.5.1 Orthogonal Polynomials

Let ω be a nonnegative, integrable weight function defined almost everywhere on $I := [-1,\,1]$. Let $L_{2,\omega}(I)$ denote the real weighted Hilbert space with the inner product

$$\langle f,\, g\rangle_{L_{2,\omega}(I)} := \int_{-1}^{1} \omega(x)\, f(x)\, g(x)\, \mathrm{d}x\,, \quad f,\, g \in L_{2,\omega}(I)\,, \tag{6.82}$$

and the related norm

$$\|f\|_{L_{2,\omega}(I)} := \sqrt{\langle f,\, f\rangle_{L_{2,\omega}(I)}}\,.$$

A sequence $(P_n)_{n=0}^{\infty}$ of real polynomials $P_n \in \mathscr{P}_n$, $n \in \mathbb{N}_0$, is called a *sequence of orthogonal polynomials* with respect to (6.82), if $\langle P_m, P_n\rangle_{L_{2,\omega}(I)} = 0$ for all distinct $m, n \in \mathbb{N}_0$ and if each polynomial P_n possesses exactly the degree $n \in \mathbb{N}_0$. If $(P_n)_{n=0}^{\infty}$ is a sequence of orthogonal polynomials, then $(c_n P_n)_{n=0}^{\infty}$ with arbitrary $c_n \in \mathbb{R}\setminus\{0\}$ is also a sequence of orthogonal polynomials. Obviously, the orthogonal polynomials P_k, $k = 0, \ldots, N$, form an orthogonal basis of $\mathscr{P}_N$ with respect to (6.82).

A *sequence of orthonormal polynomials* is a sequence $(P_n)_{n=0}^{\infty}$ of orthogonal polynomials with the property $\|P_n\|_{L_{2,\omega}(I)} = 1$ for each $n \in \mathbb{N}_0$. Starting from the sequence of monomials $M_n(x) := x^n$, $n \in \mathbb{N}_0$, a sequence of orthonormal polynomials P_n can be constructed by the known *Gram–Schmidt orthogonalization procedure*, i.e., one forms $P_0 := M_0/\|M_0\|_{L_{2,\omega}(I)}$ and then recursively for $n = 1, 2, \ldots$

$$\tilde{P}_n := M_n - \sum_{k=0}^{n-1} \langle M_n, P_k\rangle_{L_{2,\omega}(I)} P_k, \quad P_n := \frac{1}{\|\tilde{P}_n\|_{L_{2,\omega}(I)}} \tilde{P}_n .$$

For the theory of orthogonal polynomials, we refer to the books [84, 158, 406].

Example 6.58 For the weight function $\omega(x) := (1-x)^{\alpha}(1+x)^{\beta}$, $x \in (-1, 1)$, with $\alpha > -1$ and $\beta > -1$, the related orthogonal polynomials are called *Jacobi polynomial*. For $\alpha = \beta = 0$ we obtain the *Legendre polynomials*. The case $\alpha = \beta = -\frac{1}{2}$ leads to the *Chebyshev polynomials of first kind* and $\alpha = \beta = \frac{1}{2}$ to the *Chebyshev polynomials of second kind* up to a constant factor. For $\alpha = \beta = \lambda - \frac{1}{2}$ with $\lambda > -\frac{1}{2}$ we receive the *ultraspherical polynomials* which are also called *Gegenbauer polynomials*. □

For efficient computation with orthogonal polynomials, it is essential that orthogonal polynomials can be recursively calculated:

Lemma 6.59 *If $(P_n)_{n=0}^{\infty}$ is a sequence of orthogonal polynomials $P_n \in \mathscr{P}_n$, then the polynomials P_n satisfy a* three-term recurrence relation

$$P_n(x) = (\alpha_n x + \beta_n) P_{n-1}(x) + \gamma_n P_{n-2}(x), \quad n \in \mathbb{N}, \tag{6.83}$$

with $P_{-1}(x) := 0$ and $P_0(x) := 1$, where α_n, β_n, and γ_n are real coefficients with $\alpha_n \neq 0$ and $\gamma_n \neq 0$.

Proof Clearly, formula (6.83) holds for $n = 1$ and $n = 2$. We consider $n \geq 3$. If c_n and c_{n-1} are the leading coefficients of P_n and P_{n-1}, respectively, then we set $\alpha_n := \frac{c_n}{c_{n-1}} \neq 0$. Thus, $q(x) := P_n(x) - \alpha_n x P_{n-1}(x) \in \mathscr{P}_{n-1}$ can be expressed as

$$q = d_0 P_0 + \ldots + d_{n-1} P_{n-1} .$$

For $k = 0, \ldots, n-3$ this provides $d_k = 0$ by the orthogonality, since $x P_k(x) \in \mathscr{P}_{k+1}$ can be written as a linear combination of $P_0, \ldots, P_{k+1}$ and therefore

$$\langle P_k,\, q\rangle_{L_{2,\omega}(I)}=\langle P_k,\, P_n\rangle_{L_{2,\omega}(I)} - \alpha_n\, \langle x\, P_k(x),\, P_{n-1}(x)\rangle_{L_{2,\omega}(I)}=0=d_k\, \|P_k\|^2_{L_{2,\omega}(I)}\,.$$

Thus, with $\beta_n := d_{n-1}$ and $\gamma_n := d_{n-2}$, we find

$$P_n(x) = \alpha_n\, x\, P_{n-1}(x) + \beta_n\, P_{n-1}(x) + \gamma_n\, P_{n-2}(x)\,.$$

The coefficient γ_n does not vanish, since the orthogonality implies

$$0 = \langle P_n,\, P_{n-2}\rangle_{L_{2,\omega}(I)} = \frac{\alpha_n}{\alpha_{n-1}}\, \|P_{n-1}\|^2_{L_{2,\omega}(I)} + \gamma_n\, \|P_{n-2}\|^2_{L_{2,\omega}(I)}\,.$$

■

Example 6.60 The Legendre polynomials L_n normalized by $L_n(1) = 1$ satisfy the three-term recurrence relation

$$L_{n+2}(x) = \frac{2n+3}{n+2}\, x\, L_{n+1}(x) - \frac{n+1}{n+2}\, L_n(x)$$

for $n \in \mathbb{N}_0$ with $L_0(x) := 1$ and $L_1(x) := x$ (see [406, p. 81]).

The Chebyshev polynomials T_n normalized by $T_n(1) = 1$ satisfy the three-term relation

$$T_{n+2}(x) = 2x\, T_{n+1}(x) - T_n(x)$$

for $n \in \mathbb{N}_0$ with $T_0(x) := 1$ and $T_1(x) := 1$. □

Let now $(P_n)_{n=0}^{\infty}$ be a sequence of orthogonal polynomials $P_n \in \mathscr{P}_n$ with respect to (6.82). Then, every $p \in \mathscr{P}_N$ can be represented as

$$p = \sum_{k=0}^{N} \frac{\langle p,\, P_k\rangle_{L_{2,\omega}(I)}}{\langle P_k,\, P_k\rangle_{L_{2,\omega}(I)}}\, P_k\,.$$

The inner product $\langle p,\, P_k\rangle_{L_{2,\omega}(I)}$ can be exactly computed by a suitable interpolatory quadrature rule of the form

$$\langle p,\, P_k\rangle_{L_{2,\omega}(I)} = \int_{-1}^{1} \omega(x)\, p(x)\, P_k(x)\, \mathrm{d}x = \sum_{j=0}^{2N} w_j^{(2N)}\, p(x_j^{(2N)})\, P_k(x_j^{(2N)}) \tag{6.84}$$

for $k = 0, \ldots, N$, where we again employ the Chebyshev extreme points $x_j^{(2N)} = \cos\frac{j\pi}{2N}$ and where the quadrature weights $w_j^{(2N)}$ are obtained using the integrals of the Lagrange basis functions $\ell_j^{(2N)}$ related to the points $x_j^{(2N)}$, i.e.,

$$w_j^{(2N)} := \int_{-1}^{1} \omega(x)\,\ell_j^{(2N)}(x)\,\mathrm{d}x\,, \quad \ell_j^{(2N)}(x) = \prod_{\substack{k=0\\k\neq j}}^{2N} \frac{x - x_k^{(2N)}}{x_j^{(2N)} - x_k^{(2N)}}\,.$$

For the special case $\omega \equiv 1$ in (6.82), the quadrature rule in (6.84) coincides with the Clenshaw–Curtis quadrature rule of order $2N$ in Sect. 6.4, where the interpolation polynomial at the knots $x_j^{(2N)}$, $j = 0, \ldots, 2N$, has been applied. In that special case, the weights $w_j^{(2N)}$ are of the form

$$w_j^{(2N)} = \frac{\epsilon_{2N}(j)^2}{N} \sum_{\ell=0}^{N} \epsilon_{2N}(2\ell)^2\, \frac{2}{1-4\ell^2}\, \cos\frac{2\ell j\pi}{2N}\,,$$

(see (6.78)), with $\epsilon_{2N}(0) = \epsilon_{2N}(2N) = \frac{1}{\sqrt{2}}$ and $\epsilon_{2N}(j) = 1$ for $j = 1, \ldots, 2N-1$. For other weights $\omega(x)$, the expressions for $w_j^{(2N)}$ may look more complicated, but we will still be able to compute them by a fast DCT-I algorithm.

6.5.2 *Fast Evaluation of Orthogonal Expansions*

Let now $M, N \in \mathbb{N}$ with $M \geq N$ be given powers of two. In this section we are interested in efficient solutions of the following two problems (see [338]).
Problem 1: For given $a_k \in \mathbb{R}$, $k = 0, \ldots, N$, compute the *discrete polynomial transform* DPT $(N+1,\ M+1) : \mathbb{R}^{N+1} \to \mathbb{R}^{M+1}$ defined by

$$\hat{a}_j := \sum_{k=0}^{N} a_k\, P_k(x_j^{(M)})\,, \quad j = 0, \ldots, M\,, \tag{6.85}$$

where $x_j^{(M)} = \cos\frac{j\pi}{M}$, $j = 0, \ldots, M$. The corresponding transform matrix

$$\mathbf{P} := \big(P_k(x_j^{(M)})\big)_{j,k=0}^{M,\,N} \in \mathbb{R}^{(M+1)\times(N+1)}$$

is called *Vandermonde-like matrix*. This first problem addresses the evaluation of an arbitrary polynomial

$$p := \sum_{k=0}^{N} a_k\, P_k \in \mathscr{P}_N$$

on the nonuniform grid of Chebyshev extreme points $x_j^{(M)} \in I$, $j = 0, \ldots, M$. The discrete polynomial transform can be considered as a generalization of DCT-I, since for $P_k = T_k$, $k = 0, \ldots, N$, the DPT $(M+1, N+1)$ reads

$$\hat{a}_j := \sum_{k=0}^{N} a_k\, T_k(x_j^{(M)}) = \sum_{k=0}^{N} a_k \cos\frac{jk\pi}{M}, \quad j = 0, \ldots, M.$$

Transposed Problem 2: For given $b_j \in \mathbb{R}$, $j = 0, \ldots, M$, compute the *transposed discrete polynomial transform* TDPT $(M+1, N+1) : \mathbb{R}^{M+1} \to \mathbb{R}^{N+1}$ defined by

$$\tilde{b}_k := \sum_{j=0}^{M} b_j\, P_k(x_j^{(M)}), \quad k = 0, \ldots, N. \tag{6.86}$$

This transposed problem is of similar form as (6.84) for $M = 2N$ and with $b_j = w_j^{(2N)}\, p(x_j^{(2N)})$. Therefore, it needs to be solved in order to compute the Fourier coefficients of the polynomial $p \in \mathscr{P}_N$ in the orthogonal basis $\{P_k : k = 0, \ldots, N\}$.

A direct realization of (6.85) or (6.86) by the Clenshaw algorithm (see Algorithm 6.19) would require computational cost of $\mathcal{O}(MN)$. We want to derive fast algorithms to solve these two problems with only $\mathcal{O}(N(\log_2 N)^2 + M \log_2 M)$ arithmetical operations.

We will start with considering the first problem. The main idea is as follows. First, we will derive a fast algorithm for the change of basis from the polynomial expansion in the basis $\{P_k : k = 0, \ldots, N\}$ to the basis $\{T_k : k = 0, \ldots, N\}$ of Chebyshev polynomials. Then we can employ a fast DCT-I algorithm to evaluate

$$p = \sum_{k=0}^{N} a_k\, T_k, \quad a_k \in \mathbb{R}. \tag{6.87}$$

at the Chebyshev extreme points $x_j^{(M)} = \cos\frac{j\pi}{M}$, $j = 0, \ldots, M$. The values $p(x_j^{(M)})$, $j = 0, \ldots, M$, can be efficiently computed using Algorithm 6.22 involving a DCT-I $(M+1)$, where we only need to pay attention because of the slightly different normalization in (6.87) compared to (6.24). Here, we obtain

$$\big(\varepsilon_M(j)\, p(x_j^{(M)})\big)_{j=0}^{M} = \sqrt{\frac{M}{2}}\, \mathbf{C}_{M+1}^{\mathrm{I}}\, \big(\delta_M(k)\, a_k\big)_{k=0}^{M}, \tag{6.88}$$

where we set $a_k := 0$ for $k = N+1, \ldots, M$ and $\delta_M(0) := \sqrt{2}$, $\delta_M(k) := 1$ for $k = 1, \ldots, M$.

Let us now consider the problem to change the basis of a polynomial from $\{P_k : k = 0, \ldots, N\}$ to the basis $\{T_k : k = 0, \ldots, N\}$ in an efficient way. For that purpose we present in a first step a further algorithm for fast polynomial multiplication that is slightly different from the algorithm given in Theorem 6.27. Let $p \in \mathscr{P}_n$ be given in the form

$$p = \sum_{k=0}^{n} a_k\, T_k\,, \quad a_k \in \mathbb{R}\,. \tag{6.89}$$

Further, let $q \in \mathscr{P}_m$ with $m \in \mathbb{N}$ be a fixed polynomial with known polynomial values $q(x_j^{(M)})$, $j = 0, \ldots, M$, where $M = 2^s$, $s \in \mathbb{N}$, with $M/2 \le m + n < M$ is chosen. Then the Chebyshev coefficients $b_k \in \mathbb{R}$, $k = 0, \ldots, m+n$, in

$$r := p\, q = \sum_{k=0}^{n+m} b_k\, T_k$$

can be computed in a fast way by the following procedure (see [24]).

Algorithm 6.61 (Fast Polynomial Multiplication in Chebyshev Polynomial Basis)
Input: $m,\ n \in \mathbb{N}$, $M = 2^s$, $s \in \mathbb{N}$, *with* $M/2 \le m + n < M$,
polynomial values $q(x_j^{(M)}) \in \mathbb{R}$, $j = 0, \ldots, M$, *of* $q \in \mathscr{P}_m$,
Chebyshev coefficients $a_k \in \mathbb{R}$, $k = 0, \ldots, n$, *of* $p \in \mathscr{P}_n$,
$\varepsilon_M(0) = \varepsilon_M(M) := \frac{\sqrt{2}}{2}$, $\varepsilon_M(k) := 1$, $k = 1, \ldots, M-1$,
$\delta_M(0) = \delta_M(M) := \sqrt{2}$, $\delta_M(k) := 1$, $k = 1, \ldots, M-1$.

1. *Set* $a_k := 0$, $k = n+1, \ldots, M$, *and compute the values* $\varepsilon_M(j)\, p(x_j^{(M)})$, $j = 0, \ldots, M$, *by* (6.88) *using a* DCT-I $(M+1)$, *as in Algorithm* 6.28 *or Algorithm* 6.35.
2. *Evaluate the* $M+1$ *products*

$$\varepsilon_M(j)\, r(x_j^{(M)}) := \big(\varepsilon_M(j)\, p(x_j^{(M)})\big)\, q(x_j^{(M)})\,, \quad j = 0, \ldots, M\,.$$

3. *Compute*

$$(\tilde{b}_k)_{k=0}^{M-1} := \sqrt{\frac{2}{M}}\, \mathbf{C}_{M+1}^{\mathrm{I}}\, \big(\varepsilon_M(j)\, r(x_j^{(M)})\big)_{j=0}^{M}$$

by a fast algorithm of DCT-I $(M+1)$ *using Algorithm* 6.28 *or Algorithm* 6.35 *and form* $b_k := \delta_M(k)^{-1}\, \tilde{b}_k$, $k = 0, \ldots, m+n$.

Output: $b_k \in \mathbb{R}$, $k = 0, \ldots, m+n$, *Chebyshev coefficients of the product* $p\, q \in \mathscr{P}_{m+n}$.
Computational cost: $\mathscr{O}(M \log M)$.

By Theorem 6.39, the fast DCT-I (2^s+1) Algorithm 6.35 requires $2^s s - \frac{4}{3}\, 2^s + \frac{5}{2} - \frac{1}{6}(-1)^s$ multiplications and $\frac{4}{3}\, 2^s s - \frac{14}{9}\, 2^s + \frac{1}{2}\, s + \frac{7}{2} + \frac{1}{18}(-1)^s$ additions. Hence, Algorithm 6.61 realizes the multiplication of the polynomials $p \in \mathscr{P}_n$ and $q \in \mathscr{P}_m$ in the Chebyshev polynomial basis by less than $2\, M\, \log M$ multiplications and $\frac{8}{3}\, M \log M$ additions.

For the change of basis from $\{P_k : k = 0, \ldots, N\}$ to $\{T_k : k = 0, \ldots, N\}$ we want to employ a divide-and-conquer technique, where we will use so-called associated orthogonal polynomials.

Assume that the sequence $(P_n)_{n=0}^{\infty}$ of orthogonal polynomials satisfies the three-term recurrence relation (6.83). Replacing the coefficient index $n \in \mathbb{N}_0$ in (6.83) by $n + c$ with $c \in \mathbb{N}_0$, we obtain the so-called associated orthogonal polynomials $P_n(\cdot\,, c) \in \mathscr{P}_n$ defined recursively by

$$P_n(x, c) := (\alpha_{n+c}\, x + \beta_{n+c})\, P_{n-1}(x, c) + \gamma_{n+c}\, P_{n-2}(x, c)\,, \quad n \in \mathbb{N}\,, \tag{6.90}$$

with $P_{-1}(x, c) := 0$ and $P_0(x, c) := 1$. By induction one can show the following result (see [39]).

Lemma 6.62 *For all $c, n \in \mathbb{N}_0$,*

$$P_{c+n} = P_n(\cdot, c)\, P_c + \gamma_{c+1}\, P_{n-1}(\cdot, c+1)\, P_{c-1}\,. \tag{6.91}$$

Proof For $n = 0$ and $n = 1$, Eq. (6.91) is true for all $c \in \mathbb{N}_0$. Assume that (6.91) holds up to fixed $n \in \mathbb{N}$ for all $c \in \mathbb{N}_0$. We employ an induction argument. Using (6.83) and (6.90), we obtain

$$\begin{aligned}
P_{c+n+1}(x) &= (\alpha_{c+n+1}\, x + \beta_{c+n+1})\, P_{c+n}(x) + \gamma_{c+n+1}\, P_{c+n-1}(x) \\
&= (\alpha_{c+n+1}\, x + \beta_{c+n+1})\, \big(P_n(x, c)\, P_c(x) + \gamma_{c+1}\, P_{n-1}(x, c+1)\, P_{c-1}(x)\big) \\
&\quad + \gamma_{c+n+1}\, \big(P_{n-1}(x, c)\, P_c(x) + \gamma_{c+1}\, P_{n-2}(x, c+1)\, P_{c-1}(x)\big) \\
&= \big((\alpha_{c+n+1}\, x + \beta_{c+n+1})\, P_n(x, c) + \gamma_{c+n+1}\, P_{n-1}(x, c)\big)\, P_c(x) \\
&\quad + \big((\alpha_{c+n+1}\, x + \beta_{c+n+1})\, P_{n-1}(x, c+1) + \gamma_{c+n+1}\, P_{n-2}(x, c+1)\big)\, \gamma_{c+1}\, P_{c-1}(x) \\
&= P_{n+1}(x, c)\, P_c(x) + \gamma_{c+1}\, P_n(x, c+1)\, P_{c-1}(x)\,.
\end{aligned}$$

■

Lemma 6.62 implies

$$\begin{pmatrix} P_{c+n} \\ P_{c+n+1} \end{pmatrix} = \mathbf{U}_n(\cdot, c)^{\top} \begin{pmatrix} P_{c-1} \\ P_c \end{pmatrix} \tag{6.92}$$

with

$$\mathbf{U}_n(\cdot, c) := \begin{pmatrix} \gamma_{c+1} P_{n-1}(\cdot, c+1) & \gamma_{c+1} P_n(\cdot, c+1) \\ P_n(\cdot, c) & P_{n+1}(\cdot, c) \end{pmatrix}.$$

This polynomial matrix $\mathbf{U}_n(\cdot, c)$ contains polynomial entries of degree $n-1$, n, and $n+1$, respectively.

Now we describe the exchange between the bases $\{P_k : k = 0, \ldots, N\}$ and $\{T_k : k = 0, \ldots, N\}$ of $\mathscr{P}_N$, where $N = 2^t$, $t \in \mathbb{N}$. Assume that $p \in \mathscr{P}_N$ is given in the orthogonal basis $\{P_k : k = 0, \ldots, N\}$ by

$$p = \sum_{k=0}^{N} a_k \, P_k \tag{6.93}$$

with real coefficients a_k. Our goal is the fast evaluation of the related Chebyshev coefficients $\tilde{a}_k$ in the representation

$$p = \sum_{k=0}^{N} \tilde{a}_k \, T_k . \tag{6.94}$$

In an initial step, we use (6.83) and the fact that $T_1(x) = x$ to obtain

$$\begin{aligned} p(x) &= \sum_{k=0}^{N-1} a_k \, P_k(x) + a_N \left((\alpha_N x + \beta_N) P_{N-1}(x) + \gamma_N P_{N-2}(x) \right) \\ &= \sum_{k=0}^{N-1} a_k^{(0)}(x) \, P_k(x) \end{aligned}$$

with

$$a_k^{(0)}(x) := \begin{cases} a_k & k = 0, \ldots, N-3\,, \\ a_{N-2} + \gamma_N \, a_N & k = N-2\,, \\ a_{N-1} + \beta_N \, a_N \; + \alpha_N \, a_N \, T_1(x) & k = N-1\,, \end{cases} \tag{6.95}$$

where $a_{N-1}^{(0)}$ is a linear polynomial while $a_k^{(0)}$ are constants for $k = 0, \ldots, N-2$. Now, we obtain

$$\begin{aligned} p &= \sum_{k=0}^{N-1} a_k^{(0)} \, P_k = \sum_{k=0}^{N/4-1} \Big(\sum_{\ell=0}^{3} a_{4k+\ell}^{(0)} \, P_{4k+\ell} \Big) \\ &= \sum_{k=0}^{N/4-1} (a_{4k}^{(0)}, a_{4k+1}^{(0)}) \begin{pmatrix} P_{4k} \\ P_{4k+1} \end{pmatrix} + (a_{4k+2}^{(0)}, a_{4k+3}^{(0)}) \begin{pmatrix} P_{4k+2} \\ P_{4k+3} \end{pmatrix} \\ &= \sum_{k=0}^{N/4-1} \Big((a_{4k}^{(0)}, a_{4k+1}^{(0)}) + (a_{4k+2}^{(0)}, a_{4k+3}^{(0)}) \, \mathbf{U}_1(\cdot, 4k+1)^\top \Big) \begin{pmatrix} P_{4k} \\ P_{4k+1} \end{pmatrix} , \end{aligned}$$

where we have used (6.92) with $n = 1$ and $c = 4k + 1$ for $k = 0, \ldots, N/4 - 1$. This yields

$$p = \sum_{k=0}^{N/4-1} (a_{4k}^{(1)} P_{4k} + a_{4k+1}^{(1)} P_{4k+1})$$

with

$$\begin{pmatrix} a_{4k}^{(1)} \\ a_{4k+1}^{(1)} \end{pmatrix} := \begin{pmatrix} a_{4k}^{(0)} \\ a_{4k+1}^{(0)} \end{pmatrix} + \mathbf{U}_1(\cdot, 4k+1) \begin{pmatrix} a_{4k+2}^{(0)} \\ a_{4k+3}^{(0)} \end{pmatrix} . \tag{6.96}$$

Observe that the degree of polynomials in $\mathbf{U}_1(\cdot, 4k+1)$ is at most 2, and therefore the degree of the polynomials $a_{4k}^{(1)}$ and $a_{4k+1}^{(1)}$ in (6.96) is at most 3. The computation of the polynomial coefficients of $a_{4k}^{(1)}, a_{4k+1}^{(1)} \in \mathscr{P}_3$, $k = 0, \ldots, \frac{N}{4} - 1$ can be realized by employing Algorithm 6.61 with $M = 4$. However, for these polynomials of low degree, we can also compute the Chebyshev coefficients directly with $4N + 5$ multiplications and $\frac{5}{2}N + 3$ additions.

We apply the cascade summation above now iteratively (see Fig. 6.10). In the next step, we compute

$$\begin{aligned} p &= \sum_{k=0}^{N/4-1} \big(a_{4k}^{(1)} P_{4k} + a_{4k+1}^{(1)} P_{4k+1}\big) \\ &= \sum_{k=0}^{N/8-1} (a_{8k}^{(1)}, a_{8k+1}^{(1)}) \begin{pmatrix} P_{8k} \\ P_{8k+1} \end{pmatrix} + (a_{8k+4}^{(1)}, a_{8k+5}^{(1)}) \begin{pmatrix} P_{8k+4} \\ P_{8k+5} \end{pmatrix} \\ &= \sum_{k=0}^{N/8-1} \Big((a_{8k}^{(1)}, a_{8k+1}^{(1)}) + (a_{8k+4}^{(1)}, a_{8k+5}^{(1)})\, \mathbf{U}_3(\cdot, 8k+1)^\top\Big) \begin{pmatrix} P_{8k} \\ P_{8k+1} \end{pmatrix}, \end{aligned}$$

implying

$$p = \sum_{k=0}^{N/8-1} (a_{8k}^{(2)} P_{8k} + a_{8k+1}^{(2)} P_{8k+1})$$

with

$$\begin{pmatrix} a_{8k}^{(2)} \\ a_{8k+1}^{(2)} \end{pmatrix} := \begin{pmatrix} a_{8k}^{(1)} \\ a_{8k+1}^{(1)} \end{pmatrix} + \mathbf{U}_3(\cdot, 8k+1) \begin{pmatrix} a_{8k+4}^{(1)} \\ a_{8k+5}^{(1)} \end{pmatrix} .$$

Generally, in step $\tau \in \{2, \ldots, t-1\}$, we compute by (6.92) with $n = 2^\tau - 1$ the Chebyshev coefficients of the polynomials $a_{2^{\tau+1}k}^{(\tau)}, a_{2^{\tau+1}k+1}^{(\tau)} \in \mathscr{P}_{2^{\tau+1}-1}$, $k = 0, \ldots, N/2^{\tau+1} - 1$, defined by

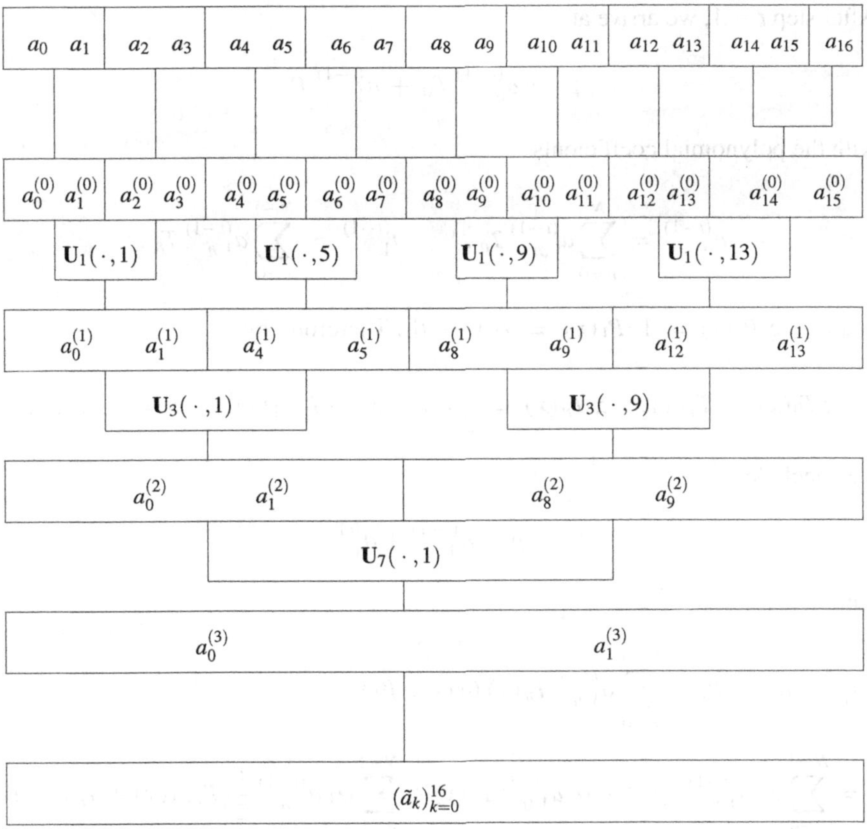

Fig. 6.10 Cascade summation for the computation of the coefficient vector $(\tilde a_k)_{k=0}^{16}$ in the case $N = 16$

$$\begin{pmatrix} a^{(\tau)}_{2^{\tau+1}k} \\ a^{(\tau)}_{2^{\tau+1}k+1} \end{pmatrix} := \begin{pmatrix} a^{(\tau-1)}_{2^{\tau+1}k} \\ a^{(\tau-1)}_{2^{\tau+1}k+1} \end{pmatrix} + \mathbf{U}_{2^\tau-1}(\cdot\,, 2^{\tau+1}k+1) \begin{pmatrix} a^{(\tau-1)}_{2^{\tau+1}k+2^\tau} \\ a^{(\tau-1)}_{2^{\tau+1}k+2^\tau+1} \end{pmatrix}, \tag{6.97}$$

where we calculate the Chebyshev coefficients of the polynomials by Algorithm 6.61 (with $M = 2^{\tau+1}$). Assume that the entries of $\mathbf{U}_{2^\tau-1}\big(x_\ell^{(2^{\tau+1})},\, 2^{\tau+1}k+1\big)$ for $k = 0, \ldots, N/2^{\tau+1}$ and $\ell = 0, \ldots, 2^{\tau+1}$ have been precomputed by Clenshaw Algorithm 6.19. Then step τ requires $4\,\frac{N}{2^{\tau+1}}$ applications of Algorithm 6.61. Therefore, we have computational costs of less than $8\,N(\tau+1)$ multiplications and $\frac{32}{3}\,N(\tau+1)+2N$ additions at step τ with the result

$$p = \sum_{k=0}^{N/2^{\tau+1}-1} \big(a^{(\tau)}_{2^{\tau+1}k} P_{2^{\tau+1}k} + a^{(\tau)}_{2^{\tau+1}k+1} P_{2^{\tau+1}k+1}\big)\,.$$

After step $t-1$, we arrive at

$$p = a_0^{(t-1)}\,P_0 + a_1^{(t-1)}\,P_1$$

with the polynomial coefficients

$$a_0^{(t-1)} = \sum_{n=0}^{N} a_{0,n}^{(t-1)}\,T_n\,, \qquad a_1^{(t-1)} = \sum_{n=0}^{N-1} a_{1,n}^{(t-1)}\,T_n\,,$$

and where $P_0(x) = 1$, $P_1(x) = \alpha_1\,x + \beta_1$. Therefore, by

$$x\,T_0(x) = T_1(x)\,, \quad x\,T_n(x) = \frac{1}{2}\big(T_{n+1}(x) + T_{n-1}(x)\big)\,, \quad n = 1, 2, \ldots\,.$$

we conclude

$$p = a_0^{(t-1)} + a_1^{(t)}$$

with

$$\begin{aligned}
a_1^{(t)} &= a_1^{(t-1)} P_1 = \sum_{n=0}^{N-1} a_{1,n}^{(t-1)} T_n(x)\,(\alpha_1 x + \beta_1) \\
&= \sum_{n=0}^{N-1} \beta_1\,a_{1,n}^{(t-1)} T_n(x) + \alpha_1 a_{1,0}^{(t-1)}\,T_1(x) + \sum_{n=1}^{N-1} \alpha_1\,a_{1,n}^{(t-1)}\,\frac{1}{2}\big(T_{n-1}(x) + T_{n+1}(x)\big) \\
&= \sum_{n=0}^{N} a_{1,n}^{(t)}\,T_n(x)\,.
\end{aligned}$$

For the coefficients we obtain

$$a_{1,n}^{(t)} := \begin{cases} \beta_1 a_{1,0}^{(t-1)} + \frac{1}{2}\alpha_1 a_{1,1}^{(t-1)} & n = 0\,, \\ \beta_1 a_{1,1}^{(t-1)} + \alpha_1 a_{1,0}^{(t-1)} + \frac{1}{2}\alpha_1 a_{1,2}^{(t-1)} & n = 1\,, \\ \beta_1 a_{1,n}^{(t-1)} + \frac{1}{2}\alpha_1 (a_{1,n-1}^{(t-1)} + a_{1,n+1}^{(t-1)}) & n = 2, \ldots, N-2\,, \\ \beta_1 a_{1,N-1}^{(t-1)} + \frac{1}{2}\alpha_1 a_{1,N-2}^{(t-1)} & n = N-1\,, \\ \frac{1}{2}\alpha_1 a_{1,N-1}^{(t-1)} & n = N\,. \end{cases} \tag{6.98}$$

The final addition of the Chebyshev coefficients of $a_0^{(t-1)}$ and $a_1^{(t)}$ yields the desired Chebyshev coefficients of p, i.e.,

$$\big(\tilde{a}_n\big)_{n=0}^{N} = \big(a_{0,n}^{(t-1)}\big)_{n=0}^{N} + \big(a_{1,n}^{(t)}\big)_{n=0}^{N}\,. \tag{6.99}$$

We summarize the discrete polynomial transform that computes the new coefficients of the Chebyshev expansion of a polynomial given in a different basis of orthogonal polynomials and solves problem 1 by evaluating the resulting Chebyshev expansion at the knots $x_j^{(M)}$.

Algorithm 6.63 (Fast Algorithm of DPT $(N+1, M+1)$**)**
Input: $N = 2^t$, $M = 2^s$ *with* $s, t \in \mathbb{N}$ *and* $s \geq t$,
$a_k \in \mathbb{R}$, $k = 0, \ldots, N$, *coefficients in* (6.93),
precomputed matrices $\mathbf{U}_{2^\tau-1}\big(x_\ell^{(2^{\tau+1})}, 2^{\tau+1}k+1\big)$ *for* $\tau = 1, \ldots, t-1$,
$k = 0, \ldots, 2^{t-\tau-1}$, *and* $\ell = 0, \ldots, 2^{\tau+1}$.

1. *Compute* $a_k^{(0)}$, $k = 0, \ldots, 2^t - 1$ *by* (6.95).
2. 2. *For* $\tau = 1, \ldots, t-1$ *do*
   Step τ. *For* $k = 0, \ldots, 2^{t-\tau-1} - 1$ *compute* (6.97) *by Algorithm* 6.61.
3. *Step* t. *Compute* $\tilde{a}_n$, $n = 0, \ldots, N$, *by* (6.98) *and* (6.99).
4. *Compute* (6.88) *by a* DCT-I $(M+1)$ *using Algorithm* 6.28 *or* 6.35.

Output: $p(x_j^{(M)})$, $j = 0, \ldots, M$.
Computational cost: $\mathcal{O}(N \log^2 N + M \log M)$.

In this algorithm, we have to store the precomputed elements of the matrices $\mathbf{U}$. Counting the arithmetic operations in each step, we verify that the complete basis exchange algorithm possesses computational costs of $\mathcal{O}(N \log^2 N)$.

Finally, using fast DCT-I $(M+1)$, the computation of (6.88) takes $M \log M$ multiplications and $\frac{4}{3} M \log M$ additions (see Theorem 6.39).

Remark 6.64 A fast algorithm for the transposed discrete polynomial transform TDPT $(N+1, M+1)$ in the transposed problem 2, i.e., the fast evaluation of

$$\hat{a}_k := \sum_{\ell=0}^{N} a_\ell \, P_k(\cos \frac{\pi \ell}{M}), \quad k = 0, \ldots, N,$$

can be obtained immediately by “reversing” Algorithm 6.63. In other words, we have simply to reverse the direction of the arrows in the flow graph of Algorithm 6.63. For the special case of spherical polynomials, this was done in [235]. See also the algorithm in [115] for the Legendre polynomials and the generalization to arbitrary polynomials satisfying a three-term recurrence relation in [116, 196]. The suggested algorithms are part of of the NFFT software (see [231, ../examples/fpt]). Furthermore, there exists a MATLAB interface (see [231, ../matlab/fpt]).

Fast algorithms based on semiseparable matrices can be found in [229, 230]. Fast algorithms based on asymptotic formulas for the Chebyshev–Legendre transform have been developed in [190, 210]. A method for the discrete polynomial transform based on diagonally scaled Hadamard products involving Toeplitz and Hankel matrices is proposed in [413]. □

Chapter 7
Fast Fourier Transforms for Nonequispaced Data

In this chapter, we describe fast algorithms for the computation of the DFT for d-variate nonequispaced data, since in a variety of applications the restriction to equispaced data is a serious drawback. These algorithms are called *nonequispaced fast Fourier transforms* (NFFT). In Sect. 7.1, we present a unified approach to the NFFT for nonequispaced data either in space or frequency domain. The NFFT is an approximate algorithm which is based on approximation of a d-variate trigonometric polynomial by a linear combination of translates of a 2π-periodic window function. For special window functions, we obtain the NFFT of A. Dutt and V. Rokhlin [121], G. Beylkin [45], and G. Steidl [396]. Section 7.2 is devoted to error estimates for special window functions. The connection between the approximation error and the arithmetic cost of the NFFT is described in Theorem 7.16. We will show that the NFFT requires asymptotically the same arithmetical cost as the FFT, since we are only interested in computing the result up to a finite precision. In Sect. 7.3, we generalize the results of Sect. 7.1. We investigate the NFFT for nonequispaced data in space and frequency domains. Based on the NFFT approach, we derive fast approximate algorithms for discrete trigonometric transforms with nonequispaced knots in Sect. 7.4. The fast summation of radial functions with a variety of applications is described in Sect. 7.5. Finally in Sect. 7.6, we develop methods for inverse nonequispaced transforms, where we distinguish between direct and iterative methods.

7.1 Nonequispaced Data Either in Space or Frequency Domain

For a given dimension $d \in \mathbb{N}$ and for large $N \in \mathbb{N}$, let

G. Plonka et al., *Numerical Fourier Analysis*, Applied and Numerical Harmonic Analysis, https://doi.org/10.1007/978-3-031-35005-4_7

$$I_N^d := \left\{ \mathbf{k} \in \mathbb{Z}^d : -\frac{N}{2}\,\mathbf{1}_d \le \mathbf{k} < \frac{N}{2}\,\mathbf{1}_d \right\}$$

be an index set, where $\mathbf{1}_d := (1, \ldots, 1)^\top \in \mathbb{Z}^d$ and the inequalities hold for each component. We use the hypercube $[-\pi, \pi)^d$ as a representative of the d-dimensional torus $\mathbb{T}^d$. The inner product of $\mathbf{x} = (x_t)_{t=1}^d$ and $\mathbf{y} = (y_t)_{t=1}^d \in \mathbb{R}^d$ is denoted by

$$\mathbf{x} \cdot \mathbf{y} := \mathbf{x}^\top \mathbf{y} = \sum_{t=1}^{d} x_t\, y_t\,.$$

First, we describe the NFFT *for nonequispaced data* $\mathbf{x}_j \in \mathbb{T}^d$, $j \in I_M^1 := \{0, \ldots, M-1\}$, *in the space domain and equispaced data in the frequency domain*, i.e., we are interested in the fast evaluation of the d-variate, 2π-periodic trigonometric polynomial

$$f(\mathbf{x}) := \sum_{\mathbf{k} \in I_N^d} \hat{f}_{\mathbf{k}}\, \mathrm{e}^{\mathrm{i}\,\mathbf{k}\cdot\mathbf{x}}\,, \quad \hat{f}_{\mathbf{k}} \in \mathbb{C}\,, \tag{7.1}$$

at arbitrary knots $\mathbf{x}_j \in \mathbb{T}^d$, $j \in I_M^1$ for given arbitrary coefficients $\hat{f}_{\mathbf{k}} \in \mathbb{C}$, $\mathbf{k} \in I_N^d$. In other words, we will derive an efficient algorithm for the fast and stable evaluation of the M values

$$f_j := f(\mathbf{x}_j) = \sum_{\mathbf{k} \in I_N^d} \hat{f}_{\mathbf{k}}\, \mathrm{e}^{\mathrm{i}\,\mathbf{k}\cdot\mathbf{x}_j}\,, \quad j \in I_M^1\,. \tag{7.2}$$

The main idea is to approximate $f(\mathbf{x})$ by a linear combination of translates of a suitable d-variate window function in a first step and to evaluate the obtained approximation at the knots $\mathbf{x}_j$, $j \in I_M^1$ in a second step. Starting with a window function $\varphi \in L_2(\mathbb{R}^d) \cap L_1(\mathbb{R}^d)$ which is well localized in space and frequency, we define the 2π-periodic function

$$\tilde{\varphi}(\mathbf{x}) := \sum_{\mathbf{r} \in \mathbb{Z}^d} \varphi(\mathbf{x} + 2\pi\,\mathbf{r}) \tag{7.3}$$

which has the uniformly convergent Fourier series

$$\tilde{\varphi}(\mathbf{x}) = \sum_{\mathbf{k} \in \mathbb{Z}^d} c_{\mathbf{k}}(\tilde{\varphi})\, \mathrm{e}^{\mathrm{i}\,\mathbf{k}\cdot\mathbf{x}} \tag{7.4}$$

with the Fourier coefficients

$$c_{\mathbf{k}}(\tilde{\varphi}) := \frac{1}{(2\pi)^d} \int_{[-\pi,\,\pi]^d} \tilde{\varphi}(\mathbf{x})\,\mathrm{e}^{-\mathrm{i}\,\mathbf{k}\cdot\mathbf{x}}\,\mathrm{d}\mathbf{x}\,, \quad \mathbf{k} \in \mathbb{Z}^d\,. \tag{7.5}$$

There is a close relation between the Fourier coefficients $c_{\mathbf{k}}(\tilde{\varphi})$ in (7.5) and the Fourier transform $\hat{\varphi}(\mathbf{k})$ of the function φ, namely

$$\hat{\varphi}(\mathbf{k}) := \int_{\mathbb{R}^d} \varphi(\mathbf{x})\,\mathrm{e}^{-\mathrm{i}\,\mathbf{k}\cdot\mathbf{x}}\,\mathrm{d}\mathbf{x} = (2\pi)^d\, c_{\mathbf{k}}(\tilde{\varphi})\,, \quad \mathbf{k} \in \mathbb{Z}^d\,, \tag{7.6}$$

which is known from the Poisson summation formula (see the proof of Theorem 4.28). Let now $\sigma \geq 1$ be an oversampling factor such that $\sigma N \in \mathbb{N}$. This factor will later determine the size of a DFT. One should choose σ such that the DFT of length σN can be efficiently realized by FFT. Now we determine the coefficients $g_{\mathbf{l}} \in \mathbb{C}$, $\mathbf{l} \in I_{\sigma N}^d$, of the linear combination

$$s_1(\mathbf{x}) := \sum_{\mathbf{l} \in I_{\sigma N}^d} g_{\mathbf{l}}\, \tilde{\varphi}\left(\mathbf{x} - \frac{2\pi \mathbf{l}}{\sigma N}\right) \tag{7.7}$$

such that the function s_1 is an approximation of the trigonometric polynomial (7.1). Computing the Fourier series of the 2π-periodic function s_1, we obtain by Lemma 4.1

$$\begin{aligned} s_1(\mathbf{x}) &= \sum_{\mathbf{k} \in \mathbb{Z}^d} c_{\mathbf{k}}(s_1)\,\mathrm{e}^{\mathrm{i}\,\mathbf{k}\cdot\mathbf{x}} = \sum_{\mathbf{k} \in \mathbb{Z}^d} \hat{g}_{\mathbf{k}}\, c_{\mathbf{k}}(\tilde{\varphi})\,\mathrm{e}^{\mathrm{i}\,\mathbf{k}\cdot\mathbf{x}} \\ &= \sum_{\mathbf{k} \in I_{\sigma N}^d} \hat{g}_{\mathbf{k}}\, c_{\mathbf{k}}(\tilde{\varphi})\,\mathrm{e}^{\mathrm{i}\,\mathbf{k}\cdot\mathbf{x}} + \sum_{\mathbf{r} \in \mathbb{Z}^d \setminus \{\mathbf{0}\}} \sum_{\mathbf{k} \in I_{\sigma N}^d} \hat{g}_{\mathbf{k}}\, c_{\mathbf{k}+\sigma N \mathbf{r}}(\tilde{\varphi})\,\mathrm{e}^{\mathrm{i}\,(\mathbf{k}+\sigma N\mathbf{r})\cdot\mathbf{x}} \end{aligned} \tag{7.8}$$

with the discrete Fourier coefficients

$$\hat{g}_{\mathbf{k}} := \sum_{\mathbf{l} \in I_{\sigma N}^d} g_{\mathbf{l}}\,\mathrm{e}^{-2\pi\mathrm{i}\,\mathbf{k}\cdot\mathbf{l}/(\sigma N)}\,. \tag{7.9}$$

Assuming that $|c_{\mathbf{k}}(\tilde{\varphi})|$ are relatively small for $\|\mathbf{k}\|_\infty \geq \sigma N - \frac{N}{2}$ and that $c_{\mathbf{k}}(\tilde{\varphi}) \neq 0$ for all $\mathbf{k} \in I_N^d$, we compare the trigonometric polynomial (7.1) with the first sum of the Fourier series (7.8) and choose

$$\hat{g}_{\mathbf{k}} = \begin{cases} \hat{f}_{\mathbf{k}}/c_{\mathbf{k}}(\tilde{\varphi}) & \mathbf{k} \in I_N^d\,, \\ 0 & \mathbf{k} \in I_{\sigma N}^d \setminus I_N^d\,. \end{cases} \tag{7.10}$$

We compute the coefficients $g_{\mathbf{l}}$ in the linear combination (7.7) by applying the d-variate inverse FFT of size $(\sigma N) \times \ldots \times (\sigma N)$ and obtain

$$g_{\mathbf{l}} = \frac{1}{(\sigma N)^d} \sum_{\mathbf{k} \in I_N^d} \hat{g}_{\mathbf{k}}\, \mathrm{e}^{2\pi \mathrm{i}\, \mathbf{k}\cdot\mathbf{l}/(\sigma N)} = \frac{1}{(\sigma N)^d} \sum_{\mathbf{k} \in I_{\sigma N}^d} \hat{g}_{\mathbf{k}}\, \mathrm{e}^{2\pi \mathrm{i}\, \mathbf{k}\cdot\mathbf{l}/(\sigma N)}\,, \quad \mathbf{l} \in I_{\sigma N}^d\,. \tag{7.11}$$

Further, we assume that the function φ is well localized in space domain and can be well approximated by its truncation $\psi := \varphi|\, Q$ on $Q := \big[-\frac{2\pi m}{\sigma N},\, \frac{2\pi m}{\sigma N}\big]^d$, where $2m \ll \sigma N$ and $m \in \mathbb{N}$. Thus, we have

$$\psi(\mathbf{x}) = \varphi(\mathbf{x})\, \chi_Q(\mathbf{x}) = \begin{cases} \varphi(\mathbf{x}) & \mathbf{x} \in Q\,, \\ 0 & \mathbf{x} \in \mathbb{R}^d \setminus Q\,, \end{cases} \tag{7.12}$$

where χ_Q denotes the characteristic function of $Q \subset [-\pi,\, \pi]^d$, since $2m \ll \sigma N$. We consider again the 2π-periodic function

$$\tilde{\psi}(\mathbf{x}) := \sum_{\mathbf{r} \in \mathbb{Z}^d} \psi(\mathbf{x} + 2\pi \mathbf{r}) \in L_2(\mathbb{T}^d) \tag{7.13}$$

and approximate s_1 by the function

$$s(\mathbf{x}) := \sum_{\mathbf{l} \in I_{\sigma N}^d} g_{\mathbf{l}}\, \tilde{\psi}\Big(\mathbf{x} - \frac{2\pi \mathbf{l}}{\sigma N}\Big) = \sum_{\mathbf{l} \in I_{\sigma N,m}(\mathbf{x})} g_{\mathbf{l}}\, \tilde{\psi}\Big(\mathbf{x} - \frac{2\pi \mathbf{l}}{\sigma N}\Big)\,. \tag{7.14}$$

Here, the index set $I_{\sigma N,m}(\mathbf{x})$ is given by

$$I_{\sigma N,m}(\mathbf{x}) := \Big\{\mathbf{l} \in I_{\sigma N}^d : \frac{\sigma N}{2\pi}\, \mathbf{x} - m\, \mathbf{1}_d \le \mathbf{l} \le \frac{\sigma N}{2\pi}\, \mathbf{x} + m\, \mathbf{1}_d\Big\}\,.$$

For a fixed knot $\mathbf{x}_j$ we see that the sum (7.14) contains at most $(2m + 1)^d$ nonzero terms. Finally, we obtain

$$f(\mathbf{x}_j) \approx s_1(\mathbf{x}_j) \approx s(\mathbf{x}_j)\,.$$

Thus, we can approximately compute the sum (7.1) for all $\mathbf{x}_j$, $j \in I_M^1$, with a computational cost of $\mathcal{O}(N^d \log N + m^d M)$ operations. The presented approach involves two approximations, s_1 and s. We will study the related error estimates in Sect. 7.2. In the following, we summarize this algorithm of NFFT as follows.

Algorithm 7.1 (NFFT)
Input: $N,\ M \in \mathbb{N}$, $\sigma > 1$, $m \in \mathbb{N}$, $\mathbf{x}_j \in \mathbb{T}^d$ *for* $j \in I_M^1$, $\hat{f}_{\mathbf{k}} \in \mathbb{C}$ *for* $\mathbf{k} \in I_N^d$.
Precomputation: (i) Compute the nonzero Fourier coefficients $c_{\mathbf{k}}(\tilde{\varphi})$ *for all* $\mathbf{k} \in I_N^d$.
(ii) Compute the values $\tilde{\psi}\big(\mathbf{x}_j - \frac{2\pi \mathbf{l}}{\sigma N}\big)$ *for* $j \in I_M^1$ *and* $\mathbf{l} \in I_{\sigma N,m}(\mathbf{x}_j)$.

1. *Let* $\hat{g}_{\mathbf{k}} := \hat{f}_{\mathbf{k}} / c_{\mathbf{k}}(\tilde{\varphi})$ *for* $\mathbf{k} \in I_N^d$.

2. *Compute the values*

$$g_{\mathbf{l}} := \frac{1}{(\sigma N)^d} \sum_{\mathbf{k}\in I_N^d} \hat{g}_{\mathbf{k}}\, \mathrm{e}^{2\pi \mathrm{i}\mathbf{k}\cdot\mathbf{l}/(\sigma N)}\,, \quad \mathbf{l} \in I_{\sigma N}^d\,.$$

using a d-variate FFT.
3. *Compute*

$$s(\mathbf{x}_j) := \sum_{\mathbf{l}\in I_{\sigma N,m}(\mathbf{x}_j)} g_{\mathbf{l}}\, \tilde{\psi}\left(\mathbf{x}_j - \frac{2\pi \mathbf{l}}{\sigma N}\right), \quad j \in I_M^1\,.$$

Output: $s(\mathbf{x}_j)$, $j \in I_M^1$, *approximating the values* $f(\mathbf{x}_j)$ *in* (7.2).
Computational cost: $\mathcal{O}(N^d \log N + m^d M)$.

Remark 7.2

1. In Sect. 7.2 we will investigate different window functions φ and ψ. Thereby, we will also use $\tilde{\psi}$, which can be very efficiently computed so that the precomputation step (ii) can be omitted.
2. For window functions, which are expensive to compute, we can apply the lookup table technique. If the d-variate window function has the form

$$\varphi(\mathbf{x}) = \prod_{t=1}^{d} \varphi_t(x_t)$$

with even univariate window functions φ_t, then the precomputation step can be performed as follows. We precompute the equidistant samples $\varphi_t(\frac{rm}{K\sigma N})$ for $r = 0, \ldots, K$ with $K \in \mathbb{N}$ and compute for the actual node $\mathbf{x}_j$ during the NFFT the values $\varphi_t((\mathbf{x}_j)_t - \frac{2\pi l_t}{\sigma N})$ for $t = 1, \ldots, d$ and $l_t \in I_{n_t,m}((\mathbf{x}_j)_t)$ by means of the linear interpolation from its two neighboring precomputed samples (see, e.g., [234] for details). □

Next we describe the NFFT *for nonequispaced data in the frequency domain and equispaced data in the space domain*. We want to compute the values

$$h(\mathbf{k}) := \sum_{j\in I_M^1} f_j\, \mathrm{e}^{\mathrm{i}\,\mathbf{k}\cdot\mathbf{x}_j}\,, \quad \mathbf{k} \in I_N^d\,, \tag{7.15}$$

with arbitrary nodes $\mathbf{x}_j \in \mathbb{T}^d$, $j \in I_M^1$, and given data $f_j \in \mathbb{C}$, $j \in I_M^1$. For this purpose we introduce the 2π-periodic function

$$\tilde{g}(\mathbf{x}) := \sum_{j\in I_M^1} f_j\, \tilde{\varphi}(\mathbf{x} + \mathbf{x}_j)$$

with $\tilde{\varphi}$ in (7.3). For the Fourier coefficients of $\tilde{g}$, we obtain by (7.5) and (7.15) the identity

$$c_{\mathbf{k}}(\tilde{g}) = \frac{1}{(2\pi)^d} \int_{[-\pi,\,\pi]^d} \tilde{g}(\mathbf{x})\, \mathrm{e}^{-\mathrm{i}\,\mathbf{k}\cdot\mathbf{x}}\, \mathrm{d}\mathbf{x} = \sum_{j\in I_M^1} f_j\, \mathrm{e}^{\mathrm{i}\,\mathbf{k}\cdot\mathbf{x}_j}\, c_{\mathbf{k}}(\tilde{\varphi})$$
$$= h(\mathbf{k})\, c_{\mathbf{k}}(\tilde{\varphi})\,, \quad \mathbf{k} \in \mathbb{Z}^d\,. \tag{7.16}$$

Thus, the unknown values $h(\mathbf{k})$, $\mathbf{k} \in I_N^d$, can be computed, if the values $c_{\mathbf{k}}(\tilde{\varphi})$ and $c_{\mathbf{k}}(\tilde{g})$ for $\mathbf{k} \in I_N^d$ are available. The Fourier coefficients (7.16) of $\tilde{g}$ can be approximated by using the trapezoidal rule

$$c_{\mathbf{k}}(\tilde{g}) \approx \frac{1}{(\sigma N)^d} \sum_{\mathbf{l}\in I_{\sigma N}^d} \sum_{j\in I_M^1} f_j\, \tilde{\varphi}\Big(\mathbf{x}_j - \frac{2\pi \mathbf{l}}{\sigma N}\Big)\, \mathrm{e}^{2\pi\mathrm{i}\,\mathbf{k}\cdot\mathbf{l}/(\sigma N)}\,.$$

Similarly, as above let φ be well-localized in the space domain, such that φ can be approximated by its truncation $\psi = \varphi|_Q$ on $Q = \big[-\frac{2\pi m}{\sigma N},\, \frac{2\pi m}{\sigma N}\big]^d$. Hence, the 2π-periodic function $\tilde{\varphi}$ can be well approximated by the 2π-periodic function $\tilde{\psi}$.

We summarize the proposed method and denote it as *nonequispaced fast Fourier transform transposed* NFFT$^\top$.

Algorithm 7.3 (NFFT$^\top$)
Input: $N \in \mathbb{N}$, $\sigma > 1$, $m \in \mathbb{N}$, $\mathbf{x}_j \in \mathbb{T}^d$ *for* $j \in I_M^1$, $\tilde{f}_{\mathbf{k}} \in \mathbb{C}$ *for* $\mathbf{k} \in I_N^d$.
Precomputation: (i) Compute the nonzero Fourier coefficients $c_{\mathbf{k}}(\tilde{\varphi})$ *for* $\mathbf{k} \in I_N^d$.
(ii) Compute the values $\tilde{\psi}\big(\mathbf{x}_j - \frac{2\pi\mathbf{l}}{\sigma N}\big)$ *for* $\mathbf{l} \in I_{\sigma N}^d$ *and* $j \in I_{\sigma N,m}^{\mathrm{T}}(\mathbf{l})$, *where* $I_{\sigma N,m}^{\mathrm{T}}(\mathbf{l}) := \{j \in I_M^1 : \mathbf{l} - m\mathbf{1}_d \le \frac{\sigma N}{2\pi}\mathbf{x}_j \le \mathbf{l} + m\mathbf{1}_d\}$.

1. *Compute*

$$\hat{g}_{\mathbf{l}} := \sum_{j\in I_{\sigma N,m}^{\mathrm{T}}(\mathbf{l})} f_j\, \tilde{\psi}\Big(\mathbf{x}_j - \frac{2\pi\mathbf{l}}{\sigma N}\Big)\,, \quad \mathbf{l} \in I_{\sigma N}^d\,.$$

2. *Compute with the d-variate* FFT

$$\tilde{c}_{\mathbf{k}}(\tilde{g}) := \frac{1}{(\sigma N)^d} \sum_{\mathbf{l}\in I_{\sigma N}^d} \hat{g}_{\mathbf{l}}\, \mathrm{e}^{2\pi\mathrm{i}\,\mathbf{k}\cdot\mathbf{l}/(\sigma N)}, \quad \mathbf{k} \in I_N^d.$$

3. *Compute* $\tilde{h}(\mathbf{k}) := \tilde{c}_{\mathbf{k}}(\tilde{g})/c_{\mathbf{k}}(\tilde{\varphi})$ *for* $\mathbf{k} \in I_N^d$.
Output: $\tilde{h}(\mathbf{k})$, $\mathbf{k} \in I_N^d$, *approximating the values* $h(\mathbf{k})$ *in* (7.15).
Computational cost: $\mathcal{O}(N^d \log N + m^d M)$.

Remark 7.4 Setting $\hat{f}_{\mathbf{k}} = \delta_{\mathbf{k}-\mathbf{m}}$ for arbitrary $\mathbf{k} \in I_N^d$ and fixed $\mathbf{m} \in I_N^d$, where $\delta_{\mathbf{k}}$ denotes the d-variate Kronecker symbol, we see that the two Algorithms 7.1 and 7.3 use the approximation

$$\mathrm{e}^{\mathrm{i}\,\mathbf{m}\cdot\mathbf{x}} \approx \frac{(2\pi)^d}{(\sigma N)^d\,\hat{\varphi}(\mathbf{m})} \sum_{\mathbf{l}\in I_{\sigma N}^d} \tilde{\psi}\left(\mathbf{x} - \frac{2\pi\mathbf{l}}{\sigma N}\right) \mathrm{e}^{2\pi\mathrm{i}\,\mathbf{m}\cdot\mathbf{l}/(\sigma N)}\,. \tag{7.17}$$

□

In some cases it is helpful to describe the algorithms as matrix–vector products. This representation shows the close relation between the Algorithms 7.1 and 7.3. In order to write the sums (7.2) and (7.15) as matrix—vector products, we introduce the vectors

$$\hat{\mathbf{f}} := (\hat{f}_{\mathbf{k}})_{\mathbf{k}\in I_N^d} \in \mathbb{C}^{N^d}\,, \quad \mathbf{f} := (f_j)_{j\in I_M^1} \in \mathbb{C}^M \tag{7.18}$$

and the *nonequispaced Fourier matrix*

$$\mathbf{A} := \left(\mathrm{e}^{\mathrm{i}\,\mathbf{k}\cdot\mathbf{x}_j}\right)_{j\in I_M^1,\mathbf{k}\in I_N^d} \in \mathbb{C}^{M\times N^d}\,. \tag{7.19}$$

Then the evaluation of the sums (7.2) for $j = -M/2, \ldots, M/2 - 1$ is equivalent to the computation of the matrix–vector product $\mathbf{A}\,\hat{\mathbf{f}}$. Thus, a naïve evaluation of $\mathbf{A}\,\hat{\mathbf{f}}$ takes $\mathcal{O}(N^d M)$ arithmetical operations.

For equispaced knots $-2\pi\,\mathbf{j}/N$, $\mathbf{j} \in I_N^d$, the matrix $\mathbf{A}$ coincides with classical d-variate Fourier matrix

$$\mathbf{F}_N^d := \left(\mathrm{e}^{-2\pi\mathrm{i}\mathbf{k}\cdot\mathbf{j}/N}\right)_{\mathbf{j},\,\mathbf{k}\in I_N^d} \in \mathbb{C}^{N^d\times N^d}\,,$$

and we can compute the matrix–vector product with the help of an FFT.

In the following, we show that Algorithm 7.1 can be interpreted as an approximate factorization of the matrix $\mathbf{A}$ in (7.19) into the product of structured matrices

$$\mathbf{B}\,\mathbf{F}_{\sigma N,N}^d\,\mathbf{D}\,. \tag{7.20}$$

Each matrix corresponds to one step in Algorithm 7.1:

1. The diagonal matrix $\mathbf{D} \in \mathbb{C}^{N^d\times N^d}$ is given by

$$\mathbf{D} := \mathrm{diag}\left(c_{\mathbf{k}}(\tilde{\varphi})^{-1}\right)_{\mathbf{k}\in I_N^d}\,. \tag{7.21}$$

2. The matrix $\mathbf{F}_{\sigma N,N}^d \in \mathbb{C}^{(\sigma N)^d\times N^d}$ is the d-variate, truncated Fourier matrix

$$\mathbf{F}_{\sigma N,N}^d := \frac{1}{(\sigma N)^d}\left(\mathrm{e}^{2\pi\mathrm{i}\,\mathbf{k}\cdot\mathbf{l}/(\sigma N)}\right)_{\mathbf{l}\in I_{\sigma N}^d,\;\mathbf{k}\in I_N^d} \tag{7.22}$$

$$= \underbrace{\mathbf{F}^1_{\sigma N,N} \otimes \ldots \otimes \mathbf{F}^1_{\sigma N,N}}_{d\text{-times}},$$

which is the Kronecker product of d truncated Fourier matrices

$$\mathbf{F}^1_{\sigma N,N} = \frac{1}{\sigma N} \left(\mathrm{e}^{2\pi \mathrm{i}\, kl/(\sigma N)}\right)_{l\in I^1_{\sigma N},k\in I^1_N}.$$

3. Finally, $\mathbf{B} \in \mathbb{R}^{M\times(\sigma N)^d}$ is a sparse multilevel band matrix

$$\mathbf{B} := \left(\tilde{\psi}\left(\mathbf{x}_j - \frac{2\pi \mathbf{l}}{\sigma N}\right)\right)_{j\in I^1_M,\mathbf{l}\in I^d_{\sigma N}}. \tag{7.23}$$

It is now obvious that we compute the values $h(\mathbf{k})$ in (7.15) by the matrix–vector multiplication with the transposed matrix of $\mathbf{A}$, i.e.,

$$(h(\mathbf{k}))_{\mathbf{k}\in I^d_N} = \mathbf{A}^\top (f_j)_{j\in I^1_M}.$$

To this end, we calculate the values $h(\mathbf{k})$ in (7.15) approximately by transposing the factorization (7.20), i.e.,

$$(h(\mathbf{k}))_{\mathbf{k}\in I^d_N} \approx \mathbf{D}^\top\,(\mathbf{F}^d_{\sigma N,N})^\top\,\mathbf{B}^\top (f_j)_{j\in I^1_M}.$$

A comparison with Algorithm 7.3 shows that this factorization describes the three steps of Algorithm 7.3. We emphasize that the Algorithms 7.1 and 7.3 compute only approximate values. Therefore, we will discuss the approximation errors in the next section.

Remark 7.5 Let $\mathbf{A} \in \mathbb{C}^{M\times N}$ with $M \le N$ be a given rectangular matrix. For $1 \le s < N$, the *restricted isometry constant* δ_s of $\mathbf{A}$ is the smallest number $\delta_s \in [0,\,1)$, for which

$$(1-\delta_s)\,\|\mathbf{x}\|_2^2 \le \|\mathbf{A}\,\mathbf{x}\|_2^2 \le (1+\delta_s)\,\|\mathbf{x}\|_2^2$$

for all *s-sparse vectors* $\mathbf{x} \in \mathbb{C}^N$, i.e., $\mathbf{x}$ possesses exactly s nonzero components. The matrix $\mathbf{A}$ is said to have the *restricted isometry property*, if δ_s is small for s reasonably large compared to M.

For a matrix $\mathbf{A}$ with restricted isometry property, the following important recovery result was shown in [78, 147, 148]: Assume that $\delta_{2s} < 0.6246$. For $\mathbf{x} \in \mathbb{C}^N$, let a noisy measurement vector $\mathbf{y} = \mathbf{A}\,\mathbf{x} + \mathbf{e}$ be given, where $\mathbf{e} \in \mathbb{C}^M$ is an error vector with small norm $\|\mathbf{e}\|_2 < \varepsilon$. Let $\mathbf{x}^* \in \mathbb{C}^N$ be the minimizer of

$$\arg\min_{\mathbf{z}\in\mathbb{C}^N} \|\mathbf{z}\|_1 \text{ subject to } \|\mathbf{A}\,\mathbf{z} - \mathbf{y}\|_2 \le \varepsilon\,.$$

Then we have

$$\|\mathbf{x} - \mathbf{x}^*\|_2 \le c_1 \frac{\sigma_s(\mathbf{x})_1}{\sqrt{s}} + c_2\,\varepsilon\,,$$

where c_1 and c_2 are positive constants depending only on δ_{2s}. In particular, if $\mathbf{x} \in \mathbb{C}^N$ is an s-sparse vector and if $\varepsilon = 0$, then the recovery in $\ell_1(\mathbb{C}^N)$ is exact, i.e., $\mathbf{x}^* = \mathbf{x}$. By $\sigma_s(\mathbf{x})_1$ we denote the best s-term approximation of $\mathbf{x} \in \mathbb{C}^N$ in $\ell_1(\mathbb{C}^N)$, i.e.,

$$\sigma_s(\mathbf{x})_1 := \inf\{\|\mathbf{y} - \mathbf{x}\|_1 : \mathbf{y} \in \mathbb{C}^N\,,\ \|\mathbf{y}\|_0 \le s\}\,,$$

where $\|\mathbf{y}\|_0$ denotes the number of nonzero components of $\mathbf{y}$. For a proof of this result, see [148, p. 44]. Several algorithms for sparse recovery such as basis pursuit, thresholding-based algorithm, and greedy algorithm are presented in [148, pp. 61–73, 141–170].

An important example of a matrix with restricted isometry property is a rectangular nonequispaced Fourier matrix

$$\mathbf{A} := \frac{1}{\sqrt{M}}\,\big(\mathrm{e}^{\mathrm{i}\,k x_j}\big)_{j,k=0}^{M-1,\,N-1}$$

where the points x_j, $j = 0, \ldots, M-1$, are chosen independently and uniformly at random from $[0,\,2\pi]$. If for $\delta \in (0,\,1)$,

$$M \ge c\,\delta^{-2}\,s\,(\ln N)^4\,,$$

then with probability at least $1 - N^{-(\ln N)^3}$, the restricted isometry constant δ_s of the nonequispaced Fourier matrix $\mathbf{A}$ satisfies $\delta_s \le \delta$, where $c > 0$ is a universal constant (see [148, p. 405]). □

7.2 Approximation Errors for Special Window Functions

In contrast to the FFT, the NFFT and NFFT$^\top$ are *approximate algorithms*. Hence, the relation between the exactness of the computed result and the computational cost of the algorithm is important.

In the following, we restrict ourselves to the one-dimensional case $d = 1$ and study the errors of NFFT for special window functions more closely. For the multidimensional case $d > 1$, we refer to [120, 129]. Further, we remark that error estimates in the norm of $L_2(\mathbb{T})$ were discussed in [213, 299].

We consider mainly the error of NFFT, since the NFFT$^\top$ produces an error of similar size (see Remark 7.8). Let f denote an arbitrary 2π-periodic trigonometric polynomial (7.1) and let s_1 as well as s be the corresponding approximating functions (7.7) and (7.14). We split the *approximation error* of Algorithm 7.1 (with $d = 1$)

$$E := \max_{x \in \mathbb{T}} |f(x) - s(x)| = \|f - s\|_{C(\mathbb{T})} \tag{7.24}$$

into the *aliasing error*

$$E_{\mathrm{a}} := \max_{x \in \mathbb{T}} |f(x) - s_1(x)| = \|f - s_1\|_{C(\mathbb{T})}$$

and the *truncation error*

$$E_{\mathrm{t}} := \max_{x \in \mathbb{T}} |s_1(x) - s(x)| = \|s_1 - s\|_{C(\mathbb{T})},$$

such that we have $E \le E_{\mathrm{a}} + E_{\mathrm{t}}$.

We are interested in an NFFT with low computational cost and high accuracy. Therefore, we use mainly a continuous window function φ with compact support $\big[-\frac{2\pi m}{\sigma N},\, \frac{2\pi m}{\sigma N}\big]$, where the oversampling factor $\sigma > 1$ and the truncation parameter $m \in \mathbb{N}\backslash\{1\}$ are low, but $N \in 2\mathbb{N}$ can be large. Often we assume that $\sigma \in \big[\frac{5}{4},\, 2\big]$ and $m \in \{2, 3, \ldots, 10\}$. Thus, the main question in this section reads: Can one obtain high accuracy of the NFFT Algorithm 7.1 for low oversampling factor $\sigma > 1$ and low truncation parameter $m \in \mathbb{N}\backslash\{1\}$? In the following, we show that this is possible for special window functions, but the argumentation requires many technical details (see [348, 349]). Note that the use of a continuous window function φ with support $\big[-\frac{2\pi m}{\sigma N},\, \frac{2\pi m}{\sigma N}\big]$ simplifies the NFFT Algorithm 7.1, since it holds $E_{\mathrm{t}} = 0$ and $E = E_{\mathrm{a}}$.

First, we consider the NFFT for special window functions constructed as follows. For fixed $m \in \mathbb{N} \setminus \{1\}$ and $\sigma > 1$, let $\Phi_{m,\sigma}$ denote the set of all functions $\varphi_0 : \mathbb{R} \to [0,\, 1]$ with following properties:

- Each function φ_0 is even, has the support $[-1,\, 1]$, and is continuous on $\mathbb{R}$.
- Each function φ_0 is decreasing on $[0,\, 1]$ with $\varphi_0(0) = 1$ and $\varphi_0(1) = 0$.
- For each function φ_0, its Fourier transform

$$\hat{\varphi}_0(\omega) = \int_{\mathbb{R}} \varphi_0(x)\, \mathrm{e}^{-\mathrm{i}\omega x}\, \mathrm{d}x = 2 \int_0^1 \varphi_0(x) \cos(\omega x)\, \mathrm{d}x\,, \quad \omega \in \mathbb{R}\,,$$

is positive and decreasing for $\omega \in \big[0,\, \frac{\pi m}{\sigma}\big]$.

Obviously, each $\varphi_0 \in \Phi_{m,\sigma}$ is of bounded variation over $[-1,\, 1]$.

For $\varphi_0 \in \Phi_{m,\sigma}$ and $N \in 2\mathbb{N}$, where $2m \ll \sigma N$, we introduce the *window function*

$$\varphi(x) := \varphi_0\Big(\frac{\sigma N x}{2\pi m}\Big)\,, \quad x \in \mathbb{R}\,. \tag{7.25}$$

By construction (7.25), the window function $\varphi : \mathbb{R} \to [0, 1]$ is even, has the compact support $\big[-\frac{2\pi m}{\sigma N}, \frac{2\pi m}{\sigma N}\big]$, and is continuous on whole $\mathbb{R}$. Further, the window function φ is decreasing on $\big[0, \frac{2\pi m}{\sigma N}\big]$ with $\varphi(0) = 1$ and $\varphi\big(\frac{2\pi m}{\sigma N}\big) = 0$. Its Fourier transform

$$\hat{\varphi}(\omega) = \int_{\mathbb{R}} \varphi(x)\, \mathrm{e}^{-\mathrm{i}\omega x}\, \mathrm{d}x = 2\int_0^{2\pi m/(\sigma N)} \varphi(x)\, \cos(\omega x)\, \mathrm{d}x$$

$$= \frac{4\pi m}{\sigma N} \int_0^1 \varphi_0(t)\, \cos\Big(\frac{2\pi m\,\omega\, t}{\sigma N}\Big)\, \mathrm{d}t = \frac{2\pi m}{\sigma N}\, \hat{\varphi}_0\Big(\frac{2\pi m \omega}{\sigma N}\Big)\,, \qquad \omega \in \mathbb{R}\,,$$

is positive and decreasing for $\omega \in \big[0, \frac{N}{2}\big]$. Further, φ is of bounded variation over $\big[-\frac{2\pi m}{\sigma N}, \frac{2\pi m}{\sigma N}\big]$, where $\big[-\frac{2\pi m}{\sigma N}, \frac{2\pi m}{\sigma N}\big]$ is contained in $[-\pi, \pi]$ by the assumption $2m \ll \sigma N$.

Example 7.6 The simplest example of such a window function can be formed by a centered cardinal B-spline. Let $M_1 : \mathbb{R} \to \mathbb{R}$ be the characteristic function of the interval $\big[-\frac{1}{2}, \frac{1}{2}\big]$. For $m \in \mathbb{N}$ let

$$M_{m+1}(x) := (M_m * M_1)(x) = \int_{-1/2}^{1/2} M_m(x-t)\, \mathrm{d}t\,, \quad x \in \mathbb{R}\,,$$

be the *centered cardinal* B-*spline of order* $m+1$. Note that M_2 is the centered hat function and that

$$N_m(x) := M_m\Big(x - \frac{m}{2}\Big)\,, \quad x \in \mathbb{R}\,,$$

is the *cardinal* B-*spline of order m*. As in [45, 396], we consider the B-*spline window function*

$$\varphi_{\mathrm{B}}(x) = \varphi_{0,\mathrm{B}}\Big(\frac{\sigma N x}{2\pi m}\Big) = \frac{1}{M_{2m}(0)}\, M_{2m}\Big(\frac{\sigma N x}{2\pi}\Big)\,, \quad x \in \mathbb{R}\,, \tag{7.26}$$

where

$$\varphi_{0,\mathrm{B}}(x) := \frac{1}{M_{2m}(0)}\, M_{2m}(mx)\,, \quad x \in \mathbb{R}\,,$$

is contained in $\Phi_{m,\sigma}$, where $\sigma > 1$ is the oversampling factor, and $2m \ll \sigma N$. Note that it holds $M_{2m}(0) > 0$ for all $m \in \mathbb{N}$, since $M_{2m}(x) > 0$ for all $x \in (-m, m)$. For instance, it holds

$$M_2(0) = 1\,, \quad M_4(0) = \frac{2}{3}\,, \quad M_6(0) = \frac{11}{20}\,, \quad M_8(0) = \frac{151}{315}\,.$$

The B-spline window function (7.26) has the compact support

$$\text{supp}\,\varphi_{\text{B}} = \left[-\frac{2\pi m}{\sigma N},\, \frac{2\pi m}{\sigma N}\right] \subset [-\pi,\, \pi]\,.$$

We compute the Fourier transform

$$\begin{aligned}\hat{\varphi}_{\text{B}}(\omega) &= \int_{\mathbb{R}} \varphi(x)\,\mathrm{e}^{-\mathrm{i}\,\omega x}\,\mathrm{d}x = \frac{1}{M_{2m}(0)} \int_{\mathbb{R}} M_{2m}\left(\frac{\sigma N x}{2\pi}\right)\,\mathrm{e}^{-\mathrm{i}\,\omega x}\,\mathrm{d}x \\ &= \frac{2\pi}{\sigma N\, M_{2m}(0)} \int_{\mathbb{R}} M_{2m}(t)\,\mathrm{e}^{-2\pi\mathrm{i}\,\omega t/(\sigma N)}\,\mathrm{d}t = \frac{2\pi}{\sigma N\, M_{2m}(0)}\,\hat{M}_{2m}\left(\frac{2\pi\,\omega}{\sigma N}\right).\end{aligned}$$

By Example 2.16, it holds

$$\hat{M}_{2m}(\omega) = \left(\text{sinc}\,\frac{\omega}{2}\right)^{2m}\,, \qquad \omega \in \mathbb{R}\,,$$

and hence

$$\hat{\varphi}_{\text{B}}(\omega) = \frac{2\pi}{\sigma N\, M_{2m}(0)}\left(\text{sinc}\,\frac{\omega\pi}{\sigma N}\right)^{2m}\,, \qquad \omega \in \mathbb{R}\,, \tag{7.27}$$

with the sinc function $\text{sinc}\,x := \frac{\sin x}{x}$ for $x \in \mathbb{R} \setminus \{0\}$ and $\text{sinc}\,0 := 1$. Note that $\hat{\varphi}_{\text{B}}(\omega) > 0$ for all $\omega \in \left[0,\, \frac{N}{2}\right]$. □

An NFFT with a window function (7.25), where $\varphi_0 \in \Phi_{m,\sigma}$, is much simpler than an NFFT with the *Gaussian window function*

$$\varphi_{\text{Gauss}}(x) := \exp\left(-\frac{\beta}{2}\left(\frac{\sigma N x}{2\pi m}\right)^2\right)\,, \qquad x \in \mathbb{R}\,,$$

with the parameter $\beta := 2\pi m\left(1 - \frac{1}{2\sigma}\right)$, since this function is supported on whole $\mathbb{R}$. Later in Theorem 7.15, we will study the error of the NFFT with the Gaussian window function.

For a window function φ of the form (7.25) with $\varphi_0 \in \Phi_{m,\sigma}$, it holds $\psi = \varphi$ by (7.12), $\tilde{\psi} = \tilde{\varphi}$ by (7.13), and hence, $s = s_1$ by (7.7) and (7.14). Thus, the truncation error E_{t} vanishes and the approximation error E coincides with the aliasing error E_{a}, i.e.,

$$E = E_{\text{a}} = \|f - s_1\|_{C(\mathbb{T})}\,.$$

For this reason, we prefer the use of window functions (7.25) with $\varphi_0 \in \Phi_{m,\sigma}$.

Theorem 7.7 *Let $\sigma > 1$ and $N \in 2\mathbb{N}$ be given, where $\sigma N \in 2\mathbb{N}$ too. Further, let $m \in \mathbb{N}\setminus\{1\}$ with $2m \ll \sigma N$. Let f be a 2π-periodic trigonometric polynomial* (7.1) *with arbitrary coefficients $\hat{f}_k \in \mathbb{C}$, $k \in I_N^1$. Let*

$$\hat{\mathbf{f}} := \left(\hat{f}_k\right)_{k \in I_N^1} \in \mathbb{C}^N$$

denote the data vector.

Then the approximation error of the NFFT *with the window function* (7.25), *where* $\varphi_0 \in \Phi_{m,\sigma}$, *can be estimated by*

$$E \le e_\sigma(\varphi) \sum_{k \in I_N^1} |\hat{f}_k| = e_\sigma(\varphi)\, \|\hat{\mathbf{f}}\|_1\,, \tag{7.28}$$

where the $C(\mathbb{T})$-error constant $e_\sigma(\varphi)$ *is defined by*

$$e_\sigma(\varphi) := \sup_{N \in 2\mathbb{N}} \left(\max_{k \in I_N^1} \Big\| \sum_{r \in \mathbb{Z}\setminus\{0\}} \frac{\hat{\varphi}(k + \sigma N r)}{\hat{\varphi}(k)}\, \mathrm{e}^{\mathrm{i}\sigma N r\,\cdot} \Big\|_{C(\mathbb{T})} \right). \tag{7.29}$$

Proof By (7.8) and (7.14), it holds

$$\begin{aligned} s_1(x) &= \sum_{k \in I_{\sigma N}^1} \hat{g}_k c_k(\tilde{\varphi})\, \mathrm{e}^{\mathrm{i}kx} + \sum_{r \in \mathbb{Z}\setminus\{0\}} \sum_{k \in I_{\sigma N}^1} \hat{g}_k c_{k+\sigma N r}(\tilde{\varphi})\, \mathrm{e}^{\mathrm{i}(k+\sigma N r)x} \\ &= \sum_{k \in I_N^1} \hat{f}_k\, \mathrm{e}^{\mathrm{i}kx} + \sum_{r \in \mathbb{Z}\setminus\{0\}} \sum_{k \in I_N^1} \hat{f}_k\, \frac{c_{k+\sigma N r}(\tilde{\varphi})}{c_k(\tilde{\varphi})}\, \mathrm{e}^{\mathrm{i}(k+\sigma N r)x}\,. \end{aligned}$$

Then from (7.1) and (7.6), it follows that

$$\begin{aligned} s_1(x) - f(x) &= \sum_{r \in \mathbb{Z}\setminus\{0\}} \sum_{k \in I_N^1} \hat{f}_k\, \frac{\hat{\varphi}(k + \sigma N r)}{\hat{\varphi}(k)}\, \mathrm{e}^{\mathrm{i}(k+\sigma N r)x} \\ &= \sum_{k \in I_N^1} \hat{f}_k\, \mathrm{e}^{\mathrm{i}kx} \left(\sum_{r \in \mathbb{Z}\setminus\{0\}} \frac{\hat{\varphi}(k + \sigma N r)}{\hat{\varphi}(k)}\, \mathrm{e}^{\mathrm{i}\sigma N r x} \right) \end{aligned}$$

and hence,

$$|s_1(x) - f(x)| \le \sum_{k \in I_N^1} |\hat{f}_k|\, \max_{k \in I_N^1} \Big| \sum_{r \in \mathbb{Z}\setminus\{0\}} \frac{\hat{\varphi}(k + \sigma N r)}{\hat{\varphi}(k)}\, \mathrm{e}^{\mathrm{i}\sigma N r x} \Big|$$

for all $x \in \mathbb{T}$. Thus, by (7.29), we obtain

$$E_{\mathrm{a}} = \|s_1 - f\|_{C(\mathbb{T})} \le e_\sigma(\varphi) \sum_{k \in I_N^1} |\hat{f}_k|\,.$$

Since $E = E_{\text{a}}$ for a window function (7.25) with $\varphi_0 \in \Phi_{m,\sigma}$, we get (7.28). ■

Remark 7.8 The error of NFFT$^\top$ with a window function (7.25) with $\varphi_0 \in \Phi_{m,\sigma}$ (see Algorithm 7.3) can be estimated by the $C(\mathbb{T})$-error constant (7.29) too. Let $\sigma > 1$ and $N, M \in 2\mathbb{N}$ be given, where also $\sigma N \in 2\mathbb{N}$. For arbitrary $f_j \in \mathbb{C}$ and $x_j \in \mathbb{T}$, $j \in I_M^1$, we consider the values $h(k)$, $k \in I_N^1$ (see (7.15)) and the related approximations $\tilde{h}(k) = \tilde{c}_k(\tilde{g})/c_k(\tilde{\varphi})$, $k \in I_N^1$. Then the error of NFFT$^\top$ can be estimated by

$$\max_{k\in I_N^1} |h(k) - \tilde{h}(k)| \le e_\sigma(\varphi) \sum_{j\in I_M^1} |f_j|\,.$$

For a proof see [349, Lemma 4]. □

Now we estimate the $C(\mathbb{T})$-error constant $e_\sigma(\varphi)$ for various special window functions φ of the form (7.25) with $\varphi_0 \in \Phi_{m,\sigma}$. First, we consider the B-spline window function (7.26).

Theorem 7.9 (see [396]) *Assume that $\sigma > 1$ and $N \in 2\mathbb{N}$ are given, where $\sigma N \in 2\mathbb{N}$ too. Further, let $m \in \mathbb{N}$ with $2m \ll \sigma N$.*

Then the $C(\mathbb{T})$-error constant (7.29) of the window function (7.26) can be estimated by

$$e_\sigma(\varphi_{\text{B}}) \le \frac{4m}{2m-1}\,(2\sigma - 1)^{-2m}\,. \tag{7.30}$$

Proof From (7.27) it follows for all $k \in I_N^1$ and $r \in \mathbb{Z} \setminus \{0\}$ that

$$\frac{|\hat{\varphi}_{\text{B}}(k + \sigma N r)|}{\hat{\varphi}_{\text{B}}(k)} = \Big(\frac{k}{k + \sigma N r}\Big)^{2m} = \Big(\frac{\frac{k}{\sigma N}}{r + \frac{k}{\sigma N}}\Big)^{2m}.$$

Setting $u = \frac{k}{\sigma N}$ for $k \in I_N^1$, it holds $|u| \le \frac{1}{2\sigma} < 1$. Now we show that

$$\sum_{r\in\mathbb{Z}\setminus\{0\}} \Big(\frac{u}{r+u}\Big)^{2m} \le \frac{4m}{2m-1}\,(2\sigma - 1)^{-2m}\,.$$

Without loss of generality, we can assume that $0 \le u \le \frac{1}{2\sigma}$. Then we have

$$\sum_{r\in\mathbb{Z}\setminus\{0\}} \Big(\frac{u}{r+u}\Big)^{2m} = \Big(\frac{u}{1-u}\Big)^{2m} + \Big(\frac{u}{1+u}\Big)^{2m} + \sum_{r=2}^{\infty} \Big[\Big(\frac{u}{r-u}\Big)^{2m} + \Big(\frac{u}{r+u}\Big)^{2m}\Big].$$

By $r + u > r - u > 0$ for each $r \in \mathbb{N}$ it follows that $\big(\frac{u}{r+u}\big)^{2m} \le \big(\frac{u}{r-u}\big)^{2m}$. Using the integral test for convergence of series, we obtain

$$\sum_{r\in\mathbb{Z}\setminus\{0\}} \Big(\frac{u}{r+u}\Big)^{2m} \le 2\big(\frac{u}{1-u}\big)^{2m} + 2\sum_{r=2}^{\infty}\big(\frac{u}{r-u}\big)^{2m}$$

$$\le 2\big(\frac{u}{1-u}\big)^{2m} + 2\int_1^{\infty}\big(\frac{u}{x-u}\big)^{2m}\,\mathrm{d}x$$

$$= 2\big(\frac{u}{1-u}\big)^{2m}\,\big(1+\frac{1-u}{2m-1}\big).$$

Since the function $\big(\frac{u}{1-u}\big)^{2m}$ increases on $[0,\,\frac{1}{2\sigma}]$, the above sum has the upper bound $\frac{4m}{2m-1}\,(2\sigma-1)^{2m}$ for each $m\in\mathbb{N}$. This implies (7.30). ■

Now we determine upper bounds of the $C(\mathbb{T})$-error constant (7.29) for some window functions (7.25) with $\varphi_0\in\Phi_{m,\sigma}$. By Hölder's inequality it follows from (7.29) that

$$e_\sigma(\varphi) \le \sup_{N\in 2\mathbb{N}}\Big(\big(\min_{k\in I_N^1}\hat\varphi(k)\big)^{-1}\max_{k\in I_N^1}\sum_{r\in\mathbb{Z}\setminus\{0\}}|\hat\varphi(k+\sigma N r)|\Big).$$

By assumption $\varphi_0\in\Phi_{m,\sigma}$, it holds

$$\min_{k\in I_N^1}\hat\varphi(k) = \hat\varphi\big(\frac{N}{2}\big).$$

Thus, we have to show that the series

$$\sum_{r\in\mathbb{Z}\setminus\{0\}}|\hat\varphi(k+\sigma N r)|$$

is uniformly bounded for all $k\in I_N^1$ and $N\in 2\mathbb{N}$. For this we have to estimate the Fourier transform $\hat\varphi(k)$ for sufficiently large frequencies $|k|\ge\sigma N-\frac{N}{2}$ very carefully. Thus, this approach results in the estimate

$$e_\sigma(\varphi) \le \sup_{N\in 2\mathbb{N}}\Big(\frac{1}{\hat\varphi\big(\frac{N}{2}\big)}\max_{k\in I_N^1}\sum_{r\in\mathbb{Z}\setminus\{0\}}|\hat\varphi(k+\sigma N r)|\Big). \tag{7.31}$$

Next we consider the *continuous Kaiser–Bessel window function* φ_{cKB} of the form (7.25) with

$$\varphi_{0,\text{cKB}}(x) := \frac{1}{I_0(\beta)-1}\cdot\begin{cases}\big(I_0(\beta\sqrt{1-x^2})-1\big) & x\in[-1,\,1]\,,\\ 0 & x\in\mathbb{R}\setminus[-1,\,1]\end{cases} \tag{7.32}$$

with $\beta = 2\pi m\,\big(1-\frac{1}{2\sigma}\big)$ and $\sigma\in\big[\frac{5}{4},\,2\big]$ (see [151, 299, 348]), where

$$I_0(x) := \sum_{k=0}^{\infty} \frac{1}{(k!)^2} \left(\frac{x}{2}\right)^{2k}, \quad x \in \mathbb{R},$$

denotes the *modified Bessel function of first kind*. Note that parameter β in (7.32) depends on m and σ. By Oberhettinger [307, 18.31], the Fourier transform of (7.32) has the form

$$\hat{\varphi}_{0,\mathrm{cKB}}(\omega) := \frac{2}{I_0(\beta) - 1} \cdot \begin{cases} \left[\frac{\sinh \sqrt{\beta^2-\omega^2}}{\sqrt{\beta^2-\omega^2}} - \mathrm{sinc}\, \omega\right] & |\omega| < \beta, \\ \left[\mathrm{sinc}\, \sqrt{\omega^2 - \beta^2} - \mathrm{sinc}\, \omega\right] & |\omega| \ge \beta. \end{cases} \tag{7.33}$$

By the scaling property of the Fourier transform (see Theorem 2.5), we obtain

$$\hat{\varphi}_{\mathrm{cKB}}(\omega) = \frac{2\pi m}{\sigma N}\, \hat{\varphi}_{0,\mathrm{cKB}}\left(\frac{2\pi m \omega}{\sigma N}\right). \tag{7.34}$$

Lemma 7.10 *If $\sigma \in \left[\frac{5}{4},\, 2\right]$ and $m \in \mathbb{N} \setminus \{1\}$ are given, then $\hat{\varphi}_{0,\mathrm{cKB}}$ is positive and decreasing on $\left[0,\, \frac{\pi m}{\sigma}\right]$.*

Proof Since $\frac{\pi m}{\sigma} < \beta = 2\pi m \left(1 - \frac{1}{2\sigma}\right)$ for $\omega \in \left[0,\, \frac{\pi m}{\sigma}\right]$, it holds

$$\hat{\varphi}_{0,\mathrm{cKB}}(\omega) = \frac{2}{I_0(\beta) - 1} \left[\frac{\sinh \sqrt{\beta^2 - \omega^2}}{\sqrt{\beta^2 - \omega^2}} - \mathrm{sinc}\, \omega\right].$$

Using the Taylor expansions of sinh and sin, we conclude

$$\begin{aligned} \hat{\varphi}_{0,\mathrm{cKB}}(\omega) &= \frac{2}{I_0(\beta) - 1} \sum_{k=0}^{\infty} \frac{1}{(2k+1)!} \left[(\beta^2 - \omega^2)^k - (-1)^k \omega^{2k}\right] \\ &= \frac{\beta^2}{3\left(I_0(\beta) - 1\right)} + \sum_{k=2}^{\infty} \frac{1}{(2k+1)!} \left[(\beta^2 - \omega^2)^k - (-1)^k \omega^{2k}\right]. \end{aligned}$$

Since $\beta = 2\pi m \left(1 - \frac{1}{2\sigma}\right)$ with $\sigma \in \left[\frac{5}{4},\, 2\right]$ and $m \in \mathbb{N} \setminus \{1\}$, all functions $(\beta^2 - \omega^2)^k - (-1)^k \omega^{2k}$ with $k \in \mathbb{N} \setminus \{1\}$ are positive and decreasing on $\left[0,\, \frac{\pi m}{\sigma}\right]$. ∎

Consequently, $\varphi_{0,\mathrm{cKB}}$ belongs to the set $\Phi_{m,\sigma}$. From Lemma 7.10 and (7.34), it follows immediately that $\hat{\varphi}_{\mathrm{cKB}}$ is positive and decreasing on $\left[0,\, \frac{N}{2}\right]$ such that

$$\min_{k \in I_N^1} \hat{\varphi}_{\mathrm{cKB}}(k) = \hat{\varphi}_{\mathrm{cKB}}\left(\frac{N}{2}\right) \ge \frac{4\pi m}{\sigma N \left(I_0(\beta) - 1\right)} \left[\frac{\sinh\left(2\pi m \sqrt{1 - \frac{1}{\sigma}}\right)}{2\pi m \sqrt{1 - \frac{1}{\sigma}}} - \frac{\sigma}{\pi m}\right]. \tag{7.35}$$

Lemma 7.11 *If $\sigma \in \left[\frac{5}{4}, 2\right]$ and $m \in \mathbb{N} \setminus \{1\}$ are given, then for $|\omega| \ge \beta = 2\pi m \left(1 - \frac{1}{2\sigma}\right)$ it holds*

$$\left|\hat{\varphi}_{0,\mathrm{cKB}}(\omega)\right| \le \frac{4\beta^2}{I_0(\beta) - 1}\, \omega^{-2}\,.$$

Proof By (7.33) we have to show that for $|\omega| \ge \beta$ it holds

$$\left|\operatorname{sinc} \sqrt{\omega^2 - \beta^2} - \operatorname{sinc} \omega\right| \le \frac{2\beta^2}{\omega^2}\,. \tag{7.36}$$

Since the sinc function is even, we consider only the case $\omega \ge \beta$. For $\omega = \beta$, the above inequality is true, since $|\operatorname{sinc} 0 - \operatorname{sinc} \beta| \le 1 + |\operatorname{sinc} \beta| < 2$. For $\omega > \beta$, we obtain

$$\begin{aligned}
&\left|\operatorname{sinc} \sqrt{\omega^2 - \beta^2} - \operatorname{sinc} \omega\right| = \left|\frac{\sin \sqrt{\omega^2 - \beta^2}}{\sqrt{\omega^2 - \beta^2}} - \frac{\sin \omega}{\omega}\right| \\
&\le \frac{1}{\omega}\left|\sin \sqrt{\omega^2 - \beta^2} - \sin \omega\right| + \left|\sin \sqrt{\omega^2 - \beta^2}\right| \Big(\frac{1}{\sqrt{\omega^2 - \beta^2}} - \frac{1}{\omega}\Big) \\
&\le \frac{2}{\omega}\left|\sin \frac{\sqrt{\omega^2 - \beta^2} - \omega}{2}\right| + \left|\sin \sqrt{\omega^2 - \beta^2}\right| \frac{\omega - \sqrt{\omega^2 - \beta^2}}{\omega \sqrt{\omega^2 - \beta^2}}\,.
\end{aligned}$$

From

$$\omega - \sqrt{\omega^2 - \beta^2} = \omega \left(1 - \sqrt{1 - \frac{\beta^2}{\omega^2}}\right) = \omega\, \frac{1 - \left(1 - \frac{\beta^2}{\omega^2}\right)}{1 + \sqrt{1 - \frac{\beta^2}{\omega^2}}} \le \frac{\beta^2}{\omega}$$

it follows that

$$\begin{aligned}
\left|\sin \frac{\sqrt{\omega^2 - \beta^2} - \omega}{2}\right| &\le \frac{1}{2}\left(\omega - \sqrt{\omega^2 - \beta^2}\right) \le \frac{\beta^2}{2\omega}\,, \\
\left|\sin \sqrt{\omega^2 - \beta^2}\right| \frac{\omega - \sqrt{\omega^2 - \beta^2}}{\omega \sqrt{\omega^2 - \beta^2}} &\le \left|\sin \sqrt{\omega^2 - \beta^2}\right| \frac{\beta^2}{\omega^2 \sqrt{\omega^2 - \beta^2}} \le \frac{\beta^2}{\omega^2}\,.
\end{aligned}$$

This completes the proof. ∎

Lemma 7.12 *For all $k \in I_N^1$ it holds*

$$\sum_{r \in \mathbb{Z}\setminus\{0\}} \left|\hat{\varphi}_{\mathrm{cKB}}(k + r\sigma N)\right| \le \frac{32 m\pi}{\sigma N \left(I_0(\beta) - 1\right)}\,. \tag{7.37}$$

Proof For $k \in I_N^1$ and $r \in \mathbb{Z}\setminus\{0\}$ it holds $|k + r\sigma N| \ge \sigma N - \frac{N}{2}$ such that by (7.33) and (7.34) the term $\hat{\varphi}_{\mathrm{cKB}}(k + r\sigma N)$ has the form

$$\frac{4\pi m}{\sigma N\,\big(I_0(\beta) - 1\big)}\,\Big[\operatorname{sinc}\sqrt{\big(\frac{2\pi m\,(k + r\sigma N)}{\sigma N}\big)^2 - \beta^2} - \operatorname{sinc}\frac{2\pi m\,(k + r\sigma N)}{\sigma N}\Big].$$

Then from (7.36) it follows that

$$\big|\hat{\varphi}_{\mathrm{cKB}}(k + r\sigma N)\big| \le \frac{8\pi m}{\sigma N\,\big(I_0(\beta) - 1\big)}\,\big(1 - \frac{1}{2\sigma}\big)^2\,\big(r + \frac{k}{\sigma N}\big)^{-2}.$$

For $k \in I_N^1$ and $r \in \mathbb{Z}\setminus\{0\}$ it holds $\frac{|k|}{\sigma N} \le \frac{1}{2\sigma}$ and $\big(r + \frac{k}{\sigma N}\big)^{-2} \le \big(|r| - \frac{1}{2\sigma}\big)^{-2}$. Hence, by the integral test for convergence of series we conclude

$$\begin{aligned}\sum_{r\in\mathbb{Z}\setminus\{0\}} \big(r + \frac{k}{\sigma N}\big)^{-2} &= \big(1 + \frac{k}{\sigma N}\big)^{-2} + \big(-1 + \frac{k}{\sigma N}\big)^{-2}\\ &\quad + \sum_{r=2}^{\infty}\Big[\big(r + \frac{k}{\sigma N}\big)^{-2} + \big(-r + \frac{k}{\sigma N}\big)^{-2}\Big]\\ &\le 2\,\big(1 - \frac{1}{2\sigma}\big)^{-2} + 2\sum_{r=2}^{\infty}\big(r - \frac{1}{2\sigma}\big)^{-2}\\ &\le 2\,\big(1 - \frac{1}{2\sigma}\big)^{-2} + 2\int_1^{\infty}\big(t - \frac{1}{2\sigma}\big)^{-2}\,\mathrm{d}t\\ &= 2\,\big(1 - \frac{1}{2\sigma}\big)^{-2} + 2\,\big(1 - \frac{1}{2\sigma}\big)^{-1} \le 4\,\big(1 - \frac{1}{2\sigma}\big)^{-2}.\end{aligned}$$

This implies the estimate (7.37). ■

We summarize as follows.

Theorem 7.13 (see [348]) *Assume that $\sigma \in \big[\frac{5}{4},\,2\big]$ and $N \in 2\mathbb{N}$ are given, where also $\sigma N \in 2\mathbb{N}$. Further, let $m \in \mathbb{N}\setminus\{1\}$ with $2m \ll \sigma N$ and $\beta = 2\pi m\,\big(1 - \frac{1}{2\sigma}\big)$.*

Then the $C(\mathbb{T})$-error constant (7.29) of the continuous Kaiser–Bessel window function φ_{cKB} can be estimated by

$$e_\sigma(\varphi_{\mathrm{cKB}}) \le 32\,m\pi\,\sqrt{1 - \frac{1}{\sigma}}\,\Big[\mathrm{e}^{2\pi m\sqrt{1-1/\sigma}} - 6\Big]^{-1}. \tag{7.38}$$

Proof From the estimate (7.31) it follows that

$$e_\sigma(\varphi_{\mathrm{cKB}}) \le \sup_{N\in 2\mathbb{N}}\Big(\frac{1}{\hat{\varphi}_{\mathrm{cKB}}\big(\frac{N}{2}\big)}\,\max_{k\in I_N^1}\sum_{r\in\mathbb{Z}\setminus\{0\}}|\hat{\varphi}_{\mathrm{cKB}}(k + \sigma N r)|\Big).$$

Applying (7.37) and (7.35), we obtain

$$
\begin{aligned}
e_\sigma(\varphi_{\mathrm{cKB}}) &\le 8 \Big[\frac{\sinh\big(2\pi m\sqrt{1-\frac{1}{\sigma}}\big)}{2\pi m\sqrt{1-\frac{1}{\sigma}}} - \frac{\sigma}{\pi m}\Big]^{-1} \\
&= 8 \Big[\frac{\mathrm{e}^{2\pi m\sqrt{1-1/\sigma}} - \mathrm{e}^{-2\pi m\sqrt{1-1/\sigma}}}{4\pi m\sqrt{1-\frac{1}{\sigma}}} - \frac{\sigma}{\pi m}\Big]^{-1} \\
&\le 32\, m\pi \sqrt{1-\frac{1}{\sigma}}\, \Big[\mathrm{e}^{2\pi m\sqrt{1-1/\sigma}} - \mathrm{e}^{-2\pi m\sqrt{1-1/\sigma}} - 4\sqrt{\sigma^2-\sigma}\Big]^{-1}.
\end{aligned}
$$

For $\sigma \in \big[\frac{5}{4},\, 2\big]$ and $m \ge 2$ it holds $2\pi\,\sqrt{1-\frac{1}{\sigma}} \ge \frac{2\pi}{\sqrt{5}}$ and $4\,\sqrt{\sigma^2-\sigma} \le 4\,\sqrt{2}$ and hence

$$
\mathrm{e}^{-2\pi m\sqrt{1-1/\sigma}} + 4\,\sqrt{\sigma^2-\sigma} \le \big(\mathrm{e}^{-2\pi/\sqrt{5}}\big)^m + 4\,\sqrt{2} < 6\,.
$$

This completes the proof. ■

As next we consider the sinh-*type window function* φ_{sinh} of the form (7.25) with

$$
\varphi_{0,\mathrm{sinh}}(x) := \frac{1}{\sinh\beta} \cdot \begin{cases} \sinh(\beta\,\sqrt{1-x^2}) & x \in [-1,\, 1]\,, \\ 0 & x \in \mathbb{R}\setminus[-1,\, 1]\,. \end{cases} \tag{7.39}
$$

Here the parameter β is defined by $\beta = 2\pi m\,\big(1-\frac{1}{2\sigma}\big)$, where $\sigma \in \big[\frac{5}{4},\, 2\big]$ and $m \in \mathbb{N}\setminus\{1\}$ (see [348]). By Oberhettinger [307, 7.58], the Fourier transform of $\varphi_{0,\mathrm{sinh}}$ reads as follows:

$$
\hat\varphi_{0,\mathrm{sinh}}(\omega) = \frac{\pi\beta}{\sinh\beta} \cdot \begin{cases} \frac{I_1(\sqrt{\beta^2-\omega^2})}{\sqrt{\beta^2-\omega^2}} & \omega \in (-\beta,\, \beta)\,, \\ \frac{1}{4} & \omega = \pm\beta\,, \\ \frac{J_1(\sqrt{\omega^2-\beta^2})}{\sqrt{\omega^2-\beta^2}} & \omega \in \mathbb{R}\setminus[-\beta,\, \beta]\,. \end{cases}
$$

Obviously, the function $\varphi_{0,\mathrm{sinh}}$ belongs to the set $\Phi_{m,\sigma}$. Using the power series expansion of the modified Bessel function I_1, we see that for $\omega \in (-\beta,\, \beta)$ it holds

$$
\frac{I_1(\sqrt{\beta^2-\omega^2})}{\sqrt{\beta^2-\omega^2}} = \frac{1}{2}\sum_{k=0}^{\infty} \frac{(\beta^2-\omega^2)^k}{4^k\, k!\,(k+1)!}
$$

such that $\hat\varphi_{0,\mathrm{sinh}}$ is positive and decreasing on $[0,\, \frac{\pi m}{\sigma}]$, since it holds $\frac{\pi m}{\sigma} < \beta = 2\pi m\,\big(1-\frac{1}{2\sigma}\big)$ for $\sigma > 1$. By the scaling property of the Fourier transform (see Theorem 2.5), we obtain

$$\hat{\varphi}_{\mathrm{sinh}}(\omega) = \frac{2\pi m}{\sigma N}\, \hat{\varphi}_{0,\mathrm{sinh}}\Big(\frac{2\pi m\omega}{\sigma N}\Big)\,.$$

Theorem 7.14 (see [348]) *Assume that $\sigma \in \big[\frac{5}{4}, 2\big]$ and $N \in 2\mathbb{N} \setminus \{2, 4, 6\}$ are given, where also $\sigma N \in 2\mathbb{N}$. Further, let $m \in \mathbb{N} \setminus \{1\}$ with $2m \ll \sigma N$ and $\beta = 2\pi m \big(1 - \frac{1}{2\sigma}\big)$.*

Then the $C(\mathbb{T})$-error constant (7.29) *of the* sinh-*type window function φ_{sinh} can be estimated by*

$$e_\sigma(\varphi_{\mathrm{sinh}}) \le \Big[40\, m^{3/2} + 3\,\big(1 - \frac{1}{2\sigma}\big)^{-3/2}\Big]\,\big(1 - \frac{1}{\sigma}\big)^{3/4}\, \mathrm{e}^{-2\pi m\sqrt{1-1/\sigma}}\,. \tag{7.40}$$

For a proof of Theorem 7.14, see [348]. Error estimates for NFFT with another window function (7.25) with $\phi_0 \in \Phi_{m,\sigma}$ can be found in [348, 349].

Finally, we analyze the error of the NFFT Algorithm 7.1 with the *Gaussian window function* (see [120, 121, 396])

$$\varphi_{\mathrm{Gauss}}(x) := \varphi_{0,\mathrm{Gauss}}\Big(\frac{\sigma N x}{2\pi m}\Big)\,, \quad x \in \mathbb{R}\,, \tag{7.41}$$

where

$$\varphi_{0,\mathrm{Gauss}}(x) := \mathrm{e}^{-\beta x^2/2}\,, \quad x \in \mathbb{R}\,, \tag{7.42}$$

with the parameter $\beta = 2\pi m \big(1 - \frac{1}{2\sigma}\big)$. We assume that $\sigma > 1$, $m \in \mathbb{N} \setminus \{1\}$, and $N \in 2\mathbb{N}$, where $\sigma N \in 2\mathbb{N}$ too. Then the window function (7.41) is supported on whole $\mathbb{R}$ such that in Algorithm 7.1 we have to apply also the truncated Gaussian window function

$$\psi_{\mathrm{Gauss}}(x) := \varphi_{\mathrm{Gauss}}(x)\, \chi_Q(x)\,, \quad x \in \mathbb{R}\,, \tag{7.43}$$

with $Q := \big[-\frac{2\pi m}{\sigma N}, \frac{2\pi m}{\sigma N}\big]$. Note that it holds

$$\psi_{\mathrm{Gauss}}\Big(\pm\frac{2\pi m}{\sigma N}\Big) = \varphi_{0,\mathrm{Gauss}}(\pm 1) = \mathrm{e}^{-\beta/2} < \mathrm{e}^{-\pi} = 0.04321\ldots\,.$$

As shown in Example 2.6, the Fourier transform of (7.42) reads as follows:

$$\hat{\varphi}_{0,\mathrm{Gauss}}(\omega) = \sqrt{\frac{2\pi}{\beta}}\, \mathrm{e}^{-\omega^2/(2\beta)}\,, \quad \omega \in \mathbb{R}\,, \tag{7.44}$$

such by the scaling property of the Fourier transform

$$\hat{\varphi}_{\mathrm{Gauss}}(\omega) = \frac{2\pi m}{\sigma N}\, \hat{\varphi}_{0,\mathrm{Gauss}}\Big(\frac{2\pi m}{\sigma N}\,\omega\Big)\,, \quad \omega \in \mathbb{R}\,. \tag{7.45}$$

Theorem 7.15 (see [396]) *Assume that $\sigma > 1$ and $N \in 2\mathbb{N}$ are given, where also $\sigma N \in 2\mathbb{N}$. Further, let $m \in \mathbb{N} \setminus \{1\}$ with $2m \ll \sigma N$ and $\beta = 2\pi m \left(1 - \frac{1}{2\sigma}\right)$. Let f be a 2π-periodic trigonometric polynomial* (7.1) *with arbitrary coefficients $\hat{f}_k \in \mathbb{C}$, $k \in I_N^1$. Denote $\hat{\mathbf{f}} := \left(\hat{f}_k\right)_{k \in I_N^1}$.*

Then the approximation error of the NFFT *Algorithm 7.1 with the Gaussian window function* (7.41) *and the truncated Gaussian window function* (7.43) *can be estimated by*

$$E \le 3\,\mathrm{e}^{-\pi m \sqrt{1-1/(2\sigma-1)}}\,\|\hat{\mathbf{f}}\|_1 . \tag{7.46}$$

Proof For the approximation error $E = \|f - s\|_{C(\mathbb{T})}$, it holds $E \le E_{\mathrm{a}} + E_{\mathrm{t}}$ with $E_{\mathrm{a}} = \|f - s_1\|_{C(\mathbb{T})}$ and $E_{\mathrm{t}} = \|s_1 - s\|_{C(\mathbb{T})}$. As shown in the proof of Theorem 7.7, the aliasing error E_{a} can be estimated by

$$E_{\mathrm{a}} \le \Big(\max_{k \in I_N^1} \sum_{r \in \mathbb{Z}\setminus\{0\}} \frac{\hat{\varphi}_{\mathrm{Gauss}}(k + \sigma N r)}{\hat{\varphi}_{\mathrm{Gauss}}(k)}\Big)\,\|\hat{\mathbf{f}}\|_1 ,$$

where by (7.44) and (7.45) it holds

$$\frac{\hat{\varphi}_{\mathrm{Gauss}}(k + \sigma N r)}{\hat{\varphi}_{\mathrm{Gauss}}(k)} = \exp\Big(-\frac{(2\pi m)^2}{2\beta}\,\big(r^2 + \frac{2kr}{\sigma N}\big)\Big) .$$

Hence, for $k \in I_N^1$ we have

$$\sum_{r \in \mathbb{Z}\setminus\{0\}} \frac{\hat{\varphi}_{\mathrm{Gauss}}(k + \sigma N r)}{\hat{\varphi}_{\mathrm{Gauss}}(k)} = \sum_{r=1}^{\infty}\Big[\exp\Big(-\frac{(2\pi m)^2}{2\beta}\,\big(r^2 + \frac{2kr}{\sigma N}\big)\Big) + \exp\Big(-\frac{(2\pi m)^2}{2\beta}\,\big(r^2 - \frac{2kr}{\sigma N}\big)\Big)\Big] .$$

Since $\mathrm{e}^x + \mathrm{e}^{-x}$ is increasing on $[0, \infty)$, for $k \in I_N^1$ it follows that

$$\begin{aligned}
&\sum_{r \in \mathbb{Z}\setminus\{0\}} \frac{\hat{\varphi}_{\mathrm{Gauss}}(k + \sigma N r)}{\hat{\varphi}_{\mathrm{Gauss}}(k)} \\
&= \sum_{r=1}^{\infty}\Big[\exp\Big(-\frac{(2\pi m)^2}{2\beta}\,\big(r^2 - \frac{r}{\sigma}\big)\Big) + \exp\Big(-\frac{(2\pi m)^2}{2\beta}\,\big(r^2 + \frac{r}{\sigma}\big)\Big)\Big] \\
&= \exp\Big(-\frac{(2\pi m)^2}{2\beta}\,\big(1 - \frac{1}{\sigma}\big)\Big) + \exp\Big(-\frac{(2\pi m)^2}{2\beta}\,\big(1 + \frac{1}{\sigma}\big)\Big) \\
&\quad + \exp\frac{(2\pi m)^2}{8\beta\sigma^2}\,\sum_{r=2}^{\infty}\Big[\exp\Big(-\frac{(2\pi m)^2}{2\beta}\,\big(r - \frac{1}{2\sigma}\big)^2\Big)
\end{aligned}$$

$$+\exp\Big(-\frac{(2\pi m)^2}{2\beta}\,\big(r+\frac{1}{2\sigma}\big)^2\Big)\Big].$$

Using the integral test for convergence of series, for all $k \in I_N^1$ we obtain

$$\sum_{r\in\mathbb{Z}\setminus\{0\}} \frac{\hat{\varphi}_{\text{Gauss}}(k+\sigma N r)}{\hat{\varphi}_{\text{Gauss}}(k)}$$

$$\le \exp\Big(-\frac{(2\pi m)^2}{2\beta}\,\big(1-\frac{1}{\sigma}\big)\Big)\Big[1+\exp\Big(-\frac{(2\pi m)^2}{\beta\sigma}\Big)\Big]$$

$$+\exp\frac{(2\pi m)^2}{8\beta\sigma^2}\int_1^\infty\Big[\exp\Big(-\frac{(2\pi m)^2}{2\beta}\,\big(x-\frac{1}{2\sigma}\big)^2\Big)$$

$$+\exp\Big(-\frac{(2\pi m)^2}{2\beta}\,\big(x+\frac{1}{2\sigma}\big)^2\Big)\Big]\,\mathrm{d}x\,.$$

Note that for $a > 0$ and $c > 0$, it holds the estimate

$$\int_a^\infty \mathrm{e}^{-cx^2}\,\mathrm{d}x = \int_0^\infty \mathrm{e}^{-c(t+a)^2}\,\mathrm{d}t \le \mathrm{e}^{-ca^2}\int_0^\infty \mathrm{e}^{-2act}\,\mathrm{d}t$$
$$= \frac{1}{2ac}\,\mathrm{e}^{-ca^2}\,, \tag{7.47}$$

since $\mathrm{e}^{-ct^2} \le 1$ for all $t \in [0,\,\infty)$. Thus, by (7.47) we obtain

$$\int_1^\infty \exp\Big(-\frac{(2\pi m)^2}{2\beta}\,\big(x-\frac{1}{2\sigma}\big)^2\Big)\,\mathrm{d}x = \int_{1-1/(2\sigma)}^\infty \exp\big(-\frac{(2\pi m)^2}{2\beta}\,t^2\big)\,\mathrm{d}t$$
$$\le \frac{2\beta\sigma}{(2\sigma-1)\,(2\pi m)^2}\,\exp\Big(-\frac{(2\pi m)^2}{2\beta}\,\big(1-\frac{1}{2\sigma}\big)^2\Big)$$

and analogously

$$\int_1^\infty \exp\Big(-\frac{(2\pi m)^2}{2\beta}\,\big(x+\frac{1}{2\sigma}\big)^2\Big)\,\mathrm{d}x = \int_{1+1/(2\sigma)}^\infty \exp\big(-\frac{(2\pi m)^2}{2\beta}\,t^2\big)\,\mathrm{d}t$$
$$\le \frac{2\beta\sigma}{(2\sigma+1)\,(2\pi m)^2}\,\exp\Big(-\frac{(2\pi m)^2}{2\beta}\,\big(1+\frac{1}{2\sigma}\big)^2\Big)\,.$$

Thus, we conclude

$$E_{\mathrm{a}} \le \|\hat{\mathbf{f}}\|_1\,\exp\Big(-\frac{(2\pi m)^2}{2\beta}\,\big(1-\frac{1}{\sigma}\big)^2\Big)\Big[1+\frac{2\beta\sigma}{(2\sigma-1)\,(2\pi m)^2}$$

$$+ \exp\Big(-\frac{(2\pi m)^2}{\beta\sigma}\Big)\Big(1 + \frac{2\beta\sigma}{(2\sigma+1)\,(2\pi m)^2}\Big)\Big].$$

Then for the special parameter $\beta = 2\pi m\,\big(1 - \frac{1}{2\sigma}\big)$, we obtain

$$E_{\mathrm{a}} \le \|\hat{\mathbf{f}}\|_1\,\mathrm{e}^{-\pi m\,(1-1/(2\sigma-1))}\Big[1 + \frac{1}{2\pi m} + \mathrm{e}^{-4\pi m/(2\sigma-1)}\Big(1 + \frac{2\sigma-1}{(2\sigma+1)\,2\pi m}\Big)\Big]. \tag{7.48}$$

Now we estimate the truncation error $E_{\mathrm{t}} = \|s_1 - s\|_{C(\mathbb{T})}$ of the NFFT Algorithm 7.1. By (7.7) and (7.14), we obtain

$$s_1(x) - s(x) = \sum_{\ell \in I_{\sigma N}^1} g_\ell\,\Big(\tilde{\varphi}_{\mathrm{Gauss}}\big(x - \frac{2\pi\ell}{\sigma N}\big) - \tilde{\psi}_{\mathrm{Gauss}}\big(x - \frac{2\pi\ell}{\sigma N}\big)\Big),$$

where the coefficients g_ℓ, $\ell \in I_{\sigma N}^1$, are given by (7.10), (7.11), and (7.6) such that

$$g_\ell = \frac{1}{\sigma N}\sum_{k \in I_{\sigma N}^1} \hat{g}_k\,\mathrm{e}^{2\pi\mathrm{i}k\ell/(\sigma N)} = \frac{1}{\sigma N}\sum_{k \in I_N^1} \frac{\hat{f}_k}{c_k(\tilde{\varphi}_{\mathrm{Gauss}})}\,\mathrm{e}^{2\pi\mathrm{i}k\ell/(\sigma N)}$$

$$= \frac{2\pi}{\sigma N}\sum_{k \in I_N^1} \frac{\hat{f}_k}{\hat{\varphi}_{\mathrm{Gauss}}(k)}\,\mathrm{e}^{2\pi\mathrm{i}k\ell/(\sigma N)}.$$

Thus, for $x \in \mathbb{T}$ it follows that the function $s_1(x) - s(x)$ can be represented as

$$\frac{2\pi}{\sigma N}\sum_{k \in I_N^1} \frac{\hat{f}_k}{\hat{\varphi}_{\mathrm{Gauss}}(k)} \sum_{\ell \in I_{\sigma N}^1} \Big(\tilde{\varphi}_{\mathrm{Gauss}}\big(x - \frac{2\pi\ell}{\sigma N}\big) - \tilde{\psi}_{\mathrm{Gauss}}\big(x - \frac{2\pi\ell}{\sigma N}\big)\Big)\,\mathrm{e}^{2\pi\mathrm{i}k\ell/(\sigma N)}.$$

Hence, $|s_1(x) - s(x)|$ has the upper bound

$$\frac{2\pi}{\sigma N}\,\|\hat{\mathbf{f}}\|_1 \max_{k \in I_N^1} \frac{1}{\hat{\varphi}_{\mathrm{Gauss}}(k)}\,\Big|\sum_{\ell \in I_{\sigma N}^1} \Big(\tilde{\varphi}_{\mathrm{Gauss}}\big(x-\frac{2\pi\ell}{\sigma N}\big) - \tilde{\psi}_{\mathrm{Gauss}}\big(x-\frac{2\pi\ell}{\sigma N}\big)\Big)\,\mathrm{e}^{2\pi\mathrm{i}k\ell/(\sigma N)}\Big|.$$

Note that it holds

$$\max_{k \in I_N^1} \frac{1}{\hat{\varphi}_{\mathrm{Gauss}}(k)} = \frac{1}{\hat{\varphi}_{\mathrm{Gauss}}\big(\frac{N}{2}\big)} = \frac{\sigma N\sqrt{\beta}}{2\pi m\sqrt{2\pi}}\,\mathrm{e}^{(\pi m)^2/(2\sigma^2\beta)}.$$

Now we simplify the sum

$$S(x) := \sum_{\ell \in I_{\sigma N}^1} \Big(\tilde{\varphi}_{\mathrm{Gauss}}\big(x - \frac{2\pi \ell}{\sigma N}\big) - \tilde{\psi}_{\mathrm{Gauss}}\big(x - \frac{2\pi \ell}{\sigma N}\big)\Big)\, \mathrm{e}^{2\pi \mathrm{i} k\ell/(\sigma N)}\,.$$

Using the functions (7.3) and (7.13), we deduce

$$S(x) = \sum_{\ell \in I_{\sigma N}^1} \sum_{r \in \mathbb{Z}} \Big(\varphi_{\mathrm{Gauss}}\big(x - \frac{2\pi \ell}{\sigma N} + 2\pi r\big)$$
$$- \varphi_{\mathrm{Gauss}}\big(x - \frac{2\pi \ell}{\sigma N} + 2\pi r\big)\, \chi_Q\big(x - \frac{2\pi \ell}{\sigma N} + 2\pi r\big)\Big)\, \mathrm{e}^{2\pi \mathrm{i} k\ell/(\sigma N)}\,.$$

Setting $n = \sigma N r - \ell$ for $r \in \mathbb{Z}$ and $\ell \in I_{\sigma N}^1$, we conclude

$$S(x) = \sum_{n \in \mathbb{Z}} \Big(\varphi_{\mathrm{Gauss}}\big(x + \frac{2\pi n}{\sigma N}\big) - \varphi_{\mathrm{Gauss}}\big(x + \frac{2\pi n}{\sigma N}\big)\, \chi_Q\big(x + \frac{2\pi n}{\sigma N}\big)\Big)\, \mathrm{e}^{-2\pi \mathrm{i} kn/(\sigma N)}$$
$$= \sum_{\left|n + \frac{\sigma N}{2\pi}\right| > m} \varphi_{\mathrm{Gauss}}\big(x + \frac{2\pi n}{\sigma N}\big)\, \mathrm{e}^{-2\pi \mathrm{i} kn/(\sigma N)}$$

and hence,

$$|S(x)| \le \sum_{\left|n + \frac{\sigma N}{2\pi}\, x\right| > m} \varphi_{\mathrm{Gauss}}\big(x + \frac{2\pi n}{\sigma N}\big)\,.$$

Thus for the truncation error E_{t}, we obtain the inequality

$$E_{\mathrm{t}} \le \frac{2\pi}{\sigma N}\, \|\hat{\mathbf{f}}\|_1\, \frac{1}{\hat{\varphi}_{\mathrm{Gauss}}\big(\frac{N}{2}\big)}\, \max_{x \in \mathbb{T}} \sum_{\left|n + \frac{\sigma N}{2\pi}\, x\right| > m} \varphi_{\mathrm{Gauss}}\big(x + \frac{2\pi n}{\sigma N}\big)$$
$$= \|\hat{\mathbf{f}}\|_1\, \frac{\sqrt{\beta}}{m\sqrt{2\pi}}\, \mathrm{e}^{(\pi m)^2/(2\sigma^2 \beta)}\, \max_{x \in \mathbb{T}} \sum_{\left|n + \frac{\sigma N}{2\pi}\, x\right| > m} \varphi_{\mathrm{Gauss}}\big(x + \frac{2\pi n}{\sigma N}\big)\,.$$

For arbitrary fixed $x \in \mathbb{T}$, we estimate the sum

$$\sum_{\left|n + \frac{\sigma N}{2\pi}\, x\right| > m} \varphi_{\mathrm{Gauss}}\big(x + \frac{2\pi n}{\sigma N}\big) = \sum_{\left|n + \frac{\sigma N}{2\pi}\, x\right| > m} \mathrm{e}^{-\frac{\beta}{2m^2}\left(\frac{\sigma N}{2\pi}\, x + n\right)^2}$$

which runs over all $n \in \mathbb{Z}$ with the property $\left|n + \frac{\sigma N}{2\pi}\, x\right| > m$ and has the upper bound

$$2 \sum_{q=m}^{\infty} \mathrm{e}^{-\frac{\beta}{2m^2} q^2} .$$

Using the integral test for convergence of series and (7.47), it follows that

$$2 \sum_{q=m}^{\infty} \mathrm{e}^{-\frac{\beta}{2m^2} q^2} \le 2\,\mathrm{e}^{-\beta/2} + 2 \int_m^{\infty} \mathrm{e}^{-\frac{\beta}{2m^2} x^2}\, \mathrm{d}x \le 2\,\mathrm{e}^{-\beta/2} \big(1 + \frac{m}{\beta}\big) .$$

Thus, we obtain the estimate

$$E_{\mathrm{t}} \le \|\hat{\mathbf{f}}\|_1 \big(\frac{2}{m} + \frac{2}{\beta}\big) \sqrt{\frac{\beta}{2\pi}}\, \mathrm{e}^{(\pi m)^2/(2\sigma^2 \beta) - \beta/2} .$$

For the special parameter $\beta = 2\pi m \big(1 - \frac{1}{2\sigma}\big)$, we deduce

$$\frac{\beta}{2} - \frac{(\pi m)^2}{2\sigma^2 \beta} = \pi m \big(1 - \frac{1}{2\sigma - 1}\big) ,$$

i.e., E_{a} and E_{t} have the *same exponential decay*. This fact explains also the special choice of the parameter β. Then it holds

$$E_{\mathrm{t}} \le \|\hat{\mathbf{f}}\|_1 \big(\frac{2}{m} + \frac{2\sigma}{\pi m\,(2\sigma - 1)}\big) \sqrt{m \big(1 - \frac{1}{2\sigma}\big)}\, \mathrm{e}^{-\pi m\,(1 - 1/(2\sigma - 1))} . \tag{7.49}$$

For $m \ge 2$ and $\sigma > 1$ it follows from (7.48) and (7.49) that

$$E \le E_{\mathrm{a}} + E_{\mathrm{t}} < 3\, \|\hat{\mathbf{f}}\|_1\, \mathrm{e}^{-\pi m\,(1 - 1/(2\sigma - 1))} .$$

This completes the proof. ■

Thus, for a fixed oversampling factor $\sigma > 1$, the approximation error of the NFFT decays exponentially with the number m of summands in (7.14). On the other hand, the computational cost of the NFFT increases with m. G. Beylkin [45, 46] used B–splines, whereas A. Dutt and V. Rokhlin [121] applied Gaussian functions as window functions. Further approaches are based on scaling vectors [302], on minimizing the Frobenius norm of certain error matrices [304], or on min–max interpolation [139]. Employing the results in [139, 304], we prefer to apply the Algorithms 7.1 and 7.3 with continuous Kaiser–Bessel window function, sinh-type window function, or *Kaiser–Bessel window function* φ_{KB} of the form (7.25) [150, 299, 300], where $\varphi_{0,\mathrm{KB}}$ is defined by

$$\varphi_{0,\mathrm{KB}}(x) := \frac{1}{I_0(\beta)} \cdot \begin{cases} I_0\big(\beta \sqrt{1 - x^2}\big) & x \in [-1,\, 1] , \\ 0 & x \in \mathbb{R} \setminus [-1,\, 1] \end{cases}$$

with $\beta := 2\pi m\left(1 - \frac{1}{2\sigma}\right)$. By Oberhettinger [307, 18.31], the Fourier transform of $\varphi_{0,\mathrm{KB}}$ reads as follows:

$$\hat{\varphi}_{0,\mathrm{KB}}(\omega) = \frac{2}{I_0(\beta)} \cdot \begin{cases} \dfrac{\sinh\sqrt{\beta^2-\omega^2}}{\sqrt{\beta^2-\omega^2}} & \omega \in (-\beta,\,\beta)\,, \\ \mathrm{sinc}\sqrt{\omega^2-\beta^2} & \omega \in \mathrm{R}\setminus[-\beta,\,\beta]\,. \end{cases} \tag{7.50}$$

In the NFFT package [234], the *modified* sinh-*type window function* φ_{msinh} of the form (7.25) is used, where $\varphi_{0,\mathrm{msinh}}$ is defined by

$$\varphi_{0,\mathrm{msinh}}(x) := \frac{1}{\sinh\beta} \cdot \begin{cases} \frac{\sinh\left(\beta\sqrt{1-x^2}\right)}{\sqrt{1-x^2}} & x \in (-1,\,1)\,, \\ 0 & x \in \mathbb{R}\setminus(-1,\,1)\,. \end{cases}$$

Then the NFFT with the modified sinh-type window function produces very small approximation errors too (see also [349] for numerical comparison). Note that the idea to consider this modified sinh-type window function comes from the structure of the Fourier transform (7.50) of the Kaiser–Bessel window function.

In [17, 18], a NFFT library with the *exponential of semicircle window function* φ_{ES} of the form (7.25) is presented, where $\varphi_{0,\mathrm{ES}}$ is defined by

$$\varphi_{0,\mathrm{ES}}(x) := \begin{cases} \mathrm{e}^{\beta\,(\sqrt{1-x^2}-1)} & x \in (-1,\,1)\,, \\ 0 & x \in \mathbb{R}\setminus(-1,\,1)\,. \end{cases} \tag{7.51}$$

Then φ_{ES} has a simpler structure than the Kaiser–Bessel window function and produces similar small approximation errors for the NFFT. Note that $\varphi_{0,\mathrm{KB}}$, $\varphi_{0,\mathrm{msinh}}$, and $\varphi_{0,\mathrm{ES}}$ possess jump discontinuities at ± 1 with relatively low size of jump.

Further, we remark that in some applications, a relatively small oversampling factor $\sigma > 1$ or even $\sigma = 1$ can be used (see [299, 300]). These papers contain error estimates related to the root mean square error as well as algorithms for tuning the involved parameter.

In summary, for the window functions $\varphi \in \{\varphi_{\mathrm{B}}, \varphi_{\mathrm{Gauss}}, \varphi_{\mathrm{cKB}}, \varphi_{\mathrm{KB}}, \varphi_{\mathrm{sinh}}, \varphi_{\mathrm{msinh}}, \varphi_{\mathrm{ES}}\}$ with the parameter $\beta = 2\pi m\left(1 - \frac{1}{2\sigma}\right)$, the approximation error of NFFT Algorithm 7.1 can be estimated as follows.

Theorem 7.16 *Assume that $\sigma > 1$ and $N \in 2\mathbb{N}$ are given, where $\sigma N \in 2\mathbb{N}$ too. Further, let $m \in \mathbb{N}$ with $2m \ll \sigma N$. Let f be a 2π-periodic trigonometric polynomial* (7.1) *with arbitrary coefficients $\hat{f}_k \in \mathbb{C}$, $k \in I_N^1$. Denote $\hat{\mathbf{f}} = \left(\hat{f}_k\right)_{k\in I_N^1}$.*

Then the approximation error of the NFFT *Algorithm* 7.1 *can be estimated by*

$$E \le C(\sigma, m)\, \|\mathbf{f}\|_1\,, \tag{7.52}$$

where the constant $C(\sigma, m)$ decays exponentially with respect to m for fixed σ.

For the NFFT Algorithm 7.1 with $\varphi = \varphi_{\mathrm{B}}$ and $\sigma > 1$, it holds by Theorems 7.7 and 7.9 that

$$C(m,\sigma) \le \frac{4m}{2m-1}\,(2\sigma-1)^{-2m}\,.$$

For the NFFT Algorithm 7.1 with $\varphi = \varphi_{\mathrm{cKB}}$ and $\sigma \in \left[\frac{5}{4},\,1\right]$, it holds by Theorems 7.7 and 7.13 that

$$C(m,\sigma) \le 32\,m\pi\,\sqrt{1-\frac{1}{\sigma}}\,\left[\mathrm{e}^{2\pi m\sqrt{1-1/\sigma}}-6\right]^{-1}.$$

For the NFFT Algorithm 7.1 with $\varphi = \varphi_{\mathrm{Gauss}}$ and $\sigma > 1$, it holds by Theorems 7.15 that

$$C(m,\sigma) \le 3\,\mathrm{e}^{-\pi m\sqrt{1-1/(2\sigma-1)}}\,.$$

For the NFFT Algorithm 7.1 with $\varphi = \varphi_{\mathrm{sinh}}$ and $\sigma \in \left[\frac{5}{4},\,1\right]$, an explicit estimate of $C(m,\sigma)$ is presented in Theorems 7.7 and 7.14. For the NFFT Algorithm 7.1 with $\varphi = \varphi_{\mathrm{ES}}$, an asymptotic estimate of $C(m,\sigma)$ for $m \to \infty$ is given in [18].

For further results on error estimates for NFFT and a numerical comparison, we refer to [348, 349]. We emphasize that the approximation error of the NFFT Algorithm 7.1 is independent on the choice of the nodes $x_j \in \mathbb{T}$, $j \in I_M^1$.

7.3 Nonequispaced Data in Space and Frequency Domain

The algorithms in Sect. 7.1 are methods for nonequispaced knots in the space/frequency domain and equispaced knots in the frequency/space domain. Now we generalize these methods to nonequispaced knots in space as well as in frequency domain. Introducing the *exponential sum* $f : [-\pi,\,\pi]^d \to \mathbb{C}$ by

$$f(\boldsymbol{\omega}) = \sum_{k\in I_{M_1}^1} f_k\,\mathrm{e}^{-\mathrm{i}N\mathbf{x}_k\cdot\boldsymbol{\omega}}\,, \quad \boldsymbol{\omega}\in[-\pi,\,\pi]^d\,,$$

we derive an algorithm for the fast evaluation of

$$f(\boldsymbol{\omega}_j) = \sum_{k\in I_{M_1}^1} f_k\,\mathrm{e}^{-\mathrm{i}N\mathbf{x}_k\cdot\boldsymbol{\omega}_j}\,, \quad j\in I_{M_2}^1\,, \tag{7.53}$$

where $\mathbf{x}_k \in [0,\,2\pi]^d$ and $\boldsymbol{\omega}_j \in [-\pi,\,\pi]^d$ are nonequispaced knots and $f_k \in \mathbb{C}$ are given coefficients. Here $N \in \mathbb{N}$ with $N \gg 1$ is called the *nonharmonic bandwidth*. We denote methods for the fast evaluation of the sums (7.53) as NNFFT. These algorithms were introduced for the first time in [129, 130] (see also [340]). The algorithms are also called nonuniform FFT of type 3 (see [269]). We will see that the NNFFT is a combination of Algorithm 7.1 and the Algorithm 7.3.

Let $\varphi_1 \in L_2(\mathbb{R}^d) \cap L_1(\mathbb{R}^d)$ be a sufficiently smooth function, and recall that its Fourier transform is given by

$$\hat{\varphi}_1(\boldsymbol{\omega}) = \int_{\mathbb{R}^d} \varphi_1(\mathbf{x})\, \mathrm{e}^{-\mathrm{i}\,\boldsymbol{\omega}\cdot\mathbf{x}}\, \mathrm{d}\mathbf{x}\,.$$

Assume that $\hat{\varphi}_1(\boldsymbol{\omega}) \neq 0$ for all $\boldsymbol{\omega} \in N\,[-\pi,\, \pi]^d$. For the function

$$G(\mathbf{x}) := \sum_{k \in I_{M_1}^1} f_k\, \varphi_1(\mathbf{x} - \mathbf{x}_k)\,, \quad \mathbf{x} \in \mathbb{R}^d\,,$$

we obtain the Fourier transformed function

$$\hat{G}(\boldsymbol{\omega}) = \sum_{k \in I_{M_1}^1} f_k\, \mathrm{e}^{-\mathrm{i}\,\mathbf{x}_k\cdot\boldsymbol{\omega}}\, \hat{\varphi}_1(\boldsymbol{\omega})\,, \quad \boldsymbol{\omega} \in \mathbb{R}^d\,,$$

and hence the relation

$$f(\boldsymbol{\omega}_j) = \frac{\hat{G}(N\boldsymbol{\omega}_j)}{\hat{\varphi}_1(N\boldsymbol{\omega}_j)}, \quad j \in I_{M_2}^1\,.$$

Using this representation, for given $\hat{\varphi}_1$ we have to compute the function $\hat{G}$ at the nodes $N\boldsymbol{\omega}_j$, $j \in I_{M_2}^1$.

For a given oversampling factor $\sigma_1 > 1$, let $N_1 := \sigma_1 N \in \mathbb{N}$. Further, let $m_1 \in \mathbb{N}$ with $2m_1 \ll N_1$ be given and choose the parameter $a = 1 + 2m_1/N_1$. Now,

$$\hat{G}(\boldsymbol{\omega}) = \sum_{k \in I_{M_1}^1} f_k \int_{\mathbb{R}^d} \varphi_1(\mathbf{x} - \mathbf{x}_k)\, \mathrm{e}^{-\mathrm{i}\,\mathbf{x}\cdot\boldsymbol{\omega}}\, \mathrm{d}\mathbf{x}$$

can be rewritten using a $2\pi a$-periodization of φ_1. We obtain

$$\hat{G}(\boldsymbol{\omega}) = \sum_{k \in I_{M_1}^1} f_k \int_{a\,[-\pi,\,\pi]^d} \sum_{\mathbf{r} \in \mathbb{Z}^d} \varphi_1(\mathbf{x} + 2\pi a\mathbf{r} - \mathbf{x}_k)\, \mathrm{e}^{-\mathrm{i}\,(\mathbf{x} + 2\pi a\mathbf{r})\cdot\boldsymbol{\omega}}\, \mathrm{d}\mathbf{x}\,. \tag{7.54}$$

We discretize this integral by the rectangular rule and find the approximation

$$\hat{G}(\boldsymbol{\omega}) \approx S_1(\boldsymbol{\omega}) := N_1^{-d} \sum_{k \in I_{M_1}^1} f_k \sum_{\mathbf{t} \in I_{aN_1}^d} \sum_{\mathbf{r} \in \mathbb{Z}^d} \varphi_1\left(\frac{2\pi\mathbf{t}}{N_1} + 2\pi a\mathbf{r} - \mathbf{x}_k\right) \times \mathrm{e}^{-\mathrm{i}\,(\frac{2\pi\mathbf{t}}{N_1} + 2\pi a\mathbf{r})\cdot\boldsymbol{\omega}}\,. \tag{7.55}$$

Similarly, as in Sect. 7.1, we assume that φ_1 is localized in space domain and can be replaced by a compactly supported function ψ_1 with $\operatorname{supp}\psi_1 \subseteq [-\frac{2\pi m_1}{N_1}, \frac{2\pi m_1}{N_1}]^d$. Then, the inner sum in (7.55) contains nonzero terms only for $\mathbf{r} = \mathbf{0}$. We change the order of summations and find

$$S_1(\boldsymbol{\omega}) \approx S_2(\boldsymbol{\omega}) := N_1^{-d} \sum_{\mathbf{t}\in I_{aN_1}^d} \Big(\sum_{k\in I_{M_1}^1} f_k\, \psi_1\big(\frac{2\pi \mathbf{t}}{N_1} - \mathbf{x}_k\big)\Big)\, \mathrm{e}^{-2\pi \mathrm{i}\,\mathbf{t}\cdot\boldsymbol{\omega}/N_1}\,.$$

After computing the inner sum over $k \in I_{M_1}^1$, we evaluate the outer sum very efficiently with the help of Algorithm 7.1. The related window function and parameters are written with the subscript 2. We summarize this approach.

Algorithm 7.17 (NNFFT)

Input: $N \in \mathbb{N}$, $\sigma_1 > 1$, $\sigma_2 > 1$, $N_1 := \sigma_1 N$, $a := 1 + \frac{2m_1}{N_1}$, $N_2 := \sigma_1\sigma_2 a\, N$,
$\mathbf{x}_k \in [0,\, 2\pi]^d$, $f_k \in \mathbb{C}$ *for* $k \in I_{M_1}^1$, $\boldsymbol{\omega}_j \in [-\pi,\, \pi]^d$ *for* $j \in I_{M_2}^1$.

Precomputation: (i) Compute the nonzero Fourier transforms $\hat{\varphi}_1(N_1\boldsymbol{\omega}_j)$ *for* $j \in I_{M_1}^1$.
(ii) Compute $\psi_1(\frac{2\pi\mathbf{t}}{N_1} - \mathbf{x}_k)$ *for* $k \in I_{N_1,m_1}^{\mathrm{T}}(\mathbf{t})$ *and* $\mathbf{t} \in I_{aN_1}^d(\mathbf{x}_k)$, *where*
$I_{N_1,m_1}^{\mathrm{T}}(\mathbf{t}) := \{k \in I_{M_1}^1 : \mathbf{t} - m_1\mathbf{1}_d \le \frac{N_1}{2\pi}\,\mathbf{x}_k \le \mathbf{t} + m_1\mathbf{1}_d\}$.
(iii) Compute the nonzero Fourier transforms $\hat{\varphi}_2(\mathbf{t})$ *for* $\mathbf{t} \in I_{aN_1}^d$.
(iv) Compute $\psi_2(\boldsymbol{\omega}_j - \frac{2\pi\mathbf{l}}{N_2})$ *for* $j \in I_{M_2}^1$ *and* $\mathbf{l} \in I_{N_2,m_2}(\boldsymbol{\omega}_j)$.

1. *Calculate*

$$F(\mathbf{t}) := \sum_{k\in I_{N_1,m}^{\mathrm{T}}} f_k\, \psi_1\big(\frac{2\pi\mathbf{t}}{N_1} - \mathbf{x}_k\big), \quad \mathbf{t} \in I_{aN_1}^d\,.$$

2. *Determine* $\hat{g}_{\mathbf{t}} := F(\mathbf{t})/\hat{\varphi}_2(\mathbf{t})$ *for* $\mathbf{t} \in I_{aN_1}^d$.
3. *Compute*

$$g_{\mathbf{l}} := N_2^{-d} \sum_{\mathbf{t}\in I_{aN_1}^d} \hat{g}_{\mathbf{t}}\, \mathrm{e}^{-2\pi\mathrm{i}\,\mathbf{t}\cdot\mathbf{l}/N_2}, \quad \mathbf{l} \in I_{N_2}^d.$$

using a d-variate FFT.

4. *Compute*

$$s(\boldsymbol{\omega}_j) := N_1^{-d} \sum_{\mathbf{l}\in I_{N_2,m_2}(\boldsymbol{\omega}_j)} g_{\mathbf{l}}\, \psi_2\big(\boldsymbol{\omega}_j - \frac{2\pi\mathbf{l}}{N_2}\big), \quad j \in I_{M_2}^1\,.$$

5. *Compute* $S(\boldsymbol{\omega}_j) := s(\boldsymbol{\omega}_j)/\hat{\varphi}_1(N_1\boldsymbol{\omega}_j)$, $j \in I_{M_2}^1$.

Output: $S(\omega_j)$ *approximate value of* $f(\omega_j)$, $j \in I_M^1$.
Computational cost: $\mathcal{O}((\sigma_1 \sigma_2 aN)^d \log(\sigma_1\sigma_2 aN) + m_1 M_1 + m_2 M_2) =$
$\mathcal{O}(N^d \log N + M_1 + M_2)$.

The approximation error of Algorithm 7.17 is estimated in [130] and [241]. An important application of the NNFFT is the fast sinh transform which is studied in [241] too.

7.4 Nonequispaced Fast Trigonometric Transforms

In this section we present fast algorithms for the discrete trigonometric transforms (see Sects. 3.5 and 6.3) at arbitrary nodes. We investigate methods for the nonequispaced fast cosine transform (NFCT) as well as methods for the nonequispaced fast sine transform (NFST). In [330], three different methods for the NDCT have been compared, where the most efficient procedure is based on an approximation of the sums by translates of a window function. We restrict ourselves to the univariate case. The generalization to the multivariate case follows similarly as for the NFFT. All presented algorithms (also in the multivariate case) are part of the software [231].

First, we develop a method for the fast evaluation of the even, 2π-periodic function

$$f^{\mathrm{c}}(x) := \sum_{k=0}^{N-1} \hat{f}_k^{\mathrm{c}} \cos(kx)\,, \quad x \in \mathbb{R}\,, \tag{7.56}$$

at the nonequidistant nodes $x_j \in [0, \pi]$, $j = 0, \ldots, M-1$, and with arbitrary real coefficients $\hat{f}_k^{\mathrm{c}}$, $k = 0, \ldots, N-1$. This evaluation can be written as matrix–vector product $\mathbf{A}\,\hat{\mathbf{f}}$ with the *nonequispaced cosine matrix*

$$\mathbf{A} := \big(\cos(kx_j)\big)_{j,k=0}^{M-1,N-1} \in \mathbb{R}^{M\times N} \tag{7.57}$$

and the vector $\hat{\mathbf{f}} = (\hat{f}_k^{\mathrm{c}})_{k=0}^{N-1}$. A fast algorithm for the *nonequispaced discrete cosine transform* (NDCT) can be deduced from the NFFT (see Algorithm 7.1).

Let an oversampling factor $\sigma > 1$ with $\sigma N \in \mathbb{N}$ be given. As in (7.3) we introduce the 2π-periodization $\tilde{\varphi}$ of an even, well-localized window function $\varphi \in L_2(\mathbb{R}) \cap L_1(\mathbb{R})$. Assume that $\tilde{\varphi}$ has a uniformly convergent Fourier series. Our goal is now to evaluate the coefficients $g_\ell \in \mathbb{R}$, $\ell = 0, \ldots, \sigma N$, in the linear combination

$$s_1(x) := \sum_{\ell=0}^{\sigma N} g_\ell\, \tilde{\varphi}\big(x - \frac{\pi \ell}{\sigma N}\big)\,, \quad x \in \mathbb{R}\,, \tag{7.58}$$

such that s_1 is an approximation of f^c. To this end, we rewrite the function f^c in (7.56) as a sum of exponentials

$$f^c(x) = f(x) := \sum_{k=-N}^{N-1} \hat{f}_k \, \mathrm{e}^{\mathrm{i}\,kx}\,, \quad x \in \mathbb{R}\,, \tag{7.59}$$

with $\hat{f}_0 = \hat{f}_0^c$, $\hat{f}_{-N} = 0$, and $\hat{f}_k = \hat{f}_{-k} = \frac{1}{2}\hat{f}_k^c$ for $k = 1, \ldots, N-1$. In other words, we have the relation $\hat{f}_k^c = 2\,(\varepsilon_N(k))^2\,\hat{f}_k$. Since $\tilde{\varphi}$ is an even 2π-periodic function, we obtain for the Fourier coefficients $c_k(\tilde{\varphi}) = c_{-k}(\tilde{\varphi})$ and with (7.10) also $\hat{g}_k = \hat{g}_{-k}$. We take into account the symmetry in step 2 of the NFFT Algorithm 7.1 and compute the coefficients g_ℓ in (7.58) by

$$g_\ell = \mathrm{Re}(g_\ell) = \frac{1}{\sigma N} \sum_{k=0}^{\sigma N} \big(\varepsilon_{\sigma N}(k)\big)^2\,\hat{g}_k\,\cos\frac{2\pi k\ell}{\sigma N}\,, \quad \ell = 0, \ldots, \sigma N\,.$$

Here we use the notation as in Lemma 3.46 with $\varepsilon_N(0) = \varepsilon_N(N) := \sqrt{2}/2$ and $\varepsilon_N(j) := 1$ for $j = 1, \ldots, N-1$. We observe that $g_\ell = g_{2\sigma Nr-\ell}$, $r \in \mathbb{Z}$, i.e., one can compute the coefficients g_ℓ in (7.58) with the help of a DCT-I of length $\sigma N + 1$ (see (3.59) and Sect. 6.3). We proceed similarly as in Sect. 7.1 and approximate s_1 by

$$s(x) := \sum_{\ell=\lfloor 2\sigma Nx\rfloor - m}^{\lceil 2\sigma Nx\rceil + m} g_\ell\,\tilde{\psi}\big(x - \frac{\pi\ell}{\sigma N}\big)\,, \quad x \in \mathbb{R}\,. \tag{7.60}$$

For a fixed node $x_j \in [0, \pi]$, the sum (7.60) contains at most $2m + 2$ nonzero summands. Hence, we approximate the sum (7.56) at the nodes x_j, $j = 0, \ldots, M-1$, due to

$$f(x) \approx s_1(x) \approx s(x)$$

by evaluation of $s(x_j)$, $j = 0, \ldots, M-1$. In summary we obtain the following algorithm.

Algorithm 7.18 (NFCT)
Input: $N, M \in \mathbb{N}$, $\sigma > 1$, $m \in \mathbb{N}$, $x_j \in [0, \pi]$ *for* $j = 0, \ldots, M-1$,
$\hat{f}_k^c \in \mathbb{R}$ *for* $k = 0, \ldots, N-1$.
Precomputation: (i) Compute the nonzero Fourier coefficients $c_k(\tilde{\varphi})$ *for all* $k = 0, \ldots, N-1$.
(ii) Compute the values $\tilde{\psi}\big(x_j - \frac{\pi\ell}{\sigma N}\big)$ *for* $j = 0, \ldots, M-1$ *and* $\ell \in I_{\sigma N,m}(x_j)$.

1. *Set*

$$\hat{g}_k := \begin{cases} \dfrac{\hat{f}_k^{\mathrm{c}}}{2(\varepsilon_{\sigma N}(k))^2\, c_k(\tilde{\varphi})} & k = 0, \ldots, N-1\,, \\ 0 & k = N, \ldots, \sigma N\,. \end{cases}$$

2. *Compute*

$$g_\ell := \frac{1}{\sigma N} \sum_{k=0}^{\sigma N} \big(\varepsilon_{\sigma N}(k)\big)^2\, \hat{g}_k \cos \frac{\pi k \ell}{\sigma N}\,, \quad \ell = 0, \ldots, \sigma N,$$

using a fast algorithm of DCT-I$(\sigma N + 1)$ *(see Algorithm* 6.28 *or* 6.35*).*
3. *Compute*

$$s(x_j) := \sum_{\ell = \lfloor 2\sigma N x_j \rfloor - m}^{\lceil 2\sigma N x_j \rceil + m} g_\ell\, \tilde{\psi}\big(x_j - \frac{\pi \ell}{\sigma N}\big)\,, \quad j = 0, \ldots, M-1\,.$$

Output: $s(x_j)$, $j = 0, \ldots, M-1$, *approximate values of* $f^{\mathrm{c}}(x_j)$ *in* (7.56).
Computational cost: $\mathcal{O}(N \log N + mM)$.

In the following we deduce a fast algorithm for the transposed problem, i.e., for the fast evaluation of

$$h(k) := \sum_{j=0}^{M-1} h_j \cos(k x_j)\,, \quad k = 0, \ldots, N-1\,, \tag{7.61}$$

with nonequispaced nodes $x_j \in [0,\, \pi]$. To this end, we write the Algorithm 7.18 in matrix–vector form, since the evaluation of the sum (7.56) at the nodes x_j is equivalent to a matrix–vector multiplication with the transposed matrix $\mathbf{A}^\top$ of the nonequispaced cosine matrix (7.57). By Algorithm 7.18, $\mathbf{A}$ can be approximated by to the matrix product $\mathbf{B}\, \mathbf{C}^{\mathrm{I}}_{\sigma N+1,\, t}\, \mathbf{D}$, where each matrix corresponds to one step of Algorithm 7.18:

1. The diagonal matrix $\mathbf{D} \in \mathbb{R}^{N \times N}$ is given by

$$\mathbf{D} := \operatorname{diag}\Big(\big(2(\varepsilon_{\sigma N}(k))^2\, c_k(\tilde{\varphi})\big)^{-1}\Big)_{k=0}^{N-1}\,.$$

2. The matrix $\mathbf{C}^{\mathrm{I}}_{\sigma N+1,\, t} \in \mathbb{R}^{\sigma N \times N}$ is a truncated cosine matrix of type I (in a nonorthogonal form)

$$\mathbf{C}^{\mathrm{I}}_{\sigma N+1,\, t} := \Big(\frac{(\varepsilon_{\sigma N}(k))^2}{\sigma N} \cos \frac{\pi k \ell}{\sigma N}\Big)_{\ell,\, k=0}^{\sigma N-1, N-1}\,.$$

3. The sparse matrix $\mathbf{B} = \big(b_{j,\ell}\big)_{j,\,\ell=0}^{M-1, \sigma N-1} \in \mathbb{R}^{M \times \sigma N}$ has the entries

$$b_{j,\ell} := \begin{cases} \tilde{\psi}\big(x_j - \frac{\pi \ell}{\sigma N}\big) & \ell \in \{\lfloor 2\sigma N x_j \rfloor - m, \ldots, \lceil 2\sigma N x_j \rceil + m\}, \\ 0 & \text{otherwise} \end{cases}$$

and possesses at most $2m + 1$ nonzero entries per row. The approximate factorization of $\mathbf{A}$ allows to derive an algorithm for the fast evaluation of (7.61), since

$$\begin{aligned} \mathbf{g} := \big(h(k)\big)_{k=0}^{N-1} &= \mathbf{A}^\top \,(h_j)_{j=0}^{M-1} \\ &\approx \mathbf{D}^\top \,(\mathbf{C}^{\mathrm{I}}_{\sigma N+1,t})^\top \,\mathbf{B}^\top \,\big(h_j\big)_{j=0}^{M-1}. \end{aligned}$$

We immediately obtain the following algorithm.

Algorithm 7.19 (NFCT$^\top$)
Input: $N \in \mathbb{N}$, $\sigma > 1$, $m \in \mathbb{N}$, $x_j \in [0, \pi]$ *for* $j = 0, \ldots, M-1$, $h_j \in \mathbb{R}$ *for* $j = 0, \ldots, M-1$.
Precomputation: (i) Compute the nonzero Fourier coefficients $c_k(\tilde{\varphi})$ *for* $k = 0, \ldots, N-1$.
(ii) Compute the values $\tilde{\psi}\big(x_j - \frac{\pi\ell}{\sigma N}\big)$ *for* $\ell \in I^1_{\sigma N}$ *and* $j \in I^{\mathrm{T}}_{\sigma N,m}(\ell)$, *where* $I^{\mathrm{T}}_{\sigma N,m}(\ell) := \{j \in \{0, \ldots, M-1\} : \ell - m \le \frac{\sigma N}{\pi}\, x_j \le \ell + m\}$.

1. *Set* $\mathbf{g} := \mathbf{B}^\top \mathbf{h}$ *by computing*

```
for ℓ = 0..., σN
    g_ℓ := 0
end
for j = 0,..., M − 1
    for ℓ = ⌊σNx_j⌋ − m,..., ⌈σNx_j⌉ + m
        g_ℓ := g_ℓ + h_j ψ̃(x_j − πℓ/(σN))
    end
end.
```

2. *Compute*

$$\hat{g}_k := \frac{1}{\sigma N} \sum_{\ell=0}^{\sigma N} (\varepsilon_{\sigma N}(\ell))^2 \, g_\ell \, \cos \frac{\pi k \ell}{\sigma N}\,, \quad k = 0, \ldots, N-1\,.$$

using a fast algorithm of DCT-I$(\sigma N + 1)$ *(see Algorithm* 6.28 *or* 6.35*).*
3. *Compute* $\tilde{h}(k) := \hat{g}_k / (2(\varepsilon_{\sigma N}(k))^2 \, c_k(\tilde{\varphi}))$ *for* $k = 0, 1, \ldots, N-1$.
Output: $\tilde{h}(k)$, $k = 0, \ldots, N-1$, *approximate values for* $h(k)$ *in* (7.61).
Computational cost: $\mathcal{O}(N \log N + mM)$.

Now we modify the NFFT in order to derive a fast algorithm for the evaluation of the odd, 2π–periodic trigonometric polynomial

$$f^{\mathrm{s}}(x) = \sum_{k=1}^{N-1} \hat{f}_k^{\mathrm{s}} \sin(kx)\,, \quad x \in \mathbb{R}\,, \tag{7.62}$$

at nonequispaced nodes $x_j \in (0, \pi)$. To this end, we rewrite f^s in (7.62) as a sum of exponentials and obtain

$$\mathrm{i} f^s(x) = f(x) = \sum_{k=-N}^{N-1} \hat{f}_k\, \mathrm{e}^{\mathrm{i}kx} = \mathrm{i} \sum_{k=1}^{N-1} 2\, \hat{f}_k \sin(kx)\,, \quad x \in \mathbb{R}$$

with $\hat{f}_0 = \hat{f}_{-N} = 0$ and $\hat{f}_k = -\hat{f}_{-k} = \frac{1}{2} \hat{f}_k^s$ for $k = 1, \ldots, N-1$. Similarly as before, we approximate $f(x)$ by a function $s_1(x)$ as in (7.58) and obtain for the coefficients g_l for $\ell = 1, \ldots, \sigma N - 1$

$$-\mathrm{i}\, g_\ell = \frac{-\mathrm{i}}{2\sigma N} \sum_{k=-\sigma N}^{\sigma N-1} \hat{g}_k\, \mathrm{e}^{\pi \mathrm{i} k\ell/(\sigma N)} = \frac{1}{\sigma N} \sum_{k=1}^{\sigma N-1} \hat{g}_k \sin \frac{\pi k \ell}{\sigma N}\,. \tag{7.63}$$

and particularly $g_0 = g_{\sigma N} = 0$. Moreover, we observe that $g_{2\sigma N r-\ell} = -g_\ell$ for all $r \in \mathbb{Z}$. Finally, we compute the sum

$$\mathrm{i}\, s(x_j) := \sum_{\ell=\lfloor 2\sigma N x_j \rfloor - m}^{\lceil 2\sigma N x_j \rceil + m} \mathrm{i}\, g_\ell\, \tilde{\psi}\big(x_j - \frac{\pi \ell}{\sigma N}\big) \tag{7.64}$$

similarly as in (7.60) and obtain the approximate values of $f^{\mathrm{s}}(x_j) = \mathrm{i}\, f(x_j) \approx \mathrm{i}\, s(x_j)$, $j = 0, \ldots, M-1$.

We summarize the fast evaluation of the nonequispaced discrete sine transform in Algorithm 7.20.

Algorithm 7.20 (NFST)
Input: $N \in \mathbb{N}$, $\sigma > 1$, $m \in \mathbb{N}$, $x_j \in (0, \pi)$ *for* $j = 0, \ldots, M-1$,
$\hat{f}_k^{\mathrm{s}} \in \mathbb{R}$ *for* $k = 1, \ldots, N-1$.
Precomputation: *(i) Compute the nonzero Fourier coefficients* $c_k(\tilde{\varphi})$ *for all* $k = 0, \ldots, N-1$.
(ii) Compute the values $\tilde{\psi}\big(x_j - \frac{\pi \ell}{\sigma N}\big)$ *for* $j = 0, \ldots, M-1$ *and* $\ell \in I_{\sigma N, m}^{\mathrm{T}}(x_j)$.

1. *Set*

$$\hat{g}_k := \begin{cases} \dfrac{\hat{f}_k^{\mathrm{s}}}{2\, c_k(\tilde{\varphi})} & k = 1, \ldots, N-1\,, \\ 0 & k = 0 \text{ and } k = N, \ldots, \sigma N\,. \end{cases}$$

2. *Compute*

$$g_\ell := \frac{1}{\sigma N} \sum_{k=1}^{\sigma N-1} \hat{g}_k \sin\frac{\pi k \ell}{\sigma N}, \quad \ell = 1, \ldots, \sigma N - 1$$

using a fast algorithm of DST-I$(\sigma N - 1)$ *(see Table* 6.1 *and Remark* 6.40*) and set* $g_0 := 0$.
3. *Compute*

$$s(x_j) := \sum_{\ell=\lfloor 2\sigma N x_j\rfloor - m}^{\lceil 2\sigma N x_j\rceil + m} g_\ell \,\tilde{\psi}\big(x_j - \frac{\pi \ell}{\sigma N}\big), \quad j = 0, \ldots, M-1\,.$$

Output: $s(x_j)$, $j = 0, \ldots, M-1$, *approximate values of* $f^{\mathrm{s}}(x_j)$ *in* (7.62).
Computational cost: $\mathcal{O}(N \log N + mM)$.

An algorithm for the fast evaluation of the values

$$h(k) := \sum_{j=0}^{M-1} h_j \sin(kx_j), \tag{7.65}$$

follows immediately by transposing the matrix–vector product as described in the case of NFCT. Note that these algorithms can also be generalized to the multivariate case. The corresponding algorithms are part of the software in [231].

Remark 7.21 Instead of (7.57) we consider the rectangular nonequispaced cosine matrix

$$\mathbf{C} := \frac{1}{\sqrt{M}} \begin{pmatrix} \sqrt{2} & \cos x_0 & \cos(2x_0) & \ldots & \cos(N-1)x_0 \\ \sqrt{2} & \cos x_1 & \cos(2x_1) & \ldots & \cos(N-1)x_1 \\ \sqrt{2} & \cos x_2 & \cos(2x_2) & \ldots & \cos(N-1)x_2 \\ \vdots & \vdots & \vdots & & \vdots \\ \sqrt{2} & \cos x_{M-1} & \cos(2x_{M-1}) & \ldots & \cos(N-1)x_{M-1} \end{pmatrix} \in \mathbb{R}^{M\times N},$$

where $x_\ell \in [0, \pi]$, $\ell = 0, \ldots, M-1$, are independent identically distributed random variables. Assume that $1 \le s < N$ and $0 < \delta < 1$. If

$$M \ge 2c\,\delta^{-2}\, s\,(\ln s)^3 \,\log N\,,$$

then with probability at least $1 - N^{-\gamma\,(\ln s)^3}$, the *restricted isometry constant* δ_s of $\mathbf{C}$ satisfies $\delta_s \le \delta$, where c and γ are positive constants. Then the nonequispaced cosine matrix $\mathbf{C}$ has the *restricted isometry property*, i.e., for all s-sparse vectors $\mathbf{x} \in \mathbb{R}^N$ with s nonzero components, it holds

$$(1-\delta_s)\,\|\mathbf{x}\|_2^2 \le \|\mathbf{C}\,\mathbf{x}\|_2^2 \le (1+\delta_s)\,\|\mathbf{x}\|_2^2\,.$$

This remarkable result is a special case of a more general issue in [364, Theorem 4.3] for the polynomial set $\{\sqrt{2},\, T_1(x)\,, \ldots,\, T_{N-1}(x)\}$ which is an orthonormal system in $L_{2,w}(I)$ by Theorem 6.3. □

7.5 Fast Summation at Nonequispaced Knots

Let K be an even, real univariate function which is infinitely differentiable at least in $\mathbb{R} \setminus \{0\}$. We form the radially symmetric, d-variate function

$$\mathscr{K}(\mathbf{x}) := K(\|\mathbf{x}\|_2)\,, \quad \mathbf{x} \in \mathbb{R}^d \setminus \{\mathbf{0}\}\,,$$

where $\|\cdot\|_2$ denotes the Euclidean norm in $\mathbb{R}^d$. If K or its derivatives have singularities at zero, then $\mathscr{K}$ is called *singular kernel function*. If K is infinitely differentiable at zero as well, then $\mathscr{K}$ is defined on $\mathbb{R}^d$ and is called *nonsingular kernel function*. For given $\alpha_k \in \mathbb{C}$ and for distinct points $\mathbf{x}_k \in \mathbb{R}^d$, $k = 1, \ldots, M_1$, we consider the d-variate function

$$f(\mathbf{y}) := \sum_{k=1}^{M_1} \alpha_k\, \mathscr{K}(\mathbf{y} - \mathbf{x}_k) = \sum_{k=1}^{M_1} \alpha_k\, K(\|\mathbf{y} - \mathbf{x}_k\|_2)\,. \tag{7.66}$$

In this section, we develop algorithms for the fast computation of the sums

$$f(\mathbf{y}_j) := \sum_{k=1}^{M_1} \alpha_k\, \mathscr{K}(\mathbf{y}_j - \mathbf{x}_k)\,, \quad j = 1, \ldots, M_2\,, \tag{7.67}$$

for given knots $\mathbf{y}_j \in \mathbb{R}^d$. In the case of a singular kernel function $\mathscr{K}$, we assume that $\mathbf{x}_k \neq \mathbf{y}_j$ for all pairs of indices.

Example 7.22 If $K(x)$ is equal to $\ln |x|$, $\frac{1}{|x|}$, or $|x|^2 \ln |x|$ for $x \in \mathbb{R} \setminus \{0\}$, then we obtain known singular kernel functions. For arbitrary $\mathbf{x} \in \mathbb{R}^d \setminus \{\mathbf{0}\}$, the singularity function of the d-variate Laplacian reads as follows:

$$\mathscr{K}(\mathbf{x}) = \begin{cases} \ln \|\mathbf{x}\|_2 & d = 2\,, \\ \|\mathbf{x}\|_2^{2-d} & d \geq 3\,, \end{cases}$$

This singular kernel function appears in particle simulation [180, 322].
The *thin-plate spline* [117]

$$\mathscr{K}(\mathbf{x}) = \|\mathbf{x}\|_2^2 \ln \|\mathbf{x}\|_2\,, \quad \mathbf{x} \in \mathbb{R}^d \setminus \{\mathbf{0}\}\,,$$

is often used for the scattered data approximation of surfaces.

For $K(x) = \sqrt{x^2 + c^2}$ with some $c > 0$, the corresponding kernel function $\mathscr{K}$ is the *multiquadrix*. For $K(x) = (x^2 + c^2)^{-1/2}$ with some $c > 0$, the corresponding kernel function $\mathscr{K}$ is the *inverse multiquadrix*. In all these cases, we obtain singular kernel functions.

For fixed $\delta > 0$, a frequently used nonsingular kernel function

$$\mathscr{K}(\mathbf{x}) = \mathrm{e}^{-\delta\, \|\mathbf{x}\|_2^2}\,, \quad \mathbf{x} \in \mathbb{R}^d\,,$$

which arises in the context of diffusion [181], image processing [131], fluid dynamics, and finance [65], is generated by the Gaussian function $K(x) = \mathrm{e}^{-\delta x^2}$. □

For equispaced knots $\mathbf{x}_k$ and $\mathbf{y}_j$, (7.67) is simply a discrete convolution, and its fast computation can be mainly realized by fast Fourier methods exploiting the basic property

$$\mathrm{e}^{\mathrm{i}\,(\mathbf{y}-\mathbf{x})} = \mathrm{e}^{\mathrm{i}\,\mathbf{y}}\, \mathrm{e}^{-\mathrm{i}\,\mathbf{x}}\,.$$

Following these lines, we propose to compute the *convolution at nonequispaced knots* (7.67) by fast Fourier transforms at nonequispaced knots, i.e., NFFT and NFFT$^\top$, as presented in Sect. 7.1. For a nonsingular kernel function $\mathscr{K}$, for example, the Gaussian kernel function, our fast summation algorithm requires $\mathscr{O}(M_1 + M_2)$ arithmetic operations for arbitrary distributed points $\mathbf{x}_k$ and $\mathbf{y}_j$. For a singular kernel function $\mathscr{K}$, we have to introduce an additional *regularization procedure* and a so-called near field correction. If either the knots $\mathbf{x}_k$ or $\mathbf{y}_j$ are "sufficiently uniformly distributed," a notation which we will clarify later, then our algorithm requires $\mathscr{O}((M_1 + M_2)\log(M_1^{1/d}))$ or $\mathscr{O}((M_1 + M_2)\log(M_2^{1/d}))$ arithmetic operations, where the big $\mathscr{O}$ constant depends on the desired accuracy of the computation.

As seen in Example 7.22, the kernel function $\mathscr{K}$ is in general a nonperiodic function, while the use of Fourier methods requires to replace $\mathscr{K}$ by a periodic version. Without loss of generality, we assume that the knots satisfy $\|\mathbf{x}_k\|_2 < \frac{\pi}{2} - \frac{\varepsilon_{\mathrm{B}}}{2}$, $\|\mathbf{y}_j\|_2 < \frac{\pi}{2} - \frac{\varepsilon_{\mathrm{B}}}{2}$, and consequently $\|\mathbf{y}_j - \mathbf{x}_k\|_2 < \pi - \varepsilon_{\mathrm{B}}$. The parameter $\varepsilon_{\mathrm{B}} \in (0, \pi)$, which we will specify later, guarantees that K has to be evaluated only at points in the interval $[-\pi + \varepsilon_{\mathrm{B}}, \pi - \varepsilon_{\mathrm{B}}]$.

First, we regularize K near zero and near the points $\pm\pi$ to obtain a 2π-periodic sufficiently smooth function K_{R}. For this purpose, we set

$$K_{\mathrm{R}}(x) := \begin{cases} T_{\mathrm{I}}(x) & |x| \le \varepsilon_{\mathrm{I}}\,, \\ K(x) & \varepsilon_{\mathrm{I}} < |x| \le \pi - \varepsilon_{\mathrm{B}}\,, \\ T_{\mathrm{B}}(|x|) & \pi - \varepsilon_{\mathrm{B}} < |x| \le \pi\,, \end{cases} \tag{7.68}$$

where $0 < \varepsilon_{\mathrm{I}} < \pi - \varepsilon_{\mathrm{B}} < \pi$. Then we extend this function 2π-periodically on $\mathbb{R}$. The functions $T_{\mathrm{I}}, T_{\mathrm{B}} \in \mathscr{P}_{2r-1}$ will be chosen such that the 2π-periodic extension of K_{R} is contained in $C^{r-1}(\mathbb{T})$ for an appropriate parameter $r \in \mathbb{N}$. This regularization of K is possible by using algebraic polynomials, but also by applying splines or trigonometric polynomials. Here we determine polynomials $T_{\rm I}$ and $T_{\rm B} \in \mathscr{P}_{2r-1}$ by *two-point Taylor interpolation*. Applying Lemma 9.35, the two-point Taylor interpolation polynomial $T_{\rm I}$ is determined by the interpolation conditions

$$T_{\rm I}^{(j)}(-\varepsilon_{\rm I}) = K^{(j)}(-\varepsilon_{\rm I}) = (-1)^j\, K^{(j)}(\varepsilon_{\rm I})\,,$$
$$T_{\rm I}^{(j)}(\varepsilon_{\rm I}) = K^{(j)}(\varepsilon_{\rm I})\,, \quad j = 0, \ldots, r-1\,.$$

Note that $T_{\rm I}$ is an even polynomial of degree $2r-2$. Analogously, we choose the two-point Taylor interpolation polynomial $T_{\rm B} \in \mathscr{P}_{2r-1}$ with the interpolation conditions

$$T_{\rm B}^{(j)}(\pi - \varepsilon_{\rm B}) = K^{(j)}(\pi - \varepsilon_{\rm B})\,, \quad T_{\rm B}^{(j)}(\pi) = \delta_j\, K(\pi)\,, \quad j = 0, \ldots, r-1\,, \tag{7.69}$$

where δ_j denotes the Kronecker symbol. Thus the 2π-periodic extension of (7.68) is contained in $C^{r-1}(\mathbb{T})$.

For $\mathbf{x} \in [-\pi,\, \pi]^d$, we introduce the function

$$\mathscr{K}_{\rm R}(\mathbf{x}) := \begin{cases} K_{\rm R}(\|\mathbf{x}\|_2) & \|\mathbf{x}\|_2 < \pi\,, \\ T_{\rm B}(\pi) & \|\mathbf{x}\|_2 \ge \pi \end{cases}$$

and extend this function 2π-periodically on $\mathbb{R}^d$.

Next we approximate the sufficiently smooth, 2π-periodic function $\mathscr{K}_{\rm R}$ by a partial sum of its Fourier series, where the Fourier coefficients are computed by a simple quadrature rule that means for sufficiently large, even $n \in \mathbb{N}$ we form

$$\mathscr{K}_{\rm RF}(\mathbf{x}) := \sum_{\mathbf{l} \in I_n^d} b_{\mathbf{l}}\, \mathrm{e}^{\mathrm{i}\,\mathbf{l}\cdot\mathbf{x}}, \tag{7.70}$$

where I_n^d denotes the index set $[-\frac{n}{2},\, \frac{n}{2} - 1]^d \cap \mathbb{Z}^d$ and where

$$b_{\mathbf{l}} := \frac{1}{n^d} \sum_{\mathbf{j} \in I_n^d} \mathscr{K}_{\rm R}\Big(\frac{2\pi \mathbf{j}}{n}\Big)\, \mathrm{e}^{-2\pi \mathrm{i}\, \mathbf{j}\cdot\mathbf{l}/n}\,, \quad \mathbf{l} \in I_n^d\,. \tag{7.71}$$

Then our original kernel function $\mathscr{K}$ splits into

$$\mathscr{K} = (\mathscr{K} - \mathscr{K}_{\rm R}) + (\mathscr{K}_{\rm R} - \mathscr{K}_{\rm RF}) + \mathscr{K}_{\rm RF} = \mathscr{K}_{\rm NE} + \mathscr{K}_{\rm ER} + \mathscr{K}_{\rm RF}\,, \tag{7.72}$$

where $\mathscr{K}_{\rm NE} := \mathscr{K} - \mathscr{K}_{\rm R}$ and $\mathscr{K}_{\rm ER} := \mathscr{K}_{\rm R} - \mathscr{K}_{\rm RF}$. Since $\mathscr{K}_{\rm R}$ is sufficiently smooth, its Fourier approximation $\mathscr{K}_{\rm RF}$ generates only a small error $\mathscr{K}_{\rm ER}$. We neglect this error and approximate f by

$$\tilde f(\mathbf{x}) := f_{\rm NE}(\mathbf{x}) + f_{\rm RF}(\mathbf{x})\,, \quad \mathbf{x} \in \mathbb{R}^d \setminus \{\mathbf{0}\}\,,$$

with the *near-field sum*

$$f_{\mathrm{NE}}(\mathbf{x}) := \sum_{k=1}^{M_1} \alpha_k \, \mathscr{K}_{\mathrm{NE}}(\mathbf{x} - \mathbf{x}_k) \tag{7.73}$$

and the *far-field sum*

$$f_{\mathrm{RF}}(\mathbf{x}) := \sum_{k=1}^{M_1} \alpha_k \, \mathscr{K}_{\mathrm{RF}}(\mathbf{x} - \mathbf{x}_k)\,. \tag{7.74}$$

Instead of $f(\mathbf{y}_j)$, $j = 1, \ldots, M_2$, we evaluate the approximate values $\tilde{f}(\mathbf{y}_j)$. If either the points $\mathbf{x}_k$ or the points $\mathbf{y}_j$ are "sufficiently uniformly distributed," this can indeed be done in a fast way as follows.

- *Near-field computation* of (7.73):
  By definition (7.68), the function $\mathscr{K}_{\mathrm{NE}}$ restricted on $[-\pi, \pi]^d$ has only values with sufficiently large magnitudes in the ball of radius ε_{I} around the origin and near the sphere $\{\mathbf{x} \in \mathbb{R}^d : \|\mathbf{x}\|_2 = \pi\}$. The second set is not of interest, since $\|\mathbf{x}_k - \mathbf{y}_j\|_2 \le \pi - \varepsilon_{\mathrm{B}}$ by assumption. To achieve the desired computational cost of our algorithm, we suppose that either the M_1 points $\mathbf{x}_k$ or the M_2 points $\mathbf{y}_j$ are *sufficiently uniformly distributed*, i.e., there exists a small constant $\nu \in \mathbb{N}$ such that every ball of radius ε_{I} contains at most ν of the points $\mathbf{x}_k$ and of the points $\mathbf{y}_j$, respectively. This implies that ε_{I} depends linearly on $M_1^{-1/d}$ or $M_2^{-1/d}$. In the following, we restrict our attention to the case

  $$\varepsilon_{\mathrm{I}} \approx \sqrt[d]{\frac{\nu}{M_2}}\,. \tag{7.75}$$

  Then the sum (7.73) contains for fixed $\mathbf{y}_j$ not more than ν summands, and its evaluation at M_2 knots requires only $\mathscr{O}(\nu\, M_2)$ arithmetic operations.
- *Far-field summation* of (7.74) by NFFT$^\top$ and NFFT:
  Substituting (7.70) for $\mathscr{K}_{\mathrm{RF}}$, we obtain

  $$f_{\mathrm{RF}}(\mathbf{y}_j) = \sum_{k=1}^{M_1} \alpha_k \sum_{\mathbf{l} \in I_n^d} b_{\mathbf{l}}\, \mathrm{e}^{\mathrm{i}\,\mathbf{l}\cdot(\mathbf{y}_j - \mathbf{x}_k)} = \sum_{\mathbf{l} \in I_n^d} b_{\mathbf{l}} \Big(\sum_{k=1}^{M_1} \alpha_k\, \mathrm{e}^{-\mathrm{i}\,\mathbf{l}\cdot\mathbf{x}_k} \Big)\, \mathrm{e}^{\mathrm{i}\,\mathbf{l}\cdot\mathbf{y}_j}\,.$$

  The expression in the inner brackets can be computed by a d-variate NFFT$^\top$ of size $n \times \ldots \times n$. This is followed by n^d multiplications with $b_{\mathbf{l}}$ and completed by a d-variate NFFT of size $n \times \ldots \times n$ to compute the outer sum with the complex exponentials. If m is the cutoff parameter and $\rho = 2$ the oversampling factor of the NFFT$^\top$ or NFFT, then the proposed evaluation of the values $f_{\mathrm{RF}}(\mathbf{y}_j)$, $j = 1, \ldots, M_2$, requires $\mathscr{O}(m^d\,(M_1 + M_2) + (\rho n)^d \log(\rho n))$ arithmetic operations.

The relation between M_1, M_2, and n is determined by the approximation error of the algorithm and is investigated in detail in [335, 336].

In summary, we obtain the following algorithm for fast summation of nonequispaced knots for a singular kernel function.

Algorithm 7.23 (Fast Summation with Singular Kernel Function)
Input: $\alpha_k \in \mathbb{C}$ *for* $k = 1, \ldots, M_1$,
$\mathbf{x}_k \in \mathbb{R}^d$ *for* $k = 1, \ldots, M_1$ *with* $\|\mathbf{x}_k\|_2 < \frac{1}{2}(\pi - \varepsilon_{\mathrm{B}})$,
$\mathbf{y}_j \in \mathbb{R}^d$ *for* $j = 1, \ldots, M_2$ *with* $\|\mathbf{y}_j\|_2 < \frac{1}{2}(\pi - \varepsilon_{\mathrm{B}})$.
Precomputation: Compute the polynomials T_{I} *and* T_{B} *by Lemma* 9.35.
Compute $(b_{\mathbf{l}})_{\mathbf{l} \in I_n^d}$ *by (7.71) and (7.68).*
Compute $\mathscr{K}_{\mathrm{NE}}(\mathbf{y}_j - \mathbf{x}_k)$ *for* $j = 1, \ldots, M_2$ *and* $k \in I_{\varepsilon_{\mathrm{I}}}^{\mathrm{NE}}(j)$,
where $I_{\varepsilon_{\mathrm{I}}}^{\mathrm{NE}}(j) := \{k \in \{1, \ldots, M_1\} : \|\mathbf{y}_j - \mathbf{x}_k\|_2 < \varepsilon_{\mathrm{I}}\}$.

1. *For each* $\mathbf{l} \in I_n^d$ *compute*
$$a_{\mathbf{l}} := \sum_{k=1}^{M_1} \alpha_k \, \mathrm{e}^{-\mathrm{i}\,\mathbf{l}\cdot\mathbf{x}_k}$$
using the d-variate NFFT$^\top$ *of size* $n \times \ldots \times n$ *(see Algorithm* 7.3*).*
2. *For each* $\mathbf{l} \in I_n^d$ *compute the products* $d_{\mathbf{l}} := a_{\mathbf{l}}\, b_{\mathbf{l}}$.
3. *For* $j = 1, \ldots, M_2$ *compute the far field sums*
$$f_{\mathrm{RF}}(\mathbf{y}_j) := \sum_{\mathbf{l} \in I_n^d} d_{\mathbf{l}}\, \mathrm{e}^{\mathrm{i}\,\mathbf{l}\cdot\mathbf{y}_j}$$
using the d-variate NFFT *of size* $n \times \ldots \times n$ *(see Algorithm* 7.1*).*
4. *For* $j = 1, \ldots, M_2$ *compute the near-field sums*
$$f_{\mathrm{NE}}(\mathbf{y}_j) := \sum_{k \in I_{\varepsilon_{\mathrm{I}}}^{\mathrm{NE}}(j)} \alpha_k\, \mathscr{K}_{\mathrm{NE}}(\mathbf{y}_j - \mathbf{x}_k)\,.$$
5. *For* $j = 1, \ldots, M_2$ *compute the near-field corrections*
$$\tilde{f}(\mathbf{y}_j) := f_{\mathrm{NE}}(\mathbf{y}_j) + f_{\mathrm{RF}}(\mathbf{y}_j).$$

Output: $\tilde{f}(\mathbf{y}_j)$, $j = 1, \ldots, M_2$, *approximate values of* $f(\mathbf{y}_j)$.
Computational cost: $\mathscr{O}\big(n^d \log n + m^d\,(M_1 + M_2)\big)$.

Remark 7.24 Algorithm 7.23 can be written in matrix–vector notation as follows. Introducing the kernel matrix
$$\mathbf{K} := \big(\mathscr{K}(\mathbf{y}_j - \mathbf{x}_k)\big)_{j,k=1}^{M_2,\,M_1} \in \mathbb{R}^{M_2 \times M_1}$$

and the coefficient vector $\boldsymbol{\alpha} := (\alpha_k)_{k=1}^{M_1} \in \mathbb{C}^{M_1}$, Algorithm 7.23 reads in matrix-vector notation

$$\mathbf{K}\boldsymbol{\alpha} \approx \big(\mathbf{A}_2\,\mathbf{D}_{\mathscr{K}_\mathrm{R}}\,\bar{\mathbf{A}}_1^\top + \mathbf{K}_\mathrm{NE}\big)\,\boldsymbol{\alpha}\,,$$

where

$$\mathbf{A}_1 := \big(\mathrm{e}^{\mathrm{i}\,\mathbf{l}\cdot\mathbf{x}_k}\big)_{k=1,\ldots,M_1,\,\mathbf{l}\in I_n^d}\,, \qquad \mathbf{D}_{\mathscr{K}_\mathrm{R}} := \mathrm{diag}\,(b_\mathbf{l})_{\mathbf{l}\in I_n^d}\,,$$
$$\mathbf{A}_2 := \big(\mathrm{e}^{\mathrm{i}\,\mathbf{l}\cdot\mathbf{y}_j}\big)_{j=1,\ldots,M_2,\,\mathbf{l}\in I_n^d}\,, \qquad \mathbf{K}_\mathrm{NE} := \big(\mathscr{K}_\mathrm{NE}(\mathbf{y}_j - \mathbf{x}_k)\big)_{j,\,k=1}^{M_2,\,M_1}\,.$$

Using the matrix factorization (7.20) of our NFFT, we have

$$\mathbf{A}_1 \approx \mathbf{B}_1\,\mathbf{F}_{\sigma n,n}^d\,\mathbf{D}_1\,, \quad \mathbf{A}_2 \approx \mathbf{B}_2\,\mathbf{F}_{\sigma n,n}^d\,\mathbf{D}_2$$

with diagonal matrices $\mathbf{D}_1$ and $\mathbf{D}_2$, sparse matrices $\mathbf{B}_1$ and $\mathbf{B}_2$ having at most $(2m+1)^d$ nonzero entries in each row and column, and the d-variate Fourier matrix given in (7.22).

$$\mathbf{F}_n^d := \big(\mathrm{e}^{-2\pi\mathrm{i}\,\mathbf{k}\cdot\mathbf{l}/n}\big)_{\mathbf{k},\,\mathbf{l}\in I_n^d}\,.$$

This can be rewritten as

$$\mathbf{K}\boldsymbol{\alpha} \approx (\bar{\mathbf{B}}_{M_2}\,\mathbf{T}\,\mathbf{B}_{M_1}^\top + \mathbf{K}_\mathrm{NE})\,\boldsymbol{\alpha}\,,$$

where $\mathbf{T} := \mathbf{F}_{\sigma n,n}\,\mathbf{D}_n\,\mathbf{D}_{\mathscr{K}_\mathrm{R}}\,\mathbf{D}_n\,\mathbf{F}_{\sigma n,n}^\top$ is a multilevel Toeplitz–Toeplitz block matrix. Note that one can avoid the complex arithmetic by using fast Toeplitz matrix–vector multiplications based on discrete trigonometric transforms (see [337, Algorithm 3]). □

Next we are interested in a nonsingular kernel function $\mathscr{K}(\mathbf{x}) = K(\|\mathbf{x}\|_2)$ for $\mathbf{x} \in \mathbb{R}^d$. For instance, the Gaussian function $K(x) = \mathrm{e}^{-\delta x^2}$ with fixed $\delta > 0$ generates such a nonsingular kernel function. Here, a regularization of K near zero is not necessary. Thus, our computation does not require a near-field correction. If $K(x)$ is very small near $x = \pm\pi$, as it happens for the Gaussian function with sufficiently large value δ, we also don't need a regularization of K near $\pm\pi$. In this case we set $K_\mathrm{R} := K$ on $[-\pi,\,\pi]$ and

$$\mathscr{K}_\mathrm{R}(\mathbf{x}) := \begin{cases} K(\|\mathbf{x}\|_2) & \|\mathbf{x}\|_2 < \pi\,, \\ K(\pi) & \mathbf{x}\in[-\pi,\,\pi]^d \text{ with } \|\mathbf{x}\|_2 \ge \pi\,. \end{cases} \tag{7.76}$$

Otherwise, we need a regularization of K near $\pm\pi$. Therefore, we choose the two-point Taylor interpolation polynomial $T_\mathrm{B} \in \mathscr{P}_{2r-1}$ with the interpolation conditions (7.69) and use the function

$$\mathscr{K}_{\mathrm{R}}(\mathbf{x}) = \begin{cases} K(\|\mathbf{x}\|_2) & \|\mathbf{x}\|_2 \le \pi - \varepsilon_{\mathrm{B}}\,, \\ T_{\mathrm{B}}(\|\mathbf{x}\|_2) & \pi - \varepsilon_{\mathrm{B}} < \|\mathbf{x}\|_2 < \pi\,, \\ T_{\mathrm{B}}(\pi) & \mathbf{x} \in [-\pi,\, \pi]^d \text{ with } \|\mathbf{x}\|_2 \ge \pi\,. \end{cases} \tag{7.77}$$

Then Algorithm 7.23 can also be applied for the fast summation with a nonsingular kernel function, and it simplifies to its first three steps. Moreover, we will see that the lack of the "near-field correction" implies that the size $n \times \ldots \times n$ of the NFFT and NFFT$^\top$ does not depend on the numbers M_1 and M_2 of the given knots. Thus, the Algorithm 7.23 with steps 1–3 requires for a nonsingular kernel only $\mathscr{O}\big((\rho n)^d \log(\rho n) + m^d(M_1 + M_2)\big) = \mathscr{O}(M_1 + M_2)$ arithmetic operations. Applying Algorithm 7.23 to the Gaussian function $K(x) = \mathrm{e}^{-\delta x^2}$ with fixed $\delta > 0$, we obtain the *fast Gauss transform*.

Algorithm 7.25 (Fast Gauss Transform)
Input: $\alpha_k \in \mathbb{C}$ *for* $k = 1, \ldots, M_1$,
$\mathbf{x}_k \in \mathbb{R}^d$ *for* $k = 1, \ldots, M_1$ *with* $\|\mathbf{x}_k\|_2 < \frac{1}{2}(\pi - \varepsilon_{\mathrm{B}})$,
$\mathbf{y}_j \in \mathbb{R}^d$ *for* $j = 1, \ldots, M_2$ *with* $\|\mathbf{y}_j\|_2 < \frac{1}{2}(\pi - \varepsilon_{\mathrm{B}})$.
Precomputation: Compute the polynomial T_{B} *by Lemma* 9.35.
Form $\mathscr{K}_{\mathrm{R}}(\mathbf{x})$ *by* (7.76) *or* (7.77) *for* $K(x) = \mathrm{e}^{-\delta x^2}$.
Compute $(b_{\mathbf{l}})_{\mathbf{l} \in I_n^d}$ *by (7.71) using* FFT *of size* $n \times \ldots \times n$.

1. *For each* $\mathbf{l} \in I_n^d$ *compute*

$$a_{\mathbf{l}} := \sum_{k=1}^{M_1} \alpha_k \, \mathrm{e}^{-\mathrm{i}\,\mathbf{l}\cdot\mathbf{x}_k}$$

using the d*-variate* NFFT$^\top$ *of size* $n \times \ldots \times n$ *(see Algorithm* 7.3*).*
2. *For each* $\mathbf{l} \in I_n^d$ *compute the products* $d_{\mathbf{l}} := a_{\mathbf{l}}\, b_{\mathbf{l}}$.
3. *For* $j = 1, \ldots, M_2$ *compute the far-field sums*

$$\tilde{f}(\mathbf{y}_j) = f_{\mathrm{RF}}(\mathbf{y}_j) := \sum_{\mathbf{l} \in I_n^d} d_{\mathbf{l}}\, \mathrm{e}^{\mathrm{i}\,\mathbf{l}\cdot\mathbf{y}_j}$$

using the d*-variate* NFFT *of size* $n \times \ldots \times n$ *(see Algorithm* 7.1*).*

Output: $\tilde{f}(\mathbf{y}_j)$, $j = 1, \ldots, M_2$, *approximate values of*

$$f(\mathbf{y}_j) = \sum_{k=1}^{M_1} \alpha_k \, \mathrm{e}^{-\delta\, \|\mathbf{y}_j - \mathbf{x}_k\|_2^2}\,.$$

Computational cost: $\mathscr{O}\big(m^d\,(M_1 + M_2)\big)$.

Remark 7.26 Fast algorithms for the discrete Gauss transforms have been introduced in [30, 182, 183]. Error estimates for fast summation at nonequispaced knots have been presented in [331, 335] and for the Gauss transform in [259]. The related software is available from [231], where a variety of different kernels is implemented. Furthermore, there exists a MATLAB interface (see [231, ./matlab/fastsum]).

7.6 Inverse Nonequispaced Discrete Transforms

Important examples of nonequispaced discrete transforms are the nonequispaced discrete Fourier transform (NDFT) (see Sect. 7.1) and the nonequispaced discrete cosine transform (NDCT) (see Sect. 7.4). As shown in Sects. 7.1 and 7.4, fast nonequispaced discrete transforms are efficient algorithms for the computation of matrix–vector products $\mathbf{A}\,\mathbf{f}$, where $\mathbf{A}$ denotes the matrix of a nonequispaced discrete transform and $\mathbf{f}$ is an arbitrary given vector. The goal of this section is to present algorithms for inverse nonequispaced discrete transforms. *Inverse nonequispaced discrete transform* of a given vector $\mathbf{f}$ means the determination of a vector $\hat{\mathbf{f}}$ as (approximate) solution of the linear system

$$\mathbf{A}\hat{\mathbf{f}} = \mathbf{f}\,.$$

First, we present direct methods for inverse NDCT and inverse NDFT in the one-dimensional case. Note that we compute the inverse nonequispaced discrete transform without knowledge of a (generalized) inverse matrix of the nonequispaced discrete transform. Instead of that, we first use a fast summation step, and then the inverse nonequispaced discrete transform can be realized as an inverse equispaced discrete transform. Later, we sketch iterative methods for inverse NDFT in the multidimensional case.

7.6.1 Direct Methods for Inverse NDCT and Inverse NDFT

We consider the one-dimensional case and study direct methods for the inverse NDCT first. We start with recalling the NDCT (7.56), where we have to evaluate the polynomial

$$p(x) = \sum_{k=0}^{N} \hat{f}_k\, T_k(x) = \sum_{k=0}^{N} \hat{f}_k\, \cos(k\ \arccos x)$$

at arbitrary distinct nodes $x_j \in [-1, 1]$. Here, $T_k(x) = \cos(k\ \arccos x)$, $x \in [-1,\ 1]$, denotes the kth Chebyshev polynomial (of first kind).

In Sect. 6.2 we have already shown that using the Chebyshev extreme points $x_j^{(N)} := \cos\frac{j\pi}{N}$, $j = 0, \ldots, N$, we can evaluate the polynomial p at the nodes $x_j^{(N)}$, $j = 0, \ldots, N$, with $\mathcal{O}(N \log N)$ arithmetic operations employing a DCT-I$(N+1)$. Vice versa, we can compute $\hat{f}_k$, $k = 0, \ldots, N$, from samples $p(x_j^{(N)})$, since the DCT-I is an orthogonal transform (see Lemma 3.46).

The *inverse* NDCT can be formulated as follows: Compute the coefficients $\hat{f}_k \in \mathbb{R}$, $k = 0, \ldots, N$, from given values

$$p(x_j) = \sum_{k=0}^{N} \hat{f}_k \, T_k(x_j)$$

at arbitrary distinct nodes $x_j \in [-1, 1]$, $j = 0, \ldots, N$. Taking the fast summation technique in Sect. 7.5, we transfer the inverse NDCT into an inverse DCT-I. To derive the inverse NDCT, we will use the following result.

Theorem 7.27 *For arbitrary distinct nodes $x_j \in [-1, 1]$, $j = 0, \ldots, N$, and $\hat{f}_k \in \mathbb{R}$, $k = 0, \ldots, N$, let*

$$f_j := \sum_{k=0}^{N} \hat{f}_k \, T_k(x_j) = \sum_{k=0}^{N} \hat{f}_k \, \cos(k \arccos x_j)\,, \quad j = 0, \ldots, N\,, \tag{7.78}$$

i.e., $(f_j)_{j=0}^N$ is the NDCT *of $(\hat{f}_k)_{k=0}^N$. Further, for the Chebyshev extreme points $x_\ell^{(N)} = \cos\frac{\ell\pi}{N}$, $\ell = 0, \ldots, N$, let*

$$g_\ell := \sum_{k=0}^{N} \hat{f}_k \, T_k(x_\ell^{(N)}) = \sum_{k=0}^{N} \hat{f}_k \, \cos\frac{\ell k \pi}{N}\,, \quad \ell = 0, \ldots, N\,, \tag{7.79}$$

i.e., $(g_\ell)_{\ell=0}^N$ is the DCT-I$(N+1)$ *(up to a normalization constant) of $(\hat{f}_k)_{k=0}^N$. Assume that $x_\ell^{(N)} \neq x_k$ for all ℓ, $k = 0, \ldots, N$.*

Then we have

$$g_\ell = c_\ell \sum_{j=0}^{N} \frac{f_j \, d_j}{x_\ell^{(N)} - x_j}\,, \quad \ell = 0, \ldots, N\,, \tag{7.80}$$

where

$$c_\ell = \prod_{k=0}^{N} \big(x_\ell^{(N)} - x_k\big)\,, \quad \ell = 0, \ldots, N\,, \tag{7.81}$$

$$d_j = \prod_{\substack{k=0 \\ k \neq j}}^{N} \frac{1}{x_j - x_k}, \qquad j = 0, \ldots, N. \tag{7.82}$$

Proof Let the polynomial p be defined by

$$p(x) = \sum_{k=0}^{N} \hat{f}_k \, T_k(x).$$

Using the Lagrange interpolation formula at the points x_j, we rewrite p in the form

$$p(x) = \sum_{j=0}^{N} p(x_j) \prod_{\substack{k=0 \\ k \neq j}}^{N} \frac{x - x_k}{x_j - x_k}.$$

Thus, for $x \neq x_j$ we obtain

$$p(x) = \prod_{n=0}^{N} (x - x_n) \sum_{j=0}^{N} \frac{p(x_j)}{x - x_j} \prod_{\substack{k=0 \\ k \neq j}}^{N} \frac{1}{x_j - x_k}, \tag{7.83}$$

and hence,

$$p\big(x_\ell^{(N)}\big) = g_\ell = c_\ell \sum_{j=0}^{N} \frac{f_j \, d_j}{x_\ell^{(N)} - x_j}.$$

This completes the proof. ■

Remark 7.28 Formula (7.83) can be considered as a special case of the *barycentric formula* (see Sect. 6.2 and [44, formula (8.1)]). □

Consequently, we can efficiently compute the values g_ℓ via (7.80) from the given values f_j by the Algorithm 7.23 using the singular kernel function $\frac{1}{x}$. Applying inverse DCT-I, we then calculate the values $\hat{f}_k$, $k = 0, \ldots, N$, from g_ℓ, $\ell = 0, \ldots, N$. We then summarize this procedure.

Algorithm 7.29 (Inverse NDCT)
Input: $x_j \in [-1, 1]$, $f_j \in \mathbb{R}$, $j = 0, \ldots, N$.
Precomputation: c_ℓ, $\ell = 0, \ldots, N$, *by* (7.81),
d_j, $j = 0, \ldots, N$, *by* (7.82) *or by Remark* 7.30.

1. *Compute the values* g_ℓ, $\ell = 0, \ldots, N$ *in* (7.80) *by Algorithm* 7.23 *with the kernel* $\frac{1}{x}$.

2. *Compute the values* $\hat{f}_k$, $k = 0, \ldots, N$ *in* (7.79) *by the inverse* DCT-I$(N+1)$ *using Algorithm* 6.28 *or* 6.35.

Output: $\hat{f}_k$, $k = 0, \ldots, N$.
Computational cost: $\mathcal{O}(N \log N)$.

Remark 7.30 The naive precomputation of c_ℓ and d_j can be improved using the relations

$$c_\ell = \prod_{k=0}^{N} \big(x_\ell^{(N)} - x_k\big) = (\mathrm{sgn}\, c_\ell)\, \exp\big(\sum_{k=0}^{N} \ln |x_\ell^{(N)} - x_k|\big)\,,$$

$$d_j = \prod_{\substack{k=0\\k\neq j}}^{N} \frac{1}{x_j - x_k} = (\mathrm{sgn}\, d_j)\, \exp\big(-\sum_{\substack{k=0\\k\neq j}}^{N} \ln |x_j - x_k|\big)$$

and applying Algorithm 7.23 with the singular kernel function $\ln |x|$. □

Based on the same ideas, we will also develop a fast algorithm for the *inverse* NDFT. In contrast to [122] we use the fast summation method of Algorithm 7.23 with the kernel $\cot x$ instead of the fast multipole method. Note that with the simple modification in (7.68), we can handle odd singular kernels as well. Taking the fast summation technique, we transfer the inverse NDFT into an inverse DFT. The inverse NDFT is based on the following result (see [122, Theorem 2.3]):

Theorem 7.31 *For* $N \in 2\mathbb{N}$, *let* $x_j \in [-\pi, \pi) \setminus \{\frac{2\pi k}{N} : k = -\frac{N}{2}, \ldots, \frac{N}{2} - 1\}$, $j = -\frac{N}{2}, \ldots, \frac{N}{2} - 1$ *be distinct nodes and let* $\hat{f}_k \in \mathbb{C}$, $k = -\frac{N}{2}, \ldots, \frac{N}{2} - 1$ *be given. Let*

$$f_j := \sum_{k=-N/2}^{N/2-1} \hat{f}_k\, \mathrm{e}^{\mathrm{i}\,k x_j}\,, \quad j = -\frac{N}{2}, \ldots, \frac{N}{2} - 1\,, \tag{7.84}$$

i.e., $(f_j)_{j=-N/2}^{N/2-1}$ *is the* NDFT *of* $(\hat{f}_k)_{k=-N/2}^{N/2-1}$. *Further, for equispaced nodes* $h_\ell^{(N)} := \frac{2\pi\ell}{N}$, $\ell = -\frac{N}{2}, \ldots, \frac{N}{2} - 1$, *let*

$$g_\ell := \sum_{k=-N/2}^{N/2-1} \hat{f}_k\, \mathrm{e}^{\mathrm{i}\,k h_\ell^{(N)}} = \sum_{k=-N/2}^{N/2-1} \hat{f}_k\, \mathrm{e}^{2\pi \mathrm{i} k\ell/N}\,, \quad \ell = -\frac{N}{2}, \ldots, \frac{N}{2} - 1\,, \tag{7.85}$$

i.e., $(g_\ell)_{\ell=-N/2}^{N/2-1}$ *is the modified* DFT *of* $(\hat{f}_k)_{k=-N/2}^{N/2-1}$.
Then we have

$$g_\ell = c_\ell \sum_{j=-N/2}^{N/2-1} f_j\, d_j \Big(\cot\big(\frac{h_\ell^{(N)} - x_j}{2}\big) - \mathrm{i}\Big) \tag{7.86}$$

with

$$c_\ell := \prod_{k=-N/2}^{N/2-1} \sin \frac{h_\ell^{(N)} - x_k}{2}, \tag{7.87}$$

$$d_j := \prod_{\substack{k=-N/2 \\ k \neq j}}^{N/2-1} \frac{1}{\sin \frac{x_j - x_k}{2}}. \tag{7.88}$$

Proof We introduce the polynomial p by

$$p(z) := \sum_{k=0}^{N-1} \hat{f}_{k-N/2}\, z^k, \quad z \in \mathbb{C}.$$

Using the Lagrange interpolation formula at the distinct nodes $z_k := \mathrm{e}^{\mathrm{i}x_k}$, $k = -\frac{N}{2}, \ldots, \frac{N}{2}$, we rewrite p in the form

$$p(z) = \sum_{j=-N/2}^{N/2-1} p(z_j) \prod_{\substack{k=-N/2 \\ k \neq j}}^{N/2-1} \frac{z - z_k}{z_j - z_k}.$$

Then for $z \neq z_j$ we obtain

$$p(z) = \prod_{n=-N/2}^{N/2-1} (z - z_n) \sum_{j=-N/2}^{N/2-1} \frac{p(z_j)}{z - z_j} \prod_{\substack{k=-N/2 \\ k \neq j}}^{N/2-1} \frac{1}{z_j - z_k}.$$

The equispaced nodes $w_N^{-\ell} := \mathrm{e}^{\mathrm{i}\,h_\ell^{(N)}} = \mathrm{e}^{2\pi\mathrm{i}\,\ell/N}$, $\ell = -\frac{N}{2}, \ldots, \frac{N}{2}$ are complex Nth roots of unity which satisfy the condition $w_N^{-\ell} \neq z_j$ for all indices ℓ and j by assumption. Hence, for $z = w_N^{-\ell}$ it follows that

$$p(w_N^{-\ell}) = \prod_{n=-N/2}^{N/2-1} (w_N^{-\ell} - z_n) \sum_{j=-N/2}^{N/2-1} \frac{p(z_j)}{w_N^{-\ell} - z_j} \prod_{\substack{k=-N/2 \\ k \neq j}}^{N/2-1} \frac{1}{z_j - z_k}. \tag{7.89}$$

By (7.84) and (7.85), we have

$$f_j = z_j^{-N/2}\, p(z_j), \quad g_j = (-1)^j\, p(w_N^{-j}), \quad j = -\frac{N}{2}, \ldots, \frac{N}{2} - 1.$$

Using (7.87), we calculate

$$\prod_{n=-N/2}^{N/2-1} (w_N^{-\ell} - z_n) = \prod_{n=-N/2}^{N/2-1} \left(\mathrm{e}^{\mathrm{i}\, h_\ell^{(N)}} - \mathrm{e}^{\mathrm{i}\, x_n}\right)$$
$$= \prod_{n=-N/2}^{N/2-1} \mathrm{e}^{\mathrm{i}\,(h_\ell^{(N)}+x_n)/2} \left(\mathrm{e}^{\mathrm{i}\,(h_\ell^{(N)}-x_n)/2} - \mathrm{e}^{-\mathrm{i}\,(h_\ell^{(N)}-x_n)/2}\right)$$
$$= (-1)^\ell\,(2\,\mathrm{i})^N \prod_{n=-N/2}^{N/2-1} \mathrm{e}^{\mathrm{i}\, x_n/2}\, \sin\frac{h_\ell^{(N)} - x_n}{2} = (-1)^\ell\,(2\,\mathrm{i})^N\, a\, c_\ell$$

with

$$a := \prod_{n=-N/2}^{N/2-1} \mathrm{e}^{\mathrm{i}\, x_n/2}\,.$$

Analogously with (7.88) we compute the product

$$\prod_{\substack{k=-N/2\\ k\neq j}}^{N/2-1} (z_j - z_k) = (2\,\mathrm{i})^{N-1}\, \mathrm{e}^{\mathrm{i}\, x_j\,(N-1)/2} \prod_{\substack{k=-N/2\\ k\neq j}}^{N/2-1} \mathrm{e}^{\mathrm{i}\, x_k/2}\, \sin\frac{x_j - x_k}{2}$$
$$= (2\,\mathrm{i})^{N-1}\, \mathrm{e}^{\mathrm{i}\, x_j\,(N-2)/2}\, \frac{a}{d_j}\,.$$

Then from (7.89) it follows that

$$p(w_N^{-\ell}) = (-1)^\ell\, g_\ell = 2\,\mathrm{i}\,(-1)^\ell\, c_\ell \sum_{j=-N/2}^{N/2-1} \mathrm{e}^{\mathrm{i}\, x_j}\, \frac{f_j\, d_j}{w_N^{-\ell} - z_j}\,.$$

With

$$\cot x = \mathrm{i}\, \frac{\mathrm{e}^{\mathrm{i}\, x} + \mathrm{e}^{-\mathrm{i}\, x}}{\mathrm{e}^{\mathrm{i}\, x} - \mathrm{e}^{-\mathrm{i}\, x}}\,, \quad x \in \mathbb{R} \setminus (\pi \mathbb{Z})\,,$$

we find

$$\cot\frac{h_\ell^{(N)} - x_j}{2} - \mathrm{i} = \frac{2\,\mathrm{i}\, z_j}{w_N^{-\ell} - z_j}\,.$$

This implies the relation (7.86). ∎

Consequently, we can efficiently compute the values g_ℓ via (7.86) from the given values f_j using a univariate variant of Algorithm 7.23 with the odd kernel $\cot x$.

Applying the modified DFT, we can calculate the values $\hat{f}_k$, $k = -\frac{N}{2}, \ldots \frac{N}{2} - 1$, from g_ℓ, $\ell = -\frac{N}{2}, \ldots \frac{N}{2} - 1$.

Algorithm 7.32 (Inverse NDFT)
Input: $N \in 2\mathbb{N}$, $x_j \in [-\pi, \pi) \setminus \{\frac{2\pi k}{N} : k = -\frac{N}{2}, \ldots, \frac{N}{2} - 1\}$, $f_j \in \mathbb{C}$, $j = -\frac{N}{2}, \ldots, \frac{N}{2} - 1$.
Precomputation: c_ℓ, $\ell = -\frac{N}{2}, \ldots, \frac{N}{2} - 1$, *by* (7.87),
d_j, $j = -\frac{N}{2}, \ldots, \frac{N}{2} - 1$, *by* (7.88).

1. *Compute the values* (7.86) *by a univariate variant of Algorithm* 7.23 *with the odd kernel* $\cot x$.
2. *Compute the values* $\hat{f}_k$ *by the inverse modified* DFT (7.85).

Output: $\hat{f}_k$, $k = -\frac{N}{2}, \ldots, \frac{N}{2} - 1$.
Computational cost: $\mathcal{O}(N \log N)$.

Remark 7.33 Algorithm 7.32 is part of the NFFT software (see [231, ./matlab/infft1D]). □

Remark 7.34 Formula (7.86) is closely related to the *barycentric formula* (see Theorem 3.9). In (7.86) we apply the Lagrange polynomials at the nonequispaced points x_k. In Theorem 3.9 we used Lagrange polynomials at the equispaced points $h_\ell^{(N)}$. Note that interchanging $w_N^{-\ell}$ and z_n in (7.89) leads to the second barycentric formula in Theorem 3.9.

Applying Lagrange interpolation and fast summation, direct methods for inverse NDFT are presented by Kircheis and Potts [239]. In the multidimensional case, a direct method for inverse NDFT with an application to image reconstruction is described in [240]. □

7.6.2 *Iterative Methods for Inverse NDFT*

Now we explain the inversion of the multidimensional NDFT using iterative methods. This approach can be applied to the mentioned NDFT variants as well.

Inversion of the NDFT means the computation of all coefficients $\hat{f}_{\mathbf{k}} \in \mathbb{C}$, $\mathbf{k} \in I_N^d$, of the d-variate, 2π-periodic trigonometric polynomial

$$f(\mathbf{x}) = \sum_{\mathbf{k} \in I_N^d} \hat{f}_{\mathbf{k}} \, \mathrm{e}^{\mathrm{i}\,\mathbf{k}\cdot\mathbf{x}},$$

if function samples $f_j = f(\mathbf{x}_j)$, $j = 0, \ldots, M - 1$, at arbitrary knots $\mathbf{x}_j \in [-\pi, \pi)^d$ are given (see (7.2)). In matrix–vector notation, this problem is equivalent to solving the linear system

$$\mathbf{A}\hat{\mathbf{f}} = \mathbf{f} \tag{7.90}$$

with the given vector $\mathbf{f} = (f_j)_{j=0}^{M-1} \in \mathbb{C}^M$ and the unknown vector $\hat{\mathbf{f}} = (\hat{f}_{\mathbf{k}})_{\mathbf{k}\in I_N^d} \in \mathbb{C}^{N^d}$, where the nonequispaced Fourier matrix $\mathbf{A}$ is given in (7.19). This linear system can be either *overdetermined*, if $N^d \le M$ (this includes the quadratic case), or *underdetermined*, if $N^d > M$. Generally, this forces us to look for a pseudo-inverse solution $\hat{\mathbf{f}}^+$ (see, e.g., [51, p. 15]). For this, we also require that the nonequispaced Fourier matrix $\mathbf{A}$ has full rank. Eigenvalue estimates in [20, 134, 258] indeed assert that this condition is satisfied, if the sampling set is uniformly dense or uniformly separated with respect to the inverse bandwidth.

In the overdetermined case, we consider a weighted least squares problem, while for the consistent underdetermined case, we look for a solution of an interpolation problem. Both problems can be iteratively solved using NFFT (see Algorithm 7.1) and adjoint NFFT (see Algorithm 7.3) to realize fast matrix–vector multiplications involving $\mathbf{A}$ or $\mathbf{A}^{\mathrm{H}}$.

If $N^d \le M$, the linear system (7.90) is overdetermined which typically implies that the given data $f_j \in \mathbb{C}$, $j = 0, \ldots, M-1$, can only be approximated up to a residual $\mathbf{r} := \mathbf{f} - \mathbf{A}\,\hat{\mathbf{f}}$. Therefore, we consider the *weighted least squares problem*

$$\min_{\hat{\mathbf{f}}} \sum_{j=0}^{M-1} w_j\,|f_j - f(\mathbf{x}_j)|^2$$

with weights $w_j > 0$. The weights might be used to compensate for clusters in the sampling set. The weighted least squares problem is equivalent to solving the *weighted normal equations of first kind*

$$\mathbf{A}^{\mathrm{H}}\,\mathbf{W}\,\mathbf{A}\,\hat{\mathbf{f}} = \mathbf{A}^{\mathrm{H}}\mathbf{W}\,\mathbf{f}$$

with the diagonal matrix $\mathbf{W} := \mathrm{diag}\,(w_j)_{j=0}^{M-1}$. This linear system can be solved using the Landweber (or Richardson) iteration, the steepest descent method, or the conjugate gradient method for least squares problems. In the NFFT library [231] all three algorithms are implemented.

If $N^d > M$, and if the linear system (7.90) is consistent, then the data $f_j \in \mathbb{C}$, $j = 0, \ldots, M-1$, can be interpolated exactly. But since there exist multiple solutions, we consider the *constrained minimization problem*

$$\min_{\hat{\mathbf{f}}} \sum_{\mathbf{k}\in I_N^d} \frac{|\hat{f}_{\mathbf{k}}|^2}{\hat{w}_{\mathbf{k}}} \quad \text{subject to} \quad \mathbf{A}\,\hat{\mathbf{f}} = \mathbf{f},$$

which incorporates "damping factors" $\hat{w}_{\mathbf{k}} > 0$. A smooth solution, i.e., a solution with rapid decay of Fourier coefficients $\hat{f}_{\mathbf{k}}$, is favored, if the damping factors $\hat{w}_{\mathbf{k}}$ are decreasing. The interpolation problem is equivalent to the *damped normal equations of second kind*

$$\mathbf{A}\,\hat{\mathbf{W}}\,\mathbf{A}^{\mathrm{H}}\tilde{\mathbf{f}} = \mathbf{f}\,, \quad \hat{\mathbf{f}} = \hat{\mathbf{W}}\,\mathbf{A}^{\mathrm{H}}\tilde{\mathbf{f}}$$

with the diagonal matrix $\hat{\mathbf{W}} := \mathrm{diag}\,(\hat{w}_{\mathbf{k}})_{\mathbf{k}\in I_N^d}$. The NFFT library [231] also contains this scheme.

We summarize the inverse d-dimensional NFFT in the overdetermined case.

Algorithm 7.35 (Inverse Iterative NDFT Using the Conjugate Gradient Method for Normal Equations of First Kind (CGNR))
Input: $N \in 2\mathbb{N}$, $\mathbf{x}_j \in [-\pi,\, \pi)^d$, $\mathbf{f} \in \mathbb{C}^M$, $\hat{\mathbf{f}}_0 \in \mathbb{C}^{N^d}$.

1. *Set* $\mathbf{r}_0 := \mathbf{f} - \mathbf{A}\,\hat{\mathbf{f}}_0$.
2. *Compute* $\hat{\mathbf{z}}_0 := \mathbf{A}^{\mathrm{H}}\,\mathbf{W}\,\mathbf{r}_0$.
3. *Set* $\hat{\mathbf{p}}_0 = \hat{\mathbf{z}}_0$.
4. *For* $\ell = 0,\, 1, \ldots$ *compute*

$$\begin{aligned}
&\mathbf{v}_\ell = \mathbf{A}\,\hat{\mathbf{W}}\,\hat{\mathbf{p}}_\ell\\
&\alpha_\ell = (\mathbf{v}_\ell^{\mathrm{H}}\,\mathbf{W}\,\mathbf{v}_\ell)^{-1}\,(\hat{\mathbf{z}}_\ell^{\mathrm{H}}\,\hat{\mathbf{W}}\,\hat{\mathbf{z}}_\ell)\\
&\hat{\mathbf{f}}_{\ell+1} = \hat{\mathbf{f}}_\ell + \alpha_\ell\,\hat{\mathbf{W}}\,\hat{\mathbf{p}}_\ell\\
&\mathbf{r}_{\ell+1} = \mathbf{r}_\ell - \alpha_\ell\,\mathbf{v}_\ell\\
&\hat{\mathbf{z}}_{\ell+1} = \mathbf{A}^{\mathrm{H}}\,\mathbf{W}\,\mathbf{r}_{\ell+1}\\
&\beta_\ell = (\hat{\mathbf{z}}_{\ell+1}^{\mathrm{H}}\,\hat{\mathbf{W}}\hat{\mathbf{z}}_{\ell+1})^{-1}\,(\hat{\mathbf{z}}_\ell^{\mathrm{H}}\,\hat{\mathbf{W}}\,\hat{\mathbf{z}}_\ell)\\
&\hat{\mathbf{p}}_{\ell+1} = \beta_\ell\,\hat{\mathbf{p}}_\ell + \hat{\mathbf{z}}_{\ell+1}.
\end{aligned}$$

Output: $\hat{\mathbf{f}}_\ell \in \mathbb{C}^{N^d}$ *ℓth approximation of the solution vector* $\hat{\mathbf{f}}$.

Remark 7.36 The algorithms presented in this chapter are part of the NFFT software [231]. The algorithmic details are described in [234]. The proposed algorithms have been implemented on top of the well-optimized FFTW software library [153, 154]. The implementation and numerical results of the multi-threaded NFFT based on OpenMP is described in [424]. Furthermore, there exist MATLAB, Octave, and Julia interfaces. We also provide Windows binaries as DLL.

Implementations on GPU are presented in [253, 446]. Parallel NFFT algorithms are developed in [322] and have been published in the PNFFT software library [320]. The implementation of these algorithms is part of the publicly available ScaFaCoS software library [11]. For the NFFT with the exponential of semicircle window function (7.51), an algorithm is presented in [16]. □

Chapter 8
High-Dimensional FFT

In this chapter, we discuss methods for the approximation of d-variate functions in high-dimension $d \in \mathbb{N}$ based on sampling along rank-1 lattices, and we derive the corresponding fast algorithms. In contrast to Chap. 4, our approach to compute the Fourier coefficients of d-variate functions is no longer based on tensor product methods. In Sect. 8.1, we introduce weighted subspaces of $L_1(\mathbb{T}^d)$ which are characterized by the decay properties of the Fourier coefficients. We show that functions in these spaces can be already well approximated by d-variate trigonometric polynomials on special frequency index sets. In Sect. 8.2, we study the fast evaluation of d-variate trigonometric polynomials on finite frequency index sets. We introduce so-called rank-1 lattices and derive an algorithm for the fast evaluation of these trigonometric polynomials at the lattice points. The special structure of the rank-1 lattice enables us to perform this computation using only a one-dimensional FFT. In order to reconstruct the Fourier coefficients of the d-variate trigonometric polynomials from the polynomial values at the lattice points exactly, the used rank-1 lattice needs to satisfy a special condition. Using so-called reconstructing rank-1 lattices, the stable computation of the Fourier coefficients of a d-variate trigonometric polynomial can be again performed by employing only a one-dimensional FFT, where the numerical effort depends on the lattice size. In Sect. 8.3, we come back to the approximation of periodic functions in weighted subspaces of $L_1(\mathbb{T}^d)$ on rank-1 lattices. Section 8.4 considers the construction of rank-1 lattices. We present a constructive component-by-component algorithm with less than $|I|^2$ lattice points, where I denotes the finite index set of nonzero Fourier coefficients that have to be computed. In particular, this means that the computational effort to reconstruct the Fourier coefficients depends only linearly on the dimension and mainly on the size of the frequency index sets of the considered trigonometric polynomials. In order to overcome the limitations of the single rank-1 lattice approach, we generalize the proposed methods to multiple rank-1 lattices in Sect. 8.5.

G. Plonka et al., *Numerical Fourier Analysis*, Applied and Numerical Harmonic Analysis, https://doi.org/10.1007/978-3-031-35005-4_8

8.1 Fourier Partial Sums of Smooth Multivariate Functions

In order to ensure a good quality of the obtained approximations of d-variate periodic functions, we need to assume that these functions satisfy certain smoothness conditions, which are closely related to the decay properties of their Fourier coefficients. As we have already seen for $d = 1$, the smoothness properties of a function strongly influence the quality of a specific approximation method (e.g., see Theorem 1.39 of Bernstein).

We consider a d-variate periodic function $f : \mathbb{T}^d \to \mathbb{C}$ with the Fourier series

$$f(\mathbf{x}) = \sum_{\mathbf{k}\in\mathbb{Z}^d} c_{\mathbf{k}}(f)\, \mathrm{e}^{\mathrm{i}\,\mathbf{k}\cdot\mathbf{x}} . \tag{8.1}$$

We will always assume that $f \in L_1(\mathbb{T}^d)$ in order to guarantee the existence of all Fourier coefficients $c_{\mathbf{k}}(f)$, $\mathbf{k} \in \mathbb{Z}^d$. For the definition of function spaces $L_p(\mathbb{T}^d)$, $1 \le p < \infty$, we refer to Sect. 4.1.

The decay properties of Fourier coefficients can also be used to characterize the smoothness of the function f (see Theorem 1.37 for $d = 1$ or Theorem 4.9 for $d > 1$). For a detailed characterization of periodic functions and suitable function spaces, in particular with respect to the decay properties of the Fourier coefficients, we refer to [378, Chapter 3].

In this section, we consider the approximation of a d-variate periodic function $f \in L_1(\mathbb{T}^d)$ using Fourier partial sums $S_I f$

$$(S_I f)(\mathbf{x}) := \sum_{\mathbf{k}\in I} c_{\mathbf{k}}(f)\, \mathrm{e}^{\mathrm{i}\,\mathbf{k}\cdot\mathbf{x}} , \tag{8.2}$$

where the finite index set $I \subset \mathbb{Z}^d$ needs to be carefully chosen with respect to the properties of the sequence of the Fourier coefficients $\big(c_{\mathbf{k}}(f)\big)_{\mathbf{k}\in\mathbb{Z}^d}$. The set I is called *frequency index set* of the Fourier partial sum. The operator $S_I : L_1(\mathbb{T}^d) \to C(\mathbb{T}^d)$ maps f to a trigonometric polynomial with frequencies supported on the finite index set I. We call

$$\Pi_I := \mathrm{span}\,\{\mathrm{e}^{\mathrm{i}\,\mathbf{k}\cdot\mathbf{x}} : \mathbf{k} \in I\}$$

the *space of trigonometric polynomials supported on* I. We will be interested in frequency index sets of type

$$I = I^d_{p,N} := \left\{\mathbf{k} = (k_s)_{s=1}^d \in \mathbb{Z}^d : \|\mathbf{k}\|_p \le N\right\} , \tag{8.3}$$

where $\|\mathbf{k}\|_p$ is the usual p-(quasi-)norm

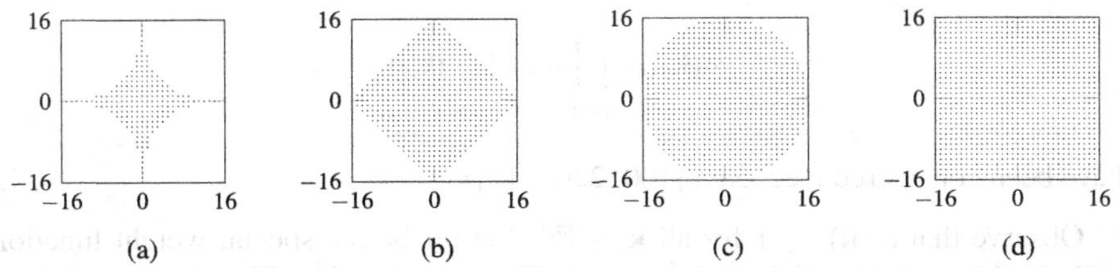

Fig. 8.1 Two-dimensional frequency index sets $I^2_{p,16}$ for $p \in \left\{\frac{1}{2},\ 1,\ 2,\ \infty\right\}$. **(a)** $I^2_{\frac{1}{2},16}$. **(b)** $I^2_{1,16}$. **(c)** $I^2_{2,16}$. **(d)** $I^2_{\infty,16}$

$$\|\mathbf{k}\|_p := \begin{cases} \left(\sum\limits_{s=1}^{d} |k_s|^p\right)^{1/p} & 0 < p < \infty\,, \\ \max\limits_{s=1,\ldots,d} |k_s| & p = \infty\,. \end{cases}$$

Figure 8.1 illustrates the two-dimensional frequency index sets $I^2_{p,16}$ for $p \in \{\frac{1}{2},\ 1,\ 2,\ \infty\}$ (see also [221, 431, 432]).

If the absolute values of the Fourier coefficients decrease sufficiently fast for growing frequency index $\mathbf{k}$, we can very well approximate the function f using only a few terms $c_{\mathbf{k}}(f)\,\mathrm{e}^{\mathrm{i}\,\mathbf{k}\cdot\mathbf{x}}$, $\mathbf{k} \in I \subset \mathbb{Z}^d$ with cardinality $|I| < \infty$. In particular, we will consider a periodic function $f \in L_1(\mathbb{T}^d)$ whose sequence of Fourier coefficients is absolutely summable. This implies by Theorem 4.9 that f has a continuous representative within $L_1(\mathbb{T}^d)$. We introduce the weighted subspace $\mathcal{A}_\omega(\mathbb{T}^d)$ of $L_1(\mathbb{T}^d)$ of functions $f : \mathbb{T}^d \to \mathbb{C}$ equipped with the norm

$$\|f\|_{\mathcal{A}_\omega(\mathbb{T}^d)} := \sum_{\mathbf{k}\in\mathbb{Z}^d} \omega(\mathbf{k})|c_{\mathbf{k}}(f)|\,, \tag{8.4}$$

if f has the Fourier expansion (8.1). Here $\omega : \mathbb{Z}^d \to [1, \infty)$ is called *weight function* and characterizes the decay of the Fourier coefficients. If ω is increasing for $\|\mathbf{k}\|_p \to \infty$, then the Fourier coefficients $c_{\mathbf{k}}(f)$ of $f \in \mathcal{A}_\omega(\mathbb{T}^d)$ have to decrease faster than the weight function ω increases with respect to $\mathbf{k} = (k_s)_{s=1}^d \in \mathbb{Z}^d$.

Example 8.1 Important examples for a weight function ω are

$$\omega(\mathbf{k}) = \omega_p^d(\mathbf{k}) := \max\{1,\ \|\mathbf{k}\|_p\}$$

for $0 < p \le \infty$. Instead of the p-norm, one can also consider a weighted p-norm. To characterize function spaces with dominating smoothness, weight functions of the form

$$\omega(\mathbf{k}) = \prod_{s=1}^{d} \max\{1,\ |k_s|\}$$

have been considered (see, e.g., [100, 220, 411]). □

Observe that $\omega(\mathbf{k}) \geq 1$ for all $\mathbf{k} \in \mathbb{Z}^d$. Let ω_1 be the special weight function with $\omega_1(\mathbf{k}) = 1$ for all $\mathbf{k} \in \mathbb{Z}^d$ and $\mathscr{A}(\mathbb{T}^d) := \mathscr{A}_{\omega_1}(\mathbb{T}^d)$. The space $\mathscr{A}(\mathbb{T}^d)$ is called *Wiener algebra*. Further, we recall that $C(\mathbb{T}^d)$ denotes the Banach space of continuous d-variate 2π-periodic functions. The norm of $C(\mathbb{T}^d)$ coincides with the norm of $L_\infty(\mathbb{T}^d)$. The next lemma (see [220, Lemma 2.1]) states that the embeddings $\mathscr{A}_\omega(\mathbb{T}^d) \subset \mathscr{A}(\mathbb{T}^d) \subset C(\mathbb{T}^d)$ are true.

Lemma 8.2 *Each function $f \in \mathscr{A}(\mathbb{T}^d)$ has a continuous representative. In particular, we obtain $\mathscr{A}_\omega(\mathbb{T}^d) \subset \mathscr{A}(\mathbb{T}^d) \subset C(\mathbb{T}^d)$ with the usual interpretation.*

Proof Let $f \in \mathscr{A}_\omega(\mathbb{T}^d)$ be given. Then the function f belongs to $\mathscr{A}(\mathbb{T}^d)$, since the following estimate holds:

$$\sum_{\mathbf{k}\in\mathbb{Z}^d} |c_{\mathbf{k}}(f)| \leq \sum_{\mathbf{k}\in\mathbb{Z}^d} \omega(\mathbf{k})\,|c_{\mathbf{k}}(f)| < \infty\,.$$

Now let $f \in \mathscr{A}(\mathbb{T}^d)$ be given. The summability of the sequence $\big(|c_{\mathbf{k}}(f)|\big)_{\mathbf{k}\in\mathbb{Z}^d}$ of the absolute values of the Fourier coefficients implies the summability of the sequence $\big(|c_{\mathbf{k}}(f)|^2\big)_{\mathbf{k}\in\mathbb{Z}^d}$ of the squared absolute values of the Fourier coefficients, and, thus, the embedding $\mathscr{A}(\mathbb{T}^d) \subset L_2(\mathbb{T}^d)$ is proved using Parseval equation (4.4).

Clearly, the function $g(\mathbf{x}) = \sum_{\mathbf{k}\in\mathbb{Z}^d} c_{\mathbf{k}}(f)\,\mathrm{e}^{\mathrm{i}\,\mathbf{k}\cdot\mathbf{x}}$ is a representative of f in $L_2(\mathbb{T}^d)$ and also in $\mathscr{A}(\mathbb{T}^d)$. We show that g is the continuous representative of f. The absolute values of the Fourier coefficients of $f \in \mathscr{A}(\mathbb{T}^d)$ are summable. So, for each $\varepsilon > 0$ there exists a finite index set $I \subset \mathbb{Z}^d$ with $\sum_{\mathbf{k}\in\mathbb{Z}^d\setminus I} |c_{\mathbf{k}}(f)| < \frac{\varepsilon}{4}$. For a fixed $\mathbf{x}_0 \in \mathbb{T}^d$, we estimate

$$\begin{aligned} |g(\mathbf{x}_0) - g(\mathbf{x})| &= \Big|\sum_{\mathbf{k}\in\mathbb{Z}^d} c_{\mathbf{k}}(f)\,\mathrm{e}^{\mathrm{i}\,\mathbf{k}\cdot\mathbf{x}_0} - \sum_{\mathbf{k}\in\mathbb{Z}^d} c_{\mathbf{k}}(f)\,\mathrm{e}^{\mathrm{i}\,\mathbf{k}\cdot\mathbf{x}}\Big| \\ &\leq \Big|\sum_{\mathbf{k}\in I} c_{\mathbf{k}}(f)\,\mathrm{e}^{\mathrm{i}\,\mathbf{k}\cdot\mathbf{x}_0} - \sum_{\mathbf{k}\in I} c_{\mathbf{k}}(f)\,\mathrm{e}^{\mathrm{i}\,\mathbf{k}\cdot\mathbf{x}}\Big| + \frac{\varepsilon}{2}\,. \end{aligned}$$

The trigonometric polynomial $(S_I f)(\mathbf{x}) = \sum_{\mathbf{k}\in I} c_{\mathbf{k}}\,\mathrm{e}^{\mathrm{i}\,\mathbf{k}\cdot\mathbf{x}}$ is a continuous function. Accordingly, for $\varepsilon > 0$ and $\mathbf{x}_0 \in \mathbb{T}^d$ there exists a $\delta_0 > 0$ such that $\|\mathbf{x}_0 - \mathbf{x}\|_1 < \delta_0$ implies $|(S_I f)(\mathbf{x}_0) - (S_I f)(\mathbf{x})| < \frac{\varepsilon}{2}$. Then we obtain $|g(\mathbf{x}_0) - g(\mathbf{x})| < \varepsilon$ for all $\mathbf{x}$ with $\|\mathbf{x}_0 - \mathbf{x}\|_1 < \delta_0$. ■

In particular for our further considerations on sampling methods, it is essential that we identify each function $f \in \mathscr{A}(\mathbb{T}^d)$ with its continuous representative in the following. Note that the definition of $\mathscr{A}_\omega(\mathbb{T}^d)$ in (8.4) using the Fourier series representation of f already comprises the continuity of the contained functions.

Considering Fourier partial sums, we will always call them exact Fourier partial sums in contrast to approximate partial Fourier sums that will be introduced later.

Lemma 8.3 *Let $I_N = \{\mathbf{k} \in \mathbb{Z}^d : \omega(\mathbf{k}) \le N\}$, $N \in \mathbb{R}$, be a frequency index set being defined by the weight function ω. Assume that the cardinality $|I_N|$ is finite.*

Then the exact Fourier partial sum

$$(S_{I_N} f)(\mathbf{x}) := \sum_{\mathbf{k} \in I_N} c_{\mathbf{k}}(f)\, \mathrm{e}^{\mathrm{i}\,\mathbf{k}\cdot\mathbf{x}} \tag{8.5}$$

approximates the function $f \in \mathscr{A}_\omega(\mathbb{T}^d)$ and we have

$$\|f - S_{I_N} f\|_{L_\infty(\mathbb{T}^d)} \le N^{-1}\, \|f\|_{\mathscr{A}_\omega(\mathbb{T}^d)}\,.$$

Proof We follow the ideas of [220, Lemma 2.2]. Let $f \in \mathscr{A}_\omega(\mathbb{T}^d)$. Obviously, $S_{I_N} f \in \mathscr{A}_\omega(\mathbb{T}^d) \subset C(\mathbb{T}^d)$ and we obtain

$$\begin{aligned}\|f - S_{I_N} f\|_{L_\infty(\mathbb{T}^d)} &= \operatorname*{ess\,sup}_{\mathbf{x}\in\mathbb{T}^d} |(f - S_{I_N} f)(\mathbf{x})| = \operatorname*{ess\,sup}_{\mathbf{x}\in\mathbb{T}^d} \Big|\sum_{\mathbf{k}\in\mathbb{Z}^d\setminus I_N} c_{\mathbf{k}}(f)\, \mathrm{e}^{\mathrm{i}\,\mathbf{k}\cdot\mathbf{x}}\Big| \\ &\le \sum_{\mathbf{k}\in\mathbb{Z}^d\setminus I_N} |c_{\mathbf{k}}(f)| \le \frac{1}{\inf\limits_{\mathbf{k}\in\mathbb{Z}^d\setminus I_N} \omega(\mathbf{k})} \sum_{\mathbf{k}\in\mathbb{Z}^d\setminus I_N} \omega(\mathbf{k})|\, c_{\mathbf{k}}(f)| \\ &\le \frac{1}{N} \sum_{\mathbf{k}\in\mathbb{Z}^d} \omega(\mathbf{k})\, |c_{\mathbf{k}}(f)| = N^{-1} \|f\|_{\mathscr{A}_\omega(\mathbb{T}^d)}.\end{aligned}$$

■

Remark 8.4 For the weight function $\omega(\mathbf{k}) = \big(\max\{1, \|\mathbf{k}\|_p\}\big)^{\alpha/2}$ with $0 < p \le \infty$ and $\alpha > 0$, we similarly obtain for the index set $I_N = I^d_{p,N}$ given in (8.3)

$$\begin{aligned}\|f - S_{I^d_{p,N}} f\|_{L_\infty(\mathbb{T}^d)} &\le N^{-\alpha/2} \sum_{\mathbf{k}\in\mathbb{Z}^d\setminus I^d_{p,N}} \big(\max\{1, \|\mathbf{k}\|_p\}\big)^{\alpha/2} |c_{\mathbf{k}}(f)| \\ &\le N^{-\alpha/2}\, \|f\|_{\mathscr{A}_\omega(\mathbb{T}^d)}\,.\end{aligned}$$

The error estimates can be also transferred to other norms. Let $H^{\alpha,p}(\mathbb{T}^d)$ denote the *periodic Sobolev space of isotropic smoothness* consisting of all $f \in L_2(\mathbb{T}^d)$ with finite norm

$$\|f\|_{H^{\alpha,p}(\mathbb{T}^d)} := \Big(\sum_{\mathbf{k}\in\mathbb{Z}^d} \big(\max\{1, \|\mathbf{k}\|_p\}\big)^{\alpha}\, |c_{\mathbf{k}}(f)|^2\Big)^{1/2}, \tag{8.6}$$

where f possesses the Fourier expansion (8.1) and where $\alpha > 0$ is the smoothness parameter. Using the Cauchy–Schwarz inequality, we obtain here

$$\begin{aligned}\|f - S_{I_{p,N}^d} f\|_{L_\infty(\mathbb{T}^d)} &\le \sum_{\mathbf{k}\in\mathbb{Z}^d\setminus I_{p,N}^d} |c_{\mathbf{k}}(f)| \\ &\le \Big(\sum_{\mathbf{k}\in\mathbb{Z}^d\setminus I_{p,N}^d} \|\mathbf{k}\|_p^{-\alpha}\Big)^{1/2} \Big(\sum_{\mathbf{k}\in\mathbb{Z}^d\setminus I_{p,N}^d} \|\mathbf{k}\|_p^{\alpha}\, |c_{\mathbf{k}}(f)|^2\Big)^{1/2} \\ &\le \Big(\sum_{\mathbf{k}\in\mathbb{Z}^d\setminus I_{p,N}^d} \|\mathbf{k}\|_p^{-\alpha}\Big)^{1/2} \|f\|_{H^{\alpha,p}(\mathbb{T}^d)}\,.\end{aligned}$$

Note that this estimate is related to the estimates on the decay of Fourier coefficients for functions $f \in C^r(\mathbb{T}^d)$ in (4.1) and Theorem 4.9. For detailed estimates of the approximation error of Fourier partial sums in these spaces, we refer to [252].

As we will see later, for efficient approximation, other frequency index sets, as, e.g., frequency index sets related to the hyperbolic crosses, are of special interest. The corresponding approximation errors have been studied in [76, 226, 227]. □

Lemma 8.5 *Let $N \in \mathbb{N}$ and the frequency index set $I_N := \{\mathbf{k} \in \mathbb{Z}^d : 1 \le \omega(\mathbf{k}) \le N\}$ with the cardinality $0 < |I_N| < \infty$ be given.*

Then the norm of the operator S_{I_N} that maps $f \in \mathcal{A}_\omega(\mathbb{T}^d)$ to its Fourier partial sum $S_{I_N} f$ on the index set I_N is bounded by

$$\frac{1}{\min_{\mathbf{k}\in\mathbb{Z}^d} \omega(\mathbf{k})} \le \|S_{I_N}\|_{\mathcal{A}_\omega(\mathbb{T}^d)\to C(\mathbb{T}^d)} \le \frac{1}{\min_{\mathbf{k}\in\mathbb{Z}^d} \omega(\mathbf{k})} + \frac{1}{N}.$$

Proof

1. Since $|I_N|$ is finite, there exists $\min_{\mathbf{k}\in I_N} \omega(\mathbf{k})$. The definition of I_N implies that $\min_{\mathbf{k}\in\mathbb{Z}^d} \omega(\mathbf{k}) = \min_{\mathbf{k}\in I_N} \omega(\mathbf{k})$. To obtain the upper bound for the operator norm, we apply the triangle inequality and Lemma 8.3

$$\begin{aligned}\|S_{I_N}\|_{\mathcal{A}_\omega(\mathbb{T}^d)\to C(\mathbb{T}^d)} &= \sup_{\substack{f\in\mathcal{A}_\omega(\mathbb{T}^d)\\ \|f\|_{\mathcal{A}_\omega(\mathbb{T}^d)}=1}} \|S_{I_N} f\|_{C(\mathbb{T}^d)} \\ &\le \sup_{\substack{f\in\mathcal{A}_\omega(\mathbb{T}^d)\\ \|f\|_{\mathcal{A}_\omega(\mathbb{T}^d)}=1}} \|S_{I_N} f - f\|_{C(\mathbb{T}^d)} + \sup_{\substack{f\in\mathcal{A}_\omega(\mathbb{T}^d)\\ \|f\|_{\mathcal{A}_\omega(\mathbb{T}^d)}=1}} \|f\|_{C(\mathbb{T}^d)} \\ &\le \sup_{\substack{f\in\mathcal{A}_\omega(\mathbb{T}^d)\\ \|f\|_{\mathcal{A}_\omega(\mathbb{T}^d)}=1}} \Big(\sum_{\mathbf{k}\in\mathbb{Z}^d} |c_{\mathbf{k}}(f)| + N^{-1}\|f\|_{\mathcal{A}_\omega(\mathbb{T}^d)}\Big)\end{aligned}$$

$$\le \sup_{\substack{f\in\mathscr{A}_\omega(\mathbb{T}^d)\\ \|f\|_{\mathscr{A}_\omega(\mathbb{T}^d)}=1}} \left(\sum_{\mathbf{k}\in\mathbb{Z}^d} \frac{\omega(\mathbf{k})}{\min_{\tilde{\mathbf{k}}\in\mathbb{Z}^d}\omega(\tilde{\mathbf{k}})} |c_{\mathbf{k}}(f)| + N^{-1}\|f\|_{\mathscr{A}_\omega(\mathbb{T}^d)} \right)$$

$$\le \frac{1}{\min_{\mathbf{k}\in\mathbb{Z}^d}\omega(\mathbf{k})} + \frac{1}{N}.$$

2. To prove the lower bound, we construct a suitable example. Let $\mathbf{k}' \in I_N$ be a frequency index with $\omega(\mathbf{k}') = \min_{\mathbf{k}\in\mathbb{Z}^d}\omega(\mathbf{k})$. We choose the trigonometric polynomial $g(\mathbf{x}) = \frac{1}{\omega(\mathbf{k}')}\,\mathrm{e}^{\mathrm{i}\,\mathbf{k}'\cdot\mathbf{x}}$ which is an element of $\mathscr{A}_\omega(\mathbb{T}^d)$ with $\|g\|_{\mathscr{A}_\omega(\mathbb{T}^d)} = 1$. Since $S_{I_N} g = g$, we find

$$\|S_{I_N}\|_{\mathscr{A}_\omega(\mathbb{T}^d)\to C(\mathbb{T}^d)} \ge \|S_{I_N} g\|_{C(\mathbb{T}^d)} = \|g\|_{C(\mathbb{T}^d)} = g(\mathbf{0}) = \frac{1}{\omega(\mathbf{k}')} = \frac{1}{\min\limits_{\mathbf{k}\in I_N}\omega(\mathbf{k})}.$$

■

Our observations in this section imply that smooth functions with special decay of their Fourier coefficients can be well approximated by d-variate trigonometric polynomials on special index sets. In the next section, we will therefore study the efficient evaluation of d-variate trigonometric polynomials on special grids, as well as the corresponding efficient computation of their Fourier coefficients.

8.2 Fast Evaluation of Multivariate Trigonometric Polynomials

As we have seen in the last section, smooth functions in $\mathscr{A}_\omega(\mathbb{T}^d)$ can be already well approximated by d-variate trigonometric polynomials on index sets $I_N = \{\mathbf{k}\in\mathbb{Z}^d : \omega(\mathbf{k}) \le N\}$. In Fig. 8.1, we have seen possible two-dimensional index sets, where $\omega(\mathbf{k}) = \max\{1,\ \|\mathbf{k}\|_p\}$. Therefore, we study trigonometric polynomials $p \in \Pi_I$ on the d-dimensional torus $\mathbb{T}^d \cong [0,\ 2\pi)^d$ of the form

$$p(\mathbf{x}) = \sum_{\mathbf{k}\in I} \hat{p}_{\mathbf{k}}\,\mathrm{e}^{\mathrm{i}\,\mathbf{k}\cdot\mathbf{x}} \tag{8.7}$$

with Fourier coefficients $\hat{p}_{\mathbf{k}} \in \mathbb{C}$ and with a fixed finite frequency index set $I \subset \mathbb{Z}^d$ of cardinality $|I|$.

Let $X \subset [0,\ 2\pi)^d$ be a finite set of sampling points with $|X|$ elements. Now we are interested in solving the following two problems:

(i) *Evaluation of trigonometric polynomials*. For given Fourier coefficients $\hat{p}_{\mathbf{k}}$, $\mathbf{k} \in I$, how to compute the polynomial values $p(\mathbf{x})$ for all $\mathbf{x} \in X$ efficiently?

(ii) *Evaluation of the Fourier coefficients.* For given polynomial values $p(\mathbf{x})$, $\mathbf{x} \in X$, how to compute $\hat{p}_{\mathbf{k}}$ for all $\mathbf{k} \in I$ efficiently?

The second problem also involves the question how the sampling set X has to be chosen such that $\hat{p}_{\mathbf{k}}$ for all $\mathbf{k} \in I$ can be uniquely computed in a stable way.

Let us consider the $|X|$-by-$|I|$ *Fourier matrix* $\mathbf{A} = \mathbf{A}(X, I)$ defined by

$$\mathbf{A} = \mathbf{A}(X, I) := \left(\mathrm{e}^{\mathrm{i}\,\mathbf{k}\cdot\mathbf{x}}\right)_{\mathbf{x}\in X,\,\mathbf{k}\in I} \in \mathbb{C}^{|X|\times|I|},$$

as well as the two vectors $\mathbf{p} := \big(p(\mathbf{x})\big)_{\mathbf{x}\in X} \in \mathbb{C}^{|X|}$ and $\hat{\mathbf{p}} := \big(\hat{p}(\mathbf{k})\big)_{\mathbf{k}\in I} \in \mathbb{C}^{|I|}$. To solve problem (i), we need to perform the matrix–vector multiplication

$$\mathbf{p} = \mathbf{A}\,\hat{\mathbf{p}}\,. \tag{8.8}$$

To compute $\hat{\mathbf{p}}$ from $\mathbf{p}$, we have to solve the inverse problem. For arbitrary polynomial $p \in \Pi_I$, this problem is only uniquely solvable if $|X| \geq |I|$ and if $\mathbf{A}$ possesses full rank $|I|$. In other words, the sampling set X needs to be large enough and the obtained samples need to contain "enough information" about p. Then $\mathbf{A}^{\mathrm{H}}\,\mathbf{A} \in \mathbb{C}^{|I|\times|I|}$ is invertible, and we have

$$\hat{\mathbf{p}} = (\mathbf{A}^{\mathrm{H}}\,\mathbf{A})^{-1}\,\mathbf{A}^{\mathrm{H}}\,\mathbf{p}\,. \tag{8.9}$$

In order to ensure stability of this procedure, we want to assume that the columns of $\mathbf{A}$ are orthogonal, i.e., $\mathbf{A}^{\mathrm{H}}\mathbf{A} = M\,\mathbf{I}_{|I|}$, where $\mathbf{I}_{|I|}$ is the $|I|$-by-$|I|$ unit matrix and $M = |X|$. Then (8.9) simplifies to

$$\hat{\mathbf{p}} = \frac{1}{M}\,\mathbf{A}^{\mathrm{H}}\,\mathbf{p}\,.$$

In the following, we will consider very special sampling sets X, so-called rank-1 lattices.

8.2.1 Rank-1 Lattices

Initially, rank-1 lattices were introduced as sampling schemes for (equally weighted) cubature formulas in the late 1950s and 1960s (see [248]). A summary of the early work on cubature rules based on rank-1 lattice sampling can be found in [303]. The recent increased interest in rank-1 lattices is particularly caused by new approaches to describe lattice rules that allow optimal theoretical error estimates for cubature formulas for specific function classes (see, e.g., [392]). We also refer to [390] for a survey on lattice methods for numerical integration. Note that lattice rules are a powerful and popular form of quasi-Monte Carlo rules [113].

In contrast to general lattices which are spanned by several vectors, we consider only sampling on so-called rank-1 lattices. This simplifies the evaluation of trigonometric polynomials essentially and allows to derive necessary and sufficient conditions for unique or stable reconstruction.

For a given vector $\mathbf{z} \in \mathbb{Z}^d$ and a positive integer $M \in \mathbb{N}$, we define the *rank*-1 *lattice*

$$X = \Lambda(\mathbf{z}, M) := \left\{ \mathbf{x}_j := \frac{2\pi}{M} (j\,\mathbf{z} \bmod M\,\mathbf{1}) \in [0, 2\pi)^d : j = 0, \ldots, M-1 \right\} \tag{8.10}$$

as spatial discretization in $[0, 2\pi)^d$. Here, $\mathbf{1} := (1)_{s=1}^d \in \mathbb{Z}^d$ and for $\mathbf{z} = (z_s)_{s=1}^d \in \mathbb{Z}^d$ the term $j\,\mathbf{z} \bmod M\,\mathbf{1}$ denotes the vector $(j\,z_s \bmod M)_{s=1}^d$. We call $\mathbf{z}$ the *generating vector* and M the *lattice size* of the rank-1 lattice $\Lambda(\mathbf{z}, M)$. To ensure that $\Lambda(\mathbf{z}, M)$ has exactly M distinct elements, we assume that M is coprime with at least one component of $\mathbf{z}$. Further, for a given rank-1 lattice $\Lambda(\mathbf{z}, M)$ with generating vector $\mathbf{z} \in \mathbb{Z}^d$, we call the set

$$\Lambda^{\perp}(\mathbf{z}, M) := \{\mathbf{k} \in \mathbb{Z}^d : \mathbf{k} \cdot \mathbf{z} \equiv 0 \bmod M\} \tag{8.11}$$

the *integer dual lattice* of $\Lambda(\mathbf{z}, M)$. The integer dual lattice $\Lambda^{\perp}(\mathbf{z}, M)$ will play an important role, when we approximate the Fourier coefficients of a function f using only samples of f on the rank-1 lattice $\Lambda(\mathbf{z}, M)$.

Example 8.6 Let $d = 2$, $\mathbf{z} = (1, 3)^\top$ and $M = 11$, and then we obtain

$$\Lambda(\mathbf{z}, M) = \frac{2\pi}{11} \left\{ \begin{pmatrix} 0 \\ 0 \end{pmatrix}, \begin{pmatrix} 1 \\ 3 \end{pmatrix}, \begin{pmatrix} 2 \\ 6 \end{pmatrix}, \begin{pmatrix} 3 \\ 9 \end{pmatrix}, \begin{pmatrix} 4 \\ 1 \end{pmatrix}, \begin{pmatrix} 5 \\ 4 \end{pmatrix}, \begin{pmatrix} 6 \\ 7 \end{pmatrix}, \begin{pmatrix} 7 \\ 10 \end{pmatrix}, \begin{pmatrix} 8 \\ 2 \end{pmatrix}, \begin{pmatrix} 9 \\ 5 \end{pmatrix}, \begin{pmatrix} 10 \\ 8 \end{pmatrix} \right\},$$

and $\Lambda^{\perp}(\mathbf{z}, M)$ contains all vectors $\mathbf{k} = (k_1, k_2)^\top \in \mathbb{Z}^2$ with $k_1 + 3k_2 \equiv 0 \bmod 11$. Figure 8.2 illustrates the construction of this two-dimensional rank-1 lattice.

A rank-1 lattice possesses the following important property.

Lemma 8.7 *Let a frequency index set $I \subset \mathbb{Z}^d$ of finite cardinality and a rank-1 lattice $X = \Lambda(\mathbf{z}, M)$ be given.*

Then two distinct columns of the corresponding M-by-$|I|$ Fourier matrix $\mathbf{A}$ are either orthogonal or equal, i.e., the $(\mathbf{h}, \mathbf{k})$th entry $(\mathbf{A}^{\mathrm{H}}\,\mathbf{A})_{\mathbf{h},\mathbf{k}} \in \{0, M\}$ for all $\mathbf{h}$, $\mathbf{k} \in I$.

Proof The matrix $\mathbf{A}^{\mathrm{H}}\,\mathbf{A}$ contains all inner products of two columns of the Fourier matrix $\mathbf{A}$, i.e., the $(\mathbf{h}, \mathbf{k})$th entry $(\mathbf{A}^{\mathrm{H}}\,\mathbf{A})_{\mathbf{h},\mathbf{k}}$ is equal to the inner product of the $\mathbf{k}$th column and the $\mathbf{h}$th column of $\mathbf{A}$. For $\mathbf{k} \cdot \mathbf{z} \not\equiv \mathbf{h} \cdot \mathbf{z} \bmod M$ we obtain

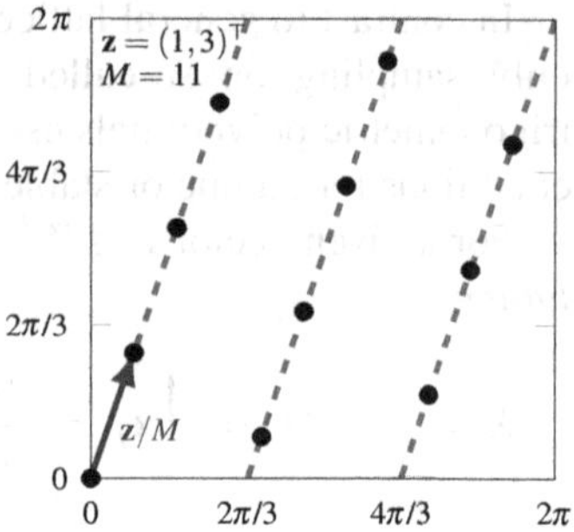

Fig. 8.2 Rank-1 lattice $\Lambda(\mathbf{z}, M)$ of Example 8.6

$$(\mathbf{A}^{\mathrm{H}}\,\mathbf{A})_{\mathbf{h},\,\mathbf{k}} = \sum_{j=0}^{M-1} \left(\mathrm{e}^{2\pi\mathrm{i}\,[(\mathbf{k}-\mathbf{h})\cdot\mathbf{z}]/M}\right)^j = \frac{\mathrm{e}^{2\pi\mathrm{i}\,(\mathbf{k}-\mathbf{h})\cdot\mathbf{z}} - 1}{\mathrm{e}^{2\pi\mathrm{i}\,[(\mathbf{k}-\mathbf{h})\cdot\mathbf{z}]/M} - 1} = 0\,,$$

since $\mathbf{k}-\mathbf{h} \in \mathbb{Z}^d$.

For $\mathbf{k}\cdot\mathbf{z} \equiv \mathbf{h}\cdot\mathbf{z} \mod M$ it follows immediately that the $\mathbf{k}$th and $\mathbf{h}$th column of $\mathbf{A}$ are equal and that $(\mathbf{A}^{\mathrm{H}}\,\mathbf{A})_{\mathbf{h},\,\mathbf{k}} = M$. ■

8.2.2 Evaluation of Trigonometric Polynomials on Rank-1 Lattice

Let us now consider the efficient evaluation of a d-variate trigonometric polynomial p supported on I on the sampling set X being a rank-1 lattice $X = \Lambda(\mathbf{z}, M)$. We have to compute $p(\mathbf{x}_j)$ for all M nodes $\mathbf{x}_j \in \Lambda(\mathbf{z}, M)$, i.e.,

$$p(\mathbf{x}_j) = \sum_{\mathbf{k}\in I} \hat{p}_{\mathbf{k}}\, \mathrm{e}^{\mathrm{i}\,\mathbf{k}\cdot\mathbf{x}_j} = \sum_{\mathbf{k}\in I} \hat{p}_{\mathbf{k}}\, \mathrm{e}^{2\pi\mathrm{i}\,j\,(\mathbf{k}\cdot\mathbf{z})/M}, \quad j = 0, \ldots, M-1.$$

We observe that $\{\mathbf{k}\cdot\mathbf{z} \mod M : \mathbf{k} \in I\} \subset \{0, \ldots, M-1\}$ and consider the values

$$\hat{g}_\ell = \sum_{\substack{\mathbf{k}\in I \\ \ell \equiv \mathbf{k}\cdot\mathbf{z} \bmod M}} \hat{p}_{\mathbf{k}}\,, \quad \ell = 0, \ldots, M-1\,. \tag{8.12}$$

Then, we can write

$$p(\mathbf{x}_j) = \sum_{\mathbf{k}\in I} \hat{p}_{\mathbf{k}}\, \mathrm{e}^{2\pi\mathrm{i}\,j\,(\mathbf{k}\cdot\mathbf{z})/M} = \sum_{\ell=0}^{M-1} \sum_{\substack{\mathbf{k}\in I \\ \ell \equiv \mathbf{k}\cdot\mathbf{z} \bmod M}} \hat{p}_{\mathbf{k}}\, \mathrm{e}^{2\pi\mathrm{i}\,j\ell/M}$$

$$= \sum_{\ell=0}^{M-1} \hat{g}_\ell \, \mathrm{e}^{2\pi \mathrm{i}\, j\ell/M} \tag{8.13}$$

for $j = 0, \ldots, M-1$. Therefore, the right-hand side of (8.13) can be evaluated using a one-dimensional FFT of length M with at most $C \cdot (M \, \log M + d\,|I|)$ arithmetic operations, where the constant C does not depend on the dimension d. Here we assume that $\hat{g}_\ell$, $\ell = 0, \ldots, M$, can be computed with $C\, d\, |I|$ arithmetic operations. The fast realization of the matrix–vector product in (8.8) or equivalently of (8.13) is presented in the following.

Algorithm 8.8 (Lattice-Based FFT (LFFT))
Input: $M \in \mathbb{N}$ *lattice size of rank-*1 *lattice* $\Lambda(\mathbf{z}, M)$,
$\mathbf{z} \in \mathbb{Z}^d$ *generating vector of* $\Lambda(\mathbf{z}, M)$,
$I \subset \mathbb{Z}^d$ *finite frequency index set,*
$\hat{\mathbf{p}} = (\hat{p}_{\mathbf{k}})_{\mathbf{k}\in I}$ *Fourier coefficients of* $p \in \Pi_I$.

1. *Set* $\hat{\mathbf{g}} := (0)_{\ell=0}^{M-1}$.
2. *For each* $\mathbf{k} \in I$ *do* $\hat{g}_{\mathbf{k}\cdot\mathbf{z} \bmod M} := \hat{g}_{\mathbf{k}\cdot\mathbf{z} \bmod M} + \hat{p}_{\mathbf{k}}$.
3. *Apply a one-dimensional* FFT *of length* M *in order to compute* $\mathbf{p} := \mathbf{F}_M^{-1}\big((\hat{g}_\ell)_{\ell=0}^{M-1}\big)$.
4. *Compute* $\mathbf{p} := M\,\mathbf{p}$.

Output: $\mathbf{p} = \mathbf{A}\,\hat{\mathbf{p}}$ *vector of values of the trigonometric polynomial* $p \in \Pi_I$.
Computational cost: $\mathcal{O}(M \log M + d|I|)$.

We immediately obtain also a fast algorithm for the matrix–vector multiplication with the adjoint Fourier matrix $\mathbf{A}^{\mathrm{H}}$.

Algorithm 8.9 (Adjoint Single Lattice-Based FFT (aLFFT))
Input: $M \in \mathbb{N}$ *lattice size of rank-*1 *lattice* $\Lambda(\mathbf{z}, M)$,
$\mathbf{z} \in \mathbb{Z}^d$ *generating vector of* $\Lambda(\mathbf{z}, M)$,
$I \subset \mathbb{Z}^d$ *finite frequency index set,*
$\mathbf{p} = \big(p(\frac{j}{M}\,\mathbf{z})\big)_{j=0}^{M-1}$ *values of the trigonometric polynomial* $p \in \Pi_I$.

1. *Apply a one-dimensional* FFT *of length* M *in order to compute* $\hat{\mathbf{g}} := \mathbf{F}_M\,\mathbf{p}$.
2. *Set* $\hat{\mathbf{a}} := (0)_{\mathbf{k}\in I}$.
3. *For each* $\mathbf{k} \in I$ *do* $\hat{a}_{\mathbf{k}} := \hat{a}_{\mathbf{k}} + \hat{g}_{\mathbf{k}\cdot\mathbf{z} \bmod M}$.

Output: $\hat{\mathbf{a}} = \mathbf{A}^{\mathrm{H}}\,\mathbf{p}$ *with the adjoint Fourier matrix* $\mathbf{A}^{\mathrm{H}}$.
Computational cost: $\mathcal{O}(M \, \log M + d\,|I|)$.

8.2.3 Evaluation of the Fourier Coefficients

Our considerations of the Fourier matrix $\mathbf{A} = \mathbf{A}(X, I)$ in (8.8) and (8.9) show that a unique evaluation of all Fourier coefficients of an arbitrary d-variate trigonometric polynomial $p \in \Pi_I$ is only possible if the $|X|$-by-$|I|$ matrix $\mathbf{A}$ has full rank $|I|$. By Lemma 8.7 we have seen that for a given frequency index set I and a rank-1 lattice $\Lambda(\mathbf{z}, M)$, two distinct columns of $\mathbf{A}$ are either orthogonal or equal. Therefore, $\mathbf{A}$ has full rank if and only if for all distinct $\mathbf{k}, \mathbf{h} \in I$,

$$\mathbf{k} \cdot \mathbf{z} \not\equiv \mathbf{h} \cdot \mathbf{z} \mod M\,. \tag{8.14}$$

If (8.14) holds, then the sums determining $\hat{g}_\ell$ in (8.12) contain only one term for each ℓ and no aliasing occurs. We define the *difference set of the frequency index set* I as

$$\mathscr{D}(I) := \{\mathbf{k} - \mathbf{l} \colon \mathbf{k},\, \mathbf{l} \in I\}\,. \tag{8.15}$$

Then the condition (8.14) is equivalent to

$$\mathbf{k} \cdot \mathbf{z} \not\equiv 0 \bmod M \quad \text{for all } \mathbf{k} \in \mathscr{D}(I) \setminus \{\mathbf{0}\}\,. \tag{8.16}$$

Therefore, we define a *reconstructing rank-*1 *lattice* to a given frequency index set I as a rank-1 lattice satisfying (8.14) or equivalently (8.16) and denote it by

$$\Lambda(\mathbf{z}, M, I) := \big\{\mathbf{x} \in \Lambda(\mathbf{z}, M) : \mathbf{k} \in \mathscr{D}(I) \setminus \{\mathbf{0}\} \text{ with } \mathbf{k} \cdot \mathbf{z} \not\equiv 0 \bmod M\big\}\,.$$

The condition (8.16) ensures that the mapping of $\mathbf{k} \in I$ to $\mathbf{k} \cdot \mathbf{z} \bmod M \in \{0, \ldots, M-1\}$ is injective. Assuming that we have a reconstructing rank-1 lattice, we will be able to evaluate the Fourier coefficients of $p \in \Pi_I$ uniquely.

If condition (8.16) is satisfied, then Lemma 8.7 implies $\mathbf{A}^{\mathrm{H}}\,\mathbf{A} = M\,\mathbf{I}_M$ for the Fourier matrix $\mathbf{A}$ such that $\hat{\mathbf{p}} = (\hat{p}_{\mathbf{k}})_{\mathbf{k} \in I} = \frac{1}{M}\,\mathbf{A}^{\mathrm{H}}\,\mathbf{p}$. Equivalently, for each Fourier coefficient, we have

$$\hat{p}_{\mathbf{k}} = \frac{1}{M} \sum_{j=0}^{M-1} p(\mathbf{x}_j)\, \mathrm{e}^{-2\pi \mathrm{i}\, j\, (\mathbf{k} \cdot \mathbf{z})/M} = \frac{1}{M} \sum_{j=0}^{M-1} p(\mathbf{x}_j)\, \mathrm{e}^{-2\pi \mathrm{i}\, j \ell/M}$$

for all $\mathbf{k} \in I$ and $\ell = \mathbf{k} \cdot \mathbf{z} \bmod M$. Algorithm 8.10 computes all Fourier coefficients $\hat{f}_{\mathbf{k}}$ using only a one-dimensional FFT of length M and the inverse mapping of $\mathbf{k} \mapsto \mathbf{k} \cdot \mathbf{z} \bmod M$ (see also [220, Algorithm 3.2]).

Algorithm 8.10 (Reconstruction via Reconstructing Rank-1 Lattice)
Input: $I \subset \mathbb{Z}^d$ *finite frequency index set,*
$M \in \mathbb{N}$ *lattice size of reconstructing rank-*1 *lattice* $\Lambda(\mathbf{z}, M, I)$,
$\mathbf{z} \in \mathbb{Z}^d$ *generating vector of reconstructing rank-*1 *lattice* $\Lambda(\mathbf{z}, M, I)$,
$\mathbf{p} = \big(p\big(\frac{2\pi}{M}\,(j\,\mathbf{z} \bmod M\,\mathbf{1})\big)\big)_{j=0}^{M-1}$ *values of* $p \in \Pi_I$.

1. *Compute* $\hat{\mathbf{a}} := \mathbf{A}^{\mathrm{H}}\,\mathbf{p}$ *using Algorithm 8.9.*
2. *Set* $\hat{\mathbf{p}} := M^{-1}\,\hat{\mathbf{a}}$.

Output: $\hat{\mathbf{p}} = M^{-1}\mathbf{A}^{\mathrm{H}}\,\mathbf{p} = (\hat{p}_{\mathbf{k}})_{\mathbf{k}\in I}$ *Fourier coefficients supported on* I.
Computational cost: $\mathcal{O}(M\,\log M + d\,|I|)$.

Example 8.11 Let $I^d_{\infty,N}$ be the full grid defined by (8.3). Then straightforward calculation shows that the rank-1 lattice $\Lambda(\mathbf{z}, M)$ with the generating vector $\mathbf{z} = (1, 2N+2, \ldots, (2N+2)^{d-1})^{\top}$ and the lattice size $M = (2N+2)^d$ is a reconstructing rank-1 lattice to the full grid $I^d_{\infty,N}$. It provides a perfectly stable spatial discretization. The resulting reconstruction algorithm is based on a one-dimensional FFT of size $(2N+2)^d$ and has similar computational cost as the usual d-dimensional tensor-product FFT (see Sect. 5.3.5). Our goal is to construct smaller reconstructing rank-1 lattices for special index sets, such that the computational cost for the reconstruction of Fourier coefficients can be significantly reduced. □

As a corollary of the observations above, we show that a reconstructing rank-1 lattice implies the following important quadrature rule (see [391]).

Theorem 8.12 *For a given finite frequency index set* I *and a corresponding reconstructing rank-*1 *lattice* $\Lambda(\mathbf{z}, M, I)$, *we have*

$$\int_{[0,\,2\pi]^d} p(\mathbf{x})\,\mathrm{d}\mathbf{x} = \frac{(2\pi)^d}{M}\sum_{j=0}^{M-1} p(\mathbf{x}_j)$$

for all trigonometric polynomials $p \in \Pi_{\mathcal{D}(I)}$, *where* $\mathcal{D}(I)$ *is defined by* (8.15).

Proof For $\mathbf{x}_j = \frac{2\pi}{M}\,(j\mathbf{z} \bmod M\,\mathbf{1}) \in \Lambda(\mathbf{z}, M, I)$ it follows that

$$\begin{aligned}\sum_{j=0}^{M-1} p(\mathbf{x}_j) &= \sum_{j=0}^{M-1}\Bigg(\sum_{\mathbf{k}\in\mathcal{D}(I)} \hat{p}_{\mathbf{k}}\,\mathrm{e}^{2\pi\mathrm{i}\,j\,(\mathbf{k}\cdot\mathbf{z})/M}\Bigg)\\ &= \sum_{\mathbf{k}\in\mathcal{D}(I)} \hat{p}_{\mathbf{k}}\Bigg(\sum_{j=0}^{M-1}\mathrm{e}^{2\pi\mathrm{i}\,j\,(\mathbf{k}\cdot\mathbf{z})/M}\Bigg).\end{aligned}$$

According to (8.16) we have $\mathbf{k}\cdot\mathbf{z} \not\equiv 0 \bmod M$ for all $\mathbf{k} \in \mathcal{D}(I)\setminus\{\mathbf{0}\}$. Therefore,

$$\sum_{j=0}^{M-1}\mathrm{e}^{2\pi\mathrm{i}\,j\,(\mathbf{k}\cdot\mathbf{z})/M} = \begin{cases} 0 & \mathbf{k}\in\mathcal{D}(I)\setminus\{\mathbf{0}\}\,,\\ M & \mathbf{k}=\mathbf{0}\,,\end{cases}$$

and the equation above simplifies to

$$\sum_{j=0}^{M-1} p(\mathbf{x}_j) = M\,\hat{p}_{\mathbf{0}} = \frac{M}{(2\pi)^d}\int_{[0,\,2\pi]^d} p(\mathbf{x})\,\mathrm{d}\mathbf{x}\,,$$

since by (8.7) it holds

$$\int_{[0,\,2\pi]^d} p(\mathbf{x})\,\mathrm{d}\mathbf{x} = \hat{p}_{\mathbf{0}}\,(2\pi)^d + 0\,.$$

■

8.3 Efficient Function Approximation on Rank-1 Lattices

Now we come back to the problem of approximation of a smooth d-variate periodic function f by a Fourier series (8.1) or by a Fourier partial sum (8.2). Let f be an arbitrary continuous function in $\mathscr{A}(\mathbb{T}^d) \cap C(\mathbb{T}^d)$. Then we determine approximate values $\hat{f}_{\mathbf{k}}$ of the Fourier coefficients $c_{\mathbf{k}}(f)$ using only the sampling values on a rank-1 lattice $\Lambda(\mathbf{z}, M)$ as given in (8.10) and obtain

$$\begin{aligned} \hat{f}_{\mathbf{k}} &:= \frac{1}{M}\sum_{j=0}^{M-1} f\Big(\frac{2\pi}{M}\,(j\,\mathbf{z} \bmod M\,\mathbf{1})\Big)\,\mathrm{e}^{-2\pi\mathrm{i}\,j\,(\mathbf{k}\cdot\mathbf{z})/M} \qquad (8.17)\\ &= \frac{1}{M}\sum_{j=0}^{M-1}\sum_{\mathbf{h}\in\mathbb{Z}^d} c_{\mathbf{h}}(f)\,\mathrm{e}^{2\pi\mathrm{i}\,j\,[(\mathbf{h}-\mathbf{k})\cdot\mathbf{z}]/M}\\ &= \sum_{\mathbf{h}\in\mathbb{Z}^d} c_{\mathbf{k}+\mathbf{h}}(f)\,\frac{1}{M}\sum_{j=0}^{M-1}\mathrm{e}^{2\pi\mathrm{i}\,j\,(\mathbf{h}\cdot\mathbf{z})/M} = \sum_{\mathbf{h}\in\Lambda^{\perp}(\mathbf{z},M)} c_{\mathbf{k}+\mathbf{h}}(f)\,, \end{aligned}$$

where the integer dual lattice $\Lambda^{\perp}(\mathbf{z}, M)$ is defined by (8.11). Obviously, we have $\mathbf{0} \in \Lambda^{\perp}(\mathbf{z}, M)$ and hence

$$\hat{f}_{\mathbf{k}} = c_{\mathbf{k}}(f) + \sum_{\mathbf{h}\in\Lambda^{\perp}(\mathbf{z},M)\setminus\{\mathbf{0}\}} c_{\mathbf{k}+\mathbf{h}}(f)\,. \qquad (8.18)$$

The absolute convergence of the series of the Fourier coefficients of f ensures that all terms in the calculation above are well-defined. We call $\hat{f}_{\mathbf{k}}$ the *approximate Fourier coefficients* of f. The formula (8.18) can be understood as an *aliasing formula for the rank-*1 *lattice* $\Lambda(\mathbf{z}, M)$. If the sum

$$\sum_{\mathbf{h}\in\Lambda^{\perp}(\mathbf{z},M)\setminus\{\mathbf{0}\}} |c_{\mathbf{k}+\mathbf{h}}(f)|$$

is sufficiently small, then $\hat{f}_{\mathbf{k}}$ is a convenient approximate value of $c_{\mathbf{k}}(f)$.

Assume that f can be already well approximated by a trigonometric polynomial p on a frequency index set I. Further, assume that we have a corresponding reconstructing rank-1 lattice $X = \Lambda(\mathbf{z}, M, I)$. Then we can compute the approximative

Fourier coefficients $\hat{f}_{\mathbf{k}}$ with $\mathbf{k} \in I$ using Algorithm 8.10 by employing M sample values $f\big(\frac{2\pi}{M}(j\,\mathbf{z} \bmod M\,\mathbf{1})\big)$ instead of the corresponding polynomial values. In this way, we obtain $\hat{f}_{\mathbf{k}}$, $\mathbf{k} \in I$, with computational cost of $\mathcal{O}(M \log M + d\,|I|)$ flops.

Now we want to study the approximation error that occurs if the exact Fourier coefficients $c_{\mathbf{k}}(f)$ are replaced by the approximate Fourier coefficients $\hat{f}_{\mathbf{k}}$ in (8.18). We consider the corresponding approximate Fourier partial sum on the frequency index set $I_N = \{\mathbf{k} \in \mathbb{Z}^d : \omega(\mathbf{k}) \le N\}$. Let $\Lambda(\mathbf{z}, M, I_N)$ be a reconstructing rank-1 lattice for I_N and $\Lambda^{\perp}(\mathbf{z}, M, I_N)$ the corresponding integer dual lattice (8.11). By definition of the reconstructing rank-1 lattice, it follows that $I_N \cap \Lambda^{\perp}(\mathbf{z}, M, I_N) = \{\mathbf{0}\}$. Generally we can show the following result.

Lemma 8.13 *Let $I \subset \mathbb{Z}^d$ be an arbitrary finite frequency index set and let $\Lambda(\mathbf{z}, M, I)$ be a reconstructing rank-1 lattice with the integer dual lattice $\Lambda^{\perp}(\mathbf{z}, M, I)$.*

Then we have

$$\big\{\mathbf{k}+\mathbf{h} : \mathbf{k} \in I, \mathbf{h} \in \Lambda^{\perp}(\mathbf{z}, M, I) \setminus \{\mathbf{0}\}\big\} \subset \mathbb{Z}^d \setminus I\,.$$

Proof Assume to the contrary that there exist $\mathbf{k} \in I$ and $\mathbf{h} \in \Lambda^{\perp}(\mathbf{z}, M, I)\setminus\{\mathbf{0}\}$ such that $\mathbf{k}+\mathbf{h} \in I$. Since $\Lambda(\mathbf{z}, M, I)$ is a reconstructing rank-1 lattice for I, it follows that $\mathbf{0} \ne \mathbf{h} = (\mathbf{k}+\mathbf{h}) - \mathbf{k} \in \mathcal{D}(I)$. Thus, $\mathbf{h} \in \mathcal{D}(I) \cap \Lambda^{\perp}(\mathbf{z}, M, I) \setminus \{\mathbf{0}\}$. But this is a contradiction, since on the one hand (8.16) implies that $\mathbf{h} \cdot \mathbf{z} \not\equiv 0 \mod M$, and on the other hand $\mathbf{h} \cdot \mathbf{z} \equiv 0 \mod M$ by definition of $\Lambda^{\perp}(\mathbf{z}, M, I)$. ■

Now we can estimate the error of the approximate Fourier sum of f as follows (see [220, Theorem 3.11]).

Theorem 8.14 *Let $f \in \mathcal{A}_{\omega}(\mathbb{T}^d)$ and let a frequency index set $I_N = \{\mathbf{k} \in \mathbb{Z}^d : \omega(\mathbf{k}) \le N\}$ of finite cardinality be given. Further, let $\Lambda(\mathbf{z}, M, I_N)$ be a reconstructing rank-1 lattice for I_N. Moreover, let the approximate Fourier partial sum*

$$(S_{I_N}^{\Lambda} f)(\mathbf{x}) := \sum_{\mathbf{k} \in I_N} \hat{f}_{\mathbf{k}}\, \mathrm{e}^{\mathrm{i}\,\mathbf{k}\cdot\mathbf{x}} \tag{8.19}$$

of f be determined by

$$\hat{f}_{\mathbf{k}} := \frac{1}{M} \sum_{j=0}^{M-1} f\left(\frac{2\pi}{M}\,(j\,\mathbf{z} \mod M\,\mathbf{1})\right)\, \mathrm{e}^{-2\pi\mathrm{i}\,j\,(\mathbf{k}\cdot\mathbf{z})/M}\,, \quad \mathbf{k} \in I_N\,, \tag{8.20}$$

that are computed using the values on the rank-1 lattice $\Lambda(\mathbf{z}, M, I_N)$.

Then we have

$$\|f - S_{I_N}^{\Lambda} f\|_{L_\infty(\mathbb{T}^d)} \le 2\,N^{-1}\,\|f\|_{\mathcal{A}_{\omega}(\mathbb{T}^d)}\,. \tag{8.21}$$

Proof Using the triangle inequality, we find

$$\|f - S_{I_N}^{\Lambda} f\|_{L_\infty(\mathbb{T}^d)} \le \|f - S_{I_N} f\|_{L_\infty(\mathbb{T}^d)} + \|S_{I_N}^{\Lambda} f - S_{I_N} f\|_{L_\infty(\mathbb{T}^d)} .$$

For the first term, Lemma 8.3 yields

$$\|f - S_{I_N} f\|_{L_\infty(\mathbb{T}^d)} \le N^{-1} \|f\|_{\mathscr{A}_\omega(\mathbb{T}^d)} .$$

For the second term, we obtain by using (8.18)

$$\begin{aligned}
\|S_{I_N}^{\Lambda} f - S_{I_N} f\|_{L_\infty(\mathbb{T}^d)} &= \operatorname*{ess\,sup}_{\mathbf{x}\in\mathbb{T}^d} \Big| \sum_{\mathbf{k}\in I_N} \big(\hat{f}_{\mathbf{k}} - c_{\mathbf{k}}(f)\big)\, \mathrm{e}^{\mathrm{i}\,\mathbf{k}\cdot\mathbf{x}} \Big| \\
&\le \sum_{\mathbf{k}\in I_N} \Big| \sum_{\mathbf{h}\in\Lambda^\perp(\mathbf{z},M)\setminus\{\mathbf{0}\}} c_{\mathbf{k}+\mathbf{h}}(f) \Big| \\
&\le \sum_{\mathbf{k}\in I_N} \sum_{\mathbf{h}\in\Lambda^\perp(\mathbf{z},M)\setminus\{\mathbf{0}\}} |c_{\mathbf{k}+\mathbf{h}}(f)| .
\end{aligned}$$

By Lemma 8.13 it follows that

$$\begin{aligned}
\|S_{I_N}^{\Lambda} f - S_{I_N} f\|_{L_\infty(\mathbb{T}^d)} &\le \sum_{\mathbf{k}\in\mathbb{Z}^d\setminus I_N} |c_{\mathbf{k}}(f)| \le \frac{1}{\inf_{\mathbf{h}\in\mathbb{Z}^d\setminus I_N} \omega(\mathbf{h})} \sum_{\mathbf{k}\in\mathbb{Z}^d} \omega(\mathbf{k})\, |c_{\mathbf{k}}(f)| \\
&\le N^{-1} \|f\|_{\mathscr{A}_\omega(\mathbb{T}^d)}
\end{aligned}$$

and hence the assertion. ∎

Theorem 8.14 states that the worst-case error of the approximation $S_{I_N}^{\Lambda} f$ in (8.19) given by the approximate Fourier coefficients computed from samples on the reconstructing rank-1 lattice $\Lambda(\mathbf{z}, M, I_N)$ is qualitatively as good as the worst-case error of the approximation $S_{I_N} f$ (see (8.5)). Improved error estimates for the approximation of functions in $\mathscr{A}_\omega(\mathbb{T}^d)$ with a special weight function ω as in Remark 8.4 can be similarly derived. The approximation error essentially depends on the considered norms. In particular, we have focused on the $L_\infty(\mathbb{T}^d)$-norm on the left-hand side and the weighted $\ell_1(\mathbb{Z}^d)$-norm of the Fourier coefficients on the right-hand side. Further results with different norms are given in [220, 425].

Remark 8.15 The idea to use special rank-1 lattices $\Lambda(\mathbf{z}, M)$ of Korobov type as sampling schemes to approximate functions by trigonometric polynomials has been already considered by V.N. Temlyakov [410]. Later, D. Li and F.J. Hickernell studied a more general setting in [271]. They presented an approximation error using an aliasing formula as (8.18) for the given rank-1 lattice $\Lambda(\mathbf{z}, M)$. But both approaches did not lead to a constructive way to determine rank-1 lattices of high quality. In contrast to their approach, we have constructed the frequency index set $I_N := \{\mathbf{k} \in \mathbb{Z}^d : \omega(\mathbf{k}) \le N\}$ with $|I_N| < \infty$ depending on the arbitrary weight function ω. The problem to find a reconstructing rank-1 lattice $\Lambda(\mathbf{z}, M, I_N)$ which is well adapted to the frequency index set I_N will be studied in the next section. Approximation properties of rank-1 lattices have been also investigated in information-based complexity and applied analysis (see, e.g., [262, 295, 451]). □

8.4 Reconstructing Rank-1 Lattices

As shown in the two last sections, we can use so-called reconstructing rank-1 lattices in order to compute the Fourier coefficients of a d-variate trigonometric polynomial in Π_I in a stable way by applying a one-dimensional FFT. The reconstructing rank-1 lattice $\Lambda(\mathbf{z}, M, I)$ for a frequency index set I is determined as a rank-1 lattice $\Lambda(\mathbf{z}, M)$ in (8.10) satisfying the condition (8.14). The computational cost to reconstruct the Fourier coefficients of the d-variate trigonometric polynomial p from its sampling values of the given rank-1 lattice mainly depends on the number M of needed sampling values. In this section we will present a deterministic procedure to obtain reconstructing rank-1 lattices using a component-by-component approach.

We start with considering the following problem: How large the number M of sampling values in $\Lambda(\mathbf{z}, M, I)$ needs to be (see also [221, 224])? For simplicity, we consider only a *symmetric frequency index set* $I \subset \mathbb{Z}^d$ satisfying the condition that for each $\mathbf{k} \in I$ also $-\mathbf{k} \in I$. For instance, all frequency index sets in Example 8.6 and Fig. 8.1 are symmetric.

Theorem 8.16 *Let I be a symmetric frequency index set with finite cardinality $|I|$ such that $I \subset [-\frac{|I|}{2}, \frac{|I|}{2}]^d \cap \mathbb{Z}^d$.*

Then there exists a reconstructing rank-1 lattice $X = \Lambda(\mathbf{z}, M, I)$ with prime cardinality M, such that

$$|I| \le M \le |\mathcal{D}(I)| \le |I|^2 - |I| + 1, \tag{8.22}$$

where $\mathcal{D}(I)$ denotes the difference set (8.15).

Proof

1. The lower bound $|I| \le M$ is obvious, since we need a Fourier matrix $\mathbf{A} = \mathbf{A}(X, I) \in \mathbb{C}^{|X| \times |I|}$ of full rank $|I|$ in (8.9) to reconstruct $\hat{\mathbf{p}}$, and this property follows from (8.14).
   Recall that $|\mathcal{D}(I)|$ is the number of all pairwise distinct vectors $\mathbf{k} - \mathbf{l}$ with $\mathbf{k}$, $\mathbf{l} \in I$. We can form at most $|I|\,(|I| - 1) + 1$ pairwise distinct vectors in $\mathcal{D}(I)$. Therefore, we obtain the upper bound $|\mathcal{D}(I)| \le |I|^2 - |I| + 1$.
2. In order to show that there exists a reconstructing rank-1 lattice with $M \le |\mathcal{D}(I)|$, we choose M as a prime number satisfying $|\mathcal{D}(I)|/2 < M \le |\mathcal{D}(I)|$ and show that there exists a generating vector $\mathbf{z}$ such that the condition (8.16)

is satisfied for $X = \Lambda(\mathbf{z}, M, I)$. The prime number M can be always chosen in $(|\mathscr{D}(I)|/2, |\mathscr{D}(I)|]$ by Bertrand's postulate.
For the special case $d = 1$ we have $I \subset [-\frac{|I|}{2}, \frac{|I|}{2}] \cap \mathbb{Z}$. Taking $z = z_1 = 1$, each $M \geq |I| + 1$ satisfies the assumption $k \cdot z = k \not\equiv 0 \bmod M$ for $k \in \mathscr{D}(I) \subset [-|I|, |I|]$. In particular, we can take M as a prime number in $(|\mathscr{D}(I)|/2, |\mathscr{D}(I)|]$, since we have $|\mathscr{D}(I)| \geq 2\,|I|$ in this case.
Let us now assume that $d \geq 2$. We need to show that there exists a generating vector $\mathbf{z}$ such that

$$\mathbf{k} \cdot \mathbf{z} \not\equiv 0 \bmod M \quad \text{for all } \mathbf{k} \in \mathscr{D}(I) \setminus \{\mathbf{0}\},$$

and want to use an induction argument with respect to the dimension d. We consider the projection of $\mathscr{D}(I)$ on the index set

$$\mathscr{D}(I_{d-1}) := \left\{ \tilde{\mathbf{k}} = (k_j)_{j=1}^{d-1} : \mathbf{k} = (k_j)_{j=1}^{d} \in \mathscr{D}(I) \right\},$$

such that each $\mathbf{k} \in \mathscr{D}(I)$ can be written as $(\tilde{\mathbf{k}}^\top, k_d)^\top$ with $\tilde{\mathbf{k}} \in \mathscr{D}(I_{d-1})$. Assume that we have found already a vector $\tilde{\mathbf{z}} \in \mathbb{Z}^{d-1}$ such that the condition

$$\tilde{\mathbf{k}} \cdot \tilde{\mathbf{z}} \not\equiv 0 \bmod M \quad \text{for all } \tilde{\mathbf{k}} \in \mathscr{D}(I_{d-1}) \setminus \{\mathbf{0}\} \tag{8.23}$$

is satisfied. We show now that there exists a vector $\mathbf{z} = (\tilde{\mathbf{z}}^\top, z_d)^\top$ with $z_d \in \{1, \ldots, M-1\}$ such that

$$\mathbf{k} \cdot \mathbf{z} = \tilde{\mathbf{k}} \cdot \tilde{\mathbf{z}} + k_d z_d \not\equiv 0 \bmod M \quad \text{for all } \mathbf{k} \in \mathscr{D}(I) \setminus \{\mathbf{0}\}. \tag{8.24}$$

For that purpose we will use a counting argument. We show that there are at most $(|\mathscr{D}(I_{d-1})| - 1)/2$ integers $z_d \in \{1, \ldots, M-1\}$ with the property

$$\mathbf{k} \cdot \mathbf{z} = \tilde{\mathbf{k}} \cdot \tilde{\mathbf{z}} + k_d z_d \equiv 0 \bmod M \qquad \text{for at least one } \mathbf{k} \in \mathscr{D}(I) \setminus \{\mathbf{0}\}. \tag{8.25}$$

Since $(|\mathscr{D}(I_{d-1})| - 1)/2 \leq (|\mathscr{D}(I)| - 1)/2 < M - 1$, we always find a z_d satisfying the desired condition (8.24).

3. We show now that for each pair of elements $\mathbf{k}$, $-\mathbf{k}$ with $\mathbf{k} = (\tilde{\mathbf{k}}^\top, k_d)^\top \in \mathscr{D}(I) \setminus \{\mathbf{0}\}$ and given $\tilde{\mathbf{z}}$ satisfying (8.23), there is at most one z_d such that (8.25) is satisfied.

   If $k_d = 0$, then (8.25) yields $\tilde{\mathbf{k}} \cdot \tilde{\mathbf{z}} \equiv 0 \bmod M$ contradicting (8.23). Thus, in this case no z_d is found to satisfy (8.25).

   If $\tilde{\mathbf{k}} = \mathbf{0}$ and $k_d \neq 0$, then (8.25) yields $k_d z_d \equiv 0 \bmod M$. Since $|k_d| \leq |I| < M$ and $z_d \in \{1, \ldots, M-1\}$, it follows that $k_d z_d$ and M are coprime such that no z_d is found to satisfy (8.25).

   If $\tilde{\mathbf{k}} \neq \mathbf{0}$ and $k_d \neq 0$, then (8.25) yields $\tilde{\mathbf{k}} \cdot \tilde{\mathbf{z}} \equiv -k_d z_d \bmod M$. Since $\tilde{\mathbf{k}} \cdot \tilde{\mathbf{z}} \neq 0$ by assumption (8.23) and k_d and M are coprime, there exists one unique solution z_d of this equation. The same unique solution z_d is found, if we replace $\mathbf{k} = (\tilde{\mathbf{k}}^\top, k_d)^\top$ by $-\mathbf{k} = (-\tilde{\mathbf{k}}^\top, -k_d)^\top$ in (8.25).

Taking into account that $\mathcal{D}(I_{d-1})$ and $\mathcal{D}(I)$ always contain the corresponding zero vector, it follows that at most $(|\mathcal{D}(I_{d-1})|-1)/2$ integers satisfy (8.25). Thus, the assertion is proved. ■

The idea of the proof of Theorem 8.16 leads us also to the so-called component-by-component Algorithm 8.17. This algorithm computes for a known lattice size M the generating vector $\mathbf{z}$ of the reconstructing rank-1 lattice (see also [221]). The component-by-component algorithm for numerical integration was presented in [95, 260].

Algorithm 8.17 (Component-by-Component Lattice Search)
Input: $M \in \mathbb{N}$ *prime, cardinality of rank-1 lattice,*
$I \subset \mathbb{Z}^d$ *finite frequency index set.*

1. *Set* $z_1 := 1$.
2. *For* $s = 2, \ldots, d$ *do*
   form the set $I_s := \big\{(k_j)_{j=1}^s : \mathbf{k} = (k_j)_{j=1}^d \in I\big\}$
   search for one $z_s \in [1,\ M-1] \cap \mathbb{Z}$ *with*

$$|\{(z_1, \ldots, z_s)^\top \cdot \mathbf{k} \bmod M :\ \mathbf{k} \in I_s\}| = |I_s|\,.$$

Output: $\mathbf{z} = (z_j)_{j=1}^d \in \mathbb{N}^d$ *generating vector.*

The construction of the generating vector $\mathbf{z} \in \mathbb{N}^d$ in Algorithm 8.17 requires at most $2d\,|I|\,M \le 2d\,|I|^3$ arithmetic operations. For each component z_s, $s \in \{2, \ldots, d\}$, of the generating vector $\mathbf{z}$ in the component-by-component step s, the tests for the reconstruction property (8.13) for a given component z_s, in step 2 of Algorithm 8.17 require at most $s\,|I|$ multiplications, $(s-1)\,|I|$ additions, and $|I|$ modulo operations. Since each component z_s, $s \in \{2, \ldots, d\}$, of the generating vector $\mathbf{z}$ can only take $M-1$ possible values, the construction requires at most $d|I|\,(M-1) \le 2\,d\,|I|\,M$ arithmetic operations in total.

Remark 8.18 The lower bound for the number M in Theorem 8.16 can be improved for arbitrary frequency index sets if we employ the exact cardinalities of the projected index sets $I_s := \{\big(k_j\big)_{j=1}^s : \mathbf{k} = \big(k_j\big)_{j=1}^d \in I\}$ (see also [221]).

The assumption on the index set can be also relaxed. In particular, the complete index set can be shifted in $\mathbb{Z}^d$ without changing the results. □

A drawback of Algorithm 8.17 is that the cardinality M needs to be known in advance. As we have shown in Theorem 8.16, M can be always taken as a prime number satisfying $|\mathcal{D}(I)|/2 < M \le |\mathcal{D}(I)|$. But this may be far away from an optimal choice. Once we have discovered a reconstructing rank-1 lattice $\Lambda(\mathbf{z}, M, I)$ satisfying for all distinct $\mathbf{k}, \mathbf{h} \in I$,

$$\mathbf{k} \cdot \mathbf{z} \not\equiv \mathbf{h} \cdot \mathbf{z} \bmod M\,,$$

we can ask for $M' < M$ such that for all distinct $\mathbf{k}, \mathbf{h} \in I$,

$$\mathbf{k} \cdot \mathbf{z} \not\equiv \mathbf{h} \cdot \mathbf{z} \bmod M'$$

is still true for the computed generating vector $\mathbf{z}$. This leads to the following simple algorithm for lattice size decreasing (see also [221]).

Algorithm 8.19 (Lattice Size Decreasing)
Input: $M \in \mathbb{N}$ *cardinality of rank-*1 *lattice,*
$I \subset \mathbb{Z}^d$ *finite frequency index set,*
$\mathbf{z} \in \mathbb{N}^d$ *generating vector of reconstructing rank-*1 *lattice* $\Lambda(\mathbf{z}, M, I)$.

1. *For* $j = |I|, \ldots, M$ *do*
   if $|\{\mathbf{k} \cdot \mathbf{z} \bmod j : \mathbf{k} \in I\}| = |I|$ *then* $M_{\min} := j$.

Output: $M_{\min}$ *reduced lattice size.*

There exist also other strategies to determine reconstructing rank-1 lattices for given frequency index sets, where the lattice size M needs not to be known a priori (see, e.g., [221, Algorithms 4 and 5]). These algorithms are also component-by-component algorithms and compute complete reconstructing rank-1 lattices, i.e., the generating vectors $\mathbf{z} \in \mathbb{N}^d$ and the lattice sizes $M \in \mathbb{N}$, for a given frequency index set I. The algorithms are applicable for arbitrary frequency index sets of finite cardinality $|I|$.

As we have seen in Theorem 8.16 the sampling size M can be bounded by the cardinality of the difference set $\mathscr{D}(I)$. Interestingly, this cardinality strongly depends on the structure of I.

Example 8.20 Let $I = I_{p,N}^d := \{\mathbf{k} \in \mathbb{Z}^d : \|\mathbf{k}\|_p \le N\}$, $N \in \mathbb{N}$, be the $\ell_p(\mathbb{Z}^d)$-ball with $0 < p \le \infty$ and the size $N \in \mathbb{N}$ (see Fig. 8.1). The cardinality of the frequency index set $I_{p,N}^d$ is bounded by $c_{p,d}\, N^d \le |I_{p,N}^d| \le C_{d,p}\, N^d$, while the cardinality of the difference set satisfies $c_{p,d}\, N^d \le |\mathscr{D}(I_{p,N}^d)| \le C_{d,p}\, 2^d\, N^d$ with the some constants $0 < c_{p,d} \le C_{d,p}$. Consequently, we can find a reconstructing rank-1 lattice of size $M \le \tilde{C}_{d,p}\, |I_{p,N}^d|$ using a component-by-component strategy, where the constant $\tilde{C}_{d,p} > 0$ only depends on p and d.

On the other hand, we obtain for $p \to 0$ the frequency index set $I := \{\mathbf{k} \in \mathbb{Z}^d : \|\mathbf{k}\|_1 = \|\mathbf{k}\|_\infty \le N\}$ with $N \in \mathbb{N}$, which is supported on the coordinate axes. In this case we have $|I| = 2d\, N + 1$, while we obtain $(2N + 1)^2 \le |\mathscr{D}(I)| \le d\,(2N + 1)^2$. Hence, there exists a positive constant $\tilde{c}_d \in \mathbb{R}$ with $\tilde{c}_d\, |I|^2 \le |\mathscr{D}(I)|$, and the theoretical upper bound on M is quadratic in $|I|$ for each fixed dimension d. In fact, reconstructing rank-1 lattices for these specific frequency index sets needs at least $\mathscr{O}(N^2)$ nodes (see [225, Theorem 3.5] and [224]). □

Example 8.21 Important frequency index sets in higher dimensions $d > 2$ are so-called (energy-norm based) hyperbolic crosses (see, e.g., [21, 72, 73, 450]). In particular, we can consider a frequency index set of the form

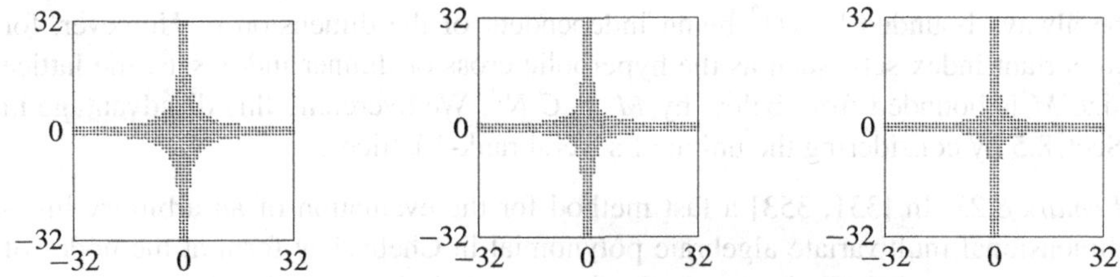

$I_{32}^{2,0}$ $I_{32}^{2,1/4}$ $I_{32}^{2,1/2}$

Fig. 8.3 Two-dimensional frequency index sets $I_{32}^{2,T}$ for $T \in \left\{0, \frac{1}{4}, \frac{1}{2}\right\}$

$$I_N^{d,T} := \left\{ \mathbf{k} \in \mathbb{Z}^d : (\max\{1, \|\mathbf{k}\|_1\})^{T/(T-1)} \prod_{s=1}^{d} (\max\{1, |k_s|\})^{1/(1-T)} \le N \right\},$$

with parameters $T \in [0, 1)$ and $N \in \mathbb{N}$ (see Fig. 8.3 for illustration). The frequency index set $I_N^{d,0}$ for $T = 0$ is a *symmetric hyperbolic cross*, and the frequency index set $I_N^{d,T}$, $T \in (0, 1)$, is called *energy-norm based hyperbolic cross*. The cardinality of $I_N^{d,T}$ can be estimated by

$$c_{d,0}\, N\, (\log N)^{d-1} \le |I_N^{d,T}| \le C_{d,0}\, N\, (\log N)^{d-1}\,, \quad \text{for } T = 0\,,$$

$$c_{d,T}\, N \le |I_N^{d,T}| \le C_{d,T}\, N\,, \quad \text{for } T \in (0, 1)$$

with some constants $0 < c_{d,T} \le C_{d,T}$, depending only on d and T (see [226, Lemma 2.6]). Since the axis cross is a subset of the considered frequency index sets, i.e., $\{\mathbf{k} \in \mathbb{Z}^d : \|\mathbf{k}\|_1 = \|\mathbf{k}\|_\infty \le N\} \subset I_N^{d,T}$ for $T \in [0, 1)$, it follows that $(2N + 1)^2 \le |\mathcal{D}(I_N^{d,T})|$. On the other hand, we obtain upper bounds of the cardinality of the difference set $\mathcal{D}(I_N^{d,T})$ of the form

$$|\mathcal{D}(I_N^{d,T})| \le \begin{cases} \tilde{C}_{d,0}\, N^2\, (\log N)^{d-2} & T = 0\,, \\ |I_N^{d,T}|^2 \le C_{d,T}^2\, N^2 & T \in (0,\, 1)\,, \end{cases}$$

(see, e.g., [219, Theorem 4.8]). Theorem 8.16 offers a constructive strategy to find reconstructing rank-1 lattices for $I_N^{d,T}$ of cardinality $M \le |\mathcal{D}(I_N^{d,T})|$. For $T \in (0, 1)$, these rank-1 lattices are of optimal order in N (see [219, Lemmata 2.1 and 2.3, and Corollary 2.4] and [220]). Reconstructing rank-1 lattices for these frequency index sets is discussed in more detail in [220]. □

Summarizing, we can construct a reconstructing rank-1 lattice $\Lambda(\mathbf{z}, M, I)$ for arbitrary finite frequency index set I. The choice of the frequency index set I always depends on the approximation properties of the considered function space. The positive statement is that the size M of the reconstructing rank-1 lattice can be always bounded by $|I|^2$ being independent of the dimension d. However, for important index sets, such as the hyperbolic cross or thinner index sets, the lattice size M is bounded from below by $M \geq C\,N^2$. We overcome this disadvantage in Sect. 8.5 by considering the union of several rank-1 lattices.

Remark 8.22 In [351, 353] a fast method for the evaluation of an arbitrary high-dimensional multivariate algebraic polynomial in Chebyshev form at the nodes of an arbitrary *rank-1 Chebyshev lattice* is suggested. An algorithm for constructing a suitable rank-1 Chebyshev lattice based on a component-by-component approach is suggested. In the two-dimensional case, the sampling points of special rank-1 Chebyshev lattice coincide with *Padua points*, (see [57]). □

8.5 Multiple Rank-1 Lattices

To overcome the limitations of the single rank-1 lattice approach, we consider now multiple rank-1 lattices which are obtained by taking a union of rank-1 lattices. For s rank-1 lattices $\Lambda(\mathbf{z}_r, M_r)$, $r = 1, \ldots, s$ as given in (8.10) we call the union

$$X = \Lambda(\mathbf{z}_1, M_1, \mathbf{z}_2, M_2, \ldots, \mathbf{z}_s, M_s) := \bigcup_{r=1}^{s} \Lambda(\mathbf{z}_r, M_r)$$

*multiple rank-*1 *lattice*.

In order to work with this multiple rank-1 lattices, we need to consider how many distinct points are contained in X. Assuming that for each r the lattice size M_r is coprime with at least one component of $\mathbf{z}_r$, the single rank-1 lattice $\Lambda(\mathbf{z}_r, M_r)$ possesses exactly M_r distinct points in $[0, 2\pi)^d$ including $\mathbf{0}$. Consequently, the number of distinct points in $\Lambda(\mathbf{z}_1, M_1, \mathbf{z}_2, M_2, \ldots, \mathbf{z}_s, M_s)$ is bounded from above by

$$|\Lambda(\mathbf{z}_1, M_1, \mathbf{z}_2, M_2, \ldots, \mathbf{z}_s, M_s)| \leq 1 - s + \sum_{r=1}^{s} M_r\,.$$

In the special case $s = 2$, we obtain the following result (see also [222, Lemma 2.1]).

Lemma 8.23 *Let $\Lambda(\mathbf{z}_1, M_1)$ and $\Lambda(\mathbf{z}_2, M_2)$ be two rank-*1 *lattices with coprime lattice sizes M_1 and M_2.*

*Then the multiple rank-*1 *lattice $\Lambda(\mathbf{z}_1, M_1, \mathbf{z}_2, M_2)$ is a subset of the rank-*1 *lattice $\Lambda(M_2\mathbf{z}_1 + M_1\mathbf{z}_2, M_1 M_2)$. Furthermore, if the cardinalities of the single rank-*1 *lattices $\Lambda(\mathbf{z}_1, M_1)$ and $\Lambda(\mathbf{z}_2, M_2)$ are M_1 and M_2, then*

$$|\Lambda(\mathbf{z}_1, M_1, \mathbf{z}_2, M_2)| = M_1 + M_2 - 1\,.$$

Proof

1. We show that $\Lambda(\mathbf{z}_1, M_1)$ is a subset of $\Lambda(M_2\mathbf{z}_1 + M_1\mathbf{z}_2, M_1M_2)$. Let

$$\mathbf{x}_j := \frac{2\pi}{M_1}(j\,\mathbf{z}_1 \bmod M_1\mathbf{1})$$

be an arbitrary point of $\Lambda(\mathbf{z}_1, M_1)$. Since M_1 and M_2 are coprime, there exists a $k \in \{0, \ldots, M_1 - 1\}$ such that $k\,M_2 \equiv j \bmod M_1$. Choose now $\ell = kM_2$, then

$$\mathbf{y}_\ell := \frac{2\pi}{M_1M_2}\big(\ell(M_2\mathbf{z}_1 + M_1\mathbf{z}_2) \bmod M_1M_2\mathbf{1}\big)$$

is a point of $\Lambda(M_2\mathbf{z}_1 + M_1\mathbf{z}_2, M_1M_2)$. Further, we find by

$$\ell\,(M_2\,\mathbf{z}_1 + M_1\,\mathbf{z}_2) \bmod M_1M_2\mathbf{1} = k\,(M_2^2\,\mathbf{z}_1 + M_1M_2\,\mathbf{z}_2) \bmod M_1M_2\mathbf{1}$$
$$= k\,M_2^2\,\mathbf{z}_1 \bmod M_1M_2\mathbf{1} = k\,M_2\,\mathbf{z}_1 \bmod M_1\mathbf{1} = j\,\mathbf{z}_1 \bmod M_1\mathbf{1}$$

that $\mathbf{x}_j = \mathbf{y}_\ell$. Analogously, we conclude that $\Lambda(\mathbf{z}_2, M_2) \subset \Lambda(M_2\mathbf{z}_1 + M_1\mathbf{z}_2, M_1M_2)$.

2. Now we prove that $\Lambda(\mathbf{z}_1, M_1) \cap \Lambda(\mathbf{z}_2, M_2) = \{\mathbf{0}\}$. For this purpose it is sufficient to show that the M_1M_2 points of $\Lambda(M_2\mathbf{z}_1 + M_1\mathbf{z}_2, M_1M_2)$ are distinct. Suppose that there is an $\ell \in \{0, \ldots, M_1M_2 - 1\}$ such that

$$\ell\,(M_2\mathbf{z}_1 + M_1\mathbf{z}_2) \equiv \mathbf{0} \bmod M_1M_2\mathbf{1}\,.$$

Then there exist $j_1, k_1 \in \{0, \ldots, M_1 - 1\}$ and $j_2, k_2 \in \{0, \ldots, M_2 - 1\}$ with $\ell = j_2\,M_1 + j_1 = k_1\,M_2 + k_2$, and we find

$$\ell\,(M_2\mathbf{z}_1 + M_1\mathbf{z}_2) \bmod M_1M_2\mathbf{1} = j_1\,M_2\mathbf{z}_1 + k_2\,M_1\mathbf{z}_2 \bmod M_1M_2\mathbf{1}\,.$$

Thus, we arrive at

$$j_1\,M_2\,\mathbf{z}_1 \equiv -k_2\,M_1\,\mathbf{z}_2 \bmod M_1M_2\mathbf{1}\,.$$

Since M_1 and M_2 are coprime, it follows that M_1 is a divisor of each component of $j_1\,\mathbf{z}_1$ and that M_2 is a divisor of each component of $-k_2\,\mathbf{z}_2$. But this can be only true for $j_1 = k_2 = 0$, since we had assumed that $\Lambda(\mathbf{z}_1, M_1)$ and $\Lambda(\mathbf{z}_2, M_2)$ have the cardinalities M_1 and M_2. This observation implies now $\ell = j_2\,M_1 = k_1\,M_2$ which is only possible for $j_2 = k_1 = 0$, since M_1 and M_2 are coprime. Thus $\ell = 0$, and the assertion is proven. ■

Lemma 8.23 can be simply generalized to the union of more than two rank-1 lattices.

Corollary 8.24 *Let the multiple rank-1 lattice $\Lambda(\mathbf{z}_1, M_1, \ldots, \mathbf{z}_s, M_s)$ with pairwise coprime lattice sizes $M_1, \ldots, M_s$ be given. Assume that $|\Lambda(\mathbf{z}_r, M_r)| = M_r$ for each $r = 1, \ldots, s$.*

Then we have

$$|\Lambda(\mathbf{z}_1, M_1, \ldots, \mathbf{z}_s, M_s)| = 1 - s + \sum_{r=1}^{s} M_r .$$

Further, let $\Lambda(\mathbf{z}, M)$ be the rank-1 lattice with the generating vector $\mathbf{z}$ and lattice size M given by

$$\mathbf{z} := \sum_{r=1}^{s} \left(\prod_{\substack{\ell=1 \\ \ell \neq r}}^{s} M_\ell \right) \mathbf{z}_r , \quad M := \prod_{r=1}^{s} M_r .$$

Then

$$\Lambda(\mathbf{z}_1, M_1, \ldots, \mathbf{z}_s, M_s) \subset \Lambda(\mathbf{z}, M).$$

Proof The proof follows similarly as for Lemma 8.23. ■

As in Sect. 8.2 we define now the *Fourier matrix* for the sampling set $X = \Lambda(\mathbf{z}_1, M_1, \mathbf{z}_2, M_2, \ldots, \mathbf{z}_s, M_s)$ and the frequency index set I

$$\begin{aligned} \mathbf{A} &= \mathbf{A}(\Lambda(\mathbf{z}_1, M_1, \mathbf{z}_2, M_2, \ldots, \mathbf{z}_s, M_s), I) \\ &:= \begin{pmatrix} \left(\mathrm{e}^{2\pi \mathrm{i}\, j\, (\mathbf{k}\cdot\mathbf{z}_1)/M_1}\right)_{j=0,\ldots,M_1-1,\, \mathbf{k}\in I} \\ \left(\mathrm{e}^{2\pi \mathrm{i}\, j\, (\mathbf{k}\cdot\mathbf{z}_2)/M_2}\right)_{j=0,\ldots,M_2-1,\, \mathbf{k}\in I} \\ \vdots \\ \left(\mathrm{e}^{2\pi \mathrm{i}\, j\, (\mathbf{k}\cdot\mathbf{z}_s)/M_s}\right)_{j=0,\ldots,M_s-1,\, \mathbf{k}\in I} \end{pmatrix}, \end{aligned} \tag{8.26}$$

where we assume that the frequency indices $\mathbf{k} \in I$ are arranged in a fixed order. Thus, $\mathbf{A}$ has $\sum_{r=1}^{s} M_r$ rows and $|I|$ columns, where the first rows of the s partial Fourier matrices coincide. We also introduce the reduced Fourier matrix

$$\widetilde{\mathbf{A}} := \begin{pmatrix} \left(\mathrm{e}^{2\pi \mathrm{i}\, j\, (\mathbf{k}\cdot\mathbf{z}_1)/M_1}\right)_{j=0,\ldots,M_1-1,\, \mathbf{k}\in I} \\ \left(\mathrm{e}^{2\pi \mathrm{i}\, j\, (\mathbf{k}\cdot\mathbf{z}_2)/M_2}\right)_{j=1,\ldots,M_2-1,\, \mathbf{k}\in I} \\ \vdots \\ \left(\mathrm{e}^{2\pi \mathrm{i}\, j\, (\mathbf{k}\cdot\mathbf{z}_s)/M_s}\right)_{j=1,\ldots,M_s-1,\, \mathbf{k}\in I} \end{pmatrix},$$

where we use beside $\left(\mathrm{e}^{2\pi \mathrm{i}\, j\, (\mathbf{k}\cdot\mathbf{z}_1)/M_1}\right)_{j=0,\ldots,M_1-1,\,\mathbf{k}\in I}$ only the partial matrices

$$\left(\mathrm{e}^{2\pi \mathrm{i}\, j\, (\mathbf{k}\cdot\mathbf{z}_r)/M_r}\right)_{j=1,\ldots,M_r-1,\,\mathbf{k}\in I}\,, \quad r = 2, \ldots, s\,,$$

such that $\widetilde{\mathbf{A}}$ has $\sum_{r=1}^{s} M_r - s + 1$ rows and $|I|$ columns. Obviously, $\mathbf{A}$ and $\widetilde{\mathbf{A}}$ have the same rank, since we have only removed redundant rows.

As in Sect. 8.2, we consider the fast evaluation of trigonometric polynomials on multiple rank-1 lattices on the one hand and the evaluation of their Fourier coefficients from samples on multiple rank-1 lattices on the other hand.

(i) Evaluation of Trigonometric Polynomials To evaluate a trigonometric polynomial at all nodes of a multiple rank-1 lattice $\Lambda(\mathbf{z}_1, M_1, \ldots, \mathbf{z}_s, M_s)$, we can apply the ideas from Sect. 8.2 and compute the trigonometric polynomial on s different rank-1 lattices $\Lambda(\mathbf{z}_1, M_1)$, …, $\Lambda(\mathbf{z}_s, M_s)$ separately. The corresponding Algorithm 8.25 applies the known Algorithm 8.8 s-times, once for each single rank-1 lattice. The computational cost of the fast evaluation at all nodes of the whole multiple rank-1 lattice $\Lambda(\mathbf{z}_1, M_1, \ldots, \mathbf{z}_s, M_s)$ is therefore $\mathcal{O}(\sum_{r=1}^{s} M_r \log M_r + s\, d\, |I|)$.

Algorithm 8.25 (Evaluation at Multiple Rank-1 Lattices)
Input: $M_1, \ldots, M_s \in \mathbb{N}$ *lattice sizes of rank-*1 *lattices* $\Lambda(\mathbf{z}_\ell, M_\ell)$, $\ell = 1, \ldots, s$,
$\mathbf{z}_1, \ldots, \mathbf{z}_s \in \mathbb{Z}^d$ *generating vectors of* $\Lambda(\mathbf{z}_\ell, M_\ell)$, $\ell = 1, \ldots, s$,
$I \subset \mathbb{Z}^d$ *finite frequency index set,*
$\hat{\mathbf{p}} = \left(\hat{p}_{\mathbf{k}}\right)_{\mathbf{k}\in I}$ *Fourier coefficients of* $p \in \Pi_I$ *in* (8.7).

1. *For* $\ell = 1, \ldots, s$ *do by Algorithm* 8.8

$$\mathbf{p}_\ell := \mathrm{LFFT}(M_\ell, \mathbf{z}_\ell, I, \hat{\mathbf{p}})\,.$$

2. *Set* $\mathbf{p} := \left(\mathbf{p}_1(1), \ldots, \mathbf{p}_1(M_1), \mathbf{p}_2(2), \ldots, \mathbf{p}_2(M_2), \ldots, \mathbf{p}_s(2), \ldots \mathbf{p}_s(M_s)\right)^\top$.

Output: $\mathbf{p} = \widetilde{\mathbf{A}}\, \hat{\mathbf{p}}$ *polynomial values of* $p \in \Pi_I$.
Computational cost: $\mathcal{O}(\sum_{\ell=1}^{s} M_\ell \log M_\ell + s\, d\, |I|)$.

The algorithm is a fast realization of the matrix–vector product with the Fourier matrix $\widetilde{\mathbf{A}}$ in (8.26). The fast computation of the matrix–vector product with the adjoint Fourier matrix $\mathbf{A}^{\mathrm{H}}$ can be realized by employing Algorithm 8.9 separately to each rank-1 lattice with a numerical effort of $\mathcal{O}(\sum_{\ell=1}^{s} M_\ell \log M_\ell + s\, d\, |I|)$.

(ii) Evaluation of the Fourier Coefficients To solve the inverse problem, i.e., to compute the Fourier coefficients of an arbitrary trigonometric polynomial $p \in \Pi_I$ as given in (8.7), we need to ensure that our Fourier matrix $\mathbf{A}$ in (8.26) has full rank $|I|$. This means that p needs to be already completely determined by the sampling set $\Lambda(\mathbf{z}_1, M_1, \ldots, \mathbf{z}_s, M_s)$. Then we can apply formula (8.9) for reconstruction. We are especially interested in a fast and stable reconstruction method.

We define a *reconstructing multiple rank-*1 *lattice* to a given frequency index set I as a multiple rank-1 lattice satisfying that

$$\mathbf{A}^{\mathrm{H}}\mathbf{A}=\mathbf{A}(\Lambda(\mathbf{z}_1, M_1, \mathbf{z}_2, M_2, \ldots, \mathbf{z}_s, M_s), I)^{\mathrm{H}}\,\mathbf{A}(\Lambda(\mathbf{z}_1, M_1, \mathbf{z}_2, M_2, \ldots, \mathbf{z}_s, M_s), I)$$

has full rank $|I|$.
In order to keep the needed number of sampling points $|\Lambda(\mathbf{z}_1, M_1, \ldots, \mathbf{z}_s, M_s)| = \sum_{r=1}^{s} M_r - s + 1$ small, we do not longer assume that each single rank-1 lattice is a reconstructing rank-1 lattice. But still, we can use Lemma 8.7 in order to compute the matrix $\mathbf{A}^{\mathrm{H}}\mathbf{A}$ in an efficient way.

Lemma 8.26 *Let* $\mathbf{A}$ *be the* $(\sum_{r=1}^{s} M_r)$*-by-*$|I|$ *Fourier matrix* (8.26) *for a frequency index set* $|I|$ *and a multiple rank-*1 *lattice* $\Lambda(\mathbf{z}_1, M_1, \mathbf{z}_2, M_2, \ldots, \mathbf{z}_s, M_s)$ *with cardinality* $1 - s + \sum_{r=1}^{s} M_r$.
Then the entries of $\mathbf{A}^{\mathrm{H}}\mathbf{A} \in \mathbb{C}^{|I|\times|I|}$ *have the form*

$$(\mathbf{A}^{\mathrm{H}}\mathbf{A})_{\mathbf{h},\mathbf{k}} = \sum_{r=1}^{s} M_r \delta_{(\mathbf{k}-\mathbf{h})\cdot\mathbf{z}_r \bmod M_r}\,,$$

where

$$\delta_{(\mathbf{k}-\mathbf{h})\cdot\mathbf{z}_r \bmod M_r} := \begin{cases} 1 & \mathbf{k}\cdot\mathbf{z}_r \equiv \mathbf{h}\cdot\mathbf{z}_r \bmod M_r\,, \\ 0 & \mathbf{k}\cdot\mathbf{z}_r \not\equiv \mathbf{h}\cdot\mathbf{z}_r \bmod M_r\,. \end{cases}$$

Proof The assertion follows directly from Lemma 8.7. The entry $(\mathbf{A}^{\mathrm{H}}\mathbf{A})_{\mathbf{h},\mathbf{k}}$ is the inner product of the $\mathbf{k}$th and the $\mathbf{h}$th column of $\mathbf{A}$. Thus, we find

$$(\mathbf{A}^{\mathrm{H}}\mathbf{A})_{\mathbf{h},\mathbf{k}} = \sum_{r=1}^{s}\sum_{j=0}^{M_r-1} \left(\mathrm{e}^{2\pi\mathrm{i}\,[(\mathbf{k}-\mathbf{h})\cdot\mathbf{z}_r]/M_r}\right)^j,$$

where the sums

$$\sum_{j=0}^{M_r-1} \left(\mathrm{e}^{2\pi\mathrm{i}\,[(\mathbf{k}-\mathbf{h})\cdot\mathbf{z}_r]/M_r}\right)^j$$

can be simply computed as in Lemma 8.7. ■

Lemma 8.26 also shows that $\mathbf{A}^{\mathrm{H}}\mathbf{A}$ can be sparse for suitably chosen rank-1 lattices. If the single rank-1 lattices are already reconstructing rank-1 lattices, then it directly follows that $\mathbf{A}^{\mathrm{H}}\mathbf{A}$ is a multiple of the identity matrix.

Now the following question remains: how to choose the parameters s as well as $\mathbf{z}_r$ and M_r, $r = 1, \ldots s$, to ensure that $\mathbf{A}^{\mathrm{H}}\mathbf{A}$ indeed possesses full rank $|I|$. The following strategy given in Algorithm 8.27 (see [223, Algorithm 1]) yields with high probability such a multiple rank-1 lattice. Here we take the lattice sizes $M_r := M$ for all $r = 1, \ldots, s$ as a prime number and choose the generating vectors $\mathbf{z}_r$ randomly in the set $[0, M-1]^d \cap \mathbb{Z}^d$. In order to determine the lattice size M large enough for the index set I, we define the *expansion of the frequency set I* by

$$N_I := \max_{j=1,\ldots,d} \left\{ \max_{\mathbf{k}\in I} k_j - \min_{\mathbf{l}\in I} \ell_j \right\}, \tag{8.27}$$

where $\mathbf{k} = (k_j)_{j=1}^d$ and $\mathbf{l} = (\ell_j)_{j=1}^d$ belong to I. The expansion N_I can be interpreted as the size of a d-dimensional cube we need to cover the index set I.

Algorithm 8.27 (Determining Reconstructing Multiple Rank-1 Lattices)
Input: $T \in \mathbb{N}$ *upper bound of the cardinality of a frequency set* I,
$d \in \mathbb{N}$ *dimension of the frequency set* I,
$N \in \mathbb{N}$ *upper bound of the expansion* N_I *of the frequency set* I,
$\delta \in (0, 1)$ *upper bound of failure probability,*
$c > 1$ *minimal oversampling factor.*

1. *Set* $c := \max\left\{c, \frac{N}{T-1}\right\}$ *and* $\lambda := c\,(T-1)$.
2. *Set* $s := \lceil \left(\frac{c}{c-1}\right)^2 \frac{\ln T - \ln \delta}{2} \rceil$.
3. *Set* $M = \operatorname{argmin}\{p > \lambda : p \in \mathbb{N} \text{ prime}\}$.
4. *For* $r = 1, \ldots, s$ *choose* $\mathbf{z}_r$ *from* $[0, M-1]^d \cap \mathbb{Z}^d$ *uniformly at random.*

Output: M *lattice size of all rank-1 lattices,*
$\mathbf{z}_1, \ldots, \mathbf{z}_s$ *generating vectors of rank-1 lattices such that*
$\Lambda(\mathbf{z}_1, M, \ldots, \mathbf{z}_s, M)$ *is a reconstructing multiple rank-1 lattice for* I *with probability at least* $1-\delta$.
Computational cost: $\mathcal{O}(\lambda \ln\ln\lambda + ds)$ *for* $c > 1$, $\lambda \sim \max\{T, N\}$, *and* $s \sim \ln T - \ln\delta$.

Due to [223, Theorem 3.4] the Algorithm 8.27 determines a reconstructing sampling set for trigonometric polynomials supported on the given frequency set I with probability at least $1-\delta_s$, where

$$\delta_s = T\,\mathrm{e}^{-2s\,(c-1)^2/c^2} \tag{8.28}$$

is an upper bound on the probability that the approach fails. There are several other strategies in the literature to find appropriate reconstructing multiple rank-1 lattices (see [222, 223, 228]). Finally, if a reconstructing multiple rank-1 lattice is found, then the Fourier coefficients of the trigonometric polynomial $p \in \Pi_I$ in (8.7) can be efficiently computed by solving the system

$$\mathbf{A}^{\mathrm{H}}\mathbf{A}\,\hat{\mathbf{p}} = \mathbf{A}^{\mathrm{H}}\mathbf{p},$$

where $\mathbf{p} := \left(p(\mathbf{x}_j)_{\mathbf{x}_j\in\Lambda(\mathbf{z}_1,M_1)}, \ldots, p(\mathbf{x}_j)_{\mathbf{x}_j\in\Lambda(\mathbf{z}_s,M_s)}\right)^{\top}$ and $\mathbf{A}^{\mathrm{H}}\mathbf{p}$ can be computed using Algorithm 8.9 for the s partial vectors.

Remark 8.28 In [228, 352] the authors suggest approximate algorithms for the reconstruction of sparse high-dimensional trigonometric polynomials, where the support in frequency domain is unknown. The main idea is the construction of the index set of frequencies belonging to the nonzero Fourier coefficients in a dimension incremental way in combination with the approximation based on rank-1 lattices. When one restricts the search space in frequency domain to a full grid $[-N, N]^d \cap \mathbb{Z}^d$ of refinement $N \in \mathbb{N}$ and assumes that the cardinality of the support of the trigonometric polynomial in frequency domain is bounded by the sparsity $s \in \mathbb{N}$, the method requires $\mathcal{O}(d\,s^2 N)$ samples and $\mathcal{O}(d\,s^3 + d\,s^2 N \log(s\,N))$ arithmetic operations in the case $c_1\sqrt{N} < s < c_2 N^d$. The number of samples is reduced to $\mathcal{O}(d\,s + d\,N)$ and the number of arithmetic operations is $\mathcal{O}(d\,s^3)$ by using a version of the Prony method. □

Remark 8.29 In this chapter we have always assumed that the user can choose the sample points for approximation and the sampling values at this points is available. This requires black-box access to the function, i.e., we need to be able to evaluate it at any point. Recently, further efficient algorithms for high-dimensional approximation based on scattered data approximation in combination with the NFFT have been developed. The main tool for this high-dimensional approximation is again based on trigonometric polynomials in combination with the analysis of variance (ANOVA) *decomposition.* For a general description of multivariate decompositions, we refer to [261] and for the classical ANOVA decomposition of periodic functions to [333]. The classical ANOVA decomposition is related to a decomposition of a multivariate polynomial in the frequency domain, where the Fourier coefficients are grouped in subsets that correspond to special subsets of coordinate indices. The ANOVA sensitivity indices then show which coordinate subsets strong correlations exist. In this way, one can find good approximations of high-dimensional polynomials by a sum of trigonometric polynomials of much lower dimensions [19]. Moreover, it is possible to decompose high-dimensional nonequispaced discrete Fourier transform into multiple transforms associated with ANOVA terms. Therefore, the grouped Fourier transform is able to handle high-dimensional frequency index sets, cf. [333]. These methods are not limited to periodic functions [334]. A very nice overview can be found in the dissertation [380]. □

Chapter 9
Numerical Applications of DFT

This chapter addresses numerical applications of DFTs. In Sect. 9.1, we describe a powerful multidimensional approximation method, the so-called cardinal interpolation by translates $\varphi(\cdot - \mathbf{k})$ with $\mathbf{k} \in \mathbb{Z}^d$, where $\varphi \in C_c(\mathbb{R}^d)$ is a compactly supported, continuous function. In this approximation method, the cardinal Lagrange function is of main interest. Applying this technique, we compute the multidimensional Fourier transform by the method of attenuation factors. Then, in Sect. 9.2, we investigate the periodic interpolation by translates on a uniform mesh, where we use the close connection between periodic and cardinal interpolation by translates. The central notion is the periodic Lagrange function. Using the periodic Lagrange function, we calculate the Fourier coefficients of a multivariate periodic function by the method of attenuation factors.

Starting with the Euler–Maclaurin summation formula, we discuss the quadrature of univariate periodic functions in Sect. 9.3. In Sect. 9.4, we present two methods for accelerating the convergence of Fourier series, namely, the Krylov–Lanczos method and the Fourier extension. Then in Sect. 9.5, we deal with fast Poisson solvers; more precisely, we solve the homogeneous Dirichlet boundary problem of the Poisson equation on the unit square by a finite difference method, where the related linear system is solved by a fast algorithm of the two-dimensional DST–I. Finally, in Sect. 9.6, we describe the Fourier analysis on the unit sphere of $\mathbb{R}^3$. We are mainly interested in fast and numerically stable algorithms for the evaluation of discrete spherical Fourier transforms.

9.1 Cardinal Interpolation by Translates

In this section, we describe a powerful approximation method of d-variate functions which can be efficiently solved by Fourier technique. The dimension $d \in \mathbb{N}$ is fixed.

G. Plonka et al., *Numerical Fourier Analysis*, Applied and Numerical Harmonic Analysis, https://doi.org/10.1007/978-3-031-35005-4_9

Let $\varphi \in C_c(\mathbb{R}^d)$ be a given complex-valued continuous *basis function* with compact support

$$\text{supp}\,\varphi := \overline{\{\mathbf{x} \in \mathbb{R}^d : \varphi(\mathbf{x}) \neq 0\}}\,.$$

Further, we assume that the d-dimensional Fourier transform

$$\hat{\varphi}(\boldsymbol{\omega}) := \int_{\mathbb{R}^d} \varphi(\mathbf{x})\,\mathrm{e}^{-\mathrm{i}\,\mathbf{x}\cdot\boldsymbol{\omega}}\,\mathrm{d}\mathbf{x}$$

belongs to $L_1(\mathbb{R}^d)$. Note that $\mathbf{x}\cdot\boldsymbol{\omega} := \sum_{\ell=1}^{d} x_\ell\,\omega_\ell$ denotes the inner product of vectors $\mathbf{x} = (x_\ell)_{\ell=1}^d$, $\boldsymbol{\omega} = (\omega_\ell)_{\ell=1}^d \in \mathbb{R}^d$. Often used basis functions are cardinal B-splines and box splines. Note that B-splines (i.e., basis splines) are splines with the smallest possible support. Cardinal B-splines are B-splines with integer knots.

Example 9.1 In the univariate case $d = 1$, let $m \in \mathbb{N} \setminus \{1\}$ be given. We choose $\varphi = N_m$ as the *cardinal* B-*spline of order* m which can be recursively defined by

$$N_m(x) := \big(N_{m-1} * N_1\big)(x) = \int_0^1 N_{m-1}(x-t)\,\mathrm{d}t\,, \quad x \in \mathbb{R}\,, \tag{9.1}$$

with

$$N_1(x) := \frac{1}{2}\,\big(\chi_{(0,\,1]}(x) + \chi_{[0,\,1)}(x)\big)\,,$$

where $\chi_{(0,\,1]}$ denotes the characteristic function of $(0,\,1]$. Then N_m is contained in $C_c^{m-2}(\mathbb{R})$ and has the compact support supp $N_m = [0,\,m]$. In the cases $m = 2,\,3,\,4$, we obtain the cardinal B-splines:

$$N_2(x) = \begin{cases} x & x \in [0,\,1)\,,\\ 2-x & x \in [1,\,2)\,,\\ 0 & x \in \mathbb{R}\setminus[0,\,2)\,,\end{cases}$$

$$N_3(x) = \begin{cases} x^2/2 & x \in [0,\,1)\,,\\ (-2x^2+6x-3)/2 & x \in [1,\,2)\,,\\ (3-x)^2/2 & x \in [2,\,3)\,,\\ 0 & x \in \mathbb{R}\setminus[0,\,3)\,,\end{cases}$$

$$N_4(x) = \begin{cases} x^3/6 & x \in [0,\,1)\,,\\ (-3x^3+12x^2-12x+4)/6 & x \in [1,\,2)\,,\\ (3x^3-24x^2+60x-44)/6 & x \in [2,\,3)\,,\\ (4-x)^3/6 & x \in [3,\,4)\,,\\ 0 & x \in \mathbb{R}\setminus[0,\,4)\,.\end{cases}$$

Further, we have $N_m \mid [k,\, k+1) \in \mathscr{P}_{m-1}$ for each $k \in \mathbb{Z}$ and $m > 1$, where $\mathscr{P}_{m-1}$ denotes the set of all algebraic polynomials up to degree $m-1$. These B-splines were introduced in [96]. The cardinal B-splines can be computed by the recurrence formula:

$$N_m(x) = \frac{x}{m-1}\, N_{m-1}(x) + \frac{m-x}{m-1}\, N_{m-1}(x-1)\,, \quad m = 2,\, 3,\, \ldots\,.$$

Obviously, we have

$$\hat{N}_1(\omega) = \int_0^1 \mathrm{e}^{-\mathrm{i}x\omega}\, \mathrm{d}x = \mathrm{e}^{-\mathrm{i}\omega/2}\, \mathrm{sinc}\, \frac{\omega}{2}\,, \quad \omega \in \mathbb{R}\,.$$

By the convolution property of the Fourier transform (see Theorem 2.5) and by (9.1), we obtain for all $\omega \in \mathbb{R}$ and $m \in \mathbb{N}$ that

$$\hat{N}_m(\omega) = \big(\hat{N}_1(\omega)\big)^m = \mathrm{e}^{-\mathrm{i}m\omega/2}\, \big(\mathrm{sinc}\, \frac{\omega}{2}\big)^m\,. \tag{9.2}$$

The *centered cardinal* B-*spline of order* $m \in \mathbb{N}$ is defined by

$$M_m(x) := N_m\big(x + \frac{m}{2}\big)\,, \quad x \in \mathbb{R}\,.$$

Then M_m is an even function with supp $M_m = [-m/2,\, m/2]$. For $m = 2,\, 3,\, 4$, the centered cardinal B-splines read as follows:

$$M_2(x) = \begin{cases} 1+x & x \in [-1,\, 0)\,, \\ 1-x & x \in [0,\, 1)\,, \\ 0 & x \in \mathbb{R} \setminus [-1,\, 1)\,, \end{cases}$$

$$M_3(x) = \begin{cases} (3+2x)^2/8 & x \in [-3/2,\, -1/2)\,, \\ (3-4x^2)/4 & x \in [-1/2,\, 1/2)\,, \\ (3-2x)^2/8 & x \in [1/2,\, 3/2)\,, \\ 0 & x \in \mathbb{R} \setminus [-3/2,\, 3/2)\,. \end{cases}$$

$$M_4(x) = \begin{cases} (x+2)^3/6 & x \in [-2,\, -1)\,, \\ (-3x^3 - 6x^2 + 4)/6 & x \in [-1,\, 0)\,, \\ (3x^3 - 6x^2 + 4)/6 & x \in [0,\, 1)\,, \\ (2-x)^3/6 & x \in [1,\, 2)\,, \\ 0 & x \in \mathbb{R} \setminus [-2,\, 2)\,. \end{cases}$$

Note that M_m is a spline on an integer grid if m is even and on a half integer grid if m is odd. The centered cardinal B-splines M_m satisfy the recurrence formula:

$$M_m(x) = \big(M_{m-1} * M_1\big)(x) = \int_{-1/2}^{1/2} M_{m-1}(x-t)\, \mathrm{d}t\,, \quad m = 2,\, 3,\, \ldots\,. \tag{9.3}$$

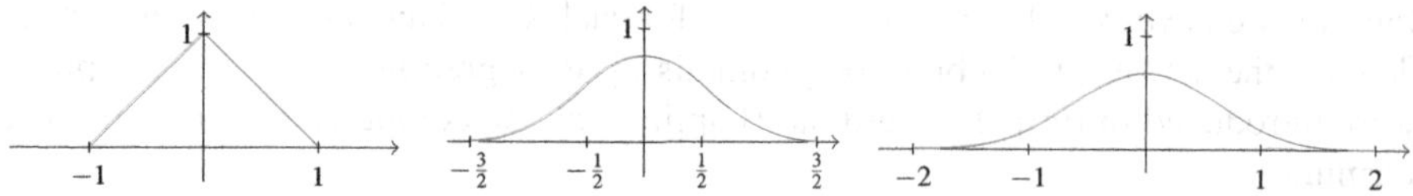

(a) M_2 (b) M_3 (c) M_4

Fig. 9.1 Centered cardinal B-splines M_2, M_3, and M_4

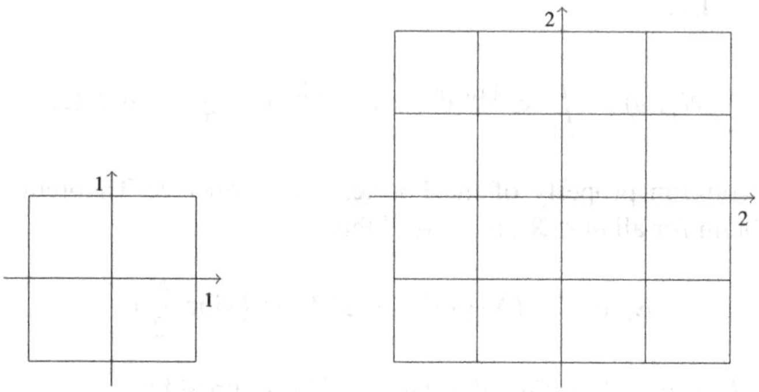

(a) support of $M^{(2,2)}$ (b) support of $M^{(4,4)}$

Fig. 9.2 Rectangular partitions of the supports of $M^{(2,2)}$ and $M^{(4,4)}$

By the convolution property of the Fourier transform and by (9.3), the Fourier transform of M_m reads as follows:

$$\hat{M}_m(\omega) = \big(\mathrm{sinc}\,\frac{\omega}{2}\big)^m\,, \quad \omega \in \mathbb{R}\,.$$

For further details, see [103] or [87, pp. 1–13] (Fig. 9.1). □

Example 9.2 A natural multivariate generalization of the univariate centered cardinal B-spline is the so-called box spline introduced in [104]. For the theory of box splines, we refer to [105] and [87, pp. 15–25]. For simplicity, we restrict us to the bivariate case $d = 2$. We choose the directions $\mathbf{d}_1 := (1,\, 0)^\top$ and $\mathbf{d}_2 := (0,\, 1)^\top$ with corresponding multiplicities $k,\, \ell \in \mathbb{N}$. The *tensor product* B-*spline* $M^{(k,\ell)}$ is defined as the tensor product of the centered cardinal B-splines M_k and M_ℓ, i.e.,

$$M^{(k,\ell)}(\mathbf{x}) := M_k(x_1)\, M_\ell(x_2)\,, \quad \mathbf{x} = (x_1,\, x_2)^\top \in \mathbb{R}^2\,.$$

Obviously, $M^{(k,\ell)}$ is supported on $[-k/2,\, k/2] \times [-\ell/2,\, \ell/2]$ and is a piecewise polynomial whose polynomial pieces are separated by a *rectangular mesh* (Fig. 9.2). The Fourier transform of $M^{(k,\ell)}$ reads as follows:

$$\hat{M}^{(k,\ell)}(\boldsymbol{\omega}) = \big(\mathrm{sinc}\,\frac{\omega_1}{2}\big)^m \big(\mathrm{sinc}\,\frac{\omega_2}{2}\big)^\ell\,, \quad \boldsymbol{\omega} = (\omega_1,\, \omega_2)^\top \in \mathbb{R}^2\,.$$

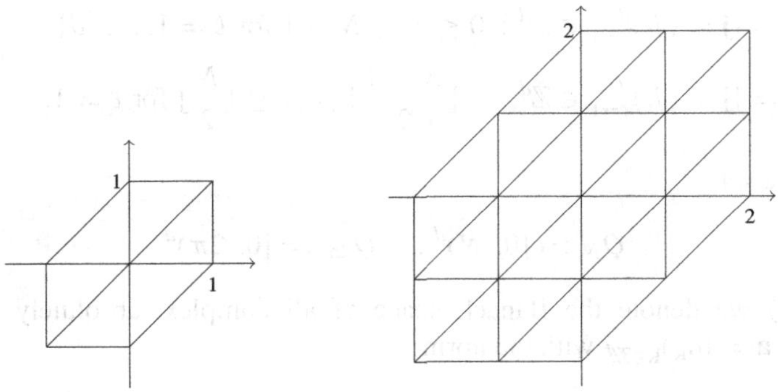

(a) support of $M^{(1,1,1)}$ (b) support of $M^{(2,2,2)}$

Fig. 9.3 Three-direction meshes of the supports of $M^{(1,1,1)}$ and $M^{(2,2,2)}$

The tensor product B-spline can be generalized by addition of the third direction $\mathbf{d}_3 := (1,\ 1)^\top$. Then for $k,\ \ell,\ m \in \mathbb{N}$, we define the *three-direction box spline*

$$M^{(k,\ell,m)}(\mathbf{x}) := \int_{-1/2}^{1/2} M^{(k,\ell,m-1)}(x_1 - t,\ x_2 - t)\,\mathrm{d}t\,, \quad \mathbf{x} = (x_1,\ x_2)^\top \in \mathbb{R}^2\,,$$

where we set $M^{(k,\ell,0)} := M^{(k,\ell)}$. Then the support of $M^{(k,\ell,m)}$ is

$$\operatorname{supp} M^{(k,\ell,m)} = \{\mathbf{x} = t_1\,k\,\mathbf{d}_1 + t_2\,\ell\,\mathbf{d}_2 + t_3\,m\,\mathbf{d}_3 : t_1,\ t_2,\ t_3 \in [-1/2,\ 1/2]\}\,,$$

which forms a hexagon with the center $(0,\ 0)^\top$ whose sides are k, ℓ, and $\sqrt{2}\,m$ long in directions $\mathbf{d}_1$, $\mathbf{d}_2$, and $\mathbf{d}_3$, respectively. The three-direction box spline $M^{(k,\ell,m)}$ is a piecewise polynomial, whose polynomial pieces are separated by a *three-direction mesh* or *type*-1 *triangulation*. Each polynomial piece is a bivariate polynomial of total degree up to $k + \ell + m - 2$. Further, $M^{(k,\ell,m)}$ possesses continuous partial derivatives up to order $r - 2$ with

$$r := k + \ell + m - \max\{k,\ \ell,\ m\}\,.$$

For example, $M^{(1,1,1)} \in C_c(\mathbb{R}^2)$ is the piecewise linear hat function with $M^{(1,1,1)}(0,0) = 1$. The three-direction box spline $M^{(2,2,1)} \in C_c^1(\mathbb{R}^2)$ consists of piecewise polynomials up to total degree 3 and $M^{(2,2,2)} \in C_c^2(\mathbb{R}^2)$ consists of piecewise polynomials up to total degree 4 (Fig. 9.3). For multivariate box splines, we refer to the literature [87, 105]. □

9.1.1 Cardinal Lagrange Function

Now we introduce some additional notations. Let $N \in \mathbb{N} \setminus \{1\}$ be fixed. By J_N and B_N we denote following sets of grid points:

$$J_N := \{\mathbf{j} = (j_\ell)_{\ell=1}^d \in \mathbb{Z}^d : 0 \le j_\ell \le N-1 \text{ for } \ell = 1, \ldots, d\},$$
$$B_N := \{\mathbf{j} = (j_\ell)_{\ell=1}^d \in \mathbb{Z}^d : -\lfloor \frac{N-1}{2} \rfloor \le j_\ell \le \lfloor \frac{N}{2} \rfloor \text{ for } \ell = 1, \ldots, d\}.$$

Further, we set

$$Q_N := [0, N)^d, \quad Q_{2\pi} := [0, 2\pi)^d.$$

By $\ell_1(\mathbb{Z}^d)$ we denote the Banach space of all complex, absolutely summable sequences $\mathbf{a} = (a_\mathbf{k})_{\mathbf{k}\in\mathbb{Z}^d}$ with the norm

$$\|\mathbf{a}\|_{\ell_1(\mathbb{Z}^d)} := \sum_{\mathbf{k}\in\mathbb{Z}^d} |a_\mathbf{k}|.$$

As usual we agree the sum of such a series by

$$\sum_{\mathbf{k}\in\mathbb{Z}^d} a_\mathbf{k} := \lim_{N\to\infty} \sum_{\mathbf{k}\in B_N} a_\mathbf{k}.$$

Let $\varphi \in C_c(\mathbb{R}^d)$ be a fixed basis function. Then we form integer translates $\varphi(\cdot - \mathbf{k})$ for $\mathbf{k} \in \mathbb{Z}^d$. A linear subspace $\mathscr{L}$ of $L_1(\mathbb{R}^d)$ is called *shift-invariant*, if for each $f \in \mathscr{L}$ all integer translates $f(\cdot - \mathbf{k})$, $\mathbf{k} \in \mathbb{Z}^d$, are also contained in $\mathscr{L}$. A special shift-invariant space is the space $\mathscr{L}(\varphi)$ of all functions s of the form

$$s = \sum_{\mathbf{k}\in\mathbb{Z}^d} a_\mathbf{k}\, \varphi(\cdot - \mathbf{k})$$

with $\left(a_\mathbf{k}\right)_{\mathbf{k}\in\mathbb{Z}^d} \in \ell_1(\mathbb{Z}^d)$. Obviously, the above series converges absolutely and uniformly on $\mathbb{R}^d$. Hence, we have $s \in L_1(\mathbb{R}^d) \cap C(\mathbb{R}^d)$, because

$$\|s\|_{L_1(\mathbb{R}^d)} \le \sum_{\mathbf{k}\in\mathbb{Z}^d} |a_\mathbf{k}|\, \|\varphi\|_{L_1(\mathbb{R}^d)} < \infty.$$

Now we study the *cardinal interpolation problem* in $\mathscr{L}(\varphi)$ and the *cardinal interpolation by translates*, respectively. For the given data $\mathbf{f} := (f_\mathbf{j})_{\mathbf{j}\in\mathbb{Z}^d} \in \ell_1(\mathbb{Z}^d)$, we determine a function $s \in \mathscr{L}(\varphi)$ with

$$s(\mathbf{j}) = f_\mathbf{j} \quad \text{for all } \mathbf{j} \in \mathbb{Z}^d. \tag{9.4}$$

In applications, one assumes often that $f_\mathbf{j} = 0$ for all $\mathbf{j} \in \mathbb{Z}^d \setminus J_N$ and $\mathbf{j} \in \mathbb{Z}^d \setminus B_N$, respectively. Thus, for $\mathbf{x} = \mathbf{j} \in \mathbb{Z}^d$, we obtain the convolution-like equation:

$$s(\mathbf{j}) = f_\mathbf{j} = \sum_{\mathbf{k}\in\mathbb{Z}^d} s_\mathbf{k}\, \varphi(\mathbf{j} - \mathbf{k}), \quad \mathbf{x} \in \mathbb{R}^d.$$

We are interested in an efficient solution of this interpolation problem by using multidimensional DFTs.

A key role in the cardinal interpolation in $\mathscr{L}(\varphi)$ plays the *symbol* σ_φ which is defined by

$$\sigma_\varphi(\boldsymbol{\omega}) := \sum_{\mathbf{k}\in\mathbb{Z}^d} \varphi(\mathbf{k})\,\mathrm{e}^{-\mathrm{i}\,\mathbf{k}\cdot\boldsymbol{\omega}}\,, \quad \boldsymbol{\omega} \in \mathbb{R}^d\,. \tag{9.5}$$

Since the basis function $\varphi \in C_c(\mathbb{R}^d)$ is compactly supported, the symbol σ_φ is a 2π-periodic, d-variate trigonometric polynomial. For the symbol, we show a property which is closely related to the Poisson summation formula (see Theorem 4.28).

Lemma 9.3 *Let $\varphi \in C_c(\mathbb{R}^d)$ be a given basis function. Assume that $\hat{\varphi}$ fulfills the condition*

$$\sum_{\mathbf{k}\in\mathbb{Z}^d} \sup\{|\hat{\varphi}(\boldsymbol{\omega} + 2\mathbf{k}\pi)| : \boldsymbol{\omega} \in Q_{2\pi}\} < \infty\,. \tag{9.6}$$

Then the symbol σ_φ can be represented in the form

$$\sigma_\varphi(\boldsymbol{\omega}) = \sum_{\mathbf{k}\in\mathbb{Z}^d} \hat{\varphi}(\boldsymbol{\omega} + 2\mathbf{k}\pi)\,.$$

Proof By condition (9.6), we see that

$$\begin{aligned}\|\hat{\varphi}\|_{L_1(\mathbb{R}^d)} &= \sum_{\mathbf{k}\in\mathbb{Z}^d} \int_{Q_{2\pi}} |\hat{\varphi}(\boldsymbol{\omega} + 2\mathbf{k}\pi)|\,\mathrm{d}\boldsymbol{\omega} \\ &\le (2\pi)^d \sum_{\mathbf{k}\in\mathbb{Z}^d} \sup\{|\hat{\varphi}(\boldsymbol{\omega} + 2\mathbf{k}\pi)| : \boldsymbol{\omega} \in Q_{2\pi}\} < \infty\end{aligned}$$

such that $\hat{\varphi} \in L_1(\mathbb{R}^d)$. By Theorem 4.21, we know that $\hat{\varphi} \in C_0(\mathbb{R}^d)$. Thus, by (9.6) the series of continuous functions

$$\sum_{\mathbf{k}\in\mathbb{Z}^d} \hat{\varphi}(\boldsymbol{\omega} + 2\mathbf{k}\pi)$$

converges uniformly on $\mathbb{R}^d$ to a 2π-periodic function $\psi \in C(\mathbb{T}^d)$. The Fourier coefficients $c_{\mathbf{j}}(\psi)$, $\mathbf{j} \in \mathbb{Z}^d$, read by Theorem 4.22 as follows:

$$\begin{aligned}c_{\mathbf{j}}(\psi) &:= \frac{1}{(2\pi)^d} \int_{Q_{2\pi}} \psi(\boldsymbol{\omega})\,\mathrm{e}^{-\mathrm{i}\,\mathbf{j}\cdot\boldsymbol{\omega}}\,\mathrm{d}\boldsymbol{\omega} = \frac{1}{(2\pi)^d} \sum_{\mathbf{k}\in\mathbb{Z}^d} \int_{Q_{2\pi}} \hat{\varphi}(\boldsymbol{\omega} + 2\mathbf{k}\pi)\,\mathrm{e}^{-\mathrm{i}\,\mathbf{j}\cdot\boldsymbol{\omega}}\,\mathrm{d}\boldsymbol{\omega} \\ &= \frac{1}{(2\pi)^d} \int_{\mathbb{R}^d} \hat{\varphi}(\boldsymbol{\omega})\,\mathrm{e}^{-\mathrm{i}\,\mathbf{j}\cdot\boldsymbol{\omega}}\,\mathrm{d}\boldsymbol{\omega} = \varphi(-\mathbf{j})\,.\end{aligned}$$

Since φ is compactly supported, the Fourier series of ψ has only finitely many nonzero summands such that

$$\psi(\boldsymbol{\omega}) = \sum_{\mathbf{j}\in\mathbb{Z}^d} \varphi(-\mathbf{j})\, \mathrm{e}^{\mathrm{i}\,\mathbf{j}\cdot\boldsymbol{\omega}} = \sum_{\mathbf{k}\in\mathbb{Z}^d} \varphi(\mathbf{k})\, \mathrm{e}^{-\mathrm{i}\,\mathbf{k}\cdot\boldsymbol{\omega}} = \sigma_\varphi(\boldsymbol{\omega})$$

for all $\boldsymbol{\omega} \in \mathbb{R}^d$. ■

Example 9.4 In the case $d = 1$, the B-splines N_m and M_m are contained in $C_c(\mathbb{R})$ for $m \in \mathbb{N} \setminus \{1\}$ and they fulfill the condition (9.6).

For the cardinal B-spline $\varphi = N_m$, the corresponding symbol reads as follows:

$$\sigma_\varphi(\omega) = \begin{cases} \mathrm{e}^{-\mathrm{i}\omega} & m = 2\,, \\ \mathrm{e}^{-3\mathrm{i}\omega/2}\, \cos\frac{\omega}{2} & m = 3\,, \\ \mathrm{e}^{-2\mathrm{i}\omega}\,(2+\cos\omega)/3 & m = 4\,. \end{cases}$$

For the centered cardinal B-spline $\varphi = M_m$, the corresponding symbol reads as follows:

$$\sigma_\varphi(\omega) = \begin{cases} 1 & m = 2\,, \\ (3+\cos\omega)/4 & m = 3\,, \\ (2+\cos\omega)/3 & m = 4\,. \end{cases}$$

Thus, the symbols of important (centered) cardinal B-splines are quite simple. □

Example 9.5 In the case $d = 2$, the three-direction box spline $\varphi = M^{(k,\ell,m)}$ with $k, \ell, m \in \mathbb{N}$ fulfills the condition (9.6). The corresponding symbol reads for $\boldsymbol{\omega} = (\omega_1, \omega_2)^\top \in \mathbb{R}^2$ as follows:

$$\sigma_\varphi(\boldsymbol{\omega}) = \begin{cases} 1 & (k,\ell,m) = (1,1,1)\,, \\ \big(7 + 2\cos\omega_1 + 2\cos\omega_2 + \cos(\omega_1+\omega_2)\big)/12 & (k,\ell,m) = (2,2,1)\,, \\ \big(3 + \cos\omega_1 + \cos\omega_2 + \cos(\omega_1+\omega_2)\big)/6 & (k,\ell,m) = (2,2,2)\,. \end{cases}$$

The symbols of often used three-direction box splines are simple trigonometric polynomials. □

A function $\lambda \in \mathscr{L}(\varphi)$ which interpolates the Kronecker data $(\delta_{\mathbf{k}})_{\mathbf{k}\in\mathbb{Z}^d}$ on the integer grid $\mathbb{Z}^d$, i.e.,

$$\lambda(\mathbf{k}) = \delta_{\mathbf{k}} := \begin{cases} 1 & \mathbf{k} = \mathbf{0}\,, \\ 0 & \mathbf{k} \in \mathbb{Z}^d \setminus \{\mathbf{0}\}\,, \end{cases}$$

is called *cardinal Lagrange function*. For a given basis function $\varphi \in C_c(\mathbb{R}^d)$ with corresponding nonvanishing symbol σ_φ, we can construct a cardinal Lagrange function as follows:

Theorem 9.6 *Let $\varphi \in C_c(\mathbb{R}^d)$ be a given basis function. Assume that $\hat{\varphi}$ fulfills the condition* (9.6) *and that $\sigma_\varphi(\boldsymbol{\omega}) \neq 0$ for all $\boldsymbol{\omega} \in Q_{2\pi}$.*

Then the function $\lambda \in \mathscr{L}(\varphi)$ defined as

$$\lambda(\mathbf{x}) := \frac{1}{(2\pi)^d} \int_{\mathbb{R}^d} \frac{\hat{\varphi}(\boldsymbol{\omega})}{\sigma_\varphi(\boldsymbol{\omega})}\, \mathrm{e}^{\mathrm{i}\,\boldsymbol{\omega}\cdot\mathbf{x}}\, \mathrm{d}\boldsymbol{\omega}\,, \quad \mathbf{x} \in \mathbb{R}^d\,, \tag{9.7}$$

is a cardinal Lagrange function of the form

$$\lambda(\mathbf{x}) = \sum_{\mathbf{k}\in\mathbb{Z}^d} \lambda_{\mathbf{k}}\, \varphi(\mathbf{x}-\mathbf{k})\,, \quad \mathbf{x} \in \mathbb{R}^d\,, \tag{9.8}$$

with the coefficients

$$\lambda_{\mathbf{k}} := \frac{1}{(2\pi)^d} \int_{Q_{2\pi}} \frac{\mathrm{e}^{\mathrm{i}\,\boldsymbol{\omega}\cdot\mathbf{k}}}{\sigma_\varphi(\boldsymbol{\omega})}\, \mathrm{d}\boldsymbol{\omega}\,, \quad \mathbf{k} \in \mathbb{Z}^d\,, \tag{9.9}$$

where $\big(\lambda_{\mathbf{k}}\big)_{\mathbf{k}\in\mathbb{Z}^d} \in \ell_1(\mathbb{Z}^d)$. The series (9.8) *converges absolutely and uniformly on $\mathbb{R}^d$.*

Proof

1. By Lemma 9.3, we know that for all $\boldsymbol{\omega} \in \mathbb{R}^d$

$$\sigma_\varphi(\boldsymbol{\omega}) = \sum_{\mathbf{n}\in\mathbb{Z}^d} \hat{\varphi}(\boldsymbol{\omega} + 2\mathbf{n}\pi) \neq 0\,.$$

   Here the series

$$\sum_{\mathbf{n}\in\mathbb{Z}^d} \hat{\varphi}(\boldsymbol{\omega} + 2\mathbf{n}\pi)$$

   converges uniformly on $\mathbb{R}^d$ by condition (9.6). By (9.7) we obtain for each $\mathbf{k} \in \mathbb{Z}^d$ that

$$\begin{aligned}\lambda(\mathbf{k}) &= \frac{1}{(2\pi)^d} \int_{\mathbb{R}^d} \frac{\hat{\varphi}(\boldsymbol{\omega})}{\sigma_\varphi(\boldsymbol{\omega})}\, \mathrm{e}^{\mathrm{i}\,\boldsymbol{\omega}\cdot\mathbf{k}}\, \mathrm{d}\boldsymbol{\omega} \\ &= \frac{1}{(2\pi)^d} \sum_{\mathbf{j}\in\mathbb{Z}^d} \int_{Q_{2\pi}} \frac{\hat{\varphi}(\boldsymbol{\omega} + 2\pi\mathbf{j})}{\sum_{\mathbf{n}\in\mathbb{Z}^d} \hat{\varphi}(\boldsymbol{\omega} + 2\pi\mathbf{n})}\, \mathrm{e}^{\mathrm{i}\,\boldsymbol{\omega}\cdot\mathbf{k}}\, \mathrm{d}\boldsymbol{\omega} = \frac{1}{(2\pi)^d} \int_{Q_{2\pi}} \mathrm{e}^{\mathrm{i}\,\boldsymbol{\omega}\cdot\mathbf{k}}\, \mathrm{d}\boldsymbol{\omega} = \delta_{\mathbf{k}}\,.\end{aligned}$$

2. Applying Lemma 9.3 to the shifted function $\psi := \varphi(\cdot + \mathbf{x})$ for arbitrary fixed $\mathbf{x} \in \mathbb{R}^d$, we obtain by Theorem 4.20

$$\hat{\psi}(\boldsymbol{\omega}) = \mathrm{e}^{\mathrm{i}\boldsymbol{\omega}\cdot\mathbf{x}}\, \hat{\varphi}(\boldsymbol{\omega})\,, \quad \boldsymbol{\omega} \in \mathbb{R}^d\,,$$

such that for all $\boldsymbol{\omega} \in \mathbb{R}^d$

$$\sigma_\psi(\boldsymbol{\omega}) = \sum_{\mathbf{k}\in\mathbb{Z}^d} \varphi(\mathbf{k}+\mathbf{x})\,\mathrm{e}^{-\mathrm{i}\,\mathbf{k}\cdot\boldsymbol{\omega}} = \sum_{\mathbf{j}\in\mathbb{Z}^d} \hat{\varphi}(\boldsymbol{\omega}+2\pi\mathbf{j})\,\mathrm{e}^{\mathrm{i}\,\boldsymbol{\omega}\cdot\mathbf{x}}\,\mathrm{e}^{2\pi\mathrm{i}\,\mathbf{j}\cdot\mathbf{x}}\,.$$

By condition (9.6), the above series on the right-hand side converges uniformly on $\mathbb{R}^d$. By definition (9.7) of the cardinal Lagrange function λ, we see that

$$\begin{aligned}\lambda(\mathbf{x}) &= \frac{1}{(2\pi)^d}\int_{\mathbb{R}^d} \frac{\hat{\varphi}(\boldsymbol{\omega})}{\sigma_\varphi(\boldsymbol{\omega})}\,\mathrm{e}^{\mathrm{i}\,\boldsymbol{\omega}\cdot\mathbf{x}}\,\mathrm{d}\boldsymbol{\omega}\\ &= \frac{1}{(2\pi)^d}\int_{Q_{2\pi}} \frac{1}{\sigma_\varphi(\boldsymbol{\omega})}\sum_{\mathbf{j}\in\mathbb{Z}^d} \hat{\varphi}(\boldsymbol{\omega}+2\pi\mathbf{j})\,\mathrm{e}^{\mathrm{i}\,\boldsymbol{\omega}\cdot\mathbf{x}}\,\mathrm{e}^{2\pi\mathrm{i}\mathbf{j}\cdot\mathbf{x}}\,\mathrm{d}\boldsymbol{\omega}\\ &= \frac{1}{(2\pi)^d}\int_{Q_{2\pi}} \frac{1}{\sigma_\varphi(\boldsymbol{\omega})}\sum_{\mathbf{k}\in\mathbb{Z}^d} \varphi(\mathbf{k}+\mathbf{x})\,\mathrm{e}^{-\mathrm{i}\,\boldsymbol{\omega}\cdot\mathbf{k}}\,\mathrm{d}\boldsymbol{\omega} = \sum_{\mathbf{k}\in\mathbb{Z}^d}\lambda_{\mathbf{k}}\,\varphi(\mathbf{x}-\mathbf{k})\,,\end{aligned}$$

where $\lambda_{\mathbf{k}}$ is given by (9.9).

3. Since $\sigma_\varphi \neq 0$ is a 2π-periodic, trigonometric polynomial, the 2π-periodic function $1/\sigma_\varphi \in C^\infty(\mathbb{T}^d)$ possesses rapidly decreasing Fourier coefficients $\lambda_{\mathbf{k}}$ by Lemma 4.6, i.e., for each $m \in \mathbb{N}_0$ it holds

$$\lim_{\|\mathbf{k}\|_2\to\infty} (1+\|\mathbf{k}\|_2)^m\,|\lambda_{\mathbf{k}}| = 0\,.$$

Since by Lemma 4.8 we have

$$\sum_{\mathbf{k}\in\mathbb{Z}^d} (1+\|\mathbf{k}\|_2)^{-d-1} < \infty$$

we obtain by Hölder's inequality that

$$\begin{aligned}\sum_{\mathbf{k}\in\mathbb{Z}^d} |\lambda_{\mathbf{k}}| &= \sum_{\mathbf{k}\in\mathbb{Z}^d} (1+\|\mathbf{k}\|_2)^{-d-1}\Big((1+\|\mathbf{k}\|_2)^{d+1}\,|\lambda_{\mathbf{k}}|\Big)\\ &\le \sum_{\mathbf{k}\in\mathbb{Z}^d} (1+\|\mathbf{k}\|_2)^{-d-1}\ \sup_{\mathbf{k}\in\mathbb{Z}^d}(1+\|\mathbf{k}\|_2)^{d+1}\,|\lambda_{\mathbf{k}}| < \infty\,.\end{aligned}$$

Hence, the series (9.8) converges absolutely and uniformly on $\mathbb{R}^d$ and $\lambda \in \mathscr{L}(\varphi)$. ■

Theorem 9.7 *Let $\varphi \in C_c(\mathbb{R}^d)$ be a given basis function. Assume that $\hat{\varphi}$ fulfills the condition* (9.6).

The cardinal interpolation problem (9.4) *in $\mathscr{L}(\varphi)$ is uniquely solvable for arbitrary given data $\mathbf{f} = \big(f_{\mathbf{k}}\big)_{\mathbf{k}\in\mathbb{Z}^d} \in \ell_1(\mathbb{Z}^d)$ if and only if the symbol σ_φ satisfies the condition $\sigma_\varphi(\boldsymbol{\omega}) \neq 0$ for all $\boldsymbol{\omega} \in Q_{2\pi}$.*

Proof

1. Assume that the cardinal interpolation problem (9.4) in $\mathscr{L}(\varphi)$ is uniquely solvable for each data $\mathbf{f} = \big(f_{\mathbf{k}}\big)_{\mathbf{k}\in\mathbb{Z}^d} \in \ell_1(\mathbb{Z}^d)$. Especially for the Kronecker data $\big(\delta_{\mathbf{j}}\big)_{\mathbf{j}\in\mathbb{Z}^d}$, there exists a function $\lambda \in \mathscr{L}(\varphi)$ of the form

$$\lambda = \sum_{\mathbf{k}\in\mathbb{Z}^d} \lambda_{\mathbf{k}}\, \varphi(\cdot - \mathbf{k})$$

with $\big(\lambda_{\mathbf{k}}\big)_{\mathbf{k}\in\mathbb{Z}^d} \in \ell_1(\mathbb{Z}^d)$ and

$$\delta_{\mathbf{j}} = \lambda(\mathbf{j}) = \sum_{\mathbf{k}\in\mathbb{Z}^d} \lambda_{\mathbf{k}}\, \varphi(\mathbf{j} - \mathbf{k})\,, \quad \mathbf{j} \in \mathbb{Z}^d\,. \tag{9.10}$$

Multiplying (9.10) by $\mathrm{e}^{-\mathrm{i}\,\mathbf{j}\cdot\boldsymbol{\omega}}$ and then summing all equations over $\mathbf{j}$, we obtain with $\mathbf{n} := \mathbf{j} - \mathbf{k}$ that

$$1 = \tau(\boldsymbol{\omega}) \sum_{\mathbf{n}\in\mathbb{Z}^d} \varphi_{\mathbf{n}}\, \mathrm{e}^{-\mathrm{i}\,\mathbf{n}\cdot\boldsymbol{\omega}} = \tau(\boldsymbol{\omega})\, \sigma_{\varphi}(\boldsymbol{\omega})$$

with

$$\tau(\boldsymbol{\omega}) := \sum_{\mathbf{k}\in\mathbb{Z}^d} \lambda_{\mathbf{k}}\, \mathrm{e}^{-\mathrm{i}\mathbf{k}\cdot\boldsymbol{\omega}}\,.$$

Using Theorem 4.7, τ is a 2π-periodic, continuous function by $\big(\lambda_{\mathbf{k}}\big)_{\mathbf{k}\in\mathbb{Z}^d} \in \ell_1(\mathbb{Z}^d)$. Hence, the symbol σ_{φ} cannot vanish.

2. Suppose that $\sigma_{\varphi} \neq 0$. By Theorem 9.6, there exists a cardinal Lagrange function $\lambda \in \mathscr{L}(\varphi)$ with the property $\lambda(\mathbf{j}) = \delta_{\mathbf{j}}$ for all $\mathbf{j} \in \mathbb{Z}^d$. For arbitrary data $\mathbf{f} = \big(f_{\mathbf{k}}\big)_{\mathbf{k}\in\mathbb{Z}^d} \in \ell_1(\mathbb{Z}^d)$, we form the function

$$s := \sum_{\mathbf{k}\in\mathbb{Z}^d} f_{\mathbf{k}}\, \lambda(\cdot - \mathbf{k})\,. \tag{9.11}$$

By

$$|f_{\mathbf{k}}\, \lambda(\mathbf{x} - \mathbf{k})| \le |f_{\mathbf{k}}|\, \sup\{|\lambda(\mathbf{u})| : \mathbf{u} \in \mathbb{R}^d\}$$

and by $\mathbf{f} \in \ell_1(\mathbb{Z}^d)$, the series in (9.11) converges absolutely and uniformly on $\mathbb{R}^d$. Further, it holds

$$\|s\|_{L_1(\mathbb{R}^d)} \le \|\mathbf{f}\|_{\ell_1(\mathbb{Z}^d)}\, \|\lambda\|_{L_1(\mathbb{R}^d)} < \infty\,.$$

Thus, $s \in L_1(\mathbb{R}^d) \cap C(\mathbb{R}^d)$ fulfills the interpolation condition $s(\mathbf{j}) = f_{\mathbf{j}}$ for all $\mathbf{j} \in \mathbb{Z}^d$.

Now we show that $s \in \mathscr{L}(\varphi)$. By Theorem 9.6, the cardinal Lagrange function λ can be represented in the form

$$\lambda = \sum_{\mathbf{j} \in \mathbb{Z}^d} \lambda_{\mathbf{j}}\, \varphi(\cdot - \mathbf{j})$$

with $\big(\lambda_{\mathbf{j}}\big)_{\mathbf{j} \in \mathbb{Z}^d} \in \ell_1(\mathbb{Z}^d)$. Then it follows that for all $\mathbf{k} \in \mathbb{Z}^d$,

$$\lambda(\cdot - \mathbf{k}) = \sum_{\mathbf{j} \in \mathbb{Z}^d} \lambda_{\mathbf{j}-\mathbf{k}}\, \varphi(\cdot - \mathbf{j})\,.$$

Thus, by (9.11) we obtain that

$$s := \sum_{\mathbf{j} \in \mathbb{Z}^d} a_{\mathbf{j}}\, \lambda(\cdot - \mathbf{j})$$

with the coefficients

$$a_{\mathbf{j}} := \sum_{\mathbf{k} \in \mathbb{Z}^d} f_{\mathbf{k}}\, \lambda_{\mathbf{j}-\mathbf{k}}\,, \quad \mathbf{j} \in \mathbb{Z}^d\,.$$

By

$$\sum_{\mathbf{j} \in \mathbb{Z}^d} |a_{\mathbf{j}}| \le \|\mathbf{f}\|_{\ell_1(\mathbb{Z}^d)}\, \|\boldsymbol{\lambda}\|_{\ell_1(\mathbb{Z}^d)} < \infty$$

we see that $\big(a_{\mathbf{j}}\big)_{\mathbf{j} \in \mathbb{Z}^d} \in \ell_1(\mathbb{Z}^d)$. Hence, $s \in \mathscr{L}(\varphi)$ is a solution of the cardinal interpolation problem (9.4).

3. Finally, we prove the unique solvability of the cardinal interpolation problem (9.4). Assume that $t \in \mathscr{L}(\varphi)$ is also a solution of (9.4), where t has the form

$$t := \sum_{\mathbf{j} \in \mathbb{Z}^d} b_{\mathbf{j}}\, \lambda(\cdot - \mathbf{j})$$

with $\big(b_{\mathbf{j}}\big)_{\mathbf{j} \in \mathbb{Z}^d} \in \ell_1(\mathbb{Z}^d)$. Then the coefficients $c_{\mathbf{j}} := a_{\mathbf{j}} - b_{\mathbf{j}}$, $\mathbf{j} \in \mathbb{Z}^d$, with the property $\big(c_{\mathbf{j}}\big)_{\mathbf{j} \in \mathbb{Z}^d} \in \ell_1(\mathbb{Z}^d)$, fulfill the equation

$$0 = \sum_{\mathbf{k} \in \mathbb{Z}^d} c_{\mathbf{k}}\, \varphi(\mathbf{j} - \mathbf{k})$$

for each $\mathbf{j} \in \mathbb{Z}^d$. Multiplying the above equation by $\mathrm{e}^{-\mathrm{i}\,\mathbf{j}\cdot\boldsymbol{\omega}}$, $\boldsymbol{\omega} \in \mathbb{R}^d$, and summing all equations over $\mathbf{j} \in \mathbb{Z}^d$, we obtain

$$0 = c(\boldsymbol{\omega})\,\sigma_\varphi(\boldsymbol{\omega})\,, \quad \boldsymbol{\omega} \in \mathbb{R}^d\,,$$

with the 2π-periodic continuous function

$$c(\boldsymbol{\omega}) := \sum_{\mathbf{k}\in\mathbb{Z}^d} c_{\mathbf{k}}\,\mathrm{e}^{-\mathrm{i}\,\mathbf{k}\cdot\boldsymbol{\omega}}\,, \quad \boldsymbol{\omega} \in \mathbb{R}^d\,.$$

Since $\sigma_\varphi(\boldsymbol{\omega}) \neq 0$ for all $\boldsymbol{\omega} \in Q_{2\pi}$ by assumption, the function $c \in C(\mathbb{T}^d)$ vanishes such that its Fourier coefficients $c_{\mathbf{k}}$ vanish too. Thus, it holds $a_{\mathbf{k}} = b_{\mathbf{k}}$ for all $\mathbf{k} \in \mathbb{Z}^d$ and hence $s = t$. ■

Remark 9.8 The proof of Theorem 9.7 is mainly based on a convolution in $\ell_1(\mathbb{Z}^d)$. For arbitrary $\mathbf{a} = \big(a_{\mathbf{k}}\big)_{\mathbf{k}\in\mathbb{Z}^d}$, $\mathbf{b} = \big(b_{\mathbf{k}}\big)_{\mathbf{k}\in\mathbb{Z}^d} \in \ell_1(\mathbb{Z}^d)$, the *convolution* in $\ell_1(\mathbb{Z}^d)$ is defined as $\mathbf{a} * \mathbf{b} := \big(c_{\mathbf{j}}\big)_{\mathbf{j}\in\mathbb{Z}^d}$ with

$$c_{\mathbf{j}} := \sum_{\mathbf{k}\in\mathbb{Z}^d} a_{\mathbf{k}}\,b_{\mathbf{j}-\mathbf{k}}\,, \quad \mathbf{j} \in \mathbb{Z}^d\,.$$

By

$$\|\mathbf{a} * \mathbf{b}\|_{\ell_1(\mathbb{Z}^d)} \le \|\mathbf{a}\|_{\ell_1(\mathbb{Z}^d)}\,\|\mathbf{b}\|_{\ell_1(\mathbb{Z}^d)} < \infty$$

we see that $\mathbf{a} * \mathbf{b} \in \ell_1(\mathbb{Z}^d)$. One can easily show that the convolution in $\ell_1(\mathbb{Z}^d)$ is a commutative, associative, and distributive operation with the unity $\big(\delta_{\mathbf{j}}\big)_{\mathbf{j}\in\mathbb{Z}^d}$. Forming the corresponding functions $a, b \in C(\mathbb{T}^d)$ by

$$a(\boldsymbol{\omega}) := \sum_{\mathbf{k}\in\mathbb{Z}^d} a_{\mathbf{k}}\,\mathrm{e}^{-\mathrm{i}\,\mathbf{k}\cdot\boldsymbol{\omega}}\,, \quad b(\boldsymbol{\omega}) := \sum_{\mathbf{k}\in\mathbb{Z}^d} b_{\mathbf{k}}\,\mathrm{e}^{-\mathrm{i}\,\mathbf{k}\cdot\boldsymbol{\omega}}\,,$$

then the convolution $\mathbf{a} * \mathbf{b} := \big(c_{\mathbf{j}}\big)_{\mathbf{j}\in\mathbb{Z}^d} \in \ell_1(\mathbb{Z}^d)$ correlates to the product $a\,b \in C(\mathbb{T}^d)$, i.e.,

$$a(\boldsymbol{\omega})\,b(\boldsymbol{\omega}) = \sum_{\mathbf{j}\in\mathbb{Z}^d} c_{\mathbf{j}}\,\mathrm{e}^{-\mathrm{i}\,\mathbf{j}\cdot\boldsymbol{\omega}}\,.$$

□

Now we show that the cardinal interpolation problem (9.4) in $\mathscr{L}(\varphi)$ can be numerically solved by multidimensional DFTs. From (9.11) and from the translation property of the d-dimensional Fourier transform (see Theorem 4.20), it follows that the Fourier transform $\hat{s}$ has the form

$$\hat{s}(\boldsymbol{\omega}) = \hat{\lambda}(\boldsymbol{\omega}) \sum_{\mathbf{k}\in\mathbb{Z}^d} f_{\mathbf{k}}\, \mathrm{e}^{-\mathrm{i}\,\mathbf{k}\cdot\boldsymbol{\omega}}\,.$$

Thus, we can estimate

$$|\hat{s}(\boldsymbol{\omega})| \le |\hat{\lambda}(\boldsymbol{\omega})| \sum_{\mathbf{k}\in\mathbb{Z}^d} |f_{\mathbf{k}}| = |\hat{\lambda}(\boldsymbol{\omega})|\, \|\mathbf{f}\|_{\ell_1(\mathbb{Z}^d)}$$

such that

$$\|\hat{s}\|_{L_1(\mathbb{R}^d)} \le \|\hat{\lambda}\|_{L_1(\mathbb{R}^d)}\, \|\mathbf{f}\|_{\ell_1(\mathbb{Z}^d)}\,,$$

i.e., $\hat{s} \in L_1(\mathbb{R}^d)$. Using the d-dimensional inverse Fourier transform of Theorem 4.22, we obtain the formula

$$s(\mathbf{x}) = \frac{1}{(2\pi)^d} \int_{\mathbb{R}^d} \hat{\lambda}(\boldsymbol{\omega}) \Big(\sum_{\mathbf{k}\in\mathbb{Z}^d} f_{\mathbf{k}}\, \mathrm{e}^{-\mathrm{i}\,\mathbf{k}\cdot\boldsymbol{\omega}}\Big)\, \mathrm{e}^{\mathrm{i}\,\mathbf{x}\cdot\boldsymbol{\omega}}\, \mathrm{d}\boldsymbol{\omega}\,. \tag{9.12}$$

We suppose again that the basis function φ fulfills the assumptions of Theorem 9.6 and that $f_{\mathbf{k}} = 0$ for all $\mathbf{k} \in \mathbb{Z}^d \setminus J_N$ with certain $N \in \mathbb{N}$. Replacing the domain of integration in (9.12) by the hypercube $[-n\pi,\, n\pi]^d$ with certain $n \in \mathbb{N}$, instead of (9.12), we compute the expression

$$\frac{1}{(2\pi)^d} \int_{[-n\pi,\, n\pi]^d} \hat{\lambda}(\boldsymbol{\omega}) \Big(\sum_{\mathbf{k}\in J_N} f_{\mathbf{k}}\, \mathrm{e}^{-\mathrm{i}\,\mathbf{k}\cdot\boldsymbol{\omega}}\Big)\, \mathrm{e}^{\mathrm{i}\,\mathbf{x}\cdot\boldsymbol{\omega}}\, \mathrm{d}\boldsymbol{\omega}$$

by a simple d-dimensional quadrature rule with step size $\frac{2\pi}{N}$, i.e., we approximate the integral (9.12) by the finite sum

$$\frac{1}{N^d} \sum_{\mathbf{m}\in B_{nN}} \hat{\lambda}\Big(\frac{2\pi\mathbf{m}}{N}\Big) \Big(\sum_{\mathbf{k}\in J_N} f_{\mathbf{k}}\, w_N^{\mathbf{k}\cdot\mathbf{m}}\Big)\, \mathrm{e}^{2\pi\,\mathrm{i}\,\mathbf{x}\cdot\mathbf{m}/N}$$

with the complex Nth root of unity $w_N := \mathrm{e}^{2\pi\,\mathrm{i}/N}$. By Theorem 9.6, we know that $\hat{\lambda} = \hat{\varphi}/\sigma_{\varphi}$. Since the symbol $\sigma_{\varphi} \ne 0$ is 2π-periodic, it holds

$$\hat{\lambda}\Big(\frac{2\pi\mathbf{m}}{N}\Big) = \frac{\hat{\varphi}\big(\frac{2\pi\mathbf{m}}{N}\big)}{\sigma_{\varphi}\big(\frac{2\pi\mathbf{m}'}{N}\big)}\,, \qquad \mathbf{m} \in \mathbb{Z}^d\,,$$

where $\mathbf{m}' = \big(m'_{\ell}\big)_{\ell=1}^d \in J_N$ denotes the *nonnegative residue of* $\mathbf{m} = \big(m_{\ell}\big)_{\ell=1}^d \in \mathbb{Z}^d$ *modulo* N, i.e., it holds $m'_{\ell} \equiv m_{\ell} \pmod N$ and $0 \le m'_{\ell} < N$ for each $\ell = 1, \ldots, d$. We will use the notation $\mathbf{m}' = \mathbf{m} \bmod N$.

Instead of the exact value $s\big(\frac{\mathbf{j}}{n}\big)$, $\mathbf{j} \in J_{nN}$, on the uniform grid $\frac{1}{n} J_{nN}$, we obtain the approximate value:

$$s_{\mathbf{j}} := \frac{1}{N^d} \sum_{\mathbf{m} \in B_{nN}} \hat{\lambda}\Big(\frac{2\pi \mathbf{m}}{N}\Big)\, w_{nN}^{-\mathbf{m}\cdot\mathbf{j}} \Big(\sum_{\mathbf{k} \in J_N} f_{\mathbf{k}}\, w_N^{\mathbf{k}\cdot\mathbf{m}}\Big), \quad \mathbf{j} \in J_{nN}. \tag{9.13}$$

Now we summarize this method, where we repeat that the given basis function $\varphi \in L_1(\mathbb{R}^d) \cap C(\mathbb{R}^d)$ with the corresponding symbol σ_φ fulfills the assumptions of Theorem 9.6.

Algorithm 9.9 (Cardinal Interpolation by Translates)
Input: n, $N \in \mathbb{N} \setminus \{1\}$, $f_{\mathbf{k}} \in \mathbb{C}$ with $\mathbf{k} \in J_N$ given data.

1. *For all $\mathbf{j} \in J_N$, compute the d-dimensional* DFT$(N \times \ldots \times N)$

$$\hat{f}_{\mathbf{j}} := \sum_{\mathbf{k} \in J_N} f_{\mathbf{k}}\, w_N^{\mathbf{j}\cdot\mathbf{k}}.$$

2. *For all $\mathbf{m} \in B_{nN}$, determine*

$$t_{\mathbf{m}} := \frac{\hat{\varphi}\big(\frac{2\pi\mathbf{m}}{N}\big)}{\sigma_\varphi\big(\frac{2\pi\mathbf{m}'}{N}\big)}\, \hat{f}_{\mathbf{m}'}$$

with $\mathbf{m}' = \mathbf{m} \bmod N$.
3. *For all $\mathbf{m} \in B_{nN}$, set $u_{\mathbf{k}} := t_{\mathbf{m}}$, where $\mathbf{k} := \mathbf{m} \bmod nN \in J_{nN}$.*
4. *For all $\mathbf{j} \in J_{nN}$, compute by d-dimensional* DFT$(nN \times \ldots \times nN)$

$$s_{\mathbf{j}} := \frac{1}{N^d} \sum_{\mathbf{k} \in J_{nN}} u_{\mathbf{k}}\, w_{nN}^{-\mathbf{j}\cdot\mathbf{k}}.$$

Output: $s_{\mathbf{j}}$, $\mathbf{j} \in J_{nN}$, approximate value of $s\big(\frac{\mathbf{j}}{n}\big)$ on the uniform grid $\frac{1}{n} J_{nN}$.
Computational cost: $\mathcal{O}\big((nN)^d \log(nN)\big)$.

9.1.2 Computation of Fourier Transforms

This subsection is devoted to the computation of the d-dimensional Fourier transform $\hat{f}$ of a given function $f \in L_1(\mathbb{R}^d) \cap C(\mathbb{R}^d)$, i.e.,

$$\hat{f}(\boldsymbol{\omega}) := \int_{\mathbb{R}^d} f(\mathbf{x})\, \mathrm{e}^{-\mathrm{i}\mathbf{x}\cdot\boldsymbol{\omega}}\, \mathrm{d}\boldsymbol{\omega}, \quad \boldsymbol{\omega} \in \mathbb{R}^d. \tag{9.14}$$

We show that the standard method for computing of (9.14) can be essentially improved without big additional work. The so-called method of attenuation factors is based on the cardinal Lagrange function in the shift-invariant space $\mathscr{L}(\varphi)$, where the basis function $\varphi \in C_c(\mathbb{R}^d)$ with the symbol σ_φ fulfills the assumptions of Theorem 9.6.

Assume that $|f(\mathbf{x})| \ll 1$ for all $\mathbf{x} \in \mathbb{R}^d \setminus [-n\pi,\, n\pi]^d$ with certain $n \in \mathbb{N}$. Replacing the domain of integration in (9.14) by the hypercube $[-n\pi,\, n\pi]^d$, we calculate the integral

$$\int_{[-n\pi,\, n\pi]^d} f(\mathbf{x})\, \mathrm{e}^{-\mathrm{i}\,\mathbf{x}\cdot\boldsymbol{\omega}}\, \mathrm{d}\boldsymbol{\omega}$$

by the simple tensor product quadrature rule with uniform step size $\frac{2\pi}{N}$ for certain sufficiently large $N \in \mathbb{N}$. Then as approximate value of $\hat{f}(\boldsymbol{\omega})$, we preserve

$$\Big(\frac{2\pi}{N}\Big)^d \sum_{\mathbf{j}\in B_{nN}} f\Big(\frac{2\pi\mathbf{j}}{N}\Big)\, \mathrm{e}^{-2\pi\mathrm{i}\,\mathbf{j}\cdot\boldsymbol{\omega}/N}\,, \qquad \boldsymbol{\omega} \in \mathbb{R}^d\,,$$

where the index set B_{nN} is equal to

$$B_{nN} = \{\mathbf{j} = (j_\ell)_{\ell=1}^d \in \mathbb{Z}^d : -\lfloor\frac{nN-1}{2}\rfloor \le j_\ell \le \lfloor\frac{nN}{2}\rfloor \text{ for } \ell = 1, \ldots, d\}\,. \tag{9.15}$$

Especially for $\boldsymbol{\omega} = \frac{\mathbf{k}}{n}$ with $\mathbf{k} \in \mathbb{Z}^d$, we get

$$\tilde{f}_{\mathbf{k}} := \Big(\frac{2\pi}{N}\Big)^d \sum_{\mathbf{j}\in B_{nN}} f\Big(\frac{2\pi\mathbf{j}}{N}\Big)\, w_{nN}^{\mathbf{j}\cdot\mathbf{k}} \tag{9.16}$$

as approximate value of $\hat{f}(\frac{\mathbf{k}}{n})$, where $w_{nN} = \mathrm{e}^{-2\pi\mathrm{i}/(nN)}$. Up to the factor $\big(\frac{2\pi}{N}\big)^d$, the expression (9.16) is a d-dimensional DFT($nN \times \ldots \times nN$). Obviously, the values $\tilde{f}_{\mathbf{k}}$, $\mathbf{k} \in \mathbb{Z}^d$, are nN-periodic, i.e., $\tilde{f}_{\mathbf{k}} = \tilde{f}_{\mathbf{k}+nN\,\mathbf{j}}$ for all $\mathbf{j}, \mathbf{k} \in \mathbb{Z}^d$. Otherwise, the Fourier transform $\hat{f}$ possesses the property: (see Lemma 4.21)

$$\lim_{\|\mathbf{k}\|_2\to\infty} \hat{f}\Big(\frac{\mathbf{k}}{n}\Big) = 0\,.$$

Thus, only for $\mathbf{k} = \big(k_\ell\big)_{\ell=1}^d \in \mathbb{Z}^d$ with $|k_\ell| < \frac{nN}{2}$, $\ell = 1, \ldots, d$, the value $\tilde{f}_{\mathbf{k}}$ can be accepted as approximation of $\hat{f}(\frac{\mathbf{k}}{n})$. Better approximations of $\hat{f}(\frac{\mathbf{k}}{n})$ can be obtained by the following *method of attenuation factors* which is mainly based on the cardinal Lagrange function $\lambda \in \mathscr{L}(\varphi)$, where $\varphi \in C_c(\mathbb{R}^d)$ is a convenient compactly supported basis function and $\mathscr{L}(\varphi)$ is the related shift-invariant space.

Theorem 9.10 (Method of Attenuation Factors for Fourier Transform) *Let n, $N \in \mathbb{N}$ be given. For $f \in L_1(\mathbb{R}^d) \cap C(\mathbb{R}^d)$, set*

$$f_{\mathbf{j}} := \begin{cases} f\big(\frac{2\pi\mathbf{j}}{N}\big) & \mathbf{j} \in B_{nN}\,, \\ 0 & \mathbf{j} \in \mathbb{Z}^d \setminus B_{nN}\,. \end{cases}$$

Let $\varphi \in C_c(\mathbb{R}^d)$ be a given compactly supported basis function. Assume that $\hat{\varphi}$ fulfills the condition (9.6) and that $\sigma_\varphi(\boldsymbol{\omega}) \neq 0$ for all $\boldsymbol{\omega} \in Q_{2\pi}$. Let $\lambda \in \mathscr{L}(\varphi)$ denote the cardinal Lagrange function (9.7).

Then the values of the Fourier transform of the function

$$s(\mathbf{x}) := \sum_{\mathbf{j}\in B_{nN}} f_{\mathbf{j}}\,\lambda\big(\frac{N\mathbf{x}}{2\pi} - \mathbf{j}\big)\,, \quad \mathbf{x} \in \mathbb{R}^d\,, \tag{9.17}$$

read as follows

$$\hat{s}\big(\frac{\mathbf{k}}{n}\big) = \hat{\lambda}\big(\frac{2\pi\mathbf{k}}{nN}\big)\,\tilde{f}_{\mathbf{k}'}\,, \quad \mathbf{k} \in \mathbb{Z}^d\,,$$

where $\mathbf{k}' := \mathbf{k} \bmod nN$ *and where* $\tilde{f}_{\mathbf{k}}$ *is defined by* (9.16). *The values*

$$\hat{\lambda}\big(\frac{2\pi\mathbf{k}}{nN}\big) = \frac{\hat{\varphi}\big(\frac{2\pi\mathbf{k}}{nN}\big)}{\sigma_\varphi\big(\frac{2\pi\mathbf{k}'}{nN}\big)}\,, \quad \mathbf{k} \in \mathbb{Z}^d\,, \tag{9.18}$$

are called attenuation factors of Fourier transform.

Proof By Theorem 9.6, there exists the cardinal Lagrange function $\lambda \in \mathscr{L}(\varphi)$. Obviously, the function (9.17) interpolates the given data $f_{\mathbf{j}}$ on the uniform grid $\frac{2\pi}{N}\,\mathbb{Z}^d$, i.e.,

$$s\big(\frac{2\pi\mathbf{j}}{N}\big) = f_{\mathbf{j}}\,, \quad \mathbf{j} \in \mathbb{Z}^d\,.$$

Further, by Theorem 9.7, we know that $s \in L_1(\mathbb{R}^d) \cap C(\mathbb{R}^d)$. Applying the d-dimensional Fourier transform, we obtain

$$\hat{s}(\boldsymbol{\omega}) = \big(\frac{2\pi}{N}\big)^d \sum_{\mathbf{j}\in B_{nN}} f_{\mathbf{j}}\,\mathrm{e}^{-2\pi\mathrm{i}\,\mathbf{j}\cdot\boldsymbol{\omega}/N}\,\hat{\lambda}\big(\frac{2\pi\,\boldsymbol{\omega}}{N}\big)\,, \quad \boldsymbol{\omega} \in \mathbb{R}^d\,.$$

By Theorem 9.6, it holds $\hat{\lambda} = \hat{\varphi}/\sigma_\varphi$. For $\boldsymbol{\omega} = \frac{\mathbf{k}}{n}$, $\mathbf{k} \in \mathbb{Z}^d$, it follows that

$$\hat{s}\big(\frac{\mathbf{k}}{n}\big) = \hat{\lambda}\big(\frac{2\pi\,\mathbf{k}}{nN}\big)\,\tilde{f}_{\mathbf{k}} = \hat{\lambda}\big(\frac{2\pi\,\mathbf{k}}{nN}\big)\,\tilde{f}_{\mathbf{k}'}\,,$$

where $\tilde{f}_{\mathbf{k}}$ is defined by (9.16) and where $\mathbf{k}' = \mathbf{k} \bmod nN$. Since the symbol σ_φ is 2π-periodic, we obtain the formula (9.18). ■

Thus, we can use $\hat{s}\big(\frac{\mathbf{k}}{n}\big)$ as approximate value of $\hat{f}\big(\frac{\mathbf{k}}{n}\big)$ for $\mathbf{k} \in \mathbb{Z}^d$. The method of attenuation factors performs two tasks. The computed values $\hat{s}\big(\frac{\mathbf{k}}{n}\big)$ *correct* the $(nN)^d$ coarse approximate values $\tilde{f}_{\mathbf{k}}$ for $\mathbf{k} \in B_{nN}$. Further, the approximate values $\tilde{f}_{\mathbf{k}}$ for $\mathbf{k} \in B_{nN}$ are *continued* to whole $\mathbb{Z}^d$ by the values $\hat{s}\big(\frac{\mathbf{k}}{n}\big)$.

The essential step for computing the Fourier transform $\hat{f}$ on a uniform grid is the d-dimensional DFT$(nN \times \ldots \times nN)$ in formula (9.16) so that we recommend to choose the positive integers n and N as powers of two.

Example 9.11 For the cardinal B-spline $\varphi = N_m$ with certain $m \in \mathbb{N} \setminus \{1\}$, the attenuation factors result from

$$\hat{\lambda}(\omega) = \begin{cases} \big(\mathrm{sinc}\,\frac{\omega}{2}\big)^2 & m = 2\,, \\ 3\,\big(\mathrm{sinc}\,\frac{\omega}{2}\big)^4\,(2 + \cos\omega)^{-1} & m = 4\,, \\ 60\,\big(\mathrm{sinc}\,\frac{\omega}{2}\big)^6\,(33 + 26\,\cos\omega + \cos 2\omega)^{-1} & m = 6 \end{cases}$$

with $\omega = \frac{2\pi k}{nN}$, $k \in \mathbb{Z}$. Note that N_3 and N_5 don't fulfill the assumptions of Theorem 9.10, because the related symbols can vanish.

For the centered cardinal B-spline $\varphi = M_m$ with certain $m \in \mathbb{N} \setminus \{1\}$, the attenuation factors account for

$$\hat{\lambda}(\omega) = \begin{cases} \big(\mathrm{sinc}\,\frac{\omega}{2}\big)^2 & m = 2\,, \\ 4\,\big(\mathrm{sinc}\,\frac{\omega}{2}\big)^3\,(3 + \cos\omega)^{-1} & m = 3\,, \\ 3\,\big(\mathrm{sinc}\,\frac{\omega}{2}\big)^4\,(2 + \cos\omega)^{-1} & m = 4 \end{cases}$$

with $\omega = \frac{2\pi k}{nN}$, $k \in \mathbb{Z}$. □

Example 9.12 For the three-direction box spline $\varphi = M^{(k,\ell,m)}$ with $k, \ell, m \in \mathbb{N}$, one obtains the attenuation factors by

$$\hat{\lambda}(\boldsymbol{\omega}) = \begin{cases} \big(\mathrm{sinc}\,\frac{\omega_1}{2}\big)\big(\mathrm{sinc}\,\frac{\omega_2}{2}\big)\big(\mathrm{sinc}\,\frac{\omega_1+\omega_2}{2}\big) & (k,\ell,m) = (1,1,1)\,, \\ \dfrac{12\,\big(\mathrm{sinc}\,\frac{\omega_1}{2}\big)^2\big(\mathrm{sinc}\,\frac{\omega_2}{2}\big)^2\big(\mathrm{sinc}\,\frac{\omega_1+\omega_2}{2}\big)}{7+2\cos\omega_1+2\cos\omega_2+\cos(\omega_1+\omega_2)} & (k,\ell,m) = (2,2,1)\,, \\ \dfrac{6\,\big(\mathrm{sinc}\,\frac{\omega_1}{2}\big)^2\big(\mathrm{sinc}\,\frac{\omega_2}{2}\big)^2\big(\mathrm{sinc}\,\frac{\omega_1+\omega_2}{2}\big)^2}{3+\cos\omega_1+\cos\omega_2+\cos(\omega_1+\omega_2)} & (k,\ell,m) = (2,2,2) \end{cases}$$

with $\boldsymbol{\omega} = (\omega_1, \omega_2)^\top = \frac{2\pi\mathbf{k}}{nN}$, $\mathbf{k} \in \mathbb{Z}^2$. □

In the univariate case with $\varphi = N_4$, we obtain the following algorithm for computing the Fourier transform of a function $f \in L_1(\mathbb{R}) \cap C(\mathbb{R})$ which fulfills the condition $|f(x)| \ll 1$ for all $|x| \ge n\pi$ with certain $n \in \mathbb{N}$. Note that the Fourier transform of the related cardinal Lagrange function $\lambda \in \mathcal{L}(\varphi)$ reads as follows:

$$\hat{\lambda}(\omega) = 3\,\big(\mathrm{sinc}\,\frac{\omega}{2}\big)^4\,(2 + \cos\omega)^{-1}\,.$$

Algorithm 9.13 (Computation of One-dimensional Fourier Transform via Attenuation Factors)
Input: n, $N \in \mathbb{N}$ *powers of two,* $f_j = f\big(\frac{2\pi j}{N}\big)$ *for* $j = -\lfloor \frac{nN-1}{2} \rfloor, \ldots, \lfloor \frac{nN}{2} \rfloor$ *given data of* $f \in L_1(\mathbb{R}) \cap C(\mathbb{R})$.

1. *Form*

$$g_j := \begin{cases} f_j & j = 0, \ldots, \lfloor \frac{nN}{2} \rfloor, \\ f_{j-nN} & j = \lfloor \frac{nN}{2} \rfloor + 1, \ldots, nN - 1. \end{cases}$$

2. *For* $k = 0, \ldots, nN - 1$ *compute the* DFT(nN)

$$\hat{g}_k := \sum_{j=0}^{nN-1} g_j\, w_{nN}^{jk}.$$

3. *With* $h := \frac{2\pi}{N}$ *form*

$$\tilde{f}_k := \begin{cases} h\, \hat{g}_k & k = 0, \ldots, \lfloor \frac{nN}{2} \rfloor, \\ h\, \hat{g}_{k+nN} & k = -\lfloor \frac{nN-1}{2} \rfloor, \ldots, -1. \end{cases}$$

4. *For* $k = -\lfloor \frac{nN-1}{2} \rfloor, \ldots, \lfloor \frac{nN}{2} \rfloor$ *calculate*

$$\hat{s}\Big(\frac{k}{n}\Big) := \hat{\lambda}\Big(\frac{2\pi k}{nN}\Big)\, \tilde{f}_k.$$

Output: $\hat{s}\big(\frac{k}{n}\big)$ *approximate value of* $\hat{f}\big(\frac{k}{n}\big)$ *for* $k = -\lfloor \frac{nN-1}{2} \rfloor, \ldots, \lfloor \frac{nN}{2} \rfloor$.
Computational cost: $\mathcal{O}\big(nN \log(nN)\big)$.

The following example shows the performance of this method of attenuation factors.

Example 9.14 We consider the even function $f \in L_1(\mathbb{R}) \cap C(\mathbb{R})$ given by $f(x) = \mathrm{e}^{-|x|}$ for $x \in \mathbb{R}$. Then the related Fourier transform reads as follows:

$$\hat{f}(\omega) = \int_{\mathbb{R}} \mathrm{e}^{-|x|}\, \mathrm{e}^{-\mathrm{i}x\omega}\, \mathrm{d}\omega = \frac{2}{1+\omega^2}, \qquad \omega \in \mathbb{R}.$$

We choose $N = 16$, $n = 1$, and $\varphi = N_4$. Instead of the exact values $\hat{f}(k)$, we obtain the coarse approximate values:

$$\tilde{f}_k = \frac{\pi}{8} \sum_{j=-7}^{8} f\Big(\frac{j\pi}{8}\Big)\, w_{16}^{jk}.$$

The method of attenuation factors creates the improved approximate values:

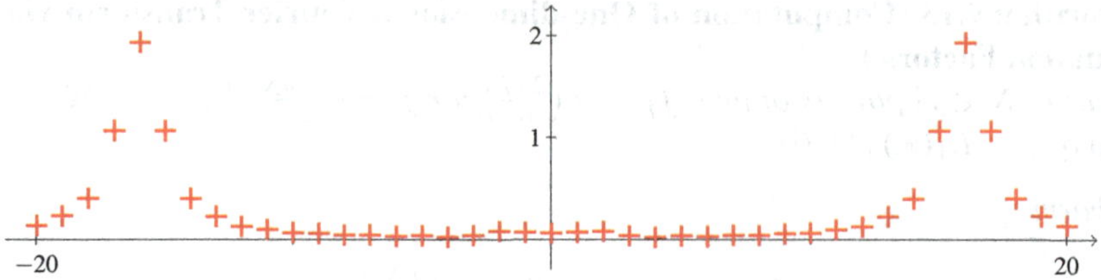

Fig. 9.4 The crosses $(k, |\tilde{f}_k - \hat{f}(k)|)$ for $k = -20, \ldots, 20$ illustrate the behavior for the classical computation of $\hat{f}(k)$ in Example 9.14

$$\hat{s}(k) = \hat{\lambda}\big(\frac{\pi k}{8}\big)\, \tilde{f}_k = \frac{3\,\big(\mathrm{sinc}\,\frac{\pi k}{16}\big)^4}{2 + \cos\frac{\pi k}{8}}\, \tilde{f}_k\,.$$

Figure 9.4 illustrates the errors for the classical computation of the values $\hat{f}(k)$. On the other hand, the method of attenuation factors produces the small maximal error:

$$\max_{|k|\le 20} |\hat{s}(k) - \hat{f}(k)| = 0.070127\,.$$

□

9.2 Periodic Interpolation by Translates

In this section, we investigate the periodic interpolation by translates on a uniform mesh. Our approach is mainly based on Sect. 9.1, since there exists a close connection between periodic and cardinal interpolation by translates.

In the following, let $N \in \mathbb{N}$ be fixed chosen. Let $\varphi \in C_c(\mathbb{R}^d)$ be a compactly supported basis function with the property (9.6). By

$$\varphi^*(\mathbf{x}) := \sum_{\mathbf{k}\in\mathbb{Z}^d} \varphi(\mathbf{x} + N\mathbf{k})\,, \quad \mathbf{x} \in \mathbb{R}^d\,, \tag{9.19}$$

we form an N-periodic function $\varphi^* \in C_N(\mathbb{R}^d)$, where $C_N(\mathbb{R}^d)$ denotes the Banach space of all N-periodic continuous functions $f : \mathbb{R}^d \to \mathbb{C}$ with the uniform norm:

$$\|f\|_{C_N(\mathbb{R}^d)} := \sup\{|f(\mathbf{x})| : \mathbf{x} \in Q_N := [0,\, N)^d\}\,.$$

Analogously, we can periodize the Fourier transform $\hat{\varphi}$, since condition (9.6) is fulfilled. Thus, we obtain the 2π-periodized Fourier transform:

$$\tilde{\varphi}(\boldsymbol{\omega}) := \sum_{\mathbf{k}\in\mathbb{Z}^d} \hat{\varphi}(\boldsymbol{\omega} + 2\pi\,\mathbf{k})\,, \quad \boldsymbol{\omega} \in \mathbb{R}^d\,. \tag{9.20}$$

By (9.6) it holds $\tilde{\varphi} \in C(\mathbb{T}^d)$.

By $\mathscr{L}_N(\varphi^*)$ we denote the space of all N-periodic continuous functions $s : \mathbb{R}^d \to \mathbb{C}$ of the form

$$s(\mathbf{x}) := \sum_{\mathbf{j}\in J_N} c_{\mathbf{j}}\, \varphi^*(\mathbf{x} - \mathbf{j})\,, \quad \mathbf{x} \in \mathbb{R}^d\,,$$

with arbitrary coefficients $c_{\mathbf{j}} \in \mathbb{C}$ and the index set

$$J_N = \{\mathbf{j} = (j_\ell)_{\ell=1}^d \in \mathbb{Z}^d : 0 \le j_\ell \le N-1 \text{ for } \ell = 1, \ldots, d\}\,.$$

If we continue the coefficients $c_{\mathbf{j}}$ by $c_{\mathbf{j}+N\mathbf{k}} := c_{\mathbf{j}}$ for all $\mathbf{j} \in J_N$ and $\mathbf{k} \in \mathbb{Z}^d$, then we get the equation

$$s(\mathbf{x}) = \sum_{\mathbf{n}\in\mathbb{Z}^d} c_{\mathbf{n}}\, \varphi(\mathbf{x} - \mathbf{n})\,,$$

where for each $\mathbf{x} \in \mathbb{R}^d$ the above sum contains only finitely many nonzero summands.

9.2.1 Periodic Lagrange Function

The *N-periodic interpolation in $\mathscr{L}_N(\varphi^*)$ on the uniform mesh J_N* means that one has to determine a function $s \in \mathscr{L}_N(\varphi^*)$ which fulfills the interpolation condition

$$s(\mathbf{j}) = f_{\mathbf{j}} \quad \text{for all } \mathbf{j} \in J_N\,, \tag{9.21}$$

where $f_{\mathbf{j}} \in \mathbb{C}$, $\mathbf{j} \in J_N$, are arbitrary given data. A function $\lambda^* \in C_N(\mathbb{R}^d)$ which interpolates the N-periodic Kronecker data $\delta_{\mathbf{k}}^{(N)}$ on the uniform mesh J_N, i.e.,

$$\lambda^*(\mathbf{k}) = \delta_{\mathbf{k}}^{(N)} = \begin{cases} 1 & \mathbf{k} = \mathbf{0}\,, \\ 0 & \mathbf{k} \in J_N \setminus \{\mathbf{0}\} \end{cases}$$

is called an *N-periodic Lagrange function*. Now we construct an N-periodic Lagrange function in the space $\mathscr{L}_N(\varphi^*)$. Similarly as in Theorem 9.6, the construction of N-periodic Lagrange function in $\mathscr{L}_N(\varphi^*)$ is based on properties of the symbol:

$$\sigma_\varphi(\boldsymbol{\omega}) = \sum_{\mathbf{k}\in\mathbb{Z}^d} \varphi(\mathbf{k})\, \mathrm{e}^{-\mathrm{i}\,\mathbf{k}\cdot\boldsymbol{\omega}}\,.$$

Since φ is compactly supported, we observe that σ_φ is a 2π-periodic trigonometric polynomial. The symbol of the shifted function $\varphi(\cdot + \mathbf{t})$ with fixed $\mathbf{t} \in \mathbb{R}^d$ is denoted by

$$\sigma_\varphi(\mathbf{t},\, \boldsymbol{\omega}) := \sum_{\mathbf{k}\in\mathbb{Z}^d} \varphi(\mathbf{k}+\mathbf{t})\,\mathrm{e}^{-\mathrm{i}\,\mathbf{k}\cdot\boldsymbol{\omega}}\,, \quad \boldsymbol{\omega} \in \mathbb{R}^d\,. \tag{9.22}$$

Obviously, we have $\sigma_\varphi(\mathbf{0},\, \boldsymbol{\omega}) = \sigma_\varphi(\boldsymbol{\omega})$.

Lemma 9.15 *For each* $\mathbf{t} \in \mathbb{R}^d$ *and all* $\mathbf{j} \in J_N$*, it holds*

$$\sigma_\varphi\big(\mathbf{t},\, \frac{2\pi\mathbf{j}}{N}\big) = \sum_{\mathbf{n}\in J_N} \varphi^*(\mathbf{n}+\mathbf{t})\, w_N^{\mathbf{j}\cdot\mathbf{n}} = \sum_{\mathbf{k}\in\mathbb{Z}^d} \hat{\varphi}\big(\frac{2\pi\mathbf{j}}{N} + 2\pi\mathbf{k}\big)\,\mathrm{e}^{2\pi\mathrm{i}\,\mathbf{j}\cdot\mathbf{t}/N}\,\mathrm{e}^{2\pi\mathrm{i}\,\mathbf{k}\cdot\mathbf{t}}\,. \tag{9.23}$$

Proof By Poisson summation formula (see Theorem 4.28) and by the translation property of the d-dimensional Fourier transform (see Theorem 4.20), we conclude that

$$\sigma_\varphi\big(\mathbf{t},\, \boldsymbol{\omega}\big) = \sum_{\mathbf{k}\in\mathbb{Z}^d} \hat{\varphi}(\boldsymbol{\omega}+2\pi\mathbf{k})\,\mathrm{e}^{\mathrm{i}\,\boldsymbol{\omega}\cdot\mathbf{t}}\,\mathrm{e}^{2\pi\mathrm{i}\,\mathbf{k}\cdot\mathbf{t}}\,.$$

The convergence of above series is ensured by condition (9.6). Especially for $\boldsymbol{\omega} = \frac{2\pi\mathbf{j}}{N}$ with $\mathbf{j} \in J_N$, we get one equation of (9.23).

Substituting $\mathbf{k} = \mathbf{n} + N\mathbf{m}$ with $\mathbf{n} \in J_N$ and $\mathbf{m} \in \mathbb{Z}^d$ in (9.22), we see that

$$\sigma_\varphi\big(\mathbf{t},\, \frac{2\pi\mathbf{j}}{N}\big) = \sum_{\mathbf{n}\in J_N}\sum_{\mathbf{m}\in\mathbb{Z}^d} \varphi(\mathbf{n}+\mathbf{t}+N\mathbf{m})\, w_N^{\mathbf{j}\cdot\mathbf{n}} = \sum_{\mathbf{n}\in J_N} \varphi^*(\mathbf{n}+\mathbf{t})\, w_N^{\mathbf{j}\cdot\mathbf{n}}\,.$$

■

Now we construct an N-periodic Lagrange function in $\mathscr{L}_N(\varphi^*)$.

Theorem 9.16 *Let* $N \in \mathbb{N}$ *be fixed. Let* $\varphi \in C_c(\mathbb{R}^d)$ *be a given basis function with the property* (9.6)*. Assume that the related symbol* σ_φ *fulfills the condition:*

$$\sigma_\varphi\big(\frac{2\pi\mathbf{j}}{N}\big) \neq 0\,, \quad \mathbf{j} \in J_N\,. \tag{9.24}$$

Then the N*-periodic function* $\lambda^* \in C_N(\mathbb{R}^d)$ *defined by the Fourier series*

$$\lambda^*(\mathbf{x}) := \sum_{\mathbf{k}\in\mathbb{Z}^d} c_{\mathbf{k}}^{(N)}(\lambda^*)\,\mathrm{e}^{2\pi\mathrm{i}\,\mathbf{k}\cdot\mathbf{x}/N} \tag{9.25}$$

with the corresponding Fourier coefficients

$$c_{\mathbf{k}}^{(N)}(\lambda^*) = \frac{\hat{\varphi}\big(\frac{2\pi\,\mathbf{k}}{N}\big)}{N^d\,\sigma_\varphi\big(\frac{2\pi\,\mathbf{k}}{N}\big)}\,, \quad \mathbf{k} \in \mathbb{Z}^d\,, \tag{9.26}$$

is an N-periodic Lagrange function in $\mathscr{L}_N(\varphi^*)$ which can be represented in the form

$$\lambda^*(\mathbf{x}) = \frac{1}{N^d} \sum_{\mathbf{n}\in J_N} \frac{\sigma_\varphi\big(\mathbf{x}, \frac{2\pi\mathbf{n}}{N}\big)}{\sigma_\varphi\big(\frac{2\pi\mathbf{n}}{N}\big)} \tag{9.27}$$

$$= \sum_{\mathbf{j}\in J_N} \lambda_{\mathbf{j}}^*\,\varphi^*(\mathbf{x}-\mathbf{j})\,, \quad \mathbf{x} \in \mathbb{R}^d \tag{9.28}$$

with the coefficients

$$\lambda_{\mathbf{j}}^* = \frac{1}{N^d} \sum_{\mathbf{n}\in J_N} \frac{w_N^{-\mathbf{j}\cdot\mathbf{n}}}{\sigma_\varphi\big(\frac{2\pi\mathbf{n}}{N}\big)}\,, \quad \mathbf{j} \in J_N\,. \tag{9.29}$$

Under the condition (9.24)*, the N-periodic Lagrange function in $\mathscr{L}_N(\varphi^*)$ is uniquely determined.*

Proof

1. By (9.6), (9.24), and (9.26), we see that

$$\sum_{\mathbf{k}\in\mathbb{Z}^d} \big|c_{\mathbf{k}}^{(N)}(\lambda^*)\big| < \infty\,.$$

Hence, the Fourier series (9.25) converges absolutely and uniformly on $\mathbb{R}^d$ such that $\lambda^* \in C_N(\mathbb{R}^d)$ (see Theorem 4.7). Especially for $\mathbf{x} = \mathbf{j} \in J_N$, we obtain by (9.25) and (9.26) that

$$\lambda^*(\mathbf{j}) = \frac{1}{N^d} \sum_{\mathbf{k}\in\mathbb{Z}^d} \frac{\hat{\varphi}\big(\frac{2\pi\mathbf{k}}{N}\big)}{\sigma_\varphi\big(\frac{2\pi\mathbf{k}}{N}\big)}\, w_N^{-\mathbf{k}\cdot\mathbf{j}}$$

with $w_N = \mathrm{e}^{-2\pi\mathrm{i}/N}$. Substituting $\mathbf{k} = \mathbf{n} + N\,\mathbf{m}$ with $\mathbf{n} \in J_N$ and $\mathbf{m} \in \mathbb{Z}^d$ in the above series, it follows from Lemma 9.3 that

$$\lambda^*(\mathbf{j}) = \frac{1}{N^d} \sum_{\mathbf{n}\in J_N} \frac{w_N^{-\mathbf{n}\cdot\mathbf{j}}}{\sigma_\varphi\big(\frac{2\pi\mathbf{n}}{N}\big)} \sum_{\mathbf{m}\in\mathbb{Z}^d} \hat{\varphi}\Big(\frac{2\pi\mathbf{n}}{N} + 2\pi\mathbf{m}\Big)$$

$$= \frac{1}{N^d} \sum_{\mathbf{n}\in J_N} w_N^{-\mathbf{n}\cdot\mathbf{j}} = \delta_{\mathbf{j}}^{(N)}\,, \quad \mathbf{j} \in J_N\,.$$

Thus, λ^* is an N-periodic Lagrange function on the uniform mesh J_N.

2. Substituting $\mathbf{k} = \mathbf{n} + N\,\mathbf{m}$ with $\mathbf{n} \in J_N$ and $\mathbf{m} \in \mathbb{Z}^d$ in the Fourier series (9.25), we receive the representation (9.27) from Lemma 9.15, since

$$\lambda^*(\mathbf{x}) = \frac{1}{N^d} \sum_{\mathbf{n}\in J_N} \frac{1}{\sigma_\varphi\big(\frac{2\pi\mathbf{n}}{N}\big)} \Big(\sum_{\mathbf{m}\in\mathbb{Z}^d} \hat{\varphi}\big(\frac{2\pi\mathbf{n}}{N} + 2\pi\mathbf{m}\big)\, \mathrm{e}^{2\pi\mathrm{i}\,\mathbf{n}\cdot\mathbf{x}/N}\, \mathrm{e}^{2\pi\mathrm{i}\,\mathbf{m}\cdot\mathbf{x}} \Big)$$

$$= \frac{1}{N^d} \sum_{\mathbf{n}\in J_N} \frac{\sigma_\varphi\big(\mathbf{x}, \frac{2\pi\mathbf{n}}{N}\big)}{\sigma_\varphi\big(\frac{2\pi\mathbf{n}}{N}\big)}\,.$$

Using Lemma 9.15, we preserve the formula (9.28) with the coefficients (9.29) such that $\lambda^* \in \mathscr{L}_N(\varphi^*)$.

3. Finally, we show the uniqueness of the N-periodic Lagrange function λ^* in $\mathscr{L}_N(\varphi^*)$. Assume that

$$\mu^*(\mathbf{x}) = \sum_{\mathbf{k}\in J_N} \mu^*_{\mathbf{k}}\, \varphi^*(\mathbf{x} - \mathbf{k})\,, \quad \mathbf{x} \in \mathbb{R}^d\,,$$

is another N-periodic Lagrange function in $\mathscr{L}_N(\varphi^*)$. Then for all $\mathbf{j} \in J_N$, we find

$$\sum_{\mathbf{k}\in J_N} (\lambda^*_k - \mu^*_k)\, \varphi^*(\mathbf{j} - \mathbf{k}) = 0\,.$$

Multiplying the above equation by $w_N^{\mathbf{j}\cdot\mathbf{n}}$ with $\mathbf{n} \in J_N$ and adding all equations over $\mathbf{j} \in J_N$, we obtain for each $\mathbf{n} \in J_N$

$$\Big(\sum_{\mathbf{k}\in J_N} (\lambda^*_k - \mu^*_k)\, w_N^{\mathbf{k}\cdot\mathbf{n}}\Big) \Big(\sum_{\mathbf{m}\in J_N} \varphi^*(\mathbf{m})\, w_N^{\mathbf{m}\cdot\mathbf{n}}\Big) = 0\,.$$

Thus, from (9.23) it follows that

$$\Big(\sum_{\mathbf{k}\in J_N} (\lambda^*_k - \mu^*_k)\, w_N^{\mathbf{k}\cdot\mathbf{n}}\Big)\, \sigma_\varphi\big(\frac{2\pi\mathbf{n}}{N}\big) = 0$$

and hence by (9.24)

$$\sum_{\mathbf{k}\in J_N} (\lambda^*_k - \mu^*_k)\, w_N^{\mathbf{k}\cdot\mathbf{n}} = 0\,, \quad \mathbf{n} \in J_N\,.$$

Since the d-dimensional DFT$(N \times \ldots \times N)$ is invertible (see Theorem 4.106), we get $\lambda^*_k = \mu^*_k$ for all $\mathbf{k} \in J_N$, i.e., both N-periodic Lagrange functions coincide. ■

The cardinal Lagrange function λ in $\mathscr{L}(\varphi)$ and the N-periodic Lagrange function λ^* in $\mathscr{L}_N(\varphi^*)$ are closely related.

Lemma 9.17 *Let $N \in \mathbb{N}$ be fixed. Let $\varphi \in C_c(\mathbb{R}^d)$ be a given basis function with the property* (9.6)*. Assume that the related symbol σ_φ fulfills the condition:*

$$\sigma_\varphi(\boldsymbol{\omega}) \neq 0\,, \quad \boldsymbol{\omega} \in Q_{2\pi}\,. \tag{9.30}$$

Then the N-periodic function λ^* in $\mathscr{L}_N(\varphi^*)$ coincides with the N-periodized cardinal Lagrange function λ, i.e.,

$$\lambda^*(\mathbf{x}) = \sum_{\mathbf{m}\in\mathbb{Z}^d} \lambda(\mathbf{x} + N\mathbf{m})\,, \quad \mathbf{x} \in \mathbb{R}^d\,. \tag{9.31}$$

For the coefficients $\lambda^*_{\mathbf{j}}$, $\mathbf{j} \in J_N$, of the N-periodic Lagrange function λ^*, it holds

$$\lambda^*_{\mathbf{j}} = \sum_{\mathbf{n}\in\mathbb{Z}^d} \lambda_{\mathbf{j}+N\mathbf{n}}\,, \quad \mathbf{j} \in J_N\,, \tag{9.32}$$

where $\lambda_{\mathbf{j}}$ denote the coefficients (9.9) *of the cardinal Lagrange function $\lambda \in \mathscr{L}(\varphi)$.*

Proof

1. From the assumption (9.30), it follows that $(\sigma_\varphi)^{-1} \in C^\infty(\mathbb{T}^d)$. By (9.9), one can interpret $\lambda_{\mathbf{j}}$ with $\mathbf{j} \in \mathbb{Z}^d$ as $(-\mathbf{j})$th Fourier coefficient of $(\sigma_\varphi)^{-1}$. As shown in Theorem 9.6, it holds

$$\sum_{\mathbf{j}\in\mathbb{Z}^d} |\lambda_{\mathbf{j}}| < \infty\,.$$

   The coefficients $\lambda^*_{\mathbf{j}}$, $\mathbf{j} \in J_N$, can be represented in the form (9.29). Using the aliasing formula (see Theorem 4.95), we conclude that it holds (9.32) for each $\mathbf{j} \in J_N$.
2. For arbitrary $\mathbf{x} \in \mathbb{R}^d$, the N-periodic Lagrange function λ^* can be written by (9.28) and (9.32) as

$$\lambda^*(\mathbf{x}) = \sum_{\mathbf{j}\in J_N} \sum_{\mathbf{n}\in\mathbb{Z}^d} \lambda_{\mathbf{j}+N\mathbf{n}}\, \varphi^*(\mathbf{x} - \mathbf{j})\,.$$

   Substituting $\mathbf{p} := \mathbf{j} + N\mathbf{n} \in \mathbb{Z}^d$ with $\mathbf{j} \in J_N$ and $\mathbf{n} \in \mathbb{Z}^d$, it follows by (9.19) that

$$\begin{aligned}\lambda^*(\mathbf{x}) &= \sum_{\mathbf{p}\in\mathbb{Z}^d} \lambda_{\mathbf{p}}\, \varphi^*(\mathbf{x} - \mathbf{p}) \\ &= \sum_{\mathbf{p}\in\mathbb{Z}^d} \sum_{\mathbf{m}\in\mathbb{Z}^d} \lambda_{\mathbf{p}}\, \varphi(\mathbf{x} - \mathbf{p} + N\mathbf{m})\,.\end{aligned}$$

Now the order of summations can be changed, since only finitely many summands don't vanish by the compact support of φ. Thus, we obtain by (9.8) that

$$\lambda^*(\mathbf{x}) = \sum_{\mathbf{m}\in\mathbb{Z}^d} \sum_{\mathbf{p}\in\mathbb{Z}^d} \lambda_{\mathbf{p}}\, \varphi(\mathbf{x} - \mathbf{p} + N\mathbf{m}) = \sum_{\mathbf{m}\in\mathbb{Z}^d} \lambda(\mathbf{x} + N\mathbf{m})\,.$$

This completes the proof. ■

Theorem 9.18 *Let $N \in \mathbb{N}$ be fixed. Let $\varphi \in C_c(\mathbb{R}^d)$ be a given basis function with the property* (9.6). *The N-periodic interpolation problem* (9.21) *is uniquely solvable in $\mathscr{L}_N(\varphi^*)$ if and only if the symbol σ_φ fulfills the condition* (9.24).

Proof For a given data $f_{\mathbf{j}} \in \mathbb{C}$ with $\mathbf{j} \in J_N$, the N-periodic interpolation problem (9.21) possesses a unique solution $s \in \mathscr{L}_N(\varphi^*)$ of the form

$$s(\mathbf{x}) = \sum_{\mathbf{k}\in J_N} c_{\mathbf{k}}\, \varphi^*(\mathbf{x} - \mathbf{k})\,, \quad \mathbf{x} \in \mathbb{R}^d\,,$$

with certain coefficients $c_{\mathbf{k}} \in \mathbb{C}$ if and only if the system of linear equations

$$f_{\mathbf{j}} = \sum_{\mathbf{k}\in J_N} c_{\mathbf{k}}\, \varphi^*(\mathbf{j} - \mathbf{k})\,, \quad \mathbf{j} \in J_N\,, \tag{9.33}$$

is uniquely solvable. Note that the right-hand side of (9.33) is equal to the $\mathbf{j}$th component of a d-dimensional cyclic convolution. Therefore, we determine the coefficients $c_{\mathbf{k}}$, $\mathbf{k} \in J_N$, using d-dimensional DFT($N \times \ldots \times N$). By the definition (9.5) of the symbol σ_φ and by (9.19), it holds for each $\mathbf{n} \in J_N$:

$$\begin{aligned}\sigma_\varphi\Big(\frac{2\pi\,\mathbf{n}}{N}\Big) &= \sum_{\mathbf{j}\in\mathbb{Z}^d} \varphi(\mathbf{j})\, w_N^{\mathbf{j}\cdot\mathbf{n}} = \sum_{\mathbf{k}\in J_N} \sum_{\mathbf{m}\in\mathbb{Z}^d} \varphi(\mathbf{k} + N\mathbf{m})\, w_N^{\mathbf{k}\cdot\mathbf{n}} \\ &= \sum_{\mathbf{k}\in J_N} \varphi^*(\mathbf{k})\, w_N^{\mathbf{k}\cdot\mathbf{n}}\,.\end{aligned}$$

Thus, we preserve by the convolution property of the d-dimensional DFT($N \times \ldots \times N$) in Theorem 4.106 that

$$\hat{f}_{\mathbf{n}} = \hat{c}_{\mathbf{n}}\, \sigma_\varphi\Big(\frac{2\pi\,\mathbf{n}}{N}\Big)\,, \quad \mathbf{n} \in J_N\,, \tag{9.34}$$

with

$$\hat{c}_{\mathbf{n}} := \sum_{\mathbf{k}\in J_N} c_{\mathbf{k}}\, w_N^{\mathbf{k}\cdot\mathbf{n}}\,, \quad \hat{f}_{\mathbf{n}} := \sum_{\mathbf{k}\in J_N} f_{\mathbf{k}}\, w_N^{\mathbf{k}\cdot\mathbf{n}}\,.$$

Thus, the unique solvability of the linear system (9.33) is equivalent to the unique solvability of (9.34). Obviously, (9.34) is uniquely solvable under the assumption (9.24). ■

Finally, we ask for an algorithm for N-periodic interpolation by translates. As before let $\varphi \in C_c(\mathbb{R}^d)$ be a given basis function which possesses the properties (9.6) and (9.24). Then the N-periodic interpolation problem (9.21) has the unique solution

$$s(\mathbf{x}) = \sum_{\mathbf{k} \in J_N} f_{\mathbf{k}}\, \lambda^*(\mathbf{x} - \mathbf{k}) \in \mathcal{L}_N(\varphi^*) \tag{9.35}$$

for arbitrary given data $f_{\mathbf{k}} \in \mathbb{C}$, $\mathbf{k} \in J_N$. Restricting the summation in the Fourier series (9.25) to the finite index set (9.15) with certain $n \in \mathbb{N}$, then in (9.35) we replace $\lambda^*(\mathbf{x} - \mathbf{k})$ for $\mathbf{k} \in J_N$ by its approximation:

$$\frac{1}{N^d} \sum_{\mathbf{m} \in B_{nN}} \frac{\hat{\varphi}\big(\frac{2\pi \mathbf{m}}{N}\big)}{\sigma_\varphi\big(\frac{2\pi \mathbf{m}}{N}\big)}\, w_N^{\mathbf{m}\cdot\mathbf{k}}\, \mathrm{e}^{2\pi \mathrm{i}\, \mathbf{m}\cdot\mathbf{x}/N} .$$

Thus, for $\mathbf{x} = \frac{\mathbf{j}}{n}$ with $\mathbf{j} \in J_{nN}$, we obtain the approximate value

$$s_{\mathbf{j}} := \frac{1}{N^d} \sum_{\mathbf{m} \in B_{nN}} \frac{\hat{\varphi}\big(\frac{2\pi \mathbf{m}}{N}\big)}{\sigma_\varphi\big(\frac{2\pi \mathbf{m}}{N}\big)}\, w_{nN}^{-\mathbf{m}\cdot\mathbf{j}} \Big(\sum_{\mathbf{k} \in J_N} f_{\mathbf{k}}\, w_N^{\mathbf{m}\cdot\mathbf{k}}\Big)$$

of the exact value $s\big(\frac{\mathbf{j}}{n}\big)$. Note that this value coincides with the approximate value (9.13) for the related cardinal interpolation problem. Thus, we can use the corresponding Algorithm 9.9 for N-periodic interpolation by translates too.

9.2.2 Computation of Fourier Coefficients

For fixed $N \in \mathbb{N}$, we calculate the Fourier coefficients

$$c_{\mathbf{k}}^{(N)}(f) := \frac{1}{N^d} \int_{Q_N} f(\mathbf{x})\, \mathrm{e}^{-2\pi \mathrm{i}\, \mathbf{x}\cdot\mathbf{k}/N}\, \mathrm{d}\mathbf{x} , \quad \mathbf{k} \in \mathbb{Z}^d , \tag{9.36}$$

of an N-periodic function $f \in C_N(\mathbb{R}^d)$, where $Q_N = [0, N)^d$ denotes the d-dimensional hypercube. Assume that the values $f_{\mathbf{j}} := f(\mathbf{j})$ on the uniform grid J_N are given. For a coarse computation of $c_{\mathbf{k}}^{(N)}(f)$, one can use the simple tensor product quadrature rule. Then one obtains the approximate value

$$\tilde{c}_{\mathbf{k}} := \frac{1}{N^d}\,\hat{f}_{\mathbf{k}}$$

with the d-dimensional DFT$(N \times \ldots \times N)$

$$\hat{f}_{\mathbf{k}} := \sum_{\mathbf{j}\in J_N} f_{\mathbf{j}}\, w_N^{\mathbf{j}\cdot\mathbf{k}}\,, \quad \mathbf{k} \in J_N\,, \tag{9.37}$$

where $w_N = \mathrm{e}^{-2\pi\mathrm{i}/N}$ means a primitive N-th root of unity. Extending $\hat{f}_{\mathbf{k}}$ onto $\mathbb{Z}^d$ by $\hat{f}_{\mathbf{k}+N\mathbf{j}} := \hat{f}_{\mathbf{k}}$ for all $\mathbf{k} \in J_N$ and $\mathbf{j} \in \mathbb{Z}^d$, we see that the sequence $\big(\tilde{c}_{\mathbf{k}}\big)_{\mathbf{k}\in\mathbb{Z}^d}$ is N-periodically. Otherwise, as shown in Lemma 4.6, we know that

$$\lim_{\|\mathbf{k}\|_2\to\infty} c_{\mathbf{k}}^{(N)}(f) = 0\,. \tag{9.38}$$

Hence, only in the case $\|\mathbf{k}\|_2 < \frac{N}{2}$, we can expect that $\tilde{c}_{\mathbf{k}}$ is a convenient approximation of $c_{\mathbf{k}}^{(N)}(f)$ (see Corollary 3.4 for $d = 1$).

A better approximate value of $c_{\mathbf{k}}^{(N)}(f)$ with the correct asymptotic behavior for $\|\mathbf{k}\|_2 \to \infty$ can be preserved by the so-called method of attenuation factors for Fourier coefficients. We choose a convenient compactly supported basis function $\varphi \in C_c(\mathbb{R}^d)$ and form the N-periodized function (9.19). Instead of calculating the integral (9.36) directly, first we determine the N-periodic interpolating function $s \in \mathscr{L}_N(\varphi^*)$ interpolating the given data $f_{\mathbf{j}}$ on the uniform grid J_N. Then we obtain the exact Fourier coefficients $c_{\mathbf{k}}^{(N)}(s)$ which are excellent approximations of the wanted Fourier coefficients (9.36).

Theorem 9.19 (Method of Attenuation Factors for Fourier Coefficients) *Let $N \in \mathbb{N}$ be fixed. Let $\varphi \in C_c(\mathbb{R}^d)$ be a given basis function with the property* (9.6). *Assume that the related symbol σ_φ fulfills the condition* (9.30). *For arbitrary given function $f \in C_N(\mathbb{R}^d)$, let $s \in \mathscr{L}_N(\varphi^*)$ be the N-periodic function interpolating $s(\mathbf{j}) = f_{\mathbf{j}} = f(\mathbf{j})$ for all $\mathbf{j} \in J_N$.*

Then the Fourier coefficients of s read as follows:

$$c_{\mathbf{k}}^{(N)}(s) = \frac{1}{N^d}\,\hat{\lambda}\Big(\frac{2\pi\mathbf{k}}{N}\Big)\,\hat{f}_{\mathbf{k}'}\,, \quad \mathbf{k} \in \mathbb{Z}^d\,, \tag{9.39}$$

where $\hat{f}_{\mathbf{k}'}$ is equal to (9.37) *and where $\mathbf{k}' := \mathbf{k} \bmod N \in J_N$ is the nonnegative residue of $\mathbf{k} \in \mathbb{Z}^d$ modulo N. The values*

$$\hat{\lambda}\Big(\frac{2\pi\mathbf{k}}{N}\Big) = \frac{\hat{\varphi}\big(\frac{2\pi\mathbf{k}}{N}\big)}{\sigma_\varphi\big(\frac{2\pi\mathbf{k}'}{N}\big)}$$

are called attenuation factors of the Fourier coefficients.

Proof By (9.35) the given data $f_{\mathbf{j}} \in \mathbb{C}$ on the uniform grid J_N will be interpolated by the N-periodic function $s \in \mathscr{L}_N(\varphi^*)$ in the form

$$s(\mathbf{x}) = \sum_{\mathbf{j} \in J_N} f_{\mathbf{j}}\, \lambda^*(\mathbf{x} - \mathbf{j})\,, \quad \mathbf{x} \in \mathbb{R}^d\,,$$

where $\lambda^* \in \mathscr{L}_N(\varphi^*)$ denotes the N-periodic Lagrange function. By the translation property of the Fourier coefficients (cf. Lemma 4.1 for the period 2π), we obtain for the $\mathbf{k}$th Fourier coefficient of the N-periodic function s:

$$c_{\mathbf{k}}^{(N)}(s) = c_{\mathbf{k}}^{(N)}(\lambda^*) \sum_{\mathbf{j} \in J_N} f_{\mathbf{j}}\, w_N^{\mathbf{j}\cdot\mathbf{k}} = c_{\mathbf{k}}^{(N)}(\lambda^*)\, \hat{f}_{\mathbf{k}}\,.$$

From Theorem 9.16, it follows that

$$c_{\mathbf{k}}^{(N)}(\lambda^*) = \frac{1}{N^d}\, \hat{\lambda}\Big(\frac{2\pi\mathbf{k}}{N}\Big) = \frac{1}{N^d}\, \frac{\hat{\varphi}\big(\frac{2\pi\mathbf{k}}{N}\big)}{\sigma_\varphi\big(\frac{2\pi\mathbf{k}}{N}\big)} = \frac{1}{N^d}\, \frac{\hat{\varphi}\big(\frac{2\pi\mathbf{k}}{N}\big)}{\sigma_\varphi\big(\frac{2\pi\mathbf{k}'}{N}\big)}\,,$$

where $\hat{\lambda}$ denotes the Fourier transform of the cardinal Lagrange function. ∎

We emphasize that the attenuation factors are independent of the given data $f_{\mathbf{j}}$ on the uniform grid J_N; they are only special values of the Fourier transform of the cardinal Lagrange function. Thus, we can use $c_{\mathbf{k}}^{(N)}(s)$ as approximate values of $c_{\mathbf{k}}^{(N)}(f)$ for all $\mathbf{k} \in \mathbb{Z}^d$. The method of attenuation factors for Fourier coefficients performs two tasks. The computed Fourier coefficients $c_{\mathbf{k}}^{(N)}(s)$ *correct* the coarse approximate values $N^{-d}\, \hat{f}_{\mathbf{k}}$ for $\mathbf{k} \in J_N$. Further, the approximate values $N^{-d}\, \hat{f}_{\mathbf{k}}$ for $\mathbf{k} \in J_N$ are *continued* to whole $\mathbb{Z}^d$ by the values $c_{\mathbf{k}}^{(N)}(s)$.

The following example shows the performance of this method of attenuation factors.

Example 9.20 We consider the even 2π-periodic function $f \in C(\mathbb{T})$ given by $f(x) := x^2$ for $x \in [-\pi, \pi)$. Then the related Fourier series

$$f(x) = \frac{\pi^2}{3} - 4\,\cos x + \cos(2x) - \frac{4}{9}\,\cos(3x) + \ldots$$

converges uniformly on $\mathbb{R}$. We choose $N = 16$ and $\varphi = N_4$. Instead of the exact Fourier coefficients

$$c_k(f) = \frac{1}{2\pi}\int_{-\pi}^{\pi} x^2\,\mathrm{e}^{-\mathrm{i}kx}\,\mathrm{d}x = \begin{cases} \frac{\pi^2}{3} & k = 0\,,\\ \frac{2\,(-1)^k}{k^2} & k \in \mathbb{Z}\setminus\{0\}\,,\end{cases}$$

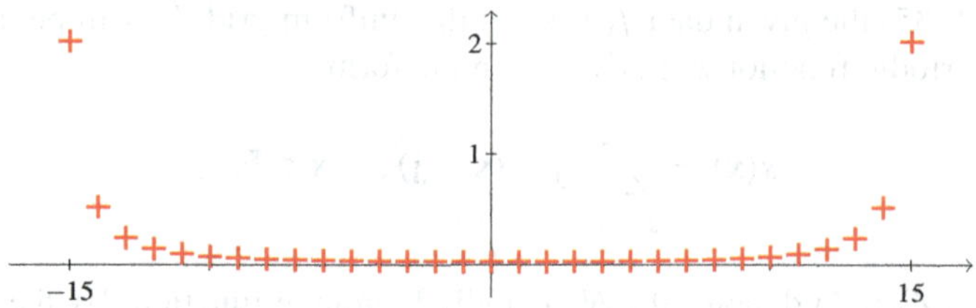

Fig. 9.5 The crosses $(k, |\frac{1}{16}\hat{f}_k - c_k(f)|)$ for $k = -15, \ldots, 15$ illustrate the error behavior for the classical computation of $c_k(f)$ in Example 9.20

we obtain the coarse approximate values:

$$\frac{1}{16}\hat{f}_k = \frac{1}{16}\sum_{j=0}^{15} f\big(\frac{j\pi}{8}\big)\, w_{16}^{jk}\,, \quad k = 0, \ldots, 15\,.$$

The method of attenuation factors creates the improved approximate values

$$c_k := \frac{1}{16}\,\hat{\lambda}\big(\frac{\pi k}{8}\big)\,\hat{f}_k$$

with the attenuation factors

$$\hat{\lambda}\big(\frac{\pi k}{8}\big) = \frac{3\,\big(\operatorname{sinc}\frac{\pi k}{16}\big)^4}{2 + \cos\frac{\pi k}{8}}\,.$$

Figure 9.5 illustrates the errors for the classical computation of the Fourier coefficients $c_k(f)$. On the other hand, the method of attenuation factors produces the small maximal error

$$\max_{|k|\le 15} |c_k - c_k(f)| = 0.026982\,.$$

□

Remark 9.21 The method of attenuation factors has a long history. Using a polynomial spline φ, the attenuation factors of Fourier coefficients were calculated first by A. Eagle [123] and later by W. Quade and L. Collatz [357] (see also [127, 164]). W. Gautschi [157] presented a general theory of attenuation factors for Fourier coefficients. He used a linear and translation invariant approximation process in order to interpolate the given data on a uniform mesh, see also [276]. Later, M.H. Gutknecht [188] extended Gautschi's approach to multivariate periodic functions. □

9.3 Quadrature of Periodic Functions

Now we consider the quadrature of univariate periodic functions. First, we derive the so-called Euler–Maclaurin summation formula, which is based on an expansion into Bernoulli polynomials. The *Bernoulli polynomial* B_n of degree n is recursively defined by $B_0(x) := 1$ and

$$B_n'(x) = n\, B_{n-1}(x)\,, \quad n \in \mathbb{N}\,, \tag{9.40}$$

with the condition

$$\int_0^1 B_n(x)\, \mathrm{d}x = 0\,, \quad n \in \mathbb{N}\,. \tag{9.41}$$

The numbers $B_n(0)$ are called *Bernoulli numbers*. Thus, the first Bernoulli polynomials read as follows:

$$B_0(x) = 1\,, \quad B_1(x) = x - \frac{1}{2}\,, \quad B_2(x) = x^2 - x + \frac{1}{6}\,,$$
$$B_3(x) = x^3 - \frac{3}{2}\,x^2 + \frac{1}{2}\,x\,, \quad B_4(x) = x^4 - 2\,x^3 + x^2 - \frac{1}{30}\,.$$

Note that by (9.40) and (9.41), it holds $B_n(0) = B_n(1)$ for all $n \in \mathbb{N} \setminus \{1\}$, since

$$B_n(1) - B_n(0) = \int_0^1 B_n'(x)\, \mathrm{d}x = n \int_0^1 B_{n-1}(x)\, \mathrm{d}x = 0\,. \tag{9.42}$$

Each Bernoulli polynomial B_n has the following symmetry property:

$$B_n(x) = (-1)^n\, B_n(1-x)\,, \tag{9.43}$$

since the polynomial $(-1)^n\, B_n(1-x)$ has the same properties (9.40) and (9.41) as B_n. For $n = 2k+1$, $k \in \mathbb{N}$, and $x = 0$, it follows that $B_{2k+1}(0) = -B_{2k+1}(1)$. Hence, by (9.42) we conclude that

$$B_{2k+1}(0) = B_{2k+1}(1) = 0\,, \quad k \in \mathbb{N}\,. \tag{9.44}$$

By b_n, $n \in \mathbb{N} \setminus \{1\}$, we denote the 1*-periodic Bernoulli function* which is defined as the 1-periodic continuation of Bernoulli polynomial B_n restricted on $[0,\ 1)$. Hence, it holds

$$b_n(x) = B_n(x - \lfloor x \rfloor)\,, \quad x \in \mathbb{R}\,, \tag{9.45}$$

where $\lfloor x \rfloor$ denotes the largest integer smaller than or equal to $x \in \mathbb{R}$. Further, b_1 is defined as the 1-periodic continuation of B_1 restricted on $(0, 1)$ with $b_1(0) = b_1(1) := 0$. Obviously, b_1 is a 1-periodic sawtooth function.

Lemma 9.22 *For each $n \in \mathbb{N}$, the 1-periodic Bernoulli function b_n can be represented as a convergent Fourier series:*

$$b_n(x) = -n! \sum_{k\in\mathbb{Z}\setminus\{0\}} \frac{1}{(2\pi\,\mathrm{i}\,k)^n}\, \mathrm{e}^{2\pi\,\mathrm{i}\,k\,x}\,. \tag{9.46}$$

Proof First, we remark that for all $k \in \mathbb{Z}$ and $n \in \mathbb{N}$,

$$c_k^{(1)}(b_n) = c_k^{(1)}(B_n) = \int_0^1 B_n(t)\,\mathrm{e}^{-2\pi\mathrm{i}kt}\,\mathrm{d}t\,.$$

The condition (9.41) leads to $c_0^{(1)}(B_n) = 0$ for each $n \in \mathbb{N}$. Now we calculate the Fourier coefficients $c_k^{(1)}(B_n)$ for $k \in \mathbb{Z} \setminus \{0\}$. For $n = 1$ we obtain

$$c_k^{(1)}(B_1) = \int_0^1 \Big(t - \frac{1}{2}\Big)\,\mathrm{e}^{-2\pi\mathrm{i}kt}\,\mathrm{d}t = -\frac{1}{2\pi\mathrm{i}k}\,. \tag{9.47}$$

By Lemma 1.6 and (9.40), we receive for $n \in \mathbb{N} \setminus \{0\}$

$$c_k^{(1)}(b_n') = 2\pi\mathrm{i}k\, c_k^{(1)}(B_n) = n\, c_k^{(1)}(B_{n-1})$$

and hence the recursion

$$c_k^{(1)}(B_n) = \frac{n}{2\pi\mathrm{i}k}\, c_k^{(1)}(B_{n-1})$$

such that by (9.47)

$$c_k^{(1)}(B_n) = -\frac{n!}{(2\pi\,\mathrm{i}\,k)^n}\,.$$

For $n \in \mathbb{N} \setminus \{1\}$, the 1-periodic Bernoulli function b_n is contained in $C_1^{(n-2)}(\mathbb{R})$ and its Fourier series (9.46) converges uniformly by Theorem 1.37. For $n = 1$, the 1-periodic Bernoulli function b_1 is piecewise linear. By Theorem 1.34 of Dirichlet–Jordan, the related Fourier series converges pointwise and uniformly on each closed interval contained in $\mathbb{R} \setminus \mathbb{Z}$. ■

Lemma 9.23 *For $n \in \mathbb{N}$ and $x \in [0, 1]$, it holds the inequality*

$$(-1)^n \big(B_{2n}(x) - B_{2n}(0)\big) \geq 0\,. \tag{9.48}$$

Proof By Lemma 9.22 we know that for each $n \in \mathbb{N}$ and $x \in [0, 1]$,

$$B_{2n}(x) = (-1)^{n+1}\,(2n)!\,\sum_{k=1}^{\infty}\frac{2\cos(2\pi kx)}{(2\pi k)^{2n}}$$

and hence

$$B_{2n}(0) = (-1)^{n+1}\,(2n)!\,\sum_{k=1}^{\infty}\frac{2}{(2\pi k)^{2n}}\,.$$

Thus, it follows that

$$(-1)^n\,\big(B_{2n}(x) - B_{2n}(0)\big) = 2\,(2n)!\,\sum_{k=1}^{\infty}\frac{1-\cos(2\pi kx)}{(2\pi k)^{2n}} \ge 0\,.$$

This completes the proof. ■

Lemma 9.24 *Let $h \in C^m[0, 1]$, $m \in \mathbb{N}$, be given. Then h can be represented in the* Bernoulli polynomial expansion*:*

$$h(x) = \sum_{j=0}^{m}\frac{B_j(x)}{j!}\int_0^1 h^{(j)}(t)\,\mathrm{d}t - \frac{1}{m!}\int_0^1 b_m(x-t)\,h^{(m)}(t)\,\mathrm{d}t$$

$$= \int_0^1 h(t)\,\mathrm{d}t + \sum_{j=1}^{m}\frac{B_j(x)}{j!}\,\big(h^{(j-1)}(1) - h^{(j-1)}(0)\big)$$

$$- \frac{1}{m!}\int_0^1 b_m(x-t)\,h^{(m)}(t)\,\mathrm{d}t\,. \qquad (9.49)$$

In the case $m = 2n + 2$, $n \in \mathbb{N}_0$, it holds for certain $\tau \in (0, 1)$

$$\int_0^1 h(t)\,\mathrm{d}t = \frac{1}{2}\,\big(h(0) + h(1)\big) - \sum_{j=1}^{n}\frac{B_{2j}(0)}{(2j)!}\,\big(h^{(2j-1)}(1) - h^{(2j-1)}(0)\big)$$

$$- \frac{B_{2n+2}(0)}{(2n+2)!}\,h^{(2n+2)}(\tau)\,. \qquad (9.50)$$

Proof

1. We show the Bernoulli polynomial expansion (9.49) by induction with respect to m. By (9.45) for b_1, we get

$$\int_0^1 b_1(x-t)\,h'(t)\,\mathrm{d}t = \int_0^x \Big(x - t - \frac{1}{2}\Big)\,h'(t)\,\mathrm{d}t + \int_x^1 \Big(x - t + \frac{1}{2}\Big)\,h'(t)\,\mathrm{d}t\,.$$

Then integration by parts leads to

$$\int_0^1 b_1(x-t)\,h'(t)\,\mathrm{d}t = \int_0^1 h(t)\,\mathrm{d}t + \big(h(1)-h(0)\big)\,B_1(x) - h(x)$$

such that (9.49) is shown for $m = 1$.

Assume that (9.49) is valid for some $m \in \mathbb{N}$. Let $h \in C^{(m+1)}[0,\,1]$ be given. By the definition of the 1-periodic Bernoulli function b_m, we obtain for the integral term in (9.49) that

$$\int_0^1 b_m(x-t)\,h^{(m)}(t)\,\mathrm{d}t = \int_0^x B_m(x-t)\,h^{(m)}(t)\,\mathrm{d}t + \int_x^1 B_m(x-t+1)\,h^{(m)}(t)\,\mathrm{d}t\,.$$

Applying integration by parts, it follows by (9.40) and (9.42) that

$$\begin{aligned}\frac{1}{m!}\int_0^1 b_m(x-t)\,h^{(m)}(t)\,\mathrm{d}t &= -\frac{B_{m+1}(x)}{(m+1)!}\,\big(h^{(m)}(1)-h^{(m)}(0)\big)\\ &\quad + \frac{1}{(m+1)!}\int_0^1 b_{m+1}(x-t)\,h^{(m+1)}(t)\,\mathrm{d}t\,,\end{aligned}$$

i.e., (9.49) is also true for $m + 1$.

2. For $x = 0$ and $m = 2n+2$, $n \in \mathbb{N}_0$, in (9.49), it follows by (9.44) and (9.43) that

$$\begin{aligned}\int_0^1 h(t)\,\mathrm{d}t &= \frac{1}{2}\,\big(h(0)+h(1)\big) - \sum_{j=1}^{n+1}\frac{B_{2j}(0)}{(2j)!}\,\big(h^{(2j-1)}(1)-h^{(2j-1)}(0)\big)\\ &\quad + \frac{1}{(2n+2)!}\int_0^1 B_{2n+2}(t)\,h^{(2n+2)}(t)\,\mathrm{d}t\\ &= \frac{1}{2}\,\big(h(0)+h(1)\big) - \sum_{j=1}^{n}\frac{B_{2j}(0)}{(2j)!}\,\big(h^{(2j-1)}(1)-h^{(2j-1)}(0)\big)\\ &\quad + \frac{1}{(2n+2)!}\int_0^1 \big(B_{2n+2}(t)-B_{2n+2}(0)\big)\,h^{(2n+2)}(t)\,\mathrm{d}t\,.\end{aligned}$$

From $h^{(2n+2)} \in C[0,\,1]$, (9.48), and (9.41), it follows by the extended mean value theorem for integrals that there exists one $\tau \in (0,\,1)$ with

$$\begin{aligned}&\int_0^1 \big(B_{2n+2}(t)-B_{2n+2}(0)\big)\,h^{(2n+2)}(t)\,\mathrm{d}t\\ &= h^{(2n+2)}(\tau)\int_0^1 \big(B_{2n+2}(t)-B_{2n+2}(0)\big)\,\mathrm{d}t\\ &= -B_{2n+2}(0)\,h^{(2n+2)}(\tau)\,.\end{aligned}$$

Thus, (9.50) is shown. ■

Corollary 9.25 (Euler–Maclaurin Summation Formula) *Let $n \in \mathbb{N}_0$ and $N \in \mathbb{N}\backslash\{1\}$ be given. Then for $h \in C^{2n+2}[0,\ N]$, it holds the* Euler–Maclaurin summation formula *:*

$$\int_0^N h(t)\mathrm{d}t = \frac{1}{2}\big(h(0) + h(N)\big) + \sum_{k=1}^{N-1} h(k) - \sum_{j=1}^{n} \frac{B_{2j}(0)}{(2j)!}\big(h^{(2j-1)}(N) - h^{(2j-1)}(0)\big) - \frac{N\, B_{2n+2}(0)}{(2n+2)!}\, h^{(2n+2)}(\sigma) \tag{9.51}$$

with one $\sigma \in (0,\ N)$.

Proof Repeated application of Lemma 9.24 to the integrals

$$\int_k^{k+1} h(t)\,\mathrm{d}t\,, \quad k = 0, \ldots, N-1\,,$$

leads to

$$\int_0^N h(t)\,\mathrm{d}t = \sum_{k=0}^{N-1} \int_k^{k+1} h(t)\,\mathrm{d}t$$
$$= \frac{1}{2}\big(h(0) + h(N)\big) + \sum_{k=1}^{N-1} h(k) - \sum_{j=1}^{n} \frac{B_{2j}(0)}{(2j)!}\big(h^{(2j-1)}(N) - h^{(2j-1)}(0)\big)$$
$$- \frac{B_{2n+2}(0)}{(2n+2)!} \sum_{k=0}^{N-1} h^{(2n+2)}(\tau_k)$$

with $\tau_k \in (k,\ k+1)$. By the intermediate value theorem of $h^{(2n+2)} \in C[0,\ N]$, there exists one $\sigma \in (0,\ N)$ with

$$\frac{1}{N}\sum_{k=0}^{N-1} h^{(2n+2)}(\tau_k) = h^{(2n+2)}(\sigma)\,.$$

■

The Euler–Maclaurin summation formula (9.51) describes a powerful connection between integrals and finite sums. By this formula, one can evaluate finite sums by integrals which can be seen as follows.

Example 9.26 For arbitrary $N \in \mathbb{N} \setminus \{1\}$, we consider the function $h(t) := t^2$ for $t \in [0,\ N]$. From (9.51) it follows that

$$\int_0^N t^2\,\mathrm{d}t = \frac{1}{2}\,N^2 + \sum_{k=1}^{N-1} k^2 - \frac{1}{6}\,N$$

and hence

$$\sum_{k=1}^{N} k^2 = \frac{1}{3}\,N^3 + \frac{1}{2}\,N^2 + \frac{1}{6}\,N = \frac{1}{6}\,N\,(N+1)\,(2N+1)\,.$$

□

Now we apply the Euler–Maclaurin summation formula (9.51) to the quadrature of a 2π-periodic smooth function g. Obviously, the trapezoidal rule with equidistant nodes $\frac{2\pi k}{N}$, $k = 0, \ldots, N-1$, coincides with the related *rectangular rule* :

$$\frac{2\pi}{N}\sum_{k=0}^{N-1} g\big(\frac{2\pi k}{N}\big)\,.$$

We estimate the quadrature error for the rectangular rule with equidistant nodes. The following lemma indicates that the simple rectangular rule with equidistant nodes is very convenient for the quadrature of 2π-periodic, $(2n+2)$-times continuously differentiable functions.

Lemma 9.27 *Let* $g \in C^{(2n+2)}(\mathbb{T})$ *with* $n \in \mathbb{N}_0$ *be given. Further, let* $N \in \mathbb{N}\setminus\{1\}$ *be fixed.*

Then the quadrature error *in the rectangular rule with equidistant nodes* $\frac{2\pi k}{N}$, $k = 0, \ldots, N-1$, *can be estimated by*

$$\Big|\int_0^{2\pi} g(x)\,\mathrm{d}x - \frac{2\pi}{N}\sum_{k=0}^{N-1} g\big(\frac{2\pi k}{N}\big)\Big| \le \frac{(2\pi)^{2n+3}\,B_{2n+2}(0)}{(2n+2)!\,N^{2n+2}}\,\|g^{(2n+2)}\|_{C(\mathbb{T})}\,.$$

Proof We apply the Euler–Maclaurin formula (9.51). The substitution $x = \frac{2\pi}{N}\,t \in [0,\,2\pi]$ for $t \in [0,\,N]$ leads to $h(t) := g\big(\frac{2\pi}{N}\,t\big)$ for $t \in [0,\,N]$. From the assumption $g \in C^{(2n+2)}(\mathbb{T})$, it follows that $h \in C^{(2n+2)}[0,\,N]$ fulfills the conditions $h^{(j)}(0) = h^{(j)}(N)$, $j = 0, \ldots, 2n+2$. Thus, by (9.51) we obtain that for certain $\xi \in (0,\,2\pi)$

$$\int_0^{2\pi} g(x)\,\mathrm{d}x - \frac{2\pi}{N}\sum_{k=0}^{N-1} g\big(\frac{2\pi k}{N}\big) = -\frac{(2\pi)^{2n+3}\,B_{2n+2}(0)}{(2n+2)!\,N^{2n+2}}\,g^{(2n+2)}(\xi)\,.$$

■

For $n = 0$, Lemma 9.27 provides:

Corollary 9.28 *For* $g \in C^2(\mathbb{T})$ *the quadrature error in the rectangular rule with equidistant nodes* $\frac{2\pi k}{N}$, $k = 0, \ldots, N-1$, *can be estimated by*

$$\Big|\int_0^{2\pi} g(x)\,\mathrm{d}x - \frac{2\pi}{N}\sum_{k=0}^{N-1} g\big(\frac{2\pi k}{N}\big)\Big| \le \frac{(2\pi)^3}{12\,N^2}\,\|g''\|_{C(\mathbb{T})}\,.$$

This result will now be used to estimate the error in the computation of the Fourier coefficients

$$c_\ell(f) = \frac{1}{2\pi}\int_0^{2\pi} f(x)\,\mathrm{e}^{-\mathrm{i}\ell x}\,\mathrm{d}x\,,\qquad \ell \in \mathbb{Z}\,,$$

of a given function $f \in C^2(\mathbb{T})$. Setting $g(x) := \frac{1}{2\pi}\,f(x)\,\mathrm{e}^{-\mathrm{i}\ell x}$, we obtain

$$g''(x) = \frac{1}{2\pi}\,\mathrm{e}^{-\mathrm{i}\ell x}\,\big(f''(x) - 2\mathrm{i}\ell\,f'(x) - \ell^2\,f(x)\big)\,.$$

Denoting

$$\hat{f}_\ell := \frac{1}{N}\sum_{k=0}^{N-1} f\big(\frac{2\pi k}{N}\big)\,w_N^{k\ell}$$

with $w_N = \mathrm{e}^{-2\pi\mathrm{i}/N}$, Corollary 9.28 leads to the estimate:

$$|c_\ell(f) - \hat{f}_\ell| \le \frac{(2\pi)^3}{12\,N^2}\max_{x\in[0,\,2\pi]}\big(|f''(x)| + 2|\ell\,f'(x)| + \ell^2\,|f(x)|\big)\,. \tag{9.52}$$

Note that for $\ell = -\frac{N}{2}$, the upper bound of the quadrature error (9.52) is essentially independent of N.

As known the Fourier coefficients $c_\ell(f)$ are well approximated by $\hat{f}_\ell$ only for $\ell = -\frac{N}{2}, \ldots, \frac{N}{2} - 1$. This follows from the aliasing formula (3.6) and Corollary 3.4. Summarizing, we can say that both 2π-periodicity and smoothness of the given function f are essentially for small quadrature errors $|c_\ell(f) - \hat{f}_\ell|$ for all $\ell = -\frac{N}{2}, \ldots, \frac{N}{2} - 1$.

9.4 Accelerating Convergence of Fourier Series

If a 2π-periodic function f is sufficiently smooth, then its Fourier series converges rapidly to f, and the related Fourier coefficients $c_k(f)$ tend to zero as $|k| \to \infty$ (see Theorem 1.39). In this case, a Fourier partial sum $S_n f$ of low-order n approximates f quite accurately, since the approximation error

$$\|f - S_n f\|_{C[0,\,2\pi]} \le \sum_{|k|>n} |c_k(f)|$$

will be small if the Fourier coefficients tend to zero rapidly enough. Otherwise, if a 2π-periodic function f is only piecewise smooth, then its Fourier partial sums $S_n f$ oscillate near a jump discontinuity of f by the Gibbs phenomenon (see Theorem 1.42) and converge very slowly to f. Can one find a rapidly convergent modified Fourier expansion of f, if f is only piecewise smooth?

In this section, we describe two methods to accelerate the convergence of a Fourier series. In the first method, we represent a 2π-periodic, piecewise smooth function f as a sum of a polynomial trend $T_m f$ and a fast convergent Fourier series of $f - T_m f$, since $f - T_m f$ is sufficiently smooth by construction.

In the second method, we consider a smooth function $\varphi \in C^\infty(I)$ defined on the interval $I := [-1, 1]$. Note that the 2-periodic extension of $\varphi|_{[-1, 1)}$ is only piecewise smooth in general. Therefore, we extend φ to a $2T$-periodic, sufficiently smooth function f with certain $T > 1$ such that f possesses a $2T$-periodic, rapidly convergent Fourier expansion.

9.4.1 Krylov–Lanczos Method

First, we consider 2π-periodic, piecewise smooth functions. A 2π-periodic function f is called *piecewise r-times continuously differentiable* or *piecewise C^r-smooth* with $r \in \mathbb{N}$, if there exist finitely many nodes x_j, $j = 1, \ldots, n$, with $0 \le x_1 < x_2 < \ldots < x_n < 2\pi$ and $x_{n+1} := x_1 + 2\pi$ so that f restricted to (x_j, x_{j+1}) belongs to $C^r[x_j, x_{j+1}]$ for each $j = 1, \ldots, n$. By $C^r[x_j, x_{j+1}]$ we mean the set of all functions f with the properties that $f, f', \ldots, f^{(r)}$ are continuous on (x_j, x_{j+1}) and have continuous extensions on $[x_j, x_{j+1}]$, i.e., there exist all one-sided finite limits $f^{(\ell)}(x_j + 0)$ and $f^{(\ell)}(x_{j+1} - 0)$ for $\ell = 0, \ldots, r$.

The Fourier series of a 2π-periodic, piecewise smooth function which is smooth on the interval $[0, 2\pi)$ has usually slow convergence due to the fact that this function has jumps at each point of $2\pi\,\mathbb{Z}$ in general. If a 2π-periodic, piecewise C^r-smooth function with $n = 1$ and $x_1 = 0$ (see Fig. 9.6) is given, then the asymptotic behavior of its Fourier coefficients:

$$c_k(f) = \frac{1}{2\pi}\int_0^{2\pi} f(t)\,\mathrm{e}^{-\mathrm{i}kt}\,\mathrm{d}t\,, \quad k \in \mathbb{Z}\,,$$

and the rate of convergence of its Fourier series depends only on the largest positive integer $m \le r$ which fulfills the condition

$$f^{(j)}(0+0) = f^{(j)}(2\pi - 0)\,, \quad j = 0, \ldots, m-1\,. \tag{9.53}$$

As known a function f with condition (9.53) possesses a uniformly convergent Fourier expansion.

Unfortunately, a 2π-periodic, piecewise C^r-smooth function with $n = 1$ and $x_1 = 0$ does not fulfill (9.53) in general. The corresponding Fourier series converges

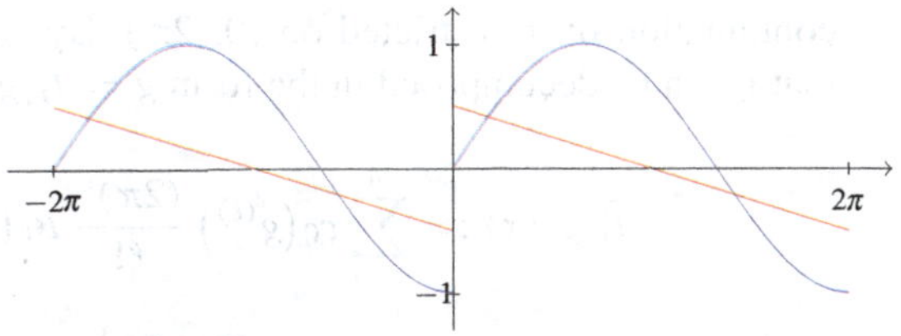

Fig. 9.6 Linear trend $T_1 f(x) = -b_1(\frac{x}{2\pi})$ (red) of the 2π-periodic, piecewise smooth function f (blue) defined by $f(x) = \sin\frac{3x}{4}$ for $x \in [0, 2\pi)$

extremely slow. In such a case, it has been proposed by A.N. Krylov and later by C. Lanczos [263] to determine a 2π-periodic, piecewise polynomial $T_m f$ such that $f - T_m f$ satisfies the condition (9.53).

Figure 9.6 shows the 2π-periodic, piecewise smooth function f with $f(x) := \sin\frac{3x}{4}$ for $x \in [0, 2\pi)$ as well as its 2π-periodic trend $T_1 f$ with $(T_1 f)(x) := \frac{1}{2} - \frac{x}{2\pi}$ for $x \in (0, 2\pi)$ and $(T_1 f)(0) := 0$. Then we have $f - T_1 f \in C(\mathbb{T})$ and $f - T_1 f \notin C^1(\mathbb{T})$.

Theorem 9.29 (Krylov–Lanczos Method of Convergence Acceleration) *For r, $m \in \mathbb{N}$ with $m \le r$, let f be a 2π-periodic, piecewise C^r-smooth function with only one node $x_1 = 0$ within $[0, 2\pi)$. Then f can be split into the sum $f = T_m f + R_m f$ on $\mathbb{R} \setminus 2\pi\,\mathbb{Z}$, where*

$$(T_m f)(x) := \sum_{\ell=1}^{m} c_0(f^{(\ell)})\, \frac{(2\pi)^\ell}{\ell!}\, b_\ell\big(\frac{x}{2\pi}\big) \tag{9.54}$$

$$= \sum_{\ell=0}^{m-1} \big(f^{(\ell)}(2\pi-0) - f^{(\ell)}(0+0)\big)\, \frac{(2\pi)^\ell}{(\ell+1)!}\, b_{\ell+1}\big(\frac{x}{2\pi}\big) \tag{9.55}$$

$$= \sum_{\ell=0}^{m-1} \big(f^{(\ell)}(0-0) - f^{(\ell)}(0+0)\big)\, \frac{(2\pi)^\ell}{(\ell+1)!}\, b_{\ell+1}\big(\frac{x}{2\pi}\big) \tag{9.56}$$

*is the 2π-*periodic trend *of f and where*

$$(R_m f)(x) := c_0(f) - \frac{(2\pi)^{m-1}}{m!} \int_0^{2\pi} b_m\big(\frac{x-t}{2\pi}\big)\, f^{(m)}(t)\, \mathrm{d}t \in C^{m-1}(\mathbb{T}) \tag{9.57}$$

possesses the uniformly convergent Fourier series

$$(R_m f)(x) = c_0(f) + \sum_{k\in\mathbb{Z}\setminus\{0\}} \frac{1}{(\mathrm{i}\,k)^m}\, c_k(f^{(m)})\, \mathrm{e}^{\mathrm{i}\,k\,x}\,. \tag{9.58}$$

Proof

1. The Krylov–Lanczos method is mainly based on the Bernoulli polynomial expansion (9.49). Let $g \in C^r[0, 2\pi]$ denote the r-times continuously differentiable

continuation of f restricted on $(0,\,2\pi)$. By substitution it follows from (9.49) that g can be decomposed in the form $g = \tilde{T}_m g + \tilde{R}_m g$ with

$$(\tilde{T}_m g)(x) := \sum_{\ell=1}^{m} c_0\big(g^{(\ell)}\big)\,\frac{(2\pi)^\ell}{\ell!}\,B_\ell\big(\frac{x}{2\pi}\big)\,,$$

$$(\tilde{R}_m g)(x) := c_0(g) - \frac{(2\pi)^{m-1}}{m!}\,\Big(\int_0^x B_m\big(\frac{x-t}{2\pi}\big)\,g^{(m)}(t)\,\mathrm{d}t + \int_x^{2\pi} B_m\big(\frac{x-t+2\pi}{2\pi}\big)\,g^{(m)}(t)\,\mathrm{d}t\Big)\,.$$

Note that it holds $c_0(g) = c_0(f)$, $g^{(j)}(0) = f^{(j)}(0+0)$, and $g^{(j)}(2\pi) = f^{(j)}(2\pi-0) = f^{(j)}(0-0)$ for $j = 0, \ldots, m$. By 2π-periodic extension of $g = \tilde{T}_m g + \tilde{R}_m g$ restricted on $(0,\,2\pi)$, we preserve the decomposition $f = T_m f + R_m f$ on $\mathbb{R} \setminus 2\pi\,\mathbb{Z}$, where $T_m f$ and $R_m f$ are defined by (9.54) and (9.57), respectively.

Obviously, $T_m f$ is a 2π-periodic, piecewise polynomial and $R_m f$ is equal to the sum of $c_0(f)$ and a convolution of 2π-periodic functions:

$$R_m f = c_0(f) - \frac{(2\pi)^{m-1}}{m!}\,b_m\big(\frac{\cdot}{2\pi}\big) * f^{(m)}\,. \tag{9.59}$$

2. Now we show that $h := f - T_m f$ restricted to $(0,\,2\pi)$ fulfills the conditions:

$$h^{(j)}(0+0) = h^{(j)}(2\pi-0)\,, \quad j = 0, \ldots, m-1\,.$$

Since $h = R_m f$, simple calculation shows that

$$h(0+0) = h(2\pi-0) = c_0(f) - \frac{(2\pi)^{m-1}}{m!}\int_0^{2\pi} B_m\big(\frac{t}{2\pi}\big)\,f^{(m)}(t)\,\mathrm{d}t\,.$$

Differentiation of $T_m f$ yields by (9.40) that

$$\begin{aligned}\frac{\mathrm{d}}{\mathrm{d}x}\,(T_m f)(x) &= \sum_{\ell=0}^{m-1}\big(f^{(\ell)}(2\pi-0) - f^{(\ell)}(0+0)\big)\,\frac{(2\pi)^{\ell-1}}{\ell!}\,B_\ell\big(\frac{x}{2\pi}\big)\\ &= c_0(f') + (T_{m-1}f')(x)\,.\end{aligned}$$

Thus, repeated differentiation provides

$$\frac{\mathrm{d}^j}{\mathrm{d}x^j}\,(T_m f)(x) = c_0(f^{(j)}) + (T_{m-j}f^{(j)})(x)\,, \quad j = 2, \ldots, m-1\,.$$

Therefore, we obtain that

$$h^{(j)}(x) = f^{(j)}(x) - (T_{m-j} f^{(j)})(x) = (R_{m-j} f^{(j)})(x)\,, \quad j = 0, \ldots, m-1\,,$$

and hence for each $j = 0, \ldots, m-1$

$$\begin{aligned} h^{(j)}(0+0) = h^{(j)}(2\pi - 0) &= c_0(f^{(j)}) \\ &- \frac{(2\pi)^{m-j-1}}{(m-j)!} \int_0^{2\pi} B_{m-j}\Big(\frac{2\pi - t}{2\pi}\Big)\, f^{(m-j)}(t)\,\mathrm{d}t\,. \end{aligned}$$

This shows that $h = R_m f \in C^{m-1}(\mathbb{T})$.

3. By the convolution property of the Fourier series (see Lemma 1.13), it follows from (9.59) that

$$c_k(R_m f) - c_0(f) = -\frac{(2\pi)^m}{m!}\, c_k\big(b_m\big(\frac{\cdot}{2\pi}\big)\big)\, c_k(f^{(m)}) \quad k \in \mathbb{Z} \setminus \{0\}\,.$$

By Lemma 9.22 we know that

$$c_k\big(b_m\big(\frac{\cdot}{2\pi}\big)\big) = -\frac{m!}{(2\pi \mathrm{i} k)^m}\,, \qquad k \in \mathbb{Z} \setminus \{0\}\,,$$

with $c_0\big(b_m\big(\frac{\cdot}{2\pi}\big)\big) = 0$. Thus, we receive the Fourier expansion (9.58) of $R_m f$. Since $f - T_m f$ is piecewise C^r-smooth by assumption and since $R_m f \in C^{m-1}(\mathbb{T})$ by step 2, the Fourier series (9.58) of $R_m f$ converges uniformly on $\mathbb{R}$ by Theorem 1.34 of Dirichlet–Jordan. ■

Example 9.30 For $m = 1$

$$(T_1 f)(x) = \big(f(2\pi - 0) - f(0+0)\big)\, b_1\big(\frac{x}{2\pi}\big)$$

is the 2π*-periodic linear trend* of f. For $m = 2$ we preserve the 2π*-periodic quadratic trend*:

$$(T_2 f)(x) = \sum_{\ell=0}^{1} \big(f^{(\ell)}(2\pi - 0) - f^{(\ell)}(0+0)\big)\, b_{\ell+1}\big(\frac{x}{2\pi}\big)\,.$$

□

Remark 9.31 Using the Krylov–Lanczos method, one can eliminate the influence of the Gibbs phenomenon, since the jumps of f are correctly represented by $T_m f + R_m f$. This idea has been widely studied for modified Fourier expansions and the rate of uniform convergence is estimated too (see [22, 407] and the references therein). The same procedure can also be applied to a highly correct computation of Fourier coefficients of a piecewise smooth function f (see [407]). The Krylov–Lanczos method is readily adapted to multivariate Fourier series in [3, 407]. □

Theorem 9.29 can be extended to an arbitrary 2π-periodic, piecewise C^r-smooth function.

Theorem 9.32 *For $r,\, m \in \mathbb{N}$ with $m \le r$, let f be a 2π-periodic, piecewise C^r-smooth function with n distinct nodes $x_j \in [0,\, 2\pi)$, $j = 1, \ldots, n$. Then f can be split into the sum $f = T_{m,n} f + R_m f$ on $\mathbb{R} \setminus \bigcup_{j=1}^{n} (\{x_j\} + 2\pi\, \mathbb{Z})$, where*

$$(T_{m,n} f)(x) := \sum_{j=1}^{n} \sum_{\ell=0}^{m-1} \big(f^{(\ell)}(x_j - 0) - f^{(\ell)}(x_j + 0)\big)\, \frac{(2\pi)^{\ell}}{(\ell+1)!}\, b_{\ell+1}\big(\frac{x - x_j}{2\pi}\big) \tag{9.60}$$

is the 2π-periodic trend of f and where $R_m f \in C^{m-1}(\mathbb{T})$ defined by (9.57) possesses the uniformly convergent Fourier series (9.58).

For a proof, see [26, 124].

Remark 9.33 The Krylov–Lanczos method is also closely related to the reconstruction of a 2π-periodic, piecewise C^r-smooth function f from given Fourier coefficients $c_k(f)$ (see [26, 124]). This recovery is based on the fact that $c_k(f) \approx c_k(T_{m,n} f)$ for large $|k|$, since the contribution of $c_k(R_m f)$ to $c_k(f)$ is negligible for large $|k|$ by the smoothness of $R_m f$. Using

$$c_k\big(b_{\ell+1}(\frac{\cdot - x_j}{2\pi})\big) = \begin{cases} -\frac{(\ell+1)!}{(2\pi \mathrm{i}\, k)^{\ell+1}}\, \mathrm{e}^{-\mathrm{i}\, k x_j} & k \in \mathbb{Z} \setminus \{0\}\,, \\ 0 & k = 0\,, \end{cases}$$

the Fourier coefficients $c_k(T_{m,n} f)$ fulfill the equations:

$$2\pi\, (\mathrm{i}\, k)^m\, c_k(T_{m,n} f) = \sum_{j=1}^{n} \mathrm{e}^{-\mathrm{i} k x_j} \sum_{\ell=0}^{m-1} (\mathrm{i}\, k)^{m-\ell-1}\, \big(f^{(\ell)}(x_j + 0) - f^{(\ell)}(x_j - 0)\big)\,.$$

Hence, the distinct nodes x_j and the associated jump magnitudes can be determined by a Prony-like method (see Sect. 10.2). □

9.4.2 Fourier Extension

Now we describe the second method for accelerating convergence of Fourier series. Let $\varphi \in C^{\infty}(I)$ with $I := [-1,\, 1]$ be given, i.e., φ is infinitely differentiable in $(-1,\, 1)$ and all one-sided limits

$$\varphi^{(j)}(-1 + 0) = \lim_{x \to -1+0} \varphi^{(j)}(x)\,, \quad \varphi^{(j)}(1 - 0) = \lim_{x \to 1-0} \varphi^{(j)}(x)$$

for each $j \in \mathbb{N}_0$ exist and are finite. In general, such a function does not fulfill the property

$$\varphi^{(j)}(-1+0) = \varphi^{(j)}(1-0)\,, \quad j = 0, \ldots, r-1\,,$$

for certain $r \in \mathbb{N}$. If $\varphi(-1+0) \neq \varphi(1-0)$, then the 2-periodic extension of the function φ restricted on $[-1,\ 1)$ is piecewise continuously differentiable with jump discontinuities at odd points. Then by the Gibbs phenomenon (see Theorem 1.42), the partial sums of the 2-periodic Fourier series oscillate near each odd point. Further, we observe a slow decay of the related Fourier coefficients and a slow convergence of the 2-periodic Fourier series.

Remark 9.34 A simple method for accelerating convergence of Fourier series is often used. Let $f : \mathbb{R} \to \mathbb{C}$ be the 4-periodic function defined on $[-1,\ 3)$ by

$$f(x) := \begin{cases} \varphi(x) & x \in [-1,\ 1]\,, \\ \varphi(2-x) & x \in (1,\ 3)\,. \end{cases}$$

Then f is continuous on whole $\mathbb{R}$ and piecewise continuously differentiable on $[-1,\ 3]$. By Theorem 1.34 of Dirichlet–Jordan, the extended function f possesses a uniformly convergent, 4-periodic Fourier series. The drawback of this method is the fact that f is not continuously differentiable in general. □

For fixed $T > 1$, let $\mathscr{T}_n^{(2T)}$ denote the linear span of the $2T$-periodic exponentials:

$$\mathrm{e}^{\mathrm{i}\,k\pi\cdot/T}\,, \quad k = -n, \ldots, n\,.$$

In the *Fourier extension*, one has to approximate a given function $\varphi \in C^\infty[-1,1]$ by a $2T$-periodic trigonometric polynomial of $\mathscr{T}_n^{(2T)}$. This Fourier extension problem was studied in [207] and the references therein. We propose the following *fast Fourier extension*.

In a first step, we approximate the one-sided finite derivatives $\varphi^{(j)}(-1+0)$, $\varphi^{(j)}(1-0)$ for $j = 0, \ldots, r-1$. Since these one-sided derivatives are very often unknown, we compute these values by interpolation at Chebyshev extreme points. Let $N \in \mathbb{N}$ be a sufficiently large power of two. Using the Chebyshev polynomials (6.1), we interpolate the given function $\varphi \in C^\infty(I)$ at the Chebyshev extreme points $x_j^{(N)} = \cos\frac{j\pi}{N}$, $j = 0, \ldots, N$. Then the interpolation polynomial $\psi \in \mathscr{P}_N$ can be expressed as

$$\psi = \frac{1}{2}\,c_0 + \sum_{k=1}^{N-1} c_k\,T_k + \frac{1}{2}\,c_N\,T_N \tag{9.61}$$

with the coefficients

$$c_k = \frac{2}{N}\Big(\frac{1}{2}\,\varphi(1) + \sum_{j=1}^{N-1} \varphi(x_j^{(N)})\,\cos\frac{jk\pi}{N} + \frac{1}{2}\,\varphi(-1)\Big), \quad k = 0, \ldots, N.$$

Since N is a power of two, the coefficients c_k can be calculated by a fast algorithm of DCT-I $(N+1)$ (by means of Algorithm 6.28 or 6.35). Then we set $c_N := 2\,c_N$. By Theorem 6.26, the coefficients d_k of the derivative

$$\psi' = \frac{1}{2}\,d_0 + \sum_{k=1}^{N-1} d_k\,T_k$$

can be recursively determined as

$$d_{N-1-k} := d_{N+1-k} + 2\,(N-k)\,c_{N-k}, \quad k = 0, \ldots, N-1,$$

with $d_{N+1} = d_N := 0$. Then

$$\psi'(1-0) = \frac{1}{2}\,d_0 + \sum_{k=1}^{N-1} d_k, \quad \psi'(-1+0) = \frac{1}{2}\,d_0 + \sum_{k=1}^{N-1} (-1)^k\,d_k$$

are approximate one-sided derivatives of φ at the endpoints ± 1. Analogously, one can calculate the higher-order one-sided derivatives $\psi^{(j)}$, $j = 2, \ldots, r-1$, at the endpoints ± 1.

In the second step, we use two-point Taylor interpolation, and we compute the unique polynomial $p \in \mathscr{P}_{2r-1}$ which fulfills the interpolation conditions:

$$p^{(j)}(1) = \psi^{(j)}(1-0), \quad p^{(j)}(2T-1) = \psi^{(j)}(-1+0), \quad j = 0, \ldots, r-1.$$

Now we describe briefly the two-point Taylor interpolation. Let $a, b \in \mathbb{R}$ be the given distinct points. Further, let $a_j, b_j \in \mathbb{R}$, $j = 0, \ldots, r-1$, be the given values for fixed $r \in \mathbb{N}$. We consider the special *Hermite interpolation problem* :

$$p^{(j)}(a) = a_j, \quad p^{(j)}(b) = b_j, \quad j = 0, \ldots, r-1, \tag{9.62}$$

for a polynomial $p \in \mathscr{P}_{2r-1}$. Then p is called *two-point Taylor interpolation polynomial* (see [272, pp. 62–67]).

Lemma 9.35 *The two-point Taylor interpolation polynomial of* (9.62) *is uniquely determined and can be expressed as*

$$p = \sum_{j=0}^{r-1} \big(a_j\,h_j^{(a,b,r)} + b_j\,h_j^{(b,a,r)}\big) \in \mathscr{P}_{2r-1}, \tag{9.63}$$

where

$$h_j^{(a,b,r)}(x) := \frac{(x-a)^j}{j!}\Big(\frac{x-b}{a-b}\Big)^r \sum_{k=0}^{r-1-j}\binom{r-1+k}{k}\Big(\frac{x-a}{b-a}\Big)^k\,, \quad j=0,\ldots,r-1\,,$$

denote the two-point Taylor basis polynomials *which fulfill the conditions:*

$$\Big(\frac{\mathrm{d}^\ell}{\mathrm{d}x^\ell}h_j^{(a,b,r)}\Big)(a) = \delta_{j-\ell}\,, \quad \Big(\frac{\mathrm{d}^\ell}{\mathrm{d}x^\ell}h_j^{(a,b,r)}\Big)(b) = 0\,, \quad j\,,\ell = 0,\ldots,r-1\,. \tag{9.64}$$

Proof First, we show the uniqueness of the two-point Taylor interpolation polynomial. Assume that $q \in \mathcal{P}_{2r-1}$ is another solution of the two-point Taylor interpolation problem (9.62). Then $p - q \in \mathcal{P}_{2r-1}$ has two distinct zeros of order r. By the fundamental theorem of algebra, this implies that $p = q$.

From the structure of $h_j^{(a,b,r)}$, it follows immediately that

$$\Big(\frac{\mathrm{d}^\ell}{\mathrm{d}x^\ell}h_j^{(a,b,r)}\Big)(b) = 0\,, \quad j\,,\ell = 0,\ldots,r-1\,.$$

By simple, but long, calculations, one can show that

$$\Big(\frac{\mathrm{d}^\ell}{\mathrm{d}x^\ell}h_j^{(a,b,r)}\Big)(a) = \delta_{j-\ell}\,, \quad j\,,\ell = 0,\ldots,r-1\,.$$

By (9.64) the polynomial (9.63) satisfies the interpolation conditions (9.62). ■

Example 9.36 For $r = 1$ we obtain that

$$h_0^{(a,b,1)}(x) = \frac{x-b}{a-b}$$

and hence

$$p(x) = a_0\,\frac{x-b}{a-b} + b_0\,\frac{x-a}{b-a} \in \mathcal{P}_1\,.$$

In the case $r = 2$, we receive the two-point Taylor basis polynomials

$$h_0^{(a,b,2)}(x) = \Big(\frac{x-b}{a-b}\Big)^2\Big(1 + 2\,\frac{x-a}{b-a}\Big)\,, \quad h_1^{(a,b,2)}(x) = \Big(\frac{x-b}{a-b}\Big)^2(x-a)$$

such that the two-point Taylor interpolation polynomial reads as follows:

$$p(x) = a_0 \left(\frac{x-b}{a-b}\right)^2 \left(1 + 2\,\frac{x-a}{b-a}\right) + a_1 \left(\frac{x-b}{a-b}\right)^2 (x-a)$$
$$+ b_0 \left(\frac{x-a}{b-a}\right)^2 \left(1 + 2\,\frac{x-b}{a-b}\right) + b_1 \left(\frac{x-a}{b-a}\right)^2 (x-b) \in \mathcal{P}_3\,.$$

□

We apply the two-point Taylor interpolation in the case $a := 1$, $b := 2T - 1$ with certain $T > 1$, and

$$a_j := \psi^{(j)}(1-0)\,, \quad b_j := \psi^{(j)}(-1+0)\,, \quad j = 0, \ldots, r-1\,.$$

Then we introduce the $2T$-periodic extension of the function ψ as

$$f(x) := \begin{cases} \psi(x) & x \in [-1,\,1]\,, \\ p(x) & x \in (1,\,2T-1)\,. \end{cases} \tag{9.65}$$

Obviously, the $2T$-periodic function f is contained in $C^{r-1}(\mathbb{R})$. For $r > 1$, f can be expressed by a rapidly convergent Fourier series.

In the third step, we choose $n \geq N$ as a power of two and use $2T$-periodic trigonometric interpolation at equidistant nodes $y_\ell := -1 + \frac{T\ell}{n}$, $\ell = 0, \ldots, 2n-1$, (see Lemma 3.7) in order to approximate f by a trigonometric polynomial

$$s_n(x) = \sum_{k=1-n}^{n-1} s_k^{(n)}\, \mathrm{e}^{\mathrm{i}k\pi\, x/T} + s_n^{(n)} \cos\frac{n\pi\, x}{T} \in \mathcal{T}_n^{(2T)}$$

with the coefficients

$$s_k^{(n)} := \frac{1}{2n} \sum_{\ell=0}^{2n-1} f(y_\ell)\, \mathrm{e}^{-\pi \mathrm{i} k\ell/n}\,, \quad k = 1-n \ldots, n\,.$$

We summarize this method:

Algorithm 9.37 (Fast Fourier extension)
Input: $N \in \mathbb{N} \setminus \{1\}$, $n \in \mathbb{N}$ *power of two with* $n \geq N$, $r \in \mathbb{N}$, $T > 1$, $\varphi \in C^\infty(I)$.

1. *By interpolation at Chebyshev extreme points* $x_j^{(N)}$, $j = 0, \ldots, N$, *determine the interpolation polynomial* $\psi \in \mathcal{P}_N$ *in the form (9.61) by a fast algorithm of* DCT-I $(N+1)$, *set* $c_N := 2\,c_N$, *and calculate the one-sided derivatives* $\psi^{(j)}(-1+0)$ *and* $\psi^{(j)}(1-0)$ *for* $j = 0, \ldots, r-1$ *recursively by Theorem 6.26.*
2. *Using two-point Taylor interpolation with the interpolation conditions*

$$p^{(j)}(1) = \psi^{(j)}(1-0)\,, \quad p^{(j)}(2T-1) = \psi^{(j)}(-1+0)\,, \quad j = 0, \ldots, r-1\,,$$

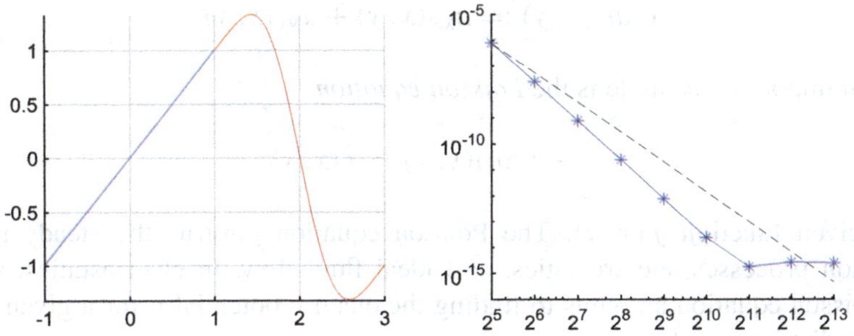

Fig. 9.7 Left: $\varphi(x) = x$ for $x \in I := [-1, 1]$ in blue and the two-point Taylor interpolation polynomial $p \in \mathscr{P}_9$ on $[1, 3]$ in red. Right: Maximum error $\|\varphi - s_n\|_{C(I)}$. The observed maximum error (with oversampling of 10) is in blue. The black dotted line illustrates the theoretical convergence rate $\mathscr{O}(\frac{\log n}{n^4})$

calculate the interpolation polynomial $p \in \mathscr{P}_{2r-1}$ *defined on the interval* $[1, 2T - 1]$.

3. *Form the* $2T$*-periodic function* (9.65) *and use* $2T$*-periodic trigonometric interpolation at equidistant nodes* $y_\ell = -1 + \frac{T\ell}{n}$, $\ell = 0, \ldots, 2n - 1$. *Compute the trigonometric polynomial* $s_n \in \mathscr{T}_n^{(2T)}$ *with the coefficients* $s_k^{(n)}$, $k = 1-n, \ldots, n$, *by a fast algorithm of* DFT$(2n)$.

Output: $s_n \in \mathscr{T}_n^{(2T)}$ *with the coefficients* $s_k^{(n)}$, $k = 1 - n, \ldots, n$.
Computational cost: $\mathscr{O}(n \log n)$.

If we apply fast algorithms of DCT-I$(N + 1)$ and DFT$(2n)$, then Algorithm 9.37 requires only $\mathscr{O}(n \log n)$ arithmetic operations for the computation of $2n$ coefficients $s_k^{(n)}$. The following example shows the performance of this fast Fourier extension.

Example 9.38 We consider the smooth function $\varphi(x) := x$ for $x \in I$. Note that the 2-periodic extension of φ is a piecewise linear sawtooth function. Choosing $T = 2$, $N = 10$, and $r = 5$, the constructed 4-periodic Fourier extension f is contained in $C^4(\mathbb{R})$. For $n = 2^t$, $t = 5, \ldots, 13$, we measure the error (Fig. 9.7):

$$\|\varphi - s_n\|_{C(I)} \approx \mathscr{O}\left(\frac{\log n}{n^4}\right).$$

□

9.5 Fast Poisson Solvers

Numerous problems of mathematical physics can be described by partial differential equations. Here we consider only elliptic partial differential equations for a bivariate function $u(x, y)$. If Δ denotes the *Laplace operator* or *Laplacian*

$$(\Delta u)(x,\, y) := u_{xx}(x,\, y) + u_{yy}(x,\, y)\,,$$

then an important example is the *Poisson equation*

$$-\,(\Delta u)(x,\, y) = f(x,\, y)$$

with given function $f(x, y)$. The Poisson equation governs the steady state in diffusion processes, electrostatics, and ideal fluid flow. In electrostatics, solving the Poisson equation amounts to finding the electric potential u for a given charge density distribution f.

In the following, we want to solve the Dirichlet boundary value problem of the Poisson equation in the open unit square $Q := (0,\, 1)^2$. Thus, we seek a function $u \in C^2(Q) \cap C(\bar{Q})$ with

$$\begin{aligned} -\,(\Delta u)(x,\, y) &= f(x,\, y)\,, \quad &(x,\, y) \in Q\,,\\ u(x,\, y) &= \varphi(x,\, y)\,, \quad &(x,\, y) \in \partial Q := \bar{Q} \setminus Q\,, \end{aligned} \tag{9.66}$$

where $f \in C(\bar{Q})$ and $\varphi \in C(\partial Q)$ are given functions.

We start our considerations with the one-dimensional boundary value problem

$$-\,u''(x) = f(x)\,, \quad x \in (0,\, 1) \tag{9.67}$$

with the boundary conditions $u(0) = \alpha$, $u(1) = \beta$ for given $f \in C[0,\, 1]$ and α, $\beta \in \mathbb{R}$. For $N \in \mathbb{N} \setminus \{1\}$, we form the uniform grid $\{x_j : j = 0, \ldots, N+1\}$ with the grid points $x_j := j\,h$ and the grid width $h := (N+1)^{-1}$. Instead of (9.67), we consider the discretization

$$-\,u''(x_j) = f(x_j)\,, \quad j = 1, \ldots, N \tag{9.68}$$

with $u(0) = \alpha$, $u(1) = \beta$. Setting $f_k := f(x_k)$ and $u_k := u(x_k)$ for $k = 0, \ldots, N+1$, we approximate $u''(x_j)$, $j = 1, \ldots, N$, by the *central difference quotient of second order* :

$$\frac{1}{h^2}\,\big(u_{j-1} - 2\,u_j + u_{j+1}\big)\,.$$

Then the corresponding discretization error can be estimated as follows:

Lemma 9.39 *Let $u \in C^4[0,\, 1]$ be given. Then for each interior grid point x_j, $j = 1, \ldots, N$, we have*

$$\Big|u''(x_j) - \frac{1}{h^2}\,\big(u_{j-1} - 2\,u_j + u_{j+1}\big)\Big| \le \frac{h^2}{12}\,\|u^{(4)}\|_{C[0,1]}\,. \tag{9.69}$$

Proof By Taylor's formula, there exist $\xi_j,\, \eta_j \in (0,\, 1)$ such that

$$u_{j-1} = u(x_j - h) = u_j - h\,u'(x_j) - \frac{h^2}{2}\,u''(x_j) - \frac{h^3}{6}\,u^{(3)}(x_j) + \frac{h^4}{24}\,u^{(4)}(x_j - \xi_j h)\,,$$

$$u_{j+1} = u(x_j + h) = u_j + h\,u'(x_j) + \frac{h^2}{2}\,u''(x_j) + \frac{h^3}{6}\,u^{(3)}(x_j) + \frac{h^4}{24}\,u^{(4)}(x_j + \eta_j h)\,.$$

Using the intermediate value theorem of $u^{(4)} \in C[0,\, 1]$, we obtain

$$\frac{1}{2}\,\big(u^{(4)}(x_j - \xi_j h) + u^{(4)}(x_j + \eta_j h)\big) = u^{(4)}(x_j + \theta_j h)$$

with some $\theta_j \in (-1,\, 1)$. Summation of the above expressions of u_{j-1} and u_{j+1} yields

$$u_{j-1} + u_{j+1} = 2\,u_j + h^2\,u''(x_j) + \frac{h^4}{12}\,u^{(4)}(x_j + \theta_j h)$$

and hence (9.69). ■

If we replace each second derivative $u''(x_j)$ by the corresponding central difference quotient, we get the following system of linear difference equations:

$$-\,u_{j-1} + 2\,u_j - u_{j+1} = h^2\,f_j\,, \quad j = 1, \ldots, N\,,$$

with $u_0 = \alpha$, $u_{N+1} = \beta$. This is the so-called finite difference method. Introducing the vectors $\mathbf{u} := (u_j)_{j=1}^N$ and $\mathbf{g} := \big(h^2\,f_j + \alpha\,\delta_{j-1} + \beta\,\delta_{N-j}\big)_{j=1}^N$ as well as the tridiagonal symmetric matrix

$$\mathbf{A}_N := \begin{pmatrix} 2 & -1 & & & \\ -1 & 2 & -1 & & \\ & \ddots & \ddots & \ddots & \\ & & -1 & 2 & -1 \\ & & & -1 & 2 \end{pmatrix} \in \mathbb{R}^{N\times N}\,,$$

we obtain the linear system:

$$\mathbf{A}_N\,\mathbf{u} = \mathbf{g}\,. \tag{9.70}$$

Obviously, $\mathbf{A}_N$ is weak diagonally dominant. Now we show that $\mathbf{A}_N$ is positive definite and therefore invertible.

Lemma 9.40 *Let $N \in \mathbb{N} \setminus \{1\}$ be given. Then for all $x \in \mathbb{R}$, we have*

$$\mathbf{A}_N\,\mathbf{s}(x) = 4\,\big(\sin\frac{x}{2}\big)^2\,\mathbf{s}(x) + \begin{pmatrix} 0 \\ \vdots \\ 0 \\ \sin(N+1)x \end{pmatrix}, \tag{9.71}$$

$$\mathbf{A}_N\,\mathbf{c}(x) = 4\,\big(\sin\frac{x}{2}\big)^2\,\mathbf{c}(x) + \begin{pmatrix} \cos x \\ 0 \\ \vdots \\ 0 \\ \cos(Nx) \end{pmatrix}, \tag{9.72}$$

with the vectors $\mathbf{s}(x) := \big(\sin(jx)\big)_{j=1}^{N}$ *and* $\mathbf{c}(x) := \big(\cos(kx)\big)_{k=0}^{N-1}$.

Proof Let $x \in \mathbb{R}$ and $j \in \{1, \ldots, N\}$ be given. From

$$\begin{aligned} \sin(j-1)x &= (\cos x)\,\sin(jx) - (\sin x)\,\cos(jx)\,, \\ \sin(j+1)x &= (\cos x)\,\sin(jx) + (\sin x)\,\cos(jx) \end{aligned}$$

it follows that

$$-\sin(j-1)x + 2\,(\cos x)\,\sin(jx) - \sin(j+1)x = 0$$

and hence

$$\begin{aligned} -\sin(j-1)x + 2\,\sin(jx) - \sin(j+1)x &= 2\,(1-\cos x)\,\sin(jx) \\ &= 4\,\big(\sin\frac{x}{2}\big)^2\,\sin(jx)\,. \end{aligned} \tag{9.73}$$

Especially for $j = 1$ and $j = N$, we obtain

$$\begin{aligned} 2\,\sin x - \sin(2x) &= 4\,\big(\sin\frac{x}{2}\big)^2\,\sin x\,, \\ -\sin(N-1)x + 2\,\sin(Nx) &= 4\,\big(\sin\frac{x}{2}\big)^2\,\sin(Nx) + \sin(N+1)x\,. \end{aligned}$$

Thus, (9.73) indicates (9.71). Analogously, (9.72) can be shown. ∎

Lemma 9.41 *For* $N \in \mathbb{N} \setminus \{1\}$, *the tridiagonal matrix* $\mathbf{A}_N$ *is positive definite and possesses the simple positive eigenvalues*

$$\sigma_j := 4\,\big(\sin\frac{j\pi}{2\,(N+1)}\big)^2\,, \quad j = 1, \ldots, N\,, \tag{9.74}$$

which are ordered in the form $0 < \sigma_1 < \ldots < \sigma_N < 4$. *An eigenvector related to* σ_j *is*

$$\mathbf{s}_j := \mathbf{s}\big(\frac{j\pi}{N+1}\big)\,, \quad j = 1, \ldots, N\,.$$

Proof For $x = \frac{j\pi}{N+1}$, $j = 1, \ldots, N$, it follows from (9.71) that

$$\mathbf{A}_N\,\mathbf{s}_j = \sigma_j\,\mathbf{s}_j\,.$$

This completes the proof. ∎

Since $\mathbf{A}_N \in \mathbb{R}^{N\times N}$ is symmetric and since the eigenvalues of $\mathbf{A}_N$ are simple, the eigenvectors $\mathbf{s}_j \in \mathbb{R}^N$ are orthogonal. By

$$\sum_{k=1}^{N} \big(\sin\,(kx)\big)^2 = \frac{N}{2} - \frac{\big(\cos\,(N+1)x\big)\,\sin\,(Nx)}{2\,\sin x}\,, \qquad x \in \mathbb{R}\setminus \pi\,\mathbb{Z}\,,$$

we obtain for $x = \frac{j\pi}{N+1}$ the equation

$$\mathbf{s}_j^\top\,\mathbf{s}_j = \sum_{k=1}^{N} \big(\sin\frac{jk\pi}{N+1}\big)^2 = \frac{N+1}{2}\,, \qquad j = 1, \ldots, N\,,$$

such that

$$\mathbf{s}_j^\top\,\mathbf{s}_k = \frac{N+1}{2}\,\delta_{j-k}\,, \quad j,\,k = 1, \ldots, N\,, \tag{9.75}$$

where δ_j denotes the Kronecker symbol. With the eigenvectors $\mathbf{s}_j$, $j = 1, \ldots, N$, we form the orthogonal *sine matrix of type* I :

$$\mathbf{S}_N^{\mathrm{I}} := \sqrt{\frac{2}{N+1}}\,\big(\mathbf{s}_1\,|\,\mathbf{s}_2\,|\,\ldots\,|\,\mathbf{s}_N\big) = \sqrt{\frac{2}{N+1}}\,\big(\sin\frac{jk\pi}{N+1}\big)_{j,k=1}^{N} \in \mathbb{R}^{N\times N}\,.$$

This matrix is symmetric and has by (9.75) the property

$$\big(\mathbf{S}_N^{\mathrm{I}}\big)^2 = \mathbf{I}_N \tag{9.76}$$

such that $\big(\mathbf{S}_N^{\mathrm{I}}\big)^{-1} = \mathbf{S}_N^{\mathrm{I}}$. We summarize:

Lemma 9.42 *For* $N \in \mathbb{N}\setminus\{1\}$, *the tridiagonal matrix* $\mathbf{A}_N$ *can be diagonalized by the sine matrix* $\mathbf{S}_N^{\mathrm{I}}$ *of type* I *in the form*

$$\mathbf{S}_N^{\mathrm{I}}\,\mathbf{A}_N\,\mathbf{S}_N^{\mathrm{I}} = \mathbf{D}_N$$

with the invertible diagonal matrix $\mathbf{D}_N := \mathrm{diag}\,(\sigma_j)_{j=1}^N$.

Applying Lemma 9.42 to the linear system (9.70), we obtain:

Theorem 9.43 *The finite difference method of the boundary value problem* (9.67) *leads to the linear system* (9.70) *which has the unique solution:*

$$\mathbf{u} = \mathbf{S}_N^{\mathrm{I}}\, \mathbf{D}_N^{-1}\, \mathbf{S}_N^{\mathrm{I}}\, \mathbf{g}\,. \tag{9.77}$$

Remark 9.44 For a real input vector $\mathbf{f} = (f_j)_{j=1}^N \in \mathbb{R}^N$, the DST–I of length N can be computed by DFT($2N+2$) (see also Table 6.1 for a similar realization). We want to compute the vector $\hat{\mathbf{f}} = \big(\hat{f}_k\big)_{k=1}^N = \mathbf{S}_N^{\mathrm{I}}\,\mathbf{f}$, i.e.,

$$\hat{f}_k = \sqrt{\frac{2}{N+1}}\, \sum_{j=1}^{N} f_j\, \sin\frac{jk\pi}{N+1}\,, \qquad k = 1, \ldots, N\,.$$

For this purpose, we form the odd vector $\mathbf{a} = (a_j)_{j=0}^{2N-1} \in \mathbb{R}^{2N}$ by

$$a_j := \begin{cases} 0 & j = 0,\ N+1\,, \\ f_j & j = 1, \ldots, N\,, \\ -f_{2N+2-j} & j = N+2, \ldots, 2N-1 \end{cases}$$

and calculate $\hat{\mathbf{a}} = \big(\hat{a}_k\big)_{k=0}^{2N+1} = \mathbf{F}_{2N+2}\,\mathbf{a}$. Simple calculation shows that $\mathrm{Re}\ \hat{a}_k = 0$, $k = 1, \ldots, N$, and

$$\hat{f}_k = -\frac{1}{\sqrt{2N+2}}\ \mathrm{Im}\ \hat{a}_k\,, \quad k = 1, \ldots, N\,.$$

Remark 9.45 The linear system (9.70) can be solved in $\mathcal{O}(N)$ arithmetic operations using the Cholesky factorization of the tridiagonal matrix $\mathbf{A}_N$. Otherwise, the solution $\mathbf{u}$ of (9.70) can be calculated by two DST–I of length N and scaling. The DST–I of length N can be realized by FFT of length $2N+2$. Thus, we need $\mathcal{O}(N \log N)$ arithmetic operations for the computation of (9.77).

A similar approach can be used for other boundary value problems of (9.67) (see [421, pp. 247–253] and [404]). □

After these preliminaries, we present a numerical solution of the boundary value problem of the Poisson equation (9.66) in the open unit square $Q = (0,\ 1)^2$.

Example 9.46 The Poisson equation $-(\Delta u)(x, y) = x^2 + y^2$ for $(x, y) \in Q$ with the boundary conditions $u(x, 0) = 0$, $u(x, 1) = -\frac{1}{2}x^2$, $u(0, y) = \sin y$, $u(1, y) = \mathrm{e}\,\sin y - \frac{1}{2}\,y^2$ for $x,\ y \in [0,\ 1]$ has the solution $u(x, y) = \mathrm{e}^x\,\sin y - \frac{1}{2}\,x^2\,y^2$. □

We use a uniform grid of $\bar{Q}$ with the grid points $(x_j,\ y_k)$, $j,\ k = 0, \ldots, N+1$, where $x_j := j\,h$, $y_k := k\,h$, and $h := (N+1)^{-1}$. In the case $j,\ k \in \{1, \ldots, N\}$,

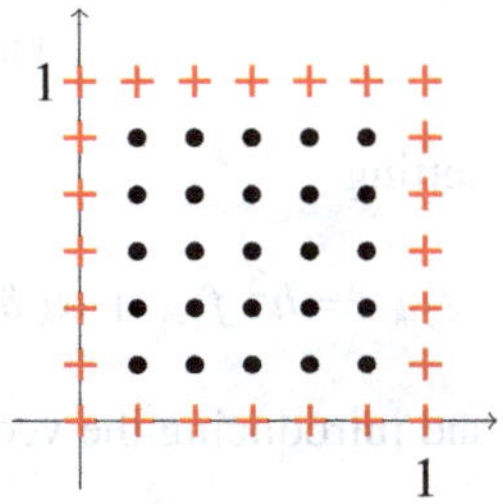

Fig. 9.8 Grid points and interior grid points in the unit square

we say that $(x_j, y_k) \in Q$ is an *interior grid point* (Fig. 9.8). Now we discretize the problem (9.66) and consider

$$\begin{aligned} &-(\Delta u)(x_j, y_k) = f(x_j, y_k)\,, \qquad j,\,k = 1,\ldots,N\,,\\ &u(0, y_k) = \alpha_k := \varphi(0, y_k)\,, \quad u(1, y_k) = \beta_k := \varphi(1, y_k)\,, \quad k = 0,\ldots,N+1\,,\\ &u(x_j, 0) = \mu_j := \varphi(x_j, 0)\,, \quad u(x_j, 1) = \nu_j := \varphi(x_j, 1)\,, \quad j,\,k = 0,\ldots,N+1\,. \end{aligned}$$

Setting $u_{j,k} := u(x_j, y_k)$ and $f_{j,k} := f(x_j, y_k)$, we approximate $(\Delta u)(x_j, y_k)$ at each interior grid point (x_j, y_k) by the *discrete Laplacian* :

$$\begin{aligned} (\Delta_h u)(x_j, y_k) &:= \frac{1}{h^2}\big(u_{j-1,k} + u_{j+1,k} - 2\,u_{j,k}\big) + \frac{1}{h^2}\big(u_{j,k-1} + u_{j,k+1} - 2\,u_{j,k}\big)\\ &= \frac{1}{h^2}\big(u_{j-1,k} + u_{j+1,k} + u_{j,k-1} + u_{j,k+1} - 4\,u_{j,k}\big) \end{aligned} \tag{9.78}$$

Obviously, the discrete Laplacian is the sum of two central partial difference quotients of second order.

Lemma 9.47 *If $u \in C^4(\bar{Q})$, then for each interior grid point (x_j, y_k), we have the estimate:*

$$\big|(\Delta u)(x_j, y_k) - (\Delta_h u)(x_j, y_k)\big| \le \frac{h^2}{12}\Big(\big\|\frac{\partial^4 u}{\partial x^4}\big\|_{C(\bar{Q})} + \big\|\frac{\partial^4 u}{\partial y^4}\big\|_{C(\bar{Q})}\Big)\,.$$

Similarly to the proof of Lemma 9.39, the above estimate can be shown by using two-dimensional Taylor's formula and intermediate value theorem. For shortness, the proof is omitted here.

If we replace the Laplacian $(\Delta u)(x_j, y_k)$ by the discrete Laplacian (9.78), we obtain the following equation at each interior grid point:

$$4\,u_{j,k} - u_{j-1,k} - u_{j+1,k} - u_{j,k-1} - u_{j,k+1} = h^2\,f_{j,k}\,, \quad j,\,k = 1,\ldots,N\,, \tag{9.79}$$

where we include the boundary conditions. For the interior grid point (x_1, y_1), this means that

$$4\,u_{1,1} - u_{2,1} - u_{1,2} = h^2\,f_{1,1} + \alpha_1 + \mu_1\,.$$

Setting

$$g_{j,k} := h^2\,f_{j,k} + \alpha_k\,\delta_{j-1} + \beta_k\,\delta_{N-j} + \mu_j\,\delta_{k-1} + \nu_j\,\delta_{N-k}\,, \qquad j,\,k = 1,\ldots,N\,,$$

and introducing the vectors

$$\mathbf{u} := \big(u_{1,1},\ldots,u_{1,N},\,\ldots\,,u_{N,1},\ldots,u_{N,N}\big)^\top \in \mathbb{R}^{N^2}\,,$$
$$\mathbf{g} := \big(g_{1,1},\ldots,g_{1,N},\,\ldots\,,g_{N,1},\ldots,g_{N,N}\big)^\top \in \mathbb{R}^{N^2}\,,$$

and the Kronecker sum

$$\mathbf{M}_{N^2} := \big(\mathbf{A}_N \otimes \mathbf{I}_N\big) + \big(\mathbf{I}_N \otimes \mathbf{A}_N\big) \in \mathbb{R}^{N^2\times N^2}\,,$$

where $\otimes$ denotes the Kronecker product of matrices (see Sect. 3.4), from (9.79) we obtain the linear system:

$$\mathbf{M}_{N^2}\,\mathbf{u} = \mathbf{g}\,. \tag{9.80}$$

Then $\mathbf{M}_{N^2}$ is a symmetric, weak diagonally dominant band matrix with bandwidth $2N-1$, where at most five nonzero entries are in each row. The *fast Poisson solver* is mainly based on the diagonalization of $\mathbf{M}_{N^2}$ by the Kronecker product $\mathbf{S}_N^{\mathrm{I}} \otimes \mathbf{S}_N^{\mathrm{I}}$.

Lemma 9.48 *The Kronecker sum $\mathbf{M}_{N^2}$ can be diagonalized by the Kronecker product $\mathbf{S}_N^{\mathrm{I}} \otimes \mathbf{S}_N^{\mathrm{I}}$, i.e.,*

$$\big(\mathbf{S}_N^{\mathrm{I}} \otimes \mathbf{S}_N^{\mathrm{I}}\big)\,\mathbf{M}_{N^2}\,\big(\mathbf{S}_N^{\mathrm{I}} \otimes \mathbf{S}_N^{\mathrm{I}}\big) = \Big(\big(\mathbf{D}_N \otimes \mathbf{I}_N\big) + \big(\mathbf{I}_N \otimes \mathbf{D}_N\big)\Big)\,, \tag{9.81}$$

where

$$\big(\mathbf{D}_N \otimes \mathbf{I}_N\big) + \big(\mathbf{I}_N \otimes \mathbf{D}_N\big) = \mathrm{diag}\,\big(\mu_{1,1},\ldots,\mu_{1,N},\,\ldots\,,\mu_{N,1},\ldots,\mu_{N,N}\big)^\top$$

is an invertible diagonal matrix with the main diagonal elements:

$$\mu_{j,k} = 4\,\big(\sin\frac{j\pi}{2\,(N+1)}\big)^2 + 4\,\big(\sin\frac{j\pi}{2\,(N+1)}\big)^2\,, \qquad j,\,k = 1,\ldots,N\,.$$

Proof Using the properties of the Kronecker product (see Theorem 3.42) and Lemma 9.42, we obtain by (9.76) that

$$\big(\mathbf{S}_N^{\mathrm{I}} \otimes \mathbf{S}_N^{\mathrm{I}}\big)\,\big(\mathbf{A}_N \otimes \mathbf{I}_N\big)\,\big(\mathbf{S}_N^{\mathrm{I}} \otimes \mathbf{S}_N^{\mathrm{I}}\big) = \big(\mathbf{S}_N^{\mathrm{I}}\,\mathbf{A}_N\,\mathbf{S}_N^{\mathrm{I}}\big) \otimes \mathbf{I}_N = \mathbf{D}_N \otimes \mathbf{I}_N\,.$$

Analogously, we see that

$$\big(\mathbf{S}_N^{\mathrm{I}} \otimes \mathbf{S}_N^{\mathrm{I}}\big)\,\big(\mathbf{I}_N \otimes \mathbf{A}_N\big)\,\big(\mathbf{S}_N^{\mathrm{I}} \otimes \mathbf{S}_N^{\mathrm{I}}\big) = \mathbf{I}_N \otimes \mathbf{D}_N\,.$$

Hence, the formula (9.81) is shown. By Lemma 9.42 we know that

$$\mathbf{D}_N = \mathrm{diag}\,\big(\sigma_j\big)_{j=1}^N\,.$$

From (9.74) we conclude that

$$\mu_{j,k} = \sigma_j + \sigma_k = 4\,\big(\sin\frac{j\pi}{2\,(N+1)}\big)^2 + 4\,\big(\sin\frac{k\pi}{2\,(N+1)}\big)^2\,, \qquad j,\,k = 1,\ldots,N\,.$$

Obviously, all $\mu_{j,k}$ are positive and fulfill $0 < \mu_{1,1} \le \mu_{j,k} < 8$. ■

By Lemma 9.48, the matrix $\mathbf{M}_{N^2}$ is invertible and its inverse reads by (9.81) as follows:

$$\mathbf{M}_{N^2}^{-1} = \big(\mathbf{S}_N^{\mathrm{I}} \otimes \mathbf{S}_N^{\mathrm{I}}\big)\;\mathrm{diag}\,\big(\mu_{1,1}^{-1},\,\ldots\,,\mu_{N,N}^{-1}\big)^\top\,\big(\mathbf{S}_N^{\mathrm{I}} \otimes \mathbf{S}_N^{\mathrm{I}}\big)\,. \tag{9.82}$$

Thus, the linear system (9.80) is uniquely solvable.

Theorem 9.49 *The finite difference method of the boundary value problem* (9.66) *leads to the linear system* (9.80) *which has the unique solution:*

$$\mathbf{u} = \mathbf{M}_{N^2}^{-1}\,\mathbf{g} = \big(\mathbf{S}_N^{\mathrm{I}} \otimes \mathbf{S}_N^{\mathrm{I}}\big)\;\mathrm{diag}\,\big(\mu_{1,1}^{-1},\,\ldots\,,\mu_{N,N}^{-1}\big)^\top\,\big(\mathbf{S}_N^{\mathrm{I}} \otimes \mathbf{S}_N^{\mathrm{I}}\big)\,\mathbf{g}\,. \tag{9.83}$$

Remark 9.50 The vector $\mathbf{u}$ can be computed by two transforms with $\mathbf{S}_N^{\mathrm{I}} \otimes \mathbf{S}_N^{\mathrm{I}}$ and scaling. We consider the transform:

$$\mathbf{h} := \big(\mathbf{S}_N^{\mathrm{I}} \otimes \mathbf{S}_N^{\mathrm{I}}\big)\,\mathbf{g} = \big(h_{1,1},\ldots,h_{1,N},\,\ldots\,,h_{N,1},\ldots,h_{N,N}\big)^\top \in \mathbb{R}^{N^2}\,.$$

Then we receive by the definition of the Kronecker product $\mathbf{S}_N^{\mathrm{I}} \otimes \mathbf{S}_N^{\mathrm{I}}$ in Sect. 3.4

$$h_{m,n} = \frac{2}{N+1}\sum_{j=1}^N\sum_{k=1}^N g_{j,k}\,\sin\frac{jm\pi}{N+1}\,\sin\frac{kn\pi}{N+1}\,, \qquad m,\,n = 1,\ldots,N\,,$$

i.e., the matrix $\big(h_{m,n}\big)_{m,n=1}^N$ is equal to the two-dimensional DST–I with size $N \times N$ which can be realized by a fast algorithm of the two-dimensional DFT of size $(2N+2) \times (2N+2)$, if $N+1$ is a power of two. Thus, the computation requires only $\mathcal{O}(N^2 \log N)$ arithmetic operations. This is now the fastest way to solve the linear system (9.80). □

We summarize:

Algorithm 9.51 (Fast Poisson Solver)
Input: $N \in \mathbb{N} \setminus \{1\}$, $N+1$ *power of two*, $h = (N+1)^{-1}$,
$x_j = j\,h$, $y_k = k\,h$, $f_{j,k} := f(x_j, y_k)$ *for* $j,\, k = 1, \ldots, N$,
$\alpha_k := \varphi(0, y_k)$, $\beta_k := \varphi(1, y_k)$ *for* $k = 1, \ldots, N$,
$\mu_j := \varphi(x_j, 0)$, $\nu_j := \varphi(x_j, 1)$ *for* $j = 1, \ldots, N$, *where* $f \in C(\bar{Q})$ *and* $\varphi \in C(\partial Q)$.

1. *Precompute the values:*

$$\mu_{m,n}^{-1} := \Big(4\,\big(\sin\frac{m\pi}{2\,(N+1)}\big)^2 + 4\,\big(\sin\frac{n\pi}{2\,(N+1)}\big)^2\Big)^{-1}, \quad m,\, n = 1, \ldots, N\,.$$

2. *Form the values:*

$$g_{j,k} := h^2\, f_{j,k} + \alpha_k\, \delta_{j-1} + \beta_k\, \delta_{N-j} + \mu_j\, \delta_{k-1} + \nu_j\, \delta_{N-k}\,, \quad j,\, k = 1, \ldots, N\,.$$

3. *Using a fast algorithm of the two-dimensional* DST–I *with size* $N \times N$, *compute*

$$\tilde{g}_{m,n} := \frac{2}{N+1} \sum_{j=1}^{N} \sum_{k=1}^{N} g_{j,k}\, \sin\frac{jm\pi}{N+1}\, \sin\frac{kn\pi}{N+1}\,, \quad m,\, n = 1, \ldots, N\,.$$

4. *Calculate*

$$h_{m,n} := \mu_{m,n}^{-1}\, \tilde{g}_{m,n}\,, \quad m,\, n = 1, \ldots, N\,.$$

5. *Using a fast algorithm of the two-dimensional* DST–I *with size* $N \times N$, *compute*

$$\tilde{u}_{j,k} := \frac{2}{N+1} \sum_{m=1}^{N} \sum_{n=1}^{N} h_{m,n}\, \sin\frac{jm\pi}{N+1}\, \sin\frac{kn\pi}{N+1}\,, \quad j,\, k = 1, \ldots, N\,.$$

Output: $\tilde{u}_{j,k}$ *approximate value of* $u(x_j, y_k)$ *for* $j,\, k = 1, \ldots, N$.
Computational cost: $\mathcal{O}(N^2 \log N)$.

Remark 9.52 This method can be extended to a rectangular domain with different step sizes in x- and y-direction. A similar approach can be used for other boundary value problems of (9.66), see [404] and [421, pp. 247 – 253]. □

Under the assumption $u \in C^4(\bar{Q})$, we present an error analysis for this method. The *local discretization error* at an interior grid point (x_j, y_k), $j,\, k = 1, \ldots, N$, is defined as

$$d_{j,k} := -\big(\Delta_h u\big)(x_j, y_k) - f_{j,k}\,.$$

By Lemma 9.47, the local discretization error $d_{j,k}$ can be estimated by

$$|d_{j,k}| \le c\,h^2\,, \quad j,\,k = 1,\ldots,N\,, \tag{9.84}$$

with the constant

$$c := \frac{1}{12}\Big(\|\frac{\partial^4 u}{\partial x^4}\|_{C(\bar{Q})} + \|\frac{\partial^4 u}{\partial y^4}\|_{C(\bar{Q})}\Big)\,.$$

Now we explore the error

$$e_{j,k} := u(x_j,\,y_k) - \tilde{u}_{j,k}\,, \quad j,\,k = 0,\ldots,N+1\,,$$

where $\tilde{u}_{j,k}$ means the approximate value of $u(x_j,\,y_k)$. By the boundary conditions in (9.66), we have

$$e_{0,k} = e_{N+1,k} = e_{j,0} = e_{j,N+1} = 0\,, \quad j,\,k = 0,\ldots,N+1\,.$$

From the definition of $d_{j,k}$, it follows that

$$-\,h^2\,(\Delta_h u)(x_j,\,y_k) - h^2\,f_{j,k} = h^2\,d_{j,k}\,, \quad j,\,k = 1,\ldots,N\,.$$

Further, we have by (9.79) that

$$4\,\tilde{u}_{j,k} - \tilde{u}_{j-1,k} - \tilde{u}_{j+1,k} - \tilde{u}_{j,k-1} - \tilde{u}_{j,k+1} - h^2\,f_{j,k} = 0\,, \quad j,\,k = 1,\ldots,N\,.$$

Subtracting both equations yields

$$4\,e_{j,k} - e_{j-1,k} - e_{j+1,k} - e_{j,k-1} - e_{j,k+1} = h^2\,d_{j,k}\,, \quad j,\,k = 1,\ldots,N\,.$$

Introducing the error vectors

$$\mathbf{e} := \big(e_{1,1},\ldots,e_{1,N},\,\ldots\,,e_{N,1},\ldots,e_{N,N}\big)^\top \in \mathbb{R}^{N^2}\,,$$
$$\mathbf{d} := \big(d_{1,1},\ldots,d_{1,N},\,\ldots\,,d_{N,1},\ldots,d_{N,N}\big)^\top \in \mathbb{R}^{N^2}\,,$$

we obtain the linear system

$$\mathbf{M}_{N^2}\,\mathbf{e} = h^2\,\mathbf{d}\,, \tag{9.85}$$

which has the unique solution

$$\begin{aligned}\mathbf{e} &= h^2\,\mathbf{M}_{N^2}^{-1}\,\mathbf{d} \\ &= h^2\,\big(\mathbf{S}_N^{\mathrm{I}} \otimes \mathbf{S}_N^{\mathrm{I}}\big)\ \mathrm{diag}\ \big(\mu_{1,1}^{-1},\,\ldots\,,\mu_{N,N}^{-1}\big)^\top\,\big(\mathbf{S}_N^{\mathrm{I}} \otimes \mathbf{S}_N^{\mathrm{I}}\big)\,\mathbf{d}\,.\end{aligned} \tag{9.86}$$

Theorem 9.53 *For a solution $u \in C^4(\bar{Q})$ of the boundary value problem (9.66), the weighted Euclidean norm of the error vector* $\mathbf{e}$ *can be estimated by*

$$\frac{1}{N}\,\|\mathbf{e}\|_2 \le \frac{h^2}{96}\Big(\big\|\frac{\partial^4 u}{\partial x^4}\big\|_{C(\bar{Q})} + \big\|\frac{\partial^4 u}{\partial y^4}\big\|_{C(\bar{Q})}\Big)\,.$$

Proof If $\|\mathbf{M}_{N^2}^{-1}\|_2$ denotes the spectral norm of $\mathbf{M}_{N^2}^{-1}$, we receive by (9.86) that

$$\|\mathbf{e}\|_2 \le h^2\, \|\mathbf{M}_{N^2}^{-1}\|_2\, \|\mathbf{d}\|_2\,.$$

Since $\mathbf{S}_N^{\mathrm{I}}$ is orthogonal and since $0 < \mu_{1,1} \le \mu_{j,k}$ for all $j, k = 1, \ldots, N$, we conclude that

$$\|\mathbf{M}_{N^2}^{-1}\|_2 = \mu_{1,1}^{-1} = \frac{1}{8}\Big(\sin\frac{\pi}{2\,(N+1)}\Big)^{-2} \le \frac{1}{8}\,(N+1)^2 = \frac{1}{8}\,h^{-2}\,.$$

By (9.84) we have $\|\mathbf{d}\|_2 \le N\,c\,h^2$ and thus we obtain that

$$\frac{1}{N}\,\|\mathbf{e}\|_2 \le \frac{c}{8}\,h^2\,.$$

■

Remark 9.54 The fast Poisson solver of Algorithm 9.51 was derived from the finite difference method. A different approach to an efficient numerical solution of (9.66) with homogeneous boundary conditions, i.e., $\varphi = 0$, follows from the *spectral method*, since the functions $\sin(\pi j x)\,\sin(\pi k y)$ for $j, k \in \mathbb{N}$ are eigenfunctions of the eigenvalue problem:

$$-\,(\Delta u)(x, y) = \lambda\, u(x, y) \quad (x, y) \in Q\,,$$
$$u(x, y) = 0\,, \quad (x, y) \in \partial Q\,.$$

Thus, we construct a numerical solution u of (9.66) with $\varphi = 0$ in the form

$$u(x, y) = \sum_{j=1}^{N}\sum_{k=1}^{N} \hat{u}_{j,k}\,\sin(\pi j x)\,\sin(\pi k y)$$

such that

$$-\,(\Delta u)(x, y) = \sum_{j=1}^{N}\sum_{k=1}^{N} \hat{u}_{j,k}\,\pi^2(j^2+k^2)\,\sin(\pi j x)\,\sin(\pi k y)\,. \tag{9.87}$$

Assume that $f \in C(\bar{Q})$ with vanishing values on ∂Q is approximated by

$$\sum_{j=1}^{N} \sum_{k=1}^{N} \hat{f}_{j,k} \, \sin(\pi j x) \, \sin(\pi k y) \, , \tag{9.88}$$

where the coefficients $\hat{f}_{j,k}$ are determined by the interpolation conditions:

$$f(x_m, y_n) = f_{m,n} = \sum_{j=1}^{N} \sum_{k=1}^{N} \hat{f}_{j,k} \, \sin \frac{\pi j m}{N+1} \, \sin \frac{\pi k n}{N+1} \, , \qquad m, n = 1, \ldots, N \, .$$

Using the orthogonal sine matrix $\mathbf{S}_N^{\mathrm{I}} = \sqrt{\frac{2}{N+1}} \big(\sin \frac{j m \pi}{N+1} \big)_{j,m=1}^{N}$, we obtain

$$\big(f_{m,n}\big)_{m,n=1}^{N} = \frac{N+1}{2} \, \mathbf{S}_N^{\mathrm{I}} \, \big(\hat{f}_{j,k}\big)_{j,k=1}^{N} \, \mathbf{S}_N^{\mathrm{I}}$$

and hence

$$\big(\hat{f}_{j,k}\big)_{j,k=1}^{N} = \frac{2}{N+1} \, \mathbf{S}_N^{\mathrm{I}} \, \big(f_{m,n}\big)_{m,n=1}^{N} \, \mathbf{S}_N^{\mathrm{I}} \, .$$

Comparing (9.87) and (9.88), we set

$$\hat{u}_{j,k} := \frac{\hat{f}_{j,k}}{\pi^2 \, (j^2 + k^2)} \, , \qquad j, k = 1, \ldots, N \, .$$

Finally, we obtain the values $u_{m,n}$ at all interior grid points $(x_m, \, y_n)$ as

$$\big(u_{m,n}\big)_{m,n=1}^{N} = \frac{N+1}{2\pi^2} \, \mathbf{S}_N^{\mathrm{I}} \, \Big(\frac{\hat{f}_{j,k}}{j^2 + k^2}\Big)_{j,k=1}^{N} \, \mathbf{S}_N^{\mathrm{I}} \, .$$

See [429] for a discussion of the different eigenvalue solution. □

A similar method based on a pseudospectral Fourier approximation and a polynomial subtraction technique is presented in [14]. Fast Poisson solvers for spectral methods based on the alternating direction implicit method are given in [146].

9.6 Spherical Fourier Transforms

The Fourier analysis on the unit sphere $\mathbb{S}^2 := \{\mathbf{x} \in \mathbb{R}^3 : \|\mathbf{x}\|_2 = 1\}$ has practical significance, for example, in tomography, geophysics, seismology, meteorology, crystallography, and applications to the geometry of convex bodies [186]. Spherical Fourier series are often used for numerical computations. They have similar remarkable properties as the Fourier series on the torus $\mathbb{T}$. Many solution methods used in numerical meteorology for partial differential equations are based on Fourier methods. The major part of the computation time is required by the calculation of the partial sums of spherical Fourier series [60, p. 402]. Numerically stable, fast algorithms for discrete spherical Fourier transforms are therefore of great interest.

Using spherical coordinates, each point $\mathbf{x} \in \mathbb{S}^2$ can be represented as

$$\mathbf{x} = (\sin\theta \, \cos\phi, \, \sin\theta \, \sin\phi, \, \cos\theta)^\top ,$$

where $\theta \in [0, \pi]$ is the *polar angle* and $\phi \in [0, 2\pi)$ is the *azimuthal angle* of $\mathbf{x}$. In other words, θ is the angle between the unit vectors $(0, 0, 1)^\top$ and $\mathbf{x} = (x_1, x_2, x_3)^\top$ and ϕ is the angle between the vectors $(1, 0, 0)^\top$ and $(x_1, x_2, 0)^\top$. Thus, any function $f : \mathbb{S}^2 \to \mathbb{C}$ can be written in the form $f(\theta, \phi)$ with $(\theta, \phi) \in [0, \pi] \times [0, 2\pi)$. By $L_2(\mathbb{S}^2)$ we denote the Hilbert space of square integrable functions f defined on $\mathbb{S}^2$, with

$$\|f\|^2_{L_2(\mathbb{S}^2)} := \frac{1}{4\pi} \int_0^{2\pi} \Big(\int_0^{\pi} |f(\theta, \phi)|^2 \sin\theta \, \mathrm{d}\theta \Big) \mathrm{d}\phi < \infty .$$

The inner product of $f, g \in L_2(\mathbb{S}^2)$ is given by

$$\langle f, g\rangle_{L_2(\mathbb{S}^2)} := \frac{1}{4\pi} \int_0^{2\pi} \Big(\int_0^{\pi} f(\theta, \phi) \, \overline{g(\theta, \phi)} \, \sin\theta \, \mathrm{d}\theta \Big) \mathrm{d}\phi .$$

First, we introduce some special functions. For $k \in \mathbb{N}_0$, the *kth Legendre polynomial* is defined as

$$P_k(x) := \frac{1}{2^k \, k!} \, \frac{\mathrm{d}^k}{\mathrm{d}x^k} (x^2 - 1)^k , \quad x \in [-1, 1] .$$

Further, the *associated Legendre function* P_k^n for $n \in \mathbb{N}_0$ and $k \geq n$ is given by

$$P_k^n(x) := \sqrt{\frac{(k-n)!}{(k+n)!}} \, (1 - x^2)^{n/2} \, \frac{\mathrm{d}^n}{\mathrm{d}x^n} P_k(x) , \quad x \in [-1, 1] . \tag{9.89}$$

Then the *spherical harmonics* Y_k^n of *degree* $k \in \mathbb{N}_0$ and *order* $n \in [-k, k] \cap \mathbb{Z}$ are of the form

$$Y_k^n(\theta, \phi) := P_k^{|n|}(\cos\theta) \, \mathrm{e}^{\mathrm{i}\,n\,\phi} . \tag{9.90}$$

Note that associated Legendre functions and spherical harmonics are not uniformly defined in the literature. The set $\{Y_k^n : k \in \mathbb{N}_0 , \; n \in [-k, k] \cap \mathbb{Z}\}$ forms an orthogonal basis of $L_2(\mathbb{S}^2)$, where we have

$$\langle Y_k^n, Y_\ell^m \rangle_{L_2(\mathbb{S}^2)} = \frac{1}{2k+1}\, \delta_{k-\ell}\, \delta_{n-m}\,.$$

An expansion of $f \in L_2(\mathbb{S}^2)$ into a Fourier series with respect to the orthogonal basis of spherical harmonics is called a *spherical Fourier series*. We say that a function $f \in L_2(\mathbb{S}^2)$ has the *bandwidth* $N \in \mathbb{N}$ if f is equal to the partial sum of the spherical Fourier series:

$$f(\theta, \phi) = \sum_{k=0}^{N} \sum_{n=-k}^{k} a_k^n(f)\, Y_k^n(\theta, \phi)\,. \tag{9.91}$$

The *spherical Fourier coefficients* of f are given by

$$\begin{aligned} a_k^n(f) &:= \langle f, Y_k^n \rangle_{L_2(\mathbb{S}^2)} \big(\langle Y_k^n, Y_k^n \rangle_{L_2(\mathbb{S}^2)}\big)^{-1} \\ &= \frac{2k+1}{4\pi} \int_0^{2\pi} \Big(\int_0^{\pi} f(\theta, \phi)\, \overline{Y_k^n(\theta, \phi)}\, \sin\theta\, \mathrm{d}\theta \Big) \mathrm{d}\phi \end{aligned} \tag{9.92}$$

with respect to the orthogonal basis of the spherical harmonics. Sometimes, the finite expansion in (9.91) is called a *spherical polynomial* of degree N.

We assume that f is given in the form (9.91). We are interested in a fast and numerically stable algorithm for the evaluation of

$$f(\theta_\ell, \phi_\ell) = \sum_{k=0}^{N} \sum_{n=-k}^{k} a_k^n(f)\, Y_k^n(\theta_\ell, \phi_\ell)\,, \quad \ell = 0, \ldots, M-1\,, \tag{9.93}$$

at arbitrary points $(\theta_\ell, \phi_\ell) \in [0, \pi] \times [0, 2\pi)$ with given spherical Fourier coefficients $a_k^n(f) \in \mathbb{C}$. Furthermore, we are interested in the "adjoint problem," in the computation of

$$\check{a}_k^n := \sum_{\ell=0}^{M-1} f_\ell\, \overline{Y_k^n(\theta_\ell, \phi_\ell)}\,, \quad k = 0, \ldots, N;\ n = -k, \ldots, k\,, \tag{9.94}$$

for given data $f_\ell \in \mathbb{C}$.

First, we develop a fast algorithm for the problem (9.93). Then we obtain immediately a fast algorithm for the adjoint problem (9.94) by writing the algorithm in matrix-vector form and forming the conjugate transpose of this matrix product. A fast algorithm for the adjoint problem will be needed for computing the spherical Fourier coefficients (9.92) of the function f by a quadrature rule (see Sect. 9.6.4).

The direct computation of the M function values $f(\theta_\ell, \phi_\ell)$ in (9.93) at arbitrarily distributed points (θ_ℓ, ϕ_ℓ) on the unit sphere $\mathbb{S}^2$ needs $\mathcal{O}(N^2 M)$ arithmetical operations and is denoted by *discrete spherical Fourier transform*, abbreviated as DSFT. For special grids of the form

$$\left(\frac{s\pi}{T}, \frac{t\pi}{T}\right) \in [0, \pi] \times [0, 2\pi), \quad s = 0, \ldots, T; \; t = 0, \ldots, 2T - 1, \tag{9.95}$$

with $N + 1 \le T \le 2^{\lceil \log_2 (N+1) \rceil}$, there exist fast realizations of the DSFT, which we denote by *fast spherical Fourier transforms*, abbreviated as FSFT. The first FSFT has been developed by J. Driscoll and D. Healy [115], where the computational cost has been reduced from $\mathcal{O}(N^3)$ to $\mathcal{O}(N^2 \log^2 N)$ flops. Further, fast algorithms for that purpose can be found in [60, 196, 289, 339, 403].

Remark 9.55 The essential drawback of the grid in (9.95) is the fact that the corresponding points on $\mathbb{S}^2$ are clustered near the north pole $(0,\,0,\,1)^\top$ and the south pole $(0,\,0,\,-1)^\top$. For many applications, such a restriction is not realistic. Therefore, it is necessary to develop algorithms for arbitrarily distributed points on the unit sphere. □

In the essential case of arbitrarily distributed points $(\theta_\ell,\, \phi_\ell) \in [0, \pi] \times [0,\, 2\pi)$, $\ell = 0, \ldots, M - 1$, we speak about a *nonequispaced discrete spherical Fourier transform*, abbreviated as NDSFT. The main idea is to combine the computation of NDSFT with the NFFT (see Chap. 7). We will suggest a fast algorithm for NDSFT, where the computational cost amounts to $\mathcal{O}(N^2\,(\log N)^2 + M)$ flops. We denote such an algorithm as *nonequispaced fast spherical Fourier transform*, abbreviated as NFSFT. We have the following relations to the transforms and algorithms on the torus:

Torus $\mathbb{T}$	Unit sphere $\mathbb{S}^2$
DFT (see Chap. 3)	DSFT
FFT (see Chap. 5)	FSFT
NDFT (see Chap. 7)	NDSFT
NFFT (see Chap. 7)	NFSFT

The fundamental idea is the fast realization of a basis exchange, such that the function in (9.91) can be approximately written in the form

$$f(\theta,\, \phi) \approx \sum_{n=-N}^{N} \sum_{k=-N}^{N} c_k^n \, \mathrm{e}^{2\mathrm{i}\,k\,\theta} \, \mathrm{e}^{\mathrm{i}\,n\,\phi} \tag{9.96}$$

with certain coefficients $c_k^n \in \mathbb{C}$. In a second step, we compute f at arbitrary points $(\theta_\ell,\, \phi_\ell) \in [0, \pi] \times [0,\, 2\pi)$, $\ell = 0, \ldots, M - 1$, by using a two-dimensional NFFT (see Algorithm 7.1).

In Sect. 9.6.1, we sketch a simple realization for an NDSFT. In Sect. 9.6.2, we present a FSFT on a special grid. Then we describe an algorithm for the fast and approximate evaluation of an NDSFT in Sect. 9.6.3. Finally, in Sect. 9.6.4, some results of fast quadrature and approximation on the unit sphere are sketched.

9.6.1 Discrete Spherical Fourier Transforms

We follow the same lines as in [257]. For given spherical Fourier coefficients $a_k^n(f) \in \mathbb{C}$ with $k = 0, \ldots, N$ and $n = -k, \ldots, k$ in (9.91), we are interested in the computation of M function values $f(\theta_\ell, \phi_\ell)$, $\ell = 0, \ldots, M-1$. To this end, we interchange the order of summation in (9.91). Using (9.90) and the function

$$h_n(x) := \sum_{k=|n|}^{N} a_k^n(f)\, P_k^{|n|}(x)\,, \quad n = -N, \ldots, N\,, \tag{9.97}$$

we obtain the sum:

$$f(\theta_\ell, \phi_\ell) = \sum_{k=0}^{N} \sum_{n=-k}^{k} a_k^n(f)\, Y_k^n(\theta_\ell, \phi_\ell) = \sum_{n=-N}^{N} h_n(\cos\theta_\ell)\, \mathrm{e}^{\mathrm{i}\, n\, \phi_\ell}\,. \tag{9.98}$$

We immediately find a first algorithm for an NDSFT, i.e., for the evaluation of f in (9.91) at arbitrarily distributed points (θ_ℓ, ϕ_ℓ) on the unit sphere $\mathbb{S}^2$.

Algorithm 9.56 (NDSFT)
Input: $N, M \in \mathbb{N}$, $a_k^n(f) \in \mathbb{C}$ *for* $k = 0, \ldots, N$ *and* $n = -k, \ldots, k$, $(\theta_l, \phi_l) \in [0,\, \pi] \times [0,\, 2\pi)$ *for* $\ell = 0, \ldots, M-1$.

1. *Compute the values:*

$$h_n(\cos\theta_\ell) := \sum_{k=|n|}^{N} a_k^n(f)\, P_k^{|n|}(\cos\theta_\ell)\,, \quad \ell = 0, \ldots, M-1\,,$$

for all $n = -N, \ldots, N$ *by the Clenshaw Algorithm* 6.19.

2. *Compute the values:*

$$f(\theta_\ell,\, \phi_\ell) := \sum_{n=-N}^{N} h_n(\cos\theta_\ell)\, \mathrm{e}^{\mathrm{i}\, n\, \phi_\ell}\,, \quad \ell = 0, \ldots, M-1\,.$$

Output: $f(\theta_\ell, \phi_\ell) \in \mathbb{C}$, $\ell = 0, \ldots, M-1$, *function values of* (9.91).
Computational cost: $\mathcal{O}(M\, N^2)$.

9.6.2 Fast Spherical Fourier Transforms

In order to use the tensor product structure of the spherical harmonics $Y_k^n(\theta,\, \phi)$ in (9.90), we develop a fast method for the computation of the discrete spherical

Fourier transform on special grid (DSFT). If we apply fast one-dimensional algorithms with respect to θ and ϕ and the row-column method (see Algorithm 5.23), then we obtain an FSFT.

We start with the task to compute the function in (9.91) for given spherical Fourier coefficients $a_k^n(f)$ for $k = 0, \ldots, N$ and $n = -k, \ldots, k$ on the special grid in (9.95). Considering h_n in (9.97) and choosing $T \in \mathbb{N}$ with $N + 1 \le T \le 2^{\lceil \log_2 (N+1) \rceil}$, we compute the values

$$h_{s,n} := h_n\big(\cos \frac{s\pi}{T}\big), \quad s = 0, \ldots, T, \tag{9.99}$$

for $n = -N, \ldots, N$ and rewrite f in (9.91) as

$$f\big(\frac{s\pi}{T}, \frac{t\pi}{T}\big) = \sum_{n=-N}^{N} h_{s,n}\, \mathrm{e}^{\mathrm{i}\pi n t/T}, \quad t = 0, \ldots, 2T - 1. \tag{9.100}$$

for all $s = 0, \ldots, T$. The computation of the function values in (9.100) can be realized by $T + 1$ DFT($2T$), and the obtained DSFT has a computational cost of $\mathcal{O}(N^3)$.

In order to speed up the computation for the DSFT, we compute for each $n = -N, \ldots, N$ the sum in (9.99) with a *fast Legendre function transform*, abbreviated as FLT. The idea for an FLT was proposed by J. Driscoll and D. Healy [115]. The associated Legendre functions P_k^n fulfill the three-term recurrence relation

$$P_{k+1}^n(x) = v_k^n\, x\, P_k^n(x) + w_k^n\, P_{k-1}^n(x), \quad k = n,\, n+1,\, \ldots, \tag{9.101}$$

with the initial conditions

$$P_{n-1}^n(x) := 0\,, \quad P_n^n(x) = \lambda_n\, (1 - x^2)^{n/2}, \quad \lambda_n := \frac{\sqrt{(2n)!}}{2^n\, n!}\,,$$

and with the coefficients

$$v_k^n := \frac{2k+1}{\sqrt{(k-n+1)(k+n+1)}}\,, \quad w_k^n := -\frac{\sqrt{(k-n)(k+n)}}{\sqrt{(k-n+1)(k+n+1)}}\,. \tag{9.102}$$

A useful idea is to define the associated Legendre functions P_k^n also for $k = 0, \ldots, n$ by means of the modified three-term recurrence relation:

$$P_{k+1}^n(x) = \big(\alpha_k^n\, x + \beta_k^n\big)\, P_k^n(x) + \gamma_k^n\, P_{k-1}^n(x) \tag{9.103}$$

for $k \in \mathbb{N}_0$ with

$$\alpha_0^n := \begin{cases} 1 & n = 0\,, \\ 0 & n \text{ odd}\,, \\ -1 & n \neq 0 \text{ even}\,, \end{cases} \qquad \alpha_k^n := \begin{cases} (-1)^{k+1} & k = 1,\, \ldots,\, n-1\,, \\ v_k^n & k = n,\, n+1,\, \ldots\,, \end{cases}$$

$$\beta_k^n := \begin{cases} 1 & k = 0,\, \ldots,\, n-1\,, \\ 0 & k = n,\, n+1,\, \ldots\,, \end{cases} \qquad \gamma_k^n := \begin{cases} 0 & k = 0,\, \ldots,\, n-1\,, \\ w_k^n & k = n,\, n+1,\, \ldots\,. \end{cases} \tag{9.104}$$

Here we set $P_{-1}^n(x) := 0$, and $P_0^n(x) := \lambda_n$ for even n and $P_0^n(x) := \lambda_n\,(1-x^2)^{1/2}$ for odd n. For $k \ge n$, this definition coincides with the recurrence (9.101). It can be easily verified by (9.89) that P_k^n is a polynomial of degree k, if n is even. Further, $(1-x^2)^{-1/2} P_k^n$ is a polynomial of degree $k-1$, if n is odd. Using the recurrence coefficients from (9.104) and introducing a shift parameter $c \in \mathbb{N}_0$, we define the *associated Legendre polynomials* $P_k^n(x,\, c)$ (see (6.90)) by

$$\begin{aligned} &P_{-1}^n(x, c) := 0\,, \quad P_0^n(x, c) := 1\,, \\ &P_{k+1}^n(x,\, c) = \big(\alpha_{k+c}^n\, x + \beta_{k+c}^n\big)\; P_k^n(x,\, c) + \gamma_{k+c}^n\; P_{k-1}^n(x,\, c)\,, \quad k \in \mathbb{N}_0\,. \end{aligned} \tag{9.105}$$

Now we can apply the discrete polynomial transform (see Sect. 6.5 and Algorithm 6.63). Note that a straightforward implementation of this transform is numerically unstable for large bandwidth. Therefore, some stabilization steps are necessary. Stabilized versions have been suggested in [235, 338, 339]. For further approaches, see [6, 190, 196, 210, 230, 289, 367, 403, 417, 418].

The stabilization method presented in [338] requires $\mathcal{O}(N^2(\log N)^2)$ arithmetical operations. For the computation of the sum in (9.100), we need for each $s = 0, \ldots, T$ an FFT of length $2T$. The proposed method is summarized in the following algorithm.

Algorithm 9.57 (FSFT on the Special Grid in (9.95)**)**
Input: $N \in \mathbb{N}$, $a_k^n(f) \in \mathbb{C}$ *for* $k = 0, \ldots, N$ *and* $n = -k, \ldots, k$,
$N + 1 \le T \le 2^{\lceil \log_2 (N+1) \rceil}$.

1. *Using* FLT, *compute the values*

$$h_{s,n} = \sum_{k=|n|}^{N} a_k^n(f)\; P_k^{|n|}\big(\cos \frac{s\pi}{T}\big)$$

   for all $s = 0, \ldots, T$ *and* $n = -N, \ldots, N$.
2. *Applying* FFT, *compute for all* $s = 0, \ldots, T-1$ *the values*

$$f\left(\frac{s\pi}{T}, \frac{t\pi}{T}\right) = \sum_{n=-N}^{N} h_{s,n}\, \mathrm{e}^{\mathrm{i}\pi n t/T}, \quad t = 0, \ldots, 2T-1\,.$$

Output: $f\big(\frac{s\pi}{T}, \frac{t\pi}{T}\big)$ *for* $s = 0, \ldots, T$ *and* $t = 0, \ldots, 2T - 1$ *function values of* f *in* (9.91) *on the grid in* (9.95).
Computational cost: $\mathscr{O}\big(N^2 (\log N)^2\big)$.

9.6.3 Fast Spherical Fourier Transforms for Nonequispaced Data

In this subsection, we develop a fast algorithm for the NDSFT improving Algorithm 9.56. The fast algorithm is denoted by NFSFT. We will start again with the fast Legendre transform (FLT) in order to compute the basis exchange. Then we will apply the two-dimensional NFFT (see Algorithm 7.1) in order to evaluate the trigonometric polynomial at arbitrarily distributed points $(\theta_\ell, \phi_\ell) \in [0, \pi] \times [0, 2\pi)$, $\ell = 0, \ldots, M - 1$.

The spherical polynomial in (9.91) can be written for arbitrary $(\theta, \phi) \in [0, \pi] \times [0, 2\pi)$ in the form

$$f(\theta, \phi) = \sum_{n=-N}^{N} h_n(\cos\theta)\, \mathrm{e}^{\mathrm{i}n\phi}\,, \tag{9.106}$$

where h_n is given in (9.97). For even $|n|$, we define the polynomial g_n of degree N by

$$g_n(x) := \sum_{k=|n|}^{N} a_k^n(f)\, P_k^{|n|}(x) \in \mathscr{P}_N\,. \tag{9.107}$$

For odd $|n|$, we introduce the polynomial g_n of degree $N - 1$ by

$$g_n(x) := \sum_{k=|n|}^{N} a_k^n(f)\, Q_k^{|n|}(x) \in \mathscr{P}_{N-1}\,, \tag{9.108}$$

where $Q_k^{|n|}$ means the polynomial factor of $P_k^{|n|}$, i.e., for $x \in (-1,\ 1)$ and $k \ge |n|$, it holds

$$Q_k^{|n|}(x) := \frac{1}{\sqrt{1-x^2}}\, P_k^{|n|}(x) \in \mathscr{P}_{|n|-1}\,.$$

As usual, $\mathscr{P}_N$ denotes the set of all algebraic polynomials of degree less or equal to N. The relation to h_n in (9.97) is given by

$$h_n(\cos\theta) = \begin{cases} g_n(\cos\theta) & n \text{ even}\,, \\ (\sin\theta)\, g_n(\cos\theta) & n \text{ odd}\,. \end{cases} \tag{9.109}$$

As in step 1 of Algorithm 9.57, we compute the data

$$g_{s,n} := g_n\big(\cos\frac{s\pi}{T}\big)\,, \quad s = 0, \ldots, T;\; n = -N, \ldots, N\,, \tag{9.110}$$

with $N + 1 \le T \le 2^{\lceil \log_2(N+1) \rceil}$ using an FLT. Then we compute the Chebyshev coefficients $\tilde{a}_k^n \in \mathbb{C}$ in

$$g_n(\cos\theta) = \begin{cases} \sum_{k=0}^{N} \tilde{a}_k^n\, T_k(\cos\theta) & n \text{ even}\,, \\ \sum_{k=0}^{N-1} \tilde{a}_k^n\, T_k(\cos\theta) & n \text{ odd}\,, \end{cases} \tag{9.111}$$

using the DCT-I Algorithm 6.28. We take into account that g_n are trigonometric polynomials of degree N or rather $N - 1$.

Applying the known relation

$$T_k(\cos\theta) = \cos(k\,\theta) = \frac{1}{2}\big(\mathrm{e}^{\mathrm{i}\,k\,\theta} + \mathrm{e}^{-\mathrm{i}\,k\,\theta}\big)\,,$$

for even n, we obtain the trigonometric polynomial

$$g_n(\cos\theta) = \sum_{k=-N}^{N} b_k^n\, \mathrm{e}^{\mathrm{i}\,k\,\theta}$$

with the Fourier coefficients

$$b_k^n := \begin{cases} \tilde{a}_0^n & k = 0\,, \\ \frac{1}{2}\,\tilde{a}_{|k|}^n & k \ne 0\,. \end{cases} \tag{9.112}$$

For odd n, it follows that $g_n(\cos\theta)$ is a trigonometric polynomial of degree $N - 1$, since

$$\begin{aligned} h_n(\cos\theta) &= (\sin\theta) \sum_{k=-N+1}^{N-1} b_k^n\, \mathrm{e}^{\mathrm{i}\,k\,\theta} \\ &= \frac{1}{2\mathrm{i}}\big(\mathrm{e}^{\mathrm{i}\theta} - \mathrm{e}^{-\mathrm{i}\theta}\big) \sum_{k=-N+1}^{N-1} b_k^n\, \mathrm{e}^{\mathrm{i}\,k\,\theta} = \sum_{k=-N}^{N} \tilde{b}_k^n\, \mathrm{e}^{\mathrm{i}\,k\,\theta} \end{aligned}$$

with

$$2\mathrm{i}\,\tilde{b}_k^n := \begin{cases} -b_{k+1}^n & k = -N,\ -N+1\,, \\ b_{k-1}^n & k = N-1,\ N\,, \\ b_{k-1}^n - b_{k+1}^n & k = -N+2, \ldots, N-2\,. \end{cases} \tag{9.113}$$

Then we can write the trigonometric polynomial in (9.109) in the form

$$h_n(\cos\theta) = \sum_{k=-N}^{N} c_k^n\, \mathrm{e}^{\mathrm{i}\,k\,\theta} \tag{9.114}$$

with the Fourier coefficients

$$c_k^n := \begin{cases} b_k^n & n \text{ even}\,, \\ \tilde{b}_k^n & n \text{ odd}\,. \end{cases} \tag{9.115}$$

Inserting the sum in (9.114) into (9.106), we arrive at the representation

$$f(\theta,\,\phi) = \sum_{n=-N}^{N} \sum_{k=-N}^{N} c_k^n\, \mathrm{e}^{\mathrm{i}\,k\,\theta}\, \mathrm{e}^{\mathrm{i}\,n\,\phi} \tag{9.116}$$

with complex coefficients $c_k^n \in \mathbb{C}$. We stress again that we have computed the discrete Fourier coefficients c_k^n in (9.116) from the given spherical Fourier coefficients $a_k^n(f)$ in (9.91) with an exact algorithm that requires $\mathcal{O}\big(N^2\,(\log N)^2\big)$ arithmetic operations independently of the chosen points $(\theta_\ell,\,\phi_\ell) \in [0,\,\pi] \times [0,\,2\pi)$.

Geometrically, this transform maps the sphere $\mathbb{S}$ to the "outer half" of the ring torus given by

$$\mathbf{x} = \big((r+\sin\theta)\,\cos\phi,\ (r+\sin\theta)\,\sin\phi,\ \cos\theta\big)^{\top}\,, \quad (\theta,\,\phi) \in [0,\,\pi] \times [0,\,2\pi)$$

with fixed $r > 1$. The "inner half" of the ring torus with the parameters $(\theta,\,\phi) \in [\pi,\,2\pi) \times [0,\,2\pi)$ is continued smoothly (see Fig. 9.9). We decompose f into an even and odd function and construct the even–odd continuation (see Fig. 9.10 for the normalized spherical harmonic):

$$\sqrt{7}\,Y_3^{-2}(\theta,\,\phi) = \sqrt{\frac{105}{8}}\,(\sin\theta)^2\,(\cos\theta)\,\mathrm{e}^{-2\mathrm{i}\,\phi}\,.$$

On the ring torus, we illustrate this normalized spherical harmonic for $(\theta,\,\phi) \in [0, 2\pi)^2$.

In a final step, we compute f at arbitrary points by using a two-dimensional NFFT (see Algorithm 7.1). We summarize:

Fig. 9.9 Topography of the earth (left) and the "outer half" of the ring torus (right), cf. Figure 5.1 in [257]

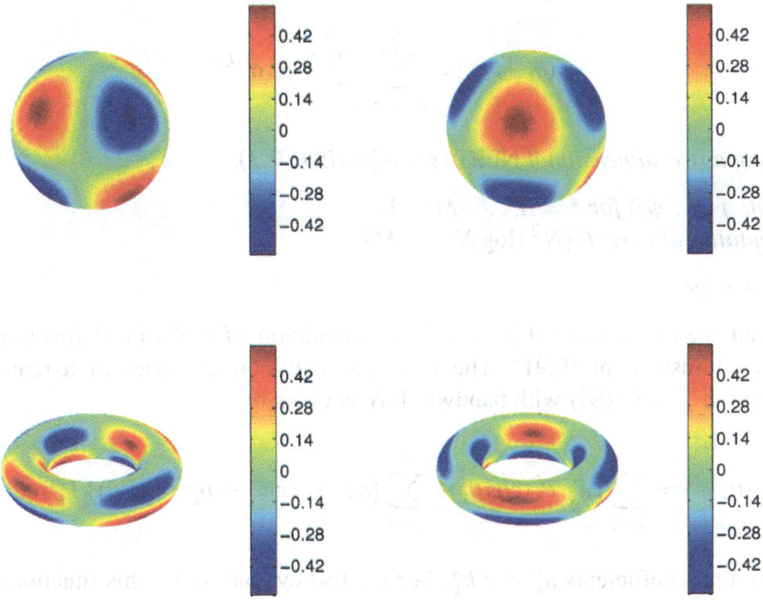

Fig. 9.10 The normalized spherical harmonic $Y_3^{-2}(\theta, \phi)$, top: real part (left) and imaginary part (right) of this normalized spherical harmonic on $\mathbb{S}^2$, down: normalized spherical harmonic $Y_3^{-2}(\theta, \phi)$ on a ring torus

Algorithm 9.58 (NFSFT)

Input: $N, M \in \mathbb{N}$, $a_k^n(f) \in \mathbb{C}$ *for* $k = 0, \ldots, N$ *and* $n = -k, \ldots, k$,
$(\theta_\ell, \phi_\ell) \in [0, \pi] \times [0, 2\pi)$ *for* $\ell = 0, \ldots, M-1$,
$N + 1 \le T \le 2^{\lceil \log_2 (N+1) \rceil}$.

1. *Using* FLT, *compute the data*

$$g_n\big(\cos\frac{s\pi}{T}\big) := \begin{cases} \sum_{k=|n|}^{N} a_k^n(f)\, P_k^{|n|}\big(\cos\frac{s\pi}{T}\big) & n \text{ even}, \\ \sum_{k=|n|}^{N} a_k^n(f)\, Q_k^{|n|}\big(\cos\frac{s\pi}{T}\big) & n \text{ odd} \end{cases}$$

for all $n = -N, \ldots, N$ *and* $s = 0, \ldots, T$.

2. *Compute the Chebyshev coefficients* $\tilde{a}_k^n$ *in* (9.111) *for* $n = -N, \ldots, N$, *by a fast* DCT-I *algorithm (see Algorithm 6.28) or 6.35.*
3. *Determine the Fourier coefficients* c_k^n *by* (9.112), (9.113), *and* (9.115) *for* $k = -N, \ldots, N$ *and* $n = -N, \ldots, N$.
4. *Compute the function values*

$$f(\theta_\ell, \phi_\ell) := \sum_{n=-N}^{N} \sum_{k=-N}^{N} c_k^n\, \mathrm{e}^{\mathrm{i}\,(k\,\theta_\ell + n\,\phi_\ell)}$$

using a two-dimensional NFFT *(see Algorithm 7.1).*

Output: $f(\theta_\ell, \phi_\ell)$ *for* $\ell = 0, \ldots, M-1$.
Computational cost: $\mathcal{O}\big(N^2\,(\log N)^2 + M\big)$.

Remark 9.59

1. Often we are interested in a real representation of real-valued function $f \in L_2(\mathbb{S}^2)$ instead of (9.91). The real spherical Fourier series of a real-valued function $f \in L_2(\mathbb{S}^2)$ with bandwidth N is given by

$$f(\theta, \phi) = \sum_{k=0}^{N} \Big(a_k^0\, P_k(\cos\theta) + \sum_{n=1}^{k} \big(a_k^n\, \cos(n\phi) + b_k^n\, \sin(n\phi)\big)\, P_k^n(\cos\theta)\Big)$$

   with real coefficients a_k^n and b_k^n. For the fast evaluation for this function in real arithmetic, one can develop similar algorithms. Instead of the NFFT, one can use the NFCT (see Algorithm 7.18) and the NFST (see Algorithm 7.20).
2. The algorithms of this section are part of the NFFT software (see [231, .../nfsft]). Furthermore, there exists a MATLAB interface (see [231, ...matlab/nfsft]). □

9.6.4 Fast Quadrature and Approximation on $\mathbb{S}^2$

Finally, we sketch some methods for the fast quadrature and approximation of functions defined on the unit sphere $\mathbb{S}^2$ being relevant for solving partial differential equations on the sphere [60, Section 18.7]. In particular, partial sums (9.91) of spherical Fourier series are used in applications. The bandwidth grows by manipulating these series, e.g., by integration or multiplication.

An important subproblem is the projection onto a space of spherical polynomials of smaller bandwidth. This task is known as *spherical filtering* (see [214]). Since the computational cost of spherical filtering amounts to $\mathcal{O}(N^3)$ flops for $\mathcal{O}(N^2)$ points on $\mathbb{S}^2$, fast algorithms are of great interest. R. Jakob–Chien and K. Alpert [214] presented a first fast algorithm for spherical filtering based on the fast multipole method, which requires $\mathcal{O}(N^2 \log N)$ flops. Later this approach was improved by N. Yarvin and V. Rokhlin [447]. M. Böhme and D. Potts suggested a method for spherical filtering based on the fast summation method (see Algorithm 7.23 with the kernel $1/x$) (see [54, 55]). This approach is easy to implement, and it is based on the NFFT and requires $\mathcal{O}(N^2 \log N)$ arithmetic operations. Furthermore, this method can be used for the fast calculation of wavelet decompositions on $\mathbb{S}^2$ (see [54, 55]).

In order to compute the spherical Fourier coefficients (9.92) by a *quadrature rule* on $\mathbb{S}^2$, we need to solve the *adjoint problem*, i.e., the fast evaluation of sums in (9.94) for given function values $f_\ell = f(\theta_\ell, \phi_\ell) \in \mathbb{C}$, $\ell = 0, \ldots, M-1$. Note that the obtained values $\check{a}_k^n$ in (9.94) do not exactly coincide with the spherical Fourier coefficients $a_k^n(f)$ from (9.91) which are given in (9.92). Good approximations of the spherical Fourier coefficients $a_k^n(f)$ can be obtained from sampled function values $f(\theta_\ell, \phi_\ell)$, $\ell = 0, \ldots, M-1$, at points $(\theta_\ell, \phi_\ell) \in [0, \pi] \times [0, 2\pi)$ provided that a quadrature rule with weights w_ℓ and sufficiently high degree of exactness is available (see also [152, 287, 288]). Then the sum (9.94) changes to

$$a_k^n(f) = \sum_{\ell=0}^{M-1} w_\ell \, f(\theta_\ell, \phi_\ell) \, \overline{Y_k^n(\theta_\ell, \phi_\ell)} \,. \tag{9.117}$$

The computation of spherical Fourier coefficients from discrete sampled data has major importance in the field of data analysis on the sphere $\mathbb{S}^2$. For special grids on the sphere, one can use the Clenshaw–Curtis quadrature with respect to $\cos\theta$ (see Sect. 6.4.2) and equidistant points with respect to ϕ. Such quadrature rules have been suggested in [115, 339].

Theorem 9.60 *Let $f \in L_2(\mathbb{S}^2)$ be a bandlimited function of the form* (9.91). *Then we can compute the spherical Fourier coefficients $a_k^n(f)$ for $k = 0, \ldots, N$ and $n = -k, \ldots, k$ by the quadrature rule*

$$a_k^n(f) = \frac{1}{2^{j+1}} \sum_{s=0}^{T} \sum_{t=0}^{T-1} \varepsilon_s w_s \, f\left(\frac{\pi s}{T}, \frac{2\pi t}{T}\right) Y_k^{-n}\left(\frac{\pi s}{T}, \frac{2\pi t}{T}\right) \tag{9.118}$$

with $2N \le T$, $\varepsilon_0 = \varepsilon_T := 2^{-1}$, $\varepsilon_s := 1$; $s = 1, \ldots, T-1$, and with the Clenshaw–Curtis weights (see Sect. 6.4.2):

$$w_s := \frac{1}{T} \sum_{u=0}^{T/2} \varepsilon_u \frac{-2}{4u^2 - 1} \cos \frac{2su\pi}{T}, \quad s = 0, \ldots, T.$$

Proof By definition of f in (9.91), it suffices to consider the basis functions $f(\theta, \phi) = Y_l^m(\theta, \phi)$, $l = 0, \ldots, N$; $m = -l, \ldots, l$. Their Fourier coefficients can be written as

$$a_k^n(f) = \frac{1}{2} \int_{-1}^{1} P_l^{|m|}(x) P_k^{|n|}(x)\, \mathrm{d}x \cdot \frac{1}{2\pi} \int_0^{2\pi} \mathrm{e}^{\mathrm{i}(m-n)\phi}\, \mathrm{d}\phi\,. \tag{9.119}$$

Now it follows for $m, n = -N, \ldots, N$ that

$$\delta_{m,n} = \frac{1}{2\pi} \int_0^{2\pi} \mathrm{e}^{\mathrm{i}(m-n)\phi}\, \mathrm{d}\phi = \frac{1}{T} \sum_{t=0}^{T-1} \mathrm{e}^{2\pi\mathrm{i}(m-n)t/T}\,. \tag{9.120}$$

Hence, for $m \neq n$, the Fourier coefficients $a_k^n(f)$ vanish. For $m = n$, we verify that $P_l^{|n|} P_k^{|n|}$ is an algebraic polynomial of degree $\leq 2N \leq T$ such that Clenshaw–Curtis quadrature gives

$$\frac{1}{2} \int_{-1}^{1} P_l^{|n|}(x) P_k^{|n|}(x)\, \mathrm{d}x = \sum_{s=0}^{T} \varepsilon_s\, w_s\, P_l^{|n|}\left(\cos \frac{\pi s}{T}\right) P_k^{|n|}\left(\cos \frac{\pi s}{T}\right)$$

with the *Chebyshev nodes* $\cos \frac{\pi s}{T}$. Together with (9.119) and (9.120), this completes the proof. ∎

In many applications, however, the distribution of the available data on the sphere is predetermined by the underlying measurement process or by data storage and access considerations. This often requires the use of techniques like spherical hyperinterpolation (see [393]) or approximate quadrature rules that differ from classical quadrature formulas.

The implementation of the algorithm for computing $a_k^n(f)$ in (9.118), which is the adjoint problem of the NFSFT in Algorithm 9.58, follows by writing the steps of Algorithm 9.58 in matrix-vector form and forming the conjugate transpose of this matrix product (see [235]). In this way, one obtains a fast algorithm for the adjoint problem which allows the efficient use of new quadrature schemes (see also [175]).

Spherical t-designs A spherical t-design is a finite point set on $\mathbb{S}^2$ which provides a quadrature rule on $\mathbb{S}^2$ with equal weights being exact for spherical polynomials up to degree t. Note that quadrature rules with equal weights are also known as quasi-Monte Carlo rules. Based on the NFSFT and the algorithm of the adjoint problem, we are able to evaluate spherical t-designs on $\mathbb{S}^2$ for high polynomial degree $t \in \mathbb{N}$ (see [173, 176]). The approach is based on computing local minimizers of a certain quadrature error. This quadrature error was also used for a variational characterization of spherical t-designs in [394].

It is commonly conjectured that there exist spherical t-designs with $M \approx \frac{1}{2}t^2$ points, but a proof is unknown up to now. Recently, a weaker conjecture was proved in [56], where the authors showed the existence of spherical t-designs with $M > c\,t^2$ points for some unknown constant $c > 0$. Moreover, in [83] it was verified that for $t = 1, \ldots, 100$, spherical t-designs with $(t+1)^2$ points exist, using the characterization of fundamental spherical t-designs and interval arithmetic. For further recent developments regarding spherical t-designs and related topics, we refer to the very nice survey article [15].

The construction of spherical t-designs is a serious challenge even for small polynomial degrees t. For the minimization problem, one can use several nonlinear optimization methods on manifolds, like Newton and conjugate gradient methods. By means of the NFSFT, evaluations of the approximate gradient and Hessian can be performed by $\mathcal{O}\big(t^2 \log t + M\,(\log \epsilon)^2\big)$ arithmetic operations, where $\epsilon > 0$ is a prescribed accuracy. Using these approaches, approximate spherical t-designs for $t \le 1000$, even in the case $M \approx \frac{1}{2}t^2$, have been presented in [171]. This method has been also generalized to Gauss-type quadratures on $\mathbb{S}^2$, where one does not require a quadrature with equal weights, but can optimize the weights as well. In this way, quadrature rules with a higher degree of exactness have been obtained using the same number of sampling points. These nonlinear optimization methods have been further generalized in order to approximate global extrema (see [174]) and to halftoning and dithering (see [177]).

Scattered Data Approximation on $\mathbb{S}^2$ Since the data collected on the surface of the earth are available as scattered nodes only, least squares approximations and interpolation of such data have attracted much attention (see, e.g., [60, 133, 152]). If we reconstruct a spherical polynomial of degree $N \in \mathbb{N}$ from sample values, we can set up a linear system with $M = (N+1)^2$ interpolation constraints which has to be solved for the unknown vector of Fourier coefficients of length $(N+1)^2$. If the nodes for interpolation are chosen such that the interpolation problem has always a unique solution, the sampling set is called a fundamental system. We can relax the condition that the number of equations M coincides with the number of unknowns $(N+1)^2$. Considering the overdetermined case $M > (N+1)^2$ or the underdetermined case $M < (N+1)^2$ leads to better distributed singular values of the system matrix. Results using fast algorithms in combination with an iterative solver were presented in [233]. For stability results, see also [58, 297]. Scattered data approximation using kernels is described in [152]. Employing radial kernels in combination with the presented fast algorithms leads to a fast summation method on the sphere as well (see [232]).

FFT *on the Rotation Group* The theory of spherical Fourier transforms can be extended to *fast Fourier transforms on the rotation group* SO(3). As known, SO(3) is the group of all rotations about the origin of $\mathbb{R}^3$. During the past years, several different techniques have been proposed for computing Fourier transforms on the rotation group SO(3) motivated by a variety of applications, like protein–protein docking [80] or texture analysis [74, 278, 376]. The spherical Fourier methods can be generalized to Fourier methods on SO(3), see [249]. The algorithm to compute an SO(3) Fourier synthesis is based on evaluating the Wigner-D functions $D_\ell^{m,n}$, which yield an orthogonal basis of $L_2(\mathrm{SO}(3))$. Using these Wigner-D functions, we expand N-bandlimited functions $f \in L_2(\mathrm{SO}(3))$ into the sum

$$f = \sum_{k=0}^{N} \sum_{m=-k}^{k} \sum_{n=-\ell}^{\ell} a_k^{m,n}\, D_k^{m,n}\,. \tag{9.121}$$

An algorithm for efficient and accurate evaluation of such N-bandlimited function $f \in L_2(\mathrm{SO}(3))$ at arbitrary samples in SO(3) has been presented in [332]. For the scattered data approximation, see [172] and for quadrature rules, see [171, 173]. These algorithms are also part of the freely available NFFT software [231, ./examples/nfsoft]. Furthermore, there exists a MATLAB interface (see [231, ./matlab/nfsoft]).

Chapter 10
Prony Method for Reconstruction of Structured Functions

The recovery of a structured function from sampled data is a fundamental problem in applied mathematics and signal processing. In Sect. 10.1, we consider the parameter estimation problem, where the classical Prony method and its relatives are described. In Sect. 10.2, we study frequently used methods for solving the parameter estimation problem, namely, MUSIC (multiple signal classification), APM (approximate Prony method), and ESPRIT (estimation of signal parameters by rotational invariance). Moreover, we briefly introduce ESPIRA (estimation of signal parameters by rational approximation). The algorithms for reconstructing exponential sums will be derived for noiseless data and then extended to the case of noisy data. The effectiveness the algorithms for noisy data depends in particular on the condition numbers of the involved Vandermonde matrices. We will deal with these stability issues in Sect. 10.3.

In Sects. 10.4 and 10.5, we consider different applications of the Prony method. We present an algorithm to recover spline functions from the given samples of its Fourier transform. Finally, we study a phase retrieval problem and investigate the question whether a complex-valued function f can be reconstructed from the modulus $|\hat{f}|$ of its Fourier transform.

10.1 Prony Method

The following problem arises quite often in electrical engineering, signal processing, and mathematical physics and is known as *parameter estimation problem* (see [313, 344] or [281, Chapter 9]):

Recover the positive integer M, distinct parameters $\phi_j \in \mathbb{C}$ with $\operatorname{Im} \phi_j \in [-\pi, \pi)$, and complex coefficients $c_j \neq 0$, $j = 1, \ldots, M$, in the *exponential sum of order M*

G. Plonka et al., *Numerical Fourier Analysis*, Applied and Numerical Harmonic Analysis, https://doi.org/10.1007/978-3-031-35005-4_10

$$h(x) := \sum_{j=1}^{M} c_j\, \mathrm{e}^{\phi_j x}\,, \quad x \in \mathbb{R}\,, \tag{10.1}$$

if noisy sampled data $h_k := h(k) + e_k$, $k = 0, \ldots, N-1$, with $N \geq 2M$ are given, where $e_k \in \mathbb{C}$ are small error terms. If $\mathrm{Re}\,\phi_j = 0$ for all j, then this problem is called *frequency analysis problem* . In many applications, h is an expansion into damped exponentials, where $\mathrm{Re}\,\phi_j \in [-\alpha, 0]$ for small $\alpha > 0$. Then the negative real part of f_j is the damping factor and the imaginary part of ϕ_j is the angular frequency of the exponential $\mathrm{e}^{\phi_j x}$.

The classical Prony method works for noiseless sampled data in the case of known order M. Following an idea of G.R. de Prony from 1795 (see [106]), we can recover all parameters of the exponential sum (10.1) from the sampled data:

$$h(k) := \sum_{j=1}^{M} c_j\, \mathrm{e}^{\phi_j k} = \sum_{j=1}^{M} c_j\, z_j^k\,, \quad k = 0, \ldots, 2M-1\,, \tag{10.2}$$

where $z_j := \mathrm{e}^{\phi_j}$ are distinct points in $\mathbb{C}$. For this purpose, we introduce the *Prony polynomial*

$$p(z) := \prod_{j=1}^{M} (z - z_j) = \sum_{k=0}^{M-1} p_k\, z^k + z^M\,, \quad z \in \mathbb{C}\,, \tag{10.3}$$

with corresponding coefficients $p_k \in \mathbb{C}$. Further, we define the *companion matrix* $\mathbf{C}_M(p) \in \mathbb{C}^{M \times M}$ of the Prony polynomial (10.3) by

$$\mathbf{C}_M(p) := \begin{pmatrix} 0 & 0 & \ldots & 0 & -p_0 \\ 1 & 0 & \ldots & 0 & -p_1 \\ 0 & 1 & \ldots & 0 & -p_2 \\ \vdots & \vdots & & \vdots & \vdots \\ 0 & 0 & \ldots & 1 & -p_{M-1} \end{pmatrix}. \tag{10.4}$$

It is known that the companion matrix $\mathbf{C}_M(p)$ has the property

$$\det\big(z\,\mathbf{I}_M - \mathbf{C}_M(p)\big) = p(z)\,,$$

where $\mathbf{I}_M \in \mathbb{C}^{M \times M}$ denotes the identity matrix. Hence, the zeros of the Prony polynomial (10.3) coincide with the eigenvalues of the companion matrix $\mathbf{C}_M(p)$. Setting $p_M := 1$, we now observe the following relation for all $m \in \mathbb{N}_0$:

$$\sum_{k=0}^{M} p_k\, h(k+m) = \sum_{k=0}^{M} p_k \left(\sum_{j=1}^{M} c_j\, z_j^{k+m}\right)$$
$$= \sum_{j=1}^{M} c_j\, z_j^m \left(\sum_{k=0}^{M} p_k\, z_j^k\right) = \sum_{j=1}^{M} c_j\, z_j^m\, p(z_j) = 0\,. \quad (10.5)$$

Using the known values $h(k)$, $k = 0, \ldots, 2M-1$, the formula (10.5) implies

$$\sum_{k=0}^{M-1} p_k\, h(k+m) = -h(M+m)\,, \quad m = 0, \ldots, M-1\,. \quad (10.6)$$

In matrix-vector notation, we obtain the linear system

$$\mathbf{H}_M(0)\,\big(p_k\big)_{k=0}^{M-1} = -\big(h(M+m)\big)_{m=0}^{M-1} \quad (10.7)$$

with the square *Hankel matrix* :

$$\mathbf{H}_M(0) := \begin{pmatrix} h(0) & h(1) & \ldots & h(M-1) \\ h(1) & h(2) & \ldots & h(M) \\ \vdots & \vdots & & \vdots \\ h(M-1) & h(M) & \ldots & h(2M-2) \end{pmatrix} = \big(h(k+m)\big)_{k,m=0}^{M-1}\,. \quad (10.8)$$

The matrix $\mathbf{H}_M(0)$ is invertible, since the special structure (10.2) of the values $h(k)$ leads to the factorization

$$\mathbf{H}_M(0) = \mathbf{V}_M(\mathbf{z})\,\big(\operatorname{diag}\mathbf{c}\big)\,\mathbf{V}_M(\mathbf{z})^\top\,,$$

where the diagonal matrix $\operatorname{diag}\mathbf{c}$ with $\mathbf{c} := (c_j)_{j=1}^M$ contains the nonzero coefficients of (10.1) in the main diagonal, and where

$$\mathbf{V}_M(\mathbf{z}) := \big(z_k^{j-1}\big)_{j,k=1}^M = \begin{pmatrix} 1 & 1 & \ldots & 1 \\ z_1 & z_2 & \ldots & z_M \\ \vdots & \vdots & & \vdots \\ z_1^{M-1} & z_2^{M-1} & \ldots & z_M^{M-1} \end{pmatrix}$$

denotes the *Vandermonde matrix* generated by the vector $\mathbf{z} := (z_j)_{j=1}^M$. Since all z_j, $j = 1, \ldots, M$, are distinct, the Vandermonde matrix $\mathbf{V}_M(\mathbf{z})$ is invertible. Note that by (10.2) we have

$$\mathbf{V}_M(\mathbf{z})\,\mathbf{c} = \big(h(k)\big)_{k=0}^{M-1} \quad (10.9)$$

to compute the coefficient vector $\mathbf{c}$. We summarize:

Algorithm 10.1 (Classical Prony Method)
Input: $M \in \mathbb{N}$, *sampled values* $h(k)$, $k = 0, \ldots, 2M-1$.

1. *Solve the linear system* (10.7).
2. *Compute all zeros* $z_j \in \mathbb{C}$, $j = 1, \ldots, M$, *of the Prony polynomial* (10.3), *i.e., calculate all eigenvalues* z_j *of the associated companion matrix* (10.4). *Form* $r_j := z_j/|z_j|$ *and* $\mathrm{Re}\,\phi_j := \ln|z_j|$, $\mathrm{Im}\,\phi_j := \mathrm{Im}(\log r_j) \in [-\pi, \pi)$, $j = 1, \ldots, M$, *where* log *is the principal value of the complex logarithm.*
3. *Solve the Vandermonde system* (10.9).

Output: $\phi_j \in \mathbb{R} + \mathrm{i}\,[-\pi, \pi)$, $\mathbf{c} = (c_j)_{j=1}^M \in \mathbb{C}^M$, $j = 1, \ldots, M$.

As shown, Prony's idea is essentially based on the separation of the unknown parameters ϕ_j from the unknown coefficients c_j. The main problem is the determination of ϕ_j, since the coefficients c_j are uniquely determined by the linear system (10.9). The following remarks explain some extensions of the Prony method and connections to related methods.

Remark 10.2 The Prony method can be also applied to the recovery of an *extended exponential sum*

$$h(x) := \sum_{j=1}^{M} c_j(x)\,\mathrm{e}^{\phi_j x}\,, \quad x \ge 0\,,$$

where c_j are polynomials of low degree. For simplicity, we sketch only the case of linear polynomials $c_j(x) = c_{j,0} + c_{j,1}x$. With distinct $z_j = \mathrm{e}^{\phi_j}$, $j = 1, \ldots, M$, the corresponding Prony polynomial reads as follows:

$$p(z) := \prod_{j=1}^{M} (z - z_j)^2 = \sum_{k=0}^{2M-1} p_k\, z^k + z^{2M}\,. \tag{10.10}$$

Assuming that the sampled values $h(k)$, $k = 0, \ldots, 4M-1$, are given, we can again derive a relation of the form

$$\sum_{k=0}^{2M} p_k\, h(k+m) = \sum_{k=0}^{2M} \sum_{j=1}^{M} c_j(k+m)\, \mathrm{e}^{\phi_j (k+m)} = 0$$

for $m \in \mathbb{Z}$ using that $p(z_j) = p'(z_j) = 0$ for $z_j = \mathrm{e}^{\phi_j}$. Thus, we have to solve the linear system

$$\sum_{k=0}^{2M-1} p_k\, h(k+\ell) = -h(2M+\ell)\,, \quad \ell = 0, \ldots, 2M-1\,,$$

and to compute all double zeros z_j of the corresponding Prony polynomial in (10.10). Introducing the *confluent Vandermonde matrix*

$$\mathbf{V}_{2M}^{c}(\mathbf{z}) := \begin{pmatrix} 1 & 0 & \dots & 1 & 0 \\ z_1 & 1 & \dots & z_M & 1 \\ z_1^2 & 2z_1 & \dots & z_M^2 & 2z_M \\ \vdots & \vdots & & \vdots & \vdots \\ z_1^{2M-1} & (2M-1)z_1^{2M-2} & \dots & z_M^{2M-1} & (2M-1)z_M^{2M-2} \end{pmatrix},$$

we finally have to solve the confluent Vandermonde system:

$$\mathbf{V}_{2M}^{c}(\mathbf{z})\,(c_{0,1},\, z_1c_{1,1},\, \ldots,\, c_{M,0},\, z_Mc_{M,1})^{\top} = \big(h(k)\big)_{k=0}^{2M-1}\,.$$

□

Remark 10.3 The Prony method can be also applied for other equidistant input values. If we know beforehand that the frequency parameters ϕ_j in (10.1) satisfy $\operatorname{Im}\phi_j \in [-K, K)$ for some $K > 0$, then we can choose a step size $\tau > 0$ such that $\tau \operatorname{Im}\phi_j \in [-\pi, \pi)$ is satisfied and use the input values $h(x_0 + \tau k)$, $k = 0, \ldots, 2M-1$, to recover $h(x)$. Indeed, the exponential sum

$$\tilde{h}(x) := h(x_0 + \tau\, x) = \sum_{j=1}^{M} (c_j\, \mathrm{e}^{\phi_j\, x_0})\, \mathrm{e}^{(\phi_j \tau) x} = \sum_{j=1}^{M} \tilde{c}_j\, \mathrm{e}^{\tilde{\phi}_j x}$$

has the same structure as h with $\tilde{c}_j = c_j\,\mathrm{e}^{\phi_j x_0}$ and $\tilde{\phi}_j = \phi_j \tau$ and can be now computed from the samples $\tilde{h}(k) = h(x_0 + \tau\, k)$, $k = 0, \ldots, 2M-1$. □

Remark 10.4 The Prony method is closely related to *Padé approximation* (see [430]). Let $(f_k)_{k\in\mathbb{N}_0}$ be a complex sequence with $\rho := \limsup_{k\to\infty} |f_k|^{1/k} < \infty$. The *z-transform* of such a sequence is the Laurent series $\sum_{k=0}^{\infty} f_k\, z^{-k}$ which converges in the neighborhood $\{z \in \mathbb{C} : |z| > \rho\}$ of $z = \infty$. Thus, the z-transform of each sequence $(z_j^k)_{k\in\mathbb{N}_0}$ is equal to $\frac{z}{z-z_j}$, $j = 1, \ldots, M$. Since the z-transform is linear, the z-transform maps the data sequence $\big(h(k)\big)_{k\in\mathbb{N}_0}$ satisfying (10.2) for all $k \in \mathbb{N}_0$ into the rational function

$$\sum_{k=0}^{\infty} h(k)\, z^{-k} = \sum_{j=1}^{M} c_j\, \frac{z}{z - z_j} = \frac{a(z)}{p(z)}\,, \tag{10.11}$$

where p is the Prony polynomial (10.3) and $a(z) := a_M\, z^M + \ldots + a_1\, z$. Now we substitute z for z^{-1} in (10.11) and form the *reverse Prony polynomial* $\operatorname{rev} p(z) := z^M\, p(z^{-1})$ of degree M with $\operatorname{rev} p(0) = 1$ as well as the reverse polynomial $\operatorname{rev} a(z) := z^M\, a(z^{-1})$ of degree at least $M-1$. Then we obtain

$$\sum_{k=0}^{\infty} h(k)\, z^k = \frac{\text{rev } a(z)}{\text{rev } p(z)} \tag{10.12}$$

converging in a neighborhood of $z = 0$. In other words, the rational function on the right-hand side of (10.12) is an $(M-1, M)$ *Padé approximant* of the power series $\sum_{k=0}^{\infty} h(k)\, z^k$ with vanishing $\mathcal{O}(z^M)$ term and we have

$$\Big(\sum_{k=0}^{\infty} h(k)\, z^k\Big)\ \text{rev } p(z) = \text{rev } a(z)$$

in a neighborhood of $z = 0$. Comparison of the coefficients of powers of z yields

$$\sum_{k=M-m}^{M} p_k\, h(k+m-M) = a_{M-m}\,, \qquad m = 0, \ldots, M-1\,,$$

$$\sum_{k=0}^{M} p_k\, h(k+m) = 0\,, \quad m \in \mathbb{N}_0\,. \tag{10.13}$$

Now the Eqs. (10.13) for $m = 0, \ldots, M-1$ coincide with (10.6). Hence, the Prony method may also be regarded as a Padé approximation. □

Remark 10.5 In signal processing, the Prony method is also known as the *annihilating filter method*, or a method to recover signals with finite rate of innovation (FRI), (see, e.g., [114, 423]). For distinct $z_j = \mathrm{e}^{\phi_j}$ and complex coefficients $c_j \neq 0$, $j = 1, \ldots, M$, we consider the discrete signal $\mathbf{h} = \big(h_n\big)_{n\in\mathbb{Z}}$ with

$$h_n := \sum_{j=1}^{M} c_j\, z_j^n, \quad n \in \mathbb{Z}\,. \tag{10.14}$$

For simplicity, we assume that M is known. Then a discrete signal $\mathbf{a} = (a_n)_{n\in\mathbb{Z}}$ is called an *annihilating filter* of the signal $\mathbf{h}$, if the discrete convolution of the signals $\mathbf{a}$ and $\mathbf{h}$ vanishes, i.e.,

$$(\mathbf{a} * \mathbf{h})_n := \sum_{\ell\in\mathbb{Z}} a_\ell\, h_{n-\ell} = 0\,, \quad n \in \mathbb{Z}\,.$$

For the construction of an annihilating filter $\mathbf{a}$, we consider

$$a(z) := \prod_{j=1}^{M} \big(1 - z_j\, z^{-1}\big) = \sum_{n=0}^{M} a_n\, z^{-n}\,, \quad z \in \mathbb{C} \setminus \{0\}\,,$$

and then $\mathbf{a} = (a_n)_{n\in\mathbb{Z}}$ with $a_n = 0, n \in \mathbb{Z}\setminus\{0, \ldots, M\}$ is an annihilating filter of $\mathbf{h}$ in (10.14). Note that $a(z)$ is the z-transform of the annihilating filter $\mathbf{a}$. Furthermore, $a(z)$ and the Prony polynomial (10.3) have the same zeros z_j, $j = 1, \ldots, M$, since $z^M\, a(z) = p(z)$ for all $z \in \mathbb{C} \setminus \{0\}$. Hence, the Prony method and the method of annihilating filters are equivalent. For details, see, e.g., [423]. Within the last years, finite rate of innovation methods have found many applications (see, e.g., [42, 310]). □

Remark 10.6 Prony methods arise also from problems of science and engineering, where one is interested in predicting future information from previous ones using a linear model. Let $\mathbf{h} = (h_n)_{n\in\mathbb{N}_0}$ be a discrete signal. The *linear prediction method* (see, e.g., [29]) aims at finding suitable predictor parameters $p_j \in \mathbb{C}$ such that the signal value $h_{\ell+M}$ can be expressed as a linear combination of the previous signal values h_j, $j = \ell, \ldots, \ell + M - 1$, i.e.,

$$h_{\ell+M} = \sum_{j=0}^{M-1} (-p_j)\, h_{\ell+j}\,, \quad \ell \in \mathbb{N}_0\,.$$

Therefore, these equations are also called *linear prediction equations*. Setting $p_M := 1$, we observe that this representation is equivalent to the homogeneous linear difference equation (10.6). Assuming that

$$h_k = \sum_{j=1}^{M} c_j\, z_j^k, \quad k \in \mathbb{N}_0\,,$$

we obtain the parameter estimation problem, i.e., the Prony polynomial (10.3) coincides with the forward predictor polynomial $\sum_{j=1}^{M}(-p_j)z^j$ up to the factor -1. The associated companion matrix $\mathbf{C}_M(p)$ in (10.4) is hence equal to the forward predictor matrix. Thus, the linear prediction method can also be considered as a Prony method. □

Remark 10.7 The Prony method has been also used for reconstruction of polygons $P \subset \mathbb{C}$ from a small number of complex moments

$$\int_P z^k\, \mathrm{d}x\, \mathrm{d}y\,, \quad k \in \mathbb{N}_0\,,$$

as, for example, from measurements of the Fourier transform of the characteristic function with support P. The idea can be also generalized to the recovery of polytopes. For more information, we refer to [91, 128, 165, 179, 187, 440]. □

Unfortunately, the classical Prony method has some numerical drawbacks. Often the order M of the exponential sum (10.1) is unknown. Further the classical Prony method is known to perform poorly if noisy sampled data are given, since the Hankel matrix $\mathbf{H}_M(0)$ and the Vandermonde matrix $\mathbf{V}_M(\mathbf{z})$ are usually badly conditioned.

We will see that one can attenuate these problems by using more sampled data. But then one has to deal with rectangular matrices.

10.2 Recovery of Exponential Sums

In this section, we present several efficient algorithms to solve the parameter estimation problem. Let $N \in \mathbb{N}$ with $N \geq 2M$ be given, where $M \in \mathbb{N}$ denotes the (unknown) order of the exponential sum in (10.1). For simplicity, we restrict ourselves to the frequency analysis problem, where $\phi_j = \mathrm{i}\varphi_j$ with $\varphi_j \in [-\pi, \pi)$. We introduce the *nonequispaced Fourier matrix* (see Chap. 7):

$$\mathbf{A}_{N,M}^{\top} := \big(\mathrm{e}^{\mathrm{i}\varphi_j(k-1)}\big)_{k,\,j=1}^{N,\,M} .$$

Note that $\mathbf{A}_{N,M}^{\top}$ coincides with the *rectangular Vandermonde matrix*

$$\mathbf{V}_{N,M}(\mathbf{z}) := \big(z_j^{k-1}\big)_{k,\,j=1}^{N,\,M}$$

with the vector $\mathbf{z} := (z_j)_{j=1}^M$, where $z_j = \mathrm{e}^{\mathrm{i}\varphi_j}$, $j = 1, \ldots, M$, are distinct nodes on the unit circle. Let $h_k = h(k) + e_k$, $k = 0, \ldots, N-1$, be the given data with small error terms $e_k \in \mathbb{C}$. Then (10.9) has the analogous matrix-vector form

$$\mathbf{V}_{N,M}(\mathbf{z})\,\mathbf{c} = \big(h(k)\big)_{k=0}^{N-1} , \tag{10.15}$$

where $\mathbf{c} = (c_j)_{j=1}^M$ is the vector of nonzero complex coefficients of the exponential sum (10.1).

In practice, the order M of the exponential sum (10.1) is often unknown. Assume that $L \in \mathbb{N}$ is a convenient upper bound of M and $M \leq L \leq N - M + 1$. In applications, such an upper bound L of M is usually known a priori. If this is not the case, then one can choose $L \approx \frac{N}{2}$. Later we will see that the choice $L \approx \frac{N}{2}$ is optimal in some sense. Often the sequence $\{h_0, h_1, \ldots, h_{N-1}\}$ of (noisy) sampled data is called a *time series of length* N. We form the L-*trajectory matrix* of this time series

$$\mathbf{H}_{L,N-L+1} := \big(h_{\ell+m}\big)_{\ell,m=0}^{L-1,\,N-L} \in \mathbb{C}^{L\times(N-L+1)} \tag{10.16}$$

with *window length* $L \in \{M, \ldots, N-M+1\}$. Obviously, $\mathbf{H}_{L,N-L+1}$ is a rectangular *Hankel matrix*.

We consider this rectangular Hankel matrix first for noiseless data $h_k = h(k)$, $k = 0, \ldots, N-1$, i.e.,

$$\mathbf{H}_{L,N-L+1} = \big(h(\ell+m)\big)_{\ell,m=0}^{L-1,\,N-L} \in \mathbb{C}^{L\times(N-L+1)} . \tag{10.17}$$

The main step to solve the parameter estimation problem is to determine the order M and to compute the parameters φ_j or alternatively the pairwise distinct nodes $z_j = \mathrm{e}^{\mathrm{i}\varphi_j}$, $j = 1, \ldots, M$. Afterward, one can calculate the coefficient vector $\mathbf{c} \in \mathbb{C}^M$ as the solution of the least squares problem:

$$\min_{\mathbf{c}\in\mathbb{C}^M} \|\mathbf{V}_{N,M}(\mathbf{z})\,\mathbf{c} - \big(h_k\big)_{k=0}^{N-1}\|_2 \,.$$

By (10.2) the L-trajectory matrix (10.17) can be factorized in the following form:

$$\mathbf{H}_{L,N-L+1} = \mathbf{V}_{L,M}(\mathbf{z})\,(\mathrm{diag}\,\mathbf{c})\,\mathbf{V}_{N-L+1,M}(\mathbf{z})^{\top} \,. \tag{10.18}$$

We denote square matrices with only one index. Additionally, we introduce the rectangular Hankel matrices

$$\mathbf{H}_{L,N-L}(s) = \big(h_{s+\ell+m}\big)_{\ell,m=0}^{L-1,\,N-L-1} \,, \quad s \in \{0,\,1\} \,, \tag{10.19}$$

for $L \in \{M, \ldots, N-M\}$, i.e., $\mathbf{H}_{L,N-L}(0)$ is obtained by removing the last column of $\mathbf{H}_{L,N-L+1}$ and $\mathbf{H}_{L,N-L}(1)$ by removing the first column of $\mathbf{H}_{L,N-L+1}$.

Lemma 10.8 *Let $N \ge 2M$ be given. For each window length $L \in \{M, \ldots, N-M+1\}$, the rank of the L-trajectory matrix* (10.17) *of noiseless data is M. The related Hankel matrices $\mathbf{H}_{L,N-L}(s)$, $s \in \{0, 1\}$, possess the same rank M for each window length $L \in \{M, \ldots, N-M\}$.*

Proof

1. As known, the square Vandermonde matrix $\mathbf{V}_M(\mathbf{z})$ is invertible. Further, we have

$$\mathrm{rank}\,\mathbf{V}_{L,M}(\mathbf{z}) = M \,, \quad \text{for} \quad L \in \{M,\,\ldots,\,N-M+1\} \,, \tag{10.20}$$

since $\mathrm{rank}\,\mathbf{V}_{L,M}(\mathbf{z}) \le \min\{L,\,M\} = M$ and since the submatrix $\big(z_k^{j-1}\big)_{j,k=1}^{M}$ of $\mathbf{V}_{L,M}(\mathbf{z})$ is invertible.
For $L \in \{M,\,\ldots,\,N-M+1\}$, we see by (10.20) that

$$\mathrm{rank}\,\mathbf{V}_{L,M}(\mathbf{z}) = \mathrm{rank}\,\mathbf{V}_{N-L+1,M}(\mathbf{z}) = M \,.$$

Since $(\mathrm{diag}\,\mathbf{c})$ is invertible, we conclude that

$$\begin{aligned}\mathrm{rank}\,\mathbf{H}_{L,N-L+1} &= \mathrm{rank}\,\Big(\mathbf{V}_{L,M}(\mathbf{z})\,\big((\mathrm{diag}\,\mathbf{c})\,\mathbf{V}_{N-L+1,M}(\mathbf{z})^{\top}\big)\Big)\\ &= \mathrm{rank}\,\mathbf{V}_{L,M}(\mathbf{z}) = M \,.\end{aligned}$$

2. By construction of the matrices $\mathbf{H}_{L,N-L}(s)$ for $s = 0, 1$, the assumption follows similarly from the corresponding factorizations:

$$\mathbf{H}_{L,N-L}(0) = \mathbf{V}_{L,M}(\mathbf{z})\,(\mathrm{diag}\,\mathbf{c})\,\mathbf{V}_{N-L,M}(\mathbf{z})^{\top}\,,$$
$$\mathbf{H}_{L,N-L}(1) = \mathbf{V}_{L,M}(\mathbf{z})\,(\mathrm{diag}\,\mathbf{c})\,(\mathrm{diag}\,\mathbf{z})\,\mathbf{V}_{N-L,M}(\mathbf{z})^{\top}\,.$$

■

Consequently, the order M of the exponential sum (10.1) coincides with the rank of the Hankel matrices in (10.17) and (10.19). Therefore, M can be computed as the numerical rank of $\mathbf{H}_{L,N-L+1}$ if it is not known beforehand.

In the next subsections, we will derive the most well-known methods to solve the parameter estimation problem, MUSIC, approximate Prony method, and ESPRIT. Further, we sketch the idea of the ESPIRA algorithm.

10.2.1 MUSIC and Approximate Prony Method

Multiple signal classification (MUSIC) [379] and the *approximate Prony method* (APM) [342] are both based on a singular value decomposition of the given Hankel matrix $\mathbf{H}_{L,N-L+1}$. The following observations also show the close connections between these approaches.

Formula (10.18) implies that the ranges of $\mathbf{H}_{L,N-L+1}$ and $\mathbf{V}_{L,M}(\mathbf{z})$ coincide in the noiseless case for $M \le L \le N-M+1$. If $L > M$, then the range of $\mathbf{V}_{L,M}(\mathbf{z})$ is a proper subspace of $\mathbb{C}^L$. This subspace is called *signal space* $\mathscr{S}_L$. The signal space $\mathscr{S}_L$ is of dimension M and is generated by the M columns $\mathbf{e}_L(\varphi_j)$, $j = 1, \ldots, M$, where

$$\mathbf{e}_L(\varphi) := \big(\mathrm{e}^{\mathrm{i}\,\ell\varphi}\big)_{\ell=0}^{L-1}\,, \quad \varphi \in [-\pi,\,\pi)\,.$$

Note that $\|\mathbf{e}_L(\varphi)\|_2 = \sqrt{L}$ for each $\varphi \in [-\pi,\,\pi)$. The *noise space* $\mathscr{N}_L$ is defined as the orthogonal complement of $\mathscr{S}_L$ in $\mathbb{C}^L$. The dimension of $\mathscr{N}_L$ is equal to $L - M$. By $\mathbf{Q}_L$ we denote the orthogonal projection of $\mathbb{C}^L$ onto the noise space $\mathscr{N}_L$. Since $\mathbf{e}_L(\varphi_j) \in \mathscr{S}_L$, $j = 1, \ldots, M$, and $\mathscr{N}_L \perp \mathscr{S}_L$, we obtain that

$$\mathbf{Q}_L\,\mathbf{e}_L(\varphi_j) = \mathbf{0}\,, \quad j = 1, \ldots, M\,.$$

For $\varphi \in [-\pi,\,\pi) \setminus \{\varphi_1, \ldots, \varphi_M\}$, the vectors $\mathbf{e}_L(\varphi_1), \ldots, \mathbf{e}_L(\varphi_M), \mathbf{e}_L(\varphi) \in \mathbb{C}^L$ are linearly independent, since the $(M+1)\times(M+1)$ Vandermonde matrix obtained by taking the first $M+1$ rows of

$$\big(\mathbf{e}_L(\varphi_1)\,|\,\ldots\,|\,\mathbf{e}_L(\varphi_M)\,|\,\mathbf{e}_L(\varphi)\big)$$

is invertible for each $L \ge M+1$. Hence, $\mathbf{e}_L(\varphi) \notin \mathscr{S}_L = \mathrm{span}\,\{\mathbf{e}_L(\varphi_1), \ldots, \mathbf{e}_L(\varphi_M)\}$, i.e., $\mathbf{Q}_L\mathbf{e}_L(\varphi) \ne \mathbf{0}$.

Thus, once the orthogonal projection $\mathbf{Q}_L$ is known, the parameters φ_j can be determined via the zeros of the *noise-space correlation function*

$$N_L(\varphi) := \frac{1}{\sqrt{L}}\, \|\mathbf{Q}_L\, \mathbf{e}_L(\varphi)\|_2\,, \quad \varphi \in [-\pi,\, \pi)\,,$$

since $N_L(\varphi_j) = 0$ for each $j = 1, \ldots, M$ and $0 < N_L(\varphi) \le 1$ for all $\varphi \in [-\pi,\, \pi)\setminus \{\varphi_1, \ldots, \varphi_M\}$. Alternatively, one can seek the peaks of the *imaging function* :

$$J_L(\varphi) := \sqrt{L}\, \|\mathbf{Q}_L\, \mathbf{e}_L(\varphi)\|_2^{-1}\,, \quad \varphi \in [-\pi,\, \pi)\,.$$

In this approach, we prefer the zeros or rather the lowest local minima of the noise-space correlation function $N_L(\varphi)$.

We determine the orthogonal projection $\mathbf{Q}_L$ of $\mathbb{C}^L$ onto the noise space $\mathscr{N}_L$ using the *singular value decomposition* (SVD) of the L-trajectory (Hankel) matrix $\mathbf{H}_{L,N-L+1}$, i.e.,

$$\mathbf{H}_{L,N-L+1} = \mathbf{U}_L\, \mathbf{D}_{L,N-L+1}\, \mathbf{W}_{N-L+1}^{\mathrm{H}}\,, \tag{10.21}$$

where

$$\mathbf{U}_L = \big(\mathbf{u}_1 \mid \ldots \mid \mathbf{u}_L\big) \in \mathbb{C}^{L\times L}\,,$$
$$\mathbf{W}_{N-L+1} = \big(\mathbf{w}_1 \mid \ldots \mid \mathbf{w}_{N-L+1}\big) \in \mathbb{C}^{(N-L+1)\times(N-L+1)}$$

are unitary and where

$$\mathbf{D}_{L,N-L+1} = \mathrm{diag}\,\big(\sigma_1, \ldots, \sigma_{\min\{L,\, N-L+1\}}\big) \in \mathbb{R}^{L\times(N-L+1)}$$

is a rectangular diagonal matrix. The diagonal entries of $\mathbf{D}_{L,N-L+1}$ are arranged in nonincreasing order:

$$\sigma_1 \ge \ldots \ge \sigma_M > \sigma_{M+1} = \ldots = \sigma_{\min\{L,N-L+1\}} = 0\,.$$

The columns of $\mathbf{U}_L$ are the *left singular vectors* of $\mathbf{H}_{L,N-L+1}$, and the columns of $\mathbf{W}_{N-L+1}$ are the *right singular vectors* of $\mathbf{H}_{L,N-L+1}$. The nonnegative numbers σ_k are called *singular values* of $\mathbf{H}_{L,N-L+1}$. The rank of $\mathbf{H}_{L,N-L+1}$ is equal to the number of positive singular values. Thus, we can determine the order M of the exponential sum (10.1) by the number of positive singular values σ_j. Practically, for noisy input data, we will have to determine the numerical rank M of $\mathbf{H}_{L,N-L+1}$. From (10.21) it follows that

$$\mathbf{H}_{L,N-L+1}\, \mathbf{W}_{N-L+1} = \mathbf{U}_L\, \mathbf{D}_{L,N-L+1}\,, \quad \mathbf{H}_{L,N-L+1}^{\mathrm{H}}\, \mathbf{U}_L = \mathbf{W}_{N-L+1}\, \mathbf{D}_{L,N-L+1}^{\top}\,.$$

Comparing the columns in the above equations, for each $k = 1, \ldots, \min\{L,\, N - L + 1\}$, we obtain

$$\mathbf{H}_{L,N-L+1}\, \mathbf{w}_k = \sigma_k\, \mathbf{u}_k\,, \quad \mathbf{H}_{L,N-L+1}^{\mathrm{H}}\, \mathbf{u}_k = \sigma_k\, \mathbf{w}_k\,.$$

Introducing the submatrices

$$\mathbf{U}_{L,M}^{(1)} := \big(\mathbf{u}_1 \,|\, \ldots \,|\, \mathbf{u}_M\big) \in \mathbb{C}^{L\times M}\,,$$

$$\mathbf{U}_{L,L-M}^{(2)} := \big(\mathbf{u}_{M+1} \,|\, \ldots \,|\, \mathbf{u}_L\big) \in \mathbb{C}^{L\times (L-M)}\,,$$

we see that the columns of $\mathbf{U}_{L,M}^{(1)}$ form an orthonormal basis of $\mathscr{S}_L$ and that the columns of $\mathbf{U}_{L,L-M}^{(2)}$ form an orthonormal basis of $\mathscr{N}_L$. Hence, the orthogonal projection onto the noise space $\mathscr{N}_L$ is of the form

$$\mathbf{Q}_L = \mathbf{U}_{L,L-M}^{(2)}\, (\mathbf{U}_{L,L-M}^{(2)})^{\mathrm{H}}\,.$$

Consequently, we obtain

$$\begin{aligned}
\|\mathbf{Q}_L\, \mathbf{e}_L(\varphi)\|_2^2 &= \langle \mathbf{Q}_L\, \mathbf{e}_L(\varphi),\, \mathbf{Q}_L\, \mathbf{e}_L(\varphi)\rangle = \langle (\mathbf{Q}_L)^2\, \mathbf{e}_L(\varphi),\, \mathbf{e}_L(\varphi)\rangle \\
&= \langle \mathbf{Q}_L\, \mathbf{e}_L(\varphi),\, \mathbf{e}_L(\varphi)\rangle = \langle \mathbf{U}_{L,L-M}^{(2)}\, (\mathbf{U}_{L,L-M}^{(2)})^{\mathrm{H}}\, \mathbf{e}_L(\varphi),\, \mathbf{e}_L(\varphi)\rangle \\
&= \langle (\mathbf{U}_{L,L-M}^{(2)})^{\mathrm{H}}\, \mathbf{e}_L(\varphi),\, (\mathbf{U}_{L,L-M}^{(2)})^{\mathrm{H}}\, \mathbf{e}_L(\varphi)\rangle = \|(\mathbf{U}_{L,L-M}^{(2)})^{\mathrm{H}}\, \mathbf{e}_L(\varphi)\|_2^2\,.
\end{aligned}$$

Hence, the noise-space correlation function can be represented by

$$\begin{aligned}
N_L(\varphi) &= \frac{1}{\sqrt{L}}\, \|(\mathbf{U}_{L,L-M}^{(2)})^{\mathrm{H}}\, \mathbf{e}_L(\varphi)\|_2 \\
&= \frac{1}{\sqrt{L}} \bigg(\sum_{k=M+1}^{L} |\mathbf{u}_k^{\mathrm{H}}\, \mathbf{e}_L(\varphi)|^2\bigg)^{1/2}\,, \qquad \varphi \in [-\pi,\, \pi)\,.
\end{aligned}$$

In MUSIC, one determines the locations of the lowest local minima of the noise-space correlation function to achieve approximations of the parameters φ_j (see, e.g., [132, 243, 281, 379]).

Algorithm 10.9 (MUSIC via SVD)
Input: $N \in \mathbb{N}$ *with* $N \ge 2\,M$, $L \approx \frac{N}{2}$ *window length,*
$h_k = h(k) + e_k \in \mathbb{C}$, $k = 0, \ldots, N-1$, *noisy sampled values in* (10.1),
$0 < \varepsilon \ll 1$ *tolerance.*

1. *Compute the singular value decomposition*

$$\mathbf{H}_{L,N-L+1} = \tilde{\mathbf{U}}_L\, \tilde{\mathbf{D}}_{L,N-L+1}\, \tilde{\mathbf{W}}_{N-L+1}^{\mathrm{H}}$$

of the rectangular Hankel matrix (10.16), *where the singular values* $\tilde{\sigma}_\ell$ *are arranged in nonincreasing order. Determine the numerical rank* M *of* (10.16) *such that* $\tilde{\sigma}_M \geq \varepsilon\,\tilde{\sigma}_1$ *and* $\tilde{\sigma}_{M+1} < \varepsilon\tilde{\sigma}_1$. *Form the matrix*

$$\tilde{\mathbf{U}}^{(2)}_{L,L-M} = \big(\tilde{\mathbf{u}}_{M+1} \,|\, \ldots \,|\, \tilde{\mathbf{u}}_L\big)$$

from the last $L - M$ *columns of* $\tilde{\mathbf{U}}_L$.

2. *Calculate the squared noise-space correlation function:*

$$\tilde{N}_L(\varphi)^2 := \frac{1}{L} \sum_{k=M+1}^{L} |\tilde{\mathbf{u}}_k^{\mathrm{H}}\, \mathbf{e}_L(\varphi)|^2$$

on the equispaced grid $\{\frac{(2k-S)\pi}{S} : k = 0, \ldots, S-1\}$ *for sufficiently large* $S \in \mathbb{N}$ *by* FFT.

3. *The* M *lowest local minima of* $\tilde{N}_L(\frac{(2k-S)\pi}{S})$, $k = 0, \ldots, S-1$, *yield the frequencies* $\tilde{\varphi}_1, \ldots, \tilde{\varphi}_M$. *Set* $\tilde{z}_j := \mathrm{e}^{\mathrm{i}\,\tilde{\varphi}_j}$, $j = 1, \ldots, M$.
4. *Compute the coefficient vector* $\tilde{\mathbf{c}} := (\tilde{c}_j)_{j=1}^M \in \mathbb{C}^M$ *as solution of the least squares problem*

$$\min_{\tilde{\mathbf{c}} \in \mathbb{C}^M} \|\mathbf{V}_{N,M}(\tilde{\mathbf{z}})\, \tilde{\mathbf{c}} - \big(\tilde{h}_k\big)_{k=0}^{N-1}\|_2\,,$$

where $\tilde{\mathbf{z}} := \big(\tilde{z}_j\big)_{j=1}^M$ *denotes the vector of computed nodes.*

Output: $M \in \mathbb{N}$, $\tilde{\varphi}_j \in [-\pi, \pi)$, $\tilde{c}_j \in \mathbb{C}$, $j = 1, \ldots, M$.

The *approximate Prony method* (APM) can be immediately derived from the MUSIC method. We start with the squared noise-space correlation function

$$\begin{aligned} N_L(\varphi)^2 &= \frac{1}{L}\, \|(\mathbf{U}^{(2)}_{L,L-M})^{\mathrm{H}}\, \mathbf{e}_L(\varphi)\|_2^2 \\ &= \frac{1}{L} \sum_{k=M+1}^{L} \big|\mathbf{u}_k^{\mathrm{H}}\, \mathbf{e}_L(\varphi)\big|^2\,, \quad \varphi \in [-\pi, \pi)\,. \end{aligned}$$

For noiseless data, all frequencies φ_j, $j = 1, \ldots, M$, are zeros of $N_L(\varphi)^2$ and hence especially zeros of

$$\big|\mathbf{u}_L^{\mathrm{H}}\, \mathbf{e}_L(\varphi)\big|^2\,.$$

Thus, we obtain $\mathbf{u}_L^{\mathrm{H}}\, \mathbf{e}_L(\varphi_j) = 0$ for $j = 1, \ldots, M$. Note that $\mathbf{u}_L^{\mathrm{H}}\, \mathbf{e}_L(\varphi)$ can have additional zeros. For noisy data, we observe small values $\big|\mathbf{u}_L^{\mathrm{H}}\, \mathbf{e}_L(\varphi)\big|^2$ near φ_j. Finally, we determine the order M of the exponential sum (10.1) by the number of sufficiently large coefficients in the reconstructed exponential sum.

Algorithm 10.10 (Approximate Prony Method (APM))
Input: $N \in \mathbb{N}$ *with* $N \geq 2\,M$, $L \approx \frac{N}{2}$ *window length,*
$h_k = h(k) + e_k \in \mathbb{C}$, $k = 0, \ldots, N-1$, *noisy sampled values of* (10.1),
$\varepsilon > 0$ *lower bound with* $|c_j| \geq 2\,\varepsilon$, $j = 1, \ldots, M$.

1. *Compute the L-th singular vector* $\mathbf{u}_L = (u_\ell)_{\ell=0}^{L-1} \in \mathbb{C}^L$ *of the rectangular Hankel matrix* (10.16).
2. *Calculate*

$$\mathbf{u}_L^{\mathrm{H}}\,\mathbf{e}_L(\varphi) = \sum_{\ell=0}^{L-1} \bar{u}_\ell\,\mathrm{e}^{\mathrm{i}\,\ell\varphi}$$

on the equispaced grid $\{\frac{(2k-S)\pi}{S} : k = 0, \ldots, S-1\}$ *for sufficiently large* $S \in \mathbb{N}$ *by* FFT.
3. *Determine the lowest local minima* ψ_j, $j = 1, \ldots, \tilde{M}$, *of* $|\mathbf{u}_L^*\,\mathbf{e}_L(\frac{(2k-S)\pi}{S})|^2$, $k = 0, \ldots, S-1$. *Set* $\tilde{w}_j := \mathrm{e}^{\mathrm{i}\tilde{\psi}_j}$, $j = 1, \ldots, \tilde{M}$.
4. *Compute the coefficients* $\tilde{d}_j \in \mathbb{C}$ *as least squares solution of the overdetermined linear system:*

$$\sum_{j=1}^{\tilde{M}} \tilde{d}_j\,\tilde{w}_j = h_k\,, \quad k = 0, \ldots, N-1\,.$$

Delete all the $\tilde{w}_k$ *with* $|\tilde{d}_k| \leq \varepsilon$ *and denote the remaining nodes by* $\tilde{z}_j$, $j = 1, \ldots, M$.
5. *Compute the coefficients* $\tilde{c}_j \in \mathbb{C}$ *as least squares solution of the overdetermined linear system:*

$$\sum_{j=1}^{M} \tilde{c}_j\,\tilde{z}_j = h_k\,, \quad k = 0, \ldots, N-1\,.$$

Output: $M \in \mathbb{N}$, $\tilde{\varphi}_j \in [-\pi, \pi)$, $\tilde{c}_j \in \mathbb{C}$, $j = 1, \ldots, M$.

10.2.2 ESPRIT

In this subsection, we sketch the frequently used ESPRIT *method* (see [344, 368]) which is also based on singular value decomposition of the rectangular Hankel matrix. First, we assume that noiseless data $h_k = h(k)$, $k = 0, \ldots, N-1$, of (10.1) are given. The set of all matrices of the form

$$z\,\mathbf{H}_{L,N-L}(0) - \mathbf{H}_{L,N-L}(1)\,, \quad z \in \mathbb{C}\,, \tag{10.22}$$

with $\mathbf{H}_{L,N-L}(0)$ and $\mathbf{H}_{L,N-L}(1)$ in (10.19) is called a rectangular *matrix pencil*. If a scalar $z_0 \in \mathbb{C}$ and a nonzero vector $\mathbf{v} \in \mathbb{C}^{N-L}$ satisfy

$$z_0\, \mathbf{H}_{L,N-L}(0)\, \mathbf{v} = \mathbf{H}_{L,N-L}(1)\, \mathbf{v}\,,$$

then z_0 is called an *eigenvalue* of the matrix pencil and $\mathbf{v}$ is called *eigenvector*. Note that a rectangular matrix pencil may not have eigenvalues in general. The ESPRIT method is based on the following result:

Lemma 10.11 *Assume that $N \in \mathbb{N}$ with $N \geq 2\,M$ and $L \in \{M, \ldots, N-M\}$ are given. In the case of noiseless data, the matrix pencil* (10.22) *has the nodes $z_j = \mathrm{e}^{\mathrm{i}\varphi_j}$, $j = 1, \ldots, M$, as eigenvalues. Further, zero is an eigenvalue of* (10.22) *with $N-L-M$ linearly independent eigenvectors.*

Proof Let p denote the Prony polynomial (10.3) and let $q(z) := z^{N-L-M}\, p(z)$. Then the companion matrix of q reads

$$\mathbf{C}_{N-L}(q) = \big(\mathbf{e}_1 \,|\, \mathbf{e}_2 \,|\, \ldots \,|\, \mathbf{e}_{N-L-1} \,|\, -\mathbf{q}\big)$$

with $\mathbf{q} := \big(0, \ldots, 0, p_0, p_1, \ldots, p_{M-1}\big)^{\top} \in \mathbb{C}^{N-L}$, where p_k are the coefficients of $p(z)$ in (10.3). Here $\mathbf{e}_k = (\delta_{k-\ell})_{\ell=0}^{N-L-1}$ denote the canonical basis vectors of $\mathbb{C}^{N-L}$. By (10.5) and (10.19), we obtain that

$$\mathbf{H}_{L,N-L}(0)\, \mathbf{q} = -\big(h(\ell)\big)_{\ell=N-L}^{N-1}$$

and hence

$$\mathbf{H}_{L,N-L}(0)\, \mathbf{C}_{N-L}(q) = \mathbf{H}_{L,N-L}(1)\,. \tag{10.23}$$

2. Thus, it follows by (10.23) that the rectangular matrix pencil in (10.22) coincides with the square matrix pencil $z\, \mathbf{I}_{N-L} - \mathbf{C}_{N-L}(q)$ up to a matrix factor:

$$z\, \mathbf{H}_{L,N-L}(0) - \mathbf{H}_{L,N-L}(1) = \mathbf{H}_{L,N-L}(0)\, \big(z\, \mathbf{I}_{N-L} - \mathbf{C}_{N-L}(q)\big)\,.$$

Now we have to determine the eigenvalues of the companion matrix $\mathbf{C}_{N-L}(q)$. By

$$\det\big(z\, \mathbf{I}_{N-L} - \mathbf{C}_{N-L}(q)\big) = q(z) = z^{N-L-M} \prod_{j=1}^{M} (z - z_j)$$

the eigenvalues of $\mathbf{C}_{N-L}(q)$ are zero and z_j, $j = 1, \ldots, M$. Obviously, $z = 0$ is an eigenvalue of the rectangular matrix pencil (10.22), which has $L - M$ linearly independent eigenvectors, since $\operatorname{rank} \mathbf{H}_{L,N-L}(0) = M$ by Lemma 10.8. For each $z = z_j$, $j = 1, \ldots, M$, we can compute an eigenvector $\mathbf{v} = (v_k)_{k=0}^{N-L-1}$ of $\mathbf{C}_{N-L}(q)$, if we set $v_{N-L-1} = z_j$. Thus, we obtain

$$\big(z_j\,\mathbf{H}_{L,N-L}(0) - \mathbf{H}_{L,N-L}(1)\big)\,\mathbf{v} = \mathbf{0}\,.$$

We have shown that the generalized eigenvalue problem of the rectangular matrix pencil (10.22) can be reduced to the classical eigenvalue problem of the square matrix $\mathbf{C}_{N-L}(q)$. ■

We start the ESPRIT method by taking the singular value decomposition (10.21) of the L-trajectory matrix $\mathbf{H}_{L,N-L+1}$ with a window length $L \in \{M, \ldots, N-M\}$. Restricting the matrices $\mathbf{U}_L$ and $\mathbf{W}_{N-L+1}$ to

$$\mathbf{U}_{L,M} := (\mathbf{u}_1 \,|\, \ldots \,|\mathbf{u}_M) \in \mathbb{C}^{L\times M}\,, \quad \mathbf{W}_{N-L+1,M} := (\mathbf{w}_1 \,|\, \ldots \,|\, \mathbf{w}_M) \in \mathbb{C}^{(N-L+1)\times M}$$

with orthonormal columns as well as the diagonal matrix $\mathbf{D}_M := \operatorname{diag}(\sigma_j)_{j=1}^M$, we obtain

$$\mathbf{H}_{L,N-L+1} = \mathbf{U}_{L,M}\,\mathbf{D}_M\,\mathbf{W}_{N-L+1,M}^{\mathrm{H}}\,.$$

Let now $\mathbf{W}_{N-L,M}(0)$ be obtained by removing the last row, and $\mathbf{W}_{N-L,M}(1)$ by removing the first row of $\mathbf{W}_{N-L+1,M}$. Then, by (10.19), the two Hankel matrices $\mathbf{H}_{L,N-L}(0)$ and $\mathbf{H}_{L,N-L}(1)$ in (10.19) can be simultaneously factorized in the form

$$\mathbf{H}_{L,N-L}(s) = \mathbf{U}_{L,M}\,\mathbf{D}_M\,\mathbf{W}_{N-L,M}(s)^{\mathrm{H}}\,, \quad s \in \{0,\,1\}\,. \tag{10.24}$$

Since $\mathbf{U}_{L,M}$ has orthonormal columns and since $\mathbf{D}_M$ is invertible, the *generalized eigenvalue problem* of the matrix pencil

$$z\,\mathbf{W}_{N-L,M}(0)^{\mathrm{H}} - \mathbf{W}_{N-L,M}(1)^{\mathrm{H}}\,, \quad z \in \mathbb{C}\,, \tag{10.25}$$

has the same nonzero eigenvalues z_j, $j = 1, \ldots, M$, as the matrix pencil in (10.22) except for additional zero eigenvalues. Therefore, we determine the nodes z_j, $j = 1, \ldots, M$, as eigenvalues of the matrix

$$\mathbf{F}_M^{\mathrm{SVD}} := \mathbf{W}_{N-L,M}(1)^{\mathrm{H}}\,\big(\mathbf{W}_{N-L,M}(0)^{\mathrm{H}}\big)^{+} \in \mathbb{C}^{M\times M}\,, \tag{10.26}$$

where $\big(\mathbf{W}_{N-L,M}(0)^{\mathrm{H}}\big)^{+}$ denotes the Moore–Penrose pseudoinverse of $\mathbf{W}_{N-L,M}(0)^{\mathrm{H}}$.

Analogously, we can handle the general case of noisy data $h_k = h(k) + e_k \in \mathbb{C}$, $k = 0, \ldots, N-1$, with small error terms $e_k \in \mathbb{C}$, where $|e_k| \le \varepsilon_1$ and $0 < \varepsilon_1 \ll 1$. For the Hankel matrix in (10.21) with the singular values $\tilde{\sigma}_1 \ge \ldots \ge \tilde{\sigma}_{\min\{L,N-L+1\}} \ge 0$, we calculate the numerical rank M of $\mathbf{H}_{L,N-L+1}$ in (10.16) taking $\tilde{\sigma}_M \ge \varepsilon\,\tilde{\sigma}_1$ and $\tilde{\sigma}_{M+1} < \varepsilon\,\tilde{\sigma}_1$ with convenient chosen tolerance ε. Using the IEEE double precision arithmetic, one can choose $\varepsilon = 10^{-10}$ for the given noiseless data. In the case of noisy data, one has to use a larger tolerance $\varepsilon > 0$.

For the rectangular Hankel matrix in (10.16) with noisy entries, we use its singular value decomposition

$$\mathbf{H}_{L,N-L+1} = \tilde{\mathbf{U}}_L \, \tilde{\mathbf{D}}_{L,N-L+1} \, \tilde{\mathbf{W}}^{\mathrm{H}}_{N-L+1}$$

and define as above the matrices $\tilde{\mathbf{U}}_{L,M}$, $\tilde{\mathbf{D}}_M := \operatorname{diag}\big(\tilde{\sigma}_j\big)_{j=1}^M$, and $\tilde{\mathbf{W}}_{N-L+1,M}$. Then

$$\tilde{\mathbf{U}}_{L,M} \, \tilde{\mathbf{D}}_M \, \tilde{\mathbf{W}}^{\mathrm{H}}_{N-L+1,M}$$

is a low-rank approximation of (10.16). Analogously to $\mathbf{W}_{N-L,M}(0)$, $\mathbf{W}_{N-L,M}(1)$ and (10.26), we introduce corresponding matrices $\tilde{\mathbf{W}}_{N-L,M}(s)$, $s \in \{0, 1\}$ and $\tilde{\mathbf{F}}_M^{\mathrm{SVD}}$. Note that

$$\tilde{\mathbf{K}}_{L,N-L}(s) := \tilde{\mathbf{U}}_{L,M} \, \tilde{\mathbf{D}}_M \, \tilde{\mathbf{W}}_{N-L,M}(s)^{\mathrm{H}} \,, \quad s \in \{0, 1\} \tag{10.27}$$

is a low-rank approximation of $\tilde{\mathbf{H}}_{L,N-L}(s)$. Thus, the SVD-based ESPRIT algorithm reads as follows:

Algorithm 10.12 (ESPRIT via SVD)
Input: $N \in \mathbb{N}$ *with* $N \gg 1$, $M \le L \le N - M$, $L \approx \frac{N}{2}$, M *unknown order of* (10.1), $h_k = h(k) + e_k \in \mathbb{C}$, $k = 0, \ldots, N-1$, *noisy sampled values of* (10.1), $0 < \varepsilon \ll 1$ *tolerance.*

1. *Compute the singular value decomposition of the rectangular Hankel matrix* $\mathbf{H}_{L,N-L+1}$ *in* (10.16). *Determine the numerical rank* M *of* $\mathbf{H}_{L,N-L+1}$ *such that* $\tilde{\sigma}_M \ge \varepsilon\,\tilde{\sigma}_1$ *and* $\tilde{\sigma}_{M+1} < \varepsilon\tilde{\sigma}_1$. *Form the matrices* $\tilde{\mathbf{W}}_{N-L,M}(s)$, $s \in \{0, 1\}$.
2. *Calculate the square matrix* $\tilde{\mathbf{F}}_M^{\mathrm{SVD}}$ *as in* (10.26) *and compute all eigenvalues* $\tilde{z}_j$, $j = 1, \ldots, M$, *of* $\tilde{\mathbf{F}}_M^{\mathrm{SVD}}$. *Replace* $\tilde{z}_j$ *by the corrected value* $\frac{\tilde{z}_j}{|\tilde{z}_j|}$, $j = 1, \ldots, M$, *and set* $\tilde{\varphi}_j := \log \tilde{z}_j$, $j = 1, \ldots, M$, *where* log *denotes the principal value of the complex logarithm.*
3. *Compute the coefficient vector* $\tilde{\mathbf{c}} := (\tilde{c}_j)_{j=1}^M \in \mathbb{C}^M$ *as solution of the least squares problem*

$$\min_{\tilde{\mathbf{c}} \in \mathbb{C}^M} \|\mathbf{V}_{N,M}(\tilde{\mathbf{z}})\,\tilde{\mathbf{c}} - \big(h_k\big)_{k=0}^{N-1}\|_2 \,,$$

where $\tilde{\mathbf{z}} := \big(\tilde{z}_j\big)_{j=1}^M$ *denotes the vector of computed nodes.*

Output: $M \in \mathbb{N}$, $\tilde{\varphi}_j \in [-\pi, \pi)$, $\tilde{c}_j \in \mathbb{C}$ *for* $j = 1, \ldots, M$.

Remark 10.13 One can avoid the computation of the Moore–Penrose pseudoinverse in (10.26). Then the second step of Algorithm 10.12 reads as follows (see [346, Algorithm 4.2]):

2′. Calculate the matrix products

$$\tilde{\mathbf{A}}_M := \tilde{\mathbf{W}}_{N-L,M}(0)^{\mathrm{H}} \, \tilde{\mathbf{W}}_{N-L,M}(0) \,, \quad \tilde{\mathbf{B}}_M := \tilde{\mathbf{W}}_{N-L,M}(1)^{\mathrm{H}} \, \tilde{\mathbf{W}}_{N-L,M}(0)$$

and compute all eigenvalues $\tilde{z}_j$, $j = 1, \ldots, M$, of the square matrix pencil $z\,\tilde{\mathbf{A}}_M - \tilde{\mathbf{B}}_M$, $z \in \mathbb{C}$, by the QZ algorithm (see [166, pp. 384–385]). Set $\tilde{\varphi}_j := \log \tilde{z}_j$, $j = 1, \ldots, M$.

The computational cost of ESPRIT is governed by the SVD of the Hankel matrix in the first step. For $L \approx \frac{N}{2}$, the SVD costs about $\frac{21}{8}N^3 + M^2(21N + \frac{91}{3}M)$ operations. In [346], a partial singular value decomposition of the Hankel matrix based on Lanczos bidiagonalization is proposed that reduces the computational cost to $18SN\log_2 N + S^2(20N + 30S) + M^2(N + \frac{1}{3}M)$ operations. Here S denotes the number of bidiagonalization steps. □

Remark 10.14 For various numerical examples as well as for a comparison between Algorithm 10.12 and a further Prony-like method, see [316]. Algorithm 10.12 is very similar to Algorithm 3.2 in [345]. Note that one can also use the QR decomposition of the rectangular Hankel matrix (10.16) instead of the singular value decomposition. In that case, one obtains an algorithm that is similar to the *matrix pencil method* [206, 374] (see also Algorithm 3.1 in [345]). The matrix pencil method has been applied to reconstruction of shapes from moments in [165] (see also Remark 10.7). In order to obtain a consistent estimation method, one can rewrite the problem of parameter estimation in exponential sums as a *nonlinear eigenvalue problem* (see, e.g., [62, 308] or the survey [452] and references therein). The obtained modification of the Prony method aims at solving the minimization problem:

$$\arg\min \left\{ \sum_{k=0}^{N-1} \Big| h_k - \sum_{j=1}^{M} c_j\, \mathrm{e}^{\mathrm{i}\varphi_j} \Big|^2 : \; c_j \in \mathbb{C},\; \varphi_j \in [-\pi, \pi),\; j = 1, \ldots, M \right\}.$$

A slightly different approach has been taken in [389], where the 1-norm of errors $\sum_{k=0}^{N-1} |h_k - \sum_{j=1}^{M} c_j\, \mathrm{e}^{\mathrm{i}\varphi_j}|$ is minimized instead of the Euclidean norm.

Remark 10.15 The numerical stability of the considered numerical methods strongly depends on the condition number of the involved Vandermonde matrix $V_{N,M}(\mathbf{z}) = (z_j^{k-1})_{k,j=1}^{N,M}$ with $z_j = \mathrm{e}^{\mathrm{i}\varphi_j}$ that appears in the factorization of the rectangular Hankel matrices $H_{L,N-L+1}$ in (10.18). Moreover, $V_{N,M}$ also occurs as the coefficient matrix in the overdetermined equation system to compute the coefficient vector (see step 4 in Algorithm 10.9, step 5 in Algorithm 10.10, or step 3 in Algorithm 10.12). In [28, 291, 347], the condition number of a rectangular Vandermonde matrix with nodes on the unit circle is estimated. It has been shown that this matrix is well conditioned, provided that the nodes z_j are not extremely close to each other and provided N is large enough. Stability issues are discussed in a more detailed manner in Sect. 10.3. □

Remark 10.16 The algorithms given in this section can be simply transferred to the general parameter estimation problem (10.1), where we only assume that $\phi_j \in \mathbb{C}$ with $\operatorname{Im}\phi_j \in [-K\pi,\, K\pi)$ with some constant $K > 0$ (see also Remark 10.3). □

Remark 10.17 The given data sequence $\{h_0, h_1, \ldots, h_{N-1}\}$ can be also interpreted as *time series*. A powerful tool of time series analysis is the *singular spectrum analysis* (see [167, 168]). Similarly as step 1 of Algorithm 10.12, this technique is based on the singular value decomposition of a rectangular Hankel matrix constructed upon the given time series h_k. By this method, the original time series can be decomposed into a sum of interpretable components such as trend, oscillatory components, and noise. For further details and numerous applications, see [167, 168]. □

Remark 10.18 The considered Prony-like method can be interpreted as a model reduction based on *low-rank approximation* of Hankel matrices (see [244, 282, 283]). The *structured low-rank approximation problem* reads as follows: For a given structure specification $\mathscr{S} : \mathbb{C}^K \to \mathbb{C}^{L\times N}$ with $L < N$, a parameter vector $\mathbf{h} \in \mathbb{C}^K$ and an integer M with $0 < M < L$, find a vector

$$\hat{\mathbf{h}}^* = \arg\min \left\{ \|\mathbf{h} - \hat{\mathbf{h}}\| : \hat{\mathbf{h}} \in \mathbb{C}^K \text{ with rank } \mathscr{S}(\hat{\mathbf{h}}) \le M \right\},$$

where $\|\cdot\|$ denotes a suitable norm in $\mathbb{C}^K$. In the special case of a Hankel matrix structure, the Hankel matrix $\mathscr{S}(\mathbf{h}) = (h_{\ell+k})_{\ell,\,k=0}^{L-1,N-1}$ is rank-deficient of order M if there exists a nonzero vector $\mathbf{p} = (p_k)_{k=0}^{M-1}$ so that

$$\sum_{k=0}^{M-1} p_k\, h(m+k) = -h(M+m)$$

for all $m = 0, \ldots, N+L-M-1$. Equivalently, the values $h(k)$ can be interpreted as function values of an exponential sum of order M in (10.1). The special kernel structure of rank-deficient Hankel matrices can already be found in [200]. □

Remark 10.19 The *d-dimensional parameter estimation problem* with fixed $d \in \mathbb{N} \setminus \{1\}$ reads as follows:

Recover the positive integer M, distinct parameter vectors $\boldsymbol{\varphi}_j \in \mathbb{R}^d + \mathrm{i}\,[-\pi,\, \pi)^d$, and complex coefficients $c_j \neq 0$, $j = 0, \ldots, M$, in the d-variate exponential sum of order M

$$h(\mathbf{x}) := \sum_{j=1}^{M} c_j\, \mathrm{e}^{\mathrm{i}\boldsymbol{\varphi}_j\cdot\mathbf{x}},$$

if noisy sampling values $h_{\mathbf{k}} := h(\mathbf{k}) + e_{\mathbf{k}}$, $\mathbf{k} \in I$, are given, where $e_{\mathbf{k}} \in \mathbb{C}$ are small error terms and where I is a suitable finite subset of $\mathbb{Z}^d$.

Up to now, there exist different approaches to the numerical solution of the d-dimensional parameter estimation problem. In [97, 98, 329, 343, 388], this problem is reduced by projections to several one-dimensional frequency analysis problems. The reconstruction of multivariate trigonometric polynomials of large sparsity is described in [350], where sampling data are given on a convenient rank-1 lattice. Direct numerical methods to solve the multivariate Prony problem are subject of very active ongoing research (see, e.g., [7, 126, 254, 256, 315, 372, 375, 388]). These approaches are, for example, based on a direct generalization of the Prony method leading to the problem of finding intersections of zero sets of multivariate polynomials (see [256, 315]) or exploit the relationship between polynomial interpolation, normal forms modulo ideals, and H-bases [254, 375]. Other ideas can be understood as direct generalization of ESPRIT or matrix pencil methods [7, 126] or are related to low-rank decomposition of Hankel matrices [372]. □

10.2.3 ESPIRA

Finally, we present a recent approach from [110] called *estimation of signal parameters by iterative rational approximation* (ESPIRA)) to reconstruct an exponential sum $h(x) = \sum_{j=1}^{M} c_j \, \mathrm{e}^{\phi_j x} = \sum_{j=1}^{M} c_j \, z_j^x$ in (10.2) from data $h_k = h(k)$ or noisy data $h_k = h(k) + e_k$, $k = 0, \ldots, N-1$, where $N > 2M$ and where $c_j \neq 0$ and $z_j = \mathrm{e}^{\phi_j}$ are pairwise distinct for $j = 1, \ldots, M$. The technique is derived for noiseless data $h_k = h(k)$, but the method works well also for noisy measurements.

For the given data vector $\mathbf{h} := (h_k)_{k=0}^{N-1}$, we consider its discrete Fourier transform $\hat{\mathbf{h}} = (\hat{h}_\ell)_{\ell=0}^{N-1}$, where

$$\hat{h}_\ell = \sum_{k=0}^{N-1} h_k \, \omega_N^{k\ell}, \qquad \ell = 0, \ldots, N-1,$$

and $\omega_N := \mathrm{e}^{-2\pi \mathrm{i}/N}$. Then we observe for $z_j^N \neq 1$ that

$$\begin{aligned} \hat{h}_\ell &= \sum_{k=0}^{N-1} \left(\sum_{j=1}^{M} c_j \, z_j^k \right) \omega_N^{k\ell} = \sum_{j=1}^{M} c_j \sum_{k=0}^{N-1} (z_j \, \omega_N^\ell)^k \\ &= \sum_{j=1}^{M} c_j \left(\frac{1 - (z_j \, \omega_N^\ell)^N}{1 - z_j \, \omega_N^\ell} \right) = \omega_N^{-\ell} \sum_{j=1}^{M} c_j \left(\frac{1 - z_j^N}{\omega_N^{-\ell} - z_j} \right). \end{aligned} \tag{10.28}$$

Using this fractional structure, the parameter estimation problem can be rephrased as a *rational interpolation problem*. We consider the rational function

$$r_M(z) := \sum_{j=1}^{M} \frac{c_j \left(1 - z_j^N\right)}{z - z_j} \tag{10.29}$$

of type $(M-1, M)$, where z_j are the wanted knots and c_j the wanted coefficients to determine $h(x)$. Then we observe from (10.28) that

$$r_M(\omega_N^{-\ell}) = \sum_{j=1}^{M} \frac{c_j\,\left(1-z_j^N\right)}{\omega_N^{-\ell}-z_j}.$$

In other words, to find M, c_j, and z_j, $j = 1, \ldots, M$, it would be also sufficient to determine a rational function r_M of type $(M-1, M)$ from the N interpolation conditions

$$r_M(\omega_N^{-\ell}) = \omega_N^{\ell}\,\hat{h}_\ell, \qquad \ell = 0, \ldots, N-1,$$

which are derived from (10.28). Then, all parameters can be simply read from the representation (10.29) of r_M. We obtain the following algorithm.

Algorithm 10.20 (ESPIRA)
Input: $N \in \mathbb{N}$ *with* $N \gg 1$, $M < N/2$, M *unknown order of* (10.1),
$\mathbf{h} = (h_k)_{k=0}^{N-1}$, $h_k = h(k) + e_k \in \mathbb{C}$, $k = 0, \ldots, N-1$, *noisy samples of* (10.1),
$0 < \varepsilon \ll 1$ *tolerance for approximation error.*

1. *Compute the* DFT *vector* $\hat{\mathbf{h}} = (\hat{h}_\ell)_{\ell=0}^{N-1}$ *using a fast algorithm for* DFT(N).
2. *Compute a rational function* $r_M(z)$ *of smallest possible type* $(M-1, M)$ *(with* $M \le N/2$*) such that*
$$|r_M(\omega_N^{-\ell}) - \omega_N^{\ell}\,\hat{h}_\ell| < \varepsilon, \qquad \ell = 0, \ldots, N-1.$$
3. *Compute a fractional decomposition of* $r_M(z)$,
$$r_M(z) = \sum_{j=1}^{M} \frac{a_j}{z-z_j},$$
i.e., compute a_j, z_j, $j = 1, \ldots, M$, *and set* $c_j := \frac{a_j}{1-z_j^N}$, $j = 1, \ldots, M$.

Output: $M \in \mathbb{N}$, z_j, $\tilde{c}_j \in \mathbb{C}$ *for* $j = 1, \ldots, M$.

For stable algorithms to perform steps 2 and 3 of Algorithm 10.20, we refer to [110]. The algorithm for rational interpolation used in [110] is based on the adaptive Antoulas–Anderson (AAA) algorithm proposed in [296]. The original AAA algorithm is intended for approximation of functions by rational functions, but in the special application in Algorithm 10.20, it acts like an interpolation algorithm in case of noiseless input data.

The complexity of Algorithm 10.20 is mainly governed by the complexity of the AAA algorithm, which is $\mathscr{O}(NM^2)$.

Remark 10.21

1. For knots z_j with $z_j^N = 1$, the fractional structure in (10.28) is not obtained. Therefore, knots of this form have to be determined in a post-processing step.
2. The ESPIRA algorithm based on the AAA algorithm can be reinterpreted as a matrix pencil method for so-called Loewner matrices instead of Hankel matrices. The ESPIRA algorithm often behaves more stably with regard to noisy input data than the other recovery methods if the number of given input data is sufficiently large.

Remark 10.22 Other numerical methods for recovery of exponential sums or extended exponential sums from a finite number of its Fourier coefficients, which are also based on rational approximation, in a fixed interval $[0, K]$ can be found in [109, 317].

There also exist special algorithms for direct recovery of cosine sums of the form $h(x) = \sum_{k=1}^{M} c_k \cos(\phi_j x)$, which completely work in real arithmetic (see, e.g., [99, 111, 323]). The approach in [345] for the reconstruction of sparse Chebyshev polynomials can also be directly generalized to the reconstruction of cosine sums with nonnegative real frequencies ϕ_j.

10.3 Stability of Exponentials

In the last section, we have derived several numerical methods for the recovery of exponential sums $h(x) = \sum_{j=1}^{M} c_j \, \mathrm{e}^{\mathrm{i}\varphi_j x}$. These methods work exactly for noiseless data. Fortunately, they can be also applied to noisy data $h_k = h(k) + e_k$, $k = 0, \ldots, N-1$, with error terms $e_k \in \mathbb{C}$ provided that the bound $\varepsilon_1 > 0$ of all $|e_k|$ is small enough. This property is based on the perturbation theory of the singular value decomposition of a rectangular Hankel matrix. Here we have to assume that the frequencies $\varphi_j \in [-\pi, \pi)$, $j = 1, \ldots, M$, are not too close to each other, that the number N of given samples is sufficiently large with $N \ge 2M$, and that the window length L satisfies $L \approx \frac{N}{2}$. We start with the following stability result (see [209], [449, pp. 162–164] or [246, pp. 59–66]).

Lemma 10.23 *Let $M \in \mathbb{N}$ and $T > 0$ be given. If the ordered frequencies $\varphi_j \in \mathbb{R}$, $j = 1, \ldots, M$, satisfy the gap condition*

$$\varphi_{j+1} - \varphi_j \ge q > \frac{\pi}{T}, \quad j = 1, \ldots, M-1, \tag{10.30}$$

then the exponentials $\mathrm{e}^{\mathrm{i}\varphi_j \cdot}$, $j = 1, \ldots, M$, are Riesz stable *in $L_2[0, 2T]$, i.e., for all vectors $\mathbf{c} = (c_j)_{j=1}^{M} \in \mathbb{C}^M$ we have the* Ingham inequalities

$$\alpha(T)\,\|\mathbf{c}\|_2^2 \le \|\sum_{j=1}^{M} c_j\,\mathrm{e}^{\mathrm{i}\varphi_j\,\cdot}\|_{L_2[0,\,2T]}^2 \le \beta(T)\,\|\mathbf{c}\|_2^2 \tag{10.31}$$

with positive constants

$$\alpha(T) := \frac{2}{\pi}\left(1 - \frac{\pi^2}{T^2 q^2}\right), \quad \beta(T) := \frac{4\sqrt{2}}{\pi}\left(1 + \frac{\pi^2}{4T^2 q^2}\right),$$

where $\|f\|_{L_2[0,2T]}$ *is given by*

$$\|f\|_{L_2[0,2T]} := \left(\frac{1}{2T}\int_0^{2T} |f(t)|^2\,\mathrm{d}t\right)^{1/2}, \quad f \in L^2[0,\,2T].$$

Proof

1. For arbitrary $\mathbf{c} = (c_j)_{j=1}^M \in \mathbb{C}^M$, let

$$h(x) := \sum_{j=1}^{M} c_j\,\mathrm{e}^{\mathrm{i}\varphi_j x}, \quad x \in [0,\,2T]. \tag{10.32}$$

Substituting $t = x - T \in [-T,\,T]$, we obtain

$$f(t) = h(t+T) = \sum_{j=1}^{M} d_j\,\mathrm{e}^{\mathrm{i}\varphi_j t}, \quad t \in [-T,\,T],$$

with $d_j := c_j\,\mathrm{e}^{\mathrm{i}\varphi_j T}$, $j = 1, \ldots, M$. Note that $|d_j| = |c_j|$ and

$$\|f\|_{L_2[-T,\,T]} = \|h\|_{L_2[0,\,2T]}\,.$$

For simplicity, we assume that $T = \pi$. If $T \ne \pi$, then we substitute $s = \frac{\pi}{T}\,t \in [-\pi, \pi]$ for $t \in [-T, T]$ such that

$$f(t) = f\left(\frac{T}{\pi}\,s\right) = \sum_{j=1}^{M} d_j\,\mathrm{e}^{\mathrm{i}\,\psi_j s}, \quad s \in [-\pi,\,\pi],$$

with $\psi_j := \frac{T}{\pi}\,\varphi_j$ and conclude from the gap condition (10.30) that

$$\psi_{j+1} - \psi_j = \frac{T}{\pi}\left(\varphi_{j+1} - \varphi_j\right) \ge \frac{T}{\pi}\,q > 1\,.$$

2. For a fixed function $k \in L_1(\mathbb{R})$ and its Fourier transform

$$\hat{k}(\omega) := \int_{\mathbb{R}} k(t)\, \mathrm{e}^{-\mathrm{i}\omega t}\, \mathrm{d}t\,, \quad \omega \in \mathbb{R}\,,$$

we see that

$$\begin{aligned}\int_{\mathbb{R}} k(t)\, |f(t)|^2\, \mathrm{d}t &= \sum_{j=1}^{M} \sum_{\ell=1}^{M} d_j\, \bar{d}_\ell \int_{\mathbb{R}} k(t)\, \mathrm{e}^{-\mathrm{i}(\psi_\ell - \psi_j)t}\, \mathrm{d}t \\ &= \sum_{j=1}^{M} \sum_{\ell=1}^{M} d_j\, \bar{d}_\ell\, \hat{k}(\psi_\ell - \psi_j)\,.\end{aligned}$$

If we choose

$$k(t) := \begin{cases} \cos\frac{t}{2} & t \in [-\pi,\, \pi]\,, \\ 0 & t \in \mathbb{R} \setminus [-\pi,\, \pi]\,, \end{cases}$$

then we obtain the Fourier transform

$$\hat{k}(\omega) = \frac{4\, \cos(\pi\omega)}{1 - 4\,\omega^2}\,, \quad \omega \in \mathbb{R} \setminus \left\{-\frac{1}{2},\, \frac{1}{2}\right\}, \tag{10.33}$$

with $\hat{k}(\pm\frac{1}{2}) = \pi$ and hence

$$\int_{-\pi}^{\pi} \cos\frac{t}{2}\, |f(t)|^2\, \mathrm{d}t = \sum_{j=1}^{M} \sum_{\ell=1}^{M} d_j\, \bar{d}_\ell\, \hat{k}(\psi_\ell - \psi_j)\,. \tag{10.34}$$

3. From (10.34) it follows immediately that

$$\int_{-\pi}^{\pi} |f(t)|^2\, \mathrm{d}t \geq \sum_{j=1}^{M} \sum_{\ell=1}^{M} d_j\, \bar{d}_\ell\, \hat{k}(\psi_\ell - \psi_j)\,.$$

Let S_1 denote that part of the above double sum for which $j = \ell$ and let S_2 be the remaining part. Clearly, by $\hat{k}(0) = 4$ we get

$$S_1 = 4 \sum_{j=1}^{M} |d_j|^2\,. \tag{10.35}$$

Since $\hat{k}$ is even and since $2\,|d_j\, \bar{d}_\ell| \leq |d_j|^2 + |d_\ell|^2$, there are constants $\theta_{j,\ell} \in \mathbb{C}$ with $|\theta_{j,\ell}| \leq 1$ and $\theta_{j,\ell} = \bar{\theta}_{\ell,j}$ such that

$$S_2 = \sum_{j=1}^{M} \sum_{\substack{\ell=1 \\ \ell \neq j}}^{M} \frac{|d_j|^2 + |d_\ell|^2}{2} \, \theta_{j,\ell} \, |\hat{k}(\psi_\ell - \psi_j)|$$

$$= \sum_{j=1}^{M} |d_j|^2 \left(\sum_{\substack{\ell=1 \\ \ell \neq j}}^{M} \mathrm{Re}\, \theta_{j,\ell} \, |\hat{k}(\psi_\ell - \psi_j)| \right).$$

Consequently, there exists a constant $\theta \in [-1, 1]$ such that

$$S_2 = \theta \sum_{j=1}^{M} |d_j|^2 \left(\sum_{\substack{\ell=1 \\ \ell \neq j}}^{M} |\hat{k}(\psi_\ell - \psi_j)| \right). \tag{10.36}$$

Since $|\psi_\ell - \psi_j| \geq |\ell - j| \, q > 1$ for $\ell \neq j$ by (10.30) (with $\pi = T$), we obtain by (10.33) that

$$\sum_{\substack{\ell=1 \\ \ell \neq j}}^{M} |\hat{k}(\psi_\ell - \psi_j)| \leq \sum_{\substack{\ell=1 \\ \ell \neq j}}^{M} \frac{4}{4\,(\ell - j)^2 \, q^2 - 1} < \frac{8}{q^2} \sum_{n=1}^{\infty} \frac{1}{4n^2 - 1}$$

$$= \frac{4}{q^2} \sum_{n=1}^{\infty} \left(\frac{1}{2n - 1} - \frac{1}{2n + 1} \right) = \frac{4}{q^2}. \tag{10.37}$$

Hence, from (10.35)–(10.37), it follows that

$$\frac{1}{2\pi} \int_{-\pi}^{\pi} |f(t)|^2 \, \mathrm{d}t \geq \alpha(\pi) \sum_{j=1}^{M} |d_j|^2$$

with $\alpha(\pi) = \frac{2}{\pi} \left(1 - \frac{1}{q^2}\right)$. In the case $T \neq \pi$, we obtain $\alpha(T) = \frac{2}{\pi} \left(1 - \frac{\pi^2}{T^2 q^2}\right)$ by the substitution in step 1 and hence

$$\|h\|^2_{L_2[0,\,2T]} \geq \alpha(T) \sum_{j=1}^{M} |c_j|^2 = \alpha(T)\, \|\mathbf{c}\|_2^2.$$

4. From (10.34)–(10.37), we conclude on the one hand

$$\int_{-\pi}^{\pi} \cos\frac{t}{2} \, |f(t)|^2 \, \mathrm{d}t \geq \int_{-\pi/2}^{\pi/2} \cos\frac{t}{2} \, |f(t)|^2 \, \mathrm{d}t \geq \frac{\sqrt{2}}{2} \int_{-\pi/2}^{\pi/2} |f(t)|^2 \, \mathrm{d}t$$

and on the other hand

$$\int_{-\pi}^{\pi} \cos\frac{t}{2}\,|f(t)|^2\,\mathrm{d}t = \sum_{j=1}^{M}\sum_{\ell=1}^{M} \hat{k}(\psi_\ell - \psi_j)\,d_j\,\bar{d}_\ell$$

$$\leq 4\sum_{j=1}^{M}|d_j|^2 + \frac{4}{q^2}\sum_{j=1}^{M}|d_j|^2 = 4\left(1+\frac{1}{q^2}\right)\sum_{j=1}^{M}|d_j|^2\,.$$

Thus, we obtain

$$\frac{1}{\pi}\int_{-\pi/2}^{\pi/2}|f(t)|^2\,\mathrm{d}t \leq \frac{4\sqrt{2}}{\pi}\left(1+\frac{1}{q^2}\right)\sum_{j=1}^{M}|d_j|^2\,. \tag{10.38}$$

5. Now we consider the function

$$g(t) := f(2t) = \sum_{j=1}^{M} d_j\,\mathrm{e}^{2\mathrm{i}\psi_j t}\,,\quad t\in\left[-\frac{\pi}{2},\,\frac{\pi}{2}\right],$$

where the ordered frequencies $2\,\psi_j$ satisfy the gap condition:

$$2\,\psi_{j+1} - 2\,\psi_j \geq 2\,q\,,\quad j = 1,\ldots,M-1\,.$$

Applying (10.38) to the function g, we receive

$$\frac{1}{2\pi}\int_{-\pi}^{\pi}|f(t)|^2\,\mathrm{d}t = \frac{1}{\pi}\int_{-\pi/2}^{\pi/2}|g(t)|^2\,\mathrm{d}t \leq \frac{4\sqrt{2}}{\pi}\left(1+\frac{1}{4q^2}\right)\sum_{j=1}^{M}|d_j|^2\,.$$

Hence, $\beta(\pi) = \frac{4\sqrt{2}}{\pi}\,(1+\frac{1}{4q^2})$ and $\beta(T) = \frac{4\sqrt{2}}{\pi}\,(1+\frac{\pi^2}{4T^2q^2})$ by the substitution in step 1. Thus, we obtain

$$\|h\|^2_{L_2[0,\,2T]} \leq \beta(T)\sum_{j=1}^{M}|d_j|^2 = \beta(T)\,\|\mathbf{c}\|_2^2\,.$$

This completes the proof. ■

Remark 10.24 The Ingham inequalities (10.31) can be considered as far-reaching generalization of the Parseval equality for Fourier series. The constants $\alpha(T)$ and $\beta(T)$ are not optimal in general. Note that these constants do not depend on M. The assumption $q > \frac{\pi}{T}$ is necessary for the existence of positive $\alpha(T)$. Compare also with [85, Theorems 9.8.5 and 9.8.6] and [254]. □

In the following, we present a discrete version of the Ingham inequalities (10.31) (see [13, 273, 290]). For sufficiently large integer $P > M$, we consider the rectangular Vandermonde matrix

$$\mathbf{V}_{P,M}(\mathbf{z}) := \big(z_j^{k-1}\big)_{k,j=1}^{P,M} = \begin{pmatrix} 1 & 1 & \ldots & 1 \\ z_1 & z_2 & \ldots & z_M \\ \vdots & \vdots & & \vdots \\ z_1^{P-1} & z_2^{P-1} & \ldots & z_M^{P-1} \end{pmatrix}$$

with $\mathbf{z} = (z_j)_{j=1}^M$, where $z_j = \mathrm{e}^{\mathrm{i}\varphi_j}$, $j = 1, \ldots, M$, are distinct nodes on the unit circle. Setting $\varphi_j = 2\pi\,\psi_j$, $j = 1, \ldots, M$, we measure the distance between distinct frequencies ψ_j, ψ_ℓ by $d(\psi_j - \psi_\ell)$, where $d(x)$ denotes the *distance of* $x \in \mathbb{R}$ *to the nearest integer*, i.e.,

$$d(x) := \min_{n\in\mathbb{Z}} |x - n| \in \Big[0, \frac{1}{2}\Big] .$$

Our aim is a good estimation of the spectral condition number of $\mathbf{V}_{P,M}(\mathbf{z})$. Therefore, we assume that ψ_j, $j = 1, \ldots, M$, satisfy the *gap condition*:

$$\min\{d(\psi_j - \psi_\ell) : \, j, \, \ell = 1, \ldots, M, \, j \neq \ell\} \geq \Delta > 0 . \tag{10.39}$$

The following discussion is mainly based on a generalization of the Hilbert inequality (see [13, 290]). Note that the Hilbert inequality reads originally as follows:

Lemma 10.25 *For all* $\mathbf{x} = (x_j)_{j=1}^M \in \mathbb{C}^M$, *we have* the Hilbert inequality:

$$\Big|\sum_{\substack{j,\ell=1\\ j\neq\ell}}^{M} \frac{x_j\,\overline{x}_\ell}{j-\ell}\Big| \leq \pi\,\|\mathbf{x}\|_2^2 .$$

Proof For an arbitrary vector $\mathbf{x} = (x_j)_{j=1}^M \in \mathbb{C}^M$, we form the trigonometric polynomial

$$p(t) := \sum_{k=1}^{M} x_k\,\mathrm{e}^{\mathrm{i}\,k\,t}$$

such that

$$|p(t)|^2 = \sum_{k,\ell=1}^{M} x_k\,\overline{x}_\ell\,\mathrm{e}^{\mathrm{i}\,(k-\ell)\,t} .$$

Using

$$\frac{1}{2\pi\,\mathrm{i}} \int_0^{2\pi} (\pi - t)\,\mathrm{e}^{\mathrm{i}nt}\,\mathrm{d}t = \begin{cases} 0 & n = 0\,, \\ \frac{1}{n} & n \in \mathbb{Z} \setminus \{0\}\,, \end{cases}$$

we obtain

$$\frac{1}{2\pi\,\mathrm{i}} \int_0^{2\pi} (\pi - t)\,|p(t)|^2\,\mathrm{d}t = \sum_{\substack{k,\ell=1 \\ k \neq \ell}}^{M} \frac{x_k\,\overline{x}_\ell}{k - \ell}\,.$$

Note that $|\pi - t| \le \pi$ for $t \in [0,\, 2\pi]$. From the triangle inequality and the Parseval equality in $L_2(\mathbb{T})$, it follows that

$$\frac{1}{2\pi} \Big| \int_0^{2\pi} (\pi - t)\,|p(t)|^2\,\mathrm{d}t \Big| \le \frac{1}{2} \int_0^{2\pi} |p(t)|^2\,\mathrm{d}t = \pi \sum_{j=1}^{M} |x_j|^2 = \pi\,\|\mathbf{x}\|_2^2\,.$$

■

To show the generalized Hilbert inequality, we need the following result:

Lemma 10.26 *For all $x \in \mathbb{R} \setminus \mathbb{Z}$, we have*

$$\big(\sin(\pi x)\big)^{-2} + 2\,\Big|\frac{\cot(\pi x)}{\sin(\pi x)}\Big| \le \frac{3}{\pi^2\,d(x)^2}\,. \tag{10.40}$$

Proof It suffices to show (10.40) for all $x \in (0,\, \frac{1}{2}]$. Substituting $t = \pi x \in (0,\, \frac{\pi}{2}]$, (10.40) means

$$3\,(\sin t)^2 \ge t^2\,(1 + 2\,\cos t)\,,$$

since $d(x)^2 = x^2 = \frac{t^2}{\pi^2}$. This inequality is equivalent to

$$3\,(\mathrm{sinc}\,t)^2 \ge 1 + 2\,\cos t\,, \quad t \in \Big[0,\, \frac{\pi}{2}\Big]\,,$$

which is true by the behaviors of the concave functions $3\,(\mathrm{sinc}\,t)^2$ and $1 + 2\,\cos t$ on the interval $[0,\, \frac{\pi}{2}]$. ■

Theorem 10.27 (See [292, Theorem 1]) *Assume that the distinct values $\psi_j \in \mathbb{R}$, $j = 1, \ldots, M$, satisfy the gap condition* (10.39) *with a constant $\Delta > 0$.*
Then the generalized Hilbert inequality

$$\Big| \sum_{\substack{j,\ell=1 \\ j\neq \ell}}^{M} \frac{x_j\,\overline{x}_\ell}{\sin\big(\pi\,(\psi_j-\psi_\ell)\big)}\Big| \le \frac{1}{\Delta}\,\|\mathbf{x}\|_2^2 \tag{10.41}$$

holds for all $\mathbf{x} = (x_j)_{j=1}^M \in \mathbb{C}^M$.

Proof

1. Setting

$$s_{j,\ell} := \begin{cases} \big[\sin\big(\pi\,(\psi_j-\psi_\ell)\big)\big]^{-1} & j\neq \ell\,, \\ 0 & j=\ell \end{cases}$$

for all $j,\,\ell = 1,\ldots,M$, we form the matrix $\mathbf{S} := -\mathrm{i}\,\big(s_{j,\ell}\big)_{j,\ell=1}^M$ which is Hermitian. Let the eigenvalues of $\mathbf{S}$ be arranged in increasing order $-\infty < \lambda_1 \le \ldots \le \lambda_M < \infty$. By the Rayleigh–Ritz theorem (see [205, pp. 234–235]), we have for all $\mathbf{x}\in\mathbb{C}^M$ with $\|\mathbf{x}\|_2 = 1$,

$$\lambda_1 \le \mathbf{x}^{\mathrm{H}}\,\mathbf{S}\,\mathbf{x} \le \lambda_M\,.$$

Suppose that $\lambda\in\mathbb{R}$ is such an eigenvalue of $\mathbf{S}$ with $|\lambda| = \max\{|\lambda_1|,\,|\lambda_M|\}$. Then we have the sharp inequality

$$|\mathbf{x}^{\mathrm{H}}\,\mathbf{S}\,\mathbf{x}| = \Big|\sum_{j,\ell=1}^{M} x_j\,\overline{x}_\ell\,s_{j,\ell}\Big| \le |\lambda|$$

for all normed vectors $\mathbf{x} = (x_j)_{j=1}^M \in \mathbb{C}^M$. Now we show that $|\lambda| \le \frac{1}{\Delta}$.

2. Related to the eigenvalue λ of $\mathbf{S}$, there exists a normed eigenvector $\mathbf{y} = (y_j)_{j=1}^M \in \mathbb{C}^M$ with $\mathbf{S}\,\mathbf{y} = \lambda\,\mathbf{y}$, i.e.,

$$\sum_{j=1}^{M} y_j\,s_{j,\ell} = \mathrm{i}\,\lambda\,y_\ell\,,\quad \ell = 1,\ldots,M\,. \tag{10.42}$$

Thus, we have $\mathbf{y}^{\mathrm{H}}\,\mathbf{S}\,\mathbf{y} = \lambda\,\mathbf{y}^{\mathrm{H}}\,\mathbf{y} = \lambda$. Applying the Cauchy–Schwarz inequality, we estimate

$$|\mathbf{y}^{\mathrm{H}}\,\mathbf{S}\,\mathbf{y}|^2 = \Big|\sum_{j=1}^{M} y_j \Big(\sum_{\ell=1}^{M} \overline{y}_\ell\,s_{j,\ell}\Big)\Big|^2 \le \|\mathbf{y}\|_2^2\,\Big(\sum_{j=1}^{M}\Big|\sum_{\ell=1}^{M}\overline{y}_\ell\,s_{j,\ell}\Big|^2\Big)$$

$$= \sum_{j=1}^{M}\Big|\sum_{\ell=1}^{M}\overline{y}_\ell\,s_{j,\ell}\Big|^2 = \sum_{j=1}^{M}\sum_{\ell,m=1}^{M}\overline{y}_\ell\,y_m\,s_{j,\ell}\,s_{j,m}$$

$$= \sum_{\ell,m=1}^{M} \overline{y}_\ell \, y_m \sum_{j=1}^{M} s_{j,\ell} \, s_{j,m} = S_1 + S_2$$

with the partial sums

$$S_1 := \sum_{\ell=1}^{M} |y_\ell|^2 \sum_{j=1}^{M} s_{j,\ell}^2 \,, \qquad S_2 := \sum_{\substack{\ell,m=1\\ \ell\neq m}}^{M} \overline{y}_\ell \, y_m \sum_{j=1}^{M} s_{j,\ell} \, s_{j,m} \,.$$

3. For distinct $\alpha, \beta \in \mathbb{R} \setminus (\pi\,\mathbb{Z})$, we have

$$\frac{1}{(\sin\alpha)\,(\sin\beta)} = \frac{\cot\alpha - \cot\beta}{\sin(\beta-\alpha)}$$

such that for all indices with $j \neq \ell$, $j \neq m$, and $\ell \neq m$, we find

$$s_{j,\ell} \; s_{j,m} = s_{\ell,m} \left[\cot\left(\pi(\psi_j - \psi_\ell)\right) - \cot\left(\pi(\psi_j - \psi_m)\right)\right].$$

Now we split the sum S_2 in the following way:

$$\begin{aligned} S_2 &= \sum_{\substack{\ell,m=1\\ \ell\neq m}}^{M} \overline{y}_\ell \, y_m \sum_{\substack{j=1\\ j\neq\ell,\, j\neq m}}^{M} s_{\ell,m} \left[\cot\left(\pi(\psi_j - \psi_\ell)\right) - \cot\left(\pi(\psi_j - \psi_m)\right)\right] \\ &= S_3 - S_4 + 2\,\mathrm{Re}\, S_5 \end{aligned}$$

with

$$S_3 := \sum_{\substack{\ell,m=1\\ \ell\neq m}}^{M} \sum_{\substack{j=1\\ j\neq\ell}}^{M} \overline{y}_\ell \, y_m \, s_{\ell,m} \, \cot\left(\pi(\psi_j - \psi_\ell)\right),$$

$$S_4 := \sum_{\substack{\ell,m=1\\ \ell\neq m}}^{M} \sum_{\substack{j=1\\ j\neq m}}^{M} \overline{y}_\ell \, y_m \, s_{\ell,m} \, \cot\left(\pi(\psi_j - \psi_m)\right),$$

$$S_5 := \sum_{\substack{j,\ell=1\\ j\neq\ell}}^{M} \overline{y}_\ell \, y_j \, s_{j,\ell} \, \cot\left(\pi(\psi_j - \psi_\ell)\right).$$

Note that $2\,\mathrm{Re}\, S_5$ is the correction sum, since S_3 contains the additional terms for $j = m$ and S_4 contains the additional terms for $j = \ell$.

4. First, we show that $S_3 = S_4$. From (10.42) it follows that

$$S_3 = \sum_{\substack{\ell,\, j=1 \\ \ell \neq j}}^{M} \overline{y}_\ell \left(\sum_{m=1}^{M} y_m \, s_{\ell,m} \right) \cot\big(\pi(\psi_j - \psi_\ell)\big)$$

$$= -\mathrm{i}\,\lambda \sum_{\substack{\ell,\, j=1 \\ \ell \neq j}}^{M} |y_\ell|^2 \cot\big(\pi(\psi_j - \psi_\ell)\big)\,.$$

Analogously, we see that

$$S_4 = \sum_{\substack{j,\, m=1 \\ j \neq m}}^{M} y_m \left(\sum_{\ell=1}^{M} \overline{y}_\ell \, s_{\ell,m} \right) \cot\big(\pi(\psi_j - \psi_m)\big)$$

$$= -\mathrm{i}\,\lambda \sum_{\substack{j,\, m=1 \\ j \neq m}}^{M} |y_m|^2 \cot\big(\pi(\psi_j - \psi_m)\big)\,.$$

Hence, we obtain the estimate:

$$|\lambda|^2 = |\mathbf{y}^{\mathrm{H}}\, \mathbf{S}\, \mathbf{y}|^2 = S_1 + S_2 = S_1 + 2\,\mathrm{Re}\, S_5 \le S_1 + 2\,|S_5|\,.$$

Using $2\,|\overline{y}_\ell\, y_j| \le |y_\ell|^2 + |y_j|^2$, we estimate

$$2\,|S_5| \le \sum_{\substack{j,\, \ell=1 \\ j \neq \ell}}^{M} 2\,|\overline{y}_\ell\, y_j|\,|s_{j,\ell} \cot\big(\pi(\psi_j - \psi_\ell)\big)|$$

$$\le 2 \sum_{\substack{j,\, \ell=1 \\ j \neq \ell}}^{M} |y_\ell|^2\,|s_{j,\ell} \cot\big(\pi(\psi_j - \psi_\ell)\big)|$$

such that

$$S_1 + 2\,|S_5| \le \sum_{\substack{j,\, \ell=1 \\ j \neq \ell}}^{M} |y_\ell|^2 \left[s_{j,\ell}^2 + 2\,|s_{j,\ell} \cot\big(\pi(\psi_j - \psi_\ell)\big)| \right].$$

By Lemma 10.26, we obtain

$$S_1 + 2\,|S_5| \le \frac{3}{\pi^2} \sum_{\ell=1}^{M} |y_\ell|^2 \sum_{\substack{j=1 \\ j \neq \ell}}^{M} d(\psi_j - \psi_\ell)^{-2} = \frac{3}{\pi^2} \sum_{\substack{j,\, \ell=1 \\ j \neq \ell}}^{M} d(\psi_j - \psi_\ell)^{-2}\,.$$

By assumption, the values ψ_j, $j = 1, \ldots, M$, are spaced from each other by at least Δ, so that

$$\sum_{\substack{j=1\\ j\neq \ell}}^{M} d(\psi_j - \psi_\ell)^{-2} < 2 \sum_{k=1}^{\infty} (k\,\Delta)^{-2} = \frac{\pi^2}{3\,\Delta^2}$$

and hence

$$|\lambda|^2 = S_1 + S_2 \le S_1 + 2\,|S_5| < \frac{1}{\Delta^2}\,.$$

■

With the natural assumption that the nodes $z_j = \mathrm{e}^{2\pi\,\mathrm{i}\,\psi_j}$, $j = 1, \ldots, M$, are well-separated on the unit circle, it can be shown that the rectangular Vandermonde matrix $\mathbf{V}_{P,M}(\mathbf{z})$ is well conditioned for sufficiently large $P > M$.

Theorem 10.28 (See [13, 273, 290, 328]) *Let $P \in \mathbb{N}$ with $P > \max\{M, \frac{1}{\Delta}\}$ be given. Assume that the frequencies $\psi_j \in \mathbb{R}$, $j = 1, \ldots, M$, satisfy the gap condition* (10.39) *with a constant $\Delta > 0$.*

Then for all $\mathbf{c} \in \mathbb{C}^M$, the rectangular Vandermonde matrix $\mathbf{V}_{P,M}(\mathbf{z})$ with $\mathbf{z} = (z_j)_{j=1}^M$ and $z_j = \mathrm{e}^{2\pi \mathrm{i} \psi_j}$ satisfies the inequalities:

$$\Big(P - \frac{1}{\Delta}\Big)\; \|\mathbf{c}\|_2^2 \le \|\mathbf{V}_{P,M}(\mathbf{z})\,\mathbf{c}\|_2^2 \le \Big(P + \frac{1}{\Delta}\Big)\; \|\mathbf{c}\|_2^2\,. \tag{10.43}$$

Further, the rectangular Vandermonde matrix $\mathbf{V}_{P,M}(\mathbf{z})$ has a uniformly bounded spectral norm condition number:

$$\mathrm{cond}_2\, \mathbf{V}_{P,M}(\mathbf{z}) \le \sqrt{\frac{P\,\Delta + 1}{P\,\Delta - 1}}\,.$$

Proof

1. Simple computation shows that

$$\|\mathbf{V}_{P,M}(\mathbf{z})\,\mathbf{c}\|_2^2 = \sum_{k=0}^{P-1} \Big|\sum_{j=1}^{M} c_j\, z_j^k\Big|^2 = \sum_{k=0}^{P-1} \sum_{j,\,\ell=1}^{M} c_j\, \overline{c}_\ell\, \mathrm{e}^{2\pi\,\mathrm{i}\,(\psi_j-\psi_\ell)\,k}$$

$$= \sum_{k=0}^{P-1} \Bigg(\sum_{j=1}^{M} |c_j|^2 + \sum_{\substack{j,\,\ell=1\\ j\neq\ell}}^{M} c_j\, \overline{c}_\ell\, \mathrm{e}^{2\pi\,\mathrm{i}\,(\psi_j-\psi_\ell)\,k}\Bigg)$$

$$= P\,\|\mathbf{c}\|_2^2 + \sum_{\substack{j,\,\ell=1\\ j\neq\ell}}^{M} c_j\,\bar{c}_\ell \left(\sum_{k=0}^{P-1} \mathrm{e}^{2\pi\,\mathrm{i}\,(\psi_j-\psi_\ell)\,k}\right).$$

Determining the sum

$$\sum_{k=0}^{P-1} \mathrm{e}^{2\pi\,\mathrm{i}\,(\psi_j-\psi_\ell)\,k} = \frac{1-\mathrm{e}^{2\pi\,\mathrm{i}\,(\psi_j-\psi_\ell)\,P}}{1-\mathrm{e}^{2\pi\,\mathrm{i}\,(\psi_j-\psi_\ell)}}$$

$$= \frac{1-\mathrm{e}^{2\pi\,\mathrm{i}\,(\psi_j-\psi_\ell)\,P}}{2\mathrm{i}\,\mathrm{e}^{\pi\,\mathrm{i}\,(\psi_j-\psi_\ell)}\,\sin\big(\pi\,(\psi_j-\psi_\ell)\big)} = -\frac{\mathrm{e}^{-\pi\,\mathrm{i}\,(\psi_j-\psi_\ell)} - \mathrm{e}^{\pi\,\mathrm{i}\,(\psi_j-\psi_\ell)\,(2P-1)}}{2\mathrm{i}\,\sin\big(\pi\,(\psi_j-\psi_\ell)\big)},$$

we obtain

$$\|\mathbf{V}_{P,M}(\mathbf{z})\,\mathbf{c}\|_2^2 = P\,\|\mathbf{c}\|_2^2 - \Sigma_1 + \Sigma_2 \tag{10.44}$$

with the sums

$$\Sigma_1 := \sum_{\substack{j,\,\ell=1\\ j\neq\ell}}^{M} \frac{c_j\,\bar{c}_\ell\,\mathrm{e}^{-\pi\,\mathrm{i}\,(\psi_j-\psi_\ell)}}{2\mathrm{i}\,\sin\big(\pi\,(\psi_j-\psi_\ell)\big)}, \qquad \Sigma_2 := \sum_{\substack{j,\,\ell=1\\ j\neq\ell}}^{M} \frac{c_j\,\bar{c}_\ell\,\mathrm{e}^{\pi\,\mathrm{i}\,(\psi_j-\psi_\ell)\,(2P-1)}}{2\mathrm{i}\,\sin\big(\pi\,(\psi_j-\psi_\ell)\big)}.$$

The nodes $z_j = \mathrm{e}^{2\pi\,\mathrm{i}\,\psi_j}$, $j = 1, \ldots, M$, are distinct, since we have (10.39) by assumption. Applying the generalized Hilbert inequality in (10.41) first with $x_k := c_k\,\mathrm{e}^{-\pi\,\mathrm{i}\,\psi_k}$, $k = 1, \ldots, M$, yields

$$|\Sigma_1| \le \frac{1}{2\,\Delta} \sum_{k=1}^{M} \big|c_k\,\mathrm{e}^{-\pi\,\mathrm{i}\,\psi_k}\big|^2 = \frac{1}{2\,\Delta}\,\|\mathbf{c}\|_2^2, \tag{10.45}$$

and then with $x_k := c_k\,\mathrm{e}^{\pi\,\mathrm{i}\,\psi_k\,(2P-1)}$, $k = 1, \ldots, M$, results in

$$|\Sigma_2| \le \frac{1}{2\,\Delta} \sum_{k=1}^{M} \big|c_k\,\mathrm{e}^{\pi\,\mathrm{i}\,\psi_k\,(2P-1)}\big|^2 = \frac{1}{2\,\Delta}\,\|\mathbf{c}\|_2^2. \tag{10.46}$$

From (10.44)–(10.46) the assertion (10.43) follows by the triangle inequality.

2. Let $\mu_1 \ge \ldots \ge \mu_M > 0$ be the ordered eigenvalues of $\mathbf{V}_{P,M}(\mathbf{z})^{\mathrm{H}}\,\mathbf{V}_{P,M}(\mathbf{z}) \in \mathbb{C}^{M\times M}$. Using the Raleigh–Ritz theorem (see [205, pp. 234–235]) and (10.43), we obtain that for all $\mathbf{c} \in \mathbb{C}^M$

$$\left(P - \frac{1}{\Delta}\right)\|\mathbf{c}\|_2^2 \le \mu_M\,\|\mathbf{c}\|_2^2 \le \|\mathbf{V}_{P,M}(\mathbf{z})\,\mathbf{c}\|_2^2 \le \mu_1\,\|\mathbf{c}\|_2^2 \le \left(P + \frac{1}{\Delta}\right)\|\mathbf{c}\|_2^2$$

and hence

$$0 < P - \frac{1}{\Delta} \le \lambda_M \le \lambda_1 \le P + \frac{1}{\Delta} < \infty\,. \tag{10.47}$$

Thus $\mathbf{V}_{P,M}(\mathbf{z})^{\mathrm{H}}\,\mathbf{V}_{P,M}(\mathbf{z})$ is positive definite and

$$\mathrm{cond}_2\,\mathbf{V}_{P,M}(\mathbf{z}) = \sqrt{\frac{\mu_1}{\mu_M}} \le \sqrt{\frac{P\,\Delta + 1}{P\,\Delta - 1}}\,.$$

■

The inequalities (10.43) can be interpreted as discrete versions of the Ingham inequalities (10.31). Now the exponentials $\mathrm{e}^{2\pi\,\mathrm{i}\,\psi_j\,\cdot}$ are replaced by their discretizations

$$\mathbf{e}_P(\psi_j) = \big(\mathrm{e}^{2\pi\,\mathrm{i}\,\psi_j\,k}\big)_{k=0}^{P-1}\,, \quad j = 1, \ldots, M\,,$$

with sufficiently large integer $P > \max\{M, \frac{1}{\Delta}\}$. Thus, the rectangular Vandermonde matrix can be written as

$$\mathbf{V}_{P,M}(\mathbf{z}) = \big(\mathbf{e}_P(\psi_1)\,|\,\mathbf{e}_P(\psi_2)\,|\,\ldots\,|\,\mathbf{e}_P(\psi_M)\big)$$

with $\mathbf{z} = (z_j)_{j=1}^M$, where $z_j = \mathrm{e}^{2\pi\,\mathrm{i}\,\psi_j}$, $j = 1, \ldots, M$, are distinct nodes on the unit circle. Then (10.43) provides the *discrete Ingham inequalities*

$$\Big(P - \frac{1}{\Delta}\Big)\,\|\mathbf{c}\|_2^2 \le \|\sum_{j=1}^M c_j\,\mathbf{e}_P(\varphi_j)\|_2^2 \le \Big(P + \frac{1}{\Delta}\Big)\,\|\mathbf{c}\|_2^2 \tag{10.48}$$

for all $\mathbf{c} = (c_j)_{j=1}^M \in \mathbb{C}^M$. In other words, (10.48) means that the vectors $\mathbf{e}_P(\varphi_j)$, $j = 1, \ldots, M$, are Riesz stable with Riesz constants $P - \frac{1}{\Delta}$ and $P + \frac{1}{\Delta}$.

Corollary 10.29 *With the assumptions of Theorem* 10.28, *the inequalities*

$$\Big(P - \frac{1}{\Delta}\Big)\,\|\mathbf{d}\|_2^2 \le \|\mathbf{V}_{P,M}(\mathbf{z})^\top\,\mathbf{d}\|_2^2 \le \Big(P + \frac{1}{\Delta}\Big)\,\|\mathbf{d}\|_2^2 \tag{10.49}$$

hold for all $\mathbf{d} \in \mathbb{C}^P$.

Proof The matrices $\mathbf{V}_{P,M}(\mathbf{z})$ and $\mathbf{V}_{P,M}(\mathbf{z})^\top$ possess the same singular values μ_j, $j = 1, \ldots, M$. By the Rayleigh–Ritz theorem, we obtain that

$$\lambda_M\,\|\mathbf{d}\|_2^2 \le \|\mathbf{V}_{P,M}(\mathbf{z})^\top\,\mathbf{d}\|_2^2 \le \lambda_1\,\|\mathbf{d}\|_2^2$$

for all $\mathbf{d} \in \mathbb{C}^P$. Applying (10.47), we obtain the inequalities in (10.49). ■

Remark 10.30 In [13, 28], the authors derive bounds on the extremal singular values and the condition number of the rectangular Vandermonde matrix $\mathbf{V}_{P,M}(\mathbf{z})$ with $P \ge M$ and $\mathbf{z} = (z_j)_{j=1}^M \in \mathbb{C}^M$, where the nodes are inside the unit disk, i.e., $|z_j| \le 1$ for $j = 1, \ldots, M$. In [328] it is investigated how the condition number of the Vandermonde matrix $\mathbf{V}_{P,M}(\mathbf{z})$ with $z_j = \mathrm{e}^{2\pi \mathrm{i}\psi_j}$ can be improved using a single shift parameter σ that transfers ψ_j to $\sigma\,\psi_j$ for $j = 1, \ldots, M$. This result is in turn applied to improve the stability of an algorithm for the fast sparse Fourier transform.

More sophisticated estimates for the condition number of Vandermonde matrices with nearly colliding nodes can be found in [25, 255]. □

Let us come back to the Hankel matrix $\mathbf{H}_{L,N-L+1}$ of rank M in (10.17), which is formed from equidistant samples $h(\ell)$ of an exponential sum $h(x) = \sum_{j=1}^M c_j\,\mathrm{e}^{2\pi \mathrm{i}\psi_j}$, where we have assumed that $c_j \ne 0$ and that $z_j = \mathrm{e}^{2\pi \mathrm{i}\psi_j}$ are pairwise distinct for $j = 1, \ldots, M$. Employing the Vandermonde decomposition of $\mathbf{H}_{L,N-L+1}$, we obtain

$$\mathbf{H}_{L,N-L+1} = \mathbf{V}_{L,M}(\mathbf{z})\,(\operatorname{diag}\mathbf{c})\,\big(\mathbf{V}_{N-L+1,M}(\mathbf{z})\big)^\top \tag{10.50}$$

as in (10.18). Therefore, we can derive the condition number of the Hankel matrix $\mathbf{H}_{L,N-L+1}$ from the condition numbers of the involved Vandermonde matrices.

Theorem 10.31 *Let $L,\, N \in \mathbb{N}$ with $M \le L \le N - M + 1$ and $\min\{L,\, N - L + 1\} > \frac{1}{\Delta}$ be given. Let $\mathbf{z} = (z_j)_{j=1}^M$ with $z_j = \mathrm{e}^{2\pi \mathrm{i}\psi_j}$, where the frequencies $\psi_j \in \mathbb{R}$, $j = 1, \ldots, M$, are well-separated at least by a constant $\Delta > 0$. Further, let $\mathbf{c} = (c_j)_{j=1}^M$ with*

$$0 < \gamma_1 \le |c_j| \le \gamma_2 < \infty\,, \quad j = 1, \ldots, M\,. \tag{10.51}$$

Then the Hankel matrix $\mathbf{H}_{L,N-L+1}$ in (10.50) satisfies

$$\gamma_1^2\,\alpha_1(L,\,N,\,\Delta)\,\|\mathbf{y}\|_2^2 \le \|\mathbf{H}_{L,N-L+1}\,\mathbf{y}\|_2^2 \le \gamma_2^2\,\alpha_2(L,\,N,\,\Delta)\,\|\mathbf{y}\|_2^2\,. \tag{10.52}$$

for all $\mathbf{y} \in \mathbb{C}^{N-L+1}$ with

$$\alpha_1(L,\,N,\,\Delta) := \Big(L - \frac{1}{\Delta}\Big)\,\Big(N - L + 1 - \frac{1}{\Delta}\Big)\,,$$

$$\alpha_2(L,\,N,\,\Delta) := \Big(L + \frac{1}{\Delta}\Big)\,\Big(N - L + 1 + \frac{1}{\Delta}\Big)\,.$$

Further, the lowest (nonzero) respectively largest singular value of $\mathbf{H}_{L,N-L+1}$ can be estimated by

$$0 < \gamma_1\,\sqrt{\alpha_1(L,\,N,\,\Delta)} \le \sigma_M \le \sigma_1 \le \gamma_2\,\sqrt{\alpha_2(L,\,N,\,\Delta)}\,. \tag{10.53}$$

The spectral norm condition number of $\mathbf{H}_{L,N-L+1}$ is bounded by

$$\text{cond}_2\, \mathbf{H}_{L,N-L+1} \le \frac{\gamma_2}{\gamma_1} \sqrt{\frac{\alpha_2(L,\, N,\, \Delta)}{\alpha_1(L,\, N,\, \Delta)}}\,. \tag{10.54}$$

Proof By the Vandermonde decomposition (10.50) of the Hankel matrix $\mathbf{H}_{L,N-L+1}$, we obtain that for all $\mathbf{y} \in \mathbb{C}^{N-L+1}$

$$\|\mathbf{H}_{L,N-L+1}\, \mathbf{y}\|_2^2 = \|\mathbf{V}_{L,M}(\mathbf{z})\, (\text{diag}\, \mathbf{c})\, \mathbf{V}_{N-L+1,M}(\mathbf{z})^\top\, \mathbf{y}\|_2^2\,.$$

The estimates in (10.43) and the assumption (10.51) imply

$$\begin{aligned}\gamma_1^2 \left(L - \frac{1}{\Delta}\right) \|\mathbf{V}_{N-L+1,M}(\mathbf{z})^\top\, \mathbf{y}\|_2^2 &\le \|\mathbf{H}_{L,N-L+1}\, \mathbf{y}\|_2^2 \\ &\le \gamma_2^2 \left(L + \frac{1}{\Delta}\right) \|\mathbf{V}_{N-L+1,M}(\mathbf{z})^\top\, \mathbf{y}\|_2^2\,.\end{aligned}$$

Using the inequalities in (10.49), we obtain (10.52). Finally, the estimates of the extremal singular values and the spectral norm condition number of $\mathbf{H}_{L,N-L+1}$ are a consequence of (10.52) and the Rayleigh–Ritz theorem. ■

Remark 10.32 For fixed N, the positive singular values as well as the spectral norm condition number of the Hankel matrix $\mathbf{H}_{L,N-L+1}$ depend strongly on $L \in \{M, \ldots, N-M+1\}$. A good criterion for the choice of an optimal window length L is to maximize the lowest positive singular value σ_M of $\mathbf{H}_{L,N-L+1}$. It was shown in [347, Lemma 3.1 and Remark 3.3] that the squared singular values increase almost monotonically for $L = M, \ldots, \lceil \frac{N}{2} \rceil$ and decrease almost monotonically for $L = \lceil \frac{N}{2} \rceil, \ldots, N-M+1$. Note that the lower bound (10.53) of the lowest positive singular value σ_M is maximal for $L \approx \frac{N}{2}$. Further, the upper bound (10.54) of the spectral norm condition number of the exact Hankel matrix $\mathbf{H}_{L,N-L+1}$ is minimal for $L \approx \frac{N}{2}$. Therefore, we prefer to choose $L \approx \frac{N}{2}$ as optimal window length. Then we can ensure that $\sigma_M > 0$ is not too small. This observation is essential for the correct detection of the order M in the first step of the MUSIC algorithm and ESPRIT algorithm. □

10.4 Recovery of Structured Functions

The reconstruction of a compactly supported, structured function from the knowledge of samples of its Fourier transform is a common problem in several scientific areas such as radio astronomy, computerized tomography, and magnetic resonance imaging.

10.4.1 Recovery from Fourier Data

Let us start with the problem of reconstruction of spline functions with arbitrary knots.

For a given $m, n \in \mathbb{N}$, let $t_j \in \mathbb{R}$, $j = 1, \ldots, m+n$, be distinct knots with $-\infty < t_1 < t_2 < \ldots < t_{m+n} < \infty$. A function $s : \mathbb{R} \to \mathbb{R}$ with compact support $\operatorname{supp} s \subseteq [t_1, t_{m+n}]$ is a *spline* of *order* m, if s restricted on $[t_j, t_{j+1})$ is a polynomial of degree $m-1$ for each $j = 1, \ldots, m+n-1$, if $s(x) = 0$ for all $x \in \mathbb{R} \setminus [t_1, t_{m+n}]$, and if $s \in C^{m-2}(\mathbb{R})$. The points t_j are called *spline knots*. Note that $C^0(\mathbb{R}) := C(\mathbb{R})$ and that $C^{-1}(\mathbb{R})$ is the set of piecewise continuous functions. Denoting with $\mathscr{S}_m[t_1, \ldots, t_{m+n}]$ the linear space of all splines of order m relative to the fixed spline knots t_j, then $\dim \mathscr{S}_m[t_1, \ldots, t_{m+n}] = n$. For $m = 1$, splines of $\mathscr{S}_1[t_1, \ldots, t_{n+1}]$ are *step functions* of the form

$$s(x) := \sum_{j=1}^{n} c_j \, \chi_{[t_j, t_{j+1})}(x)\,, \quad x \in \mathbb{R}\,, \tag{10.55}$$

where c_j are real coefficients with $c_j \neq c_{j+1}$, $j = 1, \ldots, n-1$, and where $\chi_{[t_j, t_{j+1})}$ denotes the characteristic function of the interval $[t_j, t_{j+1})$. Obviously, the piecewise constant splines

$$B_j^1(x) = \chi_{[t_j, t_{j+1})}(x)\,, \quad j = 1, \ldots, n$$

have minimal support in $\mathscr{S}_1[t_1, \ldots, t_{n+1}]$ and form a basis of the spline space $\mathscr{S}_1[t_1, \ldots, t_{n+1}]$. Therefore, B_j^1 are called B-*splines* or *basis splines* of order 1. Forming the Fourier transform

$$\hat{s}(\omega) := \int_{\mathbb{R}} s(x)\, \mathrm{e}^{-\mathrm{i} x \omega}\, \mathrm{d}x\,, \quad \omega \in \mathbb{R} \setminus \{0\}\,,$$

of $s(x)$ in (10.55), we obtain the exponential sum

$$\begin{aligned} h(\omega) := \mathrm{i}\,\omega\, \hat{s}(\omega) &= \sum_{j=1}^{n+1} (c_j - c_{j-1})\, \mathrm{e}^{-\mathrm{i}\omega t_j} \\ &= \sum_{j=1}^{n+1} c_j^1\, \mathrm{e}^{-\mathrm{i}\omega t_j}\,, \quad \omega \in \mathbb{R} \setminus \{0\}\,, \end{aligned} \tag{10.56}$$

with $c_0 = c_{n+1} := 0$, $c_j^1 := c_j - c_{j-1}$, $j = 1, \ldots, n+1$, and $h(0) := 0$. First, we consider the recovery of a real step function (10.55) by the given Fourier samples (see [329]).

Lemma 10.33 *Assume that a constant step size* $\tau > 0$ *satisfies the condition* $t_j \tau \in [-\pi, \pi)$ *for* $j = 1, \ldots, n+1$. *Then the real step function* (10.55) *can be completely recovered by given Fourier samples* $\hat{s}(\ell\tau)$, $\ell = 1, \ldots, N$ *with* $N \geq n+1$.

Proof By (10.56), the function h is an exponential sum of order $n+1$. Since s is real, we have $h(\omega) = \overline{h(-\omega)}$. The given samples $\hat{s}(\ell\tau)$, $\ell = 1, \ldots, N$, with $N \geq n+1$ therefore provide the $2N+1$ sample values $h(\ell\tau)$, $\ell = -N, \ldots, N$, of the exponential sum h, namely,

$$h(\ell\tau) = \begin{cases} \mathrm{i}\,\ell\,\tau\,\hat{s}(\ell\tau) & \ell = 1, \ldots, N\,, \\ \overline{h(-\ell\tau)} & \ell = -N, \ldots, -1\,, \\ 0 & \ell = 0. \end{cases}$$

Now we consider the function $\tilde{h}(\omega) := h((-N+\omega)\tau) = \sum_{j=1}^{n+1} (c_j^1\, \mathrm{e}^{\mathrm{i}N\tau})\, \mathrm{e}^{\mathrm{i}\omega\,(-\tau\, t_j)}$, such that $\tilde{h}(k) = h((-N+k)\tau)$ are given for $k = 0, \ldots, 2N$, and apply one of the reconstruction methods for the exponential sum $\tilde{h}$ described in Sect. 10.2 to recover all parameters of $\tilde{h}$. In this way, we determine all spline knots t_j and coefficients c_j^1, $j = 1, \ldots, n+1$. Finally, the coefficients c_j of the step function (10.55) are obtained by the recursion $c_j = c_{j-1} + c_j^1$, $j = 2, \ldots, n$, with $c_1 = c_1^1$. Hence, the step function in (10.55) can be completely reconstructed. ■

Remark 10.34 A similar technique can be applied if the support $[t_1, t_{n+1}]$ of the step function (10.55) is contained in $[-\pi, \pi]$ and some Fourier coefficients

$$c_k(s) := \frac{1}{2\pi} \int_{-\pi}^{\pi} s(x)\, \mathrm{e}^{-\mathrm{i}\,k\,x}\, \mathrm{d}x\,, \quad k \in \mathbb{Z}\,,$$

are given. If $[t_1, t_{n+1}] \not\subset [-\pi, \pi]$, we can shift and scale s properly to achieve this condition. For the step function (10.55), we obtain

$$2\pi\, \mathrm{i}\, c_k(s) = \sum_{j=1}^{n+1} (c_j - c_{j-1})\, \mathrm{e}^{-\mathrm{i}\,t_j\,k}\,, \quad k \in \mathbb{Z} \setminus \{0\}\,,$$

$$2\pi\, c_0(s) = \sum_{j=1}^{n} c_j\, (t_{j+1} - t_j)\,.$$

Thus, one can determine the breakpoints t_j and the coefficients c_j by a method of Sect. 10.2 using only the Fourier coefficients $c_k(s)$, $k = 0, \ldots, n+1$. □

This approach can be easily transferred to higher-order spline functions of the form

$$s(x) := \sum_{j=1}^{n} c_j \, B_j^m(x)\,, \quad x \in \mathbb{R}\,, \tag{10.57}$$

where B_j^m, $j = 1, \ldots, n$, is the B-spline of order m with arbitrary knots $t_j, \ldots, t_{j+m}$. The B-splines B_j^m (see [103, p. 90]) satisfy the recurrence relation

$$B_j^m(x) = \frac{x - t_j}{t_{j+m-1} - t_j}\, B_j^{m-1}(x) + \frac{t_{j+1} - x}{t_{j+m} - t_{j+1}}\, B_{j+1}^{m-1}(x)$$

with initial condition $B_j^1(x) = \chi_{[t_j,\, t_{j+1})}(x)$. The support of B_j^m is the interval $[t_j,\, t_{j+m}]$. In the case $m = 2$, we obtain the hat function as the linear B-spline:

$$B_j^2(x) = \begin{cases} \frac{x-t_j}{t_{j+1}-t_j} & x \in [t_j,\, t_{j+1})\,, \\ \frac{t_{j+1}-x}{t_{j+2}-t_{j+1}} & x \in [t_{j+1},\, t_{j+2})\,, \\ 0 & x \in \mathbb{R} \setminus [t_j,\, t_{j+2})\,. \end{cases}$$

As known, the linear B-splines B_j^2, $j = 1, \ldots, n$ form a basis of the spline space $\mathcal{S}_2[t_1, \ldots, t_{n+2}]$. For $m \geq 3$, the first derivative of B_j^m can be computed by

$$(B_j^m)'(x) = (m-1)\left(\frac{B_j^{m-1}(x)}{t_{j+m-1} - t_j} - \frac{B_{j+1}^{m-1}(x)}{t_{j+m} - t_{j+1}}\right) \tag{10.58}$$

(see [103, p. 115]). The formula (10.58) can be also applied for $m = 2$, if we replace the derivative by the right-hand derivative. Then we obtain for the kth derivative of $s(x)$ in (10.57) with $k = 1, \ldots, m-1$

$$s^{(k)}(x) = \sum_{j=1}^{n} c_j \left(B_j^m\right)^{(k)}(x) = \sum_{j=1}^{n+k} c_j^{m-k}\, B_j^{m-k}(x)\,, \tag{10.59}$$

where the real coefficients c_j^{m-k} can be recursively evaluated from c_j using (10.58). Hence, the $(m-1)$th derivative of $s(x)$ in (10.57) is a real step function

$$s^{(m-1)}(x) = \sum_{j=1}^{n+m-1} c_j^1\, B_j^1(x) = \sum_{j=1}^{n+m-1} c_j^1\, \chi_{[t_j,\, t_{j+1})}(x)\,.$$

Application of the Fourier transform yields

$$(\mathrm{i}\,\omega)^{m-1}\, \hat{s}(\omega) = \sum_{j=1}^{n+m-1} \frac{c_j^1}{\mathrm{i}\,\omega} \left(\mathrm{e}^{-\mathrm{i}\,\omega t_j} - \mathrm{e}^{-\mathrm{i}\,\omega t_{j+1}}\right) \tag{10.60}$$

$$= \frac{1}{\mathrm{i}\,\omega} \sum_{j=1}^{n+m} c_j^0 \, \mathrm{e}^{-\mathrm{i}\,\omega t_j}\,, \tag{10.61}$$

where $c_j^0 := c_j^1 - c_{j-1}^1$, $j = 1, \ldots, n+m$, with the convention $c_0^1 = c_{n+m}^1 := 0$. Thus, we obtain the exponential sum of order $n+m$:

$$(\mathrm{i}\,\omega)^m \, \hat{s}(\omega) = \sum_{j=1}^{n+m} c_j^0 \, \mathrm{e}^{-\mathrm{i}\,\omega t_j}\,. \tag{10.62}$$

Hence, we can recover a real spline function (10.57) by the given Fourier samples (see [329]).

Theorem 10.35 *Assume that* $s(x)$ *possesses the form* (10.57) *with unknown coefficients* $c_j \in \mathbb{R} \setminus \{0\}$ *and an unknown knot sequence* $-\infty < t_1 < t_2 < \ldots < t_{n+m} < \infty$. *Assume that there is a given constant* $\tau > 0$ *satisfying the condition* $t_j \tau \in [-\pi, \pi)$ *for* $j = 1, \ldots, n+m$.

Then the real spline function $s(x)$ *in* (10.57) *of order* m *can be completely recovered by the given Fourier samples* $\hat{s}(\ell\tau)$, $\ell = 1, \ldots, N$, *with* $N \geq n+m$.

Proof The Fourier transform of (10.57) satisfies the Eq. (10.62) such that $h(\omega) := (\mathrm{i}\,\omega)^m \, \hat{s}(\omega)$ is an exponential sum of order $n+m$. Using a reconstruction method of Sect. 10.2, we compute the knots t_j and the coefficients c_j^0 for $j = 1, \ldots, n+m$, as given in the proof of Lemma 10.33. Applying the formulas (10.58) and (10.59), we obtain the following recursion for the coefficients c_j^k:

$$c_j^{k+1} := \begin{cases} c_j^0 + c_{j-1}^1 & k = 0,\ j = 1, \ldots, n+m-1\,, \\ \frac{t_{m+1-k} - t_j}{m-k}\, c_j^k + c_{j-1}^{k+1} & k = 1, \ldots, m-1,\ j = 1, \ldots, n+m-k-1 \end{cases}$$

with the convention $c_0^k := 0$, $k = 1, \ldots, m$. Then c_j^m, $j = 1, \ldots, n$, are the reconstructed coefficients c_j of (10.57). ■

Remark 10.36 The proof of Theorem 10.35 is constructive. In particular, if n is unknown, but we have an upper bound of n, then the reconstruction methods in Sect. 10.2 will also find the correct knots t_j and the corresponding coefficients c_j from N Fourier samples with $N \geq n+m$, and the numerical procedure will be more stable (see, e.g., [142, 316, 342]).

In the above proof, we rely on the fact that $c_j^0 \neq 0$ for $j = 1, \ldots, n+m$. If we have the situation that $c_{j_0}^0 = 0$ for an index $j_0 \in \{1, \ldots, n+m\}$, then we will not be able to reconstruct the knot t_{j_0}. But this situation only occurs if the representation (10.57) is redundant, i.e., if the spline in (10.57) can be represented by less than n summands, so we will still be able to recover the exact function s. Observe that the above recovery procedure always results in the simplest representation of s so that the reconstructed representation of s of the form (10.57) does not possess redundant terms. □

Now we generalize this method and recover linear combinations of translates of a fixed real function $\Phi \in C(\mathbb{R}) \cap L_1(\mathbb{R})$

$$f(x) := \sum_{j=1}^{n} c_j\, \Phi(x + t_j)\,, \quad x \in \mathbb{R}\,, \tag{10.63}$$

with real coefficients $c_j \neq 0$, $j = 1, \ldots, n$, and shift parameters t_j with $-\infty < t_1 < \ldots < t_n < \infty$. Assume that Φ is a low-pass filter function with a Fourier transform $\widehat{\Phi}$ that is bounded away from zero, i.e., $|\widehat{\Phi}(\omega)| > C_0$ for $\omega \in (-T, T)$ for some positive constants C_0 and T.

Example 10.37 As a low-pass filter function Φ, we can take the *centered cardinal* B-*spline* $\Phi = M_m$ of order m; see Example 9.1 with

$$\widehat{M}_m(\omega) = \Big(\mathrm{sinc}\,\frac{\omega}{2}\Big)^m \neq 0\,, \quad \omega \in (-2\pi, 2\pi)\,,$$

the *Gaussian function* $\Phi(x) = \mathrm{e}^{-x^2/\sigma^2}$, $x \in \mathbb{R}$, with $\sigma > 0$, where the Fourier transform reads as follows:

$$\widehat{\Phi}(\omega) = \sqrt{\pi}\,\sigma\,\mathrm{e}^{-\sigma^2\omega^2/4} > 0\,, \quad \omega \in \mathbb{R}\,,$$

the *Meyer window* Φ with $T = \frac{2}{3}$ and the corresponding Fourier transform

$$\widehat{\Phi}(\omega) = \begin{cases} 1 & |\omega| \le \frac{1}{3}\,, \\ \cos\big(\frac{\pi}{2}(3|\omega| - 1)\big) & \frac{1}{3} < |\omega| \le \frac{2}{3}\,, \\ 0 & \omega \in \mathbb{R} \setminus \big[-\frac{2}{3}, \frac{2}{3}\big]\,, \end{cases}$$

or a real-valued *Gabor function* $\Phi(x) = \mathrm{e}^{-\alpha x^2} \cos(\beta x)$ with positive constants α and β, where

$$\widehat{\Phi}(\omega) = \sqrt{\frac{\pi}{4\alpha}}\big(\mathrm{e}^{-(\beta-\omega)^2/(4\alpha)} + \mathrm{e}^{-(\omega+\beta)^2/(4\alpha)}\big) > 0\,, \quad \omega \in \mathbb{R}\,.$$

□

The Fourier transform of f in (10.63) yields

$$\hat{f}(\omega) = \widehat{\Phi}(\omega) \sum_{j=1}^{n} c_j\, \mathrm{e}^{\mathrm{i}\,\omega t_j}\,, \quad \omega \in \mathbb{R}\,. \tag{10.64}$$

Theorem 10.38 *Let $\Phi \in C(\mathbb{R}) \cap L_1(\mathbb{R})$ be a given function with $|\widehat{\Phi}(\omega)| > C_0$ for all $\omega \in (-T, T)$ with some $C_0 > 0$. Assume that $f(x)$ is of the form* (10.63) *with unknown coefficients $c_j \in \mathbb{R} \setminus \{0\}$, $j = 1, \ldots, n$, and unknown shift parameters $-\infty < t_1 < \ldots < t_n < \infty$. Let $\tau > 0$ be a given constant satisfying $|\tau\, t_j| < \min\{\pi, T\}$ for all $j = 1, \ldots, n$.*

Then the function f can be uniquely recovered by the Fourier samples $\hat{f}(\ell\tau)$, $\ell = 0, \ldots, N$, with $N \geq n$.

Proof Using the assumption on $\widehat{\Phi}$, it follows from (10.64) that the function

$$h(\omega) := \frac{\hat{f}(\omega)}{\widehat{\Phi}(\omega)} = \sum_{j=1}^{n} c_j\, \mathrm{e}^{\mathrm{i}\,\omega t_j}\,, \quad \omega \in (-T, T)\,,$$

is an exponential sum of order n. Hence, we can compute all shift parameters t_j and coefficients c_j, $j = 1, \ldots, n$, by a reconstruction method given in Sect. 10.2. from the given $2N + 1$ samples $h(-\tau N + \ell\tau)$, $\ell = 0, \ldots, 2N$, similarly as in the proof of Lemma 10.33. ■

10.4.2 Recovery from Function Samples

In this section, we want to study the question of how to recover periodic structured functions directly from the given function values.

Let $\varphi \in C(\mathbb{T})$ be an even, nonnegative function with uniformly convergent Fourier expansion. Assume that all Fourier coefficients

$$c_k(\varphi) = \frac{1}{2\pi} \int_{-\pi}^{\pi} \varphi(x)\, \mathrm{e}^{-\mathrm{i}kx}\, \mathrm{d}x = \frac{1}{\pi} \int_{0}^{\pi} \varphi(x)\, \cos(kx)\, \mathrm{d}x\,, \quad k \in \mathbb{Z}\,,$$

are nonnegative and that $c_k(\varphi) > 0$ for $k = 0, \ldots, \frac{N}{2}$, where $N \in 2\,\mathbb{N}$ is fixed. Such a function φ is called a 2π-periodic *window function*.

Example 10.39 A well-known 2π-periodic window function is the 2π-*periodization*

$$\varphi(x) := \sum_{k\in\mathbb{Z}} \Phi(x + 2\pi k)\,, \quad x \in \mathbb{R}\,, \tag{10.65}$$

of the Gaussian function

$$\Phi(x) := \frac{1}{\sqrt{\pi b}}\, \mathrm{e}^{-(nx)^2/b}\,, \quad x \in \mathbb{R}\,,$$

with some $n \in \mathbb{N}$ and $b \geq 1$, where the Fourier coefficients are

$$c_k(\varphi) = \frac{1}{2\pi n}\, \mathrm{e}^{-b k^2/(4n^2)}\,, \quad k \in \mathbb{Z}\,.$$

Another window function is the 2π*-periodization* (10.65) *of the centered cardinal* B*-spline of order* $2m$

$$\Phi(x) = M_{2m}(nx)\,, \quad x \in \mathbb{R}\,,$$

with some $m, n \in \mathbb{N}$, where the Fourier coefficients are given by

$$c_k(\varphi) = \frac{1}{2\pi n}\left(\mathrm{sinc}\,\frac{k}{2n}\right)^{2m}\,, \quad k \in \mathbb{Z}\,.$$

A further 2π-periodic window function is the 2π*-periodization* (10.65) *of the Kaiser–Bessel function* (see [298, p. 80])

$$\Phi(x) = \begin{cases} \frac{\sinh(b\sqrt{m^2-n^2x^2})}{\pi\sqrt{m^2-n^2x^2}} & |x| < \frac{m}{n}\,, \\ \frac{b}{\pi} & x = \pm\frac{m}{n}\,, \\ \frac{\sin(b\sqrt{n^2x^2-m^2})}{\pi\sqrt{n^2x^2-m^2}} & |x| > \frac{m}{n} \end{cases}$$

with fixed $m, n \in \mathbb{N}$, and $b = 1 - \frac{1}{2\alpha}$, $\alpha > 1$, where the Fourier coefficients have the form

$$c_k(\varphi) = \begin{cases} \frac{1}{2\pi n}\, I_0(m\sqrt{b^2 - k^2/n^2}) & |k| \le n\,b\,, \\ 0 & |k| > n\,b\,. \end{cases}$$

Here I_0 denotes the *modified Bessel function of order zero* defined by

$$I_0(x) := \sum_{k=0}^{\infty} \frac{x^{2k}}{4^k\,(k!)^2}\,, \quad x \in \mathbb{R}\,.$$

□

Now we consider a linear combination

$$f(x) := \sum_{j=1}^{M} c_j\,\varphi(x + t_j) \tag{10.66}$$

of translates $\varphi(\cdot + t_j)$ with nonzero coefficients $c_j \in \mathbb{C}$ and distinct shift parameters:

$$-\pi \le t_1 < t_2 < \ldots < t_M \le \pi\,.$$

Then we have $f \in C(\mathbb{T})$. Let $N \in 2\mathbb{N}$ with $N > 2M+1$ be given. We introduce an *oversampling factor* $\alpha > 1$ such that $n = \alpha N$ is a power of two. Assume that perturbed, uniformly sampled data of f in (10.66),

$$f_\ell = f\left(\frac{2\pi\ell}{n}\right) + e_\ell\,, \quad \ell = 0, \ldots, n-1\,, \tag{10.67}$$

are given, where the error terms $e_\ell \in \mathbb{C}$ are bounded by $|e_\ell| \le \varepsilon_1$ with $0 < \varepsilon_1 \ll 1$. Further, we suppose that $|c_j| \gg \varepsilon_1$ for all $j = 1, \ldots, M$. Then we consider the following reconstruction problem (see [316]):

Determine the number M of translates, the shift parameters $t_j \in [-\pi, \pi)$, and the complex coefficients $c_j \ne 0$, $j = 1, \ldots, M$, in such a way that

$$f_\ell \approx \sum_{j=1}^{M} c_j\, \varphi\left(\frac{2\pi\ell}{n} + t_j\right), \quad \ell = 0, \ldots, n-1\,.$$

Note that all reconstructed values for M, t_j, and c_j depend on ε_1 and n.

This problem can be numerically solved in two steps. First, we convert the given problem into a frequency analysis problem (10.2) for an exponential sum by using Fourier techniques. Then the parameters of the exponential sum are recovered by the methods of Sect. 10.2.

For the 2π-periodic function in (10.66), the corresponding Fourier coefficients have the form

$$c_k(f) = \frac{1}{2\pi}\int_0^{2\pi} f(x)\, \mathrm{e}^{-\mathrm{i}kx}\, \mathrm{d}x = \left(\sum_{j=1}^{M} c_j\, \mathrm{e}^{\mathrm{i}kt_j}\right) c_k(\varphi) = h(k)\, c_k(\varphi) \tag{10.68}$$

with the exponential sum

$$h(x) := \sum_{j=1}^{M} c_j\, \mathrm{e}^{\mathrm{i}xt_j}\,, \quad x \in \mathbb{R}\,. \tag{10.69}$$

In applications, the Fourier coefficients $c_k(\varphi)$ of the chosen 2π-periodic window function φ are usually explicitly known (see Example 10.39), where $c_k(\varphi) > 0$ for all $k = 0, \ldots, \frac{N}{2}$. We assume that the function (10.66) is sampled on a fine grid, i.e., we have the given noisy data (10.67) on $\{\frac{2\pi\ell}{n} : \ell = 0, \ldots, n-1\}$ of $[0, 2\pi]$. Now we can approximately compute the Fourier coefficients $c_k(f)$, $k = -\frac{N}{2}, \ldots, \frac{N}{2}$, by the discrete Fourier transform:

$$c_k(f) \approx \frac{1}{n}\sum_{\ell=0}^{n-1} f\left(\frac{2\pi\ell}{n}\right) \mathrm{e}^{-2\pi\mathrm{i}k\ell/n}$$

$$\approx \tilde{c}_k := \frac{1}{n} \sum_{\ell=0}^{n-1} f_\ell \, \mathrm{e}^{-2\pi\, \mathrm{i}\, k\, \ell/n} .$$

We set

$$\tilde{h}_k := \frac{\tilde{c}_k}{c_k(\varphi)}, \quad k = -\frac{N}{2}, \ldots, \frac{N}{2} . \tag{10.70}$$

Then we obtain the following estimate of the error $|\tilde{h}_k - h(k)|$:

Lemma 10.40 *Let $\varphi \in C(\mathbb{T})$ be an even nonnegative window function with uniformly convergent Fourier expansion. Further, let $\mathbf{c} = (c_j)_{j=1}^M \in \mathbb{C}^M$ and let* (10.67) *be the given noisy sampled data of f in* (10.66). *Let h be given as in* (10.69) *and $\tilde{h}_k$ as in* (10.70).

Then, for each $k = -\frac{N}{2}, \ldots, \frac{N}{2}$, the computed approximate value $\tilde{h}_k$ of $h(k)$ satisfies the error estimate:

$$|\tilde{h}_k - h(k)| \le \frac{\varepsilon_1}{c_k(\varphi)} + \|\mathbf{c}\|_1 \max_{j=0,\ldots,N/2} \sum_{\ell \in \mathbb{Z}\setminus\{0\}} \frac{c_{j+\ell n}(\varphi)}{c_j(\varphi)} .$$

Proof The 2π-periodic function f in (10.66) has a uniformly convergent Fourier expansion. Let $k \in \{-\frac{N}{2}, \ldots, \frac{N}{2}\}$ be an arbitrary fixed index. By the aliasing formula (3.6), we have

$$\frac{1}{n} \sum_{j=0}^{n-1} f\left(\frac{2\pi j}{n}\right) \mathrm{e}^{-2\pi\, \mathrm{i}\, k\, j/n} - c_k(f) = \sum_{\ell \in \mathbb{Z}\setminus\{0\}} c_{k+\ell n}(f).$$

Observe that

$$\tilde{c}_k = \frac{1}{n} \sum_{j=0}^{n-1} f_j \, \mathrm{e}^{-2\pi\, \mathrm{i}\, k\, j/n} = \frac{1}{n} \sum_{j=0}^{n-1} (f(j) + e_j)\, \mathrm{e}^{-2\pi\, \mathrm{i}\, k\, j/n} .$$

Using the simple estimate

$$\frac{1}{n} \Big| \sum_{j=0}^{n-1} e_j \, \mathrm{e}^{-2\pi\, \mathrm{i}\, k\, j/n} \Big| \le \frac{1}{n} \sum_{j=0}^{n-1} |e_j| \le \varepsilon_1 ,$$

we conclude

$$|\tilde{c}_k - c_k(f)| \le \varepsilon_1 + \sum_{\ell \in \mathbb{Z}\setminus\{0\}} |c_{k+\ell n}(f)| .$$

From (10.68) and (10.70), it follows that

$$\tilde{h}_k - h(k) = \frac{1}{c_k(\varphi)}\,\big(\tilde{c}_k - c_k(f)\big)$$

and hence

$$|\tilde{h}_k - h(k)| \le \frac{1}{c_k(\varphi)}\,\Bigg(\varepsilon_1 + \sum_{\ell\in\mathbb{Z}\setminus\{0\}} |c_{k+\ell n}(f)|\Bigg)\,.$$

With (10.68) and

$$|h(k+\ell n)| \le \sum_{j=1}^{M} |c_j| = \|\mathbf{c}\|_1\,, \quad \ell \in \mathbb{Z}\,,$$

we obtain for all $\ell \in \mathbb{Z}$ that

$$|c_{k+\ell n}(f)| = |h(k+\ell n)|\,c_{k+\ell n}(\varphi) \le \|\mathbf{c}\|_1\,c_{k+\ell n}(\varphi)\,.$$

Thus, we receive the estimates

$$\begin{aligned}|\tilde{h}_k - h(k)| &\le \frac{\varepsilon_1}{c_k(\varphi)} + \|\mathbf{c}\|_1 \sum_{\ell\in\mathbb{Z}\setminus\{0\}} \frac{c_{k+\ell n}(\varphi)}{c_k(\varphi)}\\ &\le \frac{\varepsilon_1}{c_k(\varphi)} + \|\mathbf{c}\|_1 \max_{j=-N/2,\ldots,N/2} \sum_{\ell\in\mathbb{Z}\setminus\{0\}} \frac{c_{j+\ell n}(\varphi)}{c_j(\varphi)}\,.\end{aligned}$$

Since the Fourier coefficients of φ are even, we obtain the error estimate of Lemma 10.40. ■

Example 10.41 For a fixed 2π-periodic window function φ of Example 10.39, one can estimate the expression

$$\max_{j=0,\ldots,N/2} \sum_{\ell\in\mathbb{Z}\setminus\{0\}} \frac{c_{j+\ell n}(\varphi)}{c_j(\varphi)} \tag{10.71}$$

more precisely. Let $n = \alpha\,N$ be a power of two, where $\alpha > 1$ is the oversampling factor. For the 2π-periodized Gaussian function of Example 10.39, we obtain

$$\mathrm{e}^{-b\pi^2(1-1/\alpha)}\left[1 + \frac{\alpha}{(2\alpha-1)\,b\pi^2} + \mathrm{e}^{-2b\pi^2/\alpha}\,\Big(1 + \frac{\alpha}{(2\alpha+1)\,b\pi^2}\Big)\right]$$

as an upper bound of (10.71) (see [396]). For the 2π-periodized centered cardinal B-spline of Example 10.39, the upper bound of (10.71)

$$\frac{4m}{(2m-1)\,(2\alpha-1)^{2m}}$$

can be achieved (see [396]).
For the 2π-periodized Kaiser–Bessel function of Example 10.39, the expression (10.71) vanishes, since $c_k(\varphi) = 0$ for all $|k| \ge n$. □

Starting from the given noisy sampled data f_ℓ, $\ell = 0, \ldots, n-1$, we calculate approximate values $\tilde{h}_k$, $k = -\frac{N}{2}, \ldots, \frac{N}{2}$, of the exponential sum (10.69). In the next step, we use the ESPRIT Algorithm 10.12 in order to determine the "frequencies" t_j (which coincide with the shift parameters of (10.66)) and the coefficients c_j of (10.69).

Algorithm 10.42 (Recovery of Linear Combination of Translates)
Input: $N \in 2\,\mathbb{N}$ *with* $N > 2M + 1$, M *unknown number of translates in* (10.66),

$L \approx \frac{N}{2}$, $n = \alpha\, N$ *power of two with* $\alpha > 1$,

$f_\ell \in \mathbb{C}$, $\ell = 0, \ldots, n-1$, *noisy sampled data* (10.67),

2π*-periodic window function* φ *with* $c_k(\varphi) > 0$, $k = 0, \ldots, \frac{N}{2}$.

1. *Apply* FFT *to compute for* $k = -\frac{N}{2}, \ldots, \frac{N}{2}$

$$\tilde{c}_k := \frac{1}{n} \sum_{\ell=0}^{n-1} f_\ell\, \mathrm{e}^{2\pi \mathrm{i} k\ell/n}\,, \quad \tilde{h}_k := \frac{\tilde{c}_k}{c_k(\varphi)}\,.$$

2. *Apply Algorithm* 10.12 *to the rectangular Hankel matrix*

$$\tilde{\mathbf{H}}_{L,N-L+1} := \big(\tilde{h}_{k+\ell-N/2}\big)_{k,\,\ell=0}^{L-1,\,N-L}\,,$$

compute $M \in \mathbb{N}$, $t_j \in [-\pi,\, \pi)$, *and* $c_j \in \mathbb{C}$ *for* $j = 0, \ldots, M$.

Output: $M \in \mathbb{N}$, $t_j \in [-\pi,\, \pi)$, $c_j \in \mathbb{C}$, $j = 0, \ldots, M$.

Remark 10.43 If the 2π-periodic window function φ is well-localized, i.e., if there exists $m \in \mathbb{N}$ with $2\,m \ll n$ such that the values $\varphi(x)$ are very small for all $x \in \mathbb{R} \setminus (I_m + 2\pi\,\mathbb{Z})$ with $I_m := [-2\pi\, m/n,\, 2\pi\, m/n]$, then φ can be approximated by a 2π-periodic function ψ which is supported on $I_m + 2\pi\,\mathbb{Z}$. For a 2π-periodic window function of Example 10.39, we form its truncated version

$$\psi(x) := \sum_{k\in\mathbb{Z}} \Phi(x + 2\pi k)\, \chi_{I_m}(x + 2\pi k)\,, \quad x \in \mathbb{R}\,,$$

where χ_{I_m} denotes the characteristic function of I_m. For the 2π-periodized centered cardinal B-spline of Example 10.39, we see that $\varphi = \psi$. For each $\ell \in \{0, \ldots, n-1\}$, we define the index set:

$$J_{m,n}(\ell) := \{j \in \{1, \ldots, M\} : 2\pi\,(\ell - m) \le n\,t_j \le 2\pi\,(\ell + m)\}\,.$$

Thus, we can replace the 2π-periodic window function φ by its truncated version ψ in Algorithm 10.42. Consequently, we have only to solve the sparse linear system

$$\sum_{j \in J_{m,n}(\ell)} c_j\,\psi\Big(\frac{2\pi\,\ell}{n} + t_j\Big) = f_\ell\,, \quad \ell = 0, \ldots, n-1$$

in order to determine the coefficients c_j. For further details and examples, see [316]. For other approaches to recover special structured functions by a small number of function values, we refer to [314, 323]. □

10.5 Phase Reconstruction

In this last section, we consider the following one-dimensional phase retrieval problem. We assume that a signal f is either of the form

$$f(t) = \sum_{j=1}^{N} c_j\,\delta(t - t_j)\,, \quad t \in \mathbb{R}\,, \tag{10.72}$$

with $c_j \in \mathbb{C}$ for $j = 1, \ldots, N$ and real knots $t_1 < t_2 < \ldots < t_N$, where δ denotes the Dirac distribution (see Example 4.37), or

$$f(t) = \sum_{j=1}^{N} c_j\,\Phi(t - t_j), \quad t \in \mathbb{R}\,, \tag{10.73}$$

as in (10.66), where Φ is a known piecewise continuous function in $L_1(\mathbb{R})$. Observe that a spline function of the form (10.57) with $m \ge 1$ can also be written in the form (10.73) with $N + m$ instead of N terms using the truncated power function.

We want to study the question whether f can be reconstructed from the modulus of its Fourier transform. In other words, for given $|\mathscr{F}(f)(\omega)| = |\hat{f}(\omega)|$, $\omega \in \mathbb{R}$, we aim at reconstructing all parameters t_j and c_j, $j = 1, \ldots, N$, determining f. Applications of the phase retrieval problem occur in electron microscopy, wave front sensing, and laser optics [383, 384], as well as in crystallography and speckle imaging [361].

Unfortunately, the recovery of f is hampered by the well-known ambiguousness of the phase retrieval problem. We summarize the trivial, always occurring, ambiguities (see also [34–36]).

Lemma 10.44 *Let f be a signal of the form* (10.72) *or* (10.73). *Then*

(i) *the rotated signal* $\mathrm{e}^{\mathrm{i}\,\alpha} f$ *for* $\alpha \in \mathbb{R}$,
(ii) *the time shifted signal* $f(\cdot - t_0)$ *for* $t_0 \in \mathbb{R}$, *and*
(iii) *the conjugated and reflected signal* $\overline{f(-\cdot)}$

have the same Fourier intensity $|\mathscr{F}(f)|$ *as* f.

Proof For (i) we observe

$$|\mathscr{F}(\mathrm{e}^{\mathrm{i}\,\alpha} f)(\omega)| = |\mathrm{e}^{\mathrm{i}\,\alpha}|\,|\hat{f}(\omega)| = |\hat{f}(\omega)|\,.$$

Assertion (ii) follows from Theorem 2.5, since

$$|\mathscr{F}(f(\cdot - t_0))(\omega)| = |\mathrm{e}^{-\mathrm{i}\,t_0\,\omega}|\,|\hat{f}(\omega)| = |\hat{f}(\omega)|\,.$$

Finally,

$$|\mathscr{F}(\overline{f(-\cdot)})(\omega)| = \Big|\int_{\mathbb{R}} \overline{f(-t)}\,\mathrm{e}^{-\mathrm{i}\,\omega t}\,\mathrm{d}t\Big| = |\hat{f}(\omega)|$$

implies (iii). ■

We want to derive a constructive procedure to recover f from $|\hat{f}|$ up to the trivial ambiguities mentioned in Lemma 10.44. We observe that f in (10.72) has the Fourier transform

$$\hat{f}(\omega) = \sum_{j=1}^{N} c_j\,\mathrm{e}^{-\mathrm{i}\,\omega\,t_j}\,, \quad \omega \in \mathbb{R}\,,$$

and the known squared Fourier intensity $|\hat{f}|$ is of the form

$$|\hat{f}(\omega)|^2 = \sum_{j=1}^{N}\sum_{k=1}^{N} c_j\,\overline{c}_k\,\mathrm{e}^{-\mathrm{i}\,\omega\,(t_j - t_k)}\,. \tag{10.74}$$

Similarly, for f in (10.73), the squared Fourier intensity is a product of the exponential sum in (10.74) and $|\hat{\Phi}(\omega)|^2$.

The recovery procedure consists now of two steps. First, we will employ the Prony method to determine the parameters of the exponential sum $|\hat{f}(\omega)|^2$, i.e., the knot differences $t_j - t_k$ and the corresponding products $c_j\,\overline{c}_k$. Then, in a second step, we recover the parameters t_j and c_j, $j = 1, \ldots, N$, to obtain f. In order to be able to solve this problem uniquely, we need to assume that all knot differences $t_j - t_k$ are pairwise different for $j \neq k$ and that $|c_1| \neq |c_N|$.

First step: Recovery of the autocorrelation function $|\hat{f}(\omega)|^2$.

Since $t_j - t_k$ are distinct for $j \neq k$, the function $|\hat{f}(\omega)|^2$ can be written in the form

$$|\hat{f}(\omega)|^2 = \sum_{\ell=-N(N-1)/2}^{N(N-1)/2} \gamma_\ell \, \mathrm{e}^{-\mathrm{i}\omega\tau_\ell} = \gamma_0 + \sum_{\ell=1}^{N(N-1)/2} \big(\gamma_\ell \, \mathrm{e}^{-\mathrm{i}\omega\tau_\ell} + \overline{\gamma}_\ell \, \mathrm{e}^{\mathrm{i}\omega\tau_\ell}\big)\,, \tag{10.75}$$

where $0 < \tau_1 < \tau_2 < \ldots < \tau_{N(N-1)/2}$ and $\tau_{-\ell} = -\tau_\ell$. Then, each τ_ℓ, $\ell > 0$, corresponds to one difference $t_j - t_k$ for $j > k$ and $\gamma_\ell = c_j\,\overline{c}_k$. For $\ell = 0$ we have $\tau_0 = 0$ and $\gamma_0 = \sum_{j=1}^{N} |c_j|^2$. Thus, $|\hat{f}(\omega)|^2$ is an exponential sum with $N(N-1)+1$ terms, and all parameters τ_ℓ, γ_ℓ can be reconstructed from the equidistant samples $|\hat{f}(k\,h)|$, $k = 0, \ldots, 2(N-1)N+1$, with sampling step $0 < h < \frac{\pi}{\tau_{N(N-1)/2}}$ using one of the algorithms in Sect. 10.2.

As shown in [35], we can exploit the knowledge that $\tau_0 = 0$, $\tau_\ell = -\tau_{-\ell}$ and that $\gamma_\ell = \overline{\gamma}_{-\ell}$ for $\ell = 1, \ldots, N(N-1)/2$. Therefore, instead of $N(N-1)+1$ real values τ_ℓ and $N(N-1)+1$ complex values γ_ℓ, we only need to recover $N(N-1)/2$ real values τ_ℓ and complex values γ_ℓ for $\ell = 1, \ldots, N(N-1)/2$ as well as the real value γ_0. This can be already done using only the $3N(N-1)/2+1$ intensity values $|\hat{f}(k\,h)|$, $k = 0, \ldots, 3(N-1)N/2$. However, if more intensity values are available, these should be used to stabilize the Prony method.

Second step: Unique signal recovery.

Having determined the knot differences τ_ℓ as well as the corresponding coefficients γ_ℓ in (10.75), we aim at reconstructing the parameters t_j and c_j, $j = 1, \ldots, N$, in the second step (see [35]).

Theorem 10.45 *Let f be a signal of the form* (10.72) *or* (10.73). *Assume that the knot differences $t_j - t_k$ are distinct for $j \neq k$ and that the coefficients satisfy $|c_1| \neq |c_N|$. Further, let h be a step size satisfying $0 < h < \pi/(t_j - t_k)$ for all $j \neq k$.*

Then f can be uniquely recovered from its Fourier intensities $|\hat{f}(k\,h)|$ for all $k = 0, \ldots, 2(N-1)N+1$ up to trivial ambiguities.

Proof We follow the idea in [35]. In the first step described above, we already have obtained all parameters τ_ℓ and γ_ℓ, $\ell = 0, \ldots, N(N-1)/2$, to represent $|\hat{f}(\omega)|^2$ in (10.75). We denote by $\mathscr{T} := \{\tau_\ell : \ell = 1, \ldots, N(N-1)/2\}$ the list of positive differences ordered by size. We need to recover the mapping $\ell \to (j, k)$ such that $\tau_\ell = t_j - t_k$ and $\gamma_\ell = c_j\,\overline{c}_k$ and then extract the wanted parameters t_j and c_j, $j = 1, \ldots, N$. This is done iteratively. Obviously, the maximal distance $\tau_{N(N-1)/2}$ equals to $t_N - t_1$. Due to the shift ambiguity in Lemma 10.44 (ii), we can assume that $t_1 = 0$ and $t_N = \tau_{N(N-1)/2}$. Next, the second largest distance $\tau_{N(N-1)/2-1}$ corresponds either to $t_N - t_2$ or to $t_{N-1} - t_1$. Due to the trivial reflection and conjugation ambiguity in Lemma 10.44 (iii), we can just fix $t_{N-1} - t_1 = t_{N-1} = \tau_{N(N-1)/2-1}$. Thus, there exist a value $\tau_{\ell^*} = t_N - t_{N-1} > 0$ in $\mathscr{T}$ such that

$\tau_{\ell^*} + \tau_{N(N-1)/2-1} = \tau_{N(N-1)/2}$. Considering the corresponding coefficients, we obtain

$$c_N \, \bar{c}_1 = \gamma_{N(N-1)/2}\,, \quad c_{N-1} \, \bar{c}_1 = \gamma_{N(N-1)/2-1}\,, \quad c_N \bar{c}_{N-1} = \gamma_{\ell^*}$$

and therefore

$$|c_1|^2 = \frac{\gamma_{N(N-1)/2} \, \overline{\gamma}_{N(N-1)/2-1}}{\gamma_{\ell^*}}\,, \quad c_N = \frac{\gamma_{N(N-1)/2}}{\bar{c}_1}\,, \quad c_{N-1} = \frac{\gamma_{N(N-1)/2-1}}{\bar{c}_1}\,.$$

By Lemma 10.44 (i), f can be only recovered up to multiplication with a factor with modulus 1. Therefore, we can assume that c_1 is real and positive, and then the above equations allow us to recover c_1, c_{N-1}, and c_N in a unique way.

We proceed by considering the next largest distance $\tau_{N(N-1)/2-2}$ and notice that it corresponds either to $t_N - t_2$ or to $t_{N-2} - t_1 = t_{N-2}$. In any case, there exists a τ_{ℓ^*} such that $\tau_{\ell^*} + \tau_{N(N-1)/2-2} = \tau_{N(N-1)/2} = t_N$. We study the two cases more closely and show that they cannot be true both at the same time.

Case 1: If $\tau_{N(N-1)-2} = t_N - t_2$, then $\tau_{\ell^*} = t_2 - t_1$ and $\gamma_{\ell^*} = c_2 \, \bar{c}_1$. Further, using $\gamma_{N(N-1)/2-2} = c_N \, \bar{c}_2$, we arrive at the condition:

$$c_2 = \frac{\gamma_{\ell^*}}{\bar{c}_1} = \frac{\overline{\gamma}_{N(N-1)/2-2}}{\bar{c}_N}\,. \tag{10.76}$$

Case 2: If $\tau_{N(N-1)-2} = t_{N-2} - t_1$, then $\tau_{\ell^*} = t_N - t_{N-2}$ with coefficient $\gamma_{\ell^*} = c_N \, \bar{c}_{N-2}$. With $\gamma_{N(N-1)/2-2} = c_{N-2} \, \bar{c}_1$, we thus find the condition:

$$c_{N-2} = \frac{\overline{\gamma}_{\ell^*}}{\bar{c}_N} = \frac{\gamma_{N(N-1)/2-2}}{\bar{c}_1}\,. \tag{10.77}$$

If both conditions (10.76) and (10.77) were true, then it follows that

$$\Big|\frac{c_N}{c_1}\Big| = \Big|\frac{\gamma_{N(N-1)/2-2}}{\gamma_{\ell^*}}\Big| = \Big|\frac{c_1}{c_N}\Big|$$

contradicting the assumption $|c_1| \neq |c_N|$. Therefore, only one of the equalities (10.76) and (10.77) can be true, and we can determine either t_2 and c_2 or t_{N-2} and c_{N-2}.

We remove now all differences τ_ℓ from the set $\mathscr{T}$ that correspond to recovered knots and repeat the approach to determine the remaining knots and coefficients. ■

Remark 10.46

1. The assumptions needed for unique recovery can be checked during the algorithm. If the number N of terms in f is known beforehand, then the assumption that $t_j - t_k$ are pairwise different for $j \neq k$ is not satisfied, if the Prony method yields $|\hat{f}(\omega)|^2$ with less than $N(N-1)+1$ terms. The second assumption $|c_1| \neq |c_N|$ can be simply checked after having determined these two values.

2. The problem of recovery of the sequence of knots t_j from an unlabeled set of differences is the so-called turnpike problem that requires a backtracing algorithm with exponential complexity in the worst case [270] and is not always uniquely solvable (see [361]). □

We summarize the recovery of f from its Fourier intensities as follows:

Algorithm 10.47 (Phase Recovery from Fourier Intensities)
Input: *Upper bound* $L \in \mathbb{N}$ *of the number* N *of terms, step size* $h > 0$, *Fourier intensities* $f_k = |\hat{f}(h\,k)| \in [0, \infty)$ *for* $k = 0, \ldots, 2M$, $M > N(N-1)$, *accuracy* $\epsilon > 0$.

1. *Set* $h_k = |\hat{f}(h\,k)|^2$, *if* f *is of the form* (10.72) *and* $h_k = |\hat{f}(h\,k)/\hat{\Phi}(h\,k)|^2$, *if* f *is of the form* (10.73). *Apply Algorithm* 10.10 *to determine the knot distances* τ_ℓ *for* $\ell = -N(N-1)/2, \ldots N(N-1)/2$ *in* (10.75) *in increasing order and the corresponding coefficients* γ_ℓ.
Update the reconstructed distances and coefficients by
$$\tau_\ell := \frac{1}{2}(\tau_\ell - \tau_{-\ell}), \quad \gamma_\ell := \frac{1}{2}(\gamma_\ell + \overline{\gamma}_{-\ell}), \quad \ell = 0, \ldots, N(N-1)/2.$$
2. *Set* $t_1 := 0$, $t_N := \tau_{N(N-1)/2}$, $t_{N-1} := \tau_{N(N-1)/2-1}$. *Find the index* ℓ^* *with* $|\tau_{\ell^*} - t_N + t_{N-1}| \le \epsilon$ *and compute*
$$c_1 := \Big|\frac{\gamma_{N(N-1)/2}\,\overline{\gamma}_{N(N-1)/2-1}}{\gamma_{\ell^*}}\Big|^{1/2},$$
$$c_N := \frac{\gamma_{N(N-1)/2}}{\overline{c}_1}, \qquad c_{N-1} := \frac{\gamma_{N(N-1)/2-1}}{\overline{c}_1}.$$
Initialize the list of recovered knots and coefficients $T := \{t_1, t_{N-1}, t_N\}$ *and* $C := \{c_1, c_{N-1}, c_N\}$ *and remove the used distances from* $\mathcal{T} := \{\tau_\ell : \ell = 1, \ldots, N(N-1)/2\}$.
3. *For the maximal remaining distance* $\tau_{k^*} \in \mathcal{T}$, *determine* ℓ^* *with* $|\tau_{k^*} + \tau_{\ell^*} - t_N| \le \epsilon$.

   3.1. *If* $|\tau_{k^*} - \tau_{\ell^*}| > \epsilon$, *then compute* $d_1 = \gamma_{k^*}/\overline{c}_1$, $d_2 = \gamma_{\ell^*}/\overline{c}_1$. *If*
$$|c_N\,\overline{d}_1 - \gamma_{\ell^*}| < |c_N\,\overline{d}_2 - \gamma_{k^*}|,$$
then $T := T \cup \{\frac{1}{2}(\tau_{k^*} + t_N - \tau_{\ell^*})\}$ *and* $C := C \cup \{d_1\}$ *else* $T := T \cup \{\frac{1}{2}(\tau_{\ell^*} + t_N - \tau_{k^*})\}$ *and* $C := C \cup \{d_2\}$.

   3.2. *If* $|\tau_{k^*} - \tau_{\ell^*}| \le \epsilon$, *then the knot distance belongs to the center of the interval. Set* $T := T \cup \{t_N/2\}$, $C := C \cup \{\gamma_{k^*}/\overline{c}_1\}$.

Remove all distances between the new knot and the knots being recovered already from $\mathcal{T}$ *and repeat step* 3 *until* $\mathcal{T}$ *is empty.*

Output: knots t_j and coefficients c_j of the signal f in (10.72) *or* (10.73).

Note that this algorithm is very expensive for larger N. The computational costs are governed by Algorithm 10.10 in step 1, which is here applied to an exponential sum of the form (10.75) with $N(N-1)+1$ terms.

Example 10.48 We consider a toy example to illustrate the method. We want to recover the signal

$$f(t) = 2\,\delta(t) + (5-\mathrm{i})\,\delta(t-3) + (7+\mathrm{i})\,\delta(t-5)$$

with the Fourier transform $\hat{f}(\omega) = 2 + (5-\mathrm{i})\,\mathrm{e}^{-3\mathrm{i}\,\omega} + (7+\mathrm{i})\,\mathrm{e}^{-5\mathrm{i}\,\omega}$, i.e., we have to recover the knots $t_1 = 0$, $t_2 = 3$, $t_3 = 5$ and the coefficients $c_1 = 2$, $c_2 = 5 - \mathrm{i}$, and $c_3 = 7 + \mathrm{i}$. Thus, $N = 3$ and $\max |t_j - t_k| = 5$. We can choose a step size $h < \pi/5$. Let us take here $h = \pi/6$. Note the considered signal is already "normalized" in the sense that $t_1 = 0$ and c_1 is positive. Each other signal of the form $\mathrm{e}^{\mathrm{i}\alpha}\, f(t-t_0)$ with $\alpha, t_0 \in \mathbb{R}$ has the same Fourier intensity.

We assume that the Fourier intensities $|\hat{f}(k\pi/6)|$ for $k = 0, \ldots, L$ with $L \ge 13$ are given. Exploiting symmetry properties, also the intensities $|\hat{f}(k\,\pi/6)|$ for $k = 0, \ldots, 9$ would be sufficient. The autocorrelation function $|\hat{f}(\omega)|^2$ is of the form

$$\begin{aligned}|\hat{f}(\omega)|^2 &= (14-2\,\mathrm{i})\,\mathrm{e}^{5\mathrm{i}\,\omega} + (10+2\,\mathrm{i})\,\mathrm{e}^{3\mathrm{i}\,\omega} + (34-12\,\mathrm{i})\,\mathrm{e}^{2\mathrm{i}\,\omega} \\ &\quad + 18 + (34+12\,\mathrm{i})\,\mathrm{e}^{-2\mathrm{i}\,\omega} + (10-2\,\mathrm{i})\,\mathrm{e}^{-3\mathrm{i}\,\omega} + (14+2\,\mathrm{i})\,\mathrm{e}^{-5\mathrm{i}\,\omega}\,.\end{aligned}$$

In the first step, we recover the frequencies $\tau_0 = 0$, $\tau_1 = 2$, $\tau_2 = 3$, and $\tau_3 = 5$ as well as the coefficients $\gamma_0 = 18$, $\gamma_1 = 34 + 12\,\mathrm{i}$, $\gamma_2 = 10 - 2\,\mathrm{i}$, and $\gamma_3 = 14 + 2\,\mathrm{i}$ from the given samples using the Prony method.

In the second step, we conclude from the largest difference $\tau_3 = 5$ that $t_1 = 0$ and $t_3 = 5$. Here, we have already fixed the support of f. Indeed, any other solution with $t_1 = t_0$, $t_3 = t_0 + 5$ is also correct by Lemma 10.44 (ii). Next, from $\tau_2 = 3$ we conclude that t_2 is either 3 or 2. Both solutions are possible, and indeed $\tau_1 + \tau_2 = \tau_3 = t_3 - t_1$. For $t_2 = 3$, we find

$$c_1 := \Big|\frac{\gamma_3\overline{\gamma}_2}{\gamma_1}\Big|^{1/2} = \Big|\frac{(14+2\,\mathrm{i})\,(10+2\,\mathrm{i})}{34+12\,\mathrm{i}}\Big|^{1/2} = 2\,,$$

$$c_3 := \frac{\gamma_3}{\overline{c}_1} = \frac{14+2\,\mathrm{i}}{2} = 7+\mathrm{i}\,, \quad c_2 := \frac{\gamma_2}{\overline{c}_1} = \frac{10-2\,\mathrm{i}}{2} = 5-\mathrm{i}\,.$$

This solution recovers f.

For $t_2 = 2$, we find $c_3\,\overline{c}_1 = \gamma_3$, $c_2\,\overline{c}_1 = \gamma_1$, and $c_3\,\overline{c}_2 = \gamma_2$, and thus

$$|c_1|^2 = \Big|\frac{\gamma_3\overline{\gamma}_1}{\gamma_2}\Big| = \Big|\frac{(14+2\,\mathrm{i})\,(34-12\,\mathrm{i})}{10-2\,\mathrm{i}}\Big| = 50\,.$$

Thus, we find in this case

$$c_1 = \sqrt{50}\,, \quad c_2 = \frac{\gamma_1}{\bar{c}_1} = \frac{34 + 12\,\mathrm{i}}{\sqrt{50}}\,, \quad c_3 = \frac{\gamma_3}{\bar{c}_1} = \frac{14 + 2\,\mathrm{i}}{\sqrt{50}}\,.$$

However, this second solution

$$f_2(t) = \sqrt{50}\,\delta(t) + \frac{34 + 12\,\mathrm{i}}{\sqrt{50}}\,\delta(t-2) + \frac{14 + 2\,\mathrm{i}}{\sqrt{50}}\,\delta(t-5)$$

is indeed the conjugated and reflected signal of f, translated by 5 and multiplied with the factor $\mathrm{e}^{\mathrm{i}\,\alpha} = \frac{7-\mathrm{i}}{\sqrt{50}}$, i.e.,

$$f_2(t) = \frac{7 - \mathrm{i}}{\sqrt{50}}\,\overline{f(-t+5)}\,.$$

□

Remark 10.49 Within the last years, phase retrieval problems have been extensively studied. There exist many very different problem statements that are summarized under the term "phase retrieval" but may be quite different in nature. The applications in physics usually require a signal or image recovery from Fourier or Fresnel intensities (see, for instance, [32, 38, 40, 141]). The problem is ill-posed because of many ambiguities and can only be solved with a suitable amount of a priori knowledge about the solution signal. Often, support properties, positivity, or interference measurements can be used to reduce the solution set. For the one-dimensional discrete phase retrieval problem, we refer to the recent surveys [34, 37, 40]. The two-dimensional case remains to be not completely understood in general. Numerically, iterative projection algorithms are mainly applied in practice (see [27, 277]). □

Appendix A
List of Symbols and Abbreviations

Numbers and Related Notations

$\mathbb{N}$	Set of positive integers
$\mathbb{N}_0$	Set of nonnegative integers
$\mathbb{Z}$	Set of integers
$\mathbb{R}$	Set of real numbers
$\mathbb{R}_+$	Set of nonnegative numbers
$\mathbb{C}$	Set of complex numbers
$\mathbb{T}$	Torus of length 2π
$\mathbb{T}^d$	d-dimensional torus of length 2π
$[a,\, b]$	Closed interval in $\mathbb{R}$
$\lfloor x \rfloor$	Largest integer $\le x$ for given $x \in \mathbb{R}$
e	Euler's number
i	Imaginary unit
$\arg a$	Argument of $a \in \mathbb{C} \setminus \{0\}$
$\lvert a \rvert$	Magnitude of $a \in \mathbb{C}$
$\overline{a}$	Conjugate complex number of $a \in \mathbb{C}$
$\mathrm{Re}\, a$	Real part of $a \in \mathbb{C}$
$\mathrm{Im}\, a$	Imaginary part of $a \in \mathbb{C}$
$w_N := \mathrm{e}^{-2\pi \mathrm{i}/N}$	Primitive N-th root of unity
$\varphi(n)$	Euler totient function of $n \in \mathbb{N}$
δ_j	Kronecker symbol with $\delta_0 = 1$ and $\delta_j = 0$, $j \in \mathbb{Z} \setminus \{0\}$
$j \bmod N$	Nonnegative residue modulo N
$\delta_{j \bmod N}$	N-periodic Kronecker symbol

G. Plonka et al., *Numerical Fourier Analysis*, Applied and Numerical Harmonic Analysis, https://doi.org/10.1007/978-3-031-35005-4

$\mathcal{O}$	Landau symbol
$(\mathbb{Z}/p\,\mathbb{Z})^*$	Multiplicative group of integers modulo a prime p
$\mathbb{F}$	Set of binary floating point numbers
$\text{fl}(a)$	Floating point number of a
u	Unit roundoff in $\mathbb{F}$
$\tilde{w}_N^k$	Precomputed value of w_N^k
$\ln a$	Natural logarithm of $a > 0$ to the base e
$\log_2 N$	Binary logarithm of $N > 0$ to the base 2
$\varepsilon_N(k)$	Scaling factors with $\varepsilon_N(0) = \varepsilon_N(N) = \frac{\sqrt{2}}{2}$ and $\varepsilon_N(k) = 1$ for $k = 1, \ldots, N-1$
$\text{sgn}\ a$	Sign of $a \in \mathbb{R}$
$\alpha(t)$	Number of real additions required for an FFT for DFT(2^t)
$\mu(t)$	Number of nontrivial real multiplications required for an FFT for DFT(2^t)
$I_N,\ J_N$	Index set $\{0, \ldots, N-1\}$ if not defined differently
$k = (k_{t-1}, \ldots, k_0)_2$	t-digit binary number $k \in J_N$ with $N = 2^t$ and $k_j \in \{0,\ 1\}$
$\rho(k)$	Bit-reversed number of $k \in J_N$ with $N = 2^t$
π_N	Perfect shuffle of J_N with $N = 2^t$
I_N^d	Multivariate index set $\{\mathbf{n} = (n_j)_{j=1}^d : n_j \in I_N,\ j = 1, \ldots, d\}$
$\mathbf{n} := (n_j)_{j=1}^d$	Multivariate index
$\mathbf{1}^d$	Vector of ones, $\mathbf{1}^d = (1, \ldots, 1)^\top \in \mathbb{Z}^d$
$\mathbf{k} \bmod \mathbf{N}$	Nonnegative residue modulo $\mathbf{N}$ defined entrywise
$\mathbf{k} \circ \mathbf{N}$	Entrywise multiplication $\mathbf{k} \circ \mathbf{N} = (k_j N_j)_{j=1}^d$
$\Lambda(\mathbf{z}, M)$	Rank-1 lattice generated by $\mathbf{z} \in \mathbb{Z}^d$ and $M \in \mathbb{Z}$
$\Lambda^\perp(\mathbf{z}, M)$	Integer dual lattice of $\Lambda(\mathbf{z}, M)$

Periodic Functions and Related Notations

$f : \mathbb{T} \to \mathbb{C}$	Complex-valued 2π-periodic function
$f^{(r)}$	rth derivative of f
$C(\mathbb{T}^d)$	Banach space of continuous functions $f : \mathbb{T}^d \to \mathbb{C}$
$C^r(\mathbb{T}^d)$	Banach space of d-variate r-times continuously differentiable functions $f : \mathbb{T}^d \to \mathbb{C}$
$L_p(\mathbb{T}^d)$	Banach space of d-variate measurable functions $f : \mathbb{T}^d \to \mathbb{C}$ with integrable $\lvert f\rvert^p$, $p \ge 1$

$L_2(\mathbb{T}^d)$	Hilbert space of absolutely square-integrable functions $f : \mathbb{T}^d \to \mathbb{C}$
$\mathscr{M}(\mathbb{T}^d)$	Finite, complex-valued Borel measures on $\mathbb{T}^d$
$\mathscr{M}_+(\mathbb{T}^d)$	Finite, positive Borel measures on $\mathbb{T}^d$
$\mathscr{P}(\mathbb{T}^d)$	Probability measures on $\mathbb{T}^d$
$\mathrm{e}^{\mathrm{i}\,kx}$	kth complex exponential with $k \in \mathbb{Z}$
$\mathscr{T}_n$	Set of 2π-periodic trigonometric polynomials up to degree n
$\mathscr{T}_n^{(2T)}$	Set of $2T$-periodic trigonometric polynomials up to degree n
$c_k(f)$	kth Fourier coefficient of $f \in L_1(\mathbb{T})$ or $f \in L_2(\mathbb{T})$
$c_{\mathbf{k}}(f)$	**k**th Fourier coefficient of a d-variate function $f \in L_1(\mathbb{T}^d)$ or $f \in L_2(\mathbb{T}^d)$
$c_k^{(L)}(f)$	kth Fourier coefficient of an L-periodic function f
$\hat{c}_k(f)$	Approximate value of $c_k(f)$
$S_n f$	nth partial sum of the Fourier series of $f \in L_2(\mathbb{T})$ or $f \in L_2(\mathbb{T}^d)$
$a_k(f),\, b_k(f)$	kth real Fourier coefficients of $f : \mathbb{T} \to \mathbb{R}$
$f * g$	Convolution of $f, g \in L_1(\mathbb{T})$ or $f, g \in L_1(\mathbb{T}^d)$
D_n	nth Dirichlet kernel
F_n	nth Fejér kernel
$\sigma_n f$	nth Fejér sum
V_{2n}	nth de la Vallée Poussin kernel
$\chi_{[a,\,b]}$	Characteristic function of the interval $[a,\, b]$
$V_a^b(\varphi)$	Total variation of the function $\varphi : [a,\, b] \to \mathbb{C}$
$\tilde{S}_n f$	nth partial sum of the conjugate Fourier series of f
$f(x_0 \pm 0)$	One-sided limits of the function $f : \mathbb{T} \to \mathbb{C}$ at the point x_0
cas	Cosine-and-sine function $\cos + \sin$
$P_{2\pi}$	2π-periodization operator
$S_I f$	d-variate Fourier partial sum of $f \in L_1(\mathbb{T}^d)$ with regard to frequency index set I
Π_I	$\mathrm{Span}\{\mathrm{e}^{\mathrm{i}\mathbf{k}\cdot\mathbf{x}} : \mathbf{k} \in I\}$ space of multivariate trigonometric polynomials supported on I
$\mathscr{A}(\mathbb{T}^d)$	Weighted subspace of $L_1(\mathbb{T}^d)$
$H^{\alpha,p}(\mathbb{T}^d)$	Periodic Sobolev space of isotropic smoothness

Sequences and Related Notations

$x = (x_k)_{k\in\mathbb{Z}}$	Sequence with complex entries x_k
$\ell_\infty(\mathbb{Z})$	Banach space of bounded sequences
$\ell_p(\mathbb{Z})$	Banach space of sequences $x = (x_k)_{k\in\mathbb{Z}}$ with $\sum_{k\in\mathbb{Z}} \lvert x_k\rvert^p < \infty$, $p \ge 1$
$\ell_2(\mathbb{Z})$	Hilbert space of sequences $x = (x_k)_{k\in\mathbb{Z}}$ with $\sum_{k\in\mathbb{Z}} \lvert x_k\rvert^2 < \infty$
$V x$	Forward shift of x
$V^{-1}x$	Backward shift of x
M	Modulation filter
$\delta = (\delta_k)_{k\in\mathbb{Z}}$	Pulse sequence
$h = H\,\delta$	Impulse response of linear, time-invariant filter H
$h * x$	Discrete convolution of sequences h and x
$H(\omega)$	Transfer function of linear, time-invariant filter H

Nonperiodic Functions Defined on $\mathbb{R}^d$ and Related Notations

$f : \mathbb{R} \to \mathbb{C}$	Complex-valued function
$C_0(\mathbb{R}^d)$	Banach space of d-variate continuous functions $f : \mathbb{R} \to \mathbb{C}$ with $\lim_{\lVert\mathbf{x}\rVert\to\infty} f(\mathbf{x}) = 0$
$C_c(\mathbb{R}^d)$	Space of compactly supported, continuous functions $f : \mathbb{R}^d \to \mathbb{C}$
$C_b(\mathbb{R}^d)$	Space of bounded, continuous functions $f : \mathbb{R}^d \to \mathbb{C}$
$C^r(\mathbb{R}^d)$	Space of r-times continuously differentiable functions $f : \mathbb{R}^d \to \mathbb{C}$
$L_p(\mathbb{R}^d)$	Banach space of d-variate measurable functions $f : \mathbb{R}^d \to \mathbb{C}$ such that $\lvert f\rvert^p$ is integrable over $\mathbb{R}^d$ for $p \ge 1$
$L_2(\mathbb{R}^d)$	Hilbert space of d-variate absolutely square integrable functions $f : \mathbb{R}^d \to \mathbb{C}$
$\mathcal{M}(\mathbb{R}^d)$	Finite, complex-valued Borel measures on $\mathbb{R}^d$
$\mathcal{M}_+(\mathbb{R}^d)$	Finite, positive Borel measures on $\mathbb{R}^d$
$\mathcal{P}(\mathbb{R}^d)$	Probability measures on $\mathbb{R}^d$
$\mathrm{rba}(\mathbb{R}^d)$	Space of finitely additive set functions on $\mathbb{R}^d$
$\mathcal{S}(\mathbb{R}^d)$	Schwartz space of d-variate rapidly decreasing functions
$\mathcal{S}'(\mathbb{R}^d)$	Space of tempered distributions on $\mathcal{S}(\mathbb{R}^d)$
$\mathbb{S}^2$	Unit sphere $\mathbb{S}^2 = \{\mathbf{x} \in \mathbb{R}^3 : \lVert\mathbf{x}\rVert_2 = 1\}$
$L_2(\mathbb{S}^2)$	Hilbert space of square integrable functions f on $\mathbb{S}^2$
$\lVert f\rVert^2$	Energy of $f \in L_2(\mathbb{R})$

$\hat{f} = \mathcal{F} f$	Fourier transform of $f \in L_1(\mathbb{R})$ or $f \in L_2(\mathbb{R})$ and $f \in L_1(\mathbb{R}^d)$ or $f \in L_2(\mathbb{R}^d)$
$\|f\|^2$	Energy of $f \in L_2(\mathbb{R})$
$\hat{f} = \mathcal{F} f$	Fourier transform of $f \in L_1(\mathbb{R})$ or $f \in L_2(\mathbb{R})$ and $f \in L_1(\mathbb{R}^d)$ or $f \in L_2(\mathbb{R}^d)$
$(\hat{f})^{\vee}$	Inverse Fourier transform of $\hat{f} \in L_1(\mathbb{R})$ or $\hat{f} \in L_2(\mathbb{R})$ and $\hat{f} \in L_1(\mathbb{R}^d)$ or $\hat{f} \in L_2(\mathbb{R}^d)$
$f * g$	Convolution of $f, g \in L_1(\mathbb{R})$ or $f, g \in L_1(\mathbb{R}^d)$
sinc	Cardinal sine function
Si	Sine integral
N_m	Cardinal B-spline of order m
M_m	Centered cardinal B-spline of order m
$M^{(k,\ell)}(x_1, x_2)$	Tensor product of B-splines $M^{(k,\ell)}(x_1, x_2) = M_k(x_1)\, M_\ell(x_2)$
B_j^m	B-spline of order m with arbitrary knots $-\infty < t_j < \ldots < t_{j+m} < \infty$
H_n	Hermite polynomial of degree n
h_n	nth Hermite function
P_k	kth Legendre polynomial
P_k^n	Associated Legendre function
Y_k^n	Spherical harmonics
J_ν	Bessel function of order ν
δ	Dirac distribution
$\mathcal{H} f$	Hankel transform of $f \in L_2((0, \infty))$
Δu	Laplace operator applied to a function u
supp f	Support of $f : \mathbb{R} \to \mathbb{C}$
$\Delta_{x_0} f$	Dispersion of $f \in L_2(\mathbb{R})$ about the time $x_0 \in \mathbb{R}$
$\Delta_{\omega_0} \hat{f}$	Dispersion of $\hat{f} \in L_2(\mathbb{R})$ about the frequency $\omega_0 \in \mathbb{R}$
$(\mathcal{F}_\psi f)(b, \omega)$	Windowed Fourier transform of f with respect to the window function ψ
$\|(\mathcal{F}_\psi f)(b, \omega)\|^2$	Spectrogram of f with respect to the window function ψ
$\mathcal{F}_\alpha f$	Fractional Fourier transform for $f \in L_2(\mathbb{R})$ with $\alpha \in \mathbb{R}$
$\mathcal{L}_{\mathbf{A}} f$	Linear canonical transform for $f \in L_2(\mathbb{R})$ with a 2-by-2 matrix **A**

Vectors, Matrices, and Related Notations

$\mathbb{C}^N$	Vector space of complex column vectors $\mathbf{a} = (a_j)_{j=0}^{N-1}$
$\mathbf{a} = (a_j)_{j=0}^{N-1}$	Column vector with complex components a_j
$\mathbf{a}^\top$	Transposed vector of $\mathbf{a}$
$\overline{\mathbf{a}}$	Conjugate complex vector of $\mathbf{a}$
$\mathbf{a}^\mathrm{H}$	Transposed conjugate complex vector of $\mathbf{a}$
$\langle \mathbf{a}, \mathbf{b} \rangle$	Inner product of $\mathbf{a}, \mathbf{b} \in \mathbb{C}^N$
$\|\mathbf{a}\|_2$	Euclidean norm of $\mathbf{a} \in \mathbb{C}^N$
$\mathbf{b}_k = (\delta_{j-k})_{j=0}^{N-1}$	Standard basis vectors of $\mathbb{C}^N$ for $k = 0, \ldots, N-1$
$\mathbf{e}_k = (w_N^{jk})_{j=0}^{N-1}$	Exponential vectors of $\mathbb{C}^N$ for $k = 0, \ldots, N-1$
$\mathbf{A}_N = (a_{j,k})_{j,k=0}^{N-1}$	N-by-N matrix with complex entries $a_{j,k}$
$\mathbf{A}_N^\top$	Transposed matrix of $\mathbf{A}_N$
$\overline{\mathbf{A}}_N$	Complex conjugate matrix of $\mathbf{A}_N$
$\mathbf{A}_N^\mathrm{H}$	Transposed complex conjugate matrix of $\mathbf{A}_N$
$\mathbf{A}_N^{-1}$	Inverse matrix of $\mathbf{A}_N$
$\mathbf{A}_N^{+}$	Moore–Penrose pseudo-inverse of $\mathbf{A}_N$
$\mathbf{I}_N$	N-by-N identity matrix
$\mathbf{0}$	Zero vector and zero matrix, respectively
$\mathbf{F}_N = (w_N^{jk})_{j,k=0}^{N-1}$	N-by-N Fourier matrix with $w_N = \mathrm{e}^{-2\pi \mathrm{i}/N}$
$\frac{1}{\sqrt{N}}\,\mathbf{F}_N$	Unitary Fourier matrix
$\hat{\mathbf{a}} = (\hat{a}_k)_{k=0}^{N-1}$	Discrete Fourier transform of $\mathbf{a} \in \mathbf{C}^N$, i.e., $\hat{\mathbf{a}} = \mathbf{F}_N\,\mathbf{a}$
$\mathbf{J}_N'$	N-by-N flip matrix
$\mathbf{J}_N$	N-by-N counter-diagonal matrix
$\det \mathbf{A}_N$	Determinant of the matrix $\mathbf{A}_N$
$\mathrm{tr}\ \mathbf{A}_N$	Trace of the matrix $\mathbf{A}_N$
$\mathbf{a} * \mathbf{b}$	Cyclic convolution of $\mathbf{a}, \mathbf{b} \in \mathbb{C}^N$
$\mathbf{b}_0 = (\delta_j)_{j=0}^{N-1}$	Pulse vector
$\mathbf{V}_N$	N-by-N forward-shift matrix
$\mathbf{V}_N^{-1}$	N-by-N backward-shift matrix
$\mathbf{I}_N - \mathbf{V}_N$	N-by-N cyclic difference matrix
$\mathbf{h} = \mathbf{H}_N\,\mathbf{b}_0$	Impulse response vector of a shift-invariant, linear map $\mathbf{H}_N$
$\mathbf{M}_N$	N-by-N modulation matrix
$\mathbf{a} \circ \mathbf{b}$	Componentwise product of $\mathbf{a}, \mathbf{b} \in \mathbb{C}^N$
$\mathrm{circ}\ \mathbf{a}$	N-by-N circulant matrix of $\mathbf{a} \in \mathbb{C}^N$
$(\mathbf{a}_0 \mid \ldots \mid \mathbf{a}_{N-1})$	N-by-N matrix with the columns $\mathbf{a}_k \in \mathbb{C}^N$

diag $\mathbf{a}$	N-by-N diagonal matrix with the diagonal entries a_j, where $\mathbf{a} = (a_j)_{j=0}^{N-1}$
$\mathbf{A}_{M,N} = (a_{j,k})_{j,k=0}^{M-1,N-1}$	M-by-N matrix with complex entries $a_{j,k}$
$\mathbf{A}_{M,N} \otimes \mathbf{B}_{P,Q}$	Kronecker product of $\mathbf{A}_{M,N}$ and $\mathbf{B}_{P,Q}$
$\mathbf{a} \otimes \mathbf{b}$	Kronecker product of $\mathbf{a} \in \mathbb{C}^M$ and $\mathbf{b} \in \mathbb{C}^N$
$\mathbf{P}_N(L)$	L-stride permutation matrix with $N = L\,M$ for integers $L, M \geq 2$
$\mathbf{P}_N(2)$	Even–odd permutation matrix for even integer N
col $\mathbf{A}_{M,N}$	Vectorization of the matrix $\mathbf{A}_{M,N}$
$\mathbf{C}_{N+1}^{\mathrm{I}}$	$(N+1)$-by-$(N+1)$ cosine matrix of type I
$\mathbf{C}_N^{\mathrm{II}}, \mathbf{C}_N^{\mathrm{III}}, \mathbf{C}_N^{\mathrm{IV}}$	N-by-N cosine matrix of type II, III, and IV, respectively
$\mathbf{S}_{N-1}^{\mathrm{I}}$	$(N-1)$-by-$(N-1)$ sine matrix of type I
$\mathbf{S}_N^{\mathrm{II}}, \mathbf{S}_N^{\mathrm{III}}, \mathbf{S}_N^{\mathrm{IV}}$	N-by-N sine matrix of type II, III, and IV, respectively
$\mathbf{H}_N$	N-by-N Hartley matrix
$\mathbf{R}_N$	Bit-reversed permutation matrix for $N = 2^t$
$\mathbf{P}_N$	Perfect shuffle permutation matrix with $N = 2^t$
$\mathbf{D}_N$	Diagonal sign matrix diag $((-1)^j)_{j=0}^{N-1}$
$\mathbf{W}_{N/2}$	Diagonal matrix diag $(w_N^j)_{j=0}^{N/2-1}$ for even integer N
diag $(\mathbf{A}_N,\ \mathbf{B}_N)$	$2N$-by-$2N$ block diagonal matrix with diagonal entries $\mathbf{A}_N$ and $\mathbf{B}_N$
$\mathbf{a}^{(\ell)}$	Periodization of $\mathbf{a} \in \mathbb{C}^N$ with $N = 2^t$ and $\ell \in \{0, \ldots, t\}$
$\hat{\mathbf{A}} = \mathbf{F}_{N_1}\,\mathbf{A}\,\mathbf{F}_{N_2}$	Two-dimensional discrete Fourier transform of $\mathbf{A} \in \mathbb{C}^{N_1 \times N_2}$
$\lvert\mathbf{x}\rvert = (\lvert x_j\rvert)_{j=0}^{N-1}$	Modulus of a vector $\mathbf{x} \in \mathbb{C}^N$
$\lvert\mathbf{A}\rvert = (\lvert a_{j,k}\rvert)_{j,k=0}^{N_1-1,N_2-1}$	Modulus of matrix $\mathbf{A} \in \mathbb{C}^{N_1 \times N_2}$
$\lVert\mathbf{A}\rVert_{\mathrm{F}}$	Frobenius norm of matrix $\mathbf{A}$
$\mathbf{A} * \mathbf{B}$	Cyclic convolution of $\mathbf{A},\ \mathbf{B} \in \mathbb{C}^{N_1 \times N_2}$
$\mathbf{A} \circ \mathbf{B}$	Entrywise product of $\mathbf{A},\ \mathbf{B} \in \mathbb{C}^{N_1 \times N_2}$
$\mathbf{A} \oplus \mathbf{B}$	Block diagonal matrix diag $(\mathbf{A},\ \mathbf{B})$
$\mathbf{n} \cdot \mathbf{x}$	Inner product of $\mathbf{n} \in \mathbb{Z}^d$ and $\mathbf{x} \in \mathbb{R}^d$
$\mathbf{F}_N^d$	d-variate Fourier matrix $\mathbf{F}_N^d = (\mathrm{e}^{-2\pi \mathrm{i}\mathbf{k}\cdot\mathbf{j}/N})_{j,k \in I_N^d}$
$\mathbf{A}(X, I)$	Multivariate Fourier matrix $\mathbf{A}(X, I) = (\mathrm{e}^{\mathrm{i}\mathbf{k}\cdot\mathbf{x}})_{\mathbf{x}\in X, \mathbf{k}\in I} \in \mathbb{C}^{\lvert X\rvert \times \lvert I\rvert}$
$\lvert I\rvert$	Cardinality of finite index set I
$\mathbf{C}_M(p)$	Companion matrix to the polynomial p of degree M
$\mathbf{V}_M(\mathbf{z})$	Vandermonde matrix $\mathbf{V}_M(\mathbf{z}) = (z_k^{j-1})_{j,k=1}^M$ generated by $\mathbf{z} = (z_k)_{k=1}^M$.

$\mathbf{V}^c_{2M}(\mathbf{z})$	$2M$-by-$2M$ confluent Vandermonde matrix
$\mathbf{V}_{P,M}(\mathbf{z})$	Rectangular P-by-M Vandermonde matrix $\mathbf{V}_{P,M}(\mathbf{z}) = (z_k^{j-1})_{j,k=1}^{P,M}$
$\mathbf{H}_M(0)$	M-by-M Hankel matrix
$\mathbf{H}_{L,M}$	Rectangular L-by-M Hankel matrix
$\mathrm{cond}_2\, \mathbf{A}_{L,M}$	Spectral norm condition of an L-by-M matrix $\mathbf{A}_{L,M}$

Real-Valued Functions Defined on $[-1,\ 1]$ and Related Notations

$C(I)$	Banach space of continuous functions $f : I \to \mathbb{R}$
$C^r(I)$	Banach space of r-times continuously differentiable functions $f : I \to \mathbb{R}$
$C^\infty(I)$	Set of infinitely differentiable functions $f : I \to \mathbb{R}$
I	Closed interval $[-1,\ 1]$
$L_{2,even}(\mathbb{T})$	Subspace of even, real-valued functions of $L_2(\mathbb{T})$
$\mathscr{P}_n$	Set of real algebraic polynomials up to degree n
T_k	kth Chebyshev polynomial (of first kind)
U_k	kth Chebyshev polynomial of second kind
$w(x)$	Weight function $(1 - x^2)^{-1/2}$ for $x \in (-1,\ 1)$ (if not defined differently)
$L_{2,w}(I)$	Real weighted Hilbert space of functions $f : I \to \mathbb{R}$, where $w\,\lvert f\rvert^2$ is integrable over I
$a_k[f]$	kth Chebyshev coefficient of $f \in L_{2,w}(I)$
$C_n f$	nth partial sum of the Chebyshev series of $f \in L_{2,w}(I)$
$T_k^{[a,b]}$	kth Chebyshev polynomial with respect to the compact interval $[a,\ b]$
$x_j^{(N)}$	Chebyshev extreme points for fixed N and $j = 0, \ldots, N$
$z_j^{(N)}$	Chebyshev zero points for fixed N and $j = 0, \ldots, N-1$
ℓ_k	kth Lagrange basis polynomial
$a_k^{(N)}[f]$	kth coefficient of the interpolating polynomial of $f \in C(I)$ at Chebyshev extreme points $x_j^{(N)}$ for $j = 0, \ldots, N$
B_m	Bernoulli polynomial of degree m
b_ℓ	ℓth 1-periodic Bernoulli function
$h_j^{a,b,r}$	Two-point Taylor basis polynomials for the interval $[a,\ b]$ and $j = 0, \ldots, r-1$
λ_N	Lebesgue constant for polynomial interpolation at Chebyshev extreme points $x_j^{(N)}$ for $j = 0, \ldots, N$

Abbreviations

APM	Approximate Prony method, 580
DCT	Discrete cosine transform, 168
DFT	Discrete Fourier transform, 135
DFT(N)	Discrete Fourier transform of length N, 136
DHT	Discrete Hartley transform, 174
DSFT	Discrete spherical Fourier transform, 555
DST	Discrete sine transform, 168
ESPIRA	Estimation of signal parameters by iterative rational approximation, 586
ESPRIT	Estimation of signal parameters via rotational invariance techniques, 580
FFT	Fast Fourier transform, 265
FIR	Finite impulse response system, 55
FLT	Fast Legendre transform, 556
FRFT	Fractional Fourier transform, 116
FSFT	Fast spherical Fourier transform, 557
LFFT	Lattice based FFT, 475
LTI	Linear, time-invariant system, 53
MUSIC	Multiple signal classification, 576
NDCT	Nonequispaced discrete cosine transform, 442
NDST	Nonequispaced discrete sine transform, 442
NDSFT	Nonequispaced discrete spherical Fourier transform, 555
NFCT	Nonequispaced fast cosine transform, 443
NFFT	Nonequispaced fast Fourier transform, 416
$\text{NFFT}^\top$	Nonequispaced fast Fourier transform transposed, 418
NFSFT	Nonequispaced fast spherical Fourier transform, 561
NFST	Nonequispaced fast sine transform, 446
NNFFT	Nonequispaced FFT with nonequispaced knots in time and frequency domain, 441
SO(3)	Group of all rotations about the origin of $\mathbb{R}^3$, 566
STFT	Short time Fourier transform, 111
SVD	Singular value decomposition, 577

A.1 Table of Some Fourier Series

In this table, all functions $f : \mathbb{T} \to \mathbb{R}$ are piecewise continuously differentiable. In the left column, the 2π-periodic functions f are defined either on $(-\pi, \pi)$ or $(0, 2\pi)$. If x_0 is a point of jump discontinuity of f, then $f(x_0) := \frac{1}{2}\big(f(x_0+0) + f(x_0-0)\big)$. In the right column, the related Fourier series of f are listed. For the main properties of Fourier series, see Lemma 1.6 and Lemma 1.13. For the convergence of the Fourier series, we refer to Theorem 1.34 of Dirichlet–Jordan.

Function $f : \mathbb{T} \to \mathbb{R}$	Fourier series of f
$f(x) = x,\ x \in (-\pi, \pi)$	$2 \sum_{n=1}^{\infty} (-1)^{n+1} \frac{\sin(nx)}{n}$
$f(x) = \lvert x\rvert,\ x \in (-\pi, \pi)$	$\frac{\pi}{2} - \frac{4}{\pi} \sum_{n=1}^{\infty} \frac{\cos(2n-1)x}{(2n-1)^2}$
$f(x) = \begin{cases} 0 & x \in (-\pi, 0), \\ x & x \in (0, \pi) \end{cases}$	$\frac{\pi}{4} - \frac{2}{\pi} \sum_{n=1}^{\infty} \frac{\cos(2n-1)x}{(2n-1)^2} + \sum_{n=1}^{\infty} (-1)^{n+1} \frac{\sin(nx)}{n}$
$f(x) = x,\ x \in (0, 2\pi)$	$\pi - 2 \sum_{n=1}^{\infty} \frac{\sin(nx)}{n}$
$f(x) = x^2,\ x \in (-\pi, \pi)$	$\frac{\pi^2}{3} + 4 \sum_{n=1}^{\infty} (-1)^n \frac{\cos(nx)}{n^2}$
$f(x) = x\,(\pi - \lvert x\rvert),\ x \in (-\pi, \pi)$	$\frac{8}{\pi} \sum_{n=1}^{\infty} \frac{\sin(2n-1)x}{(2n-1)^3}$
$f(x) = \begin{cases} -1 & x \in (-\pi, 0), \\ 1 & x \in (0, \pi) \end{cases}$	$\frac{4}{\pi} \sum_{n=1}^{\infty} \frac{\sin(2n-1)x}{2n-1}$
$f(x) = \begin{cases} 0 & x \in (-\pi, 0), \\ 1 & x \in (0, \pi) \end{cases}$	$\frac{1}{2} + \frac{2}{\pi} \sum_{n=1}^{\infty} \frac{\sin(2n-1)x}{2n-1}$
$f(x) = \frac{\pi - x}{2\pi},\ x \in (0, 2\pi)$	$\frac{1}{\pi} \sum_{n=1}^{\infty} \frac{\sin(nx)}{n}$
$f(x) = \mathrm{e}^{ax},\ x \in (-\pi, \pi)$	$\frac{\sinh(a\pi)}{\pi} \sum_{n=-\infty}^{\infty} \frac{(-1)^n}{a - in}\, \mathrm{e}^{inx}, \quad a \in \mathbb{R} \setminus \{0\}$
$f(x) = \mathrm{e}^{ax},\ x \in (0, 2\pi)$	$\frac{\mathrm{e}^{2a\pi} - 1}{2\pi} \sum_{n=-\infty}^{\infty} \frac{1}{a - in}\, \mathrm{e}^{inx}, \quad a \in \mathbb{R} \setminus \{0\}$
$f(x) = \lvert \sin x\rvert,\ x \in (-\pi, \pi)$	$\frac{2}{\pi} - \frac{4}{\pi} \sum_{n=1}^{\infty} \frac{\cos(2nx)}{4n^2 - 1}$
$f(x) = \lvert \cos x\rvert,\ x \in (-\pi, \pi)$	$\frac{2}{\pi} - \frac{4}{\pi} \sum_{n=1}^{\infty} \frac{(-1)^n \cos(2nx)}{4n^2 - 1}$
$f(x) = \begin{cases} 0 & x \in (-\pi, 0), \\ \sin x & x \in (0, \pi) \end{cases}$	$\frac{1}{\pi} - \frac{2}{\pi} \sum_{n=1}^{\infty} \frac{\cos(2nx)}{4n^2 - 1} + \frac{1}{2} \sin x$
$f(x) = x \cos x,\ x \in (-\pi, \pi)$	$-\frac{1}{2} \sin x + \sum_{n=2}^{\infty} (-1)^n \frac{2n}{n^2 - 1} \sin(nx)$
$f(x) = x \sin x,\ x \in (-\pi, \pi)$	$1 - \frac{1}{2} \cos x - 2 \sum_{n=2}^{\infty} (-1)^n \frac{1}{n^2 - 1} \cos(nx)$

A.2 Table of Some Chebyshev Series

In this table, all functions $f : I \to \mathbb{R}$ are contained in $L_{2,w}(I)$, where $I := [-1, 1]$ and $w(x) := (1 - x^2)^{-1/2}$ for $x \in (-1, 1)$. In the left column, the functions f are listed. In the right column, the related Chebyshev series are listed. The basic properties of Chebyshev coefficients are described in Theorem 6.15. For the uniform convergence of Chebyshev series, see Theorems 6.12 and 6.16. Note that for $m \in \mathbb{N}_0$, the mth Bessel function and the mth modified Bessel function of first kind are defined on $\mathbb{R}$ by

$$J_m(x) := \sum_{k=0}^{\infty} \frac{(-1)^k}{k!\,(m+k)!} \left(\frac{x}{2}\right)^{m+2k}, \quad I_m(x) := \sum_{k=0}^{\infty} \frac{1}{k!\,(m+k)!} \left(\frac{x}{2}\right)^{m+2k}.$$

Further, $a \in \mathbb{R} \setminus \{0\}$ and $b \in \mathbb{R}$ are arbitrary constants.

Function $f : I \to \mathbb{R}$	Chebyshev series $\frac{1}{2}\,a_0[f] + \sum_{n=1}^{\infty} a_n[f]\,T_n(x)$
$f(x) = \lvert x \rvert$	$\frac{2}{\pi} - \frac{4}{\pi} \sum_{n=1}^{\infty} \frac{(-1)^n}{4n^2-1}\, T_{2n}(x)$
$f(x) = \operatorname{sgn} x$	$\frac{4}{\pi} \sum_{n=1}^{\infty} \frac{(-1)^{n-1}}{2n-1}\, T_{2n-1}(x)$
$f(x) = \sqrt{1+x}$	$\frac{2\sqrt{2}}{\pi} - \frac{4\sqrt{2}}{\pi} \sum_{n=1}^{\infty} \frac{(-1)^n}{4n^2-1}\, T_n(x)$
$f(x) = \sqrt{1-x}$	$\frac{2\sqrt{2}}{\pi} - \frac{4\sqrt{2}}{\pi} \sum_{n=1}^{\infty} \frac{1}{4n^2-1}\, T_n(x)$
$f(x) = \sqrt{1-x^2}$	$\frac{2}{\pi} - \frac{4}{\pi} \sum_{n=1}^{\infty} \frac{1}{4n^2-1}\, T_{2n}(x)$
$f(x) = \arcsin x$	$\frac{4}{\pi} \sum_{n=1}^{\infty} \frac{1}{(2n-1)^2}\, T_{2n-1}(x)$
$f(x) = \cos(ax)$	$J_0(a) + 2 \sum_{n=1}^{\infty} (-1)^n\, J_{2n}(a)\, T_{2n}(x)$
$f(x) = \sin(ax)$	$2 \sum_{n=1}^{\infty} (-1)^{n-1}\, J_{2n-1}(a)\, T_{2n-1}(x)$
$f(x) = \cos(ax+b)$	$J_0(a)\, \cos b + 2 \sum_{n=1}^{\infty} \cos\big(b + \frac{n\pi}{2}\big)\, J_n(a)\, T_n(x)$
$f(x) = \sin(ax+b)$	$J_0(a)\, \sin b + 2 \sum_{n=1}^{\infty} \sin\big(b + \frac{n\pi}{2}\big)\, J_n(a)\, T_n(x)$
$f(x) = \mathrm{e}^{ax}$	$I_0(a) + 2 \sum_{n=1}^{\infty} I_n(a)\, T_n(x)$
$f(x) = \cosh(ax)$	$I_0(a) + 2 \sum_{n=1}^{\infty} I_{2n}(a)\, T_{2n}(x)$
$f(x) = \sinh(ax)$	$2 \sum_{n=1}^{\infty} I_{2n-1}(a)\, T_{2n-1}(x)$
$f(x) = \mathrm{e}^{ax^2}$	$\mathrm{e}^{a/2}\, I_0(\frac{a}{2}) + 2\,\mathrm{e}^{a/2} \sum_{n=1}^{\infty} I_n(\frac{a}{2})\, T_{2n}(x)$
$f(x) = \cos(a\sqrt{1-x^2})$	$J_0(a) + 2 \sum_{n=1}^{\infty} J_{2n}(a)\, T_{2n}(x)$
$f(x) = (1+a^2x^2)^{-1}$	$\frac{1}{\sqrt{1+a^2}} + \frac{2}{\sqrt{1+a^2}} \sum_{n=1}^{\infty} \frac{(-1)^n\, a^{2n}}{(1+\sqrt{1+a^2})^{2n}}\, T_{2n}(x)$

A.3 Table of Some Fourier Transforms

In this table, all functions $f : \mathbb{R} \to \mathbb{C}$ are contained either in $L_1(\mathbb{R})$ or in $L_2(\mathbb{R})$. In the left column, the functions f are listed. In the right column, the related Fourier transforms are listed. For the main properties of Fourier transforms, see Theorems 2.5 and 2.15. See Theorems 2.10 and 2.23 for Fourier inversion formulas of functions in $L_1(\mathbb{R})$ and $L_2(\mathbb{R})$, respectively. By N_m, H_n, and h_n, we denote the cardinal B-spline of order m, the Hermite polynomial of degree n, and the nth Hermite function, respectively.

Function $f : \mathbb{R} \to \mathbb{C}$	Fourier transform $\hat{f}(\omega) = \int_{\mathbb{R}} f(x)\, \mathrm{e}^{-\mathrm{i}\omega x}\, \mathrm{d}x$
$f(x) = \begin{cases} 1 & x \in (-L,\, L)\,, \\ 0 & \text{otherwise} \end{cases}$	$2L\, \mathrm{sinc}(L\omega), \quad L > 0$
$f(x) = \begin{cases} 1 - \frac{\lvert x\rvert}{L} & x \in (-L,\, L)\,, \\ 0 & \text{otherwise} \end{cases}$	$L\, \big(\mathrm{sinc}\, \frac{L\omega}{2}\big)^2, \quad L > 0$
$f(x) = \mathrm{e}^{-x^2/2}$	$\sqrt{2\pi}\, \mathrm{e}^{-\omega^2/2}$
$f(x) = \frac{1}{\sqrt{2\pi\sigma^2}}\, \mathrm{e}^{-x^2/(2\sigma^2)}$	$\mathrm{e}^{-\sigma^2\omega^2/2}, \quad \sigma > 0$
$f(x) = \mathrm{e}^{-(a-\mathrm{i}b)\, x^2}$	$\sqrt{\frac{\pi}{a-\mathrm{i}b}}\, \exp \frac{-(a-\mathrm{i}b)\, \omega^2}{4\,(a^2+b^2)}, \quad a > 0,\ b \in \mathbb{R} \setminus \{0\}$
$f(x) = \mathrm{e}^{-a\, \lvert x\rvert}$	$\frac{2a}{a^2+\omega^2}, \quad a > 0$
$N_1(x) = \begin{cases} 1 & x \in (0,\, 1)\,, \\ 0 & \text{otherwise} \end{cases}$	$\mathrm{e}^{-\mathrm{i}\omega/2}\, \mathrm{sinc}\, \frac{\omega}{2}$
$N_m(x) = (N_{m-1} * N_1)(x)$	$\mathrm{e}^{-\mathrm{i}m\omega/2}\, \big(\mathrm{sinc}\, \frac{\omega}{2}\big)^m, \quad m \in \mathbb{N} \setminus \{1\}$
$M_m(x) = N_m(x + \frac{m}{2})$	$\big(\mathrm{sinc}\, \frac{\omega}{2}\big)^m, \quad m \in \mathbb{N}$
$h_n(x) = H_n(x)\, \mathrm{e}^{-x^2/2}$	$\sqrt{2\pi}\, (-\mathrm{i})^n\, h_n(\omega), \quad n \in \mathbb{N}_0$
$f(x) = \frac{L}{\pi}\, \mathrm{sinc}(Lx)$	$\hat{f}(\omega) = \begin{cases} 1 & \omega \in (-L,\, L)\,, \\ 0 & \text{otherwise} \end{cases} \quad L > 0$
$f(x) = \begin{cases} \mathrm{e}^{-ax}\, \cos(bx) & x > 0\,, \\ 0 & \text{otherwise} \end{cases}$	$\frac{a+\mathrm{i}\omega}{(a+\mathrm{i}\omega)^2+b^2}, \quad a > 0,\ b \geq 0$
$f(x) = \frac{a}{\pi\, (x^2+a^2)}$	$\mathrm{e}^{-a\, \lvert\omega\rvert}$

A.4 Table of Some Discrete Fourier Transforms

In the left column of this table, the components of N-dimensional vectors $\mathbf{a} = (a_j)_{j=0}^{N-1} \in \mathbb{C}^N$ are presented. In the right column, the components of the related discrete Fourier transforms $\hat{\mathbf{a}} = (\hat{a}_k)_{k=0}^{N-1} = \mathbf{F}_N\, \mathbf{a}$ of even length $N \in 2\,\mathbb{N}$ are listed, where

$$\hat{a}_k = \sum_{j=0}^{N-1} a_j\, w_N^{jk}\,, \quad k = 0, \ldots, N-1\,.$$

For the main properties of DFTs, see Theorem 3.26. By Remark 3.12, the DFT(N) of the N-periodic sequences $(a_j)_{j\in\mathbb{Z}}$ is equal to the N-periodic sequences $(\hat{a}_k)_{k\in\mathbb{Z}}$. In this table, $n \in \mathbb{Z}$ denotes an arbitrary fixed integer.

jth component $a_j \in \mathbb{C}$	kth component $\hat{a}_k$ of related DFT(N)
$a_j = \delta_{j \bmod N}$	$\hat{a}_k = 1$
$a_j = 1$	$\hat{a}_k = N\,\delta_{k \bmod N}$
$a_j = \delta_{(j-n) \bmod N}$	$\hat{a}_k = w_N^{kn} = \mathrm{e}^{-2\pi\mathrm{i}kn/N}$
$a_j = \frac{1}{2}\big(\delta_{(j+n) \bmod N} + \delta_{(j-n) \bmod N}\big)$	$\hat{a}_k = \cos\frac{2\pi nk}{N}$
$a_j = \frac{1}{2}\big(\delta_{(j+n) \bmod N} - \delta_{(j-n) \bmod N}\big)$	$\hat{a}_k = \mathrm{i}\,\sin\frac{2\pi nk}{N}$
$a_j = w_N^{jn}$	$\hat{a}_k = N\,\delta_{(k+n) \bmod N}$
$a_j = (-1)^j$	$\hat{a}_k = N\,\delta_{(k+N/2) \bmod N}$
$a_j = \cos\frac{2\pi jn}{N}$	$\hat{a}_k = \frac{N}{2}\big(\delta_{(k+n) \bmod N} + \delta_{(k-n) \bmod N}\big)$
$a_j = \sin\frac{2\pi jn}{N}$	$\hat{a}_k = \frac{\mathrm{i}N}{2}\big(\delta_{(k+n) \bmod N} - \delta_{(k-n) \bmod N}\big)$
$a_j = \begin{cases} 0 & j = 0, \\ \frac{1}{2} - \frac{j}{N} & j = 1, \ldots, N-1 \end{cases}$	$\hat{a}_k = \begin{cases} 0 & k = 0, \\ -\frac{\mathrm{i}}{2}\,\cot\frac{\pi k}{N} & k = 1, \ldots, N-1 \end{cases}$
$a_j = \begin{cases} \frac{1}{2} & j = 0, \\ \frac{j}{N} & j = 1, \ldots, N-1 \end{cases}$	$\hat{a}_k = \begin{cases} \frac{N}{2} & k = 0, \\ \frac{\mathrm{i}}{2}\,\cot\frac{\pi k}{N} & k = 1, \ldots, N-1 \end{cases}$
$a_j = \begin{cases} \frac{j}{N} & j = 0, \ldots, \frac{N}{2} - 1, \\ 0 & j = \frac{N}{2}, \\ \frac{j}{N} - 1 & j = \frac{N}{2} + 1, \ldots, N-1 \end{cases}$	$\hat{a}_k = \begin{cases} 0 & k = 0, \\ \frac{\mathrm{i}}{2}\,(-1)^k\,\cot\frac{\pi k}{N} & k = 1, \ldots, N-1 \end{cases}$
$a_j = \begin{cases} 0 & j \in \{0, \frac{N}{2}\}, \\ 1 & j = 1, \ldots, \frac{N}{2} - 1, \\ -1 & j = \frac{N}{2} + 1, \ldots, N-1 \end{cases}$	$\hat{a}_k = \begin{cases} 0 & k = 0, 2, \ldots, N-2, \\ -2\mathrm{i}\,\cot\frac{\pi k}{N} & k = 1, 3, \ldots, N-1 \end{cases}$

A.5 Table of Some Fourier Transforms of Tempered Distributions

In the left column, some tempered distributions $T \in \mathcal{S}'(\mathbb{R}^d)$ are listed. In the right column, one can see the related Fourier transforms $\hat{T} = \mathcal{F}T \in \mathcal{S}'(\mathbb{R}^d)$. The main properties of the Fourier transforms of tempered distributions are described in Theorem 4.50. For the inverse Fourier transform $\mathcal{F}^{-1}T \in \mathcal{S}'(\mathbb{R}^d)$, see Theorem 4.48. In this table, we use following notations: $\mathbf{x}_0 \in \mathbb{R}^d$, $\boldsymbol{\omega}_0 \in \mathbb{R}^d$, and $\boldsymbol{\alpha} \in \mathbb{N}_0^d$. If the Dirac distribution δ acts on functions with variable $\mathbf{x} \in \mathbb{R}^d$, then we write $\delta(\mathbf{x})$. Analogously, $\delta(\boldsymbol{\omega})$ acts on functions with variable $\boldsymbol{\omega} \in \mathbb{R}^d$.

Tempered distribution $T \in \mathcal{S}'(\mathbb{R}^d)$	Fourier transform $\mathcal{F}T \in \mathcal{S}'(\mathbb{R}^d)$
$\delta(\mathbf{x})$	1
$\delta_{\mathbf{x}_0}(\mathbf{x}) := \delta(\mathbf{x} - \mathbf{x}_0)$	$\mathrm{e}^{-\mathrm{i}\,\boldsymbol{\omega}\cdot\mathbf{x}_0}$
$(D^{\boldsymbol{\alpha}}\,\delta)(\mathbf{x})$	$\mathrm{i}^{\lvert\boldsymbol{\alpha}\rvert}\,\boldsymbol{\omega}^{\boldsymbol{\alpha}}$
$(D^{\boldsymbol{\alpha}}\,\delta)(\mathbf{x} - \mathbf{x}_0)$	$\mathrm{i}^{\lvert\boldsymbol{\alpha}\rvert}\,\boldsymbol{\omega}^{\boldsymbol{\alpha}}\,\mathrm{e}^{-\mathrm{i}\,\boldsymbol{\omega}\cdot\mathbf{x}_0}$
1	$(2\pi)^d\,\delta(\boldsymbol{\omega})$
$\mathbf{x}^{\boldsymbol{\alpha}}$	$(2\pi)^d\,\mathrm{i}^{\lvert\boldsymbol{\alpha}\rvert}\,(D^{\boldsymbol{\alpha}}\,\delta)(\boldsymbol{\omega})$
$\mathrm{e}^{\mathrm{i}\,\boldsymbol{\omega}_0\cdot\mathbf{x}}$	$(2\pi)^d\,\delta(\boldsymbol{\omega} - \boldsymbol{\omega}_0)$
$\mathbf{x}^{\boldsymbol{\alpha}}\,\mathrm{e}^{\mathrm{i}\,\boldsymbol{\omega}_0\cdot\mathbf{x}}$	$(2\pi)^d\,\mathrm{i}^{\lvert\boldsymbol{\alpha}\rvert}\,(D^{\boldsymbol{\alpha}}\,\delta)(\boldsymbol{\omega} - \boldsymbol{\omega}_0)$
$\cos(\boldsymbol{\omega}_0\cdot\mathbf{x})$	$\frac{(2\pi)^d}{2}\,\big(\delta(\boldsymbol{\omega} - \boldsymbol{\omega}_0) + \delta(\boldsymbol{\omega} + \boldsymbol{\omega}_0)\big)$
$\sin(\boldsymbol{\omega}_0\cdot\mathbf{x})$	$\frac{(2\pi)^d}{2\mathrm{i}}\,\big(\delta(\boldsymbol{\omega} - \boldsymbol{\omega}_0) - \delta(\boldsymbol{\omega} + \boldsymbol{\omega}_0)\big)$
$\mathrm{e}^{-\lVert\mathbf{x}\rVert_2^2/2}$	$(2\pi)^{d/2}\,\mathrm{e}^{-\lVert\boldsymbol{\omega}\rVert_2^2/2}$
$\mathrm{e}^{-a\,\lVert\mathbf{x}\rVert_2^2}$	$\big(\frac{\pi}{a}\big)^{d/2}\,\mathrm{e}^{-\lVert\boldsymbol{\omega}\rVert_2^2/(4a)}, \quad a > 0$

References

1. Abe, S., Sheridan, J.T.: Optical operations on wave functions as the Abelian subgroups of the special affine Fourier transformation. Opt. Lett. **19**(22), 1801–1803 (1994)
2. Abramowitz, M., Stegun, I.A. (eds.): Handbook of Mathematical Functions. National Bureau of Standards, Washington (1972)
3. Adcock, B.: Convergence acceleration of modified Fourier series in one or more dimensions. Math. Comp. **80**(273), 225–261 (2011)
4. Ahmed, N., Natarajan, T., Rao, K.R.: Discrete cosine transform. IEEE Trans. Comput. **23**, 90–93 (1974)
5. Akavia, A.: Deterministic sparse Fourier approximation via approximating arithmetic progressions. IEEE Trans. Inform. Theory **60**(3), 1733–1741 (2014)
6. Alpert, B.K., Rokhlin, V. A fast algorithm for the evaluation of Legendre expansions. SIAM J. Sci. Statist. Comput. **12**(1), 158–179 (1991)
7. Andersson, F., Carlsson, M.: ESPRIT for multidimensional general grids. SIAM J. Matrix Anal. Appl. **39**(3), 1470–1488 (2018)
8. Apostol, T.M.: Introduction to Analytic Number Theory. Springer-Verlag, New York (1976)
9. Arico, A., Serra-Capizzano, S., Tasche, M.: Fast and numerically stable algorithms for discrete Hartley transforms and applications to preconditioning. Commun. Inf. Syst. **5**(1), 21–68 (2005)
10. Arioli, M., Munthe-Kaas, H., Valdettaro, L.: Componentwise error analysis for FFTs with applications to fast Helmholtz solvers. Numer. Algor. **12**, 65–88 (1996)
11. Arnold, A., Bolten, M., Dachsel, H., Fahrenberger, F., Gähler, F., Halver, R., Heber, F., Hofmann, M., Iseringhausen, J., Kabadshow, I., Lenz, O., Pippig, M.: ScaFaCoS - Scalable fast Coloumb solvers. http://www.scafacos.de.
12. Atkinson, K., Han, W.: Theoretical Numerical Analysis. A Functional Analysis Framework. Springer-Verlag, Dordrecht (2009)
13. Aubel, C., Bölcskei, H.: Vandermonde matrices with nodes in the unit disk and the large sieve. Appl. Comput. Harmon. Anal. **47**(1), 53–86 (2019)
14. Averbuch, A., Israeli, M., Vozovoi, L.: A fast Poisson solver of arbitrary order accuracy in rectangular regions. SIAM J. Sci. Comput. **19**(3), 933–952 (1998)
15. Bannai, E., Bannai, E.: A survey on spherical designs and algebraic combinatorics on spheres. Eur. J. Comb. **30**(6), 1392–1425 (2009)
16. Barnett, A., Magland, J.: Flatiron Institute Nonuniform Fast Fourier Transform. https://finufft.readthedocs.io. Contributor: J. Magland, L. af Klinteberg, Yu-Hsuan Shih, A. Malleo, L. Lu, J. Andén

G. Plonka et al., *Numerical Fourier Analysis*, Applied and Numerical Harmonic Analysis, https://doi.org/10.1007/978-3-031-35005-4

17. Barnett: A.H. Aliasing error of the $\exp(\beta\sqrt{1-z^2})$ kernel in the nonuniform fast Fourier transform. Appl. Comput. Harm. Anal. **51**, 1–16 (2021)
18. Barnett, A.H., Magland, J., Af Klinteberg, L.: A parallel nonuniform fast Fourier transform library based on an "exponential of semicircle" kernel. SIAM J. Sci. Comput. **41**(5), C479–C504 (2019)
19. Bartel, F., Schmischke, M., Potts, D.: Grouped transformations in high-dimensional explainable ANOVA approximation. SIAM J. Sci. Comput. **44**(3), A1606–A1631 (2022)
20. Bass, R.F., Gröchenig, K.: Random sampling of multivariate trigonometric polynomials. SIAM J. Math. Anal. **36**(3), 773–795 (2004)
21. Baszenski, G., Delvos, F.-J.: A discrete Fourier transform scheme for Boolean sums of trigonometric operators. In: Multivariate Approximation Theory IV, pp. 15–24. Birkhäuser, Basel (1989)
22. Baszenski, G., Delvos, F.-J., Tasche, M.: A united approach to accelerating trigonometric expansions. Concrete analysis. Comput. Math. Appl. **30**(3–6), 33–49 (1995)
23. Baszenski, G., Schreiber, U., Tasche, M.: Numerical stability of fast cosine transforms. Numer. Funct. Anal. Optim. **21**(1–2), 25–46 (2000)
24. Baszenski, G., Tasche, M.: Fast polynomial multiplication and convolution related to the discrete cosine transform. Linear Algebra Appl. **252**, 1–25 (1997)
25. Batenkov, D., Diederichs, B., Goldman, G., Yomdin, Y.: The spectral properties of Vandermonde matrices with clustered nodes. Linear Algebra Appl. **609**, 37–72 (2021)
26. Batenkov, D., Yomdin, Y.: Algebraic Fourier reconstruction of piecewise smooth functions. Math. Comp. **81**(277), 277–318 (2012)
27. Bauschke, H.H., Combettes, P.L., Luke, D.R.: Phase retrieval, error reduction algorithm, and Fienup variants: a view from convex optimization. J. Opt. Soc. Amer. A **19**(7), 1334–1345 (2002)
28. Bazán, F.S.V.: Conditioning of rectangular Vandermonde matrices with nodes in the unit disk. SIAM J. Matrix Anal. Appl. **21**, 679–693 (2000)
29. Bazán, F.S.V., Toint, P.L.: Error analysis of signal zeros from a related companion matrix eigenvalue problem. Appl. Math. Lett. **14**(7), 859–866 (2001)
30. Beatson, R.K., Light, W.A.: Fast evaluation of radial basis functions: methods for two–dimensional polyharmonic splines. IMA J. Numer. Anal. **17**(3), 343–372 (1997)
31. Bedrosian, E.: A product theorem for Hilbert transforms. Proc. IEEE **51**(5), 868–869 (1963)
32. Beinert, R., Bredies, K.: Tensor-free proximal methods for lifted bilinear/quadratic inverse problems with applications to phase retrieval. Found. Comput. Math. **21**(5), 1181–1232 (2021)
33. Beinert, R., Jung, P., Steidl, G., Szollmann, T.: Super-resolution for doubly-dispersive channel estimation. Sampl Theory Signal Process. Data Anal. **19**(2), 1–36 (2021)
34. Beinert, R., Plonka, G. Ambiguities in one-dimensional discrete phase retrieval from Fourier magnitudes. J. Fourier Anal. Appl. **21**(6), 1169–1198 (2015)
35. Beinert, R., Plonka, G.: Sparse phase retrieval of one-dimensional signals by Prony's method. Front. Appl. Math. Stat. **3**, 5 (2017)
36. Beinert, R., Plonka, G.: Enforcing uniqueness in one-dimensional phase retrieval by additional signal information in time domain. Appl. Comput. Harmon. Anal. **45**(3), 505–525 (2018)
37. Beinert, R., Plonka, G.: One-dimensional discrete-time phase retrieval. In: Nanoscale Photonic Imaging, number 134 in Topics in Applied Physics, pp. 603–627. Springer, Cham (2020)
38. Beinert, R., Quellmalz, M.: Total variation-based reconstruction and phase retrieval for diffraction tomography. SIAM J. Imaging Sci. **15**(3), 1373–1399 (2022)
39. Belmehdi, S.: On the associated orthogonal polynomials. J. Comput. Appl. Math. **32**(3), 311–319 (1990)

40. Bendory, T., Beinert, R., Eldar, Y.C.: Fourier phase retrieval: uniqueness and algorithms. In: Compressed Sensing and Its Applications, Applied and Numerical Harmonic Analysis, pp. 55–91. Birkhäuser/Springer, Cham, 2017.
41. Benedetto, J.J.: Harmonic Analysis and Applications. CRC Press, Boca Raton (1997)
42. Berent, J., Dragotti, P.L., Blu, T.: Sampling piecewise sinusoidal signals with finite rate of innovation methods. IEEE Trans. Signal Process. **58**(2), 613–625 (2010)
43. Berg, L.: Lineare Gleichungssysteme mit Bandstruktur und ihr asymptotisches Verhalten. Deutscher Verlag der Wissenschaften, Berlin (1986).
44. Berrut, J.P., Trefethen, L.N.: Barycentric Lagrange interpolation. SIAM Rev. **46**(3), 501–517 (2004)
45. Beylkin, G.: On the fast Fourier transform of functions with singularities. Appl. Comput. Harmon. Anal. **2**(4), 363–381 (1995)
46. Beylkin, G., Cramer, R.: A multiresolution approach to regularization of singular operators and fast summation. SIAM J. Sci. Comput. **24**(1), 81–117 (2002)
47. Biswas, M.H.A., Filbir, F., Ramakrishnan, R.: New translations associated with the special affine Fourier transform and shift invariant spaces. ArXiv e-prints (2022). ArXiv:2212.05678
48. Bittens, S.: Sparse FFT for functions with short frequency support. Dolomites Res. Notes Approx. **10**, 43–55 (2017)
49. Bittens, S., Plonka, G.: Real sparse fast DCT for vectors with short support. Linear Algebra Appl. **582**, 359–390 (2019)
50. Bittens, S., Plonka, G.: Sparse fast DCT for vectors with one-block support. Numer. Algor. **82**(2), 663–697 (2019)
51. Björck, Å.: Numerical Methods for Least Squares Problems. SIAM, Philadelphia (1996)
52. Blahut, R.E.: Fast Algorithms for Digital Signal Processing. Cambridge University Press, New York (2010)
53. Bochner, S.: Vorlesungen über Fouriersche Integrale. Chelsea Publishing Company, New York (1932)
54. Böhme, M., Potts, D.: A fast algorithm for filtering and wavelet decomposition on the sphere. Electron. Trans. Numer. Anal. **16**, 70–92 (2003)
55. Böhme, M., Potts, D.: A fast algorithm for spherical filtering on arbitrary grids. In: Proceedings of SPIE. Wavelets: Applications in Signal and Image Processing X, vol. 5207 (2003)
56. Bondarenko, A., Radchenko, D., Viazovska, M.: Optimal asymptotic bounds for spherical designs. Ann. Math. (2) **178**(2), 443–452 (2013)
57. Bos, L., Caliari, M., De Marchi, S., Vianello, M., Xu, Y.: Bivariate Lagrange interpolation at the Padua points: the generating curve approach. J. Approx. Theory **143**, 15–25 (2006). Special Issue on Foundations of Computational Mathematics.
58. Böttcher, A., Kunis, S., Potts, D.: Probabilistic spherical Marcinkiewicz-Zygmund inequalities. J. Approx. Theory **157**(2), 113–126 (2009)
59. Bovik, A.: Handbook of Image and Video Processing, second edn. Academic Press (2010)
60. Boyd, J.P.: Chebyshev and Fourier Spectral Methods, second edn. Dover Press, New York (2000)
61. Bracewell, R.N.: The Hartley Transform. Clarendon Press, Oxford University Press, New York (1986)
62. Bresler, Y., Macovski, A.: Exact maximum likelihood parameter estimation of superimposed exponential signals in noise. IEEE Trans. Acoust. Speech Signal Process. **34**(5), 1081–1089 (1986)
63. Briggs, W.L., Henson, V.E.: The DFT. An Owner's Manual for the Discrete Fourier Transform. SIAM, Philadelphia (1995)
64. Brigham, E.O.: The Fast Fourier Transform. Prentice Hall, Englewood Cliffs, New York (1974)
65. Broadie, M., Yamamoto, Y.: Application of the fast Gauss transform to option pricing. Manag. Sci. **49**, 1071–1088 (2003)

66. Brown, J.L.: Analytic signals and product theorems for Hilbert transforms. IEEE Trans. Circuits Syst. **21**, 790–792 (1974)
67. Bruun, G.: z-transform DFT filters and FFT's. IEEE Trans. Acoust. Speech Signal Process. **ASSP-26**(1), 56–63 (1978)
68. Bülow, T., Sommer, G.: Hypercomplex signals—a novel extension of the analytic signal to the multidimensional case. IEEE Trans. Signal Process. **49**(11), 2844–2852 (2001)
69. Bultheel, A., Martínez, H.: An introduction to the fractional Fourier transform and friends. Cubo **7**(2), 201–221 (2005)
70. Bultheel, A., Martínez-Subaran, H.E.: Computation of the fractional Fourier transform. Appl. Comput. Harmon. Anal. **16**(3), 182–202 (2004)
71. Bultheel, A., Martínez-Sulbaran, H.: A shattered survey of the fractional Fourier transform (2003). Manuscript
72. Bungartz, H.J., Griebel, M.: A note on the complexity of solving Poisson's equation for spaces of bounded mixed derivatives. J. Complexity **15**(2), 167–199 (1999)
73. Bungartz, H.J., Griebel, M.: Sparse grids. Acta Numer. **13**, 147–269 (2004)
74. Bunge, H.J.: Texture Analysis in Material Science. Butterworths, London (1982)
75. Butzer, P.L., Nessel, R.J.: Fourier Analysis and Approximation. Volume 1: One-dimensional Theory. Academic Press, New York (1971)
76. Byrenheid, G., Kämmerer, L., Ullrich, T., Volkmer, T.: Tight error bounds for rank-1 lattice sampling in spaces of hybrid mixed smoothness. Numer. Math. **136**(4), 993–1034 (2017)
77. Calvetti, D.: A stochastic roundoff error analysis for FFT. Math. Comput. **56**(194), 755–774 (1991)
78. Candès, E.J.: The restricted isometry property and its implications for compressed sensing. C. R. Acad. Sci. Paris **346**(9–10), 589–592 (2008)
79. Candès, E.J., Fernandez-Granda, C.: Towards a mathematical theory of super-resolution. Comm. Pure Appl. Math. **67**(6), 906–956 (2014)
80. Castrillon-Candas, J.E., Siddavanahalli, V., Bajaj, C.: Nonequispaced Fourier transforms for protein-protein docking. ICES Report 05-44, Univ. Texas (2005)
81. Chandrasekaran, V., Recht, B., Parrilo, P.A., Willsky, A.S.: The convex geometry of linear inverse problems. Found. Comput. Math. **12**(6), 805–849 (2012).
82. Chandrasenkharan, K.: Classical Fourier Transforms. Spinger-Verlag, Berlin (1989)
83. Chen, X., Frommer, A., Lang, B.: Computational existence proofs for spherical t-designs. Numer. Math. **117**(2), 289–305 (2011)
84. Chihara, T.: An Introduction to Orthogonal Polynomials. Gordon and Breach Science Publishers, New York (1978)
85. Christensen, O.: An Introduction to Frames and Riesz Bases, second edn. Birkhäuser/Springer, Cham (2016)
86. Chu, C.Y.: The Fast Fourier Transform on the Hypercube Parallel Computers. Ph.D. thesis, Cornell University, Ithaca (1988)
87. Chui, C.K.: Multivariate Splines. SIAM, Philadelphia (1988)
88. Chui, C.K.: An Introduction to Wavelets. Academic Press, Boston (1992)
89. Clenshaw, C., Curtis, A.: A method for numerical integration on an automatic computer. Numer. Math. **2**, 197–205 (1960)
90. Clenshaw, C.W.: A note on the summation of Chebyshev series. Math. Tables Aids Comput. **9**, 118–120 (1955)
91. Collowald, M., Cuyt, A., Hubert, E., s. Lee, W., Celis, O.S.: Numerical reconstruction of convex polytopes from directional moments. Adv. Comput. Math. **41**, 1079–1099 (2015)
92. Condat, L.: Atomic norm minimization for decomposition into complex exponentials and optimal transport in Fourier domain. J. Approx. Theory **258**, 105456 (2020)
93. Constantin, A.: Fourier Analysis. Part I. Theory. Cambridge University Press, Cambridge (2016)
94. Cooley, J.W., Tukey, J.W.: An algorithm for machine calculation of complex Fourier series. Math. Comput. **19**, 297–301 (1965)

95. Cools, R., Nuyens, D.: Fast algorithms for component-by-component construction of rank-1 lattice rules in shift-invariant reproducing kernel Hilbert spaces. Math. Comp. **75**, 903–920 (2004)
96. Curry, H.B., Schoenberg, I.J.: On Pólya frequency functions. IV. The fundamental spline functions and their limits. J. Analyse Math. **17**, 71–107 (1966)
97. Cuyt, A., Hou, Y., Knaepkens, F., Lee, W.s.: Sparse multidimensional exponential analysis with an application to radar imaging. SIAM J. Sci. Comput. **42**(3), B675–B695 (2020).
98. Cuyt, A., Lee, W.s.: Multivariate exponential analysis from the minimal number of samples. Adv. Comput. Math. **44**(4), 987–1002 (2018)
99. Cuyt, A., Lee, W.s., Wu, M.: High accuracy trigonometric approximations of the real Bessel functions of the first kind. Comput. Math. Math. Phys. **60**(1), 119–127 (2020)
100. Dũng, D., Temlyakov, V.N., Ullrich, T.: Hyperbolic Cross Approximation. Birkhäuser, Cham (2018)
101. Daubechies, I.: Ten Lectures on Wavelets. SIAM, Philadelphia (1992)
102. Davis, P.J.: Circulant Matrices. John Wiley and Sons, New York (1979)
103. de Boor, C.: A Practical Guide to Splines, revised edn. Springer-Verlag, New York (2001)
104. de Boor, C., DeVore, R.: Approximation by smooth multivariate splines. Trans. Amer. Math. Soc. **276**(2), 775–788 (1983)
105. de Boor, C., Höllig, K., Riemenschneider, S.: Box Splines. Springer-Verlag, New York (1993)
106. de Prony, G.R.: Essai éxperimental et analytique: sur les lois de la dilatabilité des fluides élastiques et sur celles de la force expansive de la vapeur de l'eau et de la vapeur de l'alkool, à différentes températures. J. Ecole Polytechnique **1**, 24–76 (1795)
107. Deaño, A., Huybrechs, D., Iserles, A.: Computing Highly Oscillatory Integrals. SIAM, Philadelphia (2018)
108. Demeure, C.J.: Fast QR factorization of Vandermonde matrices. Linear Algebra Appl. **122/123/124**, 165–194 (1989)
109. Derevianko, N., Plonka, G.: Exact reconstruction of extended exponential sums using rational approximation of their Fourier coefficients. Anal. Appl. **20**(3), 543–577 (2022).
110. Derevianko, N., Plonka, G., Petz, M.: From ESPRIT to ESPIRA: Estimation of signal parameters by iterative rational approximation. IMA J. Numer. Anal. (2022).
111. Derevianko, N., Plonka, G., Razavi, R.: ESPRIT versus ESPIRA for reconstruction of short cosine sums and its application. Numer. Algor. (2022).
112. DeVore, R.A., Lorentz, G.G.: Constructive Approximation. Springer–Verlag, Berlin (1993)
113. Dick, J., Kuo, F.Y., Sloan, I.H.: High-dimensional integration: the quasi-Monte Carlo way. Acta Numer. **22**, 133–288 (2013).
114. Dragotti, P.L., Vetterli, M., Blu, T.: Sampling moments and reconstructing signals of finite rate of innovation: Shannon meets Strang-Fix. IEEE Trans. Signal Process. **55**, 1741–1757 (2007)
115. Driscoll, J.R., Healy, D.M.: Computing Fourier transforms and convolutions on the 2–sphere. Adv. Appl. Math. **15**(2), 202–250 (1994)
116. Driscoll, J.R., Healy, D.M., Rockmore, D.N.: Fast discrete polynomial transforms with applications to data analysis for distance transitive graphs. SIAM J. Comput. **26**(4), 1066–1099 (1997)
117. Duchon, J.: Fonctions splines et vecteurs aleatoires. Tech. rep., Seminaire d'Analyse Numerique, Universite Scientifique et Medicale, Grenoble (1975)
118. Duhamel, P., Hollmann, H.: Split-radix FFT algorithm. Electron. Lett. **20**(1), 14–16 (1984)
119. Duhamel, P., Vetterli, M.: Fast Fourier transforms: A tutorial review and a state of the art. Signal Process. **19**(4), 259–299 (1990)
120. Duijndam, A.J.W., Schonewille, M.A.: Nonuniform fast Fourier transform. Geophysics **64**, 539–551 (1999)
121. Dutt, A., Rokhlin, V.: Fast Fourier transforms for nonequispaced data. SIAM J. Sci. Stat. Comput. **14**(6), 1368–1393 (1993)
122. Dutt, A., Rokhlin, V.: Fast Fourier transforms for nonequispaced data II. Appl. Comput. Harmon. Anal. **2**(1), 85–100 (1995)

123. Eagle, A.: On the relations between the Fourier constants of a periodic function and the coefficients determined by harmonic analysis. Philos. Mag., VII. Ser. **5**, 113–132 (1928)
124. Eckhoff, K.S.: Accurate reconstructions of functions of finite regularity from truncated Fourier series expansions. Math. Comp. **64**(210), 671–690 (1995)
125. Ehler, M., Gräf, M., Neumayer, S., Steidl, G.: Curve based approximation of measures on manifolds by discrepancy minimization. Found. Comput. Math. **21**(6), 1595–1642 (2021)
126. Ehler, M., Kunis, S., Peter, T., Richter, C.: A randomized multivariate matrix pencil method for superresolution microscopy. Electron. Trans. Numer. Anal. **51**, 63–74 (2019)
127. Ehlich, H.: Untersuchungen zur numerischen Fourieranalyse. Math. Z. **91**, 380–420 (1966)
128. Elad, M., Milanfar, P., Golub, G.H.: Shape from moments—an estimation theory perspective. IEEE Trans. Signal Process. **52**(7), 145–167 (2004)
129. Elbel, B.: Mehrdimensionale Fouriertransformation für nichtäquidistante Daten. Diplomarbeit, Technische Hochschule Darmstadt (1998)
130. Elbel, B., Steidl, G.: Fast Fourier transform for nonequispaced data. In: Approximation Theory IX, pp. 39–46. Vanderbilt University Press, Nashville (1998)
131. Elgammal, A., Duraiswami, R., Davis, L.S.: Efficient non-parametric adaptive color modeling using fast Gauss transform. Technical Report, University of Maryland (2001)
132. Fannjiang, A.C.: The MUSIC algorithm for sparse objects: a compressed sensing analysis. Inverse Prob. **27**(3), 035,013 (2011)
133. Fasshauer, G.E., Schumaker, L.L.: Scattered data fitting on the sphere. In: Mathematical Methods for Curves and Surfaces II, pp. 117–166. Vanderbilt University Press, Nashville (1998)
134. Feichtinger, H.G., Gröchenig, K., Strohmer, T.: Efficient numerical methods in non-uniform sampling theory. Numer. Math. **69**(4), 423–440 (1995)
135. Feig, E.: Fast scaled-DCT algorithm. In: Proceeding SPIE 1244, Image Processing Algorithms and Techniques, vol. 2, pp. 2–13 (1990)
136. Feig, E., Winograd, S.: Fast algorithms for the discrete cosine transform. IEEE Trans. Signal Process. **40**(9), 2174–2193 (1992)
137. Feller, W.: An Introduction to Probability Theory and its Applications, vol. II, 2nd edn. Wiley, New York (1971)
138. Felsberg, M., Sommer, G.: The monogenic signal. IEEE Trans. Signal Process. **49**(12), 3136–3144 (2001)
139. Fessler, J.A., Sutton, B.P.: Nonuniform fast Fourier transforms using min-max interpolation. IEEE Trans. Signal Process. **51**(2), 560–574 (2003)
140. Févotte, C., Bertin, N., Durrieu, J.L.: Nonnegative matrix factorization with the Itakura-Saito divergence: with application to music analysis. Neural Comput. **21**(3), 793–830 (2009)
141. Filbir, F., Melnyk, O.: Image recovery for blind polychromatic ptychography. ArXiv e-prints (2022). ArXiv:2210.01626
142. Filbir, F., Mhaskar, H.N., Prestin, J.: On the problem of parameter estimation in exponential sums. Constr. Approx. **35**(2), 323–343 (2012)
143. Folland, G.B.: Fourier Analysis and its Applications. Brooks/Cole Publishing Comp., Pacific Grove (1992)
144. Folland, G.B.: Real Analysis. Modern Techniques and their Applications, second edn. John Wiley & Sons, New York (1999)
145. Förstner, W., Gülch, E.: A fast operator for detection and precise location of distinct points, corners and centres of circular features. In: Proceeding ISPRS Intercommission Conference on Fast Processing of Photogrammetric Data, pp. 281–305 (1987)
146. Fortunato, D., Townsend, A.: Fast Poisson solvers for spectral methods. IMA J. Numer. Anal. **40**(3), 1994–2018 (2020)
147. Foucart, S.: A note guaranteed sparse recovery via ℓ_1-minimization. Appl. Comput. Harmon. Anal. **29**(1), 97–103 (2010)
148. Foucart, S., Rauhut, H.: A Mathematical Introduction to Compressive Sensing. Applied and Numerical Harmonic Analysis. Birkhäuser/Springer, New York (2013)

149. Fourier, J.: The Analytical Theory of Heat. Dover Publication, New York (1955). Translated from the French
150. Fourmont, K.: Schnelle Fourier–Transformation bei nichtäquidistanten Gittern und tomographische Anwendungen. Dissertation, Universität Münster (1999)
151. Fourmont, K.: Non equispaced fast Fourier transforms with applications to tomography. J. Fourier Anal. Appl. **9**, 431–450 (2003)
152. Freeden, W., Gervens, T., Schreiner, M.: Constructive Approximation on the Sphere. Clarendon Press, Oxford University Press, New York (1998)
153. Frigo, M., Johnson, S.G.: The design and implementation of FFTW3. Proc. IEEE **93**, 216–231 (2005)
154. Frigo, M., Johnson, S.G.: FFTW, C subroutine library (2009). http://www.fftw.org
155. Gabor, D.: The theory of communication. J. IEE **93**, 429–457 (1946)
156. Gasquet, C., Witomski, P.: Fourier Analysis and Applications. Filtering, Numerical Computation, Wavelets. Springer-Verlag, Berlin (1999)
157. Gautschi, W.: Attenuation factors in practical Fourier analysis. Numer. Math. **18**, 373–400 (1971/72)
158. Gautschi, W.: Orthogonal Polynomials: Computation and Approximation. Oxford University Press, New York (2004)
159. Gentleman, W., Sande, G.: Fast Fourier transform for fun and profit. In: Fall Joint Computer Conference AFIPS, vol. 29, pp. 563–578 (1966)
160. Gilbert, A., Indyk, P., Iwen, M., Schmidt, L.: Recent developments in the sparse Fourier transform: a compressed Fourier transform for big data. IEEE Signal Process. Mag. **31**(5), 91–100 (2014)
161. Gilbert, A.C., Strauss, M.J., Tropp, J.A.: A tutorial on fast Fourier sampling. IEEE Signal Process. Mag. **25**(2), 57–66 (2008)
162. Gilbert, J.E., Murray, M.: Clifford Algebras and Dirac Operators in Harmonic Analysis. Cambridge University Press, Cambridge (1991)
163. Goerzel, G.: An algorithm for the evaluation of finite trigonometric series. Amer. Math. Monthly **65**(1), 34–35 (1958)
164. Golomb, M.: Approximation by periodic spline interpolants on uniform meshes. J. Approx. Theory **1**, 26–65 (1968)
165. Golub, G.H., Milanfar, P., Varah, J.: A stable numerical method for inverting shape from moments. SIAM J. Sci. Comput. **21**(4), 1222–1243 (1999)
166. Golub, G.H., Van Loan, C.F.: Matrix Computations, third edn. Johns Hopkins University Press, Baltimore (1996)
167. Golyandina, N., Nekrutkin, V., Zhigljavsky, A.: Analysis of Time Series Structure. SSA and Related Techniques. Chapman & Hall/CRC, Boca Raton (2001)
168. Golyandina, N., Zhigljavsky, A.: Singular Spectrum Analysis for Time Series. Springer-Verlag, Heidelberg (2013)
169. Good, I.J.: The interaction algorithm and practical Fourier analysis. J. Roy. Statist. Soc. Ser. B **20**, 361–372 (1958)
170. Gradstein, I., Ryshik, I.: Tables of Series, Products, and Integrals, vol. 2. Verlag Harri Deutsch, Thun (1981)
171. Gräf, M.: Numerical spherical designs on $\mathbb{S}^2$. http://www.tu-chemnitz.de/~potts/workgroup/graef/quadrature/index.php.en
172. Gräf, M.: An unified approach to scattered data approximation on $\mathbb{S}^3$ and SO(3). Adv. Comput. Math. **37**(3), 379–392 (2012)
173. Gräf, M.: Efficient Algorithms for the Computation of Optimal Quadrature Points on Riemannian Manifolds. Dissertation. Universitätsverlag Chemnitz (2013)
174. Gräf, M., Hielscher, R.: Fast global optimization on the torus, the sphere and the rotation group. SIAM J. Optim. **25**(1), 540–563 (2015)
175. Gräf, M., Kunis, S., Potts, D.: On the computation of nonnegative quadrature weights on the sphere. Appl. Comput. Harmon. Anal. **27**(1), 124–132 (2009)

176. Gräf, M., Potts, D.: On the computation of spherical designs by a new optimization approach based on fast spherical Fourier transforms. Numer. Math. **119**(4), 699–724 (2011)
177. Gräf, M., Potts, D., Steidl, G.: Quadrature errors, discrepancies and their relations to halftoning on the torus and the sphere. SIAM J. Sci. Comput. **34**(5), A2760–A2791 (2012).
178. Grafakos, L.: Classical Fourier Analysis, second edn. Springer-Verlag, New York (2008)
179. Gravin, N., Lasserre, J., Pasechnik, D.V., Robins, S.: The inverse moment problem for convex polytopes. Discrete Comput. Geom. **48**(3), 596–621 (2012)
180. Greengard, L.: The Rapid Evaluation of Potential Fields in Particle Systems. MIT Press, Cambridge (1988)
181. Greengard, L., Lin, P.: Spectral approximation of the free-space heat kernel. Appl. Comput. Harmon. Anal. **9**(1), 83–97 (2000)
182. Greengard, L., Strain, J.: The fast Gauss transform. SIAM J. Sci. Stat. Comput. **12**(1), 79–94 (1991)
183. Greengard, L., Sun, X.: A new version of the fast Gauss transform. In: Proceedings of the International Congress of Mathematicians (Berlin, 1998), Doc. Math., vol. 3, pp. 575–584 (1998)
184. Gröchenig, K.: An uncertainty principle related to the Poisson summation formula. Studia Math. **121**(1), 87–104 (1996)
185. Gröchenig, K.: Foundations of Time–Frequency Analysis. Birkhäuser, Boston (2001)
186. Groemer, H.: Geometric applications of Fourier series and spherical harmonics. In: Encyclopedia of Mathematics and its Applications, vol. 61. Cambridge University Press, Cambridge (1996)
187. Gustafsson, B., He, C., Milanfar, P., Putinar, M.: Reconstructing planar domains from their moments. Inverse Prob. **16**(4), 1053–1070 (2000)
188. Gutknecht, M.H.: Attenuation factors in multivariate Fourier analysis. Numer. Math. **51**(6), 615–629 (1987)
189. Hahn, S.L.: Hilbert Transforms in Signal Processing. Artech House, Boston (1996)
190. Hale, N., Townsend, A.: A fast, simple, and stable Chebyshev-Legendre transform using an asymptotic formula. SIAM J. Sci. Comput. **36**(1), A148–A167 (2014)
191. Hassanieh, H.: The Sparse Fourier Transform: Theory and Practice. ACM books, New York (2018)
192. Hassanieh, H., Indyk, P., Katabi, D., Price, E.: Simple and practical algorithm for sparse Fourier transform. In: Proceedings of the Twenty-Third Annual ACM-SIAM Symposium on Discrete Algorithms, pp. 1183–1194. ACM, New York (2012)
193. Häuser, S., Heise, B., Steidl, G.: Linearized Riesz transform and quasi-monogenic shearlets. Int.l J. Wavelets Multiresolut. Inf. Process. **12**(3), 1–25 (2014)
194. Haverkamp, R.: Approximationsfehler der Ableitungen von Interpolationspolynomen. J. Approx. Theory **30**(3), 180–196 (1980)
195. Haverkamp, R.: Zur Konvergenz der Ableitungen von Interpolationspolynomen. Computing **32**(4), 343–355 (1984)
196. Healy, D.M., Kostelec, P.J., Moore, S., Rockmore, D.N.: FFTs for the 2-sphere—improvements and variations. J. Fourier Anal. Appl. **9**(4), 341–385 (2003)
197. Healy, J.J., Kutay, M.A., Ozaktas, H.M., Sheridan, J.T.: Linear Canonical Transforms. Theory and Applications. Springer-Verlag, New York (2016)
198. Heideman, M.T., Johnson, D.H., Burrus, C.S.: Gauss and the history of the fast Fourier transform. Arch. Hist. Exact Sci. **34**(3), 265–277 (1985)
199. Heider, S., Kunis, S., Potts, D., Veit, M.: A sparse Prony FFT. In: Proceedings of 10th International Conference on Sampling Theory and Applications, vol. 9, pp. 572–575 (2013)
200. Heinig, G., Rost, K.: Algebraic Methods for Toeplitz-like Matrices and Operators. Akademie-Verlag, Berlin (1984)
201. Henrici, P.: Barycentric formulas for interpolating trigonometric polynomials and their conjugates. Numer. Math. **33**(2), 225–234 (1979)
202. Heuser, H.: Lehrbuch der Analysis. Teil 2, twelve edn. B. G. Teubner, Stuttgart (2002)

203. Hewitt, E., Ross, K.A.: Abstract Harmonic Analysis. Vol. II: Structure and analysis for compact groups. Analysis on locally compact Abelian groups. In: Die Grundlehren der mathematischen Wissenschaften, vol. 152. Springer, New York (1970)
204. Higham, N.J.: Accuracy and Stability of Numerical Algorithms, second edn. SIAM, Philadelphia (2002)
205. Horn, R.A., Johnson, C.R.: Matrix Analysis, second edn. Cambridge University Press, Cambridge (2013)
206. Hua, Y., Sarkar, T.K.: Matrix pencil method for estimating parameters of exponentially damped/undamped sinusoids in noise. IEEE Trans. Acoust. Speech Signal Process. **38**(5), 814–824 (1990)
207. Huybrechs, D.: On the Fourier extension of nonperiodic functions. SIAM J. Numer. Anal. **47**(6), 4326–4355 (2010)
208. Igari, S.: Real Analysis—With an Introduction to Wavelet Theory. American Mathematical Society, Providence (1998). Translated from Japanese
209. Ingham, A.E.: Some trigonometrical inequalities with applications to the theory of series. Math. Z. **41**(1), 367–379 (1936)
210. Iserles, A.: A fast and simple algorithm for the computation of Legendre coefficients. Numer. Math. **117**(3), 529–553 (2011)
211. Iwen, M.A.: Combinatorial sublinear-time Fourier algorithms. Found. Comput. Math. **10**(3), 303–338 (2010)
212. Iwen, M.A.: Improved approximation guarantees for sublinear-time Fourier algorithms. Appl. Comput. Harmon. Anal. **34**(1), 57–82 (2013)
213. Jacob, M.: Optimized least-square nonuniform fast Fourier transform. IEEE Trans. Signal Process. **57**(6), 2165–2177 (2009)
214. Jakob-Chien, R., Alpert, B.K.: A fast spherical filter with uniform resolution. J. Comput. Phys. **136**, 580–584 (1997)
215. Johnson, S.G., Frigo, M.: A modified split radix FFT with fewer arithmetic operations. IEEE Trans. Signal Process. **55**(1), 111–119 (2007)
216. Junghanns, P., Kaiser, R.: Collocation for Cauchy singular integral equations. Linear Algebra Appl. **439**(3), 729–770 (2013)
217. Junghanns, P., Kaiser, R., Potts, D.: Collocation–quadrature methods and fast summation for Cauchy singular integral equations with fixed singularities. Linear Algebra Appl. **491**, 187–238 (2016)
218. Junghanns, P., Rost, K.: Matrix representations associated with collocation methods for Cauchy singular integral equations. Math. Meth. Appl. Sci. **30**, 1811–1821 (2007)
219. Kämmerer, L.: Reconstructing hyperbolic cross trigonometric polynomials by sampling along rank-1 lattices. SIAM J. Numer. Anal. **51**(5), 2773–2796 (2013).
220. Kämmerer, L.: High Dimensional Fast Fourier Transform Based on Rank-1 Lattice Sampling. Dissertation. Universitätsverlag Chemnitz (2014).
221. Kämmerer, L.: Reconstructing multivariate trigonometric polynomials from samples along rank-1 lattices. In: Approximation Theory XIV: San Antonio 2013, pp. 255–271. Springer, Cham (2014)
222. Kämmerer, L.: Multiple rank-1 lattices as sampling schemes for multivariate trigonometric polynomials. J. Fourier Anal. Appl. **24**(1), 17–44 (2018)
223. Kämmerer, L.: Constructing spatial discretizations for sparse multivariate trigonometric polynomials that allow for a fast discrete Fourier transform. Appl. Comput. Harm. Anal. **47**(3), 702–729 (2019)
224. Kämmerer, L., Kunis, S., Melzer, I., Potts, D., Volkmer, T.: Computational methods for the Fourier analysis of sparse high-dimensional functions. In: Extraction of Quantifiable Information from Complex Systems, pp. 347–363. Springer, Cham (2014)
225. Kämmerer, L., Kunis, S., Potts, D.: Interpolation lattices for hyperbolic cross trigonometric polynomials. J. Complexity **28**(1), 76–92 (2012)

226. Kämmerer, L., Potts, D., Volkmer, T.: Approximation of multivariate periodic functions by trigonometric polynomials based on rank-1 lattice sampling. J. Complexity **31**(4), 543–576 (2015)
227. Kämmerer, L., Potts, D., Volkmer, T.: Approximation of multivariate periodic functions by trigonometric polynomials based on sampling along rank-1 lattice with generating vector of Korobov form. J. Complexity **31**(3), 424–456 (2015)
228. Kämmerer, L., Potts, D., Volkmer, T.: High-dimensional sparse FFT based on sampling along multiple rank-1 lattices. Appl. Comput. Harm. Anal. **51**, 225–257 (2021)
229. Keiner, J.: Gegenbauer polynomials and semiseparable matrices. Electron. Trans. Numer. Anal. **30**, 26–53 (2008)
230. Keiner, J.: Computing with expansions in Gegenbauer polynomials. SIAM J. Sci. Comput. **31**(3), 2151–2171 (2009)
231. Keiner, J., Kunis, S., Potts, D.: NFFT 3.4, C subroutine library. http://www.tu-chemnitz.de/~potts/nfft. Contributor: F. Bartel, M. Fenn, T. Görner, M. Kircheis, T. Knopp, M. Quellmalz, T. Volkmer, A. Vollrath
232. Keiner, J., Kunis, S., Potts, D.: Fast summation of radial functions on the sphere. Computing **78**(1), 1–15 (2006)
233. Keiner, J., Kunis, S., Potts, D.: Efficient reconstruction of functions on the sphere from scattered data. J. Fourier Anal. Appl. **13**(4), 435–458 (2007)
234. Keiner, J., Kunis, S., Potts, D.: Using NFFT3—a software library for various nonequispaced fast Fourier transforms. ACM Trans. Math. Software **36**, Article 19, 1–30 (2009)
235. Keiner, J., Potts, D.: Fast evaluation of quadrature formulae on the sphere. Math. Comput. **77**(261), 397–419 (2008)
236. Keriven, N., A., B., Gribonval, R., Pérez, P.: Sketching for large-scale learning of mixture models. Inf. Inference J. IMA **7**(3), 447–508 (2018)
237. King, F.W.: Hilbert Transforms, vol. I. Cambridge University Press, Cambridge (2008)
238. King, F.W.: Hilbert Transforms, vol. II. Cambridge University Press, Cambridge (2009)
239. Kircheis, M., Potts, D.: Direct inversion of the nonequispaced fast Fourier transform. Linear Algebra Appl. **575**, 106–140 (2019)
240. Kircheis, M., Potts, D.: Efficient multivariate inversion of the nonequispaced fast Fourier transform. Proc. Appl. Math. Mech. **20**(1), e202,000,120 (2021)
241. Kircheis, M., Potts, D., Tasche, M.: Nonuniform fast Fourier transforms with nonequispaced spatial and frequency data and fast sinc transforms. Numer. Algor. **23**, 1–33 (2022)
242. Kircheis, M., Potts, D., Tasche, M.: On regularized Shannon sampling formulas with localized sampling. Sampl. Theory Signal Process. Data Anal. **20**(3), 20 (2022)
243. Kirsch, A.: The MUSIC algorithm and the factorization method in inverse scattering theory for inhomogeneous media. Inverse Prob. **18**(4), 1025–1040 (2002)
244. Knirsch, H., Petz, M., Plonka, G.: Optimal rank-1 Hankel approximation of matrices: Frobenius norm and spectral norm and Cadzow's algorithm. Linear Algebra Appl. **629**, 1–39 (2021)
245. Kolmogoroff, A.: Une série de Fourier–Lebesgue divergente partout. C. R. Acad. Sci. Paris **183**, 1327–1328 (1926)
246. Komornik, V., Loreti, P.: Fourier Series in Control Theory. Springer-Verlag, New York (2005)
247. Körner, T.: Fourier Analysis, second edn. Cambridge University Press, Cambridge (1989)
248. Korobov, N.M.: Teoretiko-Chislovye Metody v Priblizhënnom Analize, second edn. Moscow (2004)
249. Kostelec, P.J., Rockmore, D.N.: FFTs on the rotation group. J. Fourier Anal. Appl. **14**(2), 145–179 (2008)
250. Kotelnikov, V.A.: On the transmission capacity of the "ether" and wire in electrocommunications. In: Modern Sampling Theory: Mathematics and Application, pp. 27–45. Birkhäuser, Boston (2001).Translated from Russian

251. Köthe, U., Felsberg, M.: Riesz-transforms versus derivatives: on the relationship between the boundary tensor and the energy tensor. In: R. Kimmel, N.A. Sochen, J. Weickert (eds.) Scale Space and PDE Methods in Computer Vision: 5th International Conference, Scale-Space 2005, pp. 179–191. Springer, Berlin (2005)
252. Kühn, T., Sickel, W., Ullrich, T.: Approximation numbers of Sobolev embeddings—sharp constants and tractability. J. Complexity **30**(2), 95–116 (2014)
253. Kunis, S., Kunis, S.: The nonequispaced FFT on graphics processing units. Proc. Appl. Math. Mech. **12**, 7–10 (2012)
254. Kunis, S., Möller, H.M., Peter, T., von der Ohe, U.: Prony's method under an almost sharp multivariate Ingham inequality. J. Fourier Anal. Appl. **24**(5), 1306–1318 (2018)
255. Kunis, S., Nagel, D.: On the condition number of Vandermonde matrices with pairs of nearly-colliding nodes. Numer. Algor. **87**, 473–496 (2021)
256. Kunis, S., Peter, T., Römer, T., von der Ohe, U.: A multivariate generalization of Prony's method. Linear Algebra Appl. **490**, 31–47 (2016)
257. Kunis, S., Potts, D.: Fast spherical Fourier algorithms. J. Comput. Appl. Math. **161**(1), 75–98 (2003)
258. Kunis, S., Potts, D.: Stability results for scattered data interpolation by trigonometric polynomials. SIAM J. Sci. Comput. **29**(4), 1403–1419 (2007)
259. Kunis, S., Potts, D., Steidl, G.: Fast Gauss transforms with complex parameters using NFFTs. J. Numer. Math. **14**(4), 295–303 (2006)
260. Kuo, F.Y.: Component-by-component constructions achieve the optimal rate of convergence for multivariate integration in weighted Korobov and Sobolev spaces. J. Complexity **19**(3), 301–320 (2003). Oberwolfach Special Issue
261. Kuo, F.Y., Sloan, I.H., Wasilkowski, G.W., Woźniakowski, H.: On decompositions of multivariate functions. Math. Comput. **79**(270), 953–966 (2010)
262. Kuo, F.Y., Sloan, I.H., Woźniakowski, H.: Lattice rule algorithms for multivariate approximation in the average case setting. J. Complexity **24**(2), 283–323 (2008)
263. Lanczos, C.: Discourse on Fourier Series, reprint of the 1966 edn. SIAM, Philadelphia (2016)
264. Larkin, K.G., Bone, D.J., Oldfield, M.A.: Natural demodulation of two-dimensional fringe patterns. I. General background of the spiral phase quadrature transform. J. Opt. Soc. America A **18**(8), 1862–1870 (2001)
265. Lasser, R.: Introduction to Fourier Series. Marcel Dekker, New York (1996)
266. Lawlor, D., Wang, Y., Christlieb, A.: Adaptive sub-linear time Fourier algorithms. Adv. Adapt. Data Anal. **5**(1), 1350,003 (2013)
267. Lebedev, N.N.: Special Functions and Their Applications. Dover Publications, New York (1972). Translated from Russian
268. Lebrat, L., de Gournay, F., Kahn, J., Weiss, P.: Optimal transport approximation of 2-dimensional measures. SIAM J. Imaging Sci. **12**(2), 762–787 (2019)
269. Lee, J.Y., Greengard, L.: The type 3 nonuniform FFT and its applications. J. Comput. Physics **206**(1), 1–5 (2005)
270. Lemke, P., Skiena, S.S., Smith, W.D.: Reconstructing sets from interpoint distances. In: Discrete and Computational Geometry, pp. 597–631. Springer-Verlag, Berlin (2003)
271. Li, D., Hickernell, F.J.: Trigonometric spectral collocation methods on lattices. In: Recent Advances in Scientific Computing and Partial Differential Equations, pp. 121–132. American Mathematical Society, Providence (2003)
272. Li, N.: 2DECOMP&FFT—Parallel FFT subroutine library. http://www.2decomp.org
273. Liao, W., Fannjiang, A.: MUSIC for single-snapshot spectral estimation: stability and super-resolution. Appl. Comput. Harmon. Anal. **40**(1), 33–67 (2016)
274. Lieb, E.H., Loss, M.: Analysis, second edn. American Mathematical Society, Providence (2014)
275. Lin, R., Zhang, H.: Convergence analysis of the Gaussian regularized Shannon sampling series. Numer. Funct. Anal. Optim. **38**(2), 224–247 (2017)
276. Locher, F.: Interpolation on uniform meshes by the translates of one function and related attenuation factors. Math. Comp. **37**(156), 403–416 (1981)

277. Luke, R.D.: Relaxed averaged alternating reflections for diffraction imaging. Inverse Prob. **21**(1), 37–50 (2005)
278. Mainprice, D., Bachmann, F., Hielscher, R., Schaeben, H.: Descriptive tools for the analysis of texture projects with large datasets using MTEX: strength, symmetry and components. Geol. Soc. London Spec. Publ. **409**(1), 251–271 (2014)
279. Majidian, H.: On the decay rate of Chebyshev coefficients. Appl. Numer. Math. **113**, 44–53 (2017)
280. Mallat, S.: A Wavelet Tour of Signal Processing. The Sparse Way, third edn. Elsevier/Academic Press, Amsterdam (2009)
281. Manolakis, D.G., Ingle, V.K., Kogon, S.M.: Statistical and Adaptive Signal Processing. McGraw-Hill, Boston (2005)
282. Markovsky, I.: Structured low-rank approximation and its applications. Automatica J. IFAC **44**(4), 891–909 (2008)
283. Markovsky, I.: Low-Rank Approximation: Algorithms, Implementation, Applications, second edn. Springer-Verlag, London (2018)
284. Maruyama, T.: Fourier Analysis and Economic Phenomena. In: Monographs in Mathematical Economics, vol. 2. Springer Nature, Singapore (2018)
285. Mason, J.C., Handscomb, D.C.: Chebyshev Polynomials. Chapman & Hall/CRC, Boca Raton (2003)
286. McClellan, J.H., Parks, T.W.: Eigenvalue and eigenvector decomposition of the discrete Fourier transform. IEEE Trans. Audio Electroacoust. **20**(1), 66–74 (1972)
287. Mhaskar, H.N., Narcowich, F.J., Ward, J.D.: Spherical Marcinkiewicz-Zygmund inequalities and positive quadrature. Math. Comput. **70**(235), 1113–1130 (2001). Corrigendum to this paper in *Math. Comput.* 71(237):453–454, 2002
288. Michel, V.: Lectures on Constructive Approximation: Fourier, Spline, and Wavelet Methods on the Real Line, the Sphere, and the Ball. Birkhäuser/Springer, New York (2013)
289. Mohlenkamp, M.J.: A fast transform for spherical harmonics. J. Fourier Anal. Appl. **5**(2–3), 159–184 (1999)
290. Moitra, A.: The threshold for super-resolution via extremal functions. Preprint, Massachusetts Institute of Technology, Cambridge (2014)
291. Moitra, A.: Super-resolution, extremal functions and the condition number of Vandermonde matrices. In: Proceedings of the Forty-Seventh Annual ACM on Symposium on Theory of Computing, pp. 821–830 (2015)
292. Montgomery, H.L., Vaughan, R.C.: Hilbert's inequality. J. London Math. Soc. **8**, 73–82 (1974)
293. Morgenstern, J.: Note on a lower bound of the linear complexity of the fast Fourier transform. J. Assoc. Comput. Mach. **20**, 305–306 (1973)
294. Morton, P.: On the eigenvectors of Schur's matrix. J. Number. Theory **12**(1), 122–127 (1980)
295. Munthe-Kaas, H., Sørevik, T.: Multidimensional pseudo-spectral methods on lattice grids. Appl. Numer. Math. **62**(3), 155–165 (2012)
296. Nakatsukasa, Y., Sète, O., Trefethen, L.N.: The AAA algorithm for rational approximation. SIAM J. Sci. Comput. **40**(3), A1494–A1522 (2018)
297. Narcowich, F.J., Sun, X., Ward, J.D., Wendland, H.: Direct and inverse Sobolev error estimates for scattered data interpolation via spherical basis functions. Found. Comput. Math. **7**(3), 369–390 (2007)
298. Natterer, F., Wübbeling, F.: Mathematical Methods in Image Reconstruction. SIAM, Philadelphia (2001)
299. Nestler, F.: Automated parameter tuning based on RMS errors for nonequispaced FFTs. Adv. Comput. Math. **42**(4), 889–919 (2016)
300. Nestler, F.: Parameter tuning for the NFFT based fast Ewald summation. Front. Phys. **4**(28) (2016)
301. Nguyen, H.Q., Unser, M., Ward, J.P.: Generalized Poisson summation formulas for continuous functions of polynomial growth. J. Fourier Anal. Appl. **23**(2), 442–461 (2017)
302. Nguyen, N., Liu, Q.H.: The regular Fourier matrices and nonuniform fast Fourier transforms. SIAM J. Sci. Comput. **21**(1), 283–293 (1999)

303. Niederreiter, H.: Quasi-Monte Carlo methods and pseudo-random numbers. Bull. Amer. Math. Soc. **84**(6), 957–1041 (1978)
304. Nieslony, A., Steidl, G.: Approximate factorizations of Fourier matrices with nonequispaced knots. Linear Algebra Appl. **366**, 337–351 (2003)
305. Nussbaumer, H.J.: Fast Fourier Transform and Convolution Algorithms, revised edn. Springer-Verlag, Berlin (1982)
306. Nyquist, H.: Certain factors affecting telegraph speed. Bell Syst. Tech. J. **3**(2), 324–346 (1924)
307. Oberhettinger, F.: Tables of Fourier Transforms and Fourier Transforms of Distributions. Springer-Verlag, Berlin (1990)
308. Osborne, M., Smyth, G.: A modified Prony algorithm for exponential function fitting. SIAM J. Sci. Comput. **16**(1), 119–138 (1995)
309. Ozaktas, H.M., Zalevsky, Z., Kutay, M.A.: The Fractional Fourier Transform with Applications in Optics and Signal Processing. John Wiley & Sons, Chichester (2001)
310. Pan, H., Blu, T., Vetterli, M.: Towards generalized FRI sampling with an application to source resolution in radioastronomy. IEEE Trans. Signal Process. **65**(4), 821–835 (2017)
311. Paszkowski, S.: Vychislitelp1nye Primeneniya Mnogochlenov i Ryadov Chebysheva. Nauka, Moscow (1983). Translated from Polish
312. Pawar, S., Ramchandran, K.: Computing a k-sparse n-length discrete Fourier transform using at most $4k$ samples and $0(k \log k)$ complexity. In: Proceedings of the IEEE International Symposium on Information Theory (ISIT), pp. 464–468 (2013)
313. Pereyra, V., Scherer, G.: Exponential Data Fitting and its Applications. Bentham Science Publishers, Sharjah (2010)
314. Peter, T., Plonka, G.: A generalized Prony method for reconstruction of sparse sums of eigenfunctions of linear operators. Inverse Problems **29**, 025,001 (2013)
315. Peter, T., Plonka, G., Schaback, R.: Prony's method for multivariate signals. Proc. Appl. Math. Mech. **15**(1), 665–666 (2015)
316. Peter, T., Potts, D., Tasche, M.: Nonlinear approximation by sums of exponentials and translates. SIAM J. Sci. Comput. **33**, 1920–1947 (2011)
317. Petz, M., Plonka, G., Derevianko, N.: Exact reconstruction of sparse non-harmonic signals from their Fourier coefficients. Sampl. Theory Signal Process. Data Anal. **19**, 7 (2021)
318. Pinsky, M.A.: Introduction to Fourier Analysis and Wavelets. American Mathematical Society, Providence (2002)
319. Pippig, M.: PFFT, Parallel FFT subroutine library (2011). http://www.tu-chemnitz.de/~potts/workgroup/pippig/software.php.en
320. Pippig, M.: PNFFT, Parallel Nonequispaced FFT subroutine library (2011). http://www.tu-chemnitz.de/~potts/workgroup/pippig/software.php.en
321. Pippig, M.: PFFT: an extension of FFTW to massively parallel architectures. SIAM J. Sci. Comput. **35**(3), C213–C236 (2013)
322. Pippig, M., Potts, D.: Parallel three-dimensional nonequispaced fast Fourier transforms and their application to particle simulation. SIAM J. Sci. Comput. **35**(4), C411–C437 (2013)
323. Plonka, G., Stampfer, K., Keller, I.: Reconstruction of stationary and non-stationary signals by the generalized Prony method. Analysis Appl. **17**(2), 179–210 (2019)
324. Plonka, G., Tasche, M.: Fast and numerically stable algorithms for discrete cosine transforms. Linear Algebra Appl. **394**, 309–345 (2005)
325. Plonka, G., von Wulffen, T.: Deterministic sparse sublinear FFT with improved numerical stability. Results Math. **76**, 53 (2021)
326. Plonka, G., Wannenwetsch, K.: A deterministic sparse FFT algorithm for vectors with small support. Numer. Algor. **71**(4), 889–905 (2016)
327. Plonka, G., Wannenwetsch, K.: A sparse fast Fourier algorithm for real non-negative vectors. J. Comput. Appl. Math. **321**, 532–539 (2017)
328. Plonka, G., Wannenwetsch, K., Cuyt, A., Lee, W.s.: Deterministic sparse FFT for m-sparse vectors. Numer. Algor. **78**(1), 133–159 (2018)
329. Plonka, G., Wischerhoff, M.: How many Fourier samples are needed for real function reconstruction? J. Appl. Math. Comput. **42**(1–2), 117–137 (2013)

330. Potts, D.: Fast algorithms for discrete polynomial transforms on arbitrary grids. Linear Algebra Appl. **366**, 353–370 (2003)
331. Potts, D.: Schnelle Fourier-Transformationen für nichtäquidistante Daten und Anwendungen. Habilitation, Universität zu Lübeck, Lübeck (2003)
332. Potts, D., Prestin, J., Vollrath, A.: A fast algorithm for nonequispaced Fourier transforms on the rotation group. Numer. Algorithms **52**(3), 355–384 (2009)
333. Potts, D., Schmischke, M.: Approximation of high-dimensional periodic functions with Fourier-based methods. SIAM J. Numer. Anal. **59(5)**, 2393–2429 (2021)
334. Potts, D., Schmischke, M.: Learning multivariate functions with low-dimensional structures using polynomial bases. J. Comput. Appl. Math. **403**, 113,821 (2022)
335. Potts, D., Steidl, G.: Fast summation at nonequispaced knots by NFFTs. SIAM J. Sci. Comput. **24**(6), 2013–2037 (2003)
336. Potts, D., Steidl, G., Nieslony, A.: Fast convolution with radial kernels at nonequispaced knots. Numer. Math. **98**(2), 329–351 (2004)
337. Potts, D., Steidl, G., Tasche, M.: Trigonometric preconditioners for block Toeplitz systems. In: Multivariate Approximation and Splines (Mannheim, 1996), pp. 219–234. Birkhäuser, Basel (1997)
338. Potts, D., Steidl, G., Tasche, M.: Fast algorithms for discrete polynomial transforms. Math. Comput. **67**(224), 1577–1590 (1998)
339. Potts, D., Steidl, G., Tasche, M.: Fast and stable algorithms for discrete spherical Fourier transforms. Linear Algebra Appl. **275/276**, 433–450 (1998)
340. Potts, D., Steidl, G., Tasche, M.: Fast Fourier transforms for nonequispaced data. A tutorial. In: Modern Sampling Theory: Mathematics and Applications, pp. 247–270. Birkhäuser, Boston (2001)
341. Potts, D., Steidl, G., Tasche, M.: Numerical stability of fast trigonometric transforms - a worst case study. J. Concr. Appl. Math. **1**(1), 1–36 (2003)
342. Potts, D., Tasche, M.: Parameter estimation for exponential sums by approximate Prony method. Signal Process. **90**, 1631–1642 (2010)
343. Potts, D., Tasche, M.: Parameter estimation for multivariate exponential sums. Electron. Trans. Numer. Anal. **40**, 204–224 (2013)
344. Potts, D., Tasche, M.: Parameter estimation for nonincreasing exponential sums by Prony-like methods. Linear Algebra Appl. **439**(4), 1024–1039 (2013)
345. Potts, D., Tasche, M.: Sparse polynomial interpolation in Chebyshev bases. Linear Algebra Appl. **441**, 61–87 (2014)
346. Potts, D., Tasche, M.: Fast ESPRIT algorithms based on partial singular value decompositions. Appl. Numer. Math. **88**, 31–45 (2015)
347. Potts, D., Tasche, M.: Error estimates for the ESPRIT algorithm. In: Large Truncated Toeplitz Matrices, Toeplitz Operators, and Related Topics, pp. 621–648. Birkhäuser/Springer, Cham (2017)
348. Potts, D., Tasche, M.: Continuous window functions for NFFT. Adv. Comput. Math. **47**(4), 34 (2021). Paper No. 53
349. Potts, D., Tasche, M.: Uniform error estimates for nonequispaced fast Fourier transforms. Sampl. Theory Signal Process. Data Anal. **19**(2), 42 (2021). Paper No. 17
350. Potts, D., Tasche, M., Volkmer, T.: Efficient spectral estimation by MUSIC and ESPRIT with application to sparse FFT. Front. Appl. Math. Stat. **2**(1) (2016)
351. Potts, D., Volkmer, T.: Fast and exact reconstruction of arbitrary multivariate algebraic polynomials in Chebyshev form. In: 11th International Conference on Sampling Theory and Applications (SampTA 2015), pp. 392–396 (2015)
352. Potts, D., Volkmer, T.: Sparse high-dimensional FFT based on rank-1 lattice sampling. Appl. Comput. Harmon. Anal. **41**(3), 713–748 (2016)
353. Potts, D., Volkmer, T.: Multivariate sparse FFT based on rank-1 Chebyshev lattice sampling. In: 12th International Conference on Sampling Theory and Applications (SampTA 2017), pp. 504–508 (2017)

354. Prestini, E.: The Evolution of Applied Harmonic Analysis. Models of the Real World, second edn. Birkhäuser/Springer, New York (2016)
355. Püschel, M., Moura, J.M.F.: The algebraic approach to the discrete cosine and sine transforms and their fast algorithms. SIAM J. Comput. **32**(5), 1280–1316 (2003)
356. Qian, L.: On the regularized Whittaker-Kotelnikov-Shannon sampling formula. Proc. Amer. Math. Soc. **131**(4), 1169–1176 (2003)
357. Quade, W., Collatz, L.: Zur Interpolationstheorie der reellen periodischen Funktionen. In: Sitzungsber. Preuß. Akad. Wiss., Phys.-Math. Kl., pp. 383–429 (1938)
358. Rader, C.: Discrete Fourier transforms when the number of data samples is prime. Proc. IEEE **56**(6), 1107–1108 (1968)
359. Ramanathan, J.: Methods of Applied Fourier Analysis. Birkhäuser, Boston (1998)
360. Ramos, G.U.: Roundoff error analysis of the fast Fourier transform. Math. Comp. **25**, 757–768 (1971)
361. Ranieri, J., Chebira, A., Lu, Y.M., Vetterli, M.: Phase retrieval for sparse signals: uniqueness conditions. ArXiv e-prints (2013). ArXiv:1308.3058v2
362. Rao, K.R., Kim, D.N., Hwang, J.J.: Fast Fourier Transforms: Algorithms and Applications. Springer, Dordrecht (2010)
363. Rao, K.R., Yip, P.: Discrete Cosine Transform: Algorithms, Advantages, Applications. Academic Press, Boston (1990)
364. Rauhut, H., Ward, R.: Sparse Legendre expansions via ℓ_1-minimization. J. Approx. Theory **164**(5), 517–533 (2012)
365. Riesz, M.: Sur les fonctions conjuguées. Math. Z. **27**(1), 218–244 (1928)
366. Rivlin, T.J.: Chebyshev Polynomials. From Approximation Theory to Algebra and Number Theory, second edn. John Wiley & Sons, New York (1990)
367. Rokhlin, V., Tygert, M.: Fast algorithms for spherical harmonic expansions. SIAM J. Sci. Comput. **27**(6), 1903–1928 (2006)
368. Roy, R., Kailath, T.: ESPRIT—estimation of signal parameters via rotational invariance techniques. In: Signal Processing, Part II, IMA Vol. Math. Appl. 23, pp. 369–411. Springer, New York (1990)
369. Runge, C.: Über empirische Funktionen und die Interpolation zwischen äquidistanten Ordinaten. Z. Math. Phys. **46**, 224–243 (1901)
370. Runge, C.: Über die Zerlegung einer empirisch gegebenen periodischen Funktion in Sinuswellen. Z. Math. Phys. **48**, 443–456 (1903)
371. Runge, C., König, H.: Vorlesungen über Numerisches Rechnen. Springer-Verlag, Berlin (1924)
372. Sahnoun, S., Usevich, K., Comon, P.: Multidimensional ESPRIT for damped and undamped signals: Algorithm, computations, and perturbation analysis. IEEE Trans. Signal Process. **65**(22), 5897–5910 (2017)
373. Salzer, H.E.: Lagrangian interpolation at the Chebyshev points $X_{n,\nu} \equiv \cos(\nu\pi/n)$, $\nu = 0(1)n$; some unnoted advantages. Comput. J. **15**, 156–159 (1972)
374. Sarkar, T.K., Pereira, O.: Using the matrix pencil method to estimate the parameters of a sum of complex exponentials. IEEE Antennas Propag. Mag. **37**(1), 48–55 (1995)
375. Sauer, T.: Prony's method in several variables: symbolic solutions by universal interpolation. J. Symbolic Comput. **84**, 95–112 (2018)
376. Schaeben, H., van den Boogaart, K.G.: Spherical harmonics in texture analysis. Tectonophysics **370**, 253–268 (2003)
377. Schatzman, J.C.: Accuracy of the discrete Fourier transform and the fast Fourier transform. SIAM J. Sci. Comput. **17**(5), 1150–1166 (1996)
378. Schmeisser, H.J., Triebel, H.: Topics in Fourier Analysis and Function Spaces. Akademische Verlagsgesellschaft Geest & Portig, Leipzig (1987)
379. Schmidt, R.: Multiple emitter location and signal parameter estimation. IEEE Trans. Antenn. Propag. **34**, 276–280 (1986)
380. Schmischke, M.: Interpretable approximation of high-dimensional data based on the ANOVA decomposition. Dissertation. Universitätsverlag Chemnitz (2022)

381. Schreiber, U.: Numerische Stabilität von schnellen trigonometrischen Transformationen. Dissertation, Universität Rostock (2000)
382. Schwartz, L.: Théorie des Distributions, nouvelle edn. Hermann, Paris (1966)
383. Seifert, B., Stolz, H., Donatelli, M., Langemann, D., Tasche, M.: Multilevel Gauss-Newton methods for phase retrieval problems. J. Phys. A **39**(16), 4191–4206 (2006)
384. Seifert, B., Stolz, H., Tasche, M.: Nontrivial ambiguities for blind frequency-resolved optical gating and the problem of uniqueness. J. Opt. Soc. Amer. B **21**(5), 1089–1097 (2004)
385. Shannon, C.E.: Communication in the presence of noise. Proc. I.R.E. **37**, 10–21 (1949)
386. Shapiro, V.L.: Fourier Series in Several Variables with Applications to Partial Differential Equations. Chapman & Hall/CRC, Boca Raton (2011)
387. Shi, H., Traonmilin, Y., Aujol, J.F.: Compressive learning for patch-based image denoising. SIAM J. Imag. Sci. **15**(3), 1184–1212 (2022)
388. Shukla, P., Dragotti, P.L.: Sampling schemes for multidimensional signals with finite rate of innovation. IEEE Trans. Signal Process. **55**(7, part 2), 3670–3686 (2007)
389. Skrzipek, M.R.: Signal recovery by discrete approximation and a Prony-like method. J. Comput. Appl. Math. **326**, 193–203 (2017)
390. Sloan, I.H., Joe, S.: Lattice Methods for Multiple Integration. Clarendon Press, Oxford University Press, New York (1994)
391. Sloan, I.H., Kachoyan, P.J.: Lattice methods for multiple integration: theory, error analysis and examples. SIAM J. Numer. Anal. **24**(1), 116–128 (1987)
392. Sloan, I.H., Reztsov, A.V.: Component-by-component construction of good lattice rules. Math. Comp. **71**(237), 263–273 (2002)
393. Sloan, I.H., Womersley, R.S.: Constructive polynomial approximation on the sphere. J. Approx. Theory **103**(1), 91–118 (2000)
394. Sloan, I.H., Womersley, R.S.: A variational characterisation of spherical designs. J. Approx. Theory **159**(2), 308–318 (2009)
395. Steidl, G.: Fast radix-p discrete cosine transform. Appl. Algebra Engrg. Comm. Comput. **3**(1), 39–46 (1992)
396. Steidl, G.: A note on fast Fourier transforms for nonequispaced grids. Adv. Comput. Math. **9**(3–4), 337–353 (1998)
397. Steidl, G., Tasche, M.: A polynomial approach to fast algorithms for discrete Fourier–cosine and Fourier–sine transforms. Math. Comp. **56**(193), 281–296 (1991)
398. Stein, E.M.: Singular Integrals and Differentiability Properties of Functions. Princeton University Press, Princeton (1970)
399. Stein, E.M., Weiss, G.: Introduction to Fourier Analysis on Euclidean Spaces. Princeton University Press, Princeton (1971)
400. Storath, M.: Directional multiscale amplitude and phase decomposition by the monogenic curvelet transform. SIAM J. Imaging Sci. **4**(1), 57–78 (2011)
401. Strang, G.: The discrete cosine transform. SIAM Rev. **41**(1), 135–147 (1999)
402. Stromberg, K.R.: An Introduction to Classical Real Analysis. AMS Chelsea Publishing, Providence (2015). Corrected reprint of the 1981 original
403. Suda, R., Takami, M.: A fast spherical harmonics transform algorithm. Math. Comp. **71**(238), 703–715 (2002)
404. Swarztrauber, P.N.: The methods of cyclic reduction, Fourier analysis and the FACR algorithm for the discrete solution of Poisson's equation on a rectangle. SIAM Rev. **19**(3), 490–501 (1977)
405. Swetz, F.J.: Mathematical Treasure: Collected Works of Chebyshev. MAA Press, Washington (2016)
406. Szegö, G.: Orthogonal Polynomials, fourth edn. American Mathematical Society, Providence (1975)
407. Tasche, M.: Accelerating convergence of univariate and bivariate Fourier approximations. Z. Anal. Anwendungen **10**(2), 239–250 (1991)

408. Tasche, M., Zeuner, H.: Roundoff error analysis for fast trigonometric transforms. In: Handbook of Analytic-Computational Methods in Applied Mathematics, pp. 357–406. Chapman & Hall/CRC Press, Boca Raton (2000)
409. Tasche, M., Zeuner, H.: Worst and average case roundoff error analysis for FFT. BIT **41**(3), 563–581 (2001)
410. Temlyakov, V.N.: Reconstruction of periodic functions of several variables from the values at the nodes of number-theoretic nets (in Russian). Anal. Math. **12**(4), 287–305 (1986)
411. Temlyakov, V.N.: Approximation of Periodic Functions. Nova Science Publishers, Commack (1993)
412. Teuber, T., Steidl, G., Gwosdek, P., Schmaltz, C., Weickert, J.: Dithering by differences of convex functions. SIAM J. Imaging Sci. **4**(1), 79–108 (2011)
413. Townsend, A., Webb, M., Olver, S.: Fast polynomial transforms based on Toeplitz and Hankel matrices. Math. Comp. **87**(312), 1913–1934 (2018)
414. Trefethen, L.N.: Is Gauss quadrature better than Clenshaw-Curtis? SIAM Rev. **50**(1), 67–87 (2008)
415. Trefethen, L.N.: Approximation Theory and Approximation Practice. SIAM, Philadelphia (2013)
416. Triebel, H.: Höhere Analysis. Deutscher Verlag der Wissenschaften, Berlin (1972)
417. Tygert, M.: Fast algorithms for spherical harmonic expansions II. J. Comput. Phys. **227**(8), 4260–4279 (2008)
418. Tygert, M.: Fast algorithms for spherical harmonic expansions, III. J. Comput. Phys. **229**(18), 6181–6192 (2010)
419. Unser, M.: Sampling—50 years after Shannon. Proc. IEEE **88**, 569–587 (2000)
420. Unser, M., Van De Ville, D.: Wavelet steerability and the higher-order Riesz transform. IEEE Trans. Image Process. **19**(3), 636–652 (2010)
421. Van Loan, C.F.: Computational Frameworks for the Fast Fourier Transform. SIAM, Philadelphia (1992)
422. Vetterli, M., Duhamel, P.: Split- radix algorithms for length-p^m DFTs. IEEE Trans. Acoust. Speech Signal Process. **37**(1), 57–64 (1989)
423. Vetterli, M., Marziliano, P., Blu, T.: Sampling signals with finite rate of innovation. IEEE Trans. Signal Process. **50**(6), 1417–1428 (2002)
424. Volkmer, T.: OpenMP parallelization in the NFFT software library. Preprint 2012-07, Faculty of Mathematics, Technische Universität Chemnitz (2012)
425. Volkmer, T.: Multivariate Approximation and High-Dimensional Sparse FFT Based on Rank-1 Lattice Sampling. Dissertation. Universitätsverlag Chemnitz (2017)
426. Wang, Z.: Fast algorithms for the discrete W transform and for the discrete Fourier transform. IEEE Trans. Acoust. Speech Signal Process **32**, 803–816 (1984)
427. Wang, Z.D.: Fast algorithms for the discrete W transform and the discrete Fourier transform. IEEE Trans. Acoust. Speech Signal Process. **32**(4), 803–816 (1984)
428. Wang, Z.D.: On computing the discrete Fourier and cosine transforms. IEEE Trans. Acoust. Speech Signal Process. **33**(5), 1341–1344 (1985)
429. Weideman, J.A.C., Trefethen, L.N.: The eigenvalues of second-order spectral differentiation matrices. SIAM J. Numer. Anal. **25**(6), 1279–1298 (1988)
430. Weiss, L., McDonough, R.N.: Prony's method, Z-transforms, and Padé approximation. SIAM Rev. **5**, 145–149 (1963)
431. Weisz, F.: Summability of Multi-dimensional Fourier Series and Hardy Spaces. Kluwer Academic Publishers, Dordrecht (2002)
432. Weisz, F.: Summability of multi-dimensional trigonometric Fourier series. Surv. Approx. Theory **7**, 1–179 (2012)
433. Wendland, H.: Scattered Data Approximation. Cambridge University Press, Cambridge (2005)
434. Werner, D.: Funktionalanalysis, third edn. Springer-Verlag, Berlin (2000)
435. Whittaker, E.T.: On the functions which are represented by the expansions of the interpolation-theory. Proc. Roy. Soc. Edinburgh **35**, 181–194 (1914)

436. Wickerhauser, M.V.: Adapted Wavelet Analysis from Theory to Software. A K Peters, Wellesley (1994)
437. Winograd, S.: Some bilinear forms whose multiplicative complexity depends on the field of constants. Math. Syst. Theory **10**, 169–180 (1977)
438. Winograd, S.: On computing the discrete Fourier transform. Math. Comp. **32**(141), 175–199 (1978)
439. Winograd, S.: Arithmetic Complexity of Computations. SIAM, Philadelphia (1980)
440. Wischerhoff, M., Plonka, G.: Reconstruction of polygonal shapes from sparse Fourier samples. J. Comput. Appl. Math. **297**, 117–131 (2016)
441. Wright, G.B., Javed, M., Montanelli, H., Trefethen, L.N.: Extension of Chebfun to periodic functions. SIAM J. Sci. Comput. **37**(5), C554–C573 (2015)
442. Wu, X., Wang, Y., Yan, Z.: On algorithms and complexities of cyclotomic fast Fourier transforms over arbitrary finite fields. IEEE Trans. Signal Process. **60**(3), 1149–1158 (2012)
443. Xiang, S., Chen, X., Wang, H.: Error bounds for approximation in Chebyshev points. Numer. Math. **116**(3), 463–491 (2010)
444. Xu, Y., Yan, D.: The Bedrosian identity for the Hilbert transform of product functions. Proc. Amer. Math. Soc. **134**(9), 2719–2728 (2006)
445. Yalamov, P.Y.: Improvements of some bounds on the stability of fast Helmholtz solvers. Numer. Algor. **26**(1), 11–20 (2001)
446. Yang, S.C., Qian, H.J., Lu, Z.Y.: A new theoretical derivation of NFFT and its implementation on GPU. Appl. Comput. Harmon. Anal. **44**(2), 273–293 (2018)
447. Yarvin, N., Rokhlin, V.: A generalized one-dimensional fast multipole method with application to filtering of spherical harmonics. J. Comput. Phys. **147**, 549–609 (1998)
448. Yavne, R.: An economical method for calculating the discrete Fourier transform. In: Proc. AFIPS Fall Joint Computer Conference, vol. 33, pp. 115–125 (1968)
449. Young, R.M.: An Introduction to Nonharmonic Fourier Series, revised first edn. Academic Press, San Diego (2001)
450. Yserentant, H.: Regularity and Approximability of Electronic Wave Functions. Springer-Verlag, Berlin (2010)
451. Zeng, X., Leung, K.T., Hickernell, F.J.: Error analysis of splines for periodic problems using lattice designs. In: Monte Carlo and Quasi-Monte Carlo Methods 2004, pp. 501–514. Springer-Verlag, Berlin (2006)
452. Zhang, R., Plonka, G.: Optimal approximation with exponential sums by maximum likelihood modification of Prony's method. Adv. Comp. Math. **45**, 1657–1687 (2019)
453. Zygmund, A.: Trigonometric Series, Vol. I, II, third edn. Cambridge University Press, Cambridge (2002)

Index

Symbols
2π-periodic function, 6
2π-periodic trend, 531

A
Accumulator, 53
Aliasing, 88
Aliasing error, 422
Aliasing formula
  for Chebyshev coefficients, 385
  for Fourier coefficients, 127
  for multivariate Fourier coefficients, 249
Analytic signal, 225
  amplitude, 225
  instantaneous phase, 225
  phase, 225
Annihilating filter, 572
Annihilating filter method, 572
ANOVA decomposition, 492
Approximate identity, 24, 73
Approximate Prony method, 576, 579
Approximation error, 422
Approximation theorem
  of Fejér, 25
  of Weierstrass, 7, 26
Array
  even, 263
  odd, 263
Associated Legendre function, 552
Associated orthogonal polynomials, 406
Attenuation factors
  of Fourier coefficients, 520
  of Fourier transform, 509
Average frequency, 108
Average time, 108
Azimuthal angle, 552

B
Backward shift, 52
Banach algebra, 72
Bandlimited function, 86
  spherical, 553
Bandwidth, 86, 553
Barycentric formula, 133, 457, 461
  for interpolating polynomial, 359
  for interpolating trigonometric polynomial, 133
Basis function, 494
Bernoulli function
  1-periodic, 523
Bernoulli numbers, 523
Bernoulli polynomial, 523
Bernoulli polynomial expansion, 525
Bernstein inequality, 26
Bessel function
  of order ν, 199
  of order zero, 198
Best approximation error, 390
Binary floating point number system, 329
Binary number, 272
Bit reversal permutation, 272
Blackman window sequence, 58
Block circulant matrix with circulant blocks, 165
Block diagonal matrix, 160
Bluestein FFT, 307

G. Plonka et al., *Numerical Fourier Analysis*, Applied and Numerical Harmonic Analysis, https://doi.org/10.1007/978-3-031-35005-4

Borel σ-algebra, 230
Bounded filter, 52
Bounded variation, 32
Box spline, 496
B-spline
of order 1, 603
of order m, 605

C
Cardinal B-spline, 71, 423, 494, 607
centered, 495
Cardinal interpolation by translates, 498, 507
Cardinal interpolation problem, 498
Cardinal Lagrange function, 500
Cardinal sine function, 61
Cascade summation, 271
Cauchy principal value, 67
Central difference quotient of second order, 540
Cesàro sum, 23
Characteristic function, 60
Characteristic polynomial, 285
Chebyshev coefficients, 348
decay, 353
Chebyshev extreme points, 343
Chebyshev polynomial
of first kind, 340
of second kind, 344
Chebyshev series, 348
Chebyshev zero points, 343
Chinese remainder theorem, 288, 307, 328
Circulant matrix, 154
basic, 156
Clenshaw algorithm, 356
Clenshaw–Curtis quadrature, 394
Comb filter, 55
Commutation property
of the Kronecker product, 163
Companion matrix, 568
Complex exponential
multivariate, 176
univariate, 7
Componentwise product of vectors, 151
Computational cost, 266
Computation of Fourier coefficients
via attenuation factors, 520
Computation of Fourier transform
via attenuation factors, 511
Computation of two-dimensional DFT
via one-dimensional transforms, 259
Confluent Vandermonde matrix, 571
Constrained minimization problem, 462
Convergence in $\mathscr{S}(\mathbb{R}^d)$, 183
Convergence in $\mathscr{S}'(\mathbb{R}^d)$, 200
Convergence theorem of Dirichlet–Jordan, 38
Convolution
of functions, 69
of measures, 236, 243
measure with function, 236, 243
in $\ell_1(\mathbb{Z}^d)$, 505
at nonequispaced knots, 449
multivariate, 191
multivariate periodic, 178
of periodic functions, 16
univariate, 69
univariate periodic, 16
Convolution property
of DFT, 151
of Fourier series, 20
of Fourier transform, 70, 82, 243
Cooley–Tukey FFT, 273
Cosine matrix
nonequispaced, 442
of type I, 168
of type II, 169
of type III, 170
of type IV, 171
Cosine vectors
of type I, 168
of type II, 169
of type III, 170
of type IV, 170
Counter-identity matrix, 142
Cyclic convolution
multidimensional, 262
one-dimensional, 147
two-dimensional, 254
Cyclic convolution property
of multidimensional DFT, 263
of one-dimensional DFT, 151
of two-dimensional DFT, 256
Cyclic correlation
of two vectors, 305

D
Damped normal equation of second kind, 462
Decimation-in-frequency FFT, 275
Decimation-in-time FFT, 275
Degree of multivariate trigonometric polynomial, 177
Digital filter, 52
Digital image, 251
Dirac comb, 215
Dirac distribution, 202
Dirichlet kernel, 179
modified, 132

Discrete convolution, 53
Discrete cosine transform
  inverse transform, 381
  inverse two-dimensional transform, 383
  nonequispaced, 442
  two-dimensional, 382
  of type I, 169
  of type II, 170
  of type III, 170
  of type IV, 172
Discrete Fourier transform
  multidimensional, 248, 260
  one-dimensional, 124
  spherical, 553
  two-dimensional, 251
Discrete Hartley transform, 174
Discrete Ingham inequalities, 600
Discrete Laplacian, 545
Discrete polynomial transform, 403
Discrete signal, 51
Discrete sine transform
  nonequispaced, 446
  of type I, 172
  of type II, 173
  of type III, 173
  of type IV, 173
Discrete trigonometric transform, 368
Dispersion of a function, 104
Distance of $x \in \mathbb{R}$ to the nearest integer, 593
Distribution
  $\mathrm{pv}(\frac{1}{x})$, 206, 213
Divide-and-conquer technique, 268

E
Energy of a signal, 103
Entrywise product
  of matrices, 254
  multidimensional, 262
Error constant of NFFT, 425
ESPIRA method, 586
ESPRIT method, 580
Euler–Maclaurin summation formula, 527
Even matrix, 258
Exponential matrix, 253
Exponential sequence, 52
Exponential sum, 439, 567
Extension
  of bounded linear operator, 193

F
Far field sum, 451
Fast Fourier extension, 535, 538
Fast Fourier transform, 265
  nonequispaced, 413, 416
  nonequispaced transposed, 418
  on the rotation group, 565
  spherical, 557
Fast Gauss transform, 454
Fast Poisson solver, 546, 548
Fejér sum, 23
Fejér summation, 49
Filon–Clenshaw–Curtis quadrature, 399
Filter coefficients, 53
Finite difference method, 541
Finitely additive set functions, 233
Finite rate of innovation, 572
FIR filter, 55
Forward shift, 52
Fourier coefficients, 8
  approximate, 478
  decay, 45
  of measures, 235
  spherical, 553
Fourier extension, 535
Fourier inversion formula, 66, 78, 189, 193
  of tempered distribution, 211
Fourier matrix, 136, 472
  for multiple rank-1 lattice and index set I, 488
  nonequispaced, 419
Fourier partial sum, 8
Fourier–Plancherel transform, 194
Fourier series
  of L-periodic function, 11
  of 2π-periodic function, 11
  real, 14
  spherical, 553
Fourier transform, 60, 64, 78, 186, 192
  discrete, 136
  inverse multivariate, 189
  kernel, 117
  of measures, 240
  modulus, 60
  on $L_2(\mathbb{R}^d)$, 194
  phase, 60
  properties, 62
  special affine, 121
  spectral decomposition, 116
  spherical, 551
  of tempered distribution, 210
Fractional Fourier transform, 117
  properties, 119
Frequency analysis problem, 568
Frequency domain, 60
  of two-dimensional DFT, 253

Frequency index set, 466
  difference set, 476
  $l_p(\mathbb{Z}^d)$ ball, 484
Frequency variance, 108
Fresnel transform, 121
Frobenius norm, 253
Function
  of bounded variation, 32
  Hölder continuous, 41
  Lipschitz continuous, 41
  piecewise continuously differentiable, 69
  piecewise C^r-smooth, 530
  piecewise r-times continuously differentiable, 530
  positive definite, 243
  rapidly decreasing, 183
  slowly increasing, 201

G
Gabor function, 607
Gabor transform, 111
Gamma function, 227
Gap condition, 593
Gaussian chirp, 64
Gaussian filter
  discrete, 255
Gaussian function, 63, 449, 607
Gaussian window, 110
Gegenbauer polynomial, 401
Gelfand triple, 214
Generalized eigenvalue problem, 582
Generating vector
  of rank-1 lattice, 473
Gentleman–Sande FFT, 290
Gibbs phenomenon, 48
Gram–Schmidt orthogonalization, 401

H
Hahn–Jordan decomposition, 234
Hamming window, 110
Hamming window sequence, 57
Hankel matrix, 569, 574
  low rank approximation, 585
Hankel transform, 198
Hanning window, 110
Hann window sequence, 57
Hartley matrix, 174
Hat function, 61
Heaviside function, 205
Heisenberg box, 113
Heisenberg's uncertainty principle, 106
Hermite function, 79, 116
Hermite interpolation problem, 536
Hermite polynomial, 79
Hilbert inequality, 593
  generalized, 594
Hilbert transform, 222
Horner scheme, 355
Hyperbolic cross
  energy-norm based, 485
  symmetric, 485

I
Ideal high-pass filter, 56
Ideal low-pass filter, 56
Imaging function, 577
Impulse response, 53
Inequality
  Bernstein, 26
  Bessel, 10
  generalized Hilbert, 594
  Heisenberg, 106
  Hilbert, 593
  Ingham, 588
  Nikolsky, 26
  Young, 17, 69
Inner product
  in $\mathbb{C}^{N_1 \times N_2}$, 253
  of multidimensional arrays, 261
In-place algorithm, 267
Integral
  highly oscillatory, 399
Interior grid point, 545
Inverse discrete Fourier transform
  multidimensional, 261
Inverse Fourier transform
  of tempered distribution, 210
Inverse multiquadrix, 449
Inverse NDCT, 456

J
Jacobi polynomial, 345, 401
Jordan's decomposition theorem, 33
Jump discontinuity, 30
Jump sequence, 52

K
Kernel, 20
  de la Vallée Poussin kernel, 23, 25
  Dirichlet kernel, 20, 25
  Fejér kernel, 22, 25
  summation kernel, 24
Kronecker product, 159

Kronecker sum, 165
Kronecker symbol, 126
  multidimensional, 248
  N-periodic, 126
Krylov–Lanczos method of convergence acceleration, 531

L

Lagrange basis polynomial, 360
Lanczos smoothing, 50
Laplace operator, 539
Laplacian filter
  discrete, 255
Largest integer smaller than or equal to n, 291
Lattice
  integer dual, 473
Lattice size
  of rank-1 lattice, 473
Lebesgue constant, 21
  for polynomial interpolation, 390
Left singular vectors, 577
Legendre function transform
  fast, 556
Legendre polynomial, 402, 552
  associated, 557
Leibniz product rule, 188
Lemma of Riemann–Lebesgue, 30
Linear canonical transform, 121
Linear difference equation with constant coefficients, 285
Linear filter, 52
Linear phase, 56
Linear prediction equations, 573
Linear prediction method, 573
Local discretization error, 548
Localized sampling, 89
L-periodic function, 11
LTI filter, 53

M

Magnitude response, 54
Matrix pencil, 581
  eigenvalue, 581
  eigenvector, 581
Matrix pencil method, 584
Matrix representation of FFT, 268
Mean value of 2π-periodic function, 11
Measure
  absolutely continuous measure, 231
  atomic measure, 231
  Borel measure, 230
  convolution of measures, 236
  Dirac measure, 231
  Fourier coefficients, 235
  Fourier transform, 240
  positive measure, 233
  Radon measure, 230
  tight, 233
  total variation measure, 230, 234
  vague convergence, 234
Mehler's formula, 118
Method of attenuation factors
  for Fourier coefficients, 520
  for Fourier transform, 508
Metric of $\mathscr{S}(\mathbb{R}^d)$, 185
Meyer window, 607
Modified Dirichlet kernel, 391
Modulus, 12
Moiré effect, 88
Monic polynomial, 344
Monogenic signal, 228
  amplitude, 228
Monte Carlo rule, 472, 564
Moore–Penrose pseudo-inverse, 157
Moving avering, 52
Multiquadrix, 449
Multivariate periodic function, 176
Multivariate trigonometric polynomial, 177
MUSIC, 576

N

n-cycle, 278
NDFT
  inverse, 458
Near field correction, 449
Near field sum, 451
Nesting method
  for two-dimensional DFT, 312
$\text{NFFT}^\top$, 418
Nikolsky inequality, 26
Node polynomial, 360
Noise space, 576
Noise-space correlation function, 577
Nonequispaced discrete transform
  inverse, 455
Nonequispaced FFT
  spherical, 561
Nonharmonic bandwidth, 439
Nonlinear eigenvalue problem, 584
Nonnegative residue modulo N, 126, 291, 506
  of an integer vector, 249
Nonsingular kernel function, 448
Normwise backward stability, 330
Normwise forward stability, 330
Nyquist rate, 88

O
Odd matrix, 258
Oversampling, 88
Oversampling factor, 610

P
Padé approximant, 572
Padua points, 486
Parameter estimation problem, 567
    multidimensional, 585
Parseval equality, 10, 77, 78, 180, 193
    for Chebyshev coefficients, 349
    for DFT, 151
    for multidimensional DFT, 263
    for two-dimensional DFT, 256
Partial sum of conjugate Fourier series, 44
Partition of unity, 132
Perfect shuffle, 276
Periodic function
    of bounded variation, 34
    piecewise continuously differentiable, 30
Periodic interpolation on uniform mesh, 513
Periodic Lagrange function, 513
Periodic signal, 52
Periodic Sobolev space
    of isotropic smoothness, 469
Periodic tempered distribution, 215
    Fourier coefficients, 219
    Fourier series, 219
Periodic window function, 608
Periodization
    of a function, 84
    of multivariate function, 194
    of a vector, 316
Periodization operator, 213
Periodized centered cardinal B-spline, 609
Periodized Gaussian function, 608
Periodized Kaiser–Bessel function, 609
Permutation
    2-stride, 161
    bit-reversed, 272
    even-odd, 161
    L-stride, 160
Permutation matrix, 160
Phase, 12
Phase recovery, 618
Phase response, 55
Pixel, 251
Poisson equation, 540
Poisson summation formula, 84, 196
    of Dirac comb, 216
Polar angle, 552
Polish space, 230
Polynomial representation of FFT, 268
Primitive Nth root of unity, 124
Principle value, 206
Pulse sequence, 52

Q
Quadratic Gauss sum, 144
Quadrature error, 394, 528
Quadrature rule
    on the unit sphere, 563
Quadrature weights, 394
Quasi-Monte Carlo rule, 472, 564

R
Rader FFT, 303
Radial function, 197
Radix-2 FFT, 269
Radix–4 FFT, 297
Radon–Nikodym derivative, 231
Rank-1 Chebyshev lattices, 486
Rank-1 lattice, 473
    multiple, 486
    reconstructing, 476
Rational interpolation problem, 586
Reconstructing multiple rank-1 lattice, 490
Rectangle function, 61
Rectangular pulse function, 16
Rectangular rule, 528
Rectangular window, 110
Rectangular window sequence, 57
Recursion, 267
Regularization procedure, 449
Regularized Shannon sampling formula, 89
Restricted isometry constant, 420, 447
Restricted isometry property, 420, 447
Reverse Prony polynomial, 571
Riemann's localization principle, 32, 68
Riesz stability
    of exponentials, 588
Riesz transform, 227
Right singular vectors, 577
Row–column method
    of multidimensional DFT, 313
    of two-dimensional DCT-II, 382
    of two-dimensional DFT, 260, 310

S
Sampling operator
    uniform, 214
Sampling period, 85
Sampling rate, 85

Sampling theorem of Shannon–Whittaker–Kotelnikov, 86
Sande–Tukey FFT, 271
Sawtooth function, 15
Schwartz space, 183
  convergence, 183
  locally convex space, 185
  metric, 185
Sequence of orthogonal polynomials, 401
Sequence of orthonormal polynomials, 401
Sequence, positive definite, 238
Shifted Chebyshev polynomial, 356
Shift-invariant filter, 52
Shift-invariant space, 498
Short-time Fourier transform, 111
Signal compression, 115
Signal flow graph, 268
Signal space, 576
Sign function, 213
Simultaneous approximation, 393
Sinc function, 61, 424
Sine integral, 47
Sine matrix
  of type I, 172, 543
  of type II, 172
  of type III, 172
  of type IV, 172
Sine vectors, 172
Singular kernel function, 448
Singular spectrum analysis, 585
Singular value decomposition, 577
Singular values, 577
Software
  discrete polynomial transform, 411
  fast Fourier transform on SO(3), 566
  fast summation, 454
  FFTW, 284
  inverse NFFT, 461
  NFFT, 463
  parallel FFT, 284
  parallel NFFT, 463
  spherical Fourier transform, 562
Space domain, 60
  of two-dimensional DFT, 253
Sparse FFT, 492
Sparse vector, 420
Spectrogram, 111
Spectrum, 12, 60
Spherical design, 564
Spherical filtering, 563
Spherical Fourier transform, 551
  fast, 554
  nonequispaced, 554
  nonequispaced discrete, 554
Spherical harmonics, 552
Spherical polynomial, 553
Spline, 603
Split–radix FFT, 300
Step function, 603
Stop band, 56
Sufficiently uniformly distributed points, 451
Sum representation of FFT, 268
Support length of a vector, 319
Symbol of cardinal interpolation, 499

T
Tempered distribution, 199
  derivative, 205
  periodic, 215
  product with smooth function, 203
  reflection, 203
  regular, 202
  scaling, 203
  translation, 203
Temporal variance, 108
Tensor product B-spline, 496
Theorem
  aliasing formula, 127
  of Banach–Steinhaus, 28
  of Bedrosian, 226
  of Bernstein, 41
  of Bochner, 244
  Chinese remainder theorem, 288
  of Dirichlet–Jordan, 38
  of Fejér, 25
  Fourier inversion, 193
  Fourier inversion formula, 66
  of Gibbs, 48
  Heisenberg's uncertainty principle, 106
  of Herglotz, 239
  Jordan decomposition, 33
  of Krylov–Lanczos, 531
  of Plancherel, 78, 194
  Poisson summation formula, 84
  of Prokhorov, 235
  of Riemann–Lebesque, 30
  Riemann's localization principle, 68
  of Shannon–Whittaker–Kotelnikov, 86
  of Weierstrass, 26
Thin-plate spline, 448
Three-direction box spline, 497
Three-direction mesh, 497
Three-term recurrence relation, 401
Time domain, 60
  of two-dimensional DFT, 253
Time-frequency analysis, 109
Time-frequency atom, 111

Time-invariant filter, 52
Time series, 574, 585
Toeplitz matrix, 154, 158
Torus, 6, 414
Total variation, 33
Total variation norm of measures, 230
Trace, 143
Trajectory matrix, 574
Transfer function, 54
Transmission band, 56
Transposed discrete polynomial transform, 404
Triangular window, 110
Trigonometric Lagrange polynomial, 132
Trigonometric polynomial, 7, 414
  supported on I, 466
Truncation error, 422
Two-point Taylor
  interpolation, 450
  interpolation polynomial, 536
Type-1 triangulation, 497

U
Ultraspherical polynomial, 401
Undersampling, 88

V
Vandermonde-like matrix, 403
Vandermonde matrix, 569, 574
Vector
  1-sparse, 316
  M-sparse, 321
  with frequency band of short support, 318
Vectorization
  of a d-dimensional array, 264
  of a matrix, 162, 259

W
Weighted least squares problem, 462
Weighted normal equation of first kind, 462
Weight function, 467
Wiener algebra, 468
Windowed Fourier transform, 111
Window function, 110, 422
  B-spline, 423
  continuous Kaiser–Bessel, 427
  exponential of semicircle, 438
  Gaussian, 424, 432
  Kaiser–Bessel, 437
  modified sinh-type, 438
  sinh-type, 431
Window length, 574
Winograd FFT, 307

Y
Young inequality, 17, 69
  generalized, 17, 70

Z
z-transform, 571

Applied and Numerical Harmonic Analysis (106 Volumes)

1. A. I. Saichev and W. A. Woyczyński: *Distributions in the Physical and Engineering Sciences* (ISBN: 978-0-8176-3924-2)
2. C. E. D'Attellis and E. M. Fernandez-Berdaguer: *Wavelet Theory and Harmonic Analysis in Applied Sciences* (ISBN: 978-0-8176-3953-2)
3. H. G. Feichtinger and T. Strohmer: *Gabor Analysis and Algorithms* (ISBN: 978-0-8176-3959-4)
4. R. Tolimieri and M. An: *Time-Frequency Representations* (ISBN: 978-0-8176-3918-1)
5. T. M. Peters and J. C. Williams: *The Fourier Transform in Biomedical Engineering* (ISBN: 978-0-8176-3941-9)
6. G. T. Herman: *Geometry of Digital Spaces* (ISBN: 978-0-8176-3897-9)
7. A. Teolis: *Computational Signal Processing with Wavelets* (ISBN: 978-0-8176-3909-9)
8. J. Ramanathan: *Methods of Applied Fourier Analysis* (ISBN: 978-0-8176-3963-1)
9. J. M. Cooper: *Introduction to Partial Differential Equations with MATLAB* (ISBN: 978-0-8176-3967-9)
10. Procházka, N. G. Kingsbury, P. J. Payner, and J. Uhlir: *Signal Analysis and Prediction* (ISBN: 978-0-8176-4042-2)
11. W. Bray and C. Stanojevic: *Analysis of Divergence* (ISBN: 978-1-4612-7467-4)
12. G. T. Herman and A. Kuba: *Discrete Tomography* (ISBN: 978-0-8176-4101-6)
13. K. Gröchenig: *Foundations of Time-Frequency Analysis* (ISBN: 978-0-8176-4022-4)
14. L. Debnath: *Wavelet Transforms and Time-Frequency Signal Analysis* (ISBN: 978-0-8176-4104-7)
15. J. J. Benedetto and P. J. S. G. Ferreira: *Modern Sampling Theory* (ISBN: 978-0-8176-4023-1)
16. D. F. Walnut: *An Introduction to Wavelet Analysis* (ISBN: 978-0-8176-3962-4)
17. A. Abbate, C. DeCusatis, and P. K. Das: *Wavelets and Subbands* (ISBN: 978-0-8176-4136-8)
18. O. Bratteli, P. Jorgensen, and B. Treadway: *Wavelets Through a Looking Glass* (ISBN: 978-0-8176-4280-80)
19. H. G. Feichtinger and T. Strohmer: *Advances in Gabor Analysis* (ISBN: 978-0-8176-4239-6)
20. O. Christensen: *An Introduction to Frames and Riesz Bases* (ISBN: 978-0-8176-4295-2)
21. L. Debnath: *Wavelets and Signal Processing* (ISBN: 978-0-8176-4235-8)
22. G. Bi and Y. Zeng: *Transforms and Fast Algorithms for Signal Analysis and Representations* (ISBN: 978-0-8176-4279-2)
23. J. H. Davis: *Methods of Applied Mathematics with a MATLAB Overview* (ISBN: 978-0-8176-4331-7)
24. J. J. Benedetto and A. I. Zayed: *Sampling, Wavelets, and Tomography* (ISBN: 978-0-8176-4304-1)

G. Plonka et al., *Numerical Fourier Analysis*, Applied and Numerical Harmonic Analysis, https://doi.org/10.1007/978-3-031-35005-4

25. E. Prestini: *The Evolution of Applied Harmonic Analysis* (ISBN: 978-0-8176-4125-2)
26. L. Brandolini, L. Colzani, A. Iosevich, and G. Travaglini: *Fourier Analysis and Convexity* (ISBN: 978-0-8176-3263-2)
27. W. Freeden and V. Michel: *Multiscale Potential Theory* (ISBN: 978-0-8176-4105-4)
28. O. Christensen and K. L. Christensen: *Approximation Theory* (ISBN: 978-0-8176-3600-5)
29. O. Calin and D.-C. Chang: *Geometric Mechanics on Riemannian Manifolds* (ISBN: 978-0-8176-4354-6)
30. J. A. Hogan: *Time-Frequency and Time-Scale Methods* (ISBN: 978-0-8176-4276-1)
31. C. Heil: *Harmonic Analysis and Applications* (ISBN: 978-0-8176-3778-1)
32. K. Borre, D. M. Akos, N. Bertelsen, P. Rinder, and S. H. Jensen: *A Software-Defined GPS and Galileo Receiver* (ISBN: 978-0-8176-4390-4)
33. T. Qian, M. I. Vai, and Y. Xu: *Wavelet Analysis and Applications* (ISBN: 978-3-7643-7777-9)
34. G. T. Herman and A. Kuba: *Advances in Discrete Tomography and Its Applications* (ISBN: 978-0-8176-3614-2)
35. M. C. Fu, R. A. Jarrow, J.-Y. Yen, and R. J. Elliott: *Advances in Mathematical Finance* (ISBN: 978-0-8176-4544-1)
36. O. Christensen: *Frames and Bases* (ISBN: 978-0-8176-4677-6)
37. P. E. T. Jorgensen, J. D. Merrill, and J. A. Packer: *Representations, Wavelets, and Frames* (ISBN: 978-0-8176-4682-0)
38. M. An, A. K. Brodzik, and R. Tolimieri: *Ideal Sequence Design in Time-Frequency Space* (ISBN: 978-0-8176-4737-7)
39. S. G. Krantz: *Explorations in Harmonic Analysis* (ISBN: 978-0-8176-4668-4)
40. B. Luong: *Fourier Analysis on Finite Abelian Groups* (ISBN: 978-0-8176-4915-9)
41. G. S. Chirikjian: *Stochastic Models, Information Theory, and Lie Groups, Volume 1* (ISBN: 978-0-8176-4802-2)
42. C. Cabrelli and J. L. Torrea: *Recent Developments in Real and Harmonic Analysis* (ISBN: 978-0-8176-4531-1)
43. M. V. Wickerhauser: *Mathematics for Multimedia* (ISBN: 978-0-8176-4879-4)
44. B. Forster, P. Massopust, O. Christensen, K. Gröchenig, D. Labate, P. Vandergheynst, G. Weiss, and Y. Wiaux: *Four Short Courses on Harmonic Analysis* (ISBN: 978-0-8176-4890-9)
45. O. Christensen: *Functions, Spaces, and Expansions* (ISBN: 978-0-8176-4979-1)
46. J. Barral and S. Seuret: *Recent Developments in Fractals and Related Fields* (ISBN: 978-0-8176-4887-9)
47. O. Calin, D.-C. Chang, and K. Furutani, and C. Iwasaki: *Heat Kernels for Elliptic and Sub-elliptic Operators* (ISBN: 978-0-8176-4994-4)
48. C. Heil: *A Basis Theory Primer* (ISBN: 978-0-8176-4686-8)
49. J. R. Klauder: *A Modern Approach to Functional Integration* (ISBN: 978-0-8176-4790-2)
50. J. Cohen and A. I. Zayed: *Wavelets and Multiscale Analysis* (ISBN: 978-0-8176-8094-7)
51. D. Joyner and J.-L. Kim: *Selected Unsolved Problems in Coding Theory* (ISBN: 978-0-8176-8255-2)
52. G. S. Chirikjian: *Stochastic Models, Information Theory, and Lie Groups, Volume 2* (ISBN: 978-0-8176-4943-2)
53. J. A. Hogan and J. D. Lakey: *Duration and Bandwidth Limiting* (ISBN: 978-0-8176-8306-1)
54. G. Kutyniok and D. Labate: *Shearlets* (ISBN: 978-0-8176-8315-3)
55. P. G. Casazza and P. Kutyniok: *Finite Frames* (ISBN: 978-0-8176-8372-6)
56. V. Michel: *Lectures on Constructive Approximation* (ISBN : 978-0-8176-8402-0)
57. D. Mitrea, I. Mitrea, M. Mitrea, and S. Monniaux: *Groupoid Metrization Theory* (ISBN: 978-0-8176-8396-2)
58. T. D. Andrews, R. Balan, J. J. Benedetto, W. Czaja, and K. A. Okoudjou: *Excursions in Harmonic Analysis, Volume 1* (ISBN: 978-0-8176-8375-7)
59. T. D. Andrews, R. Balan, J. J. Benedetto, W. Czaja, and K. A. Okoudjou: *Excursions in Harmonic Analysis, Volume 2* (ISBN: 978-0-8176-8378-8)
60. D. V. Cruz-Uribe and A. Fiorenza: *Variable Lebesgue Spaces* (ISBN: 978-3-0348-0547-6)

61. W. Freeden and M. Gutting: *Special Functions of Mathematical (Geo-)Physics* (ISBN: 978-3-0348-0562-9)
62. A. I. Saichev and W. A. Woyczyński: *Distributions in the Physical and Engineering Sciences, Volume 2: Linear and Nonlinear Dynamics of Continuous Media* (ISBN: 978-0-8176-3942-6)
63. S. Foucart and H. Rauhut: *A Mathematical Introduction to Compressive Sensing* (ISBN: 978-0-8176-4947-0)
64. G. T. Herman and J. Frank: *Computational Methods for Three-Dimensional Microscopy Reconstruction* (ISBN: 978-1-4614-9520-8)
65. A. Paprotny and M. Thess: *Realtime Data Mining: Self-Learning Techniques for Recommendation Engines* (ISBN: 978-3-319-01320-6)
66. A. I. Zayed and G. Schmeisser: *New Perspectives on Approximation and Sampling Theory: Festschrift in Honor of Paul Butzer's* 85th *Birthday* (ISBN: 978-3-319-08800-6)
67. R. Balan, M. Begue, J. Benedetto, W. Czaja, and K. A. Okoudjou: *Excursions in Harmonic Analysis, Volume 3* (ISBN: 978-3-319-13229-7)
68. H. Boche, R. Calderbank, G. Kutyniok, and J. Vybiral: *Compressed Sensing and its Applications* (ISBN: 978-3-319-16041-2)
69. S. Dahlke, F. De Mari, P. Grohs, and D. Labate: *Harmonic and Applied Analysis: From Groups to Signals* (ISBN: 978-3-319-18862-1)
70. A. Aldroubi: *New Trends in Applied Harmonic Analysis* (ISBN: 978-3-319-27871-1)
71. M. Ruzhansky: *Methods of Fourier Analysis and Approximation Theory* (ISBN: 978-3-319-27465-2)
72. G. Pfander: *Sampling Theory, a Renaissance* (ISBN: 978-3-319-19748-7)
73. R. Balan, M. Begue, J. Benedetto, W. Czaja, and K. A. Okoudjou: *Excursions in Harmonic Analysis, Volume 4* (ISBN: 978-3-319-20187-0)
74. O. Christensen: *An Introduction to Frames and Riesz Bases, Second Edition* (ISBN: 978-3-319-25611-5)
75. E. Prestini: *The Evolution of Applied Harmonic Analysis: Models of the Real World, Second Edition* (ISBN: 978-1-4899-7987-2)
76. J. H. Davis: *Methods of Applied Mathematics with a Software Overview, Second Edition* (ISBN: 978-3-319-43369-1)
77. M. Gilman, E. M. Smith, and S. M. Tsynkov: *Transionospheric Synthetic Aperture Imaging* (ISBN: 978-3-319-52125-1)
78. S. Chanillo, B. Franchi, G. Lu, C. Perez, and E. T. Sawyer: *Harmonic Analysis, Partial Differential Equations and Applications* (ISBN: 978-3-319-52741-3)
79. R. Balan, J. Benedetto, W. Czaja, M. Dellatorre, and K. A. Okoudjou: *Excursions in Harmonic Analysis, Volume 5* (ISBN: 978-3-319-54710-7)
80. I. Pesenson, Q. T. Le Gia, A. Mayeli, H. Mhaskar, and D. X. Zhou: *Frames and Other Bases in Abstract and Function Spaces: Novel Methods in Harmonic Analysis, Volume 1* (ISBN: 978-3-319-55549-2)
81. I. Pesenson, Q. T. Le Gia, A. Mayeli, H. Mhaskar, and D. X. Zhou: *Recent Applications of Harmonic Analysis to Function Spaces, Differential Equations, and Data Science: Novel Methods in Harmonic Analysis, Volume 2* (ISBN: 978-3-319-55555-3)
82. F. Weisz: *Convergence and Summability of Fourier Transforms and Hardy Spaces* (ISBN: 978-3-319-56813-3)
83. C. Heil: *Metrics, Norms, Inner Products, and Operator Theory* (ISBN: 978-3-319-65321-1)
84. S. Waldron: *An Introduction to Finite Tight Frames: Theory and Applications.* (ISBN: 978-0-8176-4814-5)
85. D. Joyner and C. G. Melles: *Adventures in Graph Theory: A Bridge to Advanced Mathematics.* (ISBN: 978-3-319-68381-2)
86. B. Han: *Framelets and Wavelets: Algorithms, Analysis, and Applications* (ISBN: 978-3-319-68529-8)
87. H. Boche, G. Caire, R. Calderbank, M. März, G. Kutyniok, and R. Mathar: *Compressed Sensing and Its Applications* (ISBN: 978-3-319-69801-4)

88. A. I. Saichev and W. A. Woyczyński: *Distributions in the Physical and Engineering Sciences, Volume 3: Random and Fractal Signals and Fields* (ISBN: 978-3-319-92584-4)
89. G. Plonka, D. Potts, G. Steidl, and M. Tasche: *Numerical Fourier Analysis* (978-3-030-04305-6)
90. K. Bredies and D. Lorenz: *Mathematical Image Processing* (ISBN: 978-3-030-01457-5)
91. H. G. Feichtinger, P. Boggiatto, E. Cordero, M. de Gosson, F. Nicola, A. Oliaro, and A. Tabacco: *Landscapes of Time-Frequency Analysis* (ISBN: 978-3-030-05209-6)
92. E. Liflyand: *Functions of Bounded Variation and Their Fourier Transforms* (ISBN: 978-3-030-04428-2)
93. R. Campos: *The XFT Quadrature in Discrete Fourier Analysis* (ISBN: 978-3-030-13422-8)
94. M. Abell, E. Iacob, A. Stokolos, S. Taylor, S. Tikhonov, J. Zhu: *Topics in Classical and Modern Analysis: In Memory of Yingkang Hu* (ISBN: 978-3-030-12276-8)
95. H. Boche, G. Caire, R. Calderbank, G. Kutyniok, R. Mathar, P. Petersen: *Compressed Sensing and its Applications: Third International MATHEON Conference 2017* (ISBN: 978-3-319-73073-8)
96. A. Aldroubi, C. Cabrelli, S. Jaffard, U. Molter: *New Trends in Applied Harmonic Analysis, Volume II: Harmonic Analysis, Geometric Measure Theory, and Applications* (ISBN: 978-3-030-32352-3)
97. S. Dos Santos, M. Maslouhi, K. Okoudjou: *Recent Advances in Mathematics and Technology: Proceedings of the First International Conference on Technology, Engineering, and Mathematics, Kenitra, Morocco, March 26–27, 2018* (ISBN: 978-3-030-35201-1)
98. Á. Bényi, K. Okoudjou: *Modulation Spaces: With Applications to Pseudodifferential Operators and Nonlinear Schrödinger Equations* (ISBN: 978-1-0716-0330-7)
99. P. Boggiato, M. Cappiello, E. Cordero, S. Coriasco, G. Garello, A. Oliaro, J. Seiler: *Advances in Microlocal and Time-Frequency Analysis* (ISBN: 978-3-030-36137-2)
100. S. Casey, K. Okoudjou, M. Robinson, B. Sadler: *Sampling: Theory and Applications* (ISBN: 978-3-030-36290-4)
101. P. Boggiatto, T. Bruno, E. Cordero, H. G. Feichtinger, F. Nicola, A. Oliaro, A. Tabacco, M. Vallarino: *Landscapes of Time-Frequency Analysis: ATFA 2019* (ISBN: 978-3-030-56004-1)
102. M. Hirn, S. Li, K. Okoudjou, S. Saliana, Ö. Yilmaz: *Excursions in Harmonic Analysis, Volume 6: In Honor of John Benedetto's 80th Birthday* (ISBN: 978-3-030-69636-8)
103. F. De Mari, E. De Vito: *Harmonic and Applied Analysis: From Radon Transforms to Machine Learning* (ISBN: 978-3-030-86663-1)
104. G. Kutyniok, H. Rauhut, R. J. Kunsch, *Compressed Sensing in Information Processing* (ISBN: 978-3-031-09744-7)
105. P. Flandrin, S. Jaffard, T. Paul, B. Torresani, *Theoretic Physics, Wavelets, Analysis, Genomics: An Indisciplinary Tribute to Alex Grossmann* (ISBN: 978-3-030-45846-1)
106. G. Plonka-Hoch, D. Potts, G. Steidl, M. Tasche, *Numerical Fourier Analysis, Second Edition* (ISBN: 978-3-031-35004-7)

For an up-to-date list of ANHA titles, please visit **http://www.springer.com/series/4968**